Review Formulas

$$\frac{a}{d} + \frac{b}{d} = \frac{a+b}{d}. \qquad \frac{a}{b}\frac{c}{d} = \frac{ac}{bd}. \qquad \frac{a}{b} = \frac{ad}{bd}. \qquad \frac{a}{b} + \frac{c}{d} = \frac{ad+bc}{bd}. \qquad \frac{a/b}{c/d} = \frac{ad}{bc}.$$

$$\frac{-a}{b} = \frac{a}{-b} = -\frac{a}{b}. \qquad\qquad ax^2 + bx + c = 0 \quad \text{has roots} \quad r = \frac{-b \pm \sqrt{b^2 - 4ac}}{2a}.$$

$$a^2 - b^2 = (a-b)(a+b). \qquad a^3 - b^3 = (a-b)(a^2 + ab + b^2).$$

$$a^n - b^n = (a-b)(a^{n-1} + a^{n-2}b + \cdots + a^{n-1-k}b^k + \cdots + b^{n-1}).$$

$$(a+b)^n = a^n + \binom{n}{1}a^{n-1}b + \binom{n}{2}a^{n-2}b^2 + \cdots + \binom{n}{k}a^{n-k}b^k + \cdots + b^n,$$

$$\text{where} \quad \binom{n}{k} = \frac{n!}{k!(n-k)!}.$$

$$a^u a^v = a^{u+v}. \quad a^u b^u = (ab)^u. \quad (a^u)^v = a^{uv}. \quad a^{u/v} = \sqrt[v]{a^u} = (\sqrt[v]{a})^u. \quad a^{-u} = \frac{1}{a^u}. \quad a^{u-v} = \frac{a^u}{a^v}.$$

$$\frac{\sin\theta}{\cos\theta} = \tan\theta. \qquad \frac{\cos\theta}{\sin\theta} = \cot\theta. \qquad \frac{1}{\cos\theta} = \sec\theta. \qquad \frac{1}{\sin\theta} = \csc\theta.$$

$$\sin^2\theta + \cos^2\theta = 1. \qquad \sec^2\theta - \tan^2\theta = 1. \qquad \csc^2\theta - \cot^2\theta = 1.$$

$$\sin(A+B) = \sin A \cos B + \cos A \sin B. \qquad\qquad \sin\left(\frac{\pi}{2} - A\right) = \cos A.$$

$$\sin(A-B) = \sin A \cos B - \cos A \sin B.$$

$$\cos(A+B) = \cos A \cos B - \sin A \sin B. \qquad\qquad \cos\left(\frac{\pi}{2} - A\right) = \sin A.$$

$$\cos(A-B) = \cos A \cos B + \sin A \sin B. \qquad\qquad \tan\left(\frac{\pi}{2} - A\right) = \cot A.$$

$$\tan(A+B) = \frac{\tan A + \tan B}{1 - \tan A \tan B}. \qquad\qquad \sin 2A = 2\sin A \cos A.$$

$$\tan(A-B) = \frac{\tan A - \tan B}{1 + \tan A \tan B}. \qquad\qquad \cos 2A = \cos^2 A - \sin^2 A.$$

$$\tan 2A = \frac{2\tan A}{1 - \tan^2 A}.$$

$$\cos A \cos B = \frac{1}{2}[\cos(A+B) + \cos(A-B)].$$

$$\sin A \sin B = \frac{1}{2}[\cos(A-B) - \cos(A+B)]. \qquad\qquad \cos^2 A = \frac{1 + \cos 2A}{2}.$$

$$\sin A \cos B = \frac{1}{2}[\sin(A+B) + \sin(A-B)]. \qquad\qquad \sin^2 A = \frac{1 - \cos 2A}{2}.$$

CALCULUS:
Concepts and Calculations

Have you obtained your *Study Guide for Goodman and Saff's Calculus: Concepts and Calculations?*

It can help you with this course by providing:
1. A summary of the important formulas for each chapter.
2. A review of the important ideas of each chapter.
3. Detailed solutions of many of the problems in the exercises for each chapter, including a few of the earlier problems and most of the really difficult problems.

You can get a copy of the *Study Guide for Goodman and Saff's Calculus: Concepts and Calculations,* Vols. I and II, at your College Bookstore. If your bookstore doesn't have it in stock, please ask the bookstore manager to order you a copy.

CALCULUS:
Concepts and Calculations

A. W. Goodman & E. B. Saff

THE UNIVERSITY OF SOUTH FLORIDA

Macmillan Publishing Co., Inc.

NEW YORK

Collier Macmillan Publishers

LONDON

Macmillan Publishing Co., Inc.
866 Third Avenue, New York, New York 10022

Collier Macmillan Canada, Ltd.

Library of Congress Cataloging in Publication Data

Goodman, A. W.   (date)
   Calculus, concepts and calculations.

   Includes index.
   1. Calculus. I. Saff, E. B.   (date)   joint author.
II. Title.
QA303.G6262   515   79-26449
ISBN 0-02-344740-0

Printing: 1 2 3 4 5 6 7 8   Year: 1 2 3 4 5 6 7

Dedication

To our parents, whose affection and guidance will always be an inspiration

Hannah and William Goodman   *Rose and Irving Saff*

Preface

Every calculus text has its special slant, and ours, as the subtitle indicates, is "concepts and calculations."

A glance at the table of contents shows that all the "concepts" typically covered in a three-semester course are, indeed, present. But we have attempted to tailor their explanations to be within ready grasp of today's student, and we have designed the subject matter so as to afford the instructor with real flexibility in the order of presentation.

To be more specific, limits and continuity are presented in an intuitive manner, with the more subtle aspects explained by means of a substantial set of examples. The more difficult work with ϵ and δ has been placed in an appendix, where a detailed discussion (including exercises) is present for the teacher (or student) who wishes to use it.

Many instructors argue that the notion of the "limit of a sequence" is more fundamental and more readily understood than the "limit of a function." Whatever the case, the instructor has the option of presenting sequences and their limits earlier in the course, for they are covered in great detail in Section 4 of Chapter 3. However, the teacher who prefers to discuss limits of functions first can ignore this section and go directly from Section 3 of Chapter 3 to Chapter 4 (Differentiation) without sacrificing the logical development.

Where appropriate, we have included, for the aid of the student, step-by-step procedures for working problems (such as graphing functions and maximum–minimum problems). The exercises contain problems ranging from the routine to the more challenging (indicated by a *) and were designed to reinforce concepts and to explore further directions. We have also included a substantial number of review problems at the end of the chapters.

Many of the theorems in the text are proved in great detail, but those proofs which are either "obvious" or require the more sophisticated notions of real analysis have been deleted. In this regard we have attempted to be more honest with the student by labeling appropriate theorems with **PWO** (the proof will be omitted) or **PLE** (the proof is left as an exercise).

Calculus has its own aesthetic beauty and some will argue that it can stand aloof from the sciences and engineering. But "applications" can provide gratifying reinforcement for the student and give him a clearer picture of the roots of the subject. We have attempted, therefore, to maintain a reasonable balance between the theory and the applications. Prospective scientists or engineers will find particularly useful tools in the sections that cover:

1. Solving equations using (a) the bisection method, (b) Newton's method, and (c) successive substitutions.
2. Approximation methods for integrals (Simpson's Rule and the Trapezoidal Rule).
3. Use of tables of integrals.
4. Equations of growth and decay.
5. The regression line (least squares).
6. Computation of forces due to gravitational attraction.
7. Lagrange multipliers.
8. Green's Theorem, the Divergence Theorem, and Stokes' Theorem.

Today's inexpensive hand-held calculator is another tool that places many fascinating applications at the fingertips of the student. When properly guided, its use can also significantly reinforce many concepts of calculus as well as provide the student with a sophisticated discovery tool. The authors recognize, however, that some course schedules hardly allow sufficient time to complete the required syllabus, much less to include new material involving the calculator. With this in mind, we have designed the "calculations" portions of the text as being purely optional. Those problems which require a calculator are placed at the end of the exercise set and are marked with a "C" so that they are easy to identify. There are over 300 such calculator problems as well as optional sections on Calculator Number Systems and Polynomial Evaluation.

The ideal calculator for this book should have the usual function buttons together with y^x, $\sqrt{x}$, $\ln x$, e^x, the trigonometric functions and the inverse trigonometric functions. However, a programmable calculator is not necessary.

Any course outline must be framed to fit the ability and previous preparation of the students. For the students who require it, we have included a review of inequalities and a brief review of trigonometry. A well-prepared class should skip Chapter 1 and the first 8 sections of Chapter 2, which give the elements of analytic geometry. Such a class can start with the conic sections and within six lessons begin the differential calculus.

The following schedule can be covered in three semesters by a class meeting 4 hours a week for 16 weeks in each semester (64 lessons each semester).

First Semester		Second Semester		Third Semester	
Chapter	Hours	Chapter	Hours	Chapter	Hours
2	9	8	6	16	12
3	6	9	6	17	8
4	11	10	8	18	14
5	12	11	6	19	13
6	9	12	7	20	8
7	7	13	5	Exams	9
Exams	10	14	6	or Review	
or Review		15	10		
Total	64	Exams	10	Total	64
		or Review			
		Total	64		

This schedule is intended only as a guide and certainly should be modified to suit the local calendar and conditions. If the class meets only three times a week for four semesters then the breaks might best be made after Chapters 5, 11, and 16. A school on the quarter system, with 10 weeks in each quarter, could also break after Chapters 5, 11, and 16 if the class meets 5 hours a week (50 lessons) in each of the four quarters.

We are grateful for the assistance of our many colleagues who offered helpful suggestions. Our thanks also go to Miss Denisse R. Thompson who carefully read the entire manuscript, worked all the problems, and corrected our slips and oversights. We are pleased to acknowledge our thanks to Mrs. Loretta Saff, who cheerfully and accurately typed the entire manuscript. We are also indebted to Mrs. Elaine Wetterau, production supervisor of the Macmillan Publishing Co., Inc., who so diligently and carefully guided this book through its final stages to completion.

Finally, each of the authors owes a great debt to his teachers and this obvious debt should be mentioned. Those teachers who have had the greatest influence on our education are (for A. W. Goodman) Paul Erdös, Hans Rademacher, and Otto Szasz, and (for E. B. Saff) W. E. Kirwan and J. L. Walsh.

Tampa, Florida   A. W. G.
                                                                          E. B. S.

Contents

Preliminaries 1

1

Introduction to Analytic Geometry 2

22

Integration 6
236

Applications of Integration 7
295

The Trigonometric Functions 8
329

Vectors in the Plane 12
470

Polar Coordinates 13
522

Indeterminate Forms and Improper Integrals 14
546

Infinite Series 15
565

16 *Vectors and Solid Analytic Geometry*

636

17 *Moments and Centroids*

703

Partial Differentiation 18
733

Multiple Integrals 19
804

Line and Surface Integrals 20
865

Appendix 1 Sequences: The ϵ, N Definition of Limit

910

Appendix 2 Functions: The ϵ, δ Definition of Limit

918

Appendix 3 Determinants

926

Appendix 4 Units: The British and Metric Systems

936

Appendix 5 Evaluation of Polynomial Functions

940

Tables

Answers to Odd-Numbered Problems

Index

CALCULUS:
Concepts and Calculations

1

Preliminaries

The Real Numbers 1

The foundation for the development of calculus is the real number system. We assume that the reader is familiar with the algebraic properties of the real numbers and we shall not pause here to give a complete or detailed account of them. Instead, we mention some facts about real numbers that are particularly relevant to the study of calculus.

The real numbers consist of the positive numbers, the negative numbers, and zero. The building blocks of the real number system are the *integers:*

$$\ldots, \quad -3, \quad -2, \quad -1, \quad 0, \quad 1, \quad 2, \quad 3, \quad \ldots.$$

By taking ratios of integers, we obtain the *rational* numbers. For example,

$$\frac{2}{3}, \quad \frac{-100}{47}, \quad \frac{75}{238}, \quad \frac{10^5}{37^{21}}, \quad \frac{-14}{1}$$

are each rational numbers. The decimal representation of a rational number will either be a *terminating decimal,* such as

$$\frac{1}{8} = 0.125, \qquad \frac{-237}{10} = -23.7,$$

or a *nonterminating repeating decimal,* such as

$$\frac{2}{3} = 0.6666\ldots, \qquad \frac{215}{99} = 2.171717\ldots.$$

There are indeed real numbers that are not rational. These are the numbers that have a *nonrepeating nonterminating decimal* expansion, such as

$$\sqrt{2} = 1.4142\ldots, \qquad \sqrt{7} = 2.6457\ldots, \qquad \pi = 3.1415\ldots.$$

A real number that is not rational is said to be *irrational*. There are infinitely many rational numbers and infinitely many irrationals. In fact, between any two distinct rationals there is an irrational number, and between any two distinct irrationals there is a rational number.

The operations of addition and multiplication of real numbers are governed by the following laws, which hold for arbitrary real numbers a, b, c:

Commutative Laws: $\qquad a + b = b + a, \quad ab = ba$

Associative Laws: $\qquad a + (b + c) = (a + b) + c, \quad a(bc) = (ab)c$

Distributive Law: $\qquad a(b + c) = ab + ac$

Existence of Identities: $\qquad a + 0 = a, \quad a \cdot 1 = a$

Existence of Inverses: $\qquad a + (-a) = 0, \quad a \cdot \left(\dfrac{1}{a}\right) = 1 \quad$ for $a \neq 0$.

Many properties of the real numbers can be rigorously derived from the preceding laws. One such familiar fact is that if $ab = 0$, then $a = 0$ or $b = 0$.

★2 *Calculator Number Systems*

Any calculator, and indeed any computing device built by human beings, can handle only *finitely many* decimal places. This means that the number system of a calculator is different from the real number system, which contains nonterminating decimals. It is important to keep this distinction in mind when using the calculator for purposes of computation or numerical experimentation. We shall emphasize this point by considering a typical calculator which displays 10 significant digits and uses scientific notation with two-digit exponents.

Such a calculator is capable of displaying numbers of the form

(1) $$\pm(a_1.a_2a_3a_4a_5a_6a_7a_8a_9a_{10}) \times 10^S,$$

where $a_1, a_2, \ldots, a_{10}$ are each digits $(0, 1, \ldots, 9)$ and the power S of 10 is an integer between -99 and $+99$. For nonzero numbers, the first digit a_1 is always taken to be 1 or greater. We call the portion $(a_1.a_2a_3 \ldots a_{10})$ the *mantissa*.

As far as the calculator is concerned, the numbers (1) are the only numbers that exist.[1] Thus, although there is no largest *real* number, there is a largest calculator number, namely

$$9.999999999 \times 10^{99}.$$

Adding a positive number to this quantity will either leave it unchanged or result in a warning from your calculator that you have exceeded its capacity.

We can also see from (1) that the calculator deals only with *rational* numbers. It is likely that your device has a button for the irrational number π. However, this number is not actually part of your calculator system—what is displayed is an *approximation* to the real number π, accurate to perhaps nine decimal places. It is again clear from (1) that there are only *finitely many* rational numbers in the calculator system. The rational number 2/3 is certainly not present because its decimal expansion is nonterminating. Of course, you can perform the division $2 \div 3$ on your calculator, but the answer will not be 0.666 . . . ; it will probably be 0.666666667, which represents 0.666 . . . rounded to nine decimal places.

Another important difference in the calculator number system is that some of the basic laws of real number arithmetic fail to hold because of roundoff. For example, on our typical calculator we find that

(2) $$10^4 + (8 \times 10^{-7}) = 10^4$$

[write out the decimal form of the sum on the left in (2) to see why this happens]. Thus the equation $a + x = a$ does not necessarily imply that $x = 0$. For many calculators it is also easy to show that the associative law of addition is violated. This means that adding a list of numbers from left to right may not result in the same answer as adding them from right to left.

The differences we have mentioned thus far point out that the equality symbol of a calculator is not the same as the equality symbol for the real number system. To avoid this ambiguity, it is often advisable to use the *approximation symbol* $\approx$ when computing an expression involving given real numbers. For example, with the aid of a calculator we find that

(3) $$\frac{4 + \pi}{7} \approx 1.020227522.$$

In many of the calculator examples and exercises throughout the text we use the notion of *rounding to k decimal places*. Consider the decimal $n.b_1b_2b_3 \ldots$, where n is an integer and $b_1, b_2, \ldots$ are digits. To round off this number to the kth decimal place, we find the closest possible approximation having k decimal places. Thus, if the first digit to

---

[1] Some calculators displaying 10 significant digits will utilize 12- or 13-digit internal registers for the purpose of minimizing roundoff error.

be dropped, b_{k+1}, is under 5, we get $n.b_1 b_2 \ldots b_k$, whereas if b_{k+1} is 5 or greater, we get $n.b_1 b_2 \ldots b_k + 10^{-k}$.

For example, the number $8.24993 \ldots$ rounded

to one decimal place is 8.2,

to two decimal places is 8.25,

to three decimal places is 8.250,

to four decimal places is 8.2499.

Rounding off is a common occurrence in the workings of a calculator. If we could look inside our typical calculator, we would find that the computations are carried out with a fixed number of digits (typically 13). But once a computation is completed, the answer is automatically rounded to the maximum number of display digits (usually 10). For example, internally the calculator finds the product

$$(2.987825) \times (1.111111) = 3.319805223575$$

and then displays the answer rounded to 3.319805224.

In working with numbers expressed in scientific notation, it is often required to *round the computation to r significant digits,* that is, to find the closest number with an r-digit mantissa. We obtain such an approximation to the number $\pm(a_1.a_2 a_3 a_4 \ldots) \times 10^S$ by simply rounding its mantissa to $(r - 1)$ decimal places.

For example, the number $(6.0638\ldots) \times 10^{-8}$ rounded

to one significant digit is 6×10^{-8},

to two significant digits is 6.1×10^{-8},

to three significant digits is 6.06×10^{-8}.

As a rule of thumb, the reader should keep in mind that when computing with approximate numbers, the answer should generally not be given with more significant digits than is possessed by the approximate number having the *fewest* significant digits.

3 *Some Unusual Symbols*

The reader is already familiar with the standard symbols, such as $+, -, \cdot, \times, \sqrt{}$, and so forth. Other symbols will be explained as they appear for the first time. A few special symbols will be treated here.

∎ The proof is completed. It is convenient to have a mark to signal the end of a proof. Thus, if the reader has trouble understanding the proof, he can at least locate the place where the proof is completed and then reread the proof until it does become clear. In the past this place was often indicated by the letters Q.E.D., which abbreviate "quod erat demonstrandum," the Latin phrase for "which was to be demonstrated." In recent times it has become the custom to use the symbol ▌ with exactly the same meaning. In this book we will use ▌.

● The solution of the example is completed. It is also convenient to have a symbol marking the end of the solution of an example. Now, an example may also be a theorem, and a theorem may also serve as an example. Nevertheless, there may be some advantage in distinguishing between the two, and hence we will use the two different symbols ▌ and ●.

PWO. The proof will be omitted. Frequently, there are good reasons for omitting a proof. Perhaps the proof is (1) long, (2) uninteresting, or (3) uses advanced methods and concepts beyond the scope of this text. In any one of these cases we label the theorem PWO to advise the reader that the proof will be omitted.

PLE. The proof is left as an exercise. We also label a theorem PLE to indicate that the proof will be omitted. But in this case the proof of the theorem is well within the capacity of the student and he is expected to supply the proof for himself.

To avoid the burden of too many symbols, we will make little use of certain popular ones that are often used in other texts. However, your teacher may wish to use them in his lectures or you may meet these symbols when you read other books, so we list them here for your convenience:

$\mathscr{R} \equiv$ set of all real numbers		$\mathscr{I} \equiv$ set of all integers	
$\in \equiv$ is an element of the set		$\wedge \equiv$ and	
$\ni \equiv$ such that		$\vee \equiv$ or	
$\Longrightarrow \equiv$ implies		$\exists \equiv$ there exists	
$\Longleftrightarrow \equiv$ implies and is implied by		$\forall \equiv$ for all (for every)	

Example 1. Translate the following symbolic statements into English.

(a) $x \in \mathscr{R} \Longrightarrow x^2 \geq 0$.   (b) $x < 0 \Longrightarrow x^3 < 0$.

(c) $x^3 < 0 \wedge x \in \mathscr{R} \Longrightarrow x < 0$.   (d) $x = 0 \vee y = 0 \Longleftrightarrow xy = 0$.

Solution. (a) "x is a real number implies that x^2 is either zero or positive." This sentence may seem a little awkward at first. A smoother translation which has exactly the same meaning is: "If x is a real number, then x^2 is either zero or positive."

(b) "If x is less than zero, then x^3 is less than zero."

(c) "If x^3 is less than zero and x is a real number, then x is less than zero."

(d) "If x is zero or y is zero, then their product is zero. Conversely, if the product of x and y is zero, then either x is zero or y is zero."   ●

EXERCISE 1

1. Find the decimal expansions of each of the following rational numbers: (a) $16/15$, (b) $37/22$, (c) $20/10^5$, and (d) $-314/999$.
2. How can you tell when a rational number a/b has a terminating decimal expansion?
3. Which of the following statements are always true?
   (a) The product of two rational numbers is rational.
   (b) The product of two irrational numbers is irrational.
   (c) If x and y are rational numbers and $y \neq 0$, then x/y is rational.
   (d) A nonzero rational times an irrational is irrational.
*4. Find an irrational number between 0.213 and 0.214.
5. Which of the decimals 6.85983478 and 6.85982978 is larger?
6. Which of the decimals -14.03952667 and -14.040533589 is larger?

In Problems 7 through 13, a statement is given with symbols. Translate the statement into English.

7. $a = b \wedge c = d \implies ac = bd.$
8. $ac = bd \wedge a = b \wedge a \neq 0 \implies c = d.$
9. $a < c \wedge b < d \implies a + b < c + d.$
10. $a < c \wedge b < d \nRightarrow a + c < b + d.$
11. $n \geq 2 \implies \exists$ prime $p \ni n < p < 2n.$
12. $a \in \mathscr{R} \wedge b \in \mathscr{R} \implies a + b \in \mathscr{R} \wedge ab \in \mathscr{R}.$
13. $a + c < b + c \iff a < b.$

CALCULATOR PROBLEMS

C1. With the aid of a calculator, find each of the following numbers rounded to four decimal places.

    (a) $\dfrac{13}{71}.$ (b) $\dfrac{16.23598}{11.06045}.$ (c) $\sqrt{5}.$ (d) $\pi^2 + \dfrac{0.3336}{6.8394}.$

C2. For your own calculator, find the smallest (positive) nonzero number. What happens when you divide this number by 2? Is there a smallest (positive) real number?

C3. With the aid of a calculator, find each of the following rounded to three significant digits.

    (a) $(2.102 \times 10^6) \cdot (3.581 \times 10^4).$ (b) $\dfrac{6.238 \times 10^{-2}}{8.973 \times 10^3}.$

    (c) $\dfrac{(4.837 \times 10^2) - (2.134 \times 10^3)}{9.637}.$ (d) $(25.23 + 30.025)^4.$

C4. Computations that involve differences of nearly equal numbers will often result in a loss of significant digits, and hence poor approximations. Sometimes this difficulty can be avoided by rewriting the expression. For example, consider the computation of $(\sqrt{x+1}-1)/x$ for small values of x. By rationalizing the numerator,

$$\frac{\sqrt{x+1}-1}{x} = \frac{\sqrt{x+1}-1}{x} \cdot \frac{\sqrt{x+1}+1}{\sqrt{x+1}+1} = \frac{x}{x(\sqrt{x+1}+1)} = \frac{1}{\sqrt{x+1}+1};$$

that is,

$$\frac{\sqrt{x+1}-1}{x} = \frac{1}{\sqrt{x+1}+1} \qquad \text{for } x \neq 0.$$

(a) Compute $(\sqrt{x+1}-1)/x$ for $x = 10^{-1}, 10^{-2}, 10^{-3}, \ldots, 10^{-10}$ and arrange the results in a table.

(b) Compute $1/(\sqrt{x+1}+1)$ for $x = 10^{-1}, 10^{-2}, 10^{-3}, \ldots, 10^{-10}$ and arrange the results in a table.

(c) Compare the tables in (a) and (b) and explain the discrepancies. Which formula gives the better approximations?

C5. Calculate the value of $\sqrt{10-3(10/3)}$ in the manner displayed (start with $10 \div 3$). Did you obtain the answer 0? Explain.

Inequalities 4

Many of the proofs and problems in calculus require a knowledge of inequalities. Here we give a brief review of this topic.

DEFINITION 1. If a and b are real numbers, we write

(4) $\qquad\qquad\qquad a < b \qquad$ (read "a is less than b")

if and only if the difference $b - a$ is positive. Under these conditions we also write

(5) $\qquad\qquad\qquad b > a \qquad$ (read "b is greater than a").

The relations (4) and (5) are called *inequalities*. By definition, the two inequalities (4) and (5) are equivalent. It also follows from this definition that $0 < b$ if and only if b is a positive number. As trivial examples, we have

$$86 < 99, \qquad \text{because } 99 - 86 = 13, \text{ a positive number,}$$
$$-1000 < 1, \qquad \text{because } 1 - (-1000) = 1001, \text{ a positive number,}$$

$$\frac{55}{89} > \frac{21}{34}, \qquad \text{because } \frac{55}{89} - \frac{21}{34} = \frac{1}{3026}, \text{ a positive number,}$$

$$2.0376 > 2.0369, \qquad \text{because } 2.0376 - 2.0369 = 0.0007, \text{ a positive number.}$$

THEOREM 1. If a and b are any two real numbers, then exactly one of the three relations

$$\text{(A) } a < b, \qquad \text{(B) } a = b, \qquad \text{(C) } b < a$$

holds.

Proof. For any two numbers a and b we have either $b - a$ is positive, or $b - a = 0$, or $b - a$ is negative, and these three cases are mutually exclusive. In the first case $a < b$, in the second case $a = b$, and in the third case if $b - a$ is negative, then $-(b - a) = a - b$ is positive, and hence $b < a$. ∎

THEOREM 2. If $a < b$ and c is any positive number, then $ca < cb$.

In other words, *a true inequality remains true when multiplied on both sides by the same positive number.*

Proof. Since $a < b$, then $b - a$ is a positive number. Since the product of two positive numbers is again positive, we have that

$$c(b - a) = cb - ca$$

is positive. By Definition 1, this gives $ca < cb$. ∎

THEOREM 3 (PLE). If $a < b$ and c is any negative number, then $ca > cb$.

In other words, *when an inequality is multiplied on both sides by the same negative number, the inequality sign is reversed.*

THEOREM 4. If $a < b$ and $b < c$, then $a < c$.

Proof. By hypothesis, $b - a$ and $c - b$ are positive numbers. Thus the sum $(b - a) + (c - b)$ is a positive number. But this sum is $c - a$. Since $c - a$ is positive, we have by Definition 1 that $a < c$. ∎

By a similar type of argument the student can prove

THEOREM 5 (PLE). If $a < c$ and $b < d$, then $a + b < c + d$.

In other words, *two inequalities can be added termwise to give a true inequality.* Of course, the inequality sign must be in the same direction in all three inequalities.

THEOREM 6. If $a < b$ and c is any number, then $a + c < b + c$.

Proof. Clearly, $(b + c) - (a + c) = b - a$ is positive. ∎

Thus *a true inequality remains true when the same number is added to both sides.* Note that c can be a negative number, so that this theorem includes subtraction.

THEOREM 7. If $0 < a < b$, then

(6)
$$\frac{1}{a} > \frac{1}{b} > 0.$$

Thus *reciprocation reverses the inequality sign, when both members are positive.*

Proof. Multiply both sides of $a < b$ by the positive number $1/ab$ and use Theorem 2. ∎

THEOREM 8 (PLE). If $0 < a < b$ and $0 < c < d$, then

$$ac < bd.$$

Thus *multiplication of the corresponding terms of an inequality preserves the inequality.* Of course, all terms should be positive, and the inequality sign must be in the same direction in all three inequalities.

THEOREM 9. If $0 < a < b$ and n is any positive integer, then

(7) $a^n < b^n$.

Proof. Apply Theorem 8, $n - 1$ times with $c = a$, and $d = b$. ∎

THEOREM 10. If $0 < a < b$ and n is any positive integer, then

$$\sqrt[n]{a} < \sqrt[n]{b},$$

where, if n is even, the symbol $\sqrt[n]{}$ means the positive nth root.

Proof. The proof is a little complicated because it uses the method of contradiction. By Theorem 1 there are only three possibilities:

   (A) $\sqrt[n]{a} < \sqrt[n]{b}$,   (B) $\sqrt[n]{a} = \sqrt[n]{b}$,   (C) $\sqrt[n]{b} < \sqrt[n]{a}$.

We prove that the first case must hold by showing that (B) and (C) are impossible. In each of the latter two cases we take the nth power of both sides. In case (B) we find obviously that $a = b$. But this is impossible, because by hypothesis $a < b$. In case (C) we apply Theorem 9 to $\sqrt[n]{b} < \sqrt[n]{a}$ and find that $b < a$. Again this is contrary to the hypothesis that $a < b$. Since each of the cases (B) and (C) leads to a contradiction, the only case that can occur is (A). ∎

DEFINITION 2. If a and b are real numbers, we write

$$a \leqq b \qquad \text{(read "a is less than or equal to b")}$$

if and only if $a < b$ or $a = b$. Under these conditions we also write $b \geqq a$.

Thus $14 \leqq 14$, $-2 \leqq 0.38$, and $0.38 \geqq -2$.

In most of the theorems we can allow the equality sign to occur in the hypotheses as long as we make suitable modifications in the conclusions. In this way we obtain a large number of new theorems that vary only slightly from the ones already proved. There is no need to list them all because they are really obvious. Merely as an illustration of the type of theorem to be expected, we give the following variations on Theorems 6 and 8.

THEOREM 6★ (PLE). If $a \leqq b$ and c is any number, then $a + c \leqq b + c$.

THEOREM 8★ (PLE). If $0 < a \leqq b$ and $0 < c < d$, then $ac < bd$.

In establishing that a certain inequality is valid, it is sometimes not clear where to begin. One strategy is to start with the given inequality and see if we can deduce one that we *know* to be true. This procedure of working backwards is called the *analysis* of the problem. The official justification of the given inequality then amounts to showing that the steps in the analysis can be reversed.

Example 1. Without using a calculator, prove that

(8) $$\sqrt{2} + \sqrt{6} < \sqrt{3} + \sqrt{5}.$$

Solution. We give the analysis. Squaring both sides of (8) yields

(9) $$2 + 2\sqrt{2}\sqrt{6} + 6 < 3 + 2\sqrt{3}\sqrt{5} + 5,$$

or, on subtracting 8 from both sides and dividing by 2,

(10) $$\sqrt{12} < \sqrt{15}.$$

But since $12 < 15$, the inequality (10) is obviously true (Theorem 10). To prove the inequality (8), we start with the remark that $12 < 15$ and reverse the above steps.   ●

The theorems we have presented enable us to solve inequalities involving an unknown x in much the same manner as we do equations.

Example 2. Find all values of x that satisfy

(11) $$2x - 3 < 6x + 2.$$

Solution. We proceed to isolate x on one side of the inequality. Adding $-2x$ to both sides of (11), we get

(12) $$-3 < 4x + 2,$$

and adding -2 to both sides of (12) yields

(13) $$-5 < 4x.$$

Finally, multiplying this last inequality by 1/4 we arrive at the solution, that is, the set of all x such that

$$\frac{-5}{4} < x. \quad \bullet$$

Example 3. Solve the inequality

(14)
$$\frac{4x - 1}{x + 2} < 1.$$

Solution. We wish to simplify the problem by multiplying both sides by $x + 2$, but we do not know whether $x + 2$ is positive or negative. Thus we consider two cases:

(I) $x + 2 > 0$ (i.e., $x > -2$)   or   (II) $x + 2 < 0$ (i.e., $x < -2$).

For case (I), multiplication of (14) by $x + 2$ yields the inequality

$$4x - 1 < x + 2,$$

from which we obtain $3x < 3$ or $x < 1$. Hence this case gives the solution $x < 1$, *provided that* $x > -2$. For case (II), multiplication of (14) by $x + 2$ yields

$$4x - 1 > x + 2,$$

which gives $x > 1$. But here x must also satisfy $x < -2$, and since there is no number that is simultaneously greater than 1 and less than -2, case (II) gives no solution.

Consequently, the answer to the problem consists of all those numbers x that satisfy both of the inequalities

$$-2 < x \quad \text{and} \quad x < 1. \quad \bullet$$

The solution to Example 3 can be expressed in streamlined form by writing

$$-2 < x < 1.$$

More generally, $a < b < c$ means that $a < b$ *and* $b < c$.

Example 4. Solve the inequality

(15)
$$2x - 3 < 4x < -x + 1.$$

Solution. We need to find those numbers x that satisfy both of the inequalities

(16)
$$2x - 3 < 4x,$$

(17)
$$4x < -x + 1.$$

The first inequality (16) gives $-3 < 2x$ or $-3/2 < x$, and the second inequality (17) implies that $5x < 1$ or $x < 1/5$. Hence the solution set is

$$-\frac{3}{2} < x < \frac{1}{5}. \quad \bullet$$

EXERCISE 2

In Problems 1 through 4, determine which of the two given numbers is the larger without using a calculator.

1. $\sqrt{19} + \sqrt{21}, \ \sqrt{17} + \sqrt{23}$.

2. $\sqrt{11} - \sqrt{8}, \ \sqrt{17} - \sqrt{15}$.

3. $\sqrt{17} + 4\sqrt{5}, \ 5\sqrt{7}$.

4. $2\sqrt{2}, \ \sqrt[3]{23}$.

5. Which of the numbers -0.08, -0.083, -0.08351, and -0.08362 satisfy the inequality $-0.08360 < x < -0.08350$?

6. Prove that for any two numbers a, b,

$$2ab \leq a^2 + b^2$$

and that the equality sign occurs if and only if $a = b$. HINT: Use the fact that $(a - b)^2 \geq 0$.

In Problems 7 through 14, prove the given inequality under the assumption that all the quantities involved are positive. Determine the conditions under which the equality sign occurs.

7. $a + \dfrac{1}{a} \geq 2$.

8. $\dfrac{a}{5b} + \dfrac{5b}{4a} \geq 1$.

9. $\sqrt{\dfrac{c}{d}} + \sqrt{\dfrac{d}{c}} \geq 2$.

10. $(c + d)^2 \geq 4cd$.

11. $\dfrac{a + b}{2} \geq \sqrt{ab} \geq \dfrac{2ab}{a + b}$.

12. $(a + 5b)(a + 2b) \geq 9b(a + b)$.

13. $x^2 + 4y^2 \geq 4xy$.

14. $x^2 + y^2 + z^2 \geq xy + yz + zx$.

15. Prove Theorems 3, 5, and 8.

In Problems 16 through 45, find the set of values x for which the given inequality is true.

16. $5x - 3 < 7$.

17. $6x - 3 > 8x$.

18. $9x - 5 < 2x + 2$.

19. $2x + 1 > 5 - x$.

20. $1 + 3x \leq 8x + 8$.

21. $4x - 3 \geq 2x + 7$.

22. $1.5x - 10^9 \geq 10^9 + 0.5x$.

23. $6x - 6 \leq 6x + 1$.

24. $5x - 8 \geq 5x + 2$.

25. $-3 < x + 2 < 3$.

26. $-1 \leq x - 2 \leq 5$.

27. $1 \leq 2x + 1 \leq 4$.

28. $-83 < 19x - 37 < 0$.

29. $3x - 5 < x < 9x + 2$.

30. $x - 1 < 2x + 6 < 8x.$

31. $5 - x < 2x + 1 < 4x + 3.$

32. $1 + x < 3x + 6 < 4x + 8.$

33. $x - 5 < 3x + 1 < -x + 2.$

34. $x \leq 1 + 4x \leq 2x - 1.$

35. $\dfrac{1 + 2x}{x - 1} < 1.$

36. $\dfrac{x - 3}{x + 2} < 2.$

37. $\dfrac{4x + 3}{x - 2} > 10.$

38. $\dfrac{5 - 4x}{x + 3} \leq 1.$

39. $\dfrac{x - 6}{x + 6} < -1.$

40. $(x - 2)(x + 3) < 0.$

41. $(x + 8)(x - 1) > 0.$

42. $x^2 - x - 2 > 0.$

43. $x^2 + 2x \geq 0.$

44. $\sqrt{19 - x} > 2.$

45. $\sqrt{x - 3} < 6.$

CALCULATOR PROBLEMS

In Problems C1 through C6, use a calculator to determine which of the two given numbers is larger.

C1. $\dfrac{60}{51}, \dfrac{159}{135}.$

C2. $\pi^2, \dfrac{484}{49}.$

C3. $\sqrt{2} + \sqrt{5}, \sqrt{6 + (1.8)^2}.$

C4. $\dfrac{4}{3 + \pi}, \sqrt{\dfrac{15}{31}}.$

C5. $(1.023)^3, 2 - (0.99)^4.$

C6. $21^4 + 63, 19^4 + 32^3.$

In Problems C7 through C14, give the solution set rounded to the third decimal place.

C7. $16x - \dfrac{4}{17} > \dfrac{2}{3}x + \dfrac{15}{37}.$

C8. $\sqrt{47.325x} + 4 < (8.6)^2 - 3x.$

C9. $7x + \sqrt{93} < 2\pi x - \sqrt{0.651}.$

C10. $\dfrac{x}{8.314} + 1 < 2x < (6.231)^3.$

C11. $0.376x + (428)^{1/4} < x < 8 - \dfrac{x}{64.221}.$

C12. $\dfrac{9x - \pi}{17 - 6x} < \dfrac{1}{(2.364)^2}.$

C13. $\dfrac{2\pi x - 0.391}{x - \sqrt{5}} > 1.$

C14. $\sqrt{8 - x} \geq \cos 36°.$

C15. For what integer values of n is it true that $\left(\dfrac{1}{3}\right)^n < 10^{-5}$?

5 *Absolute Value*

It is convenient to have a special symbol $|x|$ (read "the absolute value of x" or "the magnitude of x") that denotes x if $x \geq 0$, and denotes $-x$ (which is positive) if $x < 0$.

DEFINITION 3 (Absolute Value). For each real number x, the absolute value of x is defined by

(18)
$$|x| \equiv \begin{cases} x, & \text{if } x \geq 0, \\ -x, & \text{if } x < 0. \end{cases}$$

For example, if $x = -7$, then it falls in the second case of (18). Consequently, we find that $|x| = |-7| = -(-7) = 7$. By the nature of the definition, we have

THEOREM 11 (PLE). For all x

(19)
$$-|x| \leq x \leq |x| \quad \text{and} \quad |x| \geq 0.$$

Further, $|x| = 0$ if and only if $x = 0$.

THEOREM 12 (PLE). If x and y are any pair of real numbers, then

(20)
$$|xy| = |x|\,|y|,$$

(21)
$$|x - y| = |y - x|,$$

(22)
$$\sqrt{x^2} = |x|,$$

and, if $y \neq 0$, then

(23)
$$\left| \frac{x}{y} \right| = \frac{|x|}{|y|}.$$

As examples of equation (22) we mention

$$\sqrt{(-3)^2} = |-3| = 3, \qquad \sqrt{(2.5)^2} = |2.5| = 2.5, \qquad \sqrt{1 - \cos^2 \theta} = \sqrt{\sin^2 \theta} = |\sin \theta|.$$

THEOREM 13. If r is any nonnegative number, then $|x| \leq r$ if and only if

(24)
$$-r \leq x \leq r.$$

Proof. Suppose first that $|x| \leq r$. If $x > r$ or $x < -r$, then $|x| > r$, which is false. Hence $-r \leq x \leq r$ if $|x| \leq r$. Now assume that $-r \leq x \leq r$. If $0 \leq x \leq r$, then $|x| = x \leq r$; if $-r \leq x < 0$, then $|x| = -x \leq r$. Thus $|x| \leq r$ if $-r \leq x \leq r$. ∎

An extremely useful property of absolute value is the *triangle inequality,* which asserts that the absolute value of the sum of two numbers is less than or equal to the sum of their absolute values. This result is included in the next theorem.

THEOREM 14 (Triangle Inequality). If x and y are any pair of real numbers, then

(25) $$|x + y| \leq |x| + |y|,$$

(26) $$|x| - |y| \leq |x - y|.$$

Proof. By Theorem 11 we have $-|x| \leq x \leq |x|$ and $-|y| \leq y \leq |y|$. If we add these two inequalities, we obtain

(27) $$-(|x| + |y|) \leq x + y \leq |x| + |y|.$$

Hence by Theorem 13 (with $r = |x| + |y|$) we get the desired inequality (25).

If we apply (25) to the identity $x = y + (x - y)$, we find that

$$|x| = |y + (x - y)| \leq |y| + |x - y|,$$

and subtracting $|y|$ from both sides, we obtain (26). ∎

We leave it for the reader to show that (26) is equivalent to the more complicated looking inequality

$$||x| - |y|| \leq |x - y|.$$

Example 1. Solve the equation $|2x - 1| = 3.2$.

Solution. The only numbers with absolute value 3.2 are ± 3.2. Hence we consider two cases:

$$2x - 1 = 3.2 \quad \text{or} \quad 2x - 1 = -3.2.$$

The solutions are $x = (1 + 3.2)/2 = 2.1$ and $x = (1 - 3.2)/2 = -1.1$. ●

Example 2. Solve the equation $|x^2 + 3x| = |x - 1|$.

Solution. Two numbers have the same absolute value if and only if one is equal to plus or minus the other. Hence there are two cases:

(28) $$x^2 + 3x = x - 1 \quad \text{or} \quad x^2 + 3x = -(x - 1).$$

This gives the two quadratic equations

(29) $x^2 + 2x + 1 = 0$ [using the first equation in (28)],

(30) $x^2 + 4x - 1 = 0$ [using the second equation in (28)].

Equation (29) has the single solution $x = -1$, and using the quadratic formula, we can solve (30):

$$x = \frac{-4 \pm \sqrt{16 + 4}}{2} = -2 \pm \sqrt{5}.$$

Thus $x = -1$, $-2 + \sqrt{5}$, or $-2 - \sqrt{5}$. ●

Example 3. Solve the inequality

(31) $$|2x - 1| \leqq 4.$$

Solution. By Theorem 13, inequality (31) is equivalent to

(32) $$-4 \leqq 2x - 1 \leqq 4.$$

Adding 1 to each part of (32) and then dividing each part by 2 gives the solution set

$$-\frac{3}{2} \leqq x \leqq \frac{5}{2}. \quad ●$$

Example 4. Solve the inequality

(33) $$|x - 1| > 3.$$

Solution. The only numbers whose absolute value exceeds 3 are those numbers greater than 3 or less than -3. Hence inequality (33) gives $x - 1 > 3$ or $x - 1 < -3$. The former inequality implies that $x > 4$, and the latter asserts that $x < -2$. Thus x satisfies (33) if and only if $x > 4$ *or* $x < -2$. ●

WARNING: Do not write nonsense such as $4 < x < -2$.

Example 5. Show that if $|x| < 1$, then $|3x - 5| < 8$.

Solution. By the triangle inequality (25), we have

$$|3x - 5| \leqq |3x| + |-5| = 3|x| + 5,$$

and since $|x| < 1$, it follows that $3|x| + 5 < 3(1) + 5 = 8$. Thus $|3x - 5| < 8$. ●

Example 6. Prove that if $|x - 5| < 1$ and $y \equiv x^2 - 3x + 11$, then y satisfies

(34) $9 < y < 35.$

Solution. If $|x - 5| < 1$, then $-1 < x - 5 < 1$, and consequently $4 < x < 6$. We then have

$$16 < x^2 \qquad\qquad < 36$$
$$-18 < \quad -3x \qquad < -12$$

Adding, $\dfrac{11 = \qquad\qquad 11 = 11}{9 < x^2 - 3x + 11 < 35}$

which is the same as inequality (34).   ●

EXERCISE 3

In Problems 1 through 21, solve for x.

1. $|x - 1| = 4.$
2. $|4x - 6| = 3.$
3. $|9x - 6| = 0.$
4. $|3 - 2x| = 4.$
5. $|7 + 3x| = 1.$
6. $|x + 2| = |2 - 4x|.$
7. $|5x - 1| = |3x - 2|.$
8. $|x - 1| = |1 - x|.$
9. $|x^2 + 2x| = |x|.$
10. $|1 - x^2| = |3x + 1|.$
11. $|x^2 + 2| = |1 - 2x|.$
12. $\left|\dfrac{x - 2}{x - 4}\right| = 2.$
13. $\left|\dfrac{2 + x}{3x - 4}\right| = 1.$
14. $|3x - 1| < 5.$
15. $|x - 8| < 1.$
16. $8 \geq |3x - 2|.$
17. $|7x - 4| < 3.$
18. $|3 - x| \geq 6.$
19. $|5 - 3x| \geq 2.$
20. $|6x - 1| \leq 0.$
21. $|4x - 3| \geq 1.$

22. Show that if $|x| < 6$, then $\left|1 - \dfrac{x}{2}\right| < 4.$

23. Show that if $|x - 1| < 2$, then $|2x - 3| < 5.$
24. Show that if $|x| < 2$, then $|x^3 + x| < 10.$
25. Show that if $|x| < 1$, then $|x^2 - 2x + 2| < 5.$
26. Show that if $|x| < 2$, then $|x^4 - x^3 + x - 16| < 42.$
27. Prove that if $|x - 3| < 2$ and $y \equiv x^3 - 2x^2 + 3x - 4$, then $-50 < y < 134.$ (Using calculus, we will be able to obtain the sharper inequality $-2 < y < 86.$)
28. Prove that if $|x - 5| < 1$ and $y \equiv x/(x + 10)$, then $1/4 < y < 3/7.$ (With a little effort, the sharper inequality $2/7 < y < 3/8$ can be obtained.)

29. Prove that if x, y, and z are any real numbers, then
   (a) $|xyz| = |x| \cdot |y| \cdot |z|$.
   (b) $|x + y + z| \le |x| + |y| + |z|$.
   (c) $|x - y| \le |x - z| + |y - z|$.
30. Prove that if $|x| < |y|$, then $-|y| < x < |y|$.
31. Let $r > 0$. Prove that $|x - a| < r$ if and only if $a - r < x < a + r$.

CALCULATOR PROBLEMS

In Problems C1 through C9, solve for x by using a calculator. Give the solution set rounded to the third decimal place.

C1. $|3.264 - (2.119)x| = \sqrt{5}$.   **C2.** $\left| \dfrac{x}{16} - \pi \right| = \dfrac{97}{376}$.   **C3.** $|\pi x - 1| = \left(\dfrac{47}{39} \right)^2$.

C4. $|6x - \pi^{-1}| = |x|$.   **C5.** $|(0.863)x - 1| = |\sqrt{2} - x|$.

C6. $|(2.03)x - \sqrt{2}| < \sqrt{5}$ **C7.** $\left| \dfrac{x}{2.685} - 3 \right| < \sqrt{6}$.

C8. $|6x - 8| \ge \dfrac{\sqrt{\pi}}{2}$.   **C9.** $|4.301 - x| \ge \sqrt{14}$.

Radian Measure   6

The selection of a unit for measuring any quantity is quite arbitrary. The reader may reflect on the source of the foot, inch, and centimeter as units for measuring length. In a rational approach, we try to select a unit that will be convenient or at least offer some advantages.

Use of the degree as a unit for measuring angles is an inheritance from ancient times. This unit is defined by assigning a measure of 360 degrees (360°) to the angle associated with one revolution (see Fig. 1). For most purposes the degree is quite satisfactory, but for the calculus it is much more convenient to use the *radian* as a unit for measuring angles. The definition of a radian uses the length of the arc of a circle, and the concept of the length of an arc is not treated carefully until Chapter 7. But we may proceed on an intuitive basis to formulate the definition.

Let $A_1 B_1$ and $A_2 B_2$ be arcs of concentric circles subtending the same central angle, as indicated in Fig. 2. If s_1 and s_2 are the lengths of these arcs and r_1 and r_2 are the radii of the corresponding circles, then from the similarity of the two sectors

(35)
$$\frac{s_1}{r_1} = \frac{s_2}{r_2}.$$

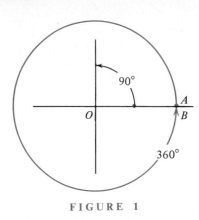

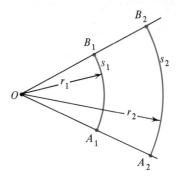

FIGURE 1   FIGURE 2

It follows from (35) that for a given angle, the ratio s/r does not depend on the radius of the circle. Hence the ratio is a suitable measure for the angle, and it is this number that is taken as the radian measure of the angle. We note that since s/r is the ratio of two lengths, it is dimensionless.

It is advantageous to give a direction to our angles, indicated by attaching a sign to the radian measure. In mathematics a counterclockwise direction of rotation is taken as positive, and the opposite direction of rotation (clockwise) is negative.

DEFINITION 4 (Radian Measure). The radian measure of an angle AOB is given by

(36) $$\theta = \frac{s}{r},$$

where $|OA| = |OB| = r$ and s is the directed distance traveled by the point A as the side OA is rotated about O into the side OB. Here $s > 0$ if the rotation is counter-clockwise, and $s < 0$ if the rotation is clockwise.

By the definition of the number π, the circumference of a circle of radius 1 is 2π. Hence by equation (36) the radian measure of the revolution is 2π. Thus $360°$ and 2π radians are measures for the same angle. This gives the formula

(37) $$180° = \pi \text{ radians,}$$

from which it follows that

$$1° = \frac{\pi}{180} \text{ radians} \quad \text{and} \quad 1 \text{ radian} = \frac{180}{\pi} \text{ degrees.}$$

Example 1. Express 200° in radians.

Solution. $200° = 200 \times 1° = 200 \times \pi/180 = 10\pi/9$ radians. Therefore, an angle of 200° is equal to an angle of $10\pi/9$ radians. ●

Example 2. Express $5\pi/6$ radians in degrees.

Solution. $5\pi/6$ radians $= 5\pi/6 \times 1$ radian $= 5\pi/6 \times 180°/\pi = 150°$. Therefore, an angle of $5\pi/6$ radians is the same as an angle of 150°. ●

To facilitate the change from degrees to radians, we will use both units of measure interchangeably at first. When we are concerned with the differentiation and integration of the trigonometric functions (Chapter 8), we must use radian measure if the formulas are to be correct.

EXERCISE 4

1. Find an approximate value for the radian measure of an angle of 1°.
2. Find an approximate value for the degree measure of an angle of 1 radian.
3. Find the radian measure for each of the following angles: **(a)** 60°, **(b)** 240°, **(c)** 720°, **(d)** −135°, **(e)** 12°, **(f)** 132°, **(g)** 36°, **(h)** −150°, and **(i)** 85°.
4. Find the degree measure for each of the following angles: **(a)** $\pi/4$, **(b)** $-9\pi/4$, **(c)** $5\pi/2$, **(d)** $11\pi/120$, **(e)** $-2\pi/9$, **(f)** $19\pi/36$, **(g)** 7π, and **(h)** -8π.
5. Find each of the following: **(a)** $\sin(\pi/6)$, **(b)** $\cos(-2\pi/3)$, **(c)** $\tan(3\pi/4)$, **(d)** $\sec 3\pi$, **(e)** $\cot(-\pi/2)$, **(f)** $\csc 0$, **(g)** $\sin(7\pi/2)$, and **(h)** $\cos 7\pi$.

CALCULATOR PROBLEMS
C1. Find each of the following rounded to four decimal places: **(a)** $\sin 32°$, **(b)** $\cos(\pi/7)$, **(c)** $\tan 2$, **(d)** $\cos(-8.2397°)$, **(e)** $\sec(73.2)$, and **(f)** $\cot(-1.02385)$.
C2. A student who had developed an unhealthy addiction to calculators was asked by the instructor for the numerical value of $\sin \pi$. The student's answer was 0.0548 .... What went wrong?

2
Introduction to Analytic Geometry

1 Objective

Before the sixteenth century, algebra and geometry were regarded as separate subjects. It was René Descartes (1596–1650) who first noticed that these two subjects could be united, and that each subject could contribute to the development of the other. This union, which we now call analytic geometry, has been fruitful far beyond the wildest dreams of Descartes. Our objective in this chapter is to see just how algebra and geometry are brought together. The first step in this process is the interpretation of real numbers as points on a line.

2 Coordinates on a Line

Everyone is familiar with coordinates on a line in an intuitive way. A ruler, or a scale on a thermometer are common examples. We now make this concept precise. Let $\mathcal{L}$ be a line (see Fig. 1) and on this line we select a unit for measurement, an initial point marked O and called the *origin,* and a positive direction indicated by an arrow. Such a line will henceforth be called a *directed line.*

Let us recall the standard method for assigning numbers to all the points on this line. The origin is assigned the number zero, and it divides $\mathcal{L}$ into two portions; the portion in the direction of the arrow is called the positive side, while the portion in the direction opposed to the arrow is the negative side. If a point P lies on the positive portion, we assign

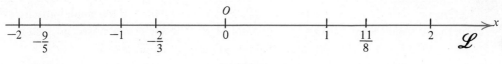

FIGURE 1

it the positive number d, where d is the distance of P from the origin. If a point P lies on the negative side of $\mathscr{L}$, it is assigned the negative number $-d$, where again d is the distance from P to O. The assigned real number x is called the *coordinate* of the point P.

In the manner just described, every point on the directed line $\mathscr{L}$ is assigned a unique real number (its coordinate); conversely, for each real number there corresponds a unique point on the line $\mathscr{L}$.

When we wish to distinguish between numbers and their corresponding points, we will use lowercase letters a, b, c, . . . for the numbers, and capitals A, B, C, . . . for the corresponding points. Frequently, there is no need to make a distinction, and we will often say "the point b" instead of "the point B that corresponds to the number b."

Inequalities are particularly easy to visualize on the number line. For example, the inequality $a < b$ means that the point a lies to the left of the point b on the line $\mathscr{L}$ of Fig. 1. Also, a number x will satisfy $a < x < b$ if and only if it lies strictly between the points a and b.

Example 1. Draw a picture showing the sets of points for which

    (a) $-3 < x < -1$.   **(b)** $1 \leqq x \leqq 2$.   **(c)** $4 < x$.

Solution. These sets, $\mathscr{A}$, $\mathscr{B}$, and $\mathscr{C}$, are shown in color in Fig. 2. ●

FIGURE 2

Sets like $\mathscr{A}$, $\mathscr{B}$, and $\mathscr{C}$ occur so frequently in mathematics that it is worthwhile to have special names and symbols for them.

DEFINITION 1 (Intervals). The set of points x for which $a \leqq x \leqq b$ is called a *closed* interval and is denoted by $[a, b]$. The set of points x for which $a < x < b$ is called an *open* interval and is denoted by (a, b). The points a and b are called the *end points* of the interval. The points in (a, b) are called *interior points* of the interval.

Referring to Fig. 2, we see that the set $\mathscr{A}$ is an open interval, and the small half-circles indicate that the end points -3 and -1 are not in $\mathscr{A}$. The set $\mathscr{B}$ is a closed interval and the solid dots indicate that $\mathscr{B}$ contains its end points $x = 1$ and $x = 2$.

An interval may be *half open* (or *half closed*). Thus the set $a \leq x < b$ is indicated by writing $[a, b)$, and the set $a < x \leq b$ is indicated by writing $(a, b]$. Any one of these four types of sets is called an *interval* and may be denoted by $\mathscr{I}$.

The set $\mathscr{C}$ of Fig. 2 is called a *ray*. Following the conventions for an interval, this ray is indicated by writing $\mathscr{C} = (4, \infty)$. Any set of the form (a, ∞), $(-\infty, a)$, $[a, \infty)$, or $(-\infty, a]$ is called a *ray*. The first two types are *open rays* because they do not contain the end point $x = a$. The last two types are *closed rays* because they contain $x = a$.

Intervals and rays are often quite helpful in finding and visualizing the solutions of conditional inequalities.

Example 2. Solve the conditional inequality

(1)
$$x^2 - x > 30.$$

Solution. The inequality (1) is equivalent to

$$x^2 - x - 30 > 0,$$

and factoring the quadratic, we obtain

(2)
$$(x + 5)(x - 6) > 0.$$

Now the left side of (2) is zero if $x = -5$ or $x = 6$. These two points divide the line naturally into the three sets indicated in Fig. 3. To solve (2), we merely examine the

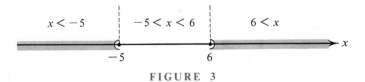

FIGURE 3

sign of each factor on these three sets. The results are arranged systematically in Table 1, with the last column giving the sign of the product $(x + 5)(x - 6)$.

TABLE 1

Value of x	Sign of the Factor		Sign of the Product
	$x + 5$	$x - 6$	
$x < -5$	$-$	$-$	$+$
$-5 < x < 6$	$+$	$-$	$-$
$6 < x$	$+$	$+$	$+$

This table shows clearly that (2) is true if and only if x is in the open ray $(-\infty, -5)$ or in the open ray $(6, \infty)$. Thus (1) is true if and only if x is in $(-\infty, -5) \cup (6, \infty)$. This set is shown in color in Fig. 3. ●

From our construction of the number line, we know that if a point P has the coordinate x, then the distance from P to the origin is given by $|x|$, the *absolute value* of x. More generally, if P_1 and P_2 are two points with coordinates x_1 and x_2, respectively, then it is easy to see that the distance between P_1 and P_2 is given by $|x_2 - x_1|$. This formula is true regardless of the ordering of x_1 and x_2 (see Fig. 4).

FIGURE 4

Using the symbol $|P_1P_2|$ to denote the distance between P_1 and P_2, we therefore have

(3) $$|P_1P_2| = |x_2 - x_1|.$$

For example, if P_1 and P_2 have the respective coordinates $x_1 = 4$ and $x_2 = -6$, then by (3),

$$|P_1P_2| = |-6 - 4| = |-10| = 10.$$

Example 3. Give a geometric interpretation of the inequality

(4) $$|x - 1| \leqq 3$$

and find the solution set.

Solution. The left-hand side of (4) is the distance between x and 1. Hence $|x - 1| \leqq 3$ means that the distance between x and 1 cannot exceed 3. We can find all such numbers x by starting at the point 1 and proceeding 3 units in either direction. Referring to Fig. 5, we see that the solution set of (4) is the closed interval $[-2, 4]$. ●

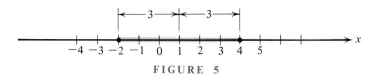

FIGURE 5

We can obtain the same result algebraically. If $|x - 1| \leqq 3$, then $-3 \leqq x - 1 \leqq 3$. Adding one to both ends and the middle, we have $-2 \leqq x \leqq 4$.

Calculus deals with changes in variable quantities, and it is therefore convenient to have at hand a special symbol to represent change. Suppose that a particle P moves along a line, and let x denote its varying coordinate. If P starts at the point $x_1 = 3$ and stops at the point $x_2 = 8$ (see Fig. 6), what is the change in x during motion?

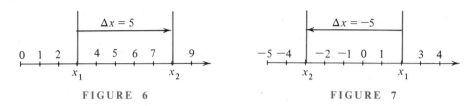

FIGURE 6   FIGURE 7

Obviously, the change in x is 5, and this can be computed from x_1 and x_2 by taking the difference $x_2 - x_1 = 8 - 3 = 5$. The symbol universally used to denote this change is Δx (read "delta x") and by definition[1] if x changes from x_1 to x_2, then

(5)
$$\Delta x \equiv x_2 - x_1.$$

If the particle is moving in the negative direction on the line, then the x-coordinate is decreasing and we would expect that the change Δx will be negative. This is indeed the case because in this case x_2 will be less than x_1 in equation (5). For example, suppose that the moving particle starts at $x_1 = 2$ and stops at $x_2 = -3$, as indicated in Fig. 7. Then by definition

$$\Delta x = x_2 - x_1 = -3 - 2 = -5$$

and Δx is negative, just as we expected.

To summarize, we see that Δx is positive if the direction from x_1 to x_2 coincides with that of the given line, and Δx is negative if the direction from x_1 to x_2 is opposite to that of the given line. We also note that the absolute value of Δx gives the distance between the points x_1 and x_2, and so Δx gives the *directed distance* from x_1 to x_2.

EXERCISE 1

1. Draw a picture showing each of the following sets.

   (a) $-2 \leqq x < 3$.   (b) $x < 3$.   (c) $\dfrac{3}{2} \geqq x > \dfrac{1}{2}$.

   (d) $|x| \leqq 5$.   (e) $|x + 1| < 4$.   (f) $|x - 2| \geqq 1$.

[1] The symbol Δ (delta) is the Greek d and is selected because it suggests the word *difference*. Indeed, as equation (5) indicates, Δx, the change in x, is just the difference between the initial value of x and the final value of x. In words, Δx is the final value of x minus the initial value of x. Although Δx is composed of two distinct symbols Δ and x, the compound symbol represents a single number (e.g., Δx represents 5 in Fig. 6).

2. Suppose that x is in the interval $[-1, 3]$. Prove that
   (a) $4 + x$ is in the interval $[3, 7]$.   (b) $4 - x$ is in the interval $[1, 5]$.

In Problems 3 through 11, find the set of values of x for which the given inequality is true.

3. $x(x - 1) > 0$.

4. $(x - 8)(x + 1) < 0$.

5. $(2x - 1)(x + 3) \leq 0$.

6. $2x - 19 \leq 11(x - 2)$.

7. $x^2 - 8x + 24 \leq 9$.

8. $4x^2 - 13x + 4 < 1$.

9. $x^3 - 16x \geq 0$.

10. $x^3 + 3x^2 \geq 13x + 15$.

11. $x^4 + 36 \geq 13x^2$.

12. **Midpoint.** Let P_1 and P_2 have coordinates x_1 and x_2. Prove that the point M with coordinate $(x_1 + x_2)/2$ is the midpoint of the line segment joining P_1 and P_2 by showing that $|P_1M| = |MP_2|$.

13. Find the midpoint for the line segment joining P_1 and P_2 with coordinates x_1 and x_2 if:
   (a) $x_1 = 4$, $x_2 = 6$.
   (b) $x_1 = -3$, $x_2 = 11$.
   (c) $x_1 = -9$, $x_2 = -3$.
   (d) $x_1 = 7/13$, $x_2 = 13/7$.
   (e) $x_1 = 4 - \sqrt{3}$, $x_2 = 11 + \sqrt{3}$.
   (f) $x_1 = 2 - \sqrt{18}$, $x_2 = 18 + \sqrt{2}$.

14. To find the points that divide an interval $[x_1, x_2]$ into n subintervals of equal length, first note that the common length l must be $(x_2 - x_1)/n$. Show that the desired points are then given by

$$x_1 + l, \quad x_1 + 2l, \quad x_1 + 3l, \ldots, \quad x_1 + (n - 1)l.$$

15. Using the result of Problem 14, find the two points that divide each of the following intervals $[x_1, x_2]$ into three equal parts.

   (a) $[4, 13]$.   (b) $[-7, 32]$.   (c) $[2/5, 5/2]$.   (d) $[-14 - 3\sqrt{2}, -5 + 6\sqrt{2}]$.

16. Using the result of Problem 14, find the points that divide the interval $[2/3, 33/2]$ into five equal parts.

17. Which one of the following points is farthest from the origin: $-3, \sqrt{7}, 4.32, \dfrac{-20}{3}$, and $-\sqrt{15}$? Which one is closest to the origin?

18. Find the distance between the points x_1 and x_2 in each of the following cases:

   (a) $x_1 = 10$, $x_2 = -10$.   (b) $x_1 = \dfrac{13}{21}$, $x_2 = \dfrac{8}{13}$.   (c) $x_1 = \sqrt[3]{2}$, $x_2 = \sqrt{3}$.

19. Give a geometric interpretation of the inequality $|x + 2| \leq 5$ and find the solution set.

20. Give a geometric interpretation of the equation $|x - 3| = |x + 1|$ and find x.

21. Give a geometric interpretation of the inequality $|x - 3| < |x - 4|$ and find the solution set.

In Problems 22 through 27, find the set of values x for which the given inequality is true.

22. $\dfrac{x - 3}{x + 1} < 0$.

23. $\dfrac{x - 2}{x - 4} \geq 0$.

24. $\dfrac{1}{x - 2} > 100$.

25. $\dfrac{1}{|x - 2|} > 100$.

26. $\dfrac{x(x - 1)}{x + 2} \leq 0$.

27. $\dfrac{x - 4}{(x - 3)(x + 1)} > 0$.

CALCULATOR PROBLEMS

C1. Find the distance (to three decimal places) between the points x_1 and x_2 in each of the following cases.

(a) $x_1 = \sqrt{4.263}$, $x_2 = \dfrac{\pi}{2}$.

(b) $x_1 = \dfrac{270}{143}$, $x_2 = \dfrac{13}{22}$.

(c) $x_1 = 0.2359$, $x_2 = (1.0318)^2$.

C2. Using the result of Problem 14, find the two points that divide the interval $[-60.012, 1.987]$ into three equal parts. Give your answers to three decimal places.

C3. Using the result of Problem 14, find the six points that divide the interval $[\sqrt{2}, 3.2145]$ into seven equal parts. Give your answers to four decimal places.

3 *The Rectangular Coordinate System*

The old familiar rectangular coordinate system is just two directed lines meeting at right angles (see Fig. 8). The point of intersection is called the *origin* and is usually lettered O. It is customary to make one of these lines horizontal and to take the direction to the right of O as the positive direction on this line. This horizontal line is called the *x-axis,* or the *horizontal axis.* The other directed line that is perpendicular to the x-axis is called the *y-axis,* or the *vertical axis,* and the positive direction on this axis is upward from O. These two axes divide the plane into four quadrants, which are labeled Q. I, Q. II, Q. III, and Q. IV for convenience, as indicated in Fig. 8.

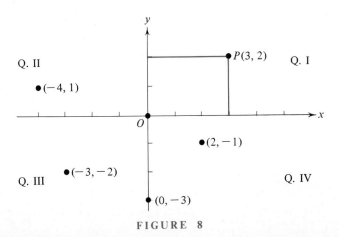

FIGURE 8

Once a rectangular coordinate system has been chosen, any point in the plane can be located with respect to it. To specify the location of a point P, we draw horizontal and vertical lines through P (see Fig. 8). The value of x where the vertical line intersects the

x-axis is called the *x-coordinate* of P, or the *abscissa* of P. The value of y at the intersection of the horizontal line with the y-axis is called the *y-coordinate* of P, or the *ordinate* of P. For example, in Fig. 8, the abscissa of P is 3 and the ordinate is 2. It is customary to enclose this pair of numbers in parentheses, thus: (x, y), or in our specific case $(3, 2)$, and these numbers are called the *coordinates* of the point P.[1]

Figure 8 shows a number of other points with their coordinates. The reader should check each point to see if its coordinates appear to be consistent with the position of the point in the figure.

Of course, this procedure can be reversed. Given the coordinates $A(5, -8)$, for example, the point A can be located by moving five units to the right of O on the x-axis and then proceeding downward eight units along a line parallel to the y-axis.

The discussion we have just given proves

THEOREM 1. With a given rectangular coordinate system each point P in the plane has a uniquely determined pair of coordinates (x, y), where x and y are real numbers. Conversely, for each pair (x, y) of real numbers there is exactly one point P which has this pair for its coordinates.

The rectangular coordinate system is frequently called the *Cartesian* coordinate system in honor of René Descartes. A brief but highly entertaining account of the life of this genius can be found in *Men of Mathematics* by E. T. Bell (New York: Simon and Schuster, 1937).

Given two points P_1 and P_2 in the plane, with respective coordinates (x_1, y_1) and (x_2, y_2), it is quite easy to determine the distance between the two points. Indeed, if we draw a line segment joining the two points and then make this segment the hypotenuse of a right triangle $P_1 Q P_2$ as indicated in Fig. 9 by drawing suitable lines parallel to the axes, the distance $|P_1 P_2|$ can be computed by the Pythagorean Theorem:

(6)
$$|P_1 P_2| = \sqrt{a^2 + b^2},$$

where a is the length of the line segment $P_1 Q$ and b is the length of the line segment $Q P_2$.

Now along the line $P_1 Q$, the height above the x-axis does not change, so that $a = |\Delta x|$. Similarly, along the line $Q P_2$ the x-coordinate does not change, so that $b = |\Delta y|$, the absolute value of the change in y. In the case pictured in Fig. 9, both Δx and Δy are positive, but in other cases they may not be, so that absolute value signs are necessary. However, on squaring, these absolute value signs may be dropped because for any number A, we have $(A)^2 = (-A)^2$. Substituting for a and b in equation (6) yields

---

[1] The reader will recall that when $a < b$ we use the same symbol (a, b) for an open interval. The appropriate meaning of (a, b) will therefore be dictated by the context.

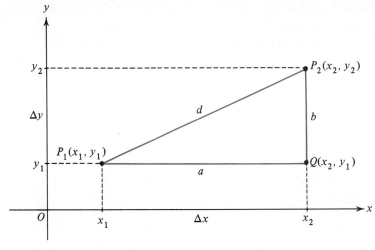

FIGURE 9

THEOREM 2. The distance between the points $P_1(x_1, y_1)$ and $P_2(x_2, y_2)$ is given by

(7)
$$d = \sqrt{(\Delta x)^2 + (\Delta y)^2} = \sqrt{(x_2 - x_1)^2 + (y_2 - y_1)^2}.$$

In formula (7) we assume that the unit length on the coordinate axes are the same. We also remark that since distance is a nonnegative quantity, the positive square root is to be taken in (7). Further, given two points in the plane, the distance between them does not depend on the letter assigned, so either could be called P_1 and the other one P_2. Thus, either logically or by inspection of equation (7), the quantity d is unchanged if the subscripts 1 and 2 are interchanged.

Formula (7) is easy to memorize because it is just a disguised form of the Pythagorean Theorem.

Example 1. Find the distance between $(-9, 3)$ and $(3, -2)$.

Solution. We select $(-9, 3)$ as P_1 and then $(3, -2)$ is P_2. Therefore,

$$\Delta x = x_2 - x_1 = 3 - (-9) = 12, \qquad \Delta y = y_2 - y_1 = -2 - 3 = -5,$$

and so, by equation (7),

(8)
$$d = \sqrt{12^2 + (-5)^2} = \sqrt{144 + 25} = \sqrt{169} = 13. \quad \bullet$$

Example 2. Let $\mathscr{L}$ be the set of all points P for which $|PA| = |PB|$, where A is the fixed point $(4, 1)$ and B is the fixed point $(2, 3)$. Find a simple equation that the coordinates (x, y) of P must satisfy.

Solution. By equation (7), we have for the distances

(9) $|PA| = \sqrt{(x - 4)^2 + (y - 1)^2}, \qquad |PB| = \sqrt{(x - 2)^2 + (y - 3)^2}.$

By the conditions of the problem, $|PA| = |PB|$. Squaring both sides in (9), we have

$$x^2 - 8x + 16 + y^2 - 2y + 1 = x^2 - 4x + 4 + y^2 - 6y + 9.$$

Dropping the terms x^2 and y^2 which appear on both sides, and collecting like terms, gives

(10) $4y = 4x - 4,$

or

(11) $y = x - 1.$ ●

Conversely, it can be shown that if $y = x - 1$, then the point $P(x, y)$ is equidistant from $A(4, 1)$ and $B(2, 3)$ because the steps that lead from (9) to (11) can be reversed to show that $|PA| = |PB|$.

We shall see shortly that an equation like (11), in which x and y appear only to the first degree, always describes a straight line, and this is naturally what we expect to obtain for the perpendicular bisector of the line segment AB.

EXERCISE 2

1. Using coordinate paper, plot the following points: $(3, 4)$, $(5, 4\sqrt{6})$, $(-5, 12)$, $(-4, -3)$, $(\sqrt{5}, -2)$, $(0, -7)$, and $(\sqrt{2}, \sqrt{3})$. Find the distance of each point from the origin.

2. In each of the following, compute the distance $|AB|$ using formula (7). Then make a careful drawing to scale, and check your answer by measuring the distance with a ruler.

   (a) $A(-9, -1)$, $B(3, 4)$.   (b) $A(-2, 9)$, $B(1, 5)$.
   (c) $A(7, 1)$, $B(3, 3)$.   (d) $A(7, 11)$, $B(-9, -5)$.

3. What figure is formed by the set of all points that have: **(a)** y-coordinate equal to 6, and **(b)** x-coordinate equal to -3?

4. Show that the triangle with vertices at $P(-1, 4)$, $Q(-4, 1)$, and $R(-5, 5)$ is isosceles.

In Problems 5 through 10, use the distance formula to determine whether the points P, Q, and R are the vertices of a right triangle. For this purpose recall the converse of the Pythagorean

Theorem: If the square of the length of one side of a triangle is equal to the sum of the squares of the lengths of the other sides, then the triangle is a right triangle.

5. $P(1, 1)$, $Q(3, 2)$, $R(0, 8)$.   6. $P(-1, 4)$, $Q(2, 2)$, $R(6, 8)$.
7. $P(-4, 1)$, $Q(-1, -1)$, $R(1, 4)$.   8. $P(2, -6)$, $Q(-5, 1)$, $R(-1, 5)$.
9. $P(5, 4)$, $Q(-3, -5)$, $R(-8, -1)$.   10. $P(-13, 9)$, $Q(3, 2)$, $R(-3, -3)$.

In Problems 11 through 18, find a simple equation satisfied by the coordinates of a point $P(x, y)$ subject to the given conditions.

11. $P(x, y)$ is the same distance from $(-1, -3)$ and $(3, 5)$.
12. $P(x, y)$ is the same distance from $(-1, 2)$ and $(-5, 7)$.
13. $P(x, y)$ is 5 units from the point $(3, -4)$.
14. $P(x, y)$ is 9 units from the point $(-4, 5)$.
15. The distance from $P(x, y)$ to the x-axis is equal to the distance from $P(x, y)$ to $(1, 3)$.
16. The distance from $P(x, y)$ to the y-axis is equal to the distance from $P(x, y)$ to the point $(4, 1)$.
17. The distance from $P(x, y)$ to the point $(3, 2)$ is twice its distance from the point $(-4, 1)$.
18. The distance from $P(x, y)$ to the point $(0, 1)$ is half its distance from the point $(2, 0)$.

19. Using formula (7), prove that the lengths of the diagonals of any rectangle are equal. HINT: For convenience, choose the coordinate axes so that the origin is at one vertex, the x-axis is along one side, and the y-axis is along another side.
20. Prove that if each coordinate of the points P_1 and P_2 is doubled to form the coordinates of new points Q_1 and Q_2, then $|Q_1Q_2| = 2|P_1P_2|$.

CALCULATOR PROBLEMS

C1. On a map of Florida, a coordinate system was set up with its origin at Tampa. Orlando was located at coordinates $(104.6, 80.5)$ in kilometers and Jacksonville at coordinates $(56.2, 289.8)$. If an airplane flew a straight path from Orlando to Jacksonville averaging 324.8 km/hr, how long would the flight take?
C2. A hurricane was traced from a point 132 miles due east and 217 miles due south of the National Hurricane Center. This hurricane traveled along a fairly straight path for 7 hours. After that time it was located 104 miles due east and 171 miles due south of the Center. What distance had the hurricane traveled in those 7 hours? After 7 hours, how far was the hurricane from the Center?
C3. Ms. A. Sokol, a saleswoman who flies her own plane, must start from her home at A, visit three cities, B, C, and D, and return home. Using her home as the origin in a coordinate system, the other cities are at the points $B(3.1, 0.2)$, $C(1.2, 5.8)$, and $D(4.3, 3.2)$, where each unit represents 100 km. Find one route that is shortest, and the length of this route.

4 Graphs of Equations and Equations of Graphs

Let us consider all the pairs of numbers (x, y) that satisfy the equation

(12) $$y = \frac{1}{3}x - 1.$$

It is easy to find such pairs. We merely select a particular value for x, say $x = 15$, and using it in (12) we find that $y = \frac{1}{3} \times 15 - 1 = 4$. Thus the pair $(15, 4)$ satisfies equation (12). Corresponding to this pair there is a point in the plane with this pair $(15, 4)$ as coordinates. We can continue to find more pairs, and to mark out more points in the plane. It is desirable to introduce some system into the computations by arranging the work in a little table, selecting values for x for which the computation is easy, and finding the corresponding y. Such a table is given below, and the corresponding points are marked in Fig. 10.

x	-6	-3	0	3	6	9	12	15	18
$y = \dfrac{1}{3}x - 1$	-3	-2	-1	0	1	2	3	4	5

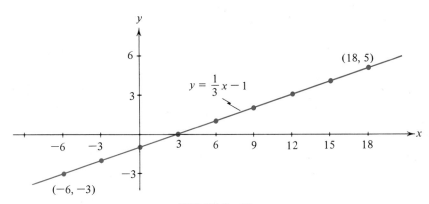

FIGURE 10

The points all seem to lie on a straight line, and we feel reasonably confident that if we continue to select values for x, compute y, and plot the corresponding points, these new points will also fall on the line determined by the points already plotted in Fig. 10. Of course, this must be proved, and we will give a proof in Section 6. But first our ideas must be made precise.

DEFINITION 2 (Graph). The *graph* of an equation involving x and y is the set of all points (x, y) whose coordinates satisfy that equation. The graph is also called a curve.[1]

It is obviously impossible to compute all such pairs and mark the corresponding points in the plane, since for most equations there will be infinitely many such points. In the case of equation (12), however, the matter is simplified because we suspect (and it will be proved later) that the graph is a straight line. So we can plot just two of the points and then draw the straight line that passes through those two points.

Example 1. Sketch the graph of the equation

(13)
$$y = \frac{x^2}{4} + 1.$$

Solution. In this case we have infinitely many points on the curve, so it is impossible to obtain the full graph. So we compute the coordinates of enough points to enable us to form a good guess as to the appearance of the graph, and then "sketch" the rest of the graph by assuming that the curve is nice and smooth. The table of values can be condensed by observing that x is squared in (13), so that both x and $-x$ will lead to the same value for y. For example, if $x = 1$ or $x = -1$, we find $y = 5/4$. The table is given below and the sketch is shown in Fig. 11. ●

x	0	± 1	± 2	± 3	± 4	± 5
y	1	5/4	2	13/4	5	29/4

Example 2. Sketch the graph of the equation $y^2 = x^2$.

Solution. Taking square roots of both sides of the equation, we find that either $y = x$ or $y = -x$. We leave it to the reader to make a table of values. The graph of $y = x$ appears to be a straight line through the origin that makes an angle of $45°$ with the

---

[1] Strictly speaking, the words *graph* and *curve* do not have the same meaning. The distinction is rather complicated, however, and the reader need not worry about it at present. The graph is also called the *locus* of an equation.

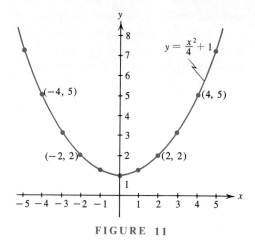

FIGURE 11

positive x-axis, and the graph of $y = -x$ appears to be a line perpendicular to this line at the origin. Then the graph of $y^2 = x^2$ is the pair of straight lines shown in Fig. 12. ●

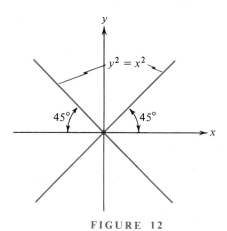

FIGURE 12

We have seen how each equation in x and y leads to a unique graph, the graph of the equation. One might expect that conversely each graph would lead to a uniquely determined equation. Unfortunately, this is not true.

Let us consider the straight line of Fig. 10. Certainly, equation (12), $y = (1/3)x - 1$, is an equation for this straight line, since the straight line was plotted from this equation. If we add 1 to both sides of this equation, we have

$$(14) \qquad\qquad y + 1 = \frac{1}{3}x.$$

But equation (14) is satisfied whenever equation (12) is, and equation (12) is satisfied whenever equation (14) is. Hence both equation (12) and (14) have the *same* graph. So if we are given the graph, its equation is *not unique*. Indeed, we can obtain a large variety of equations, each having the straight line of Fig. 10 for its graph. A few of these are $3y + 3 = x$, $3(y - 1) = x - 6$, and $27y^3 = x^3 - 9x^2 + 27x - 27$.

Because of the multiplicity of equations that can arise from a given graph, we cannot say "the" equation of a graph, but rather we must say "an" equation of a graph. However, we will always try to find among the set of all such equations a suitably attractive and simple one to call (erroneously) "the" equation of the graph.

DEFINITION 3 (Equation for a Graph). An equation is called an *equation for a graph* if it is satisfied by the coordinates of those, and only those, points that lie on the graph.

Example 3. Find an equation for the graph consisting of all points P such that $|PF| = |PD|$, where F is the point $(0, 2)$ and $|PD|$ denotes the distance from P to the x-axis.

Solution. Let (x, y) be the coordinates of a point on the graph. Then the distance $|PD|$ from the x-axis is $|y|$, and $|PF| = \sqrt{(x - 0)^2 + (y - 2)^2}$. Hence P is on the graph if and only if

(15) $$|y| = \sqrt{x^2 + y^2 - 4y + 4}.$$

This is an equation for the graph. To find a simpler one, we square both sides and obtain

$$y^2 = x^2 + y^2 - 4y + 4,$$
$$4y = x^2 + 4,$$

(16) $$y = \frac{x^2}{4} + 1. \quad \bullet$$

Equation (16) stands as the solution because it cannot be simplified further. Observe that all the steps in going from the graph to equation (16) can be reversed, so that a point that satisfies equation (16) must be on the graph. Note also that equation (16) is identical with equation (13), so that we already have a sketch of this graph in Fig. 11. This type of graph is called a *parabola,* .and we will study it in detail later.

EXERCISE 3

In Problems 1 through 6, the graph of the equation is a straight line. Find two points satisfying the given equation, draw the line, and check that other points satisfying the equation also seem to lie on the line.

1. $y = 2x + 5$.
2. $y = -3x + 7$.
3. $y = x - 4$.
4. $y = -2x - 5$.
5. $3y = x + 6$.
6. $4y = -x - 8$.

7. Sketch the graphs of the equations (all with the same set of coordinate axes):
   (a) $y = x^2$. (b) $y = x^2/4$. (c) $y = x^2/10$. (d) $y = -x^2$.
8. Sketch the graphs of the equations (all with the same set of coordinate axes):
   (a) $y = x^2$. (b) $y = x^3$. (c) $y = x^4$.
9. Sketch the graphs of the equations (on separate coordinate systems):
   (a) $y = \sqrt{x}$. (b) $y = -\sqrt{x}$. (c) $y^2 = x$.
10. Starting with the graph for $y = x^2/4 + 1$ in Fig. 11, explain how it can be shifted to obtain the graphs of the following equations:

   (a) $y = \dfrac{x^2}{4} + 2$. (b) $y = \dfrac{x^2}{4} - 2$. (c) $y = \dfrac{(x-1)^2}{4} + 1$.

In Problems 11 through 24, sketch the graph of the given equation.

11. $x^2 + y^2 = 25$.
12. $y^2 = 9x^2$.
13. $y = |x|$.
14. $y = -3$.
15. $x = 5$.
16. $xy = 1$.
17. $y = 2x^3$.
18. $4x = y^2$.
19. $y = 2\sqrt{x} - 2$.
20. $y = 2\sqrt{x} + 4$.
21. $y = |x - 3|$.
22. $|y| = x - 3$.
23. $3x^2 + 4y^2 = 0$.
24. $(y - x)(y - 3) = 0$.

25. Write an equation whose graph is (a) the x-axis, (b) the y-axis, and (c) all points on either the x-axis or the y-axis.

In Problems 26 through 30, the point $P(x, y)$ is subject to certain conditions. In each case find a simple equation for the graph consisting of all points P that satisfy the given condition.

26. P is equidistant from $A(2, 5)$ and $B(4, 3)$.
27. P is on a circle with center at the origin and radius 3.
28. P has an ordinate of 7.
29. P is on a line parallel to the y-axis and 3 units to the right of the y-axis.
30. P is 7 units from the point $(1, 2)$.

31. Determine the constants a and b so that the graph of $y = ax + b$ contains the points $(1, 5)$ and $(3, 11)$.
32. Do Problem 31 if the two points are
   (a) $(9, -2)$ and $(-3, 6)$. (b) $(-1, -6)$ and $(4, 5)$.

⋆**33.** Determine the constants a, b, and c so that the curve $y = ax^2 + bx + c$ passes through the points $(1, 0)$, $(2, 0)$, and $(3, 2)$.

34. Explain why the graph of the equation $y - 6x^8 - 3x^2 - 1 = 0$ lies above the x-axis.

CALCULATOR PROBLEM

C1. Sketch the graphs of the given equations.

    (a) $y = 4x^{3/2}$.   (b) $y = \dfrac{3}{x^2 + 1}$.   (c) $4y = x^4 - 4x^2$.

5   *The Slope of a Line*

Let $\mathscr{L}$ be a line in the plane and let α be the angle from the x-axis to the line, that is, the smallest angle through which the x-axis must rotate in a positive direction (counterclockwise) in order that it coincide with the given line (see Fig. 13). If the line is falling as we

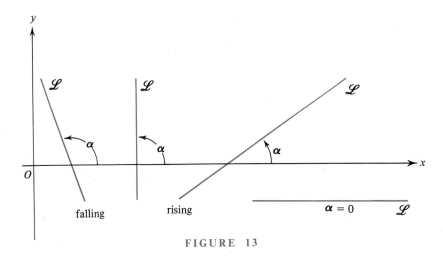

falling   rising $\alpha = 0$

FIGURE 13

progress from left to right, then clearly $90° < \alpha < 180°$ and $\tan \alpha$ is negative. If the line is rising, then $0 < \alpha < 90°$ and $\tan \alpha$ is positive. If the line is horizontal, then the line need not intersect the x-axis. In this case we put $\alpha = 0$. If the line is vertical, then $\alpha = 90°$ and $\tan \alpha$ is undefined. However, no harm is done if we use the symbol $\tan 90° = \infty$, as long as we keep in mind that ∞ is not a number and do not try to do algebraic manipulations with it. The key idea is that $\tan \alpha$ is a convenient measure for describing the behavior of the line: $\tan \alpha$ is positive for a rising line, negative for a falling line, zero for a horizontal line, and large values for $|\tan \alpha|$ indicate that the line is very steep.

DEFINITION 4 (Slope). The *slope* of a line is denoted by m and is defined by

(17) $$m \equiv \tan \alpha,$$

where α is the smallest positive angle from the x-axis to the line. If the line is horizontal, then $m = 0$, and if the line is vertical, then the slope is undefined. The angle α is called the *angle of inclination* of the line.

It is easy to compute the slope of a line if we know the coordinates of two points on the line. Indeed, let $P_1(x_1, y_1)$ and $P_2(x_2, y_2)$ be two distinct points on $\mathscr{L}$. We can select the subscripts so that P_2 lies to the right of P_1 and hence $x_2 > x_1$. We construct a right triangle P_1QP_2 (as shown in Figs. 14 and 15) by drawing lines through P_1 and P_2 parallel to the x-axis and y-axis, respectively.

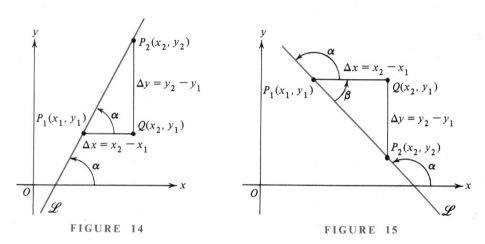

FIGURE 14   FIGURE 15

In the case of a rising line, as indicated in Fig. 14, it is obvious that the angle $\alpha = \angle QP_1P_2$ of the right triangle, so that $\tan \alpha = |QP_2|/|P_1Q| = \Delta y/\Delta x$, and hence

(18) $$m = \frac{\Delta y}{\Delta x} = \frac{y_2 - y_1}{x_2 - x_1}.$$

In the case of a falling line, formula (18) is still valid, but a little twist is needed in the proof to take care of the negative sign. Referring to Fig. 15, we see that α is now the *exterior* angle of the right triangle, so we first compute the tangent of its supplementary angle β. From Fig. 15, we have

$$\tan \beta = \frac{|P_2Q|}{|P_1Q|} = \frac{|\Delta y|}{\Delta x},$$

where we used $|\Delta y|$ because in this case $\Delta y < 0$. Since $\alpha + \beta = 180°$, we know from elementary trigonometry that $\tan \alpha = -\tan \beta$. Thus

$$\tan \alpha = \frac{-|\Delta y|}{\Delta x} = \frac{-(-\Delta y)}{\Delta x} = \frac{\Delta y}{\Delta x},$$

that is, equation (18) holds.

If the line is horizontal, $\Delta y = 0$ and $m = 0$. If the line is vertical, $x_1 = x_2$, so that $\Delta x = 0$. In this case formula (18) involves a division by zero, and this is consistent with our agreement that for a vertical line the slope is undefined. This proves

---

THEOREM 3. If $P_1(x_1, y_1)$ and $P_2(x_2, y_2)$ are any two distinct points on a line (that is not vertical), then the slope m is given by equation (18).

Example 1. Find the slope of a line through the points $(13, 3)$ and $(-5, 7)$.

Solution. Let $(13, 3)$ be the point P_2 and $(-5, 7)$ be the point P_1. Then, by equation (18),

$$m = \frac{y_2 - y_1}{x_2 - x_1} = \frac{3 - 7}{13 - (-5)} = \frac{-4}{18} = -\frac{2}{9}. \quad \bullet$$

Since the slope is negative, we know that the line is falling. Further, we know that for every increase in x of 9 units, the line falls 2 units.

We conclude this section by deriving a formula for the *midpoint* of the line segment joining two points $P_1(x_1, y_1)$ and $P_2(x_2, y_2)$. Let $M(x, y)$ be this midpoint. Referring to Fig. 16 for the case of a rising segment, we see that the triangles P_1QM and MRP_2 are congruent. Hence $|P_1Q| = |MR|$ so that $x - x_1 = x_2 - x$. Solving this last equation for x gives $x = (x_1 + x_2)/2$. Similarly, $|QM| = |RP_2|$ so that $y - y_1 = y_2 - y$ and thus $y = (y_1 + y_2)/2$. The reader can easily verify that the same formulas hold for the case of a falling segment, a horizontal segment, or a vertical segment. Consequently, we have

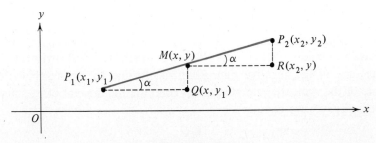

FIGURE 16

THEOREM 4. If M is the midpoint of the line segment joining $P_1(x_1, y_1)$ and $P_2(x_2, y_2)$, then the coordinates of M are given by

(19)
$$x_m = \frac{x_1 + x_2}{2} \quad \text{and} \quad y_m = \frac{y_1 + y_2}{2}.$$

Just for fun we use the formulas in (19) to verify by means of the slope that the three points P_1, M, P_2 are collinear. The slope of P_1M is, by (18),

$$\frac{y_m - y_1}{x_m - x_1} = \frac{\dfrac{y_1 + y_2}{2} - y_1}{\dfrac{x_1 + x_2}{2} - x_1} = \frac{y_2 - y_1}{x_2 - x_1},$$

which is indeed the same as the slope of P_1P_2.

Example 2. Find the midpoint of the segment joining $P_1(-3, -1)$ and $P_2(5, 4)$.

Solution. According to equations (19), we take the average of the x-coordinates and the average of the y-coordinates of the end points P_1, P_2. Thus

$$\left(\frac{-3 + 5}{2}, \frac{-1 + 4}{2} \right) = \left(1, \frac{3}{2} \right)$$

gives the coordinates of the midpoint. ●

EXERCISE 4

In each of Problems 1 through 6, plot the points P and Q and compute the slope of the line through P and Q.

1. $P(2, -1)$, $Q(7, 4)$.
2. $P(13, -3)$, $Q(-7, -4)$.
3. $P(-5, 7)$, $Q(1, -11)$.
4. $P(0, 0)$, $Q(a, b)$, $a \neq 0$.
5. $P(a, b)$, $Q(3a, 7b)$, $a \neq 0$.
6. $P(a, 0)$, $Q(0, b)$, $a \neq 0$.

In each of Problems 7 through 10, plot the points P, Q, and R and determine whether they are collinear by computing the slopes of the lines PQ and PR. Check your work by computing the slope of the line QR.

7. $P(-1, -2)$, $Q(5, -5)$, $R(-5, 0)$.
8. $P(2, 4)$, $Q(-3, 2)$, $R(-13, -2)$.

9. $P(-6, 8)$, $Q(4, 2)$, $R(12, -3)$.
10. $P(-2, 5)$, $Q(-12, -11)$, $R(6, 18)$.

11. Find the midpoint of the line segment PQ in each of the following cases.
    (a) $P(-2, 5)$, $Q(4, -1)$.   (b) $P(0, 4)$, $Q(-2, 8)$.
    (c) $P(5, 2)$, $Q(-6, 2)$.   (d) $P(-1, -3)$, $Q(-6, -4)$.
12. The end points of the diameter of a circle are $P(-7, 2)$ and $Q(1, 6)$. Find the coordinates of the center of the circle. What is the radius of the circle?
13. A line segment has one end point at $P_1(3, 6)$ and midpoint at $M(2, 1)$. Find the other end point.
14. Find the slopes of the medians of a triangle having vertices $P(1, 2)$, $Q(2, -4)$, and $R(-2, 2)$.
15. Using formulas (19), verify that $|P_1 M| = |MP_2|$.
16. Find the coordinates of the three points that divide the line segment from $P_1(1, 2)$ to $P_2(3, 10)$ into four equal parts.
17. The *law of reflection* states that for a flat mirror the angle of incidence of a ray of light equals the angle of reflection. If the mirror is the x-axis of a Cartesian coordinate system, and the slope of the incidence ray is m, show that the slope of the reflected ray is $-m$. (The incidence and reflection angles are both measured from a perpendicular to the mirror.)
18. Prove analytically that the midpoint of the hypotenuse of a right triangle is equidistant from the three vertices. HINT: Put the right angle at the origin.
19. Prove analytically that the diagonals of a parallelogram bisect each other. HINT: Take one vertex at the origin and one side along the x-axis. Then prove that the diagonals have the same midpoint.
20. Prove that if A and B are two distinct points on a straight line, then no matter which one is selected to be P_1 (with the other being P_2) in Theorem 3, the slope computed by equation (18) is the same. HINT: Let the coordinates of A be (x_A, y_A) and the coordinates of B be (x_B, y_B). Then observe that

$$\frac{y_B - y_A}{x_B - x_A} = \frac{y_A - y_B}{x_A - x_B}.$$

*21. Let $\mathscr{L}_1$ and $\mathscr{L}_2$ be two straight lines that meet at a point P and have slopes m_1 and m_2, respectively. Let φ be the least positive angle that $\mathscr{L}_1$ must be turned to bring it into coincidence with $\mathscr{L}_2$. The angle φ is called *the angle from $\mathscr{L}_1$ to $\mathscr{L}_2$*. Find and prove a formula for $\tan \varphi$ in terms of m_1 and m_2.

CALCULATOR PROBLEM

C1. The slope of a line is 1.328. What is its angle of inclination?

Equations for the Line 6

Our first task is to find an equation for a line when the slope and one point on the line are given. If m is the given slope, and $P_1(x_1, y_1)$ is the given point on the line, then the variable point $P(x, y)$ is on this line if and only if the slope of the line P_1P is equal to m. If we use equation (18), this gives immediately

$$(20) \qquad\qquad \frac{y - y_1}{x - x_1} = m.$$

The point $P_1(x_1, y_1)$ is an exceptional point for equation (20) because when we place $x = x_1$ and $y = y_1$, both the numerator and denominator are zero so that the left side is really meaningless. However, if we clear fractions, we obtain

$$(21) \qquad\qquad y - y_1 = m(x - x_1),$$

and the exceptional case disappears. This equation is satisfied if and only if $P(x, y)$ is on the line.

If the line is vertical so that $m = \infty$, equations (20) and (21) are both meaningless, but if we divide (21) on both sides by m and then let m grow indefinitely so that the left side becomes zero, we obtain the meaningful equation $0 = x - x_1$, or

$$(22) \qquad\qquad x = x_1.$$

But a quick inspection of (22) shows that it is indeed the equation of a vertical line through the point $P_1(x_1, y_1)$, even if it was obtained by rather questionable means. This proves

---

THEOREM 5. Suppose that a line is given by specifying its slope and one point on the line. If the line is vertical, then equation (22) is an equation for the line. In all other cases equation (21) is an equation for the line. Conversely, the graph of equation (21) is a straight line that has slope m and passes through P_1.

For obvious reasons, (21) is called the *point-slope form* of the equation of the line with slope m through the point (x_1, y_1).

Example 1. Find an equation for the line with slope $3/4$, passing through the point $(5, -6)$.

Solution. Substituting the given numbers in equation (21), the point-slope form, we find that

$$y - (-6) = \frac{3}{4}(x - 5),$$

or

$$4y + 24 = 3x - 15,$$

or

$$4y = 3x - 39. \quad \bullet$$

Example 2. Prove that the graph of $3x + 2y = 7$ is a straight line, and find its slope and one point on the line.

Solution. By transposition and division by 2, the given equation is equivalent to

$$(23) \qquad\qquad y - \frac{7}{2} = -\frac{3}{2}x.$$

But (23) has just the form of (21) with $y_1 = 7/2$, $m = -3/2$, and $x_1 = 0$. Therefore, by Theorem 5, the graph is a straight line through the point $(0, 7/2)$ with slope $-3/2$. $\bullet$

If two points are given (instead of the slope and one point), then one can use the two given points to compute the slope. Using this in equation (20) yields

$$(24) \qquad\qquad \frac{y - y_1}{x - x_1} = \frac{y_2 - y_1}{x_2 - x_1}.$$

The point $x = x_1$, $y = y_1$ is also an exceptional point for (24) just as it is for (20). But this exceptional case disappears after appropriate clearing of fractions. Despite this slight defect, equation (24) is to be preferred because its symmetry makes it easier to memorize. Thus we have

---

THEOREM 6. Suppose that the points $P_1(x_1, y_1)$ and $P_2(x_2, y_2)$ are given with $x_1 \neq x_2$. Then equation (24) is an equation for the line passing through P_1 and P_2.

For obvious reasons, (24) is called the *two-point form* of the equation of the straight line.

Example 3. Find an equation for the line passing through $(1, -3)$ and $(-4, 5)$.

Solution. Let the given points be P_1 and P_2, respectively. Equation (24) then gives

$$\frac{y - (-3)}{x - 1} = \frac{5 - (-3)}{-4 - 1}.$$

Simplification gives $-5(y + 3) = 8(x - 1)$, or finally $5y + 8x + 7 = 0$, as a suitably simple form for the equation of this line. ●

Observe that had we selected $(-4, 5)$ as P_1 and $(1, -3)$ as P_2, equation (24) would then give

$$\frac{y - 5}{x - (-4)} = \frac{-3 - 5}{1 - (-4)}.$$

But simplification of this equation also leads to $5y + 8x + 7 = 0$.

DEFINITION 5 (Intercept). If a line intersects the x-axis at the point $(a, 0)$, then a is called the *x-intercept* of the line. If the line intersects the y-axis at $(0, b)$, then b is called the *y-intercept* of the line.

If we are given the slope and the y-intercept of a line, then using Theorem 5, equation (21), we can immediately write an equation for the line. Since $(0, b)$ is a point on the line, we have $y - b = m(x - 0)$, or transposing,

(25)
$$y = mx + b.$$

For obvious reasons, (25) is called the *slope-intercept form* of the equation of a straight line.

THEOREM 7. If m is the slope and b is the y-intercept of a line, then equation (25) is an equation for the line.

Example 4. Find an equation for the line with slope 3 that meets the y-axis 5 units below the origin.

Solution. Here $m = 3$ and $b = -5$ so that equation (25) gives $y = 3x - 5$. ●

Using any one of Theorems 5, 6, or 7, we see that any straight line has an equation of the form $Ax + By + C = 0$. We are now in a good position to prove the converse, which we have always suspected to be true anyway.

THEOREM 8. If A and B are not simultaneously zero, then the graph of the equation

(26)
$$Ax + By + C = 0$$

is a straight line.

Proof. Suppose that $B = 0$. Then $A \neq 0$, so (26) becomes $Ax + C = 0$, or $x = -C/A$. But this is just the equation of a vertical line through the point $(-C/A, 0)$.

If $B \neq 0$, then (26) can be written in the form

$$y = -\frac{Ax}{B} - \frac{C}{B}.$$

But this is just the slope-intercept form, so the graph is a line with slope $-A/B$ and y-intercept $-C/B$. ∎

Example 5. What is the graph of $-x + 3y + 7 = 0$?

Solution. This equation is equivalent to $y = x/3 - 7/3$ so that the graph is a straight line with slope $1/3$ and y-intercept $-7/3$. ●

An equation of the form (26), which contains only the first power of x and y, is called a *linear* equation. We can sloganize our results by saying that *every straight line has a linear equation, and every linear equation has a straight line for its graph.*

EXERCISE 5

In Problems 1 through 6, find a simple equation for the line satisfying the given conditions.

1. Slope 3, passing through the point $(-1, 2)$.
2. Slope $-1/4$, passing through the point $(5, -3)$.
3. Passing through the given pair of points.
    (a) $(5, 6)$, $(-2, -1)$.   (b) $(-1, 6)$, $(13, -1)$.
    (c) $(5/4, 1/2)$, $(-3/4, -7/2)$.   (d) $(1 - \sqrt{3}, 3)$, $(1 + \sqrt{3}, 5)$.
4. Slope 10, y-intercept 5.

5. Slope $-1/3$, y-intercept $7/6$.

6. Slope 2, x-intercept -3.

7. Find the slope and y-intercept for each of the following lines.
   - (a) $2x + 3y + 4 = 0$.
   - (b) $5x - y - 7 = 0$.
   - (c) $x = 3y + 9$.
   - (d) $57x + 19y = 114$.
   - (e) $y = 10$.
   - (f) $3x - \sqrt{3}\,y + 12 = 0$.

8. Find the angle of inclination for each of the following lines.
   - (a) $y = x + 2$.
   - (b) $y = x + \pi$.
   - (c) $y = -x + \sqrt{15}$.
   - (d) $y = \sqrt{3}\,x - 11$.
   - (e) $\sqrt{3}\,y = x - 11$.
   - (f) $x = 100$.

9. Find the x-intercept for each of the following lines.
   - (a) $y = 2x - 4$.
   - (b) $3x - 2y = 14$.
   - (c) $2x - y - 7 = 0$.

10. Prove that if a line has x-intercept $a \neq 0$ and y-intercept $b \neq 0$, then

$$\frac{x}{a} + \frac{y}{b} = 1$$

is an equation for the line. This is called the *intercept form* of the equation of a straight line. Use this to find a simple equation for the line with x- and y-intercepts: (a) 5, -4, (b) 1, 1, (c) 2, 7, and (d) $-1/3$, $1/6$.

In Problems 11 through 14, sketch the graph of the given equation.

11. $x + 3 = 2y - 1$.

12. $(y - 2x + 1)(y + x) = 0$.

13. $y = 1 + x + |x|$.

14. $y = x - |x|$.

In Problems 15 through 17, find the point of intersection (if there is one) for the two lines $\mathcal{L}_1$ and $\mathcal{L}_2$. HINT: *Solve the pair of equations by first combining them to eliminate one variable.*

15. $\mathcal{L}_1\!:y = 2x + 7$, $\mathcal{L}_2\!:y = x - 5$.

16. $\mathcal{L}_1\!:y = 5x + 17$, $\mathcal{L}_2\!:10x - 2y = 13$.

17. $\mathcal{L}_1\!:2x + 3y = 9$, $\mathcal{L}_2\!:3x + 2y = 11$.

18. Find the coordinates of the vertices of the triangle formed by the lines $y = 2$, $y = 2x$, and $y = -x + 9$.

19. The relationship between temperatures measured in degrees Fahrenheit, F, and in degrees Celsius (centigrade), C, is a linear one. Water freezes at $32°$F or $0°$C, and it boils at $212°$F or $100°$C. Find an equation expressing the linear relationship between the two measures. Draw a graph of this equation using degrees Fahrenheit for the x-axis and degrees Celsius for the y-axis. Express each of the following in the alternative scale.
   - (a) $95°$F.
   - (b) $125°$C.
   - (c) $98.6°$F.
   - (d) $0°$F.
   - (e) $60°$C.

*20. A line $\mathscr{L}:y = mx + b$ divides the plane into two regions: the set of all points above the line, and the set of all points below the line. Show that for the point $P_1(x_1, y_1)$, if $y_1 > mx_1 + b$, then P_1 lies above the line; and if $y_1 < mx_1 + b$, then P_1 lies below the line.

*21. Without making a drawing, determine which of the following points lie above the line $y = -x/2 + 5/3$ and which lie below the line.
   (a) $(10, -3)$.   (b) $(50, -23)$.   (c) $(-40, 21)$.   (d) $(5/2, 2/5)$.

In applications of mathematics it frequently happens that the graph of two related quantities is not a straight line, but a straight line is used as a first approximation. This is the case in Problems 22 through 24. In each problem assume (erroneously?) that a linear equation relates the two quantities.

22. The United States population to the nearest million was 76,000,000 in 1900 and 106,000,000 in 1920. Using t as the number of years after 1900, find a linear equation that gives the population in terms of t. Use this equation to compute the population: (a) in 1910, (b) in 1960, and (c) in 1880.

23. A standard life insurance table gives the life expectancy E for a man of age A. This means that at age A a man may expect to live E years longer. One particular table gives $E = 20$ when $A = 54$, and $E = 14$ when $A = 63$. Based on these data, find a linear equation that gives E in terms of A. Compute: (a) the age at which a man may expect to live 28 years more, (b) E for a newly born male child, and (c) the age at which $E = 0$.

24. A certain spring had a length L of 12 in. when supporting a weight W of 12 lb. When $W = 24$ lb, $L = 16$ in. Compute L when: (a) $W = 18$ lb, (b) $W = 0$ lb, and (c) $W = 99$ lb.

CALCULATOR PROBLEMS

C1. A laboratory equipment company pays its first-year salesmen \$85 per week plus a $5\frac{3}{4}$ percent commission. Express the weekly income I of a salesman in terms of the dollar amount S of sales he makes in a week. If a salesman sells \$3284 in equipment in 1 week, how much money will he make?

C2. A salesman working for the company in Problem C1 is offered another job, which pays a straight 18 percent commission. How much must he sell each week in order that the new job pay as much as his present job?

C3. Find the equation of the line through $P(\sqrt{3}, \sqrt{5})$ having angle of inclination $70°$. Give the constants in the equation to three decimal places.

Parallel and Perpendicular Lines 7

Given two straight lines, can we look at their equations and merely by inspection determine whether the lines are parallel or perpendicular? The answer is yes, and the method is given in

---

THEOREM 9. Let $\mathcal{L}_1$ be a line with slope m_1 and let $\mathcal{L}_2$ be a line with slope m_2. Then

  (i) The lines are parallel if and only if $m_1 = m_2$.

  (ii) The lines are mutually perpendicular if and only if

(27)
$$m_1 m_2 = -1.$$

Proof. Two lines are parallel if and only if they make the same angle with the x-axis, and consequently if and only if they have the same slopes.

   The perpendicularity criterion is a little harder to prove. We will use a slight amount of trigonometry. Assume first that $\mathcal{L}_1$ and $\mathcal{L}_2$ intersect in a point P that lies above the x-axis, as indicated in Fig. 17. Then from the figure

$$\alpha_2 = \varphi + \alpha_1.$$

If $\varphi = 90°$, then

$$m_2 = \tan \alpha_2 = \tan (90° + \alpha_1) = -\cot \alpha_1 = -\frac{1}{\tan \alpha_1} = -\frac{1}{m_1},$$

and hence equation (27) is satisfied. Conversely, if (27) is satisfied, then obviously $\tan \alpha_2 = -\cot \alpha_1$, and hence α_1 and α_2 must differ by 90°.

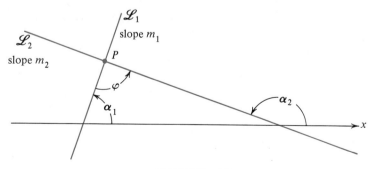

FIGURE 17

If P lies on or below the x-axis, we shift the lines upward so that P lies above the x-axis (or lower the axis) without altering either the slopes or the angle of intersection of the two lines. The details are supplied in Problems 19 and 20 of Exercise 6.

If m_1 or m_2 is undefined, then one of the lines is vertical. For parallelism, both lines must be vertical, and for perpendicularity one line must be vertical and the other horizontal. ∎

Example 1. Find an equation of the line through the point $(5, 1)$ that is **(a)** parallel to the line $y = 3x + 7$; **(b)** perpendicular to the line $y = 3x + 7$.

Solution. **(a)** Since the line is to be parallel, it must have slope $m = 3$ so that we can write

$$y = 3x + b,$$

where b is unknown. Since $(5, 1)$ is on the line, its coordinates must satisfy this equation so that we have

$$1 = 3 \times 5 + b.$$

Therefore, $b = -14$ and the equation is $y = 3x - 14$.

(b) By equation (27), $3m_2 = -1$ or $m_2 = -1/3$. Hence the sought equation is

$$y = -\frac{1}{3}x + b,$$

where again b is an unknown. Again the point $(5, 1)$ is on the line so that

$$1 = -\frac{1}{3} \times 5 + b,$$

and consequently $b = 8/3$. Therefore, the perpendicular line has as its equation $y = -x/3 + 8/3$, or $3y + x = 8$. ●

Example 2. A line segment is drawn from the point $(0, 5)$ perpendicular to the line $\mathscr{L}: y = x + 3$. Find the point where the segment meets $\mathscr{L}$.

Solution. We first find the equation of the line through $(0, 5)$ perpendicular to $\mathscr{L}$. Since $\mathscr{L}$ has slope 1, the perpendicular line must have slope -1, and it must pass through $(0, 5)$. Hence $y = -x + 5$ is an equation of the perpendicular line. Now we find the point of intersection of the two lines

$$y = x + 3, \qquad y = -x + 5.$$

Setting $x + 3 = -x + 5$, we obtain $x = 1$ as the abscissa and $y = 1 + 3 = 4$ as the ordinate. Hence $(1, 4)$ is the point where the perpendicular meets $\mathscr{L}$. ●

We remark that in Example 2, the distance from $(0, 5)$ to $(1, 4)$, namely

$$\sqrt{(0 - 1)^2 + (5 - 4)^2} = \sqrt{2},$$

gives the (shortest) distance of the point $(0, 5)$ to the line $\mathcal{L}: y = x + 3$.

EXERCISE 6

In Problems 1 through 6, a point P and a line $\mathcal{L}$ are given. Find an equation for the line through P and (a) parallel to $\mathcal{L}$; (b) perpendicular to $\mathcal{L}$.

1. $(5, -5)$, $y = x + 10$.
2. $(0, 11)$, $2y = x - 7$.
3. $(0, 0)$, $3y + x = \pi$.
4. $(-1, -1)$, $5y - 2x = 9$.
5. $(100, 200)$, $x - 37 = 0$.
6. $(3/4, 5/6)$, $7x - 6y = \sqrt{5}$.

In Problems 7 and 8, determine whether PQRS is a parallelogram.

7. $P(-3, -2)$, $Q(8, 1)$, $R(13, 10)$, $S(2,7)$.
8. $P(13, 7)$, $Q(8, -3)$, $R(-7, 0)$, $S(-1, 12)$.

9. Are the two lines $3y - 4x + 2 = 0$ and $4y + 3x - 7 = 0$ perpendicular?
10. Are the two lines $4y - x - 1 = 0$ and $8y + 4x + 9 = 0$ parallel?
11. Find an equation for the perpendicular bisector of the line segment from $P(3, 1)$ to $Q(-2, 0)$.
12. Given the line $\mathcal{L}: 3y - 2x = 8$ and the point $P(1, -1)$, find:
    (a) An equation of the line through P perpendicular to $\mathcal{L}$.
    (b) The distance from P to $\mathcal{L}$.
13. Given the line $\mathcal{L}: y = 2x - 1$ and the point $P(1, 2)$, find:
    (a) An equation of the line through P perpendicular to $\mathcal{L}$.
    (b) The distance from P to $\mathcal{L}$.
14. Find the distance between the parallel lines $2y - x - 7 = 0$, $2y - x + 1 = 0$.
15. Find the distance between the parallel lines $4y + 2x - 30 = 0$, $2y + x + 10 = 0$.
16. Three consecutive vertices of a parallelogram are $(-5, 0)$, $(1, 2)$, and $(7, 8)$. Find the coordinates of the fourth vertex.
17. Prove by means of slopes that the three points $P(5, 1)$, $Q(6, 4)$, and $R(8, 0)$ are the vertices of a right triangle. Find the area of the triangle.
18. Prove that the two lines

$$Ax + By + C = 0,$$
$$Ax + By + D = 0$$

are parallel. Prove that the two lines

$$Ax + By + C = 0,$$
$$Bx - Ay + D = 0$$

are mutually perpendicular.

19. Let $\mathscr{L}_1$ and $\mathscr{L}_2$ be the lines $y = mx + b$ and $y = mx + B$, with $B \neq b$. Show that these lines are parallel and that the line $\mathscr{L}_2$ lies above $\mathscr{L}_1$ if and only if $B > b$.

*20. Suppose that the two lines $\mathscr{L}_1 : y = m_1 x + b_1$ and $\mathscr{L}_2 : y = m_2 x + b_2$ intersect in a point below the x-axis. Then by increasing the constants b_1 to B_1 and b_2 to B_2 the two lines can be moved upward so that they intersect above the x-axis. But this does not change m_1, or m_2, or the angle between the lines $\mathscr{L}_1$ and $\mathscr{L}_2$. Theorem 9 was proved only in case the lines intersect above the x-axis. Show that the above argument extends Theorem 9 to all cases.

21. Let $A(1, -3)$, $B(3, 11)$, and $C(5, -9)$ be the vertices of a triangle and let M_1 and M_2 be the midpoints of the segments AB and BC. Prove that the line segments $M_1 M_2$ and AC are parallel by showing that both have the slope $-3/2$.

*22. Convert Problem 21 into a theorem and give a proof. HINT: Replace the numerical coordinates for A, B, and C by letters.

*23. Find an equation for the perpendicular bisector of the segment joining $A(a, b)$ and $B(c, d)$ when $a \neq c$. Discuss the case $a = c$. Find two different methods for working this problem.

24. Let A, B, C, and D be any four points in the plane and let P_1, P_2, P_3, and P_4 be midpoints of the line segments AB, BC, CD, and DA, respectively. Prove that $P_1 P_2 P_3 P_4$ is a parallelogram. HINT: Draw the line BD and use Problem 22.

8 The Circle

The set of all points that have a given fixed distance r from a given fixed point C is called a *circle*. The distance r is the *radius* and C is the *center*. Using our distance formula, it is easy to find a nice equation for a circle.

> **THEOREM 10.** If (h, k) is the center and r is the radius, then
>
> (28) $$r^2 = (x - h)^2 + (y - k)^2$$
>
> is an equation for the circle. Conversely, if $r^2 > 0$, the graph of (28) is a circle.

Proof. In Theorem 2, formula (7), let $d = r$, let P_2 be the point (x, y) on the circle, and let P_1 be the center (h, k). Then square both sides. ∎

| **Example 1.** Find an equation for the circle of radius 5 and center $(-3, 4)$.

Solution. From equation (28), we have

$$25 = (x - (-3))^2 + (y - 4)^2$$
$$= x^2 + 6x + 9 + y^2 - 8y + 16.$$

The constants can be combined and the terms rearranged to give

$$x^2 + y^2 + 6x - 8y = 0. \quad \bullet$$

Obviously, this circle passes through the origin. Why?

This procedure can be reversed, as illustrated in

Example 2. Describe the graph of the equation $x^2 + y^2 - 4x + 6y + 9 = 0$.

Solution. Since this is a quadratic equation and the coefficients of x^2 and y^2 are the same, this suggests that the graph is a circle. Rearranging the terms and completing the squares, we have

$$
\begin{array}{ll}
x^2 - 4x \quad\quad + y^2 + 6y \quad\quad = -9 \\
\quad\quad + 4 \quad\quad\quad\quad\quad + 9 = \quad\quad + 4 + 9 \\
\hline
x^2 - 4x + 4 + y^2 + 6y + 9 = \quad 4 \\
(x - 2)^2 \quad\quad + (y - (-3))^2 = \quad 4.
\end{array}
$$

Using Theorem 10, we find that our suspicions are confirmed: The graph is a circle with center $(2, -3)$ and radius 2. $\bullet$

This example suggests

THEOREM 11. The graph of the equation

(29)
$$x^2 + y^2 + Ax + By + C = 0$$

is either a circle, a point, or has no points at all. When it is a circle, the center is $(-A/2, -B/2)$ and the radius is

(30)
$$r = \frac{1}{2}\sqrt{A^2 + B^2 - 4C}.$$

Proof. Completing the square in (29) just as in the example, we have

$$x^2 + Ax + \frac{A^2}{4} + y^2 + By + \frac{B^2}{4} = \frac{A^2}{4} + \frac{B^2}{4} - C,$$

or

$$\left(x + \frac{A}{2}\right)^2 + \left(y + \frac{B}{2}\right)^2 = \frac{1}{4}(A^2 + B^2 - 4C).$$

If the right side is positive, the graph is obviously the circle described in the theorem. Since the square of a real number can never be negative, the left side cannot be negative for any real point (x, y). Consequently, if $A^2 + B^2 - 4C < 0$, there are no points on the graph. If $A^2 + B^2 - 4C = 0$, the only possible point is the one with coordinates $x = -A/2$, $y = -B/2$, and the graph consists of a single point (a circle of radius zero).   ∎

It does not pay to memorize formula (30) because the process of completing the square is quite simple and leads to r^2 on the right side of the equation.

The circle has always been an interesting figure because it has so many axes of symmetry. Two points P_1 and P_2 are said to be *symmetric with respect to a line* $\mathscr{L}$ if the line $\mathscr{L}$ is the perpendicular bisector of the segment P_1P_2. Each of the points is said to be the *reflection* of the other in the line $\mathscr{L}$. A curve is said to be *symmetric with respect to the line* $\mathscr{L}$ if for every point P_1 on the curve, its reflection in the line $\mathscr{L}$ is also on the curve. In this case the line $\mathscr{L}$ is called an *axis of symmetry* for the curve. Physically, symmetry about a line means that if the curve is drawn carefully on a piece of paper, and if the paper is folded on the line of symmetry, then the two halves of the curve will coincide.

A curve is said to be *symmetric with respect to a point* C if for each point P of the curve there is a corresponding point Q on the curve such that C bisects the line segment PQ. If a curve is symmetric with respect to a point C, then C is called a *center of symmetry for the curve* or, more briefly, a *center of the curve.*

It is intuitively obvious that a circle is symmetric with respect to any line through its center and it is easy to prove this using congruent triangles. It is also clear that the center of the circle is a center of symmetry for that curve. In due time we will develop algebraic methods for checking the symmetry of any curve from an equation for the curve. However, this is best postponed until we have become experts in the use of function notation. The impatient reader should consult Chapter 11, Section 3. In the meantime, geometric methods (congruent triangles) will be sufficient for our problems.

EXERCISE 7

In *Problems 1 through 6, find an equation for the circle with the given center and given radius.*

1. $C(5, 12)$, $r = 13$.
2. $C(0, 0)$, $r = 7$.
3. $C(1, -1)$, $r = 2$.
4. $C(-4, -5)$, $r = 6$.
5. $C(a, 2a)$, $r = \sqrt{5}a$.
6. $C(3, b)$, $r = b$.

In Problems 7 through 10, find an equation for the circle satisfying the given conditions.

7. Tangent to the x-axis, center at $(3, 2)$.
8. Tangent to the y-axis, center at $(-5/2, -3/2)$.
9. Tangent to the line $y = 7$, center at $(5, -2)$.
10. Center at $(1, 1)$, passing through the point $(3, 2)$.

In Problems 11 through 18, describe the graph of the given equation.

11. $x^2 + y^2 - 4x + 2y - 20 = 0$.
12. $x^2 + y^2 + 6x + 8y + 24 = 0$.
13. $x^2 + y^2 - 6x - 16y + 73 = 0$.
14. $4x^2 + 4y^2 - 4x + 20y + 36 = 0$.
15. $5x^2 + 5y^2 - 8x - 4y - 121 = 0$.
16. $2x^2 + 2y^2 - 2x + 6y - 3 = 0$.
17. $2x^2 + 2y^2 + 12x - 8y + 31 = 0$.
18. $9x^2 + 9y^2 + 6x - 6y + 5 = 0$.

19. Find an equation for the circle with center at the point $(6, 1)$ and tangent to the line $\mathscr{L}: y - x - 1 = 0$. HINT: To find r, first find the point of tangency of $\mathscr{L}$ with the circle. The radial line drawn to the point of tangency must be perpendicular to $\mathscr{L}$.

20. Find an equation for the circle with center at the origin and tangent to the line $y = -3x + 5$.

*21. Find the equations of the lines that pass through the point $(3, 1)$ and are tangent to the circle $x^2 + y^2 = 2$.

22. Sketch the graph of the equation $y = \sqrt{4 - (x - 1)^2}$.

*23. Find an equation for the circle through the three points $(0, -2)$, $(8, 2)$, and $(3, 7)$.

*24. Find an equation for the line that joins the points of intersection of the two circles $x^2 + y^2 = 4x + 4y - 4$ and $x^2 + y^2 = 2x$. HINT: If a point lies on two circles, its coordinates satisfy the equation of each circle, and hence they satisfy the difference of the two equations.

*25. Find the points of intersection of the two circles of Problem 24.

*26. The equation $x^2 + y^2 = 2x$ is the equation of a certain circle $\mathscr{C}$. Prove that if x_1, y_1 are real numbers such that $x_1{}^2 + y_1{}^2 < 2x_1$, then the point $P(x_1, y_1)$ lies inside the circle $\mathscr{C}$. What can you say about the position of P if $x_1{}^2 + y_1{}^2 > 2x_1$?

*27. Prove that if the line $y = mx + b$ is tangent to the circle $x^2 + y^2 = r^2$, then we have $b^2 = r^2 + m^2r^2$.

28. Prove analytically that an angle inscribed in a semicircle is a right angle. HINT: Place the center of the circle at the origin.

29. Consider the chords of the circle $x^2 + y^2 = 1$ that have one end point fixed at $(1, 0)$ and the other end point varying over the circle. Find an equation for the locus of midpoints $M(x_0, y_0)$ of these chords. Describe the locus.

30. Prove that a circle is symmetric with respect to any diameter.

31. Prove that a square has four axes of symmetry.

32. How many axes of symmetry does a rectangle have if it is not a square?

33. Find the number of axes of symmetry for: (a) an equilateral triangle, (b) a regular pentagon, and (c) a regular hexagon.

34. Does the figure consisting of two circles have an axis of symmetry if the circles are not concentric, and the circles have different radii?
35. Use the definition of a circle to prove that it has a center of symmetry.
36. How many centers of symmetry does a rectangle have? How many for an equilateral triangle?

CALCULATOR PROBLEMS

C1. Find the circumference of the circle $x^2 + y^2 + 2.38x - 6.04y - 4.71 = 0$.
C2. Does the point $P(17, 19)$ lie inside or outside the circle

$$89x^2 + 89y^2 - 107x - 378y = 48,800?$$

9 *The Conic Sections*

The calculus provides us with a systematic method of studying curves. Hence it would be wise to postpone our study of the conic sections until after we have developed some portions of the calculus. But in this very development, it is advantageous to have at hand numerous examples. For this reason we now give a brief introduction to conic sections. We will return to a more detailed study of these important curves in Chapter 11.

As the name implies, these curves are obtained by cutting a cone with a plane as shown in Figs. 18a–18c. Three main types of curves occur: the *ellipse,* the *parabola,* and the

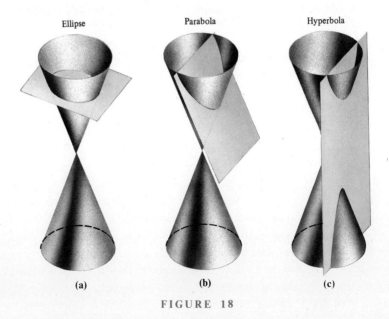

Ellipse   Parabola   Hyperbola

(a)   (b)   (c)

FIGURE 18

hyperbola, depending upon the inclination of the cutting plane to the axis of the cone.[1] However, this geometric definition has fallen from favor because the analytic definitions (given below) are easier to use.

DEFINITION 6 (Parabola). The set of all points P that are equidistant from a fixed point F (the *focus*) and a fixed line $\mathscr{D}$ (the *directrix*) is called a *parabola.*

In order to find a simple equation for a parabola, we locate the focus and the directrix in a suitably chosen position. Or if the focus and the directrix are not to be moved, then we will place our coordinate axes in an appropriate position. Whatever point of view we adopt, let us agree that the x-axis is parallel to the directrix and runs midway between the focus and the directrix, and that the focus in on the positive y-axis (see Fig. 19). Then the focus F is at $(0, p)$, where p is some positive number, and the equation of the directrix is $y = -p$.

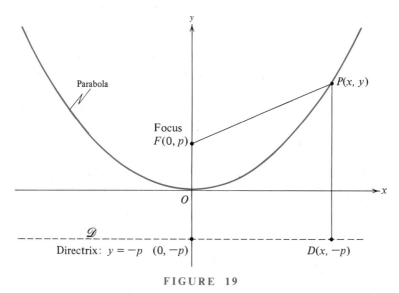

FIGURE 19

Now by definition $P(x, y)$ is a point of the parabola if and only if

(31) $$|PF| = |PD|.$$

Using our distance formula, equation (7), we have

$$\sqrt{(x - 0)^2 + (y - p)^2} = \sqrt{(x - x)^2 + (y + p)^2}.$$

---

[1] As special cases we can also obtain circles, straight lines, and a point.

Squaring both sides and simplifying yields

$$x^2 + y^2 - 2py + p^2 = y^2 + 2py + p^2,$$

or

(32)
$$x^2 = 4py.$$

Each step is reversible so that if (32) is satisfied, so is (31) and $P(x, y)$ is a point of the parabola. We have proved

---

THEOREM 12. The graph of equation (32) is a parabola with focus at $(0, p)$ and directrix $y = -p$.

The line through the focus and perpendicular to the directrix is called the *axis* of the parabola. The point of intersection of the axis with the parabola is called the *vertex* of the parabola. It is obvious from Fig. 19 that with our special choice of the position of the coordinate axes, the vertex turns out to be at the origin.

Example 1. Give an equation for the parabola with focus at $(0, 4)$ and directrix $y = -4$.

Solution. Here $p = 4$, so using (32), we obtain $x^2 = 16y$. ●

Example 2. What is the distance from the focus to the directrix for the parabola $y = 3x^2$?

Solution. To match equation (32), we must write

$$x^2 = \frac{1}{3}y = 4 \times \frac{1}{12}y.$$

Hence $p = 1/12$, and the desired distance is twice this, or $1/6$. ●

If we interchange the x-axis and the y-axis in our discussion of the parabola, then the focus is at $(p, 0)$ and the directrix is the vertical line having the equation $x = -p$. The resulting equation for this parabola is

(33)
$$y^2 = 4px.$$

We therefore have

THEOREM 13. The graph of equation (33) is a parabola with focus at $(p, 0)$ and directrix $x = -p$.

Example 3. Find the focus and the equation of the directrix for the parabola $y^2 = 3x$. Draw a sketch of the parabola.

Solution. The given equation has the form of equation (33). Hence $4p = 3$, or $p = 3/4$. Thus the focus is at $(3/4, 0)$ and the directrix is $x = -3/4$. A sketch of the parabola is given in Fig. 20. ●

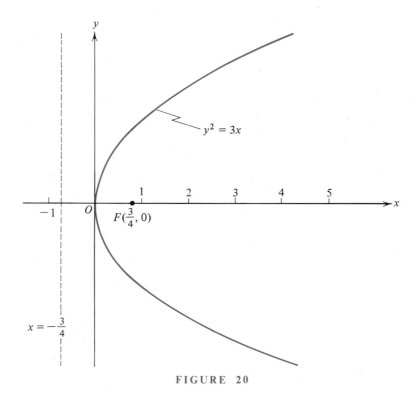

FIGURE 20

DEFINITION 7 (Ellipse). The set of all points P, such that the sum of its distances, $|PF_1| + |PF_2|$, from two fixed points F_1 and F_2 is a constant, is called an *ellipse*. The points F_1 and F_2 are called the *foci* of the ellipse.

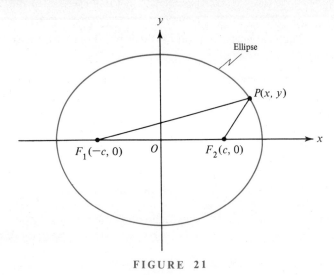

FIGURE 21

As before, the task of finding an equation for an ellipse is simplified if the coordinate system is selected judiciously. Let the x-axis pass through the two foci F_1 and F_2 (as indicated in Fig. 21) and let the y-axis bisect the segment F_1F_2. If we denote the distance between the two foci by $2c$, then the coordinates for the foci will be $(-c, 0)$ and $(c, 0)$. Further, it is convenient to denote the constant sum $|PF_1| + |PF_2|$ by $2a$. Then according to the definition of an ellipse the point P is a point of the ellipse if and only if

(34) $$|PF_1| + |PF_2| = 2a.$$

Our distance formula, equation (7), gives

$$\sqrt{(x + c)^2 + y^2} + \sqrt{(x - c)^2 + y^2} = 2a.$$

We transpose the second radical to the right side and then square both sides, obtaining

$$x^2 + 2xc + c^2 + y^2 = x^2 - 2xc + c^2 + y^2 - 4a\sqrt{(x - c)^2 + y^2} + 4a^2.$$

If we drop the common terms x^2, y^2, and c^2 from both sides, and then transpose, we obtain

(35) $$4a\sqrt{(x - c)^2 + y^2} = 4a^2 - 4xc.$$

After dividing through by 4 and again squaring, we have

$$a^2(x^2 - 2xc + c^2 + y^2) = a^4 - 2a^2xc + x^2c^2.$$

The terms $-2a^2xc$ drop out, and on transposing and grouping, we arrive at

$$x^2(a^2 - c^2) + a^2y^2 = a^2(a^2 - c^2),$$

and on dividing by $a^2(a^2 - c^2)$, we have

(36)
$$\frac{x^2}{a^2} + \frac{y^2}{a^2 - c^2} = 1.$$

This suggests that we introduce a new quantity $b > 0$, defined by the equation $b^2 = a^2 - c^2$. When this is done, equation (36) assumes the very simple form

(37)
$$\frac{x^2}{a^2} + \frac{y^2}{b^2} = 1.$$

It is not completely obvious that all the steps above are reversible. Indeed, in taking square roots we must assure ourselves that we have taken the positive square root on both sides. Once this subtle point is settled, the proof will be completed for

---

THEOREM 14. If $a > b > 0$, then equation (37) is an equation for the ellipse

$$|PF_1| + |PF_2| = 2a,$$

where the foci are $F_1(-c, 0)$ and $F_2(c, 0)$, and

(38)
$$c = \sqrt{a^2 - b^2}.$$

Equation (37) is called the *standard form* for the equation of the ellipse. A sketch of the ellipse $x^2/25 + y^2/16 = 1$ is shown in Fig. 21.

Example 4. Describe the graph of

$$\frac{x^2}{169} + \frac{y^2}{25} = 1.$$

Solution. By Theorem 14, the graph is an ellipse with $a = 13$ and $b = 5$. Further, for this ellipse, $|PF_1| + |PF_2| = 2a = 2 \times 13 = 26$. From equation (38) we have $c^2 = a^2 - b^2 = 13^2 - 5^2 = 169 - 25 = 144$, and hence $c = 12$. Therefore, the foci of this ellipse are at $(-12, 0)$ and $(12, 0)$. ●

If we set $y = 0$ in equation (37), we find that $x = \pm a$. Thus the points $V_1(a, 0)$ and $V_2(-a, 0)$ are points of the ellipse and are indeed the intersection points of the ellipse with the x-axis. Similarly, if we set $x = 0$ in (37), we find that $y = \pm b$ so that the points $V_3(0, b)$ and $V_4(0, -b)$ are also on the ellipse and are the intersection points of the ellipse with the

y-axis. It is intuitively obvious that the ellipse (37) is symmetric with respect to the x- and y-axis (see Fig. 22a).

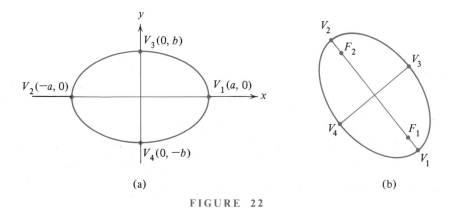

(a)   (b)

FIGURE 22

The chord V_1V_2 of length $2a$ is called the *major axis* of the ellipse. The chord V_3V_4 of length $2b$ is called the *minor axis* of the ellipse. The end points of the major axis, V_1 and V_2, are called the *vertices* of the ellipse.

If the ellipse is in general position as shown in Fig. 22b, then by definition the *major axis* is the chord through the foci, and it can be proved to be the largest chord of the ellipse. The *minor axis* is the chord cut from the perpendicular bisector of the major axis by the ellipse. The *vertices* of an ellipse are the end points of the major axis. These are labeled V_1 and V_2 in Fig. 22b.

The point of intersection of the major and minor axes is called the *center* of the ellipse. For the ellipse described in Theorem 14, the center is at the origin. It is intuitively clear that an ellipse is symmetric with respect to its center.

As was the case for the parabola, the roles of x and y can be interchanged to yield an equation for an ellipse whose foci are on the y-axis.

THEOREM 15 (PLE). If $a > b > 0$, then

(39)
$$\frac{x^2}{b^2} + \frac{y^2}{a^2} = 1$$

is an equation for the ellipse $|PF_1| + |PF_2| = 2a$ with foci at $F_1(0, -c)$ and $F_2(0, c)$, where $c = \sqrt{a^2 - b^2}$.

For example, the equation

$$\frac{x^2}{9} + \frac{y^2}{16} = 1$$

is an equation for an ellipse with major axis along the y-axis. In this case $a = 4, b = 3$, and $c = \sqrt{16 - 9} = \sqrt{7}$. Hence the foci of the ellipse are at $(0, \pm\sqrt{7})$ and the vertices are $(0, \pm 4)$.

DEFINITION 8 (Hyperbola). The set of all points P, such that the difference of its distances from two fixed points F_1 and F_2 is a constant, is called a *hyperbola*. The points F_1 and F_2 are called the foci of the hyperbola.

Here we should observe that in the definition, the order of the difference is not specified, so there are two possibilities. To be specific, suppose that the constant difference is $2a$. Then the point P is a point of the hyperbola if either

(40)
$$|PF_1| - |PF_2| = 2a,$$

or

(41)
$$|PF_2| - |PF_1| = 2a.$$

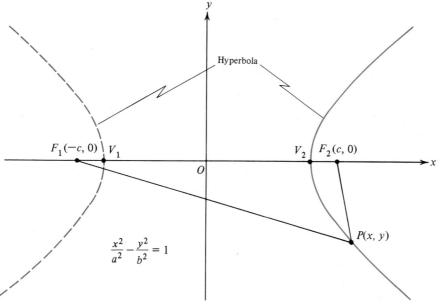

FIGURE 23

To find a simple form for the equation of the hyperbola, we again run the x-axis through the foci F_1 and F_2, and let the y-axis be the perpendicular bisector of the segment F_1F_2 (see Fig. 23). If the distance between the two foci is $2c$, then the coordinates of the foci will be $(-c, 0)$ and $(c, 0)$. Let $P(x, y)$ be a point of the hyperbola that satisfies equation (40). Then by the distance formula

(42)
$$\sqrt{(x + c)^2 + y^2} - \sqrt{(x - c)^2 + y^2} = 2a.$$

We transpose the second radical of (42) to the right side and then square both sides, obtaining

$$x^2 + 2xc + c^2 + y^2 = x^2 - 2xc + c^2 + y^2 + 4a\sqrt{(x - c)^2 + y^2} + 4a^2.$$

If we drop the common terms x^2, y^2, and c^2 from both sides and then transpose, we obtain

$$-4a\sqrt{(x - c)^2 + y^2} = 4a^2 - 4xc.$$

Now this equation is almost identical with equation (35) obtained in the derivation of the equation of the ellipse. The only difference is the negative sign in front of the radical, and on squaring, this difference disappears. Therefore, we can follow that work and arrive again at

(36)
$$\frac{x^2}{a^2} + \frac{y^2}{a^2 - c^2} = 1.$$

But for the ellipse the condition $|PF_1| + |PF_2| = 2a$ implied that $a \geq c$, and as a result $a^2 - c^2$ was positive (or zero) and we could set $b^2 = a^2 - c^2$. This time we are dealing with the hyperbola, and $|PF_1| - |PF_2| = 2a$. But since $|F_1F_2| = 2c$, it is obvious that $2a \leq 2c$, and if the equality sign does not hold, then $a^2 - c^2$ is negative. Hence, in introducing b^2, we turn the quantity around and write

(43)
$$b^2 = c^2 - a^2$$

(instead of $b^2 = a^2 - c^2$ for the ellipse). Using (43) in (36), we find

(44)
$$\frac{x^2}{a^2} - \frac{y^2}{b^2} = 1$$

as the *standard form* for the equation of the hyperbola with foci at $(\pm c, 0)$.

If the point P satisfies equation (41) instead of equation (40), a little computation will show that its coordinates still satisfy (44). Conversely, it can be proved that if (x, y) satisfies (44), then P satisfies either equation (40) or (41), but we omit the details. Thus we have

THEOREM 16. Equation (44) is an equation for the hyperbola

$$\pm(|PF_1| - |PF_2|) = 2a,$$

where the foci are $F_1(-c, 0)$ and $F_2(c, 0)$, and $c = \sqrt{a^2 + b^2}$.

The hyperbola $x^2/16 - y^2/9 = 1$ is shown in Fig. 23. The curve falls into two pieces called *branches*. The right-hand branch, shown solid, is the branch for which $|PF_1| - |PF_2| = 2a = 8$. On the left-hand branch, shown dashed, $|PF_2| - |PF_1| = 8$.

Example 5. Find an equation for the curve described by a point moving so that the difference of its distances from $(-5, 0)$ and $(5, 0)$ is 8. Sketch this curve.

Solution. The curve is a hyperbola with foci at $(-5, 0)$ and $(5, 0)$. From equation (40) we have $2a = 8$ and hence $a = 4$. Further, $c = 5$. Then equation (43) gives $b^2 = c^2 - a^2 = 5^2 - 4^2 = 25 - 16 = 9$. Hence $b = 3$ and an equation for this hyperbola is

$$\frac{x^2}{16} - \frac{y^2}{9} = 1.$$

The sketch is shown in Fig. 23. ●

The line through the foci is called the *transverse axis* of the hyperbola. It is intuitively obvious that the hyperbola is symmetric with respect to this axis. It is also symmetric with respect to the perpendicular bisector of the segment F_1F_2. This line, which coincides with the y-axis in Fig. 23, is called the *conjugate axis* of the hyperbola. The intersection points of the line F_1F_2 and the hyperbola are called the *vertices* of the hyperbola. These points are labeled V_1 and V_2 in Fig. 23.

The point of intersection of the transverse axis and the conjugate axis is called the *center* of the hyperbola. For the hyperbola described in Theorem 16, the center is at the origin. It is intuitively clear that a hyperbola is symmetric with respect to its center.

As with the parabola and the ellipse, the roles of x and y can be reversed to obtain an equation for the hyperbola with foci along the y-axis.

THEOREM 17 (PLE). The equation

(45) $$\frac{y^2}{a^2} - \frac{x^2}{b^2} = 1$$

is an equation for the hyperbola $\pm(|PF_1| - |PF_2|) = 2a$ with foci at $F_1(0, -c)$ and $F_2(0, c)$, where $c = \sqrt{a^2 + b^2}$.

For example, the equation

(46)
$$\frac{y^2}{36} - \frac{x^2}{64} = 1$$

describes a hyperbola having a vertical transverse axis. Here $a = 6$, $b = 8$, and $c = \sqrt{36 + 64} = 10$. Hence the foci of the hyperbola (46) are $(0, \pm 10)$ and the vertices are $(0, \pm 6)$. A sketch of this hyperbola is given in Fig. 24.

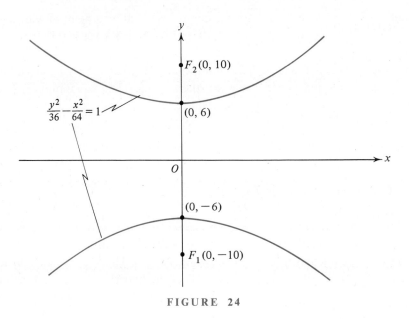

FIGURE 24

EXERCISE 8

THE PARABOLA

1. Sketch the graph of each of the following parabolas and give the coordinates of the focus and the equation of the directrix.
   (a) $x^2 = 4y$.　　(b) $y = x^2$.　　(c) $y = 4x^2$.　　(d) $y = x^2/32$.
   (e) $y^2 = 2x$.　　(f) $x = 16y^2$.　　(g) $x = 3y^2$.　　(h) $x = y^2/9$.

2. Suppose that the focus is on the y-axis, but below the x-axis instead of above it. We can still use $(0, p)$ for the focus, but then p is negative. Or we may agree that p is

always positive and let $(0, -p)$ be the focus when it lies below the x-axis. In this book we adopt the latter convention and assume that $p > 0$. Prove that $x^2 = -4py$ is an equation for the parabola with focus at $(0, -p)$ and directrix $y = p$.

3. Using the result of Problem 2, sketch each of the following parabolas and give the focus and directrix.

   (a) $y^2 = -2x$.   (b) $x^2 = -4y$.   (c) $y^2 + 4x = 0$.

   (d) $y = -x^2/8$.   (e) $x + y^2 = 0$.   (f) $7y = -5x^2$.

4. Prove that the point of a parabola which is closest to the focus is the vertex.

5. Find an equation for each parabola described.

   (a) Focus at $(0, 3)$;   directrix $y = -3$.

   (b) Focus at $(-2, 0)$;   directrix $x = 2$.

   (c) Vertex at $(0, 0)$;   directrix $y = 4$.

   (d) Vertex at $(0, 0)$;   focus at $(8, 0)$.

6. Derive a simple equation for the parabola with focus at $(4, 5)$ and directrix $y = 1$. Sketch the parabola from the equation.

7. Repeat Problem 6 with the focus at $(-2, -7)$ and directrix the y-axis.

8. Derive the general equation $(x - h)^2 = 4p(y - k)$ for a parabola with focus at $(h, k + p)$ and directrix the horizontal line $y = k - p$. The vertex of this parabola is at (h, k).

9. Using the result of Problem 8, prove that the graph of the equation $y = Ax^2 + Bx + C$ is a parabola if $A \neq 0$. HINT: Complete the square.

10. The parabola $y = x^2$ divides the plane into two regions, one above the parabola and one below the parabola. Show that if $y_1 > x_1^2$, then the point $P(x_1, y_1)$ lies in the region above the parabola $y = x^2$.

THE ELLIPSE

11. Write the standard form for the equation of each of the following ellipses and sketch.

   (a) Distance sum is 10, and foci at $(\pm 3, 0)$.

   (b) Distance sum is 20, and foci at $(0, \pm 6)$.

   (c) Distance sum is 10, and foci at $(\pm 2\sqrt{6}, 0)$.

   (d) Distance sum is 10, and foci at $(0, \pm 1)$.

12. Prove Theorem 15.

13. Find the foci and distance sum for each of the following ellipses, and sketch the graph.

   (a) $\dfrac{x^2}{25} + \dfrac{y^2}{9} = 1$.   (b) $\dfrac{x^2}{25} + \dfrac{y^2}{24} = 1$.   (c) $\dfrac{x^2}{9} + \dfrac{y^2}{25} = 1$.

   (d) $9x^2 + 4y^2 = 36$.   (e) $\dfrac{x^2}{3} + \dfrac{y^2}{4} = 1$.   (f) $\dfrac{x^2}{25} + 4y^2 = 1$.

14. Prove that an ellipse is symmetric with respect to its major axis, and also with respect to its minor axis.

15. Find the ellipse in standard form that has one focus at $(0, -5)$ and one vertex at $(0, 13)$.

16. Find the ellipse in standard form that passes through the points $(4, -1)$ and $(-2, -2)$.

THE HYPERBOLA

17. Write the standard form for the equation of the hyperbola with:
    (a) $\pm(|PF_1| - |PF_2|) = 8$, and foci at $(-5, 0)$ and $(5, 0)$.
    (b) $\pm(|PF_1| - |PF_2|) = 6$, and foci at $(-5, 0)$ and $(5, 0)$.
    (c) $\pm(|PF_1| - |PF_2|) = 4$, and foci at $(-5, 0)$ and $(5, 0)$.
    (d) $\pm(|PF_1| - |PF_2|) = 2$, and foci at $(0, -3)$ and $(0, 3)$.
    (e) $\pm(|PF_1| - |PF_2|) = 1$, and foci at $(0, -2)$ and $(0, 2)$.
    (f) $\pm(|PF_1| - |PF_2|) = 8$, and foci at $(0, -5)$ and $(0, 5)$.

18. Sketch each of the hyperbolas in Problem 17. Put the first three on the same coordinate system.

19. Find the foci for each of the following hyperbolas.

    (a) $\dfrac{x^2}{25} - \dfrac{y^2}{144} = 1.$ (b) $\dfrac{x^2}{2} - \dfrac{y^2}{2} = 1.$ (c) $\dfrac{x^2}{144} - \dfrac{y^2}{25} = 1.$

    (d) $\dfrac{y^2}{4} - \dfrac{x^2}{5} = 1.$ (e) $\dfrac{y^2}{25} - \dfrac{x^2}{144} = 1.$ (f) $x^2 - 2y^2 = 6.$

20. Prove that a hyperbola is symmetric with respect to its transverse axis.

21. Prove that a hyperbola is symmetric with respect to its conjugate axis.

*22. Let $K > 0$. Show that the graph of $xy = K^2/2$ is a hyperbola, by showing that this equation is the equation of the hyperbola with foci at (K, K) and $(-K, -K)$ and distance difference $2K$.

23. Sketch the hyperbola of Problem 22 when $K = \sqrt{2}$.

24. Find the equation of the hyperbola that goes through the point $(2, 3)$ and has foci at $(\pm 2, 0)$.

25. Find the equation of the hyperbola that goes through the points $(\sqrt{6}, 4)$ and $(4, 6)$ if its foci are on the y-axis and the hyperbola has the x-axis as an axis of symmetry.

26. Find the equation of the hyperbola described in Problem 25 if it goes through the points $(3\sqrt{2}, 2)$ and $(12, 5)$.

INTERSECTIONS

In Problems 27 through 32, find the intersection of the given pair of curves. Then sketch the curves and check your work geometrically.

27. $y = 3 + 4x - x^2$ and $y + x = -3$.

28. $y = 9 + 2x - x^2$ and $y = -3x + 13$.

29. $y = 1 + x^2$ and $y = 1 + 4x - x^2$.

30. $\dfrac{x^2}{18} + \dfrac{y^2}{8} = 1$ and $\dfrac{x^2}{3} - \dfrac{y^2}{2} = 1.$

31. $x + y = 1$ and $x^2 + y^2 = 1.$

32. $x + y = 10$ and $x^2 + y^2 = 10.$

REVIEW PROBLEMS

1. Find the solution set for each of the following inequalities.

   (a) $x^2 - 3x + 2 < 0.$ (b) $\dfrac{x}{x-1} \geq 0.$ (c) $|x + 1| \geq 4.$

   (d) $x(x-1)(x-2) \geq 0.$ (e) $|x - 3| \leq |x - 5|.$ (f) $\dfrac{1}{x} + \dfrac{1}{x^2} \leq 2.$

2. A particle moving in the plane starts at the point $(2, -1)$ and stops at the point $(-3, -6)$. Find Δx and Δy.

3. Determine the longest side of the triangle with vertices $P(1, 1)$, $Q(0, 0)$, and $R(\sqrt{3}, -\sqrt{3})$.

4. Are the three points $P(0, -2)$, $Q(1, 0)$, and $R(62, 122)$ collinear?

5. Find the slope and y-intercept of the following lines.

   (a) $3x - 2y = 6.$ (b) $x = -3y + 9.$ (c) $4x - 2y - 8 = 0.$

6. Sketch the graphs of the lines in Problem 5.

In Problems 7 through 10, find the equation of the line that satisfies the given conditions.

7. The line through the two points:

   (a) $P(-5, 6)$, $Q(3, -10).$ (b) $P(-5, -8)$, $Q(7, -4).$
   (c) $P(6, -9)$, $Q(-3, 11).$ (d) $P(4, 7)$, $Q(7, 10).$

8. The perpendicular bisector of the line segment PQ, for the pairs of points in Problem 7.

9. The line through P perpendicular to the line $\mathcal{L}$:

   (a) $P(2, -3)$, $\mathcal{L}:2y = 3x + 1776.$
   (b) $P(4, -5)$, $\mathcal{L}:4y - 5x = 1984.$

10. The line through P with angle of inclination α:

    (a) $P(\sqrt{3}, 5)$, $\alpha = 60°.$ (b) $P(3, \sqrt{5} - \sqrt[3]{17})$, $\alpha = 90°.$
    (c) $P(\sqrt{7}, 2\sqrt{7})$, $\alpha = 135°.$ (d) $P(\sqrt[3]{19} + \sqrt[5]{81}, -7)$, $\alpha = 0°.$

11. Let M be the midpoint of the line segment joining $P_1(-1, -3)$ and $P_2(2, 5)$.

    (a) Find the coordinates of M.

    (b) What is the equation of the circle with center M that passes through P_1 and P_2?

12. Find the equation of the circle with center at $(-1, -1)$ that has $\mathcal{L}:y = 5 - x$ as a tangent line.

13. Let A and B be the points $(0, 0)$ and $(0, 12)$, respectively. Find an equation for the set of all points $P(x, y)$ such that:

    (a) $2|AP| = |BP|$.   (b) $3|AP| = |BP|$.

*14. Prove the following theorem. If k is a positive constant and A and B are two fixed points, then the set of all points $P(x, y)$ such that $k|AP| = |BP|$ is a circle if $k \neq 1$, and is a straight line if $k = 1$.

In Problems 15 through 20, graph the given equation.

15. $x^2 = -16y$.

16. $y^2 = -8x$.

17. $x^2 + y^2 + 10x - 4y - 7 = 0$.

18. $\dfrac{x^2}{3} - \dfrac{y^2}{6} = 1$.

19. $6x^2 + 9y^2 - 54 = 0$.

20. $4y^2 - 12x^2 = 36$.

21. Find an equation for the set of all points $P(x, y)$ such that the difference of its distances from the two points $A(0, 6)$ and $B(0, -6)$ is 3.

22. Find an equation for the ellipse with foci at $(\pm 4, 0)$ and vertices at $(\pm 6, 0)$.

In Problems 23 through 26, find the point (or points) of intersection of the two given curves and check your solutions by sketching the graphs.

23. $5y + 7x = 1$ and $y - 6x = -22$.

24. $y = 2x^2 - 4x + 5$ and $y = 2x^2 + 6x + 15$.

25. $y = x^2 + 2x + 2$ and $y = -x^2 + 2x + 1$.

26. $x^2 + y^2 - 8x + 6 = 0$ and $x^2 + y^2 + 2x - 10y + 6 = 0$.

3

Functions and Sequences

Objective 1

The concept of *function* plays a leading role in mathematics. There are basically two different ways of defining a function: the classical definition and the modern ordered-pair one. In this chapter we will emphasize the classical approach because it is more intuitive and lends itself more readily to physical applications. We will also discuss the idea of *limit* which will be needed in defining the basic operations of differentiation and integration in later chapters. The limit concept is a distinguishing feature of calculus and is largely responsible for the many powerful problem-solving techniques of the subject.

Functions and Function Notation 2

We have already met functions in an informal way. The equation of a straight line

$$(1) \qquad\qquad y = 3x + 7$$

is an example of a function because for each number x, this equation gives an associated value of y. Other natural examples of functions are

$$(2) \qquad\qquad y = x^2,$$

$$(3) \qquad\qquad y = 7\sqrt{25 - x^2},$$

$$(4) \qquad\qquad y = \sqrt{x(x - 1)(x - 2)}.$$

71

In each of these examples we have a rule that tells us how to find the real number y whenever the real number x is given. In these simple cases, the rule is given by a certain formula involving algebraic operations. For the general concept of a function we admit a much wider latitude, and in fact the rule may be quite wild. Further, it is not necessary for the rule to relate numbers. Indeed, a function may relate any two sets. For example, one set might consist of automobiles, and the second set may consist of brand names. Then with each car there is associated some definite brand name: Chevrolet, Ford, Plymouth, and so on. Thus the brand name is a function of the car. As another example we may consider a particular automobile race. With each car entered in that race, there is associated a particular person, the driver of the car in that race. Of course, these last two examples are not very interesting to a mathematician, but they illustrate the general nature of a function. Thus a function may relate elements from any two well-defined sets.

DEFINITION 1 (Function). Let $\mathscr{A}$ and $\mathscr{B}$ be any two nonempty sets, and let x represent an element from $\mathscr{A}$, and let y represent an element from $\mathscr{B}$. A *function* from $\mathscr{A}$ to $\mathscr{B}$ is a rule (method, or procedure) that assigns a unique element y of $\mathscr{B}$ to each x in $\mathscr{A}$. When y is related to x by a rule of this type, we say that y is a function of x. A function is also called a *mapping* from $\mathscr{A}$ to $\mathscr{B}$.

Frequently, we use the symbol f to represent a function. The value y that f assigns to an element x in $\mathscr{A}$ is called the *image* of x and is denoted by $f(x)$. This notation is read "f of x" or "f at x." For the function f given by equation (1), we have

$$f(x) = 3x + 7,$$

so that in particular $f(0) = 3 \cdot 0 + 7 = 7$, and $f(-1) = 3(-1) + 7 = 4$. In other words, this function assigns the value of $y = 7$ to the number $x = 0$ and assigns the value $y = 4$ to $x = -1$.

Example 1. Compute $f(3)$ for each of the functions defined by equations (2), (3), and (4).

Solution. Whenever a function f is given by a formula involving the variable x, we merely replace x by the particular number 3 and compute in accordance with the formula.

If $f(x) = x^2$,   then $f(3) = 3^2 = 9$.

If $f(x) = 7\sqrt{25 - x^2}$,   then $f(3) = 7\sqrt{25 - 9} = 28$.

If $f(x) = \sqrt{x(x - 1)(x - 2)}$,   then $f(3) = \sqrt{3 \cdot 2 \cdot 1} = \sqrt{6}$. ●

Example 2. If $f(x) = x^2$, find $f(w + 3z)$.

Solution. This collection of symbols tells us that we are to replace x by $w + 3z$ in $f(x) = x^2$. When we do this, we obtain

$$f(w + 3z) = (w + 3z)^2 = w^2 + 6wz + 9z^2. \quad \bullet$$

In Definition 1 we call x the *independent variable* of the function because it can be selected arbitrarily from the set $\mathscr{A}$. We call y the *dependent variable* because its value depends on the particular x selected. When we wish to emphasize that x is the independent variable of a function f, we will say for brevity "the function $f(x)$." Similarly, we speak of "the function $y = f(x)$" in order to emphasize that x and y are, respectively, the independent and dependent variables for f.

DEFINITION 2 (Domain and Range). If f is a function from $\mathscr{A}$ to $\mathscr{B}$, the set $\mathscr{A}$ is called the *domain* of f and is denoted by $\mathscr{D}(f)$. The set $\mathscr{C}$ consisting of all y such that $y = f(x)$ for some x in $\mathscr{A}$ is called the *range* of f. If $\mathscr{C} = \mathscr{B}$, we say that f maps $\mathscr{A}$ onto $\mathscr{B}$.

It is sometimes helpful to think of a function f as being a machine or, in the terminology of design engineers, a *black box* (see Fig. 1). When we push x into the machine, out pops its corresponding y. The set of permissible inputs to the machine is precisely the domain of f (i.e., the set $\mathscr{A}$). The collection of all possible outputs of the box constitutes the range of f.

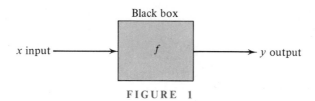

FIGURE 1

Example 3. Discuss the domain and range of each of the four functions defined by equations (1) through (4).

Solution. A function is not properly given unless the domain is stated in advance. Hence the question about the domain is an improper one, since this information has not been supplied. But in any book on the calculus of real functions of a real variable, it is reasonable to suppose that if a function is given by a formula [as in equations (1) through (4)], then the domain is the largest set $\mathscr{D}$ of real numbers for which

$f(x)$ is real if x is in $\mathscr{D}$. If we make this agreement, then for the functions

$$y = 3x + 7 \quad \text{and} \quad y = x^2,$$

the domain is $\mathscr{R}$, the set of all real numbers.

On the other hand, the formula $y = 7\sqrt{25 - x^2}$ does not determine y as a real number when $x^2 > 25$. If we adopt the agreement just mentioned, then the domain of this function is the interval $-5 \leq x \leq 5$. The formula $y = 7\sqrt{25 - x^2}$ shows that if $-5 \leq x \leq 5$, then $0 \leq y \leq 35$.

The function $y = \sqrt{x(x - 1)(x - 2)}$ may be treated similarly. If the quantity under the radical is negative, then either $x < 0$ or $1 < x < 2$. By our agreement, the domain of this function is the complement of this set; that is, $\mathscr{D}(f) = [0, 1] \cup [2, \infty)$.

Since we can solve the equation $y = 3x + 7$ for x in terms of y [obtaining $x = (y - 7)/3$], it is clear that y may be any real number. Hence the range of the function $y = 3x + 7$ is $\mathscr{R}$. On the other hand, for real x, we have $x^2 \geq 0$, so the range of the function $y = x^2$ is the ray $[0, \infty)$. We leave it for the student to argue that the range of the function $y = \sqrt{x(x - 1)(x - 2)}$ is the same ray. ●

Whenever a function is presented for study, an explicit statement of the domain of the function should be given at the same time. This is usually done by writing the restriction just after the formula that gives the function. Thus, in the two examples just studied, we would write

(3) $$y = 7\sqrt{25 - x^2}, \quad -5 \leq x \leq 5,$$

and

(4) $$y = \sqrt{x(x - 1)(x - 2)}, \quad x \in [0, 1] \cup [2, \infty),$$

to indicate the domain. We will avoid the burden of always specifying the domain by adopting the agreement mentioned earlier. *Whenever a function is defined by a formula (or several formulas) and no domain is specified, then the domain of the function is the largest set for which the formula (or formulas) gives a real-valued function of a real variable.* With this agreement the functions defined by (3) and (4) automatically have the domains already displayed.

A function need not be given by a formula, or it may be given by a combination of several formulas, using different formulas for different subsets of the domain. As an example of the first type, consider the statement "y is 1 if x is irrational, and y is zero if x is rational." This defines y as a function of x because it gives a rule that allows us to find a particular y for each x. We might display this function by writing

(5) $$y = f(x) = \begin{cases} 0, & \text{if } x \text{ is a rational number,} \\ 1, & \text{if } x \text{ is an irrational number.} \end{cases}$$

Here the range of the function has only two elements: $y = 0$ and $y = 1$. This interesting function is often called the *Dirichlet function.*

To illustrate that a function may be given by several formulas, we consider the function defined as follows:

$$(6) \qquad y = \begin{cases} -3x - 4, & \text{if } x < -1, \\ x, & \text{if } -1 \le x \le 1, \\ (x - 2)^2, & \text{if } 1 < x. \end{cases}$$

The meaning of this collection of symbols should be clear. Given a particular number x_1 we examine it to see if $x_1 < -1$, or if $-1 \le x_1 \le 1$, or if $x_1 > 1$. At least one of these cases must occur, and x_1 cannot be in two of these sets simultaneously. We can then compute the associated $y_1 = f(x_1)$ by using the proper one of the three formulas in (6). If $x_1 = -3$, we use $y = -3x - 4$ and we find that $y_1 = 5$. If $x_1 = 0$, we use $y = x$ and find that $y_1 = 0$. If $x_1 = 5$, we use $y = (x - 2)^2$ and find that $y_1 = 9$.

The function defined by equation (6) may look artificial, but in truth many functions encountered in everyday affairs are defined in pieces just as in (6). If y is the cost in dollars of a telephone call from Tampa to Los Angeles that lasts x minutes, then y is a function of x for $x > 0$. According to present direct-dial weekday rates, we have

$$(7) \qquad y = \begin{cases} 0.54, & \text{if } 0 < x \le 1, \\ 0.92, & \text{if } 1 < x \le 2, \\ 1.30, & \text{if } 2 < x \le 3, \\ 1.68, & \text{if } 3 < x \le 4, \quad \text{etc.} \end{cases}$$

The reader will recall other functions that are defined by different formulas for different intervals. The cost of mailing a first-class letter as a function of the letter's weight is one such example. Income tax due as a function of taxable income, and the value of a savings bond as a function of time, are also natural examples of this type of function.

So far we have exclusively used the symbol f in the notation for a function. In practice, any letter may be used, although some have become standard favorites. The most popular ones for representing a function are f, g, h, φ, ψ, F, G, and H. We can also distinguish different functions by using letters with subscripts. Thus g_1, g_2, and g_3 may well represent three different functions. Similarly, we can use symbols other than x or y to denote the variables of a function.

Example 4. Let g be the function defined for $r > 0$ whose value $g(r)$ is the circumference of a circle of radius r. Compute $g(2.3)$.

Solution. The values of $g(r)$ are given by the formula $g(r) = 2\pi r$. To find $g(2.3)$, we merely replace r by 2.3 in the formula:

$$g(2.3) = 2\pi(2.3) = 4.6\pi \approx 14.4513. \quad \bullet$$

We remark that the domain of the function g in Example 4 is given as the set of all positive numbers r. Of course, the expression $2\pi r$ can be evaluated for nonpositive numbers r as well. But when we allow this possibility, we are actually speaking of a new function with domain $\mathcal{R}$. This new function is considered to be different from g, since its domain is different.

Example 5. Let $F(x) = 4x + 5$. Is it true that

(8)
$$F(x_1 + x_2) = F(x_1) + F(x_2)$$

for all pairs of real numbers x_1, x_2?

Solution. The left side of (8) looks like a multiplication problem in which F multiplies the sum of x_1 and x_2. Since the Distributive Law of Multiplication,

(9)
$$A(B + C) = AB + AC,$$

is true for all real numbers A, B, C, a beginner might easily believe that (8) is also true for any function F. But in general, (8) is **false.** To settle this, we need only one *counterexample*. We need to find one pair of numbers for which (8) is false. Selecting at random, we let $x_1 = -1$ and $x_2 = 3$. Then the left side of equation (8) gives

$$F(x_1 + x_2) = F(-1 + 3) = F(2) = 4 \cdot 2 + 5 = 13.$$

For the right side we find

$$F(x_1) + F(x_2) = F(-1) + F(3) = [4(-1) + 5] + [4 \cdot 3 + 5] = 1 + 17 = 18.$$

Since $13 \neq 18$, the assertion (8) is false, when $F(x) = 4x + 5$.   ●

EXERCISE 1

1. If $f(x) = x^3 - 2x^2 + 3x - 4$, show that $f(1) = -2, f(2) = 2, f(3) = 14, f(0) = -4$, and $f(-2) = -26$.
2. If $f(x) = 2^x$, show that $f(1) = 2$, $f(5) = 32$, $f(0) = 1$, and $f(-2) = 1/4$.
3. If $g(x) = 2x/(x^2 + 1)$, find $g(1)$, $g(-2)$, and $g(1/2)$.
4. A function h is defined for $x > 1/2$ by $h(x) = (2x - 1)^{-3/2}$. Find $h(1)$, $h(5/2)$, and $h(5)$.
5. A function f is defined for all x and its values are given by

$$f(x) = \begin{cases} 2 - x, & \text{if } x < -1, \\ x^2, & \text{if } -1 \leq x < 2, \\ \sqrt{x - 2}, & \text{if } 2 \leq x. \end{cases}$$

Compute $f(0)$, $f(-2)$, and $f(11)$.

6. A function g is defined for $t > 2$ and its values are given by

$$g(t) = \begin{cases} \dfrac{1}{t}, & \text{if } 2 < t \leq 7, \\[2mm] 13, & \text{if } 7 < t. \end{cases}$$

     Find $g(16/5)$, $g(507)$, and $g(20/3)$.

7. If $f(x) = 2x - x^2$, find $f(3 + h)$.

8. If $f(x) = 6 + 5x$, find $f(w - 2z)$.

In Problems 9 through 16, find the domain of the given function, that is, the largest set $\mathscr{D}$ of real numbers for which the formula defines a mapping of $\mathscr{D}$ into the real numbers.

9. $f(x) = \dfrac{2x + 3}{(x - 1)(x + 2)}.$ 10. $g(x) = \dfrac{6}{x^2 + x - 6}.$

11. $h(x) = 3 + \sqrt{4 - x}.$ 12. $F(x) = (x + 1)^{-3/2}.$

13. $G(x) = (9 - x^2)^{5/2}.$ 14. $H(x) = \sqrt{x^2 - 25}.$

15. $K(x) = \sqrt{(x^2 - 1)(x - 4)}.$ 16. $L(x) = \dfrac{1}{\sqrt{x(x - 1)(x - 2)}}.$

In Problems 17 through 25, find the range of the function defined by the given formula.

17. $y = 1 - x.$ 18. $y = -x^2.$ 19. $y = x^2 + 1.$

20. $y = 5 - 2x^2.$ 21. $y = \dfrac{1}{x}.$ 22. $y = x^3.$

23. $y = \sin x.$ 24. $y = 2 \cos x.$ 25. $y = 11 - 7 \sin x.$

26. The variables s and t are related by the equation $s^2 + t = 1$. Is s a function of t? Is t a function of s?

27. Let x be the length of one edge of a cube. The surface area of the cube is then a function of x. Find a formula for this function.

28. The perimeter of a square is a function of its area. Find a formula for this function.

29. A farmer uses 100 meters of fencing to enclose a rectangular area of length L and width w. Give a formula for L as a function of w. What is the domain and range of this function?

30. Given that one leg of a right triangle is 10 cm, express the length l of the other leg as a function of the length h of the hypotenuse. What is the domain of this function?

31. Two men start walking at the same time from the same point along perpendicular paths. One of them walks at the rate of 2.8 miles/hr, the other walks 3 miles/hr. Express the distance between the two men as a function of the hours t they have spent walking.

32. A line is drawn from the point $(-1, 1)$ to a second point on the parabola $y = x^2$.

Express the slope of the line as a function of the x-coordinate of the second point on the parabola.

⋆**33.** Most numbers have a unique decimal representation, but 1/4 has two different ones, namely 0.2500 . . . and 0.2499 . . . , where the second set of dots indicates an infinite sequence of nines. If we exclude this latter type of decimal representation, then each number has only one decimal representation. Let $d_2(x)$ be the function that gives, for each x, the second digit after the decimal point in the decimal representation of x. Find $d_2(1/2)$, $d_2(\sqrt{2})$, $d_2(\pi)$, $d_2(4)$, and $d_2(1/6)$. What is the range of this function?

34. If $f(x) = 11x$, prove that $f(x + y) = f(x) + f(y)$ for all x, y.

35. If $f(x) = 3x + 5$, prove that $f(4x) = 4f(x) - 15$ for all x.

36. If $f(x) = x^2$, prove that $f(x + h) - f(x) = h(2x + h)$ for all x, h.

37. If $F(x) = \sqrt{x}$, prove that

$$\frac{F(x + h) - F(x)}{h} = \frac{1}{\sqrt{x + h} + \sqrt{x}}, \qquad x > 0, \quad h > 0.$$

FUNCTIONS OF SEVERAL VARIABLES

Function notation extends to functions of several variables. Thus $f(x, y)$ is used to denote a function of the two variables x and y. Then $f(5, 3)$ is the value of the function when $x = 5$ and $y = 3$. Similarly, we can write $f(x, y, z)$ for a function of three variables, $f(x, y, z, u)$ for a function of four variables, and so on.

38. If $f(x, y) = x^2 + xy - 2y^2$, show that $f(1, 3) = -14$, $f(3, -2) = -5$, $f(5, 3) = 22$, $f(t, t) \equiv 0$, and $f(x - y, y) = x^2 - xy - 2y^2$.

39. If $f(x, y) = 2x + 3y^2 - 1$, find: **(a)** $f(-1, 2)$, **(b)** $f(3, 1)$, and **(c)** $f(1, 3)$.

40. If $F(x, y) = (xy - 3x)^2$, find: **(a)** $F(1/3, 0)$, **(b)** $F(-2, 1)$, and **(c)** $F(1, -2)$.

41. If $h(x, y) = 2x/(x + y)$, find: **(a)** $h(0, 3)$, **(b)** $h(1, 2)$, and **(c)** $h(2, 1)$.

42. If $G(x, y) = xy/(x^2 + y^2)$, show that $G(x, x) = 1/2$ for $x \neq 0$.

43. If $g(x, y) = (x - y)/(x + y)$,   find: **(a)** $g(1, 1)$,   **(b)** $g(2, 3)$,   **(c)** $g(3, -2)$,   **(d)** $g(2t, -t)$, and **(e)** $g(x + y, x - y)$.

44. For the function given in Problem 43, prove that

$$g(x, y) \equiv \frac{1}{g(x, -y)} \equiv -g(y, x) \equiv g(tx, ty),$$

(assuming that none of the denominators is zero).

45. If $H(x, y, z) = (x - y)(y - z)(z - x)$,   find: **(a)** $H(3, 2, 1)$,   **(b)** $H(5, 4, 3)$, **(c)** $H(2, 0, -3)$, and **(d)** $H(4t, t, 0)$.

46. For the function defined in Problem 45, prove that

$$H(x, y, z) \equiv H(y, z, x) \equiv H(z, x, y).$$

Further, prove that $H(x^2, y^2, z^2) = (x + y)(y + z)(z + x)H(x, y, z)$.

CALCULATOR PROBLEMS

C1. On many calculators there are special keys for computing functions such as $y = 1/x$, $y = \sqrt{x}$, $y = x^2$, $y = 10^x$, and so on. When an input is not in the domain of the function, the calculator will display a warning. For your own calculator see what happens when: **(a)** $x = 0$ is the input for $y = 1/x$, **(b)** $x = -3$ is the input for $y = \sqrt{x}$, and **(c)** $x = 90°$ is the input for $y = \tan x$.

C2. The sum of $100 is deposited in a bank that pays r percent interest per year, compounded quarterly. After 1 year the balance $B(r)$ is given by the formula

$$B(r) = 100\left(1 + \frac{r}{400}\right)^4.$$

Compute $B(5)$ and $B(5\frac{3}{4})$.

C3. Let f be the function defined by

$$f(x) = 5\left(\sqrt{x} + \frac{1}{\sqrt{x}}\right), \qquad \text{for } 0.8 \leqq x \leqq 1.3.$$

Compute the values of $f(x)$ for $x = 0.8, 0.9, 1.0, 1.1, 1.2$, and 1.3, and arrange your answers in a table. Based on this data, what would you *guess* is the minimum (smallest) value assumed by $f(x)$ over its *whole* domain $[0.8, 1.3]$? Prove algebraically that your guess is correct.

The Graph of a Function **3**

> **DEFINITION 3.** Let $y = f(x)$ be a real-valued function of a real variable. The *graph of f* is the collection of all points (x, y) in the plane for which x is in the domain of f and $y = f(x)$.

This is reminiscent of Definition 2, Chapter 2, for the graph of an equation. In fact, the ideas are quite similar. The essential difference is this: If y is a function of x, then for each x in the domain of the function, there is *exactly one y,* but for an arbitrary equation relating x and y, there may be more than one y. For example, the equation $y^2 = x^2$ gives two values of y for each $x \neq 0$. The graph of this equation is shown in Fig. 12 of Chapter 2.

Stated geometrically, each vertical line intersects the graph of a *function* in at most one point. However, each vertical line may intersect the graph of an equation in n points, where n is any integer greater than or equal to zero; it may even intersect the graph in infinitely many points. For the graphs in Fig. 2, only (b) is the graph of a function because every vertical line has at most one intersection with the graph. This is not the case for Figs. 2(a) and 2(c).

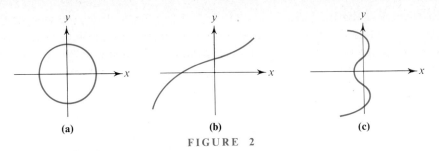

FIGURE 2

Example 1. Sketch the graph of the function f given by $f(x) = x + 1$.

Solution. Here $y = x + 1$, so the graph is a straight line with slope 1 and y-intercept 1 (see Fig. 3). ●

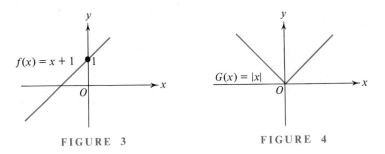

FIGURE 3   **FIGURE 4**

Example 2. Graph the function $G(x) = |x|$.

Solution. From the definition of absolute value, we have

$$(10) \qquad y = |x| = \begin{cases} x, & \text{if } x \geq 0, \\ -x, & \text{if } x < 0. \end{cases}$$

Thus we draw the portion of the line $y = x$ for which $x \geq 0$ and draw the part of the line $y = -x$ for $x < 0$ (see Fig. 4). ●

Example 3. Graph the function

$$(11) \qquad H(x) = \begin{cases} -1, & \text{if } x < 0, \\ x^2, & \text{if } 0 \leq x \leq 1, \\ \dfrac{x+2}{3}, & \text{if } x > 1. \end{cases}$$

Solution. Considering the separate formulas for H, we draw the portion of the horizontal line $y = -1$ for $x < 0$, the portion of the parabola $y = x^2$ for $0 \leq x \leq 1$, and the portion of the line $y = (x + 2)/3$ for $x > 1$ (see Fig. 5). From the definition of H it is clear that its domain is $\mathscr{R}$, the set of all real numbers. From Fig. 5 we can also see that the range of H, which is the set of all the y-coordinates of the points on the graph, is $[0, +\infty) \cup \{-1\}$. ●

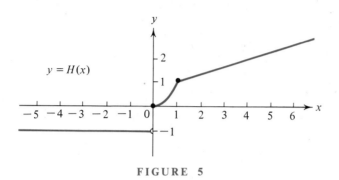

$y = H(x)$

FIGURE 5

The graph of $H(x)$, shown in Fig. 5, has a break or jump at $x = 0$. We therefore say that this function is *discontinuous* at $x = 0$. We will give a careful definition of continuous and discontinuous functions in Chapter 4.

For the construction of certain special types of discontinuous functions it is convenient to introduce the symbol $[x]$ (read "square bracket of x" or "the integer of x").

DEFINITION 4 (The Integer of x). The symbol $[x]$ is defined to be the greatest integer that is less than or equal to x.

In other words, if for a fixed x we let n be the unique integer such that

(12)
$$n \leq x < n + 1,$$

then

(13)
$$[x] = n.$$

Example 4. Find $[x]$ for $x = 2$, 3.78, $-1/3$, and $-\pi$.

Solution. From equations (12) and (13), we have $[2] = 2$, $[3.78] = 3$, $[-1/3] = -1$, and $[-\pi] = -4$. ●

THEOREM 1 (PLE). If $[x] = n$, then $x = n + \theta$, where θ is some number such that $0 \leq \theta < 1$. Conversely, if $x = n + \theta$, with $0 \leq \theta < 1$, then $[x] = n$.

Example 5. Graph the functions **(a)** $f(x) = [x]$ and **(b)** $g(x) = x - [x]$.

Solution. **(a)** Using equations (12) and (13), we find that the graph of $f(x) = [x]$ consists of the horizontal segments $y = n$ for $n \leq x < n + 1$, $n = 0, \pm 1, \pm 2, \ldots$ (see Fig. 6).
(b) Similarly, to graph $g(x) = x - [x]$, we sketch the line segments $y = x - n$ for $n \leq x < n + 1$, $n = 0, \pm 1, \pm 2, \ldots$ (see Fig. 7).   ●

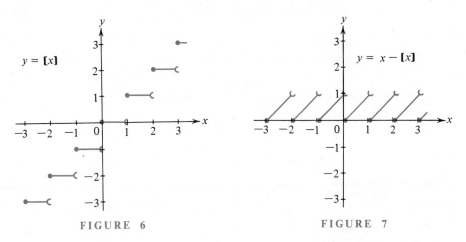

FIGURE 6   FIGURE 7

Example 6. The graph of a function f is given in Fig. 8. Find $f(1)$ and $f(-2)$. What is the domain and range of f?

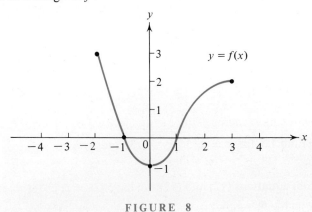

FIGURE 8

Solution. We remark that in reading off function values from a given graph, we can generally give only *approximate* answers. The accuracy of these answers will depend on the scale used for the coordinate system. If we read from the graph in Fig. 8, it appears that $f(1) = 0$ and $f(-2) = 3$.

The domain of f is the set of all abscissas of the points on the graph, that is, the closed interval $[-2, 3]$, and its range is the set of all ordinates on the graph, namely the closed interval $[-1, 3]$. ●

Frequently, functions have certain characteristics that are worthy of a special name. Two such are covered in

---

DEFINITION 5 (Even and Odd Functions). Let $f(x)$ be defined in a set $\mathscr{D}$. The function f is called an *even function* in $\mathscr{D}$ if

(14)
$$f(-x) = f(x)$$

whenever x and $-x$ are both in $\mathscr{D}$. The function f is called an *odd function* in $\mathscr{D}$ if

(15)
$$f(-x) = -f(x)$$

whenever x and $-x$ are both in $\mathscr{D}$.

THEOREM 2 (PLE). Let $f(x) = x^n$, where n is an integer. If n is an even integer, then $f(x)$ is an even function. If n is an odd integer, then $f(x)$ is an odd function.

Clearly, it is this theorem that accounts for the names "even" and "odd" for the two types of functions. Note that by our agreement, $\mathscr{D} = \mathscr{R}$ if n is a positive integer, and $\mathscr{D} = \mathscr{R} - \{0\}$ if n is a negative integer.

If f is an *even* function on an interval such as $[-a, a]$, then its graph will be symmetric with respect to the y-axis. Indeed, if the point (x_0, y_0) is on the graph of f, then so is the point $(-x_0, y_0)$ because $f(-x_0) = f(x_0) = y_0$ (see Fig. 9).

The graph of an *odd* function f on $[-a, a]$ is symmetric with respect to the origin. Whenever the point (x_0, y_0) lies on the graph, so does the point $(-x_0, -y_0)$ (see Fig. 10). Notice that the origin is the midpoint of the line segment joining (x_0, y_0) and $(-x_0, -y_0)$.

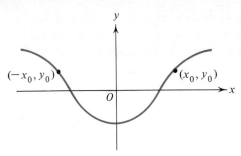

FIGURE 9 **(Even function)**

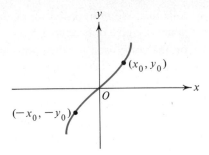

FIGURE 10 **(Odd function)**

EXERCISE 2

1. Graph each of the following functions (all with the same set of coordinate axes) for the interval $0 \leq x \leq 3$.
   (a) $y = x$.   (b) $y = \sqrt{x}$.   (c) $y = \sqrt[3]{x}$.   (d) $y = \sqrt[4]{x}$.
2. Graph the function f defined by

$$f(x) = \begin{cases} 2 + \sqrt{4 - x^2}, & \text{if } -2 \leq x \leq 2, \\ 2, & \text{for all other } x. \end{cases}$$

   What is the range of f?
3. Graph each of the following functions.
   (a) $y = |x + 2|$.   (b) $y = 1 + x - |x|$.
   (c) $y = \begin{cases} 2, & \text{if } x \geq 2, \\ 3x - 4, & \text{if } 1 < x < 2, \\ -1, & \text{if } x \leq 1. \end{cases}$ (d) $y = \begin{cases} x^2, & \text{if } -2 < x < 2, \\ 2x, & \text{for all other } x. \end{cases}$
4. Graph each of the following functions.
   (a) $y = -|x|$.   (b) $y = |x^2 - 1|$.
   (c) $y = \begin{cases} 1, & \text{if } x \geq 1, \\ x, & \text{if } -1 < x < 1, \\ -1, & \text{if } x \leq -1. \end{cases}$ (d) $y = \begin{cases} 4 - x^2, & \text{if } -1 \leq x \leq 1, \\ 2, & \text{for all other } x. \end{cases}$
5. Which of the functions in Problem 3 are discontinuous? For what values of x do the discontinuities occur?

In Problems 6 through 13, sketch the graph of the given function f.

6. $f(x) = [2x], x \in [0,2)$.   7. $f(x) = [x + 3]$.
8. $f(x) = 4 - [x]$.   9. $f(x) = [x] - x$.
10. $f(x) = \left[\dfrac{1}{1 + x^2} \right]$.   11. $f(x) = \left[\dfrac{3}{1 + x^2} \right]$.
★12. $f(x) = x^2 - [x^2], x \in [-\sqrt{3}, \sqrt{3}]$.   ★13. $f(x) = |x + 1| + |x - 1| - 2|x|$.

*14. For the greatest integer function, prove that for all $x, y \in \mathcal{R}$:

    (a) $[x + y] \geqq [x] + [y]$.

    (b) $[2x] = [x] + [x + 1/2]$.

    (c) $[x + n] = [x] + n$, for any integer n.

    HINT: Use Theorem 1.

*15. Prove Theorem 2.

16. Let f and g be two functions both defined in $\mathcal{D}$. Prove that:

    (a) If f and g are even in $\mathcal{D}$, then $f + g$ is even in $\mathcal{D}$.

    (b) If f and g are even in $\mathcal{D}$, then fg is even in $\mathcal{D}$.

    (c) If f and g are odd in $\mathcal{D}$, then $f + g$ is odd in $\mathcal{D}$.

    (d) If f and g are odd in $\mathcal{D}$, then fg is *even* in $\mathcal{D}$.

    What is the situation if f is even in $\mathcal{D}$ and g is odd?

17. Let f be an odd function in $\mathcal{D}$ and suppose that 0 is in $\mathcal{D}$. Prove that $f(0) = 0$.

18. For each of the following, determine whether the given function is even, odd, or neither.

    (a) $f(x) = \dfrac{x}{1 + x^2}$.   (b) $f(x) = 3x^2 + 1$.   (c) $f(x) = x + 3$.

    (d) $f(x) = x^{1/3}$.   (e) $f(x) = \sin x$.

19. For each of the following, determine whether the given function is even, odd, or neither.

    (a) $f(x) = |x|$.   (b) $f(x) = 2x^3 - x$.   (c) $f(x) = 4x^2 + x$.

    (d) $f(x) = [x]$.   (e) $f(x) = \cos x$.

*20. Criticize the definition of an even and odd function on the following grounds. Let $\mathcal{D}$ be a set that contains only positive numbers. Then the requirement that x and $-x$ are both in $\mathcal{D}$ is never satisfied. Hence f satisfies (vacuously) the definition of an even function (or an odd function). This proves the following theorem. *If $\mathcal{D}$ contains only positive numbers, then every function is both an odd function and an even function in $\mathcal{D}$.* Should we accept this theorem, or alter the definition?

21. Graph each of the following functions f. Then with the aid of your graph determine (i) the maximum value of $f(x)$, and (ii) the minimum value of $f(x)$ over its domain.

    (a) $f(x) = x^2 - 3$ for $x \in [-3, 2]$.   (b) $f(x) = \begin{cases} \sqrt{x} & \text{if } 0 \leq x < 1, \\ 2 - x, & \text{if } 1 \leq x \leq 4. \end{cases}$

CALCULATOR PROBLEMS

C1. Find $[x]$ for (a) $x = -4{,}178/369$, (b) $x = \sqrt{763}$, (c) $x = 9\pi^2 - 45$, and (d) $x = -\sqrt[4]{521/13}$.

C2. Evaluate the function $f(x) = (3 - x)/(1 + x)$ on the interval $[0, 2]$ starting with $x = 0$ and using increments of 0.25 [i.e., compute $f(0), f(0.25), f(0.50), \ldots, f(2)$]. Use these results to help sketch the graph of f on $[0, 2]$.

C3. Do Problem C2 for $f(x) = 3x/(x^2 - 4)$ on $[1, 3]$, using increments of 0.25. Reexamine this function near $x = 2$.

C4. Do Problem C2 for $f(x) = x^4 - 1.8x$ on $[-1, 1]$, using increments of 0.25.

C5. Sketch the graph of $f(x) = 1.3x^3 - 2.6x + 0.9$ on $[-1, 1]$ by first evaluating the function starting with $x = -1$ and using increments of 0.25. Observe from your graph that $f(x) = 0$ for some x between 0.25 and 0.50. Now magnify the graph on the interval $[0.25, 0.50]$ by starting with $x = 0.25$ and evaluating the function using increments of 0.05. It should now be apparent that $f(x) = 0$ for some x between 0.35 and 0.40.

C6. Sketch the graph of $f(x) = x^3 + x - 1$ on $[0, 1]$ by first evaluating the function starting with $x = 0$ and using increments of 0.1. Observe from your graph that $f(x) = 0$ for some x between 0.6 and 0.7. Now magnify the graph on the interval $[0.6, 0.7]$ by starting with $x = 0.6$ and evaluating the function using increments of 0.01. It should now be apparent that $f(x) = 0$ for some x between 0.68 and 0.69.

★4 Sequences and Limits [1]

The word *sequence* is often used to mean a list or succession of quantities. Using the definition of a function of a single variable, we can make this meaning more precise.

DEFINITION 6 (Sequence). A sequence is a function whose domain is the set of all positive integers.[2]

If S denotes such a function, then we can display its values in the form

(16) $S(1),$ $S(2),$ $S(3),$ $\ldots,$ $S(n),$ $\ldots,$

where the final three dots indicate that the list continues indefinitely. In dealing with sequences it is customary to depart from the ordinary function notation and use subscripts, writing S_1 for $S(1)$, S_2 for $S(2)$, and so on. With this agreement the display (16) becomes

(17) $S_1,$ $S_2,$ $S_3,$ $\ldots,$ $S_n,$ $\ldots.$

The values in (17) are called the *terms* of the sequence; S_1 is the first term, S_2 the second term, and, more generally, S_n is the nth term. Since a sequence is completely specified

---

[1] The teacher who prefers can defer Sections 4 and 5 until later. However, these sections will be particularly helpful for the calculator applications.

[2] Some authors refer to such a function as an *infinite* sequence.

when all its terms are known, we are justified in saying that *any unending list of numbers is a sequence.*

Sequences are often defined by an explicit formula, such as

(18) $$S_n = n^2: \qquad 1, \, 4, \, 9, \, 16, \dots$$

(19) $$S_n = \frac{n}{n+1}: \qquad \frac{1}{2}, \, \frac{2}{3}, \, \frac{3}{4}, \, \frac{4}{5}, \dots$$

(20) $$S_n = \frac{(-1)^n}{n}: \qquad -1, \, \frac{1}{2}, \, -\frac{1}{3}, \, \frac{1}{4}, \dots$$

In each of the preceding examples the formula is on the left, and the first four terms of the sequence are on the right. These terms were obtained by simply substituting $n = 1, 2, 3,$ and 4 in the given formula.

As with ordinary functions we can use letters different from S for a sequence.

Example 1. Find **(a)** the 19th term and **(b)** the $(n-1)$st term of the sequence

(21) $$a_n = \frac{2n}{3n^2 + 1}.$$

Solution. **(a)** To compute the 19th term, we put $n = 19$ in formula (21):

$$a_{19} = \frac{2 \cdot 19}{3(19)^2 + 1} = \frac{38}{1084}.$$

(b) Since equation (21) gives the nth term of the sequence, the $(n-1)$st term is obtained by replacing n by $n - 1$:

$$a_{n-1} = \frac{2(n-1)}{3(n-1)^2 + 1} = \frac{2n-2}{3n^2 - 6n + 4}. \quad \bullet$$

In some important applications a sequence will be defined by means of a *recurrence relation* that expresses the nth term of the sequence as a function of its preceeding terms. An example of this type of sequence is

(22) $$b_1 = 2,$$

(23) $$b_n = nb_{n-1} + 1, \qquad n = 2, 3, 4, \dots.$$

Substituting $n = 2$ in the recurrence relation (23), we find that

$$b_2 = 2b_1 + 1 = 2 \cdot 2 + 1 = 5,$$

where we have used the given fact that $b_1 = 2$. Having $b_2 = 5$, we can compute b_3 by again using (23), now with $n = 3$:

$$b_3 = 3b_2 + 1 = 3 \cdot 5 + 1 = 16.$$

Proceeding in this stepwise manner, we can generate each term of the sequence by knowing the value of the previous term. The first five terms of the sequence defined by (22) and (23) are

(24) $2, 5, 16, 65, 326, \ldots .$

We turn now to the important notion of *limit* of a sequence. By way of example, let us examine the sequence

(25) $s_n = \dfrac{1}{n}$: $1, \dfrac{1}{2}, \dfrac{1}{3}, \dfrac{1}{4}, \dfrac{1}{5}, \ldots .$

Note that as we proceed down the list, the terms are getting closer and closer to zero. Indeed, if we plot the terms on a real number line as shown in Fig. 11, we see that they are

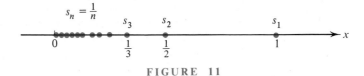

FIGURE   11

crowding toward $x = 0$. For this reason we say that the sequence $s_n = 1/n$ *converges* to zero, or has *limit* zero. We indicate this by writing

(26) $\displaystyle\lim_{n \to \infty} \frac{1}{n} = 0$ or $\dfrac{1}{n} \to 0$ as $n \to \infty,$

which is read "the limit as n approaches infinity of $1/n$ is zero."
     For the sequence

(27) $s_n = 2 - \dfrac{1}{n^2}$: $1, \dfrac{7}{4}, \dfrac{17}{9}, \dfrac{31}{16}, \dfrac{49}{25}, \ldots$

the terms are getting closer and closer to the number 2 as we look way down the list. Indeed, for large integers n the quantity $1/n^2$ is small (i.e., near zero) so that $2 - (1/n^2)$ will be very close to 2. Hence we say that the sequence (27) converges to 2 and we write

$$\lim_{n \to \infty} \left(2 - \frac{1}{n^2} \right) = 2.$$

With the above examples as background, we now give

DEFINITION 7 (Limit of a Sequence). A sequence s_1, s_2, s_3, . . . is said to have *limit L,* or to *converge* to the number L, if the s_n eventually (i.e., for all large values of n) stay arbitrarily close to L. In such a case we write

(28) $$\lim_{n \to \infty} s_n = L \quad \text{or} \quad s_n \to L \text{ as } n \to \infty.$$

Geometrically, the limit of a sequence is L if the terms of the sequence eventually cluster at $x = L$ (see Fig. 12).

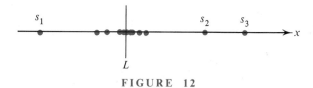

FIGURE 12

In Definition 7, the statement that the terms of the sequence *eventually stay arbitrarily close to L* means that any open interval centered at L (no matter how small in length) will contain all the terms s_n from some value of n onward. This interpretation is developed further in Appendix 1.

We wish to emphasize that not all sequences converge. If $s_n = (-1)^n$, then the first few terms are $-1, 1, -1, 1, -1, 1, \ldots$. Clearly, the terms oscillate between -1 and $+1$. Hence there is no *single* number L that the terms approach as n becomes large.

DEFINITION 8 (Divergent Sequence). A sequence that does not converge is said to *diverge.*

Another example of a divergent sequence is

(29) $$s_n = n: \quad 1, 2, 3, 4, 5, \ldots.$$

Here the terms grow without bound and therefore do not get close to any (finite) number L.

Example 2. Does the sequence

(30) $$a_n = \frac{3n^2}{2n^2 + n + 1}$$

converge or diverge? If it converges, find the limit.

Solution. It is helpful to rephrase the problem in the following way. If we look at the terms of the sequence $a_1, a_2, a_3, \ldots$ having large subscripts, are they all getting closer to some specific number L? To gain some insight, we have calculated (to five decimal places) several terms of the sequence (30) and arranged their values in Table 1. It appears from this table that the terms are "zeroing in" on the number 1.5. Indeed, it is easy to give a convincing argument of this fact.

TABLE 1

n	10	100	1000	10,000	100,000
$\dfrac{3n^2}{2n^2 + n + 1}$	1.42180	1.49246	1.49925	1.49992	1.49999

Dividing the numerator and denominator of (30) by n^2, we find that

$$(31) \qquad a_n = \frac{3}{2 + \dfrac{1}{n} + \dfrac{1}{n^2}}.$$

Now as n becomes large, $1/n$ and $1/n^2$ are each getting closer to zero so that the denominator

$$2 + \frac{1}{n} + \frac{1}{n^2}$$

is a number close to 2. Hence the fraction (31) gets closer and closer to $3/2 = 1.5$. In symbols,

$$\lim_{n \to \infty} \frac{3n^2}{2n^2 + n + 1} = \frac{3}{2}. \quad \bullet$$

In the example above we tacitly used some facts about sequences that are worthy of explicit mention. First, if p is any fixed positive number, then

$$(32) \qquad \lim_{n \to \infty} \frac{1}{n^p} = 0, \qquad p > 0;$$

that is, as n becomes large, $1/n^p$ is getting closer and closer to zero. For example,

$$\lim_{n \to \infty} \frac{1}{\sqrt{n}} = 0, \qquad \lim_{n \to \infty} \frac{1}{n^2} = 0, \qquad \text{and} \qquad \lim_{n \to \infty} \frac{1}{n^{25/7}} = 0.$$

Next, if we have a constant sequence

$$s_n = c: \qquad c, c, c, c, \ldots,$$

where c is any fixed number, then the terms certainly approach (in fact, equal) c. Hence

(33)
$$\lim_{n \to \infty} c = c.$$

Now suppose that $a_1, a_2, a_3, \ldots$ and $b_1, b_2, b_3, \ldots$ are two convergent sequences with

(34)
$$\lim_{n \to \infty} a_n = L \quad \text{and} \quad \lim_{n \to \infty} b_n = M.$$

If by termwise addition we form the new sequence

(35)
$$a_1 + b_1, \quad a_2 + b_2, \quad a_3 + b_3, \quad \ldots, \quad a_n + b_n, \quad \ldots,$$

then what can be said about its convergence? From (34) we know that for large values of n, the terms a_n are close to the number L and the terms b_n are close to M. Thus the sum $a_n + b_n$ will be a number close to $L + M$. In other words, *the limit of the sum of two convergent sequences is the sum of the limits*:

$$\lim_{n \to \infty} (a_n + b_n) = \lim_{n \to \infty} a_n + \lim_{n \to \infty} b_n = L + M.$$

Similarly, we can form new sequences by taking termwise products or termwise quotients, to obtain

(36)
$$a_1 b_1, \quad a_2 b_2, \quad a_3 b_3, \quad \ldots, \quad a_n b_n, \ldots,$$

and

(37)
$$\frac{a_1}{b_1}, \quad \frac{a_2}{b_2}, \quad \frac{a_3}{b_3}, \quad \ldots, \quad \frac{a_n}{b_n}, \ldots.$$

The convergence behavior of these sequences is as one would expect; *the limit of the product sequence* (36) *is the product of the limits $L \cdot M$,* and *the limit of the quotient sequence* (37) *is the quotient of the limits L/M, provided that $M \neq 0$.* For convenience in reference we collect these facts in

---

THEOREM 3. If $a_1, a_2, a_3, \ldots$ and $b_1, b_2, b_3, \ldots$ are two convergent sequences, then

(38)
$$\lim_{n \to \infty} (a_n \pm b_n) = \lim_{n \to \infty} a_n \pm \lim_{n \to \infty} b_n,$$

(39)
$$\lim_{n \to \infty} a_n b_n = \left(\lim_{n \to \infty} a_n \right) \cdot \left(\lim_{n \to \infty} b_n \right),$$

(40)
$$\lim_{n \to \infty} \frac{a_n}{b_n} = \frac{\lim_{n \to \infty} a_n}{\lim_{n \to \infty} b_n} \quad \text{if } \lim_{n \to \infty} b_n \neq 0.$$

A formal proof of Theorem 3 can be found in Appendix 1. Here are some examples of how this theorem can be applied:

(41)
$$\lim_{n\to\infty} \left(3 + \frac{1}{n}\right) = \lim_{n\to\infty} 3 + \lim_{n\to\infty} \frac{1}{n} = 3 + 0 = 3,$$

(42)
$$\lim_{n\to\infty} \left(2 - \frac{1}{\sqrt{n}}\right) = \lim_{n\to\infty} 2 - \lim_{n\to\infty} \frac{1}{\sqrt{n}} = 2 - 0 = 2,$$

(43)
$$\lim_{n\to\infty} \left(3 + \frac{1}{n}\right)\left(2 - \frac{1}{\sqrt{n}}\right) = \lim_{n\to\infty} \left(3 + \frac{1}{n}\right) \cdot \lim_{n\to\infty} \left(2 - \frac{1}{\sqrt{n}}\right) = 3 \cdot 2 = 6,$$

(44)
$$\lim_{n\to\infty} \frac{2 - \dfrac{1}{\sqrt{n}}}{3 + \dfrac{1}{n}} = \frac{\lim_{n\to\infty}\left(2 - \dfrac{1}{\sqrt{n}}\right)}{\lim_{n\to\infty}\left(3 + \dfrac{1}{n}\right)} = \frac{2}{3}.$$

Note, however, that we cannot use equation (40) of Theorem 3 to find

(45)
$$\lim_{n\to\infty} \frac{7}{\left(\dfrac{1}{n} + \dfrac{1}{n^2}\right)},$$

because the limit of the denominator is zero. In fact, the sequence in (45) is divergent, since as *n* becomes large its terms are of the form

$$7 \div (\text{small positive number}) = \text{large positive number},$$

where the large positive number grows without bound. Thus the limit (45) does not exist.

Example 3. For the sequence

(46)
$$s_n = \frac{4n^3 + n}{n^5 - 1},$$

find the limit if it exists.

Solution. On dividing the numerator and denominator in (46) by n^5, and using Theorem 3, we can write

$$\lim_{n\to\infty} \frac{4n^3 + n}{n^5 - 1} = \lim_{n\to\infty} \frac{\dfrac{4}{n^2} + \dfrac{1}{n^4}}{1 - \dfrac{1}{n^5}} = \frac{\lim_{n\to\infty}\left(\dfrac{4}{n^2} + \dfrac{1}{n^4}\right)}{\lim_{n\to\infty}\left(1 - \dfrac{1}{n^5}\right)}$$

$$= \frac{\lim_{n\to\infty}\dfrac{4}{n^2} + \lim_{n\to\infty}\dfrac{1}{n^4}}{\lim_{n\to\infty}1 - \lim_{n\to\infty}\dfrac{1}{n^5}} = \frac{0 + 0}{1 - 0} = \frac{0}{1} = 0. \quad \bullet$$

The astute reader may well balk at Example 3 because when $n = 1$ the denominator in (46) is zero so that the first term s_1 is not defined. Thus for Example 3 to be meaningful we need to slightly generalize the definition of "sequence." Before doing so, we wish to emphasize an important point: If we alter the first 89 terms of a sequence $s_1, s_2, s_3, \ldots$, this will *not* affect the convergence or divergence of the sequence. Indeed, recall from Definition 7 that when dealing with limits we only care about what happens "eventually," that is, way down the list where the subscript n is large. Thus in questions of convergence we can always ignore the first 89 terms. Of course, this statement is not only true for the first 89 terms, but also for the first 329 or 11,226 terms; more generally, *we can ignore any fixed number of terms in dealing with limits.* By way of illustration, consider the sequence

$$(47) \qquad s_n = \begin{cases} 3, & \text{if } 1 \leq n \leq 81 \\ 6, & \text{if } n > 81 \end{cases} : \quad 3, 3, \ldots, 3, 6, 6, 6, 6, 6, \ldots .$$

Here the first 81 terms equal 3, while all the remaining terms equal 6. Thus, ignoring the first 81 terms, we have

$$\lim_{n \to \infty} s_n = 6.$$

Returning to the sequence of Example 3, we see that even though the first term is undefined, the limit is not affected. It is in this spirit that we often modify the definition of sequence to include a function whose domain is the set of all positive integers greater than some arbitrary number.

Example 4. Decide whether the sequence

$$(48) \qquad s_n = \frac{30n}{2^n}$$

converges or diverges. If it converges, find the limit.

Solution. Again some insight is provided by calculating the terms s_n for several positive integers n. This is done in Table 2, where the values are rounded to five decimal places.

TABLE 2

n	1	5	10	15	20	25
$\dfrac{30n}{2^n}$	15	4.68750	0.29297	0.01373	0.00057	0.00002

From Table 2 we are lead to guess that

$$(49) \qquad \lim_{n \to \infty} \frac{30n}{2^n} = 0.$$

That this is in fact the case can be seen from the following argument.
According to the Binomial Theorem of algebra,

$$2^n = (1 + 1)^n = 1 + n + \frac{n(n-1)}{2} + \cdots + 1 > \frac{n(n-1)}{2} \quad \text{for } n \geqq 2.$$

Thus

(50) $$0 < \frac{30n}{2^n} < \frac{30n}{\dfrac{n(n-1)}{2}} = \frac{60}{n-1} \quad \text{for } n \geqq 2.$$

Now as n becomes large, the sequence $60/(n-1)$ approaches zero. Since from inequality (50) the terms $30n/2^n$ are "squeezed" between 0 and $60/(n-1)$, there is no alternative but for $30n/2^n$ to approach zero also. ●

We remark that by using an argument similar to the one in Example 4, it can be shown that for any fixed number r in the interval $-1 < r < 1$,

(51) $$\lim_{n \to \infty} r^n = 0,$$

while if $|r| > 1$, the sequence r^n diverges. For example,

$$\lim_{n \to \infty} \left(\frac{2}{3}\right)^n = 0, \qquad \lim_{n \to \infty} (-0.68)^n = 0, \qquad \text{and} \qquad \lim_{n \to \infty} \frac{1}{3^n} = 0,$$

but the limits of the sequences $(1.01)^n$ and $(-7)^n$ do not exist.
For some sequences $s_1, s_2, s_3, \ldots$ the question of convergence or divergence is not an easy matter. Furthermore, even if the sequence is known to converge, the value of its limit may not always be obvious.

Example 5. Find

(52) $$\lim_{n \to \infty} \left(1 + \frac{2}{n}\right)^n$$

if it exists.

Solution. Choosing various values for n, we calculate (to five decimals) some terms of the sequence. These values are arranged in Table 3.

TABLE 3

n	10	50	100	10^4	10^6	10^8
$\left(1 + \dfrac{2}{n}\right)^n$	6.19174	7.10668	7.24465	7.38758	7.38904	7.38906

From this table, the terms appear to be very slowly converging to a number L with the numerical value $L = 7.389 \ldots$. Unfortunately, this does not seem to be any familiar number. Furthermore, because we can only calculate to finitely many decimal places, it is not at all clear whether the limit L is a rational or irrational number. Thus at this point in our study of calculus all we can cautiously say is that the sequence *appears* to be converging to a number having 7.389 as its first few decimals. ●

We wish to stress that the question of convergence or divergence of a sequence can never be settled by computing the values of several terms. Indeed, we can only calculate a finite number of terms, and, as previously mentioned, the limiting behavior of a sequence is not influenced by such values. Nonetheless, computations can often lend insight and help us make an educated *guess* as to convergence of the sequence.

EXERCISE 3

1. Find the first four terms of each of the following sequences.

   (a) $s_n = \dfrac{n}{n+7}$.   (b) $s_n = (-2)^n$.   (c) $s_n = n^2 - n$.

2. Find the first four terms of each of the following sequences.

   (a) $s_n = (-1)^{n-1}$.   (b) $s_n = (n+1)^2$.   (c) $s_n = \dfrac{5n}{n+3}$.

3. What is the 25th term of the sequence $a_n = (-1)^n/(n+5)$?
4. What is the 100th term of the sequence $b_n = 3n^2/100$?
5. For the sequence $b_1 = 0$, $b_n = n + b_{n-1}^2$, $n = 2, 3, \ldots$, find the first four terms.
6. For the sequence $a_1 = -1$, $a_n = 3a_{n-1} + 7$, $n = 2, 3, \ldots$, find the first four terms.
7. Find the first five terms of the sequence given by $s_1 = 0$, $s_2 = 1$, and $s_n = ns_{n-1} - s_{n-2}$, $n = 3, 4, 5, \ldots$.
8. The *Fibonacci sequence* is defined by $F_1 = 1$, $F_2 = 1$, and $F_n = F_{n-1} + F_{n-2}$, $n = 3, 4, 5, \ldots$. Find the first eight terms of this sequence.
9. The *Fibonacci ratio sequence* is given by $R_n = F_{n+1}/F_n$, $n = 1, 2, 3, \ldots$, where the F_n are the Fibonacci numbers defined in Problem 8. Find the first seven terms of the ratio sequence.
10. Plot several terms of the sequence $s_n = (2 - n)/(1 + n)$ on the real number line. Do they appear to be clustering toward a single point?

In Problems 11 through 16, verify the given limit.

11. $\displaystyle\lim_{n\to\infty} \left(\frac{1}{n^2} - \frac{3}{\sqrt{n}} \right) = 0$.   12. $\displaystyle\lim_{n\to\infty} \frac{n+8}{n} = 1$.

13. $\lim\limits_{n\to\infty} \dfrac{1-n^2}{1-2n^2} = \dfrac{1}{2}$.

14. $\lim\limits_{n\to\infty} \left(\dfrac{2n}{n+1} + \dfrac{n-4}{3n}\right) = \dfrac{7}{3}$.

15. $\lim\limits_{n\to\infty} \left(\dfrac{n}{n+2}\right)\left(\dfrac{n+6}{3n+2}\right) = \dfrac{1}{3}$.

16. $\lim\limits_{n\to\infty} \left(\dfrac{2}{3}\right)^n = 0$.

In Problems 17 through 22, explain why each of the given sequences diverge.

17. $s_n = 3n^2$.

18. $s_n = 6 - n$.

19. $s_n = 2 + (-1)^n$.

20. $s_n = \dfrac{n^2}{n+1}$.

21. $s_n = 3^n$.

22. $s_n = \begin{cases} 0, & \text{if } n \text{ is even,} \\ \dfrac{n}{n+1}, & \text{if } n \text{ is odd.} \end{cases}$

In Problems 23 through 37, decide whether the given sequence converges or diverges. If it converges, find the limit.

23. $\dfrac{6n}{n+1}$.

24. $5 + \dfrac{3}{n^2}$.

25. $9 - \dfrac{4}{n} + \dfrac{1}{n^3}$.

26. $7 - n^2$.

27. $\dfrac{(-1)^n}{n}$.

28. $\dfrac{5n^3 + 8}{n^3 + 1}$.

29. $\dfrac{n^2 - 3n + 7}{2n^2 + 1}$.

30. $\dfrac{2 - \sqrt{n}}{7 + \sqrt{n}}$.

31. $\dfrac{n - 6}{n^{1/3} + 1}$.

32. $n^2\left(\dfrac{n+1}{n-9}\right)$.

33. $\left(\dfrac{4n+1}{n-1}\right)\left(\dfrac{n+1}{2n-3}\right)$.

34. $\dfrac{n}{2^n}$.

35. $5 - \dfrac{2}{3^n}$.

36. $3 + \dfrac{8}{n^4} - (0.6)^n$.

37. $(-2.3)^n$.

38. Write down the first several terms of the sequence

$$s_n = \begin{cases} 0, & \text{if } n \text{ is even,} \\ \dfrac{1}{n}, & \text{if } n \text{ is odd.} \end{cases}$$

Is this sequence convergent or divergent?

39. Find the limits of each of the following sequences.

(a) $s_n = \begin{cases} 1/n, & \text{if } 1 \leqq n \leqq 29, \\ 7, & \text{if } n > 29. \end{cases}$

(b) $s_n = \begin{cases} 3, & \text{if } 1 \leqq n \leqq 439, \\ \dfrac{2n}{n^2 + 1}, & \text{if } n > 439. \end{cases}$

40. Suppose that $s_1, s_2, s_3, \ldots$ is a convergent sequence with $\lim\limits_{n\to\infty} s_n = L$. Explain why

(a) $\lim\limits_{n\to\infty} s_{n-1} = L$.
(b) $\lim\limits_{n\to\infty} 5s_n = 5L$.
(c) $\lim\limits_{n\to\infty} s_n^2 = L^2$.

Suppose that $a_1, a_2, a_3, \ldots$ is a convergent sequence with $\lim\limits_{n\to\infty} a_n = -2$. Find

(a) $\lim\limits_{n\to\infty} (2 + a_n)$.　　　(b) $\lim\limits_{n\to\infty} 7a_n^2$.　　　(c) $\lim\limits_{n\to\infty} \dfrac{a_n}{2a_n - 3}$.

42. If $f(x) = 2 - x$, find

(a) $\lim\limits_{n\to\infty} f\left(\dfrac{1}{n}\right)$.　　　(b) $\lim\limits_{n\to\infty} f\left(\dfrac{n}{n + 1}\right)$.　　　(c) $\lim\limits_{n\to\infty} f\left(3 - \dfrac{1}{n}\right)$.

43. Let $h_1, h_2, h_3, \ldots$ be a sequence of nonzero numbers with $\lim\limits_{n\to\infty} h_n = 0$. If $f(x) = x^2$, find

(a) $\lim\limits_{n\to\infty} \dfrac{f(h_n)}{h_n}$.　　　　　　　　　(b) $\lim\limits_{n\to\infty} \dfrac{f(1 + h_n) - f(1)}{h_n}$.

44. Let $h_1, h_2, h_3, \ldots$ be a sequence of nonzero numbers with $\lim\limits_{n\to\infty} h_n = 0$. If $g(x) = 1/x$, find

(a) $\lim\limits_{n\to\infty} \dfrac{g(1 + h_n) - g(1)}{h_n}$.　　　　　(b) $\lim\limits_{n\to\infty} \dfrac{g(3 + h_n) - g(3)}{h_n}$.

45. Using the Binomial Theorem, show that $(1.01)^n > (0.01)n$ for $n \geq 1$, and thereby conclude that the sequence $s_n = (1.01)^n$ is divergent.

46. Using the Binomial Theorem, show that $0 < n/(1.1)^n < 200/(n - 1)$ for $n \geq 2$, and thereby conclude that the sequence $s_n = n/(1.1)^n$ converges to zero.

CALCULATOR PROBLEMS

C1. Illustrate each of the following limits by computing the terms of the sequence for several large values of n. Arrange your answers in tabular form.

(a) $\lim\limits_{n\to\infty} \dfrac{2n}{n + 315} = 2$.　　　　　　　(b) $\lim\limits_{n\to\infty} \dfrac{1000}{n} = 0$.

(c) $\lim\limits_{n\to\infty} \dfrac{n^2 - 40}{n^2 + 83} = 1$.　　　　　　(d) $\lim\limits_{n\to\infty} \sqrt[n]{n} = 1$.

(e) $\lim\limits_{n\to\infty} (0.82)^n = 0$.　　　　　　　(f) $\lim\limits_{n\to\infty} \dfrac{50n^2}{2^n} = 0$.

C2. In Example 5, the sequence $(1 + 2/n)^n$ does in fact converge to the number $L = 7.389 \ldots$. However, when $n = 10^{13}, 10^{14}, 10^{15}$, and so on, were used as inputs on a calculator, the value obtained for $(1 + 2/n)^n$ was 1. (Try this for your own calculator.) Explain why the calculator gives this misleading answer.

C3. The sequence $\sqrt{3}, \sqrt{\sqrt{3}}, \sqrt{\sqrt{\sqrt{3}}}, \ldots$ has the general formula $s_n = 3^{1/2^n}$. Starting with s_1, calculate the successive terms of this sequence until you are convinced that $\lim\limits_{n\to\infty} s_n = 1$. How large must n be in order that $|s_n - 1| < 0.001$?

C4. Let $s_n = 2^n/n!$, where $n! = 1 \cdot 2 \cdots (n - 1)n$ is the product of all positive integers up to n. Show that $s_1 = 2$ and that $s_n = 2s_{n-1}/n$ for $n = 2, 3, 4, \ldots$. Starting with s_1 and using the recurrence relation, calculate the successive terms of the sequence until you are convinced that $\lim_{n \to \infty} s_n = 0$.

C5. The sequence given by $a_1 = 4$, $a_n = 6 - a_{n-1}^2$ for $n \geq 2$ diverges. By calculating the successive terms of the sequence, explain the nature of this divergence.

C6. The sequence given by $b_1 = 0.2$, $b_n = 5/(b_{n-1}^2 + 1)$ for $n \geq 2$ diverges. By calculating the successive terms of this sequence, explain the nature of this divergence.

In Problems C7 through C14, calculate several terms of the given sequence so that you can make a reasonable guess as to convergence or divergence. If the sequence appears to be converging, give the limit to at least three decimal places.

C7. $\sqrt{\dfrac{2n}{n+1}}$.

C8. $\dfrac{n^4}{2^n}$.

C9. $\sqrt{n+1} - \sqrt{n}$.

C10. $\dfrac{n}{\log_{10} n}$.

C11. $\left(1 + \dfrac{1}{n}\right)^n$.

C12. $(0.9)^{1/n}$.

C13. $n \sin(1/n)$, where $1/n$ is the radian measure of an angle.

C14. $a_1 = 1$, $a_n = \dfrac{a_{n-1}}{2} + \dfrac{2}{a_{n-1}}$ for $n = 2, 3, \ldots$.

C15. Each of the sequences $1/2n$, $75/n^2$, $(0.98)^n$, and $1/\log_{10} n$ converges to zero. Evaluate the terms of these sequences for $n = 100, 500, 1000, 5000$, and $10{,}000$ and arrange your answers in tabular form. Based on your table, which of the sequences appears to be converging "most rapidly" to zero? What are the second, third, and fourth most rapidly converging sequence?

C16. Calculate several successive terms of the sequence $b_1 = 200$, $b_n = b_{n-1} + (1/b_{n-1})$, $n = 2, 3, \ldots$. Note that the terms are getting larger but that the increase is rather slow. Thus it is difficult to guess whether or not the sequence converges. In fact, the sequence diverges. To see this, write the recurrence relation in the form $b_n b_{n-1} = b_{n-1}^2 + 1$. Then argue that if the sequence b_n *does* converge to a number L, we must have $L^2 = L^2 + 1$, which is absurd.

★5 *Solving Equations*

Many mathematical and physical problems can be ultimately reduced to solving an equation. If we are lucky enough to have a quadratic equation, then we can use the known formula to express the exact solutions. Formulas also exist for solving third- and fourth-degree equations, but they are much more complicated. And for equations of fifth degree or higher, no such formulas can even be found. Thus, for many algebraic equations and for certain equations of nonalgebraic type such as

(53) $$\tan x = x + 4,$$

we generally cannot write down exact solutions. In these cases our only alternative is to approximate the solution.

    In this section we present two different methods for constructing sequences of approximations. The first method deals with equations of the form

(54) $$f(x) = 0,$$

where f is some given function. A solution of (54) is called a *zero* of f. It is also called a *root* of the equation (54). Although many equations are not immediately of this form, it is an easy matter to transform them. For example, the equations

$$\tan x = x + 4 \qquad \text{and} \qquad 2x^2 = \frac{5}{x^3 + 1}$$

can be written as

$$\tan x - x - 4 = 0 \qquad \text{and} \qquad 2x^2(x^3 + 1) - 5 = 0.$$

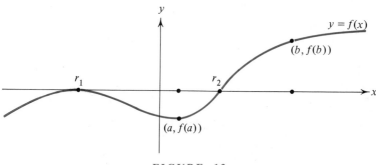

FIGURE 13

    Geometrically, solving equation (54) is equivalent to finding the points where the graph of $y = f(x)$ crosses (or touches) the x-axis (see Fig. 13 where r_1 and r_2 are roots). The procedure that we now describe for approximating solutions of (54) is called the *bisection method*.

    If the graph of a function f defined on an interval $[a, b]$ is an unbroken curve, and if f is negative at one end point and positive at the other, then the graph of f must cross the x-axis somewhere between a and b. This is illustrated in Fig. 13, where $f(a)$ is negative, $f(b)$ is positive, and a zero of f lies between a and b.

    Here is how the bisection method works. As the first step, we experiment with our function f to find two numbers, a_1 and b_1 ($a_1 < b_1$), such that the function values $f(a_1)$ and $f(b_1)$ have opposite signs. By our previous remark, a zero of f must reside somewhere between a_1 and b_1, provided that $y = f(x)$ is an unbroken curve. Next we compute the midpoint m_1 of the interval $[a_1, b_1]$:

$$(55) \qquad\qquad m_1 = \frac{a_1 + b_1}{2}$$

and evaluate $f(m_1)$. If $f(m_1) = 0$, then m_1 is the desired solution and our work is done. If $f(m_1) \neq 0$, then we consider two separate cases.

CASE 1. If the sign of $f(m_1)$ is opposite to that of $f(a_1)$, then f must have a zero in the smaller interval $[a_1, m_1]$. In this case we set

$$(56) \qquad\qquad a_2 = a_1 \quad \text{and} \quad b_2 = m_1.$$

CASE 2. If the sign of $f(m_1)$ is the same as $f(a_1)$, and hence opposite to that of $f(b_1)$, then f will have a zero in the interval $[m_1, b_1]$. In this case we set

$$(57) \qquad\qquad a_2 = m_1 \quad \text{and} \quad b_2 = b_1.$$

At this stage we have produced an interval $[a_2, b_2]$, having half the length of our original interval $[a_1, b_1]$, in which $f(x) = 0$ has a solution. Now we repeat the whole procedure using a_2 and b_2 instead of a_1 and b_1. In other words, we calculate the midpoint

$$m_2 = \frac{a_2 + b_2}{2}$$

and examine the sign of $f(m_2)$, provided, of course, that $f(m_2) \neq 0$. Again we consider two cases, $f(m_2) > 0$ and $f(m_2) < 0$. One of these two cases must give a third interval $[a_3, b_3]$ containing our sought-after zero of f.

Repeating the procedure again and again (assuming that we did not stumble upon an actual zero), we generate two sequences of numbers,

$$(58) \qquad\qquad a_1, a_2, a_3, a_4, \ldots, a_n, \ldots,$$

and

$$(59) \qquad\qquad b_1, b_2, b_3, b_4, \ldots, b_n, \ldots,$$

with the property that a zero of f always lies between a_n and b_n for each $n = 1, 2, 3, \ldots$. Furthermore, because the length of the intervals $[a_n, b_n]$ are reduced by one half at each step in the procedure, the numbers a_n and b_n are getting closer and closer together. Since the desired solution is trapped between them, each of the sequences (58) and (59) will converge to the solution.

Example 1. Use the bisection method to approximate a solution of the equation

$$(60) \qquad\qquad x^3 + 4x - 1 = 0.$$

Solution. Here $f(x) = x^3 + 4x - 1$. First we notice that $f(0) = -1 < 0$ and $f(1) = 4 > 0$ so that a solution of (60) lies between 0 and 1. Setting $a_1 = 0$, $b_1 = 1$, we find the midpoint:

$$m_1 = \frac{0 + 1}{2} = 0.5.$$

Since $f(0.5) = (0.5)^3 + 4(0.5) - 1 = 1.125$ is positive, we know that f has a zero in the interval $[0, 0.5]$, and we put $a_2 = 0$ and $b_2 = 0.5$. Next, we find the midpoint $m_2 = (a_2 + b_2)/2 = (0 + 0.5)/2 = 0.25$. Then we examine the sign of $f(0.25)$, and so on. In Table 4 we have carried on the computations for 11 steps. To simplify matters, the values of a_n, b_n, m_n, and $f(m_n)$ were rounded to five decimal places at each stage of the calculations. Note that once the eleventh step is completed, we know that equation (60) has a solution x^* lying between 0.24610 and 0.24659. Thus we know its first three digits after the decimal point: $x^* = 0.246\ldots$. We remind the reader that the sequence of left end points, the sequence of right end points, and, of course, the sequence of midpoints will each converge to x^*. Furthermore, the sequence of function values $f(m_1), f(m_2), \ldots$ will converge to zero, as indeed seems the trend from the fifth column of Table 4. ●

TABLE 4

Bisection Method for $x^3 + 4x - 1 = 0$

n	a_n	b_n	m_n	$f(m_n)$	Sign $f(a_n)$	Sign $f(b_n)$	Sign $f(m_n)$
1	0.0	1.0	0.5	1.125	−	+	+
2	0.0	0.5	0.25	0.01563	−	+	+
3	0.0	0.25	0.125	−0.49805	−	+	−
4	0.125	0.25	0.1875	−0.24341	−	+	−
5	0.1875	0.25	0.21875	−0.11453	−	+	−
6	0.21875	0.25	0.23438	−0.04960	−	+	−
7	0.23438	0.25	0.24219	−0.01703	−	+	−
8	0.24219	0.25	0.24610	−0.00069	−	+	−
9	0.24610	0.25	0.24805	0.00746	−	+	+
10	0.24610	0.24805	0.24708	0.00340	−	+	+
11	0.24610	0.24708	0.24659	0.00135	−	+	+
12	0.24610	0.24659			−	+	

The good news about the bisection method is its reliability: It always works provided that the graph of $y = f(x)$ is an unbroken curve. The bad news is apparent from Example 1: It is a lot of work for only three-decimal-place accuracy! In short, the approximating sequences are converging rather slowly. Starting with a smaller interval $[a_1, b_1]$ will certainly help matters, but experimenting to find two such nearby points where $f(a_1)$ and $f(b_1)$ have opposite signs may significantly add to the number of computations. It is thus worthwhile to consider alternative methods for constructing sequences of approximations.

Our next method deals with equations written in the form

(61) $$x = f(x).$$

The solutions of (61) are called the *fixed points* of the given function f. Geometrically, solving equation (61) is equivalent to finding the points where the graph of $y = f(x)$ intersects the line $y = x$. For the *specific* graph depicted in Fig. 14, we can easily devise a geometric procedure which, starting with some fixed point P_1 on the curve $y = f(x)$, will eventually lead to the desired intersection point P.

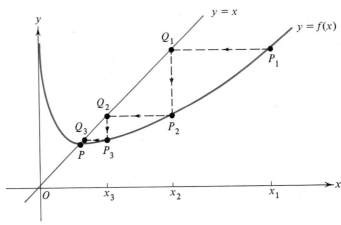

FIGURE 14

First we draw a horizontal line through P_1 to obtain the point Q_1 on the line $y = x$ (see Fig. 14). Next we draw a vertical line through Q_1, obtaining the point P_2 on the curve $y = f(x)$. From P_2 we again draw a horizontal line that intersects $y = x$ in the point Q_2. We then draw a vertical line from Q_2, obtaining the third point P_3 on the curve $y = f(x)$. Continuing with this "staircase" procedure, we generate a sequence of points $P_1, P_2, \ldots,$ $P_n, \ldots,$ which geometrically appears to be getting closer and closer to the intersection point P.

Now let us translate the above scheme into algebraic language. Let (x_1, y_1) be the coordinates of P_1, (x_2, y_2) the coordinates of P_2, and, in general, let (x_n, y_n) be the coordinates of P_n. Since each of these points lies on the graph of f, we have $y_1 = f(x_1), y_2 = f(x_2),$ and so on. Now recall from Fig. 14 that Q_1 and P_1 lie on the same horizontal line. Hence these points have the same ordinate y_1; that is, (y_1, y_1) are the coordinates of point Q_1. Since P_2 and Q_1 lie on the same vertical line, their abscissas must be identical: $x_2 = y_1$. As

$y_1 = f(x_1)$, we therefore have

(62) $$x_2 = f(x_1),$$

and repetition of this reasoning gives $x_3 = f(x_2)$, $x_4 = f(x_3)$, $x_5 = f(x_4)$, and so forth. So, in general, the sequence of numbers $x_1, x_2, x_3, \ldots$ satisfies the recurrence relation

(63) $$x_{n+1} = f(x_n), \qquad n = 1, 2, 3, \ldots.$$

Of course, our hope here is that for a reasonable starting guess x_1, the sequence x_1, x_2, x_3, $\ldots$, generated by (63) will converge to a solution of our original equation $x = f(x)$. This procedure for obtaining approximations is called the method of *successive substitutions*. It is a process based on *iteration*, that is, repeated use of the same formula.

Example 2. Use the method of successive substitutions to approximate a solution of the equation

(64) $$x = \frac{1}{2}\left(x + \frac{3}{x}\right).$$

Solution. A portion of the graph of $y = (1/2)(x + 3/x)$ is shown in Fig. 14. According to the recurrence relation (63), we replace x on the left side of (64) by x_{n+1} and replace x on the right side of the equation by x_n. This gives

(65) $$x_{n+1} = \frac{1}{2}\left(x_n + \frac{3}{x_n}\right), \qquad n = 1, 2, 3, \ldots.$$

If we use $x_1 = 3$ as our initial guess, we find from (65) that

$$x_2 = \frac{1}{2}\left(x_1 + \frac{3}{x_1}\right) = \frac{1}{2}\left(3 + \frac{3}{3}\right) = 2.$$

We then substitute the computed value for x_2 into equation (65) to find x_3:

$$x_3 = \frac{1}{2}\left(x_2 + \frac{3}{x_2}\right) = \frac{1}{2}\left(2 + \frac{3}{2}\right) = 1.75.$$

In Table 5 we give x_n for $n = 1, 2, \ldots, 7$.

TABLE 5

$x_1 = 3$	$x_4 = 1.732142857\ldots$	$x_6 = 1.732050808\ldots$
$x_2 = 2$	$x_5 = 1.732050810\ldots$	$x_7 = 1.732050808\ldots$
$x_3 = 1.75$		

The computations can stop after we find x_7 because with the hand calculator used, x_7 is indistinguishable from x_6. The same will be true for x_8 and x_7, and so on. This is, in fact, encouraging news because if $x_6 \approx x_7$, then

$$x_6 \approx \frac{1}{2}\left(x_6 + \frac{3}{x_6}\right),$$

which means that $x_6 = 1.732050808\ldots$ is our desired approximation to a solution of (64). ●

If the number x_6 in Example 2 looks vaguely familiar, it is because to nine decimal places it agrees with the irrational number $\sqrt{3}$. Indeed, the reader can show that equation (64) is equivalent to $x^2 = 3$.

If, instead of using a positive number x_1 as our initial guess, we had used a negative number, say $x_1 = -5$, then the method of successive substitutions would lead to the approximation $-1.732050808\ldots$ for $-\sqrt{3}$.

Example 3. Use the method of successive substitutions to approximate a solution of the equation of Example 1:

(60) $$x^3 + 4x - 1 = 0.$$

Solution. Note that our given equation (60) is not of the "fixed-point form," $x = f(x)$. Thus our first task is to rewrite (60) so that x is isolated on one side of the equation. This can be done in a variety of ways:

(66) $x = \frac{1}{4}(1 - x^3)$, $x = \frac{1}{x^2 + 4}$,   and $x = \sqrt[3]{1 - 4x}$.

Let us try working with the first equation in (66). Then the recurrence relation (63) becomes

(67) $$x_{n+1} = \frac{1}{4}(1 - x_n^3), \qquad n = 1, 2, 3, \ldots.$$

Choosing $x_1 = 0.5$ as our initial guess, we obtain from (67) the computed values that are listed in Table 6. The successive approximation procedure gives

TABLE 6

$x_1 = 0.5$	$x_4 = 0.246215137\ldots$	$x_7 = 0.246266177\ldots$
$x_2 = 0.21875$	$x_5 = 0.246268493\ldots$	$x_8 = 0.246266172\ldots$
$x_3 = 0.247383118\ldots$	$x_6 = 0.246266067\ldots$	$x_9 = 0.246266172\ldots$

$x_8 = 0.246266172\ldots$ as our approximate solution. This answer should be compared with the approximations obtained by means of the bisection method in Example 1. As a further check we substitute x_8 in our original equation (60) and find that

$$x_8{}^3 + 4x_8 - 1 = -7.0 \times 10^{-10},$$

which is indeed a number close to zero.   ●

Although the method of successive substitutions appears to be a highly efficient procedure, it, too, has a significant drawback—it does not always converge! To illustrate this, we return to the recurrence relation (67) and use $x_1 = 5$ as our initial guess. The results of our computations are given in Table 7. From this table we see that the sequence is diverg-

TABLE 7

$x_1 = 5$	$x_4 = -1.032901748 \times 10^{11}$	$x_6 = -5.227465172 \times 10^{96}$
$x_2 = -31.0$	$x_5 = 2.754971094 \times 10^{32}$	$x_7 = $ overflow
$x_3 = 7448.0$		

ing. Thus the success of the method of successive substitutions depends on choosing a good starting value x_1. Further, there are equations for which the method will fail for every starting value x_1 other than the actual solution. This is illustrated in Fig. 15 and discussed further in Section 12 of Chapter 5.

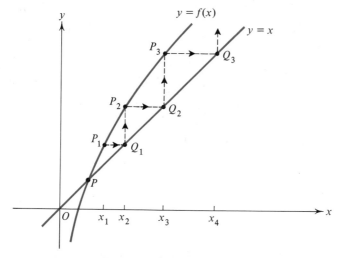

FIGURE 15 (Divergence)

EXERCISE 4

CALCULATOR PROBLEMS

C1. Using the bisection method, find to three decimal places the solution of $x^2 - 5.6 = 0$ that lies between $a_1 = 2$ and $b_1 = 3$. Compare your answer with the calculator's value for $\sqrt{5.6}$.

C2. Using the bisection method, find to three decimal places the solution of $x^5 - 37.4 = 0$ that lies between $a_1 = 2$ and $b_1 = 2.5$.

C3. Using the bisection method, find to three decimal places the solution of $x^3 + 3x = 1$ that lies between $a_1 = 0.3$ and $b_1 = 0.4$.

C4. In the bisection method, if $|a_1 - b_1| = 1$, show that $|a_n - b_n| = 1/2^{n-1}$. How large must n be so that $|a_n - b_n| < 0.001$?

C5. The equation

$$f(x) = x^4 + 8x^2 - 4x - 1 = 0$$

has two real solutions. Use the bisection method to find both of them correct to two decimal places. HINT: Since $f(-1) > 0, f(0) < 0$, and $f(1) > 0$, one solution lies in $(-1, 0)$ and the other in $(0, 1)$.

C6. The equation

$$f(x) = x^4 + 6x^2 - 4x - 1 = 0$$

has two real solutions. Use the bisection method to find both of them correct to two decimal places.

C7. Use the bisection method to find the negative root of $x^5 + x + 1 = 0$ correct to two decimal places.

C8. Use the bisection method to find the solution of $x \cdot 10^x = 13$ correct to two decimal places.

C9. Use the recurrence relation (65) starting with $x_1 = 1.8$ to find an approximation for $\sqrt{3}$.

C10. Show that $x = \sqrt{a}, a > 0$, is a solution of the equation

$$x = \frac{1}{2}\left(x + \frac{a}{x}\right).$$

Then write down the recurrence relation for solving this equation by the method of successive substitutions. This is known as the *divide and average* algorithm for approximating square roots.

C11. Using Problem C10, generate approximations for **(a)** $\sqrt{2}$, starting with $x_1 = 1.5$, and **(b)** $\sqrt{32.65}$, starting with $x_1 = 6$.

C12. To solve equation (60) in Example 3, use the equivalent form

$$x = \frac{1}{x^2 + 4}$$

to generate successive approximations starting with $x_1 = 0.25$. Also sketch the graphs of $y = x$ and $y = 1/(x^2 + 4)$ on the same coordinate system. It should be apparent from these graphs that there is only one solution to equation (60).

C13. Use the method of successive substitutions to approximate a solution of the equation $x = (\cos x)/2$, where x is in radians. Start with $x_1 = 0.45$.

C14. Use the method of successive substitutions to approximate a solution of the equation $x = 1 + \sin x$, where x is in radians. Start with $x_1 = 2$.

C15. Show that the equation $3x^3 - 7x - 1 = 0$ can be written in the fixed-point form

$$x = \frac{1}{3x^2 - 7} \qquad \text{for } x \neq \pm\sqrt{\frac{7}{3}},$$

and then use the method of successive substitutions to approximate a solution starting with $x_1 = 0$. Check your answer by substituting back into the original equation.

C16. Show that the equation $x^4 - 8x + 3 = 0$ can be written in the fixed-point form

$$x = \frac{3}{8 - x^3} \qquad \text{for } x \neq 2,$$

and then use the method of successive substitutions to approximate a solution starting with $x_1 = 0$. Check your answer by substituting back into the original equation.

C17. The equation $x^2 - 4x + 3 = 0$ has the solutions $x = 1$ and $x = 3$. This equation can be written in the fixed-point form $x = (x^2 + 3)/4$. Sketch the graph of the line $y = x$ and the parabola $y = (x^2 + 3)/4$ on the same coordinate system. Show graphically why the method of successive substitutions will *fail* to produce the solution at $x = 3$ even if one uses a starting value x_1 close to 3, say $x_1 = 2.9$ or $x_1 = 3.1$ (starting with $x_1 = 3$ is not fair). Then numerically corroborate this claim by attempting to use the recurrence relation $x_{n+1} = (x_n^2 + 3)/4$ starting with $x_1 = 3.1$. Finally, rewrite the original equation in the form $x = (x^2 - 3)/(2x - 4)$ and try a starting value close to 3.

C18. The equation $3x^4 + x - 5 = 0$ has a solution near $x = 1$. Try to obtain this solution by using the fixed-point form

$$x = \frac{5}{3x^3 + 1},$$

to generate approximations starting with $x_1 = 1$. If you are not successful in reaching the solution near $x = 1$, try rewriting the equation in the form

$$x = \sqrt[4]{\frac{5 - x}{3}}$$

and use successive substitutions starting with $x_1 = 1$.

C19. Show graphically that if the method of successive substitutions is used to solve for the intersection of the two lines $y = x$ and $y = mx + b$ via the recurrence relation $x_{n+1} = mx_n + b$, then the method will converge for any starting value provided that the slope m satisfies $|m| < 1$. On the other hand, illustrate that if $|m| > 1$, the method will diverge for any starting value (other than the actual solution).

4

Differentiation of Algebraic Functions

1 Objective

In Chapter 2 we introduced the concept of the slope of a straight line as a measure of the change in y for a given change in x [see page 39, equation (18)]. Can a curve have a slope? At first glance, the answer would seem to be no because the change in y for a given change in x would vary from point to point on the curve. However, if we fix on one point on the curve and draw the line tangent to the curve at this point (if there is such a line), we might call the slope of the tangent line, the slope of the curve. This is indeed the definition.

Our objective is to learn how to compute the slope of a curve, that is, to learn how to compute the slope of a line tangent to a given curve at a given point on the curve.

The method for computing the slope is surprisingly easy when we consider the difficulty of the problem. Further, the method has applications that are very important and go far beyond the geometric problem of finding tangent lines for an arbitrary curve.

2 An Example

Let us consider a specific curve, the parabola $y = x^2/4$ shown in Fig. 1, and let us concentrate our attention on the fixed point $P(1, 1/4)$ lying on the curve. If we select a neighboring point Q on the parabola, it is an easy matter to compute the slope of the line PQ. Now we can imagine a succession of different positions $Q_1, Q_2, Q_3, \ldots, Q_n$ for Q, each on the

108

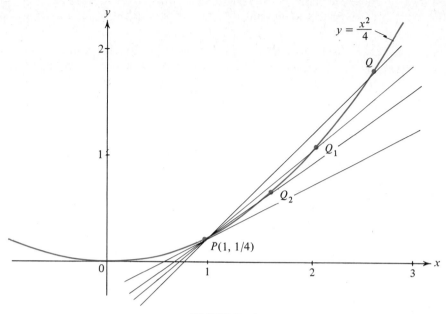

FIGURE 1

parabola and each closer to P than the preceding one. For each point Q_n we can draw the line PQ_n and compute its slope. Or we may imagine the point Q sliding along the curve toward P. In either case the lines seem to tend to a limiting position as Q approaches P, and whenever this occurs we call this limiting position the *tangent* to the curve at the point P. Before going further with the theory, let us show how easy it is to find the slope of this tangent line for the particular curve of Fig. 1. We note that at P, $x = 1$ and $y = 1/4$. Let us suppose that for the neighboring point Q, the x-coordinate has increased by an amount Δx, so that Q has $1 + \Delta x$ for its x-coordinate. Since $y = x^2/4$, the y-coordinate of Q is $(1 + \Delta x)^2/4$. Then for the slope m_{PQ} of the line PQ, we have

$$m_{PQ} = \frac{y_2 - y_1}{x_2 - x_1} = \frac{\dfrac{(1 + \Delta x)^2}{4} - \dfrac{1^2}{4}}{(1 + \Delta x) - 1} = \frac{1}{4} \frac{(1 + \Delta x)^2 - 1}{\Delta x},$$

(1) $$m_{PQ} = \frac{1 + 2\,\Delta x + (\Delta x)^2 - 1}{4\,\Delta x} = \frac{2\,\Delta x + (\Delta x)^2}{4\,\Delta x} = \frac{1}{2} + \frac{\Delta x}{4}.$$

Now as Q approaches P, the difference in the x-coordinates, Δx, gets closer and closer to zero. Hence in equation (1) the slope of the line PQ gets closer and closer to 1/2. We have proved that:

The line tangent to the curve $y = x^2/4$ at the point $(1, 1/4)$ has slope $m = 1/2$.

The simple process that we have just illustrated provides us with the means of computing the slope of the tangent line at any point on any reasonably decent curve. The differential calculus is just a systematic exploitation of the procedure illustrated above. However, before we can develop this key idea, we need to lay a firm foundation for the limiting process that we used in going from equation (1) to the conclusion that $m = 1/2$ for the tangent line.

EXERCISE 1

1. Let m be the slope of the line tangent to the curve $y = x^2/4$ at the point $(x, x^2/4)$. Follow the example and prove that:
   (a) $m = 1$ at $(2, 1)$.   (b) $m = 2$ at $(4, 4)$.
   (c) $m = 4$ at $(8, 16)$.   (d) $m = -1$ at $(-2, 1)$.

2. Make a careful sketch of the curve of Problem 1, draw the tangent line, and measure the slope at each of the four points named. If the same scale is used on both axes, then the measured value should be very close to the value computed in Problem 1.

3. Let m be the slope of the line tangent to the curve $y = x^2$ at the point (x, x^2). Follow the example and prove that:
   (a) $m = 0$ at $(0, 0)$.   (b) $m = 2$ at $(1, 1)$.
   (c) $m = 4$ at $(2, 4)$.   (d) $m = 6$ at $(3, 9)$.

4. Make a careful sketch of the curve of Problem 3, draw the tangent line, and measure the slope at each of the four points named. If the same scale is used on both axes, then the measured value should be very close to the value computed in Problem 3.

5. Prove that if (x_1, y_1) is any point on the curve $y = x^2/4$, then $m = x_1/2$. HINT: Set $x_2 = x_1 + \Delta x$. Then $x_2 - x_1 = \Delta x$, and

$$m_{PQ} = \frac{y_2 - y_1}{x_2 - x_1} = \frac{1}{\Delta x}\left[\frac{(x_1 + \Delta x)^2}{4} - \frac{x_1^2}{4}\right] = \frac{2x_1 + \Delta x}{4}.$$

6. Using the method of Problem 5, show that the line tangent to the curve $y = x^2$ at the point (x_1, y_1) has slope $2x_1$.

7. The line tangent to the curve $y = x^2/4$ at the point (x_1, y_1) has slope $m = 2$. Find x_1 and y_1. HINT: From Problem 5, the slope of the line is $x_1/2$.

8. Find an equation for the tangent line to the curve of the example at the point $(1, 1/4)$.

9. Find an equation for the tangent line to the curve of Problem 3 at the point $(0, 0)$.

10. Find an equation for the tangent line to the curve of Problem 3 that is parallel to the line $y = 2x + 3$.

11. Find an equation for the line perpendicular at $(1, 1/4)$ to the line obtained in Problem 8. This new line is called the *normal* line to the curve at the point $(1, 1/4)$.

12. Let c be a constant and let $y = cx^2$. Prove that for a tangent line to this curve at the

point (x, cx^2) we have $m = 2cx$. HINT: First consider the point $(x_1, cx_1{}^2)$ and prove that $m = 2cx_1$. Then drop the subscript.

13. Find an equation for the tangent line to the curve of Problem 12 at the point $(2, 4c)$.
14. Find the formula for the slope of the line tangent to the curve $y = cx^3$ at the point (x, cx^3).
15. Find an equation for the tangent line to the curve of Problem 14 at the point $(1, c)$.
16. Repeat Problem 14 for the curve $y = c/x$.
$\star$17. On the basis of the results obtained in Problems 12, 14, and 16, guess at a formula for m when $y = cx^n$, where n is any integer.

Limits of Functions 3

The concept of a *limit* of a sequence of numbers is discussed in Section 4 of Chapter 3. Related ideas are used in defining the limit of a function.

Let $y = 3x + 7$. When $x = 2$, it is easy to see that $y = 13$. But at this moment we are interested in how y behaves when x is *close* to 2. Is y close to 13? This certainly seems to be the case, and the accompanying two tables of values for x and y support this belief (Tables 1 and 2).

TABLE 1

x	1.5	1.8	1.9	1.99	1.999	1.9999
$y = 3x + 7$	11.5	12.4	12.7	12.97	12.997	12.9997

TABLE 2

x	2.5	2.2	2.1	2.01	2.001	2.0001
$y = 3x + 7$	14.5	13.6	13.3	13.03	13.003	13.0003

In Table 1 the variable x is approaching 2 through values that are less than 2. In this case we say that x is approaching 2 *from the left* and in symbols we write $x \to 2^-$ (read "x tends to 2 minus" or "x approaches 2 from the left").

In Table 2 the variable x is approaching 2 through values that are greater than 2. In this case we say that x is approaching 2 *from the right* and in symbols we write $x \to 2^+$ (read "x tends to 2 plus" or "x approaches 2 from the right").

If we wish to indicate that x may approach 2 without restricting its direction of approach, we use the symbol $x \to 2$ (read "x approaches 2"), omitting the $\pm$ signs. It is the latter situation that is most common.

Now how is y behaving as $x \to 2$? The tables illustrate that as x gets closer to 2, then y gets closer to 13 (in symbols, $y \to 13$). How close to 13 does y get? Answer: As close as we wish. But to make y close to 13, we must insist that x be close to 2. In other words, the two variables x and y are related (in this example by $y = 3x + 7$) and y can be made close to 13 by restricting x to be close to 2.

Summarizing the preceding discussion, we say that if $y = 3x + 7$, then y approaches 13 as x approaches 2 and we write this in symbols as

$$(2) \qquad\qquad \lim_{x \to 2} (3x + 7) = 13$$

(read "the limit of $3x + 7$, as x approaches 2, is 13").

The discussion above applies to any function $f(x)$, where the variable x may approach any suitable constant a, and as $x \to a$, the function values get closer and closer to some particular number L.

DEFINITION 1 (Limit of a Function). Let f be a function that is defined at every number in some open interval $\mathscr{I}$ containing a, except possibly at the number a itself. If the function values $f(x)$ approach a specific number L as x approaches a, then we say that *L is the limit of $f(x)$ as x approaches a,* and we write

$$(3) \qquad\qquad \lim_{x \to a} f(x) = L,$$

or

$$(4) \qquad\qquad f(x) \to L \qquad \text{as} \qquad x \to a.$$

Another way to express this definition is to say that the function values $f(x)$ stay arbitrarily close to L whenever x is sufficiently near (but not equal to) a. This interpretation is discussed further in Appendix 2.

Example 1. Find $\lim_{x \to 3} (x^2 - 7)$.

Solution. Here $f(x) = x^2 - 7$ and the variable x is approaching 3 from either side. It should be apparent that when x is close to 3 its square is close to 9, and hence $f(x) = x^2 - 7$ is getting closer to $9 - 7 = 2$. Thus

$$(5) \qquad\qquad \lim_{x \to 3} (x^2 - 7) = 2.$$

In Tables 3 and 4 we illustrate this fact by giving (to five decimal places) the values of $x^2 - 7$ for several numbers x near 3. ●

TABLE 3

x	2.8	2.9	2.99	2.999	2.9995	2.99999
$f(x) = x^2 - 7$	0.84	1.41	1.9401	1.99400	1.99700	1.99994

TABLE 4

x	3.2	3.1	3.01	3.001	3.0005	3.00001
$f(x) = x^2 - 7$	3.24	2.61	2.0601	2.00600	2.00300	2.00006

The top row of Table 3 (or Table 4) can be regarded as displaying the first few terms of a sequence $x_1, x_2, x_3, \ldots$ that converges to 3. The bottom row then shows the corresponding terms of the sequence of function values $f(x_1), f(x_2), f(x_3), \ldots$. In the case of Example 1, this latter sequence will converge to 2.

In general, if

$$(3) \qquad \lim_{x \to a} f(x) = L,$$

then for *any* sequence $x_1, x_2, x_3, \ldots$ of numbers (different from a) that converges to a, the corresponding sequence of function values $f(x_1), f(x_2), f(x_3), \ldots$ converges to L.[1]

Example 2. Find the limit of

$$(6) \qquad f(x) = \frac{\sqrt{x} - 2}{x - 4}$$

as x approaches 4.

Solution. It is helpful to rephrase the problem in the following way. When x is near and on either side of 4, is $f(x) = (\sqrt{x} - 2)/(x - 4)$ near some specific number L? To gain some insight, we have calculated in Tables 5 and 6 the values of $f(x)$ for several numbers x near 4. (The computations are shown rounded to six decimal places.) From these tables it appears that $f(x)$ is getting closer and closer to 0.25 as

TABLE 5

x	3.7	3.8	3.9	3.99	3.999	3.9999
$f(x) = \dfrac{\sqrt{x} - 2}{x - 4}$	0.254872	0.253206	0.251582	0.250156	0.250016	0.250002

[1] The converse of this statement is also true and can be used as an alternative form of Definition 1.

TABLE 6

x	4.3	4.2	4.1	4.01	4.001	4.0001
$f(x) = \dfrac{\sqrt{x} - 2}{x - 4}$	0.245480	0.246951	0.248457	0.249844	0.249984	0.249998

x gets closer and closer to 4, and so we are lead to guess that

(7)
$$\lim_{x \to 4} \frac{\sqrt{x} - 2}{x - 4} = 0.25.$$

We now give a convincing argument of this fact. Note that $f(x)$ is not defined at $x = 4$. Indeed, if we set $x = 4$ in (6), we obtain the meaningless expression $0/0$. However, for all numbers x different from 4, we can rationalize the numerator to obtain

$$f(x) = \frac{\sqrt{x} - 2}{x - 4} = \frac{(\sqrt{x} - 2)(\sqrt{x} + 2)}{(x - 4)(\sqrt{x} + 2)} = \frac{x - 4}{(x - 4)(\sqrt{x} + 2)} = \frac{1}{\sqrt{x} + 2};$$

that is,

(8)
$$f(x) = \frac{1}{\sqrt{x} + 2} \qquad \text{for all } x \neq 4, x \geq 0.$$

Now when x is close to 4, its square root will be a number close to 2 and hence the fraction in (8) approaches

$$\frac{1}{2 + 2} = \frac{1}{4} = 0.25. \quad \bullet$$

Example 2 illustrates an important feature of the limit concept; namely, in defining

$$\lim_{x \to a} f(x) = L$$

no reference is made to the value of the function $f(x)$ *at* $x = a$. The limiting value of $f(x)$ depends only on its behavior for x *near* but not equal to a. It may turn out that $f(a)$ is undefined (as in Example 2) or, if f is defined at a, its limiting value may be different from $f(a)$. This is illustrated in

Example 3. Sketch the graph of the function

(9)
$$f(x) = \begin{cases} 2x^2, & \text{if } x < 1, \\ 4, & \text{if } x = 1, \\ 3 - x, & \text{if } x > 1. \end{cases}$$

Then find the limit of $f(x)$ as x approaches 1.

Solution. The graph of f is shown in Fig. 2, and consists of the portion of the parabola $y = 2x^2$ for $x < 1$, the isolated point $(1, 4)$, and the portion of the line $y = 3 - x$ for $x > 1$.

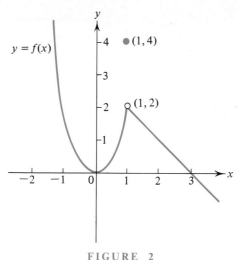

FIGURE 2

Note that as x approaches 1 from the left, the corresponding point $(x, f(x))$ on the graph of f slides along the parabolic portion toward $(1, 2)$. If x approaches 1 from the right, then $(x, f(x))$ slides along the linear portion of the graph, again toward $(1, 2)$. Thus as x approaches 1 from either side, $f(x)$ approaches 2; that is,

$$\lim_{x \to 1} f(x) = 2.$$

Note here that the limiting value 2 is different from $f(1) = 4$. ●

We wish to emphasize that the limit of a function as $x \to a$ *may not exist.* For example, consider the function

(10)
$$g(x) = \begin{cases} 2x, & \text{if } x < 1, \\ 3, & \text{if } x \geq 1, \end{cases}$$

whose graph is shown in Fig. 3. Both from the graph and the definition of this function, it is obvious that as x approaches 1 from the right $(x \to 1^+)$, the function $g(x)$ approaches (in fact, equals) 3. But as x tends to 1 from the left $(x \to 1^-)$, $g(x)$ equals $2x$ and hence approaches the number 2. Thus there is no *single* number L that $g(x)$ approaches as $x \to 1$, and we write

$$\lim_{x \to 1} g(x) \text{ does not exist.}$$

For this function g we can say that if x approaches 1 from *one side only,* then a limit does

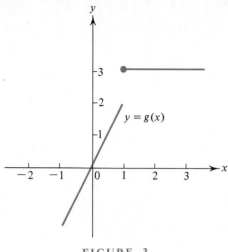

FIGURE 3

exist. We indicate this by writing

(11)
$$\lim_{x \to 1^-} g(x) = 2 \quad \text{and} \quad \lim_{x \to 1^+} g(x) = 3.$$

The first relation in (11) is read "the limit of $g(x)$ as x approaches 1 from the left is 2." The second relation is read "the limit of $g(x)$ as x approaches 1 from the right is 3."

Example 4. Let f be defined by the equation

(12)
$$f(x) = \sqrt{x - 2}.$$

Find **(a)** $\lim_{x \to 2^+} f(x)$ if it exists and **(b)** $\lim_{x \to 2^-} f(x)$ if it exists.

Solution. A sketch of the graph of f is shown in Fig. 4. From this graph it is apparent that as x tends to 2 from the right, $f(x)$ approaches 0; that is,

$$\lim_{x \to 2^+} \sqrt{x - 2} = 0.$$

However,

$$\lim_{x \to 2^-} \sqrt{x - 2} \text{ does not exist}$$

because if x is less than 2, then $x - 2$ is negative and $\sqrt{x - 2}$ is not a real number. ●

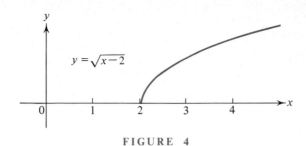

$$y = \sqrt{x-2}$$

FIGURE 4

It is important to keep in mind that the *two-sided limit*

$$\lim_{x \to a} f(x)$$

exists and equals L if and only if the one-sided limits

$$\lim_{x \to a^-} f(x) \qquad \text{and} \qquad \lim_{x \to a^+} f(x)$$

both exist and both equal L.

EXERCISE 2

In Problems 1 through 9, find the indicated limit.

1. $\lim\limits_{x \to 0} (4x - 6)$.

2. $\lim\limits_{x \to -2} (7 - 3x)$.

3. $\lim\limits_{x \to -1} (2x^2 - 8)$.

4. $\lim\limits_{x \to 3} \dfrac{10}{x + 2}$.

5. $\lim\limits_{x \to 1^-} [x]$.

6. $\lim\limits_{x \to 9} \dfrac{\sqrt{x} - 3}{x - 9}$.

7. $\lim\limits_{y \to -2} (y^3 + 1)$.

8. $\lim\limits_{t \to 0^-} \dfrac{|t|}{t}$.

9. $\lim\limits_{x \to 2} \dfrac{x^2 - 4}{x - 2}$.

In Problems 10 through 17, sketch the graph of the given function and find the indicated limits (if they exist).

10. $f(x) = \begin{cases} 2x - 1, & \text{if } x < 0 \\ -1, & \text{if } x \geq 0 \end{cases}$;   (a) $\lim\limits_{x \to 0} f(x)$,   (b) $\lim\limits_{x \to -2} f(x)$.

11. $g(x) = \begin{cases} x^2, & \text{if } x < -1 \\ 2 + x, & \text{if } x \geq -1 \end{cases}$;   (a) $\lim\limits_{x \to -1} g(x)$,   (b) $\lim\limits_{x \to 0} g(x)$.

12. $h(x) = \begin{cases} |x|, & \text{if } x \neq 0 \\ 3, & \text{if } x = 0 \end{cases}$;   (a) $\lim\limits_{x \to 0} h(x)$,   (b) $\lim\limits_{x \to -5} h(x)$.

13. $F(x) = \begin{cases} x + 1, & \text{if } x < 1 \\ -1, & \text{if } x = 1 \\ 2x, & \text{if } x > 1 \end{cases}$;   (a) $\lim\limits_{x \to 1} F(x)$,   (b) $\lim\limits_{x \to 3} F(x)$.

14. $G(x) = \begin{cases} -x, & \text{if } x < 0 \\ 3, & \text{if } x = 0; \\ x^2, & \text{if } x > 0 \end{cases}$ (a) $\lim_{x \to 0} G(x)$,   (b) $\lim_{x \to 2} G(x)$.

15. $H(x) = \begin{cases} 1, & \text{if } x < 0 \\ x - 3, & \text{if } 0 < x < 2; \\ -1, & \text{if } x > 2 \end{cases}$ (a) $\lim_{x \to 0^-} H(x)$,   (b) $\lim_{x \to 0^+} H(x)$,   (c) $\lim_{x \to 0} H(x)$,   (d) $\lim_{x \to 2} H(x)$.

16. $f(x) = \begin{cases} 1 - x^2, & \text{if } x < -1 \\ 0, & \text{if } -1 < x < 1; \\ 3x, & \text{if } x > 1 \end{cases}$ (a) $\lim_{x \to -1^-} f(x)$,   (b) $\lim_{x \to -1^+} f(x)$,   (c) $\lim_{x \to -1} f(x)$,   (d) $\lim_{x \to 1} f(x)$.

17. $g(x) = \begin{cases} -x^2, & \text{if } 1 < x \le 2 \\ -4, & \text{if } 2 < x < 3 \end{cases};$ (a) $\lim_{x \to 2^-} g(x)$,   (b) $\lim_{x \to 2^+} g(x)$,   (c) $\lim_{x \to 2} g(x)$,   (d) $\lim_{x \to 1^-} g(x)$.

18. If $\lim_{x \to 1} f(x) = 7$ and $g(x) = f(x)$ for all $x \ne 1$, what can be said about $\lim_{x \to 1} g(x)$?

19. Given a function $f(x)$ for which $\lim_{x \to 0} f(x) = 2$ and $\lim_{x \to 1} f(x) = 3$, find the limits of each of the following sequences s_n.

   (a) $s_n = f\left(\dfrac{1}{n}\right)$.   (b) $s_n = f\left(\dfrac{1}{2^n}\right)$.   (c) $s_n = f\left(1 - \dfrac{1}{n^2}\right)$.

CALCULATOR PROBLEMS

C1. Illustrate each of the following limits as $x \to 1$ by evaluating the function for $x = 0.9$, 0.99, 0.999, and 0.9999 and for $x = 1.1$, 1.01, 1.001, and 1.0001.

   (a) $\lim_{x \to 1} \dfrac{4}{x^2 + 1} = 2.$

   (b) $\lim_{x \to 1} \sqrt{\dfrac{x}{x + 3}} = \dfrac{1}{2}.$

   (c) $\lim_{x \to 1} (3 - x)^2 = 4.$

   (d) $\lim_{x \to 1} \dfrac{x^3 - 1}{x - 1} = 3.$

In Problems C2 through C7, guess the limit by computing the function values at several suitable numbers x.

C2. $\lim_{x \to 0} \dfrac{\sqrt{x^2 + 4} - 2}{x}.$

C3. $\lim_{x \to 2} \dfrac{256 - x^8}{x^2 - 4}.$

C4. $\lim_{x \to 2} \dfrac{x^4 - x^2 - 12}{x^3 - 5x + 2}.$

C5. $\lim_{x \to 0} \dfrac{(x + 2)^3 - 8}{(x - 2)^3 + 8}.$

C6. $\lim_{x \to -1} \dfrac{\sqrt[3]{x} + 1}{x^3 + 1}.$

C7. $\lim_{x \to 4} \dfrac{\sqrt{x} - 2}{\sqrt{7 + \sqrt{x}} - 3}.$

C8. In using computations to guess at the limit of a function $f(x)$ as $x \to a$, we have often used values of x that differ from a by successive powers of $1/10$, that is, $x = a \pm 0.1$, $a \pm 0.01$, $a \pm 0.001$, and so on. This practice can occasionally backfire. For example, consider the limit as $x \to 0^+$ of

$$f(x) = \sin\left(\frac{\pi}{x}\right),$$

where the quantity in parentheses represents the radian measure of an angle. If $x_n = 10^{-n}$, $n = 1, 2, 3, \ldots$, then $f(x_n) = \sin(10^n \pi) = 0$ for all n, so we might guess that $f(x) \to 0$ as $x \to 0^+$. Show, however, that if $x_n = 2/(2n + 1)$ (a sequence that also approaches zero), then $f(x_n) = (-1)^n$. This means that $\lim_{x \to 0^+} f(x)$ does not exist.

Some Limit Theorems 4

If $f(x)$ and $g(x)$ are two functions having the same domain, then we can construct new functions by taking their sum, product, or quotient. For example, if

(13) $$f(x) = 3x + 7 \quad \text{and} \quad g(x) = 5x + 1,$$

then

(14) $$f(x) + g(x) = (3x + 7) + (5x + 1) = 8x + 8,$$

(15) $$f(x)g(x) = (3x + 7)(5x + 1) = 15x^2 + 38x + 7,$$

(16) $$\frac{f(x)}{g(x)} = \frac{3x + 7}{5x + 1}.$$

In general, if the limits of $f(x)$ and $g(x)$ as $x \to a$ are known, what can be said about the limits of these new functions? The answer is contained in

---

THEOREM 1. If

$$\lim_{x \to a} f(x) = L \quad \text{and} \quad \lim_{x \to a} g(x) = M,$$

then

(17) $$\lim_{x \to a} [f(x) + g(x)] = \lim_{x \to a} f(x) + \lim_{x \to a} g(x) = L + M,$$

(18) $$\lim_{x \to a} f(x)g(x) = \left[\lim_{x \to a} f(x)\right] \cdot \left[\lim_{x \to a} g(x)\right] = LM,$$

and *if the limit M of g(x) is not zero, then*

(19)
$$\lim_{x \to a} \frac{f(x)}{g(x)} = \frac{\lim\limits_{x \to a} f(x)}{\lim\limits_{x \to a} g(x)} = \frac{L}{M}.$$

Equation (17) asserts that *the limit of the sum of two functions is the sum of the limits of the functions.* For example, if we start with the obvious relations

$$\lim_{x \to 2} (3x + 7) = 13 \quad \text{and} \quad \lim_{x \to 2} (5x + 1) = 11,$$

then since $8x + 8 = (3x + 7) + (5x + 1)$, we have

$$\lim_{x \to 2} (8x + 8) = \lim_{x \to 2} (3x + 7) + \lim_{x \to 2} (5x + 1) = 13 + 11 = 24.$$

Similarly, equation (18) states that *the limit of the product of two functions is the product of the limits of the functions.* In particular, using (15) we find that

$$\lim_{x \to 2} (15x^2 + 38x + 7) = \left\{ \lim_{x \to 2} (3x + 7) \right\} \cdot \left\{ \lim_{x \to 2} (5x + 1) \right\} = 13 \cdot 11 = 143.$$

Finally, equation (19) states that *the limit of the quotient of two functions is the quotient of the limits of the two functions, provided that the limit of the denominator is not zero.* For example,

$$\lim_{x \to 2} \frac{3x + 7}{5x + 1} = \frac{\lim\limits_{x \to 2} (3x + 7)}{\lim\limits_{x \to 2} (5x + 1)} = \frac{13}{11}.$$

Theorem 1 should be apparent from a commonsense understanding of limits (the proof of this result is given in Appendix 2). Along with Theorem 1 we mention two other rather obvious facts.

THEOREM 2. $\lim\limits_{x \to a} x = a$; that is, *the limit of x, as x approaches a, is a.*

THEOREM 3. If $g(x) \equiv c$, a constant function, then

$$\lim_{x \to a} g(x) = c.$$

Note that by combining Theorem 3 with equation (18) of Theorem 1, we have

$$(20) \qquad \lim_{x \to a} cf(x) = \left\{ \lim_{x \to a} c \right\} \cdot \left\{ \lim_{x \to a} f(x) \right\} = c \cdot \left\{ \lim_{x \to a} f(x) \right\}.$$

In words, *the limit of a constant times a function is the constant times the limit of the function.* For example, since $18x + 42 = 6(3x + 7)$, we have

$$\lim_{x \to 2} (18x + 42) = 6 \lim_{x \to 2} (3x + 7) = 6 \cdot 13 = 78.$$

Example 1. Use Theorems 1, 2, and 3 to show that

$$(21) \qquad \lim_{x \to 2} (7x^3 - 5x^2) = 36.$$

Solution. By Theorem 2, $\lim\limits_{x \to 2} x = 2$. Using equation (18) of Theorem 1 [with $f(x) = x$ and $g(x) = x$], we have

$$\lim_{x \to 2} x^2 = 2 \cdot 2 = 4.$$

Again using equation (18) [this time with $f(x) = x^2$ and $g(x) = x$], we obtain

$$\lim_{x \to 2} x^3 = 4 \cdot 2 = 8.$$

Applying equation (20), first with $c = 7$ and second with $c = -5$, we have

$$\lim_{x \to 2} 7x^3 = 7 \cdot 8 = 56 \qquad \text{and} \qquad \lim_{x \to 2} -5x^2 = -5 \cdot 4 = -20.$$

Finally, combining these two by means of Theorem 1, equation (17), we find that

$$\lim_{x \to 2} (7x^3 - 5x^2) = \lim_{x \to 2} 7x^3 + \lim_{x \to 2} -5x^2 = 56 - 20 = 36. \quad \bullet$$

Often we are interested only in the numerical value of a limit. In the case of Example 1, we can find the limit by replacing x by 2 in $7x^3 - 5x^2$ and computing $7 \cdot 2^3 - 5 \cdot 2^2 = 56 - 20 = 36$. The purpose of Example 1 was to carefully justify that this computation leads to the correct value of the limit.

Starting with the simple function $f(x) = x$ together with the constant functions, we can take sums and products to build up any *polynomial* function, that is, any function of the form

$$(22) \qquad P(x) = a_n x^n + a_{n-1} x^{n-1} + a_{n-2} x^{n-2} + \cdots + a_2 x^2 + a_1 x + a_0,$$

where n is a nonnegative integer.[1] The function $7x^3 - 5x^2$ of Example 1 is a polynomial, and the same reasoning used to justify its limit can be applied to establish

---

[1] An efficient method for calculating the values of a polynomial function is given in Appendix 5.

THEOREM 4. If $P(x)$ is any polynomial [cf. (22)], then

(23) $$\lim_{x \to a} P(x) = P(a).$$

Example 2. Find $\lim_{x \to 3} (6x^3 - 2x^2 - 5x)$.

Solution. By Theorem 4,

$$\lim_{x \to 3} (6x^3 - 2x^2 - 5x) = 6 \cdot 3^3 - 2 \cdot 3^2 - 5 \cdot 3 = 162 - 18 - 15 = 129. \quad \bullet$$

Combining Theorem 4 with equation (19) of Theorem 1 immediately gives

THEOREM 5. If $f(x)$ is any *rational* function, that is, if

$$f(x) = \frac{N(x)}{D(x)},$$

where $N(x)$ and $D(x)$ are polynomials, and if $D(a) \neq 0$, then

$$\lim_{x \to a} f(x) = \frac{N(a)}{D(a)}.$$

Example 3. Find

$$\lim_{x \to -1} \frac{x^3 + 5x^2 + 3x}{2x^5 + 3x^4 - 2x^2 - 1}.$$

Solution. By Theorem 5,

$$\lim_{x \to -1} \frac{x^3 + 5x^2 + 3x}{2x^5 + 3x^4 - 2x^2 - 1} = \frac{(-1)^3 + 5(-1)^2 + 3(-1)}{2(-1)^5 + 3(-1)^4 - 2(-1)^2 - 1} = -\frac{1}{2}. \quad \bullet$$

Example 4. Find

$$\lim_{x \to 4} \frac{x^2 - 16}{x - 4}.$$

Solution. At $x = 4$, both the numerator and the denominator are zero, so that Theorem 5 is not applicable here. But we can do some preliminary algebra that will be helpful. Indeed, when $x \neq 4$, we have

$$\frac{x^2 - 16}{x - 4} = \frac{(x - 4)(x + 4)}{x - 4} = x + 4$$

so that

$$\lim_{x \to 4} \frac{x^2 - 16}{x - 4} = \lim_{x \to 4} (x + 4) = 4 + 4 = 8. \quad \bullet$$

EXERCISE 3

In Problems 1 through 6, use Theorems 1, 2, and 3 to carefully justify the given limit.

1. $\lim_{x \to 3} (2x^2 + 7x) = 39.$

2. $\lim_{x \to 0} (4 - x^2 - 2x^3) = 4.$

3. $\lim_{x \to 4} \dfrac{x^2}{3x - 2} = \dfrac{8}{5}.$

4. $\lim_{x \to -1} \dfrac{2x - 1}{x^4 - x^3} = -\dfrac{3}{2}.$

5. If $\lim_{x \to 0} f(x) = -4$ and $\lim_{x \to 0} g(x) = 2$, then $\lim_{x \to 0} [f(x) - 3f(x)g(x)] = 20.$

6. If $\lim_{x \to 1} F(x) = 6$ and $\lim_{x \to 1} G(x) = -5$, then $\lim_{x \to 1} \dfrac{2F(x)}{G(x) + x} = -3.$

In Problems 7 through 20, find the indicated limit.

7. $\lim_{x \to 2} (4x^3 - x^2 - 3x + 1).$

8. $\lim_{x \to -1} (8x^9 - 5x^4 + 2).$

9. $\lim_{x \to 0} (x^{10} - 8x^5 + 3x - 47).$

10. $\lim_{x \to 8} \sqrt{4^3 + \pi^2}.$

11. $\lim_{x \to 1} \dfrac{9x^5 - 14x^2}{2x^3 + x^2 + 2}.$

12. $\lim_{x \to -2} \dfrac{5x^2 + 3x + 4}{x^4 + 2x - 3}.$

13. $\lim_{x \to 3} \dfrac{x^2 + 2x - 15}{x - 3}.$

14. $\lim_{x \to -5} \dfrac{x^2 - 25}{x + 5}.$

15. $\lim_{t \to 3} \dfrac{t^2 - 1}{4t^2 - t - 6}.$

16. $\lim_{z \to -1} \dfrac{z^2 + 2z - 4}{z + 5}.$

17. $\lim_{s \to 1} \dfrac{s - 1}{s^3 - 1}.$

18. $\lim_{w \to 1} \dfrac{w^4 - 1}{w - 1}.$

19. $\lim_{h \to 0} \dfrac{(3 + h)^2 - 9}{h}.$

20. $\lim_{h \to 0} \dfrac{(1 + h)^3 - 1}{h}.$

21. Given that $\lim_{x \to 8} f(x) = -2$ and $\lim_{x \to 8} g(x) = 3$, find

    (a) $\lim_{x \to 8} [4f(x) - 5]$.   (b) $\lim_{x \to 8} \left[\dfrac{f(x) - g(x)}{x + g(x)} \right]$.   (c) $\lim_{x \to 8} \left[\dfrac{f(x)g(x)}{1 + f(x)} \right]$.

22. Given that $\lim_{x \to 0} F(x) = 1$ and $\lim_{x \to 0} G(x) = 9$, find

    (a) $\lim_{x \to 0} [xF(x) - G(x)]$.   (b) $\lim_{x \to 0} \left[\dfrac{F(x) + G(x)}{F(x) - G(x)} \right]$.   (c) $\lim_{x \to 0} [F(x) + G(x)]^2$.

23. If $\lim_{x \to -4} f(x) = 2$, find $\lim_{x \to -4} [6 + f(x)^2 - f(x)^3]$.

24. If $\lim_{x \to 10} g(x) = -1$, find $\lim_{x \to 10} \left[\dfrac{2g(x) + 3}{g(x)^2 - x} \right]$.

CALCULATOR PROBLEMS

C1. For the two functions $f(x) = 2x + 1$ and $g(x) = \pi x^2$, we have $\lim_{x \to 1} f(x) = 3$ and $\lim_{x \to 1} g(x) = \pi$. According to equation (18) of Theorem 1, $\lim_{x \to 1} f(x)g(x) = 3\pi$. Illustrate this fact by making a table of values for $f(x)g(x)$ for $x = 0.99, 0.9987, 0.99994$, and 0.999999; and for $x = 1.02, 1.001, 1.00025$, and 1.000001. Compare your answers with the calculator's value for 3π.

C2. According to equation (19) of Theorem 1, we have for the two functions f and g in Problem C1 that $\lim_{x \to 1} f(x)/g(x) = 3/\pi$. Illustrate this fact by making a table of values for $f(x)/g(x)$ for several numbers x near 1. Compare your answers with the calculator's value for $3/\pi$.

In Problems C3 through C6, find the indicated limit. Give your answers to three decimal places.

C3. $\lim_{x \to \sqrt{2}} (5x^3 - 3x + \sqrt{7})$.   C4. $\lim_{x \to \pi} (6x^2 - \pi x + 1)$.

C5. $\lim_{x \to 1.28} \left(\dfrac{8.3x^2}{6.2x^3 + 5.1x^2 + 4.7} \right)$.   C6. $\lim_{x \to -2.4} \left(\dfrac{x^3 + 6.1}{1.3x^2 - 7.2} \right)^2$.

⋆5 *Limits Involving Infinity*

The notation

(3) $$\lim_{x \to a} f(x) = L$$

can be used with $a = \pm\infty$, or $L = \pm\infty$, or any combination. Since there are three possibilities for a (finite, $+\infty$, or $-\infty$) and a similar set for L, there are $3 \times 3 = 9$ special cases for equation (3). Only the case in which a and L are both finite was covered by Definition 1. Hence we need eight more definitions, one for each of the remaining cases. However, we can spare ourselves this labor, because the ideas involved are so simple that the reader could easily guess the definitions for himself. Instead, we will illustrate the ideas involved with a few examples.

The notation $x \to \infty$ means that x grows without bound; that is, given any M, no matter how large, x eventually becomes and remains larger than M. The notation $x \to -\infty$ means that x is negative but that $|x|$ grows without bound. Similarly, $\lim\limits_{x \to a} f(x) = \infty$ means that as $x \to a$, $f(x)$ grows without bound. Some of the possibilities are illustrated in the following examples.

Example 1. Find

$$(24) \qquad \qquad \lim_{x \to \infty} \frac{2x^2}{x^2 + 1}.$$

Solution. In (24) we first divide both the numerator and the denominator by x^2. This gives us

$$(25) \qquad \lim_{x \to \infty} \frac{2x^2}{x^2 + 1} = \lim_{x \to \infty} \frac{2}{1 + \dfrac{1}{x^2}}.$$

If x is a large positive number, then $1/x^2$ is close to zero, and as x tends to plus infinity, $1/x^2 \to 0$. Thus $1 + 1/x^2 \to 1$ as $x \to \infty$, so it is obvious that the limit in (25) equals $2/1 = 2$.

Note that as $x \to -\infty$, we again have $1/x^2 \to 0$, and hence we can also write

$$\lim_{x \to -\infty} \frac{2x^2}{x^2 + 1} = 2.$$

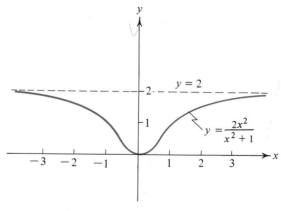

FIGURE 5

Geometrically, these results mean that the graph of $f(x) = 2x^2/(x^2 + 1)$, shown in Fig. 5, approaches the horizontal line $y = 2$ as it moves off to the left and to the right. Such a line is called a *horizontal asymptote*[1] of the graph. ●

Example 2. Find

$$\text{(26)} \qquad \lim_{x \to \infty} (2 + 3x^2 - x^3).$$

Solution. Factoring out the term containing the highest power of x in (26), we obtain

$$\lim_{x \to \infty} (2 + 3x^2 - x^3) = \lim_{x \to \infty} - x^3 \left(\frac{-2}{x^3} - \frac{3}{x} + 1 \right).$$

As $x \to \infty$, the first factor $-x^3$ approaches $-\infty$ and the second factor approaches 1. Hence

$$\text{(27)} \qquad \lim_{x \to \infty} (2 + 3x^2 - x^3) = -\infty.$$

If we let $x \to -\infty$, then $-x^3$ approaches plus infinity and the second factor again tends to 1. Thus

$$\text{(28)} \qquad \lim_{x \to -\infty} (2 + 3x^2 - x^3) = \infty.$$

The geometric meaning of the relations (27) and (28) is illustrated in Fig. 6. The graph drops down infinitely far as x moves off to the right and it rises up infinitely far as x moves off to the left. ●

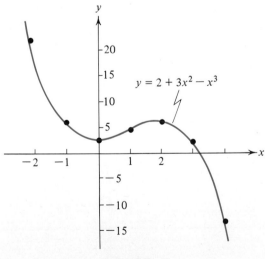

FIGURE 6

[1] Asymptotes are discussed in detail in Section 2 of Chapter 11.

We remark that the method of Example 2 can be used to show that the limit of any polynomial function

$$P(x) = a_n x^n + a_{n-1} x^{n-1} + \cdots + a_1 x + a_0, \qquad a_n \neq 0,$$

as $x \to \infty$ or as $x \to -\infty$ is the same as the limit of its term containing the highest power of x; that is,

$$\lim_{x \to \infty} P(x) = \lim_{x \to \infty} a_n x^n \qquad \text{and} \qquad \lim_{x \to -\infty} P(x) = \lim_{x \to -\infty} a_n x^n.$$

In particular,

$$\lim_{x \to -\infty} (3x^6 - 5x^4 + 17x^2 + \pi) = \lim_{x \to -\infty} 3x^6 = \infty.$$

Example 3. Find

(29)
$$\lim_{x \to -\infty} \frac{x^2 - 3x}{2x + 10}.$$

Solution. This problem can be worked using the same manipulation of factoring out the highest powers of x from the numerator and the denominator. We write

(30)
$$\lim_{x \to -\infty} \frac{x^2 - 3x}{2x + 10} = \lim_{x \to -\infty} \frac{x^2 \left(1 - \dfrac{3}{x}\right)}{x \left(2 + \dfrac{10}{x}\right)} = \lim_{x \to -\infty} x \left(\frac{1 - \dfrac{3}{x}}{2 + \dfrac{10}{x}}\right).$$

Now as x tends to minus infinity, the second factor in the last expression tends to $1/2$ and the first factor x approaches $-\infty$. Hence the product tends to minus infinity, that is,

$$\lim_{x \to -\infty} \frac{x^2 - 3x}{2x + 10} = -\infty.$$

Observe that if we let $x \to \infty$, the limit is changed in sign so that

$$\lim_{x \to \infty} \frac{x^2 - 3x}{2x + 10} = \infty. \quad \bullet$$

Example 4. Find

$$\lim_{x \to 3} \frac{4}{(x - 3)^2}.$$

Solution. If x is near and on either side of 3, then $(x - 3)^2$ is a small positive number.

Thus $4/(x - 3)^2$ is a large positive number that grows without bound. Consequently,

$$\lim_{x \to 3} \frac{4}{(x - 3)^2} = \infty.$$

Geometrically, this means that the graph of $f(x) = 4/(x - 3)^2$, shown in Fig. 7, goes up on either side of the vertical line $x = 3$. This line is called a *vertical asymptote* of the graph. ●

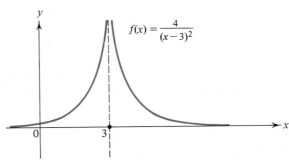

$$f(x) = \frac{4}{(x-3)^2}$$

FIGURE 7

Example 5. Find

(31)
$$\lim_{x \to 2} \frac{3x^2 + 4x + 5}{x^2 + 8x - 20}.$$

Solution. In (31) we factor the denominator and write

$$\lim_{x \to 2} \frac{3x^2 + 4x + 5}{x^2 + 8x - 20} = \lim_{x \to 2} \frac{1}{x - 2} \cdot \frac{3x^2 + 4x + 5}{x + 10}.$$

The second factor tends to $25/12$ as $x \to 2$. The first factor approaches ∞ if $x \to 2^+$, but it reverses sign and tends to $-\infty$ if $x \to 2^-$. Hence (31) must be separated into two cases,

$$\lim_{x \to 2^+} \frac{3x^2 + 4x + 5}{x^2 + 8x - 20} = \infty \quad \text{and} \quad \lim_{x \to 2^-} \frac{3x^2 + 4x + 5}{x^2 + 8x - 20} = -\infty.$$

In the first case we say that the right-hand limit at $x = 2$ is ∞, and in the second case we say that the left-hand limit at $x = 2$ is $-\infty$.

Geometrically, these results imply that the graph of

$$f(x) = \frac{3x^2 + 4x + 5}{x^2 + 8x - 20}$$

goes up just to the right of the vertical line $x = 2$, and goes down just to the left of this line. Again we call $x = 2$ a vertical asymptote of the graph. ●

EXERCISE 4

In Problems 1 through 24, find the indicated limit.

1. $\lim\limits_{x \to \infty} \left(3 + \dfrac{2}{x^2} - \dfrac{1}{x} \right).$

2. $\lim\limits_{x \to -\infty} \left(-6 + \dfrac{\pi}{x^8} - \dfrac{\sqrt{2}}{x^4} \right).$

3. $\lim\limits_{x \to -\infty} (5x^3 + 2x^2 - 17).$

4. $\lim\limits_{x \to \infty} (18 - 4x^2 + 9x^6).$

5. $\lim\limits_{x \to \infty} \dfrac{5x^3 + 1}{20x^3 - 8000x}.$

6. $\lim\limits_{x \to 20^+} \dfrac{5x^3 + 1}{20x^3 - 8000x}.$

7. $\lim\limits_{x \to \infty} \dfrac{50x^{10} + 100}{x^{11} + x^6 + 1}.$

8. $\lim\limits_{x \to -\infty} \dfrac{x^{25} + x}{x^{10}(2x^{15} + \pi)}.$

9. $\lim\limits_{x \to -1} \left[2 - \dfrac{6}{(x + 1)^4} \right].$

10. $\lim\limits_{x \to 0} \left(16x + \dfrac{3}{x^2} \right).$

11. $\lim\limits_{x \to \infty} 2^x.$

12. $\lim\limits_{x \to -\infty} 2^x.$

13. $\lim\limits_{x \to \infty} 8 \dfrac{10 + 3^x}{20 - 3^x}.$

14. $\lim\limits_{x \to -\infty} 8 \dfrac{10 + 3^x}{20 - 3^x}.$

15. $\lim\limits_{x \to 2^-} \dfrac{5x^2 + 3x - 1}{x^2 + x - 6}.$

16. $\lim\limits_{x \to 2^+} \dfrac{5x^2 + 3x - 1}{x^2 + x - 6}.$

17. $\lim\limits_{x \to 4^+} \left[8x - \dfrac{1}{(x - 4)^3} \right].$

18. $\lim\limits_{x \to \pi^-} \left[\dfrac{17}{(x - \pi)^5} + 1000 \right].$

19. $\lim\limits_{x \to 1^+} \dfrac{x^2 - 2x + 1}{x^3 - 3x^2 + 3x - 1}.$

20. $\lim\limits_{x \to 5^-} \dfrac{x^{100} - 4x^{99}}{x - 5}.$

21. $\lim\limits_{x \to 3^+} \dfrac{x^2 |x - 3|}{x - 3}.$

22. $\lim\limits_{x \to 3^-} \dfrac{x^2 |x - 3|}{x - 3}.$

23. $\lim\limits_{x \to 3^+} \dfrac{x^2 [x - 3]}{x - 3}.$

24. $\lim\limits_{x \to 3^-} \dfrac{x^2 [x - 3]}{x - 3}.$

★★25. Assuming that A, B, C, and D are all different from 0, find

$$\lim_{y \to \infty} \left(\lim_{x \to \infty} \frac{Ax + By}{Cx + Dy} \right) - \lim_{x \to \infty} \left(\lim_{y \to \infty} \frac{Ax + By}{Cx + Dy} \right).$$

26. Suppose that $N(x) = a_n x^n + a_{n-1} x^{n-1} + \cdots + a_1 x + a_0$ is a polynomial of nth degree so that $a_n \neq 0$. Suppose further that

$$D(x) = b_m x^m + b_{m-1} x^{m-1} + \cdots + b_1 x + b_0, \qquad \text{with } b_m \neq 0.$$

Prove that

$$\lim_{x \to \infty} \frac{N(x)}{D(x)} = \begin{cases} 0, & \text{if } n < m, \\ \dfrac{a_n}{b_m}, & \text{if } n = m, \\ \pm\infty, & \text{if } n > m, \text{ and } a_n/b_m \gtrless 0. \end{cases}$$

27. Given that $\lim\limits_{x \to \infty} f(x) = 5$, what can you say about $\lim\limits_{x \to 0^+} f(1/x)$?

28. Suppose that $\lim\limits_{x \to a} f(x) = \infty$. Show that $\lim\limits_{x \to a} 1/f(x) = 0$. Give an example to show that the converse of this assertion is false.

CALCULATOR PROBLEMS

In Problems C1 through C4, illustrate the limit by evaluating the function for the given numbers x.

C1. $\lim\limits_{x \to \infty} \dfrac{3x^2 + \pi}{x^2 + x} = 3$; $x = 20, 50, 200, 3000.$

C2. $\lim\limits_{x \to -\infty} \dfrac{6 + 2^x}{3 - 2^x} = 2$; $x = -4, -8, -20, -35.$

C3. $\lim\limits_{x \to 2^-} \dfrac{5x - 6}{(x - 2)(x + 1)} = -\infty$; $x = 1.5, 1.9, 1.985, 1.99996.$

C4. $\lim\limits_{x \to 1^+} \dfrac{1 - 2x}{x^2(x - 1)} = -\infty$; $x = 1.5, 1.01, 1.0025, 1.00001.$

6 *Continuous Functions*

Let us consider the function defined by

(32) $$y = f(x) = \frac{x^2 - 3x}{2x - 6}, \qquad x \neq 3.$$

This function is defined for all x, except $x = 3$. At $x = 3$, a computation gives $f(3) = 0/0$, and this is meaningless. A proper graph of this function must omit a point corresponding to $x = 3$ (as we have tried to indicate in Fig. 8). Of course, it is easy to fill this gap in the graph. For when $x \neq 3$, we have (factoring and canceling)

$$y = \frac{x^2 - 3x}{2x - 6} = \frac{x(x - 3)}{2(x - 3)} = \frac{x}{2}$$

so that the graph of the function (32) is just the straight line $y = x/2$ minus the single point $(3, 3/2)$. To fill the gap, we would merely define $f(3)$ to be $3/2$. But we could fill the gap (in the definition) by being contrary and making the definition $f(3) = 4$ (it is our function to define as we please). With the first definition, $f(3) = 3/2$, the function becomes *continuous,* and the graph has no breaks. With the second definition, $f(3) = 4$, the function becomes discontinuous at $x = 3$, and the graph has a break or jump at $x = 3$. This illustrates the following definition of a continuous function.

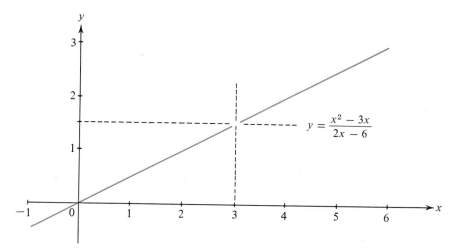

FIGURE 8

DEFINITION 2 (Continuity at a Point). A function $f(x)$ is said to be *continuous at a point* $x = a$ if

(a) $f(x)$ is defined at $x = a$.

(b) $\lim\limits_{x \to a} f(x)$ exists (as a real number).

(c) $\lim\limits_{x \to a} f(x) = f(a)$.

Otherwise, the function is said to be *discontinuous* at $x = a$.

It is customary in writing the definition of a continuous function to write only **(c)** because the symbols themselves imply that **(a)** and **(b)** are satisfied [otherwise **(c)** would not make any sense]. We have listed all three conditions for clarity.

The function (32) is not continuous at $x = 3$ because condition **(a)** fails. But observe that condition **(b)** is satisfied, namely,

$$\lim_{x \to 3} \frac{x^2 - 3x}{2x - 6} = \lim_{x \to 3} \frac{x(x - 3)}{2(x - 3)} = \lim_{x \to 3} \frac{x}{2} = \frac{3}{2}.$$

If we make the definition $f(3) = 4$, then conditions **(a)** and **(b)** are satisfied at $x = 3$, but **(c)** is not. Then the function is not continuous at $x = 3$. But if we make the definition $f(3) = 3/2$, then **(a)**, **(b)**, and **(c)** are satisfied and hence the revised $f(x)$ is continuous at $x = 3$.

Example 1. A function f is given by

(33)
$$f(x) = \begin{cases} -x^2, & \text{if } x \leq 0, \\ 2/x, & \text{if } 0 < x < 1, \\ 3, & \text{if } x = 1, \\ 2, & \text{if } 1 < x. \end{cases}$$

Sketch the graph of f and determine the values of x at which it is discontinuous.

Solution. The graph of f is shown in Fig. 9. The function is discontinuous at $x = 0$ and $x = 1$, where its graph is obviously broken. More precisely, the function is not continuous at $x = 0$ because $\lim_{x \to 0} f(x)$ does not exist, violating condition **(b)** of Definition 2. It is discontinuous at $x = 1$ because $\lim_{x \to 1} f(x) = 2$ is different from $f(1) = 3$, which violates condition **(c)**. ●

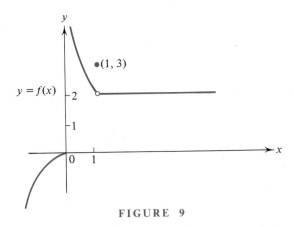

FIGURE 9

DEFINITION 3 (Continuity in an Interval). A function $f(x)$ is said to be *continuous in*[1] *an open interval* if it is continuous at every point of the interval. A function is said to be *continuous in the closed interval* $a \leq x \leq b$ if it is continuous in the open interval $a < x < b$ and if at the end points

(34)
$$\lim_{x \to a^+} f(x) = f(a) \qquad \text{and} \qquad \lim_{x \to b^-} f(x) = f(b).$$

[1] Here custom permits a variety of words. A function is continuous in (on, over, throughout) a set if it is continuous at every point of the set.

For example, the function $f(x)$ given in equation (33) is continuous in the intervals $[-4, -1]$, $(0, 1)$, $[2, 7]$, and, more generally, in any interval that omits the points 0 and 1. It is also continuous in the closed interval $[-3, 0]$ because $\lim_{x \to -3^+} f(x) = -9 = f(-3)$ and $\lim_{x \to 0^-} f(x) = 0 = f(0)$.

Definitions 2 and 3 may appear to be so much excess luggage, since "Everyone knows that a continuous function is one whose graph has no breaks or jumps." Furthermore, in the examples given it is very easy to locate the points of discontinuity by inspection. But we want to be precise in our work, and to be precise we need to give correct definitions.

Each of the fundamental theorems on limits (Theorems 1, 2, 3, 4, and 5) has a counterpart in a theorem on continuous functions. For example, the sum and product of two continuous functions is a continuous function. The quotient of two continuous functions is also a continuous function if the denominator is not zero. For the convenience of the student who wishes to pursue this matter, Appendix 2 contains precise statements of these theorems and their proofs. We pause here only to mention the immediate consequences of Theorems 4 and 5 concerning polynomial and rational functions.

THEOREM 6. Any polynomial function is continuous at all values of x. A rational function is continuous wherever it is defined (i.e., at all points except those which make a denominator zero).

Example 2. Find the points of discontinuity for the functions

(a) $9x^5 - 8x^4 + 2$.   **(b)** $\dfrac{3x^2 + \pi}{(x - 1)(x - 2)}$.   **(c)** $\dfrac{19x^5}{x^2 + 1}$.

Solution. The function in **(a)** is a polynomial and therefore has no points of discontinuity. The rational function **(b)** is continuous except at $x = 1$ and $x = 2$, where its denominator is zero. The denominator of the rational function **(c)** is never zero, so it has no points of discontinuity. ●

An important property possessed by continuous functions is described in

---

THEOREM 7 PWO (Intermediate Value Theorem). If $f(x)$ is continuous in an interval that contains the points a and b ($a < b$), then for each number K between $f(a)$ and $f(b)$, there is at least one number c in the interval $a \leqq x \leqq b$ for which $f(c) = K$.

In short, Theorem 7 states that a continuous function f in $[a, b]$ assumes every value between $f(a)$ and $f(b)$. This is illustrated in Fig. 10, where there are in fact, three points, labeled c_1, c_2, and c_3, where $f(x) = K$.

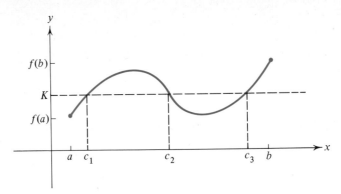

FIGURE 10

The proof of Theorem 7 involves rather sophisticated notions (including a firm grasp of the properties of the real number system) and therefore we omit the details. Nonetheless, it should be apparent to the reader that if one is to draw an *unbroken* curve from the point $(a, f(a))$ to the point $(b, f(b))$, then sooner or later it will have to cross the horizontal line $y = K$ for every number K between $f(a)$ and $f(b)$.

Example 3. Prove that the equation

(35) $$x^4 + x^3 + 2x = 3.26$$

has at least one solution in the interval [0, 1].

Solution. The function $f(x) = x^4 + x^3 + 2x$ is a polynomial and therefore is continuous for all x. Its value at $x = 0$ is 0 and its value at $x = 1$ is 4. Since 3.26 is a number between 0 and 4, the Intermediate Value Theorem asserts that there is at least one solution of (35) in [0, 1]. ●

A method, based on Theorem 7, for actually computing the solutions of equation (35) is described in (optional) Section 5 of Chapter 3.

EXERCISE 5

In Problems 1 through 10, state whether the given function is continuous for all x, and if it is not continuous locate the points of discontinuity.

1. $f(x) = x^3 + 1000x$.

2. $f(x) = 50x^{10}$.

3. $f(x) = \dfrac{5x^4 - 3x^2}{x^2 - 25}$.

4. $f(x) = \dfrac{5x^4 - 3x^2}{x^2 + 25}$.

5. $f(x) = 3 - \dfrac{1}{x} + \dfrac{x^3}{x-4}.$

6. $f(x) = \dfrac{x^2 - 2}{x^2 - 4} + \dfrac{\pi}{x^3}.$

7. $f(x) = \begin{cases} \dfrac{x^4 - 16}{x - 2}, & \text{if } x \neq 2, \\ 16, & \text{if } x = 2. \end{cases}$

8. $f(x) = \begin{cases} \dfrac{x^2 - 16}{x - 4}, & \text{if } x \neq 4, \\ 8, & \text{if } x = 4. \end{cases}$

9. $f(x) = \begin{cases} \dfrac{x^2 + 5x}{10x + 50}, & \text{if } x \neq -5, \\ -1/2, & \text{if } x = -5. \end{cases}$

10. $f(x) = \begin{cases} \dfrac{x^3 - x^2 + 2x - 2}{x - 1}, & \text{if } x \neq 1, \\ 4, & \text{if } x = 1. \end{cases}$

11. Let $f(x) = 1/x$, if $x \neq 0$. Is it possible to define $f(0)$ so that $f(x)$ becomes continuous at $x = 0$?

★**12.** Let $f(x)$ be the Dirichlet function, defined in Chapter 3, equation (5), page 74. Discuss the continuity of this function.

In Problems 13 through 16, determine the value of the constant so that the given function is everywhere continuous. Sketch a graph of the function.

13. $f(x) = \begin{cases} Ax^2, & \text{if } x \leq 2, \\ 3, & \text{if } x > 2. \end{cases}$

14. $f(x) = \begin{cases} mx + 5, & \text{if } x \leq 2, \\ x - 1, & \text{if } x > 2. \end{cases}$

15. $f(x) = \begin{cases} -Bx^3, & \text{if } x < 4, \\ -6x + 16, & \text{if } x \geq 4. \end{cases}$

16. $f(x) = \begin{cases} C(x^2 - 2x), & \text{if } x < 0, \\ \cos x, & \text{if } x \geq 0. \end{cases}$

17. Let

$$f(x) = \begin{cases} 1, & \text{if } x \leq 3, \\ ax + b, & \text{if } 3 < x < 5, \\ 7, & \text{if } 5 \leq x. \end{cases}$$

Determine the constants a and b so that this function is everywhere continuous, and graph this function.

18. Use the Intermediate Value Theorem to prove that the equation $x^5 - 2x^2 = -1.39$ has at least one solution in the interval $[-1, 0]$.

19. Prove that the equation $(x^4 - 2x)/(x^2 + 1) = 3$ has at least one solution in $[-2, -1]$.

20. Prove that the equation $ax^3 + bx^2 + cx + d = 0$ has at least one real solution for any choice of the constants a, b, c, and d with $a \neq 0$. HINT: Consider the limits of the polynomial function as $x \to \infty$ and as $x \to -\infty$.

21. Generalize Problem 20 by showing that for any polynomial $P(x)$ of *odd* degree, the equation $P(x) = 0$ has at least one real solution. Is the analogous statement true for polynomials of even degree?

CALCULATOR PROBLEMS

C1. Let x^* be the unique solution of the equation $x^3 + 3x - 1 = 0$. Use the Intermediate Value Theorem to prove that the approximate solution $a = 0.322$ satisfies the inequality $|x^* - a| \leq 0.001$. HINT: Compute $x^3 + 3x - 1$ at 0.322 and at 0.323.

C2. Let x^* be the unique solution of the equation $x^5 + x = -1$. Prove that the approximate solution $a = -0.75$ satisfies $|x^* - a| \leq 0.01$.

C3. In Problem 19, find the indicated solution to two decimal places.

7 *The Derivative*

Geometrically, the derivative of a function $y = f(x)$ is a second function that gives the slope of the tangent line to the graph of the first function at each point. But first we will give the analytic definition and postpone the geometric interpretation of the derivative until the next section.

DEFINITION 4 (Derivative). Let f be a function defined in an open interval containing the point x_1. If the limit

$$(36) \qquad \lim_{h \to 0} \frac{f(x_1 + h) - f(x_1)}{h}$$

exists,[1] then that limit is called the *derivative of $f(x)$ at $x = x_1$* and is written $f'(x_1)$ (read "f prime of x sub-one"). When the limit (36) exists, the function is said to be *differentiable* at x_1.

The fraction in (36) is called the *difference quotient* for f at x_1. Note that this quotient (which is a function of the variable h) is not defined at $h = 0$ but is defined in an open interval to the left and to the right of $h = 0$.

Example 1. Find the derivative of $f(x) = 3x^2$ at $x = x_1$.

Solution. We have

$$f(x_1 + h) = 3(x_1 + h)^2 \qquad \text{and} \qquad f(x_1) = 3x_1^2.$$

Thus the difference quotient for f at x_1 is

$$\frac{f(x_1 + h) - f(x_1)}{h} = \frac{3(x_1 + h)^2 - 3x_1^2}{h},$$

---

[1] This means that the ratio in (36) does tend to some finite limiting value as h approaches zero. There are weird functions for which the ratio oscillates violently as h approaches zero and hence does not have a limit.

and applying Definition 4, we obtain

$$f'(x_1) = \lim_{h \to 0} \frac{3(x_1 + h)^2 - 3x_1^2}{h} = \lim_{h \to 0} \frac{3x_1^2 + 6x_1 h + 3h^2 - 3x_1^2}{h}$$

$$= \lim_{h \to 0} \frac{6x_1 h + 3h^2}{h} = \lim_{h \to 0} (6x_1 + 3h) = 6x_1 + 3 \times 0 = 6x_1.$$

Hence $f'(x_1)$, the derivative of this function at x_1, is $6x_1$. ●

The subscript on x serves to remind us that x is fixed at x_1 and that it is h that is varying in the limit process. Once these ideas are clear, we can drop the subscript and write: If $f(x) = 3x^2$, then $f'(x) = 6x$.

The definition of a derivative can be stated in a number of different ways. First, we can condense Definition 4 by writing

(37)
$$f'(x_1) = \lim_{h \to 0} \frac{f(x_1 + h) - f(x_1)}{h}$$

whenever the limit on the right side exists. In this form x_1 is fixed and the variable is h.

If we wish to let x be the variable, we set $x = x_1 + h$ so that $h = x - x_1$ and $x \to x_1$ as $h \to 0$. Then (37) takes the form

(38)
$$f'(x_1) = \lim_{x \to x_1} \frac{f(x) - f(x_1)}{x - x_1}.$$

For example, using this last formula, we find for the function $f(x) = 3x^2$,

$$f'(x_1) = \lim_{x \to x_1} \frac{3x^2 - 3x_1^2}{x - x_1} = \lim_{x \to x_1} \frac{3(x + x_1)(x - x_1)}{x - x_1}$$

$$= \lim_{x \to x_1} 3(x + x_1) = 3(x_1 + x_1) = 6x_1,$$

which is indeed the same as the answer obtained in Example 1.

Let us look again at the expression (36). In the numerator we evaluate the function at $x_1 + h$ and also at x_1. Thus h is really the change in x, as x goes from x_1 to $x_1 + h$. It is convenient to use the symbol Δx to denote this change in x, so we set

$$h = \Delta x.$$

Further, the numerator in (36), $f(x_1 + h) - f(x_1)$, is just the value of the function when x is $x_1 + h$, minus the value of the function at x_1, and this is just the change in $y = f(x)$, so we are justified in calling it Δy. Then the expression (36) takes on the simple form

$$\lim_{\Delta x \to 0} \frac{\Delta y}{\Delta x}.$$

This form suggests an alternative notation for the derivative which turns out to be very convenient. Indeed in the limiting process, the Greek letters are replaced by their English equivalents and we have (by definition)

(39)
$$\frac{dy}{dx} = \lim_{\Delta x \to 0} \frac{\Delta y}{\Delta x}.$$

This new symbol $\dfrac{dy}{dx}$ is just another symbol for the derivative and is read "the derivative of y with respect to x" or "dy over dx."

Although $\dfrac{dy}{dx}$ looks like a fraction, it is not. It is the limiting value of a fraction, and as such it enjoys many properties of fractions. The use of this symbol makes many otherwise difficult manipulations become childishly simple. For this reason, we want to employ the defining equation (39) as frequently as possible. But whenever there is a suspicion that the notation (39) may be leading us into error, because it looks like a fraction, then we should return to the definition[1]

(40)
$$f'(x) = \lim_{h \to 0} \frac{f(x + h) - f(x)}{h},$$

where there is no temptation to treat the derivative $f'(x)$ as a fraction.

In computing derivatives from equation (39), the procedure can be divided into four steps.

(I) In $f(x)$, replace x by $x + \Delta x$ to find

(41)
$$f(x + \Delta x).$$

(II) Subtract $f(x)$ from (41) to obtain

(42)
$$\Delta y = f(x + \Delta x) - f(x).$$

(III) Divide both sides of (42) by Δx, obtaining the difference quotient

(43)
$$\frac{\Delta y}{\Delta x} = \frac{f(x + \Delta x) - f(x)}{\Delta x}.$$

---

[1] Note that in going from equation (37) to equation (40) we have dropped the subscript on x. The subscript merely reminds us that x is fixed at x_1 as $h \to 0$. Once this is clear, the subscript may be deleted.

(IV) Take the limit in (43) as the variable Δx approaches zero,

$$\frac{dy}{dx} = \lim_{\Delta x \to 0} \frac{\Delta y}{\Delta x} = \lim_{\Delta x \to 0} \frac{f(x + \Delta x) - f(x)}{\Delta x}.$$

Example 2. Find the derivative of $y = \dfrac{x^2}{2} - 2x + 9$ using the notation of equation (39).

Solution. Following the four steps just outlined, we have

(I) $\quad f(x + \Delta x) = \dfrac{(x + \Delta x)^2}{2} - 2(x + \Delta x) + 9.$

(II) $\quad \Delta y = \left[\dfrac{(x + \Delta x)^2}{2} - 2(x + \Delta x) + 9 \right] - \left(\dfrac{x^2}{2} - 2x + 9 \right)$

$\qquad\qquad = \left[\dfrac{x^2}{2} + x\,\Delta x + \dfrac{(\Delta x)^2}{2} - 2x - 2\,\Delta x + 9 \right] - \left(\dfrac{x^2}{2} - 2x + 9 \right).$

$\qquad\quad \Delta y = x\,\Delta x + \dfrac{(\Delta x)^2}{2} - 2\,\Delta x.$

(III) $\quad \dfrac{\Delta y}{\Delta x} = x + \dfrac{\Delta x}{2} - 2, \qquad \text{for } \Delta x \neq 0.$

(IV) $\quad \dfrac{dy}{dx} = \lim_{\Delta x \to 0} \dfrac{\Delta y}{\Delta x} = \lim_{\Delta x \to 0} \left(x + \dfrac{\Delta x}{2} - 2 \right) = x + 0 - 2.$

$\qquad\quad \dfrac{dy}{dx} = x - 2. \quad \bullet$

To avoid excessive writing, we will hereafter streamline the four-step procedure as illustrated in

Example 3. If $y = \sqrt{x}$ and $x > 0$, find $\dfrac{dy}{dx}$.

Solution. We immediately write

$$\frac{\Delta y}{\Delta x} = \frac{\sqrt{x + \Delta x} - \sqrt{x}}{\Delta x}.$$

Letting $\Delta x \to 0$ in the preceding expression yields $0/0$, or no information. We must first prepare the way with some clever algebra, which allows the cancellation of the

Δx. To accomplish our aim, we rationalize the numerator.

$$\frac{\Delta y}{\Delta x} = \frac{\sqrt{x + \Delta x} - \sqrt{x}}{\Delta x} \cdot \frac{\sqrt{x + \Delta x} + \sqrt{x}}{\sqrt{x + \Delta x} + \sqrt{x}} = \frac{x + \Delta x - x}{\Delta x(\sqrt{x + \Delta x} + \sqrt{x})}$$

$$= \frac{\Delta x}{\Delta x(\sqrt{x + \Delta x} + \sqrt{x})} = \frac{1}{\sqrt{x + \Delta x} + \sqrt{x}}.$$

Now we can let $\Delta x \to 0$ comfortably. We find that

$$\frac{dy}{dx} = \lim_{\Delta x \to 0} \frac{1}{\sqrt{x + \Delta x} + \sqrt{x}} = \frac{1}{\sqrt{x} + \sqrt{x}} = \frac{1}{2\sqrt{x}}.$$

Thus if $y = \sqrt{x}$, then $\dfrac{dy}{dx} = \dfrac{1}{2\sqrt{x}}$. ●

In Example 1 we saw that if $y = f(x) = 3x^2$, then the derivative is $6x$. This fact can be written with symbols in a variety of ways, all of which are equivalent:

$$f'(x) = 6x, \qquad y' = 6x, \qquad \frac{dy}{dx} = 6x, \qquad Df(x) = 6x,$$

$$\frac{d}{dx}y = 6x, \qquad \frac{d}{dx}(3x^2) = 6x.$$

In the last two expressions the symbol $\dfrac{d}{dx}$ (read "the derivative with respect to x of") may be thought of as an *operator* or machine that produces a new function, the derivative, by operating on the original or primitive function.

When we wish to evaluate the derivative at a particular point x_1, we can use any of the equivalent forms

$$f'(x_1) = y'(x_1) = \frac{dy}{dx}\bigg|_{x=x_1} = Df\bigg|_{x=x_1}.$$

Thus, if $y = f(x) = 3x^2$, we have

$$f'(2) = 6 \cdot 2 = 12, \qquad y'(2) = 12, \qquad \frac{dy}{dx}\bigg|_{x=2} = 12, \qquad Df\bigg|_{x=2} = 12.$$

Computing derivatives from the definition can be a very tedious process if the given $f(x)$ is complicated. In Sections 9, 10, and 11, we will prove some formulas for finding derivatives that will greatly reduce the labor. But first the student must serve his apprenticeship by finding derivatives in the manner illustrated above. One formula (which will be proved later) is that if

$$y = cx^n,$$

then

$$\frac{dy}{dx} = ncx^{n-1}.$$

It is clear that this formula does give the correct answers in the three examples just worked.

One useful property of a differentiable function is given in

THEOREM 8. If the derivative $f'(x_1)$ exists, then the function f is continuous at x_1.

Proof. First we write, for $x \neq x_1$, the identity

$$f(x) - f(x_1) = \left[\frac{f(x) - f(x_1)}{x - x_1}\right](x - x_1)$$

or

(44) $$f(x) = f(x_1) + \left[\frac{f(x) - f(x_1)}{x - x_1}\right](x - x_1).$$

Recalling the formula (38) for $f'(x_1)$, we take the limit as $x \to x_1$ in (44) and find that

$$\lim_{x \to x_1} f(x) = \lim_{x \to x_1} f(x_1) + \lim_{x \to x_1}\left[\frac{f(x) - f(x_1)}{x - x_1}\right] \cdot \lim_{x \to x_1}(x - x_1)$$

$$= f(x_1) + f'(x_1) \cdot 0 = f(x_1).$$

Thus $\lim_{x \to x_1} f(x) = f(x_1)$, which means that f is continuous at x_1. ∎

It is possible for a function f not to have a derivative at a point x_1 even though it is continuous there. For example, $f(x) = |x|$ is continuous at $x_1 = 0$ but is not differentiable at $x_1 = 0$ because the difference quotient

$$\frac{f(0 + \Delta x) - f(0)}{\Delta x} = \frac{|\Delta x| - 0}{\Delta x} = \frac{|\Delta x|}{\Delta x}$$

does not have a two-sided limit as $\Delta x \to 0$. Indeed,

$$\lim_{\Delta x \to 0^+} \frac{|\Delta x|}{\Delta x} = \lim_{\Delta x \to 0^+} \frac{\Delta x}{\Delta x} = 1, \quad \text{but} \quad \lim_{\Delta x \to 0^-} \frac{|\Delta x|}{\Delta x} = \lim_{\Delta x \to 0^-} \frac{-\Delta x}{\Delta x} = -1.$$

For this example, we say that the function has a *right-hand derivative* and a *left-hand derivative,* but not a derivative at $x = 0$.

EXERCISE 6

In each of Problems 1 through 12, compute the derivative, either by the method of Example 1 or the method of Example 2. Then check your answer by the formula $\dfrac{d}{dx}cx^n = ncx^{n-1}$, *wherever this formula is applicable.*

1. $y = 2x.$

2. $y = 5x - \pi.$

3. $y = x + x^2.$

4. $y = 4 - x^2.$

5. $y = \dfrac{1}{x}, \quad x \neq 0.$

6. $y = \dfrac{3}{x + 1}, \quad x \neq -1.$

7. $y = -x^3.$

8. $y = x^3 - 12x.$

9. $y = \sqrt{5x}, \quad x > 0.$

10. $y = \sqrt{x + 5}, \quad x > -5.$

11. $y = \dfrac{x + 2}{x - 5}, \quad x \neq 5.$

12. $y = \dfrac{6}{1 + x^2}.$

13. For $f(x) = 2/x$, find $f'(1)$ by using the form (38).

14. If $f(x) = |x|$, show that $f'(x) = 1$ for $x > 0$ and $f'(x) = -1$ for $x < 0$.

15. Graph the function

$$y = \begin{cases} x, & \text{if } x \leq 1, \\ 2 - x, & \text{if } 1 < x. \end{cases}$$

Prove that at $x_1 = 1$ this function does not have a derivative. Is the function continuous at $x_1 = 1$?

16. Let

$$f(x) = \begin{cases} x^2, & \text{if } x \neq 1, \\ 2, & \text{if } x = 1. \end{cases}$$

Is f differentiable at $x = 1$?

CALCULATOR PROBLEMS

In Problems C1 through C6, guess the value of the derivative of the given function $f(x)$ at the indicated value of x_1 by computing the difference quotient $[f(x_1 + \Delta x) - f(x_1)]/\Delta x$ for $\Delta x = 0.1, 0.01, 0.001, 0.0001$, and for $\Delta x = -0.1, -0.01, -0.001, -0.0001.$

C1. $f(x) = x^{1/4}, \quad x_1 = 1.$

C2. $f(x) = 5x^4, \quad x_1 = 2.$

C3. $f(x) = \dfrac{x^2}{x - 1}, \quad x_1 = 2.$

C4. $f(x) = \sqrt{x^2 + 3}, \quad x_1 = 1.$

C5. $f(x) = 8\sqrt{2 + \sqrt{x}}, \quad x_1 = 4.$

C6. $f(x) = (x^2 - 8)^{1/4}, \quad x_1 = 3.$

C7. For the function $f(x) = x^{1/4}$ in Problem C1, the derivative $f'(1) = 0.25$. However, if $\Delta x = 10^{-11}$ is used in computing the difference quotient $[f(1 + \Delta x) - f(1)]/\Delta x$, the calculator gives the misleading answer 0. Try this for your own calculator and explain why the value 0 is obtained. Finally, rewrite the difference quotient in a form that gives a more accurate value when $\Delta x = 10^{-11}$. HINT: Rationalize the numerator.

The Tangent Line to a Curve 8

We select a point $P(x_1, y_1)$ on the curve $y = f(x)$, and a neighboring point $Q(x_2, y_2)$, also on the curve, and consider the secant line PQ as shown in Fig. 11.

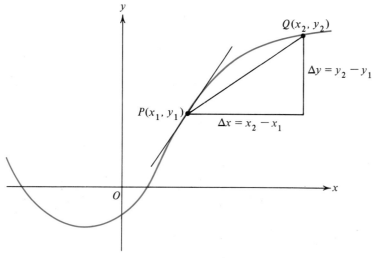

FIGURE 11

DEFINITION 5 (Tangent Line). The tangent line to the curve $y = f(x)$ at the point $P(x_1, y_1)$ is defined to be the line that has the limiting position of the secant PQ as the point Q approaches the point P along the curve.

Of course the curve may be so bumpy near the point P that the secant line may jiggle violently as Q approaches P, and it may not have a limit position. If this does occur, then we say that the curve does not have a tangent line at P.

Figure 11 shows the point Q approaching the point P from the right ($x_2 > x_1$ and Δx is positive). It could also approach from the left ($x_2 < x_1$ and Δx is negative). For a curve to have a tangent line, the limiting position of the secant must be the same, whether Q approaches P from one side or the other.

We can find the equation of the tangent line if we know one point on the line and its slope. We have already selected the point (x_1, y_1) on the curve so that all that remains is to compute the slope. Now let Δx be the change in x in going from x_1 to x_2. Then $x_2 = x_1 + \Delta x$. For the y-coordinate of Q we have $y_2 = f(x_2) = f(x_1 + \Delta x)$, and the change in the y-coordinate is just

$$\Delta y = y_2 - y_1 = f(x_1 + \Delta x) - f(x_1).$$

Then the slope of the secant PQ is given by

$$(45) \qquad m_{PQ} = \frac{\Delta y}{\Delta x} = \frac{f(x_1 + \Delta x) - f(x_1)}{\Delta x}.$$

If we take the limit as $Q \to P$, namely, as $\Delta x \to 0$, the middle term in (45) is identical with the right side of (39) in the definition of the derivative. This proves

---

THEOREM 9. Let $y = f(x)$ be the equation of a curve. Then the derivative $f'(x_1)$ is the slope of the line tangent to the curve at the point $P(x_1, y_1)$ on the curve.

Using the point-slope formula for a line, an equation of the tangent line to the curve $y = f(x)$ at $P(x_1, y_1)$ is therefore given by

$$(46) \qquad \boxed{y - y_1 = f'(x_1)(x - x_1),} \qquad \text{where } y_1 = f(x_1).$$

Example 1. Give an equation for the tangent line to the graph of

$$(47) \qquad y = \frac{x^2}{2} - 2x + 9$$

at the point $(4, 9)$.

Solution. First we should check that the given point is on the curve, for otherwise the problem would be meaningless. At $x = 4$, equation (47) gives

$$y = \frac{4^2}{2} - 2 \cdot 4 + 9 = 8 - 8 + 9 = 9.$$

Hence $(4, 9)$ is on the curve.

Next we need the derivative of $f(x) = x^2/2 - 2x + 9$ at $x = 4$. In Example 2 of Section 7 we found that

$$(48) \qquad f'(x) = \frac{dy}{dx} = x - 2.$$

Therefore, $f'(4) = 4 - 2 = 2$ is the slope of the tangent line at the point $(4, 9)$. Finally, formula (46) with $x_1 = 4$, $y_1 = 9$ gives the equation

$$y - 9 = 2(x - 4) \qquad \text{or} \qquad y = 2x + 1$$

for the tangent line. ●

Example 2. Sketch the curve $y = x^2/2 - 2x + 9$ of Example 1 by plotting a few points and the tangent lines at those points.

Solution. From equation (48) we have $f'(x) = x - 2$. For each x we can compute the corresponding y and also the slope of the tangent at the point (x, y). The results of several such computations using equations (47) and (48) are shown in Table 7. These data are used to plot the points and the tangent lines shown in Fig. 12. From these points and lines the curve is easily visualized, because at each point plotted we now know the direction of the curve. ●

TABLE 7

x	-4	-2	0	2	4	6	8
$y = \dfrac{x^2}{2} - 2x + 9$	25	15	9	7	9	15	25
$m = f'(x) = x - 2$	-6	-4	-2	0	2	4	6

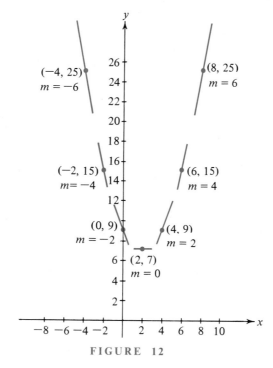

FIGURE 12

Example 3. Find the low point on the curve of Example 1.

Solution. Since $f'(x) = x - 2$, we know that $f'(x)$ is negative for $x < 2$. This means that the slope of the tangent line is negative, and this in turn means that the curve is falling as x increases. On the other hand, $f'(x)$ is positive for $x > 2$. This means that the slope of the tangent line is positive, and this means that the curve is rising as x increases. Since the curve falls until $x = 2$ and then rises, it is obvious that $(2, 7)$ is a low point on the curve. Such a point is called a minimum point on the curve, and the value for y (in this case 7) is called a minimum for the function. ●

We notice that the minimum occurs, in this example, at $x = 2$, where the derivative is zero. A value of x for which the derivative $f'(x)$ is zero is called a *critical value* or a *critical point* for the function $f(x)$. The reader should plot a few more points and complete the sketch of the curve of Fig. 12, and convince himself that the function

$$y = \frac{x^2}{2} - 2x + 9$$

is never less than 7.

Similarly, a curve can have a high point or a maximum. We will discuss these high and low points more carefully and more thoroughly in Chapter 5.

EXERCISE 7

In Problems 1 through 6, a function $f(x)$ and its derivative $f'(x)$ are given. Sketch the curve $y = f(x)$ by plotting a few points together with the tangent lines at those points. Find all the critical values for the function. Find high and low points (if any) on the curve $y = f(x)$.

1. $f(x) = 2 - x^2$; $f'(x) = -2x$.   2. $f(x) = x^2 - 4x$; $f'(x) = 2x - 4$.

3. $f(x) = x^3 - 12x + 2$; $f'(x) = 3x^2 - 12$.   4. $f(x) = 3x - x^3$; $f'(x) = 3 - 3x^2$.

5. $f(x) = \dfrac{6}{1 + x^2}$; $f'(x) = \dfrac{-12x}{(1 + x^2)^2}$.   6. $f(x) = -x^3$; $f'(x) = -3x^2$.

In Problems 7 through 9, use the definition to find the derivative and give an equation of the tangent line to the curve at the indicated point. Finally, sketch the curve and the tangent line.

7. The parabola $y = 4 - 2x^2$ at $(2, -4)$.

8. The curve $y = 2/x^2$ at $(1, 2)$.

9. The hyperbola $y = 1 - 2/x$ at $(1, -1)$.

10. **Normal Line.** For a curve $y = f(x)$, the line through $(x_1, f(x_1))$ and perpendicular to the tangent line to the curve at that point is called the *normal line* to the curve at that point. Show that the normal line to the curve $y = f(x)$ at $(x_1, f(x_1))$ has the equation

$$y - f(x_1) = -\frac{1}{f'(x_1)}(x - x_1) \qquad \text{if } f'(x_1) \neq 0,$$

and has the equation $x = x_1$ if $f'(x_1) = 0$.

11. Give an equation of the normal line to the graph $y = 2 - x^2$ at $(3, -7)$ (see Problem 1).

12. Give an equation of the normal line to the graph of $y = x^3 - 12x + 2$ at $(1, -9)$ (see Problem 3).

Differentiation Formulas 9

We have just seen how helpful the derivative is in sketching curves. But this is only one of the many important uses of the derivative. However, before we examine some of the other applications, we want to simplify the actual computation of the derivative by proving some useful formulas.

THEOREM 10. The derivative of a constant is zero,

$$\frac{d}{dx} c = 0.$$

Proof. Let $y = c$. If x is increased by an amount Δx, then y, being constant, remains unchanged. Hence $\Delta y = 0$. Consequently, $\Delta y / \Delta x = 0$ and

$$\frac{dy}{dx} = \lim_{\Delta x \to 0} \frac{\Delta y}{\Delta x} = \lim_{\Delta x \to 0} 0 = 0. \quad \blacksquare$$

THEOREM 11 (Power Rule). If n is a positive integer, then

(49) $$\frac{d}{dx} x^n = nx^{n-1}.$$

For example,

$$\frac{d}{dx} x^7 = 7x^6 \quad \text{and} \quad \frac{d}{dx} x^{285} = 285x^{284}.$$

Formula (49) is easy to memorize. In differentiating, the exponent n becomes a multiplier in front, and the exponent on x is decreased by 1. A nice feature of this formula is that it is true for all values of n, whether an integer or not. Later we will prove that this formula holds for any rational n.

Proof of Theorem 11. Let $y = x^n$. Then

(50)
$$\frac{\Delta y}{\Delta x} = \frac{(x + \Delta x)^n - x^n}{\Delta x}.$$

To prepare the way for taking the limit as $\Delta x \to 0$ in (50), we first expand the numerator. According to the Binomial Theorem of algebra,

$$(x + \Delta x)^n = x^n + nx^{n-1}(\Delta x) + \frac{n(n-1)}{2}x^{n-2}(\Delta x)^2 + \cdots + (\Delta x)^n.$$

In other words,

(51) $(x + \Delta x)^n = x^n + nx^{n-1}(\Delta x) + $ [terms involving $(\Delta x)^2, (\Delta x)^3, \ldots, (\Delta x)^n$].

Consequently, if we subtract x^n from both sides of (51) and then divide both sides by Δx, we obtain

$$\frac{(x + \Delta x)^n - x^n}{\Delta x} = nx^{n-1} + \text{[terms involving } \Delta x, (\Delta x)^2, \ldots, (\Delta x)^{n-1}].$$

Since the terms involving Δx and its higher powers tend to zero as $\Delta x \to 0$, we have

$$\frac{dy}{dx} = \lim_{\Delta x \to 0} \frac{(x + \Delta x)^n - x^n}{\Delta x} = nx^{n-1}. \quad \blacksquare$$

In the statement of the next theorem we depart from the usual notation by choosing $u(x)$ to denote any differentiable function.

THEOREM 12. The derivative of a constant times a function is the constant times the derivative of the function,

$$\frac{d}{dx}(cu) = c\frac{du}{dx} \qquad \text{for any constant } c.$$

Proof. Let $y = cu(x)$. Then

$$\frac{\Delta y}{\Delta x} = \frac{cu(x + \Delta x) - cu(x)}{\Delta x} = c\left[\frac{u(x + \Delta x) - u(x)}{\Delta x}\right] = c\frac{\Delta u}{\Delta x}.$$

Hence from the basic properties of limits, we have

$$\frac{dy}{dx} = \lim_{\Delta x \to 0} c\frac{\Delta u}{\Delta x} = c\lim_{\Delta x \to 0}\frac{\Delta u}{\Delta x} = c\frac{du}{dx}. \quad \blacksquare$$

Combining Theorems 11 and 12, we see that for any positive integer n and any constant c,

(52)
$$\frac{d}{dx}cx^n = cnx^{n-1}.$$

Example 1. Find dy/dx if **(a)** $y = 10x^7$, **(b)** $y = -13x^{101}$, **(c)** $y = \pi^3 x$, and **(d)** $y = [(17^\pi - \sin \sqrt{7} \tan \sqrt[3]{11})/(\pi^5 - 41)]\sqrt{3}$.

Solution. Theorems 10, 11, and 12 give immediately

(a) $\dfrac{dy}{dx} = 10\dfrac{d}{dx}x^7 = 10(7x^6) = 70x^6.$ **(b)** $\dfrac{dy}{dx} = -13(101x^{100}) = -1313x^{100}.$

(c) $\dfrac{dy}{dx} = \pi^3 \dfrac{d}{dx}x = \pi^3 \cdot 1 = \pi^3.$ **(d)** $\dfrac{dy}{dx} = 0.$ ●

THEOREM 13. The derivative of the sum of two differentiable functions is the sum of the derivatives of the two functions,

(53)
$$\frac{d}{dx}(u + v) = \frac{du}{dx} + \frac{dv}{dx}.$$

Proof. Let $y = u(x) + v(x)$. Then the difference quotient for y is

$$\frac{\Delta y}{\Delta x} = \frac{[u(x + \Delta x) + v(x + \Delta x)] - [u(x) + v(x)]}{\Delta x}$$

$$= \left[\frac{u(x + \Delta x) - u(x)}{\Delta x}\right] + \left[\frac{v(x + \Delta x) - v(x)}{\Delta x}\right] = \frac{\Delta u}{\Delta x} + \frac{\Delta v}{\Delta x}.$$

Thus as $\Delta x \to 0$ in this equation, we obtain

$$\frac{dy}{dx} = \lim_{\Delta x \to 0}\left(\frac{\Delta u}{\Delta x} + \frac{\Delta v}{\Delta x}\right) = \lim_{\Delta x \to 0}\frac{\Delta u}{\Delta x} + \lim_{\Delta x \to 0}\frac{\Delta v}{\Delta x} = \frac{du}{dx} + \frac{dv}{dx}. \quad ∎$$

Example 2. Find $\dfrac{dy}{dx}$ if $y = 10x^7 + \pi x^3$.

Solution. $\dfrac{dy}{dx} = \dfrac{d}{dx}(10x^7) + \dfrac{d}{dx}(\pi x^3) = 70x^6 + 3\pi x^2.$ ●

By the use of mathematical induction,[1] Theorem 13 can be extended to any finite number of functions, as stated in

---

THEOREM 14 (PLE). The derivative of the sum of any finite number of differentiable functions is the sum of the derivatives of the functions,

$$\frac{d}{dx}(u_1 + u_2 + \cdots + u_n) = \frac{du_1}{dx} + \frac{du_2}{dx} + \cdots + \frac{du_n}{dx}.$$

Example 3. If $y = 3x^4 - 8x^3 + 6x^2 - 5x + 18$, find $\dfrac{dy}{dx}$.

Solution. Observe that a subtraction such as $-8x^3$ can be written as $+(-8)x^3$ so that Theorem 14 can be applied to this sum. This together with the earlier theorems gives

$$\frac{dy}{dx} = \frac{d}{dx}3x^4 + \frac{d}{dx}(-8)x^3 + \frac{d}{dx}6x^2 + \frac{d}{dx}(-5)x + \frac{d}{dx}18$$

$$= 12x^3 + (-8)3x^2 + 12x + (-5) = 12x^3 - 24x^2 + 12x - 5. \quad \bullet$$

Example 4. If $y = (2x - 5)^3$, find $\dfrac{dy}{dx}$.

Solution. None of the theorems proved so far handles this function directly, although later we will prove a formula that will be applicable here. In the meantime we must first expand by the Binomial Theorem. This gives

$$y = 8x^3 - 60x^2 + 150x - 125.$$

Then by our differentiation formulas

$$\frac{dy}{dx} = 24x^2 - 120x + 150. \quad \bullet$$

Example 5. If $s = 16t^2 + 64t$, find $\dfrac{ds}{dt}$.

Solution. Although x and y are standard and convenient letters for the independent

[1]Mathematical induction is explained in detail in the *Study Guide*.

and dependent variables, respectively, there is no law that says these letters, and these only, must be used. In computing the derivative we are computing the limiting value of the ratio of the change in the dependent variable to the change in the independent variable. Hence the formulas hold no matter what letters are used. Thus if $s = 16t^2 + 64t$, then

$$\frac{ds}{dt} = 32t + 64. \quad \bullet$$

EXERCISE 8

In Problems 1 through 12, use the formulas to compute the derivative of the dependent variable with respect to the independent variable.

1. $y = 3x^8 - 4x^6$.

2. $y = \pi x^4 - 6x$.

3. $y = x(5x^4 + x^2 + 1)$.

4. $y = x^{12} - 2x^6 + 4x^3 - 6x^2 + 12x$.

5. $y = \dfrac{-x}{3} + 1{,}000{,}000$.

6. $y = \dfrac{x^4 - 2x}{x}$.

7. $s = t^{84} - 2t^{36} + 14$.

8. $u = 3v^8 - 4v^6$.

9. $V = \dfrac{4}{3}\pi r^3$.

10. $w = (2z - 3)^2$.

11. $y = (2x - 3)^3$.

12. $y = (2x - 1)^4$.

In Problems 13 through 16, find the indicated derivative.

13. $f'(\sqrt{2})$, where $f(x) = 3x^5 - 2x^3$.

14. $F'(-1)$, where $F(x) = 4G(x)$ and $G'(-1) = 3$.

15. $f'(0)$, where $f(t) = 5g(t) - 4h(t)$ and $g'(0) = -1$, $h'(0) = 2$.

16. $F'(\pi)$, where $F(t) = \dfrac{G(t)}{2} + \dfrac{H(t)}{3}$ and $G'(\pi) = 0$, $H'(\pi) = 6$.

In Problems 17 through 20, compute the derivative and use the information to sketch the graph of the given equation. Find all the critical values of the function and the coordinates of high and low points (if any) on the graph.

17. $y = x^2 + 6x + 5$.

18. $y = 9 + 8x - x^2$.

19. $y = x(x^2 - 9)$.

20. $y = \dfrac{x^3}{9} - \dfrac{x^2}{2} + 5$.

In Problems 21 through 24, find an equation for the indicated line.

21. Tangent line to $y = 4x^3 - x$ at $(1, 3)$.

22. Tangent line to $y = (x + 1)^2$ at $(-3, 4)$.

23. Normal line to $y = x^{10}/2 + x^5/5$ at $(1, 7/10)$. (See Problem 10 of Exercise 7.)

24. Normal line to $y = x^5 - 4x^4 + 3$ at $(0, 3)$.

25. Find the points on the curve $y = 2x^3 + 3x^2 - 36x + 10$ where the tangent line is horizontal.

26. Find the points on the curve $y = x^3 + x^2 + x$ where the tangent line is parallel to $y = 2x + 3$.

27. A minicar is traveling at night up a hill having the equation $y = x^3 - 15x^2 + 76x$. Find the angle of inclination of the car's headlights at the moment the car reaches the point $(5, 130)$ on the hill. (The car is traveling left to right.)

★28. Determine the constants a, b, and c so that the curve $y = ax^2 + bx + c$ will go through the origin and the point $(1, 1)$, and have slope 3 at the point $(1, 1)$.

29. Two curves are said to be *tangent* at a point where they meet if they have the same tangent line at the point. Show that the two curves $y = 1 - 2x + 2x^2$ and $y = x^2$ are tangent at the point $(1, 1)$.

★30. Determine the constants a, b, and c so that the two curves $y = x^2 + ax + b$ and $y = cx - x^2$ are tangent at the point $(1, 3)$.

31. Using the theorems of this section, prove that for any polynomial function $P(x) = a_n x^n + a_{n-1} x^{n-1} + \cdots + a_2 x^2 + a_1 x + a_0$, we have

$$P'(x) = na_n x^{n-1} + (n-1)a_{n-1} x^{n-2} + \cdots + 2a_2 x + a_1.$$

CALCULATOR PROBLEMS

C1. Find the angle of inclination (to five significant digits) of the tangent line to the graph of $y = 3.1x^3 - 2.8x^2 + 1.9x$ at the point $(1, 2.2)$.

C2. Find the angle of inclination (to five significant digits) of the tangent line to the graph of $y = \sqrt{2}x^5 + \sqrt{6}x - \sqrt{17}$ at the point $(0, -\sqrt{17})$.

In Problems C3 through C5, find all the critical points of the given function to five significant digits.

C3. $f(x) = \pi x^3 - 8x^2 + 3x - 1$.   **C4.** $f(x) = \sqrt{2}x^3 + \sqrt{13}x^2 + \sqrt{2}x + \sqrt{6}$.

C5. $f(x) = x^4 + x^2 + 5x + 9$. HINT: $f'(x) = 0$ has only one real root.

10 *The Product and Quotient Rules*

Functions that can be expressed as the product or quotient of simpler functions can be differentiated by using the Product or Quotient Rules, which we discuss in this section.

> **THEOREM 15 (Product Rule).** The derivative of the product of two differentiable functions is the first times the derivative of the second plus the second times the

derivative of the first,

(54)
$$\frac{d}{dx} uv = u\frac{dv}{dx} + v\frac{du}{dx}.$$

Proof. Let $y = u(x)v(x)$. Then the difference quotient for y is

$$\frac{\Delta y}{\Delta x} = \frac{u(x + \Delta x)v(x + \Delta x) - u(x)v(x)}{\Delta x},$$

and in order to facilitate taking the limit we use the trick of adding and subtracting the term $u(x + \Delta x)v(x)$ in the numerator. We then obtain

$$\frac{\Delta y}{\Delta x} = \frac{u(x + \Delta x)v(x + \Delta x) - u(x + \Delta x)v(x) + u(x + \Delta x)v(x) - u(x)v(x)}{\Delta x}$$

$$= u(x + \Delta x)\left[\frac{v(x + \Delta x) - v(x)}{\Delta x}\right] + v(x)\left[\frac{u(x + \Delta x) - u(x)}{\Delta x}\right].$$

(55)
$$\frac{\Delta y}{\Delta x} = u(x + \Delta x)\frac{\Delta v}{\Delta x} + v(x)\frac{\Delta u}{\Delta x}.$$

Since u is differentiable at x, we know from Theorem 8 that it is continuous there. Hence $u(x + \Delta x)$ tends to $u(x)$ as $\Delta x \to 0$. Keeping in mind that x is held fixed throughout our discussion, we can now take the limit as $\Delta x \to 0$ in (55):

$$\frac{dy}{dx} = \lim_{\Delta x \to 0}\left[u(x + \Delta x)\frac{\Delta v}{\Delta x} + v(x)\frac{\Delta u}{\Delta x}\right]$$

$$= \lim_{\Delta x \to 0} u(x + \Delta x) \cdot \lim_{\Delta x \to 0}\frac{\Delta v}{\Delta x} + \lim_{\Delta x \to 0} v(x) \cdot \lim_{\Delta x \to 0}\frac{\Delta u}{\Delta x}$$

$$= u(x)\frac{dv}{dx} + v(x)\frac{du}{dx}. \quad \blacksquare$$

Example 1. If $y = (x^4 + 3)(3x^3 + 1)$, find $\dfrac{dy}{dx}$.

Solution. By the formula just proved,

$$\frac{dy}{dx} = (x^4 + 3)\frac{d}{dx}(3x^3 + 1) + (3x^3 + 1)\frac{d}{dx}(x^4 + 3),$$

$$\frac{dy}{dx} = (x^4 + 3)(9x^2) + (3x^3 + 1)(4x^3) = 9x^6 + 27x^2 + 12x^6 + 4x^3,$$

(56) $\qquad \dfrac{dy}{dx} = 21x^6 + 4x^3 + 27x^2,$

when the terms are arranged with decreasing powers of x. ●

Note that we could first multiply the two factors and then differentiate. This method gives

(57) $\qquad\qquad y = (x^4 + 3)(3x^3 + 1) = 3x^7 + x^4 + 9x^3 + 3.$

Then using the right side of (57), we have

(58) $\qquad\qquad\qquad \dfrac{dy}{dx} = 21x^6 + 4x^3 + 27x^2.$

We observe that equations (56) and (58) are identical, and hence our formulas are consistent. This is certainly what we must expect (and indeed demand).

Since this problem could be solved without formula (54), this formula may seem to be useless (although it is certainly interesting). At present this is indeed so, but further on we will see that (54) is an extremely useful formula.

THEOREM 16 (Quotient Rule). The derivative of the quotient of two differentiable functions is the denominator times the derivative of the numerator minus the numerator times the derivative of the denominator, all divided by the square of the denominator, if the denominator is not zero,

(59) $$\frac{d}{dx}\frac{u}{v} = \frac{v\dfrac{du}{dx} - u\dfrac{dv}{dx}}{v^2}, \qquad v \neq 0.$$

The proof of the Quotient Rule (59) is similar to the proof of the Product Rule and is left to the Exercises (see Problem 28).

Example 2. If $y = \dfrac{x}{x^2 - 3x + 5}$, find $\dfrac{dy}{dx}$.

Solution. By the Quotient Rule, equation (59),

$$\frac{dy}{dx} = \frac{(x^2 - 3x + 5)\dfrac{d}{dx}x - x\dfrac{d}{dx}(x^2 - 3x + 5)}{(x^2 - 3x + 5)^2}$$

$$= \frac{(x^2 - 3x + 5)1 - x(2x - 3)}{(x^2 - 3x + 5)^2} = \frac{x^2 - 3x + 5 - 2x^2 + 3x}{(x^2 - 3x + 5)^2}$$

$$= \frac{-x^2 + 5}{(x^2 - 3x + 5)^2} = -\frac{x^2 - 5}{(x^2 - 3x + 5)^2}. \quad \bullet$$

Example 3. Prove that the formula for differentiating $y = x^n$ (Theorem 11) holds when n is a negative integer.

Solution. Since we are assuming that n is negative, we want to bring out this negative character where it can be seen. So we let $n = -m$, where m is now a positive integer. Hence $y = x^n = x^{-m}$. Using (59) on the latter expression gives

$$\frac{dy}{dx} = \frac{d}{dx} x^{-m} = \frac{d}{dx} \frac{1}{x^m} = \frac{x^m \cdot 0 - 1(mx^{m-1})}{(x^m)^2}$$

$$= \frac{-mx^{m-1}}{x^{2m}} = (-m)x^{-m-1}.$$

But $n = -m$, so substituting in the last expression, we obtain

$$\frac{dy}{dx} = nx^{n-1}. \quad \bullet$$

For example, if $y = 5/x^3$, then $y = 5x^{-3}$ and $\dfrac{dy}{dx} = 5(-3)x^{-3-1} = -15x^{-4} = -\dfrac{15}{x^4}$.

Example 4. Sketch the graph of $y = \dfrac{8x}{x^2 + 4}$.

Solution. The derivative will yield valuable information. Indeed, for this function

$$\frac{dy}{dx} = \frac{(x^2 + 4)8 - 8x \cdot 2x}{(x^2 + 4)^2} = \frac{8x^2 + 32 - 16x^2}{(x^2 + 4)^2}$$

$$= \frac{-8x^2 + 32}{(x^2 + 4)^2} = -8\frac{x^2 - 4}{(x^2 + 4)^2}.$$

We first remark that in this last expression the denominator $(x^2 + 4)^2$ is always positive (in fact, it is always greater than or equal to 16). Therefore, the sign of the derivative is the sign of $-8(x^2 - 4)$. We see immediately that:

If $x < -2$, then $x^2 - 4 > 0$, and hence $\dfrac{dy}{dx}$ is negative,

If $-2 < x < 2$, then $x^2 - 4 < 0$, and hence $\dfrac{dy}{dx}$ is positive,

If $x > 2$, then $x^2 - 4 > 0$, and hence $\dfrac{dy}{dx}$ is negative.

Consequently, as x steadily increases from $-\infty$ to $+\infty$, the curve is first falling $\left(\dfrac{dy}{dx}\text{ is negative}\right)$ until x reaches -2, then the curve begins to rise $\left(\dfrac{dy}{dx}\text{ is positive}\right)$ until x reaches $+2$, and then the curve again falls. At $x = -2$, the equation $y = 8x/(x^2 + 4)$ gives $y = -2$, so the point $(-2, -2)$ is a low point on the curve.

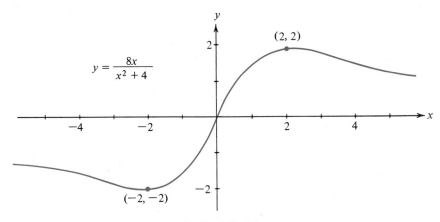

FIGURE 13

At $x = 2$, we have $y = 2$, so the point $(2, 2)$ is a high point on the curve. To find the behavior of y as x becomes very large, we note that

$$\lim_{x \to \infty} y = \lim_{x \to \infty} \frac{8x}{x^2 + 4} = \lim_{x \to \infty} \frac{8}{x + \dfrac{4}{x}} = 0,$$

and the same result when $x \to -\infty$. These facts suggest that the curve must have the form shown in Fig. 13, and this is easily confirmed by plotting a few more points. ●

EXERCISE 9

In Problems 1 through 20, use the appropriate formulas to compute the derivative of the dependent variable with respect to the independent variable. Simplify the result as much as possible.

1. $y = (x^2 - 1)(x^3 + 3)$.

2. $y = (x - 1)(x^4 + x^3 + x^2 + x + 1)$.

3. $y = (x^2 + 2x + 1)(x^2 - 2x + 1)$.

4. $y = 5x^{-3} + x^{-1}$.

5. $y = \dfrac{2}{x^2} - \dfrac{3}{x}$.

6. $y = \dfrac{x^3 - 1}{x^5}$.

7. $y = (x^2 + 5)x^{-3}$.

8. $y = \left(x + \dfrac{1}{x^2}\right)(2 + 5x)$.

9. $y = \dfrac{x - 1}{x + 1}$.

10. $y = \dfrac{ax + b}{cx + d}$.

11. $y = \dfrac{x^2 - 2x + 3}{x - 4}$.

12. $y = \dfrac{x + x^{-1}}{x - x^{-2}}$.

13. $y = \dfrac{x^2 + 2x + 1}{x^2 - 2x + 1}$.

14. $y = \dfrac{ax^2 + bx + c}{dx + e}$.

15. $y = (2x^2 + 5)(3x^{-2} - 3)$.

16. $u = v^{-2} + 13 + v^2$.

17. $r = \dfrac{\theta}{\theta^2 + 1}$.

18. $u = \dfrac{3t + 5}{7t + 9}$.

19. $v = \dfrac{u + u^{-3}}{u - u^{-1}}$.

20. $v^{-1} = \dfrac{u - u^{-1}}{u + u^{-3}}$.

In Problems 21 through 26, compute y' and use it to sketch the curve. Find all the critical values of the function and high and low points on the curve, if any.

21. $y = 2x + \dfrac{1}{x^2}$.

22. $y = \dfrac{2 + 5x^2}{1 + x^2}$.

23. $y = \dfrac{2x + 3}{x - 1}$.

24. $y = \dfrac{1}{x^2} + x^2$.

25. $y = \dfrac{3x}{1 + 2x^2}$.

★26. $y = -\dfrac{21 - 20x + 5x^2}{x^2 - 4x + 5}$.

★27. Test the consistency of the product formula with the formula for differentiating $y = cx^n$, by computing the derivative of $y = x^7 = x^2 \cdot x^5$ in two ways and showing that both ways give the same result.

28. Prove the Quotient Rule (Theorem 16). HINT: Write the difference quotient for $y = u(x)/v(x)$, simplify this compound fraction, and then add and subtract the term $u(x)v(x)$ in the numerator.

29. Find $f'(2)$, where $f(x) = (x^2 + 1)g(x)$, $g(2) = -1$, and $g'(2) = 3$.

30. Find $F'(1)$, where $F(x) = G(x)/x$, $G(1) = \pi$, and $G'(1) = \pi$.

31. Find $f'(0)$, where $f(x) = \dfrac{g(x) + x^2}{g(x) + 1}$, $g(0) = 0$, and $g'(0) = 1$.

32. Extend Theorem 15 to the product of three differentiable functions by proving that

$$\frac{d}{dx}\, uvw = uv\,\frac{dw}{dx} + uw\,\frac{dv}{dx} + vw\,\frac{du}{dx}.$$

*33. State the formula for differentiating the product $y = tuvw$ of four functions of x.

*34. There are two points P on the curve $y = \dfrac{x - 4}{x + 5}$ such that the line from the origin to P is tangent to the curve. Find the coordinates of these two points and sketch the curve.

*35. Find all those points P on the curve $y = \dfrac{2}{1 + x^2}$ such that the line OP from the origin to P is normal to the curve. Sketch the curve.

11 Composite Functions and the Chain Rule

If we wanted to compute the derivative of the function

(60) $$y = (x^2 - 3x + 5)^{25},$$

we could do so with the formulas developed so far. We would just compute the twenty-fifth power of $x^2 - 3x + 5$, and then differentiate the resulting polynomial expression, using the methods of Section 9. But what an unpleasant task! Fortunately, the Chain Rule (Theorem 17) will give us the derivative,

$$\frac{dy}{dx} = 25(x^2 - 3x + 5)^{24}(2x - 3),$$

immediately.

Let us examine the structure of the function (60). We see that it is made up of two parts; first we have a function (let us call it u)

(61) $$u = x^2 - 3x + 5$$

and then y is an appropriate function of u, namely

(62) $$y = u^{25}.$$

The original function (60) is obtained by composing (61) and (62); that is, substituting the expression for u given in (61) into (62). Such a function is called a *composite function*. In general, if y is some function of u, say $y = f(u)$, and if u is simultaneously some function of x, say $u = g(x)$, then y is a function of x. For each time x is given, $u = g(x)$ determines a unique value of u and then this particular value of u determines a corresponding value of y through the function $y = f(u)$. The two functions together, $y = f(u)$ and $u = g(x)$, form the composite function[1]

---

[1] Some authors use the circle notation $f \circ g$ to denote the composite function $f(g(x))$.

(63) $$y = f(g(x)) \qquad \text{(read ``}f\text{ of }g\text{ of }x\text{'')}$$

obtained by substituting $g(x)$ for u in $y = f(u)$. In our particular example, $f(u)$ is u^{25} and $g(x)$ is $x^2 - 3x + 5$, and $f(g(x)) = (x^2 - 3x + 5)^{25}$.

We can visualize the composite function (63) as a chain of two "black boxes," as shown in Fig. 14. Of course, for the linkup to make sense the outputs $g(x)$ of the black box g must be permissible inputs for the black box f. In other words, the domain of $f(g(x))$ is the set of all numbers x in the domain of g such that $g(x)$ is in the domain of f.

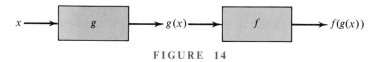

FIGURE 14

For owners of scientific hand calculators, forming composite functions simply amounts to a succession of button pushing. As a simple example, suppose that one wishes to calculate $1/x^2$ on a device that has a button for squaring and a button for taking reciprocals. Then $1/x^2$ can be computed by entering the nonzero number x, pushing the "square button," and then pushing the "reciprocal button," as shown in Fig. 15. In this

FIGURE 15

particular example it makes no difference if the order of operations is reversed by first taking the reciprocal and then squaring. However, in most situations the order of operations is *crucial*. This is illustrated in

Example 1. If

$$f(u) = \frac{1}{u + 4} \qquad \text{and} \qquad g(x) = x^2,$$

find the composite functions **(a)** $f(g(x))$ and **(b)** $g(f(u))$.

Solution. **(a)** $f(g(x)) = \dfrac{1}{g(x) + 4} = \dfrac{1}{x^2 + 4}$.

(b) $g(f(u)) = [f(u)]^2 = \left(\dfrac{1}{u + 4}\right)^2 = \dfrac{1}{u^2 + 8u + 16}$.   ●

In part **(b)**, if we replace u by the variable x, we find that

$$g(f(x)) = \frac{1}{x^2 + 8x + 16},$$

which is quite different from the function $f(g(x))$ found in part **(a).** Thus *composition is not a commutative operation.*

Let us return to the problem of differentiating the composite function

$$y = f(u) \qquad \text{with} \qquad u = g(x).$$

We wish to find the derivative of y *with respect to x,* so we give x an increment of the amount Δx. This causes a change in u of amount Δu, and this in turn induces a change in y, which we denote as usual by Δy. By definition,

$$\frac{dy}{dx} = \lim_{\Delta x \to 0} \frac{\Delta y}{\Delta x}.$$

If Δu is not zero, we can multiply on the right side by $\Delta u/\Delta u$, since $\Delta u/\Delta u = 1$. A slight shuffle then gives

(64)
$$\frac{dy}{dx} = \lim_{\Delta x \to 0} \frac{\Delta y}{\Delta x} \frac{\Delta u}{\Delta u} = \lim_{\Delta x \to 0} \frac{\Delta y}{\Delta u} \frac{\Delta u}{\Delta x}.$$

If y is a differentiable function of u, then

(65)
$$\lim_{\Delta u \to 0} \frac{\Delta y}{\Delta u} = \frac{dy}{du},$$

and if u is a differentiable function of x, then

(66)
$$\lim_{\Delta x \to 0} \frac{\Delta u}{\Delta x} = \frac{du}{dx}.$$

Finally, we must remark that as Δx approaches zero, the induced change in u, Δu, also approaches zero. Using (65) and (66) in (64), we obtain

$$\frac{dy}{dx} = \lim_{\Delta x \to 0} \frac{\Delta y}{\Delta u} \frac{\Delta u}{\Delta x} = \left(\lim_{\Delta u \to 0} \frac{\Delta y}{\Delta u} \right)\left(\lim_{\Delta x \to 0} \frac{\Delta u}{\Delta x} \right) = \frac{dy}{du} \frac{du}{dx}.$$

We have almost proved

THEOREM 17 (The Chain Rule). If $y = f(u)$ and $u = g(x)$ are two differentiable functions, then the derivative of the composite function $y = f(g(x))$ is given by

(67)
$$\frac{dy}{dx} = \frac{dy}{du} \frac{du}{dx}.$$

In terms of the functions f and g, the Chain Rule can be equivalently expressed as

$$\frac{d}{dx}f(g(x)) = f'(g(x))g'(x),$$

where f' stands for df/du and g' stands for dg/dx.

Unfortunately, the argument that led us to Theorem 17 has a slight error that would surely go unnoticed by anyone but an expert. Indeed, in this procedure, we may have inadvertently divided by zero, and such an operation is not allowed. In computing the derivative through the definition

$$\frac{dy}{dx} = \lim_{\Delta x \to 0} \frac{\Delta y}{\Delta x},$$

the denominator Δx is small and getting close to zero, but it is *never equal to zero*. On the other hand, Δy, the change in y, could be zero without any harm, since it is in the numerator.

Now let us look at equation (64). The change Δx may induce no change in u so that $\Delta u = 0$. Then equation (64) falls because it contains a forbidden division by zero. This pulls down with it equation (65) and indeed the entire proof. But Theorem 17 is true. The reader may accept this fact on intuitive grounds, or read through the following proof.

Proof of Theorem 17. From equation (65), the difference quotient $\Delta y/\Delta u$ is close to the derivative dy/du when Δu is close to zero. Hence we can write

$$(68) \qquad\qquad \frac{\Delta y}{\Delta u} = \frac{dy}{du} + \epsilon, \qquad \Delta u \neq 0,$$

where ϵ (Greek lowercase letter epsilon) is a function of Δu and

$$\lim_{\Delta u \to 0} \epsilon = \lim_{\Delta u \to 0} \left(\frac{\Delta y}{\Delta u} - \frac{dy}{du} \right) = \frac{dy}{du} - \frac{dy}{du} = 0.$$

We shall make ϵ continuous at $\Delta u = 0$ by defining $\epsilon(0) = 0$.

If we multiply equation (68) by Δu, we have

$$(69) \qquad\qquad \Delta y = \frac{dy}{du}\Delta u + \epsilon\, \Delta u.$$

Although equation (68) is meaningless when $\Delta u = 0$, equation (69) is meaningful and correct even when $\Delta u = 0$; for if u does not change (i.e., $\Delta u = 0$), then y also does not change ($\Delta y = 0$) and equation (69) merely states that $0 = 0$ in this case.

Next we divide both sides of equation (69) by Δx ($\Delta x \neq 0$) to obtain

$$\frac{\Delta y}{\Delta x} = \frac{dy}{du}\frac{\Delta u}{\Delta x} + \epsilon\frac{\Delta u}{\Delta x}.$$

Taking the limit of both sides of this last equation as Δx approaches zero gives

$$\lim_{\Delta x \to 0} \frac{\Delta y}{\Delta x} = \frac{dy}{du} \cdot \lim_{\Delta x \to 0} \frac{\Delta u}{\Delta x} + \lim_{\Delta x \to 0} \epsilon \cdot \lim_{\Delta x \to 0} \frac{\Delta u}{\Delta x}$$

or

$$(70) \qquad \frac{dy}{dx} = \frac{dy}{du}\frac{du}{dx} + \lim_{\Delta x \to 0} \epsilon \cdot \frac{du}{dx}.$$

But since $u = g(x)$ is differentiable, it follows from Theorem 8 that u is also continuous, so $\Delta x \to 0$ implies that $\Delta u \to 0$. Consequently,

$$\lim_{\Delta x \to 0} \epsilon = \lim_{\Delta u \to 0} \epsilon = 0,$$

and substituting this value in (70) gives

$$\frac{dy}{dx} = \frac{dy}{du}\frac{du}{dx} + 0 \cdot \frac{du}{dx} = \frac{dy}{du}\frac{du}{dx}. \quad \blacksquare$$

Example 2. Find $\dfrac{dy}{dx}$ if $y = (x^4 - 5x^3 + 3)^{50}$.

Solution. The given function becomes a composite function if we let $u = x^4 - 5x^3 + 3$, and $y = u^{50}$. Then by (67),

$$\frac{dy}{dx} = \frac{dy}{du}\frac{du}{dx} = 50u^{49}(4x^3 - 15x^2).$$

Since $u = x^4 - 5x^3 + 3$, this is equivalent to

$$\frac{dy}{dx} = 50(x^4 - 5x^3 + 3)^{49}(4x^3 - 15x^2). \quad \bullet$$

In working problems of this type, the student may find it helpful at first to put in the intermediate steps. But after a certain amount of practice, he should be able to write the answer immediately.

Sometimes the Chain Rule must be used in conjunction with other formulas, as illustrated in

Example 3. Find $\dfrac{dy}{dx}$ if $y = \left(\dfrac{x^2 - 1}{x^2 + 1}\right)^5$.

Solution. Here we have $y = u^5$, where u is a quotient of two functions of x. Then

$$\frac{dy}{dx} = \frac{dy}{du}\frac{du}{dx} = 5u^4 \frac{d}{dx}\left(\frac{x^2 - 1}{x^2 + 1}\right)$$

$$= 5\left(\frac{x^2 - 1}{x^2 + 1}\right)^4 \frac{(x^2 + 1)2x - (x^2 - 1)2x}{(x^2 + 1)^2} = \frac{20x(x^2 - 1)^4}{(x^2 + 1)^6}. \quad \bullet$$

If we apply the Chain Rule to $y = u^n$, we obtain

THEOREM 18. If n is any integer and $u(x)$ is any differentiable function, then

(71)
$$\frac{d}{dx}u^n = nu^{n-1}\frac{du}{dx}.$$

COROLLARY. If n is any integer, then

$$\frac{d}{dx}(x - a)^n = n(x - a)^{n-1}.$$

EXERCISE 10

1. Let $f(u) = (u - 1)/(u + 1)$ and $g(x) = 2/x$. Find the composite functions (a) $f(g(x))$, (b) $g(f(u))$, and (c) $g(g(x))$, and state the domain of these functions.
2. Let $f(u) = u^2 - 2$ and $g(x) = \sqrt{x}$. Find the composite functions (a) $f(g(x))$, (b) $g(f(u))$, and (c) $g(g(x))$, and state the domain of these functions.

In Problems 3 through 6, find $\dfrac{dy}{dx}$ in terms of x alone. HINT: *See Theorem 17.*

3. $y = u^2,$ $u = x^3 + x.$ 4. $y = u^{-1},$ $u = x^2 - x.$

5. $y = \dfrac{u}{u + 1},$ $u = x^2.$ 6. $y = u^{-1} + u^{-2},$ $u = 3x - 2.$

In Problems 7 through 28, compute the derivative of the given function.

7. $y = (3x + 5)^{10}.$ 8. $y = (x^2 + x - 2)^4.$

9. $y = (2x^3 - 3x^2 + 6x)^5.$ 10. $y = (x - 5)^{-3}.$

11. $y = (x^3 + x^2 - 1)^{-2}.$ 12. $y = (x^4 - 4x - 11)^{-3}.$

13. $y = \left(\dfrac{x+2}{x-3}\right)^3.$

14. $y = \left(\dfrac{ax+b}{cx+d}\right)^7.$

15. $y = (7x+3)^4(4x+1)^7.$

16. $y = (x^2-1)^{10}(x^2+1)^{15}.$

17. $y = (2x+1)^{-1}(x+3)^{-2}.$

18. $y = (x^3+2)^2(x^2+1)^{-3}.$

19. $y = \left(\dfrac{x^2+5}{x^3-1}\right)^2.$

20. $y = \left(\dfrac{x+2}{x^2+2x-1}\right)^3.$

21. $y = \dfrac{(x^2+1)^3}{(x^3-1)^2}.$

22. $y = \dfrac{(x+1)^5}{(x-1)^{-7}}.$

23. $y = \dfrac{(x-1)^{-1}}{(x+1)^{-2}}.$

24. $y = \dfrac{(x^{-1}+1)^2}{(x+1)^{-1}}.$

25. $u = (v^3+17)^{12}.$

26. $s = (t^3-3t)^{11}.$

27. $r = (\theta+1)^6(\theta-1)^8.$

28. $w = \dfrac{(u^2-1)^5}{(u^5+3)^2}.$

In Problems 29 through 32, find $\dfrac{dy}{dx}$ *in terms of x alone.* HINT: *Select the easiest way to work the problem.*

29. $y = \dfrac{u^2+1}{u^2-1},$ where $u = \sqrt{x^2+1}.$

30. $y = \dfrac{u^4+u^2+4}{u^4-6u^2+9},$ where $u = \sqrt{x^2+3}.$

31. $y = \dfrac{3u+5}{4u+7},$ where $u = \dfrac{7x^2-5}{-4x^2+3}.$

32. $y = (u^3+6)^4(u^3+4)^6,$ where $u = \sqrt[3]{x^2-5}.$

33. Find $\dfrac{d}{dx}[f(x^3)]$ at $x = 2,$ where $f'(8) = -5.$

34. Find $\dfrac{d}{dx}[F(x^2-3x)]$ at $x = 1,$ where $F'(-2) = 4.$

35. Find $\dfrac{d}{dx}[f^3(x)]$ at $x = -1,$ where $f(-1) = 2, f'(-1) = -3.$

36. Find $\dfrac{d}{dx}[F(x)]^{-5}$ at $x = 0,$ where $F(0) = -1, F'(0) = 6.$

37. The function $y = (x+3)^8(x-4)^8$ can be differentiated by two methods: first by considering it as the product of two composite functions, and second by performing the multiplication and differentiating the result $y = (x^2-x-12)^8.$ Show that both methods give the same result for the derivative.

★38. A polynomial $P(x)$ is said to have an nth-order *zero* at $x = a$ if it can be written in the form

$$P(x) = (x-a)^n Q(x), \qquad n \geq 1,$$

where $Q(x)$ is a polynomial and $Q(a) \neq 0$. Prove that if $P(x)$ has an nth-order zero at $x = a$, then its derivative has a zero of order $n - 1$ at $x = a$.

★39. Prove that if $y = \dfrac{P(x)}{(x - a)^n}$, $n \geq 1$, where $P(x)$ is a polynomial with $P(a) \neq 0$, then

$$\frac{dy}{dx} = \frac{Q(x)}{(x - a)^{n+1}},$$

where $Q(x)$ is an appropriate polynomial and $Q(a) \neq 0$.

Problems 38 and 39 show that differentiation decreases the order of a zero by 1, but it increases the order of an "infinity" by 1.

Implicit Functions 12

Suppose that in the equation

(72) $$y^7 + x^7 = 4x^5 y^5 - 2$$

we select a fixed value x_1 for x. Then (72) becomes a seventh-degree polynomial in y. A seventh-degree polynomial always has seven roots, although some of the roots may be repeated roots and some may be complex (involve $\sqrt{-1}$). If at least one of these roots is real, we may select it as the image of x_1, in an attempt to construct a real-valued function based on equation (72). Suppose that we follow this procedure for each x in a given interval, always selecting a corresponding y such that the pair (x, y) satisfies (72). Then we can say that the equation (72) generates a function $y = f(x)$. If there is such a function, then when we use it in place of y in equation (72), we obtain

$$f^7(x) + x^7 = 4x^5 f^5(x) - 2,$$

for each x in the given interval.[1]

    If we can solve equation (72) and give a formula for y, that formula determines y as an *explicit* function of x. Otherwise, we say that equation (72) determines y as an *implicit* function of x. In the specific example presented in equation (72), there is no formula that gives y as an explicit function of x and that is composed of a finite number of the algebraic operations (addition, subtraction, multiplication, division, and root extraction). And yet, in accordance with the method described above, equation (72) does determine y implicitly as a function of x.

    Is this function differentiable? This is a rather difficult question and we postpone it until the end of this section. Assuming that the function determined implicitly by (72) is differentiable, can we compute its derivative? This is a much simpler question and the answer is yes. In fact, the computation is very easy. Assuming that $f(x)$ is differentiable, we

---

[1] The notation $f^7(x)$ means $f(x)$ raised to the seventh power, that is, $[f(x)]^7$.

merely differentiate both sides of equation (72) with respect to x, keeping in mind that y is a function of x. We will need our formula for differentiating a composite function, since y^7 is now a composite function of x. For this derivative, we find that

$$\frac{d}{dx} y^7 = 7y^6 \frac{dy}{dx}.$$

On the right side of (72), we have a product in which one of the factors is the composite function y^5. For the derivative of this product, we have

$$\frac{d}{dx} 4x^5 y^5 = 4x^5 \left(5y^4 \frac{dy}{dx} \right) + 20x^4 y^5 = 20x^5 y^4 \frac{dy}{dx} + 20x^4 y^5.$$

Hence if we differentiate both sides of (72) with respect to x, we find that

(73) $$7y^6 \frac{dy}{dx} + 7x^6 = 20x^5 y^4 \frac{dy}{dx} + 20x^4 y^5.$$

Solving for $\dfrac{dy}{dx}$, we obtain

$$\frac{dy}{dx} (7y^6 - 20x^5 y^4) = 20x^4 y^5 - 7x^6,$$

(74) $$\frac{dy}{dx} = \frac{20x^4 y^5 - 7x^6}{7y^6 - 20x^5 y^4},$$

and this is the derivative of y with respect to x when y and x are related by equation (72). The procedure used in going from (72) to (73) is called *implicit differentiation*.

What right did we have to differentiate equation (72) with respect to x? The work looks correct, but is it? If we recall the old slogan "Equals added to equals gives equals," this suggests the corresponding new slogan "Equals differentiated with respect to the same variable gives equals." Making this slogan precise, we have

---

THEOREM 19. If two differentiable functions are equal in an interval, then their derivatives with respect to the same variable are equal in that interval.

Proof. The difficulty here is that the proof of this theorem is too simple. Indeed, differentiation is a uniquely defined operation, so that to each differentiable function there is exactly one corresponding function, its derivative. So if two functions are equal, they are perhaps different forms for the same function, and hence the derivative of either form must be the same function. Hence the derivatives are equal. ∎

For example, the equation (72) really states that $y^7 + x^7$ and $4x^5y^5 - 2$ are merely different forms of one and the same function, so the two different forms for the derivative of this function, given by the two sides of equation (73), are equal.

Example 1. Find the slope of the tangent line to the graph of $y^7 + x^7 = 4x^5y^5 - 2$ at the point $(1, 1)$.

Solution. Note that $x = 1, y = 1$ satisfy the given equation. We compute the derivative just as before and obtain equation (74). Putting $x = 1$, $y = 1$ in (74) yields

$$m = \frac{dy}{dx} = \frac{20 - 7}{7 - 20} = \frac{13}{-13} = -1,$$

so the slope is -1. ●

Is it not surprising that we can find the tangent to this curve so easily when the task of sketching this curve is very difficult?

How about the slope at the point $(2, 2)$? Equation (74) again yields $m = -1$ when $x = 2$ and $y = 2$. This computation is merely a manipulation with numbers and has *no meaning*. The point $(2, 2)$ is not on the curve. For if $x = 2$ and $y = 2$, the left side of (72) is $2^7 + 2^7 = 256$, while the right side is $4 \times 2^{10} - 2 = 4094$, and these are not equal. Summarizing: *The expression for the derivative found in equation (74) is true only for pairs (x, y) that satisfy (72).*

Example 2. Find the derivative of $y = \sqrt{1 - x^2}$, for $-1 < x < 1$.

Solution. By squaring and transposing this leads to $x^2 + y^2 = 1$. Hence by implicit differentiation

$$2x + 2y \frac{dy}{dx} = 0,$$

$$\frac{dy}{dx} = -\frac{x}{y}.$$

But $y = \sqrt{1 - x^2}$, so that

$$\frac{dy}{dx} = -\frac{x}{\sqrt{1 - x^2}}. ●$$

We can use the technique of implicit differentiation to prove that the Power Rule (Theorem 11) is true when n is any rational number.

THEOREM 20. If n is any rational number, then

(49)
$$\frac{d}{dx} x^n = nx^{n-1}.$$

Proof. We have already proved (49) if n is any integer (see Example 3 of Section 10). Now let $n = p/q$, where p and q are integers and $q \neq 0$. Set

(75)
$$y = x^n = x^{p/q}.$$

Then raising both sides to the qth power, we have

(76)
$$y^q = x^p.$$

Implicit differentiation gives

$$qy^{q-1} \frac{dy}{dx} = px^{p-1},$$

or

(77)
$$\frac{dy}{dx} = \frac{p}{q} \frac{x^{p-1}}{y^{q-1}}.$$

But $y^q = x^p$ and $y^{-1} = 1/x^{p/q}$. Substitution in (77) yields

$$\frac{dy}{dx} = \frac{p}{q} \frac{x^{p-1}}{x^p} x^{p/q} = \frac{p}{q} x^{(p/q)-1} = nx^{n-1}. \quad \blacksquare$$

Theorem 20 and the Chain Rule give immediately

THEOREM 21. If n is any rational number and $u(x)$ is any differentiable function, then

(78)
$$\frac{d}{dx} cu^n = cnu^{n-1} \frac{du}{dx}, \qquad \text{where } c \text{ is a constant.}$$

Example 3. Solve Example 2 without implicit differentiation.

Solution. Set $u = 1 - x^2$; then $y = \sqrt{1 - x^2} = (1 - x^2)^{1/2} = u^{1/2}$,

$$\frac{dy}{dx} = \frac{1}{2}u^{(1/2)-1}\frac{du}{dx} = \frac{1}{2\sqrt{u}}(-2x) = \frac{-x}{\sqrt{u}} = -\frac{x}{\sqrt{1-x^2}}. \quad \bullet$$

Let us return to equation (72) and the assertion that it determines y as a differentiable function of x. This very plausible and innocent-sounding statement is in fact false, unless the proper interpretation is given to the words.

To clarify the matter, we consider the simpler equation

$$(79) \qquad\qquad\qquad x^2 + y^2 = 1.$$

One might assume that equation (79) determines the function

$$(80) \qquad\qquad\qquad y = \sqrt{1-x^2},$$

already treated in Examples 2 and 3. But the function

$$(81) \qquad\qquad y = f(x) \equiv \begin{cases} \sqrt{1-x^2}, & \text{if } x \text{ is rational, } |x| \leq 1, \\ -\sqrt{1-x^2}, & \text{if } x \text{ is irrational, } |x| \leq 1, \end{cases}$$

also has the property that

$$x^2 + f^2(x) = 1$$

is an identity for all x in $-1 \leq x \leq 1$.

Now the function (81) is not differentiable; in fact, it is not even continuous in the open interval $-1 < x < 1$. Hence any assertion that "(79) *determines y as a differentiable function of x*" is erroneous. The true situation is this: Equation (79) can be satisfied by infinitely many different functions, and (80) and (81) give only two of the many possible functions. Equation (79) does not by itself make any selection. It is the mathematician who selects a nice function, that is, one that is differentiable.

Generally speaking, an equation such as (79) or (72) gives rise to many functions that satisfy the equation. Among these there will be at least one that is differentiable. In what follows we will always assume there is at least one such differentiable function. When we say that the equation implicitly determines y as a function of x, we mean that we have selected one of these differentiable functions. It is rather difficult to prove that there is always such a function, and in the interest of brevity we omit this item.

EXERCISE 11

In Problems 1 through 14, find the derivative of the given function.

1. $y = x^{3/2} + x^{1/2}$.   2. $y = 4x^{5/2} - 10x^{3/2}$.   3. $y = (x^2 - 1)^{4/7}$.

4. $y = (\sqrt{x} - 1)^{3/2}$.   5. $y = x\sqrt{1-x^2}$.   6. $y = (x^2 + 1)^{1/3}(x^3 - 1)^{1/2}$.

7. $y = \dfrac{x^2}{\sqrt{1 - x^2}}.$

8. $y = \dfrac{\sqrt{x} + 1}{\sqrt{x} - 1}.$

9. $y = \dfrac{x}{9\sqrt{9 - x^2}}.$

10. $y = \dfrac{\sqrt{10 - x^2}}{x}.$

11. $y = \left(\dfrac{x + 9}{x - 9}\right)^{5/3}.$

★12. $y = \sqrt{4 + \sqrt{4 - x}}.$

★13. $y = \sqrt{4x + \dfrac{1}{\sqrt{x}}}.$

★14. $y = \sqrt[3]{x^2 + \sqrt{x^3}}.$

In Problems 15 through 26, find $\dfrac{dy}{dx}$ by implicit differentiation.

15. $x^2 + y^2 = 14.$

16. $xy^2 - 1 = x.$

17. $x^3 + y^3 = 6xy.$

18. $x^3 - 2xy^2 + 3y^3 = 7.$

19. $x^4y^4 = x^4 + y^4.$

20. $x^4 + x^3y + y^4 = 3.$

21. $y + \sqrt{y} = 2x^2.$

22. $\sqrt{x} + \sqrt{y} = 4.$

23. $xy^{3/2} + 4 = 2x + y.$

24. $y^{1/3} + y^{1/5} = x.$

25. $(x + y)^4 = x^4 + y^4.$

26. $\sqrt{x + y} + xy = 12.$

27. Find an equation for the tangent line to the curve $y + y^{2/3} = x$ at $(12, 8)$.

28. Find an equation for the tangent line to the curve $2y + 2x - xy^{3/2} = 8$ at $(0, 4)$.

29. Show that the two curves $y = x^{4/3} + x$ and $x = y + y^5$ are tangent at the origin.

30. In elementary geometry it is proved that a tangent line to a circle is perpendicular to the radial line at the point of contact. Give a second proof of this theorem using implicit differentiation on the equation $x^2 + y^2 = r^2$.

★31. The graph of $25x^2 + 25y^2 - 14xy = (24)^2$ is an ellipse with center at the origin. The end points of the major and minor axes are $(4, 4)$, $(-4, -4)$, $(3, -3)$, and $(-3, 3)$. One would expect that these four points would be the only points at which the line from the origin is normal to the curve. Use implicit differentiation to prove this fact.

13 *Higher-Order Derivatives*

The derivative of $y = x^5$ is $5x^4$. But we can differentiate $5x^4$. If we do, we obtain $20x^3$, and this new function is called the *second derivative* of y with respect to x. Repeating the process again, we obtain $60x^2$, the *third derivative* of y with respect to x. We may continue this process indefinitely. All that is needed is a notation for these derivatives of higher order. A variety of notations are in current use. If $y = f(x)$, the derivatives can be written as follows:

First derivative: $\qquad \dfrac{dy}{dx} \qquad \dfrac{d}{dx}y \qquad y' \qquad f'(x) \qquad Df(x).$

Second derivative: $\dfrac{d^2y}{dx^2}$ $\dfrac{d^2}{dx^2}y$ y'' $f''(x)$ $D^2f(x)$.

Third derivative: $\dfrac{d^3y}{dx^3}$ $\dfrac{d^3}{dx^3}y$ y''' $f'''(x)$ $D^3f(x)$.

nth derivative: $\dfrac{d^ny}{dx^n}$ $\dfrac{d^n}{dx^n}y$ $y^{(n)}$ $f^{(n)}(x)$ $D^nf(x)$.

The reader should observe the peculiar location of the superscripts in the first two columns. The reason for this choice of notation is that the second derivative is the derivative of the first derivative, and hence it is natural to write

$$\frac{d}{dx}\left(\frac{dy}{dx}\right) = \frac{d^2y}{dx^2}.$$

In other words, in the numerator it is the differentiation that is repeated twice, so we expect d^2, but in the denominator it is the variable x that is repeated; that is, we have differentiated twice with respect to x.

The entries in the third column are read y prime, y double prime, y triple prime, and y upper n, and similarly the entries in the fourth column are read f prime of x, f double prime of x, and so on.

It may well appear that there are too many different notations for the derivative, but this is not the case. Each notation has its own particular appeal, and each is very useful in certain situations, as the reader will discover as he goes further and deeper into mathematics.

We will postpone a study of the uses of the higher-order derivatives until later in the book. For the present we will practice the technique of computing them.

Example 1. Find the derivatives of all orders for $y = x^4$.

Solution. From our power formula, we have

$$\frac{dy}{dx} = 4x^3 \qquad \frac{d^2y}{dx^2} = 12x^2 \qquad \frac{d^3y}{dx^3} = 24x$$

$$\frac{d^4y}{dx^4} = 24 \qquad \frac{d^5y}{dx^5} = 0 \qquad \frac{d^ny}{dx^n} = 0 \qquad \text{for } n \geqq 5. \quad \bullet$$

Example 2. Discover and prove a formula for the nth derivative of $y = \dfrac{1}{1-x}$.

Solution. To discover a general formula, we must examine a few of the earlier cases. Computation gives

$$\frac{dy}{dx} = \frac{d}{dx}(1-x)^{-1} = -1(1-x)^{-2}(-1) \quad = (1-x)^{-2} = \frac{1}{(1-x)^2}.$$

$$\frac{d^2y}{dx^2} = \frac{d}{dx}(1-x)^{-2} = -2(1-x)^{-3}(-1) \quad = 2(1-x)^{-3} = \frac{2}{(1-x)^3}.$$

$$\frac{d^3y}{dx^3} = \frac{d}{dx}2(1-x)^{-3} = -2\cdot 3(1-x)^{-4}(-1) = 6(1-x)^{-4} = \frac{6}{(1-x)^4}.$$

A study of the first and last columns, and the method of obtaining the derivatives, suggests the general formula

(82)
$$\frac{d^n y}{dx^n} = \frac{n!}{(1-x)^{n+1}}, \quad n \geqq 1.$$

Indeed, a little thought shows that this formula is really obvious, but if a proof is demanded, we must use mathematical induction. We have already proved that (82) is correct when $n = 1$. In fact, it checks when $n = 2$ and $n = 3$ as well. Assuming now that (82) is true when $n = k$, and computing the next higher derivative, we have

$$\frac{d^{k+1}y}{dx^{k+1}} = \frac{d}{dx}\left(\frac{d^k y}{dx^k}\right) = \frac{d}{dx}\frac{k!}{(1-x)^{k+1}} = \frac{d}{dx}k!(1-x)^{-(k+1)}$$

$$= -(k+1)k!(1-x)^{-(k+1)-1}(-1) = \frac{(k+1)!}{(1-x)^{k+2}}.$$

But this is just equation (82) for the index $n = k + 1$. ●

Example 3. Use implicit differentiation to obtain a simple form for y'' for the ellipse $b^2x^2 + a^2y^2 = a^2b^2$.

Solution. For the first derivative we have $2b^2x + 2a^2y\dfrac{dy}{dx} = 0$, or

(83)
$$\frac{dy}{dx} = -\frac{b^2x}{a^2y}.$$

Using the Quotient Rule and remembering that y is a function of x,

$$\frac{d^2y}{dx^2} = -\frac{a^2yb^2 - b^2xa^2\dfrac{dy}{dx}}{a^4y^2}.$$

Substituting from (83), we have

$$\frac{d^2y}{dx^2} = -\frac{a^2b^2y - a^2b^2x\left(-\dfrac{b^2x}{a^2y}\right)}{a^4y^2} = -\frac{a^4b^2y^2 + a^2b^4x^2}{a^6y^3}$$

$$= -\frac{b^2(a^2y^2 + b^2x^2)}{a^4y^3}.$$

But (x, y) must lie on the ellipse $a^2y^2 + b^2x^2 = a^2b^2$. Hence

$$\frac{d^2y}{dx^2} = -\frac{b^2(a^2b^2)}{a^4y^3} = -\frac{b^4}{a^2y^3}. \quad \bullet$$

EXERCISE 12

In Problems 1 through 6, find the second derivative of the dependent variable with respect to the independent variable.

1. $y = x^{10} + 3x^6$.

2. $y = \sqrt{x^3 + 1}$.

3. $y = \dfrac{x}{(1 - x)^2}$.

4. $u = \dfrac{(1 - v)^3}{v}$.

5. $s = t^2(t - 1)^5$.

6. $w = (z^2 + 2)^{5/2}$.

In Problems 7 through 14, find the indicated derivative.

7. $\dfrac{d^2y}{dx^2}$, where $y = x^{2/3} + 5x$.

8. $f''(x)$, where $f(x) = \dfrac{2}{x}$.

9. $f^{(3)}(x)$, where $f(x) = (2x - 1)^5$.

10. $\dfrac{d^3y}{dx^3}$, where $y = x^4 - 3x^3 - 2x$.

11. $D^3f(x)$, where $f(x) = \dfrac{x}{(1 - x)^2}$.

12. $f^{(4)}(x)$, where $f(x) = (x - 2)^3$.

13. $\dfrac{d^4s}{dt^4}$, where $s = \dfrac{1}{1 + t}$.

14. $\dfrac{d^3w}{dz^3}$, where $w = \sqrt{4z + 1}$.

In Problems 15 through 20, use implicit differentiation to find a simple form for y''.

15. $4x^2 - 9y^2 = 36$.

16. $x^2 + y^2 = r^2$.

17. $x^3 + y^3 = 1$.

18. $y^2 = 4px$.

19. $x^{1/2} + y^{1/2} = a^{1/2}$.

20. $b^2x^2 - a^2y^2 = a^2b^2$.

21. The notation $f''(2)$ means the second derivative $f''(x)$ computed when $x = 2$. Determine the constants in the function $f(x) = ax^3 + bx^2$ so that $f'(1) = 5$ and $f''(2) = 32$.

22. Determine the constant k in the function $f(x) = k - 2kx^3 + k^2x^4$ so that $f'''(0) = 24$.

In Problems 23 through 30, find $f^{(n)}(x)$ for the given function.

23. $y = \dfrac{1}{2 + x}$.

24. $y = \dfrac{1}{1 - 2x}$.

25. $y = \dfrac{1}{2 + 3x}$.

26. $y = \dfrac{2}{3 + 2x}$.

27. $y = \dfrac{a}{b + cx}$.

28. $y = \dfrac{a}{(b + cx)^2}$.

$\star$29. $y = \sqrt{1 - x}$.

$\star$30. $y = \sqrt{a - bx}$.

31. Show that Problem 27 contains Problems 23 and 25 as special cases and check the answers to these problems by selecting a, b, and c properly.

$\star$32. Find a formula for $\dfrac{d^2}{dx^2}(uv)$ and $\dfrac{d^3}{dx^3}(uv)$, where u and v are functions of x.

CALCULATOR PROBLEMS

C1. Prove that if f is differentiable at x, then

$$f'(x) = \lim_{\Delta x \to 0} \frac{f(x + \Delta x) - f(x - \Delta x)}{2 \Delta x}.$$

HINT: First use Definition 4 with $h = -\Delta x$ to show that

$$f'(x) = \lim_{\Delta x \to 0} \frac{f(x) - f(x - \Delta x)}{\Delta x}.$$

Then write

$$\frac{f(x + \Delta x) - f(x - \Delta x)}{2 \Delta x} = \frac{1}{2}\left[\frac{f(x + \Delta x) - f(x)}{\Delta x} + \frac{f(x) - f(x - \Delta x)}{\Delta x}\right]$$

and let $\Delta x \to 0$.

C2. Give a geometric interpretation for the quotient $[f(x + \Delta x) - f(x - \Delta x)]/2 \Delta x$ in Problem C1.

C3. The purpose of this problem is to compare the accuracy in approximating $f'(x)$ by the ordinary difference quotient and by the quotient $[f(x + \Delta x) - f(x - \Delta x)]/2 \Delta x$ of Problem C1. For the specific function $f(x) = (x - 3)^4$, complete the following table of approximations to $f'(1) = -32$.
   Which of the quotients in this table gives the more accurate approximations?

Δx	0.2	0.1	0.01	0.001
$[f(1 + \Delta x) - f(1)]/\Delta x$				
$[f(1 + \Delta x) - f(1 - \Delta x)]/2 \Delta x$				

C4. Prove that if $p(x) = ax^2 + bx + c$ is any polynomial of degree 2, then $p'(x) = [p(x + \Delta x) - p(x - \Delta x)]/2\,\Delta x$ for all $\Delta x \neq 0$.

C5. Let $f(x) = 10(5x^2 + x + \sqrt{x})^{1/4}/(11x + \sqrt{x} + 17)^{1/2}$. Approximate $f'(2)$ by computing $[f(x + \Delta x) - f(x - \Delta x)]/2\,\Delta x$ for $x = 2$ and $\Delta x = 0.1, 0.01$, and 0.001.

C6. As is suggested by Problem C4, the quotient $[f(x + \Delta x) - f(x - \Delta x)]/2\,\Delta x$ can be shown to equal the derivative at x of a quadratic polynomial p which fits the graph of f at the three points $(x - \Delta x, f(x - \Delta x))$, $(x, f(x))$, and $(x + \Delta x, f(x + \Delta x))$. For such a polynomial it can be further verified that $p''(x)$ equals

$$Q \equiv \frac{f(x - \Delta x) - 2f(x) + f(x + \Delta x)}{(\Delta x)^2},$$

and hence this last quotient approximates the second derivative $f''(x)$. For the function $f(x) = x^4 - 3\sqrt{x}$, estimate $f''(1)$ by computing the quotient Q for $x = 1$ and $\Delta x = 0.1, 0.01$, and 0.001. [Compare your answers with the exact value for $f''(1)$.] What happens when Q is computed for $\Delta x = 10^{-6}$?

C7. For $f(x) = \sqrt{x^3 + 8}$, estimate $f''(2)$ by computing the quotient Q in Problem C6 for $x = 2$ and $\Delta x = 0.1, 0.05$, and 0.01. [Compare your answers with the exact value for $f''(2)$.] Experiment with using smaller values for Δx, say $\Delta x = 10^{-4}$, to see if the approximations improve or worsen.

REVIEW PROBLEMS

In Problems 1 through 15, compute the derivative of the given function.

1. $y = 3x^2 - 6x$.

2. $y = 2x - \dfrac{1}{x^2}$.

3. $u = 3v^4 - 12v$.

4. $w = \dfrac{az + b}{cz + d}$.

5. $w = \dfrac{z}{(1 - z)^2}$.

6. $w = \sqrt{2}^{\sqrt{3}\sqrt{5} - \pi}$.

7. $y = (7x + 1)^5(5x - 1)^7$.

8. $r = (\theta^{3/2} + 2)^4$.

9. $v = \left(u^2 + \dfrac{1}{u}\right)^5$.

10. $y = (x^3 + 3x)^{4/3}(2x^2 - 5)^{7/4}$.

11. $y = (x^2 - 1)^{3/2}(x^4 + 5)^{5/4}$.

12. $s = \dfrac{t^2 + 3t}{t^3 - 1}$.

13. $r = \dfrac{s^3 - 3s}{s^2 + 1}$.

*14. $y = x^4(x - 1)^5(x + 3)^6$.

*15. $y = x^{3/2}(x - 1)^{5/2}(x + 3)^{7/2}$.

In Problems 16 through 21, use implicit differentiation to find $\dfrac{dy}{dx}$.

16. $x^2 + 3xy = 17$.

17. $y^2 = (x + 3)(x + 6)$.

18. $3y^4 - 4y^3x^2 + 5x^3 = 4$.

19. $2y^5 - y^3x^2 + 3x^4 = 4$.

20. $x^3 = \dfrac{x^2 - y}{y^2 - x}$.

21. $x = \dfrac{ax + by}{cx + dy}$.

22. Show that Problems 16, 17, and 21 are easy to solve without implicit differentiation.

23. Sketch the curve $12y = x^3 + x^2 - 12x$ by drawing the tangent line to the curve at each of the following points on the curve: $(-4, 0)$, $(-3, 3/2)$, $(-1, 1)$, $(0, 0)$, $(1, -5/6)$, $(2, -1)$, $(3, 0)$, and $(4, 8/3)$.

24. Do Problem 23 for the curve $y = x^2/(x^2 + 1)$ using the points $(0, 0)$, $(\pm 1, 1/2)$, $(\pm 3, 9/10)$, and $(\pm 5, 25/26)$.

25. Find an equation for the line tangent to the given curve at the given point.

(a) $y = \dfrac{4}{x}$, $(8, 1/2)$.   (b) $x^2 + y^2 = 6x + 4y$, $(6, 4)$.

(c) $y = \dfrac{2x}{1 - x^2}$, $(0, 0)$.   (d) $y = \sqrt{x^3 - 2}$, $(3, 5)$.

(e) $x^2y + y^2x = 12$, $(3, 1)$ and $(1, 3)$.   (f) $x^3y + y^3x = 10$, $(2, 1)$ and $(1, 2)$.

26. Find an equation for the line normal to the curve at the given point for each of the curves in Problem 25.

27. For each of the following curves, find the points where the tangent line is parallel to the x-axis.

(a) $y = x^2 + 4x - 7$.   (b) $y = x^3 - 3x^2 + 5$.

(c) $y = 2x^3 - 3x^2 - 36x - 11$.   (d) $y = (x + 3)^3(x - 5)^2$.

(e) $y = \dfrac{x}{x^2 + 4}$.   (f) $y = \dfrac{x + 3}{x - 1}$.

28. Find an equation for the line (or lines) tangent to the curve $y = x^2$ and passing through the point: (a) $(5, 9)$, and (b) $(0, -3)$.

29. Find the second derivative for each of the following functions.

(a) $y = \sqrt{x^2 + 4}$.   (b) $v = u^2 + \dfrac{2}{u}$.

(c) $y = x\sqrt{x^3 - 1}$.   (d) $y = \dfrac{t}{\sqrt{t^2 + 1}}$.

(e) $s = \dfrac{\sqrt{t} + 1}{\sqrt{t} - 1}$.   (f) $x = \left(\dfrac{at + b}{ct + d}\right)^{3/2}$.

30. Prove that the Product Rule applied to uv^{-1} and the Quotient Rule applied to u/v give the same result.

31. For each of the following functions, find $f'(x)$ at each point where the derivative exists.

    (a) $f(x) = x + |x|$.   (b) $f(x) = x - [x]$.

★32. Let $f(x)$ be an odd function that is twice differentiable in an interval $(-a, a)$. Prove that in that interval $f'(x)$ is an even function and $f''(x)$ is an odd function. HINT: Use the Chain Rule.

5

Applications of the Derivative

1 Objective

The derivative has been introduced as an aid to curve sketching, namely as a method of finding the line tangent to a given curve at a fixed point on the curve. But the derivative is useful in other problems. Some of these uses will be considered in this chapter.

2 Motion on a Straight Line

We suppose that a particle P is moving on a straight line. For simplicity we assume that the line is horizontal. The motion of the particle is completely determined if we specify where the particle is at each instant of time. This is easy to do if we select some point on the line as the origin and introduce a coordinate system. In place of the usual x, we use the letter s to denote the directed distance on this line. Let $s = f(t)$ be the particular function that gives the directed distance from the origin to the moving particle P at time t. Whenever such a function is given, the motion of the particle is completely specified.

For example, let $s = t^2 - 4$ in some convenient units. To be specific, let s be in meters and t in seconds. This equation states that when $t = 0$ then $s = -4$, so the particle is 4 meters to the left of the origin. We indicate this in Fig. 1, where the lower line is the directed line on which the particle moves, and the upper line indicates the motion of the particle but is drawn distinct from the lower line for clarity. The time values are on the upper line, and the corresponding s values are on the lower line. When $t = 2$, $s = 0$, so the particle passes through the origin 2 sec after starting. When $t = 5$, the particle is 21 meters to the right of the origin.

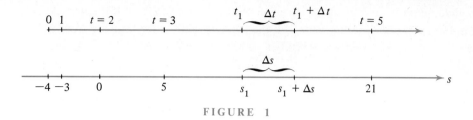

FIGURE   1

The velocity of a moving particle is usually found by taking the directed distance it travels and dividing by the time required to cover that distance. This is quite satisfactory if the particle is moving uniformly; that is, if it does not go faster or slower at different times. If the motion is not uniform, as is the case in our example, then this ratio gives only an *average velocity*, and this average need not be a constant. To illustrate this statement, suppose again that $s = t^2 - 4$. In the 2-sec interval between $t = 1$ and $t = 3$ the particle moves from the point $s = -3$ to the point $s = 5$ and thus travels a distance $\Delta s = 5 - (-3) = 8$ m. The average velocity during this 2-sec interval is $\Delta s/\Delta t = 8/2 = 4$ m/sec. In the 2-sec interval between $t = 3$ and $t = 5$, the particle moves from the point $s = 3^2 - 4 = 5$ to the point $s = 5^2 - 4 = 21$, a distance of 16 m. The average velocity during this 2-sec interval is $\Delta s/\Delta t = 16/2 = 8$ m/sec. If we moved the time interval or changed its length, we would obtain still different values for the average velocity. To obtain a unique velocity at a specified time, we take the limit $\Delta s/\Delta t$ as $\Delta t \to 0$. This gives the *instantaneous velocity* at that time.

DEFINITION 1 (Velocity). If $s = f(t)$ gives the location of a particle moving on a straight line, then $v = v(t_1)$, the instantaneous velocity at $t = t_1$ is defined by the equation

(1)
$$v(t_1) = \lim_{\Delta t \to 0} \frac{\Delta s}{\Delta t} = \lim_{\Delta t \to 0} \frac{f(t_1 + \Delta t) - f(t_1)}{\Delta t}.$$

For brevity, the instantaneous velocity is called the *velocity*. The *speed* of the particle is the absolute value of its velocity.

It is clear that the velocity is just the derivative of $f(t)$, and indeed we may write

(2)
$$v = v(t) = f'(t) = \frac{ds}{dt} = s'(t),$$

selecting whichever notation seems most suitable.

The words *speed* and *velocity* are frequently confused. The difference is that velocity is a signed quantity, and hence on occasion may be negative. Thus, if the particle is moving from left to right on the line of Fig. 1, then $v \geq 0$, while if it is moving from right to left, then s is decreasing and $v \leq 0$. But in either case the speed is $|v|$, and hence is never negative.

By definition the *acceleration* of a particle is just the instantaneous rate of change of the velocity. Using $a = a(t)$ to denote this quantity, we have by the definition of acceleration,

$$(3) \qquad a = a(t) = \lim_{\Delta t \to 0} \frac{v(t + \Delta t) - v(t)}{\Delta t} = \lim_{\Delta t \to 0} \frac{\Delta v}{\Delta t} = \frac{dv}{dt}.$$

Since $v = \dfrac{ds}{dt}$, it follows that

$$(4) \qquad a = \frac{dv}{dt} = \frac{d}{dt}\left(\frac{ds}{dt}\right) = \frac{d^2s}{dt^2}.$$

In words, the acceleration of a particle moving on a straight line is the second derivative with respect to time of the function that gives the directed distance of the particle from some fixed origin.

Example 1. Discuss the motion of a particle moving on a straight line if its position is given by $s = t^2 - 12t$, where s is in meters and t is in seconds.

Solution. At $t = 0$, the equation gives $s = 0$, so the particle starts at the origin. The velocity at any time is given by[1]

$$(5) \qquad v = \frac{ds}{dt} = \frac{d}{dt}(t^2 - 12t) = 2t - 12 = 2(t - 6).$$

Hence at $t = 0, v = -12$ m/sec, so the particle is moving to the left. These facts are recorded schematically in Fig. 2. Although the particle is moving on the s-axis, the lower line on the figure, its motion is indicated on the upper curve for clarity. The arrow indicates the direction of motion, and just as before we place the t values on the upper curve.

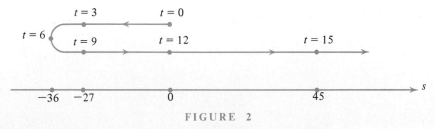

FIGURE 2

[1] During the computation, it is convenient to drop the physical units such as meters and seconds, and to restore them when we have the final result.

If $t < 6$, the second factor on the right side of (5) is negative and hence the velocity is negative. Thus in the interval $0 \leq t < 6$, the particle is moving to the left. At $t = 6$, $v = 0$ and the particle is momentarily at rest. For $t > 6$, the velocity is positive, and for these values of t the particle is moving to the right. Consequently, at $t = 6$, the particle attains its extreme position to the left of the origin. In other words, s is a minimum at $t = 6$. An easy computation gives $s = 6^2 - 12 \times 6 = -36$ for this minimum value of s. To find the time when the particle is again at the origin, we solve the equation $s = t^2 - 12t = 0$. This gives $t = 0$ and $t = 12$.

Finally, we compute the acceleration. From (4) and (5),

$$a = \frac{dv}{dt} = \frac{d}{dt}(2t - 12) = 2.$$

In other words, the velocity is steadily increasing at the rate of 2 m/sec each second (written 2 m/sec^2). Thus the velocity starts at -12 m/sec; in 6 sec it has increased to 0, and in 6 more seconds it has increased to 12 m/sec. During this same period the speed has at first decreased from 12 m/sec down to 0, and then increased back to 12 m/sec. ●

Motion Under Gravity **3**

It is an experimental fact that near the surface of the earth a free-falling object has a downward acceleration that is constant, provided that we neglect air resistance. If we select the positive direction upward, then the basic equation governing the motion of a body falling under the influence of gravity is

(6) $$a = -g,$$

where g denotes this constant acceleration. The value of g depends on the units used. In the British system $g \approx 32$ ft/sec^2 (more accurately 32.17 ft/sec^2) and in the MKS (meter, kilogram, second) system $g \approx 9.8$ m/sec^2 (more accurately 9.805 m/sec^2). Furthermore, this constant is quite different on the moon or on planets other than the earth.

Since acceleration is the derivative of the velocity with respect to time, equation (6) leads to

(7) $$v = -gt + C_1,$$

where C_1 is some constant. The process of going from equation (6) to equation (7) is the inverse of differentiation. Instead of differentiating a given function, we are given its derivative and asked to find its primitive, that is, the original function. Such a process is called *integration* or *antidifferentiation*. This process will be studied systematically in the next chapter. For the present we note that the derivative of v given by (7) does yield (6), no matter what value is assigned to the constant C_1.

To find an expression for s, we recall that $v = ds/dt$, so we look for a function whose derivative gives the right side of (7). Inspection shows that

(8)
$$s = -\frac{g}{2}t^2 + C_1 t + C_2$$

is such a function, where C_2 is any constant. For the derivative of (8) does give (7), and the derivative of (7) does give (6). It is intuitively clear that any function $s(t)$ for which $s''(t) = -g$ must have the form (8). In due time we will be able to prove this, as a corollary of Theorem 10 (page 201).

What can be said about the constants C_1 and C_2? If we put $t = 0$ in equation (7), we see that $v = C_1$ at the time $t = 0$. We call this velocity the *initial velocity,* and symbolize it with v_0. Therefore, $C_1 = v_0$, the initial velocity. Similarly, if we put $t = 0$ in equation (8), we find $s = C_2$. We call this the *initial position* and symbolize it by s_0. Therefore, $C_2 = s_0$, the location of the moving object at $t = 0$. Then equation (8) has the form

(9)
$$s = -\frac{g}{2}t^2 + v_0 t + s_0.$$

Summarizing: *Equation (9) is the equation of an object falling under the influence of gravity, when air resistance is neglected. In this equation s_0 is a constant that gives the initial position of the object, and v_0 is a constant that gives the initial velocity.*

In the British system, with $g = 32$, equation (9) becomes

(9a)
$$s = -16t^2 + v_0 t + s_0 \qquad \text{feet.}$$

In the MKS system, with $g = 9.8$, equation (9) becomes

(9b)
$$s = -4.9t^2 + v_0 t + s_0 \qquad \text{meters.}$$

Example 1. A stone is thrown upward from the top of a building 24.5 meters high with an initial velocity of 19.6 m/sec. Find the maximum height the stone attains. At what time does it pass the top of the building on the way down? When does it hit the ground?

Solution. With the positive direction upward and the origin on the ground, we have $s_0 = 24.5$ and $v_0 = 19.6$. Equation (9b) gives

(10)
$$s = -4.9t^2 + 19.6t + 24.5.$$

Differentiating gives for the velocity

$$v = \frac{ds}{dt} = -9.8t + 19.6 = -9.8(t - 2).$$

For $t < 2$, the velocity is positive, and the stone is traveling upward. At $t = 2$, the velocity is zero, and the stone is stationary for an instant. Then it starts to descend. Therefore, the maximum height of the stone is obtained by putting $t = 2$ into equation (10).

$$s = -4.9(4) + 19.6(2) + 24.5 = 44.1.$$

Hence the maximum height is 44.1 meters.

The stone is at the top of the building when $s = 24.5$. Equation (10) then gives

$$24.5 = -4.9t^2 + 19.6t + 24.5$$

$$0 = -4.9t^2 + 19.6t$$

$$0 = -4.9t(t - 4).$$

Hence $t = 0$ or $t = 4$. So the stone passes the top of the building 4 sec after it is thrown.

The stone hits the ground when $s = 0$. Equation (10) gives

$$0 = -4.9t^2 + 19.6t + 24.5$$

$$0 = -4.9(t^2 - 4t - 5)$$

$$0 = -4.9(t - 5)(t + 1).$$

Thus $s = 0$ when $t = 5$, or $t = -1$. The second answer is physically meaningless and may be rejected. Therefore, the stone hits the ground 5 sec after being thrown. ●

EXERCISE 1

In Problems 1 through 4, a particle is moving on a horizontal line in accordance with the given equation. Find the velocity and acceleration and determine any extreme positions for $t \geqq 0$. Make a graph similar to Fig. 2 showing the motion for $t \geqq 0$.

1. $s = 10 + 6t - t^2$.   2. $s = 2t^2 - 20t + 5$.
3. $s = t^3 - 9t - 7$.   4. $s = t^3 + 3t^2 - 45t + 8$.

5. A stone is thrown upward from the top of a building 48 ft high with an initial velocity of 32 ft/sec. Find the maximum height of the stone, and when it hits the ground. Where is the stone when the velocity is −64 ft/sec? HINT: Use equation (9a).

6. Suppose that in Problem 5, the building is 14.7 meters high and the initial velocity is 9.8 m/sec. Find the maximum height of the stone, and when it hits the ground. Where is the stone when the velocity is −19.6 m/sec?

7. A bomb is dropped from a plane 4410 meters above the ground. How much time does it take to reach the ground, and what is the speed of the bomb just before it hits the ground? Neglect the air resistance and any horizontal motion of the bomb due to

the motion of the plane. Actually, the bomb will travel along a parabola (as we will learn in Chapter 12) but assume here that the drop is vertical.

8. Do Problem 7 if the plane is 1600 ft above the ground.

9. A man standing on the ground throws a rock vertically upward. Find a formula giving the maximum height in meters of the stone in terms of the initial velocity v_0. Neglect the height of the man. What is the least value of v_0 that will suffice for the stone to land on top of a 40-meter building?

10. Do Problem 9 if the height of the building is 100 ft.

11. A stone is dropped from the roof of a building 144 ft high. One second later a second stone is thrown downward from the top of the same building with an initial speed $|v_0|$. What must be the value of $|v_0|$ in order that both stones hit the ground at the same time?

12. Do Problem 11 if the height of the building is 78.4 meters.

13. A ball is dropped from a height of H feet above the ground. Show that it strikes the ground in $\sqrt{H}/4$ sec.

14. If the ball of Problem 13 is thrown downward with an initial speed of $|v_0|$ ft/sec, show that it hits the ground in $(\sqrt{v_0^2 + 64H} - |v_0|)/32$ sec.

15. Find the formulas corresponding to those in Problems 13 and 14 if the metric (MKS) system is used.

⋆16. Is it possible to have a particle move on a horizontal line from left to right (s always increasing), and yet have $v(t_0) = 0$ for some particular t_0?

⋆17. Is it possible to have a motion as described in Problem 16 with $v(t_0) = 0$ for two different values of t_0?

18. An astronaut stands on the edge of a cliff and drops a rock. He observes that it takes 3 sec for the rock to land. How high is the cliff (above the landing point of the rock) if: (a) the astronaut is on Venus, where $g = 28$ ft/sec²; (b) he is on Mars, where $g = 12$ ft/sec²; and (c) he is on the moon, where $g = 5.5$ ft/sec²?

19. An associate astronaut stands at the bottom of the cliff in Problem 18 and returns the rock so that the experiment can be repeated. In each case find the minimum velocity with which he must throw the rock in order that it lands on top of the cliff.

20. Find the value of g in the MKS system to three significant figures for: (a) Venus, (b) Mars, and (c) the moon. HINT: Use g given in Problem 18.

21. An astronaut throws a rock directly upward and catches it exactly 4 sec later. Find the initial velocity of the rock in m/sec if the astronaut is: (a) on Earth, (b) on Venus, (c) on Mars, and (d) on the moon.

22. An astronaut develops his technique so that he can throw a stone upward with a velocity of exactly 80 ft/sec. On Earth he catches the stone after 5 sec; on Venus, after 5.7 sec; on Mars after 13.3 sec; and on the moon, after 29 sec. Use these data to compute g for each of these bodies.

23. The position of a particle moving on the x-axis is given by $x = t^3 - 6t^2 + 15t - 5$, where t denotes time in seconds. Find the location of the particle when the velocity is 6 units/sec. HINT: Two answers are possible.

24. Do Problem 23 if $x = t^3 - 6t^2 + 6t + 11$.
25. For the motion described in Problem 4, find the location of the particle and the velocity when: (a) the acceleration is 12, and (b) when the acceleration is 30.
26. Suppose that $x = t^4 - 12t^3 + 60t^2 - 11t - \sqrt{29}$ gives the position of a particle moving on the x-axis. Find the velocity of the particle when: (a) the acceleration is 24, (b) the acceleration is 60, (c) the acceleration is 12, and (d) the acceleration is zero.

Related Rates 4

The location of a particle is not the only quantity that can depend on time. Indeed, any physical quantity Q that is changing gives rise to a function $Q(t)$, and the derivative $Q'(t)$ gives the *instantaneous rate of change of Q with respect to t*. For example, Q might be the volume of a balloon into which gas is being pumped, or Q might be the concentration of acid in a vat in which a chemical reaction is taking place, or Q might be the quantity of electrical charge on a condenser, or Q might be the stress in a certain steel beam in a bridge, over which a truck is passing. It may be difficult or even impossible to determine $Q(t)$ and hence its derivative, but in all cases the procedure is the same. We set up an equation relating Q, the quantity in question, and other quantities for which we know the rate of change. We then differentiate this equation with respect to t and solve the resulting equation for $Q'(t)$. By definition, this is the (instantaneous) rate of change of Q with respect to t.

Example 1. Gas is being pumped into a spherical balloon at the rate of 8 cm³/sec. Find the rate of change of the radius of the balloon when the radius of the balloon is 10 cm.

Solution. The known formula for the volume of a sphere is

(11) $$V = \frac{4}{3}\pi r^3.$$

Differentiating both sides of (11) with respect to t (Theorem 18 of Chapter 4 and the Chain Rule), we obtain

$$\frac{dV}{dt} = 4\pi r^2 \frac{dr}{dt} \qquad \text{or} \qquad \frac{dr}{dt} = \frac{1}{4\pi r^2}\frac{dV}{dt}.$$

We are given that $dV/dt = 8$ and that $r = 10$. Hence

$$\frac{dr}{dt} = \frac{8}{4\pi \cdot 100} = \frac{1}{50\pi} \approx 0.006366 \text{ cm/sec.} \quad \bullet$$

In working problems of this type, many students try to substitute $r = 10$ directly in equation (11) and then differentiate. When this is done, the right side of equation (11) becomes a constant and then $dV/dt = 0$. This is, of course, a ridiculous answer. We leave it to the reader to explain what is erroneous about this computation.

Example 2. A man is standing on the top rung of a 13-ft ladder, which is leaning against a wall, when a scientific-minded joker starts to pull the bottom of the ladder away from the wall steadily at the rate of 6 ft/min. At what rate is the man on the ladder descending (he remains standing on the top rung) when the bottom of the ladder is 5 ft from the wall? When the bottom is 12 ft from the wall?

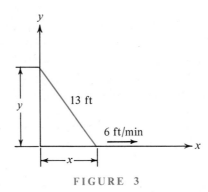

FIGURE 3

Solution. We introduce a coordinate system as indicated in Fig. 3. Then no matter what the position of the ladder,

(12) $$x^2 + y^2 = 13^2 = 169.$$

Differentiating both sides of this equation with respect to t gives

$$2x\frac{dx}{dt} + 2y\frac{dy}{dt} = 0,$$

or

(13) $$\frac{dy}{dt} = -\frac{x}{y}\frac{dx}{dt}.$$

When $x = 5$, equation (12) gives $y = \sqrt{169 - 25} = 12$. Further, we are given that $dx/dt = 6$ ft/min. Using these values for x, y, and dx/dt in equation (13), we find that

$$\frac{dy}{dt} = -\frac{5}{12}6 = -2.5 \text{ ft/min},$$

or the man is descending at the rate of 2.5 ft/min. When $x = 12$, then $y = \sqrt{169 - 144} = 5$, and in this case equation (13) gives

$$\frac{dy}{dt} = -\frac{12}{5} 6 = -14.4 \text{ ft/min,}$$

and this is somewhat faster. ●

EXERCISE 2

1. A snowball is melting at the rate of 1 ft³/hr. If it is always a sphere, how fast is the radius changing when the snowball is 18 in. in diameter?
2. How fast is the surface area changing for the snowball of Problem 1?
3. At noon a certain ship A is 35 kilometers due north of a ship B. Ship A is traveling south with a speed of 14 km/hr, and ship B is traveling east with a speed of 20 km/hr. Find a general expression for the distance between these two ships at any time t. How fast is this distance increasing at 1:00 P.M.?
4. A conical tank full of water is 20 meters high and 10 meters in diameter at the top. If the water is flowing out at the bottom at the rate of 2 m³/min, find the rate at which the water level is falling: (a) when the water level is 16 meters above the bottom, and (b) when the water level is 2 meters above the bottom.
5. Sand is being poured onto a conical pile at the rate of 9 meters³/min. Friction forces in the sand are such that the slope of the sides of the conical pile is always 2/3. How fast is the altitude increasing when the radius of the base of the pile is 6 meters?
6. A man 6 ft tall walks away from the base of a street light at a rate of 3.5 ft/sec. If the light is 20 ft above the ground, find the rate of change of the length of the man's shadow: (a) when he is 10 ft from the base of the light, and (b) when he is 50 ft from the base of the light.
7. In Problem 6, how fast is the farther end of his shadow moving?
8. A boat is fastened to a rope that is wound about a windlass 20 ft above the level at which the rope is attached to the boat. If the boat is moving away from the dock at the rate of 5 ft/sec, how fast is the rope unwinding when the boat is 40 ft from the point at water level directly under the windlass?
9. The surface area of a cube is changing at the rate of 8 cm²/sec. How fast is the volume changing when the surface area is 60 cm²?
10. A particle P moves on the curve $y = x^3$ in such a way that the x-coordinate changes 5 units/sec.
    (a) How fast is the y-coordinate of P changing?
    (b) What is the rate of change of the slope of the curve at P?
★11. A particle P moves on the line $y = x + 5$ in such a way that the x-coordinate

changes 3 units/sec. Find the rate of change of the distance between P and the point $(2, 0)$ when P is (a) at $(-5, 0)$, (b) at $(-1, 4)$, and (c) at $(7, 12)$. Find the limit as $x \to \infty$ of the rate of change of this distance.

$\star$12. A particle P moves on the parabola $y = x^2$ in such a way that the x-coordinate changes 2 units/sec. Find the rate of change of the distance between P and the point $(3, 0)$ when P is (a) at $(-1, 1)$, (b) at $(1, 1)$, and (c) at $(3, 9)$. Find the limit as $x \to \infty$ of the rate of change of this distance.

13. A particle P moves on the curve $y = x^2 + 1$ in such a way that $dx/dt = 4$. If s is the distance of P from the point $(0, 6)$, find ds/dt: (a) in general as a function of x, and (b) when P is at $(2, 5)$.

$\star$14. A particle P moves to the right on the part of the curve $y = 1/x$ in the first quadrant in such a way that $dy/dt = -1/s$, where s is the distance from P to the origin. Find dx/dt: (a) in general as a function of x, (b) when P is at $(1, 1)$, and (c) when P is at $(3, 1/3)$.

15. A particle moves on the curve $y = x^2$. At what point (or points) on the curve will the x-coordinate and the y-coordinate be changing at the same rate? Assume that the particle is never stationary.

16. Do Problem 15 for the curve: (a) $12y = x^3$, (b) $y = x^3 - x^2 + 2$, (c) $y = \sqrt{16 - x^2}$, and (d) $y = 1/x$.

5 *Increasing and Decreasing Functions*

We have already met increasing and decreasing functions in a casual way in Section 8 of Chapter 4. We now study them more carefully.

The function graphed in Fig. 4 is not increasing everywhere, but it is increasing in the interval $a \leqq x \leqq b$.

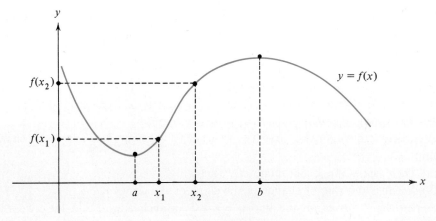

FIGURE 4

> **DEFINITION 2 (Increasing Function).** A function $f(x)$ is said to be *increasing in the interval* $\mathscr{I}$ if, whenever x_1 and x_2 are in $\mathscr{I}$ and $x_1 < x_2$, then
>
> (14) $$f(x_1) < f(x_2).$$

If a function f has a positive derivative at each point in an interval $\mathscr{I}$, then we know geometrically that its graph has a rising tangent line at every point. In such a case it is reasonably clear that the function f must be increasing in $\mathscr{I}$. This result is stated in Theorem 1, with the official proof being deferred to Section 7.

> **THEOREM 1.** If $f'(x) > 0$ for every point in an interval $\mathscr{I}$, then $f(x)$ is an increasing function in $\mathscr{I}$.

The definition of a *decreasing* function is similar to the definition of an increasing function—only the inequality sign is reversed in (14). We leave the task of writing this definition and the analogue of Theorem 1 to the reader. It is important to observe that $g(x)$ *is a decreasing function if and only if* $f(x) \equiv -g(x)$ *is an increasing function.*

Example 1. In what intervals is the function $f(x) = 17 - 15x + 9x^2 - x^3$ increasing? In what intervals is this function decreasing?

Solution. Computing the derivative, we find that

$$f'(x) = -15 + 18x - 3x^2 = -3(5 - 6x + x^2),$$

(15) $$f'(x) = -3(x - 5)(x - 1).$$

Clearly, the derivative is zero when $x = 1$ and $x = 5$, and only for those values. These two points, where the derivative is zero, break the x-axis into three sets:

$$\mathscr{I}_1: -\infty < x < 1, \qquad \mathscr{I}_2: 1 < x < 5, \qquad \mathscr{I}_3: 5 < x < \infty.$$

In each of these sets we may expect the derivative to have constant sign. Let us look first at $\mathscr{I}_3$. When $x > 5$, the last two factors in (15) are positive. Hence $f'(x) < 0$ and the function is decreasing in $\mathscr{I}_3$. A similar type of discussion gives the sign of the derivative in the other two sets. The work may be arranged systematically as in the table given on the next page.

The graph of this function is shown in Fig. 5, where for convenience we have used different scales on the two axes. It is clear from the picture that the function increases and decreases as predicted by the entries in the table. ●

	Domain of x	Sign of $-3(x-5)(x-1)$	Sign of $f'(x)$	Function Is
$\mathcal{I}_1$	$-\infty < x < 1$	$(-)(-)(-)$	$-$	decreasing
$\mathcal{I}_2$	$1 < x < 5$	$(-)(-)(+)$	$+$	increasing
$\mathcal{I}_3$	$5 < x < \infty$	$(-)(+)(+)$	$-$	decreasing

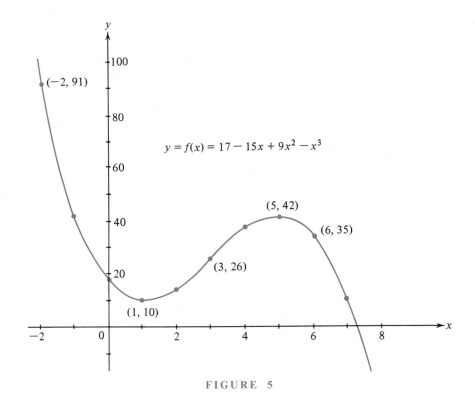

FIGURE 5

6 *Extreme Values of a Function*

Briefly, the maximum value of a function is its highest value, and the minimum value is its lowest. But to be precise we must specify the interval we are considering. In other words, it must be clear which values of x are allowed to enter the competition for making $f(x)$ a maximum or a minimum.

DEFINITION 3 (Maximum, Minimum). The function $f(x)$ is said to have a *maximum value* in an interval $\mathscr{I}$ at x_0 if x_0 is in the interval and if $f(x_0) \geq f(x)$ for all x in $\mathscr{I}$. The number $f(x_0)$ is said to be the *maximum* of $f(x)$ in that interval. Similarly, $f(x_0)$ is the *minimum* if $f(x_0) \leq f(x)$ for all x in the interval $\mathscr{I}$.

The graph of a function may have several high points, one of which may be higher than all the others. To distinguish among such high points, the highest one is called the *maximum* or the *absolute maximum,* and each of the others is called a *relative maximum.* Precisely, we have

DEFINITION 4 (Relative Maximum). The function $f(x)$ is said to have a relative maximum at x_0 if there is an interval with x_0 in the interior, such that in that interval, $f(x_0)$ is the maximum value of the function.

Similarly, we may define a *relative minimum* for $f(x)$. All such values, whether maximum or minimum (absolute or relative), are called *extreme values*. One expects the extreme values occur where the derivative is zero. The precise statement is

THEOREM 2. If $f(x)$ has a relative maximum at $x = x_0$, and if $f(x)$ is differentiable at x_0, then $f'(x_0) = 0$.

Proof. Let us compute the derivative of $y = f(x)$ at x_0. By definition,

$$(16) \qquad f'(x_0) = \lim_{h \to 0} \frac{f(x_0 + h) - f(x_0)}{h}.$$

By hypothesis, $f(x_0)$ is a relative maximum value, so $f(x_0) \geq f(x_0 + h)$ when $|h|$ is sufficiently small, whence the numerator of (16) is always negative or at most zero. Now, if h approaches 0 through positive values, the ratio on the right in (16) is always negative (or zero), and hence the limit process gives

$$(17) \qquad f'(x_0) \leq 0.$$

On the other hand, if h is negative and approaching zero in (16), then the quotient on the right side of (16) is always positive (or zero) and hence the limit process gives

$$(18) \qquad f'(x_0) \geq 0.$$

Now the two conditions on the derivative, (17) and (18), are incompatible unless $f'(x_0) = 0$. ∎

THEOREM 3 (PLE). If $f(x)$ has a relative minimum at x_0, and if $f(x)$ is differentiable at x_0, then $f'(x_0) = 0$.

Example 1. Find the extreme values of the function

(19) $$f(x) = 17 - 15x + 9x^2 - x^3$$

(a) for all real x, (b) in the interval $-2 \leqq x \leqq 6$, and (c) in the interval $3 \leqq x \leqq 6$.

Solution. (a) This function is the same as the one treated in the example of the preceding section and the graph is shown in Fig. 5. To locate all the relative extreme points, we first find the values of x for which the derivative is zero. From the preceding section we recall that

(15) $$f'(x) = -3(x - 5)(x - 1).$$

Therefore, relative extreme values can occur only at $x = 1$ and $x = 5$. Since $f(x)$ is decreasing for $x < 1$, and increasing if $1 < x < 5$, it is clear that the curve descends, stops at $x = 1$, then rises, so that a relative minimum occurs at $x = 1$, and this minimum value is $f(1) = 17 - 15 + 9 - 1 = 10$. Similarly, the function is increasing as x runs from 1 to 5, stops at $x = 5$, and thereafter is decreasing; hence a relative maximum occurs at $x = 5$, and this maximum value is

$$f(5) = 17 - 15 \times 5 + 9 \times 25 - 125 = 42.$$

Thus $(1, 10)$ is a relative minimum point and $(5, 42)$ is a relative maximum point.

Is 10 the *absolute* minimum value of the function? No! For if we put $x = 20$, we find that $f(20) = 17 - 300 + 3600 - 8000 = -4683$, and this is less than 10. But this is not the minimum either, because

$$f(100) = 17 - 1500 + 90,000 - 1,000,000 = -911,483,$$

and this is still less than -4683. Indeed, we have seen that for $x > 5$ the function $f(x)$ is steadily decreasing as x increases, and hence there is no absolute minimum. Thus 10 is a relative minimum but not a minimum.

Similarly, by letting x approach $-\infty$, we see that 42 is a relative maximum but not an absolute maximum value for the function. For example, $f(-20) = 17 + 300 + 3600 + 8000 = 11,917$ and $11,917 > 42$. Since $f(x)$ is decreasing as x increases for $x < 1$, we can reverse the direction of x and say that $f(x)$ is increasing as x decreases toward $-\infty$. Hence $f(x)$ has no maximum.

(b) The situation changes when we restrict x to lie in some interval. Suppose now that x must lie in the interval $-2 \leqq x \leqq 6$. It is clear that we only need to compare the relative maximum and minimum with the values of $f(x)$ at the end points. Now $f(-2) = 17 + 30 + 36 + 8 = 91 > 42$. Hence the maximum value of $f(x)$ in the interval $-2 \leqq x \leqq 6$ is 91 and it occurs at $x = -2$. At the other end point, $x = 6$, we have $f(6) = 17 - 90 + 324 - 216 = 35$. Hence 10 is the minimum value of $f(x)$ in this interval. In this case the relative minimum and the absolute minimum are the same.

(c) In the interval $3 \leqq x \leqq 6$, the point $(5, 42)$ is the only relative extreme point. At the end points we have $f(3) = 17 - 45 + 81 - 27 = 26$ and $f(6) = 35$. Comparing these two values with $f(5) = 42$, we can conclude that in the interval $3 \leqq x \leqq 6$, the minimum value of $f(x)$ is 26 and the maximum value is 42. ●

Example 2. Find the extreme values of the function $y = x^3$ for all real x.

Solution. First we find where the derivative vanishes:

(20) $$\frac{dy}{dx} = 3x^2 = 0, \qquad \text{at } x = 0.$$

At $x = 0$, $y = 0$, and this is the only possible relative maximum or minimum. But if x is positive, then y is positive, and if x is negative, y is negative, so near $x = 0$ this function assumes values greater than zero and also values that are less than zero.

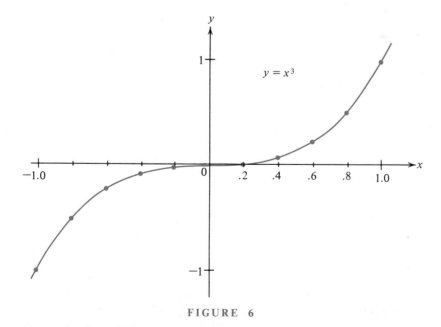

FIGURE 6

Therefore, at $x = 0$ there is neither a relative maximum nor a relative minimum. Indeed, the derivative is always positive, except at $x = 0$, and hence the function is everywhere increasing. Therefore, $y = x^3$ has neither an absolute maximum, nor an absolute minimum, nor a relative maximum, nor a relative minimum (see Fig. 6). ●

This second example shows that the vanishing of the derivative is a *necessary* condition for a relative extreme point, but it is not a *sufficient* condition. The meaning of these new technical terms *necessary* and *sufficient* has already been illustrated in the preceding work. However, for clarity we give a formal definition.

A condition C is said to be *necessary* for a property P if P implies C.
A condition C is said to be *sufficient* for a property P if C implies P.

In the present situation the condition is "$f'(x_0) = 0$," and the property is "$f(x_0)$ is a relative extreme."

The condition is necessary. If $f(x_0)$ is a relative extreme, then (necessarily) the derivative must vanish at x_0. This is the content of Theorems 2 and 3.

However, the *condition is not sufficient.* For as shown in Example 2, the derivative of the function $y = x^3$ vanishes at $x = 0$, but the function does not have a relative extreme at $x = 0$.

It is easy to guess at a set of sufficient conditions for a relative extreme. These are given in the next two theorems.

THEOREM 4 (PLE) (First-Derivative Test for Relative Minimum). Suppose that $f'(x_0) = 0$ and there is an interval $c \leqq x \leqq d$ about x_0 such that

(21) $$f'(x) < 0, \qquad \text{if } c \leqq x < x_0$$

and

(22) $$f'(x) > 0, \qquad \text{if } x_0 < x \leqq d.$$

Then $f(x_0)$ is a relative minimum for the function $f(x)$, and, moreover, $f(x_0)$ is the absolute minimum of $f(x)$ for x in the interval $c \leqq x \leqq d$.

THEOREM 5 (PLE) (First-Derivative Test for Relative Maximum). Suppose that $f'(x_0) = 0$ and there is an interval $c \leqq x \leqq d$ about x_0 such that

(23) $$f'(x) > 0, \quad \text{if } c \leqq x < x_0$$

and

(24) $$f'(x) < 0, \quad \text{if } x_0 < x \leqq d.$$

Then $f(x_0)$ is a relative maximum for $f(x)$, and, moreover, $f(x_0)$ is the absolute maximum of $f(x)$ for x in the interval $c \leqq x \leqq d$.

We leave it to the reader to make a sketch showing the situation described in these two theorems. The truth of these theorems will then be immediately obvious.

By way of summary we list the steps for determining the relative extreme values of a differentiable function f.

STEP 1. Find $f'(x)$.

STEP 2. Find the *critical numbers* of f, that is, the points where $f'(x) = 0$. These points furnish the *candidates* for the location of relative extremes.

STEP 3. Test each critical point to see if the conditions of Theorems 4 or 5 hold.

We remark that if the sign of f' does *not* change in going from one side of a critical point to the other, then f fails to have a relative extreme value at this critical point.

To find the absolute maximum and absolute minimum of $f(x)$ on a closed interval $a \leqq x \leqq b$, we must also check the end points. Hence we have

STEP 4. Compute $f(a)$ and $f(b)$ and compare these values with the relative extreme values.

EXERCISE 3

In Problems 1 through 14, find all of the intervals (or rays) in which the given function is increasing, and find all of the relative maximum and minimum points.

1. $y = x^2 + 2x - 5$.
2. $y = 10 - 6x - 2x^2$.
3. $y = 6 + 3x - x^2$.
4. $y = 2x^2 + 8x + 17$.
5. $y = 2x^3 - 3x^2 - 36x + 7$.
6. $y = 1 - 12x - 9x^2 - 2x^3$.
7. $y = 3 + 6x^2 - 2x^3$.
8. $y = x^3 - 3x + 11$.
★9. $y = 5 - (x + 2)^3(x - 3)^2$.
★10. $y = (x + 1)^3(x - 3)^3$.
★11. $y = 20 - 6x + 9x^2 - 5x^3$.
★12. $y = x^3 + 6x^2 + 12x + 8$.
★13. $y = \dfrac{9 + 2x - x^2}{1 + x}$.
★14. $y = \dfrac{5 - 3x}{1 - x}$.

In Problems 15 through 20, find the absolute minimum and the absolute maximum for the given function in the given interval.

15. $f(x) = x^3 - 2x^2, \quad -1 \leqq x \leqq 1$.
16. $f(x) = x^2 + \dfrac{16}{x}, \quad 1 \leqq x \leqq 3$.

17. $f(x) = \dfrac{3x + 1}{x^2 + x + 3}$, $-4 \leqq x \leqq -1$.

18. $f(x) = \dfrac{x^2}{1 + x^2}$, $9 \leqq x \leqq 10$.

19. $f(x) = \dfrac{2 - x}{5 - 4x + x^2}$, $-100 \leqq x \leqq 2$.

20. $f(x) = \dfrac{1 + x + x^2 + x^3}{1 + x^3}$, $2 \leqq x \leqq 5$.

*21. Find the minimum value of $y = x^4 + 4x^3 + 6x^2 + 4x + 12$ without using the calculus.

*22. Find the maximum value of the derivative of the function $y = x/(1 + x^2)$.

23. Find constants a, b, and c so that the curve $y = ax^2 + bx + c$ goes through the point $(0, 3)$ and has a relative extreme at $(1, 2)$.

*24. Let $f(x) = x$ if $-1 < x < 1$ and let $f(-1) = f(1) = 0$. Does this function have a relative maximum or a relative minimum in the closed interval $-1 \leqq x \leqq 1$?

*25. Find the maximum and minimum points (if there are any) for
    (a) $f(x) = |x|$.   (b) $g(x) = 1 - |x|$.   (c) $h(x) = |x - 1|$.

**26. Do Problem 25 for:
    (a) $F(x) = 3 + |x - 5|$.   (b) $G(x) = 2 - |x + 3|$.   (c) $H(x) = |x| + |x - 1|$.

27. Show that the polynomial

$$P(x) = x^5 + 4x^3 + 6x^2 - 13x + 1$$

has at least one critical point in the interval $(0, 1)$ by examining the sign of $P'(0)$ and $P'(1)$. Then prove that there is exactly one critical point of $P(x)$ in $(0, 1)$ by examining the sign of $P''(x)$ on $(0, 1)$.

28. Prove that if a differentiable function $f(x)$ is increasing on the interval $a < x < b$, then $f'(x) \geqq 0$ for $a < x < b$. HINT: Use equation (16) and let $h \to 0$ through positive values.

7 Rolle's Theorem and Some of Its Consequences

We know that the derivative of a constant is zero. How about the converse? If the derivative of a function is zero for all x in a certain interval, does it follow that the function is constant in that interval? This seems to be obviously true, and we could simplify matters by merely stating "this is an obvious fact." But we can also give a proof, and we shall do so shortly (Theorem 9). The real question at issue is this: Which assertions can be accepted as obvious, and which assertions demand a proof? Unfortunately, we cannot stop to prove all of the theorems required for the calculus, because to do so starting from a bare minimum of axioms would require so much time and energy that the reader would lose all interest before arriving at the calculus.

As a practical way out of this dilemma, we start somewhere in the middle rather than at the beginning.[1] Thus we take, without proof, certain theorems that are obviously true anyway and use them (without apology) whenever they are needed. One such theorem that will be particularly useful in this section is

---

THEOREM 6 (PWO).[2] If $f(x)$ is a continuous function of x in the closed interval $a \leqq x \leqq b$, then it has a maximum value at some point of the interval. It also has a minimum value at some (other) point of the interval.

If the function is not continuous, then it is possible that the function does not have a maximum or a minimum. One example of this type of function was given as Problem 24 of Exercise 3. If $f(x)$ is defined only in an *open* interval, then it is again possible that $f(x)$ has no maximum or minimum. But if the conditions of Theorem 6 are satisfied, then it must have a maximum point (at least one) and a (different) minimum point.

We need Theorem 6 to prove the fundamental

THEOREM 7 (Rolle's Theorem). Suppose that the function $f(x)$ is defined in the closed interval $a \leqq x \leqq b$ and that:

1. $f(a) = f(b) = 0$,
2. $f(x)$ is continuous in $a \leqq x \leqq b$,
3. $f(x)$ is differentiable in $a < x < b$.

Then there is some point ξ (Greek lowercase letter xi) in the open interval $a < \xi < b$ such that $f'(\xi) = 0$.

Proof. If $f(x) = 0$ for all x in $a \leqq x \leqq b$, then $f'(x) = 0$ for all x in that interval, and then any ξ will do. Suppose now that $f(x)$ is not zero for some x in $a \leqq x \leqq b$. By Theorem 6,

---

[1] In the authors' opinion, the beginning consists of Peano's five axioms for the positive integers. Anyone who wants to start at the beginning can find an excellent presentation in the book by E. Landau, *Foundations of Analysis* (Chelsea Publishing Co., New York, 1951). However, we recommend that the reader defer this material until he has mastered at least the calculus.

[2] The interested reader can find a proof of this theorem in any advanced calculus or real variables text.

$f(x)$ has a maximum and a minimum in that interval. Let M and m be the maximum and minimum values, respectively, of $f(x)$. Then either $M > 0$ or $m < 0$, or perhaps both (see Fig. 7). Suppose that $M > 0$, and $M = f(\xi_1)$. Then $a < \xi_1 < b$ because at the end points the function is zero. Now the conditions of Theorem 2 are satisfied at $x = \xi_1$ and hence $f'(\xi_1) = 0$. If the maximum M is 0, then the minimum $m < 0$. We apply the same argument to ξ_2, where $f(\xi_2) = m$, and find (using Theorem 3) that $f'(\xi_2) = 0$. ∎

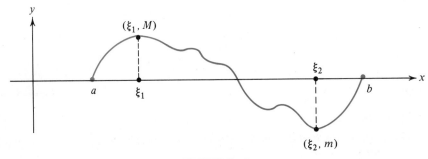

FIGURE 7

We may picture Rolle's Theorem in the following physical way. Suppose that $f(t)$ gives the position of a particle traveling along a straight line. If the motion of the particle is smooth (differentiable) and the particle makes a round trip during the interval of time $a \leq t \leq b$, [i.e., $f(a) = f(b)$], then at some time $t = \xi$ during the trip, this particle must have reversed its direction. In other words, there will be a time ξ when the velocity $f'(\xi)$ is zero.

Example 1. Prove that the equation $g(x) = 4x^3 + 3x^2 - 6x + 1 = 0$ has a root between 0 and 1.

Solution. The given equation is a cubic and can be solved by Cardan's formula (really Tartaglia's formula). When this is done, it is clear that $g(x) = 0$ has a root in the specified interval. But nowadays Tartaglia's formula is seldom taught, and only a few scholars are familiar with the general solution of the cubic. Hence another approach is desirable. By inspection we see that if

(25) $$f(x) = x^4 + x^3 - 3x^2 + x,$$

then the derivative of $f(x)$ is $g(x)$; that is,

(26) $$f'(x) = 4x^3 + 3x^2 - 6x + 1 = g(x).$$

Now $f(0) = 0$ and $f(1) = 1 + 1 - 3 + 1 = 0$. Thus condition (1) of Rolle's Theorem is satisfied with $a = 0$ and $b = 1$. The other two conditions are always satisfied for any polynomial. Therefore, by Rolle's theorem, $f'(x)$ has at least one zero in the open interval $0 < x < 1$. ●

In solving this problem, it was necessary to reverse the process of differentiation; that is, given $g(x)$, we found $f(x)$ such that $f'(x) = g(x)$. We have already encountered this situation in Section 3. The function $f(x)$ is called an *integral* of $g(x)$, and the process of going from $g(x)$ to $f(x)$ is called *integration*. Thus integration is the inverse operation of differentiation. In Chapter 6 we will introduce a suitable notation for the process of integration and we will make a systematic study of this new operation. For the present we will need only the following formula: If $g(x) = ax^n$ and $n \neq -1$, then its integral is

$$f(x) = \frac{ax^{n+1}}{n + 1} + C,$$

where C is some arbitrary constant. This is easily checked by differentiating $f(x)$ to obtain $f'(x) = ax^n = g(x)$.

THEOREM 8 (The Mean Value Theorem). Suppose that:

1. $f(x)$ is continuous in the closed interval $a \leq x \leq b$,
2. $f(x)$ is differentiable in the open interval $a < x < b$.

Then there is a point ξ in the open interval $a < \xi < b$ such that

(27)
$$f'(\xi) = \frac{f(b) - f(a)}{b - a}.$$

It is frequently convenient to write equation (27) in the equivalent form

(28)
$$f(b) - f(a) = (b - a)f'(\xi), \qquad a < \xi < b.$$

Before proving this theorem, we try to picture its meaning. If we graph the function $y = f(x)$, then the ratio

$$\frac{f(b) - f(a)}{b - a}$$

is just the slope of the line segment joining the two points $(a, f(a))$ and $(b, f(b))$ (see Fig. 8). Now $f'(\xi)$ is the slope of the line tangent to the curve at $(\xi, f(\xi))$. So equation (27) asserts that there is some point on the curve where the tangent line is parallel to the chord line. When equation (27) is regarded in this light, it appears obvious that some such point must exist on the curve.

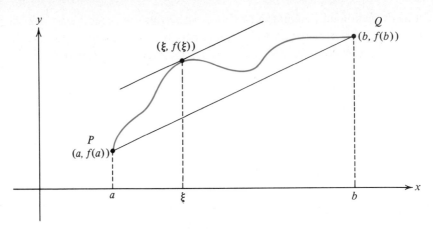

F I G U R E   8

We can picture the Mean Value Theorem in another interesting way. Suppose that a car moves on a straight line and the distance from its starting point is given by $f(t)$. Since the velocity is the derivative $f'(t)$, the Mean Value Theorem tells us that there will be some time ξ during the trip when the instantaneous velocity $f'(\xi)$ is exactly the same as the average velocity, $[f(b) - f(a)]/(b - a)$.

Proof of Theorem 8. We intend to apply Rolle's Theorem, and so we want to use $f(x)$ to construct a new function $F(x)$ that is zero at $x = a$ and $x = b$. We do this by subtracting the equation $y = l(x)$ of the straight line that passes through P and Q, the end points of the graph of $y = f(x)$. These points $P(a, f(a))$ and $Q(b, f(b))$ are indicated in Fig. 8.

It is easy to see that

$$(29) \qquad l(x) = (x - a)\frac{f(b) - f(a)}{b - a} + f(a)$$

is the equation of a straight line for which $l(a) = f(a)$ and $l(b) = f(b)$ so that $y = l(x)$ passes through P and Q.

Now consider

$$(30) \qquad F(x) \equiv f(x) - l(x) = f(x) - \left[(x - a)\frac{f(b) - f(a)}{b - a} + f(a)\right].$$

Then $F(a) = 0$, $F(b) = 0$, and

$$(31) \qquad F'(x) = f'(x) - \frac{f(b) - f(a)}{b - a}.$$

By Rolle's Theorem, there is a ξ in the interval $a < x < b$ for which $F'(\xi) = 0$. For this ξ, equation (31) gives equation (27). ∎

Although these two theorems, Rolle's Theorem and the Mean Value Theorem, look innocent and ineffective, they are in fact fundamental in the development of calculus. We now present some theoretical consequences of the Mean Value Theorem.

THEOREM 9. If the derivative of $f(x)$ is zero in an interval $a \leqq x \leqq b$, then the function is a constant in that interval.

Proof. Since a function that is differentiable in $a \leqq x \leqq b$ is also continuous in that interval, $f(x)$ satisfies the conditions of the Mean Value Theorem. We rewrite equation (27) in the form

$$(28) \qquad f(b) - f(a) = (b - a)f'(\xi), \qquad a < \xi < b,$$

from which we have

$$(32) \qquad f(b) = f(a) + (b - a)f'(\xi), \qquad a < \xi < b.$$

We now let b be variable. To emphasize this change, we replace b in (32) by x, where x is any number in the interval $a < x \leqq b$. In this situation ξ is also a variable depending on x, but we always have $a < \xi < x$. Then (32) becomes

$$(33) \qquad f(x) = f(a) + (x - a)f'(\xi), \qquad a < \xi < x.$$

By the hypothesis of Theorem 9, $f'(\xi) = 0$ for every ξ in $a < \xi < b$. Then (33) simplifies to

$$(34) \qquad f(x) = f(a),$$

for every x in $a < x \leqq b$. This means that $f(x)$ is the constant $f(a)$. ∎

THEOREM 10. If $F(x)$ and $G(x)$ are two functions each defined in $a \leqq x \leqq b$ and having the same derivative there, then they differ by a constant. That is, there is a constant C such that

$$(35) \qquad F(x) \equiv G(x) + C, \qquad a \leqq x \leqq b.$$

Proof. We are given that

$$(36) \qquad \frac{d}{dx}F(x) = \frac{d}{dx}G(x), \qquad a \leqq x \leqq b$$

and this means that for all x in that interval

$$\frac{d}{dx}(F(x) - G(x)) = 0.$$

Now apply Theorem 9 to the function $f(x) \equiv F(x) - G(x)$, whose derivative is zero. It follows from Theorem 9 that $F(x) - G(x) \equiv C$, and hence the identity (35). ∎

COROLLARY. If n is a rational number, $n \neq -1$, and if

(37)
$$\frac{d}{dx} F(x) = ax^n$$

in some interval, then in that interval

(38)
$$F(x) = \frac{ax^{n+1}}{n+1} + C,$$

where C is a constant.

Proof. We already know one function $G(x)$ whose derivative is ax^n. This is

(39)
$$G(x) = \frac{ax^{n+1}}{n+1}.$$

By Theorem 10, if $F(x)$ is any other function with the same derivative, then $F(x) \equiv G(x) + C$. ∎

We conclude this section by supplying the promised proof of Theorem 1 in Section 5. First we restate this result in a slightly generalized form.

THEOREM 1′. If $f(x)$ is continuous on the closed interval $\mathscr{I}$ and if $f'(x)$ is positive for all x in the interior of $\mathscr{I}$, then f is increasing in $\mathscr{I}$.

Proof. Let x_1 and x_2 be arbitrary points in $\mathscr{I}$ with $x_1 < x_2$. Since f is continuous for $x_1 \leq x \leq x_2$ and differentiable for $x_1 < x < x_2$, we can apply the Mean Value Theorem to obtain

$$f(x_2) - f(x_1) = f'(\xi)(x_2 - x_1), \qquad \text{where } x_1 < \xi < x_2.$$

But $f'(\xi) > 0$ and $(x_2 - x_1) > 0$, so the right-hand side of this last equation must be positive. Therefore

$$f(x_2) - f(x_1) > 0 \qquad \text{or} \qquad f(x_2) > f(x_1). \quad ∎$$

EXERCISE 4

In Problems 1 through 4, check that the given function satisfies the conditions of Rolle's Theorem for the given interval. Then find the ξ predicted by Rolle's Theorem.

1. $f(x) = x^2 - 13x + 30$, $\qquad 3 \leq x \leq 10$.

2. $f(x) = \dfrac{x(x-2)}{x^2 - 2x + 2}$, $\qquad 0 \leq x \leq 2$.

3. $f(x) = x(x-1)(x-2)$, $\qquad 0 \leq x \leq 1$.

4. $f(x) = x(x-1)(x-2)$, $\qquad 1 \leq x \leq 2$.

5. If $B^2 - 4AC > 0$, and $A > 0$, then the quadratic function $f(x) = Ax^2 + Bx + C$ vanishes at the real points

$$a = \frac{-B - \sqrt{B^2 - 4AC}}{2A} \qquad \text{and} \qquad b = \frac{-B + \sqrt{B^2 - 4AC}}{2A}.$$

According to Rolle's Theorem, $f'(x)$ must vanish once in the open interval $a < x < b$. Show that this is really the case. Describe ξ geometrically.

**6. Let $f(x) = x(x-a)(x-b)$, where $0 < a < b$. Find the roots of $f'(x) = 0$ and prove that they lie in the intervals predicted by Rolle's Theorem. See Problems 3 and 4.

In Problems 7 through 12, find an integral (an antiderivative) for the given function.

7. $g(x) = 6x^2$. $\quad$ 8. $g(x) = 15x^4$. $\qquad$ 9. $g(x) = 3/x^2$.

10. $g(x) = 9\sqrt{x}$. $\quad$ 11. $g(x) = 20(x^3 - x^9)$. $\quad$ 12. $g(x) = ax^n + bx^m$, $m \neq -1$, $n \neq -1$.

13. Prove that the equation

$$6x^5 + 5x^4 + 4x^3 - 9x^2 - 2x + 1 = 0$$

has at least one root in $0 < x < 1$.

The Mean Value Theorem states that for a suitable ξ the tangent line at $(\xi, f(\xi))$ is parallel to the chord. In each of Problems 14 through 18, a curve and the end points of a chord are given. Find a value of ξ satisfying the requirements of the Mean Value Theorem. Sketch the curve subtended by the given chord.

14. $y = x^2$, $\qquad (2, 4)$, $\qquad (3, 9)$. $\qquad$ 15. $y = \sqrt{x}$, $\qquad (25, 5)$, $\qquad (36, 6)$.

16. $y = x^3 - 9x$, $\qquad (-3, 0)$, $\qquad (4, 28)$. $\qquad$ 17. $y = \dfrac{1}{x-7}$, $\qquad (7.1, 10)$, $\qquad (7.2, 5)$.

18. $y = \dfrac{1}{x-7}$, $\qquad (7.01, 100)$, $\qquad (7.02, 50)$.

19. Use Theorem 10 to prove that the two functions

$$F(x) = \frac{3x + 5}{x - 7} \quad \text{and} \quad G(x) = \frac{-x + 33}{x - 7}$$

differ by a constant, by proving that $F'(x) \equiv G'(x)$. Find the constant.

20. Do Problem 19 for

$$F(x) = -\frac{2 + 5x - 10x^2}{1 + 3x - 5x^2 + 4x^3} \quad \text{and} \quad G(x) = \frac{x(1 + 8x^2)}{1 + 3x - 5x^2 + 4x^3}.$$

*21. If $y = mx + b$, then $y'' = 0$. Prove conversely that if $y'' = 0$ for all x in an interval $\mathscr{I}$, then $y = mx + b$ for all x in $\mathscr{I}$.

*22. Use the Mean Value Theorem to prove that under suitable conditions

$$f'(x_2) - f'(x_1) = (x_2 - x_1)f''(\xi), \qquad x_1 < \xi < x_2.$$

What are the conditions?

23. Use Rolle's Theorem to prove that if $a > 0$, then the cubic equation

$$x^3 + ax + b = 0$$

cannot have more than one real root no matter what value is assigned to b.

24. Is the following assertion correct? A necessary and sufficient condition that a given function $f(x)$ is a constant for all points in an interval is that $f'(x)$ is zero throughout that interval.

25. The function

$$f(x) = \begin{cases} x, & \text{if } 0 \leq x \leq 1, \\ 2 - x, & \text{if } 1 \leq x \leq 2 \end{cases}$$

is zero at $x = 0$ and $x = 2$. Therefore, by Rolle's Theorem the derivative should be zero at least once in the interval $0 < x < 2$. But it is not zero at any point of this interval. Where lies the trouble? HINT: Make a graph of this function.

26. Repeat Problem 25 for the function

$$f(x) = \begin{cases} 0, & \text{if } x = 1, \\ x - 2, & \text{if } 1 < x < 3, \\ 0, & \text{if } x = 3, \end{cases}$$

and the interval $1 < x < 3$.

CALCULATOR PROBLEMS

In Problems C1 through C4, give answers to four decimal places.

C1. Show that $f(x) = x^4 - 15x^3 + 51x^2 - 15x + 50$ satisfies Rolle's Theorem for the interval $5 \leq x \leq 10$ and find a root of $f'(x) = 0$ in that interval.

C2. Do Problem C1 for $f(x) = \dfrac{x^2 - 10x}{x^4 + 1}$ in the interval $0 \leq x \leq 10$.

C3. A sketch of the graph of the function in Problem C2 shows that $f'(x) = 0$ should have a second root ξ_2, with $\xi_2 > 10$, and a third root $\xi_3 < 0$. Find ξ_2.

C4. Find ξ_3 as described in Problem C3.

Concave Curves and Inflection Points 8

Let us consider a curve, the graph of $y = f(x)$ in an interval $a \leq x \leq b$, and let us pick two values of x, say x_1 and x_2, such that $a \leq x_1 < x_2 \leq b$. Then the line segment joining the points $(x_1, f(x_1))$ and $(x_2, f(x_2))$ on the curve is called a *chord* of the curve (or a chord of the function) in that interval (Fig. 9). Now a chord may either cut the curve in a third point, or it may lie entirely above the curve (except for its end points), or it may lie entirely below the curve (except for its end points). This suggests

---

DEFINITION 5 (Concave Curve). A curve (or function) is said to be *concave upward* in an interval if every chord in that interval lies above the curve except for its end points (see Fig. 9). The function is said to be *concave downward* if every such chord, except for its end points, lies below the curve.

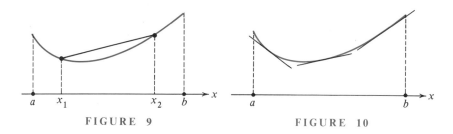

FIGURE 9   FIGURE 10

If we look for an analytic test for curves that are concave upward, it is easy to find one. Suppose that we draw the tangent lines to the curve at a sequence of points along the curve, as shown in Fig. 10. Then it is obvious from the graph that the slope of the tangent line is increasing. For the three lines shown in Fig. 10, the slope of the first is negative, the slope for the second one is positive but small, and the slope of the third one is positive and larger. But if the slope is increasing, then the derivative of the slope,

$$\frac{dm}{dx} = \frac{d}{dx}\left(\frac{dy}{dx}\right) = \frac{d^2y}{dx^2},$$

should be positive. These geometrical considerations suggest

THEOREM 11 (PWO). If $f''(x) > 0$ in an interval $a \leq x \leq b$, then the curve $y = f(x)$ is concave upward in that interval, if $f''(x) < 0$ in $a \leq x \leq b$, then the curve $y = f(x)$ is concave downward in that interval.

This result is quite easy to remember using the following device:

Second derivative positive.	Second derivative negative.
Bowl will hold water.	Bowl won't hold water.
Curve is concave upward.	Curve is concave downward.

Actually, we can go a little further and allow the second derivative to be zero at isolated points in the interval, and still further refinement is possible. Also, Theorem 11 has a partial converse: If the curve $y = f(x)$ is concave upward and if $f(x)$ is twice differentiable, then $f''(x) \geq 0$. Similarly, if the curve is concave downward, then $f''(x) \leq 0$.

Although Fig. 9 shows a nice smooth curve, Definition 5 is formulated so that it applies to an arbitrary curve (the function might not have a second derivative at some points, and yet the graph might still be concave).

Example 1. Discuss the concavity of the curve

(40) $$y = f(x) = \frac{x^3 - 18x^2 + 81x + 36}{18}.$$

Solution. Differentiating, we find that

$$\frac{dy}{dx} = f'(x) = \frac{1}{18}(3x^2 - 36x + 81),$$

(41) $$f'(x) = \frac{1}{6}(x^2 - 12x + 27) = \frac{1}{6}(x - 3)(x - 9).$$

So the critical points are $x = 3$ and $x = 9$. Differentiating again, we obtain

(42) $$\frac{d^2y}{dx^2} = f''(x) = \frac{1}{6}(2x - 12) = \frac{1}{3}(x - 6).$$

For $x > 6$, we have $f''(x) > 0$, and hence for $x > 6$ the curve is concave upward. Similarly, for $x < 6$, $f''(x) < 0$, and the curve is concave downward.

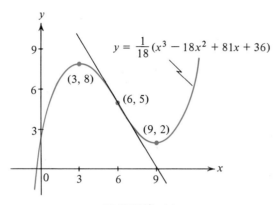

FIGURE 11

The curve is shown in Fig. 11. For the critical points we have

$$f(3) = \frac{1}{18}(3^3 - 6 \times 3^3 + 9 \times 3^3 + 36) = \frac{1}{18}(4 \times 3^3 + 4 \times 3^2) = 8,$$

$$f(9) = \frac{1}{18}(9^3 - 2 \times 9^3 + 9^3 + 36) \qquad = \frac{1}{18}(36) = 2. \; \bullet$$

What can we say about the point $(6, 5)$? From the figure it appears that the tangent line at $(6, 5)$ actually cuts through the curve, part of the curve lying on one side of the tangent line and part on the other side. It is clear that this occurs because for $x < 6$, the slope is decreasing; that is, the tangent line is turning in a clockwise direction as x increases, while for $x > 6$, the slope is increasing; that is, the tangent line is turning in a counterclockwise direction as x increases. Such a point is called an *inflection point*. More precisely, we have

---

DEFINITION 6 (Inflection Point). A point (x_1, y_1) on the curve $y = f(x)$ is called an inflection point of the curve if there is an interval $a \leqq x \leqq b$ around x_1 such that in that interval the derivative is increasing on one side of x_1, and decreasing on the other side.

Two cases are possible:

(I) $f'(x)$ is increasing for $a \leqq x < x_1$, and
$f'(x)$ is decreasing for $x_1 < x \leqq b$.

(II) $f'(x)$ is decreasing for $a \leqq x < x_1$, and
$f'(x)$ is increasing for $x_1 < x \leqq b$.

THEOREM 12. If (x_1, y_1) is an inflection point of the curve $y = f(x)$ and if the second derivative $f''(x)$ is continuous at x_1, then $f''(x_1) = 0$.

Proof.[1] By definition, $f'(x)$ is increasing on one side of x_1, so $f''(x) \geqq 0$ on that side of x_1; and $f'(x)$ is decreasing on the other side of x_1, so $f''(x) \leqq 0$ on that side of x_1. Since $f''(x)$ is continuous at x_1, it must be zero at x_1. ∎

In Example 1, we suspect that the point $(6, 5)$ is an inflection point. It is clear from equation (42) that $f'(x)$ is decreasing for $x < 6$ and increasing for $x > 6$. Thus by Definition 6, this point is indeed an inflection point. But then Theorem 12 asserts that $f''(6) = 0$, and this is certainly the case by inspection of equation (42).

The condition that $f''(x_1) = 0$ is a necessary condition for an inflection point, but it is not a sufficient one. For example, consider the function $y = x^4$, and its graph, shown in Fig. 12. Here we have $f'(x) = 4x^3$, $f''(x) = 12x^2$, and hence $f''(x) = 0$ if $x = 0$. But $(0, 0)$

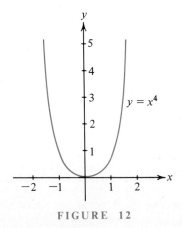

$y = x^4$

FIGURE 12

---

[1] We use the fact that a differentiable function that is increasing in an interval has a nonnegative derivative in that interval (see Problem 28, Exercise 3).

is not an inflection point on the curve because the tangent line in this case is just the x-axis, and it lies entirely on one side of the curve, except of course for the point of tangency.

This same curve illustrates another subtle point: The second derivative may vanish at points in an interval of concavity. Clearly, the curve $y = x^4$ is concave upward for all x. But we cannot say that therefore $f''(x) > 0$ because in this example $f''(0) = 0$.

THEOREM 13. If the curve $y = f(x)$ is concave upward in $a \leq x \leq b$ and if the second derivative $f''(x)$ is continuous in $a \leq x \leq b$, then $f''(x) \geq 0$ for $a \leq x \leq b$. If the curve is concave downward, then $f''(x) \leq 0$ in that interval.

Proof. First assume that the curve $y = f(x)$ is concave upward. If there is a point c in $\mathscr{I}: a \leq x \leq b$ for which $f''(c) < 0$, then, by continuity, $f''(x)$ would be negative on some interval $\mathscr{I}^\star$ containing c. Then by Theorem 11 the curve $y = f(x)$ is concave downward in $\mathscr{I}^\star$. This is a contradiction to the assumption that the curve is concave upward. Hence $f''(x) \geq 0$ in $\mathscr{I}$.

Similarly, the assumption that the curve $y = f(x)$ is concave downward and that $f''(c) > 0$ leads to a contradiction, so that if the curve $y = f(x)$ is concave downward in $\mathscr{I}$, then $f''(x) \leq 0$ in $\mathscr{I}$. ∎

The concepts developed in this section are quite helpful in sketching the graph of a function. Perhaps the following outline may be useful as a guide.

A. Given $f(x)$, compute (carefully) $f'(x)$ and $f''(x)$. Note the domain in which $f(x)$, $f'(x)$, and $f''(x)$ are defined and continuous.

B. Find the zeros of $f'(x)$ and use these zeros (the critical points) to find the intervals in which $f(x)$ is increasing and the intervals in which $f(x)$ is decreasing.

C. Use the information from B to determine all relative maximum points and all relative minimum points.

D. Find all zeros of $f''(x)$ and use these zeros to find the intervals in which $f(x)$ is concave upward, and the intervals in which $f(x)$ is concave downward.

E. Use the information from D to determine the inflection points.

F. Determine the limits of $f(x)$ as $x \to \infty$ and $x \to -\infty$.

G. Find all points where the curve crosses the axes.

From the information developed above, it should be easy to sketch the graph of $y = f(x)$. One can always compute a few points to check the work.

EXERCISE 5

In Problems 1 through 20, determine the intervals in which the curve is concave upward, and locate all the points of inflection.

1. $y = x^2 + 6x - 13$.

2. $y = 25 - 8x - x^2$.

3. $y = x(x^2 - 9)$.

4. $y = \dfrac{1}{x}$.

5. $y = x^{2/3}$.

6. $y = x^3 - 6x^2 + 17$.

7. $y = x^5$.

8. $y = x^6$.

9. $y = x^3 - 3x^2 - 9x$.

10. $y = x^4 - 12x^3 + 48x^2$.

11. $y = \dfrac{4}{1 + x^2}$.

12. $y = \dfrac{10x}{1 + 3x^2}$.

13. $y = x + \dfrac{1}{x}$.

14. $y = 3x^2 - \dfrac{16}{x^2}$.

★15. $y = x^5(x - 6)^5$.

★16. $y = (x + 1)^3(x - 5)^6$.

17. $y = \dfrac{x^3}{x^2 + 3}$.

18. $y = \dfrac{4}{\sqrt{x}} + \dfrac{\sqrt{x}}{3}$.

19. $y = x - \dfrac{4}{x^2}$.

★20. $y = x^{2/3}(x - 40)$.

21. Sketch the graphs of some of the functions given in Problems 1 through 20. Note that it may be helpful to use different scales on the two axes.

★22. Determine the constants so that $y = Ax^3 + Bx^2 + Cx + D$ has a relative maximum point at $(1, 16)$ and a relative minimum point at $(5, -16)$. Where is the inflection point?

★★23. Let (x_1, y_1) and (x_2, y_2) be relative extreme points for an arbitrary cubic: $y = Ax^3 + Bx^2 + Cx + D$. If (x_3, y_3) is the inflection point, prove that x_3 bisects the segment $[x_1, x_2]$ and y_3 bisects the segment $[y_1, y_2]$. HINT: Is it possible to move the curve (or the x-axis) so that $x_1 = -a$ and $x_2 = a$?

9 The Second Derivative Test

The second derivative furnishes a convenient method for determining which of the critical points are relative maximum points, and which are relative minimum points. For if a curve is concave upward in a neighborhood of (x_1, y_1) and if $f'(x_1) = 0$, it is obvious that (x_1, y_1) is a relative minimum point. If the curve is concave downward, then (x_1, y_1) is obviously a relative maximum point. Stated in terms of the second derivative, we have

---

THEOREM 14. Suppose that $f''(x)$ is continuous in a neighborhood of x_1 and that $f'(x_1) = 0$. If $f''(x_1)$ is positive, then the point (x_1, y_1) is a relative minimum point for the curve, $y = f(x)$, where $y_1 = f(x_1)$. If $f''(x_1)$ is negative, then the point (x_1, y_1) is a relative maximum point.

This criterion is easy to recall if we keep in mind that $f''(x_1)$ positive means the bowl will hold water, and that $f''(x_1)$ negative means the bowl will not hold water.

Example 1. Locate the extreme points for the curve

$$(43) \qquad\qquad y = x^4 - 4x^3 - 2x^2 + 12x - 5.$$

Solution. Differentiating equation (43), we obtain

$$(44) \qquad \frac{dy}{dx} = f'(x) = 4x^3 - 12x^2 - 4x + 12 = 4(x^3 - 3x^2 - x + 3).$$

Differentiating again, we have

$$(45) \qquad\qquad \frac{d^2y}{dx^2} = f''(x) = 4(3x^2 - 6x - 1).$$

The critical values for x are obtained by solving

$$f'(x) = 0 \qquad \text{or} \qquad x^3 - 3x^2 - x + 3 = 0.$$

Factoring, we find that

$$f'(x) = x^2(x - 3) - (x - 3) = (x^2 - 1)(x - 3) = (x - 1)(x + 1)(x - 3).$$

Thus the critical values for x are $x = -1, 1,$ and 3. The corresponding y values are $-14, 2,$ and -14, respectively. For the second derivative at $x = -1, 1,$ and 3, equation (45) gives

$$f''(-1) = 4[3 \times (-1)^2 - 6 \times (-1) - 1] = \quad 32 > 0,$$
$$f''(1) \ = 4[3 \times (1)^2 \quad - 6 \times 1 \quad - 1] = -16 < 0,$$
$$f''(3) \ = 4[3 \times (3)^2 \quad - 6 \times 3 \quad - 1] = \quad 32 > 0.$$

Therefore, the two points $(-1, -14)$ and $(3, -14)$ are relative minimum points, and $(1, 2)$ is a relative maximum point. ●

Example 2. Locate the extreme points for the curve $y = x^4$ shown in Fig. 12.

Solution. Since x^4 is always positive or zero, and is zero only at $x = 0$, the curve has an absolute minimum at $x = 0$, as well as a relative minimum there.

If we apply the calculus, we find that $f'(x) = 4x^3$ and this is zero only at $x = 0$. Hence the only extreme point is the minimum at $(0, 0)$. ●

What does the Second Derivative Test tell us? Since $f''(x) = 12x^2$, and this is zero at $x = 0$, the test gives no information, even though the point $(0, 0)$ is obviously a minimum point.

We conclude from this example that $f''(x_1) > 0$ and $f'(x_1) = 0$ together form a sufficient condition for a relative minimum at x_1, but not a necessary condition.

EXERCISE 6

Use the second derivative test, whenever it is applicable, to locate the relative maximum points and relative minimum points, for each of the functions given in Exercise 5.

10 *Applications of the Theory of Extremes*

So far we have been concerned with maximum and minimum points on a curve. Now there are many quite natural problems that arise in the physical world that can be reduced to the problem of finding a maximum or minimum point on a curve. Thus we are in a position to solve such problems quite simply. We first give a number of concrete examples, and then we will formulate some general principles based on the experience gained through the examples.

Example 1. Find two numbers whose sum is 24 and whose product is as large as possible.

Solution. Let x and y be the numbers. Then the conditions of the problem state that

(46) $$x + y = 24.$$

If we let P denote the product, then we are to maximize

(47) $$P = xy.$$

One might expect to begin by differentiating P, since we seek its maximum. But be careful! P depends on *two* variables, and so far our calculus has all been developed for a *single* independent variable. So we must first alter P in some way so that it will depend on just one variable. Equation (46) gives us the means to do this. Solving (46) for y, we have $y = 24 - x$. Substituting in (47), we obtain

(48) $$P = x(24 - x) = 24x - x^2,$$

a function of a single variable. Differentiating twice, we find that

$$\frac{dP}{dx} = 24 - 2x, \qquad \frac{d^2P}{dx^2} = -2.$$

Clearly, the derivative is zero when $24 - 2x = 0$, or when $x = 12$. Since the second derivative is negative, we have located a maximum point. When $x = 12$, equation (46) gives $y = 12$. Thus the point (12, 144) is a relative maximum point on the curve of equation (48). But more than that it is an absolute maximum. We leave the proof of this last statement to the student. Thus the maximum product is 144, and it occurs when 24 is split into 12 and 12. ●

Example 2. A farmer with a field adjacent to a straight river wishes to fence a rectangular region for grazing. If no fence is needed along the river, and the farmer has available 1600 meters of fencing, what should be the dimensions of the field in order that it have a maximum area?

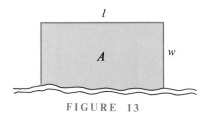

FIGURE 13

Solution. Let l and w denote the length and width of the field, respectively, and let A be the area (see Fig. 13). Using all the fencing available gives

(49) $$l + 2w = 1600.$$

We are to maximize

(50) $$A = lw.$$

This is a function of two variables, but solving equation (49) for l we find that $l = 1600 - 2w$, and using this in (50), we obtain

$$A = w(1600 - 2w) = 1600w - 2w^2.$$

Then

$$\frac{dA}{dw} = 1600 - 4w, \qquad \frac{d^2A}{dw^2} = -4.$$

The derivative is zero at $w = 400$, and the second derivative is negative, so we have a maximum. The dimensions of the field should be 400 by 800 meters and the maximum area is $400 \times 800 = 320{,}000$ m². ●

Example 3. The strength of a rectangular beam varies directly as the width and the square of the depth. What are the dimensions of the strongest rectangular beam that can be cut from a cylindrical log of radius r?

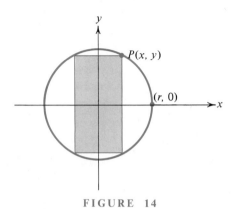

FIGURE 14

Solution. We place the cross section of the log on a rectangular coordinate system as shown in Fig. 14. Let (x, y) denote the coordinates of the point P in the first quadrant, where the corner of the rectangle lies on the circle. Then the width of the beam is $2x$ and the depth is $2y$. If S denotes the strength of the beam, then

(51) $$S = C(2x)(2y)^2 = 8Cxy^2,$$

where C is a positive constant that depends on the type of wood (small for balsa and larger for oak) but is of no interest here. Since P lies on the circle $x^2 + y^2 = r^2$, we have

(52) $$y^2 = r^2 - x^2,$$

and substituting this in equation (51), we find that

$$S = 8Cx(r^2 - x^2) = 8C(xr^2 - x^3).$$

Differentiating twice gives

(53) $$\frac{dS}{dx} = 8C(r^2 - 3x^2), \qquad \frac{d^2S}{dx^2} = -48Cx.$$

The first derivative is zero for $x = \pm r/\sqrt{3}$. The negative value for x may be rejected because it has no physical meaning in our problem, since the width $2x$ must be positive. At $x = r/\sqrt{3}$, the second derivative is negative, so this value gives the maximum strength. For this x we find that

$$y = \sqrt{r^2 - x^2} = \sqrt{r^2 - \frac{r^2}{3}} = \sqrt{\frac{2}{3}}r \approx 0.8165r.$$

The width is therefore $2r/\sqrt{3} \approx 1.155r$ and the depth is $2r\sqrt{2/3} \approx 1.633r$, for a beam of maximum strength. ●

Example 4. A piece of wire of length L is to be cut into two pieces, and each piece bent so as to form a square. How should the wire be cut if the sum of the areas enclosed by the two squares is to be a maximum?

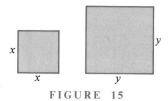

FIGURE 15

Solution. Let x and y be the lengths of a side of the two squares (Fig. 15). Then for the total area, we have

(54) $$A = x^2 + y^2.$$

Since all the material comes from the given wire, the perimeter of the two figures must give L so that

(55) $$L = 4x + 4y.$$

Solving (55) for x, we find $x = L/4 - y$, so equation (54) becomes

(56) $$A = \left(\frac{L}{4} - y\right)^2 + y^2 = \frac{L^2}{16} - \frac{L}{2}y + 2y^2.$$

Differentiating twice, we obtain

(57) $$\frac{dA}{dy} = -\frac{L}{2} + 4y, \qquad \frac{d^2A}{dy^2} = 4.$$

The derivative is zero at $y = L/8$, but this time the second derivative is positive, so we have a *minimum* point. In other words, if we cut the wire in half and form a square from each piece, the squares will each have sides of length $L/8$ and the total area enclosed will be as *small* as possible. This minimum area is $2(L/8)^2$ or $L^2/32$. To find a maximum value, we observe that under the conditions of the problem $0 \leq y \leq L/4$. By symmetry it suffices to check one end point, say $y = L/4$. Then $A = (L/4)^2 = L^2/16$. This is the maximum value of A given by (56) for y in this interval. But the problem states that the wire is to be cut into *two* pieces. If we adhere strictly to the statement of the problem and insist on two pieces, then the problem has *no solution*. For no matter how small we make the x square, we can increase the total area by making x still smaller and y still larger because for y near $L/4$ the derivative dA/dy is positive. ●

Let us summarize in a general form the procedure used in the first four examples.

STEP 1. If Q is the quantity to be maximized or minimized, find an expression for Q involving one or possibly more variables.

In our examples this expression is given by equations (47), (50), (51), and (54).

STEP 2. If the expression for Q involves more than one variable, search for other equations relating the variables. Use these equations to reduce Q to a function of a single variable.

In our examples, equations (46), (49), (52), and (55) play the role of auxiliary equations which are used to simplify the expressions for P, A, S, and A, respectively.

STEP 3. If $Q = Q(x)$, compute the first and second derivatives $Q'(x)$ and $Q''(x)$. Find the critical values of x, that is, the values of x for which $Q'(x) = 0$. Then use the Second Derivative Test on each critical point to determine if it gives a relative minimum or a relative maximum.

In all these physical problems there is a *natural interval* for each of the variables, and it is understood that the variables assume only such values as are physically meaningful. In most problems it is physically obvious that the extreme values exist, and occur for special values inside the natural interval of the variable. But we must be careful, because as we saw in Example 4, an extreme value may occur at the end point of an interval. Hence in all problems one should examine the end points.

To illustrate these vague and general remarks about the natural interval, let us reexamine our examples.

In Example 1, $x + y = 24$ and we are to maximize the product $P = xy$. Here x and y could be any real numbers, but since the product is negative if either x or y are negative (they cannot both be negative), it is obvious that no harm is done if we insist that x and y both be nonnegative. Then the natural interval is $0 \leq x \leq 24$ for x, and the same for y. At the end points, the product $P = 0$, so the end points cannot furnish a maximum.

In Example 2, the length and width of the field must both be positive, so the natural intervals are $0 < l < 1600$, and $0 < w < 800$. At either end point of these two intervals, the area, as given by equation (50), is zero.

In Example 3, the natural intervals for the width $2x$ and depth $2y$ of the beam are $0 < 2x < 2r$ and $0 < 2y < 2r$, respectively. At either end point of these two intervals, $S = 0$.

In Example 4, the natural intervals for the variables x and y are $0 < x < L/4$ and $0 < y < L/4$. But this time the expression to be maximized, $x^2 + y^2$, has a larger value at either end point of these intervals than at any interior point. But since the end points are not admitted in the problem, the problem as stated has no solution.

STEP 4. Determine the natural intervals for the variables involved in the problem, from the physical meaning of the variables. Check the values of Q at the end points of these natural intervals.

EXERCISE 7

1. A square piece of tin 18 in. on each side is to be made into a box, without a top, by cutting a square from each corner and folding up the flaps to form the sides (see Fig. 16). What size corners should be cut in order that the volume of the box be as large as possible?

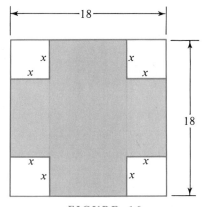

FIGURE 16

2. Solve Problem 1 if the given piece of tin is a rectangle 24 by 45 cm.
3. Prove that of all rectangles inscribed in a given fixed circle (see Fig. 14), the square has the largest area.
4. Prove that if we maximize the perimeter of the rectangle instead of the area in Problem 3, the solution is still the square.
*5. A generalization of Problem 4. Let l and w be the length and width of a rectangle inscribed in a circle of fixed radius, and suppose that we are to find the rectangle that maximizes the quantity

$$P = l^n + w^n,$$

where n is some fixed positive rational number. Prove that if $n < 2$, then the square makes P a maximum, but if $n > 2$, then the maximum is given by the degenerate rectangle, in which l or w is the diameter of the circle while the other dimension is zero. Note that if $n = 2$, then P is the same for all rectangles.

6. The stiffness of a rectangular beam varies directly with the width and the cube of the depth. Find the dimensions of the stiffest beam that can be cut from a cylindrical log of radius R.

7. Three planks, each 50 cm wide, are made into a trough. If the cross section of the trough is in the form of a trapezoid, how far apart should the top of the planks be set in order that the trapezoid have a maximum area?

8. Find the maximum of the area for all circular sectors of given fixed perimeter P. HINT: Use radian measure for the angle of the sector.

9. Find two positive numbers whose product is 100 and whose sum is as small as possible.

*10. A closed cylindrical can is to contain a certain fixed volume V. What should be the ratio of the height to the radius of the can in order that the can requires the least amount of material, that is, in order that the surface area be a minimum?

11. A closed rectangular box has a square base and has a fixed volume. What is the ratio of the height to a side of the base in order that the surface area of the box be a minimum?

12. Solve Problem 11 if the box is open on top.

13. Find the altitude of the cylinder of maximum volume that can be inscribed in a right circular cone of height 12 and radius of base 7 (see Fig. 17).

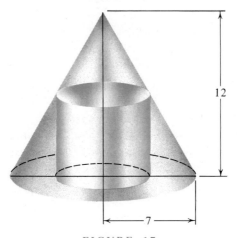

FIGURE 17

14. Solve Problem 13 if the height of the cone is H and the radius of the base is R.

15. Find the altitude of the cylinder of maximum volume that can be inscribed in a sphere of radius R.

16. Find the altitude of the cone of maximum volume that can be inscribed in a sphere of radius R.

*17. Prove that if a right circular cone is circumscribed about a sphere, the volume of the cone is greater than or equal to twice the volume of the sphere.

18. Find two positive numbers x and y such that their sum is 60 and the product xy^3 is a maximum.

19. Find two positive numbers x and y such that their sum is 35 and the product x^2y^5 is a maximum.

★20. Generalization of Problems 18 and 19. Let p and q be fixed positive rational numbers. Show that among all pairs of positive numbers x and y such that $x + y = S$, the maximum of the product $x^p y^q$ is obtained when $x = pS/(p + q)$ and $y = qS/(p + q)$.

21. Find two positive numbers whose sum is 16 and the sum of whose cubes is a maximum.

22. Solve Problem 21 if we are to minimize the sum of the cubes.

23. Find two numbers whose sum is 18 and for which the sum of the fourth power of the first and the square of the second is a minimum.

24. Find the point on the parabola $y = x^2$ that is closest to the point $(10, 2)$.

25. Find the points on the parabola $8y = x^2 - 40$ that are closest to the origin. What is the radius of a circle with center at the origin that is tangent to this parabola?

26. Plans for a new supermarket require a floor area of 900 m². The supermarket is to be rectangular in shape with three solid brick walls and a very fancy all-glass front. If glass costs 1.88 times as much as the brick wall per linear meter, what should be the dimensions of the building so that the cost of materials for the walls is a minimum?

27. Suppose that in the supermarket of Problem 26 the heat loss across the glass front is seven times as great as the heat loss across the brick per square meter. Neglecting the heat loss across the roof and through the floor, what should be the dimensions of the building so that the heat loss is a minimum?

★28. Suppose that the architects for the supermarket of Problems 26 and 27 wish to take account of the heat losses across the floor and ceiling and that the ceiling is to be 6 m high. Suppose further that the rate of heat loss per square meter of ceiling is K_1 times the rate of heat loss per square meter of brick wall, and that for the floor the multiplier is K_2. Find the dimensions of the supermarket that minimize the heat loss.

★29. Let us call space in an attic *livable* if a 6-ft man can stand upright without bumping his head. As indicated in Fig. 18, let s be the slant height of the roof and suppose that the cost of building the roof is proportional to the material used and this in turn is proportional to s. If the base is 36 ft, find the slope of the roof that minimizes the cost per square foot of livable attic space. HINT: Show that this amounts to minimizing s/x.

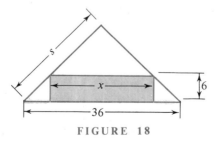

FIGURE 18

30. A certain handbill requires 600 cm² for the printed message and must have a 6-cm margin at the top and bottom and a 4-cm margin on each side. Find the dimensions of this handbill if the amount of paper used is a minimum.

*31. A certain stained glass window consists of a rectangle together with a matching semicircle set on the upper base of the rectangle. If the perimeter of the window is fixed at P, find the height of the rectangle and the radius of the semicircle for the window that will let in the most light (have maximum area).

32. A man at a point A on one shore of a lake 6 km wide with parallel shore lines wishes to reach a point C on the other side 13 km along the bank of the lake from a point B directly opposite the point A. If he can row 4 km/hr and walk 5 km/hr and if he sets out by boat, find how far from B he should land in order to make the trip as quickly as possible. How long does the trip take? How much longer does it take if he rows first to B and then walks to C?

33. If light travels from a point A to a point P on a plane mirror and is then reflected to a point B, the most careful measurements seem to show that the angle of incidence equals the angle of reflection. With the lettering of Fig. 19, this states that $\angle CPA = \angle DPB$. Make the assumption that light always takes the shortest path (in air) and then prove this law by showing that the path APB is shortest when $a/x = b/(l - x)$.

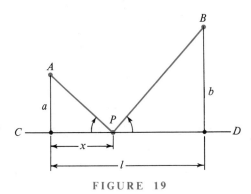

FIGURE 19

34. The illumination at a point P due to a light source is directly proportional to the strength of the source and inversely proportional to the square of the distance of P from the source. Two light sources of strength A and B, respectively, are distance L apart. Find that point on the line segment joining the two sources where the total illumination is a minimum.

35. A ship A is 40 km due west of a ship B and A is sailing east at 12 km/hr. At the same time, B is sailing north at 16 km/hr. Find the minimum distance between the two ships.

36. Among all rectangles with sides parallel to the axes and inscribed in the ellipse

$$\frac{x^2}{a^2} + \frac{y^2}{b^2} = 1,$$

find the dimensions of the one with the largest area.

37. Find the minimum value of $x^2 + 2x + 7$ without using calculus. HINT: Use the identity $x^2 + 2x + 7 = (x + 1)^2 + 6$.

In Problems 38 through 41, find the extreme values of the given function without using calculus.

38. $x^2 + 6x - 11$.

39. $20 + 10x - x^2$.

40. $\dfrac{10}{x^2 - 2x + 3}$.

41. $\dfrac{18}{x^2 + 5x + 7}$.

CALCULATOR PROBLEMS

In Problems C1 through C4, use a calculator to estimate the answer to three significant figures.

C1. Find the point on the curve $y = x^2$ that is closest to: **(a)** $P(10, 0)$, **(b)** $P(10, 5)$, and **(c)** $P(3, 1)$. HINT: The critical points of $f(x) = \sqrt{(x - 10)^2 + x^4}$ are the same as those of $F(x) = (x - 10)^2 + x^4$.

C2. Do Problem C1 for the curve $y = x^3$.

C3. Do Problem 2 of this exercise if the piece of tin is 1 meter by 50 cm.

C4. Do Problem 29 of this exercise if the base of the house is 40 ft rather than 36 ft.

Differentials 11

We have already mentioned that the symbol

(58) $$\frac{dy}{dx},$$

introduced for the derivative, looks like a fraction but is not one. It has the appearance of a fraction, and in certain circumstances actually acts like one. For example, if we use cancellation (as in a fraction) in

$$\frac{dy}{du}\frac{du}{dx} = \frac{dy}{du}\frac{du}{dx} = \frac{dy}{dx},$$

we obtain the correct formula for the derivative of a composite function.

Our present objective is to give a meaning to the pieces of (58), namely, a meaning to dy and dx so that their quotient is indeed the derivative $f'(x)$.

Since the derivative $f'(x)$ is the limit value of the ratio $\Delta y/\Delta x$, the difference of these two quantities is tending to zero as $\Delta x \to 0$. Let ϵ (Greek lowercase letter epsilon) denote this difference. In other words, let

(59) $$\frac{\Delta y}{\Delta x} = f'(x) + \epsilon, \qquad \Delta x \neq 0,$$

where $\epsilon \to 0$ as $\Delta x \to 0$. Multiplying through by Δx gives

(60) $$\Delta y = f'(x)\,\Delta x + \epsilon\,\Delta x.$$

If we drop the last term in (60) and replace the Greek Δ by the corresponding English *d*, we obtain

(61) $$dy = f'(x)\,dx,$$

where the meaning of dx and dy is still to be explained. During the first century of the development of the calculus, the quantities dx and dy (called *differentials*) were regarded as "vanishingly small" or "infinitesimal" and the term $\epsilon\,\Delta x$ could be dropped in (60) because it was "an infinitesimal of higher order." Although these terms are vague and mystic, they served remarkably well in guiding mathematicians of those days toward the solution of difficult problems and the discovery of new and important results.

When we try to frame a definition of dx and dy, we must reject the idea of a vanishingly small number, and consequently we are forced to admit that dx may be any number. We thus arrive at

---

DEFINITION 7 (The Differential of *x*). If x is an independent variable, then dx is a second variable and consequently may be any real number. The quantity dx is called the *differential of x*.

Although dx is just another real variable, in our applications we will use it as a small change in x, and under these conditions we have $dx = \Delta x$.

DEFINITION 8 (The Differential of *y*). If y is a differentiable function of x, then dy, the *differential of y*, is defined by the equation $dy = f'(x)\,dx$.

Why make these two definitions? Answer: Now dy and dx each have a meaning, and if $dx \neq 0$, we see from (61) that if dy is divided by dx, we do get $f'(x)$, the derivative of y with respect to x.

Further, if dx is reasonably small (the size here depends on the problem under consideration), then dy is very close to Δy, the actual change in y. For if $\Delta x = dx$, then equations (60) and (61) yield

(62)
$$\Delta y = dy + \epsilon \, \Delta x.$$

From (62) it is clear that as $\Delta x \to 0$, the difference $\Delta y - dy$ is tending to zero much more rapidly than either Δy or dy because $\epsilon \, \Delta x$ is the product of two quantities, ϵ and Δx, that are both approaching zero.

It may be helpful to picture these quantities on the graph of $y = f(x)$ at some fixed point P. Referring to Fig. 20, we let Δx be represented by the directed line segment PR.

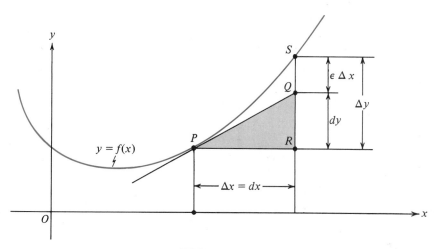

FIGURE 20

Then Δy is represented by the directed line segment RS, while $dy = f'(x) \, \Delta x$ is just the directed line segment RQ, and $\epsilon \, \Delta x$ is the directed line segment QS. Briefly,

$\Delta y = RS$ is the change in y along the curve,
$dy = RQ$ is the change in y along the tangent line to the curve at the point $P(x, f(x))$.

Although in Fig. 20, $\epsilon \, \Delta x$ appears to be of the same order of magnitude as Δx or Δy, it actually tends to zero much more rapidly than Δx. This will be illustrated in Example 1.

Differentials are frequently useful in approximations. Let $y = f(x)$ and suppose that $f(x_0) = y_0$ is known and we wish to compute $f(x_0 + \Delta x)$. We obtain an approximation for $y_0 + \Delta y$, when we replace Δy by dy. When Δx is small, the approximation is usually quite close. The error is just

(63)
$$\Delta y - dy = \epsilon \, \Delta x$$

and the *absolute error is* $|\Delta y - dy|$. As we have already remarked, the right side of (63) is usually much smaller than either term on the left side of (63).

Example 1. For the function $y = x^2$, find Δy and dy **(a)** in general, and **(b)** at $x_0 = 3$. **(c)** At $x_0 = 3$, compute Δy and dy for $\Delta x = 0.1, 0.01$, and 0.001. Find the error when Δy is replaced by dy.

Solution. **(a)** If $y = x^2$, then at $x + \Delta x$ we have

$$\Delta y = (x + \Delta x)^2 - x^2 = x^2 + 2x\, \Delta x + (\Delta x)^2 - x^2$$
$$= 2x\, \Delta x + (\Delta x)^2.$$

By contrast,

$$dy = f'(x)\, dx = 2x\, dx = 2x\, \Delta x.$$

Consequently,

$$\Delta y - dy = 2x\, \Delta x + (\Delta x)^2 - 2x\, \Delta x = (\Delta x)^2.$$

Hence in this case

$$\epsilon\, \Delta x = (\Delta x)^2.$$

(b) In particular, at $x_0 = 3$, we have $\Delta y = 6\, \Delta x + (\Delta x)^2$, $dy = 6\, \Delta x$, and hence $\Delta y - dy = (\Delta x)^2$.

(c) The results for specific values of Δx are presented in the following table:

Δx	$\Delta y = 6\, \Delta x + (\Delta x)^2$	$dy = 6\, \Delta x$	Error, $\Delta y - dy = \epsilon\, \Delta x$
0.1	0.61	0.6	0.01
0.01	0.0601	0.06	0.0001
0.001	0.006001	0.006	0.000001

It is clear that the entries in the fourth column are much smaller than the corresponding entries in the second or third columns, just as we expected they would be. This shows that in this example it is quite satisfactory to replace Δy by dy if Δx is small. ●

Example 2. Find an approximate value for $\sqrt[3]{26.5}$, without using tables or a calculator.

Solution. We observe that 26.5 is close to 27, a perfect cube, so this suggests the use of differentials. This will tell us approximately how much $\sqrt[3]{x}$ is changing as x changes from $x_0 = 27$ to $x_0 + dx = 26.5$. Since x is decreasing, we will have $dx = -0.5$ in this case. Set

(64) $$y = f(x) = \sqrt[3]{x} = x^{1/3}.$$

Then at x_0,

$$\frac{dy}{dx} = f'(x_0) = \frac{1}{3}x_0^{-2/3} = \frac{1}{3\sqrt[3]{x_0^2}},$$

$$dy = f'(x_0)\,dx = \frac{dx}{3\sqrt[3]{x_0^2}}.$$

Setting $x_0 = 27$, and $dx = -0.5$, we have

$$dy = \frac{-0.5}{3\sqrt[3]{27^2}} = \frac{-0.5}{3 \times 9} = -\frac{1}{54}.$$

We use the symbol $\approx$ to denote approximate equality. Then

$$\sqrt[3]{x_0 + \Delta x} = \sqrt[3]{26.5} = \sqrt[3]{27} + \Delta y \approx \sqrt[3]{27} + dy$$

$$\approx 3 - \frac{1}{54} \approx 3 - 0.0185 = 2.9815. \quad \bullet$$

A calculator or tables gives $\sqrt[3]{26.5} = 2.98137$ to six significant figures. The method of differentials gives the first four figures correctly, and this is quite good. The absolute error here is less than 0.0002.

In many cases, the magnitude of the error is not too important. It is more important to know the ratio of the error to the quantity being measured or computed. This leads to

---

DEFINITION 9 (Relative Error). In the estimation or measurement of any quantity,

$$\text{relative error} = \left| \frac{\text{error}}{\text{correct value of the quantity}} \right|.$$

The *percent error* is $100 \times$ (relative error).

Of course, Definition 9 is difficult to use because often we do not know the denominator, and hence we do not know the numerator. But in practical applications we are content with an estimate for the relative error.

Example 3. Estimate to three significant figures the relative error and the percent error for the quantities computed in Examples 1 and 2.

Solution. We arrange the results in a table.

Example	Δx	$y + \Delta y$	$y + dy$	Error, $\epsilon \Delta x = \Delta y - dy$	Relative error, $\lvert \epsilon \Delta x/(y + \Delta y)\rvert$	Percent error
1	0.1	9.61	9.6	0.01	0.00104	0.104
1	0.01	9.0601	9.06	0.0001	0.0000110	0.00110
1	0.001	9.006001	9.006	0.000001	0.000000111	0.0000111
2	−0.5	2.981366	2.981481	−0.000115	0.0000386	0.00386

From this table it is clear that the relative error is extremely small, when we replace $y + \Delta y$ by $y + dy$. ●

Now that differentials have been defined, we can have differential formulas as well as derivative formulas. For example, the formula

$$\frac{d}{dx}x^2 = 2x$$

can now be written as

$$d(x^2) = 2x \, dx.$$

Each formula for differentiating an expression gives rise to a corresponding differential formula, by multiplying through by dx. Here are the differentiation formulas obtained so far in this book together with the corresponding differential formulas.

(65) $\qquad \dfrac{dc}{dx} = 0.$ $\qquad\qquad\qquad dc = 0.$

(66) $\qquad \dfrac{d(x^n)}{dx} = nx^{n-1}.$ $\qquad\qquad d(x^n) = nx^{n-1} \, dx.$

(67) $\qquad \dfrac{d(au)}{dx} = a\dfrac{du}{dx}.$ $\qquad\qquad d(au) = a \, du.$

(68) $\qquad \dfrac{d(u^n)}{dx} = nu^{n-1}\dfrac{du}{dx}.$ $\qquad\quad d(u^n) = nu^{n-1} \, du.$

(69) $\qquad \dfrac{d(u + v)}{dx} = \dfrac{du}{dx} + \dfrac{dv}{dx}.$ $\qquad d(u + v) = du + dv.$

(70) $\qquad \dfrac{d(uv)}{dx} = u\dfrac{dv}{dx} + v\dfrac{du}{dx}.$ $\qquad d(uv) = u \, dv + v \, du.$

(71)
$$\frac{d\left(\dfrac{u}{v}\right)}{dx} = \frac{v\dfrac{du}{dx} - u\dfrac{dv}{dx}}{v^2}. \qquad d\left(\frac{u}{v}\right) = \frac{v\,du - u\,dv}{v^2}.$$

EXERCISE 8

In Problems 1 through 6, obtain expressions for: (a) Δy, and (b) $\Delta y - dy$. Show that in each case $(\Delta x)^2$ is a factor of $\Delta y - dy$.

1. $y = 2x^3$.

2. $y = 3x - x^2$.

3. $y = x^3 - 2x$.

4. $y = 3x^2 + 6x + 15$.

5. $y = \dfrac{1}{x^2}$.

6. $y = \dfrac{x}{10 + x}$.

In Problems 7 through 14, use differentials to find an approximate value to four significant figures. Then check your answer with that given by tables or by a calculator.

7. $\sqrt{65}$.

8. $\sqrt[3]{65}$.

9. $\sqrt[6]{65}$.

10. $\sqrt[3]{999}$.

11. $\sqrt{141}$.

12. $\sqrt{26.5}$.

13. $\sqrt{4.12}$.

14. $\sqrt{0.037}$.

15. The approximations are good only if Δx is small. Find $\sqrt{111}$ first by regarding 111 as near 100, and second by regarding 111 as near 121, and compare the results. Tables give $\sqrt{111} = 10.5356.\ldots$.

16. The derivative of the formula $A = \pi r^2$ for the area of a circle with respect to r gives $2\pi r$, the circumference of the circle. Explain this on the basis of differentials. In the same way, explain why the derivative of $4\pi r^3/3$ gives $4\pi r^2$, the surface area of a sphere. Can this method be used to obtain the surface area of a cone from the volume?

17. The side of a cube is measured and found to be 8 cm long. If this measurement is subject to an error of ± 0.05 cm on each edge, find an approximation for the maximum error that is made in computing the volume of the cube. What is the approximate relative error?

18. A stone dropped from a bridge landed in the water 4 sec later. Using $s = 16t^2$, compute the height of the bridge. If the time measured may be off by as much as $1/5$ sec, find an approximation for the maximum error in the computed height. Find the approximate relative error.

19. The range of a gun is given by the formula $x = (V_0{}^2 \sin 2\alpha)/32$, where V_0 is the muzzle velocity in ft/sec, x is in feet, and α is the angle of elevation. To hit a certain target, the powder charge was designed to give $V_0 = 640$ ft/sec. If $\alpha = 15°$, compute the theoretical range. Find an approximation for the error in the range if the initial velocity is off by 1 percent. Find the approximate relative error.

20. Suppose that in Problem 19 the initial velocity is correct as given. Estimate the error made by using $g = 32$ in place of $g = 32.17$. What is the relative error?

21. A circular plate expands under heating so that its radius increases 2 percent. Find an approximation for the change in area if $r = 10$ cm before heating.

CALCULATOR PROBLEMS

C1. If the MKS system is used, the range of a gun is given by the formula $x = (V_0^2 \sin 2\alpha)/9.8$ in meters. Do Problem 19 if $\alpha = 20°$ and $V_0 = 300$ m/sec.

C2. Suppose that in Problem C1 the initial velocity is exactly 300 m/sec. Estimate the error made by using $g = 9.8$ in place of the more accurate $g = 9.8054$. What is the approximate relative error?

C3. The period for one vibration of a pendulum is given by the formula

$$t = \pi \sqrt{\frac{l}{g}},$$

where t is in seconds, g is the gravitational constant, and l is the length. Find t if $l = 50$ cm. Here use $\pi = 3.1416$ and $g = 9.8$.

C4. Use differentials to estimate the error made in Problem C3 if $l = 51$ cm rather than 50 cm. Find the percent error.

C5. In Problem C3, assume that l is exactly 50 cm. What error was made by using $g = 9.8$ in place of $g = 9.8054$?

$\star 12$ Newton's Method and the Banach Fixed-Point Theorem

To solve an equation $f(x) = 0$, we must find the point (or points) where the graph of $y = f(x)$ meets the x-axis (see Fig. 21). We can use calculus (just as Newton did) to find a sequence of points $x_1, x_2, x_3, \ldots$ that get closer and closer to the solution $x^\star$.

To find the root $x^\star$ we do not need to draw the graph of the function—but we do need the picture in Fig. 21 to follow the logic behind the steps.

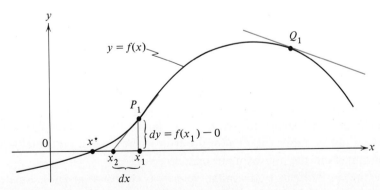

FIGURE 21

1. We guess at a solution and we call our guess x_1. Of course, we make a reasonable guess rather than a wild one. This item will be discussed later.

2. We compute $f(x_1)$. If $f(x_1) = 0$, we are done. If $f(x_1)$ is not zero, then we are off by a (small) amount, and we call this dy (see Fig. 21). Then

$$(72) \qquad \frac{dy}{dx} = f'(x_1)$$

and hence

$$(73) \qquad dx = \frac{dy}{f'(x)}.$$

Now (see Fig. 21) $dx = x_1 - x_2$ or $x_2 = x_1 - dx$ so that

$$(74) \qquad x_2 = x_1 - \frac{dy}{f'(x_1)} = x_1 - \frac{f(x_1)}{f'(x_1)}.$$

3. All the quantities on the right side of equation (74) can be computed from the given function. Use equation (74) to find x_2. Then x_2 is the next approximation to $x^\star$, a root of $f(x) = 0$.

4. We repeat the process as often as we wish. We use x_2 in place of x_1 on the right side of (74) to find x_3. Then we use x_3 on the right side of (74) to find x_4, and so on. To indicate this process, we rewrite equation (74) as

$$(75) \qquad x_{n+1} = x_n - \frac{f(x_n)}{f'(x_n)}, \qquad n = 1, 2, 3, 4, \ldots,$$

The geometric nature of this process is illustrated in Fig. 22.

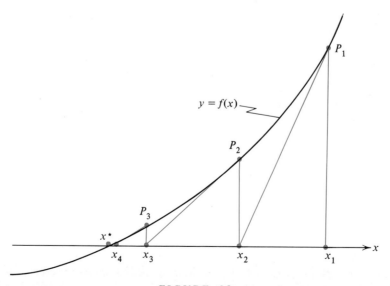

FIGURE 22

If we make a bad guess for x_1, the process may lead away from $x^\star$. For example, if we select x_1 to be the x-coordinate of Q_1 in Fig. 21, then it is clear that x_2 will be still farther from $x^\star$ than x_1.

Example 1. Find one root of

(76)
$$2x^3 + x = 10.$$

[This is the equation for Problem C1(a) in Exercise 7.]

Solution. Setting $f(x) = 2x^3 + x - 10$, we first note that $f'(x) = 6x^2 + 1$ is positive for all x, so that $f(x)$ is everywhere increasing. Thus there is only one root for equation (76). Furthermore, since $f(1) = 2 + 1 - 10 = -7$ and $f(2) = 16 + 2 - 10 = 8$, this root must lie between 1 and 2. Thus we begin Newton's process with the initial guess $x_1 = 1.5$. Using $f(x) = 2x^3 + x - 10$, equation (75) becomes

(77)
$$x_{n+1} = x_n - \frac{2x_n^3 + x_n - 10}{6x_n^2 + 1}, \qquad n = 1, 2, 3, \ldots.$$

We use this equation repeatedly, recording the results in Table 1. With $x_1 = 1.5$, equation (77) gives

$$x_2 = 1.5 - \frac{2(1.5)^3 + 1.5 - 10}{6(1.5)^2 + 1} = 1.5 - \frac{-1.75}{14.5} = 1.620689 \ldots \approx 1.62069.$$

To simplify matters, we only carry five decimal places in our work. Thus we put $x_2 = 1.62069$ and use equation (77) to find x_3. The further outcomes are shown in Table 1.

TABLE 1

x_n	$f(x_n)$ $2x_n^3 + x_n - 10$	$f'(x_n)$ $6x_n^2 + 1$	$Q \equiv \dfrac{f(x_n)}{f'(x_n)}$	$x_{n+1} = x_n - Q$
$x_1 = 1.5$	-1.75	14.5	-0.12069	$x_2 = 1.62069$
$x_2 = 1.62069$	0.13462	16.75982	0.00803	$x_3 = 1.61266$
$x_3 = 1.61266$	0.00066	16.60403	0.00004	$x_4 = 1.61262$
$x_4 = 1.61262$	-0.000004	16.60326	-0.0000002	$x_5 = 1.61262$

Since we introduced uncertainties in the fifth decimal place (because of round-off), we surmise from Table 1 that the unique root $x^\star$ of $2x^3 + x - 10 = 0$ agrees with 1.6126 to four decimal places. Indeed,

$$f(1.6126) = -0.0003 \ldots \qquad \text{and} \qquad f(1.6127) = +0.0013 \ldots$$

so that $1.6126 < x^\star < 1.6127$. Consequently, $x^\star = 1.6126 \ldots$. In fact, $x^\star$ agrees with 1.61262 to five decimal digits.   ●

In general, how can we be certain that the sequence $x_1, x_2, x_3, \ldots$ defined by (75) will converge to a zero of $f(x)$? If we examine Fig. 22, we can easily guess at a set of sufficient conditions, namely

---

THEOREM 15 (PWO). Suppose that an interval $\mathcal{I} : a < x < b$, contains a zero $x^\star$ of $f(x)$ and that in that interval

$$(78) \qquad\qquad f'(x) > 0 \qquad \text{and} \qquad f''(x) > 0.$$

If we select x_1 so that $x^\star < x_1 < b$, then the sequence generated by Newton's method [equation (75)] will steadily decrease to $x^\star$.

By a geometric argument, based on Fig. 22, the student can easily persuade himself that Theorem 15 is true.

Other special cases can be handled by a similar geometric argument. Suppose that in $\mathcal{I} : a < x < b$ we have

$$(79) \qquad\qquad f'(x) < 0 \qquad \text{and} \qquad f''(x) > 0.$$

Then we should select x_1 so that $a < x_1 < x^\star < b$. If we do, then the sequence $x_1, x_2, x_3, \ldots$ will increase steadily to $x^\star$.

We leave it for the student to consider the case $f''(x) < 0$ in $\mathcal{I}$ (the curve is concave downward).

The real danger occurs if $f''(x)$ changes sign in $\mathcal{I}$. In this case the sequence may diverge.

We now turn to the general method of successive substitutions (also called *iteration*) introduced in Section 5 of Chapter 3. In this method the equation to be solved is put in the form

$$(80) \qquad\qquad x = f(x)$$

and then the sequence of successive approximates is

$$(81) \qquad\qquad x_2 = f(x_1), \quad x_3 = f(x_2), \quad \ldots, x_{n+1} = f(x_n), \quad \ldots.$$

Is there a set of conditions under which we can be certain that this sequence converges to a solution of (80)? The answer is yes and is contained in

> **THEOREM 16 (PWO) The Banach Fixed-Point Theorem.**[1] Suppose that there is some closed interval $\mathscr{I}$ such that in $\mathscr{I}$, we have $|f'(x)| \leqq k < 1$. If $f(x)$ lies in $\mathscr{I}$ for all x in $\mathscr{I}$, then the sequence (81) converges to the unique solution of (80) in $\mathscr{I}$ for any choice of the starting value x_1 in $\mathscr{I}$.

One can easily visualize this theorem by thinking of the interval $\mathscr{I}$ as pliable, and that $f(x)$ squashes the interval into a subset of itself [$k < 1$ and $f(x)$ carries $\mathscr{I}$ into $\mathscr{I}$]. Then in this squashing process there is always one point x that is fixed [the solution of (80)].

Example 2. Apply the Banach Fixed-Point Theorem to $2x^3 + x = 10$, the equation of Example 1.

Solution. If we write this equation in the form

$$x = 10 - 2x^3 \quad \text{or} \quad x = \frac{10 - x}{2x^2},$$

the function on the right side does not satisfy the conditions of Theorem 16. But if we write

(82)
$$x = \sqrt[3]{\frac{10 - x}{2}},$$

then $f(x) = \sqrt[3]{(10 - x)/2}$ does satisfy the conditions of Theorem 16. First $f(1) = \sqrt[3]{4.5} \approx 1.65$ and $f(2) = \sqrt[3]{4} \approx 1.58$. Since $f(x)$ is steadily decreasing, the interval $\mathscr{I} : 1 \leqq x \leqq 2$ goes into a subset of $\mathscr{I}$ under $f(x)$. Further, we have

$$|f'(x)| = \left| \frac{1}{3} \frac{1}{2} \left(\frac{2}{10 - x} \right)^{2/3} \right| \leqq \frac{1}{6} \left(\frac{1}{4} \right)^{2/3} < 1 \quad \text{in } \mathscr{I}.$$

We start with $x_1 = 1.5$, and generate the sequence (81) with $f(x) = \sqrt[3]{(10 - x)/2}$. Keeping five decimal places, we find that

$$
\begin{aligned}
x_2 &= f(1.5) = 1.61981, \\
x_3 &= f(1.61981) = 1.61216, \\
x_4 &= f(1.61216) = 1.61265, \\
x_5 &= f(1.61265) = 1.61262, \\
x_6 &= f(1.61262) = 1.61262.
\end{aligned}
$$

---

[1] This theorem is really a very special case of the theorem proved by Banach. The more general theorem is one of the truly beautiful and important results created in the twentieth century by the Polish School of Mathematicians.

Hence (to five decimal places) $x^\star = 1.61262$ is the fixed point—and is also the solution of equation (76). This answer, of course, agrees with the one found in Example 1 by Newton's method. ●

Actually, Newton's method is a special case of successive substitution. To solve the equation $f(x) = 0$ by Newton's technique, we rewrite it in the form

$$x = x - \frac{f(x)}{f'(x)} \qquad \text{for } f'(x) \neq 0,$$

and find the fixed points of the function $g(x) = x - f(x)/f'(x)$ using iteration.

EXERCISE 9

CALCULATOR PROBLEMS

In Problems C1 through C8, use a calculator and Newton's method [equation (75)] to find the indicated root of the given equation to five significant figures.

C1. $x^2 = 7, \quad x > 0.$

C2. $x^5 = 3.456.$

C3. $x^3 + 2x - 4 = 0.$

C4. $x^4 + 3x - 5 = 0, \quad x > 0.$

C5. $x^5 - 7x - 50 = 0, \quad x > 0.$

C6. $x^4 - x^3 - 75 = 0, \quad x > 0.$

C7. $x^4 - x^3 - 75 = 0, \quad x < 0.$

C8. $x^2 + 2\sqrt{x} = 1000.$

C9. Show that the equation

$$\frac{x^2 + 2}{3x^2 + 7} = \frac{4}{x + 12}$$

leads to the equation of Problem C3 and hence has the same solution.

C10. The equation

$$\frac{x + 1}{x^3 + 2x^2 + 3x + 9} = \frac{x}{2x^2 + x + 5}$$

is equivalent to one of those in Problems C1 through C8. Which one is it?

C11. Find the absolute minimum of the function $f(x) = x^4 + 6x^2 - 28x + 60.$

C12. Find the absolute maximum and absolute minimum of the function $f(x) = 4x^5 - 5x^4 - 600x + 30$ on the interval $[2.5, 3.5]$.

C13. Find the absolute maximum and absolute minimum of the function $f(x) = x^4 - 15x^2 + 3x + 6$ on $[0, 0.5]$.

C14. For the function in Problem C13, find the absolute maximum and absolute minimum on the interval $[2, 3]$.

In Problems C15 through C18, show that the given equation has been rewritten in a form suitable for using the Banach Fixed-Point Theorem. Then solve by using the sequence (81).

C15. $x^3 + 2x - 4 = 0$; $\quad x = f(x) = \sqrt[3]{4 - 2x}$, $\quad 1 \le x \le 1.5$ (see Problem C3).

C16. $x^4 + 3x - 5 = 0$; $\quad x = f(x) = \sqrt[4]{5 - 3x}$, $\quad 1 \le x \le 4/3$ (see Problem C4).

C17. $x^5 - 7x - 50 = 0$; $\quad x = f(x) = \sqrt[5]{50 + 7x}$, $\quad 1 \le x \le 3$ (see Problem C5).

C18. $x^2 + 2\sqrt{x} = 1000$; $\quad x = f(x) = \sqrt{1000 - 2\sqrt{x}}$, $\quad 9 \le x \le 100$ (see Problem C8).

In Problems C19 and C20, transform the given equation into a form suitable for using the Banach Fixed-Point Theorem. In each case state $f(x)$ and the interval $\mathscr{I}$. Then solve by using the sequence (81).

C19. $x^5 - 60x^3 + 1 = 0$, $x > 0$.

C20. $1 + \sqrt{x} + \sqrt[3]{x} = 10\sqrt[5]{x}$.

REVIEW PROBLEMS

In Problems 1 through 4, find the extreme points and the inflection points (if there are any) and sketch the graph.

1. $y = 10 + 9x + 3x^2 - x^3$.

2. $y = 2x^3 - 3x^2 - 36x + 7$.

3. $y = x^2\sqrt{5 - x}$.

4. $y = \dfrac{x^2}{x^2 - 16}$.

In Problems 5 and 6, find the intervals in which the function is increasing. Find the intervals in which the graph is concave upward.

5. $f(x) = x^4 - 4x^3 - 2x^2 + 12x + \pi\sqrt{19}$.

6. $f(x) = 3x^5 - 15x^4 - 40x^3 + \sqrt[3]{11}$.

7. A man standing on a bridge throws a stone upward. Exactly 2 sec later the stone passes the man on the way down, and 3 sec later it hits the water. Find the initial velocity of the stone (in ft/sec) and the height of the bridge (in ft) above the water. HINT: Use equation (9a).

8. Find the point on the curve $8y = 40 - x^2$ that is closest to the origin.

9. Square corners are cut from a rectangular piece of tin 16 ft by 10 ft, and the edges are turned up to make an open rectangular box. Find the size of the corners that must be removed to maximize the volume of the box.

10. Do Problem 9 if the piece of tin is 16 ft by 6 ft.

11. Find two positive numbers whose sum is 20 and such that x^3y^7 is as large as possible.

12. Let P be a point on the x-axis, let s_1 be the distance from P to $(0, 4)$, and let s_2 be the distance from P to $(6, 1)$. Where is P when (a) $s_1^2 + s_2^2$ is a minimum, (b) $s_1^2 + 2s_2^2$ is a minimum, and (c) $s_1^2 - 2s_2^2$ is a maximum?

13. Using differentials, estimate to three significant figures: (a) $\sqrt[3]{8.5}$, (b) $\sqrt[4]{17}$, (c) $\sqrt[3]{7.5}$, (d) $\sqrt[4]{15}$, and (e) $\sqrt{10}$.

14. Among all rectangles with sides parallel to the axes and inscribed in the ellipse

$$\frac{x^2}{6} + \frac{y^2}{2} = 1,$$

find the dimensions of the one that has the largest perimeter.

15. Do Problem 14 for the ellipse $x^2/11 + y^2/14 = 1$.

16. Let a^2 and b^2 be positive constants. Solve Problem 14 for the ellipse

$$\frac{x^2}{a^2} + \frac{y^2}{b^2} = 1.$$

Use your general formula to check your answers to Problems 14 and 15.

In Problems 17 through 20, the motion of a particle on the x-axis is governed by the given equation. In each case find the indicated quantity.

17. $x = t^3 - 9t^2 + 25t - 19$. Find the location of the particle when the velocity is a minimum.

18. $x = -11 + 12t + 3t^2 + 4t^3 - t^4$. Find the location of the particle and the velocity when the acceleration is a maximum.

★19. $x = (1 + t)^{3/2} + 15(1 + t)^{1/2}$, $t \geq 0$. Find the location of the particle when the velocity is a minimum.

★20. $x = 48t/(t + 12)^{3/2}$. Find the location of the particle and the acceleration when the particle is as far to the right as possible.

21. A particle P moves on the ellipse $9x^2 + 16y^2 = 144$ in such a way that $dx/dt = -4y$. Find dy/dt. At what point is dy/dt a maximum?

22. Do Problem 21 if the particle moves on the right branch of $9x^2 - 16y^2 = 144$.

23. For the particle of Problem 21, find ds/dt, where s is the distance of P from the origin.

24. A particle moves steadily on the curve $y = x^2 - 2x + 5$. Find the point (or points) where the y-coordinate is changing twice as fast as the x-coordinate.

25. Do Problem 24 for the curve: (a) $14y = x^3 + x$, (b) $y = 8/x^2$, (c) $y = 12\sqrt{x + 4}$, and (d) $y = 3x + 5$.

26. The area of an equilateral triangle is increasing at the rate of 4 in.^{2}/sec. Find the rate of change of the perimeter.

6

Integration

1 Objective

Starting from the fact that the area of a rectangle is the product of the lengths of two adjacent sides, we learn in elementary geometry how to find the area of a parallelogram, a triangle, and then the area of any polygon by decomposing it into the sum of a number of triangles. Thus we are able to find the area of any plane figure, as long as it is bounded by a finite number of straight-line segments (e.g., the shaded region shown in Fig. 1).

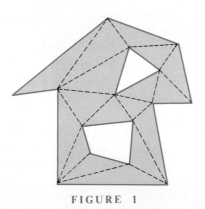

FIGURE 1

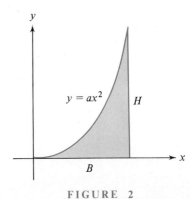

FIGURE 2

But what shall we do if a part of the boundary of the figure is a curve? For example, what is the area of the shaded region shown in Fig. 2, which is bounded by the parabola $y = ax^2$, the x-axis, and the vertical line $x = B$? Fantastic as it may seem, this problem was solved more than 2000 years ago by Archimedes (287–212 B.C.) by a very ingenious

236

method. In fact, he showed that $A = BH/3$. Unfortunately, the method of Archimedes could not be applied to figures bounded by other curves, and so this general problem of finding the area of a region bounded wholly or in part by curves continued to plague mathematicians for the next 1800 years. One can well imagine the tremendous excitement that was generated when almost simultaneously Newton (1642–1727) and Leibniz (1646–1716) showed how calculus can be used in a systematic way to obtain the area bounded by any curve or set of curves. Is it any wonder then that in 1669 Isaac Barrow resigned his position as a professor of mathematics at Cambridge University in favor of his student, Isaac Newton?

Our purpose in this chapter is to learn just how calculus is used to find areas. Further we will see that finding areas of regions bounded by curves is more than just a game. Many very practical problems can be reduced to equivalent problems in finding the area of a region bounded by curves. Such, for example, is the problem of finding the total force on the side of a dam due to the water in the reservoir.

The key to the solution of the area problem is the integral. In this chapter we will introduce two different types of integrals: (I) the indefinite integral and (II) the definite integral. These two integrals are symbolized[1] by

(I) $$\int f(x)\,dx \qquad [\text{read “the integral of } f(x)\,dx\text{”}]$$

and

(II) $$\int_a^b f(x)\,dx \qquad [\text{read “the integral from } a \text{ to } b \text{ of } f(x)\,dx\text{”}].$$

The reader is warned in advance not to confuse these different integrals merely because the family name is the same and the symbols are so similar. These mathematical cousins are distinct entities, each with its own characteristics. The exact nature of their relationship will be investigated after we have become better acquainted with them.

The Indefinite Integral, or the Antiderivative 2

We have already mentioned in Chapter 5, Sections 2 and 7, that integration is the inverse of differentiation. In other words, given

(1) $$\frac{dy}{dx} = f(x),$$

[1] The integral sign is really an elongated S and this symbol was actually used for an S several hundred years ago. The letter S or $\int$ is intended to remind us of the word sum, and as we will learn in Section 5, the definite integral (II) is the limit of a certain sum.

we are to find a function $y = F(x)$ such that on differentiating $F(x)$ we obtain $f(x)$. If such a function $F(x)$ exists, it is called an *indefinite integral of $f(x)$*. It is also called an *antiderivative*, because it is obtained by reversing the process of differentiation.

To introduce a suitable notation for this process, we first write (1) in the differential form

$$(2) \qquad dy = f(x)\, dx$$

and then prefix an integral sign $\int$ in front of both sides, obtaining

$$(3) \qquad \int dy = \int f(x)\, dx.$$

Here the integral sign indicates that we are to find a function whose differential is the quantity standing after the sign. The function $f(x)$ in (3) is called the *integrand*.

For example, if

$$dy = x^3\, dx,$$

then on integrating both sides, we would have

$$(4) \qquad \int dy = \int x^3\, dx,$$

$$(5) \qquad y = \frac{x^4}{4} + C.$$

The left side of (5) is y because the differential of y is dy and the right side of (5) is $x^4/4 + C$ because the derivative with respect to x of this function is x^3 so that its differential is $x^3\, dx$.

Now the differential of any constant C is zero so that the integral is not uniquely determined. To obtain the most general integral, we must add an arbitrary constant C. Because of the indefiniteness of this constant, the integral in (3) is called an *indefinite integral*.

When we perform the indicated integration in equation (4), why not add a constant to each side? The two constants to be added, one on each side, may be different so that we must use different symbols for these two constants. Let C_1 and C_2 denote these constants. Then we obtain from (4)

$$(6) \qquad y + C_1 = \frac{x^4}{4} + C_2,$$

or

$$(7) \qquad y = \frac{x^4}{4} + (C_2 - C_1) = \frac{x^4}{4} + C,$$

where $C = C_2 - C_1$ is just another constant. Thus (6) is equivalent to (5), and so there is no advantage or generality to be gained by adding a constant on both sides.

Example 1. It is known that a certain curve passes through the point $(3, 38)$ and has slope $m = 2x^3 - x + 5$ for each x. Find the equation of the curve.

Solution. From the conditions of the problem

$$m = \frac{dy}{dx} = 2x^3 - x + 5.$$

Then, in differential form,

$$dy = (2x^3 - x + 5)\,dx.$$

Integrating both sides of this equation, we obtain

$$\int dy = \int (2x^3 - x + 5)\,dx,$$

(8)
$$y = \frac{x^4}{2} - \frac{x^2}{2} + 5x + C.$$

Since the curve passes through $(3, 38)$, these values must satisfy the equation. Substituting $x = 3$ and $y = 38$ in (8), we find that

$$38 = \frac{81}{2} - \frac{9}{2} + 15 + C = 51 + C,$$

whence $C = 38 - 51 = -13$. Using this in (8), we find that the solution to our problem is

$$y = \frac{x^4}{2} - \frac{x^2}{2} + 5x - 13. \quad \bullet$$

The side condition that the curve goes through the point $(3, 38)$ enables us to determine the value of C. Thus we can select from the infinity of indefinite integrals (8), the particular one fitting the conditions of the problem. In any natural problem such side conditions always appear. Frequently, these side conditions are called *initial conditions*. The idea is that the curve "starts" from the point $(3, 38)$, even though in reality it may "start" somewhere else, or "start" may have no real meaning. Initial conditions are also called *boundary conditions*.

We have already proved (Theorem 10 of Chapter 5, page 201) that if two functions have the same derivative, then they differ by a constant. This means that if we are seeking the indefinite integral of a particular function $f(x)$ and find one suitable solution $F(x)$, then any other indefinite integral must differ from $F(x)$ by at most a constant. Thus we obtain the most general integral merely by adding an arbitrary constant C to $F(x)$. These remarks give

---

THEOREM 1. If $F'(x) = f(x)$, then

(9)
$$\int f(x)\,dx = F(x) + C.$$

Since integration is the inverse of differentiation, every differentiation formula may be written as an integral formula. For ready reference we now collect the important ones. It may seem like a burdensome task to be asked to memorize all of these new formulas, equations (10) through (15), but the perceptive student will quickly recognize that these are all old friends, just dressed up in a new disguise so that memorization is hardly any trouble at all.

(10)
$$\int dx = x + C.$$

(11)
$$\int du = u + C.$$

(12)
$$\int cf(x)\,dx = c\int f(x)\,dx.$$

(13)
$$\int (f(x) + g(x))\,dx = \int f(x)\,dx + \int g(x)\,dx.$$

(14)
$$\int x^n\,dx = \frac{x^{n+1}}{n+1} + C, \qquad \text{if } n \neq -1.$$

(15)
$$\int u^n\,du = \frac{u^{n+1}}{n+1} + C, \qquad \text{if } n \neq -1.$$

The formulas (10) and (11) are really identical. So also are (14) and (15). First, we want to emphasize that there is nothing magic about the letter x. Any letter is suitable to indicate a variable quantity. Another reason for replacing x by u to obtain (11) and (15) will appear in Example 2.

Formulas (12) and (13) are just the integral forms of the differentiation formulas

$$\frac{d}{dx}cF(x) = c\frac{d}{dx}F(x)$$

and

$$\frac{d}{dx}[F(x) + G(x)] = \frac{d}{dx}F(x) + \frac{d}{dx}G(x),$$

where $\dfrac{d}{dx}F(x) = f(x)$ and $\dfrac{d}{dx}G(x) = g(x)$.

The student will observe that in (14) and (15) the value $n = -1$ is forbidden. For if $n = -1$, then the denominator on the right side is zero, and the formula is meaningless. We will see that the case $n = -1$; that is, the determination of the integral

$$\int \frac{dx}{x},$$

is one of the most fascinating little chapters in calculus. But this must be reserved for Chapter 9.

Example 2. Find the indefinite integrals

(a) $\displaystyle\int \left(7x^3 - \frac{3}{x^2}\right) dx.$ (b) $\displaystyle\int (x^2 + 6)^{3/2}x \, dx.$

Solution. For **(a)** using formulas (13), (12), and (14) in turn, we have

$$\int \left(7x^3 - \frac{3}{x^2}\right) dx = \int 7x^3 \, dx + \int -\frac{3}{x^2} \, dx = 7 \int x^3 \, dx - 3 \int x^{-2} \, dx$$

$$= 7\frac{x^4}{4} - 3\frac{x^{-2+1}}{-2+1} + C = \frac{7}{4}x^4 + \frac{3}{x} + C.$$

The integral **(b)** does not fit any of the standard formulas. However, if we select u properly, it turns out that **(b)** can be made to fit formula (15). To do this, set $u = x^2 + 6$. Then $du = 2x \, dx$. In **(b)** we have a term $x \, dx$, and this is almost du. In fact, it is $du/2$. With these substitutions we have

$$\int (x^2 + 6)^{3/2}x \, dx = \int u^{3/2}\frac{du}{2} = \frac{1}{2} \int u^{3/2} \, du.$$

Using formula (15), we find that

$$\frac{1}{2} \int u^{3/2} \, du = \frac{1}{2}\left(\frac{u^{5/2}}{5/2}\right) + C = \frac{1}{2}\left(\frac{2}{5}u^{5/2}\right) + C.$$

Replacing u by $x^2 + 6$, we then have

(16) $$\int (x^2 + 6)^{3/2}x \, dx = \frac{1}{5}(x^2 + 6)^{5/2} + C. \quad \bullet$$

The perceptive reader will observe that the work that leads to equation (16) is just the inverse of the Chain Rule for differentiating a composite function $f(u)$, where u is a function of x. The cautious or critical reader might demand the proof of some theorem that covers all substitutions of the type used in solving Example 2**(b).** Such theorems are easy to prove once we have acquired the proper background. In the meantime, we contend that we did *not* violate any standards of rigor. To prove that our results are correct, it is sufficient to differentiate the right side of (16) and show that this yields the integrand on the left side. Indeed,

$$\frac{d}{dx}\left[\frac{1}{5}(x^2 + 6)^{5/2} + C\right] = \frac{5}{2}\frac{1}{5}(x^2 + 6)^{3/2}(2x) = (x^2 + 6)^{3/2}x.$$

EXERCISE 1

In Problems 1 through 18, find the indefinite integral.

1. $\displaystyle\int 18x^5\, dx.$

2. $\displaystyle\int 160x^9\, dx.$

3. $\displaystyle\int \sqrt{y}\, dy.$

4. $\displaystyle\int 4x^{-5/3}\, dx.$

5. $\displaystyle\int \frac{21\, dt}{\sqrt{t^9}}.$

6. $\displaystyle\int -88y^{8/3}\, dy.$

7. $\displaystyle\int (1000x - 5x^4)\, dx.$

8. $\displaystyle\int (\pi x^2 + \sqrt{x})\, dx.$

9. $\displaystyle\int \left(\sqrt{2}x^7 - \frac{3}{2}x^2 + \frac{1}{x^5}\right) dx.$

10. $\displaystyle\int (2x + 7)^3\, dx.$

11. $\displaystyle\int (2x + 7)^{513}\, dx.$

12. $\displaystyle\int (1 - 7x)^{1/2}\, dx.$

13. $\displaystyle\int (3 + 11x^2)^{5/2}x\, dx.$

14. $\displaystyle\int \frac{t\, dt}{(11 + 7t^2)^2}.$

15. $\displaystyle\int \frac{u^2\, du}{(1 + u^3)^{5/2}}.$

16. $\displaystyle\int \frac{\sqrt{z^6 + 5z^4}}{z}\, dz, \qquad z \neq 0.$

17. $\displaystyle\int (y^3 + y + 55)^{7/2}(3y^2 + 1)\, dy.$

18. $\displaystyle\int \frac{\sqrt{3}(w + 1)\, dw}{\sqrt{5w^2 + 10w + 11}}.$

In Problems 19 through 22, find the equation of the curve, given the derivative and one point P on the curve.

19. $\dfrac{dy}{dx} = 3x^2 + 2x, \qquad P(0, 0).$

20. $\dfrac{dy}{dx} = \dfrac{2}{x^3} - \dfrac{3}{x^4}, \quad P(1, 2).$

21. $\dfrac{dy}{dx} = x\sqrt{4 + 5x^2}, \quad P\left(-1, \dfrac{1}{5}\right).$

22. $\dfrac{dy}{dx} = mx + b, \qquad P(-1, 0).$

CALCULATOR PROBLEMS

In Problems C1 through C4, find the indefinite integral. Give coefficients and exponents to four significant figures.

C1. $\int 25.79(1 + 14x)^{7.531}\,dx.$

C2. $\int 8.492x^2(x^3 + \pi)^{-0.567}\,dx.$

C3. $\int \dfrac{2594\,dx}{(x + 1)^{0.9813}}.$

C4. $\int 5(2x + 1)(x^2 + x)^{0.123}\,dx.$

Differential Equations 3★

A differential equation is an equation that relates a function and some of its derivatives in a given set. The set is usually an interval, a ray, or the set of all real numbers. If the differential equation involves the *n*th derivative, but no derivative of higher order, then the equation is called an *nth-order differential equation*. In this section we will look briefly at first-order differential equations.

> **Example 1.** Solve the differential equation
>
> (17)
> $$x^2\frac{dy}{dx} = x^4y^2 + y^2.$$
>
> **Solution.** In this case the variables x and y can be separated, by factoring the right side and dividing by x^2. This gives
>
> $$\frac{dy}{dx} = y^2\left(x^2 + \frac{1}{x^2}\right),$$
>
> $$\frac{dy}{y^2} = \left(x^2 + \frac{1}{x^2}\right)dx.$$
>
> Integrating both sides of this equation, we obtain
>
> $$-\frac{1}{y} = \frac{x^3}{3} - \frac{1}{x} + C,$$
>
> (18)
> $$y = -\frac{1}{\dfrac{x^3}{3} - \dfrac{1}{x} + C} = \frac{3x}{3 - x^4 - 3Cx}.$$
>
> Observing that since C is an arbitrary constant, we could replace C by $-C/3$, which is just as arbitrary. Then we could write the solution of our differential equation as
>
> (19)
> $$y = \frac{3x}{3 - x^4 + Cx}.$$
>
> Equation (19) is just as much a solution as (18), because each formula generates the same set of functions as C runs through the set of all real numbers. ●

Before we can check that (19) is a solution, we need to state precisely what we mean by a solution.

DEFINITION 1 (Solution of a Differential Equation). The function $f(x)$ defined in an interval $\mathscr{I}$ is said to be a solution of the first-order differential equation

(20) $$\varphi(x, y, y') = 0$$

in $\mathscr{I}$, if

(21) $$\varphi(x, f(x), f'(x)) \equiv 0$$

for all x in $\mathscr{I}$.

Equation (20), which looks a little strange, is merely a symbol that represents[1] an equation in the three variables x, y, and y'. For example, we can put (17) in the form (20) by transposing the right side to obtain

(22) $$x^2 \frac{dy}{dx} - x^4 y^2 - y^2 = 0.$$

Equation (21) merely states that equation (20) becomes an identity when y is replaced by $f(x)$ and y' is replaced by $f'(x)$ in the function φ.

Let us apply this definition to the function $f(x)$ defined by (19) to test whether we really have a solution. From (19) we easily find that

$$f'(x) = \frac{(3 - x^4 + Cx)3 - 3x(-4x^3 + C)}{(3 - x^4 + Cx)^2}$$

(23) $$f'(x) = \frac{9 + 9x^4}{(3 - x^4 + Cx)^2}.$$

Then, using (19) and (23) in (22), we find that the left side of (22) is

(24) $$\frac{x^2(9 + 9x^4)}{(3 - x^4 + Cx)^2} - x^4 \left(\frac{3x}{3 - x^4 + Cx}\right)^2 - \left(\frac{3x}{3 - x^4 + Cx}\right)^2.$$

Now this is obviously zero for all x for which the expression (24) has meaning, that is, for all x in $\mathscr{R} - \mathscr{N}$, where $\mathscr{N}$ is merely the set of points for which $3 - x^4 + Cx = 0$. Thus we have proved that (19) is a solution of the differential equation (17) for a domain that consists of all real numbers except for at most four points.

---

[1] For an explanation of the notation for a function of several variables, see Chapter 3, Exercise 1, Problems 38–46, page 78.

Example 2. Solve the differential equation (17) subject to the initial conditions
(a) $y = 6$ when $x = 2$, **(b)** $y = 1$ when $x = 0$, and **(c)** $y = 0$ when $x = 0$.

Solution. **(a)** Using $y = 6$ and $x = 2$ in (19), we see that

$$6 = \frac{6}{3 - 16 + 2C}.$$

Hence $1 = 2C - 13$ or $C = 7$. It follows that

(25)
$$y = \frac{3x}{3 - x^4 + 7x}$$

is the one solution from the family (19) that satisfies the given initial condition.
(b) If we set $y = 1$ and $x = 0$ in (19), we obtain $1 = 0/3$ for every value of C. Hence
the family of solutions (19) does not contain one that satisfies the initial condition. This is no surprise, for if we look at the differential equation (17), the left
side is zero when $x = 0$ while the right side is 1. Hence the equation cannot have
a solution in this case.
(c) In contrast to this lack of a solution in **(b)**, we observe that every curve of the
family defined by (19) passes through the point $(0, 0)$. Hence every solution of
(17) satisfies the initial conditions $y = 0$ when $x = 0$. ●

EXERCISE 2

*In Problems 1 through 8, solve the given differential equation subject to the given initial
condition.*

1. $\dfrac{ds}{dt} = 32t + 5$, $s = 100$ when $t = 0$.

2. $\dfrac{dy}{dx} = xy^2$, $y = 6$ when $x = 1$.

3. $y\dfrac{dy}{dx} = x(y^4 + 2y^2 + 1)$, $y = 1$ when $x = -3$.

4. $\dfrac{dy}{dx} = \dfrac{x}{y}$, $y = 5$ when $x = 2\sqrt{6}$.

5. $\dfrac{du}{dv} = \sqrt{uv}$, $u = 100$ when $v = 9$.

6. $\dfrac{dy}{dx} = \dfrac{1 + 3x^2}{2 + 2y}$, $y = 1$ when $x = 2$.

7. $\dfrac{dy}{dx} = \dfrac{x(1 + y^2)^2}{y(1 + x^2)^2},$ $y = 3$ when $x = 1.$

8. $\dfrac{dw}{dz} = \sqrt{wz - 2w - 3z + 6},$ $w = 12$ when $z = 6.$

In Problems 9 through 16, prove that the given function is a solution of the given differential equation for any selection of the constants A, B, and C.

9. $x\dfrac{dy}{dx} = y + x^2,$ $y = Ax + x^2.$

10. $x\dfrac{dy}{dx} = 3y - 2x,$ $y = x + Ax^3.$

11. $x\dfrac{dy}{dx} = y - \dfrac{1}{\sqrt{x^2 + 1}},$ $y = \sqrt{x^2 + 1} + Ax.$

12. $(x^2 + 1)\dfrac{dy}{dx} = xy + 1,$ $y = A\sqrt{x^2 + 1} + x.$

★13. $x^2\dfrac{d^2y}{dx^2} - 2x\dfrac{dy}{dx} + 2y = 0,$ $y = Ax + Bx^2.$

★14. $(x^2 + 1)\dfrac{d^2y}{dx^2} + x\dfrac{dy}{dx} - y = 0,$ $y = Ax + B\sqrt{x^2 + 1}.$

★★15. $x^3\dfrac{d^3y}{dx^3} + x^2\dfrac{d^2y}{dx^2} - 2x\dfrac{dy}{dx} + 2y = 0,$ $y = \dfrac{A}{x} + Bx + Cx^2.$

★★16. $4x^2\dfrac{d^3y}{dx^3} + 12x\dfrac{d^2y}{dx^2} + 3\dfrac{dy}{dx} = 0,$ $y = Ax^{1/2} + Bx^{-1/2} + C.$

4 *The Summation Notation*

The definition of the *definite* integral requires the use of sums involving many terms. The work can be simplified tremendously if we have available a shorthand notation for writing these sums. This section will be devoted to explaining and illustrating a very convenient notation for sums. The student who exercises enough with this new notation to feel at home with it will find the subsequent work with sums rather easy.

The symbol $\sum$ is a capital sigma in the Greek alphabet and corresponds to our English S. Thus it naturally reminds us of the word *sum*. The symbol

$$\sum f(k)$$

means that we are to sum the numbers $f(k)$ for various integer values of k. The range for the integers is indicated by placing them below and above the $\sum$. For example,

$$\sum_{k=1}^{4} f(k) \qquad \text{means} \qquad f(1) + f(2) + f(3) + f(4)$$

and is read "the sum from $k = 1$ to 4 of $f(k)$." Thus we substitute in $f(k)$ successively all the integers between and including the lower and the upper limits of summation, in this case $k = 1, 2, 3,$ and 4, and then add the results.

The sum need not start at 1 or end at 4. Further, any letter can be used instead of k. The following examples should indicate the various possibilities. The new shorthand notation is on the left side, and its meaning is on the right side in each of these equations.

$$\sum_{k=1}^{7} k^2 = 1 + 4 + 9 + 16 + 25 + 36 + 49.$$

$$\sum_{j=1}^{4} f(j) = f(1) + f(2) + f(3) + f(4).$$

$$\sum_{k=2}^{6} g(k) = g(2) + g(3) + g(4) + g(5) + g(6).$$

$$\sum_{j=1}^{8} 1 = 1 + 1 + 1 + 1 + 1 + 1 + 1 + 1 = 8.$$

$$\sum_{k=1}^{n} f(k) = f(1) + f(2) + f(3) + \cdots + f(n).$$

Sometimes the terms to be added involve subscripts, or combinations of functions with subscripts. These possibilities are illustrated below.

$$\sum_{n=3}^{7} a_n = a_3 + a_4 + a_5 + a_6 + a_7.$$

$$\sum_{n=1}^{5} n b_n = b_1 + 2b_2 + 3b_3 + 4b_4 + 5b_5.$$

$$\sum_{k=1}^{n} \frac{a_k}{k} = a_1 + \frac{a_2}{2} + \frac{a_3}{3} + \cdots + \frac{a_n}{n}.$$

In order to see that this is really a nice notation, the student should consider the task of writing the sum of the squares of the first 100 positive integers. This would take quite a lot of time and energy. But with our new notation we can write the same sum as

$$\sum_{k=1}^{100} k^2$$

in just a few seconds. To assist the student to master this new notation, we will frequently use both the new and the old notation together.

Example 1. Show by direct computation that $\sum_{k=1}^{5} k^2 = 55$.

Solution. Writing out the left side, we have

$$\sum_{k=1}^{5} k^2 = 1^2 + 2^2 + 3^2 + 4^2 + 5^2 = 1 + 4 + 9 + 16 + 25 = 55. \quad \bullet$$

Example 2. Prove that the sum of the first n positive integers is $n(n + 1)/2$.

Solution. We are to prove that

(26)
$$\sum_{k=1}^{n} j = 1 + 2 + 3 + \cdots + n = \frac{n(n + 1)}{2}.$$

We use mathematical induction.[1] When $n = 1$, equation (26) gives

$$\sum_{j=1}^{1} j = 1 = \frac{1(1 + 1)}{2} = 1.$$

Assume that (26) is true when $n = k$. Thus we assume that

$$\sum_{j=1}^{k} j = 1 + 2 + 3 + \cdots + k = \frac{k(k + 1)}{2}.$$

[1]Mathematical induction is explained in detail in Volume 1 Appendix A of the *Study Guide* for this book.

Adding $(k + 1)$ to both sides of this equation, we obtain

$$\sum_{j=1}^{k+1} j = 1 + 2 + 3 + \cdots + k + (k + 1) = \frac{k(k + 1)}{2} + (k + 1)$$

$$= \frac{k(k + 1) + 2(k + 1)}{2}$$

$$= \frac{(k + 1)(k + 2)}{2}.$$

But this is equation (26) when $n = k + 1$. ●

Example 3. Prove that if $x \neq 1$ and $n \geq 1$, then

(27) $$\sum_{j=0}^{n} x^j = 1 + x + x^2 + x^3 + \cdots + x^n = \frac{1 - x^{n+1}}{1 - x}.$$

Solution. We can prove (27) by mathematical induction, but for variety we select a different method. Let s denote the sum on the left side of (27). Thus

(28) $$s = 1 + x + x^2 + \cdots + x^n.$$

We multiply both sides of (28) by x and obtain

(29) $$sx = x + x^2 + x^3 + \cdots + x^n + x^{n+1}.$$

If we subtract equation (29) from (28), most of the terms drop out. We find that $s - sx = 1 - x^{n+1}$ or, on factoring,

(30) $$s(1 - x) = 1 - x^{n+1}.$$

If $x \neq 1$, we can divide both sides of (30) by $1 - x$. This gives (27). ●

EXERCISE 3

1. Show by direct computations that each of the following assertions is true.

(a) $\displaystyle\sum_{k=1}^{6} k^2 = 91.$

(b) $\displaystyle\sum_{k=1}^{5} k^3 = 225.$

(c) $\displaystyle\sum_{n=1}^{10} 2n = 110.$

(d) $\displaystyle\sum_{s=0}^{5} \frac{1}{2} s(s - 1) = 20.$

(e) $\displaystyle\sum_{d=1}^{5} \frac{1}{d} = \frac{137}{60}$.

$\star$(f) $\displaystyle\sum_{t=1}^{5} \frac{1}{t(t+1)} = \frac{5}{6}$.

In Problems 2 through 16, a number of assertions are given with the summation notation. In each case write out both sides of the equation in full and decide whether the given assertion is always true, or sometimes may be false.

2. $c \displaystyle\sum_{k=1}^{n} k^4 = \sum_{k=1}^{n} ck^4$.

3. $c \displaystyle\sum_{k=1}^{n} a_k = \sum_{k=1}^{n} ca_k$.

4. $\displaystyle\sum_{k=1}^{N} b_k = \sum_{j=1}^{N} b_j$.

5. $\displaystyle\sum_{k=1}^{n} f(k) = \sum_{k=2}^{n+1} f(k)$.

6. $\displaystyle\sum_{k=1}^{n} f(k) = \sum_{k=2}^{n+1} f(k-1)$.

7. $\displaystyle\left(\sum_{k=1}^{n} a_k\right)\left(\sum_{k=1}^{n} b_k\right) = \sum_{k=1}^{n} a_k b_k$.

8. $\displaystyle\sum_{k=1}^{n} b_k + \sum_{k=n+1}^{N} b_k = \sum_{k=1}^{N} b_k$, $\quad 1 < n < N$.

9. $\displaystyle\left(\sum_{k=1}^{n} a_k\right)^2 = \sum_{k=1}^{n} a_k^2 + \sum_{k=1}^{n} 2a_k + \sum_{k=1}^{n} 1$.

10. $\displaystyle\sum_{k=1}^{n} a_k + \sum_{k=1}^{n} b_k = \sum_{k=1}^{n} (a_k + b_k)$.

11. If $a_k \leqq b_k$ for each positive integer k, then $\displaystyle\sum_{k=1}^{n} a_k \leqq \sum_{k=1}^{n} b_k$.

12. $\displaystyle\sum_{k=0}^{n} a_k = \sum_{k=0}^{n} a_{n-k}$.

13. $\displaystyle\frac{d}{dx}\left(\sum_{k=1}^{n} x^k\right) = \sum_{k=1}^{n} kx^{k-1}$.

14. $\displaystyle\int\left(\sum_{k=1}^{n} x^k\right) dx = C + \sum_{k=1}^{n} \frac{x^{k+1}}{k+1}$.

15. $\displaystyle\sum_{k=1}^{n} (a_{k+1} - a_k) = a_{n+1} - a_1$.

16. $\displaystyle\sum_{k=1}^{n} [(k+1)^2 - k^2] = \sum_{k=1}^{n} (2k+1) = n + 2\sum_{k=1}^{n} k$.

$\star$17. Combine the results of Problems 15 and 16 to get a new proof of formula (26) of Example 2.

In Problems 18 through 23, write out the given assertion in full and then use mathematical induction to prove that the assertion is true for every positive integer n.

18. $\sum_{j=1}^{n} (2j - 1) = n^2.$

19. $\sum_{j=1}^{n} (3j - 1) = \dfrac{n(3n + 1)}{2}.$

20. $\sum_{j=1}^{n} j(j + 1) = \dfrac{n(n + 1)(n + 2)}{3}.$

21. $\sum_{k=1}^{n} k^2 = \dfrac{n(n + 1)(2n + 1)}{6}.$

22. $\sum_{j=1}^{n} \dfrac{1}{j(j + 1)} = \dfrac{n}{n + 1}.$

23. $\sum_{j=1}^{n} j^3 = \dfrac{n^2(n + 1)^2}{4}.$

24. Combine the results of Problem 23 and Example 2 [equation (26)] to prove that

$$\sum_{j=1}^{n} j^3 = \left(\sum_{j=1}^{n} j \right)^2.$$

State the result in words.

25. Prove that for any function $f(x)$

$$\sum_{k=1}^{n} [f(x_k) - f(x_{k-1})] = f(x_n) - f(x_0).$$

*26. Solve the equation $\displaystyle\sum_{k=1}^{5} k^2 = \sum_{k=1}^{5} t^2$ for t.

*27. Use mathematical induction to prove that for any positive integer n,

$$(x + y)^n = \sum_{k=0}^{n} \binom{n}{k} x^{n-k} y^k,$$

where (by definition)

$$\binom{n}{k} = \dfrac{n!}{k!(n - k)!}.$$

**28. Let u and v be functions of x and let $D^k u$ denote the kth derivative of u with respect to x, where $D^0 u$ is defined to be u. Prove that for each positive integer n,

$$D^n(uv) = \sum_{k=0}^{n} \binom{n}{k} D^{n-k} u \, D^k v.$$

This formula is known as the *Leibniz rule* for the nth derivative of a product.

CALCULATOR PROBLEMS

In Problems C1 through C8, find the indicated sum to five significant figures.

C1. $\displaystyle\sum_{k=1}^{6} \sqrt{k}.$

C2. $\displaystyle\sum_{k=1}^{10} \frac{1}{k}.$

C3. $\displaystyle\sum_{k=1}^{10} \frac{1}{\sqrt{k}}.$

C4. $\displaystyle\sum_{k=1}^{10} \frac{1}{k!}.$

C5. $\displaystyle\sum_{k=0}^{5} (0.345)^k.$

C6. $\displaystyle\sum_{k=0}^{10} (0.345)^k.$

C7. $\displaystyle\sum_{k=0}^{10} \frac{k}{2^k}.$

C8. $\displaystyle\sum_{k=0}^{10} \frac{k}{3^k}.$

C9. Use the formula of Example 3 to check your answers in Problems C5 and C6.

C10. Give to five significant figures:

(a) $\displaystyle\sum_{k=0}^{100} \left(\frac{1}{2}\right)^k,$ (b) $\displaystyle\sum_{k=0}^{100} \left(\frac{1}{5}\right)^k,$ (c) $\displaystyle\sum_{k=0}^{100} (0.8)^k.$

5 *The Definition of the Definite Integral*

The definite integral originates with the problem of computing the area of a region bounded by one or more curves. We begin with a very simple case.

Example 1. Find the area of the region $\mathscr{F}$ bounded above by $y = x^2$, below by the segment $0 \le x \le 1$ of the x-axis, and on the side by the vertical line $x = 1$ (see Fig. 2, 3, or 4). For brevity we call $\mathscr{F}$ the region under the curve $y = x^2$ from $x = 0$ to $x = 1$ and denote the area by $A(\mathscr{F})$ (read "A of $\mathscr{F}$").

Solution. Our plan is to estimate this area using a finite number of rectangles and then to make the estimate better and better by letting the number of rectangles used approach infinity.

As indicated in Fig. 3, one set of rectangles lies entirely within the region $\mathscr{F}$. In Fig. 4 we show a second set of rectangles whose union contains $\mathscr{F}$. Consequently,

(31)   area of rectangles in Fig. 3 $\le A(\mathscr{F}) \le$ area of rectangles in Fig. 4.

Before proceeding further with this example, we need to set up a rather elaborate notation so that we can express our relations [such as (31)] more accurately.

We divide the interval $0 \le x \le 1$ into n subintervals all of the same length, and we indicate the division points by

$$0 = x_0, x_1, x_2, \ldots, x_n = 1.$$

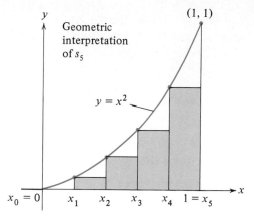

FIGURE 3

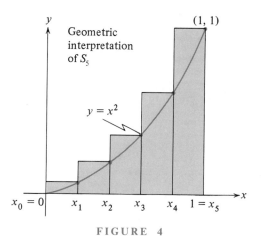

FIGURE 4

In Figs. 3 and 4, we have selected $n = 5$, and with this choice of n, we have $x_k = k/5$, for $k = 0, 1, 2, 3, 4, 5$.

For arbitrary n, we refer to the interval $x_{k-1} \leqq x \leqq x_k$ as the kth subinterval. To describe accurately the rectangles shown in Figs. 3 and 4, we set

(32 $$m_k = \text{minimum of } x^2 \text{ in the } k\text{th subinterval}$$

and

(33) $$M_k = \text{maximum of } x^2 \text{ in the } k\text{th subinterval.}$$

With the kth subinterval as a base we build a rectangle with height m_k for $k = 1, 2, \ldots, n$. This gives the set of rectangles shown in Fig. 3. If we use M_k as the height in place of m_k, then we obtain the rectangles in Fig. 4.

Now the area of a rectangle is the base $x_k - x_{k-1}$ times the height. Consequently, the inequality (31) can be written as

$$(34) \qquad \sum_{k=1}^{n} m_k(x_k - x_{k-1}) \leq A \leq \sum_{k=1}^{n} M_k(x_k - x_{k-1}).$$

We use s_n to denote the sum of the areas of the (smaller) rectangles on the left side, and we use S_n to denote the sum of the areas of the (larger) rectangles on the right side. Then (34) can be condensed to

$$(35) \qquad s_n \leq A \leq S_n.$$

Let us actually do these computations when $n = 5$. Since $f(x) = x^2$ is an increasing function, the minimum value of x^2 in the kth subinterval always occurs at the left end point and the maximum value of x^2 always occurs at the right end point. Hence

$$(36) \qquad m_k = \text{minimum of } x^2 \text{ for } \frac{k-1}{5} \leq x \leq \frac{k}{5} \quad \text{is} \quad \frac{(k-1)^2}{25}$$

and

$$(37) \qquad M_k = \text{maximum of } x^2 \text{ for } \frac{k-1}{5} \leq x \leq \frac{k}{5} \quad \text{is} \quad \frac{k^2}{25}.$$

Since $x_k - x_{k-1} = 1/5$ for every k, the sums s_5 and S_5 in (34) and (35) are

$$s_5 = \frac{0^2}{25} \cdot \frac{1}{5} + \frac{1^2}{25} \cdot \frac{1}{5} + \frac{2^2}{25} \cdot \frac{1}{5} + \frac{3^2}{25} \cdot \frac{1}{5} + \frac{4^2}{25} \cdot \frac{1}{5} = \frac{30}{125} = \frac{6}{25}$$

and

$$S_5 = \frac{1^2}{25} \cdot \frac{1}{5} + \frac{2^2}{25} \cdot \frac{1}{5} + \frac{3^2}{25} \cdot \frac{1}{5} + \frac{4^2}{25} \cdot \frac{1}{5} + \frac{5^2}{25} \cdot \frac{1}{5} = \frac{55}{125} = \frac{11}{25}.$$

We have proved that

$$(38) \qquad \frac{6}{25} < A < \frac{11}{25}.$$

Although this in itself is not very exciting, it points the way to the construction of better and better estimates of A, that will eventually lead to the exact value of A.

In place of $n = 5$, we now let n be an arbitrary positive integer. Then (36) and (37) are replaced by

$$(39) \qquad m_k = \text{minimum of } x^2 \text{ for } \frac{k-1}{n} \leq x \leq \frac{k}{n} \quad \text{is} \quad \left(\frac{k-1}{n}\right)^2$$

and

(40) $M_k = \text{maximum of } x^2 \text{ for } \dfrac{k-1}{n} \leqq x \leqq \dfrac{k}{n}$ is $\left(\dfrac{k}{n}\right)^2.$

Furthermore,

$$s_n = \sum_{k=1}^{n} \left(\frac{k-1}{n}\right)^2 (x_k - x_{k-1}) = \sum_{k=1}^{n} \frac{(k-1)^2}{n^2} \cdot \frac{1}{n} = \frac{1}{n^3} \sum_{k=1}^{n} (k-1)^2$$

and

$$S_n = \sum_{k=1}^{n} \left(\frac{k}{n}\right)^2 (x_k - x_{k-1}) = \sum_{k=1}^{n} \frac{k^2}{n^2} \cdot \frac{1}{n} = \frac{1}{n^3} \sum_{k=1}^{n} k^2.$$

We can apply here the formula proved in Problem 21 of Exercise 3, namely

(41) $$\sum_{k=1}^{n} k^2 = \frac{n(n+1)(2n+1)}{6}.$$

When we apply this to s_n, we must replace n by $n-1$ in (41). Then

$$s_n = \frac{1}{n^3} \frac{(n-1)n(2n-1)}{6} = \frac{1}{6}\left(1 - \frac{1}{n}\right)\left(2 - \frac{1}{n}\right),$$

and

$$S_n = \frac{1}{n^3} \frac{n(n+1)(2n+1)}{6} = \frac{1}{6}\left(1 + \frac{1}{n}\right)\left(2 + \frac{1}{n}\right).$$

Since $s_n \leqq A \leqq S_n$ for every positive integer n, we have

(42) $$\frac{1}{6}\left(1 - \frac{1}{n}\right)\left(2 - \frac{1}{n}\right) \leqq A \leqq \frac{1}{6}\left(1 + \frac{1}{n}\right)\left(2 + \frac{1}{n}\right).$$

Now take the limit in (42) as $n \to \infty$. Then A is squashed between two convergent sequences both of which have the same limit, $1/3$.

We have computed the area under the parabola $y = x^2$ between $x = 0$ and $x = 1$ (just as Archimedes did over 2000 years ago) and obtained the same exact value, namely $1/3$. ●

We now go from this specific example to the general theory. We let $f(x)$ be an arbitrary bounded function defined on the interval $a \leqq x \leqq b$. Although such a function may still be rather wild, we will picture the function as positive and continuous (see Fig. 5). (When $f(x)$ is continuous and positive, we let $\mathscr{F}$ be the region bounded above by the curve $y = f(x)$, below by the x-axis, and on the two sides by suitable segments of the lines $x = a$ and $x = b$. A typical region of this type is shown in Fig. 5. To avoid lengthy descriptions, we will henceforth describe this $\mathscr{F}$ as *the region under the curve* $y = f(x)$ *between a and b*).

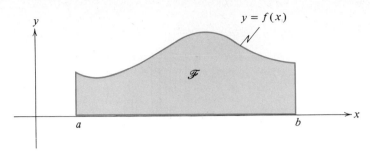

FIGURE 5

We divide the interval $a \leqq x \leqq b$ into n subintervals using $n + 1$ distinct points x_0, $x_1, x_2, \ldots, x_n$. The notation is selected so that $x_0 = a$, $x_n = b$, and all the points are in order on the line, namely

$$(43) \qquad\qquad a = x_0 < x_1 < x_2 < \cdots < x_n = b.$$

Any set $\mathscr{P}_n = \{x_0, x_1, \ldots, x_{n-1}, x_n\}$ that satisfies the inequalities (43) is called a *partition* of the interval $a \leqq x \leqq b$. We number these subintervals in the natural way; that is, the first subinterval is the interval $x_0 \leqq x \leqq x_1$, the second subinterval is $x_1 \leqq x \leqq x_2$, and in general the kth subinterval is the interval $x_{k-1} \leqq x \leqq x_k$, where $k = 1, 2, 3, \ldots, n$. Thus the nth subinterval is the interval $x_{n-1} \leqq x \leqq x_n = b$ (see Fig. 6).

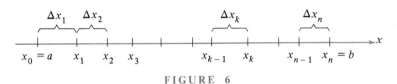

FIGURE 6

We let Δx_k denote the length of the kth subinterval so that

$$(44) \qquad \Delta x_1 = x_1 - x_0, \qquad \Delta x_2 = x_2 - x_1, \qquad \ldots, \qquad \Delta x_n = x_n - x_{n-1},$$

and we let $\mu(\mathscr{P}_n)$ denote the maximum of these lengths. The number $\mu(\mathscr{P}_n)$ is called the *mesh* of the partition, or the *norm* of the partition $\mathscr{P}_n$.

Turning now to the function $f(x)$, for each of these n subintervals, we let

$$(45) \qquad\qquad m_k = \text{minimum of } f(x) \text{ in the interval } x_{k-1} \leqq x \leqq x_k,$$

$$(46) \qquad\qquad M_k = \text{maximum of } f(x) \text{ in the interval } x_{k-1} \leqq x \leqq x_k.$$

Using these numbers, we form two sums: a lower sum s_n and an upper sum S_n defined by

$$(47) \qquad\qquad s_n = m_1 \Delta x_1 + m_2 \Delta x_2 + \cdots + m_n \Delta x_n = \sum_{k=1}^{n} m_k \Delta x_k$$

and

(48) $$S_n = M_1 \Delta x_1 + M_2 \Delta x_2 + \cdots + M_n \Delta x_n = \sum_{k=1}^{n} M_k \Delta x_k.$$

Now consider the limit of s_n and S_n as the number of subdivisions of $a \leq x \leq b$ approaches infinity and the maximum of the lengths of the subintervals approaches zero. This is symbolized by writing

(49) $$\lim_{\substack{n \to \infty \\ \mu(\mathscr{P}_n) \to 0}} s_n \quad \text{and} \quad \lim_{\substack{n \to \infty \\ \mu(\mathscr{P}_n) \to 0}} S_n.$$

If we select the partitions in a malicious manner, we can find a sequence of partitions $\mathscr{P}_1$, $\mathscr{P}_2, \ldots, \mathscr{P}_n, \ldots$ for which $n \to \infty$ but $\mu(\mathscr{P}_n) \not\to 0$. In the reverse direction, if $\mu(\mathscr{P}_n) \to 0$ for a sequence of partitions, then the number of subintervals must approach infinity. Consequently, we can replace (49) with

(50) $$\lim_{\mu(\mathscr{P}_n) \to 0} s_n \quad \text{and} \quad \lim_{\mu(\mathscr{P}_n) \to 0} S_n$$

with exactly the same meaning. Frequently, one writes (erroneously)

(51) $$\lim_{n \to \infty} s_n \quad \text{and} \quad \lim_{n \to \infty} S_n,$$

but this is permissible if we understand that $\mu(\mathscr{P}_n) \to 0$ in (51).

The definite integral is just the common limit in (49) or (51) whenever there is one. We summarize the above in

---

DEFINITION 2 (The Definite Integral). Let $f(x)$ be continuous in $a \leq x \leq b$. Suppose there is a number L such that for every sequence of partitions of $a \leq x \leq b$, for which $\mu(\mathscr{P}_n) \to 0$, we have

(52) $$\lim_{\mu(\mathscr{P}_n) \to 0} s_n = L \quad \text{and} \quad \lim_{\mu(\mathscr{P}_n) \to 0} S_n = L.$$

Then L is called the *definite integral* of $f(x)$ from a to b. This is symbolized by writing either

(53) $$L = \int_a^b f(x)\, dx \quad \text{or} \quad L = \int_a^b f.$$

The numbers a and b are called the lower and upper limits[1] of the integral and $f(x)$ is called the *integrand*.

---

[1] Do not confuse "limits of the integral" with the "limits" we have studied earlier. We are now using the *same* word with a *new* meaning.

Many writers prefer the second form in (53) because the integral does not depend on x, only on the function f and the end points a and b. But the first form is more useful, so we will retain it.

It is not necessary to require that $f(x)$ be continuous in this definition, but it is a convenience that simplifies matters. If $f(x)$ is not continuous, then it may not have a minimum or maximum in certain of the subintervals, and if this is the case the symbols m_k and M_k in (45) and (46) are meaningless. In the general theory, the minimum in (45) is replaced by the "greatest lower bound of $f(x)$," and the maximum in (46) is replaced by "the least upper bound of $f(x)$." But we wish to avoid these more sophisticated concepts.

We have already seen a geometric interpretation for the sums s_n and S_n in Figs. 3 and 4. We now go one step further and look for a geometric interpretation of the definite integral. Suppose that in addition to being continuous in $a \leqq x \leqq b$, the function $f(x)$ is also positive[1] in $a < x < b$.

We assume (temporarily) that the region $\mathscr{F}$ has an area which we denote by $A(\mathscr{F})$. We can estimate this area from below and above by considering the areas of two sets of rectangles $\mathscr{R}_1, \mathscr{R}_2, \ldots, \mathscr{R}_n$ and $\mathscr{T}_1, \mathscr{T}_2, \ldots, \mathscr{T}_n$ as pictured in Figs. 7 and 8. For $k = 1, 2, \ldots, n$, the rectangle $\mathscr{R}_k$ has for its base the interval $x_{k-1} \leqq x \leqq x_k$, and has height m_k.

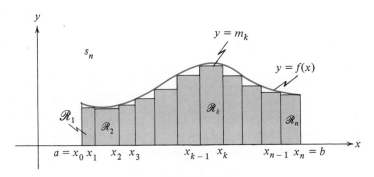

FIGURE 7

FIGURE 8

---

[1] We specify an open interval here because we want to admit the possibility that $f(x)$ is zero at the end points of the interval.

The rectangle $\mathcal{T}_k$ has the same base but has height M_k. For the areas of these rectangles we have $A(\mathcal{R}_k) = m_k \, \Delta x_k$ and $A(\mathcal{T}_k) = M_k \, \Delta x_k$, $k = 1, 2, \ldots, n$. Then the lower sum defined by (47) gives

(54) $$s_n = A(\mathcal{R}_1) + A(\mathcal{R}_2) + \cdots + A(\mathcal{R}_n) \leqq A(\mathcal{F}),$$

(see Fig. 7). The upper sum (48) gives

(55) $$S_n = A(\mathcal{T}_1) + A(\mathcal{T}_2) + \cdots + A(\mathcal{T}_n) \geqq A(\mathcal{F})$$

(see Fig. 8). It is now intuitively obvious that these sums s_n and S_n give approximations to $A(\mathcal{F})$ and that these approximations become closer as $\mu(\mathcal{P}_n) \to 0$. Consequently, we expect that the limits in (52) exist and that

(56) $$\int_a^b f(x) \, dx = A(\mathcal{F}).$$

Even if $f(x)$ is not always positive in $a < x < b$, the above considerations strongly suggest

---

THEOREM 2 (PWO). If $f(x)$ is continuous in $a \leqq x \leqq b$, then the definite integral exists.

This means that there is a number L such that $s_n \to L$ and $S_n \to L$ for every sequence of partitions for which $\mu(\mathcal{P}_n) \to 0$.

A correct proof of Theorem 2 is long and tedious, and so we omit it.[1] However, a little discussion is in order. We first reiterate that in Definition 2 and Theorem 2 the function may be negative, but in (56) we are assuming that $f(x) \geqq 0$. This is not a major stumbling block in the proof of Theorem 2. Indeed, if $f(x) < 0$, we can always add some constant C such that $F(x) \equiv f(x) + C$ is positive in $a \leqq x \leqq b$. Consequently, if we could prove Theorem 2 for positive functions, then it would be an easy matter to deduce Theorem 2 for an arbitrary continuous function.

Now suppose that $f(x) > 0$ in $a < x < b$. We have already remarked that in this case

$$\lim_{\mu(\mathcal{P}_n) \to 0} s_n = A(\mathcal{F}) \qquad \text{and} \qquad \lim_{\mu(\mathcal{P}_n) \to 0} S_n = A(\mathcal{F}).$$

Doesn't this prove that the two limits exist, because both give the area of the region $\mathcal{F}$ shown in Fig. 5? If so, then Theorem 2 is proved. There is only a "small" error in this approach—an error in logic. Indeed, the area of a region bounded by curves *has not been defined* and we do not know that $\mathcal{F}$ has an area. If we grant that a region of the type shown in Fig. 5 always has an area, that is, a number $A(\mathcal{F})$ that satisfies the inequalities (54) and (55) for every partition, then it is almost a trivial matter to prove Theorem 2. Unfortu-

---

[1] See footnote 2, page 197.

nately, the correct order is just the reverse. We must first prove Theorem 2 and then we can make

---

DEFINITION 3 (Area). Let $f(x)$ be positive in $a < x < b$ and continuous in $a \leq x \leq b$. Let $\mathscr{F}$ be the region under the curve $y = f(x)$ between $x = a$ and $x = b$. Then

(56)
$$A(\mathscr{F}) \equiv \int_a^b f(x)\, dx.$$

SUMMARY. *In view of the above discussion, Theorem 2 is intuitively obvious. We omit the proof and henceforth assume* (*in this book*) *that Theorem 2 is true.*[1]

Example 2. Show that

(57)
$$\int_0^2 (x - 1)\, dx = 0.$$

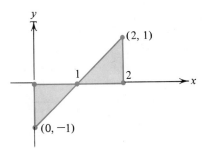

FIGURE 9

Solution. The graph of $y = x - 1$ is shown in Fig. 9. We divide the interval $0 \leq x \leq 2$ into n equal parts. Then each subinterval has length $2/n$ and the points of the partition are $0, 2/n, 4/n, 6/n, \ldots, 2(n-1)/n, 2n/n = 2$. To establish (57), it is sufficient to compute either the limit of the lower sums s_n or the limit of the upper sums S_n, since by Theorem 2 these two limits must be the same. In this example we compute the lower sum s_n. For this particular function the minimum m_k always occurs at the left end point of the kth subinterval. Since this is $x_{k-1} = 2(k-1)/n$,

---

[1] See the discussion at the beginning of Section 7 of Chapter 5 (pages 196–197).

we have

$$m_k = x_{k-1} - 1 = \frac{2(k-1)}{n} - 1.$$

Then the lower sum, equation (47), for this specific function is

$$s_n = \sum_{k=1}^{n} \left[\frac{2(k-1)}{n} - 1 \right] \frac{2}{n} = \sum_{k=1}^{n} \frac{4(k-1)}{n^2} - \sum_{k=1}^{n} \frac{2}{n}$$

$$= \frac{4}{n^2}[0 + 1 + 2 + \cdots + (n-1)] - \frac{2}{n}(1 + 1 + \cdots + 1).$$

The second sum is merely $2n/n = 2$. By Example 2 of Section 4,

$$\frac{4}{n^2}[0 + 1 + 2 + \cdots + (n-1)] = \frac{4}{n^2} \cdot \frac{(n-1)n}{2}.$$

Hence

$$s_n = \frac{4}{n^2} \cdot \frac{(n-1)n}{2} - 2 = \frac{2(n-1)}{n} - 2 = 2 - \frac{2}{n} - 2 = -\frac{2}{n}.$$

Consequently, as $n \to \infty$ we see that $s_n \to 0$, as required. ●

We leave it to the reader to compute the area of the shaded region in Fig. 9 and to explain why this is not given by the definite integral (57).

In Examples 1 and 2 we used partitions $\mathscr{P}_n$ consisting of equally spaced points. However by Definition 2, any choice of partitions $\mathscr{P}_n$ for which $\mu(\mathscr{P}_n) \to 0$ would suffice. The advantage of equal subintervals is that it greatly facilitates the computation of upper and lower sums.

EXERCISE 4

1. Let $f(x) = x$, $a = 0$, and $b = 1$. Using a partition of the interval $0 \leq x \leq 1$ into n equal parts, compute s_n and S_n for (a) $n = 4$, (b) $n = 8$, and (c) $n = 10$.

2. Arrange the numbers s_4, S_4, s_8, S_8, s_{10}, and S_{10} found in Problem 1 in an increasing sequence. Compute $S_n - s_n$ for $n = 4$, 8, and 10.

3. Use the method of this section to compute

$$\int_0^1 x\, dx.$$

HINT: If the interval $0 \leq x \leq 1$ is divided into n equal parts, then

$$s_n = \frac{0}{n} \cdot \frac{1}{n} + \frac{1}{n} \cdot \frac{1}{n} + \frac{2}{n} \cdot \frac{1}{n} + \cdots + \frac{n-1}{n} \cdot \frac{1}{n}$$

and

$$S_n = \frac{1}{n} \cdot \frac{1}{n} + \frac{2}{n} \cdot \frac{1}{n} + \frac{3}{n} \cdot \frac{1}{n} + \cdots + \frac{n}{n} \cdot \frac{1}{n}.$$

Use Example 2 of Section 4 to evaluate either one of these sums and then let $n \to \infty$.

4. Follow the methods used in Problem 3 to compute

(a) $\displaystyle\int_0^2 x \, dx.$ (b) $\displaystyle\int_0^1 (3 + x) \, dx.$ (c) $\displaystyle\int_1^5 (6 - x) \, dx.$

In each case draw a figure showing the function and the associated region. Check your work by computing the area using formulas from geometry.

5. Use the methods of this section to obtain the formula $A = bh/2$ for the area of a right triangle with base b and height h. HINT: Let $y = hx/b$ in the interval $0 \leq x \leq b$. If this interval is divided into n equal parts, then

$$S_n = \frac{h}{n} \cdot \frac{b}{n} + \frac{2h}{n} \cdot \frac{b}{n} + \frac{3h}{n} \cdot \frac{b}{n} + \cdots + \frac{nh}{n} \cdot \frac{b}{n} = \frac{bh}{n^2} \cdot \frac{n(n+1)}{2} = \frac{bh}{2}\left(1 + \frac{1}{n}\right).$$

6. Use the methods of this section to verify that

(a) $\displaystyle\int_0^2 x^2 \, dx = \frac{8}{3}.$ (b) $\displaystyle\int_0^1 3x^2 \, dx = 1.$ (c) $\displaystyle\int_0^2 Cx^2 \, dx = \frac{8C}{3}.$

$\star$7. Compute the definite integral $\displaystyle\int_a^b x^2 \, dx$, where $0 \leq a < b$ and interpret the result as an area. HINT: If the n subdivisions are equal, then $\Delta x_k = (b - a)/n \equiv \Delta x$, since they are all equal. Hence

$$S_n = \sum_{k=1}^n (a + k\,\Delta x)^2 \, \Delta x = \Delta x\left[\sum_{k=1}^n a^2 + 2a\,\Delta x \sum_{k=1}^n k + (\Delta x)^2 \sum_{k=1}^n k^2\right].$$

$\star\star$8. Compute the definite integral $\displaystyle\int_a^b x^3 \, dx$, where $0 \leq a < b$. HINT: Follow the method of Problem 7, and then use the formula proved in Problem 23 of Exercise 3.

9. Look carefully at the results of Problems 7 and 8 and form a conjecture (guess) for the value of $\displaystyle\int_a^b x^n \, dx$ when n is a positive integer. Is the conjecture correct when $n = 0$?

*10. Give an example of a sequence of partitions of $0 \leq x \leq 2$ for which $n \to \infty$ but $\mu(\mathscr{P}_n) \not\to 0$.

*11. Let $f(x)$ be the Dirichlet function: $f(x) = 0$ if x is rational and $f(x) = 1$ if x is irrational. For this function and the interval $0 \leq x \leq 3$, prove that $s_n = 0$ and $S_n = 3$ for every partition. Does this result show that Theorem 2 is false?

CALCULATOR PROBLEMS

In Problems C1 through C8, divide the given interval into 10 equal parts and compute s_{10} and S_{10} to five significant figures. Then compute $S_{10} - s_{10}$ in order to estimate the maximum error made in using either s_{10} or S_{10} for

$$\int_a^b f(x)\, dx.$$

C1. $f(x) = \sqrt{x}$, $a = 1$, $b = 2$.   C2. $f(x) = \sqrt{x}$, $a = 4$, $b = 5$.

C3. $f(x) = \dfrac{1}{x}$, $a = 1$, $b = 2$.   C4. $f(x) = \dfrac{1}{x}$, $a = 10$, $b = 11$.

C5. $f(x) = \dfrac{1}{x^2 + 1}$, $a = 0$, $b = 1$.   C6. $f(x) = \dfrac{1}{x^2 + 1}$, $a = 5$, $b = 6$.

C7. $f(x) = x^4$, $a = 0$, $b = 1$.   C8. $f(x) = x^4$, $a = 1$, $b = 2$.

The Fundamental Theorem of the Calculus   6

Suppose that in forming a sum, as in equations (47) and (48), we did not select the minimum m_k nor the maximum M_k of the function $f(x)$ in the kth subinterval. Instead, let us select in each subinterval an arbitrary point $x_k^{\star}$, where $x_{k-1} \leq x_k^{\star} \leq x_k$, and use as our multiplier $f(x_k^{\star})$ rather than m_k or M_k. By the definition of m_k and M_k as minimum and maximum values of $f(x)$ in the kth interval, we always have

$$m_k \leq f(x_k^{\star}) \leq M_k, \qquad k = 1, 2, \ldots, n,$$

and hence multiplying by the positive number Δx_k (the length of the kth subinterval),

(58) $m_k \Delta x_k \leq f(x_k^{\star}) \Delta x_k \leq M_k \Delta x_k, \qquad k = 1, 2, \ldots, n.$

Let $S_n^{\star}$ denote the sum with these new multipliers, namely,

(59) $S_n^{\star} = f(x_1^{\star}) \Delta x_1 + f(x_2^{\star}) \Delta x_2 + \cdots + f(x_n^{\star}) \Delta x_n = \displaystyle\sum_{k=1}^{n} f(x_k^{\star}) \Delta x_k.$

The sums in (59) are known as *Riemann sums*. The inequality (58) implies that for any Riemann sum, we have

$$\sum_{k=1}^{n} m_k \, \Delta x_k \leqq \sum_{k=1}^{n} f(x_k^{\star}) \, \Delta x_k \leqq \sum_{k=1}^{n} M_k \, \Delta x_k,$$

Using (47), (59), and (48) in turn, we find that this is equivalent to

(60) $$s_n \leqq S_n^{\star} \leqq S_n$$

for each n.

By Theorem 2, if $f(x)$ is continuous in $a \leqq x \leqq b$, then as $\mu(\mathscr{P}_n) \to 0$, both s_n and S_n tend to the same limit L. But the inequality (60) shows that $S_n^{\star}$ is "squeezed" between s_n and S_n and hence must approach the same limit. Thus

(61) $$\lim_{\mu(\mathscr{P}_n) \to 0} [f(x_1^{\star}) \, \Delta x_1 + f(x_2^{\star}) \, \Delta x_2 + \cdots + f(x_n^{\star}) \, \Delta x_n] = \int_a^b f(x) \, dx.$$

We have proved

THEOREM 3. Let $f(x)$ be continuous in $a \leqq x \leqq b$. Let $\mathscr{P}_1, \mathscr{P}_2, \ldots, \mathscr{P}_n, \ldots$ be a sequence of partitions for which $\mu(\mathscr{P}_n) \to 0$ as $n \to \infty$. Let $x_k^{\star}$ be an arbitrary point in the kth subinterval of $\mathscr{P}_n$, for $k = 1, 2, \ldots, n$, and form the approximating sum $S_n^{\star}$ defined by (59). Then equation (61) holds.

This theorem supplies the oil for a slick proof of

THEOREM 4 (The Fundamental Theorem of the Calculus). Let $f(x)$ be continuous in $a \leqq x \leqq b$ and let $F(x)$ be such that $F'(x) = f(x)$ for each point in $a \leqq x \leqq b$. Then

(62) $$\int_a^b f(x) \, dx = F(b) - F(a).$$

This is the theorem that permits us to compute many integrals without ever touching the sums s_n, S_n, or $S_n^{\star}$. Thus to compute a definite integral we only need to find an antiderivative $F(x)$ for the integrand $f(x)$.

Example 1. Compute $\displaystyle\int_1^2 15x^2 \, dx$.

Solution. Here $f(x) = 15x^2$. According to Theorem 4, we search for an antiderivative of $15x^2$. Clearly, $F(x) = 5x^3$ is one such. Then equation (62) gives

$$\int_1^2 15x^2\, dx = F(2) - F(1) = 5 \cdot 2^3 - 5 \cdot 1^3 = 40 - 5 = 35. \quad \bullet$$

Proof of Theorem 4. We partition the interval $a \leqq x \leqq b$ into n subintervals in the usual manner. By the Mean Value Theorem (Theorem 8, Chapter 5), there is in each subinterval $x_{k-1} \leqq x \leqq x_k$, an $x_k^\star$ such that for the antiderivative $F(x)$ we have

$$F(x_k) \quad - F(x_{k-1}) = F'(x_k^\star)(x_k - x_{k-1})$$
$$= f(x_k^\star)\, \Delta x_k.$$

We write one such equality for each subinterval:

$$F(x_1) \quad - F(a) \quad = f(x_1^\star)\, \Delta x_1$$
$$F(x_2) \quad - F(x_1) \quad = f(x_2^\star)\, \Delta x_2$$
$$F(x_3) \quad - F(x_2) \quad = f(x_3^\star)\, \Delta x_3$$
$$\vdots$$
$$F(x_{n-1}) - F(x_{n-2}) = f(x_{n-1}^\star)\, \Delta x_{n-1}$$
$$F(b) \quad - F(x_{n-1}) = f(x_n^\star)\, \Delta x_n.$$

Now add these n equations. On the left side, all the terms cancel pairwise except $F(a)$ and $F(b)$, and we have

(63) $$F(b) - F(a) = f(x_1^\star)\, \Delta x_1 + f(x_2^\star)\, \Delta x_2 + \cdots + f(x_n^\star)\, \Delta x_n.$$

Now take the limit in (63) as $n \to \infty$ and $\mu(\mathscr{P}_n) \to 0$. The right side is an approximating sum for the definite integral. But the left side is a constant, $F(b) - F(a)$, for each partition of $a \leqq x \leqq b$ into n subintervals, and for each positive integer n. Hence from Theorem 3 and equation (63),

$$F(b) - F(a) = \lim_{\mu(\mathscr{P}_n) \to 0} \sum_{k=1}^{n} f(x_k^\star)\, \Delta x_k = \int_a^b f(x)\, dx. \quad \blacksquare$$

COROLLARY. If $n \neq -1$ is any rational number for which x^n is continuous in $a \leqq x \leqq b$, then

$$\int_a^b x^n\, dx = \frac{1}{n+1}(b^{n+1} - a^{n+1}).$$

For example, if $n < 0$ and the interval $a \leq x \leq b$ contains $x = 0$, then x^n is not continuous, so the corollary does not apply. Another exceptional case occurs if $n = p/q$ in lowest terms and q is even. If $a < 0$, then x^n will require the computation of an even root of a negative number and this does not give a real number.

Proof of the Corollary. If $F(x) = \dfrac{1}{n+1} x^{n+1}$, then $F'(x) = x^n$.   ∎

Example 2. Compute the definite integral

$$\int_{-1}^{4} (20x^4 - 64x^3 - 3)\, dx.$$

Solution. According to Theorem 4, we need to find an antiderivative of $20x^4 - 64x^3 - 3$. There are many such, but any one will do. By inspection (or from the formulas of Section 2),

(64) $$F(x) = 4x^5 - 16x^4 - 3x$$

is one such function. Then by Theorem 4,

$$\int_{-1}^{4} (20x^4 - 64x^3 - 3)\, dx = F(4) - F(-1),$$

where F is defined by (64). It is easy to see that

$$F(4) = 4 \cdot 4^5 - 16 \cdot 4^4 - 12 = -12$$

and

$$F(-1) = 4(-1)^5 - 16(-1)^4 - 3(-1) = -17.$$

Hence

$$\int_{-1}^{4} (20x^4 - 64x^3 - 3)\, dx = -12 - (-17) = 5.   \bullet$$

When we have found an indefinite integral for $f(x)$, we carry along the limits a and b by writing

(65) $$\int_{a}^{b} f(x)\, dx = F(x)\,\Big|_{a}^{b},$$

where by definition the right side of (65) is just $F(b) - F(a)$. The technique for using this new notation is illustrated in

Example 3. Evaluate each of the given integrals.

(a) $\displaystyle\int_{1}^{2} x^4\, dx.$ (b) $\displaystyle\int_{-1}^{2} x^4\, dx.$ (c) $\displaystyle\int_{5}^{6} (x-5)^7\, dx.$

Solution. In each case an antiderivative for $f(x)$ is easy to find.

(a) $\displaystyle \int_1^2 x^4 \, dx = \frac{x^5}{5} \bigg|_1^2 = \frac{2^5}{5} - \frac{1^5}{5} = \frac{32}{5} - \frac{1}{5} = \frac{31}{5}.$

(b) $\displaystyle \int_{-1}^2 x^4 \, dx = \frac{x^5}{5} \bigg|_{-1}^2 = \frac{2^5}{5} - \frac{(-1)^5}{5} = \frac{32}{5} + \frac{1}{5} = \frac{33}{5}.$

(c) $\displaystyle \int_5^6 (x - 5)^7 \, dx = \frac{(x - 5)^8}{8} \bigg|_5^6 = \frac{1}{8} - 0 = \frac{1}{8}.$ ●

Example 4. Find the area of the region under the parabola

$$y = \frac{12 + 2x - x^2}{4}$$

between $x = -2$ and $x = 4$.

Solution. The parabola and the region $\mathscr{F}$ are shown in Fig. 10. By Definition 3 and Theorem 4,

$$A(\mathscr{F}) = \int_{-2}^4 \frac{12 + 2x - x^2}{4} \, dx = \frac{1}{4} \left(12x + x^2 - \frac{x^3}{3} \right) \bigg|_{-2}^4$$

$$= \frac{1}{4} \left(48 + 16 - \frac{64}{3} \right) - \frac{1}{4} \left(-24 + 4 - \frac{-8}{3} \right)$$

$$= \frac{64}{4} - \frac{16}{3} + \frac{20}{4} - \frac{2}{3} = 21 - \frac{16 + 2}{3} = 15. \ ●$$

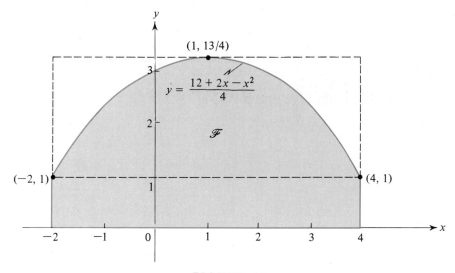

$(1, 13/4)$

$y = \dfrac{12 + 2x - x^2}{4}$

$\mathscr{F}$

$(-2, 1)$ $(4, 1)$

FIGURE 10

Of course, this answer is correct. But it is also reasonable because the smallest rectangle containing $\mathscr{F}$ has area $6 \times 13/4 = 19.5$, while the longest rectangle (not largest) contained in $\mathscr{F}$ has area $6 \times 1 = 6$ (see the dashed lines in Fig. 10). Clearly, we must have $6 < A(\mathscr{F}) < 19.5$, and $A(\mathscr{F}) = 15$ satisfies this condition. Such checks as these are useful in finding errors in computations.

Example 5. Evaluate

$$\int_{-2}^{3} x(x^2 - 1)^{1/2}\, dx.$$

Solution. An eager and trusting student will undoubtedly write

$$\int_{-2}^{3} x(x^2 - 1)^{1/2}\, dx = \frac{1}{2}\int_{-2}^{3}(x^2 - 1)^{1/2}(2x\, dx) = \frac{1}{2}\frac{(x^2 - 1)^{3/2}}{3/2}\Big|_{-2}^{3}$$

$$= \frac{1}{3}(9 - 1)^{3/2} - \frac{1}{3}(4 - 1)^{3/2} = \frac{16}{3}\sqrt{2} - \sqrt{3}.$$

However, this result does not even have the respectability of being wrong. Rather it is pure nonsense because the "function" $f(x) = x(x^2 - 1)^{1/2}$ is not even defined in a portion of the interval of integration. Indeed, if x is in the open interval $-1 < x < 1$, then $x^2 - 1 < 0$, and $(x^2 - 1)^{1/2}$ is not a real number. Hence the problem posed is actually meaningless. ●

This example is intended to show the need for phrases "Let f be a well-defined function," and "Let f be continuous." Annoying as they may be, they serve somewhat as "flashing yellow lights" to help us avoid an error. But they also tend to slow us down and prevent the pleasure of a fast trip to our objective.

EXERCISE 5

In Problems 1 through 8, evaluate the given definite integral.

1. $\displaystyle\int_{1}^{5}(x + 2)\, dx.$

2. $\displaystyle\int_{1}^{5}(u + 2)\, du.$

3. $\displaystyle\int_{-2}^{2} x^3\, dx.$

4. $\displaystyle\int_{-2}^{2} x^4\, dx.$

5. $\displaystyle\int_{0}^{5}\frac{x\, dx}{\sqrt{144 + x^2}}.$

6. $\displaystyle\int_{1}^{3}\left(x - \frac{2}{x}\right)^2 dx.$

7. $\displaystyle\int_{-1}^{1}(ax^2 + bx + c)\, dx.$

8. $\displaystyle\int_{0}^{B} ax^2\, dx.$

In Problems 9 through 15, find the area of the figure bounded by the x-axis, the given curve, and the two given vertical lines. In each case first sketch the figure and guess an approximate value for the area by a consideration of suitable rectangles.

9. $y = x^2$, $x = 0$, $x = 1$.   10. $y = x^4$, $x = 0$, $x = 1$.

11. $y = 2x + 5$, $x = -1$, $x = 6$.   12. $y = x^2 + 1$, $x = -2$, $x = 2$.

13. $y = x\sqrt{6x^2 + 1}$, $x = 0$, $x = 2$.   14. $y = \dfrac{1}{x^2}$, $x = 1$, $x = 100$.

15. $y = \dfrac{x}{\sqrt{x^2 + 1}}$, $x = 1$, $x = \sqrt{7}$.

*16. The figure bounded by the x-axis, the line $y = mx + b$, and the lines $x = 0$ and $x = a$ is a trapezoid if the constants a, b, and m are all positive. It is known that the area of such a trapezoid is its width a times the average of the two vertical sides b and $ma + b$. Prove this rule by using the calculus to find the area. Show that if $m = 0$, integration gives the area of a rectangle.

*17. Prove the theorem of Archimedes mentioned in Section 1. In other words, prove that the area of the figure bounded by the parabola $y = ax^2$, the x-axis, and the lines $x = 0$ and $x = B$ is $BH/3$, where $H = aB^2$ is the height of the parabola.

*18. Prove that if n is a positive even integer, then

$$\int_{-a}^{a} x^n \, dx = 2 \int_{0}^{a} x^n \, dx,$$

and if n is an odd positive integer,

$$\int_{-a}^{a} x^n \, dx = 0.$$

CALCULATOR PROBLEMS

C1. Up to this point we do not have an antiderivative for $1/x$, and hence to compute

$$\int_{a}^{b} \frac{1}{x} \, dx$$

we must use approximations. For $a = 1$ and $b = 2$, divide the interval into $n = 20$ equal parts and compute s_n and S_n to five significant figures.

C2. Do Problem C1 if $a = 2$ and $b = 4$.

C3. Do Problem C1 if $a = 4$ and $b = 8$.

C4. Can you guess at a relation among the integrals estimated in Problems C1, C2, and C3?

7 *Properties of the Definite Integral*

We recall that we have defined the definite integral only for continuous functions, but the definition can be modified to include discontinuous functions as well. If $f(x)$ is continuous, then the two limits, $\lim s_n$ and $\lim S_n$, always exist and are equal. If $f(x)$ is not continuous, these limits may still exist and be equal. If $\lim s_n = \lim S_n$ as $\mu(\mathscr{P}_n) \to 0$, then we say that $f(x)$ is *integrable over the interval* $a \leqq x \leqq b$. Otherwise, we say that $f(x)$ is *not integrable*. We have already had one example of a function that is not integrable (see Problem 11 of Exercise 4).

The definite integral has certain important properties that are very helpful.

(A) *The homogeneous property.* If c is constant, and $f(x)$ is integrable over $a \leqq x \leqq b$, then

$$(66) \qquad \int_a^b cf(x)\,dx = c\int_a^b f(x)\,dx.$$

This result states that "We can pass through the integral sign with a multiplicative constant."

(B) *The additive property.* If $f(x)$ and $g(x)$ are both integrable over $a \leqq x \leqq b$, then

$$(67) \qquad \int_a^b [f(x) + g(x)]\,dx = \int_a^b f(x)\,dx + \int_a^b g(x)\,dx.$$

This result states that "We can integrate a sum by adding the integrals of each term in the sum."

(C) *The linear property.* If c_1 and c_2 are any two constants, and $f(x)$ and $g(x)$ are integrable over $a \leqq x \leqq b$, then

$$(68) \qquad \int_a^b [c_1 f(x) + c_2 g(x)]\,dx = c_1\int_a^b f(x)\,dx + c_2\int_a^b g(x)\,dx.$$

Obviously, (68) is merely a combination of (66) and (67), so it is sufficient to prove the first two properties.

(D) *The interval sum property.* If $f(x)$ is integrable over $a \leqq x \leqq b$, and $a < c < b$, then

$$(69) \qquad \int_a^b f(x)\,dx = \int_a^c f(x)\,dx + \int_c^b f(x)\,dx.$$

If we assume that $f(x) > 0$ in $a \leqq x \leqq b$, then equation (69) has a geometric interpretation that makes it easy to understand. Indeed, if we examine Fig. 11, we see that the

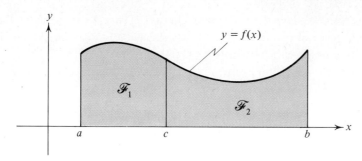

FIGURE 11

terms on the right side of (69) give $A(\mathscr{F}_1) + A(\mathscr{F}_2)$, while the term on the left side is $A(\mathscr{F}_1 \cup \mathscr{F}_2)$. When viewed in this way, equation (69) is obviously true.

We are not quite prepared to prove such general results. However, this is no great loss, because we will use these properties only when the functions involved are continuous. This assumption of continuity simplifies the proofs of the formulas (66), (67), (68), and (69).

THEOREM 5 (PLE). Let $f(x)$ and $g(x)$ be continuous in $a \le x \le b$ and let c_1 and c_2 be arbitrary constants. Then equations (66) through (69) hold.

As an illustration of the method used in proving Theorem 5, we will prove equation (67) which is the same as (68) when $c_1 = c_2 = 1$.

Let $\mathscr{P}_n$ be a partition of $a \le x \le b$. Let $S_n^{\star}(f)$ refer to an approximating sum for $f(x)$ and let $S_n^{\star}(g)$ refer to an approximating sum for $g(x)$. Then

$$(70) \qquad S_n^{\star}(f) = \sum_{k=1}^{n} f(x_k^{\star})\,\Delta x_k \qquad \text{and} \qquad S_n^{\star}(g) = \sum_{k=1}^{n} g(x_k^{\star})\,\Delta x_k.$$

Adding the equations in (70), we obtain

$$(71) \qquad S_n^{\star}(f) + S_n^{\star}(g) = \sum_{k=1}^{n} [f(x_k^{\star}) + g(x_k^{\star})]\,\Delta x_k,$$

an approximating sum for the integral of the function $[f(x) + g(x)]$. Taking the limit in (71) as $\mu(\mathscr{P}_n) \to 0$ gives

$$(67) \qquad \int_a^b f(x)\,dx + \int_a^b g(x)\,dx = \int_a^b [f(x) + g(x)]\,dx.$$

Example 1. Evaluate the definite integral

$$(72) \qquad I \equiv \int_5^7 [3x\sqrt{x^2 - 9} + 24(x - 5)^2 + 666{,}000(x - 6)^5]\,dx.$$

Solution. Here we have the form of equation (67) except that we have three terms in the sum in place of two terms. But clearly this equation can be extended to the sum of any finite number of terms.

An indefinite integral is obvious for the last two terms in (72). For the first term we set $u = x^2 - 9$ and hence $du = 2x\, dx$. Then

$$\int 3x \sqrt{x^2 - 9}\, dx = 3 \int \sqrt{x^2 - 9}\, x\, dx = \frac{3}{2} \int \sqrt{x^2 - 9}\, (2x\, dx)$$

$$= \frac{3}{2} \int u^{1/2}\, du = \frac{3}{2} \frac{u^{3/2}}{3/2} + C = u^{3/2} + C = (x^2 - 9)^{3/2} + C.$$

Consequently, by Theorem 4,

$$I_1 \equiv \int_5^7 3x \sqrt{x^2 - 9}\, dx = (x^2 - 9)^{3/2} \Big|_5^7 = 40^{3/2} - 16^{3/2} = 80\sqrt{10} - 64,$$

$$I_2 \equiv \int_5^7 24(x - 5)^2\, dx = 24\frac{(x - 5)^3}{3} \Big|_5^7 = 8(7 - 5)^3 - 0 = 64,$$

$$I_3 \equiv \int_5^7 666{,}000(x - 6)^5\, dx = 666{,}000 \frac{(x - 6)^6}{6} \Big|_5^7$$

$$= 111{,}000 - 111{,}000 = 0.$$

Hence $I = I_1 + I_2 + I_3 = 80\sqrt{10} - 64 + 64 + 0 = 80\sqrt{10}.$ ●

For simplicity, we divided the computation into three separate parts. With a little practice the student will learn to combine the parts into one, and thus save time and space. This is illustrated in

Example 2. Evaluate $I \equiv \int_1^2 [19x(x^2 - 1)^{1/2} - 10x(x^2 - 1)^{3/2}]\, dx.$

Solution. Set $u = x^2 - 1$ and hence $du = 2x\, dx$. Then

$$I = \frac{19}{2} \int_1^2 (x^2 - 1)^{1/2}(2x\, dx) - 5 \int_1^2 (x^2 - 1)^{3/2}(2x\, dx)$$

$$= \frac{19}{2} \int_{x=1}^{x=2} u^{1/2}\, du - 5 \int_{x=1}^{x=2} u^{3/2}\, du = \left(\frac{19}{2} \frac{u^{3/2}}{3/2} - 5 \frac{u^{5/2}}{5/2}\right) \Big|_{x=1}^{x=2}$$

$$= \left(\frac{19}{3} (x^2 - 1)^{3/2} - 2(x^2 - 1)^{5/2}\right) \Big|_1^2$$

$$= \frac{19}{3} 3\sqrt{3} - 2 \cdot 9\sqrt{3} - 0 = \sqrt{3}.$$ ●

So far we have defined the quantities that appear in equation (69) only if $a < c < b$. What happens to Theorem 5 if a, c, and b do not occur in that order on the real line? We want to define the definite integral so that equation (69) holds for any three points.

DEFINITION 4. Let $f(x)$ be integrable over an interval $\mathscr{I}$. If a is in $\mathscr{I}$, then

$$(73) \qquad \int_a^a f(x)\, dx \equiv 0.$$

If $b < a$ and both are in $\mathscr{I}$, then

$$(74) \qquad \int_a^b f(x)\, dx \equiv - \int_b^a f(x)\, dx.$$

THEOREM 6 (PLE). Let $f(x)$ be continuous in an interval $\mathscr{I}$. If a, b, and c are any three points in $\mathscr{I}$, then

$$(69) \qquad \int_a^b f(x)\, dx = \int_a^c f(x)\, dx + \int_c^b f(x)\, dx.$$

The complete proof requires a consideration of a large number of cases (see Exercise 6). It is sufficient to illustrate the method with just one case. Suppose that $b < a < c$ and all are in $\mathscr{I}$. By Theorem 5,

$$(75) \qquad \int_b^c f(x)\, dx = \int_b^a f(x)\, dx + \int_a^c f(x)\, dx.$$

By Definition 4, equation (74) applied to (75) yields

$$(76) \qquad - \int_c^b f(x)\, dx = - \int_a^b f(x)\, dx + \int_a^c f(x)\, dx.$$

Simple algebra converts (76) into (69).

EXERCISE 6

In Problems 1 through 8, compute the given definite integral.

1. $\displaystyle \int_0^4 (24x^{1/2} - 10x^{3/2})\, dx.$

2. $\displaystyle \int_1^4 (3x^{1/2} - 4x^{-1/2})\, dx.$

3. $\displaystyle\int_0^4 2x^{1/2}(12 - 5x)\,dx.$

4. $\displaystyle\int_1^4 \frac{3x - 4}{\sqrt{x}}\,dx.$

5. $\displaystyle\int_0^1 (6x\sqrt{x^2 + 1} + 4x)\,dx.$

6. $\displaystyle\int_0^4 6x(\sqrt{x^2 + 9} - 10)\,dx.$

7. $\displaystyle\int_0^1 \frac{10x^3 - 20x^{5/2} + 9x^2}{x^{3/2}}\,dx.$

8. $\displaystyle\int_0^4 15\sqrt{x}\,(\sqrt{x} - 1)^2\,dx$

9. Prove Theorem 6 in each of the following cases: (1) $a = c < b$, (2) $a < c = b$, and (3) $a = b = c$.

10. Three distinct points, a, b, and c, may be placed on a line in six different ways as far as order is concerned. We already know that equation (69) is true in the two cases $a < c < b$ and $b < a < c$. List the four remaining cases.

11. Prove that equation (69) is true in each of the remaining cases listed in Problem 10.

12. Assume that $c = -1$ in equation (66). In this case how would you interpret equation (66) as a relation among areas? Draw a picture to illustrate the relationship.

13. Suppose that $f(x)$ is continuous and $m \leq f(x) \leq M$ in $a \leq x \leq b$. Prove that

(77)
$$m(b - a) \leq \int_a^b f(x)\,dx \leq M(b - a).$$

HINT: Consider the Riemann sums for f.

14. Suppose that $f(x)$ and $g(x)$ are continuous and $g(x) \leq f(x)$ in $a \leq x \leq b$. Prove that

(78)
$$\int_a^b g(x)\,dx \leq \int_a^b f(x)\,dx.$$

15. Prove that if $f(x)$ is continuous in $a \leq x \leq b$, then

(79)
$$\left|\int_a^b f(x)\,dx\right| \leq \int_a^b |f(x)|\,dx.$$

HINT: Use Problem 14 and the inequalities $-|f(x)| \leq f(x) \leq |f(x)|$.

*16. Prove that:

(a) $\displaystyle\int_a^b \cos^2 x\,dx = b - a - \int_a^b \sin^2 x\,dx.$

(b) $\displaystyle\int_a^b \cos^2 x\,dx = \int_a^b \cos 2x\,dx + \int_a^b \sin^2 x\,dx.$

(c) $\displaystyle\int_0^{\pi/4} \sec^2 x\,dx = \frac{\pi}{4} + \int_0^{\pi/4} \tan^2 x\,dx.$

Areas 8

We have already considered the area of a region under the curve $y = f(x)$ between $x = a$ and $x = b$, when $f(x) > 0$ in $a < x < b$. We now suppose that the curve touches the x-axis or perhaps crosses it. Some of the possibilities are shown in Figs. 12 and 13. Under these circumstances we would hesitate to speak of the shaded set as the region *under* the curve, but we can describe it as the *figure bounded by the curve $y = f(x)$ and the x-axis, between $x = a$ and $x = b$.*

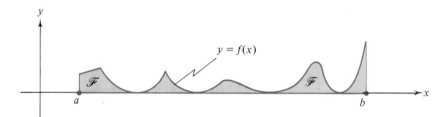

FIGURE 12

We first consider the case illustrated in Fig. 12 in which $f(x) \geq 0$ in $a \leq x \leq b$. We assume that $f(x)$ has zeros at $r_1, r_2, \ldots, r_{m-1}$ in the open interval $a < x < b$. These points divide $a \leq x \leq b$ into m subintervals. The area of the region $\mathscr{F}_k$ bounded by the curve and the x-axis, between $x = r_{k-1}$ and $x = r_k$, is (by Definition 3)

(80)
$$A(\mathscr{F}_k) = \int_{r_{k-1}}^{r_k} f(x)\, dx, \qquad k = 2, 3, \ldots, m - 1.$$

We can include the two end subintervals ($k = 1$ and $k = m$) in formula (80) if we set $r_0 = a$ and $r_m = b$, so we do this.

Clearly, we want the area function to be *additive*. This means that if $A(\mathscr{F})$ denotes the area of the figure bounded by the curve $y = f(x)$ and the x-axis, between $x = a$ and $x = b$, then

(81)
$$A(\mathscr{F}) = A(\mathscr{F}_1) + A(\mathscr{F}_2) + \cdots + A(\mathscr{F}_m).$$

Using (80), we have

(82)
$$A(\mathscr{F}) = \int_{a}^{r_1} f(x)\, dx + \int_{r_1}^{r_2} f(x)\, dx + \cdots + \int_{r_{m-1}}^{b} f(x)\, dx.$$

By equation (69) applied $m - 1$ times, equation (82) yields

(83)
$$A(\mathscr{F}) = \int_a^b f(x)\, dx.$$

We now take up the case in which the curve may cross the x-axis. Before stating a general result, let us examine

Example 1. Find the area of the figure $\mathscr{F}$ bounded by the curve

$$y = \frac{3x^2 - 18x + 15}{5}$$

and the x-axis, between $x = 0$ and $x = 6$.

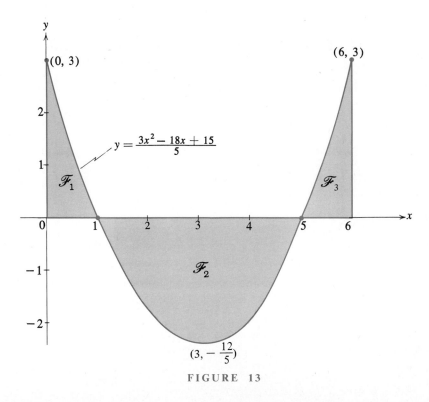

FIGURE 13

Solution. The set in question is shown shaded in Fig. 13. Since $f(x) = 0$ when $x = 1$ or $x = 5$, the set is the union of the three regions $\mathscr{F}_1$, $\mathscr{F}_2$, and $\mathscr{F}_3$, indicated in the figure. It is easy to compute $A(\mathscr{F}_1)$ and $A(\mathscr{F}_3)$. Indeed,

$$A(\mathscr{F}_1) = \frac{1}{5} \int_0^1 (3x^2 - 18x + 15)\, dx = \frac{1}{5}(x^3 - 9x^2 + 15x)\Big|_0^1 = \frac{7}{5},$$

$$A(\mathscr{F}_3) = \frac{1}{5} \int_5^6 (3x^2 - 18x + 15)\, dx = \frac{1}{5}(x^3 - 9x^2 + 15x)\Big|_5^6$$

$$= \frac{1}{5}(216 - 324 + 90) - \frac{1}{5}(125 - 225 + 75) = \frac{7}{5}.$$

We observe in passing that $A(\mathscr{F}_1) = A(\mathscr{F}_3)$, and it is clear from Fig. 13 that we should have expected this.

To find $A(\mathscr{F}_2)$, we turn the curve "upside down" and create a new region $\mathscr{F}_2^{\bigstar}$ as shown in Fig. 14. We want the area function to be invariant under any rigid motion.

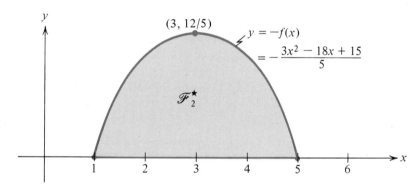

FIGURE 14

This means $A(\mathscr{F}_2) = A(\mathscr{F}_2^{\bigstar})$. On the other hand, the curve is turned up-side down by taking $y = -f(x)$ as the equation of the new boundary curve. Therefore,

$$A(\mathscr{F}_2) = \int_1^5 -f(x)\, dx = -\frac{1}{5} \int_1^5 (3x^2 - 18x + 15)\, dx$$

$$= -\frac{1}{5}(x^3 - 9x^2 + 15x)\Big|_1^5$$

$$= -\frac{1}{5}(125 - 225 + 75) + \frac{1}{5}(1 - 9 + 15) = \frac{32}{5}.$$

These three results give

(84) $$A(\mathscr{F}) = \frac{7}{5} + \frac{32}{5} + \frac{7}{5} = \frac{46}{5}. \quad \bullet$$

For contrast, let us integrate $f(x)$ from 0 to 6. We find that

$$I = \int_0^6 \frac{1}{5}(3x^2 - 18x + 15)\,dx = \frac{1}{5}(x^3 - 9x^2 + 15x)\,\Big|_0^6$$

(85)

$$= \frac{1}{5}(216 - 324 + 90) - 0 = -\frac{18}{5}.$$

We can observe that a change of sign in (84) also yields $-18/5$; indeed,

(86)
$$\frac{7}{5} - \frac{32}{5} + \frac{7}{5} = -\frac{18}{5} = I.$$

It is clear then that the definite integral in (85) also measures the area of $\mathscr{F}$, but in so doing, it gives the area of regions below the x-axis with a negative sign. This example suggests

---

THEOREM 7. Let $f(x)$ be continuous in $a \leq x \leq b$ and have a finite number of zeros in $a \leq x \leq b$. The figure $\mathscr{F}$ bounded by the curve $y = f(x)$, the x-axis, and the lines $x = a$ and $x = b$ is the union of several regions, some below the x-axis and others above the x-axis. If T_1 and T_2 denote the total areas of these two types of regions, respectively, then

(87)
$$A(\mathscr{F}) = \int_a^b |f(x)|\,dx = T_2 + T_1$$

and

(88)
$$\int_a^b f(x)\,dx = T_2 - T_1.$$

In our example, (87) is given by (84), where $T_2 = 14/5$ and $T_1 = 32/5$. Equation (88) is given by either (85) or (86). To distinguish between the two items in (87) and (88), we introduce some terminology. The quantity $T_1 + T_2$ given by (87) is the *area,* which may also be called the *geometric area* or the *absolute area.* The quantity $T_2 - T_1$ given by (88) will be called the *algebraic area* because it is a signed area, counting the areas of regions above the x-axis with a positive sign, and the areas of regions below the x-axis with a negative sign.

We omit a formal proof of Theorem 7 because the result is certainly obvious. The student who wants to run through a proof can start by writing

$$\int_a^b |f(x)|\,dx = \sum_{k=1}^m \int_{r_{k-1}}^{r_k} |f(x)|\,dx = \int_a^{r_1} |f(x)|\,dx + \cdots + \int_{r_{m-1}}^b |f(x)|\,dx,$$

where $r_0 = a$, $r_m = b$, and $r_1 < r_2 < \cdots < r_{m-1}$ are the distinct zeros of $f(x)$ in $a < x < b$.

EXERCISE 7

In Problems 1 through 10, compute (a) the definite integral of the given function over the given interval, and (b) the area of the figure bounded by the curve $y = f(x)$, the x-axis, and the vertical lines at the end points of the interval.

1. $f(x) = 4x - x^2$, $0 \leq x \leq 5$.   2. $f(x) = 4 - 4x^3$, $0 \leq x \leq 3$.

3. $f(x) = x^2 - 2x$, $1 \leq x \leq 4$.   4. $f(x) = x^2 + 3x$, $-1 \leq x \leq 2$.

5. $f(x) = x^2 - 4x + 3$, $0 \leq x \leq 4$.   *6. $f(x) = 9 - 3x^2$, $0 \leq x \leq 3$.

7. $f(x) = x^4 - x^2$, $-2 \leq x \leq 2$.   *8. $f(x) = x\sqrt{x^2 + 1}$, $-1 \leq x \leq 2$.

9. $f(x) = x - \dfrac{8}{\sqrt{x}}$, $1 \leq x \leq 16$.   10. $f(x) = x - \dfrac{8}{x^2}$, $1 \leq x \leq 4$.

Average Value of a Function. The average value of a function $f(x)$ over the interval $a \leq x \leq b$ is denoted by $AV(f(x))$ and is defined by

$$\frac{1}{b-a}\int_a^b f(x)\,dx.$$

If $f(x)$ is positive, then $AV(f(x))$ gives the height of a rectangle with base $b - a$ that has the same area as the area under the curve $y = f(x)$ between $x = a$ and $x = b$.

In Problems 11 through 18, find the average value of the given function over the given interval. Make a sketch showing the function and the average value for Problems 11, 12, 16, 17, and 18.

11. $f(x) = x$, $5 \leq x \leq 11$.   12. $f(x) = x^2$, $0 \leq x \leq 4$.

13. $f(x) = x$, $0 \leq x \leq b$.   14. $f(x) = x^2$, $0 \leq x \leq b$.

15. $f(x) = x^3$, $-c \leq x \leq c$.   16. $f(x) = x\sqrt{x^2 + 9}$, $0 \leq x \leq 4$.

17. $f(x) = \dfrac{x}{\sqrt{x^2 + 9}}$, $0 \leq x \leq 4$.   18. $f(x) = \dfrac{x}{(x^2 + 1)^2}$, $0 \leq x \leq 2$.

19. If $f(x)$ is an odd function, what can you say about $AV(f(x))$ over $-c \leq x \leq c$?

20. Prove that $AV(c + f(x)) = c + AV(f(x))$ when both averages are taken over the same interval.

21. Prove that $AV(cf(x)) = cAV(f(x))$ when both averages are taken over the same interval.

22. Suppose that $a < c < b$. Is it true that

$$AV(f(x)) \Big|_a^b = AV(f(x)) \Big|_a^c + AV(f(x)) \Big|_c^b$$

for every integrable function?

23. On a certain morning in the town of Chotshpot, Iceland, the temperature (in degrees centigrade) was given by the equation

$$y = -30 + 4t - \frac{t^2}{12},$$

where t is the time in hours. Find **(a)** the average temperature from midnight (where $t = 0$) to noon ($t = 12$) and **(b)** the extreme values of the temperature for that interval.

CALCULATOR PROBLEMS

In Problems C1 and C2, compute to five significant figures: (a) the definite integral of $f(x)$ over $a \leq x \leq b$, and (b) the area of the figure bounded by the curve $y = f(x)$ and the x-axis between $x = a$ and $x = b$.

C1. $f(x) = 1.357x - x^2$, $a = 0$, $b = 2.468$.
C2. $f(x) = x^3 - x^2 - 1$, $a = 0$, $b = 1.5$.

C3. Find $AV(f(x))$ for the function and interval in Problem C1.
C4. Find $AV(f(x))$ for the function and interval in Problem C2.

★9 The Integral as a Function of the Upper Limit

We recall that the two sums

$$(89) \qquad \sum_{k=1}^{5} k^2 \quad \text{and} \quad \sum_{t=1}^{5} t^2$$

are identical because each sum gives $1 + 4 + 9 + 16 + 25 = 55$. Since the sums in (89) do not depend upon k or t, these letters do not represent variables. When symbols are used in this way, they are called *dummy variables* (or *dummy indices*).

The same phenomenon also occurs in integration. For example,

$$(90) \qquad \int_1^4 x^2 \, dx = \int_1^4 t^2 \, dt = \int_1^4 z^2 \, dz = \int_1^4 \theta^2 \, d\theta$$

because each of the expressions in (90) is equal to

$$\frac{x^3}{3}\Big|_1^4 = \frac{64-1}{3} = 21.$$

Since the number 21 does not depend on x, t, z, or θ, these are dummy variables in (90).

To emphasize this point, many mathematicians prefer to use the more accurate notation

(91)
$$\int_a^b f,$$

rather than the more convenient notation

(92)
$$\int_a^b f(x)\,dx.$$

If we insist on the form (91) and try to be consistent, then the sums in (89) should be written

(93)
$$\sum_1^5 (\ )^2,$$

and the integrals in (90) should be written

(94)
$$\int_1^4 (\ )^2.$$

It is clear that the forms (89) and (90) are more attractive and satisfactory than the corresponding (93) and (94).

In most situations the choice of notation is really a trivial matter. However, suppose that we wish to regard the upper limit as a variable. Then the integral defines a new function $F(x)$. Our first impulse might be to write

(95)
$$F(x) = \int_a^x f(x)\,dx$$

for this new function. However, this notation is definitely confusing because x is used with two different meanings in one equation. First, it appears as an upper limit on the integral sign and there x is a true variable. Second, x is a dummy variable in the integrand. To be correct, one should use a different letter for the dummy variable and write

(96)
$$F(x) \equiv \int_a^x f(t)\,dt.$$

Many authors use (95) when they mean (96). This is acceptable as long as the reader is clear about the meaning of (95). The important fact about functions of the form (96) is stated in

---

THEOREM 8. Let f be continuous in an interval $\mathscr{I}$ and let a be any fixed point in $\mathscr{I}$. Let F be the function defined in $\mathscr{I}$ by

(96)
$$F(x) \equiv \int_a^x f(t)\, dt.$$

Then F is differentiable in $\mathscr{I}$, and for each x in $\mathscr{I}$,

(97)
$$\frac{d}{dx} F(x) = f(x).$$

This theorem provides a method of solving difficult problems in an easy way.

Example 1. Find a function F which has for its derivative the function

$$f(x) = (x^7 + 111x^4 - 5x^2 + 13x + 33)^{1/3}.$$

Solution. The problem of finding antiderivatives will be studied in a systematic way in Chapter 10, after we have learned how to differentiate the trigonometric and exponential functions and their inverse functions. But even with this added power we will not be able to find a suitable function $F(x)$ that can be obtained as a finite combination of our elementary functions, using the algebraic operations.[1]
   Nevertheless, we can solve this problem. It suffices to define $F(x)$ by

$$F(x) = \int_0^x (t^7 + 111t^4 - 5t^2 + 13t + 33)^{1/3}\, dt.$$

By Theorem 8 this function has the required derivative. Further, we really know every solution to the problem, for if $G(x)$ is any other function with the same derivative, then for some suitably selected constant C, we have $G(x) = F(x) + C$ for all x (see Theorem 10 of Chapter 5). ●

Proof of Theorem 8. By definition, the derivative of $F(x)$ is

$$\frac{dF(x)}{dx} = \lim_{h \to 0} \frac{F(x + h) - F(x)}{h},$$

---

[1] The algebraic operations are addition, subtraction, multiplication, division, raising to a power, and extracting a root.

where it is understood that h is always selected so that $x + h$ is also in the given interval. From equation (96),

$$(98) \qquad \frac{F(x + h) - F(x)}{h} = \frac{1}{h} \left[\int_a^{x+h} f(t)\, dt - \int_a^x f(t)\, dt \right] = \frac{1}{h} \int_x^{x+h} f(t)\, dt,$$

by Theorem 6 with $c = x$ and $b = x + h$. Let $\mathcal{K}$ be the interval $x \leq t \leq x + h$ if $h > 0$, and let $\mathcal{K}$ be the interval $x + h \leq t \leq x$ if $h < 0$. Let m and M be the minimum and maximum values, respectively, of $f(t)$ for t in $\mathcal{K}$.

We assume first that $h > 0$. Then by Exercise 6, Problem 13,

$$(99) \qquad\qquad mh \leq \int_x^{x+h} f(t)\, dt \leq Mh.$$

Dividing by the positive number h, and using the result in (98) gives

$$(100) \qquad\qquad m \leq \frac{F(x + h) - F(x)}{h} \leq M.$$

Now as $h \to 0^+$ both m and M approach $f(x)$, since f is continuous. This proves that $F'(x) = f(x)$ in the case that $h \to 0$ through positive values.

If h is negative, there is no change in equation (98), but (99) is replaced by

$$m|h| \leq \int_{x+h}^x f(t)\, dt \leq M|h|, \qquad |h| = -h > 0.$$

Dividing by $|h|$ yields

$$m \leq \frac{1}{|h|} \int_{x+h}^x f(t)\, dt = \frac{1}{h} \int_x^{x+h} f(t)\, dt \leq M$$

and when this is used in (98), we get (100), just as before. As $h \to 0^-$, both m and M approach $f(x)$ and this completes the proof that $F'(x) = f(x)$.   ∎

EXERCISE 8

1. Prove that $\dfrac{d}{dx} \displaystyle\int_x^b f(t)\, dt = -f(x)$.

2. Prove that if u is any differentiable function of x, then

$$\frac{d}{dx} \int_a^u f(t)\, dt = f(u)\, \frac{du}{dx}.$$

In Problems 3 through 14, compute the indicated derivative.

3. $\dfrac{d}{dx}\displaystyle\int_0^x \sin t\, dt.$

4. $\dfrac{d}{dx}\displaystyle\int_0^x \sin\theta\, d\theta.$

5. $\dfrac{d}{dt}\displaystyle\int_0^t \sin x\, dx.$

6. $\dfrac{d}{dx}\displaystyle\int_\pi^x \sin^3 s\, ds.$

★7. $\dfrac{d}{dx}\displaystyle\int_1^{x^2} u\sin u\, du.$

★8. $\dfrac{d}{dx}\displaystyle\int_1^{x^3} \sqrt{1+u^4}\, du.$

★9. $\dfrac{d}{dt}\displaystyle\int_1^{t^2} r\sin r\, dr.$

★10. $\dfrac{d}{dt}\displaystyle\int_{t^2}^1 \theta^3\cos^2(\theta^4)\, d\theta.$

11. $\dfrac{d}{dx}\displaystyle\int_3^7 \sqrt{t^9+5}\, dt.$

12. $\dfrac{d}{dx}\left[\displaystyle\int_1^{x^3} 5^t\, dt + \int_{x^3}^4 5^t\, dt\right].$

★13. $\dfrac{d}{dx}\displaystyle\int_x^{x^2} \sqrt{1+t^5}\, dt.$

14. $\dfrac{d}{dy}\displaystyle\int_{y^2}^{y^3} \sin u\, du.$

15. Find the interval in which the graph of $y = \displaystyle\int_0^x \dfrac{s\,ds}{\sqrt{128+s^6}}$ is concave upward.

★10 Approximation Methods for Integrals

Consider for a moment the problem of computing the definite integrals:

$$I_1 \equiv \int_1^2 \frac{dx}{x}, \qquad I_2 \equiv \int_0^{\pi/4} \sin x\, dx,$$

$$I_3 \equiv \int_1^3 2^x\, dx, \qquad I_4 \equiv \int_0^\pi \sqrt{1+\sin^2 x}\, dx.$$

In each case, evaluation would be an easy matter if we could find an antiderivative $F(x)$, for then we could apply Theorem 4. In the first three integrals, such antiderivatives can be found—and we will learn how this is done in Chapters 8 and 9. But there is no antiderivative for $\sqrt{1+\sin^2 x}$ that can be expressed as a finite combination of elementary functions (see Example 1 of Section 9). Consequently, we should have available methods of computing integrals such as I_4.

We already have one such method, namely the approximation by rectangles. This was explored in Section 5, and in fact those approximations s_n and S_n formed the basis for the definition of the definite integral. Now approximations by rectangles give only crude estimates, and we can obtain much better estimates if we use trapezoids, as indicated in Figs. 15 and 16.

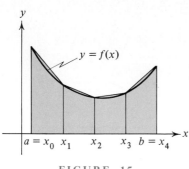

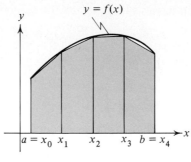

FIGURE 15   FIGURE 16

As usual, we divide the interval $a \leq x \leq b$ into n equal parts and let $x_0, x_1, \ldots, x_n$ be the points of the partition. Let $I_T(n)$ be the approximation obtained when trapezoids are used (as indicated in the Figs. 15 and 16). Then we have

---

THEOREM 9 (PWO). If $_k = f(x_k)$, $k = 0, 1, 2, \ldots, n$, then the approximation of

$$I \equiv \int_a^b f(x)\, dx$$

by means of n trapezoids is given by

(101)
$$I_T(n) = \frac{b - a}{2n} \left(y_0 + y_n + 2 \sum_{k=1}^{n-1} y_k \right).$$

If $f(x)$ is concave upward, then $I_T(n) > I$ (see Fig. 15). If $f(x)$ is concave downward, then $I_T(n) < I$ (see Fig. 16).

We will not prove Theorem 9 because there is a better formula, known as *Simpson's Rule*, which we now derive. Simpson's Rule is based on the approximation of $f(x)$ by a set of parabolas.

To begin the process, we suppose that we have only three points on the curve $y = f(x)$: (x_0, y_0), (x_1, y_1), and (x_2, y_2).

THEOREM 10. Given three points (x_0, y_0), (x_1, y_1), and (x_2, y_2), with $x_0 < x_1 < x_2$, there is always a unique parabola

$$y = A + Bx + Cx^2$$

which passes through these three points. If

$$x_1 = x_0 + h \qquad \text{and} \qquad x_2 = x_0 + 2h,$$

then for this parabola

(102)
$$\int_{x_0}^{x_2} (A + Bx + Cx^2)\, dx = \frac{h}{3}(y_0 + 4y_1 + y_2).$$

Proof. To find the parabola that passes through the three given points, it is sufficient to solve the system of three equations

(103)
$$\begin{aligned} y_0 &= A + Bx_0 + Cx_0^2, \\ y_1 &= A + Bx_1 + Cx_1^2, \\ y_2 &= A + Bx_2 + Cx_2^2, \end{aligned}$$

for the three unknowns A, B, and C. Of course, if $C = 0$, the parabola degenerates into a straight line, and if in addition $B = 0$, the line is horizontal.

To compute the definite integral, we observe that it is invariant under a horizontal translation of the points and the curve. Hence we can set $x_0 = -h$, $x_1 = 0$, and $x_2 = h$. Then the set of equations (103) simplifies to

(104)
$$\begin{aligned} y_0 &= A - Bh + Ch^2, \\ y_1 &= A, \\ y_2 &= A + Bh + Ch^2. \end{aligned}$$

Using these equations, we find that

$$\int_{-h}^{h} (A + Bx + Cx^2)\, dx = \left(Ax + \frac{1}{2}Bx^2 + \frac{1}{3}Cx^3 \right)\Bigg|_{-h}^{h}$$

$$= 2Ah + \frac{2}{3}Ch^3 = \frac{h}{3}(6A + 2Ch^2)$$

$$= \frac{h}{3}(y_0 + 4y_1 + y_2). \quad \blacksquare$$

In order to apply Theorem 10 over a large region, we divide it into an *even* number of panels and use equation (102) on successive pairs of panels (see Fig. 17). To emphasize that the number of panels must be even, we let $2n$ denote the number of panels. If h is the width of each panel and $a \leq x \leq b$ is partitioned by the set $\mathcal{P} = \{x_0, x_1, x_2, \ldots, x_{2n}\}$, where $x_0 = a$ and $x_{2n} = b$, then

$$x_k = a + kh = a + k\frac{b-a}{2n}, \qquad k = 0, 1, 2, \ldots, 2n.$$

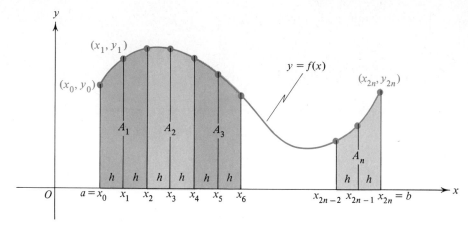

FIGURE 17

In each of the subintervals $[x_0, x_2]$, $[x_2, x_4]$, ..., $[x_{2n-2}, x_{2n}]$, we approximate the curve $y = f(x)$ by a suitable parabola (as described in Theorem 10) and we approximate the integral by using the integral of the approximating parabola. We let $I_P(2n)$ denote this approximation. Then repeated application of equation (102) to each successive triple of points gives

$$I_P(2n) = \sum_{k=1}^{n} \frac{h}{3}(y_{2k-2} + 4y_{2k-1} + y_{2k})$$

$$= \frac{b-a}{6n}(y_0 + 4y_1 + y_2$$
$$+ y_2 + 4y_3 + y_4$$
$$+ y_4 + 4y_5 + y_6 + \cdots$$
$$+ y_{2n-2} + 4y_{2n-1} + y_{2n}).$$

On combining these terms, we have *Simpson's Rule* (or the *parabolic rule*):

THEOREM 11. If the curve is approximated by parabolas and the interval $a \leq x \leq b$ is divided into $2n$ equal subintervals, then

(105) $$\int_a^b f(x)\, dx \approx I_P(2n),$$

where

$$(106) \quad I_P(2n) = \frac{b-a}{6n}(y_0 + 4y_1 + 2y_2 + 4y_3 + 2y_4 + \cdots + 4y_{2n-1} + y_{2n})$$

$$= \frac{b-a}{6n}\left(y_0 + y_{2n} + 4\sum_{k=1}^{n} y_{2k-1} + 2\sum_{k=1}^{n-1} y_{2k}\right).$$

Although at first glance this formula may look frightening, it is quite easy to use, as we will learn in the next two examples.

Example 1. Use Simpson's Rule with $2n = 8$ ($n = 4$) to estimate

$$(107) \qquad\qquad I \equiv \int_0^4 x^2 \, dx.$$

Solution. To compute the sum on the right side of (106), we arrange the work as shown in Table 1.

TABLE 1

k	x_k	$y_k = x_k^2$	Multiplier	Term
0	0.0	0.00	1	0.0
1	0.5	0.25	4	1.0
2	1.0	1.00	2	2.0
3	1.5	2.25	4	9.0
4	2.0	4.00	2	8.0
5	2.5	6.25	4	25.0
6	3.0	9.00	2	18.0
7	3.5	12.25	4	49.0
8	4.0	16.00	1	16.0
				$\Sigma = 128.0$

By Simpson's Rule an approximate value for I is

$$I_P(8) = \frac{4-0}{6(4)} \Sigma = \frac{1}{6}(128) = 21\frac{1}{3}.$$

The interested reader should observe that this approximate value is really the exact value of the integral (107). Why should we have expected this? ●

Example 2. Use Simpson's rule with 10 panels ($n = 5$) to approximate

$$I_1 \equiv \int_1^2 \frac{1}{x} \, dx.$$

Solution. The work is given in Table 2, where the quotients are rounded off to six decimal places.

TABLE 2

k	x_k	$y_k = 1/x_k$	Multiplier	Term
0	1.0	1.000000	1	1.000000
1	1.1	0.909091	4	3.636364
2	1.2	0.833333	2	1.666666
3	1.3	0.769231	4	3.076924
4	1.4	0.714286	2	1.428572
5	1.5	0.666667	4	2.666668
6	1.6	0.625000	2	1.250000
7	1.7	0.588235	4	2.352940
8	1.8	0.555556	2	1.111112
9	1.9	0.526316	4	2.105264
10	2.0	0.500000	1	0.500000
				$\Sigma \approx 20.794510$

The error in $\sum$ due to rounding off is at most 0.000010. Hence

$$\int_1^2 \frac{1}{x}\, dx \approx \frac{b-a}{6n} \sum = \frac{2-1}{30}(20.794510) \approx 0.6931503,$$

where now the error due to rounding off is at most 0.0000004.   ●

In any serious work we would like to know how far the approximate value is from the correct value. One answer to the question is given in

---

THEOREM 12 (PWO). Let $M = \max |f^{(4)}(x)|$ for x in the interval $a \leq x \leq b$ and let h be the width of each panel, $h = (b-a)/2n$. If

$$\int_a^b f(x)\, dx = I_P + E,$$

where I_P is the approximation supplied by Simpson's Rule [the right side of equation (106)] and E is the error, then

(108) $$|E| \leq \frac{nh^5}{90} M.$$

Example 3. Estimate the error made in Example 2.

Solution. Since $f(x) = 1/x$, we have $M = \max |4!/x^5| = 24$ if $1 \leqq x \leqq 2$. In this example $h = 1/10$ and hence, from (108),

$$|E| \leqq \frac{nh^5}{90} \times 24 = \frac{5\left(\dfrac{1}{10}\right)^5}{90} \times 24 \approx 0.0000133. \quad \bullet$$

EXERCISE 9

CALCULATOR PROBLEMS

In Problems C1 through C6, use Simpson's Rule (with $n = 5$) to estimate the value of the given integral to four significant figures. Then check your work by using the calculus to find the true value of the integral.

C1. $\displaystyle\int_1^2 x^3 \, dx.$ **C2.** $\displaystyle\int_3^4 \frac{dx}{x^2}.$ **C3.** $\displaystyle\int_1^4 x^3 \, dx.$

★C4. $\displaystyle\int_3^4 \frac{dx}{\sqrt{x}}.$ **★C5.** $\displaystyle\int_0^2 \sqrt{1 + x} \, dx.$ **★C6.** $\displaystyle\int_0^5 x \sqrt{1 + x^2} \, dx.$

C7. By using inverse trigonometric functions, it can be shown that

$$\pi = 4 \int_0^1 \frac{dx}{1 + x^2}.$$

Use Simpson's Rule with $n = 5$ to estimate π to four decimal places.

C8. The gas tank of an airplane is often placed inside the wing, as shown in Figure 18. The dimensions (in meters) of a typical cross section of the wing are shown in the figure. Use Simpson's Rule to estimate the area of this cross section of the gas tank. From this area, it is an easy matter to compute the volume of the gas tank.

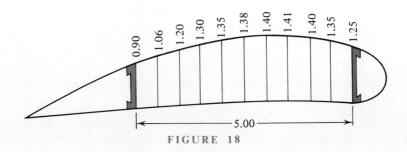

FIGURE 18

C9. The dimensions (in miles) of a certain lake were scaled from an aerial photograph and the results are indicated in Fig. 19. Estimate the area of this lake to the nearest hundredth of a square mile.

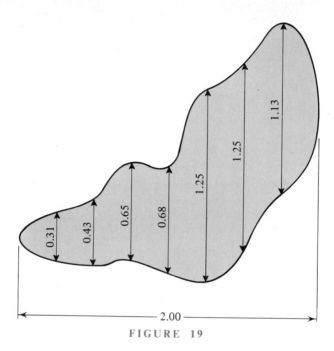

FIGURE 19

C10. Measurements (in km) from an aerial photograph of Lake Newton are shown in Figure 20. Use Simpson's Rule to estimate the surface area of this lake to three significant figures.

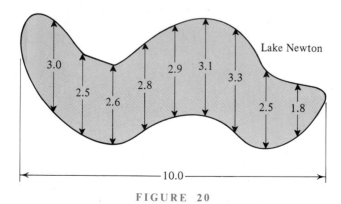

FIGURE 20

C11. Ms. Sheila Brown is trying to sell the waterfront property shown in Figure 21. She claims that the area is approximately one third of an acre. Use the dimensions shown to check her claim. Note: One acre is 43,560 ft².

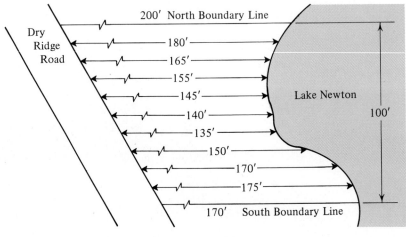

FIGURE 21

REVIEW PROBLEMS

In Problems 1 through 4, find the indefinite integral (antiderivative).

1. $\int (1 + 2x^2 + x^5)^{5/4}(5x^4 + 4x)\, dx.$

2. $\int \sqrt[3]{x^2 + 2x + 1}\, dx.$

3. $\int \dfrac{r^4}{\sqrt{1 - 3r^5}}\, dr.$

4. $\int (1 + \sqrt{2t - 1})^3\, dt.$

In Problems 5 through 8, solve the given differential equation, subject to the given initial conditions.

5. $\dfrac{dy}{dx} = \sqrt[3]{\dfrac{y}{x}}, \qquad y = 1$ when $x = 4.$

6. $\dfrac{dy}{dx} = \dfrac{y^3(1 + x)}{x^3(1 + y)}, \qquad y = \sqrt{11}$ when $x = \sqrt{11}.$

7. $u^3 \dfrac{dv}{du} = v^2(u^4 - 1), \quad v = -8$ when $u = 1/2.$

8. $r \dfrac{dr}{d\theta} = \theta \sqrt{1 + 4r^2}, \quad r = \sqrt{2}$ when $\theta = 0.$

9. Show that the formal solution of Problem 6 can be factored into the product of two solutions $y = x$ and $y = -x/(1 + 2x)$, but one of these does not satisfy the initial conditions.

10. It is easy to check that for any constant A, the function $y = Ax$ is a solution of the differential equation

$$\frac{dy}{dx} = \frac{y}{x}.$$

Explain why the methods learned so far are insufficient for the solution of this differential equation.

In Problems 11 through 16, evaluate the given sum using the formulas of Section 4.

11. $\displaystyle\sum_{k=1}^{n} 6k.$

12. $\displaystyle\sum_{k=1}^{n} 6k^2.$

13. $\displaystyle\sum_{j=1}^{n} (4j + 1)^2.$

14. $\displaystyle\sum_{j=1}^{n} (aj + b)^2.$

15. $\displaystyle\sum_{t=1}^{n} t(t - 8).$

16. $\displaystyle\sum_{k=1}^{n} (2k + 1)(k + 6).$

*17. Prove that $\dfrac{1}{k} - \dfrac{1}{k + 1} = \dfrac{1}{k(k + 1)}$. Use this to prove that

$$\sum_{k=1}^{n} \frac{1}{k(k + 1)} = 1 - \frac{1}{n + 1} = \frac{n}{n + 1}.$$

*18. Prove that

$$\frac{1}{k(k + 1)(k + 2)} = \frac{1}{2}\left(\frac{1}{k} - \frac{2}{k + 1} + \frac{1}{k + 2}\right).$$

Use this to evaluate the sum

$$\sum_{k=1}^{n} \frac{1}{k(k + 1)(k + 2)}.$$

In Problems 19 through 28, compute the area of the figure bounded by the given curve and the x-axis between the given values of a and b.

19. $y = \dfrac{1}{\sqrt{5x}},$ $a = 1,\quad b = 2.$

20. $y = \dfrac{1}{\sqrt{3x + 13}},$ $a = 1,\quad b = 4.$

21. $y = (\sqrt{x} - 2)^2,$ $a = 0,\quad b = 9.$

*22. $y = 1 + |x|,$ $a = -1,\quad b = 2.$

23. $y = x - \sqrt{x},$ $a = 4,\quad b = 9.$

24. $y = x - \sqrt{x},$ $a = 0,\quad b = 9.$

25. $y = \sqrt{x} - 2$, $a = 0$, $b = 9$. *26. $y = \sqrt{x}\sqrt{1 + x\sqrt{x}}$, $a = 1$, $b = 2$.

*27. $y = (x - c)(x - 2c)$, $a = 0$, $b = 3c$. 28. $y = x^4 - 2x^3 + x^2$, $a = -1$, $b = 1$.

In Problems 29 through 34, find the average value of the given function over the given interval. (See Exercise 7, page 279.)

29. $f(x) = x^n$, $0 \leq x \leq b$, $n > 0$.   30. $f(x) = x^n + x^m$, $0 \leq x \leq b$, $m, n > 0$.

31. $f(x) = \dfrac{1}{\sqrt{x + 1000}}$, $0 \leq x \leq b$.   32. $f(x) = \dfrac{x}{\sqrt{x^2 + 16}}$, $0 \leq x \leq b$.

33. $f(x) = \dfrac{x^2}{\sqrt[4]{x^3 + 1}}$, $0 \leq x \leq b$.   34. $f(x) = \sqrt{x} - \sqrt[3]{x}$, $0 \leq x \leq b$.

*35. Find the limit of the average value as $b \to \infty$ for the function and the interval given in: **(a)** Problem 31, **(b)** Problem 32, **(c)** Problem 33, and **(d)** Problem 34.

7

Applications of Integration

In Chapter 6 we approximated the "area" of a region under a curve by a certain sum, and in a natural way we arrived at the definition of the area as a definite integral. There are many other quantities that can be handled in much the same way. Among these we find volume, pressure, and work. In each case the quantity in question is estimated by lower sums s_n and upper sums S_n, and the quantity is defined through a common limit. Thus, if Q denotes the quantity under investigation, then

$$\lim_{n \to \infty} s_n \leqq Q \qquad \text{and} \qquad Q \leqq \lim_{n \to \infty} S_n,$$

and if the two limits are equal, Q is *defined* as the common limit. If the sums s_n and S_n are generated by a function $f(x)$ that is continuous in $a \leqq x \leqq b$, then $\lim s_n = \lim S_n = \lim S_n^\star$, and Q is easily computed as a definite integral:

$$Q = \lim_{n \to \infty} S_n^\star = \lim_{n \to \infty} \sum_{k=1}^{n} f(x_k^\star)\,\Delta x_k = \int_a^b f(x)\,dx,$$

where $x_{k-1} \leqq x_k^\star \leqq x_k$ for $k = 1, 2, \ldots, n$ and $\mu(\mathscr{P}_n) \to 0$ as $n \to \infty$.

But once we have witnessed this process in operation for the area under a curve, it seems unnecessary and unwise to repeat the description of this process for each new quantity that we meet. All that is necessary is the recognition that such a treatment is possible. In this chapter we will slide over the details and rely heavily on our intuition. We can now solve many fascinating problems with very little effort.

2 *The Area Between Two Curves*

Suppose that we have two curves $y = f_1(x)$ and $y = f_2(x)$ and we want to find the area A of the region $\mathscr{F}$ bounded by these two curves, and two vertical lines $x = a$ and $x = b$. The situation is illustrated in Fig. 1. Naturally, we select the notation so that the curve $y = f_2(x)$ lies above the curve $y = f_1(x)$, and this means that $f_2(x) \geqq f_1(x)$ for $a \leqq x \leqq b$. To distinguish between the ordinates on the two curves, we use y_2 for the upper curve and y_1 for the lower curve.

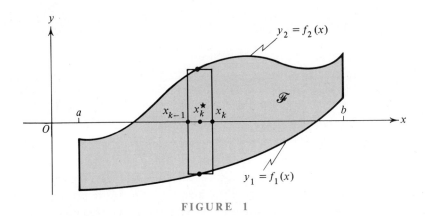

FIGURE 1

We approximate the area by summing the areas of suitably selected rectangles. A representative one is shown in the figure. In each subinterval an $x_k^{\bigstar}$ is selected, and the height h_k of the approximating rectangle is just

$$h_k = y_2 - y_1 \equiv f_2(x_k^{\bigstar}) - f_1(x_k^{\bigstar}).$$

The area of the rectangle is $h_k \, \Delta x_k$ and hence an approximation for the area of the region bounded by the given curves between $x = a$ and $x = b$ is

$$\sum_{k=1}^{n} [f_2(x_k^{\bigstar}) - f_1(x_k^{\bigstar})] \, \Delta x_k.$$

In fact, by the definition of area

(1)
$$A = \lim_{\mu(\mathscr{P}) \to 0} \sum_{k=1}^{n} [f_2(x_k^{\bigstar}) - f_1(x_k^{\bigstar})] \, \Delta x_k$$

and by Theorem 3 of Chapter 6 this is just the definite integral

(2)
$$A = \int_a^b [f_2(x) - f_1(x)]\, dx.$$

This same procedure will be used repeatedly, so in order to speed up the work, we introduce some shortcuts in our notation and in our thought process.

We have already seen that the limit of the sum given in equation (1) does not depend on the particular $x_k^\star$ selected, just as long as each $x_k^\star$ is in the kth subinterval, and the functions involved are continuous. So we can drop the superscript $\star$ in equation (1). Next we can think of Δx_k as a differential of x and regard each rectangle as a thin rectangle of width dx and height h. Each term in the sum (1) then has the form $h\, dx$, and we call this the *differential element of area*. With this simplified view and simplified notation, we can write

(3)
$$dA = h\, dx,$$

read "the differential of the area is the height times the differential width." But for each x, $h = f_2(x) - f_1(x)$, so substituting this in (3) and integrating both sides leads immediately to equation (2), which we already know to be correct.

Example 1. Find the area of the region bounded by the curves $y = 3 - x^2$ and $y = -x + 1$, between $x = 0$ and $x = 2$.

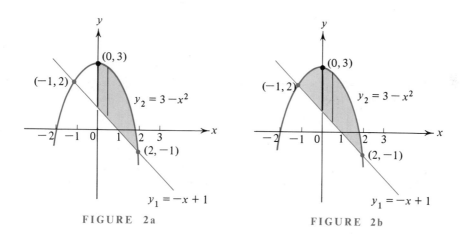

FIGURE 2a   FIGURE 2b

Solution. The region is shown shaded in Fig. 2a. From the graph it is clear that in the interval $0 \le x \le 2$, the parabola $y = 3 - x^2$ lies above (or meets) the straight line $y = -x + 1$. Hence the height of the rectangle is

$$h = y_2 - y_1 = 3 - x^2 - (-x + 1) = 2 + x - x^2.$$

The differential element of area is $dA = h\,dx = (2 + x - x^2)\,dx$ and hence

$$A = \int_0^2 (2 + x - x^2)\,dx = \left(2x + \frac{x^2}{2} - \frac{x^3}{3}\right)\Big|_0^2$$

$$= 4 + \frac{4}{2} - \frac{8}{3} - 0 = \frac{10}{3} = 3\frac{1}{3}. \quad \bullet$$

Example 2. Find the area of the region bounded by the curves $y = 3 - x^2$ and $y = -x + 1$.

Solution. Here the terminology implies that the two curves intersect in at least two points, and if they intersect in exactly two points these two points determine a and b, the extreme values of x for the figure. The case in which there are more than two intersection points for the two curves is illustrated in the next example.

Solving $y = 3 - x^2$ and $y = -x + 1$ leads to $3 - x^2 = -x + 1$ or $x^2 - x - 2 = 0$. The roots are $x = 2$ and $x = -1$, and the points of intersection are $(2, -1)$ and $(-1, 2)$, just as indicated in Fig. 2b. Then, just as in Example 1,

$$A = \int_{-1}^2 h\,dx = \int_{-1}^2 (y_2 - y_1)\,dx = \int_{-1}^2 [3 - x^2 - (-x + 1)]\,dx$$

$$= \int_{-1}^2 (2 + x - x^2)\,dx = \left(2x + \frac{x^2}{2} - \frac{x^3}{3}\right)\Big|_{-1}^2$$

$$= 4 + \frac{4}{2} - \frac{8}{3} - \left(-2 + \frac{1}{2} + \frac{1}{3}\right) = \frac{27}{6} = 4\frac{1}{2}. \quad \bullet$$

Example 3. Find the area of the figure bounded by the curves $y = x$ and $y = x^5/16$.

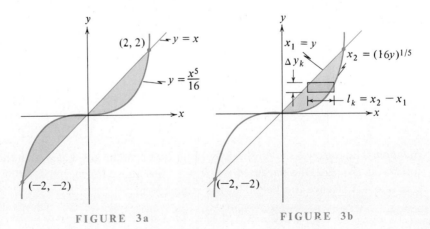

FIGURE 3a   FIGURE 3b

Solution. It is easy to prove that these two curves intersect at the points $(-2, -2)$, $(0, 0)$, and $(2, 2)$ and nowhere else. The graphs are shown in Figs. 3a and 3b.

In the interval $0 < x < 2$, the straight line $y = x$ lies above the curve $y = x^5/16$, but in the interval $-2 < x < 0$, their roles are reversed. Hence

$$A = \int_{-2}^{0} \left(\frac{x^5}{16} - x\right) dx + \int_{0}^{2} \left(x - \frac{x^5}{16}\right) dx$$

$$= \left(\frac{x^6}{96} - \frac{x^2}{2}\right)\Big|_{-2}^{0} + \left(\frac{x^2}{2} - \frac{x^6}{96}\right)\Big|_{0}^{2}$$

$$= 0 - \left(\frac{64}{96} - \frac{4}{2}\right) + \left(\frac{4}{2} - \frac{64}{96}\right) = 2\left(2 - \frac{2}{3}\right) = \frac{8}{3}. \quad \bullet$$

In this situation, one could also appeal to the symmetry of the figure and say that A is just twice the area of the region bounded by the two curves in the first quadrant. Thus

$$A = 2\int_{0}^{2} \left(x - \frac{x^5}{16}\right) dx = 2\left(\frac{x^2}{2} - \frac{x^6}{96}\right)\Big|_{0}^{2} = 2\left(2 - \frac{2}{3}\right) = \frac{8}{3}.$$

Example 4. Find the area in Example 3 using "horizontal strips" rather than "vertical strips" (see Fig. 3b).

Solution. By symmetry we can restrict ourselves to the region in the first quadrant. If this region is divided into horizontal strips, the thickness of each strip is represented by dy and its length by $x_2 - x_1$, so $dA = (x_2 - x_1)\, dy$. For the particular curves under consideration, $x_2 = (16y)^{1/5}$ and $x_1 = y$. Hence

$$A = 2\int_{y=0}^{y=2} [(16y)^{1/5} - y]\, dy = 2\left(16^{1/5}\frac{5}{6}y^{6/5} - \frac{y^2}{2}\right)\Big|_{0}^{2}$$

$$= 2\left(16^{1/5} \cdot \frac{5}{6} \cdot 2^{6/5} - \frac{4}{2}\right) = 2\left(2^{4/5} \cdot \frac{5}{6} \cdot 2^{6/5} - 2\right)$$

$$= 2\left(4 \cdot \frac{5}{6} - 2\right) = \frac{8}{3}. \quad \bullet$$

EXERCISE 1

In Problems 1 through 14, find the area of the figure bounded by the given pair of curves.

1. $y = 4x - x^2 - 3$, $y = -3$.
2. $y = 2x - x^2$, $y = x - 2$.
3. $x = 4y - y^2 - 3$, $x = -3$.
4. $x = 2y - y^2$, $y = x + 2$.
5. $y = x^2$, $y = x^4$.
6. $y = x$, $y = x^3$.

7. $y = 4 - x^2$, $y = x + 2$.   8. $y = x^2 + 1$, $y = 2x^2 - 8$.
★9. $y = x^3$, $y = 3x + 2$.   10. $y = 6x - 2 - x^2$, $y = 3$.
11. $y = 9 + 2x - x^2$, $y = -3x + 13$.   12. $y = 1 + 4x - x^2$, $y = 1 + x^2$.
★13. $y = x^3 - x^2 - 2x + 2$, $y = 2$.   ★14. $y = x^3 - 3x^2 + x + 5$, $y = x + 5$.

15. Find the area of the region lying in the first quadrant and bounded by the curves $y = -2x + 13$ and $y = 2x + 9/x^2$.

16. Find the area of the region above $y = 1/x^2$ that is bounded by the parabola $y = \sqrt{x}$ and the line $x = 9$.

17. Find the area of the region bounded by the curves $y = x^2$ and $y = x^3$ in two different ways: first, by using vertical strips, and second, by using horizontal strips.

18. Find the area of the region bounded by the curves $x = 9y$ and $x = y^2$ in two different ways: first, by using vertical strips, and second, by using horizontal strips.

19. Find the area of the triangular-shaped region bounded by the parabola $y = x^2$ and the two straight lines $y = 4$ and $y = x + 2$.

20. Find the area of the region bounded by the curve $y = 4/x^2$ and the two lines $y = 1$ and $y = 4$.

21. Find the area of the region bounded by the parabola $x = y^2$ and by the lines $y = x + 2$, $y = -1$, and $y = 1$.

22. Find the area of the region bounded by the curves $x = y^3 + 9$, $x = y^3$, $y = 2$, and $y = -2$.

3   *Volumes of Solids of Revolution*

If the region below a curve $y = f(x)$ between $x = a$ and $x = b$ is rotated about the x-axis, it generates a solid figure called a *solid of revolution*. Because of the symmetrical way in which it is generated (by a rotation of a plane figure), it is easy to compute its volume.

The situation is pictured in Fig. 4, where in order to aid visualization, one fourth of the solid has been removed. To compute the volume we first cut the solid into disks by a number of planes each perpendicular to the x-axis. We notice that each disk is very nearly a right circular cylinder and is like a coin standing on its edge. The height h of the cylinder is the thickness of the slice, and the radius r of the cylinder is the y-coordinate of an appropriate point on the curve. Thus an approximation for the volume of the solid of revolution is obtained by adding the volumes of the approximating right circular cylinders, and the *exact* volume is obtained by taking the limit of this sum.

To carry out this program in detail, we form a partition:

$$a = x_0 < x_1 < x_2 \cdots < x_n = b$$

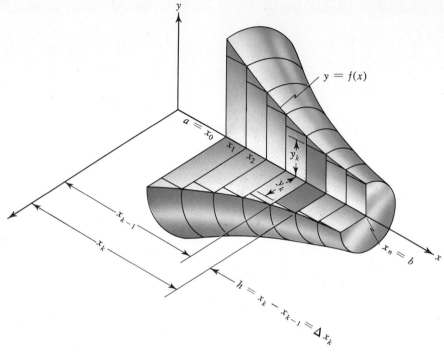

FIGURE   4

and at each point x_k pass a plane perpendicular to the x-axis. If ΔV_k denotes the volume of that portion of the solid contained between the two planes at $x = x_{k-1}$ and $x = x_k$, then

$$\Delta V_k \approx \pi r_k^2 h = \pi y_k^2 \Delta x_k.$$

Here $h = \Delta x_k$ is the thickness of the disk (the height of the cylinder) and $r_k = y_k$ is a suitably selected radius. Any value for $y = f(x)$, for x in the subinterval $x_{k-1} \leq x \leq x_k$, can be taken for r_k. Figure 4 shows the case in which the minimum value of $f(x)$ was selected in each subinterval. Thus an approximation for V is obtained by adding the volume of each of the n disks,

$$V \approx \sum_{k=1}^{n} \pi f^2(x_k)\,\Delta x_k = \pi[f^2(x_1)\,\Delta x_1 + f^2(x_2)\,\Delta x_2 + \cdots + f^2(x_n)\,\Delta x_n].$$

By taking the limit as $\mu(\mathscr{P}) \to 0$, the sum tends to the *exact* value of V on the one hand, and on the other hand the limit is the corresponding definite integral. Hence

(4) $$V = \int_a^b \pi f^2(x)\,dx = \int_a^b \pi y^2\,dx.$$

The student is advised *not* to memorize this formula. Rather it is better to recall that the volume of a right circular cylinder is $\pi r^2 h$, and hence the approximate volume of one slice is the *differential element of volume*

(5) $$dV = \pi r^2 \, dx.$$

Then we can proceed directly from (5) to (4). Because the differential element of volume is a disk, this method of computing volumes is called the *disk method*.

Example 1. Find the volume of the solid of revolution obtained by rotating about the x-axis the region under the curve $y = x^{2/3}$ between $x = 0$ and $x = 8$.

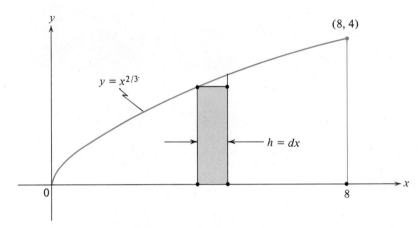

FIGURE 5

Solution. It is not always necessary to make a perspective drawing of the solid. As indicated in Fig. 5, it is sufficient to draw the plane region that is rotated to generate the solid. A representative rectangular strip is shown in Fig. 5, and on rotation this generates a cylindrical disk of volume

$$dV = \pi r^2 h = \pi y^2 \, dx.$$

Since $y = x^{2/3}$ and hence $y^2 = (x^{2/3})^2 = x^{4/3}$, we have

$$V = \int_0^8 \pi y^2 \, dx = \int_0^8 \pi x^{4/3} \, dx = \pi \frac{3}{7} x^{7/3} \Big|_0^8 = \frac{384}{7} \pi = 54\frac{6}{7}\pi. \quad \bullet$$

In computing volumes, it is frequently convenient to "slice" the solid into "shells" rather than disks. This situation arises when we rotate the plane region around the y-axis instead of around the x-axis. Such a solid is shown in Fig. 6, where one fourth of the resulting solid has been removed to aid visualization. A representative rectangular strip between $x = x_{k-1}$ and $x = x_k$ is shown shaded in the figure, and during the rotation this rectangular strip generates a cylindrical shell. If y_k is the height of this shell, then the

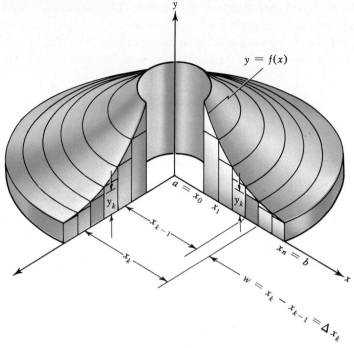

FIGURE 6

precise volume of the cylindrical shell is

(6) $$V_k = \pi x_k^2 y_k - \pi x_{k-1}^2 y_k = \pi(x_k + x_{k-1})(x_k - x_{k-1}) y_k,$$

obtained by taking the volume of a solid cylinder, $\pi r^2 y_k$, where $r = x_k$, and subtracting the volume of the hole, $\pi r^2 y_k$, where $r = x_{k-1}$. But x_k and x_{k-1} are approximately equal, so we can replace $x_k + x_{k-1}$ by $2x_k$. Thus (6) becomes

(7) $$V_k \approx 2\pi x_k y_k \, \Delta x_k.$$

This formula is easy to remember if we realize that $2\pi x_k$ is the perimeter of the outer circle, y_k is the height of the shell, and Δx_k is its width. So formula (7) has the form "length × height × width." On adding, we have

$$V \approx \sum_{k=1}^{n} 2\pi x_k y_k \, \Delta x_k$$

and finally, on taking a limit in the usual manner,

(8) $$V = \int_a^b 2\pi x f(x) \, dx = \int_a^b 2\pi x y \, dx.$$

Again the student should *not* memorize this formula. This formula is very similar to (4), and the student who just memorizes will almost certainly confuse the two formulas and probably will use the wrong one. It is much better to recall that the volume of a shell is "length × height × width" when the shell is "cut and flattened out." Then we can write immediately that

$$(9) \qquad\qquad dV = 2\pi rhw = 2\pi xy\,dx$$

and then proceed from (9) directly to (8) by integration.

Just as our first method of computing volumes was called the disk method, this procedure is called the *shell method,* or the method of cylindrical shells. It is helpful to keep in mind that for the disk method the rectangular strips are taken *perpendicular* to the axis of revolution, while in the shell method the rectangular strips are *parallel* to the axis of revolution.

Example 2. Check the solution to Example 1 by computing the volume by the shell method.

Solution. This time the region is to be divided into horizontal strips, as indicated in Fig. 7. Then when the region is rotated about the *x*-axis, each horizontal strip gener-

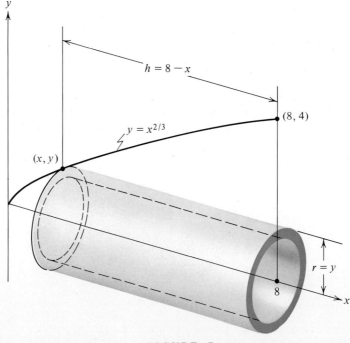

FIGURE 7

ates a cylindrical shell. The height of each shell is $8 - x$, the radius is y, and the width (or thickness) is dy. Hence

$$dV = 2\pi y(8 - x)\,dy$$

and

(10)
$$V = \int_0^4 2\pi y(8 - x)\,dy.$$

The reader should compare equation (10) with equation (8) and observe how much they differ, although both arise from the method of cylindrical shells. This should be a convincing argument against pure memorization of formulas for this type of problem. A firm grasp of the basic principles is highly recommended.

To complete the computation, we observe that the equation $y = x^{2/3}$ for the curve yields $x = \sqrt{y^3}$. Hence (10) becomes

$$V = \int_0^4 2\pi y(8 - y^{3/2})\,dy = 2\pi\left(4y^2 - \frac{2}{7}y^{7/2}\right)\Big|_0^4 = \frac{2\pi}{7}(64 \cdot 7 - 2 \cdot 128)$$

$$= \frac{2\pi}{7}(448 - 256) = \frac{2\pi}{7} \cdot 192 = \frac{384}{7}\pi. \quad \bullet$$

Observe that both the disk and shell methods give the same volume, $384\pi/7$.

Example 3. The region bounded by the two curves $y = x^{1/3}$ and $y = x^2$ is rotated about the y-axis. Find the volume of the solid of revolution by using **(a)** the shell method, and **(b)** the disk method.

Solution. **(a)** A sketch of the region is shown in Fig. 8a, together with a typical strip that generates a shell. The height h of the shell is just the difference in y-coordinates, namely, $h = x^{1/3} - x^2$. Furthermore, the radius of the shell is x and its thickness is

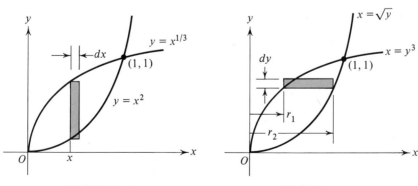

FIGURE 8a   FIGURE 8b

dx. Hence

$$dV = 2\pi xh\,dx = 2\pi x(x^{1/3} - x^2)\,dx$$

so that

$$V = \int_0^1 2\pi x(x^{1/3} - x^2)\,dx.$$

We leave it to the reader to verify that the value of this integral is $5\pi/14$.

(b) For the disk method we take horizontal strips as illustrated in Fig. 8b. In this case rotation about the y-axis will produce a *washer* (a disk with a hole). The inner radius of the washer is $r_1 = y^3$; its outer radius is $r_2 = \sqrt{y}$; and the thickness (height) is dy. Hence the volume of the washer is given by

$$dV = \pi r_2^2\,dy - \pi r_1^2\,dy = \pi(r_2^2 - r_1^2)\,dy = \pi(y - y^6)\,dy,$$

so

$$V = \int_0^1 \pi(y - y^6)\,dy = \pi\left(\frac{y^2}{2} - \frac{y^7}{7}\right)\Big|_0^1 = \pi\left(\frac{1}{2} - \frac{1}{7}\right) = \frac{5\pi}{14}. \quad \bullet$$

EXERCISE 2

In Problems 1 through 4, the region under the given curve for the given interval is rotated about the x-axis. Use the disk method to find the volume of the solid of revolution so obtained.

1. $y = 2x$, $\qquad 0 \leq x \leq 5$.
2. $y = x^3$, $\qquad 0 \leq x \leq 2$.
3. $y = \sqrt{x(2 - x)}$, $\quad 0 \leq x \leq 2$.
4. $y = 4x - x^2$, $\quad 0 \leq x \leq 4$.

In Problems 5 through 8, the region under the curve for the given interval is rotated about the y-axis. Use the shell method to find the volume of the solid of revolution so obtained.

5. $y = 2x$, $\qquad 3 \leq x \leq 5$.
6. $y = x^3$, $\qquad 2 \leq x \leq 3$.
7. $y = 4x - x^2$, $\quad 0 \leq x \leq 4$.
8. $y = 4x - x^2 - 3$, $\quad 1 \leq x \leq 3$.

In Problems 9 through 11, the region bounded by the given pair of curves is rotated about the x-axis. Find the volume of the solid (a) by the shell method and (b) by the disk method and show that the results are the same. Observe that in the disk method each disk will have a circular hole.

9. $y = x^2$, $y = 2x$.   10. $y = \sqrt{x}$, $y = x^3$.   11. $y = x^2$, $y = x^5/8$.

12. Use the disk method to prove that the volume of a sphere of radius R is $4\pi R^3/3$.

13. The ellipse $\dfrac{x^2}{a^2} + \dfrac{y^2}{b^2} = 1$ is rotated about the *x*-axis. Find the volume of the region bounded by the surface of revolution so generated.

14. Find the volume if the ellipse of Problem 13 is rotated about the *y*-axis. Find all pairs *a* and *b* for which this volume is equal to the volume of the solid of Problem 13.

15. The region bounded by the hyperbola $x^2 - y^2 = 1$ and the lines $y = -2$ and $y = 2$ is rotated about the *y*-axis. Find the volume of the resulting solid.

16. The region bounded by the parabola $y = x^2$ and the line $y = 4$ is rotated about the line $y = 5$. Find the volume of the solid generated.

17. The region bounded by the curve $y = x^3$ and the lines $y = -1$ and $x = 1$ is rotated about the line $y = -1$. Find the volume of the solid generated.

18. The triangle formed by the lines $y = x - 2$, $y = 0$, and $x = 3$ is rotated about the line $x = 1$. Find the volume of the solid generated.

19. Find the volume of the solid that is generated by revolving about the line $x = -2$ the region in the first quadrant that is above the parabola $y = x^2$ and below the parabola $y = 2 - x^2$.

Volumes of Other Solids   **4**

The technique of *slicing* used in the disk method can also be applied to find volumes of solids other than solids of revolution. The method can be employed whenever parallel cross sections of the solid have a simple geometric shape, such as all squares, all triangles, and so on. This is illustrated in

> **Example 1.** Compute the volume of the pyramid *OABC* shown in Fig. 9 (see next page). The line segments *OA, OB,* and *OC* are mutually perpendicular, and have lengths 1, 2, and 3, respectively.
>
> **Solution.** We slice the solid by planes perpendicular to the edge *OC*, obtaining triangular slices. If we regard *OC* as on the *y*-axis with *O* the origin, then the slices have thickness *dy* and the differential element of volume is
>
> $$dV = \frac{1}{2} bh \, dy,$$
>
> where *b* and *h* are the base and height of the triangular face of the slice. By similar triangles
>
> $$\frac{b}{3 - y} = \frac{1}{3} \quad \text{and} \quad \frac{h}{3 - y} = \frac{2}{3}.$$

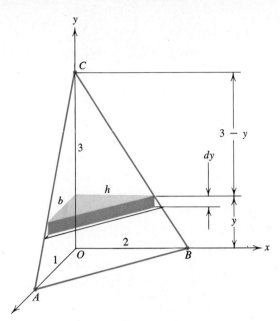

FIGURE 9

Hence $b = (3 - y)/3$ and $h = 2(3 - y)/3$. Then

$$V = \int_0^3 \frac{1}{2} bh \, dy = \int_0^3 \frac{1}{2} \left(\frac{3 - y}{3} \right) \left(\frac{2(3 - y)}{3} \right) dy = \frac{1}{9} \int_0^3 (3 - y)^2 \, dy$$

$$= -\frac{1}{9} \frac{(3 - y)^3}{3} \bigg|_0^3 = 0 - \left(-\frac{1}{9} \frac{3^3}{3} \right) = 1. \quad \bullet$$

Observe that this is consistent with the formula for the volume of any pyramid, namely $V = AH/3$, where A is the area of the base and H is the altitude of the pyramid.

Example 2. A certain solid has a circular base in the xy-plane described by the inequality $x^2 + y^2 \leq r^2$. Each plane perpendicular to the x-axis that meets this solid cuts the solid in a square. Find the volume of the solid.

Solution. Figure 10 illustrates cross sections of the described solid by three different planes drawn perpendicular to the x-axis. Because of symmetry we need only compute the volume of the solid to the right of the y-axis, and then double the answer.

For a typical slice of thickness dx perpendicular to the x-axis, the square sides have length $2y = 2\sqrt{r^2 - x^2}$ so that we have

$$dV = (2y)^2 \, dx = 4y^2 \, dx = 4(r^2 - x^2) \, dx.$$

Hence

$$V = 2 \int_0^r 4(r^2 - x^2)\, dx = 8 \int_0^r (r^2 - x^2)\, dx = 8 \left(r^2 x - \frac{x^3}{3} \right) \Big|_0^r$$

$$= 8 \left(r^3 - \frac{r^3}{3} \right) = \frac{16 r^3}{3}. \quad \bullet$$

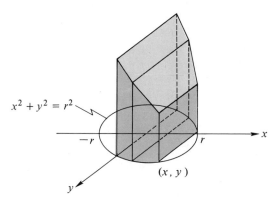

FIGURE 10

EXERCISE 3

1. A certain solid has a circular base in the xy-plane described by the inequality $x^2 + y^2 \leq 1$. Each plane perpendicular to the x-axis that meets this solid cuts the solid in an equilateral triangle. Find the volume of this solid.

2. In the pyramid of Fig. 9, suppose as before that the lines OA, OB, and OC are mutually perpendicular but this time have lengths α, β, and γ, respectively. Prove the general formula for the volume of such a pyramid, namely,

$$V = \frac{1}{6}\alpha\beta\gamma = \frac{1}{3}AH,$$

where A is the area of the base and H is the altitude.

3. A certain solid has its base in the xy-plane and the base is the region bounded by the parabola $y^2 = 4x$ and the line $x = 9$. Each plane perpendicular to the x-axis that meets this solid cuts it in a square. Find the volume of this solid.

4. Find the volume of the solid described in Problem 3 if each of the planes cuts the solid in an isosceles triangle of height 5.

5. Find the volume of the solid described in Problem 4 if the height of the isosceles triangle is x^2.

6. Locate the plane perpendicular to the x-axis that will cut in half the solid of Problem 3 (divide the volume in half).

7. A solid whose base is the region bounded by the two curves $y = x^2$ and $y = 1 - x^2$ in the xy-plane has the property that cross sections perpendicular to the x-axis are semicircles with diameter in the xy-plane. Find the volume of this solid.

8. A solid figure has as its base an isosceles triangle with base and altitude both equal to 2. Each cross section perpendicular to the triangle's altitude is a semicircle with its diameter in the triangle. Compute the volume of the solid.

$\star$9. A wedge is cut from a cylinder of radius 10 cm by two half-planes, one perpendicular to the axis of the cylinder. The second plane meets the first plane at an angle of $45°$ along a diameter of the circular cross section made by the first plane. Find the volume of the wedge.

$\star$10. A "hole" is drilled in a solid cylindrical rod of radius r, by a drill also of radius r. The axes of the rod and the drill meet at right angles, and because the radii are equal the "hole" just separates the rod into two pieces. Find the volume of the material cut out by the drill.

5 *The Length of a Plane Curve*

Physically speaking, the length of a plane curve is quite a simple concept. But mathematically, it is a little more complicated. From a physical point of view, we merely take a piece of wire, bend it to fit the curve, snip off the excess, if there is any, straighten out the wire, and measure it with a ruler. What we now want is a mathematical definition that will give us just the number which our feelings about the physical nature of the problem demand that we should get.

We assume that the curve $\mathscr{C}$ is given by an equation $y = f(x)$ for $a \leqq x \leqq b$, and our problem is to define and compute the length of $\mathscr{C}$. We partition the interval $a \leqq x \leqq b$ into n subintervals with points x_k: $a = x_0 < x_1 < x_2 < \cdots < x_n = b$. Let P_k be the point (x_k, y_k) on the curve so that $y_k = f(x_k)$. Let s_n be the length of the polygonal path $P_0 P_1 P_2 P_3 \cdots P_{n-1} P_n$ formed by joining the successive points P_k on $\mathscr{C}$ with straight-line segments as indicated in Fig. 11. If n is large, then clearly s_n should be close to the length of $\mathscr{C}$. This is the basis for

---

DEFINITION 1 (Length of a Curve). If the limit of s_n exists as $\mu(\mathscr{P}) \to 0$, this limit is the length of $\mathscr{C}$.

If we use s to denote this length, then by definition

(11)
$$s = \lim_{\mu(\mathscr{P}) \to 0} s_n = \lim_{\mu(\mathscr{P}) \to 0} \sum_{k=1}^{n} |P_{k-1} P_k|.$$

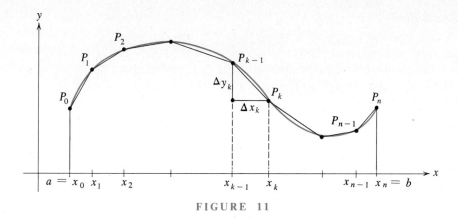

FIGURE 11

The general theory is a little more complicated than one might suppose. Indeed, there is a function $f(x)$ that is continuous in $a \leq x \leq b$, and for which the curve $\mathscr{C}$ does *not* have finite length. But if $f'(x)$ is continuous in $a \leq x \leq b$, then $\mathscr{C}$ has a (finite) length. When $f'(x)$ is continuous in $a \leq x \leq b$, the curve $\mathscr{C}$ is called a *smooth curve.*

THEOREM 1. If $f'(x)$ is continuous in $a \leq x \leq b$, then the length of the curve $y = f(x)$ between $x = a$ and $x = b$ is given by

(12)
$$s = \int_a^b \sqrt{1 + [f'(x)]^2}\, dx.$$

At first glance this formula may appear to be hard to memorize. But if we use dy/dx for $f'(x)$, we can write

(13)
$$\int \sqrt{1 + [f'(x)]^2}\, dx = \int \sqrt{1 + \left(\frac{dy}{dx}\right)^2}\, dx = \int \sqrt{dx^2 + dy^2}.$$

Hence (12) appears as a highly disguised form of the Pythagorean Theorem. Interchanging the role of x and y in Theorem 1 or manipulating with (13), we arrive at

(14)
$$s = \int_c^d \sqrt{1 + \left(\frac{dx}{dy}\right)^2}\, dy.$$

In (14), y is the independent variable and the equation of the curve is $x = g(y)$ for y in the interval $c \leq y \leq d$.

Proof of Theorem 1. The length of the line segment joining P_{k-1} and P_k is given by the distance formula

(15) $$|P_{k-1}P_k| = \sqrt{(x_k - x_{k-1})^2 + (y_k - y_{k-1})^2}, \qquad k = 1, 2, \ldots, n.$$

Since $f(x)$ is differentiable in the interval $a \leq x \leq b$, we can apply the Mean Value Theorem (Chapter 5, Theorem 8) to $f(x)$ and find an $x_k^{\star}$ such that $x_{k-1} < x_k^{\star} < x_k$ and such that

(16) $$y_k - y_{k-1} = f'(x_k^{\star})(x_k - x_{k-1}) = f'(x_k^{\star}) \Delta x_k.$$

Using (16) in (15), we obtain

$$|P_{k-1}P_k| = \sqrt{(\Delta x_k)^2 + [f'(x_k^{\star}) \Delta x_k]^2} = \sqrt{1 + [f'(x_k^{\star})]^2} \, \Delta x_k.$$

Then on adding the lengths of the individual line segments, we have

(17) $$s_n = \sum_{k=1}^{n} \sqrt{1 + [f'(x_k^{\star})]^2} \, \Delta x_k.$$

Since $f'(x)$ is continuous, this sum approaches a limit as $\mu(\mathscr{P}) \to 0$, and this limit is just the definite integral on the right side of (12). ∎

> **Example 1.** Find the length of the arc of the curve $y = 2\sqrt{x^3}$ between $x = 1/3$ and $x = 5/3$.
>
> **Solution.** We use the word *arc* to denote a piece of a curve. In this example the curve is given by the equation $y = 2\sqrt{x^3}$ for all $x \geq 0$, and we are to find the length of a piece of the curve, the arc for which $1/3 \leq x \leq 5/3$. By equation (12),
>
> $$s = \int_{1/3}^{5/3} \sqrt{1 + \left(\frac{dy}{dx}\right)^2}\, dx = \int_{1/3}^{5/3} \sqrt{1 + (3\sqrt{x})^2}\, dx$$
>
> $$= \frac{1}{9} \int_{1/3}^{5/3} (1 + 9x)^{1/2} \, 9\, dx = \frac{1}{9}\frac{2}{3}(1 + 9x)^{3/2} \Big|_{1/3}^{5/3}$$
>
> $$= \frac{2}{27}(64 - 8) = \frac{112}{27} = 4\frac{4}{27}. \quad \bullet$$

If the reader will select a curve at random and try to compute its length, he will see that equation (12) frequently leads to integrals that are hard to evaluate. We must select our curves very carefully in order to have $\sqrt{1 + [f'(x)]^2}$ simplify nicely. This difficulty will partially disappear when we learn more about integration in Chapter 10.

EXERCISE 4

In each of Problems 1 through 7, find the length of the arc of the given curve between the given limits.

    1. $y = x^{3/2}$, $0 \leqq x \leqq 4$.   2. $3y = 2(1 + x^2)^{3/2}$, $1 \leqq x \leqq 4$.

3. $3y = (x^2 + 2)^{3/2}, \quad 0 \leq x \leq 3.$

4. $y = \dfrac{x^3}{3} + \dfrac{1}{4x}, \qquad 1 \leq x \leq 4.$

5. $y = \dfrac{x^3}{6} + \dfrac{1}{2x}, \quad 1 \leq x \leq 3.$

6. $y = (a^{2/3} - x^{2/3})^{3/2}, \quad 0 \leq x \leq a.$

7. $x = 2\sqrt{7}y^{3/2}, \qquad 0 \leq y \leq 1.$

8. Prove that the length of the arc in Problem 7 is greater than the length of the chord joining the end points of the arc.

★9. Let A be any positive constant. Show that finding the arc length for the curve

$$y = \frac{1}{3\sqrt{A}}(2 + Ax^2)^{3/2}$$

leads to

$$s = \int_a^b (1 + Ax^2)\, dx.$$

Show that when $A = 2$, this is the curve of Problem 2, and that when $A = 1$, this is the curve of Problem 3.

★10. Let A and B be positive constants. Show that finding the arc length of the curve

$$y = Ax^3 + \frac{B}{x}$$

will lead to the integral

$$s = \int_a^b \left(3Ax^2 + \frac{B}{x^2}\right) dx$$

if A and B satisfy the condition $12AB = 1$. Show that when $A = 1/3$ and $B = 1/4$, this is the curve of Problem 4, and when $A = 1/6$ and $B = 1/2$, this is the curve of Problem 5.

★11. Prove that for the curve of Problem 6 the derivative y' is not continuous in $0 \leq x \leq a$. In computing the length of the arc we used the formula given in Theorem 1. But in this theorem we assume that y' is continuous. Give an argument which you believe would justify this misuse of the formula.

In Problems 12 through 16, set up the integral for computing the length of the arc of the given curve between the given limits, but do not attempt to evaluate the integral.

12. $y = x + x^2, \quad -1 \leq x \leq 1.$

13. $y = x^3, \qquad 0 \leq x \leq 4.$

14. $y = x + 1/x, \quad 4 \leq x \leq 17.$

15. $x = 2y^{5/2}, \quad 4 \leq y \leq 9.$

16. $b^2x^2 + a^2y^2 = a^2b^2, \quad a > b > 0, \, -a/2 \leq x \leq a/2, \, y \geq 0.$

CALCULATOR PROBLEMS

C1. Estimate the length of the arc $y = x^3$, $0 \leq x \leq 4$, of Problem 13 by calculating the sum of the lengths of the chords obtained by partitioning the interval $[0, 4]$ into: **(a)** $n = 1$, **(b)** $n = 2$, and **(c)** $n = 4$ equal subintervals (see Fig. 11).

C2. Approximate the length of the arc in Problem 13 by using Simpson's Rule (Chapter 6, Section 10) with $2n = 8$ ($n = 4$) to estimate the integral expression for s. Compare with your answers to Problem C1.

6 *The Area of a Surface of Revolution*

If a curve is rotated about an axis, it generates a surface, called a *surface of revolution*. Our problem is to find the area of such a surface. If we consider in particular the curve of Fig. 11 and rotate this curve around the x-axis, then at the same time the polygonal path $P_0 P_1 P_2 \cdots P_n$ is also rotated about the same axis, and generates a second surface whose area σ_n (Greek letter sigma) is very close to the area of the surface generated by the curve. Each line segment $P_{k-1} P_k$ generates a section or frustum of a cone and the area of the frustum of a cone is known from solid geometry. Thus we can obtain σ_n by adding the areas A_k of each of these sections of a cone.

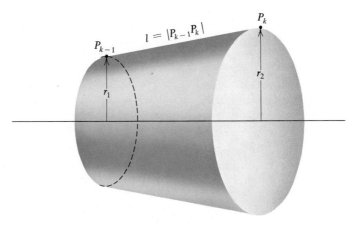

FIGURE 12

DEFINITION 2 (Area of Surface of Revolution). If σ denotes the area of the surface of revolution, then

(18)
$$\sigma = \lim_{\mu(\mathscr{P}) \to 0} \sigma_n.$$

Now the area for the frustum of a cone is

$$A = 2\pi \frac{r_1 + r_2}{2} l = \pi(r_1 + r_2)l,$$

where the symbols have the meaning shown in Fig. 12. Applying this to the frustum generated by the chord $P_{k-1}P_k$ of the curve of Fig. 11, we have

(19)
$$A_k = \pi[f(x_{k-1}) + f(x_k)]\sqrt{(\Delta x_k)^2 + (\Delta y_k)^2}.$$

Just as in the derivation of the formula for the length of arc, the Mean Value Theorem permits us to write that

(20)
$$\Delta y_k = f'(x_k^\star)\,\Delta x_k,$$

where $x_k^\star$ is a suitably chosen point in the interval $x_{k-1} < x < x_k$. Using (20) in (19) and summing, we have

$$\sigma_n = \sum_{k=1}^{n} \pi[f(x_{k-1}) + f(x_k)]\sqrt{1 + [f'(x_k^\star)]^2}\,\Delta x_k,$$

(21)
$$\sigma_n = \sum_{k=1}^{n} \pi f(x_{k-1})\sqrt{1 + [f'(x_k^\star)]^2}\,\Delta x_k + \sum_{k=1}^{n} \pi f(x_k)\sqrt{1 + [f'(x_k^\star)]^2}\,\Delta x_k.$$

Let us examine the first sum in (21). This looks very much like a Riemann sum for the function $\pi f(x)\sqrt{1 + [f'(x)]^2}$, so we would suspect that

(22)
$$\lim_{\mu(\mathscr{P})\to 0} \sum_{k=1}^{n} \pi f(x_{k-1})\sqrt{1 + [f'(x_k^\star)]^2}\,\Delta x_k = \int_a^b \pi f(x)\sqrt{1 + [f'(x)]^2}\,dx.$$

The fly in the ointment is this: Theorem 3 of Chapter 6, which gives the definite integral as a limit of Riemann sums, requires that the *same* $x_k^\star$ replace x wherever it occurs in the integrand. But for the sum in (22), x_{k-1} appears in the first factor while $x_k^\star$ appears in the second factor, and although both lie in the interval $x_{k-1} \le x \le x_k$, they are not always the same. Nevertheless, equation (22) is still valid—a fact that we shall not take time here to prove.[1] On intuitive grounds, the reader may convince himself of the validity of (22) by arguing that if $\mu(\mathscr{P})$ is small, then $|x_{k-1} - x_k^\star|$ is small, and hence x_{k-1} and $x_k^\star$ are very nearly equal.

Similar remarks apply as well to the second sum in (21); that is,

$$\lim_{\mu(\mathscr{P})\to 0} \sum_{k=1}^{n} \pi f(x_k)\sqrt{1 + [f'(x_k^\star)]^2}\,\Delta x_k = \int_a^b \pi f(x)\sqrt{1 + [f'(x)]^2}\,dx.$$

[1] For a proof, see A. W. Goodman's *Analytic Geometry and the Calculus*, 4th ed. (New York: Macmillan Publishing Co., Inc., 1980), Theorem 2, pp. 318–319.

Putting these facts together with Definition 2 gives

THEOREM 2 (PWO). Suppose that $f(x) \geq 0$ and $f'(x)$ is continuous in the interval $a \leq x \leq b$. Then the area of the surface of revolution generated by rotating the curve $y = f(x)$ between $x = a$ and $x = b$ about the x-axis is given by

(23) $$\sigma = \int_a^b 2\pi f(x) \sqrt{1 + [f'(x)]^2} \, dx = \int_a^b 2\pi y \sqrt{1 + \left(\frac{dy}{dx}\right)^2} \, dx.$$

Naturally, if the curve $x = g(y)$ between $y = c$ and $y = d$ is rotated about the y-axis, then the area is given by

(24) $$\sigma = \int_c^d 2\pi g(y) \sqrt{1 + [g'(y)]^2} \, dy = \int_c^d 2\pi x \sqrt{1 + \left(\frac{dx}{dy}\right)^2} \, dy.$$

Finally, the curve of Theorem 2 might be rotated about an axis parallel to the x-axis, say the line $y = y_0$. In this case the formula is

(25) $$\sigma = \int_a^b 2\pi (y - y_0) \sqrt{1 + \left(\frac{dy}{dx}\right)^2} \, dx.$$

In formulas (23) and (25) the curve must lie above the axis of rotation, and in (24) it must lie to the right of this axis. We leave it to the reader to discover the reason for this additional restriction (see Exercise 5, Problem 12).

Example 1. The arc of the curve $y = x^3$ lying between $x = 0$ and $x = 2$ is rotated about the x-axis. Find the area of the surface generated.

Solution. By formula (23),

$$\sigma = \int_a^b 2\pi y \sqrt{1 + \left(\frac{dy}{dx}\right)^2} \, dx = \int_0^2 2\pi x^3 \sqrt{1 + (3x^2)^2} \, dx$$

$$= \frac{2\pi}{36} \int_0^2 (1 + 9x^4)^{1/2} 36x^3 \, dx = \frac{2\pi}{36} \frac{2}{3} (1 + 9x^4)^{3/2} \Big|_0^2$$

$$= \frac{\pi}{27} [(145)^{3/2} - 1]. \quad \bullet$$

EXERCISE 5

In Problems 1 through 7, find the area of the surface generated by rotating the given arc about the x-axis.

1. $y = 2\sqrt{x}$, $0 \leq x \leq 3$.   2. $y = 4\sqrt{x}$, $5 \leq x \leq 8$.

3. $y = \dfrac{x^3}{3}$, $\quad 0 \leq x \leq \sqrt[4]{15}$. $\qquad$ 4. $y = mx$, $m > 0$, $a \leq x \leq b$, $a \geq 0$.

5. $y = \dfrac{x^3}{3} + \dfrac{1}{4x}$, $1 \leq x \leq 2$. $\qquad$ 6. $3y = \sqrt{x}(3 - x)$, $0 \leq x \leq 3$.

7. $8B^2 y^2 = x^2(B^2 - x^2)$, $0 \leq x \leq B$.

8. Prove that the area of the surface of a sphere of radius r is $4\pi r^2$.

9. The arc of $x = 2\sqrt{15 - y}$ lying in the first quadrant is rotated about the y-axis. Find the area of the surface generated.

10. A zone on a sphere is the portion of the sphere lying between two parallel planes that intersect the sphere. The altitude of the zone is the distance between the two parallel planes. Prove that for a zone of altitude h on the sphere of radius r, the surface area is $2\pi r h$. Notice that this states that the area does not depend on the location of the zone on the sphere.

11. The arc of Problem 5 is rotated about the line $y = -C$, $C > 0$. Find the area of the surface generated.

12. If the line segment $y = x - 3$, $1 \leq x \leq 5$ is rotated about the x-axis, it generates a piece of a cone. Show that formal manipulations, using equation (23) to compute the surface area, lead to the integral

$$ I = \int_1^5 2\pi(x - 3)\sqrt{2}\,dx, $$

but that $I = 0$. Find the surface area and explain why it is not given by I.

Fluid Pressure 7

Everyone who swims is familiar with the fact that the pressure of the water on the body increases as one goes deeper under the water. This pressure is most noticeable on the ears, which are rather sensitive to such external forces. Careful measurements indicate that for any fluid, the pressure at a point is directly proportional to the distance of the point below the surface of the fluid. Furthermore, the constant of proportionality is just the product $g\rho$, where g is the gravitational constant and ρ (Greek letter rho) is the mass density of the fluid (mass per unit volume). The product $g\rho$ is the weight density[1] δ (Greek lowercase letter delta) of the fluid in suitable units. Thus, if the distance h is measured in feet and the gravitational constant is in ft/sec[2], then the mass density[2] would be measured in slugs per cubic foot and the weight density in pounds per cubic foot. The pressure, given by the formula

---

[1] Just as weight is the force exerted by a mass on earth, the weight density is the force per unit volume exerted by the fluid.

[2] One slug is the unit of mass that would have a gravitational force (weight) of 32 lb at the earth's surface. For more information on units, see Appendix 4.

(26)
$$P = \rho g h = \delta h,$$

would be in pounds per square foot. Similarly, if h is in centimeters, g in cm/sec^2, ρ in grams/cm^3, or δ in dynes/cm^3, then (26) gives P in dynes/cm^2. If one does not change to a planet with a different value for the gravitational constant, then one can easily use the weight density.

Since pressure is the force on a unit surface area, formula (26) is completely reasonable and one could attempt a theoretical proof of (26) by arguing that a flat plate of unit area should support the weight of the column of fluid directly over it. Since most fluids are incompressible over a reasonable depth, the density is a constant, so equation (26) gives just this weight for the unit area.

What is not at all obvious is the fact that this pressure is the same in all directions. This means that if we take a flat metal plate, then no matter how that plate is oriented in the fluid, formula (26) still gives the pressure normal (perpendicular) to the face of the plate at each point. Of course, this pressure will usually vary as we go from point to point on the plate because h will usually vary. If h is constant, as it is for a horizontal plate, then the total force F on one face of the plate is just the pressure times the area.

(27)
$$F = PA = \delta h A.$$

When the plate is not horizontal, h is a variable and simple arithmetic fails, but this is just the situation in which the definite integral is useful.

Example 1. A vertical dam across a certain small stream has (roughly) the shape of a parabola. It is 36 ft across the top and is 9 ft deep at the center. Find the maximum force that the water can exert on the face of this dam.

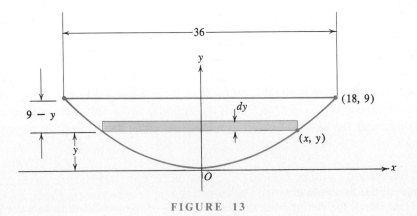

FIGURE   13

Solution. Obviously, the maximum force occurs during flood time when the water is at the top of the dam and just about to overflow. The density of water is a variable, depending on the type and amount of dissolved material as well as the temperature and the depth of the water, but we will suppose for simplicity that for water $\delta = 62.5$ lb/ft^3. We place the river and the dam on a coordinate system, as indicated in Fig. 13. The equation of the parabolic bottom of the dam has the form

$$y = cx^2$$

and since the curve passes through $(18, 9)$ we have $9 = c(18)^2$, and hence

$$c = \frac{9}{(18)^2} = \frac{1}{2 \times 18} = \frac{1}{36}.$$

Therefore, in the first quadrant $x = \sqrt{y/c} = 6\sqrt{y}$. We divide the face of the dam into thin horizontal strips, as indicated in the figure. The representative strip shown is at distance $9 - y$ from the surface of the water. The pressure there is $P = \delta h = \delta(9 - y)$. The total force on the strip is

$$dF = P\,dA = \delta(9 - y)\,dA = \delta(9 - y)2x\,dy$$

and hence on integrating (summing over all such strips, and taking a limit)

(28)
$$F = \int_0^9 \delta(9 - y)2x\,dy.$$

For the particular case in hand, $x = 6\sqrt{y}$, so we find that

$$F = 12\delta \int_0^9 (9 - y)\sqrt{y}\,dy = 12\delta \left(\frac{2}{3}9y^{3/2} - \frac{2}{5}y^{5/2}\right)\Big|_0^9$$

$$= 12\delta \left(\frac{2}{3}3^5 - \frac{2}{5}3^5\right) = 24 \times 3^5\delta \left(\frac{1}{3} - \frac{1}{5}\right)$$

$$= 24 \times 3^5 \times \frac{2}{15}\delta = \frac{2^4 \times 3^5\delta}{5} \text{ lb.}$$

Computation with $\delta = 62.5$ gives $F = 48{,}600$ lb, or 24.3 tons. ●

We could develop a general formula similar to equation (28) for this type of problem. But it is better not to burden the mind with such a formula. The student should master the general principles as embodied in equations (26) and (27), and apply these general principles to each particular problem.

Example 2. Suppose that in Example 1 the face of the dam on the water side was slanted at a 45° angle. Find the total force of the water on the face of the dam.

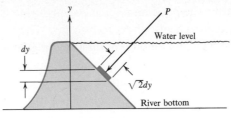

FIGURE 14

Solution. A cross section of the dam is shown in Fig. 14. The pressure on a representative strip is just the same as before, but now the width of the strip is $\sqrt{2}\, dy$, because of the slant of the face; hence the area of the strip, and consequently each subsequent equation in the solution of Example 1, must be multiplied by $\sqrt{2}$. Then the force of the water on the face of the dam is $\sqrt{2} \times 24.3$ tons, or 34.4 tons approximately. ●

The force of 34.4 tons is perpendicular to the face of the dam. If we resolve this force into horizontal and vertical components, we find that there is a horizontal force of 24.3 tons tending to push the dam down the river, just as before. The vertical component is also 24.3 tons, and this component is pressing the dam downward against its foundations. This force would be added to the weight of the concrete in the computations for the design of the foundation.

EXERCISE 6

1. Find the force of the water on one end of a tank if the end is a rectangle 4 ft wide and 6 ft high, and the tank is full of water.
2. Solve Problem 1 if the end is an inverted triangle 8 ft wide at the top and 6 ft high.
3. Solve Problem 1 if the end is a trapezoid 5 ft wide at the top, 3 ft wide at the bottom, and 6 ft high. Note that the area of the end of the tank is the same in Problems 1, 2, and 3. Compare the three forces on the end.
4. A vertical dam has the form of a segment of a parabola 800 ft wide at the top and 100 ft high at the center. Find the maximum force that the water behind the dam can exert on the dam.
5. Find the force on one side of the vertical triangular plate shown in Fig. 15 when the plate is submerged in water.

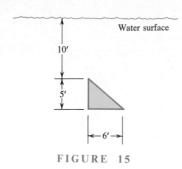

FIGURE 15

6. A gate for a dam is in the form of an inverted isosceles triangle. The base is 6 ft, the altitude 10 ft, and the base is 10 ft below the surface of the water. Find the force of the water on the gate.

7. A cylindrical drum lying with its axis horizontal is half full of oil. If the weight density of the oil is 50 lb/ft^3 and the radius of the drum is 9 in., find the force of the oil on one end of the drum.

8. The cross section of a cylindrical gasoline tank is an ellipse with major axis 3 ft and minor axis 1 ft. Naturally, the tank is placed so that the axis of the cylinder and the major axis of the elliptical cross section are horizontal. Find the force of the gasoline on one end of the tank when it is half full of gasoline with a weight density of 50 lb/ft^3.

9. Solve Problem 4 if the dam has the form of half an ellipse, that is, the portion of an ellipse lying below the major axis.

10. Solve Problem 5 if the base of the triangular plate is B ft instead of 6 ft, all other dimensions being the same.

11. A trough is 8 ft long and 2 ft high. Vertical cross sections are isosceles right triangles with the hypotenuse horizontal. Find the force on one (inclined) side if the trough is filled with water.

$\star$12. Find a general formula for the force on one side of a vertical triangular plate sub-merged in water as shown in Fig. 15, when the base is B ft, the altitude is A ft, and the top is H ft below the surface of the water.

$\star$13. Suppose that in Problem 8 the major and minor axes of the ellipse are $2a$ and $2b$, respectively. Prove that the force of the fluid on the end of the cylinder is $2\delta ab^2/3$ whenever the tank is half full of a liquid of weight density δ.

$\star$14. Suppose that in Problem 4 the parabolic segment has width $2a$ and depth b. Show that the force on the dam is $8\delta ab^2/15$.

$\star$15. The bow of a landing barge consists of a rectangular flat plate A ft wide and B ft long. When the barge is floating, this plate makes an angle of 30° with the surface of the water. Show that the maximum normal force of the water on this plate is $\delta AB^2/4$.

8 Work

When a constant force F acts in a straight line through a distance s, the product Fs is called the *work* done by the force and is denoted by W. This concept of work is introduced in physics, and it turns out to be a very useful one. If the force is a variable, then the arithmetic definition $W = Fs$ is no longer available. If the variable force is acting along a straight line, we make this line an x-axis, and if the force acts from $x = a$ to $x = b$, that is, if it is pushing some object from $x = a$ to $x = b$, then by *definition* the work done is

$$(29) \qquad W \equiv \int_a^b F(x)\, dx,$$

where $F(x)$ gives the force for each x in $a \leq x \leq b$. There is nothing to prove here because this is just a definition of W. But it is clear that the underlying motivation for the definition is the fact that for a small displacement of the object through a distance Δx_k the force is nearly constant, and hence $F(x_k)\, \Delta x_k$ is a good approximation to the work done during the small displacement. Then W should be the limit of the sum

$$\sum_{k=1}^n F(x_k)\, \Delta x_k$$

as $n \to \infty$ and $\mu(\mathscr{P}) \to 0$. But if $F(x)$ is continuous in $a \leq x \leq b$, then this limit exists and is just the definite integral (29).

The compression of a spring furnishes a good illustration of these principles. Each spring has a *natural length L*, the length of the spring when no external forces are applied other than the gravitational forces. The force required to compress or extend this spring an amount x is directly proportional to x, and this proportionality factor is called the *spring constant*. Thus

$$(30) \qquad\qquad F(x) = cx,$$

where $F(x)$ is the force on the spring and x is the difference in length from the natural length. The situation is shown in Fig. 16.

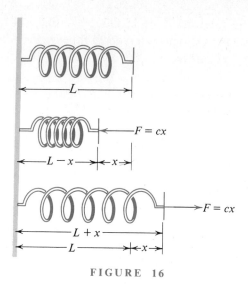

Example 1. A spring has a natural length of 20 in. and a 40-lb force is required to compress it to a length of 18 in. How much work is done on the spring in compressing it to a length of 17 in. starting from its natural length?

Solution. We must first find the spring constant c from the given data. We know $F = 40$ lb when $x = 20 - 18 = 2$ in., whence equation (30) gives $40 = c2$ or $c = 20$ lb/in. Then, to find the work done when the spring is compressed to a length of 17 in. from 20 in., equation (29) yields

$$W = \int_a^b F(x)\, dx = \int_0^3 20x\, dx = 10x^2 \Big|_0^3 = 90 \text{ in.-lb.} \quad \bullet$$

If this same spring were stretched from a length of 20 in. to a length of 23 in., the work done on the spring would be the same. In each case the work in compressing or extending the spring appears to be stored in the spring and can be recovered. Thus if the spring that has been compressed to a length of 17 in. now pushes some object 2 in., so that the spring extends to a length of 19 in., then the spring does work on the object, and the amount is the same as the work done earlier on the spring to compress it from 19 in. to 17 in. This amount is given by

$$\int_1^3 F(x)\, dx = \int_1^3 20x\, dx = 10x^2 \Big|_1^3 = 90 - 10 = 80 \text{ in.-lb.}$$

Example 2. Find the work done in filling a cylindrical tank (Fig. 17) with oil of weight density 50 lb/ft³ from a reservoir 15 ft below the bottom of the tank. The tank is 10 ft high, 8 ft in diameter, and the oil is pumped in through a hole in the bottom.

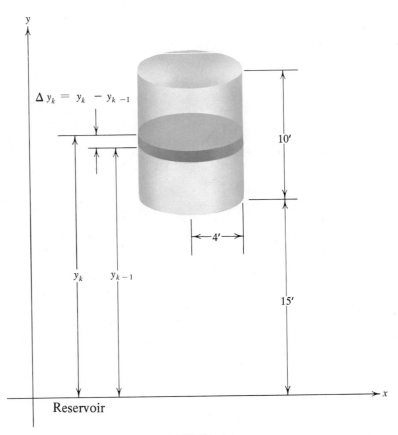

$$\Delta y_k = y_k - y_{k-1}$$

10′

4′

y_k y_{k-1}

15′

Reservoir

FIGURE   17

Solution. Equation (29) is not directly applicable, because we do not have a variable force acting over an interval. Hence we must start with the basic principle and obtain a new formula (or definition) for work that will cover this situation.

As shown in Fig. 17, we introduce a y-axis. Let $\mathscr{P}$ be a partition of the interval $15 \leqq y \leqq 25$, and through each point y_k of the partition we pass a horizontal plane. These planes intersect the cylindrical tank to form disks, and the volume of the kth disk is $\pi r^2 (y_k - y_{k-1}) = \pi r^2 \Delta y_k$. The weight of the oil in this disk is $\delta \Delta V_k = \pi r^2 \delta \Delta y_k$, where δ is the weight density of the oil. Now each particle in the kth disk must be raised at least through a distance y_{k-1} and at most through a distance y_k. Since the force required is acting against gravity, then it is just the weight of the disk.

Thus ΔW_k, the work necessary to lift this disk of oil from the reservoir to its place in the tank, must satisfy the inequality

$$(\pi r^2 \delta \, \Delta y_k) y_{k-1} \leqq \Delta W_k \leqq (\pi r^2 \delta \, \Delta y_k) y_k.$$

Consequently, on summing we have

$$\sum_{k=1}^{n} \pi r^2 \delta y_{k-1} \, \Delta y_k \leqq W \leqq \sum_{k=1}^{n} \pi r^2 \delta y_k \, \Delta y_k.$$

If we now let $\mu(\mathcal{P}) \to 0$, both of these sums approach the definite integral

$$\int_{15}^{25} \pi r^2 \delta y \, dy,$$

and hence this is W. Since $r = 4$ ft and $\delta = 50 \, \text{lb/ft}^3$, we find that

$$W = \int_{15}^{25} \pi r^2 \delta y \, dy = \pi r^2 \delta \frac{y^2}{2} \Big|_{15}^{25} = \pi r^2 \delta \, 200$$

$$= \pi \times 16 \times 50 \times 200 = 160{,}000\pi \ \text{ft-lb.} \quad \bullet$$

One should observe that we have really computed a minimum value for W. For in actual fact there may be friction losses. Further, in a practical case it might be necessary to carry the oil all the way to the top and then discharge it into the tank. In this latter situation all of the oil is lifted 25 ft and the work done on the oil is simply $10\pi r^2 \delta \times 25 = 10 \times 16 \times 50 \times 25\pi = 200{,}000\pi$ ft-lb. This is larger than $160{,}000\pi$ ft-lb, as we know it should be.

EXERCISE 7

1. A tug boat that is pulling a barge along a straight-line path exerts a force of $2x + 1$ tons when the barge is x feet away from the dock. Find the work done by the tug boat in towing the barge 30 ft from the dock.
2. If the spring constant is 100 lb/in., find the work done in compressing a spring of natural length 20 in. from a length of 19 in. to a length of 15 in.
3. A force of 40 lb is required to compress a spring of natural length 20 in. to a length of 18 in. Find the work done in compressing this spring: (a) from 20 to 19 in., (b) from 19 to 18 in., and (c) from 18 to 17 in.
4. A force of 5 dynes is required to stretch a spring 8 cm from its natural length. How much work is done in stretching the spring from 2 cm to 4 cm beyond its natural length?

5. It took 7 dyne-cm of work to stretch a spring 3 cm from its natural length. Find the force (in dynes) that the spring exerts when it is stretched 2 cm from its natural length.

6. A chain 30 ft long and weighing 1 lb/ft, is dangling from the top of a cliff. How much work is required to pull the chain to a horizontal position at the top of the cliff?

7. A chain that is lying on the floor is 20 ft long and weighs 0.5 lb/ft. How much work is required to lift one end of the chain to a height of 30 ft?

8. According to Coulomb's Law, the magnitude of the repulsive force acting on each of two negative point charges is inversely proportional to the square of the distance between them. If this force is 6 dynes when the charges are 3 cm apart, how much work is required to move one of the charges from a distance of 8 cm to a distance of 4 cm from the other?

9. A cylindrical tank of radius 5 ft and height 8 ft is filled with oil of weight density 40 lb/ft^3. Find the work required to lift all the oil to a height 3 ft above the top of the tank.

10. Find the amount of work required to empty a hemispherical reservoir 10 ft deep if it is full of a liquid of weight density δ lb/ft^3 and the liquid must be pumped to the top.

11. Find the work required to empty a conical reservoir of radius 6 ft at the top and height 8 ft if it is full of a liquid of weight density δ lb/ft^3 and if the liquid must be lifted 4 ft above the top of the reservoir.

12. Solve Problem 11 if the reservoir is filled only to a depth of 4 ft.

13. A bucket weighing 2 lb is filled with 1 ft^3 of water. Suppose that the bucket is lifted at the constant rate of 3 ft/sec for a period of 4 sec. If the bucket leaks water at the rate of 0.25 ft^3/sec, how much work is done in lifting the bucket? (Assume that 1 ft^3 of water weighs 62.5 lb.)

REVIEW PROBLEMS

In Problems 1 through 8, find the area of the figure bounded by the given pair of curves.

1. $y = x^2,$ $y = -x^2 + 6x.$

2. $y = x + 6,$ $y = x^2 - x - 2.$

3. $y = x - 4,$ $x = y^2 + 2y + 2.$

4. $y = x^2,$ $y = 2 + \dfrac{x^2}{2}.$

★5. $y = \sqrt{3x},$ $y = x^2 - 2x.$

★6. $y = x - 1,$ $y = \dfrac{27(x-1)}{x^3}.$

★7. $y = \dfrac{x^3}{2},$ $y = \dfrac{x^5}{8}.$

★8. $6 = x^2 y,$ $21 = 9x + 2y.$

9. Find the area of the region bounded by the curve $y = x^3 - 3x^2 + 2x$ and the line tangent to this curve at the origin.

★10. Find the area of the figure bounded by the *x*-axis and the curve $y = x(x - b)(x + a)$, where $a > 0$, $b > 0$.

In Problems 11 through 14, the region under the given curve for the given interval is rotated about the x-axis. Compute the volume of the solid generated.

11. $y = x^2$, $0 \leq x \leq 2$.

12. $y = x(a - x)$, $0 \leq x \leq a$.

13. $y = \dfrac{a}{x}$, $1 \leq x \leq b$, $a > 0$.

14. $y = x^2(a - x)$, $0 \leq x \leq a$.

15. Each of the regions described in Problems 11 through 14 is rotated about the y-axis. In each case compute the volume generated.

16. Consider the region under the curve $y = x^2$ for x between 0 and b, where b is a fixed positive number. The area of this region is bisected by some line $x = x_0$. Find x_0.

17. Suppose that the region in Problem 16 is rotated about the x-axis. Some line $x = x_1$ will generate a plane that bisects the volume of the solid. Find x_1.

18. Referring to Problems 16 and 17, guess which is the larger number, x_0 or x_1. Then prove that your guess is correct (or wrong).

In Problems 19 through 22, the region bounded by the given pair of curves is rotated about the x-axis. Find the volume of the solid generated.

19. $y = \sqrt{x}$, $y = \sqrt[3]{2x}$.

20. $y = \sqrt{ax}$, $y = \dfrac{x^2}{a}$, $a > 0$.

21. $y = \dfrac{2}{x}$, $x + y = 3$.

*22. $y = 4x - x^2$, $y = 3$.

23. A swimming pool is 10 ft wide, 30 ft long, and has a uniform depth of 8 ft. Find the force of the water: (a) on the bottom, (b) on one side, and (c) on one end.

24. Check your answers in Problems 19, 20, and 21 by finding the volume by a different method.

25. Find the formula for the work done on a spring in compressing it a distance A from its natural length. Use k for the spring constant.

26. If an attractive force between two objects varies inversely as the square of the distance between them ($F = k/r^2$), find the work done in moving one object from a separation distance a to a separation distance b, where $b > a$. What is the work done if $b < a$?

27. A 10-ft length of chain weighing 30 lb is hanging over the side of a boat, but does not quite touch the water. How much work is required to raise the chain to the deck of the boat?

28. Prove that the integral formula gives $s = \sqrt{(x_2 - x_1)^2 + (y_2 - y_1)^2}$ for the length of the line segment joining $P_1(x_1, y_1)$ and $P_2(x_2, y_2)$.

29. Prove that if A and B are positive constants such that $32AB = 1$, then the length of arc of the curve $y = Ax^4 + B/x^2$ for $0 < a \leq x \leq b$ is given by

$$\int_a^b \left(4Ax^3 + \frac{2B}{x^3} \right) dx = \left(Ax^4 - \frac{B}{x^2} \right)\bigg|_a^b.$$

30. Find the length of arc of the curve $y = (2x^6 + 1)/8x^2$ for $1 \leqq x \leqq 2$.
31. Find the length of arc of the curve $y = x^4/64 + 2/x^2$ for $2 \leqq x \leqq 4$.
32. Prove that under the conditions of Problem 29, the curve always has a local minimum at $x = \sqrt[3]{4B}$.
33. Find the area of the surface generated by rotating the arc

$$y = \frac{x^3}{6} + \frac{1}{2x}, \qquad 1 \leqq x \leqq 3,$$

about the x-axis.

8

The Trigonometric Functions

Brief Review of Trigonometry *1*

Our objective is to obtain formulas for differentiating and integrating the trigonometric functions and to look at a few of the many applications of these formulas. For this purpose we first review trigonometry.

Originally, the trigonometric functions were defined as ratios of the sides in a right triangle. Thus, referring to Fig. 1, we have the six definitions

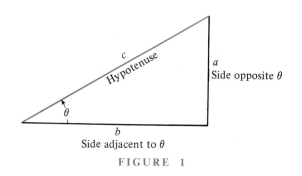

FIGURE 1

(1) $$\sin \theta = \frac{\text{side opposite } \theta}{\text{hypotenuse}} = \frac{a}{c}, \qquad \csc \theta = \frac{\text{hypotenuse}}{\text{side opposite } \theta} = \frac{c}{a},$$

(2) $$\cos \theta = \frac{\text{side adjacent to } \theta}{\text{hypotenuse}} = \frac{b}{c}, \qquad \sec \theta = \frac{\text{hypotenuse}}{\text{side adjacent to } \theta} = \frac{c}{b},$$

(3) $$\tan \theta = \frac{\text{side opposite } \theta}{\text{side adjacent to } \theta} = \frac{a}{b}, \qquad \cot \theta = \frac{\text{side adjacent to } \theta}{\text{side opposite } \theta} = \frac{b}{a}.$$

329

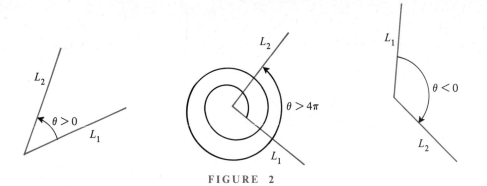

FIGURE 2

Although this classical definition is often useful, it is meaningful only if θ is an acute angle. If we regard θ as a measure of the rotation of one line L_1 to the position of a second line L_2, as indicated in Fig. 2 (positive for a counterclockwise rotation, and negative for a clockwise rotation), then the classical definition fails. In this case we place the "angle" on a rectangular coordinate system (see Fig. 3) with L_1 along the positive x-axis and vertex at the origin. We select a point $P(x, y)$ on L_2 and let $r = |OP| > 0$. With these three numbers, x, y, and r, the six trigonometric functions are defined by

(4) $$\sin \theta \equiv \frac{y}{r}, \qquad \csc \theta \equiv \frac{r}{y},$$

(5) $$\cos \theta \equiv \frac{x}{r}, \qquad \sec \theta \equiv \frac{r}{x},$$

(6) $$\tan \theta \equiv \frac{y}{x}, \qquad \cot \theta \equiv \frac{x}{y}.$$

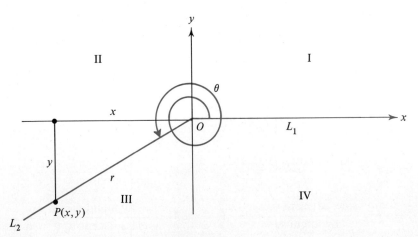

FIGURE 3

Clearly, when θ is an acute angle, the ratios in equations (4), (5), and (6) coincide with the ratios given in equations (1), (2), and (3), respectively.

In certain cases it is convenient to select P on the unit circle (as indicated in Fig. 4). In this case $r = |OP| = 1$ in equations (4) and (5), and we have

(7)
$$\sin \theta = y \quad \text{and} \quad \cos \theta = x.$$

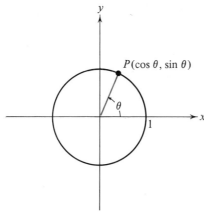

FIGURE 4

When P is on the unit circle, the coordinates of P are $(\cos \theta, \sin \theta)$. For this reason the trigonometric functions are often called the *circular functions*.

From these definitions it is easy to compute the trigonometric functions for certain "popular" angles, and the results are listed in Table 1 (see next page). Here und. (undefined) means that the corresponding function does not exist because the definition requires an illegal division by zero. Values of the trigonometric functions for other acute angles may be found in Table A at the end of the book.

If L_2, the terminal side of the angle θ, lies in some quadrant other than Q.I (see Fig. 3), then the trigonometric functions may be negative, because x or y may be negative. The sign ($\pm$) of each of the trigonometric functions is given in Table 2. Observe that r is always positive in equations (4) through (6).

The trigonometric functions satisfy the following elementary identities:

(8)
$$\sec \theta = \frac{1}{\cos \theta}, \quad \sec \theta \cos \theta = 1, \quad \cos \theta = \frac{1}{\sec \theta}.$$

(9)
$$\csc \theta = \frac{1}{\sin \theta}, \quad \csc \theta \sin \theta = 1, \quad \sin \theta = \frac{1}{\csc \theta}.$$

(10)
$$\cot \theta = \frac{1}{\tan \theta}, \quad \cot \theta \tan \theta = 1, \quad \tan \theta = \frac{1}{\cot \theta}.$$

TABLE 1

θ Radians	θ Degrees	$\sin \theta$	$\cos \theta$	$\tan \theta$	$\cot \theta$	$\sec \theta$	$\csc \theta$
0	0	0	1	0	und.	1	und.
$\dfrac{\pi}{6}$	30°	$\dfrac{1}{2}$	$\dfrac{\sqrt{3}}{2}$	$\dfrac{\sqrt{3}}{3}$	$\sqrt{3}$	$\dfrac{2\sqrt{3}}{3}$	2
$\dfrac{\pi}{4}$	45°	$\dfrac{\sqrt{2}}{2}$	$\dfrac{\sqrt{2}}{2}$	1	1	$\sqrt{2}$	$\sqrt{2}$
$\dfrac{\pi}{3}$	60°	$\dfrac{\sqrt{3}}{2}$	$\dfrac{1}{2}$	$\sqrt{3}$	$\dfrac{\sqrt{3}}{3}$	2	$\dfrac{2\sqrt{3}}{3}$
$\dfrac{\pi}{2}$	90°	1	0	und.	0	und.	1

TABLE 2

Quadrant	sin	cos	tan	cot	sec	csc
I	+	+	+	+	+	+
II	+	−	−	−	−	+
III	−	−	+	+	−	−
IV	−	+	−	−	+	−

Equations (8), (9), and (10) follow immediately from (5), (4), and (6), respectively. Using the Pythagorean Theorem, we can also derive the identities

(11) $\qquad \sin^2 \theta + \cos^2 \theta = 1, \qquad \sin^2 \theta = 1 - \cos^2 \theta, \qquad \cos^2 \theta = 1 - \sin^2 \theta.$

(12) $\qquad \sec^2 \theta - \tan^2 \theta = 1, \qquad \sec^2 \theta = 1 + \tan^2 \theta, \qquad \tan^2 \theta = \sec^2 \theta - 1.$

(13) $\qquad \csc^2 \theta - \cot^2 \theta = 1, \qquad \csc^2 \theta = 1 + \cot^2 \theta, \qquad \cot^2 \theta = \csc^2 \theta - 1.$

Formulas for the trigonometric functions of $\alpha \pm \beta$ are

(14) $\qquad\qquad \sin (\alpha + \beta) = \sin \alpha \cos \beta + \cos \alpha \sin \beta,$

(15) $\qquad\qquad \sin (\alpha - \beta) = \sin \alpha \cos \beta - \cos \alpha \sin \beta,$

(16) $\qquad\qquad \cos (\alpha + \beta) = \cos \alpha \cos \beta - \sin \alpha \sin \beta,$

(17) $\qquad\qquad \cos (\alpha - \beta) = \cos \alpha \cos \beta + \sin \alpha \sin \beta,$

(18)
$$\tan (\alpha + \beta) = \frac{\tan \alpha + \tan \beta}{1 - \tan \alpha \tan \beta},$$

(19)
$$\tan (\alpha - \beta) = \frac{\tan \alpha - \tan \beta}{1 + \tan \alpha \tan \beta}.$$

In all these formulas, the unit of measure for the angles is unimportant in trigonometry. However, in calculus, certain formulas become much simpler if we use radian measure for the angles, and so henceforth we will use only radian measure unless otherwise indicated (see Chapter 1, Section 6).

Two angles are *complementary* if their sum is $\pi/2$. For complementary angles, we have the identities

(20)
$$\sin \left(\frac{\pi}{2} - \theta \right) = \cos \theta, \qquad \cos \left(\frac{\pi}{2} - \theta \right) = \sin \theta,$$

(21)
$$\tan \left(\frac{\pi}{2} - \theta \right) = \cot \theta, \qquad \cot \left(\frac{\pi}{2} - \theta \right) = \tan \theta.$$

The trigonometric functions have certain other interesting properties: sine, cosine, secant, and cosecant are periodic functions with period 2π, but tangent and cotangent have period π. By this we mean that for all θ,

(22)
$$\sin (\theta + 2\pi) = \sin \theta, \qquad \cos (\theta + 2\pi) = \cos \theta,$$

(23)
$$\sec (\theta + 2\pi) = \sec \theta, \qquad \csc (\theta + 2\pi) = \csc \theta.$$

But (replacing 2π by π)

(24)
$$\tan (\theta + \pi) = \tan \theta, \qquad \cot (\theta + \pi) = \cot \theta.$$

Further, sine, tangent, cotangent, and cosecant are odd functions, but cosine and secant are even functions. By this we mean that for all θ,

(25)
$$\sin (-\theta) = -\sin \theta, \qquad \csc (-\theta) = -\csc \theta,$$

(26)
$$\tan (-\theta) = -\tan \theta, \qquad \cot (-\theta) = -\cot \theta.$$

But (dropping the minus sign on the right side)

$$\cos (-\theta) = \cos \theta, \qquad \sec (-\theta) = \sec \theta.$$

All the trigonometric functions are continuous at every point where they are defined.

The graph of $y = \sin x$ is the familiar wave shown in Fig. 5 on the next page. We will have more to say about this graph (and graphs of the other trigonometric functions) when we consider the inverse functions in Section 5.

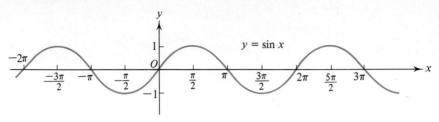

FIGURE 5

EXERCISE 1

In Problems 1 through 12, find sin θ, cos θ, and tan θ for the given angle. Observe that it is then easy to find cot θ, sec θ, and csc θ, by taking the reciprocals [see equations (8), (9), and (10)]. HINT: *Use Tables 1 and 2.*

1. $\dfrac{9\pi}{4}$

2. $\dfrac{7\pi}{6}$

3. $-\dfrac{2\pi}{3}$

4. 4π

5. $\dfrac{11\pi}{6}$

6. $\dfrac{3\pi}{4}$

7. $\dfrac{7\pi}{3}$

8. $-\dfrac{3\pi}{2}$

9. $0°$

10. $120°$

11. $-420°$

12. $330°$

13. Prove the first identity in each of equations (8), (9), and (10). The other two follow immediately.

14. Prove the first identity in each of equations (11), (12), and (13). The other two follow immediately.

In Problems 15 through 22, prove the given identity, using equations (8) through (13).

15. $\cos A \tan A = \sin A$.

16. $\cos B \csc B \tan B = 1$.

17. $\cos^2 C \csc^2 C = \csc^2 C - 1$.

18. $\sec D \csc D = \tan D + \cot D$.

19. $\dfrac{\sin^4 \theta - \cos^4 \theta}{\sin^2 \theta - \cos^2 \theta} = 1$.

20. $\dfrac{\tan \theta \sin \theta}{1 + \cos \theta} = \sec \theta - 1$.

21. $\dfrac{2 - \cos^2 \alpha}{1 - \sin^2 \alpha} = \sec^4 \alpha - \tan^4 \alpha$.

★22. $\dfrac{1 + \sin \beta + \cos \beta}{1 + \sin \beta - \cos \beta} = \dfrac{\sin \beta}{1 - \cos \beta}$.

In Problems 23 through 30, use the addition formulas [the identities given by equations (14) through (19)].

23. Use the known values of the trigonometric functions of $\pi/4$ and $\pi/6$ to compute (a) $\sin (\pi/12)$, (b) $\cos (\pi/12)$, and (c) $\tan (\pi/12)$. HINT: $\pi/12 = 15° = 45° - 30°$.

24. Compute (a) $\sin (5\pi/12)$, (b) $\cos (5\pi/12)$, and (c) $\tan (5\pi/12)$.

In Problems 25 through 29, prove the given identity.

25. $\cos(\theta + \pi) = -\cos\theta$.

26. $\sin(\pi - \theta) = \sin\theta$.

27. $\tan\left(\dfrac{\pi}{4} + \theta\right) = \dfrac{1 + \tan\theta}{1 - \tan\theta}$.

28. $\dfrac{\cos(A + B)}{\cos(A - B)} = \dfrac{1 - \tan A \tan B}{1 + \tan A \tan B}$.

29. $\dfrac{\cos(C + D) - \cos(C - D)}{\cos(C + D) + \cos(C - D)} = -\tan C \tan D$.

30. Prove the double-angle formulas

$$\sin 2A = 2\sin A \cos A,$$

$$\cos 2A = \cos^2 A - \sin^2 A,$$

$$\tan 2A = \frac{2\tan A}{1 - \tan^2 A}.$$

31. Prove the half-angle formulas

$$\sin\frac{\theta}{2} = \pm\sqrt{\frac{1 - \cos\theta}{2}},$$

$$\cos\frac{\theta}{2} = \pm\sqrt{\frac{1 + \cos\theta}{2}},$$

$$\tan\frac{\theta}{2} = \frac{\sin\theta}{1 + \cos\theta}.$$

CALCULATOR PROBLEMS

C1. Fill in the blanks in the following table to four significant figures.

θ	1°	1 radian	3°	3 radians	6.5°	6.5 radians
sin						
cos						
tan						

In Problems C2 through C7, first state the value of the given expression by using an appropriate trigonometric identity. Then check your answer by computing the expression.

C2. $\sin^2(2.9) + \cos^2(2.9)$.

C3. $\sec^2(0.85) - \tan^2(0.85)$.

C4. $\sin\left(4\pi + \dfrac{\pi}{6}\right)$.

C5. $\tan\left(\pi + \dfrac{\pi}{4}\right)$.

C6. $\sin(4.2) - 2\sin(2.1)\cos(2.1)$.

C7. $\cos(8.35) - \cos(-8.35)$.

2 *The Derivative of the Sine and Cosine*

We are looking for a function $f(x)$ which is the derivative of $\sin x$. We begin experimentally by using our calculators to approximate the derivative of $\sin x$ at $x = 0$. Since

$$(27) \qquad \frac{d}{dx} \sin x \bigg|_{x=0} = \lim_{\Delta x \to 0} \frac{\sin(0 + \Delta x) - \sin 0}{\Delta x} = \lim_{\Delta x \to 0} \frac{\sin \Delta x}{\Delta x},$$

we will compute the ratio on the right for small values of Δx. The results are listed (to five significant figures in Table 3).

TABLE 3

Δx (radians)	0.10	0.08	0.06	0.04	0.02
$\sin \Delta x$	0.099833	0.079915	0.059964	0.039989	0.019999
$\dfrac{\sin \Delta x}{\Delta x}$	0.99833	0.99893	0.99940	0.99973	0.99993

The last line of Table 3 seems to indicate that $(\sin \Delta x)/\Delta x \to 1$ as $\Delta x \to 0$. Of course, there are many functions $f(x)$ for which $f(0) = 1$ that could be candidates for the derivative of $\sin x$, but the derivative of a periodic function should be periodic, and $\cos x$ is a periodic function for which $\cos 0 = 1$. This leads us to suspect that

$$(28) \qquad \frac{d}{dx} \sin x = \cos x.$$

Continuing with our calculator, we might try $x = \pi/3$ in (28). If we select $\Delta x = 0.01$, we have (keeping six significant figures)

$$(29) \qquad \frac{d}{dx} \sin x \bigg|_{x=\pi/3} \approx \frac{\sin\left(\dfrac{\pi}{3} + 0.01\right) - \sin \dfrac{\pi}{3}}{0.01}$$

$$\approx \frac{\sin(1.05720) - \dfrac{\sqrt{3}}{2}}{0.01} \approx \frac{0.870982 - 0.866025}{0.01} = 0.4957.$$

This is certainly close enough to $1/2 = \cos(\pi/3)$.

To prove formula (28), we need two preliminary results. We begin with the sector shown in Fig. 6, and we recall that if θ is the radian measure of the vertex angle and r is the radius $(r = |OP|)$, then

(30)
$$\text{area of the sector } SOP = \frac{r^2\theta}{2}.$$

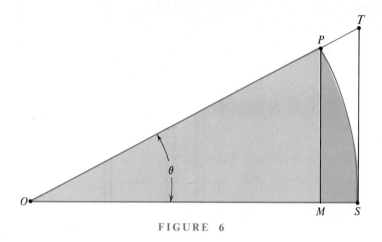

FIGURE 6

If we select $r = 1$, then $|OM| = r\cos\theta = \cos\theta$. Further $|MP| = r\sin\theta = \sin\theta$, and $|ST| = r\tan\theta = \tan\theta$. A comparison of the areas of the triangles MOP and SOT, the first inside the sector and the second containing the sector, gives

$$\text{area of } \triangle MOP < \text{area of sector } SOP < \text{area of } \triangle SOT,$$

or

$$\frac{1}{2}|OM| \times |MP| < \frac{\theta}{2} < \frac{1}{2}|OS| \times |ST|,$$

or

(31)
$$\frac{1}{2}\cos\theta\sin\theta < \frac{\theta}{2} < \frac{1}{2}\tan\theta = \frac{1}{2}\frac{\sin\theta}{\cos\theta}.$$

Dividing by $\sin\theta$ and multiplying by 2, we obtain

(32)
$$\cos\theta < \frac{\theta}{\sin\theta} < \frac{1}{\cos\theta}.$$

Now let $\theta \to 0^+$ in (32). Since $\cos\theta \to 1$ and $1/\cos\theta \to 1$, it follows that

$$\lim_{\theta \to 0^+} \frac{\theta}{\sin\theta} = 1.$$

Taking reciprocals, we see that $\displaystyle\lim_{\theta \to 0^+} \frac{\sin\theta}{\theta} = 1$.

THEOREM 1. If θ is the radian measure of an angle, then

(33)
$$\lim_{\theta \to 0} \frac{\sin \theta}{\theta} = 1.$$

We have given the proof for positive θ. To complete the proof, we observe that if θ is negative and small, then $\sin \theta$ is also negative. Since for all θ we have $\sin \theta = -\sin(-\theta)$, it follows that

$$\frac{\sin \theta}{\theta} = \frac{-\sin(-\theta)}{\theta} = \frac{\sin(-\theta)}{-\theta} = \frac{\sin |\theta|}{|\theta|}.$$

Hence the limit of this ratio is also 1, as $\theta \to 0^-$. ∎

From (33) it is easy to prove that

(34)
$$\lim_{\theta \to 0} \frac{\cos \theta - 1}{\theta} = 0.$$

Indeed, multiplying the numerator and denominator on the left side by $\cos \theta + 1$, we find that

$$\frac{\cos \theta - 1}{\theta} = \frac{(\cos \theta - 1)(\cos \theta + 1)}{(\cos \theta + 1)\theta} = \frac{\cos^2 \theta - 1}{(\cos \theta + 1)\theta} = \frac{-\sin^2 \theta}{(\cos \theta + 1)\theta} = -\frac{\sin \theta}{\cos \theta + 1} \cdot \frac{\sin \theta}{\theta}.$$

Now as $\theta \to 0$, the first factor tends to $-0/2$ and the second factor tends to 1. This proves (34). Equations (33) and (34) will now give the main result,

(28)
$$\frac{d}{dx} \sin x = \cos x.$$

To prove (28), we recall that by definition

$$\frac{d}{dx} \sin x = \lim_{\Delta x \to 0} \frac{\sin(x + \Delta x) - \sin x}{\Delta x}.$$

Since $\sin(x + \Delta x) = \sin x \cos \Delta x + \cos x \sin \Delta x$, we have

$$\frac{d}{dx} \sin x = \lim_{\Delta x \to 0} \frac{\sin x \cos \Delta x + \cos x \sin \Delta x - \sin x}{\Delta x},$$

(35)
$$\frac{d}{dx} \sin x = \lim_{\Delta x \to 0} \left[\frac{\sin x (\cos \Delta x - 1)}{\Delta x} + \cos x \frac{\sin \Delta x}{\Delta x} \right].$$

If we apply (33) and (34) with θ replaced by Δx, we obtain (28) from (35).

Combining the Chain Rule and equation (28) immediately gives

THEOREM 2. If u is any differentiable function of x, then

(36)
$$\frac{d}{dx}\sin u = \cos u \frac{du}{dx}.$$

For the cosine function we use the identity $\cos u = \sin\left(\frac{\pi}{2} - u\right)$. Hence

$$\frac{d}{dx}\cos u = \frac{d}{dx}\sin\left(\frac{\pi}{2} - u\right) = \cos\left(\frac{\pi}{2} - u\right)\frac{d}{dx}\left(\frac{\pi}{2} - u\right)$$

$$= \cos\left(\frac{\pi}{2} - u\right)\left(-\frac{du}{dx}\right) = -\cos\left(\frac{\pi}{2} - u\right)\frac{du}{dx} = -\sin u \frac{du}{dx}.$$

Hence we have proved

THEOREM 3. If u is any differentiable function of x, then

(37)
$$\frac{d}{dx}\cos u = -\sin u \frac{du}{dx}.$$

Example 1. Find the derivative for each of the following functions: **(a)** $y = \sin 5x$, **(b)** $y = \sin(x^3 + 3x - 5)$, **(c)** $y = \cos^7 t$, and **(d)** $y = (5 + \sin 2x)/\cos^4 3x$.

Solution

(a) $\dfrac{dy}{dx} = \cos 5x \dfrac{d}{dx}5x = (\cos 5x)\,5 = 5\cos 5x.$

(b) $\dfrac{dy}{dx} = \cos(x^3 + 3x - 5)\dfrac{d}{dx}(x^3 + 3x - 5) = (3x^2 + 3)\cos(x^3 + 3x - 5).$

(c) $\dfrac{dy}{dt} = 7(\cos t)^6 \dfrac{d}{dt}\cos t = 7\cos^6 t\,(-\sin t) = -7\sin t \cos^6 t.$

(d) $\dfrac{dy}{dx} = \dfrac{\cos^4 3x\,(2\cos 2x) - (5 + \sin 2x)\,4\cos^3 3x\,(-\sin 3x)3}{\cos^8 3x}$

$\qquad = \dfrac{2\cos 2x \cos 3x + 12(5 + \sin 2x)\sin 3x}{\cos^5 3x}.$ ●

EXERCISE 2

In Problems 1 through 15, find the derivative of the given function with respect to the indicated independent variable.

1. $y = \sin 3x$.   2. $y = \cos 7x$.   3. $y = \sin^2 x$.   4. $y = \cos^2 5x$.

5. $y = \cos^2 x^3$.   6. $y = \sin^3 x^2$.   7. $y = \sin 2x \cos 3x$.   8. $y = \sin^2 x \cos^3 x$.

9. $y = (\sin 2x + \cos 2x)^3$.   10. $y = \dfrac{1}{\sin x}$.   11. $y = \dfrac{\sin^2 3x}{\cos^3 2x}$.

★12. $y = \sqrt{\dfrac{1 - \sin^3 x}{1 - \cos^3 x}}$.   13. $z = t \sin^2 (3t^2 + 5)$.   14. $r = (\theta^2 + 2) \sin^2 (5\theta - 1)$.

15. $v = \dfrac{1}{u} \sin \dfrac{1}{u^2}$.

16. Prove that $\sin^2 u + \cos^2 u$ is a constant by showing that its derivative is zero.

17. Show that the two functions $(\cos^2 u - \sin^2 u)^2$ and $-4 \sin^2 u \cos^2 u$ differ by a constant, by showing that they both have the same derivative.

18. If $f(x) = \sin 3x$, find a formula for $f^{(2n)}(x)$, the $2n$th derivative, valid for each positive integer n.

19. If $g(x) = \sin 5x$, find $g^{(2n-1)}(x)$.   20. Find $f^{(2n)}(x)$ if $f(x) = \cos (-2x)$.

In Problems 21 through 24, find the extreme values of the given function for $0 \le x \le 2\pi$. Sketch the curve.

21. $y = \sin 2x$.   22. $y = 2 + \cos 3x$.

23. $y = \sin x + \sin^2 x$.   24. $y = \sin^2 x - \cos x$.

25. Find all the critical points of $y = x + \sin x$. Prove that this function is always increasing and hence has neither a relative maximum nor a relative minimum point. Sketch the curve.

★26. Show that in the interval $0 \le x \le \pi/2$, the curve $y = \sin x$ is concave downward and hence always lies above a certain line, joining its end points. Use this fact to prove that in the interval $0 \le x \le \pi/2$, we have the inequality

$$\sin x \ge \frac{2}{\pi}x.$$

27. Prove that the function $y = A \sin kt + B \cos kt$ satisfies the differential equation

$$\frac{d^2y}{dt^2} + k^2y = 0.$$

In Problems 28 through 31, find all the relative maximum and minimum points and sketch the curve.

28. $y = 3 \sin x + 4 \cos x$.

29. $y = \sin^4 x$.

30. $y = x - 2 \sin x$.

31. $y = \dfrac{\sin^2 x}{1 + \cos^2 x}$.

*★32.** Show that for $y = \sin^4 x$ (the curve of Problem 29) the second derivative is zero at $x = n\pi/3$, where n is any integer. Prove that the points $(n\pi, 0)$ are not inflection points on the curve, but the points corresponding to the other multiples of $\pi/3$ *are* inflection points.

33. Using differentials, find an approximate value for: **(a)** $\sin 32°$, **(b)** $\sin 44°$, and **(c)** $\cos 59°$. Recall that the differentiation formulas are valid only for radian measure and that $1° = \pi/180$ radian. Give your answer to three decimal places.

*★34.** Find the angle from the curve $y = \sin x$ to the curve $y = \sin 2x$ at each point P of intersection of these two curves. By definition, the angle of intersection at P is the angle between the two tangent lines to the two curves at P.

CALCULATOR PROBLEMS

In Problems C1 through C4, solve the given equation to three significant figures. Note that x is in radians. Give the answer: **(a)** *in radians, and* **(b)** *in degrees.*

C1. $x - 2 \sin x = 0$, $x > 0$.   **C2.** $x - 2 \cos x = 0$.

C3. $2x^3 - \sin x = 0$, $x > 0$.   **C4.** $x^3 = 4 \cos x$, $x > 0$.

C5. Experiment with your calculator to guess the value of the limit

$$\lim_{\theta \to 0} \frac{\sin 3\theta}{\sin 2\theta}, \qquad \text{where } \theta \text{ is in radians.}$$

Then prove that your guess is correct.

The Integral of the Sine and Cosine   3

Since each differentiation formula gives rise to a corresponding integral formula, equations (37) and (36) yield

(38) $$\int \sin u \, du = -\cos u + C,$$

(39) $$\int \cos u \, du = \sin u + C.$$

Example 1. Compute each of the following integrals.

(a) $\displaystyle\int \cos (3x + 11) \, dx.$ **(b)** $\displaystyle\int x^4 \sin (x^5 + 7) \, dx.$

(c) $\displaystyle\int_0^{\pi/4} \frac{\sin x}{\cos^3 x}\, dx.$ 　　　(d) $\displaystyle\int_0^{\pi/2} \sin t \cos t (\sin t + \cos t)\, dt.$

Solution. In **(a)** let $u = 3x + 11$; then $du = 3\, dx$.

$$\int \cos (3x + 11)\, dx = \frac{1}{3}\int \cos(3x + 11)(3\, dx) = \frac{1}{3}\int \cos u\, du$$

$$= \frac{1}{3}\sin u + C = \frac{1}{3}\sin(3x + 11) + C.$$

(b) $\displaystyle\int x^4 \sin(x^5 + 7)\, dx = \frac{1}{5}\int \sin(x^5 + 7)5x^4\, dx$

$$= -\frac{1}{5}\cos(x^5 + 7) + C.$$

(c) $\displaystyle\int_0^{\pi/4} \frac{\sin x}{\cos^3 x}\, dx = -\int_0^{\pi/4} \cos^{-3} x(-\sin x\, dx) = -\int_{x=0}^{x=\pi/4} u^{-3}\, du$

$$= -\frac{u^{-2}}{-2}\Big|_{x=0}^{x=\pi/4} = \frac{1}{2}\frac{1}{\cos^2 x}\Big|_0^{\pi/4} = \frac{1}{2}\left(\frac{1}{1/2} - \frac{1}{1}\right) = \frac{1}{2}.$$

(d) $\displaystyle\int_0^{\pi/2} \sin t \cos t (\sin t + \cos t)\, dt = \int_0^{\pi/2} (\sin^2 t \cos t + \cos^2 t \sin t)\, dt$

$$= \left(\frac{\sin^3 t}{3} - \frac{\cos^3 t}{3}\right)\Big|_0^{\pi/2} = \left(\frac{1}{3} - 0\right) - \left(0 - \frac{1}{3}\right) = \frac{2}{3}. \quad \bullet$$

EXERCISE 3

In Problems 1 through 18, compute the given integral.

1. $\displaystyle\int 10 \sin 2x\, dx.$ 　　　　　　　2. $\displaystyle\int 9 \cos 3x\, dx.$

3. $\displaystyle\int 8 \sin x \cos x\, dx.$ 　　　　　　4. $\displaystyle\int 18x^2 \sin x^3\, dx.$

5. $\displaystyle\int 20 t^3 \cos t^4\, dt.$ 　　　　　　6. $\displaystyle\int_0^{\pi/8} \sin 4x\, dx.$

7. $\displaystyle\int_0^{\pi/15} \cos 5t\, dt.$ 　　　　　8. $\displaystyle\int_0^{\sqrt[3]{\pi}} t^2 \sin t^3\, dt.$

9. $\displaystyle\int (2t + 1) \cos (t^2 + t + 5)\, dt.$

*10. $\displaystyle\int \sin^3 \theta\, d\theta.$

*11. $\displaystyle\int (\sin \theta + \cos \theta)^3\, d\theta.$

12. $\displaystyle\int \sqrt{x}\, \cos x^{3/2}\, dx.$

13. $\displaystyle\int \frac{5 \sin 2\sqrt{x}}{\sqrt{x}}\, dx.$

14. $\displaystyle\int \cos^n ax \sin ax\, dx, \quad n \neq -1.$

15. $\displaystyle\int \frac{\cos cx\, dx}{(a + b \sin cx)^n}, \quad n \neq 1.$

16. $\displaystyle\int (x + \sin x)^3 (1 + \cos x)\, dx.$

17. $\displaystyle\int (x^2 + \cos 6x)^9 (x - 3 \sin 6x)\, dx.$

18. $\displaystyle\int 15 \sin x \cos x (\cos^3 x - \sin^3 x)\, dx.$

19. Find the area of the region under one arch of the curve $y = \sin 3x$.

20. Find the area of the region bounded by the curves $y = \cos x$ and $y = \cos^3 x$ between $x = 0$ and $x = \pi/2$.

21. The region under $y = \sin x$ between $x = 0$ and $x = \pi/2$ is rotated about the x-axis. Find the volume of the solid generated. HINT: Use the trigonometric identity $\sin^2 x = (1 - \cos 2x)/2$.

22. Find the area of the region bounded by the curves $y = \sin x$ and $y = \cos x$ between any pair of successive intersection points of the two curves.

*23. Anticipate the results of the next section by proving formulas for the derivative of: (a) $\tan x$, (b) $\cot x$, (c) $\sec x$, and (d) $\csc x$. HINT: $\tan x = \sin x/\cos x$, and so on.

*24. State the integration formulas that correspond to the formulas you obtained in Problem 23.

The Other Trigonometric Functions 4

The differentiation formulas for all six of the trigonometric functions are given in

---

THEOREM 4. If u is a differentiable function of x, then:

(40) $\quad \dfrac{d}{dx} \sin u = \cos u \dfrac{du}{dx},$

(41) $\quad \dfrac{d}{dx} \cos u = -\sin u \dfrac{du}{dx},$

(42) $\quad \dfrac{d}{dx} \tan u = \sec^2 u \dfrac{du}{dx},$

(43) $\quad \dfrac{d}{dx} \cot u = -\csc^2 u \dfrac{du}{dx},$

(44) $\dfrac{d}{dx}\sec u = \sec u \tan u \dfrac{du}{dx},$ (45) $\dfrac{d}{dx}\csc u = -\csc u \cot u \dfrac{du}{dx}.$

Proof. We have already proved (40) and (41); see formulas (36) and (37) of Section 2. To prove (42), we use the quotient formula in conjunction with (40) and (41). Indeed,

$$\frac{d}{dx}\tan u = \frac{d}{dx}\frac{\sin u}{\cos u} = \frac{\cos u \dfrac{d}{dx}\sin u - \sin u \dfrac{d}{dx}\cos u}{\cos^2 u}$$

$$= \frac{\cos u \cos u - \sin u(-\sin u)}{\cos^2 u}\frac{du}{dx} = \frac{1}{\cos^2 u}\frac{du}{dx} = \sec^2 u\frac{du}{dx}.$$

To prove (44), we have

$$\frac{d}{dx}\sec u = \frac{d}{dx}(\cos u)^{-1} = -1(\cos u)^{-2}(-\sin u)\frac{du}{dx}$$

$$= \frac{\sin u}{\cos^2 u}\frac{du}{dx} = \frac{1}{\cos u}\frac{\sin u}{\cos u}\frac{du}{dx} = \sec u \tan u\frac{du}{dx}.$$

The proofs of (43) and (45) are similar and are left for the student. ∎

Example 1. Find the derivative for each of the following functions:

(a) $y = \tan 7x.$ **(b)** $y = \cot x^3.$ **(c)** $y = \tan \sqrt{1 + x^2}.$

(d) $y = \tan x \sec^2 x + 2 \tan x.$ **(e)** $y = \csc (\sin x).$ **(f)** $y = \dfrac{1 - \tan^2 t}{1 + \tan^2 t}.$

Solution

(a) $\dfrac{dy}{dx} = 7 \sec^2 7x.$ **(b)** $\dfrac{dy}{dx} = -3x^2 \csc^2 x^3.$

(c) $\dfrac{dy}{dx} = \sec^2 \sqrt{1 + x^2}\dfrac{d}{dx}\sqrt{1 + x^2} = \dfrac{x}{\sqrt{1 + x^2}}\sec^2 \sqrt{1 + x^2}.$

(d) $\dfrac{dy}{dx} = \sec^2 x \sec^2 x + \tan x(2 \sec x)\sec x \tan x + 2 \sec^2 x$

$= \sec^4 x + 2 \sec^2 x(\tan^2 x + 1) = \sec^4 x + 2 \sec^2 x \sec^2 x = 3 \sec^4 x.$

(e) $\dfrac{dy}{dx} = -\csc (\sin x) \cot (\sin x)\dfrac{d}{dx}\sin x = -\cos x \csc (\sin x) \cot (\sin x).$

Observe that this expression cannot be simplified.

(f) $\dfrac{dy}{dt} = \dfrac{(1 + \tan^2 t)(-2 \tan t \sec^2 t) - (1 - \tan^2 t)2 \tan t \sec^2 t}{(1 + \tan^2 t)^2}$

$$= \frac{-4 \tan t \sec^2 t}{(1 + \tan^2 t)^2} = \frac{-4 \tan t \sec^2 t}{(\sec^2 t)^2}$$

$$= -4 \frac{\sin t}{\cos t} \cos^2 t = -4 \sin t \cos t = -2 \sin 2t.$$

But the given expression could be simplified before differentiation. Indeed,

$$y = \frac{1 - \tan^2 t}{1 + \tan^2 t} = \frac{1 - \tan^2 t}{\sec^2 t} = \cos^2 t - \frac{\sin^2 t}{\cos^2 t} \cos^2 t$$

$$= \cos^2 t - \sin^2 t = \cos 2t.$$

Hence $dy/dt = -2 \sin 2t$. This example shows that it is sometimes advantageous to try to simplify an expression before differentiating it.   ●

With the four new differentiation formulas we have immediately the four new integration formulas

(46) $\displaystyle\int \sec^2 u \, du = \tan u + C,$ **(47)** $\displaystyle\int \csc^2 u \, du = -\cot u + C,$

(48) $\displaystyle\int \sec u \tan u \, du = \sec u + C,$ **(49)** $\displaystyle\int \csc u \cot u \, du = -\csc u + C.$

Example 2. Compute each of the following integrals.

(a) $\displaystyle\int 10 \sec^2 5x \, dx.$ **(b)** $\displaystyle\int \tan 3x \sec^2 3x \, dx.$ **(c)** $\displaystyle\int_{\pi/4}^{\pi/2} \csc^8 x \cot x \, dx.$

Solution

(a) $\displaystyle\int 10 \sec^2 5x \, dx = \int 2 \sec^2 5x \, d(5x) = 2 \tan 5x + C.$

(b) Since $d(\tan 3x) = 3 \sec^2 3x \, dx$, we can write

$$\int \tan 3x \sec^2 3x \, dx = \frac{1}{3} \int \tan 3x \, d(\tan 3x) = \frac{1}{6} \tan^2 3x + C.$$

(c) $\displaystyle\int_{\pi/4}^{\pi/2} \csc^8 x \cot x \, dx = -\int_{\pi/4}^{\pi/2} \csc^7 x(-\csc x \cot x) \, dx$

$\displaystyle = -\frac{\csc^8 x}{8}\Big|_{\pi/4}^{\pi/2} = -\frac{1}{8}\left(\csc^8 \frac{\pi}{2} - \csc^8 \frac{\pi}{4}\right)$

$\displaystyle = -\frac{1}{8}[1 - (\sqrt{2})^8] = -\frac{1}{8}(1 - 16) = \frac{15}{8}. \quad\bullet$

EXERCISE 4

In Problems 1 through 12, find the derivative of the given function with respect to the indicated independent variable.

1. $y = \tan^2 x$.
2. $y = \sec^2 5x$.
3. $y = \csc^3 (2x + 1)$.
4. $y = \tan x^2 - x^2$.
5. $y = \sin x \tan x$.
6. $y = 4 \cos x \sin 4x - \sin x \cos 4x$.
7. $s = \tan^3 t + 3 \tan t$.
8. $y = 3\theta + 3 \cot \theta - \cot^3 \theta$.
9. $y = \sec^2 x \csc^3 x$.
10. $r = \tan^3 \theta \cot^4 \theta$.
★11. $y = \dfrac{x + \sec 2x}{x + \tan 2x}$.
★12. $y = \sqrt{\dfrac{\tan x + \sin x}{\tan x - \sin x}}$.

In Problems 13 through 16, find the third derivative of the given function.
★13. $y = x^2 \tan x$.　★14. $y = \sec x^2$.　★15. $y = \sin 2x \tan 2x$.　★16. $y = \tan x \csc x$.

17. Explain why the two functions $y = \tan^2 x$ and $y = \sec^2 x$ both have the same derivative, $y' = 2 \sec^2 x \tan x$.

In Problems 18 through 25, compute the given integral.

18. $\displaystyle\int \sec^2 5x \tan^3 5x \, dx$.
19. $\displaystyle\int \csc^5 6x \cot 6x \, dx$.
20. $\displaystyle\int x^2 \sec^2 (5x^3 + 7) \, dx$.

21. $\displaystyle\int (\sec^2 x + \tan^2 x) \, dx$.
22. $\displaystyle\int \frac{\theta \, d\theta}{\sin^2 4\theta^2}$.
23. $\displaystyle\int \sin y \sec^3 y \, dy$.

24. $\displaystyle\int (\sec z + \tan z)^2 \, dz$.
25. $\displaystyle\int \sec^4 \theta \, d\theta$.

26. Find the minimum point on the curve $y = \tan x + \cot x$ in the interval $0 < x < \pi/2$.

27. Solve Problem 26 without using the calculus by first proving the identity

$$\tan x + \cot x = \frac{2}{\sin 2x}.$$

28. Prove that if $A > 0$, then the minimum of $f(x) = \tan x + A \cot x$ is $2\sqrt{A}$ for x in the interval $0 < x < \pi/2$.

29. Use differentials to find approximate values for: **(a)** $\tan 46.8°$, **(b)** $\tan 44.1°$, **(c)** $\csc 31°$, and **(d)** $\sec 29°$. Give answers to three decimal places.

30. Find the area of the region under the curve $y = \sec^2 x$ between $x = 0$ and $x = \pi/4$.

31. The region under the curve $y = \sec x$ between $x = 0$ and $x = \pi/4$ is rotated about the x-axis. Find the volume of the solid generated.

32. Repeat Problem 31 for the curve $y = \tan x$.

33. Repeat Problem 31 for the curve $y = \sec^2 x$.

34. The region under the curve $y = \sec^2 \pi x^2$ between $x = 0$ and $x = 1/2$ is rotated about the y-axis. Find the volume of the solid generated.

In Problems 35 through 38, use implicit differentiation to find $\dfrac{dy}{dx}$.

35. $y + x = \sin y \cos x$.

37. $yx = \sin y + \cos x$.

36. $y \tan x + x \sec y = 1$.

38. $\sin (x + y) = \tan (x + y)$.

In Problems 39 through 42, determine the extreme values of the given function in the interval $-\pi/2 < x < \pi/2$.

39. $y = \tan x + \sec x$.

41. $y = \tan x - 8 \sin x$.

40. $y = 2 \tan x + \sec^2 x$.

42. $y = \tan x - 2x$.

43. Prove that the graph of the equation of Problem 38 consists of the collection of straight lines $y = -x + n\pi$, where n is any integer.

★44. A ladder 27 ft long rests against a wall 8 ft high. On the other side of the wall, 12 ft away, is a tall building. Prove that with the ladder touching the ground at one end, the other end cannot rest against the building. How close to the wall must the building be in order that the ladder may just barely reach the building?

★45. Find the error (if there is one) in the computation

$$\int_0^\pi \sec^2 x \, dx = \tan x \Big|_0^\pi = \tan \pi - \tan 0 = 0.$$

⋆**46.** By using implicit differentiation, it is an easy matter to compute $\dfrac{dy}{dx}$ and find the slope of the tangent line for points on the graph of

$$\cos(x + y) = 1 + \tan^4(x + 3y).$$

Would such a computation be meaningful?

CALCULATOR PROBLEMS

C1. Compute the absolute minimum of the function $f(x) = (x - 1)\tan x$ in the interval $[0, 1]$.

C2. Use Newton's method to find the root (to four significant figures) of the equation $\sec x + x^2 - 2 = 0$ in the interval $0 < x < \pi/2$. Start with $x_1 = 1$.

C3. Use Simpson's Rule (Chapter 6, Section 10) with $2n = 10(n = 5)$ to approximate the length of the curve $y = \tan x$ from $x = 0$ to $x = 1$.

C4. Use Simpson's Rule with $2n = 8$ $(n = 4)$ to approximate the length of the curve $y = \sec x$ from $x = 0$ to $x = 1/2$.

5 *The Inverse Trigonometric Functions*

A function $y = f(x)$ generates a set of ordered pairs (a, b). Thus, for each $x = a$, the corresponding $y = b$ is obtained by computing $f(a)$ to determine b.

The inverse function (if it exists) is the set obtained by interchanging a and b, to obtain a set of pairs (b, a). Thus the roles of independent variable and dependent variable are interchanged. This new set of ordered pairs (b, a) is not always a function, so we call the new set the *inverse relation* of $y = f(x)$.

To make a picture of the inverse relation, we observe that the two points (a, b) and (b, a) are symmetric with respect to the line $y = x$ (mirror images of each other in the line $y = x$). Thus we arrive at

> **THEOREM 5.** Let $\mathscr{F}$ be the graph of $y = f(x)$ and let $\mathscr{G}$ be the graph of the inverse relation. Then $\mathscr{G}$ can be obtained from $\mathscr{F}$ by reflecting $\mathscr{F}$ in the line $y = x$. Conversely $\mathscr{F}$ can be obtained from $\mathscr{G}$ in the same way.

Example 1. Sketch the graph of the function $f(x) = x^2 + 2x + 2$ and its inverse relation.

Solution. These two graphs are shown in Fig. 7. It is easy to sketch the parabola, because the minimum occurs at $x = -1$ and is $y = 1$. Then $\mathcal{G}$ is obtained by reflection of $\mathcal{F}$ in the line $y = x$. ●

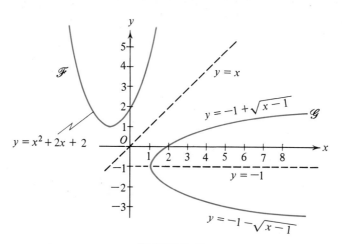

FIGURE 7

It is clear that in this case the inverse relation is *not* a function because for each $x > 1$, a vertical line meets $\mathcal{G}$ in two points (there are two distinct values of y that correspond to the given x), and this violates the definition of a function.

A little algebra will show that if $y = x^2 + 2x + 2$, then $x^2 + 2x + 2 - y = 0$ gives $x = -1 \pm \sqrt{y - 1}$. If we interchange x and y, we have a formula for the inverse relation that can be written in two pieces:

(50) $$y = -1 + \sqrt{x - 1} \quad \text{and} \quad y = -1 - \sqrt{x - 1}.$$

Each equation in (50) defines a function, and each piece is called a *branch of the inverse relation,* or a *branch of the inverse function.* In the graph of $\mathcal{G}$ shown in Fig. 7, the upper portion is the graph of $y = -1 + \sqrt{x - 1}$ and the lower portion is the graph of $y = -1 - \sqrt{x - 1}$.

When confronted with a specific function, we usually settle on some particular piece of the inverse relation and call it the *inverse function,* or the *principal branch of the inverse function.* If $y = G(x)$ is the designated inverse function of $y = F(x)$, then for each fixed x_0, the corresponding $y_0 = G(x_0)$ is called a *principal value* of the inverse function.

In this section we are interested in the trigonometric functions and their inverses. The graph of $y = \sin x$ is shown in Fig. 8. To obtain the graph of the inverse relation, we reflect the curve in Fig. 8 about the line $y = x$. The result is shown in Fig. 9.

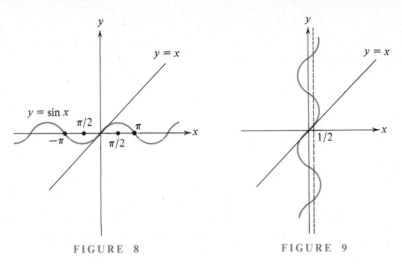

FIGURE 8   FIGURE 9

It is customary to write $y = \sin^{-1} x$ (read "y is the angle whose sine is x" or "y is the inverse sine of x") for this inverse relation.

Clearly, the graph in Fig. 9 is not the graph of a function because y is not uniquely determined for each x. For example, if $x = 1/2$, then $y = \pi/6 + 2n\pi$ and $y = 5\pi/6 + 2n\pi$ (n any integer) all have the property that $\sin y = 1/2$. As soon as we select a specific value of y from among the many possible ones, then $y = \sin^{-1} x$ becomes a function. We do this by requiring that y lie in the interval $-\pi/2 \leqq y \leqq \pi/2$. Thus the conditions

(51)
$$\sin y = x \quad \text{and} \quad -\frac{\pi}{2} \leqq y \leqq \frac{\pi}{2}$$

suffice to *define* the inverse sine function. To distinguish the function defined by (51) from the inverse relation, we use capital S. Thus

(52)
$$y = \text{Sin}^{-1} x$$

represents the function defined by (51) and $y = \sin^{-1} x$ represents the inverse relation shown in Fig. 9.

The function $\text{Sin}^{-1} x$ defined by (51) is called the *principal branch* of the inverse sine function. The word *principal* is used to emphasize the fact that a definite choice of y has been made, but strictly speaking the word is unnecessary because the function $y = \text{Sin}^{-1} x$ will always mean this principal branch. The graph of the function (52) is shown in Fig. 10. Observe that the function is defined only for x in the interval $-1 \leqq x \leqq 1$.

In a similar manner, we can define inverse functions for the remaining five trigonometric functions, by selecting suitable principal branches. For the cosine, tangent, and cotangent functions, we make the following definitions:

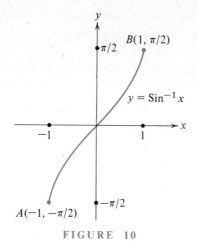

FIGURE 10

(53) $y = \text{Cos}^{-1} x$ (read "y is the inverse cosine of x")
                    if $x = \cos y$ and $0 \leq y \leq \pi$.

(54) $y = \text{Tan}^{-1} x$ (read "y is the inverse tangent of x")

                    if $x = \tan y$ and $-\dfrac{\pi}{2} < y < \dfrac{\pi}{2}$.

(55) $y = \text{Cot}^{-1} x$ (read "y is the inverse cotangent of x")
                    if $x = \cot y$ and $0 < y < \pi$.

The inverse cosine function is defined only for $-1 \leq x \leq 1$. The inverse tangent and cotangent functions are defined for all real x. Any number y lying in the indicated intervals in equations (51), (53), (54), and (55) is called a *principal value* of the inverse function.

The symbol -1 in equations (52), (53), (54), and (55) is not an exponent, but is

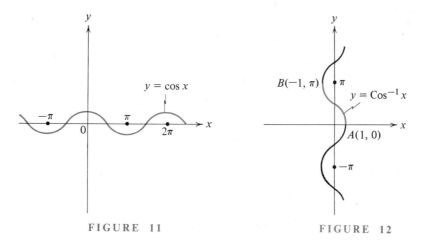

FIGURE 11   FIGURE 12

selected in agreement with the general plan of indicating the inverse function of $f(x)$ by $f^{-1}(x)$. Some authors use Arc sin x in place of Sin^{-1} x to avoid confusing Sin^{-1} x with the reciprocal of sin x, namely $(\sin x)^{-1} = 1/\sin x$.

The graphs of the trigonometric functions cosine, tangent, and cotangent and their inverse relations are shown in Figs. 11 through 16. In Figs. 12, 14, and 16 the inverse functions (the principal branches) are shown in color.

It is quite natural to look for definitions of Sec^{-1} x and Csc^{-1} x, and in fact principal branches can be selected for these functions. However, there is no universal agreement on this selection, so we prefer to omit these functions completely. There is no practical need for defining Sec^{-1} x and Csc^{-1} x, since any natural problem that can be solved using these functions can be solved just as easily without them.

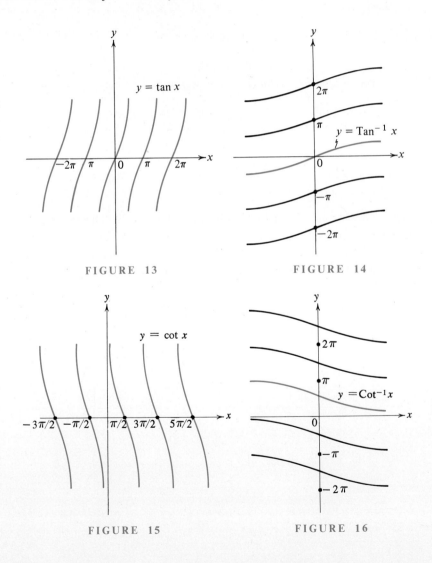

FIGURE 13   FIGURE 14

FIGURE 15   FIGURE 16

DEFINITION 1 (Inverse Trigonometric Functions). The inverse trigonometric functions are

(51) $\qquad y = \operatorname{Sin}^{-1} x, \quad$ if $x = \sin y \quad$ and $\quad -\dfrac{\pi}{2} \leqq y \leqq \dfrac{\pi}{2}.$

(53) $\qquad y = \operatorname{Cos}^{-1} x, \quad$ if $x = \cos y \quad$ and $\quad 0 \leqq y \leqq \pi.$

(54) $\qquad y = \operatorname{Tan}^{-1} x, \quad$ if $x = \tan y \quad$ and $\quad -\dfrac{\pi}{2} < y < \dfrac{\pi}{2}.$

(55) $\qquad y = \operatorname{Cot}^{-1} x, \quad$ if $x = \cot y \quad$ and $\quad 0 < y < \pi.$

Example 2. Find **(a)** $\operatorname{Sin}^{-1} 1$, **(b)** $\operatorname{Tan}^{-1}(-1)$, and **(c)** $\operatorname{Cot}^{-1}(-1)$.

Solution. **(a)** From trigonometry, $\sin\left(\dfrac{\pi}{2} + 2n\pi\right) = 1$, and these are the only values for y such that $\sin y = 1$. But among these only $y = \pi/2$ lies in the required interval $-\pi/2 \leqq y \leqq \pi/2$. Hence $\operatorname{Sin}^{-1} 1 = \pi/2$.
(b) $\tan(3\pi/4 + 2n\pi) = -1$ and $\tan(-\pi/4 + 2n\pi) = -1$. But

$$-\frac{\pi}{2} < -\frac{\pi}{4} < \frac{\pi}{2},$$

whence $\operatorname{Tan}^{-1}(-1) = -\pi/4$.
(c) $\cot y = -1$ for the same values of y for which $\tan y = -1$ [see part **(b)**]. But now the principal value must lie in the interval $0 < y < \pi$. Hence $\operatorname{Cot}^{-1}(-1) = 3\pi/4$. ●

Example 3. Find $\sin[\operatorname{Cos}^{-1} 2/3 + \operatorname{Sin}^{-1}(-3/4)]$.

Solution. Let $A = \operatorname{Cos}^{-1} 2/3$ and let $B = \operatorname{Sin}^{-1}(-3/4)$. Then we are to compute $\sin(A + B)$. Of course, we have

(56) $\qquad\qquad \sin(A + B) = \sin A \cos B + \cos A \sin B.$

Since $\cos A = 2/3 > 0$, A is a first quadrant angle. But $\sin B = -3/4 < 0$, so B is a fourth quadrant angle; that is, $-\pi/2 < B < 0$, since B is a principal value. For convenience these angles are shown in Fig. 17 on the next page. From this figure we see that $\sin A = \sqrt{5}/3$, $\cos B = \sqrt{7}/4$, and substituting in (56), we find that

$$\sin(A + B) = \frac{\sqrt{5}}{3}\frac{\sqrt{7}}{4} + \frac{2}{3}\frac{(-3)}{4} = \frac{\sqrt{35} - 6}{12}. \ ●$$

We observe that $\sin(A + B)$ is negative but very small. From this we infer that $|B|$ is slightly larger than A.

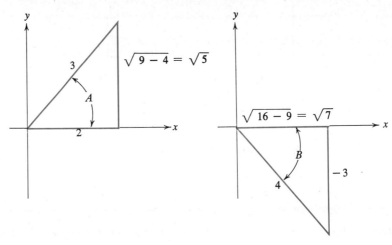

EXERCISE 5

In Problems 1 through 9, give the indicated angle in radians.

1. $\text{Tan}^{-1} 0$.
2. $\text{Sin}^{-1}(-1/2)$.
3. $\text{Cos}^{-1} 0$.
4. $\text{Cos}^{-1}(-\sqrt{2}/2)$.
5. $\text{Tan}^{-1} \sqrt{3}$.
6. $\text{Tan}^{-1}(-1/\sqrt{3})$.
7. $\text{Cot}^{-1}(-\sqrt{3}/3)$.
8. $\text{Cos}^{-1}(-\sqrt{3}/2)$.
9. $\text{Sin}^{-1}(-\sqrt{3}/2)$.

10. Give a numerical value for:
    (a) $\sin(\text{Cos}^{-1}(-3/5))$.
    (b) $\cos(\text{Sin}^{-1}(-3/5))$.
    (c) $\sin(\text{Tan}^{-1}(-\sqrt{3}))$.
    (d) $\sin(\text{Cot}^{-1}(-\sqrt{3}))$.

11. Give a numerical value for:
    (a) $\tan[\text{Sin}^{-1}(1/2) + \text{Sin}^{-1}(-2/3)]$.
    (b) $\cos[\text{Tan}^{-1} 1 + \text{Cos}^{-1}(-3/4)]$.
    (c) $\sin[\text{Cot}^{-1}(2/5) + \text{Tan}^{-1}(3/7)]$.

In Problems 12 through 22, identify the statement as true or false in the given domain.

12. $\sin(\text{Cos}^{-1} x) = \sqrt{1 - x^2}$, $-1 \leq x \leq 1$.

13. $\cos(\text{Sin}^{-1} x) = \sqrt{1 - x^2}$, $-1 \leq x \leq 1$.

14. $\text{Tan}^{-1} \dfrac{\sqrt{1 - x^2}}{x} = \text{Cos}^{-1} x$, $0 < x \leq 1$.

15. $\text{Tan}^{-1} \dfrac{\sqrt{1 - x^2}}{x} = \text{Cos}^{-1} x$, $0 < |x| \leq 1$.

16. $\text{Tan}^{-1} u = \text{Cot}^{-1}(1/u)$, $0 < |u|$.

17. $\text{Sin}^{-1}(-v) = -\text{Sin}^{-1} v$, $-1 \leq v \leq 1$.

18. $\text{Cos}^{-1}(-w) = \text{Cos}^{-1} w,$ $\qquad\qquad -1 \leq w \leq 1.$

19. $\text{Cos}^{-1}(-w) = \pi - \text{Cos}^{-1} w,$ $\qquad -1 \leq w \leq 1.$

20. $\text{Cos}^{-1} x + \text{Sin}^{-1} x = \pi/2,$ $\qquad -1 \leq x \leq 1.$

21. $2 \text{Sin}^{-1} y = \text{Cos}^{-1}(1 - 2y^2),$ $\qquad -1 \leq y \leq 1.$

22. $\text{Tan}^{-1} m + \text{Tan}^{-1} n = \text{Tan}^{-1} \dfrac{m+n}{1-mn},$ $\qquad mn \neq 1.$

23. For what values of y is the equation of Problem 21 true?

24. Find some domain for m and n such that the equation of Problem 22 is true.

CALCULATOR PROBLEMS

In Problems C1 through C9, use a calculator to find the indicated angle in radians to five significant figures.

C1. $\text{Sin}^{-1}(0.2).$ $\qquad$ C2. $\text{Cos}^{-1}(0.3).$ $\qquad$ C3. $\text{Sin}^{-1}(-0.3).$

C4. $\text{Cos}^{-1}(-0.3).$ $\qquad$ C5. $\text{Cos}^{-1}(-0.8).$ $\qquad$ C6. $\text{Cos}^{-1}(-0.9).$

C7. $\text{Tan}^{-1}(0.08).$ $\qquad$ C8. $\text{Tan}^{-1} 100.$ $\qquad$ C9. $\text{Tan}^{-1}(-100).$

C10. Although most scientific calculators have buttons for Sin^{-1}, Cos^{-1}, and Tan^{-1}, there probably will not be a button for Cot^{-1}. Nonetheless, it is easy to compute Cot^{-1} by using the button for Tan^{-1}. First prove that

$$\text{Cot}^{-1} x = \frac{\pi}{2} - \text{Tan}^{-1} x.$$

Now use your calculator to find each of the following indicated angles (in radians) to five significant figures:

(a) $\text{Cot}^{-1}(0.123).$ $\qquad$ (b) $\text{Cot}^{-1}(-5).$ $\qquad$ (c) $\text{Cot}^{-1}(-1/7).$

In Problems C11 through C16, find the indicated root of the equation to five significant digits.

C11. $\sin[\pi(x+1)] = 0.4,$ $\qquad -1 < x < -0.5.$

C12. $\sin(\pi x) = 0.7,$ $\qquad 0.7 < x < 0.8.$

C13. $\tan(\pi x) = 23,$ $\qquad 0.5 < x < 1.5.$

C14. $\tan(1 - 3x) = 25,$ $\qquad -0.19 < x < -0.15.$

C15. $5\cos(x^2 - 1) = -4,$ $\qquad 1.8 < x < 1.9.$

C16. $9\cos(4x^3) = 7,$ $\qquad 0.5 < x < 0.6.$

6 Calculus of the Inverse Trigonometric Functions

A general formula for the derivative of an inverse function can be derived.[1] However, we do not need such a formula because we can obtain the desired results using implicit differentiation.

Let $y = \text{Sin}^{-1} x$. Then, by Definition 1,

(51) $$x = \sin y, \qquad -\frac{\pi}{2} \leq y \leq \frac{\pi}{2}.$$

Differentiating both sides of (51) with respect to x gives

(57) $$1 = \cos y \frac{dy}{dx} \qquad \text{or} \qquad \frac{dy}{dx} = \frac{1}{\cos y},$$

as long as $\cos y \neq 0$. But $\cos y = \pm\sqrt{1 - \sin^2 y} = \pm\sqrt{1 - x^2}$, from equation (51). In the interval $-\pi/2 \leq y \leq \pi/2$, $\cos y$ is never negative, so $\cos y = \sqrt{1 - x^2}$ and (57) becomes

(58) $$\frac{dy}{dx} = \frac{1}{\sqrt{1 - x^2}}, \qquad -1 < x < 1.$$

Finally, since $y = \text{Sin}^{-1} x$, we have

(59) $$\frac{d}{dx} \text{Sin}^{-1} x = \frac{1}{\sqrt{1 - x^2}}, \qquad -1 < x < 1.$$

Differentiation formulas for the other inverse trigonometric functions can be obtained in the same way.

Let $\quad y = \text{Cos}^{-1} x$.	Let $\quad y = \text{Tan}^{-1} x$.
Then $\quad x = \cos y$.	Then $\quad x = \tan y$.

Differentiating with respect to x gives

$$1 = -\sin y \frac{dy}{dx}$$

Differentiating with respect to x gives

$$1 = \sec^2 y \frac{dy}{dx}$$

---

[1] If $y = f(x)$ and $x = g(y)$ are inverse functions, then the derivatives are related by $g'(y_0) = 1/f'(x_0)$, where $y_0 = f(x_0)$. This relation is often written as $\dfrac{dx}{dy} = 1 \left/ \dfrac{dy}{dx} \right.$.

$$\frac{dy}{dx} = \frac{-1}{\sin y} = \frac{-1}{\sqrt{1 - \cos^2 y}}$$

$$\frac{dy}{dx} = -\frac{1}{\sqrt{1 - x^2}}.$$

Hence, if $-1 < x < 1$, then

(60) $\qquad \dfrac{d}{dx} \operatorname{Cos}^{-1} x = -\dfrac{1}{\sqrt{1 - x^2}}.$

$$\frac{dy}{dx} = \frac{1}{\sec^2 y} = \frac{1}{1 + \tan^2 y}$$

$$\frac{dy}{dx} = \frac{1}{1 + x^2}.$$

Hence for all x,

(61) $\qquad \dfrac{d}{dx} \operatorname{Tan}^{-1} x = \dfrac{1}{1 + x^2}.$

We leave it as an exercise for the reader to prove that for all x

(62) $$\frac{d}{dx} \operatorname{Cot}^{-1} x = \frac{-1}{1 + x^2}.$$

Of course, these formulas are valid only for suitable values of x. Thus, as already indicated in (59) and (60), x is restricted to the interval $-1 < x < 1$, but in (61) and (62), x can be any real number. Putting $x = \pm 1$ in (59) or (60) yields infinity for the derivative. Of course, this is not a number, but the occurrence of the zero in the denominator is consistent with the fact that the curves shown in Figs. 10 and 12 for the principal branches are vertical at the end points A and B. It is also worthwhile to note that the derivatives for $\operatorname{Sin}^{-1} x$ and $\operatorname{Tan}^{-1} x$ are always positive, so these functions are increasing functions of x. Similarly, the derivatives for $\operatorname{Cos}^{-1} x$ and $\operatorname{Cot}^{-1} x$ are always negative, so these functions are decreasing functions of x. These facts are consistent with the curves shown in Figs. 10, 12, 14, and 16.

If in our differentiation formulas we replace x by a differentiable function $u(x)$, then the Chain Rule gives the more general formulas of

---

THEOREM 6. If u is any differentiable function of x, then

(63) $\qquad \dfrac{d}{dx} \operatorname{Sin}^{-1} u = \dfrac{1}{\sqrt{1 - u^2}} \dfrac{du}{dx}.$

(64) $\qquad \dfrac{d}{dx} \operatorname{Cos}^{-1} u = \dfrac{-1}{\sqrt{1 - u^2}} \dfrac{du}{dx}.$

(65) $\qquad \dfrac{d}{dx} \operatorname{Tan}^{-1} u = \dfrac{1}{1 + u^2} \dfrac{du}{dx}.$

(66) $\qquad \dfrac{d}{dx} \operatorname{Cot}^{-1} u = \dfrac{-1}{1 + u^2} \dfrac{du}{dx}.$

Of course, in formulas (63) and (64), u must be restricted to lie in the interval $-1 < u < 1$.

Example 1. Find $\dfrac{dy}{dx}$ for the following functions.

(a) $y = \text{Sin}^{-1}(3x - 1)$.　　**(b)** $y = \text{Cos}^{-1} x^3$.　　**(c)** $y = x - \dfrac{1}{2}(x^2 + 4)\,\text{Tan}^{-1}\dfrac{x}{2}$.

Solution. **(a)** $\dfrac{dy}{dx} = \dfrac{1}{\sqrt{1 - (3x-1)^2}} \dfrac{d}{dx}(3x - 1) = \dfrac{3}{\sqrt{6x - 9x^2}}$.

(b) $\dfrac{dy}{dx} = \dfrac{-1}{\sqrt{1 - (x^3)^2}} \dfrac{d}{dx} x^3 = \dfrac{-3x^2}{\sqrt{1 - x^6}}$.

(c) $\dfrac{dy}{dx} = 1 - x\,\text{Tan}^{-1}\dfrac{x}{2} - \dfrac{1}{2}(x^2 + 4)\dfrac{1}{1 + \dfrac{x^2}{4}}\dfrac{1}{2}$

$$= 1 - x\,\text{Tan}^{-1}\dfrac{x}{2} - \dfrac{x^2 + 4}{x^2 + 4} = -x\,\text{Tan}^{-1}\dfrac{x}{2}. \quad \bullet$$

Each of the four differentiation formulas (63)–(66) leads to a corresponding integration formula, but because the differentiation formulas occur in pairs that differ just by a minus sign [(63), (64) and (65), (66)], only two of the four integration formulas are needed for practical purposes. These are

$$(67) \qquad \int \frac{du}{\sqrt{1 - u^2}} = \text{Sin}^{-1} u + C, \qquad -1 < u < 1,$$

and

$$(68) \qquad \int \frac{du}{1 + u^2} = \text{Tan}^{-1} u + C.$$

Example 2. Find each of the following integrals.

(a) $\displaystyle\int \frac{x^2\, dx}{1 + 4x^6}$.　　**(b)** $\displaystyle\int_0^{1/\sqrt{2}} \frac{y\, dy}{\sqrt{1 - 2y^4}}$.　　**(c)** $\displaystyle\int \frac{dx}{\sqrt{6x - x^2}}$.

Solution. **(a)** Set $u = 2x^3$. Then $du = 6x^2\, dx$; hence

$$\int \frac{x^2\, dx}{1 + 4x^6} = \frac{1}{6}\int \frac{6x^2\, dx}{1 + (2x^3)^2} = \frac{1}{6}\int \frac{du}{1 + u^2}$$

$$= \frac{1}{6}\text{Tan}^{-1} u + C = \frac{1}{6}\text{Tan}^{-1}(2x^3) + C.$$

(b) Set $u = \sqrt{2}\,y^2$. Then $du = 2\sqrt{2}\,y\,dy$. Consequently,

$$\int \frac{y\,dy}{\sqrt{1-2y^4}} = \frac{1}{2\sqrt{2}} \int \frac{2\sqrt{2}\,y\,dy}{\sqrt{1-(\sqrt{2}y^2)^2}} = \frac{1}{2\sqrt{2}} \int \frac{du}{\sqrt{1-u^2}}$$

$$= \frac{1}{2\sqrt{2}} \operatorname{Sin}^{-1}(\sqrt{2}\,y^2) + C.$$

Then for the definite integral, we have

$$\int_0^{1/\sqrt{2}} \frac{y\,dy}{\sqrt{1-2y^4}} = \frac{1}{2\sqrt{2}} \operatorname{Sin}^{-1}(\sqrt{2}\,y^2) \Big|_0^{1/\sqrt{2}}$$

$$= \frac{1}{2\sqrt{2}} \left(\operatorname{Sin}^{-1} \frac{\sqrt{2}}{2} - \operatorname{Sin}^{-1} 0 \right) = \frac{\pi}{8\sqrt{2}}.$$

(c) It is not easy to see that this integral fits the form of either (67) or (68). But if we complete the square under the radical, we have

$$6x - x^2 = 9 - 9 + 6x - x^2 = 9 - (x^2 - 6x + 9) = 9 - (x-3)^2.$$

Whence we can match this integral with (67) thus:

$$\int \frac{dx}{\sqrt{6x-x^2}} = \int \frac{dx}{\sqrt{9-(x-3)^2}} = \frac{1}{3} \int \frac{dx}{\sqrt{1-\left(\frac{x-3}{3}\right)^2}}$$

$$= \operatorname{Sin}^{-1} \frac{x-3}{3} + C. \quad \bullet$$

We should observe that $\operatorname{Sin}^{-1}[(x-3)/3]$ is not defined if $|(x-3)/3| > 1$. This means that x must be restricted to a domain in which $-1 \leq (x-3)/3 \leq 1$ or $0 \leq x \leq 6$. This is reasonable because if x is not in this interval, the quantity under the radical in the integrand of **(c)** is negative. Actually, we should also exclude the end points $x = 0$ and $x = 6$ because for these values of x the denominator of the integrand is zero. We shall return to this point and discuss it fully when we consider improper integrals in Chapter 14.

EXERCISE 6

In Problems 1 through 10, find the derivative of the given function with respect to the independent variable.

1. $y = \operatorname{Cos}^{-1} 5x.$

2. $y = \operatorname{Tan}^{-1} t^4.$

3. $y = \operatorname{Sin}^{-1} \sqrt{x}.$

4. $z = t \operatorname{Cot}^{-1}(1 + t^2).$

5. $z = \operatorname{Cot}^{-1} \dfrac{y}{1-y^2}.$

6. $w = \operatorname{Tan}^{-1} \dfrac{1}{t}.$

7. $x = \text{Sin}^{-1} \sqrt{1 - t^4}$.

8. $s = \dfrac{t}{\sqrt{1 - t^2}} + \text{Cos}^{-1} t$.

9. $y = (x^2 - 2) \text{Sin}^{-1} \dfrac{x}{2} + \dfrac{x}{2} \sqrt{4 - x^2}$.

10. $y = \sqrt{Ax - B^2} - B \text{Tan}^{-1} \dfrac{\sqrt{Ax - B^2}}{B}$, $A > 0$, $B \neq 0$.

11. Prove that $\text{Sin}^{-1} x + \text{Cos}^{-1} x$ is a constant by showing that the derivative is zero. What is the constant?

12. Repeat Problem 11 for $\text{Tan}^{-1} x + \text{Cot}^{-1} x$.

13. Sketch the curve $y = \text{Tan}^{-1}(1/x)$. Does the curve have any inflection points?

*14. Prove that the derivative of the function $y = \text{Tan}^{-1} x + \text{Tan}^{-1}(1/x)$ is zero if $x \neq 0$. But y is not a constant. Compare this result with Problems 11 and 12. Sketch the graph of this function.

15. A picture 6 ft in height is hung on a wall so that its lower edge is 2 ft above the eye of an observer. How far from the wall should a person stand so that the picture subtends the largest angle at the person's eye?

16. A road sign 20 ft high stands on a slight rise so that the bottom of the sign is 20 ft above the horizontal plane of the road. If the eye of the driver of an automobile is 4 ft above the road, at what horizontal distance from the sign will the sign appear to be largest to the driver (subtend the largest angle)?

17. An airplane is flying level 2000 ft above ground level at 180 miles/hr. Let θ be the angle between the line of sight from an observer to the plane and a vertical line. How fast is θ changing **(a)** when the plane is directly overhead, and **(b)** when θ is $\pi/4$? When is the rate of change a maximum?

18. Find the angle of intersection of the two curves $y = \text{Sin}^{-1} x$ and $y = \text{Cos}^{-1} x$. (See Problem 34 of Exercise 2.)

19. Prove that if a is any positive constant, then

(69)
$$\int \frac{du}{\sqrt{a^2 - u^2}} = \text{Sin}^{-1} \frac{u}{a} + C, \qquad -a < u < a$$

and

(70)
$$\int \frac{du}{a^2 + u^2} = \frac{1}{a} \text{Tan}^{-1} \frac{u}{a} + C.$$

These are simple generalizations of (67) and (68) and may be more convenient for practical applications.

In Problems 20 through 27, find the given integral.

20. $\displaystyle\int \frac{dx}{\sqrt{25 - 4x^2}}$.

21. $\displaystyle\int \frac{dy}{36 + 4y^2}$.

22. $\displaystyle\int \frac{z \, dz}{5 + 2z^4}$.

23. $\int \dfrac{\sin x \, dx}{\sqrt{10 - \cos^2 x}}$.

24. $\int \dfrac{dx}{\sqrt{5 + 4x - x^2}}$.

25. $\int \dfrac{7 \, dx}{25 - 12x + 4x^2}$.

★26. $\int \dfrac{5 \, dt}{9\sqrt{t} + \sqrt{t^3}}$.

27. $\int \dfrac{3 \, dy}{\sqrt{5 - 12y - 9y^2}}$.

★28. Sketch the curve $y = \dfrac{1}{1 + x^2}$. Find the area of the region under this curve between $x = 0$ and $x = M$, where M is a positive constant. Find the limit of this area as $M \to \infty$.

29. Sketch the curve $y = 1/\sqrt{1 - x^2}$ for the interval $0 \leq x < 1$. Find the area of the region under this curve between $x = 0$ and $x = M$, where $0 < M < 1$. Find the limit of this area as $M \to 1^-$.

CALCULATOR PROBLEMS

C1. This problem concerns solving the equation

$$2x = \tan x, \qquad \frac{\pi}{4} < x < \frac{\pi}{2}.$$

(a) Prove that the given equation has a unique root in the interval $(\pi/4, \pi/2)$. HINT: Show that $f(x) = 2x - \tan x$ is strictly decreasing on the interval and examine its end-point behavior.

(b) Attempt solving the equation for the indicated root by using successive substitutions (iteration) with the equation written in the fixed-point form $x = (1/2) \tan x$. Explain why this procedure fails to produce the desired root.

(c) Solve the given equation (to five significant figures) by using successive substitutions with the equation written in the fixed-point form $x = \text{Tan}^{-1}(2x)$. Show that the Banach Fixed-Point Theorem (Chapter 5, Section 12) assures convergence of the iterations to the desired root for any starting value in $(\pi/4, \pi/2)$.

(d) Find the root of the equation $2x = \tan x$ (to five significant figures) in the interval $\pi/2 < x < 3\pi/2$. HINT: Write the equation in the form $x = \pi + \text{Tan}^{-1}(2x)$.

C2. The equation $\sin x = 1/x$ has a unique root in the interval $\pi/2 < x < \pi$. Explain why the fixed point form $x = \text{Sin}^{-1}(1/x)$ is inappropriate for generating successive approximations that converge to this root. Then find the root (to five significant figures) using the fixed-point form $x = \pi - \text{Sin}^{-1}(1/x)$.

In Problems C3 through C6, find the indicated root to four significant figures.

C3. $1 + \sin x = \tan x, \quad 0 < x < \pi/2$.
C4. $2x = \tan x, \quad 5\pi < x < 11\pi/2$.
C5. $\text{Cos}^{-1}(2x) = \text{Tan}^{-1} x$.
C6. $x + \sin x = \cot x, \quad 0 < x < \pi/2$.

C7. Compute the absolute maximum of the function $f(x) = x \cos x + 3$ on the interval $[0, \pi/2]$.

C8. Compute the absolute minimum of the function $g(x) = \sin x + x \cos x$ on the interval $[\pi/2, 3\pi/2]$.

REVIEW PROBLEMS

In Problems 1 through 12, find the derivative of the given function.

1. $y = \cot 2x$.

2. $y = \sec x^2$.

3. $y = \sqrt{\csc x^3}$.

4. $v = \cos \sqrt{u}$.

5. $w = \text{Sin}^{-1} z^2$

6. $r = \tan^2 \theta$.

7. $s = \dfrac{1 + \cot t}{1 - \cot t}$.

8. $r = \dfrac{\text{Sin}^{-1} s}{s^2}$.

9. $w = \dfrac{\sqrt{1 - t^2}}{\text{Cos}^{-1} t}$.

10. $y = t^2 \text{Tan}^{-1} t$.

11. $y = \text{Sin}^{-1} \dfrac{x - 1}{x + 1}$.

12. $\text{Tan}^{-1} y = \text{Cos}^{-1} x$.

In Problems 13 through 18, find the given integral.

13. $\displaystyle\int x \sin (5x^2)\, dx$.

14. $\displaystyle\int \sqrt{x} \sec^2 x^{3/2}\, dx$.

15. $\displaystyle\int \cos \theta \csc^5 \theta\, d\theta$.

16. $\displaystyle\int (5 \tan^2 w + 7 \sec^2 w)\, dw$.

17. $\displaystyle\int \dfrac{dx}{\sqrt{6x - x^2 - 1}}$.

18. $\displaystyle\int \dfrac{dx}{10 - 6x + x^2}$.

19. Find the average value over the interval $0 \leqq x \leqq \pi$ for each of the functions: (a) $\sin x$, (b) $\cos x$, (c) $\sin 3x$, (d) $\sin mx$ (m a positive integer), and (e) $\sec^2 (x/3)$. HINT: The average value is the integral of the function over an interval divided by the length of the interval.

In Problems 20 through 25, find the extreme points of the given function in the given interval. Sketch the graph of the function.

20. $y = 2 \cos x + \sin 2x$, $0 \leqq x \leqq \pi$.

21. $y = x + 3 \sin x$, $0 \leqq x \leqq \pi$.

22. $y = x + \text{Tan}^{-1} x$, $-\infty < x < \infty$.

23. $y = x - 5 \text{Tan}^{-1} x$, $-\infty < x < \infty$.

24. $y = 2x + \text{Cos}^{-1} x$, $-1 \leqq x \leqq 1$.

*25. $y = 4 \sin^2 x - \tan^2 x$, $-\pi \leqq x \leqq \pi$.

*26. Prove that the graph of $y = x \sin (1/x)$ has an infinite number of relative extreme points in the interval $0 < x < \pi$.

*27. Find the area of the region under the curve:

(a) $y = \cos x$ from $-\pi/4$ to $\pi/4$.

(b) $y = \cos^2 x$ from $-\pi/4$ to $\pi/4$.

(c) $y = \cos^3 x$ from $-\pi/4$ to $\pi/4$.

Which of these three regions has the largest area and why?

28. The region under the curve $y = 1/\sqrt{1 + x^2}$ between $x = 0$ and $x = M$ is rotated about the x-axis. Find the volume of the solid generated. Find the limit of this volume as $M \to \infty$.

9

The Logarithmic and Exponential Functions

Objective 1

The exponential and logarithmic functions arise quite often in a variety of natural problems, for example, the shape of a suspended cable, the rate of a chemical reaction, and the growth or decay of living organisms. Our natural objective is to obtain differentiation formulas for the functions $y = a^u$ and $y = \log_a u$, where u is a differentiable function of x.

Review 2

The reader is already familiar with the exponential function a^u in an intuitive way. Let us proceed on this basis to list the important properties of the exponential function. After a brief examination of this list we will discuss the question of *proving* these properties.

If $a > 0$, $b > 0$, and u and v are any pair of real numbers, then

(1) $\qquad a^u > 0.$

(2) $\qquad a^u a^v = a^{u+v}.$

(3) $\qquad \dfrac{a^u}{a^v} = a^{u-v}.$

(4) $\qquad a^{-u} = \dfrac{1}{a^u}.$

(5) $\qquad (a^u)^v = a^{uv}.$

(6) $\qquad a^0 = 1.$

(7) $\qquad 1^u = 1.$

(8) $\qquad a^u b^u = (ab)^u.$

(9) $\qquad a^{u/v} = \sqrt[v]{a^u} = (a^u)^{1/v} = (\sqrt[v]{a})^u = (a^{1/v})^u, \quad v \neq 0.$

363

(10)   If $a > 1$, then $\lim_{x \to \infty} a^x = \infty$.

(11)   If $a > 1$, then $\lim_{x \to -\infty} a^x = 0$.

(12) If $a > 1$, the function a^x is increasing for $-\infty < x < \infty$;

thus if $u < v$, then $a^u < a^v$.

(13)   The function $f(x) = a^x$ is a continuous function for $-\infty < x < \infty$.

Of course, these properties are not all independent. For example, (6) can be proved from (1) and (2), and then (4) follows from (2) and (6). Then (3) is an immediate consequence of (2) and (4). Further, (1) and (8) are sufficient to prove (7). On the other hand, it is impossible to deduce (7) without using (8) or some equivalent.

To gain a feeling for the behavior of the exponential function, it is worthwhile to sketch its graph for some fixed base. In Fig. 1 we show a few points on the graph of $y = 2^x$, and the curve is sketched on the assumption that $f(x)$ is continuous and increasing. The curve obtained illustrates properties (10) and (11). The points indicated in Fig. 1 correspond to integer values of x. The reader may use a calculator to check a few other places on the curve. Thus, if $x = 1.3$, then $2^x = 2.462\ldots$.

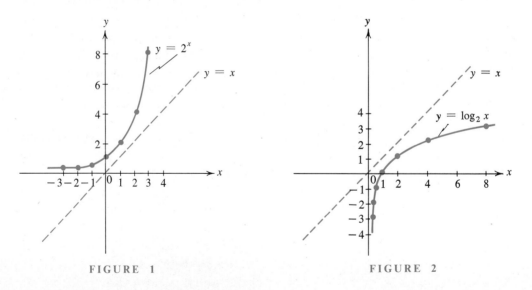

FIGURE 1   FIGURE 2

We define the logarithmic function (to the base a) as the inverse of the exponential function (with the same base).

DEFINITION 1 (Logarithm). The logarithm to the base a ($a > 0$, $a \neq 1$) of the number N ($N > 0$) is the number L such that $a^L = N$. Thus

(14) $$L = \log_a N \quad \text{if and only if} \quad a^L = N.$$

For example,

(15) $$\log_2 8 = 3 \quad \text{because} \quad 2^3 = 8.$$

Since the functions $y = a^x$ and $y = \log_a x$ are inverse functions, the graph of either can be obtained from the other by a reflection in the line $y = x$. In this way we can obtain the graph of $y = \log_2 x$ shown in Fig. 2 by a reflection of the graph shown in Fig. 1.

If $a > 0$ and $a \neq 1$, the domain of the function $f(x) = a^x$ consists of all real numbers, but the range is only the set of all positive numbers (a^x is never zero or negative). Consequently, the inverse function $g(x) = \log_a x$ has the domain $0 < x < \infty$, and has the set of all real numbers as its range.

There is a logical difficulty in proving the assertions (1) through (13) that might pass unnoticed if we did not call attention to it. As long as the exponent u is an integer, and the base a is a rational number, the computation of a^u is reasonably easy. For example, we have $(2/5)^6 = 64/15{,}625$. But what can we say about $(\sqrt{2})^{\sqrt{3}}$? Since both $\sqrt{2}$ and $\sqrt{3}$ are irrational numbers, does $(\sqrt{2})^{\sqrt{3}}$ really have some meaning, in the sense that it is a certain number? The answer is yes, but this is by no means obvious. The proof is long and tedious. A rigorous treatment of the exponential and logarithmic functions requires that this point be settled first; namely, it must be proved that a^u is well defined for every $a > 0$ and every real number u. Then equations (1) through (9) must be proved, not only when u and v are integers (this case is easy) but for all u and v. Once this sticky point is passed, it is easy to deduce the following properties of the logarithmic function. Suppose that $a > 0$; $a \neq 1$; M and N are any positive numbers; and n is any real number. Then

(16) $$\log_a MN = \log_a M + \log_a N,$$

(17) $$\log_a \frac{M}{N} = \log_a M - \log_a N,$$

(18) $$\log_a M^n = n \log_a M,$$

(19) $$\log_a 1 = 0, \quad \log_a a = 1.$$

We will base our treatment in Sections 4 and 5 on the assumption that the properties (1) through (13) and (16) through (19) have already been proved for all real numbers in the domain of the functions a^x and $\log_a x$. We will also assume that

$$\lim_{h \to 0} (1 + h)^{1/h}$$

exists. This limit is a transcendental number approximately equal to 2.71828 and denoted by e. We will consider this limit intuitively in the next section. A rigorous treatment of these assumptions can be deferred to a more advanced mathematics course.

EXERCISE 1

1. Sketch the graph of $y = a^x$ for (a) $a = 1/2$, (b) $a = 1$, (c) $a = \sqrt{2}$, (d) $a = 2$, and (e) $a = 3$, all with the same coordinate system.

2. Sketch the graph of $y = \log_a x$ for (a) $a = \sqrt{2}$, (b) $a = 2$, (c) $a = 3$, and (d) $a = 10$, all with the same coordinate system.

3. Explain why $\log_a N$ is meaningless if $a = 1$.

4. Explain why a^u is not defined for all u if a is *negative*.

5. Find each of the following:
   (a) $\log_3 9$.
   (b) $\log_2 (1/4)$.
   (c) $\log_{10} (0.001)$.
   (d) $5^{\log_5 \pi}$,
   (e) $2^{\log_2 7 + \log_2 6}$.

6. Find each of the following:
   (a) $\log_\pi \pi^5$.
   (b) $\log_5 (1/25)$.
   (c) $\log_{\sqrt{2}} 1$.
   (d) $3^{-\log_3 2}$,
   (e) $4^{\log_4 3 - \log_4 7}$.

7. If $0 < a < 1$, what is $\lim\limits_{x \to \infty} a^x$?

8. If u is a positive integer and a^u is defined by $a^u = aaa \cdots a$, with u factors, prove that (2) and (3) are true for all positive integers u and v.

9. Derive property (11) from property (10) and equation (4).

10. Prove that if $b \neq 0$, then $(1/b)^u = 1/b^u$, using equations (7) and (8).

11. Prove that $(a/b)^u = a^u/b^u$, if $b \neq 0$.

*12. Assuming the "laws of exponents," equations (1) through (9), prove the "laws of logarithms," equations (16) through (19).

13. In changing the base of the logarithms from b to a, log M is changed in accordance with the formula

$$\log_a M = \log_a b \log_b M.$$

Prove this equation. HINT: Set $x = \log_a M$ and $y = \log_b M$. Then $M = a^x = b^y$. Now take the logarithm of both sides to the base a.

14. Prove that $\log_A B \log_B A = 1$.

In Problems 15 through 22, solve the given equations for x. Assume that $a > 1$ and log means logarithm to the base 10.

15. $3^{x(x-1)} = 9$.

16. $a^x a^{4-x^2} = a^2$.

17. $a^x a^{x^2} a^{x^3} = a^{13x}$.

18. $\dfrac{a^{x^2}}{a} = a^5$.

19. $\log x + \log (x - 10) = \log 24$.

20. $\log 2x + \log (3 + x) = \log 20$.

21. $\log x^5 - \log (5x + 14) = 3 \log x$.

22. $\log (x - 1) + \log (x + 2) + \log (x - 3) = \log 6$.

23. Relative gains or losses in electrical power are measured in *decibels* (dB) by the formula

$$dB = 10 \log_{10} \frac{P_2}{P_1},$$

where P_2 is the larger power and P_1 is the smaller power. If the power input to an amplifier is 0.01 watt and its output is 10 watts, find the amplifier gain.

24. An amplifier has a 50-dB gain (see Problem 23) and a 100-watt output. What input power does it have?

CALCULATOR PROBLEMS

In Problems C1 and C2, use a calculator to estimate the given quantity to four decimal places.

C1. (a) $\sqrt{2}^{\sqrt{3}}$.   (b) $\sqrt{3}^{\sqrt{2}}$.   (c) $\pi^{1.111}$.   (d) $\sqrt{3}^{-\sqrt{1.2}}$.

C2. (a) $\sqrt{2}^{\sqrt{5}}$.   (b) $\sqrt{5}^{\sqrt{2}}$.   (c) $\pi^{-1.111}$.   (d) $\sqrt{3}^{\sqrt{1.2}}$.

C3. Use the numbers obtained in Problems C1 and C2 to check (approximately) the following statements.

     (a) $\sqrt{2}^{\sqrt{3}} \sqrt{2}^{\sqrt{5}} = \sqrt{2}^{\sqrt{3}+\sqrt{5}}$.   (b) $\pi^{1.111}\pi^{-1.111} = 1$.   (c) $\sqrt{3}^{-\sqrt{1.2}}\sqrt{3}^{\sqrt{1.2}} = 1$.

C4. What happens when you enter $x = -3$ and hit the log button on your calculator? Explain.

C5. Use a calculator to evaluate each of the following expressions in the form given (log means logarithm to the base 10). Then use the properties of the logarithm to *prove* that your answers are (approximately) correct.

     (a) $\log 3 + \log (1/3)$.   (b) $\log 49 - 2 \log 7$.   (c) $10^{\log 5 - \log 2}$.

The Number e   3

In arithmetic work, choosing the base number of the logarithm equal to 10 offers certain computational advantages. In calculus, however, the most convenient choice for the base is the number e, which we define as follows.

DEFINITION 2

(20)
$$e \equiv \lim_{h \to 0} (1 + h)^{1/h}.$$

What is the decimal value of this limit? More important, how do we even know that the limit of $(1 + h)^{1/h}$ as $h \to 0$ *exists?* The answers to these questions are by no means obvious. Letting $h \to 0^+$, we observe that while the quantity $(1 + h)$ gets closer and closer to 1, this quantity is being raised to the higher and higher powers $1/h$, and the outcome of the competition is unclear.

Some insight is furnished if we consider the special values $h = 1/n$, where n is an integer. For such values it is easy to compute $y = (1 + h)^{1/h}$. The results are recorded in Table 1 to four decimal places, and the corresponding points are plotted in Fig. 3.

TABLE 1

h	$n = 1/h$	$y = (1 + h)^{1/h}$	h	$n = 1/h$	$y = (1 + h)^{1/h}$
1	1	2.0000	—	—	—
1/2	2	2.2500	−1/2	−2	4.0000
1/3	3	2.3704	−1/3	−3	3.3750
1/4	4	2.4414	−1/4	−4	3.1605
1/5	5	2.4883	−1/5	−5	3.0518
1/10	10	2.5937	−1/10	−10	2.8680

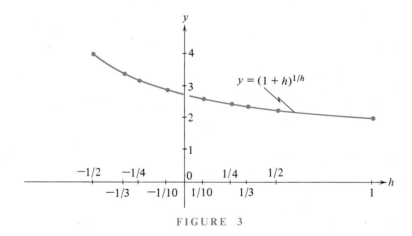

FIGURE 3

From Table 1 and the curve in Fig. 3, it appears that as $h \to 0$, the quantity $(1 + h)^{1/h}$ tends to a limit that is roughly 2.7. This limit is the number we call e. In Chapter 15 we will learn an easy method for computing e to any required degree of accuracy, but for most practical purposes it is sufficient to use $e \approx 2.71828$.

Just like π, the number e is a *transcendental number;* that is, it is not the root of any polynomial with integer coefficients. The proof that e is transcendental is very difficult, and in fact for many years the question of the nature of e was an unsolved problem. The first proof that e is transcendental was given by Charles Hermite in 1873.

EXERCISE 2

In Problems 1 and 2, find the limit in terms of the number e.

1. (a) $\lim\limits_{n\to\infty} \left(1 + \dfrac{1}{n}\right)^n$.
   (b) $\lim\limits_{h\to 0} \left(1 + \dfrac{h}{2}\right)^{1/h}$.
   (c) $\lim\limits_{h\to 0} (1 + h)^{-1/h}$.

2. (a) $\lim\limits_{n\to\infty} \left(1 + \dfrac{2}{n}\right)^n$.
   (b) $\lim\limits_{h\to 0} (1 + 3h)^{1/h}$.
   (c) $\lim\limits_{h\to 0} h(1 + h)^{1/h}$.

CALCULATOR PROBLEMS

C1. Use your calculator to find $(1 + h)^{1/h}$ to four decimal places when
   (a) $h = 1/100$.
   (b) $h = -1/100$.
   (c) $h = 10^{-4}$.
   (d) $h = 10^{-7}$.

C2. Try to evaluate $(1 + h)^{1/h}$ when $h = 10^{-14}$. Does your calculator give a misleading answer? Explain.

The Derivative of the Logarithmic Function **4**

Let $y = \log_a x$, and let x be changed by an amount Δx, giving a change of Δy in y. Then we have

$$\Delta y = \log_a (x + \Delta x) - \log_a x.$$

Using the fact that the difference of logarithms is the logarithm of the quotient, we obtain

$$\Delta y = \log_a \left(\frac{x + \Delta x}{x}\right).$$

Dividing by Δx and then inserting the factor x/x yields

$$\frac{\Delta y}{\Delta x} = \frac{1}{\Delta x} \log_a \left(1 + \frac{\Delta x}{x}\right) = \frac{1}{x}\frac{x}{\Delta x} \log_a \left(1 + \frac{\Delta x}{x}\right).$$

Applying here equation (18) with $n = x/\Delta x$ gives

(21)
$$\frac{\Delta y}{\Delta x} = \frac{1}{x} \log_a \left(1 + \frac{\Delta x}{x}\right)^{x/\Delta x}.$$

We are to let $\Delta x \to 0$ in (21). For simplicity, let us replace $\Delta x/x$ by h in equation (21). Since x is fixed, and $\Delta x \to 0$, then $h \to 0$, and (21) gives

$$\frac{dy}{dx} = \lim_{\Delta x\to 0} \frac{1}{x} \log_a \left(1 + \frac{\Delta x}{x}\right)^{x/\Delta x} = \lim_{h\to 0} \frac{1}{x} \log_a (1 + h)^{1/h} = \frac{1}{x} \log_a e,$$

by the definition of e in (20). We have proved that if $y = \log_a x$, then

$$\frac{dy}{dx} = \frac{1}{x} \log_a e.$$

Using the Chain Rule for differentiating a function of a function, we have

---

THEOREM 1. If u is any differentiable function of x, and $u > 0$, then

(22)
$$\frac{d}{dx} \log_a u = \frac{1}{u} \frac{du}{dx} \log_a e.$$

Equation (22) can be simplified by selecting a suitable number for the base a. Indeed, if we select e as the base, then $\log_e e = 1$, and the nuisance factor in (22) can be dropped.

Logarithms to the base e are called *natural logarithms,* although at first glance such a base as $2.71828\ldots$ may seem to be most unnatural. But equation (22) does simplify nicely when the base is e. As the student pursues his scientific studies, he will find more reasons for regarding e as a natural base, and logarithms to the base e as natural logarithms. A table of natural logarithms is given at the end of the book (Table B).

In order to avoid the subscript a, we will use the symbol $\ln u$ to denote the natural logarithm of u, and $\log u$ to denote the logarithm to the base 10 of u; that is,

$$\log_{10} u = \log u \qquad \text{and} \qquad \ln u = \log_e u.$$

With this notation we have the two special cases of equation (22):

(23)
$$\frac{d}{dx} \log u = \frac{1}{u} \frac{du}{dx} \log_{10} e = \frac{1}{u} \frac{du}{dx} \times 0.434\ldots$$

and

(24)
$$\frac{d}{dx} \ln u = \frac{1}{u} \frac{du}{dx}.$$

Example 1. Find the derivative of:

(a) $y = \ln (x^2 + 1)$, (b) $y = \ln \cos x$, (c) $y = \ln [x^4(x^2 + 4)^{3/2}]$.

Solution. Using equation (24), we have

(a) $\dfrac{dy}{dx} = \dfrac{1}{x^2 + 1} \dfrac{d}{dx} (x^2 + 1) = \dfrac{2x}{x^2 + 1}.$

(b) $\dfrac{dy}{dx} = \dfrac{1}{\cos x}\dfrac{d}{dx}\cos x = \dfrac{-\sin x}{\cos x} = -\tan x.$

We observe that $\ln \cos x$ is well defined only if $\cos x$ is positive, so it is understood that the manipulations and the results are valid only if x lies in one of the intervals $-\pi/2 + 2n\pi < x < \pi/2 + 2n\pi$, where n is some integer.

We will frequently meet similar situations in the future. Henceforth, whenever an expression of the form $\ln u(x)$ is under consideration, we assume that x is restricted to lie in some set for which $u(x) > 0$. By this agreement we can avoid a lengthy and pointless investigation into the precise composition of the domain of x.

(c) The computations are simplified if we first use the laws of logarithms, equations (16) and (18), to simplify the expression for y. Indeed,

$$y = \ln [x^4(x^2 + 4)^{3/2}] = \ln x^4 + \ln (x^2 + 4)^{3/2} = 4 \ln x + \frac{3}{2} \ln (x^2 + 4).$$

Then

$$\frac{dy}{dx} = \frac{4}{x} + \frac{3}{2}\frac{2x}{x^2 + 4} = \frac{4}{x} + \frac{3x}{x^2 + 4} = \frac{4x^2 + 16 + 3x^2}{x(x^2 + 4)} = \frac{7x^2 + 16}{x(x^2 + 4)}. \quad \bullet$$

The properties of logarithms can be used to simplify an otherwise complicated problem in differentiation. The method is illustrated in

Example 2. Find $\dfrac{dy}{dx}$ if $y = \dfrac{(x^2 + 1)^{1/2}(6x + 5)^{1/3}}{(x^2 - 1)^{1/2}}.$

Solution. We first take the natural logarithm of both sides and use equations (16), (17), and (18). This gives

$$\ln y = \frac{1}{2} \ln (x^2 + 1) + \frac{1}{3} \ln (6x + 5) - \frac{1}{2} \ln (x^2 - 1).$$

Differentiating both sides with respect to x yields

$$\frac{1}{y}\frac{dy}{dx} = \frac{x}{x^2 + 1} + \frac{2}{6x + 5} - \frac{x}{x^2 - 1}$$

$$= \frac{x(6x + 5)(x^2 - 1) + 2(x^2 + 1)(x^2 - 1) - x(x^2 + 1)(6x + 5)}{(x^2 + 1)(6x + 5)(x^2 - 1)}$$

$$= \frac{2(x^4 - 6x^2 - 5x - 1)}{(x^2 + 1)(6x + 5)(x^2 - 1)}.$$

Multiplying through by y and using the expression for y in terms of x, we have

$$\frac{dy}{dx} = \frac{(x^2 + 1)^{1/2}(6x + 5)^{1/3}}{(x^2 - 1)^{1/2}} \cdot \frac{2(x^4 - 6x^2 - 5x - 1)}{(x^2 + 1)(6x + 5)(x^2 - 1)}$$

$$= \frac{2(x^4 - 6x^2 - 5x - 1)}{(x^2 + 1)^{1/2}(6x + 5)^{2/3}(x^2 - 1)^{3/2}}. \quad \bullet$$

The procedure just illustrated is called *logarithmic differentiation.*

Each differentiation formula leads to an integral formula. Thus equation (24) gives $d \ln u = (1/u) \, du$, and hence

(25)
$$\int \frac{du}{u} = \ln u + C.$$

There is a slight difficulty with (25) because the function $\ln u$ has meaning only if u is positive. Suppose that u is negative. Then $-u = |u|$ is positive, and using (25) on $-u$, we can write

$$\int \frac{du}{u} = \int \frac{-du}{-u} = \int \frac{d(-u)}{-u} = \ln(-u) + C = \ln |u| + C.$$

Combining this last formula with (25), we have

THEOREM 2. If u is not zero, then

(26)
$$\int \frac{du}{u} = \ln |u| + C.$$

Of course, if $u = 0$, the integrand $1/u$ becomes infinite. In any natural problem this exceptional case will not arise.

Example 3. Find the area of the region under the curve $y = 1/x$ between $x = 1$ and $x = 5$.

Solution. Using (26) and Table B in the appendix, we find that

$$A = \int_1^5 y \, dx = \int_1^5 \frac{1}{x} \, dx = \ln x \Big|_1^5 = \ln 5 - \ln 1$$

$$= 1.609 \ldots - 0 \approx 1.609.$$

A calculator that has an ln function will give $A = \ln 5 = 1.60943791 \ldots . \quad \bullet$

Example 4. Find the indefinite integrals.

(a) $\displaystyle\int \frac{x^3\,dx}{2x^4 + 1}$.   (b) $\displaystyle\int \tan x\,dx$.   (c) $\displaystyle\int \sec x\,dx$.

Solution. (a) Let $u = 2x^4 + 1$. Then $du = 8x^3\,dx$.

$$\int \frac{x^3\,dx}{2x^4 + 1} = \frac{1}{8}\int \frac{8x^3\,dx}{2x^4 + 1} = \frac{1}{8}\int \frac{du}{u} = \frac{1}{8}\ln(2x^4 + 1) + C.$$

Note that we have dropped the absolute value signs in $\ln|2x^4 + 1|$ because $2x^4 + 1$ is always positive.

(b) $\displaystyle\int \tan x\,dx = \int \frac{\sin x}{\cos x}\,dx = -\int \frac{d(\cos x)}{\cos x} = -\ln|\cos x| + C.$

(c) We multiply the integrand by $1 = (\sec x + \tan x)/(\sec x + \tan x)$. Then we have

$$\int \sec x\,dx = \int \frac{(\sec x + \tan x)\sec x}{\sec x + \tan x}\,dx = \int \frac{(\sec^2 x + \sec x \tan x)\,dx}{\sec x + \tan x}$$

$$= \int \frac{d(\sec x + \tan x)}{\sec x + \tan x} = \ln|\sec x + \tan x| + C. \quad \bullet$$

These last two examples suggest that we expand our list of fundamental integration formulas by adjoining the following:

(27)
$$\int \tan u\,du = -\ln|\cos u| + C.$$

(28)
$$\int \cot u\,du = \ln|\sin u| + C.$$

(29)
$$\int \sec u\,du = \ln|\sec u + \tan u| + C.$$

(30)
$$\int \csc u\,du = -\ln|\csc u + \cot u| + C.$$

We have just proved (27) and (29). We leave the proofs of (28) and (30) to the reader.

EXERCISE 3

In Problems 1 through 15, find dy/dx.

1. $y = \ln(x^6 + 3x^2 + 1)$.   2. $y = \ln(x + 1)^3$.

3. $y = 3\ln(5x + 5)$.   4. $y = \ln x^2$.

5. $y = \ln^2 x$.   6. $y = \ln \sec x^2$.

7. $y = x^2 \ln x$.   8. $y = x \ln x^2 - 2x$.

9. $y = \ln[2x\sqrt{x^2 + 4}]$.   10. $y = [\ln x][\ln(1 - x)]$.

11. $y = \ln \tan x + \ln \cot x$.   12. $y = \ln \dfrac{1 + x^2}{1 - x^2}$.

13. $y = 4x\,\text{Tan}^{-1} 2x - \ln(4x^2 + 1)$.   14. $y = x(\sin \ln x + \cos \ln x)$.

15. $y = x\sqrt{x^2 - 5} - 5\ln(x + \sqrt{x^2 - 5})$.

Use logarithmic differentiation in Problems 16 through 19 to find dy/dx.

16. $y = \sqrt{(x^2 - 1)(x^2 + 2)}$.   17. $y = \sqrt[3]{(x - 1)(x + 2)(x + 5)}$.

18. $y = 6\dfrac{(3x + 2)^{1/2}}{(2x + 1)^{1/3}}$.   19. $y = \dfrac{\sqrt{x^2 - 5}}{x^6\sqrt{x^2 + 7}}$.

20. Explain why the answers in Problems 2 and 3 are the same.

21. Explain why the answer in Problem 11 is zero.

In Problems 22 through 33, find the given integral.

22. $\displaystyle\int \dfrac{x\,dx}{x^2 + 4}$.   23. $\displaystyle\int \dfrac{\sin x\,dx}{5 - 3\cos x}$.   24. $\displaystyle\int x\tan x^2\,dx$.

25. $\displaystyle\int \sec 5x\,dx$.   26. $\displaystyle\int \dfrac{x + 3}{x^2 + 4}\,dx$.   ★27. $\displaystyle\int \dfrac{dx}{\sqrt{x}(1 + x)}$.

★28. $\displaystyle\int \dfrac{x\,dx}{\sin x^2}$.   ★29. $\displaystyle\int \dfrac{x^3\,dx}{\tan x^4}$.   ★30. $\displaystyle\int \dfrac{\ln x\,dx}{x}$.

★31. $\displaystyle\int \dfrac{dx}{x\ln x}$.   32. $\displaystyle\int_0^1 \dfrac{dx}{1 + 5x}$.   33. $\displaystyle\int_0^{\pi/4} \tan x\,dx$.

34. Find the area of the region under the curve:
    (a) $y = 1/x$ between $x = 1$ and $x = 7$.
    (b) $y = 1/x$ between $x = 4$ and $x = 28$.
    (c) $y = 2x/(x^2 + 3)$ between $x = 1$ and $x = 5$.

35. The region bounded by the y-axis and the curve $y = 1/x^2$, between the lines $y = 5$ and $y = 35$, is rotated about the y-axis. Find the volume of the solid generated.

36. Find the length of arc of the curve:

    (a) $y = \ln \cos x,$ between $x = 0$ and $x = \pi/4,$

    (b) $y = \dfrac{x^2}{2} - \dfrac{1}{4}\ln x,$ between $x = 1$ and $x = 16.$

37. Sketch the curve $y = x^2 - 8\ln x$ for $x > 0$ and locate all extreme points and inflection points on the curve.

38. Formula (22) is not in a useful form because $\log_a e$ is not available in tables. Use Problem 14 of Exercise 1 to show that this factor can be replaced by $1/\ln a$. This factor is easy to compute using Table B, or a calculator.

39. Prove that for $0 < x < \pi/2$ the functions $\ln(\csc 2x - \cot 2x)$ and $\ln \tan x$ have the same derivative. Consequently, in that interval $\ln(\csc 2x - \cot 2x) = \ln \tan x + C$. Find C.

*40. Let k be a positive constant. Prove that the equation $kx + \ln x = 0$ always has exactly one solution. HINT: Prove that the left side is an increasing function.

*41. Prove that the equation $x = \ln x$ has no solution. HINT: Find the minimum value of $f(x) \equiv x - \ln x$.

In Problems 42 through 47, find $f'(x)$ for the given $f(x)$.

42. $\ln(\ln x)$.

43. $\ln(\ln(\ln x))$.

44. $\ln(x \ln x)$.

45. $\ln(\tan^7 x)$.

46. $\dfrac{\ln(\sin^4 x)}{\ln(\cos^4 x)}$.

47. $\dfrac{\sqrt{5}\ln(\cos^7 x)}{\ln(1 + \tan^2 x)}$.

CALCULATOR PROBLEMS

C1. Use the definition of the derivative

$$\frac{d}{dx}\ln x = \lim_{\Delta x \to 0}\frac{\ln(x + \Delta x) - \ln x}{\Delta x} = \lim_{\Delta x \to 0}\frac{1}{\Delta x}\ln\left(1 + \frac{\Delta x}{x}\right)$$

to estimate the derivative at $x = 2$, by computing (to five decimal places) the right side when: (a) $\Delta x = 0.1$, (b) $\Delta x = 0.01$, and (c) $\Delta x = 0.001$.

C2. Do Problem C1 when: (a) $\Delta x = -0.1$, (b) $\Delta x = -0.01$, and (c) $\Delta x = -0.001$.

C3. Since $\displaystyle\int_1^2 \frac{dx}{x} = \ln 2 - \ln 1 = \ln 2$, we may regard $\ln 2$ as the area under the curve $y = 1/x$ between $x = 1$ and $x = 2$. Use Simpson's Rule with 10 subintervals to compute $\ln 2$. Carry six decimal places and round off answer to five decimal places.

C4. Find to five decimal places: (a) $\log 10$, (b) $\ln 10$, (c) $\log 100$, (d) $\ln 100$, (e) $\log e$, (f) $\log e^2$, (g) $\ln e^3$, (h) $\ln(-5)$, and (i) $(\ln 10)(\log e)$.

C5. Find each root of $\ln x^3 = x$ to five significant figures.

C6. Find the smallest positive root of $\ln x = \dfrac{100}{x}$.

5 *The Exponential Function*

In order to find a differentiation formula for the function

(31) $$y = a^u$$

we first take the natural logarithm of both sides. This yields

(32) $$\ln y = \ln a^u = u \ln a.$$

If we differentiate this equation with respect to x we have, by (24),

$$\frac{1}{y} \frac{dy}{dx} = \frac{du}{dx} \ln a.$$

Multiplying both sides by y and using (31), we obtain

(33) $$\frac{dy}{dx} = y \frac{du}{dx} \ln a = a^u \frac{du}{dx} \ln a.$$

Hence, if u is any differentiable function of x,

(34) $$\frac{d}{dx} a^u = a^u \frac{du}{dx} \ln a.$$

The most important case occurs when the base a is e. Since $\ln e = \log_e e = 1$, the nuisance factor in a, in (34), becomes 1 when $a = e$. Hence

(35) $$\frac{d}{dx} e^u = e^u \frac{du}{dx},$$

(36) $$\frac{d}{dx} e^x = e^x.$$

Note that e^x is a function that is its own derivative. The simplicity of this formula lends weight to the feeling that e is a natural base. Just as logarithms to the base e are natural logarithms, so e^x is a "natural" exponential function.[1] A table of values for e^x is given in Table C, at the end of the book. It is important to keep in mind that e^x and $\ln x$ are inverse functions so that

$$e^{\ln x} = x, \; x > 0 \quad \text{and} \quad \ln e^x = x \text{ for all } x.$$

Each of the three differentiation formulas (34), (35), and (36) yields an equivalent integration formula. These are

(37) $$\int a^u \, du = \frac{a^u}{\ln a} + C,$$

---

[1] To avoid the use of superscripts, some authors write exp (x) instead of e^x.

(38) $$\int e^u \, du = e^u + C,$$ (39) $$\int e^x \, dx = e^x + C.$$

Example 1. Find the derivative of:

(a) $y = e^{3x}$.   (b) $y = e^{\tan x^2}$.   (c) $y = (e^{2x} - 1)/(e^{2x} + 1)$.

Solution. (a) $\dfrac{dy}{dx} = e^{3x} \dfrac{d}{dx} 3x = 3e^{3x}$.

(b) $\dfrac{dy}{dx} = e^{\tan x^2} \dfrac{d}{dx} \tan x^2 = e^{\tan x^2} 2x \sec^2 x^2 = 2x \, e^{\tan x^2} \sec^2 x^2$.

(c) $\dfrac{dy}{dx} = \dfrac{(e^{2x} + 1)2e^{2x} - (e^{2x} - 1)2e^{2x}}{(e^{2x} + 1)^2} = \dfrac{4e^{2x}}{(e^{2x} + 1)^2}$.   ●

Example 2. Find each of the integrals:

(a) $\displaystyle\int_0^2 \dfrac{dx}{e^x}$.   (b) $\displaystyle\int e^x \sin e^x \, dx$.   (c) $\displaystyle\int \dfrac{e^x \, dx}{1 + 5e^x}$.   (d) $\displaystyle\int_2^5 4e^{\ln x} \, dx$.

Solution. (a) $\displaystyle\int_0^2 \dfrac{dx}{e^x} = \int_0^2 e^{-x} \, dx = -\int_0^2 e^{-x} d(-x)$

$$= -e^{-x} \Big|_0^2 = e^0 - e^{-2} = 1 - e^{-2} = 1 - 0.1353 \ldots \approx 0.865.$$

(b) $\displaystyle\int e^x \sin e^x \, dx = \int \sin u \, du$ (where $u = e^x$)

$$= -\cos u + C = -\cos e^x + C.$$

(c) $\displaystyle\int \dfrac{e^x \, dx}{1 + 5e^x} = \dfrac{1}{5} \int \dfrac{5e^x \, dx}{1 + 5e^x} = \dfrac{1}{5} \int \dfrac{du}{u}$ (where $u = 1 + 5e^x$)

$$= \dfrac{1}{5} \ln u + C = \dfrac{1}{5} \ln (1 + 5e^x) + C.$$

(d) $\displaystyle\int_2^5 4e^{\ln x} \, dx = \int_2^5 4x \, dx = 2x^2 \Big|_2^5 = 2(25 - 4) = 42.$ ●

Example 3. Find the derivative of $y = x^{x^2}$.

Solution. Since both the base and the exponent are variables, none of our formulas cover this case. But logarithmic differentiation will help us to bypass this difficulty. Taking natural logarithms of both sides gives

$$\ln y = x^2 \ln x$$

and now we have a product to differentiate. Hence

$$\frac{1}{y}\frac{dy}{dx} = 2x \ln x + x^2\frac{1}{x},$$

$$\frac{dy}{dx} = y(2x \ln x + x) = (x + 2x \ln x)x^{x^2}. \quad \bullet$$

EXERCISE 4

In Problems 1 through 11, find the derivative.

1. $y = x^2 e^{-3x}$.

2. $y = e^{1/x^2}$.

3. $y = e^{\sin^2 5x}$.

4. $y = \ln(1 + 5e^x)$.

5. $y = \dfrac{x^2}{e^x + x}$.

6. $y = \ln\dfrac{1 + e^{3x}}{1 - e^{3x}}$.

7. $y = x^{\sin x}$.

8. $y = (\sin x)^x$.

9. $y = (1 + 3x)^{1/x}$.

10. $y = (\cos x^2)^{x^3}$.

11. $y = (24 + 24x + 12x^2 + 4x^3 + x^4)e^{-x}$.

In Problems 12 through 17, find a formula for the nth derivative.

12. $y = \ln x$.

13. $y = \ln(1 + x)^5$.

14. $y = e^x$.

15. $y = (e^x)^7$.

$\star$16. $y = xe^x$.

$\star$17. $y = x^2 \ln x$.

In Problems 18 through 23, find the indicated integral.

18. $\displaystyle\int e^{-4x}\, dx$.

19. $\displaystyle\int 14xe^{x^2}\, dx$.

20. $\displaystyle\int e^{\tan x}\sec^2 x\, dx$.

21. $\displaystyle\int \frac{6e^x\, dx}{1 + e^{2x}}$.

22. $\displaystyle\int \frac{9e^{3x}\, dx}{1 + e^{3x}}$.

23. $\displaystyle\int \frac{e^x + e^{-x}}{e^x - e^{-x}}\, dx$.

In Problems 24 through 29, sketch the graph of the given function and find all the relative maximum, relative minimum, and inflection points.

24. $y = e^{-x^2}$.

25. $y = e^x + e^{-x}$.

26. $y = e^x - e^{-x}$.

27. $y = xe^{x/3}$.

$\star$28. $y = e^{-x}\cos x$.

29. $y = xe^{-x^2}$.

30. Find the sides of the largest rectangle that can be drawn with two vertices on the x-axis, and two vertices on the curve $y = e^{-x^2}$.

31. From the point (a, e^a) on the curve $y = e^x$ a line is drawn normal to this curve. Find the x-intercept of this line.

32. Prove that $2x - \ln(3 + 6e^x + 3e^{2x}) = C - 2\ln(1 + e^{-x})$ by showing that both sides have the same derivative. What is C?

33. Find the length of the arc of $y = \dfrac{e^x + e^{-x}}{2}$ between $x = 0$ and $x = a$.

34. The arc of Problem 33 is rotated about the x-axis. Find the area of the surface generated.

★35. Prove that if $x > 0$, then $e^x > 1 + x$. HINT: Integrate the inequality $e^t \geq 1$ over the interval $0 \leq t \leq x$.

★36. Prove that the inequality of Problem 35 also holds if $x < 0$.

★37. By integrating the inequality of Problem 35, prove that if $x > 0$, then

$$e^x > 1 + x + \frac{x^2}{2}.$$

★38. Prove that if $x > 0$ and n is a positive integer, then

$$e^x > 1 + \sum_{k=1}^{n} \frac{x^k}{k!}.$$

39. Find all the real roots of $e^{5x} + e^{4x} - 6e^{3x} = 0$.

40. Find all the real roots of

$$3e^{\mathrm{Cos}^{-1} 1} \sin^2 x - 5x^2 e^{-\ln(x^2+1)} + (7 + 4\cos \pi) \cos^2 x = 0.$$

CALCULATOR PROBLEMS

In Problems C1 through C4, find the indicated root to five significant figures.

C1. $e^x = x + 20, \quad x > 0$.   C2. $e^x = x + 3, \quad x < 0$.
C3. $e^x = 10 - x^3$.   C4. $xe^x = 5$.

Growth and Decay 6

Quantities that change with time are visible wherever we turn. The population of a city grows with time; the amount of pollution is increasing (unfortunately) in most places. Money left on deposit in a well-run bank increases. The intensity of light decreases as it passes through a body of water. The amount of a radioactive substance decreases with time.

Most of the quantities mentioned obey the same differential equation, and a very simple one at that. If Q measures the amount of the item at time t, then in many cases the rate of change of Q is directly proportional to Q itself, namely

(40)
$$\frac{dQ}{dt} = kQ,$$

where k is the constant of proportionality. The rate of increase in the population of any species (bacteria, rabbits, or humans) will be proportional to the number of the species present—unless the growth is restricted by other factors.

The amount of a radioactive substance decreases at a rate proportional to the amount present. Thus it obeys the differential equation (40), but in this case k is negative.

By solving equation (40), we obtain at one stroke the equation that governs the behavior of a large variety of items. From (40) we easily obtain

$$\frac{dQ}{Q} = k\, dt,$$

(41)
$$\ln Q = \int \frac{dQ}{Q} = \int k\, dt = kt + \ln C,$$

where we added a constant in the form $\ln C$ for convenience. If we exponentiate the first and last terms in (41), we have

$$Q = e^{\ln Q} = e^{kt+\ln C} = e^{\ln C}e^{kt} = Ce^{kt}.$$

Thus $Q = Ce^{kt}$. To determine C, we assume the initial condition $Q = Q_0$ when $t = t_0$. Thus $Q_0 = Ce^{kt_0}$, and hence $C = Q_0 e^{-kt_0}$. Using this value of C, we have

$$Q = Q_0 e^{-kt_0}e^{kt} = Q_0 e^{k(t-t_0)}.$$

THEOREM 3. If $Q(t)$ satisfies the differential equation (40) and $Q(t_0) = Q_0$, then

(42)
$$Q(t) = Q_0 e^{k(t-t_0)}.$$

This is the equation for simple growth or decay. It is growth if $k > 0$, and decay if $k < 0$. Of course, if $k = 0$, then Q is constant. In using this equation, we need two pieces of information (two points on the curve). First, we need to know the initial conditions. Here it is often convenient to set $t_0 = 0$ (i.e., start measuring time when $Q = Q_0$). Second, we need one other piece of information to determine k in equation (42).

Example 1. The rate of growth of a certain colony of bacteria is proportional to the number present. If there are 4000 bacteria at 1:00 P.M. and 8000 at 3:00 P.M., how many will there be at 4:30 P.M.?

Solution. Despite the fact that bacteria come in units and are not continually divisible, there are so many present that we may assume that the growth process is continuous.

We let $Q(t)$ denote the population of the colony at the time t hours after 1:00 P.M. Then, by the conditions stated, Q satisfies the differential equation (40), and $Q_0 = 4000$ when $t_0 = 0$. Thus, from equation (42), we have

(43) $$Q(t) = 4000e^{kt}.$$

In order to solve for the constant k, we use the additional fact that 8000 are present at 3:00 P.M. Thus

(44) $$8000 = 4000e^{2k} \quad \text{or} \quad 2 = e^{2k},$$

and on taking the natural logarithms, we get

(45) $$\ln 2 = 2k \quad \text{or} \quad k = \frac{1}{2}\ln 2 = \ln \sqrt{2}.$$

Thus equation (43) becomes

(46) $$Q(t) = 4000e^{t \ln \sqrt{2}} = 4000(e^{\ln \sqrt{2}})^t = 4000(\sqrt{2})^t.$$

When $t = 3.5$, we obtain

$$Q(3.5) = 4000(\sqrt{2})^{3.5} \approx 13{,}500$$

for the approximate number present at 4:30 P.M. ●

For future reference, it is worthwhile to develop a general formula for finding the constant k. If $Q = Q_1$ when $t = t_1$ ($t_1 \neq t_0$), then equation (42) gives $Q_1 = Q_0 e^{k(t_1 - t_0)}$. Hence

$$\ln \frac{Q_1}{Q_0} = \ln e^{k(t_1 - t_0)} = k(t_1 - t_0)$$

and

(47) $$k = \frac{1}{t_1 - t_0} \ln \frac{Q_1}{Q_0}.$$

Example 2. When listing properties of radioactive elements, it is customary to ignore k in equation (42) and to give the half-life instead. In other words, the tables give the value of t (assume that $t_0 = 0$) for which $Q(t) = Q_0/2$. If the half-life of radium is 1620 years, find k to three significant figures. If we start with 1 gram of radium, how much will be left after **(a)** 200 years and **(b)** 400 years?

Solution. The half-life of radium tells us that $t_1 - t_0 = 1620$ years when $Q_1/Q_0 = 1/2$. Hence, from (47),

$$k = \frac{\ln (1/2)}{1620} = -\frac{\ln 2}{1620} \approx -\frac{0.69315}{1620} \approx -0.000428.$$

Thus, for radium, equation (42) gives the general law

(48) $$Q = Q_0 e^{-0.000428t}.$$

(a) When $t = 200$, the exponent is -0.0856. Suitable tables or a calculator gives $Q = e^{-0.0856} = 0.918$ gram (to three significant figures).

(b) When $t = 400$, the exponent in (48) is -0.1712. Then $Q = 0.843$ gram. Observe that the loss of 0.157 gram in 400 years is *not* twice 0.082 gram, the loss in 200 years. ●

Example 3. H. F. Libby took the simple computations of Example 2, replaced radium by an isotope of carbon, and showed how the same method could be used to assign very accurate dates to some of the items found by archeologists. For this excellent work, Professor Libby won the Nobel prize.

Carbon-14 is a radioactive isotope of the stable form of carbon (carbon-12). The ratio of ^{14}C to ^{12}C in the atmosphere is constant, and the ratio in all *living* plants and animals is the same because of continuous respiration. When a plant or animal dies, however, the ^{12}C remains but the ^{14}C undergoes radioactive decay. Thus the ratio of ^{14}C to ^{12}C begins to decrease. By chemical analysis that requires great care and precision, this ratio can be determined, and such a determination will tell what fraction of the radioactive ^{14}C has disintegrated. If the half-life of ^{14}C is 5560 years, find k.

Archeologists digging near Tel-Aviv in Israel found the remains of some wood used for heating or cooking. Analysis showed that 25 percent of the carbon ^{14}C had disintegrated. Find approximately how many years ago the wood was cut for the fire.

Solution. To find k, we use equation (47). Then

$$k = \frac{-\ln 2}{5560} \approx -\frac{0.69315}{5560} \approx -0.000125 = \frac{-1}{8000}.$$

If we solve equation (42) for t, we find that

(49) $$t - t_0 = \frac{1}{k} \ln \frac{Q}{Q_0}.$$

In our case, $t_0 = 0$ and $k = -1/8000$. Hence

(50) $$t = \frac{1}{\dfrac{-1}{8000}} \ln \frac{Q}{Q_0} = 8000 \ln \left(\frac{Q}{Q_0}\right)^{-1} = 8000 \ln \frac{Q_0}{Q}.$$

Now set $Q_0/Q = 1/0.75 = 4/3$. Then equation (50) and a calculator give

$$t = 8000 \ln \left(\frac{4}{3}\right) \approx 8000(0.2877) \approx 2300 \text{ years.}$$

Observe that the chemical analysis was given with only two significant figures, so we round off the answer to two significant figures. ●

Under suitable conditions money obeys the same growth law as bacteria and disintegrating atoms. We recall that if we deposit P dollars in a well-run bank that pays simple interest at the rate i, then after 1 year the amount A on deposit is given by $A = P(1 + i)$.

If the yearly interest rate i is *compounded quarterly*, this means there are four interest periods per year and, at each of these times (every 3 months), we receive 1/4th of the yearly rate on the accumulated amount. Hence, after 1 year,

$$A = P\left(1 + \frac{i}{4}\right)\left(1 + \frac{i}{4}\right)\left(1 + \frac{i}{4}\right)\left(1 + \frac{i}{4}\right) = P\left(1 + \frac{i}{4}\right)^4.$$

More generally, if the yearly interest rate i is *compounded n times per year,* then after 1 year,

$$A = P\left(1 + \frac{i}{n}\right)^n,$$

and *after t years,* the amount is given by

(51)
$$A = P\left(1 + \frac{i}{n}\right)^{tn}.$$

Quite recently, many savings institutions have begun paying interest that is *compounded continuously*. This means that the number n of interest periods per year tends to infinity. Now as $n \to \infty$ in (51), we have

$$A = \lim_{n \to \infty} P\left(1 + \frac{i}{n}\right)^{tn} = \lim_{n \to \infty} P\left[\left(1 + \frac{i}{n}\right)^{n/i}\right]^{ti} = \lim_{h \to 0} P[(1 + h)^{1/h}]^{ti},$$

where $h = i/n$ and $h \to 0$ as $n \to \infty$. By equation (20), the limit of the quantity in brackets is e. We have therefore proved

THEOREM 4. If P dollars earn interest compounded continuously for t years at the yearly rate i, then the amount A is given by

(52)
$$A = Pe^{ti}.$$

This equation deserves comment. First, t can be a fraction. We can deposit our money for 39 days and then withdraw it. In this case $t = 39/365$. Second, equation (52) makes A a continuous function of t (of course, no one leaves money on deposit for $\sqrt{5}$ years). Finally, equation (52) has exactly the form of equation (42) for exponential growth.

Example 4. $100 is deposited in a bank where interest is compounded continuously. Find the amount on deposit after 1 year if the yearly rate is: **(a)** 5 percent, **(b)** 6 percent, and **(c)** 7.5 percent. Compare this with the amount on deposit if interest is compounded quarterly.

Solution. If interest is compounded continuously, we use equation (52) with $P = 100$ and $t = 1$. Observe that 5 percent means $i = 0.05$. If interest is compounded quarterly, we use equation (51) with $n = 4$. Rounding off to the nearest cent, we have

Continuously   *Quarterly*

(a) $A = 100\, e^{0.05} = \$105.13,$ $100\left(1 + \dfrac{0.05}{4}\right)^4 = \$105.09,$

(b) $A = 100\, e^{0.06} = \$106.18,$ $100\left(1 + \dfrac{0.06}{4}\right)^4 = \$106.14,$

(c) $A = 100\, e^{0.075} = \$107.79,$ $100\left(1 + \dfrac{0.075}{4}\right)^4 = \$107.71.$ ●

EXERCISE 5

CALCULATOR PROBLEMS

C1. Using the constant k determined in Example 2 for radium, find the amount of radium left from 1 gram after: **(a)** 100 years, **(b)** 500 years, and **(c)** 1000 years.

C2. The team of archeologists in Example 3 dug through the remains of one city and below found evidence of a second city and more half-burned wood. In this second city careful analysis of a sample showed that 35 percent of the ^{14}C had disintegrated. How many years ago (approximately) would this date the second city?

C3. Suppose that in Example 3, the chemical analysis was off by 1 percent. Find approximately the error made in dating the wood. Use either arithmetic or differentials.

C4. Repeat Problem C1 for the data in Problem C2.

C5. Suppose that the half-life of ^{14}C is 5700 years instead of 5560 as used in Example 3. Find the new value of k to four significant figures. What number would replace 8000 in equation (50)?

C6. Suppose that the analysis of material found buried with a mummy in an Egyptian pyramid showed that 40 percent of the ^{14}C had disintegrated. How long ago (approximately) did the interment take place? Use the value for k in Example 3.

C7. The half-life of ^{99}Mo (molybdenum-99) is 66 hours. As it disintegrates, it forms a new radioactive element, ^{99}Tc (technetium-99), that has a half-life of 6.04 hours. Technetium is extremely useful in medical research, and at present it is the most widely used radioactive element in medical diagnosis. The molybdenum is delivered to hospitals in the form of molybdic oxide buried in a matrix of lead or alumina, all nicely packaged in a cylinder prepared for "milking." The technetium that is formed appears as a compound called pertechnetate, and "milking" consists of washing it from the cylinder with physiological saline solution as it is needed.

Suppose that a cylinder contains 10 grams of ^{99}Mo when it is packed at noon on Friday. Find the number of grams of ^{99}Mo that will be in the cylinder at noon the next week: (a) on Monday, (b) on Wednesday, and (c) on Friday.

C8. Mr. Goodbettor received an injection of a solution containing 1 milligram of ^{99}Tc (see Problem C7) at 10:00 A.M. We assume that the only loss is through disintegration (all of the ^{99}Tc is absorbed in the tissues). Find the number of milligrams of ^{99}Tc still in him at 6:00 P.M. that day, when he starts cooking dinner for his wife. What is his ^{99}Tc content the next day at the same time? When will his ^{99}Tc content be only 10 percent of the original injection? For simplicity, use $k = -0.115$.

C9. According to Newton's law of cooling, if an object at temperature T is immersed in a medium having the constant temperature M, then the rate of change of T is proportional to the difference of temperatures $T - M$. This gives the differential equation $dT/dt = k(T - M)$.

(a) Solve the differential equation for T.

(b) A thermometer reading 80° is placed in a medium having a constant temperature of 30°. After 10 minutes the thermometer reads 75°. What will it read after 25 minutes?

C10. Use Newton's law of cooling (Problem C9) to answer the following. A pie is removed from an oven at 450° and left to cool at a room temperature of 70°. After 30 minutes the pie is at 200°. How many minutes after being removed from the oven will the pie be at the ideal serving temperature of 100°?

C11. **Dating the Garden of Eden.** Suppose that the world population was 250,000,000 in the year A.D. 1500, and that prior to this date the population grew at a rate proportional to its size with $k = 0.0025$. If we use $Q_0 = 2$ (for Adam and Eve), what (approximate) date does that assign to their existence?

C12. **The Banker's Definition of e.** Find the amount A if $1 is left on deposit for 1 year at 100 percent interest compounded continuously.

C13. Suppose that $100 is deposited in a bank. Find the amount after 3 years: (a) if interest is compounded annually at the rate of 5 percent, and (b) if the annual rate is 5 percent but it is compounded continuously.

C14. A bank lends money at 12 percent, compounded continuously. How much would the bank lend a person who agreed to pay back $100 after 3 years? HINT: Solve equation (52) for P. This is the present value of an amount A, due t years later.

C15. How much could the person in Problem C14 borrow if he or she agreed to pay back $1000: (a) 10 years later and (b) 20 years later?

C16. In 1626, Peter Minuit paid the Indians $24 for the land in New York City that is now known as Manhattan. If he could have invested the money at 4 percent interest compounded continuously, find the amount this would have given in 1981. Find the amount if the interest rate was 6 percent, and the amount if the interest rate was 7 percent. Give answers to three significant figures.

C17. A certain slowly growing colony of bacteria increases only 5 percent every hour. A laboratory technician left a 1-gram colony with ample food at 5:00 P.M. on Friday. If the growth is unrestricted, how many grams of these bacteria will she find when she shows up for work at 9 A.M. the next Monday?

C18. If we let $k = 0.05$ in Problem C17, find the time at which the colony of bacteria had doubled in size.

C19. If a product is not advertised, then the sales will very likely fall off at a rate proportional to the amount of the product sold. If a product follows the law of decay, equation (40), with a constant $k = -0.10$ when t is in years, how long will it take for the volume of sales to be: (a) cut in half, and (b) reduced to 20 percent of the original amount?

*C20. Suppose that brine containing 2 lb of salt per gallon runs into a tank initially filled with 100 gal of water containing 25 lb of salt. If the brine enters at 5 gal/min, the concentration is kept uniform by stirring, and the mixture flows out at the same rate, find the amount of salt in the tank after 10 minutes. HINT: Let Q denote the pounds of salt in the tank at t minutes after the process begins, and use the fact that

$$\text{rate of increase in } Q = \text{rate of input} - \text{rate of exit.}$$

C21. If light hits the surface of a body of water vertically, part of it is absorbed as it passes through the water, and the amount remaining follows the same decay equation [equation (42)], except that now t is replaced by the distance x that the light has traveled. The coefficient k depends on the type of impurities present in the water and the units used. Suppose that $k = -0.014$ when x is in centimeters. If light hits the water vertically with "unit" intensity, find the intensity at a depth of: (a) 50 cm, (b) 100 cm, (c) 200 cm, and (d) 10 meters.

C22. In general, any radiation is reduced as it passes through some barrier. In place of the half-life of a radioactive element, we use the term *half-value layer* for the thickness necessary to reduce the intensity of the rays 50 percent. This figure for light passing through water is 50 cm. In the case of x rays, the half-value layer depends on the wavelength of the x ray. Half-value layers of various materials, in centimeters, for x rays of wavelength 2×10^{-9} cm are given in Table 2.

TABLE 2

Water	Aluminum	Copper	Lead
3.9	0.92	0.051	0.013

We wish to surround an x-ray machine with a barrier that will remove 90 percent of the radiation. Find the minimum thickness of the barrier if we construct it using (a) aluminum, (b) copper, and (c) lead.

C23. Let x_0 be the half-value layer (see Problem C22). If $x^\star$ is the thickness needed to remove 90 percent of the same type of radiation with the same type of barrier, prove that

$$x^\star = \frac{\ln 10}{\ln 2} x_0 \approx 3.322 \, x_0.$$

Find the equation for the thickness needed to remove 95 percent of the radiation.

The Hyperbolic Functions 7

Certain combinations of the exponential function appear so frequently, both in the applications of mathematics and in the theory, that it is worthwhile to give them special names. We shall see that these functions satisfy identities that are quite similar to the standard trigonometric identities. It is this similarity with the trigonometric functions that accounts for the names attached to the functions.

The function $(e^x - e^{-x})/2$ is called the *hyperbolic sine* of x and is abbreviated $\sinh x$. The function $(e^x + e^{-x})/2$ is called the *hyperbolic cosine* of x and is abbreviated $\cosh x$. The remaining four hyperbolic functions are then defined in terms of $\sinh x$ and $\cosh x$, in just the same way that the remaining four trigonometric functions can be defined in terms of $\sin x$ and $\cos x$. Precisely, we have

---

DEFINITION 3 (Hyperbolic Functions). The six hyperbolic functions are

$$\sinh x \equiv \frac{e^x - e^{-x}}{2}, \qquad\qquad \cosh x \equiv \frac{e^x + e^{-x}}{2},$$

(53)
$$\tanh x \equiv \frac{\sinh x}{\cosh x} = \frac{e^x - e^{-x}}{e^x + e^{-x}}, \qquad \coth x \equiv \frac{\cosh x}{\sinh x} = \frac{e^x + e^{-x}}{e^x - e^{-x}},$$

$$\operatorname{sech} x \equiv \frac{1}{\cosh x} = \frac{2}{e^x + e^{-x}}, \qquad \operatorname{csch} x \equiv \frac{1}{\sinh x} = \frac{2}{e^x - e^{-x}}.$$

For each identity among the trigonometric functions, there is a corresponding identity among the hyperbolic functions. We will prove a few of these as our first example and reserve the rest for Exercise 6.

Example 1. Prove that for all x,

(54) $\cosh^2 x - \sinh^2 x = 1,$ (55) $\operatorname{sech}^2 x + \tanh^2 x = 1.$

The reader will note that (54) and (55) are similar to $\cos^2 x + \sin^2 x = 1$ and $\sec^2 x - \tan^2 x = 1$, respectively, but there is a change of sign. From (54) we see that the point with coordinates $(\cosh x, \sinh x)$ lies on the hyperbola $u^2 - v^2 = 1$ in the uv-plane while $(\cos x, \sin x)$ lies on the circle $u^2 + v^2 = 1$.

Solution. In proving (54), we will use the fact that $e^x e^{-x} = e^{x-x} = e^0 = 1$. By the definition of the hyperbolic functions, the left side gives

$$\cosh^2 x - \sinh^2 x = \left(\frac{e^x + e^{-x}}{2}\right)^2 - \left(\frac{e^x - e^{-x}}{2}\right)^2$$

$$= \frac{e^{2x} + 2e^0 + e^{-2x}}{4} - \frac{e^{2x} - 2e^0 + e^{-2x}}{4} = \frac{2+2}{4} = 1.$$

We can prove (55) in a similar fashion, but now that (54) has been established, we can use it to give a second and quicker proof. Indeed, if we divide both sides of (54) by $\cosh^2 x$, we have

$$\frac{\cosh^2 x - \sinh^2 x}{\cosh^2 x} = \frac{1}{\cosh^2 x}.$$

Then using the definitions of $\tanh x$ and $\operatorname{sech} x$, this gives

$$1 - \tanh^2 x = \operatorname{sech}^2 x,$$

and this is equivalent to (55). ●

The graphs of the hyperbolic functions are easy to sketch using the values for e^x from Table C at the end of the book, and these are shown in Figs. 4 through 9. But in the graphs the similarity between the trigonometric functions and the hyperbolic functions breaks down. For one thing the hyperbolic functions are not periodic. Further, $\sin x$ and $\cos x$ are bounded functions; that is, for all x we have $|\sin x| \leq 1$ and $|\cos x| \leq 1$. But $\sinh x$ varies from $-\infty$ to $+\infty$ and $\cosh x$ varies between $+1$ and $+\infty$. On the other hand, $|\tanh x| < 1$ and $0 < \operatorname{sech} x \leq 1$, while it is their trigonometric counterpart that is unbounded. We leave it for the reader to prove these fundamental properties of the hyperbolic functions.

Now let us find the derivatives of $\sinh x$ and $\cosh x$. Using the defining equations (53) and the known derivatives $d(e^x)/dx = e^x$ and $d(e^{-x})/dx = -e^{-x}$, we have

$$\frac{d}{dx} \sinh x = \frac{d}{dx} \frac{e^x - e^{-x}}{2} = \frac{e^x + e^{-x}}{2} = \cosh x$$

and

$$\frac{d}{dx} \cosh x = \frac{d}{dx} \frac{e^x + e^{-x}}{2} = \frac{e^x - e^{-x}}{2} = \sinh x.$$

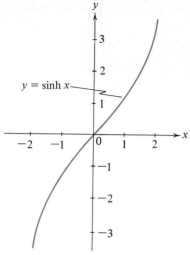

FIGURE 4

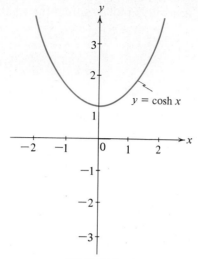

FIGURE 5

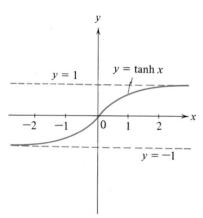

FIGURE 6

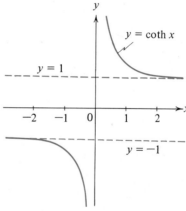

FIGURE 7

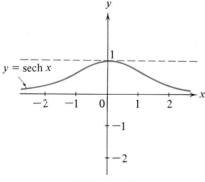

FIGURE 8

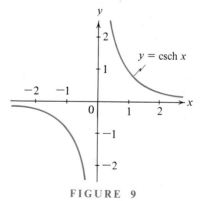

FIGURE 9

More generally, it follows from the chain rule that if u is any differentiable function of x, then

(56)
$$\frac{d}{dx}\sinh u = \cosh u\,\frac{du}{dx}, \qquad \frac{d}{dx}\cosh u = \sinh u\,\frac{du}{dx}.$$

These differentiation formulas lead to the integration formulas

(57)
$$\int \sinh u\,du = \cosh u + C, \qquad \int \cosh u\,du = \sinh u + C.$$

Example 2. Find dy/dx for $y = \sinh^2(3x^3 + 1)$.

Solution. We have

$$\frac{dy}{dx} = 2\sinh(3x^3 + 1)\cdot\frac{d}{dx}\sinh(3x^3 + 1) = 2\sinh(3x^3 + 1)\cosh(3x^3 + 1)\cdot 9x^2$$

$$= 18x^2\sinh(3x^3 + 1)\cosh(3x^3 + 1). \quad\bullet$$

Example 3. Find the following integrals:

$$\textbf{(a)}\ \int x^2\sinh x^3\,dx, \qquad \textbf{(b)}\ \int \cosh^3 x\sinh x\,dx.$$

Solution. For **(a)**, we let $u = x^3$ so that $du = 3x^2\,dx$. Then

$$\int x^2\sinh x^3\,dx = \frac{1}{3}\int \sinh u\,du = \frac{1}{3}\cosh u + C = \frac{1}{3}\cosh x^3 + C.$$

For **(b)**, we have

$$\int \cosh^3 x\sinh x\,dx = \int \cosh^3 x\,d(\cosh x) = \frac{\cosh^4 x}{4} + C. \quad\bullet$$

The formulas for the derivatives of the remaining four hyperbolic functions are also easy to derive and are left as an exercise for the reader:

(58)
$$\frac{d}{dx}\tanh u = \operatorname{sech}^2 u\,\frac{du}{dx}, \qquad \frac{d}{dx}\coth u = -\operatorname{csch}^2 u\,\frac{du}{dx},$$

(59)
$$\frac{d}{dx}\operatorname{sech} u = -\operatorname{sech} u\tanh u\,\frac{du}{dx}, \qquad \frac{d}{dx}\operatorname{csch} u = -\operatorname{csch} u\coth u\,\frac{du}{dx}.$$

As a memory aid, observe that the derivatives of the first three hyperbolic functions (sinh u, cosh u, and tanh u) carry a positive sign, while the derivatives of the last three carry a negative sign. Otherwise, they are identical with the formulas for differentiating the trigonometric functions.

Since the hyperbolic functions are defined in terms of exponential functions, it is reasonable to expect that the *inverse hyperbolic functions* (sinh^{-1} x, cosh^{-1} x, tanh^{-1} x, etc.) can be expressed in terms of the natural logarithm.

Example 4. Find a formula for sinh^{-1} x in terms of the natural logarithm. Also find the derivative of sinh^{-1} x.

Solution. By definition

(60) $$y = \sinh^{-1} x \qquad \text{if and only if} \qquad x = \sinh y.$$

Writing the last equation in the form

$$x = \frac{e^y - e^{-y}}{2},$$

we multiply both sides by $2e^y$ and obtain $2xe^y = e^{2y} - 1$, or

$$(e^y)^2 - 2xe^y - 1 = 0.$$

This can be solved as a quadratic in e^y, giving

$$e^y = \frac{2x \pm \sqrt{4x^2 + 4}}{2} = x \pm \sqrt{x^2 + 1}.$$

Here we must take the plus sign, since e^y is positive. Therefore,

(61) $$y = \sinh^{-1} x = \ln (x + \sqrt{x^2 + 1}).$$

There are two possible methods for finding the derivative of sinh^{-1} x. One is to differentiate the equation $x = \sinh y$ *implicitly* with respect to x; the other is to simply differentiate the formula in (61). We shall illustrate the latter approach:

$$\frac{d}{dx} \sinh^{-1} x = \frac{d}{dx} \ln (x + \sqrt{x^2 + 1}) = \frac{\frac{d}{dx}(x + \sqrt{x^2 + 1})}{x + \sqrt{x^2 + 1}}$$

$$= \frac{1 + \frac{1}{2}(x^2 + 1)^{-1/2} 2x}{x + \sqrt{x^2 + 1}} = \frac{\sqrt{x^2 + 1} + x}{\sqrt{x^2 + 1}(x + \sqrt{x^2 + 1})} = \frac{1}{\sqrt{x^2 + 1}}.$$

Thus

(62)
$$\frac{d}{dx} \sinh^{-1} x = \frac{1}{\sqrt{x^2 + 1}}. \quad \bullet$$

EXERCISE 6

1. Derive the differentiation formulas (58) and (59).
2. Prove each of the following assertions about the hyperbolic functions, and check that the graphs illustrate these assertions.
   - (a) $\cosh x \geqq 1$.
   - (b) $-1 < \tanh x < 1$.
   - (c) $|\coth x| > 1$.
   - (d) $0 < \operatorname{sech} x \leqq 1$.
   - (e) $\lim_{x \to \infty} \tanh x = 1$.
   - (f) $\lim_{x \to \infty} \operatorname{sech} x = 0$.

3. Prove the following assertions.
   - (a) $y = \sinh x$ is an increasing function for all x.
   - (b) $y = \cosh x$ is an increasing function for $x > 0$.
   - (c) $y = \cosh x$ is concave upward for all x.
   - (d) $y = \sinh x$ is concave upward for $x > 0$.
   - (e) $y = \tanh x$ is an increasing function for all x.

In Problems 4 through 10, prove the given identity and state the corresponding trigonometric identity.

4. $\coth^2 x - \operatorname{csch}^2 x = 1$.
5. $\sinh 2x = 2 \sinh x \cosh x$.
6. $\cosh 2x = \cosh^2 x + \sinh^2 x$.
7. $2 \cosh^2 x = \cosh 2x + 1$.
8. $2 \sinh^2 x = \cosh 2x - 1$.
9. $\cosh (x + y) = \cosh x \cosh y + \sinh x \sinh y$.
10. $\sinh (x + y) = \sinh x \cosh y + \cosh x \sinh y$.

In Problems 11 through 16, find the derivative of the given function.

11. $y = \ln (\sinh x^3)$.
12. $y = e^x \cosh x$.
13. $y = \cosh^4 (x^{1/4})$.
14. $y = \sqrt{1 + \sinh^2 x}$.
15. $y = \tanh e^{-x}$.
16. $y = \operatorname{sech} (2x^2 - 1)$.

17. Given that $\sinh x_0 = 4/3$, use the identities involving hyperbolic functions to give the exact values for the other five hyperbolic functions at x_0.
18. Show that $y = \cosh x$ satisfies the differential equation $y'' - y = 0$ and the initial conditions $y(0) = 1$, $y'(0) = 0$.

In Problems 19 through 26, find the given integral.

19. $\displaystyle\int \sinh (5x + 1) \, dx$.

20. $\displaystyle\int \frac{\cosh \sqrt{x}}{\sqrt{x}} \, dx$.

21. $\displaystyle\int \cosh x \sinh^2 x \, dx.$

⋆22. $\displaystyle\int \sinh^3 x \cosh^3 x \, dx.$

23. $\displaystyle\int \tanh x \, dx.$

24. $\displaystyle\int \frac{\sinh x}{1 + \cosh x} \, dx.$

⋆25. $\displaystyle\int \cosh^2 x \, dx.$

26. $\displaystyle\int \coth (5x) \, dx.$

27. Derive formula (62) for the derivative of $y = \sinh^{-1} x$ by using the method of implicit differentiation.

28. Use the results of Example 4 to derive the integration formula

$$\int \frac{du}{\sqrt{u^2 + a^2}} = \ln\left(u + \sqrt{u^2 + a^2}\right) + C.$$

29. The inverse hyperbolic cosine is defined by $y = \cosh^{-1} x$ if and only if $x = \cosh y$ *and $y \geq 0$*. Derive the formula

$$\cosh^{-1} x = \ln\left(x + \sqrt{x^2 - 1}\right), \qquad x \geq 1,$$

and find the derivative of $\cosh^{-1} x$.

30. Use the results of Problem 29 to derive the integration formula

$$\int \frac{du}{\sqrt{u^2 - a^2}} = \ln\left(u + \sqrt{u^2 - a^2}\right) + C.$$

31. The inverse hyperbolic tangent is defined by $y = \tanh^{-1} x$ if and only if $x = \tanh y$. Derive the formula

$$\tanh^{-1} x = \frac{1}{2} \ln \frac{1 + x}{1 - x}, \qquad -1 < x < 1,$$

and find the derivative of $\tanh^{-1} x$.

32. Find the absolute minimum of the function $y = \sqrt{5} \cosh x + \sinh x$.

33. Find the absolute maximum of the function $y = 2 \sinh x - 3 \cosh x$.

34. A hanging cable of uniform density ρ lb/ft and horizontal tension T lb takes the shape of the graph of a solution of the differential equation

$$\frac{d^2 y}{dx^2} = \frac{\rho}{T} \sqrt{1 + \left(\frac{dy}{dx}\right)^2}.$$

This curve is called a *catenary*. Show that $y = a \cosh(x/a) + b$ with $a = T/\rho$ satisfies the differential equation.

CALCULATOR PROBLEMS

In Problems C1 through C4, it is assumed that the student's calculator does not have hyperbolic function keys, but does have exponential and log keys.

C1. Compute to three decimal places:

    (a) $\sinh(2.75)$.   **(b)** $\cosh(-0.5)$.   **(c)** $\sinh^{-1}(1.5)$.

C2. Compute to three decimal places:

    (a) $\tanh(8.3)$.   **(b)** $\operatorname{sech}(2.5)$.   **(c)** $\sinh^{-1}(-3.6)$.

C3. Solve the equation $4 - \sinh(2x^2 - 1) = 0$ for x. Give answer to three decimal places.

C4. Solve the equation $2x - \cosh x = 0$ for x. Give answers to three decimal places.

REVIEW PROBLEMS

In Problems 1 through 4, solve the given equation.

1. $3^{x+6} = 9^{-x}$.

2. $\ln(1 + x) - \ln(2 - x) = 0$.

3. $\ln(x^2 + x) = \ln(15 - x^2)$.

4. $\log_2(x - 3) = 4$.

In Problems 5 through 16, find the derivative of the given function.

5. $\ln[x^2(x^2 - 4)]$.

6. $\ln(\cos^2 x)$.

7. $\ln[x^5/(x - 2)^3]$.

8. $x^2 e^{x^2+2x+3}$.

9. $x2^{x^2+1}$.

10. x^{x^3}.

11. $(2x + 1)^x$.

12. $\sinh x^3$.

13. $\cosh^3(4x - 1)$.

14. $\ln(\ln^3 x)$.

15. $\ln|\sin x|$.

16. $e^{2x}\sinh 4x$.

In Problems 17 through 24, find the given integral.

17. $\displaystyle\int e^{\cos x}\sin x\,dx$.

18. $\displaystyle\int(e^x + 1)^2\,dx$.

19. $\displaystyle\int\frac{(x^3 + 2)\,dx}{x^4 + 8x}$.

20. $\displaystyle\int\frac{(1 + \sec^2 x)\,dx}{5 + x + \tan x}$.

21. $\displaystyle\int 4x\sec x^2\,dx$.

22. $\displaystyle\int e^x\tan e^x\,dx$.

23. $\displaystyle\int\operatorname{sech} 3x\tanh 3x\,dx$.

24. $\displaystyle\int x^2\sinh x^3\,dx$.

25. Find the area of the region bounded by the curves $y = e^x$ and $y = e^{3x}$, between $x = 0$ and $x = 1$.

26. Find the area of the region bounded by $y = \ln x$ and $x = 0$, between $y = -1$ and $y = 2$.

27. The region under the curve $y = 1/\sqrt{x}$ between $x = 1$ and $x = M > 1$ is rotated about the x-axis. Find the volume of the solid generated. What is the limit of the volume as $M \to \infty$?

28. Find the length of the arc of the curve $y = \cosh x$ between $x = -1$ and $x = 1$.

29. The arc in Problem 28 is rotated about the x-axis. Find the area of the surface generated. HINT: $2 \cosh^2 x = \cosh 2x + 1$.

★30. In Chapter 14 we will see that if n is any positive integer, then $x^n \ln x \to 0$ as $x \to 0^+$. Assume this fact, and sketch each of the following curves.
   (a) $y = x \ln x$.   (b) $y = x^2 \ln x$.   (c) $y = x^3 \ln x$.
   In each case find the extreme points on the curve, where the curve is increasing, and the inflection points.

★31. In Chapter 14 we will see that if n is any positive integer, then $x^n e^{-x} \to 0$ as $x \to \infty$. Assume this fact and sketch each of the following curves for $x \geq 0$.
   (a) $y = xe^{-x}$.   (b) $y = x^2 e^{-x}$.   (c) $y = x^3 e^{-x}$.
   In each case find the extreme points on the curve, where the curve is increasing, and the inflection points.

In Problems 32 through 34, prove that for arbitrary values of the constants A and B, the given function satisfies the given differential equation.

32. $y = Ax^2 + \ln x$, $x \dfrac{dy}{dx} = 2y - 2 \ln x + 1$.

33. $y = Ax^3 + e^x$, $x \dfrac{dy}{dx} = 3y + (x - 3)e^x$.

34. $y = A \sinh x + B \cosh x$, $\dfrac{d^2 y}{dx^2} = y$.

35. A radioactive material has a half-life of 1 year. How long will it take for 10 grams of this material to decay to 1 gram?

36. It is found that a certain colony of bacteria triples in size in 1 hour. In how many hours will it be 50 times the original number?

37. Find the solution of the differential equation $y' = (10 - y)/2$ that satisfies $y(0) = 5$.

10

Methods of Integration

1 Objective

We have seen the importance of integration, and we have learned a few tricks for computing indefinite integrals. But at this moment we are not able to find any of the following indefinite integrals:

(1)   **(a)** $\displaystyle\int \frac{x^3}{\sqrt{1+x}}\,dx.$ **(b)** $\displaystyle\int x \sin x \, dx.$

(2)   **(c)** $\displaystyle\int \frac{x^2}{(x^2-1)(x-2)}\,dx.$ **(d)** $\displaystyle\int \sqrt{1+\cos^2 x}\,dx.$

This chapter is devoted to learning a few new tricks that will greatly enlarge the number of functions we can integrate. Of course, we can always search in a table of integrals, a procedure that we will discuss in Section 5, and this is often the simplest and best method. But we should learn some new tricks of integration for several reasons: (i) we may not find the particular integral we need in the table, (ii) using one of our tricks may be quicker, and (iii) we might want to know how the various integrals in the table were obtained.

What tricks can we use on the integrals just given? For (a) we can use algebraic substitution (see Section 3). For (b) we have available integration by parts (see Section 4). The integral in (c) will yield to partial fractions (see Section 8). But try as we may, no trick is sufficient to give (d) in a finite number of terms. By this we mean that if we have available the elementary functions (rational functions, trigonometric functions, inverse trigonometric functions, exponential functions, and logarithmic functions) and permit the use of the standard operations—addition, subtraction, multiplication, division, taking roots, and making composite functions—a finite number of times, we cannot[1] find an antiderivative for $\sqrt{1 + \cos^2 x}$.

We will not stop to prove this fact. Instead, our purpose is to learn how to recognize the integrals that can be expressed in a finite number of terms and to develop methods for computing them.

Summary of Basic Formulas **2**

For reference purposes we summarize the important integration formulas covered so far in this book. Before proceeding, the student should be certain that he has memorized the first 20 formulas [(3) through (22)].

(3)
$$\int af(u)\,du = a \int f(u)\,du.$$

(4)
$$\int (f(u) + g(u))\,du = \int f(u)\,du + \int g(u)\,du.$$

(5)
$$\int u^n\,du = \frac{u^{n+1}}{n+1} + C, \qquad n \neq -1.$$

(6)
$$\int \frac{du}{u} = \ln|u| + C.$$

(7)
$$\int e^u\,du = e^u + C.$$

(8)
$$\int a^u\,du = \frac{a^u}{\ln a} + C, \qquad a > 0, \qquad a \neq 1.$$

(9)
$$\int \sin u\,du = -\cos u + C.$$

(10)
$$\int \cos u\,du = \sin u + C.$$

(11)
$$\int \tan u\,du = -\ln|\cos u| + C.$$

---

[1] This is proved in the little book by J. F. Ritt, *Integration in Finite Terms* (New York: Columbia University Press, 1948).

(12)
$$\int \cot u \, du = \ln |\sin u| + C.$$

(13)
$$\int \sec u \, du = \ln |\sec u + \tan u| + C.$$

(14)
$$\int \csc u \, du = -\ln |\csc u + \cot u| + C.$$

(15)
$$\int \sec^2 u \, du = \tan u + C.$$

(16)
$$\int \csc^2 u \, du = -\cot u + C.$$

(17)
$$\int \sec u \tan u \, du = \sec u + C.$$

(18)
$$\int \csc u \cot u \, du = -\csc u + C.$$

(19)
$$\int \sinh u \, du = \cosh u + C.$$

(20)
$$\int \cosh u \, du = \sinh u + C.$$

(21)
$$\int \frac{du}{a^2 + u^2} = \frac{1}{a} \operatorname{Tan}^{-1} \frac{u}{a} + C.$$

(22)
$$\int \frac{du}{\sqrt{a^2 - u^2}} = \operatorname{Sin}^{-1} \frac{u}{a} + C.$$

(23)
$$\int \frac{du}{\sqrt{u^2 + a^2}} = \ln (u + \sqrt{u^2 + a^2}) + C.$$

(24)
$$\int \frac{du}{\sqrt{u^2 - a^2}} = \ln (u + \sqrt{u^2 - a^2}) + C, \quad u > a > 0.$$

We do not insist that the student memorize (23) and (24) because an alternative method for these two integrals will be presented in Section 7.

Algebraic Substitutions 3

In many problems the integral does not fit directly any of the formulas we have learned so far. In such cases a substitution will frequently help.

Example 1. Find

(25) $$I_1 = \int x\sqrt{2x+1}\,dx.$$

Solution. Clearly, this does not fit any of the standard formulas listed in Section 2. One possible attack is to find some new variable that on substitution will make (25) more attractive. We let $u = \sqrt{2x+1}$. Then we have

(26) $$u^2 = 2x+1, \qquad x = \frac{u^2-1}{2}, \qquad dx = u\,du.$$

If we use these expressions in (25), we find that in terms of the new variable u,

(27) $$\int x\sqrt{2x+1}\,dx = \int \frac{u^2-1}{2}\,u(u\,du) = \frac{1}{2}\int (u^4 - u^2)\,du$$

$$= \frac{1}{2}\left(\frac{u^5}{5} - \frac{u^3}{3}\right) + C = \frac{u^3}{2}\left(\frac{u^2}{5} - \frac{1}{3}\right) + C$$

$$= \frac{u^3}{30}(3u^2 - 5) + C.$$

Returning to the original variable x, via $u = \sqrt{2x+1}$, we find that

$$I_1 = \frac{(2x+1)^{3/2}}{30}[3(2x+1) - 5] + C$$

(28) $$I_1 = \frac{(2x+1)^{3/2}}{30}(6x-2) + C = \frac{(2x+1)^{3/2}}{15}(3x-1) + C. \quad \bullet$$

We remark that the substitution $u = \sqrt{2x+1}$ is not the only substitution that will allow us to compute the integral in (25). Another substitution that works just as well is $u = 2x+1$.

In the case of a definite integral, the work may be simpler if we make a corresponding change in the limits along with the change in the variable. This is illustrated in

Example 2. Evaluate $I_2 = \displaystyle\int_0^3 \frac{x^2+2x}{\sqrt{1+x}}\,dx.$

Solution. Again the radical seems to be a source of trouble, so we try to remove it by the substitution

(29)
$$y = \sqrt{1 + x},$$

or $y^2 = 1 + x$. This substitution requires that $2y\,dy = dx$ and $x = y^2 - 1$.
Hence for the indefinite integral we have

(30)
$$\int \frac{(x^2 + 2x)\,dx}{\sqrt{1 + x}} = \int \frac{(y^2 - 1)^2 + 2(y^2 - 1)}{y} 2y\,dy$$

$$= \int (y^4 - 2y^2 + 1 + 2y^2 - 2)2\,dy$$

$$= 2\int (y^4 - 1)\,dy.$$

Now as x increases from 0 to 3, equation (29) dictates that y increases from $\sqrt{1} = 1$ to $\sqrt{1 + 3} = 2$. Putting these limits on the integrals in (30), we have

$$\int_{x=0}^{x=3} \frac{(x^2 + 2x)\,dx}{\sqrt{1 + x}} = 2\int_{y=1}^{y=2} (y^4 - 1)\,dy = 2\left(\frac{y^5}{5} - y\right)\Big|_1^2$$

$$= 2\left[\frac{2^5}{5} - 2 - \left(\frac{1}{5} - 1\right)\right] = \frac{52}{5}. \quad \bullet$$

Example 3. Find $I_3 = \displaystyle\int x^3(1 + x^2)^{3/2}\,dx.$

Solution. Let $u = 1 + x^2$. Then $du = 2x\,dx$ and $x^2 = u - 1$. Hence

$$I_3 = \int x^3(1 + x^2)^{3/2}\,dx = \int x^2(1 + x^2)^{3/2}(x\,dx)$$

$$= \int (u - 1)u^{3/2}\frac{du}{2} = \frac{1}{2}\int (u^{5/2} - u^{3/2})\,du$$

$$= \frac{1}{2}\left(\frac{2}{7}u^{7/2} - \frac{2}{5}u^{5/2}\right) + C$$

$$= \frac{u^{5/2}}{35}(5u - 7) + C = \frac{(x^2 + 1)^{5/2}(5x^2 - 2)}{35} + C. \quad \bullet$$

Example 4. Find

(31)
$$I_4 = \int \frac{(2x + 3)\,dx}{\sqrt{x}(1 + \sqrt[3]{x})}.$$

Solution. Here the difficulty lies in the terms $x^{1/2}$ and $x^{1/3}$ that appear in the denominator. Since 6 is the least common multiple of 2 and 3, we select the new variable

$$(32) \qquad\qquad\qquad u = x^{1/6}.$$

Then $x^{1/3} = u^2$ and $x^{1/2} = u^3$ and the denominator in (31) will become a polynomial in u. Continuing with this substitution, we have, from (32),

$$du = \frac{1}{6}x^{-5/6}\,dx = \frac{1}{6x^{5/6}}\,dx = \frac{dx}{6u^5},$$

or $dx = 6u^5\,du$. Using this and (32) in (31), we find that

$$(33) \qquad I_4 = \int \frac{(2u^6 + 3)6u^5\,du}{u^3(1 + u^2)} = 6\int \frac{2u^8 + 3u^2}{1 + u^2}\,du.$$

We now have a rational function to integrate. The general rule for a rational function is to divide the numerator by the denominator, obtaining a polynomial for the quotient, and a remainder, which is also a polynomial with degree less than the degree of the denominator. Following this rule, we find that

$$\frac{2u^8 + 3u^2}{1 + u^2} = 2u^6 - 2u^4 + 2u^2 + 1 - \frac{1}{1 + u^2}.$$

Using this in (33) and integrating, we obtain

$$I_4 = 6\int \frac{2u^8 + 3u^2}{1 + u^2}\,du = 6\left(\frac{2u^7}{7} - \frac{2u^5}{5} + \frac{2u^3}{3} + u - \text{Tan}^{-1}u + C\right).$$

Returning to the original variable x by $u = x^{1/6}$, we find that

$$(34) \qquad I_4 = 6\left(\frac{2}{7}x^{7/6} - \frac{2}{5}x^{5/6} + \frac{2}{3}x^{1/2} + x^{1/6} - \text{Tan}^{-1}x^{1/6} + C\right). \quad \bullet$$

The key idea in Example 4 is to find the common denominator for the fractional exponents that occur in the integrand. If we were asked to compute

$$(35) \qquad\qquad\qquad I_5 = \int \frac{x^{1/3}}{x^2 + x^{2/5}}\,dx,$$

then we would let $u = x^{1/15}$ because 15 is the common denominator of $1/3$ and $2/5$.

The proof that the method of substitution gives a correct result is a little bit lengthy, and in the interest of simplicity we omit it. The method of substitution is essentially the inverse of the chain rule for finding the derivative of a composite function.

However, in the case of an indefinite integral, no theoretical justification is needed. We can always differentiate the answer, and if this gives the integrand, then the answer must be correct.

EXERCISE 1

In Problems 1 through 24, find the indicated indefinite integral.

1. $\displaystyle\int x\sqrt{x+1}\,dx.$

2. $\displaystyle\int 2x\sqrt{x-3}\,dx.$

3. $\displaystyle\int x\sqrt{2x-1}\,dx.$

4. $\displaystyle\int \frac{x}{\sqrt{x+2}}\,dx.$

5. $\displaystyle\int \frac{x}{\sqrt{x-3}}\,dx.$

6. $\displaystyle\int x^2\sqrt{2-x}\,dx.$

7. $\displaystyle\int \frac{x^2}{\sqrt{x+1}}\,dx.$

8. $\displaystyle\int (x-3)\sqrt{2x+1}\,dx.$

9. $\displaystyle\int (x+2)\sqrt{1-x}\,dx.$

★10. $\displaystyle\int x\sqrt[3]{1-2x}\,dx.$

★11. $\displaystyle\int x\sqrt[3]{x+1}\,dx.$

★12. $\displaystyle\int x^3\sqrt{x^2+1}\,dx.$

★13. $\displaystyle\int x^3\sqrt{2x^2+1}\,dx.$

14. $\displaystyle\int \frac{x^2-x}{\sqrt{x+1}}\,dx.$

15. $\displaystyle\int \frac{8\,dx}{1+4\sqrt{x}}.$

16. $\displaystyle\int \frac{x}{1+\sqrt{x}}\,dx.$

17. $\displaystyle\int \frac{x^{1/3}}{x^{1/3}+1}\,dx.$

18. $\displaystyle\int \frac{dx}{x^{1/2}(1+x^{1/4})}.$

19. $\displaystyle\int \frac{dx}{x^{1/2}+x^{1/3}}.$

20. $\displaystyle\int \frac{x^{1/2}\,dx}{x+x^{4/5}}.$

21. $\displaystyle\int \frac{5x^2+20x-24}{\sqrt{x+5}}\,dx.$

22. $\displaystyle\int \frac{x^5-8x^3}{\sqrt{x^2-4}}\,dx.$

23. $\displaystyle\int \frac{2x+3}{\sqrt{x-1}}\,dx.$

24. $\displaystyle\int \frac{8x+21\sqrt{2x-5}}{4+\sqrt{2x-5}}\,dx.$

In Problems 25 through 34, compute the definite integral.

25. $\displaystyle\int_0^2 x\sqrt{6x+4}\,dx.$

26. $\displaystyle\int_0^4 x^2\sqrt{2x+1}\,dx.$

27. $\displaystyle\int_0^7 x\sqrt[3]{x+1}\,dx.$

28. $\displaystyle\int_2^5 \frac{x+1}{\sqrt{x-1}}\,dx.$

29. $\displaystyle\int_1^5 \frac{x+3}{\sqrt{2x-1}}\,dx.$

30. $\displaystyle\int_0^1 \frac{x^{3/2}}{1+x}\,dx.$

31. $\displaystyle\int_0^8 \frac{dx}{4+x^{1/3}}.$

32. $\displaystyle\int_6^{32} \frac{(x-5)^{2/3}\,dx}{(x-5)^{2/3}+3}.$

33. $\displaystyle\int_0^1 \frac{t^8\,dt}{\sqrt{1+t^3}}.$

34. $\displaystyle\int_0^1 \frac{\sqrt{x}}{1+\sqrt[4]{x}}\,dx.$

35. Find $\displaystyle\int \frac{e^{2x}}{\sqrt{e^x-1}}\,dx.$ HINT: Let $u=e^x-1.$

36. Find the area of the region under the curve $y=1/(1+\sqrt{x})$ between $x=0$ and $x=M$, where $M>0$.

★37. Find the volume of the solid generated when the region of Problem 36 is rotated about the x-axis.

38. Find the volume of the solid generated when the region of Problem 36 is rotated about the y-axis.

★39. Let V be the volume of the solid generated when the region of Problem 36 is rotated about the line $y = -R$, where $R \geq 0$. Prove that $V = V_0 + 2\pi RA$, where V_0 is the volume of the solid of Problem 37 and A is the area of the region of Problem 36. Find V when $R = 1/2$.

★40. Find the length of arc of the curve $y = \frac{4}{5}x^{5/4}$ from $x = 0$ to $x = 1$.

Integration by Parts   **4**

If u and v are each differentiable functions of x, then by the rule for differentiating a product,

$$(36) \qquad \frac{d}{dx}(uv) = u\frac{dv}{dx} + v\frac{du}{dx} = uv' + vu'.$$

Integrating both sides of this equation, we obtain

$$(37) \qquad uv = \int uv'\, dx + \int vu'\, dx,$$

or on transposing, we have the fundamental formula

$$(38) \qquad \int uv'\, dx = uv - \int vu'\, dx.$$

Since $dv = v'\, dx$ and $du = u'\, dx$, equation (38) can be put in the more compact form

$$(39) \qquad \int u\, dv = uv - \int v\, du.$$

Equation (39) is the basic equation for integration by parts. The idea is that the integrand on the left side may be very complicated but if we select the part for u and the part for dv quite carefully, the integral on the right side may be much easier.

Example 1. Find the indefinite integral $I_1 = \int x \ln x\, dx$.

Solution. Obviously, this fits none of the formulas we have had so far, so we try to use

(39). There are two natural ways of selecting u and dv so that the product is $x \ln x \, dx$.

First possibility: let $u = x$, $dv = \ln x \, dx$.

Second possibility: let $u = \ln x$, $dv = x \, dx$.

With the first selection we cannot find v, the integral of $\ln x$, so we come to a dead end. In the second case we can find v. We arrange the work as follows. Let

$$u = \ln x, \qquad dv = x \, dx.$$

Then
$$du = \frac{1}{x} \, dx, \qquad v = \frac{x^2}{2}.$$

Making these substitutions in (39), we have for the left side

$$\int u \, dv = \int \ln x (x \, dx) = \int x \ln x \, dx,$$

and for the right side

(40) $$uv - \int v \, du = \frac{x^2}{2} \ln x - \int \frac{x^2}{2} \frac{1}{x} \, dx = \frac{x^2}{2} \ln x - \frac{1}{2} \int x \, dx.$$

Therefore, by (39),

(41) $$I_1 = \int x \ln x \, dx = \frac{x^2}{2} \ln x - \frac{1}{4} x^2 + C. \quad \bullet$$

Note that we used $v = x^2/2$ rather than $v = x^2/2 + C$. It is clear that equation (39) is correct for every value of C and in most cases it is simpler to set $C = 0$. The arbitrary constant that we need can be added in the last step, as we have done in going from equation (40) to (41).

Example 2. Find the indefinite integral $I_2 = \int x \cos x \, dx$.

Solution. Let $u = x, \longrightarrow dv = \cos x \, dx.$

Then $du = dx, \longleftarrow v = \sin x.$

The reader may find that the arrows are helpful in carrying out the steps. The top line gives the original integral. The arrow from right to left may serve to remind us of the negative sign in (39). Following this diagram [or equation (39)], we have

$$I_2 = x \sin x - \int \sin x \, dx = x \sin x - (-\cos x) + C$$

$$= x \sin x + \cos x + C. \quad \bullet$$

Interested readers might try setting $u = \cos x$ and $dv = x\, dx$. They will find that "matters get worse" instead of better.

Example 3. Find the indefinite integral $I_3 = \int \operatorname{Tan}^{-1} 3x\, dx$.

Solution. At first it seems as though this integrand cannot be split into two parts, but we can always consider "1" as a factor of any expression.

Let
$$u = \operatorname{Tan}^{-1} 3x, \longrightarrow dv = 1\, dx.$$

Then
$$du = \frac{3\, dx}{1 + 9x^2}, \longleftarrow v = x.$$

Following the arrows, or using equation (39), we have

$$I_3 = \int u\, dv = \int \operatorname{Tan}^{-1} 3x\, dx = x \operatorname{Tan}^{-1} 3x - \int \frac{x\, 3\, dx}{1 + 9x^2}$$

$$= x \operatorname{Tan}^{-1} 3x - \frac{1}{6} \ln (1 + 9x^2) + C. \quad \bullet$$

Sometimes it will be necessary to apply integration by parts more than once in order to evaluate a particular integral.

Example 4. Find the indefinite integral $I_4 = \int x^2 e^x\, dx$.

Solution. Let
$$u = x^2, \longrightarrow dv = e^x\, dx.$$

Then
$$du = 2x\, dx, \longleftarrow v = e^x.$$

Equation (39) gives

(42)
$$I_4 = x^2 e^x - \int e^x (2x)\, dx = x^2 e^x - 2 \int x e^x\, dx.$$

We can apply integration by parts to the last integral in (42).

Let
$$u = x, \longrightarrow dv = e^x\, dx.$$

Then
$$du = dx, \longleftarrow v = e^x.$$

This time equation (39) gives

$$\int x e^x\, dx = x e^x - \int e^x\, dx = x e^x - e^x + C.$$

Using this in (42), we find that

$$I_4 = x^2 e^x - 2[xe^x - e^x + C] = x^2 e^x - 2xe^x + 2e^x - 2C$$

$$I_4 = x^2 e^x - 2xe^x + 2e^x + C_1. \quad \bullet$$

Example 5. Find the indefinite integral

(43) $$I_5 = \int e^{2x} \cos x \, dx.$$

Solution. Here there is no clear reason for selecting either factor as u. Let us select (at random)

$$u = e^{2x}, \xrightarrow{\hspace{2cm}} dv = \cos x \, dx.$$

Then $$du = 2e^{2x} \, dx, \xleftarrow{\hspace{2cm}} v = \sin x.$$

From (39),

(44) $$I_5 = e^{2x} \sin x - 2 \int e^{2x} \sin x \, dx.$$

But this new integral looks just as difficult as the original one. We try integration by parts on this new integral.

Let $$u = e^{2x}, \xrightarrow{\hspace{2cm}} dv = \sin x \, dx.$$

Then $$du = 2e^{2x} \, dx, \xleftarrow{\hspace{2cm}} v = -\cos x.$$

(45) $$\int e^{2x} \sin x \, dx = -e^{2x} \cos x + 2 \int e^{2x} \cos x \, dx$$

and we are back to the integral that we started with. But there is still hope. Using (45) in (44), we find that

$$I_5 = e^{2x} \sin x - 2(-e^{2x} \cos x + 2I_5) + C_1,$$

where I_5 has the meaning of (43). But the coefficients of I_5 on the two sides are different. Transposition gives

$$I_5 + 4I_5 = e^{2x} \sin x + 2e^{2x} \cos x + C_1$$

and dividing by 5 yields

$$I_5 = \frac{e^{2x}}{5}(\sin x + 2 \cos x) + C. \quad \bullet$$

Example 6. Find $I_6 = \int \sec^3 x\, dx$.

Solution. Let $\qquad u = \sec x, \qquad\qquad dv = \sec^2 x\, dx.$

Then $\qquad\qquad du = \sec x \tan x\, dx, \qquad v = \tan x.$

$$\int \sec^3 x\, dx = \sec x \tan x - \int \sec x \tan^2 x\, dx$$

$$= \sec x \tan x - \int \sec x (\sec^2 x - 1)\, dx$$

$$= \sec x \tan x - \int \sec^3 x\, dx + \int \sec x\, dx.$$

Therefore, by transposition,

$$2 \int \sec^3 x\, dx = \sec x \tan x + \int \sec x\, dx,$$

$$I_6 = \int \sec^3 x\, dx = \frac{1}{2}(\sec x \tan x + \ln |\sec x + \tan x|) + C. \quad \bullet$$

We remark that any positive odd power of sec x can be integrated by the method used in Example 6. [See formula (73) in Table D at the end of the book.]

EXERCISE 2

In Problems 1 through 27, find the indicated indefinite integral.

1. $\int x^2 \ln x\, dx.$ 2. $\int x \sin x\, dx.$ 3. $\int \theta \cos 3\theta\, d\theta.$

4. $\int xe^{-4x}\, dx.$ 5. $\int xe^{3x}\, dx.$ 6. $\int x^5 \ln x\, dx.$

7. $\int \frac{\ln x}{x^3}\, dx.$ 8. $\int \ln x\, dx.$ 9. $\int \ln x^2\, dx.$

10. $\int x \sin 5x\, dx.$ 11. $\int \text{Sin}^{-1} 2x\, dx.$ 12. $\int \text{Tan}^{-1} 5x\, dx.$

13. $\int x \ln (x + 3)\, dx.$ 14. $\int \frac{x^2}{e^x}\, dx.$ 15. $\int y^2 e^{5y}\, dy.$

16. $\displaystyle\int x^2 \cos x \, dx.$

17. $\displaystyle\int y^2 \sin 2y \, dy.$

18. $\displaystyle\int x \, \text{Tan}^{-1} x \, dx.$

19. $\displaystyle\int x^2 \, \text{Tan}^{-1} x \, dx.$

$\star$20. $\displaystyle\int e^x \sin 2x \, dx.$

$\star$21. $\displaystyle\int \frac{\sin x}{e^{3x}} \, dx.$

22. $\displaystyle\int \cos x \ln (\sin x) \, dx.$

$\star$23. $\displaystyle\int x^3 \cos 4x \, dx.$

24. $\displaystyle\int (\ln x)^2 \, dx.$

$\star$25. $\displaystyle\int \sec^5 x \, dx.$

26. $\displaystyle\int x \sinh x \, dx.$

27. $\displaystyle\int x \cosh ax \, dx.$

28. The region under the curve $y = e^{-x}$ between $x = 0$ and $x = 2$ is rotated about the y-axis. Find the volume of the solid.

29. The region under the curve $y = \sin x$ between $x = 0$ and $x = \pi$ is rotated about the y-axis. Find the volume of the solid.

5 *Tables of Integrals*

The rich variety of functions makes a systematic arrangement of all functions impossible. However, the functions that one meets in applications are usually not too complicated and for these we can give a reasonable classification, together with references to the applicable formulas in Table D at the end of the book.

Integrand a Form Involving:	Use Formulas:
1. $a + bu$	25–32
2. $\sqrt{a + bu}$	33–40
3. $\sqrt{a^2 - u^2}$	41–47
4. $\sqrt{u^2 \pm a^2}$	48–60
5. $\sqrt{Ax^2 + Bx + C}$	See page 957.
6. Trigonometric functions	61–81
7. Inverse trigonometric functions	82–89
8. Exponential and trigonometric functions	90–95
9. Natural logarithm function	96–101

Example 1. Find $I_1 = \displaystyle\int \frac{5u^2 \, du}{(2 + 3u)^2}$.

Solution. Except for the factor 5, this fits formula 27 in Table D with $a = 2$ and $b = 3$. Therefore,

$$I_1 = \frac{5}{27}\left(2 + 3u - \frac{4}{2 + 3u} - 4 \ln|2 + 3u|\right) + C. \quad \bullet$$

Example 2. Find $I_2 = \displaystyle\int \frac{dx}{(x + 3)^2 \sqrt{x^2 + 6x + 58}}$.

Solution. Since the integrand involves a factor of the form $\sqrt{Ax^2 + Bx + C}$, we use a substitution of the type described on page 957. Here $A = 1$, $B = 6$, and $C = 58$, so that $B^2 - 4AC = 6^2 - 4(58) < 0$. This fits case 2 so we should make the substitution $u = \sqrt{A}x + B/2\sqrt{A} = x + 3$. Then $du = dx$ and

$$x^2 + 6x + 58 = x^2 + 6x + 9 + 49 = (x + 3)^2 + 7^2 = u^2 + 7^2.$$

Hence, by formula 55 of Table D,

$$I_2 = \int \frac{du}{u^2 \sqrt{u^2 + 7^2}} = -\frac{\sqrt{u^2 + 7^2}}{7^2 u} + C = -\frac{\sqrt{(x + 3)^2 + 49}}{49(x + 3)} + C. \quad \bullet$$

Example 3. Find $I_3 = \displaystyle\int \sec^4 5x \, dx$.

Solution. We use formula 73 of Table D with $n = 4$. Here $5x = u$, so $dx = du/5$. Then

$$I_3 = \frac{1}{5}\int \sec^4 u \, du = \frac{1}{5}\left[\frac{\tan u \sec^2 u}{3} + \frac{2}{3}\int \sec^2 u \, du\right]$$

$$= \frac{\tan u \sec^2 u}{15} + \frac{2}{15}\tan u + C = \frac{\tan 5x \sec^2 5x}{15} + \frac{2}{15}\tan 5x + C. \quad \bullet$$

In each of these problems, we can easily check the work. We merely differentiate the answer, and if no errors have been made, then the derivative of the answer will be the integrand. Similarly, we can check any indefinite integral formula that may look suspicious in Table D. If we differentiate the right side, we should obtain the integrand on the left side of the formula. Occasionally, a formula from Table D must be used more than once. This is illustrated in

Example 4. Find $I_4 = \displaystyle\int \sec^5 x \, dx$.

Solution. We do not find this particular integral in Table D, but we do find in formula 73,

(46)
$$\int \sec^n u \, du = \frac{\tan u \sec^{n-2} u}{n - 1} + \frac{n - 2}{n - 1} \int \sec^{n-2} u \, du.$$

An equation such as (46) is called a *reduction formula*. It expresses an integral involving the integer n in terms of an integral, where n has been decreased. (In this case n is replaced by $n - 2$.) If we use (46) with $n = 5$ and $u = x$, we find that

$$I_4 = \int \sec^5 x \, dx = \frac{\tan x \sec^3 x}{4} + \frac{3}{4} \int \sec^3 x \, dx.$$

We can now use (46) again to find the integral[1] of $\sec^3 x$. From (46) with $n = 3$ and $u = x$,

$$\int \sec^3 x \, dx = \frac{\tan x \sec x}{2} + \frac{1}{2} \int \sec x \, dx$$

$$= \frac{\tan x \sec x}{2} + \frac{1}{2} \ln |\sec x + \tan x| + C.$$

If we combine this with our earlier expression for I_4, we obtain

$$\int \sec^5 x \, dx = \frac{1}{4} \tan x \sec^3 x + \frac{3}{8} \tan x \sec x + \frac{3}{8} \ln |\sec x + \tan x| + C. \quad \bullet$$

EXERCISE 3

In Problems 1 through 24, use the proper formula from Table D (or any other table) to find the indefinite integral. In some cases a preliminary substitution may be helpful.

1. $\displaystyle\int \frac{6 \, dx}{(2 - x)x^2}.$

2. $\displaystyle\int \frac{5e^x \, dx}{\sqrt{e^{2x} - 3}}.$

3. $\displaystyle\int 20 \, \mathrm{Tan}^{-1} (2x + 1) \, dx.$

4. $\displaystyle\int \cos (\ln x) \, dx.$

5. $\displaystyle\int \frac{dt}{t^3 \sqrt{t^2 - 4}}.$

6. $\displaystyle\int \frac{ds}{(9 - s^2)^{3/2}}.$

7. $\displaystyle\int \frac{8 \, dz}{z(4 + 13z)}.$

8. $\displaystyle\int \frac{10 \, dy}{y \sqrt{4 + 13y}}.$

9. $\displaystyle\int \frac{12x^2 \, dx}{\sqrt{5 - x^2}}.$

10. $\displaystyle\int \frac{\sqrt{x^2 + 2x + 3}}{(x + 1)^2} \, dx.$

11. $\displaystyle\int \sin^2 4x \, dx.$

12. $\displaystyle\int x \cos^2 (4x^2) \, dx.$

13. $\displaystyle\int \cos 3v \cos 5v \, dv.$

14. $\displaystyle\int \cos 2\theta \sin 8\theta \, d\theta.$

15. $\displaystyle\int \frac{dx}{(6x - x^2 - 5)^{3/2}}.$

[1] This antiderivative can also be obtained from formula 67 of Table D. Also see Example 6, page 407.

16. $\displaystyle\int z^2 \cos z \, dz.$

17. $\displaystyle\int x^3 \cos x \, dx.$

18. $\displaystyle\int 16x \operatorname{Sin}^{-1} 2x \, dx.$

19. $\displaystyle\int 52 e^{2x} \cos 3x \, dx.$

20. $\displaystyle\int \frac{u^2 \, du}{(1 + 3u)^5}.$

21. $\displaystyle\int \frac{\sqrt{x^2 + 4x - 6}}{x + 2} \, dx.$

22. $\displaystyle\int \tan^3 (x - 2) \, dx.$

23. $\displaystyle\int \frac{x \, dx}{\sqrt{2x - 3}}.$

24. $\displaystyle\int \frac{\cos x \, dx}{(\sin x)\sqrt{2 \sin x + 5}}.$

In Problems 25 through 30, use a suitable reduction formula from Table D to find the integral.

25. $\displaystyle\int \sin^5 x \, dx.$

26. $\displaystyle\int x^3 e^x \, dx.$

27. $\displaystyle\int \tan^6 x \, dx.$

28. $\displaystyle\int \tan^4 3x \, dx.$

29. $\displaystyle\int x^3 e^{5x} \, dx.$

30. $\displaystyle\int x^5 (\ln x)^2 \, dx.$

Trigonometric Integrals   *6*

We already know how to integrate each of the trigonometric functions [see Section 2, formulas (9) through (14)]. Here we consider integrals involving powers of these functions.

THEOREM 1 (PWO). If either m or n is a positive odd integer, then

(47) $$\int \sin^m x \cos^n x \, dx$$

can be integrated in finite terms.

We can omit the formal proof of this theorem because a simple example will suffice to show the general method.

Example 1. Find $I_1 = \displaystyle\int \sin^{2/5} x \cos^3 x \, dx.$

Solution. Here $n = 3$ and is odd. We recall that $\cos^2 x = 1 - \sin^2 x$. Then we can write $\cos^3 x = \cos^2 x \cos x = (1 - \sin^2 x) \cos x$, and hence

$$I_1 = \int \sin^{2/5} x (1 - \sin^2 x) \cos x \, dx$$

$$= \int \sin^{2/5} x \cos x \, dx - \int \sin^{12/5} x \cos x \, dx$$

$$= \frac{5}{7} \sin^{7/5} x - \frac{5}{17} \sin^{17/5} x + C. \quad \bullet$$

Suppose that in (47) m is odd. Indeed, let $m = 2p + 1$, where p is a positive integer. Then we use the identity

$$\sin^m x = \sin^{2p+1} x = \sin^{2p} x \sin x = (1 - \cos^2 x)^p \sin x$$

and proceed just as in Example 1, obtaining a sum of terms of the form

$$\int \cos^q x \sin x \, dx = - \int u^q \, du = - \frac{\cos^{q+1} x}{q + 1} + C.$$

THEOREM 2 (PWO). If m and n are both even integers, then (47) can be integrated in finite terms.

All that is needed is to alter the powers by a suitable trick so that either m or n or both become odd. This is done using the two trigonometric identities

(48)
$$\sin^2 \theta = \frac{1 - \cos 2\theta}{2}, \qquad \cos^2 \theta = \frac{1 + \cos 2\theta}{2}.$$

Example 2. Find $I_2 = \displaystyle\int \sin^2 3x \cos^2 3x \, dx$.

Solution. We apply (48) with $\theta = 3x$ and have

$$I_2 = \int \frac{1 - \cos 6x}{2} \cdot \frac{1 + \cos 6x}{2} \, dx = \frac{1}{4} \int (1 - \cos^2 6x) \, dx.$$

Applying (48) again, this time with $\theta = 6x$, we have

$$I_2 = \frac{1}{4} \int \left(1 - \frac{1 + \cos 12x}{2} \right) dx$$

$$= \frac{1}{8} \int (1 - \cos 12x) \, dx = \frac{1}{8} x - \frac{1}{96} \sin 12x + C. \quad \bullet$$

This same example could also be worked using the identity

(49)
$$\sin \theta \cos \theta = \frac{1}{2} \sin 2\theta.$$

With this identity we find that

$$I_2 \equiv \int \sin^2 3x \cos^2 3x \, dx = \int \left(\frac{1}{2}\sin 6x\right)^2 dx$$

$$= \frac{1}{4}\int \sin^2 6x \, dx = \frac{1}{4}\int \frac{1 - \cos 12x}{2}\, dx = \frac{1}{8}x - \frac{1}{96}\sin 12x + C.$$

A number of similar theorems can be proved about the integrals of $\tan^m \theta \sec^n \theta$ and $\cot^m \theta \csc^n \theta$. We are content with the following one, selected as representative.

THEOREM 3 (PWO). If m is a positive odd integer or n is a positive even integer, then

(50) $$\int \tan^m x \sec^n x \, dx$$

can be integrated in finite terms.

A simple example covering each case will suffice to show the general method, which depends on the identity $\sec^2 x = 1 + \tan^2 x$.

Example 3. Find

(51)   **(a)** $\displaystyle\int \tan^4 x \sec^6 x \, dx$ and   **(b)** $\displaystyle\int \tan^3 x \sqrt{\sec x}\, dx.$

Solution. **(a)** Here n is even. We note that $d(\tan x) = \sec^2 x \, dx$. Hence for **(a)** we have

$$\int \tan^4 x \sec^6 x \, dx = \int \tan^4 x \sec^4 x \sec^2 x \, dx$$

$$= \int \tan^4 x (\tan^2 x + 1)^2 \sec^2 x \, dx$$

$$= \int (\tan^8 x + 2\tan^6 x + \tan^4 x)\, d(\tan x)$$

(52) $$\int \tan^4 x \sec^6 x \, dx = \frac{1}{9}\tan^9 x + \frac{2}{7}\tan^7 x + \frac{1}{5}\tan^5 x + C.$$

(b) Here m is odd. We use the formula $d(\sec x) = \sec x \tan x \, dx$. Then

$$\int \tan^3 x \sqrt{\sec x} \, dx = \int \tan x (\sec^2 x - 1) \sqrt{\sec x} \, dx$$

$$= \int (\sec^2 x - 1) \frac{1}{\sec^{1/2} x} \sec x \tan x \, dx$$

$$= \int (\sec^{3/2} x - \sec^{-1/2} x) \, d(\sec x)$$

(53) $$\int \tan^3 x \sqrt{\sec x} \, dx = \frac{2}{5} \sec^{5/2} x - 2 \sec^{1/2} x + C. \quad \bullet$$

We summarize Theorems 1, 2, and 3 in the following table.

Theorem Number	Integrand Involves	Condition	Use the Identity
1	$\sin^m x \cos^n x$	m odd positive	$\sin^2 x = 1 - \cos^2 x$
1	$\sin^m x \cos^n x$	n odd positive	$\cos^2 x = 1 - \sin^2 x$
2	$\sin^m x \cos^n x$	m, n even positive	$\sin^2 x = \dfrac{1 - \cos 2x}{2}$
			$\cos^2 x = \dfrac{1 + \cos 2x}{2}$
3	$\tan^m x \sec^n x$	m odd positive	$\tan^2 x = \sec^2 x - 1$
3	$\tan^m x \sec^n x$	n even positive	$\sec^2 x = \tan^2 x + 1$

A perceptive student will observe that in Theorems 1, 2, and 3 the same argument "x" occurs in both terms of the product. Suppose that this does not occur. Then it may be helpful to recall the trigonometric identities

(54) $$\sin Ax \cos Bx = \frac{1}{2}[\sin (A + B)x + \sin (A - B)x],$$

(55) $$\cos Ax \cos Bx = \frac{1}{2}[\cos (A + B)x + \cos (A - B)x],$$

(56) $$\sin Ax \sin Bx = \frac{1}{2}[\cos (A - B)x - \cos (A + B)x].$$

Example 4. Find $I_4 = \displaystyle\int \sin 7x \sin 3x \, dx$.

Solution. Using (56), we can write

$$I_4 = \int \frac{1}{2}\Big[\cos (7-3)x - \cos (7+3)x\Big] dx = \frac{1}{8}\sin 4x - \frac{1}{20}\sin 10x + C. \quad \bullet$$

EXERCISE 4

In Problems 1 through 18, find the indicated indefinite integral.

1. $\displaystyle\int \sin^3 x \, dx.$

2. $\displaystyle\int \sin^5 \theta \, d\theta.$

3. $\displaystyle\int \sin^2 y \cos^3 y \, dy.$

4. $\displaystyle\int \sin^4 (2x) \, dx.$

5. $\displaystyle\int \sin^2 x \cos^4 x \, dx.$

6. $\displaystyle\int \tan^2 5x \cos^4 5x \, dx.$

7. $\displaystyle\int \frac{\cos^3 x \, dx}{(\sin x)^{3/2}}.$

8. $\displaystyle\int \tan^5 x \cos x \, dx.$

9. $\displaystyle\int \cot^2 3x \csc^4 3x \, dx.$

10. $\displaystyle\int \tan^5 x \sec^3 x \, dx.$

11. $\displaystyle\int \sin 3x \cos 2x \, dx.$

*12. $\displaystyle\int \sin^2 3x \sin^2 5x \, dx.$

13. $\displaystyle\int \tan^4 (x - 5) \, dx.$

14. $\displaystyle\int \frac{dx}{\sin^2 6x}.$

15. $\displaystyle\int (\sec 5x + \csc 5x)^2 \, dx.$

16. $\displaystyle\int \sec^4 \theta \, d\theta.$

17. $\displaystyle\int \cot^3 x \, dx.$

18. $\displaystyle\int \sqrt[3]{\cos \theta} \, \sin^3 \theta \, d\theta.$

In Problems 19 through 22, the region below the given curve between the given values for x is rotated about the x-axis. Find the volume of the solid generated.

19. $y = \sin x, \quad x = 0$ and $x = \pi$.

20. $y = \sec x, \quad x = 0 \quad$ and $x = \pi/4$.

21. $y = \tan 2x, \quad x = 0$ and $x = \pi/8$.

22. $y = \cos^2 x, \quad x = \pi/2$ and $x = \pi$.

23. Let $m \geq 0$ and $n \geq 0$ be any two distinct integers. Prove that

$$\int_0^{2\pi} \sin mx \sin nx \, dx = 0 \quad \text{and} \quad \int_0^{2\pi} \cos mx \cos nx \, dx = 0.$$

24. Evaluate the integrals in Problem 23, when $m = n$ an integer.

25. Prove that for any two integers m and n,

$$\int_0^{2\pi} \sin mx \cos nx \, dx = 0.$$

*26. Show that the formal substitution $1 - \cos 2\theta = 2 \sin^2 \theta$ gives

$$\int_0^{2\pi} (1 - \cos 2\theta)^{3/2}\, d\theta = \int_0^{2\pi} 2\sqrt{2} \sin^3 \theta\, d\theta = 0.$$

But the first integral is not zero. Where lies the trouble?

27. Evaluate the first integral in Problem 26.

7 Trigonometric Substitutions

The integration of certain algebraic expressions is simplified by introducing suitable trigonometric functions. If the integrand[1] involves:

$$\text{(I)} \quad \sqrt{a^2 - u^2}, \qquad \text{set } u = a \sin \theta,$$

$$\text{(II)} \quad \sqrt{a^2 + u^2}, \qquad \text{set } u = a \tan \theta,$$

$$\text{(III)} \quad \sqrt{u^2 - a^2}, \qquad \text{set } u = a \sec \theta.$$

These substitutions are easy to remember if we associate with each of the cases a right triangle whose sides are a, u, and a suitable radical. The labeling of the triangle depends on the particular case at hand. For example, in (II) it is clear that the hypotenuse must be $\sqrt{a^2 + u^2}$, while in (I) the hypotenuse must be a, and in (III) the hypotenuse is u. The three triangles and the related quantities are shown in Figs. 1, 2, and 3.

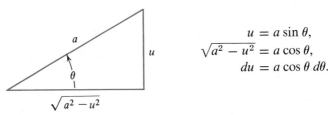

$$u = a \sin \theta,$$
$$\sqrt{a^2 - u^2} = a \cos \theta,$$
$$du = a \cos \theta\, d\theta.$$

FIGURE 1

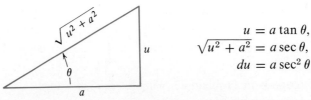

$$u = a \tan \theta,$$
$$\sqrt{u^2 + a^2} = a \sec \theta,$$
$$du = a \sec^2 \theta\, d\theta.$$

FIGURE 2

[1] The radical sign may not be visible. For example, if the integrand is $1/(a^2 + u^2)^5$, this comes under case (II), because $1/(a^2 + u^2)^5 = 1/(\sqrt{a^2 + u^2})^{10}$.

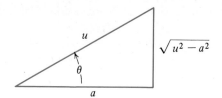

$$u = a \sec \theta,$$
$$\sqrt{u^2 - a^2} = a \tan \theta,$$
$$du = a \sec \theta \tan \theta \, d\theta.$$

FIGURE 3

Example 1. Find the indefinite integral $I_1 = \displaystyle\int \frac{x^2 \, dx}{\sqrt{9 - x^2}}$.

Solution. This integral is of type (I) with $a = 3$ and $u = x$. We first construct a triangle as shown in Fig. 4.

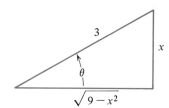

$$x = 3 \sin \theta,$$
$$\sqrt{9 - x^2} = 3 \cos \theta,$$
$$dx = 3 \cos \theta \, d\theta.$$

FIGURE 4

From this triangle we see immediately that

$$I_1 = \int \frac{x^2 \, dx}{\sqrt{9 - x^2}} = \int \frac{(3 \sin \theta)^2 \, 3 \cos \theta \, d\theta}{3 \cos \theta} = 9 \int \sin^2 \theta \, d\theta,$$

(57)
$$I_1 = \frac{9}{2} \int (1 - \cos 2\theta) \, d\theta = \frac{9}{2}\theta - \frac{9}{4} \sin 2\theta + C$$

$$= \frac{9}{2}\theta - \frac{9}{2} \sin \theta \cos \theta + C.$$

Returning to the original variable x, we can obtain the expressions for θ, $\sin \theta$, and $\cos \theta$ directly from the triangle in Fig. 4. We find that

$$\theta = \mathrm{Sin}^{-1} \frac{x}{3}, \qquad \sin \theta = \frac{x}{3}, \qquad \cos \theta = \frac{\sqrt{9 - x^2}}{3}.$$

Hence

$$I_1 = \frac{9}{2} \mathrm{Sin}^{-1} \frac{x}{3} - \frac{9}{2} \frac{x}{3} \frac{\sqrt{9 - x^2}}{3} + C,$$

(58)
$$I_1 = \frac{9}{2} \mathrm{Sin}^{-1} \frac{x}{3} - \frac{1}{2} x \sqrt{9 - x^2} + C. \quad \bullet$$

As a check, the student is advised to differentiate I_1 [given by equation (58)] and show that the derivative is indeed $x^2/\sqrt{9 - x^2}$.

Example 2. Compute $I_2 = \int_{-1.5}^{1.5} \dfrac{x^2\, dx}{\sqrt{9 - x^2}}$.

Solution. The integrand here is the same as in Example 1, so we could just substitute the proper limits in equation (58). Our purpose, however, is to illustrate the method of changing the limits along with the variable. Referring to Fig. 4, we see that when $x = 1.5$, then $\theta = 30°$ or $\theta = \pi/6$. When $x = -1.5$, the relation $\theta = \operatorname{Sin}^{-1}(x/3)$ requires θ to be in the fourth quadrant, or $\theta = -\pi/6$. Thus as x varies continuously from -1.5 to $+1.5$, the angle θ also varies continuously from $-\pi/6$ to $\pi/6$. Whence from equation (57), we have

$$\int_{-1.5}^{1.5} \frac{x^2\, dx}{\sqrt{9 - x^2}} = \left(\frac{9}{2}\theta - \frac{9}{4}\sin 2\theta\right)\Bigg|_{\theta = -\pi/6}^{\theta = \pi/6}$$

$$= \frac{9}{2}\frac{\pi}{6} - \frac{9}{4}\frac{\sqrt{3}}{2} - \left[\frac{9}{2}\left(-\frac{\pi}{6}\right) - \frac{9}{4}\left(-\frac{\sqrt{3}}{2}\right)\right]$$

$$= \frac{3}{2}\pi - \frac{9}{4}\sqrt{3}. \quad \bullet$$

Example 3. Find $I_3 = \displaystyle\int \frac{dx}{(x^2 + 4)^{3/2}}$.

Solution. The integrand is of type (II), so we construct the triangle shown in Figure 5.

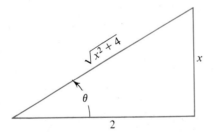

$$x = 2\tan\theta,$$
$$\sqrt{x^2 + 4} = 2\sec\theta,$$
$$dx = 2\sec^2\theta\, d\theta.$$

FIGURE 5

From this triangle it is easy to see that

$$I_3 = \int \frac{dx}{(x^2 + 4)^{3/2}} = \int \frac{2\sec^2\theta\, d\theta}{(2\sec\theta)^3} = \frac{1}{4}\int \frac{d\theta}{\sec\theta}$$

$$= \frac{1}{4}\int \cos\theta\, d\theta = \frac{1}{4}\sin\theta + C = \frac{1}{4}\frac{x}{\sqrt{x^2 + 4}} + C. \quad \bullet$$

Example 4. Find $I_4 = \int \dfrac{\sqrt{9x^2 - 1}}{x}\, dx$.

Solution. If we let $u = 3x$, this integral is of type (III). We construct the triangle shown in Fig. 6.

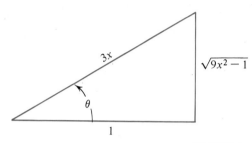

$$3x = \sec\theta,$$
$$\sqrt{9x^2 - 1} = \tan\theta,$$
$$3dx = \sec\theta\tan\theta\, d\theta.$$

FIGURE 6

From this triangle, it is easy to see that

$$I_4 = \int \frac{\tan\theta\, \tfrac{1}{3}\sec\theta\tan\theta\, d\theta}{\tfrac{1}{3}\sec\theta} = \int \tan^2\theta\, d\theta$$

$$= \int (\sec^2\theta - 1)\, d\theta = \tan\theta - \theta + C = \sqrt{9x^2 - 1} - \mathrm{Cos}^{-1}\frac{1}{3x} + C. \quad \bullet$$

If the integrand contains $\sqrt{Ax^2 + Bx + C}$, then one must complete the square before making the appropriate trigonometric substitution. This is illustrated in

Example 5. Find $I_5 = \int (4x^2 + 8x + 13)^{-1/2}\, dx$.

Solution. To complete the square, we write

$$4x^2 + 8x + 13 = 4(x^2 + 2x) + 13 = 4(x^2 + 2x + 1) - 4 + 13$$
$$= 4(x + 1)^2 + 9 = 4(x + 1)^2 + 3^2 = (2x + 2)^2 + 3^2.$$

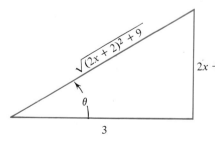

$$2x + 2 = 3\tan\theta,$$
$$\sqrt{4x^2 + 8x + 13} = 3\sec\theta,$$
$$2dx = 3\sec^2\theta\, d\theta.$$

FIGURE 7

From the triangle shown in Fig. 7, we find that

$$I_5 = \int (3 \sec \theta)^{-1} \left(\frac{3}{2} \sec^2 \theta \, d\theta \right) = \frac{1}{2} \int \sec \theta \, d\theta$$

$$= \frac{1}{2} \ln |\sec \theta + \tan \theta| + C$$

$$= \frac{1}{2} \ln \left(\frac{\sqrt{4x^2 + 8x + 13}}{3} + \frac{2x + 2}{3} \right) + C$$

$$= \frac{1}{2} \ln \left(\sqrt{4x^2 + 8x + 13} + 2x + 2 \right) + C_1. \quad \bullet$$

EXERCISE 5

In Problems 1 through 15, find the indicated integral.

1. $\displaystyle \int \frac{x^3 \, dx}{\sqrt{x^2 + 9}}$.

2. $\displaystyle \int \frac{dx}{(4x^2 - 1)^{3/2}}$.

3. $\displaystyle \int \sqrt{2 - y^2} \, dy$.

4. $\displaystyle \int \frac{dx}{(9 - x^2)^{3/2}}$.

5. $\displaystyle \int \frac{dy}{y^2 \sqrt{y^2 - 6}}$.

6. $\displaystyle \int \frac{dy}{(25 + y^2)^2}$.

7. $\displaystyle \int \frac{\sqrt{9 - x^2}}{x} \, dx$.

8. $\displaystyle \int \frac{dx}{(x^2 + 5)^{3/2}}$.

9. $\displaystyle \int \frac{\sqrt{4 - y^2}}{y^2} \, dy$.

10. $\displaystyle \int \frac{du}{u^4 \sqrt{a^2 - u^2}}$.

11. $\displaystyle \int \frac{du}{u^4 \sqrt{u^2 - a^2}}$.

12. $\displaystyle \int \frac{\sqrt{x^2 + 2x - 3}}{x + 1} \, dx$.

★13. $\displaystyle \int \frac{(3x + 7) \, dx}{\sqrt{x^2 + 4x + 5}}$.

★14. $\displaystyle \int \frac{\sqrt{10x - 21 - x^2}}{x - 5} \, dx$.

★15. $\displaystyle \int \sqrt{x^2 + 4} \, dx$.

16. Use the methods of this section to derive formulas (23) and (24) (Section 2, page 398).
17. Find the length of the curve $y = \ln 3x$ between $x = 1$ and $x = \sqrt{8}$.
18. Prove by integration that the area of a circle of radius r is πr^2.
19. In a circle of radius r, a chord b units from the center divides the circle into two parts. Use calculus to find a formula for the area of the smaller part. Observe that if we know the area of the full circle (Problem 18), then this problem can be solved without integration.
20. The region under the curve $y = x^{3/2}/\sqrt{x^2 + 4}$ between $x = 0$ and $x = 4$ is rotated about the x-axis. Find the volume.
21. The region under the curve $y = (4 - x^2)^{1/4}$ between $x = -1$ and $x = 2$ is rotated about the x-axis. Find the volume.

*22. If a circle of radius r is rotated about an axis R units from the center of the circle (where $R > r$), the solid generated is called a *torus,* or anchor ring, and resembles a smooth doughnut. Find the volume of the torus.

Partial Fractions, Linear Factors **8**

Given the choice of integrating the rational function

(59)
$$\frac{7x^2 + 10x - 1}{x^3 + 3x^2 - x - 3}$$

or integrating the sum

(60)
$$\frac{2}{x - 1} + \frac{1}{x + 1} + \frac{4}{x + 3},$$

which would you select? No doubt (60) is the easier one. Actually, the two functions given in (59) and (60) are identical—and this can be checked by combining the simple fractions in (60) to obtain (59). Thus, if we are faced with the problem of integrating a complicated-looking rational function such as (59), we want to be able to write it as a sum of simpler rational functions such as those in (60). The method of proceeding from (59) to (60) is the *method of partial fractions.*

THEOREM 4 (PWO). If $R(x)$ is any rational function, then

$$\int R(x)\, dx$$

can be integrated in finite terms.

We will not actually prove this theorem. Instead, we will outline the method and present a series of examples.

Let $R(x) = N(x)/D(x)$, where $N(x)$ and $D(x)$ are polynomials. If the degree of $N(x)$ is greater than or equal to the degree of $D(x)$, then division gives

(61)
$$\frac{N(x)}{D(x)} = P(x) + \frac{Q(x)}{D(x)},$$

where $P(x)$ and $Q(x)$ are polynomials and $Q(x)$ has degree less than that of $D(x)$.

Example 1. Write each of the rational functions

$$R_1(x) = \frac{x^2}{x^2 - 4} \quad \text{and} \quad R_2(x) = \frac{x^4 - 7x^2 - 5x + 5}{x^2 + 3x + 2}$$

in the form (61).

Solution. Long division (see any algebra text) gives

$$\frac{x^2}{x^2 - 4} = 1 + \frac{4}{x^2 - 4}, \qquad \frac{x^4 - 7x^2 - 5x + 5}{x^2 + 3x + 2} = x^2 - 3x + \frac{x + 5}{x^2 + 3x + 2}.$$

In each case the right side has the form $P(x) + Q(x)/D(x)$ and the degree of $Q(x)$ is less than the degree of $D(x)$. ●

Since the integration of the polynomial $P(x)$ in (61) is a trivial matter, we now concentrate on the integration of the remainder term $Q(x)/D(x)$.

According to a theorem first proved by Gauss, any polynomial such as $D(x)$ with real coefficients can be factored into a product of linear and quadratic factors with real coefficients. There are four cases to consider for the factors of $D(x)$:

(I) Nonrepeated linear factors.

(II) Repeated linear factors.

(III) Nonrepeated quadratic factors.

(IV) Repeated quadratic factors.

In this section we discuss cases (I) and (II). The remaining cases will be treated in the next section.

Example 2. Find the indefinite integral $I_2 = \displaystyle\int \frac{(2x + 41)\, dx}{x^2 + 5x - 14}$.

Solution. We can find the factors of the denominator by inspection. Indeed,

$$x^2 + 5x - 14 = (x - 2)(x + 7).$$

We now ask, are there numbers A and B such that

(62) $$\frac{2x + 41}{x^2 + 5x - 14} = \frac{A}{x - 2} + \frac{B}{x + 7}$$

is an identity in x? If we can find these unknowns A and B, then the right side of equation (62) is called the *partial fraction decomposition* of the left side of (62). That such numbers can always be found is the statement of a rather complicated algebraic theorem. We now give two methods for finding A and B in (62).

FIRST METHOD. Assuming that numbers A and B exist so that (62) is an identity, multiply both sides of (62) by $(x - 2)(x + 7)$. This gives us

(63) $$2x + 41 = A(x + 7) + B(x - 2).$$

Put $x = -7$ in (63). This gives us

$$-14 + 41 = A(0) + B(-9)$$

and hence $B = 27/(-9) = -3$. Put $x = 2$ in (63). Then

$$4 + 41 = A(9) + B(0)$$

and hence $A = 45/9 = 5$.

SECOND METHOD. We first obtain equation (63), just as before, and then rearrange the right side, grouping together the constants, the terms in x, and so on. Thus (63) is equivalent to

(64) $$2x + 41 = (A + B)x + 7A - 2B.$$

In order that (64) be true for all x, the corresponding coefficients must be the same

Equating coefficients of x yields $\quad 2 = A + B.$

Equating coefficients of 1 yields $\quad 41 = 7A - 2B.$

Thus we have a system of two equations in the two unknowns A and B. We leave it to the reader to solve this system and show that $A = 5$ and $B = -3$.

Now returning to equation (62) and using these values for A and B, we have

$$\frac{2x + 41}{x^2 + 5x - 14} = \frac{5}{x - 2} - \frac{3}{x + 7},$$

and hence

$$I_2 = \int \frac{5\,dx}{x - 2} - \int \frac{3\,dx}{x + 7} = 5 \ln |x - 2| - 3 \ln |x + 7| + C$$

$$= \ln \left| \frac{(x - 2)^5}{(x + 7)^3} \right| + C. \quad \bullet$$

We may have more than two factors in the denominator. The method is still the same, but the algebraic manipulations become more complicated.

Example 3. Find the indefinite integral $I_3 = \displaystyle\int \frac{(3x^2 + 11x + 4)\,dx}{x^3 + 4x^2 + x - 6}$.

Solution. The denominator factors into $(x - 1)(x + 2)(x + 3)$. Therefore, we search for three numbers A, B, and C such that

(65) $$\frac{3x^2 + 11x + 4}{x^3 + 4x^2 + x - 6} = \frac{A}{x - 1} + \frac{B}{x + 2} + \frac{C}{x + 3}$$

is an identity in x. Multiplying both sides of (65) by the common denominator $(x - 1)(x + 2)(x + 3)$, we have

(66) $3x^2 + 11x + 4 = A(x + 2)(x + 3) + B(x - 1)(x + 3) + C(x - 1)(x + 2).$

FIRST METHOD. Putting $x = 1$ in (66) gives

$$3 + 11 + 4 = A(3)(4),$$

and hence $A = 18/12 = 3/2$. Putting $x = -2$ in (66) gives

$$12 - 22 + 4 = B(-3)(1),$$

and hence $B = -6/(-3) = 2$. Putting $x = -3$ in (66) gives

$$27 - 33 + 4 = C(-4)(-1),$$

and hence $C = -2/4 = -1/2$.

SECOND METHOD. We regroup the terms on the right side of (66), in descending powers of x.

$$3x^2 + 11x + 4 = A(x^2 + 5x + 6) + B(x^2 + 2x - 3) + C(x^2 + x - 2)$$
$$= (A + B + C)x^2 + (5A + 2B + C)x + (6A - 3B - 2C).$$

This is an identity in x if and only if the corresponding coefficients are equal.

Equating coefficients of x^2 yields $A + B + C = 3.$

Equating coefficients of x^1 yields $5A + 2B + C = 11.$

Equating coefficients of x^0 yields $6A - 3B - 2C = 4.$

Solving these three simultaneous equations by any of the standard methods gives $A = 3/2$, $B = 2$, and $C = -1/2$. Naturally, these values for A, B, and C are the same as those obtained by the first method. To compute I_3, we use (65) with the known values for A, B, and C and have

$$I_3 = \frac{3}{2} \int \frac{dx}{x - 1} + 2 \int \frac{dx}{x + 2} - \frac{1}{2} \int \frac{dx}{x + 3}.$$

$$= \frac{3}{2} \ln|x - 1| + 2 \ln|x + 2| - \frac{1}{2} \ln|x + 3| + C$$

$$= \frac{1}{2} \ln \left| \frac{(x - 1)^3(x + 2)^4}{x + 3} \right| + C. \quad \bullet$$

In general, if $D(x)$ can be factored into a product of *distinct linear factors,*

(67) $$D(x) = (x - r_1)(x - r_2)\cdots(x - r_m),$$

then we set

(68) $$\frac{Q(x)}{D(x)} = \frac{A_1}{x - r_1} + \frac{A_2}{x - r_2} + \cdots + \frac{A_m}{x - r_m},$$

where $A_1, A_2, \ldots, A_m$ are constants to be determined. Either of the methods illustrated in Examples 2 and 3 can be used to determine these constants—but in case the factors are all distinct (the case under consideration), the first method is quicker. We multiply both sides of (68) by $D(x)$ and then substitute $r_1, r_2, \ldots, r_m$ in the resulting equation to obtain $A_1, A_2, \ldots, A_m$.

If the denominator of the integrand has a *factor $(x + b)$ repeated n times,* then in the partial fraction decomposition the corresponding term is

(69) $$\frac{A_1}{x + b} + \frac{A_2}{(x + b)^2} + \cdots + \frac{A_n}{(x + b)^n},$$

where $A_1, A_2, \ldots, A_n$ are the unknowns to be determined.

Example 4. Find the indefinite integral $I_4 = \displaystyle\int \frac{3x^3 + 3x^2 + 3x + 2}{x^3(x + 1)}\,dx$.

Solution. Since the factor x is repeated three times in the denominator, our partial fraction decomposition must be

(70) $$\frac{3x^3 + 3x^2 + 3x + 2}{x^3(x + 1)} = \frac{A}{x + 1} + \frac{B}{x} + \frac{C}{x^2} + \frac{D}{x^3},$$

where A, B, C, and D are unknowns to be determined. Multiplying both sides of (70) by $x^3(x + 1)$, we have

(71) $$3x^3 + 3x^2 + 3x + 2 = Ax^3 + Bx^2(x + 1) + Cx(x + 1) + D(x + 1).$$

Here the first method will give A and D directly. For if we set $x = 0$ in (71), we obtain $2 = D$, and if we set $x = -1$ in (71), we obtain $-3 + 3 - 3 + 2 = A(-1)$ or $A = 1$. But B and C are not obtained so readily. Other values of x used in (71) give equations in which both B and C appear. If we use the second method, we equate corresponding coefficients

The coefficients of x^3 yield	$A + B$	$= 3.$
The coefficients of x^2 yield	$B + C$	$= 3.$
The coefficients of x^1 yield	$C + D$	$= 3.$
The coefficients of x^0 yield	D	$= 2.$

Solving this system of four linear equations in four unknowns gives $A = 1$, $B = 2$, $C = 1$, and $D = 2$. Hence

$$I_4 = \int \left(\frac{1}{x+1} + \frac{2}{x} + \frac{1}{x^2} + \frac{2}{x^3} \right) dx = \ln |(x+1)x^2| - \frac{1}{x} - \frac{1}{x^2} + C. \quad \bullet$$

EXERCISE 6

In Problems 1 through 26, find the indicated indefinite integral.

1. $\displaystyle\int \frac{(4x+5)\,dx}{x^2+3x+2}.$

2. $\displaystyle\int \frac{dx}{x^2+5x+6}.$

3. $\displaystyle\int \frac{2\,dx}{x^2-1}.$

4. $\displaystyle\int \frac{(5x+4)\,dx}{x^2+x-2}.$

5. $\displaystyle\int \frac{2\,dx}{x^2+8x+15}.$

6. $\displaystyle\int \frac{(2x+7)\,dx}{x^2+4x-5}.$

7. $\displaystyle\int \frac{7x\,dx}{x^2+x-12}.$

8. $\displaystyle\int \frac{(x^2-37)\,dx}{x^2+x-12}.$

9. $\displaystyle\int \frac{x^2\,dx}{x^2-4}.$

10. $\displaystyle\int \frac{2x^3-2x^2-7x-5}{x^2-x-6}\,dx.$

11. $\displaystyle\int \frac{12x^2+4x-8}{x^3-4x}\,dx.$

12. $\displaystyle\int \frac{(3x^2-6x-12)\,dx}{x^3+x^2-4x-4}.$

13. $\displaystyle\int \frac{(x^2-x+1)\,dx}{x^3+6x^2+11x+6}.$

★14. $\displaystyle\int \frac{(11+10x-x^2)\,dx}{x^3+3x^2-13x-15}.$

15. $\displaystyle\int \frac{(7x+4)\,dx}{x^3-4x^2}.$

16. $\displaystyle\int \frac{18-3x-2x^2}{x(x-3)^2}\,dx.$

17. $\displaystyle\int \frac{x^2-2x-8}{x^3-4x^2+4x}\,dx.$

18. $\displaystyle\int \frac{8x^2+4x-2}{(x^2-x)^2}\,dx.$

19. $\displaystyle\int \frac{(x^3+12x+14)\,dx}{(x+2)^3(x-1)}.$

20. $\displaystyle\int \frac{(x+1)(x+9)}{(x^2-9)^2}\,dx.$

21. $\displaystyle\int \frac{4x+8}{4x^2-1}\,dx.$

22. $\displaystyle\int \frac{e^x\,dx}{(e^x-1)(e^x-3)}.$

23. $\displaystyle\int \frac{\cos x\,dx}{4-\sin^2 x}.$

24. $\displaystyle\int \frac{19x^2-3x+10}{x^4+5x^3}\,dx$

★25. $\displaystyle\int \frac{2x^5+17x^4+40x^3-3x^2-92x-58}{(x+1)(x+2)(x+3)(x+4)}\,dx.$

★26. $\displaystyle\int \frac{2x^4-2x^3+x^2+13x-66}{x^4-13x^2+36}\,dx.$

27. Find the area of the region under the curve $y = 1/(x^2+4x+3)$ between $x = 0$ and $x = 5$.

28. Find the area of the region under the curve $y = 1/(x^2+4x+3)$ between $x = 0$ and $x = M(M > 0)$. What is the limit of this area as $M \to \infty$?

29. The region under the curve $y = 1/\sqrt{x^2+6x+8}$ between $x = -1$ and $x = 4$ is rotated about the x-axis. Find the volume of the solid generated.

30. The region of Problem 27 is rotated about the y-axis. Find the volume of the solid generated.

*31. If we make a formal computation for the area of the region under the curve $y = (4x - 10)/(x^2 - 5x + 6)$ between $x = 1$ and $x = 5$, we obtain

$$A = \int_1^5 \frac{(4x - 10)\,dx}{x^2 - 5x + 6} = \ln(x^2 - 5x + 6)^2 \Big|_1^5 = \ln \frac{6^2}{2^2} = \ln 9.$$

Why is this answer incorrect?

*32. Use partial fractions to derive the formula

$$\int \frac{du}{a^2 - u^2} = \frac{1}{2a} \ln \frac{a + u}{a - u} + C, \qquad |u| < a.$$

Partial Fractions, Quadratic Factors 9

Suppose that the denominator has a factor such as $x^2 + 2x + 5$. Since

$$x^2 + 2x + 5 = x^2 + 2x + 1 + 4 = (x + 1)^2 + 2^2$$

is the sum of two squares, the equation $x^2 + 2x + 5 = 0$ has no real roots. Hence factorization into linear factors with real coefficients is impossible. To test a general quadratic equation $ax^2 + bx + c = 0$ for real roots, we examine the *discriminant $b^2 - 4ac$*. If $b^2 - 4ac \geq 0$, then $ax^2 + bx + c$ has real linear factors, but if $b^2 - 4ac < 0$, we cannot decompose it further. In this latter case of a *nonrepeated quadratic factor,* the partial fraction decomposition has the corresponding term

(72) $$\frac{Ax + B}{ax^2 + bx + c},$$

where A and B are the unknowns.

Example 1. Find the indefinite integral $I_1 = \displaystyle\int \frac{(x^2 + 4x + 1)\,dx}{x^3 + 3x^2 + 4x + 2}$.

Solution. Factoring the denominator yields

(73) $$D(x) = x^3 + 3x^2 + 4x + 2 = (x + 1)(x^2 + 2x + 2).$$

The discriminant of the second factor gives $b^2 - 4ac = 2^2 - (4)(1)(2) = -4 < 0$. Hence the partial fraction decomposition is

(74) $$\frac{x^2 + 4x + 1}{(x + 1)(x^2 + 2x + 2)} = \frac{A}{x + 1} + \frac{Bx + C}{x^2 + 2x + 2}.$$

Multiplying both sides of (74) by $(x + 1)(x^2 + 2x + 2)$, we have

$$x^2 + 4x + 1 = A(x^2 + 2x + 2) + (Bx + C)(x + 1)$$
$$= (A + B)x^2 + (2A + B + C)x + (2A + C).$$

The second method leads to the set of equations

$$
\begin{array}{llll}
x^2: & A + B & & = 1. \\
x^1: & 2A + B & + C & = 4. \\
x^0: & 2A & + C & = 1.
\end{array}
$$

Solving this set gives $A = -2$, $B = 3$, and $C = 5$. Hence

$$I_1 = \int \left(\frac{-2}{x + 1} + \frac{3x + 5}{x^2 + 2x + 2} \right) dx = -2 \ln |x + 1| + \int \frac{3x + 5}{x^2 + 2x + 2} \, dx.$$

To compute the last integral, we note that $d(x^2 + 2x + 2) = (2x + 2) \, dx$ and hence we write the numerator in the form

$$3x + 5 = \frac{3}{2}(2x + 2) - 3 + 5$$

$$= \frac{3}{2}(2x + 2) + 2.$$

Then

$$I_1 = -2 \ln |x + 1| + \frac{3}{2} \int \frac{(2x + 2) \, dx}{x^2 + 2x + 2} + \int \frac{2 \, dx}{(x + 1)^2 + 1}$$

$$= -2 \ln |x + 1| + \frac{3}{2} \ln (x^2 + 2x + 2) + 2 \, \mathrm{Tan}^{-1} (x + 1) + C. \quad \bullet$$

The only case left to discuss is that of a *repeated quadratic factor*. If the factor is $(ax^2 + bx + c)^n$, then in the partial fraction decomposition the corresponding term[1] is

(75) $$\frac{A_1 x^{2n-1} + A_2 x^{2n-2} + \cdots + A_{2n-1} x + A_{2n}}{(ax^2 + bx + c)^n},$$

where $A_1, A_2, \ldots, A_{2n}$ are unknowns. The integration of a term such as (75) is then carried out by a suitable trigonometric substitution (see Section 7).

Example 2. Find the indefinite integral $I_2 = \displaystyle\int \frac{x^3 \, dx}{(x^2 + 4x + 13)^2}$.

Solution. Here the decomposition into partial fractions is already accomplished because the integrand has the form (75). Since

$$x^2 + 4x + 13 = (x + 2)^2 + 3^2,$$

---

[1] The procedure followed here is slightly different from that of most textbooks.

the trigonometric substitution is dictated by the triangle of Fig. 8. We have

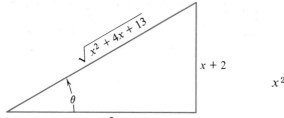

$$x + 2 = 3 \tan \theta,$$
$$dx = 3 \sec^2 \theta \, d\theta,$$
$$x^2 + 4x + 13 = 9 \sec^2 \theta.$$

FIGURE 8

$$I_2 = \int \frac{(3 \tan \theta - 2)^3 \, 3 \sec^2 \theta \, d\theta}{(9 \sec^2 \theta)^2}$$

$$= \int \frac{27 \tan^3 \theta - 54 \tan^2 \theta + 36 \tan \theta - 8}{27 \sec^2 \theta} \, d\theta$$

$$= \int \left(\frac{\sin^3 \theta}{\cos \theta} - 2 \sin^2 \theta + \frac{4}{3} \sin \theta \cos \theta - \frac{8}{27} \cos^2 \theta \right) d\theta$$

$$= \int \left[(1 - \cos^2 \theta) \frac{\sin \theta}{\cos \theta} - (1 - \cos 2\theta) + \frac{4}{3} \sin \theta \cos \theta - \frac{4}{27} (1 + \cos 2\theta) \right] d\theta$$

$$= \int \left(\frac{\sin \theta}{\cos \theta} + \frac{1}{3} \sin \theta \cos \theta - \frac{31}{27} + \frac{23}{27} \cos 2\theta \right) d\theta$$

$$= -\ln |\cos \theta| - \frac{1}{6} \cos^2 \theta - \frac{31}{27} \theta + \frac{23}{54} \sin 2\theta + C.$$

Now from Fig. 8 we see that

$$\cos \theta = \frac{3}{\sqrt{x^2 + 4x + 13}}, \qquad \theta = \mathrm{Tan}^{-1} \frac{x + 2}{3},$$

$$\sin 2\theta = 2 \sin \theta \cos \theta = 2 \left(\frac{x + 2}{\sqrt{x^2 + 4x + 13}} \right) \left(\frac{3}{\sqrt{x^2 + 4x + 13}} \right) = \frac{6(x + 2)}{x^2 + 4x + 13},$$

and so

$$I_2 = -\ln \frac{3}{\sqrt{x^2 + 4x + 13}} - \frac{3}{2(x^2 + 4x + 13)} - \frac{31}{27} \mathrm{Tan}^{-1} \frac{x + 2}{3}$$

$$+ \frac{23}{27} \left[\frac{3(x + 2)}{x^2 + 4x + 13} \right] + C$$

$$I_2 = \frac{1}{2} \ln (x^2 + 4x + 13) + \frac{46x + 65}{18(x^2 + 4x + 13)} - \frac{31}{27} \mathrm{Tan}^{-1} \frac{x + 2}{3} + C_1. \quad \bullet$$

**Example 3. Find the indefinite integral

$$I_3 = \int \frac{4x^5 + 5x^4 + 15x^3 - 28x^2 - 2x - 3}{x^2(x - 3)(x^2 + x + 1)^2} \, dx.$$

Solution. Here the corresponding partial fraction is

$$\frac{Q(x)}{D(x)} = \frac{A}{x^2} + \frac{B}{x} + \frac{C}{x - 3} + \frac{Dx^3 + Ex^2 + Fx + G}{(x^2 + x + 1)^2}.$$

If we multiply both sides by $D(x)$ and equate coefficients of powers of x, we obtain seven equations in the seven unknowns $A, \ldots, G$:

$$
\begin{array}{lrcl}
x^6: & B + C + D & = & 0, \\
x^5: & A - B + 2C - 3D + E & = & 4, \\
x^4: & -A - 3B + 3C - 3E + F & = & 5, \\
x^3: & -3A - 7B + 2C - 3F + G = & 15, \\
x^2: & -7A - 5B + C - 3G = & -28, \\
x: & -5A - 3B & = & -2, \\
1: & -3A & = & -3.
\end{array}
$$

After some labor, the very careful student will find that $A = 1$, $B = -1$, $C = 1$, $D = 0$, $E = 0$, $F = 0$, and $G = 9$. Then

$$I_3 = \int \left[\frac{1}{x^2} - \frac{1}{x} + \frac{1}{x - 3} + \frac{9}{(x^2 + x + 1)^2} \right] dx.$$

The trigonometric substitution $2x + 1 = \sqrt{3} \tan \theta$ will eventually lead to

$$I_3 = -\frac{1}{x} + \ln \left| \frac{x - 3}{x} \right| + 4\sqrt{3} \, \text{Tan}^{-1} \left(\frac{2x + 1}{\sqrt{3}} \right) + \frac{3(2x + 1)}{x^2 + x + 1} + C. \quad \bullet$$

EXERCISE 7

In Problems 1 through 16, find the indicated indefinite integral.

1. $\displaystyle \int \frac{dx}{x(x^2 + 1)}.$

2. $\displaystyle \int \frac{x^3 + 2x^2 + 8}{x^2(x^2 + 4)} \, dx.$

3. $\displaystyle \int \frac{2x^2 - 3x + 8}{(x^2 + 1)(x^2 + 4)} \, dx.$

4. $\displaystyle \int \frac{3x + 3}{x^3 - 1} \, dx.$

5. $\displaystyle \int \frac{4x^2 + 2x + 1}{16x^4 - 1}.$

6. $\displaystyle \int \frac{7x^3 - 9x^2 + x - 1}{x^2(9x^2 + 1)} \, dx.$

7. $\displaystyle\int \frac{4x\,dx}{(x^2+9)^2}.$

8. $\displaystyle\int \frac{6x^2\,dx}{(x^2+9)^2}.$

9. $\displaystyle\int \frac{6x^3\,dx}{(x^2+9)^2}.$

10. $\displaystyle\int \frac{8x^3\,dx}{(x^2+1)^3}.$

11. $\displaystyle\int \frac{(4x+7)\,dx}{x^2+1}.$

12. $\displaystyle\int \frac{(10x+36)\,dx}{x^2+2x+5}.$

13. $\displaystyle\int \frac{2x^2-13x+18}{x(x^2+9)}\,dx.$

14. $\displaystyle\int \frac{(x^6+3x^4+4x^3+3x^2+1)}{(x^2+1)^3}\,dx.$

15. $\displaystyle\int \frac{(x^3-8x-8)\,dx}{x^2(x^2+2x+2)}.$

⋆16. $\displaystyle\int \frac{x^5-2x^4-x^3+x-2}{x^6+2x^4+x^2}\,dx.$

REVIEW PROBLEMS

Find each of the following integrals, using any method you wish.

1. $\displaystyle\int \frac{5x^2-3x+1}{x^2(x^2+1)}\,dx.$

2. $\displaystyle\int \frac{2\,dx}{x(x^2+1)^2}.$

3. $\displaystyle\int \frac{dx}{\sin x \cos x}.$

4. $\displaystyle\int \frac{x\,dx}{1+\sqrt{x}}.$

5. $\displaystyle\int \frac{dx}{e^{2x}-1}.$

6. $\displaystyle\int \frac{\sin x\,dx}{\sqrt{1+\cos x}}.$

7. $\displaystyle\int \frac{\sin x\,dx}{1+\cos^2 x}.$

8. $\displaystyle\int \cosh^3 x\,dx.$

9. $\displaystyle\int \frac{\sin \sqrt{x}}{\sqrt{x}}\,dx.$

10. $\displaystyle\int \frac{d\theta}{\cot \theta - \tan \theta}.$

11. $\displaystyle\int \frac{d\theta}{2\csc\theta - \sin\theta}.$

12. $\displaystyle\int \frac{x^2}{e^{5x}}\,dx.$

13. $\displaystyle\int \mathrm{Cot}^{-1} 2x\,dx.$

14. $\displaystyle\int (\ln x)^2\,dx.$

15. $\displaystyle\int \frac{(\ln x)^7}{x}\,dx.$

16. $\displaystyle\int \frac{5\cos x\,dx}{6+\sin x - \sin^2 x}.$

17. $\displaystyle\int \frac{dx}{1+e^x}.$

18. $\displaystyle\int \frac{(1+\tan^2 x)\,dx}{1+9\tan^2 x}.$

19. $\displaystyle\int \sinh^3 2x\,dx.$

20. $\displaystyle\int x\sinh x\,dx.$

21. $\displaystyle\int \sin\frac{x}{2}\cos\frac{5x}{2}\,dx.$

22. $\displaystyle\int \frac{dx}{x\ln x}.$

23. $\displaystyle\int e^{\tan y}\sec^2 y\,dy.$

24. $\displaystyle\int \sin^5 x\,\sqrt{\cos x}\,dx.$

25. $\displaystyle\int \frac{e^{3x}\,dx}{1+e^{6x}}.$

26. $\displaystyle\int \frac{18x\,dx}{9x^2+6x+5}.$

27. $\displaystyle\int \sqrt{1+\sin\theta}\,d\theta.$

28. $\displaystyle\int 36x^5 e^{2x^3}\,dx.$

29. $\displaystyle\int \ln(x^2+a^2)^5\,dx.$

30. $\displaystyle\int \frac{\sin t\,dt}{\sec t + \cos t}.$

31. $\displaystyle\int \sqrt{\sec y}\,\tan y\,dy.$

32. $\displaystyle\int \sin\sqrt{x}\,dx.$

33. $\displaystyle\int \cos 3x \cos 7x\,dx.$

★34. $\displaystyle\int \frac{dz}{1 - \tan^2 z}$.

35. $\displaystyle\int \frac{6e^{4x}\, dx}{1 - e^x}$.

36. $\displaystyle\int \frac{8e^x\, dx}{3 + 2e^x - e^{2x}}$.

37. $\displaystyle\int \theta \sec^2 \theta\, d\theta$.

38. $\displaystyle\int \sqrt{1 - \cos \theta}\, d\theta$.

39. $\displaystyle\int \frac{\tan x\, dx}{\ln (\cos x)}$.

40. $\displaystyle\int \sin \frac{x}{3} \sin \frac{5x}{3}\, dx$.

41. $\displaystyle\int \sinh x \sin x\, dx$.

42. $\displaystyle\int 125\, x^4 (\ln x)^2\, dx$.

43. $\displaystyle\int (\tan \theta + \cot \theta)^2\, d\theta$.

44. $\displaystyle\int e^{ax} \sin bx\, dx$.

45. $\displaystyle\int \frac{24\, dx}{x(x^2 - 1)(x^2 - 4)}$.

46. $\displaystyle\int \frac{(\mathrm{Tan}^{-1} y)^2}{1 + y^2}\, dy$.

47. $\displaystyle\int \cos (\ln x)\, dx$.

48. $\displaystyle\int \frac{\sin \theta \cos^3 \theta\, d\theta}{1 + \sin^2 \theta}$.

★49. $\displaystyle\int \tanh (\ln x)\, dx$.

50. $\displaystyle\int e^x \sinh x \cos x\, dx$.

51. $\displaystyle\int \frac{1 - \sqrt{x}}{1 + \sqrt{x}}\, dx$.

★52. $\displaystyle\int \frac{\cos^3 x\, dx}{1 + \cos^2 x}$.

★53. $\displaystyle\int \frac{d\theta}{9 \cos^2 \theta - 4 \sin^2 \theta}$.

54. $\displaystyle\int \frac{1 + \sin x}{1 - \sin x}\, dx$.

★55. $\displaystyle\int \frac{\sin \theta + 7 \cos \theta}{\sin \theta + 2 \cos \theta}\, d\theta$.

56. $\displaystyle\int \cos 6\theta \sin 8\theta\, d\theta$.

57. $\displaystyle\int \frac{\sqrt{t + 1} + 1}{\sqrt{t + 1} - 1}\, dt$.

58. $\displaystyle\int \frac{\cos^9 t \tan^2 t \sec^2 t}{\csc^4 t}\, dt$.

★59. $\displaystyle\int \frac{\sin 2\theta \cos 5\theta\, d\theta}{\sec 4\theta}$.

★60. $\displaystyle\int \sqrt{\frac{1 + s}{1 - s}}\, ds$.

11

More Analytic Geometry

Translation of Axes 1

In all the curves that we have met so far, the interesting features of the curves were close to the origin. For example, the curve $y = x^2 - 2x + 3$ can be written $y = (x - 1)^2 + 2$, from which the minimum point is $(1, 2)$, and this is a "reasonable" location for a minimum point. The student soon learns from experience to expect such nice behavior, and therefore in making a table of values from which to sketch the curve, the student starts by using $x = 0, \pm 1, \pm 2$, and so on. And the quizmasters usually keep faith by proposing only "decent" curves for the student to plot. But suppose that we are confronted with a monstrosity such as

(1) $$y = x^2 - 100x + 3000$$

to graph. We certainly have no desire to put $x = 0, \pm 1, \pm 2$, and so on. We can use the calculus to determine the extreme point on the curve. Indeed, $y' = 2x - 100$ and $y'' = 2$, so the curve has a minimum point when $x = 50$ and the minimum value for y is $(50)^2 - (100)(50) + 3000 = 500$.

Shall we now graph this curve in the usual way? Of course not. We would first move our coordinate axes so that the new origin is somewhere near this minimum point, and while we are in the moving business we may as well put the new origin right at the minimum point. To do this, we split the constant into a sum of two terms so that we have a perfect square on the right side. Thus

$$y = x^2 - 100x + 3000 = x^2 - 100x + 2500 + 500$$

$$y = (x - 50)^2 + 500$$

433

and hence

(2) $$y - 500 = (x - 50)^2.$$

We introduce two new variables X and Y defined by

(3) $$X = x - 50 \qquad Y = y - 500;$$

and when these new variables are used in equation (2) we have

(4) $$Y = X^2,$$

an old friend.

Let us now analyze just what the substitution (3) does geometrically. Consider two rectangular coordinate systems placed with the corresponding axes parallel and similarly directed, as shown in Fig. 1. Let O' be the origin for the XY-system, and suppose that (h, k) are the coordinates for O' in the xy-system.

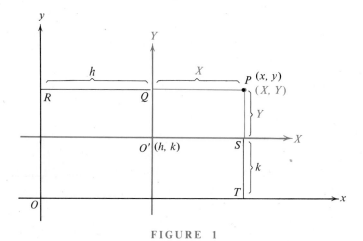

FIGURE 1

Now each point P in the plane has two sets of coordinates: (x, y) when referred to the xy-system, and (X, Y) when referred to the XY-system. From the figure it is obvious that $x = X + h$ and $y = Y + k$, and hence

(5) $$X = x - h \qquad \text{and} \qquad Y = y - k.$$

These equations relate the coordinates in the one system to the coordinates in the other. It is convenient to think of the xy-system as the original or primitive system, and to regard the XY-system as the new system obtained from the original system by shifting (translating) the two axes from their original position to the new position. Then the equations (5) are the equations for the translation of the axes. Keeping in mind that only the axes and the coordinates are changed but that the points remain just where they were, we have

> **THEOREM 1.** If the xy-system is translated so that the origin of the new system (the XY-system) is at the point (h, k) in the original system, then the coordinates of a point are related by the equation set
>
> (5) $$X = x - h \quad \text{and} \quad Y = y - k.$$

Proof. The equation set (5) is certainly true when P and O' have the special position shown in Fig. 1. To complete the proof, we must consider all possible locations for P and O'. Now coordinates are directed distances and, with the lettering of Fig. 1, the equation $x = X + h$ is equivalent to the equation

(6) $$RP = RQ + QP.$$

Further, the equation $y = Y + k$ is equivalent to

(7) $$TP = TS + SP.$$

But equations (6) and (7) are always true no matter how the three points are distributed on a directed line. ∎

Example 1. Discuss the graph of equation (1).

Solution. The substitution $X = x - 50$ and $Y = y - 500$ changes equation (1) into $Y = X^2$. The graph of $Y = X^2$ in the XY-system is a parabola that opens upward, has vertex at $(0, 0)$, focus at $(0, 1/4)$, and the directrix is the line $Y = -1/4$.

    Returning to the original system, we find that the graph of $y = x^2 - 100x + 3000$ is the same parabola. The vertex is at $(50, 500)$, the focus is at $(50, 500.25)$, and the directrix is the line $y = 499.75$. ●

In Chapter 2 we found standard forms for the equations of certain curves:

(8) $$y^2 = 4px \qquad \text{Parabola,}$$

(9) $$x^2 = 4py \qquad \text{Parabola,}$$

(10a) $$\frac{x^2}{a^2} + \frac{y^2}{b^2} = 1 \qquad \text{Ellipse,} \qquad a > b > 0,$$

(10b) $$\frac{x^2}{b^2} + \frac{y^2}{a^2} = 1 \qquad \text{Ellipse,} \qquad a > b > 0,$$

(11)
$$\frac{x^2}{a^2} - \frac{y^2}{b^2} = 1 \qquad \text{Hyperbola,}$$

(12)
$$\frac{y^2}{a^2} - \frac{x^2}{b^2} = 1 \qquad \text{Hyperbola.}$$

For the parabola (8) the vertex is at the origin, the focus is at $(p, 0)$, and the directrix is $x = -p$. For the parabola (9) the focus is at $(0, p)$ and the directrix is $y = -p$.

For the ellipse, set $c = \sqrt{a^2 - b^2}$. For (10a), the foci are at $(\pm c, 0)$. For (10b), the foci are at $(0, \pm c)$.

For the hyperbola (11), the foci are at $(\pm c, 0)$, where now $c = \sqrt{a^2 + b^2}$. For the hyperbola (12), the foci are at $(0, \pm c)$.

> **Example 2.** Discuss the graph of
>
> (13)
> $$25x^2 - 4y^2 - 150x - 16y + 109 = 0.$$
>
> *Solution.* Just as in Example 1, we complete the squares by suitably grouping the terms and adding appropriate constants to both sides. Equation (13) can be put in the form
>
> (14)
> $$25(x^2 - 6x \qquad) - 4(y^2 + 4y \qquad) = -109.$$
>
> We complete the squares inside the parentheses by adding 9 and 4, respectively. This amounts to adding 25×9 and subtracting 4×4 on the left side, and hence the same additions and subtractions must be made on the right side. Equation (14) then gives
>
> $$25(x^2 - 6x + 9) - 4(y^2 + 4y + 4) = -109 + 25 \times 9 - 4 \times 4$$
>
> $$25(x - 3)^2 - 4(y + 2)^2 = -109 + 225 - 16 = 100$$
>
> $$\frac{(x - 3)^2}{4} - \frac{(y + 2)^2}{25} = 1.$$
>
> The translation of axes, $X = x - 3$, $Y = y + 2$ yields
>
> (15)
> $$\frac{X^2}{2^2} - \frac{Y^2}{5^2} = 1, \qquad c = \sqrt{a^2 + b^2} = \sqrt{29}.$$
>
> We know that the graph of (15) is a hyperbola with foci at $(\pm\sqrt{29}, 0)$ in the XY-system. Returning to the original system, the graph is still a hyperbola, but now the foci are at $(3 \pm \sqrt{29}, -2)$. Since the X- and Y-axes are axes of symmetry for the hyperbola (15), it follows that the lines $x = 3$, $y = -2$ are axes of symmetry for the graph of (13). Similarly, the points $(1, -2)$ and $(5, -2)$ are vertices of this hyperbola in the original coordinate system. The graph of (13) is shown in Fig. 2. ●

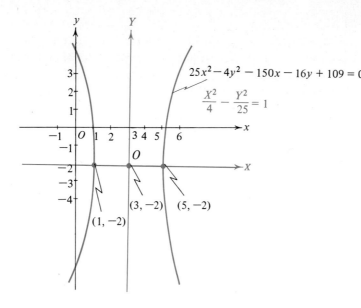

25x²−4y²−150x−16y+109=0

$$\frac{X^2}{4} - \frac{Y^2}{25} = 1$$

(3, −2)   (5, −2)

(1, −2)

FIGURE 2

Example 3. Find an equation for the ellipse with vertices at $(-1, 0)$ and $(9, 0)$ and foci at $(2, 0)$ and $(6, 0)$.

Solution. The center of the ellipse will be at the midpoint of the line segment joining the vertices. It can also be found as the midpoint of the line segment joining the foci. In either case we find that the center of the ellipse is at $(4, 0)$. Then $a = 9 - 4 = 5$ and $c = 6 - 4 = 2$. Hence $b = \sqrt{a^2 - c^2} = \sqrt{25 - 4} = \sqrt{21}$. Thus an equation for the ellipse is

$$\frac{(x - 4)^2}{25} + \frac{y^2}{21} = 1. \quad \bullet$$

EXERCISE 1

In Problems 1 through 10, find an equation for the conic section satisfying the given conditions. HINT: *It may help to sketch the curve before finding the equation.*

1. A parabola with focus at $(0, 5)$ and directrix $y = 1$.
2. A parabola with focus at $(3, 5)$ and directrix $y = 7$.
3. A parabola with focus at $(-1, 3)$ and directrix the line $x = 1$.

4. A parabola with focus at $(-2, -2)$ and directrix $x = -8$.
5. An ellipse with vertices at $(1, 0)$ and $(11, 0)$ and foci at $(3, 0)$ and $(9, 0)$.
6. An ellipse with vertices at $(0, 3)$ and $(10, 3)$ and foci at $(1, 3)$ and $(9, 3)$.
7. An ellipse with vertices at $(-2, 2 \pm \sqrt{2})$ and foci at $(-2, 1)$ and $(-2, 3)$.
8. A hyperbola with vertices at $(3, -2)$ and $(-5, -2)$ and foci at $(4, -2)$ and $(-6, -2)$.
9. A hyperbola with vertices at $(3, \pm \sqrt{2})$ and foci at $(3, \pm 2)$.
10. A hyperbola with vertices at $(-2, 3)$ and $(-2, 1)$ and foci at $(-2, 2 \pm \sqrt{5})$.

In Problems 11 through 20, put the given equation into standard form by a translation of the axes, and identify the graph. If it is a parabola, give its focus, vertex, and directrix. If it is an ellipse or hyperbola, give its foci, vertices, and axes of symmetry.

11. $x^2 + 2x + 16y + 33 = 0$.   12. $x + 20y^2 + 40y + 27 = 0$.
13. $y^2 - 4y + 12 = 8x$.   14. $2y + 180x = x^2 + 7950$.
15. $9x^2 + 25y^2 - 90x - 150y + 225 = 0$. 16. $25x^2 + 16y^2 + 100x - 192y + 276 = 0$.
17. $16x^2 + y^2 - 32x + 4y + 16 = 0$.   18. $16y^2 - 9x^2 - 64y - 54x = 161$.
19. $y^2 + 20 = x^2 + 10y + 4x$.   20. $6x^2 + 84x + 69 = 15y^2 + 90y$.

In Problems 21 through 26, find h and k so that the translation $x = X + h$, $y = Y + k$ changes the given equation into one of the indicated form.

21. $xy + 5x - 3y = 17$ into $XY = C$.
22. $xy - 7x + 4y = 1$ into $XY = C$.
23. $xy + 2x + 5y = 9$ into $XY = C$.
24. $xy - 6x - y = 4$ into $XY = C$.
25. $y = x^3 - 12x^2 + 7x$ into $Y = X^3 + CX$.
*26. $y = ax^3 + bx^2 + cx$ into $Y = AX^3 + CX$.

*27. Prove that if $A \neq 0$, the curve $y = Ax^2 + Bx + C$ is a parabola. Find a formula for the focus of this parabola.
*28. Consider the family of parabolas $y = Ax^2 + C$, where C is a constant and A is a variable. Let $(0, F)$ be the focus. Compute dF/dA. What is the limit of F as $A \to 0^+$; as $A \to \infty$? Sketch a few parabolas from this family when $C = 1$.
*29. Consider the family of ellipses $x^2 + B^2y^2 = 1$, where B varies. Let $(\pm F, 0)$ be the foci when $B > 1$, with $F > 0$. Compute dF/dB. What is the limit position of the foci as $B \to 0^+$; as $B \to \infty$? Sketch a few ellipses from this family.
*30. Consider the family of hyperbolas $x^2 - y^2 = K$, with $K > 0$. With the notation of Problem 29, find dF/dK. What is the limit position of the foci as $K \to 0^+$; as $K \to \infty$? Do any two distinct members of this family have points in common? Sketch a few hyperbolas from this family.

Asymptotes **2**

The graph of the simple equation

(16) $$y = \frac{1}{x}$$

will supply us with suitable examples of asymptotes. The graph can be sketched quite quickly by computing the coordinates of a few points, and the curve obtained is shown in Fig. 3. The important features of this curve are

(a) When $x = 0$, there is no corresponding y, so the graph does not meet the y-axis. For every other value of x, (16) gives a corresponding y, and the function is continuous, so the graph falls into two pieces (called *components*), separated by the y-axis.

(b) If x is small and positive, the corresponding y is very large. For example, if $x = 0.001$, then $y = 1000$. Since y can be made arbitrarily large by taking x sufficiently small, this means that

(17) $$\lim_{x \to 0^+} y = \infty.$$

(c) If x is small and negative, the corresponding y has large absolute value, but is negative. For example, if $x = -0.0001$, then $y = -10,000$. Since $|y|$ can be made arbitrarily large by taking x sufficiently small and negative, this means that

(18) $$\lim_{x \to 0^-} y = -\infty.$$

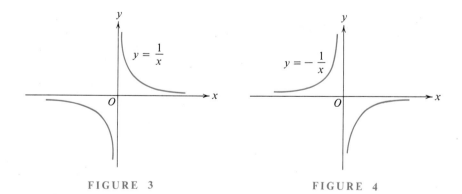

FIGURE 3   FIGURE 4

(d) If x is very large, the corresponding y is positive and very small. For example, if $x = 100,000$, then $y = 0.00001$. Clearly, we have

$$\lim_{x \to \infty} y = 0. \tag{19}$$

(e) Similar considerations show that

$$\lim_{x \to -\infty} y = 0. \tag{20}$$

The graph of $y = -1/x$ is shown in Fig. 4. Since the only new item is the factor -1, the graph in Fig. 4 can be obtained from that in Fig. 3 either **(i)** by rotating the graph about the x-axis through an angle of $180°$ or **(ii)** by reflecting it in the x-axis.

Let us observe that a point P on the graph in Fig. 3 draws closer to the y-axis as the point travels upward, receding farther from the origin. If the point travels to the right going farther from the origin, it draws closer to the x-axis. These lines, the x- and y-axes, are called *asymptotes* of the graph (or curve). Similar considerations show that the graph of Fig. 4 has these same lines as asymptotes.

DEFINITION 1 (Asymptote). Let P be a point on a graph $\mathscr{C}$, let s be the distance of P from a line $\mathscr{L}$, and let r be the distance of P from the origin. If there is a curve $\mathscr{C}_0$, contained in $\mathscr{C}$ such that

$$\lim_{r \to \infty} s = 0, \tag{21}$$

for P on $\mathscr{C}_0$, then the line $\mathscr{L}$ is called an asymptote of the graph $\mathscr{C}$.

It is clear that with this formal definition the x- and y-axes are asymptotes for the graphs of $y = 1/x$ and $y = -1/x$. The asymptote $\mathscr{L}$ does not need to be one of the coordinate axes. This is illustrated in

Example 1. Find the asymptotes for $\mathscr{C}$, the graph of

$$y = \frac{3x - 11}{x - 5}. \tag{22}$$

Solution. Taking the limit as $x \to \infty$, we have

$$\lim_{x\to\infty} y = \lim_{x\to\infty} \frac{3x-11}{x-5} = \lim_{x\to\infty} \frac{3 - \dfrac{11}{x}}{1 - \dfrac{5}{x}} = \frac{3}{1} = 3.$$

Therefore, the line $y = 3$ is a horizontal asymptote. Looking for a vertical asymptote, we observe that

$$\lim_{x\to 5^+} y = \lim_{x\to 5^+} \frac{3x-11}{x-5} = \infty \qquad \text{and} \qquad \lim_{x\to 5^-} y = \lim_{x\to 5^-} \frac{3x-11}{x-5} = -\infty.$$

At $x = 5$, y is undefined, so the vertical line $x = 5$ divides the graph into two curves, and each curve has this line as a vertical asymptote. The curves are shown in Fig. 5. ●

Alternative Solution. If we divide $x - 5$ into $3x - 11$, we find that

$$y = \frac{3x-11}{x-5} = \frac{4}{x-5} + 3,$$

(23) $$y - 3 = \frac{4}{x-5}.$$

The substitutions $X = x - 5$, $Y = y - 3$ reduce (23) to the form

(24) $$Y = 4\frac{1}{X},$$

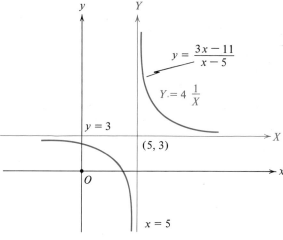

FIGURE 5

and this is essentially equation (16) except for the factor 4. Hence the graph of $y = (3x - 11)/(x - 5)$ is essentially that of Fig. 3 except that the two curves have been stretched by a factor of 4 in the vertical direction, and the origin of the XY-system is at $(5, 3)$ in the xy-system. The graph is shown in Fig. 5. Since the X- and Y-axes are asymptotes for the graph of (24), it follows from the theorem on the translation of axes that the lines $x = 5$ and $y = 3$ are asymptotes of the graph of (22). ●

This example suggests that it is easy to locate horizontal and vertical asymptotes by virtue of

---

THEOREM 2 (PLE). Let $\mathscr{C}$ be the graph of $y = f(x)$. If

$$(25) \qquad \lim_{x \to \infty} f(x) = L_1,$$

then the line $y = L_1$ is a horizontal asymptote of $\mathscr{C}$ as the point P on $\mathscr{C}$ recedes to the right. If

$$(26) \qquad \lim_{x \to -\infty} f(x) = L_2,$$

then the line $y = L_2$ is a horizontal asymptote of $\mathscr{C}$ as the point P on $\mathscr{C}$ recedes to the left.

If $f(x)$ can be written in the form

$$(27) \qquad y = \frac{g(x)}{(x - a)^n},$$

where n is positive, $g(a) \neq 0$, and $g(x)$ is continuous at $x = a$, then the line $x = a$ is a vertical asymptote of $\mathscr{C}$.

In some applications the limits L_1 and L_2 in equations (25) and (26) will be the same. Frequently, the exponent n in equation (27) will be 1. When $n = 1$ and $g(a)$ is positive, the graph resembles Fig. 3 with a vertical asymptote at $x = a$, instead of the y-axis. If $g(a)$ is negative, then the curve resembles Fig. 4, with a vertical asymptote at $x = a$.

Example 2. Locate the horizontal and vertical asymptotes for the graph of

$$(28) \qquad y = \frac{4x^3}{(x - 4)^2(x + 2)}.$$

Solution. For a horizontal asymptote we divide the numerator and denominator by x^3. This gives

$$\lim_{x \to \infty} \frac{4x^3}{(x-4)^2(x+2)} = \lim_{x \to \infty} \frac{4}{\left(1 - \dfrac{4}{x}\right)^2\left(1 + \dfrac{2}{x}\right)} = 4.$$

The limit is also 4 when $x \to -\infty$. Therefore, the line $y = 4$ is a horizontal asymptote both to the right and to the left. The graph of the function (28) is shown in Fig. 6. Theorem 2, equation (27), makes it easy to locate the vertical asymptotes. We only need to find the zeros of the denominator. These are obviously $x = 4$ and $x = -2$, so the lines $x = 4$ and $x = -2$ are vertical asymptotes. For $x = 4$, the function $g(x)$ of the theorem is $4x^3/(x + 2)$. Since $g(4) > 0$, $\lim_{x \to 4^+} y = \infty$, as indicated in Fig. 6.

But $(x - 4)$ occurs to an even power; hence as $x \to 4^-$, we still have $y \to \infty$. For $x = -2$, the function $g(x)$ of the theorem is $4x^3/(x - 4)^2$. Since $g(-2) < 0$ and since $n = 1$, the graph is somewhat similar to Fig. 4 in the neighborhood of $x = -2$. All these facts are illustrated in Fig. 6. ●

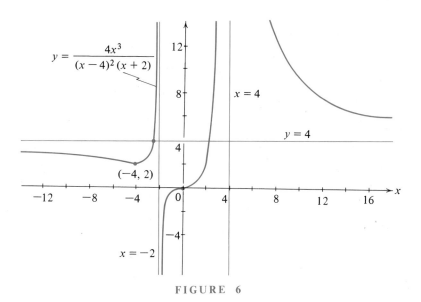

FIGURE 6

Does the curve cut the x-axis? We determine this by setting $y = 0$, and this leads to the equation $4x^3 = 0$. So the curve cuts the x-axis only at the origin, and since $x = 0$ is a triple root of the equation $4x^3 = 0$, we suspect that the curve has a point of inflection at $(0, 0)$. Does the curve meet the horizontal asymptote $y = 4$? The condition $y = 4$ leads to the equation

$$4 = \frac{4x^3}{(x-4)^2(x+2)}.$$

Hence

$$(x-4)^2(x+2) = x^3$$
$$x^3 - 6x^2 + 32 = x^3$$
$$6x^2 = 32.$$

Therefore, the curve crosses the horizontal asymptote at $x = \pm 4/\sqrt{3} \approx \pm 2.309$.

What are the extreme points on this curve? Computing the derivative for (28) gives us

$$\frac{dy}{dx} = \frac{-4x^2(6x+24)}{(x-4)^3(x+2)^2}.$$

Then $x = -4$ is a critical value, and for this value $y = 2$. Since as x decreases from -2 to -4, y changes from $+\infty$ to $+2$, and since as $x \to -\infty$, y approaches 4, it is clear that the point $(-4, 2)$ is a relative minimum and that the curve approaches the line $y = 4$ from below as $x \to -\infty$. Similarly, the curve approaches the line $y = 4$ from above as $x \to \infty$.

How about asymptotes that are neither vertical nor horizontal? These can be located using

---

THEOREM 3. Let $\mathscr{C}$ be the graph of $y = f(x)$. If $f(x)$ can be written in the form

(29) $$f(x) = Ax + B + g(x),$$

where $\lim\limits_{x \to \infty} g(x) = 0$, then the line $y = Ax + B$ is an asymptote of $\mathscr{C}$ as the point P recedes to the right. If $\lim\limits_{x \to -\infty} g(x) = 0$, then the line is an asymptote as P recedes to the left.

Proof. We must distinguish between the coordinates on the curve and on the line. For a given x, let $y_C = f(x)$ be the corresponding y-coordinate on the curve of $y = f(x)$, and let $y_L = Ax + B$ be the corresponding y-coordinate on the line. Let P_C and P_L be the points (x, y_C) and (x, y_L), respectively.

From equation (29) and the meaning of y_C and y_L, we have

$$y_C - y_L = f(x) - (Ax + B) = g(x).$$

Suppose now that $g(x) \to 0$ as $x \to \infty$. Then $y_C - y_L \to 0$ as $x \to \infty$. But (as illustrated in Fig. 7, the curve for Example 3) the distance s from the point P_C to the line $\mathscr{L}$ is less than $|y_C - y_L|$. Therefore, $s \to 0$ as $x \to \infty$.

The proof is just the same, in case $g(x) \to 0$ as $x \to -\infty$. ∎

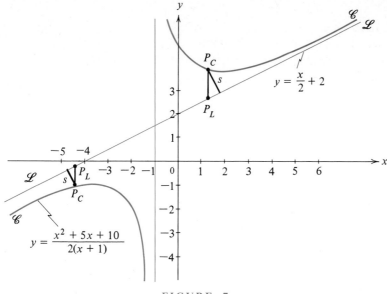

FIGURE 7

If $y_C - y_L$ is positive, then obviously the curve lies above the line. If $y_C - y_L$ is negative, then the curve lies below the line. But $y_C - y_L$ is just $g(x)$, so it is a simple matter to determine when the curve lies above the asymptote and when it lies below.

Example 3. Find the asymptotes for the graph of

$$(30) \qquad y = \frac{x^2 + 5x + 10}{2(x + 1)}.$$

Solution. By Theorem 2 there is a vertical asymptote $x = -1$, since the denominator has a zero at $x = -1$. If we perform the indicated division we find a quotient of $\frac{1}{2}x + 2$ and a remainder of 6. Therefore,

$$y = \frac{x^2 + 5x + 10}{2(x + 1)} = \frac{1}{2}x + 2 + \frac{3}{x + 1}.$$

Comparing this equation with equation (29), we see that $Ax + B = \frac{1}{2}x + 2$ and $g(x) = 3/(x + 1)$. Since $\lim\limits_{x \to \pm\infty} 3/(x + 1) = 0$, it follows from Theorem 3 that the line $y = \frac{1}{2}x + 2$ is an asymptote. ●

For $x > -1$, $g(x)$ is positive, so that for $x > -1$ the curve lies above the asymptote. When $x < -1$, $g(x)$ is negative and the curve lies below the asymptote. The graph of (30) is shown in Fig. 7.

EXERCISE 2

In Problems 1 through 24, find all the asymptotes, and sketch the graph.

1. $y = \dfrac{x + 1}{x - 1}$.

2. $y = \dfrac{x - 3}{x + 4}$.

3. $y = \dfrac{x - 2}{2x - 5}$.

4. $y = \dfrac{2x - 3}{x + 1}$.

5. $y = \dfrac{x^2 + 1}{x^2}$.

6. $y = x + \dfrac{1}{x}$.

7. $y = x + \dfrac{1}{x^2}$.

8. $y = \dfrac{1 - e^x}{1 + e^x}$.

9. $y = \dfrac{2e^{-x} + 3}{e^{-x} - 1}$.

10. $y = x^2 - \ln x$.

11. $y = \dfrac{x^2 - 1}{2x^2}$.

12. $y = \dfrac{2x^2}{x^2 - 1}$.

13. $y = \text{Tan}^{-1}(5x)$.

14. $y = 2x + \text{Tan}^{-1} x$.

15. $y = \dfrac{x^2 + \sin x}{2x}$.

16. $y = 5\dfrac{(x + 1)^2}{x(x + 2)}$.

17. $y = \dfrac{4(x^3 - 7x)}{(x - 2)^2(x + 5)}$.

18. $y = \dfrac{-6(x + 2)^4}{(x^2 + 4x)^2}$.

19. $y = \dfrac{2x^2 - x - 8}{2x - 3}$.

20. $y = \dfrac{x^3}{2(x^2 - 1)}$.

21. $y = \dfrac{x^3 - 6x^2 + 6}{2x^2}$.

22. $y = \dfrac{-x^3 + 2x^2 - x}{x^2 + 1}$.

23. $y = \dfrac{x^4}{x^2 + 1}$.

24. $y = \dfrac{x^4}{2x^3 + 16}$.

25. Prove that the line $y = ax/b$ is an asymptote for the hyperbola

$$\frac{y^2}{a^2} - \frac{x^2}{b^2} = 1.$$

HINT: This amounts to proving that $\lim\limits_{x \to \infty} \left(\dfrac{ax}{b} - \dfrac{a}{b} \sqrt{x^2 + b^2} \right) = 0$.

Note that $\dfrac{a}{b}(x - \sqrt{x^2 + b^2})(x + \sqrt{x^2 + b^2}) = \dfrac{a}{b}(-b^2) = -ab$.

Hence $\lim\limits_{x \to \infty} \left(\dfrac{ax}{b} - \dfrac{a}{b} \sqrt{x^2 + b^2} \right) = \lim\limits_{x \to \infty} \dfrac{-ab}{x + \sqrt{x^2 + b^2}} = 0$.

26. Prove that $y = -ax/b$ is also an asymptote for the hyperbola of Problem 25.
27. Prove that the lines $y = \pm bx/a$ are both asymptotes for the hyperbola

$$\frac{x^2}{a^2} - \frac{y^2}{b^2} = 1.$$

28. Does the parabola $y = x^2$ have an asymptote?
29. Does the straight line $y = mx + b$ have an asymptote?
*30. Prove that if $D(x)$ is a polynomial of degree m and $N(x)$ is a polynomial of degree

$m + 1$, then the graph of $y = N(x)/D(x)$ always has one asymptote that is not vertical. (See Problems 19, 20, 21, 22, and 24.)

★31. In Definition 1, the introduction of a curve $\mathscr{C}_0$ contained in a graph $\mathscr{C}$ seems unnatural. Explain the need for such a phrase by considering the graph of

$$(y - x^2)\left(y - \frac{1}{x}\right) = y^2 - y\left(x^2 + \frac{1}{x}\right) + x = 0.$$

32. Find all relative extreme points on the curve of Problem 20.

33. Do Problem 32 for curve of Problem 23.

CALCULATOR PROBLEMS

In Problems C1 and C2, find all extreme points to four significant figures for the given curve.

C1. The curve of Problem 21.

C2. The curve of Problem 24.

Symmetry 3

We recall from Chapter 2 (p. 54) some facts about symmetry. Two points P_1 and P_2 are
said to be *symmetric with respect to a line* $\mathscr{L}$ if the line $\mathscr{L}$ is the perpendicular bisector of
the segment P_1P_2. Each of the points is said to be the *reflection* of the other point in the line
$\mathscr{L}$. We can think of $\mathscr{L}$ as a mirror, and each of the points is an image of the other point in
the mirror. A curve (or graph) is said to be *symmetric with respect to the line* $\mathscr{L}$ if for every
point P_1 on the curve its reflection P_2 in the line $\mathscr{L}$ is also on the curve. In this case the line
$\mathscr{L}$ is called an *axis of symmetry for the curve.*

We may also have symmetry with respect to a point. Two points P_1 and P_2 are said to
be *symmetric with respect to a point* C if the point C is the bisector of the segment P_1P_2.
Each of the points is said to be the *reflection* of the other in the point C. Of course, in this
case the physical picture of C acting as a mirror is missing. A curve (or graph) is said to be
symmetric with respect to a point C if for every point P on the curve its reflection in the
point C is also on the curve. In this case the point C is called a *center of symmetry for the
curve.* As an example, the center of a circle is a center of symmetry for that curve because
every diameter of a circle is bisected by the center.

Our objective is to give an algebraic test for the symmetry of curves. Since we can
always translate the coordinate axes, we can assume that the axis of symmetry is a line
through the origin, and the center of symmetry is at the origin.

Any equation in x and y can be put in the form

(31) $$F(x, y) = 0,$$

where $F(x, y)$ denotes a suitable function of two variables (see Chapter 3, Exercise 1,

Problems 38 to 46). Indeed, to obtain (31) we merely transpose all the terms of the given equation to the left side. For example, the parabola $y^2 = 16x$ can be written in the form

$$(32) \qquad\qquad y^2 - 16x = 0,$$

where on comparison with (31) we see that the notation $F(x, y)$ represents the function $y^2 - 16x$ in this case.

Actually, the graph of (31) may consist of several curves. For example, the graph of

$$(33) \qquad\qquad y^3 - 16xy = 0$$

consists of the parabola $y^2 = 16x$ and the x-axis, since the given function can be factored into the product $y(y^2 - 16x) = 0$, so that either $y = 0$ (the x-axis) or $y^2 - 16x = 0$ (the parabola).

Two equations, $F(x, y) = 0$ and $G(x, y) = 0$, are said to be *equivalent* if they have the same graphs. The only fact that we will need is the obvious one that if $G(x, y) \equiv CF(x, y)$, where C is some nonzero constant, then the two equations are equivalent.

THEOREM 4. If $F(x, -y) = 0$ is equivalent to $F(x, y) = 0$, then the graph of $F(x, y) = 0$ is symmetric with respect to the x-axis.

If $F(-x, y) = 0$ is equivalent to $F(x, y) = 0$, then the graph of $F(x, y) = 0$ is symmetric with respect to the y-axis.

In simple terms this means that we are to replace one of the variables by its negative and examine the resulting equation. For example, if we replace y by $-y$ in $y^3 - 16xy = 0$, we have $(-y)^3 - 16x(-y) = 0$ or

$$(34) \qquad\qquad -y^3 + 16xy = 0.$$

Since (33) can be obtained from (34) by multiplying by -1, the two equations are equivalent. Hence by Theorem 4 the graph of (33) is symmetric with respect to the x-axis.

If we replace x by $-x$ in (33), we obtain $y^3 + 16xy = 0$. Clearly, this equation is not equivalent to (33), so the graph of (33) is not symmetric with respect to the y-axis.

Proof of Theorem 4. Let $P_1(a, b)$ be a point on the graph of $F(x, y) = 0$. If P_2 is the image of P_1 in the x-axis, then P_2 has the coordinates $(a, -b)$ (see Fig. 8). Since P_1 is on the given graph, $F(a, b) = 0$. Then the coordinates of P_2 will satisfy the equation $F(x, -y) = 0$. If $F(x, -y) = 0$ is equivalent to $F(x, y) = 0$, this means that P_2 is on the original graph. Hence the original graph is symmetric with respect to the x-axis.

Similarly, the point $P_3(-a, b)$ is the image of $P_1(a, b)$ in the y-axis. So if $F(-x, y) = 0$ is equivalent to $F(x, y) = 0$, both P_1 and P_3 lie on the graph together, and the graph is symmetric with respect to the y-axis. ∎

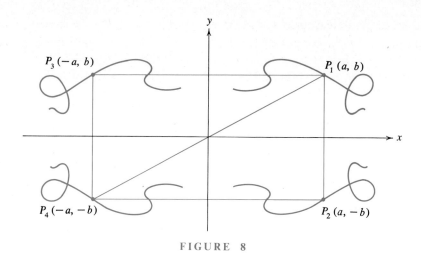

FIGURE 8

Referring to Fig. 8, we also notice that the point $P_4(-a, -b)$ is the image of $P_1(a, b)$ in the origin. Hence we have

---

THEOREM 5. If $F(-x, -y) = 0$ is equivalent to $F(x, y) = 0$, then the graph of $F(x, y) = 0$ is symmetric with respect to the origin.

Example 1. Discuss the symmetry of the graph of

$$(35) \qquad\qquad y = \frac{x^3}{2(x^2 - 1)}.$$

Solution. It is not necessary to transpose the terms and obtain the form $F(x, y) = 0$. We can substitute directly in (35). Replacing y by $-y$ gives

$$(36) \qquad\qquad -y = \frac{x^3}{2(x^2 - 1)}.$$

Replacing x by $-x$ gives

$$(37) \qquad\qquad y = \frac{-x^3}{2(x^2 - 1)}.$$

Making both replacements, x by $-x$ and y by $-y$, we obtain

$$(38) \qquad\qquad -y = \frac{-x^3}{2(x^2 - 1)}.$$

Equations (36) and (37) are *not* equivalent to (35), but (38) is equivalent to (35).

Hence the graph is symmetric with respect to the origin but not with respect to either axis. ●

The graph of equation (35), along with its asymptotes, is shown in Fig. 9.

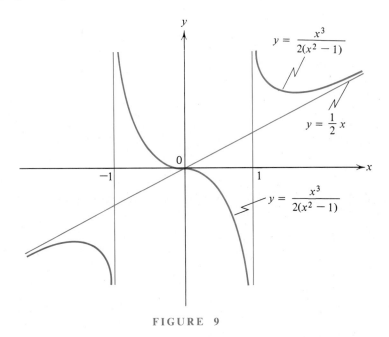

FIGURE 9

EXERCISE 3

1. Prove that for each of the following equations, the graph has the indicated axis of symmetry or center of symmetry.

   (a) $\dfrac{x^2}{a^2} + \dfrac{y^2}{b^2} = 1$, *x*-axis, *y*-axis, origin.   (b) $y^2 = 4px$, *x*-axis.

   (c) $\dfrac{x^2}{a^2} - \dfrac{y^2}{b^2} = 1$, *x*-axis, *y*-axis, origin.   (d) $x^2 = 4py$, *y*-axis.

   (e) $xy = C$,   origin.

2. Find the axes of symmetry (if there are any) for each of the curves.

   (a) $\dfrac{(x-h)^2}{a^2} + \dfrac{(y-k)^2}{b^2} = 1$.   (b) $\dfrac{(y-k)^2}{b^2} - \dfrac{(x-h)^2}{a^2} = 1$.

   (c) $(x-h)^2 = 4p(y-k)$.   (d) $(y-k)^2 = 4p(x-h)$.

3. Do any of the curves in Problem 2 have a center of symmetry?

In Problems 4 through 9, examine the given equation for symmetry and sketch the graph of the given equation.

4. $x^2 y^2 = 16$.

5. $y^2 = 4 + x^3$.

6. $y^2 = x^2(1 - x^2)$.

7. $y^2 = 4\dfrac{x - 2}{x - 5}$.

8. $y^2 = \dfrac{x^2 - 6x}{x - 3}$.

9. $y^2 = \dfrac{x^2 + 1}{x^2 - x}$.

*10. Prove that if the equation $F(x, y) = 0$ is equivalent to the equation $F(y, x) = 0$, obtained by interchanging the variables, then the graph of $F(x, y) = 0$ is symmetric with respect to the line $y = x$.

11. Test each of the following equations for symmetry about the line $y = x$ (see Problem 10).

(a) $x^n + y^n = r^n$.

(b) $xy = 1$.

(c) $y = \dfrac{x^2 y^2}{x^3 - y^3}$.

(d) $x^2 + y^2 = x^3 + y^3$.

*12. By a suitable translation of the coordinate axes, prove that the graph of $y = (3x - 2)/(x - 1)$ is symmetric with respect to the line $y = x + 2$.

*13. Use implicit differentiation to show that the graph of $x^2 + y^2 = x^3 + y^3$ is falling (decreasing function) for those points for which $x > 1$ and $y > 1$. Then show that there are no points on the graph that lie in the quarter-plane $x > 1, y > 1$.

14. Prove that the graph of $y = x^3 - 4x$ is symmetric with respect to the origin. Using a suitable translation, find a center of symmetry for the graph of $y = (x - 7)(x - 9)(x - 11)$.

15. Let $y = (x - a)(x - b)(x - c)$, and suppose that $b = (a + c)/2$. What can you say about a center of symmetry of this graph?

The Conic Sections Again 4

In Chapter 2 the parabola, ellipse, and hyperbola were treated individually. But there is one property that brings these three curves together in a single family, and this property is the *eccentricity*. Further we will see that the ellipse and hyperbola also have a directrix (in fact, each of these curves has two directrices).

The starting point for our discussion is the following algebraic definition for a general conic section.

DEFINITION 2. Let $e > 0$ be a fixed number, let $\mathscr{D}$ be a fixed line, and let F be a fixed point not on the line $\mathscr{D}$. Let $|PD|$ denote the distance of the point $P(x, y)$ from the line $\mathscr{D}$, and let $|PF|$ denote the distance from P to F. Then the collection $\mathscr{G}$ of all points P such that

(39)
$$\frac{|PF|}{|PD|} = e$$

is called a *conic section*. The point F is a *focus* of the graph $\mathscr{G}$, and the line $\mathscr{D}$ is called a *directrix*. The number e in equation (39) is called the *eccentricity* of $\mathscr{G}$.

That Definition 2 truly includes all our old friends is stated in

THEOREM 6 (PLE). If $0 < e < 1$, then the graph of (39) is an ellipse. If $e = 1$, equation (39) describes a parabola, and if $e > 1$, the graph is a hyperbola.

Notice that when $e = 1$ in (39), then this equation is identical with the defining equation for a parabola, so Theorem 6 is certainly true in this case. For the other values of e we leave the detailed justification of Theorem 6 to the reader. Our concern, instead, is to give the specific formulas for computing e from the familiar equations for the ellipse and hyperbola.

ELLIPSE. If the ellipse has the standard form

(10a) $\quad \dfrac{x^2}{a^2} + \dfrac{y^2}{b^2} = 1, \quad a > b > 0 \quad$ or $\quad$ (10b) $\quad \dfrac{x^2}{b^2} + \dfrac{y^2}{a^2} = 1, \quad a > b > 0,$

set

(40)
$$c = \sqrt{a^2 - b^2}, \qquad e = \frac{c}{a}, \qquad d = \frac{a}{e}.$$

In either case, $e = c/a$ gives the eccentricity.

For equation (10a), if F is the point $(c, 0)$ and $\mathscr{D}$ is the vertical line $x = d$, with c and d given in (40), then it is not difficult to show (see Problem 33) that the points of the ellipse (10a) satisfy equation (39) with $e = c/a$. Since the graph of (10a) is symmetric with respect to the y-axis, we see immediately that the point $(-c, 0)$ also acts as a focus for this conic section with the line $x = -d$ as the associated directrix.

For the ellipse (10b), the foci are $(0, \pm c)$ and the directrices are the horizontal lines $y = \pm d$, where c and d are again given in (40).

We observe that $c < a$ so that we always have $0 < e < 1$. If $a = b$, the ellipse is a circle and the eccentricity is zero. The effect of changing e is indicated by the three ellipses shown in Figs. 10a to 10c. The foci and directrices of a typical ellipse are shown in Fig. 11. The important fact is that for all points on the ellipse, we have $|PF|/|PD| = e$.

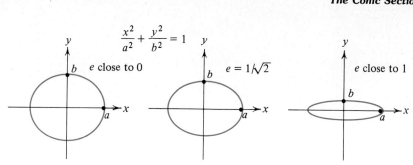

FIGURE 10a   FIGURE 10b   FIGURE 10c

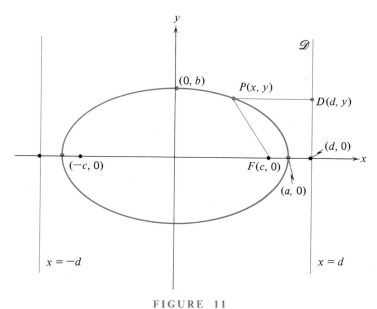

FIGURE 11

Example 1. Find the eccentricity and directrices for the ellipse $x^2/5 + y^2/9 = 1$.

Solution. Since $c = \sqrt{a^2 - b^2} = \sqrt{9 - 5} = 2$, equation (40) gives

$$e = \frac{c}{a} = \frac{2}{3} \quad \text{and} \quad d = \frac{a}{e} = \frac{3}{2/3} = \frac{9}{2}.$$

Thus the directrices are the two horizontal lines $y = \pm 4.5$. ●

HYPERBOLA. If the hyperbola has the standard form

(11)
$$\frac{x^2}{a^2} - \frac{y^2}{b^2} = 1$$

or

(12)
$$\frac{y^2}{a^2} - \frac{x^2}{b^2} = 1,$$

set

(41)
$$c = \sqrt{a^2 + b^2}, \qquad e = \frac{c}{a}, \qquad d = \frac{a}{e}.$$

In either case e gives the eccentricity and this is always greater than 1. For equation (11) the two foci are $(\pm c, 0)$ and the two directrices are the vertical lines $x = \pm d$. For equation (12) the foci are $(0, \pm c)$ and the directrices are the horizontal lines $y = \pm d$.

The effect of changing e is indicated by the three hyperbolas shown in Figs. 12a to 12c.

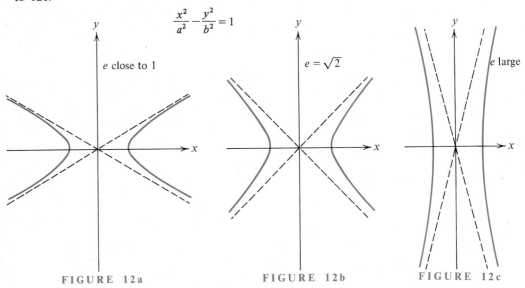

FIGURE 12a   FIGURE 12b   FIGURE 12c

The foci and directrices of a typical hyperbola are shown in Fig. 13. The important fact is that for all points on the hyperbola, equation (39) is again true: $|PF|/|PD| = e$.

Example 2. Discuss the graph of

(42)
$$3x^2 - y^2 - 18x + 2y = 22.$$

Solution. Completing the square in the usual way gives

$$3(x^2 - 6x + 9) - (y^2 - 2y + 1) = 22 + 27 - 1 = 48,$$

or

(43)
$$\frac{(x - 3)^2}{16} - \frac{(y - 1)^2}{48} = 1.$$

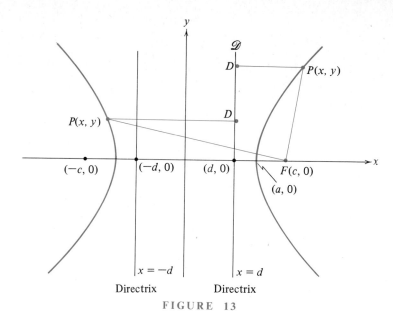

FIGURE 13

The change of variables $X = x - 3$, $Y = y - 1$ reduces (43) to the standard form

(44)
$$\frac{X^2}{16} - \frac{Y^2}{48} = 1$$

and amounts to a translation of the coordinate axes. Thus the graph of equation (42) is a hyperbola. Here $a = 4$, $b = \sqrt{48}$, and consequently

$$c = \sqrt{a^2 + b^2} = \sqrt{16 + 48} = 8, \qquad e = \frac{c}{a} = \frac{8}{4} = 2,$$

and $d = a/e = 4/2 = 2$. Hence the foci are $(\pm 8, 0)$ in the XY-system and the directrices are $X = \pm 2$. By Problem 27 of Exercise 2, the asymptotes are $Y = \pm\sqrt{3}\,X$.

We can now transfer all these data back to the original xy-system, through the substitutions $X = x - 3$, $Y = y - 1$. We find that in the xy-system, the foci are at $(11, 1)$ and $(-5, 1)$, the directrices are $x = 5$ and $x = 1$, and the vertices of the hyperbola are at $(7, 1)$ and $(-1, 1)$. Further, we find that the asymptotes are $y = \pm\sqrt{3}(x - 3) + 1$. These two asymptotes intersect at $(3, 1)$, the center of the hyperbola, and make angles of $60°$ and $120°$, respectively, with the positive x-axis. The curve, together with the important items, is shown in Fig. 14 (see next page).   ●

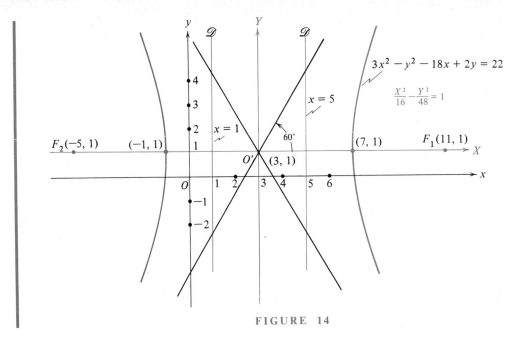

FIGURE 14

EXERCISE 4

In Problems 1 through 16, find the eccentricity, foci, and directrices of the given conic section. If the curve is a hyperbola find its asymptotes. It may be helpful to sketch the curve.

1. $\dfrac{x^2}{25} + \dfrac{y^2}{16} = 1.$

2. $\dfrac{x^2}{169} + \dfrac{y^2}{25} = 1.$

3. $\dfrac{x^2}{7} + \dfrac{y^2}{11} = 1.$

4. $\dfrac{x^2}{2} + \dfrac{y^2}{11} = 1.$

5. $y^2 - x^2 = 1.$

6. $3y^2 - x^2 = 3.$

7. $3x^2 - y^2 = 3.$

8. $5y^2 - 4x^2 = 20.$

9. $25x^2 + 9y^2 = 225.$

10. $25x^2 - 9y^2 = 225.$

11. $3x^2 + 4y^2 - 16y = 92.$

12. $25x^2 + 16y^2 + 200x + 400 = 160y.$

13. $8x^2 + 32x + 23 = y(y + 2).$

14. $9y^2 + 96x = 16x^2 + 72y + 144.$

15. $2(y^2 - 6y + 3) = x(x + 4).$

16. $4x(x - 2) + 3y(y + 2) = 41.$

17. Prove that the ellipse of Problem 12 is tangent to both the *x*- and *y*-axes.

18. Prove that the hyperbola of Problem 14 is tangent to the *x*-axis.

19. Let *r* be the ratio of the minor axis to the major axis in an ellipse. Prove that

$r = \sqrt{1 - e^2}$ and hence as $e \to 0$, $r \to 1$. Prove that if $e = 0$, then (10a) is the equation of a circle. Prove that if $r = 1$, then (10a) is also the equation of a circle.

In Problems 20 through 28, find the equation of the conic section with center of symmetry at the origin and satisfying the given conditions.

20. Major axis 6, focus at $(2, 0)$.

21. Eccentricity $1/9$, focus at $(1, 0)$.

22. Focus at $(10, 0)$, directrix $x = 8$.

23. Eccentricity 5, directrix $y = 2$.

24. Focus at $(3, 0)$, eccentricity 1.5.

25. Directrix $y = 13$, eccentricity $12/13$.

26. Focus at $(4, 0)$, directrix $x = -9$.

27. Ellipse passing through $(2, 1)$ and $(1, 3)$.

28. Focus at $(\sqrt{5}, 0)$, asymptotes $2y = \pm x$.

$\star$29. Prove that for a hyperbola with foci on the x-axis $e = \sqrt{1 + m^2}$, where $\pm m$ is the slope of the asymptotes. Thus the eccentricity measures the deviation of the hyperbola from the x-axis.

$\star$30. Prove that for each fixed $a > 1$, the ellipse

$$\frac{x^2}{a^2} + \frac{y^2}{a^2 - 1} = 1$$

has foci at $(\pm 1, 0)$. If a is large, this ellipse resembles a circle $x^2 + y^2 = a^2$. What curve (or figure) does this ellipse approach as $a \to 1^+$?

$\star\star$31. Prove that for each fixed $M > 2$, the ellipse

$$\frac{(x - M)^2}{M^2} + \frac{y^2}{2M} = 1$$

has center at $(M, 0)$ and vertices at $(0, 0)$, $(2M, 0)$. Show that $e = \sqrt{1 - 2/M}$ and the foci are at $(M \pm \sqrt{M^2 - 2M}, 0)$. Use the vertices to sketch a few of these ellipses for large values of M, for example, $M = 50$, $M = 5000$, and so on.

$\star\star$32. Solve the equation of Problem 31 for y^2 and show that as $M \to \infty$, this equation approaches $y^2 = 4x$. Therefore, under certain appropriate conditions the ellipse becomes a parabola as $e \to 1$. As $M \to \infty$, one focus approaches $(1, 0)$ and one directrix approaches the line $x = -1$. This is somewhat difficult to prove at present, but becomes easy with L'Hospital's Rule covered in Chapter 14.

$\star\star$33. Prove that if $a > b > 0$ and d and e are defined by equation set (40), then $|PF|/|PD| = e$ for every point on the ellipse defined by equation (10a) (see Fig. 11).

$\star\star$34. Do Problem 33 for the hyperbola defined by equation (11) (see Fig. 13).

CALCULATOR PROBLEMS

In Problems C1 through C4, give your answer to four significant figures.

C1. Each planet travels around the sun in an elliptical orbit with the sun as one of the foci. For the earth, the major semiaxis is 92,910,000 miles and the eccentricity is 0.01673.

Find the minor semiaxis. Find the maximum and minimum distance of the earth from the sun.

C2. Do Problem C1 for Pluto if the major semiaxis is 3,664,000,000 miles and $e = 0.2502$.

C3. If the orbit of Mars has $a = 141,570,000$ and $b = 140,950,000$ miles, find the eccentricity of the path traveled by Mars.

C4. Do Problem C3 for Mercury, where $a = 35,960,000$ miles and $b = 35,192,000$ miles.

5 *The Angle Between Two Curves*

If $\mathscr{L}_1$ and $\mathscr{L}_2$ are two distinct straight lines that are not parallel, they form four angles at their point of intersection. When we wish to specify φ, the angle of intersection of two lines, we take the least positive angle that the line $\mathscr{L}_1$ must be turned in a counterclockwise direction about the point of intersection to bring it into coincidence with $\mathscr{L}_2$ (see Fig. 15) and we call this angle *the angle from $\mathscr{L}_1$ to $\mathscr{L}_2$*. Thus φ is uniquely determined and, under these conditions, φ will lie in the range $0 < \varphi < 180°$. If the lines are given by their equations, it is easy to find φ by computing $\tan \varphi$, using the formula of Theorem 7.

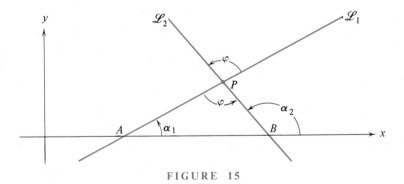

FIGURE 15

THEOREM 7. If $\mathscr{L}_1$ and $\mathscr{L}_2$ have slopes m_1 and m_2, respectively, and φ is the angle from $\mathscr{L}_1$ to $\mathscr{L}_2$, then

(45)
$$\tan \varphi = \frac{m_2 - m_1}{1 + m_1 m_2}.$$

Proof. If $\mathscr{L}_1$ and $\mathscr{L}_2$ are parallel or coincide, then $m_1 = m_2$ and equation (45) gives $\tan \varphi = 0$. In this case we agree to take $\varphi = 0$ as the angle from $\mathscr{L}_1$ to $\mathscr{L}_2$.

If $\mathcal{L}_1$ and $\mathcal{L}_2$ are not parallel, they will meet at a point P. Suppose that (as shown in Fig. 15) P lies above the x-axis and that φ is an interior angle of the triangle ABP formed by $\mathcal{L}_1$, $\mathcal{L}_2$, and the x-axis. Then it is clear that $\alpha_2 = \alpha_1 + \varphi$, or

(46) $$\varphi = \alpha_2 - \alpha_1.$$

Taking the tangent of both sides of (46), we find that

(47) $$\tan \varphi = \tan (\alpha_2 - \alpha_1) = \frac{\tan \alpha_2 - \tan \alpha_1}{1 + \tan \alpha_1 \tan \alpha_2} = \frac{m_2 - m_1}{1 + m_1 m_2}.$$

If we interchange the labels on the lines $\mathcal{L}_1$ and $\mathcal{L}_2$ in Fig. 15, then φ, the angle from $\mathcal{L}_1$ to $\mathcal{L}_2$, is an exterior angle of the triangle ABP. The energetic student can show that in this case equation (45) still holds.

Finally, suppose that P, the point of intersection, lies on or below the x-axis. Then by a suitable translation of the axes, P could be made to appear above the x-axis. But in this translation neither φ nor the slopes m_1 and m_2 of the lines $\mathcal{L}_1$ and $\mathcal{L}_2$ are changed. Therefore, formula (45), which has been proved when P lies above the x-axis, is true in all cases. ∎

Example 1. Find the angle between the lines

$$\mathcal{L}_1 : 2y = x + 5099 \qquad \text{and} \qquad \mathcal{L}_2 : 3y = -x + \sqrt{17\pi}.$$

Solution. We have $m_1 = 1/2$ and $m_2 = -1/3$. If φ denotes the angle from $\mathcal{L}_1$ to $\mathcal{L}_2$, equation (45) gives

$$\tan \varphi = \frac{-\dfrac{1}{3} - \dfrac{1}{2}}{1 + \left(\dfrac{1}{2}\right)\left(-\dfrac{1}{3}\right)} = \frac{-2 - 3}{6 - 1} = \frac{-5}{5} = -1.$$

Hence $\varphi = 135°$. A similar computation for θ, the angle from $\mathcal{L}_2$ to $\mathcal{L}_1$, will give $\tan \theta = +1$ or $\theta = 45°$. Since the problem did not specify a preferred direction, it is reasonable to present the smaller of the two angles, $45°$, as the answer. ●

DEFINITION 3 (Angle Between Curves). The angle between two curves at a point of intersection P is the angle between the tangent lines to the two curves at P.

Example 2. Find the angle of intersection of the two curves $y = x^3$ and $y = 12 - x^2$.

Solution. One point of intersection is $(2, 8)$. We leave it to the student to sketch these two curves and prove that there are no other points of intersection. For the curve $y = x^3$, $m_1 = 3x^2 = 12$ at $(2, 8)$, and for the curve $y = 12 - x^2$, $m_2 = -2x = -4$ at $(2, 8)$. Then

$$\tan \varphi = \frac{m_2 - m_1}{1 + m_1 m_2} = \frac{-4 - 12}{1 + (12)(-4)} = \frac{-16}{-47} = \frac{16}{47}.$$

Hence $\tan \varphi \approx 0.340$, and a calculator gives $\varphi \approx 18°47'$.   ●

EXERCISE 5

In each of Problems 1 through 12, find $\tan \varphi$, where φ is the angle from the first curve to the second curve, at each point of intersection of the two curves. In each case make a reasonably accurate sketch of the two curves near the point of intersection.

1. $y = x$,	$y = 2x$.	2. $y = 2x$,	$y = 3x$.
3. $y = x + \pi^2$,	$y = 2x - \sqrt{37}$.	4. $y = x$,	$y = 3x$.
5. $3y = x$,	$2y = x$.	6. $y = 5x$,	$y = -2x$.
7. $3y = x$,	$3y = -x$.	8. $y = -2x + 3$,	$y = x^3$.
9. $y = x^2$,	$y = 2x(x - 1)$.	10. $y = x^2$,	$y = x^3$.
★★11. $xy = 8$,	$y = 10 - x - x^2$.	12. $2y = x$,	$y = x^3$.

★13. Let P be an arbitrary point on the parabola $y^2 = 4px$ and let F be the focus $(p, 0)$

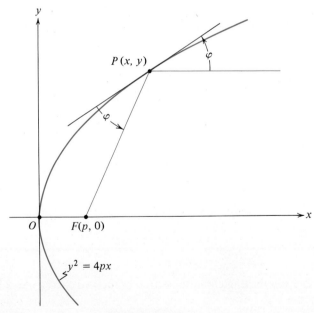

FIGURE 16

with $p > 0$. Prove that the line segment PF and the ray through P parallel to the x-axis make equal angles with the tangent line to the parabola at P (see Fig. 16). Thus a ray of light issuing from F will be reflected by the parabola along a line parallel to the x-axis. This explains why headlights and searchlights have the form of a paraboloid of revolution with the light source at the focus. Reflecting telescopes have the same type of design.

$\star$**14.** Prove that a ray of light issuing from one focus of the ellipse $x^2/a^2 + y^2/b^2 = 1$ will pass through the other focus. In other words, if P is any point on the ellipse, prove that the line segments PF_1 and PF_2 make equal angles with the tangent to the ellipse at P. HINT: The tangent of each of the angles is b^2/yc.

$\star$**15.** Prove that the line $y = x$ bisects the angle from the line $y = m_1 x$ to the line $y = m_2 x$ if and only if $m_1 m_2 = 1$.

$\star$**16.** Prove that the angle from $y = m_1 x + b_1$ to $y = m_2 x + b_2$ is $45°$ if and only if $m_2 = (1 + m_1)/(1 - m_1)$.

Rotation of Axes 6

In Section 1 we obtained the equations for the change in the coordinates of a point when the axes were translated. We now obtain a similar set of equations when the axes are rotated. Suppose that as indicated in Fig. 17, two rectangular coordinate systems have the same origin and that the $x'y'$-system can be obtained by rotating the xy-system about the origin in a counterclockwise direction through an angle α. If α is negative, then the rota-

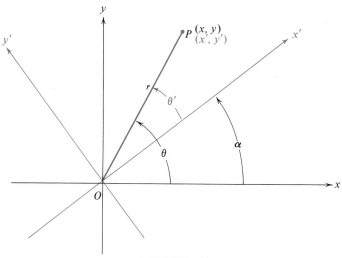

FIGURE 17

tion is understood to be in a clockwise direction through an angle $|\alpha|$. For brevity we merely say that the axes have been rotated through an angle α.

If P is any point in the plane, it has two sets of coordinates (x, y) and (x', y'), one for each coordinate system. To obtain equations relating these four quantities, we recall from trigonometry that if $r = |OP|$ is the length of the line segment joining P with the origin and if this line segment makes an angle θ with the positive x-axis, then

$$(48) \qquad\qquad\qquad x = r \cos \theta, \qquad y = r \sin \theta.$$

Similarly, if the line OP makes an angle θ' with the x'-axis, then

$$(49) \qquad\qquad\qquad x' = r \cos \theta', \qquad y' = r \sin \theta'.$$

Now $\theta = \theta' + \alpha$, so using this in (48), we find that

$$x = r \cos (\theta' + \alpha) = r \cos \theta' \cos \alpha - r \sin \theta' \sin \alpha.$$

$$y = r \sin (\theta' + \alpha) = r \sin \theta' \cos \alpha + r \cos \theta' \sin \alpha.$$

Substituting from (49), we find that these equations become

$$(50) \qquad\qquad \boxed{\begin{aligned} x &= x' \cos \alpha - y' \sin \alpha. \\ y &= x' \sin \alpha + y' \cos \alpha. \end{aligned}}$$

These equations can be solved inversely for x' and y' in the usual way. It is more instructive, however, to remark that if the $x'y'$-system is obtained from the xy-system by a rotation through an angle α, then the xy-system is obtained from the $x'y'$-system by a rotation through an angle $-\alpha$. Then the inverse equations can be obtained by replacing α by $-\alpha$, and shifting the primes to the unprimed letters. Since $\cos (-\alpha) = \cos \alpha$ and $\sin (-\alpha) = -\sin \alpha$, this gives

$$(51) \qquad\qquad \boxed{\begin{aligned} x' &= x \cos \alpha + y \sin \alpha, \\ y' &= -x \sin \alpha + y \cos \alpha. \end{aligned}}$$

THEOREM 8. If the xy-system is rotated through an angle α to obtain the $x'y'$-system, then the coordinates of a fixed point P are related by equations (50) and (51).

The rotation of axes is useful for putting certain equations into recognizable form.

Example 1. Find the new equation for the curve $xy = A^2$ if the axes are rotated through an angle of $45°$. Use this new equation to discuss the curve.

Solution. We use equation set (50) with $\sin \alpha = \cos \alpha = \sin 45° = \sqrt{2}/2$. Replacing x and y in $xy = A^2$ by their equivalent expressions (50), we have

$$\left(x'\frac{\sqrt{2}}{2} - y'\frac{\sqrt{2}}{2}\right)\left(x'\frac{\sqrt{2}}{2} + y'\frac{\sqrt{2}}{2}\right) = A^2,$$

$$\frac{x'^2}{2} - \frac{y'^2}{2} = A^2.$$

But this is just the equation of a hyperbola with $a^2 = b^2 = 2A^2$ and hence the hyperbola has foci at $(\pm 2A, 0)$ in the $x'y'$-system. The eccentricity is $c/a = 2A/\sqrt{2}A = \sqrt{2}$, and the directrices are the lines $x' = \pm A$. Then a rotation back to the original coordinate system makes it obvious that the curve $xy = A^2$ is a hyperbola with foci at $(\sqrt{2}A, \sqrt{2}A)$ and $(-\sqrt{2}A, -\sqrt{2}A)$. The eccentricity is, of course, still $\sqrt{2}$, but the directrices are now the lines $x + y = \pm\sqrt{2}A$. ●

The methods used in this example are suitable for removing the xy term in any quadratic expression. Indeed, suppose that

(52) $$Ax^2 + Bxy + Cy^2 + Dx + Ey + F = 0$$

is the equation of some curve in the xy-plane. Rotating the axes through an angle α leaves the curve unchanged but changes its equation to

(53) $$A'x'^2 + B'x'y' + C'y'^2 + D'x' + E'y' + F' = 0,$$

where (53) is obtained by using (50) in (52). We leave it to the student to carry out this substitution and show that the new coefficients A', B', C', D', E', and F' are given in terms of the old coefficients by the equations

(54)
$$\begin{cases} A' = A\cos^2\alpha + B\cos\alpha\sin\alpha + C\sin^2\alpha, \\ B' = B(\cos^2\alpha - \sin^2\alpha) + 2(C - A)\sin\alpha\cos\alpha, \\ C' = A\sin^2\alpha - B\sin\alpha\cos\alpha + C\cos^2\alpha, \\ D' = D\cos\alpha + E\sin\alpha, \\ E' = -D\sin\alpha + E\cos\alpha, \\ F' = F. \end{cases}$$

To remove the xy term, we merely set $B' = 0$. This gives

(55) $$B(\cos^2\alpha - \sin^2\alpha) = -2(C - A)\sin\alpha\cos\alpha,$$

$$\frac{\cos 2\alpha}{\sin 2\alpha} = \frac{A - C}{B} \qquad \text{if } B \neq 0.$$

We have proved

THEOREM 9. The xy term in $Ax^2 + Bxy + Cy^2 + Dx + Ey + F = 0$ can be removed by a rotation of the coordinate axes through an angle α, where

(56)
$$\cot 2\alpha = \frac{A - C}{B}, \quad \text{if } B \neq 0.$$

We remark for future reference that if we return to equation (55) and divide both sides by $B \cos^2 \alpha$, we find that $\tan \alpha$ is a positive root of

$$\tan^2 \alpha + \frac{2(A - C)}{B} \tan \alpha - 1 = 0, \quad B \neq 0.$$

Hence

(57)
$$\tan \alpha = \frac{C - A}{B} + \sqrt{\left(\frac{C - A}{B}\right)^2 + 1}$$

gives $\tan \alpha$ in terms of A, B, and C. In some cases it may be simpler to use equation (56) to determine $\cos 2\alpha$ and $\sin 2\alpha$. Then we can determine α from the relation

(58)
$$\tan \alpha = \frac{1 - \cos 2\alpha}{\sin 2\alpha}.$$

It now seems to be obvious that the curve defined by (52) is always a conic section, but, unfortunately, such a simplified statement of the result is false because certain special degenerate cases may arise. For example, (52) may be factorable and hence represent two lines. Or if $A = B = C = 0$, then (52) may represent a single line. Other possibilities can occur, and we leave these to the next exercise. Stated precisely, we have

---

THEOREM 10 (PLE). The graph of (52) is one of the following figures: an ellipse, a circle, a parabola, a hyperbola, two lines, a single line, a point, or no points.

It is convenient to have at hand a test that can be applied to the quadratic expression (52) in order to determine the type of conic, without actually performing the rotation. For simplicity, we ignore the various degenerate cases.

THEOREM 11 (PWO). If the graph of (52) is a hyperbola, parabola, or ellipse, then:

(1) It is a hyperbola if $B^2 - 4AC > 0$.

(2) It is a parabola if $B^2 - 4AC = 0$.

(3) It is an ellipse if $B^2 - 4AC < 0$.

The quantity $B^2 - 4AC$ is called the *discriminant* of the quadratic expression (52). The proof of Theorem 11 depends upon the fact that $B^2 - 4AC$ is *invariant* under the transformation equations (54), that is, under a rotation of the axes. The term *invariant* means in this case that for every value of α in (54), we have

(59) $$B'^2 - 4A'C' = B^2 - 4AC.$$

Other expressions may also be invariant, and the concept of invariance may be applied to other types of transformations.

Example 2. Discuss the graph of

(60) $$2x^2 + 3xy + 2y^2 = 7.$$

Solution. Since $B^2 - 4AC = 9 - 16 = -7 < 0$, the curve is an ellipse. To obtain more information about this ellipse, we must rotate the axes through an angle α, where $\cot 2\alpha = (A - C)/B = (2 - 2)/3 = 0$, whence $2\alpha = 90°$ and $\alpha = 45°$. The transformation equations (50) give

$$x = (x' - y')\sqrt{2}/2, \qquad y = (x' + y')\sqrt{2}/2.$$

Either direct substitution in (60) or using the equation set (54) yields

$$\frac{x'^2}{2} + \frac{y'^2}{14} = 1.$$

Here $a = \sqrt{14}$ and $b = \sqrt{2}$, and this equation has the form (10b) (see page 452). From equation set (40),

$$c = \sqrt{a^2 - b^2} = \sqrt{14 - 2} = 2\sqrt{3},$$

$$e = \frac{c}{a} = \frac{2\sqrt{3}}{\sqrt{14}} = \sqrt{\frac{6}{7}} < 1,$$

$$d = \frac{a}{e} = \frac{\sqrt{14}}{\sqrt{6/7}} = \frac{7}{\sqrt{3}}.$$

It follows that in the $x'y'$-system the foci are $(0, \pm 2\sqrt{3})$ and the directrices are $y' = \pm 7/\sqrt{3}$.

Using equations (50) and (51) to return to the original coordinate system, we find that the curve is still an ellipse, but now the foci are $(\sqrt{6}, -\sqrt{6})$ and

$(-\sqrt{6}, \sqrt{6})$ and the directrices are $y - x = \pm 14/\sqrt{6}$. The vertices at the ends of the major axis are $(\sqrt{7}, -\sqrt{7})$ and $(-\sqrt{7}, \sqrt{7})$, and the end points of the minor axis are $(1, 1)$ and $(-1, -1)$. ●

If the original quadratic equation has linear terms in x and y, then it may be necessary to perform both a rotation of the axes and a translation of the axes to transform the equation into one of the standard forms. The actual computation may become very complicated and tedious, but no new ideas are involved, so we will omit further discussion. The curious reader can find an example of this type and more problems in Goodman's *Analytic Geometry and the Calculus,* 4th. ed. (New York: Macmillan Publishing Co., Inc., 1980).

EXERCISE 6

In Problems 1 through 10, identify the conic sections, and then by a rotation of the axes, find the eccentricity, the foci, and directrices (or focus and directrix if the curve is a parabola). Find the asymptotes if the curve is a hyperbola. Sketch the graph in each case.

1. $x^2 - xy + y^2 = 12$.
2. $x^2 + xy + y^2 = 24$.
3. $x^2 + 3xy + y^2 = 15$.
4. $x^2 - 3xy + y^2 = 10$.
★5. $x^2 + 2xy + y^2 = 8y - 8x$.
★6. $4x^2 + 4xy + y^2 = 60x - 120y$.
7. $13x^2 - 2\sqrt{3}\, xy + 15y^2 = 192$.
8. $7x^2 - 18xy + 7y^2 = 16$.
9. $5x^2 + 6xy - 3y^2 = 24$.
10. $7x^2 + 6\sqrt{3}\, xy + 13y^2 = 64$.

11. Prove that the equation $x^2 + y^2 = r^2$ is unchanged under a rotation of the axes through any angle.

★12. Show by specific examples that the graph of the equation $Ax^2 + Bxy + Cy^2 + Dx + Ey + F = 0$ may be: **(a)** two parallel lines, **(b)** two intersecting lines, **(c)** a single line, **(d)** a point, or **(e)** no points.

★13. Derive equation set (54) by using (50) in (52).

In Problems 14 through 18, use equation set (5) as the definition of a translation, and equation set (50) as the definition of a rotation.

14. Prove that under a translation, the equation of a straight line goes into the equation of another straight line with the same slope. Thus m in $y = mx + b$ is an invariant under a translation.

15. Prove that m in Problem 14 is *not* an invariant under a rotation.

★16. Prove that the distance between two points is an invariant under a translation of the coordinate axes.

★17. Prove that the distance between two points is an invariant under a rotation of the axes.

*18. Prove that under a rotation of the axes: **(a)** F is an invariant, **(b)** $A + C$ is an invariant, and **(c)** $D^2 + E^2$ is an invariant.

*19. Suppose that the coordinate axes are *fixed,* but the *point* $P(x, y)$ is *moved* to a new position $P'(x', y')$ by a rotation of the point through an angle α about the origin. Naturally, in this rotation the point is always on a fixed circle with center at the origin. Prove that

$$x' = x \cos \alpha - y \sin \alpha,$$

$$y' = x \sin \alpha + y \cos \alpha.$$

*20. Any transformation of the plane that takes each point $P(x, y)$ into the new point $P'(x', y')$, defined by the equations of Problem 19, is called a *rotation of the plane about the origin through an angle* α.

   Prove that as far as the correspondence $(x, y) \leftrightarrow (x', y')$ is concerned, a rotation of the plane about the origin through an angle α is identical with the rotation of the coordinate axes through an angle $-\alpha$.

   It follows from this that the quantities mentioned in Problems 17 and 18 are invariant under any rotation of the plane.

*21. Prove that under a rotation of the plane right angles go into right angles. HINT: If in a triangle, $a^2 + b^2 = c^2$, then the angle opposite c is a right angle. But distance is invariant under a rotation.

THE INVARIANCE OF $B^2 - 4AC$

22. Deduce from equation set (54) the relations

$$B' = B \cos 2\alpha + (C - A) \sin 2\alpha,$$

$$A' + C' = A + C,$$

$$A' - C' = (A - C) \cos 2\alpha + B \sin 2\alpha.$$

23. Prove that for any A', B', C'

$$B'^2 - 4A'C' = B'^2 + (A' - C')^2 - (A' + C')^2.$$

24. Use the formulas from Problems 22 and 23 to prove that

$$B'^2 - 4A'C' = B^2 - 4AC.$$

REVIEW PROBLEMS

In Problems 1 through 12, sketch a graph of the given equation. It may be helpful to examine the equation for symmetries and asymptotes.

1. $y = \dfrac{x^2 + 12}{x^2 + 4}$.

2. $y = \dfrac{x^2 + 12}{x^2 - 4}$.

3. $y = 18 \dfrac{x + 4}{x^2 + 9}$.

4. $y = 18 \dfrac{x + 5}{x^2 - 9}$.

5. $y = \dfrac{x^3}{9 - x^2}$.

6. $y = \dfrac{6x}{\sqrt{x^2 - x - 12}}$.

7. $y = \dfrac{x^2 - 4}{\sqrt{x^3 - 16x}}$.

8. $y = \dfrac{\sqrt{x^4 - 16}}{x}$.

9. $y^2(x^2 - 9) = x$.

10. $y^2 = x^4 - 16x^2$.

11. $y^2 x^2 = y^2 x^4 + 1$.

★12. $12 y^2 x^2 = y^2 x^4 + x$.

13. Prove that the graph of $y = (ax + b)/(cx + d)$ is always a straight line or a hyperbola. Is it possible for the graph to be a horizontal straight line?

★14. Prove that the graph of $y = \sqrt{(x - a)(x - b)}$ is a portion of a hyperbola if $a \neq b$. Without any computation, give the center and vertices of this hyperbola.

★15. Prove that the graph of $\sqrt{x} + \sqrt{y} = k$, $k > 0$, is a portion of a parabola. HINT: It is not necessary to find a focus or directrix.

★16. Find the slope of the line tangent to $\sqrt{x} + \sqrt{y} = k$ at $(k^2, 0)$. Note that this is a one-sided tangent and you are computing a left derivative.

★17. For the graph of $y^2 = x^2(1 - x^2)$ find the extreme values of y. The graph decomposes into two curves near the origin. Find the slope of the tangent line to each of these curves at the origin. (See Exercise 3, Problem 6.)

In Problems 18 through 23, find tan φ, *where φ is the angle from the first curve to the second curve at each point of intersection of the two curves.*

18. $2y = x^2$, $xy = 4$.

19. $y^2 = 4x$, $x^2 + y^2 = 5$.

20. $y = x^2$, $y = 2x^2 + x - 2$.

21. $x^2 + y^2 = 10$, $y^2 = 3x$.

22. $y^2 = x^3 + 1$, $3x^2 + 2y^2 = 7$.

23. $y = 3x$, $x^2 - xy + 2y^2 = 16$.

★24. Two one-parameter families of curves $\mathscr{F}_1$ and $\mathscr{F}_2$ are said to be *mutually orthogonal* if each curve from $\mathscr{F}_1$ meets each curve from $\mathscr{F}_2$ in a right angle at every point of intersection of the two curves. Show that

$$\left(\frac{dy}{dx}\right)_1 \left(\frac{dy}{dx}\right)_2 = -1$$

(where the subscripts refer to the curves from the two families) is a criterion for the orthogonality of the families.

★25. Use the criterion of Problem 24 to show that the two families $y = mx$ and $x^2 + y^2 = r^2$ are mutually orthogonal. HINT: Show that

$$\left(\frac{dy}{dx}\right)_1 = \frac{y}{x} \quad \text{and} \quad \left(\frac{dy}{dx}\right)_2 = -\frac{x}{y}.$$

In Problems 26 through 29, prove that the two given families are mutually orthogonal. HINT: *In each case express the derivative in a form that is free of the parameter (as in Problem 25).*

$\star$**26.** $xy = a$, $\qquad y^2 - x^2 = b$.

$\star$**27.** $x^2 - 2y^2 = a$, $\qquad y = \dfrac{b}{x^2}$.

$\star$**28.** $y = ax^2$, $\qquad x^2 + 2y^2 = b^2$.

$\star$**29.** $y^2 = ax^3$, $\qquad 2x^2 + 3y^2 = b^2$.

$\star\star$**30.** Sketch the graph of each of the following equations.

(a) $|y + x^4| = 4$.

(b) $y^2 = [x]$.

(c) $[y^2] = x$.

(d) $[x^2 + y^2] = 1$.

(e) $[x^2 + 4y^2] = 4$.

(f) $[y] = |x|$.

In Problems 31 through 36, rotate the axes through a suitable angle to remove the xy term, and put the resulting equation in standard form.

31. $5x^2 - 6xy + 5y^2 = 32$.

32. $5x^2 + 4xy + 2y^2 = 6$.

33. $5x^2 - 3xy + 9y^2 = 18$.

34. $4x^2 + 10xy + 4y^2 = 9$.

35. $2x^2 - \sqrt{3}xy + y^2 = 50$.

36. $11x^2 - 24xy + 4y^2 + 40 = 0$.

12

Vectors in the Plane

1 Objective

Certain quantities in nature possess both a magnitude and a direction. Force is such a quantity. Similarly, the velocity of a moving particle has a magnitude (called its *speed*) and a direction, the direction in which the particle is moving.

Initially, the theory of vectors was constructed to handle problems involving forces and velocities. It is quite natural to represent a force (or a velocity) by a directed line segment (an arrow). The length of the line segment is the magnitude of the force (or the speed of the moving particle). The arrow and the position of the line segment give the direction of the force (or of the moving particle). Consequently, in the first organization of the theory, a vector was defined to be a directed line segment. Later the theory was generalized and refined. At an intermediate stage, a vector is defined algebraically as an ordered *n*-tuple of numbers, written in the form $[a_1, a_2, \ldots, a_n]$. Today a vector is any element of a vector space, and a vector space is any collection of elements that satisfy a certain set of axioms.

Although the algebraic definition of a vector has many advantages in simplicity and efficiency, the student who is introduced to vectors this way may not realize the source of the definitions and may find it difficult to relate the theory to either geometric problems or physical problems. Consequently, we prefer to use the geometric definition. The reader who is aware of the various possibilities in defining a vector will have no difficulty in adjusting to different definitions as he continues his study of mathematics.

Our objective in this chapter is to study the algebra and calculus of directed line segments (vectors) in the plane. The theory of three-dimensional vectors is considered in

Chapter 16. The student who wishes to study Chapter 16 at the conclusion of this chapter can easily do so, since the material in the intervening chapters is not used in Chapter 16.

The Algebra of Vectors 2

DEFINITION 1 (Vector). A vector is a directed line segment.

In creating a mathematical theory, we must start by deciding just what is meant by equality, addition, subtraction, multiplication, and so on, for our new quantities. In keeping with this program, we state

---

DEFINITION 2 (Equality of Vectors). Two vectors are said to be equal if they have the same length and the same direction.

Thus two vectors may be equal without being collinear.

To distinguish vectors from numbers, we may use boldface type: Thus **A** is a vector and A is a number. Frequently, it is convenient to use letters with arrows to indicate a vector, or double letters giving the beginning point and ending point of the directed line segment. Thus, in Fig. 1, **B**, **PQ**, $\vec{B}$, and $\overrightarrow{PQ}$ all denote the same vector. The point P is called the *initial point* and Q is called the *terminal point* of the vector **PQ.**

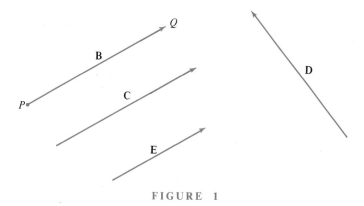

FIGURE 1

As an illustration of Definition 2, we see that the vectors **B** and **C** of Fig. 1 are equal (**B** = **C**) because they have the same length and the same direction. *Thus a given vector is* *equal to any vector obtained by shifting the given vector parallel to itself.* The vectors **B** and **D** of Fig. 1 have the same length but do not have the same direction, so **B** $\neq$ **D**. Similarly, the vectors $\overrightarrow{PQ}$ and $\overrightarrow{QP}$ are not equal because one has exactly the opposite direction of the other. The vectors **B** and **E** of Fig. 1 have the same direction, but **B** $\neq$ **E** because the lengths are different.

The length, or magnitude, of a vector **B** is frequently denoted by |**B**| and is always a nonnegative number. Sometimes it is convenient to use the corresponding letter in ordinary type for the length; for example, $B = $ |**B**| is the length of the vector **B**. We also use |**PQ**| to denote the length of the vector **PQ** from the point P to the point Q.

A point can be regarded as a vector of zero length. Because a point has no particular direction it is not a vector by our previous definition. The casual reader would not be disturbed by this slight defect. But rigor requires that we amend our previous definition. Hence to our collection of vectors covered in Definition 1 we adjoin a particular new vector denoted by **0** and called the *zero vector*. By definition this vector is the only one that has zero length. Whenever we speak of a vector from the point P to the point Q, it is understood that this vector is the zero vector if and only if the points P and Q coincide.

DEFINITION 3 (Addition of Vectors). To add the vectors **AB** and **CD,** place **CD** so that its beginning point C falls on the end point B of **AB.** The sum **AB** + **CD** is then the vector **AD.**

This definition is illustrated in Fig. 2. Note that two vectors do *not* need to be parallel or perpendicular in order to form their sum. The sum of two vectors is frequently called the *resultant,* and each of the vectors in the sum is called a *vector component* of the resultant.

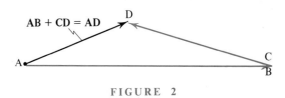

FIGURE 2

Why did we define the sum of two vectors in this way? Because our objective is to develop a theory that will be useful in studying forces and velocities, and it is a fact that forces and velocities do indeed combine in just the manner described in Definition 3. The student is probably already familiar with the *parallelogram law* for adding two forces. This rule is illustrated in Fig. 3. In this definition for the sum of two vectors, the two vectors to be added are placed so that their beginning points coincide. The sum or *resultant* is then the diagonal of the parallelogram shown in Fig. 3.

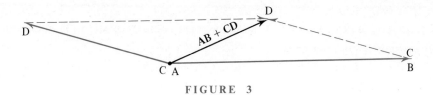

FIGURE 3

It is clear from Fig. 3 that adding two vectors by the parallelogram method gives the same result as adding the same two vectors by the "tail-to-head" method of Definition 3. Hence the two definitions are equivalent and we may use whichever definition convenience dictates.

The definition of addition gives a number of important algebraic properties of vectors. Some of these are stated in

---

THEOREM 1 (PLE). Vector addition is commutative, associative, and the zero vector is the additive identity.

By this we mean that if **A, B,** and **C** are any three vectors, then

(1) $$\mathbf{A} + \mathbf{B} = \mathbf{B} + \mathbf{A} \qquad \text{(Commutative Law)}$$

(2) $$\mathbf{A} + (\mathbf{B} + \mathbf{C}) = (\mathbf{A} + \mathbf{B}) + \mathbf{C} \qquad \text{(Associative Law)}$$

(3) $$\mathbf{0} + \mathbf{A} = \mathbf{A} + \mathbf{0} = \mathbf{A}.$$

Diagrams for the proofs of (1) and (2) are given in Figs. 4 and 5. For example, to prove the associative law, equation (2), Fig. 5 shows that the vector **S** is the sum of **A** and **(B + C)** (examine $\triangle OPR$), and it is also the sum of **(A + B)** and **C** (examine $\triangle OQR$). Therefore, both sides of equation (2) give the same vector, **S.**

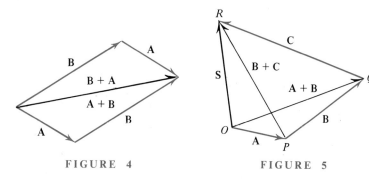

FIGURE 4   FIGURE 5

DEFINITION 4 (The Negative of a Vector). The vector **B** is said to be the negative of the vector **A,** and we write

(4) $$\mathbf{B} = -\mathbf{A}$$

if

(5) $$\mathbf{B} + \mathbf{A} = \mathbf{0}.$$

THEOREM 2. If $\mathbf{A} = \overrightarrow{PQ}$, then the negative of the vector **A** is the vector $\overrightarrow{QP}$.

Proof. This is illustrated in Fig. 6. The sum of the two vectors

$$\overrightarrow{PQ} + \overrightarrow{QP}$$

is the vector from P to P and clearly this is the zero vector. ∎

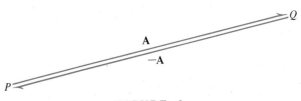

FIGURE 6

If we add the vector **A** to itself, we obtain a vector in the same direction but twice as long. Such a vector would naturally be written as $\mathbf{A} + \mathbf{A} = 2\mathbf{A}.$ This suggests

---

DEFINITION 5 (Scalar Times a Vector). If $\mathbf{A} \neq \mathbf{0}$ is a vector and c is a number, then the product $c\mathbf{A}$ is a vector whose magnitude is $|c|\,|\mathbf{A}|$. The product has the direction of **A** if $c > 0$, and the opposite direction if $c < 0$. If $c = 0$, or $\mathbf{A} = \mathbf{0}$, then $c\mathbf{A}$ is the zero vector.

This definition is illustrated in Fig. 7. It is clear that the product $(-1)\mathbf{A}$ is the same as the negative of the vector as described in Definition 4. In symbols $(-1)\mathbf{A} = -\mathbf{A}.$ Thus the two definitions are consistent.

A number is frequently called a *scalar* to distinguish it from a vector. Definition 5 gives the rules for multiplying a scalar and a vector. The rules for multiplying two vectors will be given in Chapter 16.

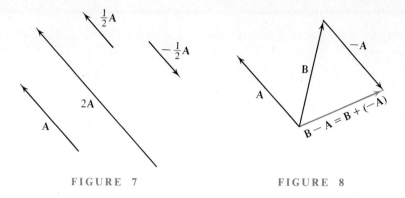

FIGURE 7   FIGURE 8

DEFINITION 6 (Subtraction). To subtract the vector **A** from the vector **B** add the negative of **A** to **B**. In symbols,

$$\mathbf{B} - \mathbf{A} \equiv \mathbf{B} + (-1)\mathbf{A}.$$

This definition is illustrated in Fig. 8.

If both the vectors **A** and **B** have the same initial point, then (as illustrated in Fig. 9) the parallelogram law gives the very important

---

THEOREM 3. If both **A** and **B** have the same initial point, then **B** − **A** is the vector drawn from the terminal point of **A** to the terminal point of **B**.

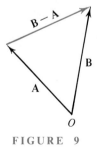

FIGURE 9

The preceding definitions and theorems give a large number of important relations, most of which are easy to prove but nevertheless must be stated. These are collected in

> **THEOREM 4 (PLE).** If **A** and **B** are any two vectors and c and d are any two numbers, then
>
> $$1\mathbf{A} = \mathbf{A}, \qquad\qquad (-1)\mathbf{A} = -\mathbf{A},$$
>
> $$(\mathbf{A} - \mathbf{B}) + \mathbf{B} = \mathbf{A}, \qquad\qquad \mathbf{A} - \mathbf{B} = -(\mathbf{B} - \mathbf{A}),$$
>
> (6) $$c(d\mathbf{A}) = (cd)\mathbf{A}, \qquad\qquad (c + d)\mathbf{A} = c\mathbf{A} + d\mathbf{A},$$
>
> (7) $$c(\mathbf{A} + \mathbf{B}) = c\mathbf{A} + c\mathbf{B}.$$

We would continue the development of vector algebra, using purely geometric methods (see Figs. 1 through 9), but the time has come to introduce the *algebraic method* for representing vectors. This algebraic form is ideal for computations with vectors.

We suppose, for the rest of this chapter, that all the vectors lie in a plane. We introduce into this plane the usual x- and y-coordinate axes. Along with this coordinate system we introduce two new elements, namely, two vectors **i** and **j** each of unit length (i.e., of length 1). The vector **i** points in the direction of the positive x-axis, and the vector **j** points in the direction of the positive y-axis. Since a vector may be shifted parallel to itself, it is convenient to think of **i** as going from $(0, 0)$ to $(1, 0)$. Similarly, **j** can be realized as the vector from $(0, 0)$ to $(0, 1)$ (see Fig. 10). It is now obvious that if P is the point $(7, 5)$, then the vector **OP** from the origin O to the point P is just $7\mathbf{i} + 5\mathbf{j}$ (see the definition for the sum of two vectors).

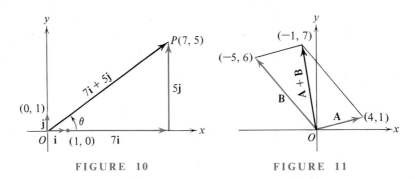

FIGURE 10   FIGURE 11

Any vector in the plane can be written using these unit vectors **i** *and* **j**. Indeed, if **A** is the vector, we merely shift it parallel to itself so that the initial point of **A** falls on the origin. If, after the shift, the terminal point of **A** falls on (a_1, a_2), then $\mathbf{A} = a_1\mathbf{i} + a_2\mathbf{j}$. The quantities a_1 and a_2 are called the *components* of the vector along the x- and y-axis, respectively. For example, in Fig. 10 the vector **OP** has the components $a_1 = 7$ and $a_2 = 5$. These components are scalars and should be distinguished from the vector components, which for the vector **OP** are 7**i** and 5**j**. A component can be negative; for example, the components of the vector **B** of Fig. 11 are -5 and 6. Sometimes it is convenient to use letter subscripts for components. Thus we might write $\mathbf{A} = A_x\mathbf{i} + A_y\mathbf{j}$. With this notation the vector **A** of Fig. 11 has the components $A_x = 4$ and $A_y = 1$.

THEOREM 5. If $\mathbf{A} = a_1\mathbf{i} + a_2\mathbf{j}$ and $\mathbf{B} = b_1\mathbf{i} + b_2\mathbf{j}$, then

$$(8) \qquad \mathbf{A} + \mathbf{B} = (a_1 + b_1)\mathbf{i} + (a_2 + b_2)\mathbf{j}.$$

Proof. Using Theorems 1 and 4, we can write

$$\mathbf{A} + \mathbf{B} = (a_1\mathbf{i} + a_2\mathbf{j}) + (b_1\mathbf{i} + b_2\mathbf{j})$$
$$= a_1\mathbf{i} + b_1\mathbf{i} + a_2\mathbf{j} + b_2\mathbf{j}$$
$$= (a_1 + b_1)\mathbf{i} + (a_2 + b_2)\mathbf{j}. \quad \blacksquare$$

In other words, to add two vectors, just add the corresponding components. For example, the vectors shown in Fig. 11 are $\mathbf{A} = 4\mathbf{i} + \mathbf{j}$ and $\mathbf{B} = -5\mathbf{i} + 6\mathbf{j}$, and by Theorem 5,

$$\mathbf{A} + \mathbf{B} = (4 - 5)\mathbf{i} + (1 + 6)\mathbf{j} = -\mathbf{i} + 7\mathbf{j}.$$

In a similar manner, Theorem 4 also gives

THEOREM 6. If $\mathbf{A} = a_1\mathbf{i} + a_2\mathbf{j}$, then $c\mathbf{A} = (ca_1)\mathbf{i} + (ca_2)\mathbf{j}$.

Example 1. If $\mathbf{A} = 4\mathbf{i} + \mathbf{j}$ and $\mathbf{B} = -5\mathbf{i} + 6\mathbf{j}$, find $3\mathbf{A}$, $-4\mathbf{B}$, $3\mathbf{A} - 4\mathbf{B}$, and $5\mathbf{A} + 4\mathbf{B}$.

Solution. Using Theorems 5 and 6, we see that

$$3\mathbf{A} = 3(4\mathbf{i} + \mathbf{j}) = 12\mathbf{i} + 3\mathbf{j},$$
$$-4\mathbf{B} = -4(-5\mathbf{i} + 6\mathbf{j}) = 20\mathbf{i} - 24\mathbf{j},$$
$$3\mathbf{A} - 4\mathbf{B} = 12\mathbf{i} + 3\mathbf{j} + (20\mathbf{i} - 24\mathbf{j}) = 32\mathbf{i} - 21\mathbf{j},$$
$$5\mathbf{A} + 4\mathbf{B} = 5(4\mathbf{i} + \mathbf{j}) + 4(-5\mathbf{i} + 6\mathbf{j})$$
$$= 20\mathbf{i} + 5\mathbf{j} - 20\mathbf{i} + 24\mathbf{j} = 0\mathbf{i} + 29\mathbf{j} = 29\mathbf{j}. \quad \bullet$$

The Pythagorean Theorem gives immediately the formula

(9) $$|\mathbf{A}| = |a_1\mathbf{i} + a_2\mathbf{j}| = \sqrt{a_1^2 + a_2^2}$$

for the length of the vector **A**.

We can specify the direction of a nonzero vector, by giving θ, the angle that the vector makes with the positive x-axis. In order that θ be uniquely determined, we require that $0 \leq \theta < 2\pi$. A simple formula for θ is not available, but if $\mathbf{A} = a_1\mathbf{i} + a_2\mathbf{j}$, then

(10) $$\tan \theta = \frac{a_2}{a_1}.$$

Of course, (10) fails if $a_1 = 0$, but this causes no difficulty because $\theta = \pi/2$ or $\theta = 3\pi/2$ according as a_2 is positive or negative.

One may be tempted to transform equation (10) into

(11) $$\theta = \text{Tan}^{-1}\frac{a_2}{a_1},$$

but strictly speaking (10) and (11) are not equivalent because (11) requires that $-\pi/2 < \theta < \pi/2$, while we have demanded that $0 \leq \theta < 2\pi$. For example, suppose that $\mathbf{C} = -5\mathbf{i} - 5\mathbf{j}$. Then $\tan/\theta = 1$, and formula (11) gives $\theta = \pi/4$. But it is obvious that this vector **C** makes an angle $\theta = 5\pi/4$ with the positive x-axis. If $\mathbf{D} = -3\mathbf{j}$, the x-component d_1 is zero, but obviously $\theta = 3\pi/2$.

THEOREM 7. Any vector **A** can be written in the form

(12) $$\mathbf{A} = |\mathbf{A}| \cos \theta \mathbf{i} + |\mathbf{A}| \sin \theta \mathbf{j} = |\mathbf{A}| (\cos \theta \mathbf{i} + \sin \theta \mathbf{j}),$$

where $|\mathbf{A}|$ is the length of the vector **A** and θ is the angle that the vector makes with the positive x-axis.

Proof. The x-component of any vector **A** is just the projection of the vector on the x-axis and this is just $|\mathbf{A}| \cos \theta$. Similarly, the y-component is the projection on the y-axis and clearly this is $|\mathbf{A}| \sin \theta$. ∎

If $\mathbf{A} = a_1\mathbf{i} + a_2\mathbf{j}$, then equation (12) takes the form

$$\mathbf{A} = \sqrt{a_1^2 + a_2^2} (\cos \theta \mathbf{i} + \sin \theta \mathbf{j}).$$

A *unit vector* is a vector of length 1. It is always possible to reduce any given vector to a unit vector with the same direction as the given vector, merely by dividing the given vector by its length (providing the length is not zero).

Example 2. Find a unit vector with the same direction as $\mathbf{E} = 5\sqrt{3}\,\mathbf{i} - 5\mathbf{j}$.

Solution. By (9), $|\mathbf{E}| = \sqrt{75 + 25} = 10$. Then for a unit vector, we have

$$\mathbf{e} = \frac{1}{10}\mathbf{E} = \frac{\mathbf{E}}{10} = \frac{\sqrt{3}}{2}\mathbf{i} - \frac{1}{2}\mathbf{j}. \quad \bullet$$

Since $\theta = 11\pi/6$, we could also write for this vector

$$\mathbf{e} = \cos\frac{11\pi}{6}\mathbf{i} + \sin\frac{11\pi}{6}\mathbf{j}.$$

DEFINITION 7 (Position Vector). The vector from the origin of the coordinate system to a point P is called the position vector of the point P.

If P has coordinates (p_1, p_2), it is clear that the position vector **OP** has these as components, and indeed $\mathbf{OP} = p_1\mathbf{i} + p_2\mathbf{j}$.

It is easy to find the vector from the point A to the point B by using the position vectors to A and B. This is the content of

---

THEOREM 8. For any two points A and B,

(13) $$\mathbf{AB} = \mathbf{OB} - \mathbf{OA}.$$

Proof. This is merely a restatement of Theorem 3, where now the two vectors **B** and **A** both have their initial point at the origin, and hence are named **OB** and **OA,** and **AB** is the vector from A to B. ∎

Example 3. Find the length of the vector **AB** from the point $A(-3, 5)$ to the point $B(7, 9)$. Find a unit vector with the direction of **AB** (see Fig. 12 on the next page).

Solution. The position vectors are $\mathbf{OB} = 7\mathbf{i} + 9\mathbf{j}$ and $\mathbf{OA} = -3\mathbf{i} + 5\mathbf{j}$. Therefore, by equation (13), $\mathbf{AB} = \mathbf{OB} - \mathbf{OA}$, or

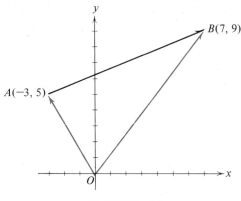

FIGURE 12

$$\mathbf{AB} = 7\mathbf{i} + 9\mathbf{j} - (-3\mathbf{i} + 5\mathbf{j}) = 7\mathbf{i} + 9\mathbf{j} + 3\mathbf{i} - 5\mathbf{j} = 10\mathbf{i} + 4\mathbf{j}.$$

By equation (9) the length is $|\mathbf{AB}| = \sqrt{10^2 + 4^2} = 2\sqrt{5^2 + 2^2} = 2\sqrt{29}$. If $\mathbf{u}$ denotes a unit vector with the direction of $\mathbf{AB}$, then

$$\mathbf{u} = \frac{1}{2\sqrt{29}}(10\mathbf{i} + 4\mathbf{j}) = \frac{5}{\sqrt{29}}\mathbf{i} + \frac{2}{\sqrt{29}}\mathbf{j}. \quad \bullet$$

It is clear that two vectors

$$\mathbf{A} = a_1\mathbf{i} + a_2\mathbf{j} \quad \text{and} \quad \mathbf{B} = b_1\mathbf{i} + b_2\mathbf{j}$$

are equal if and only if their corresponding components are equal; that is, $\mathbf{A} = \mathbf{B}$ if and only if

$$a_1 = b_1 \quad \text{and} \quad a_2 = b_2.$$

Since a plane vector is completely specified by giving its components a_1 and a_2, one may regard the unit vectors $\mathbf{i}$ and $\mathbf{j}$ as unnecessary adornments in the equation $\mathbf{A} = a_1\mathbf{i} + a_2\mathbf{j}$. In view of this, many authors use the notation $[a_1, a_2]$ for the vector $\mathbf{A}$. Since the symbol $[a_1, a_2]$ is also used for the closed interval $a_1 \leq x \leq a_2$, some confusion may occur and to avoid this mishap other writers use $\langle a_1, a_2 \rangle$ for the vector $a_1\mathbf{i} + a_2\mathbf{j}$. Of course, any clear symbolism is satisfactory, but we will continue to use $\mathbf{i}$ and $\mathbf{j}$.

EXERCISE 1

1. Let $\mathbf{A} = \mathbf{i} + 2\mathbf{j}, \mathbf{B} = 3\mathbf{i} - 5\mathbf{j}, \mathbf{C} = -8\mathbf{i} + 7\mathbf{j}$, and $\mathbf{D} = -3\mathbf{i} - 9\mathbf{j}$. Compute each of the following vectors.
   - (a) $\mathbf{A} + \mathbf{B}$.
   - (b) $\mathbf{A} + \mathbf{C}$.
   - (c) $\mathbf{A} + \mathbf{D}$.
   - (d) $\mathbf{A} + \mathbf{B} + \mathbf{C}$.
   - (e) $\mathbf{A} + 2\mathbf{B}$.
   - (f) $\mathbf{A} - 3\mathbf{C}$.

2. Using the vectors from Problem 1, compute:
   (a) $\mathbf{A} - \mathbf{B}$.   (b) $\mathbf{B} - \mathbf{A}$.   (c) $\mathbf{A} - 2\mathbf{D}$.
   (d) $2\mathbf{A} - 3\mathbf{B}$.   (e) $\mathbf{C} + \mathbf{A} + \dfrac{1}{3}\mathbf{D}$.   (f) $8\mathbf{A} - 5\mathbf{B} - 2\mathbf{C} + 3\mathbf{D}$.

3. Compute the length of the vectors $\mathbf{A}$, $\mathbf{B}$, $\mathbf{C}$, and $\mathbf{D}$ of Problem 1.

4. Determine the angle between the positive x-axis and each of the following vectors:
   $\mathbf{A} = 2\mathbf{i} - 2\mathbf{j}$, $\mathbf{B} = -4\mathbf{i} + 4\sqrt{3}\,\mathbf{j}$, $\mathbf{C} = -25\mathbf{i}$, $\mathbf{D} = \pi\mathbf{j}$, and $\mathbf{E} = -(\pi/2)\mathbf{i}$. Then write
   each of the vectors in the form $r(\cos\theta\mathbf{i} + \sin\theta\mathbf{j})$.

5. In each of the following, find the vector from the first-named point to the second-
   named point and find the length of the vector.
   (a) $A(11, 7)$, $B(3, 13)$.   (b) $C(111, 59)$, $D(141, 99)$.

   (c) $E(-5, -7)$, $F(-17, -12)$.   (d) $G\!\left(-\dfrac{11}{2}, \dfrac{13}{3}\right)$, $H\!\left(\dfrac{13}{2}, -\dfrac{11}{3}\right)$.

6. For each of the following vectors, find a unit vector with the same direction:
   $\mathbf{A} = 3\mathbf{i} + 4\mathbf{j}$, $\mathbf{B} = 8\mathbf{i} - 15\mathbf{j}$, $\mathbf{C} = -21\mathbf{i} + 20\mathbf{j}$, $\mathbf{D} = 4\mathbf{i} - 7\mathbf{j}$, and $\mathbf{E} = -6\mathbf{i} - 3\mathbf{j}$.

7. Find two vectors that are parallel to the line:
   (a) $y = 3x + 2$.   (b) $y = -2x + 11$.   (c) $y = \sqrt{3}x + \sqrt{17}$.

8. Express each of the following vectors in the form $a_1\mathbf{i} + a_2\mathbf{j}$.
   (a) A vector of length 2 having the same direction as $3\mathbf{i} - 4\mathbf{j}$.
   (b) A vector of length 6 that makes an angle of $60°$ with the positive x-axis.
   (c) A vector with 4 times the length of $-\mathbf{i} + 2\mathbf{j}$ and in the opposite direction.

9. Prove that two vectors $\mathbf{A} = a_1\mathbf{i} + a_2\mathbf{j}$ and $\mathbf{B} = b_1\mathbf{i} + b_2\mathbf{j}$ are equal if and only if
   $a_1 = b_1$ and $a_2 = b_2$. In other words, two vectors are equal if and only if their
   corresponding components are equal.

★10. Let $\mathbf{OP}$ be the position vector to P, the midpoint of the line segment AB, where A
   is (a_1, a_2) and B is (b_1, b_2). Prove that

$$\mathbf{OP} = \frac{1}{2}\mathbf{OA} + \frac{1}{2}\mathbf{OB}.$$

From this deduce the formula for the coordinates of the midpoint,

$$\left(\frac{a_1 + b_1}{2}, \frac{a_2 + b_2}{2}\right). \qquad \text{HINT: } \mathbf{OP} = \mathbf{OA} + \frac{1}{2}\mathbf{AB}.$$

This formula for the coordinates of the midpoint of a given line segment is important and should be memorized.

11. Find the midpoint of the line segment AB for each of the following pairs of points.
    (a) $A(3, 2)$, $B(11, 20)$.   (b) $A(5, -9)$, $B(9, -5)$.
    (c) $A(-6, 10)$, $B(8, -6)$.   (d) $A(\sqrt{2} + 7, \pi + 3e)$, $B(\sqrt{2} - 7, \pi - e)$.

12. Do Problem 11 for the following pairs of points.
    (a) $A(5, -7)$, $B(-9, 23)$.   (b) $A(-8, -17)$, $B(4, 11)$.
    (c) $A(6, 18)$, $B(-5, -3)$.   (d) $A(\sqrt{2}, -3\sqrt{5})$, $B(5\sqrt{2}, 11\sqrt{5})$.

★★13. Following the methods of Problem 10, prove that the point

$$P\left(\frac{2a_1 + b_1}{3}, \frac{2a_2 + b_2}{3}\right)$$

is one of the trisection points of the line segment AB, where A is (a_1, a_2) and B is (b_1, b_2). In fact, P is that trisection point that is nearer to A. Find a formula for the trisection point nearer to B.

14. Prove that if the vector $\mathbf{A} = a_1\mathbf{i} + a_2\mathbf{j}$ is not the zero vector, then the vector $\mathbf{B} = -a_2\mathbf{i} + a_1\mathbf{j}$ is perpendicular to $\mathbf{A}$. HINT: $\mathbf{B}$ is $90°$ in advance of $\mathbf{A}$.

15. Find scalars x and y such that

$$x(\mathbf{i} + 2\mathbf{j}) + y(3\mathbf{i} + 4\mathbf{j}) = 7\mathbf{i} + 9\mathbf{j}.$$

16. Repeat Problem 15 for

$$x(3\mathbf{i} + \mathbf{j}) + y(2\mathbf{i} + 4\mathbf{j}) = -\mathbf{i} - 12\mathbf{j}.$$

★17. Find all integers x, y, and z such that

$$x(3\mathbf{i} - 4\mathbf{j}) + y(-\mathbf{i} + 3\mathbf{j}) + z(2\mathbf{i} + 4\mathbf{j}) = \mathbf{0}.$$

★★18. Repeat Problem 17 for

$$x(5\mathbf{i} + 3\mathbf{j}) + y(-\mathbf{i} + 4\mathbf{j}) + z(-8\mathbf{i} - 2\mathbf{j}) = \mathbf{0}.$$

19. Give a geometric argument to prove that for any two vectors, $|\mathbf{A} + \mathbf{B}| \leq |\mathbf{A}| + |\mathbf{B}|$. Express this inequality in algebraic terms.

20. Use the result of Problem 19 to prove that for any two vectors $|\mathbf{A} - \mathbf{B}| \geq |\mathbf{A}| - |\mathbf{B}|$. Express this in algebraic form.

21. Two forces $\mathbf{F}_1 = 2\mathbf{i} - \mathbf{j}$ and $\mathbf{F}_2 = 3\mathbf{i} + 6\mathbf{j}$ with magnitude given in Newtons are acting on an object at the same point.
   (a) Find the single force that would have the same effect when applied at the same point.
   (b) Find the single force that would cancel the effect of the forces $\mathbf{F}_1$ and $\mathbf{F}_2$ when applied at the same point.

22. Do Problem 21 if three forces $\mathbf{F}_1 = 3\mathbf{i} + 4\mathbf{j}$, $\mathbf{F}_2 = \mathbf{i} - 5\mathbf{j}$, and $\mathbf{F}_3 = -6\mathbf{i} - 2\mathbf{j}$ are acting at the same point on a body.

23. A train is traveling 80 km/hr in the direction of the vector $\mathbf{i}$ when Jane throws a ball out the window in the direction of the vector $\mathbf{j}$ with a speed (relative to the train) of 10 km/hr. Find the velocity vector of the ball relative to the ground just after Jane releases the ball. Find the speed of the ball relative to the ground.

24. Do Problem 23 if John throws the ball at 20 km/hr from a train that is only going 60 km/hr.

CALCULATOR PROBLEMS

C1. Find the components (to four significant figures) of the vector:
  (a) $\mathbf{A} = \sqrt{13}\,(\cos 37°\mathbf{i} + \sin 37°\mathbf{j})$.
  (b) $\mathbf{B} = \sqrt{19}\,(\cos 129°\mathbf{i} + \sin 129°\mathbf{j})$.
  (c) $\mathbf{A} + \mathbf{B}$.

C2. Do Problem C1 for:
  (a) $\mathbf{C} = 123.4(\cos 201°\mathbf{i} + \sin 201°\mathbf{j})$.
  (b) $\mathbf{D} = 341.2(\cos 307°\mathbf{i} + \sin 307°\mathbf{j})$.
  (c) $\mathbf{C} + \mathbf{D}$.

C3. A river 2.08 km wide is flowing at the rate of 1.61 km/hr in the direction of $\mathbf{i}$. A man who can swim 2.45 km/hr wishes to cross the river in the direction of the vector $\mathbf{j}$. Find: **(a)** the direction in which he should swim, and **(b)** the time needed for him to cross.

Vectors and Parametric Equations   *3*

Vectors can be very useful in describing curves. Let us suppose that a particle P moving in the plane describes some curve. To be specific, let t denote the time, and suppose that for each value of t, the particle P has a definite location. Let $\mathbf{R}$ be the position vector of P, that is, the vector from the origin to the particle P. Then for each value of t, the vector $\mathbf{R}$ is specified so that $\mathbf{R}$ is a function, a *vector function,* of the scalar t. We may write $\mathbf{R} = R(t)$ to indicate this dependence of the vector $\mathbf{R}$ on the scalar t. Of course, a vector is specified whenever its components are specified. So the function $\mathbf{R}(t)$ really consists of two scalar functions and we can write

(14) $$\mathbf{R}(t) = f(t)\mathbf{i} + g(t)\mathbf{j},$$

where $f(t)$ is some function that gives the x-component of $\mathbf{R}$ and $g(t)$ is another function that gives the y-component of $\mathbf{R}$.

Actually, equation (14) is just a shorthand way of writing the pair of equations

(15) $$x = f(t), \qquad y = g(t).$$

The equations in (15) are called the *parametric equations* of the curve and t is called the *parameter.*

In order to find a point on the curve, one uses some fixed value t in (15) and computes for that t the coordinate pair (x, y). Thus (15) is a set of simultaneous equations. If we can eliminate t from this pair of equations, by solving simultaneously, we will obtain one equation in the two variables x and y,

(16) $$F(x, y) = 0 \quad \text{or} \quad y = f(x),$$

an equation for the curve in its customary form. To distinguish between these various ways of describing a curve, (16) is called a *Cartesian equation* for the curve and (14) is called a *vector equation* for the curve.

The vector equation (14) and the parametric equations (15) have an advantage over the Cartesian equation, because they are more flexible and they give more information, as we shall see later in this chapter. For one thing, (16) merely tells which points are on the curve, but (14) and (15) tell us where the moving particle P is at any given time t, that is, how the curve was described.

Further, the vector equation or the parametric equations give the curve a direction, namely, the direction in which P is traveling as the parameter t increases. Henceforth when we speak of a positive direction along a curve we will mean the direction in which P moves as t increases. Of course, a curve by itself has no direction. The direction only arises when we give a set of parametric equations. By changing the functions $f(t)$ and $g(t)$ it is possible to reverse the positive direction on the curve.

Example 1. Discuss the curve whose vector equation is $\mathbf{R} = (4 - t^4)\mathbf{i} + t^2\mathbf{j}$.

Solution. This vector equation is equivalent to the pair of parametric equations

(17) $x = 4 - t^4, \qquad y = t^2.$

To sketch this curve we make a table, selecting convenient values of t and computing the corresponding coordinates of P, the end point of the position vector $\mathbf{R}$.

TABLE 1

t	0	1	$\sqrt{2}$	$\sqrt{3}$	2	$\sqrt{5}$	3	4
x	4	3	0	-5	-12	-21	-77	-252
y	0	1	2	3	4	5	9	16

Table 1 shows the results of such computations for $t = 0, 1, \sqrt{2}, \sqrt{3}, 2, \sqrt{5}, 3, 4$. From this table it is clear that the points $(4, 0), (3, 1), (0, 2), \ldots, (-252, 16)$ all lie on the curve (see Fig. 13).

We can obtain more information about this curve if we eliminate t from the pair of equations (17) and examine the resulting Cartesian equation. Squaring the second equation in (17) and substituting in the first, we find that

$$x = 4 - y^2,$$

and hence the tip of the vector $\mathbf{R}$ describes a parabola. Note, however, that the vector equation does not give the whole parabola, just the part on and above the x-axis. For, given any real number t, its square is either positive or zero, so from (17) we

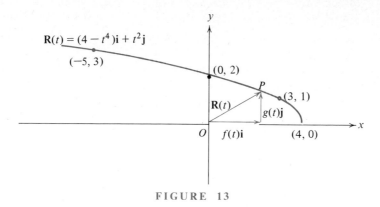

FIGURE 13

have $y \geq 0$ for all t. If we start our moving particle P at time $t = 0$, it is at the point $(4, 0)$, and as time passes, the particle moves along the upper half of the parabola toward the left, steadily getting farther away from the vertex $(4, 0)$. Thus for $t \geq 0$, the vector equation $\mathbf{R} = (4 - t^4)\mathbf{i} + t^2\mathbf{j}$ describes the upper half of the parabola $x = 4 - y^2$ and the positive direction on the parabola is to the left. For $t \leq 0$, the equation describes the same half of the parabola, but now the positive direction on the parabola is to the right. ●

It is not necessary to think of the parameter t as time. This is merely a convenient interpretation that aids visualization and understanding. Of course, when we are concerned with the trajectory of some moving object such as a shell or a planet, then it is only natural and fitting to let t denote the time. But we are always free to use any letter we wish as a parameter. Thus the equations

(18) $$x = 4 - \theta^4, \qquad y = \theta^2$$

can be regarded as parametric equations of a curve with θ as the parameter, and when so regarded (18) gives exactly the same curve as (17). The variable θ may have some geometric significance as an angle, but this is not necessary, and in the particular case of the parabola (18) there is no angle in Fig. 13 that is related to the parameter θ.

Example 2. A wheel of radius a rolls on a straight line without slipping. Let P be a fixed point on the wheel, at distance b from the center of the wheel. The curve described by the point P is called a *trochoid*. If $b = a$, the curve is called a *cycloid*. Find parametric equations for the trochoid, and sketch a portion of the curve.

Solution. For convenience we select the x-axis to be the straight line on which the wheel rolls. Further, we select the initial position of the wheel so that the center C of the wheel is on the y-axis and the point P is on the y-axis below C (see Fig. 14). Let

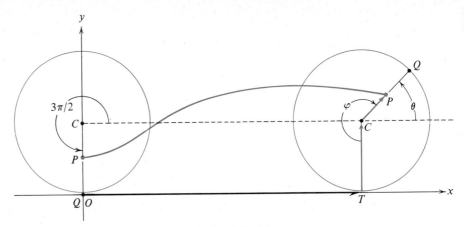

FIGURE 14

Q be that point on the rim of the wheel that lies on the radial line CP. Then, in the initial position of the wheel, the point Q coincides with the origin O.

Let us suppose now that the wheel turns through an angle φ (the radial line CQ turns through an angle φ) in going from its initial position to some general position, as shown in Fig. 14. It is easy to write the vector equation for the point P. Indeed,

(19) $\mathbf{OP} = \mathbf{OT} + \mathbf{TC} + \mathbf{CP}.$

Since the wheel rolls without slipping, the distance OT is just the length of the circular arc TQ, namely, $a\varphi$, when φ is measured in radians. Hence $\mathbf{OT} = a\varphi\mathbf{i}.$ Clearly, $\mathbf{TC} = a\mathbf{j}.$ Finally, if θ denotes the angle that the radial line CQ makes with the positive x-axis, then by Theorem 7 we have $\mathbf{CP} = b(\cos\theta\mathbf{i} + \sin\theta\mathbf{j}).$ Then equation (19) becomes

(20) $\mathbf{OP} = a\varphi\mathbf{i} + a\mathbf{j} + b\cos\theta\mathbf{i} + b\sin\theta\mathbf{j}$
                $= (a\varphi + b\cos\theta)\mathbf{i} + (a + b\sin\theta)\mathbf{j}.$

But equation (20) contains two parameters, φ and θ, and there should be only one. To eliminate the excess parameter, we observe that for any position of the wheel, $\varphi + \theta = 3\pi/2$. Therefore,

$\cos\theta = \cos(3\pi/2 - \varphi) = -\sin\varphi$ and $\sin\theta = \sin(3\pi/2 - \varphi) = -\cos\varphi.$

Equation (20) now simplifies to

(21) $\mathbf{OP} = (a\varphi - b\sin\varphi)\mathbf{i} + (a - b\cos\varphi)\mathbf{j},$

the vector equation of the trochoid. Consequently,

(22) $x = a\varphi - b\sin\varphi, \qquad y = a - b\cos\varphi,$

form a set of parametric equations for the trochoid. A portion of a trochoid is shown in Fig. 14. ●

If $b = a$, the curve is called a *cycloid*. Obviously, the cycloid will meet the x-axis at intervals of length $2\pi a$. When $b = a$, the equation set (22) becomes

(23) $$x = a(\varphi - \sin \varphi), \qquad y = a(1 - \cos \varphi),$$

the parametric equations for the cycloid. This curve is shown in Fig. 15.

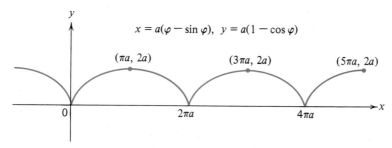

$$x = a(\varphi - \sin \varphi), \quad y = a(1 - \cos \varphi)$$

$(\pi a, 2a)$ $(3\pi a, 2a)$ $(5\pi a, 2a)$

$2\pi a$ $4\pi a$

FIGURE 15

It is clear from Fig. 15 that y is a function of x, but it is also clear from equation set (23) that it is not possible[1] to find a simple expression for this function. Thus for many curves, and in particular the cycloid, the parametric equations are simpler than the Cartesian equations.

EXERCISE 2

In each of Problems 1 through 10, a curve is given by a set of parametric equations. Sketch the curve for the indicated range of the parameter. In Problems 1 through 6, obtain a Cartesian equation for the curve by eliminating the parameter, and identify the curve. We will consider some of these curves in more detail in Exercise 3.

1. $x = t$, $y = 1 - t$, $0 \leq t \leq 1$.
2. $x = 5 \cos t$, $y = 5 \sin t$, $0 \leq t \leq 2\pi$.
3. $x = 5 \cos^2 t$, $y = 5 \sin^2 t$, $0 \leq t \leq 2\pi$.
4. $x = 3 \cos \theta$, $y = 5 \sin \theta$, $0 \leq \theta \leq \pi$.
5. $x = \sin \alpha$, $y = \cos 2\alpha$, $0 \leq \alpha \leq \pi$.
6. $x = \cosh u$, $y = \sinh u$, $-\infty < u < \infty$.
7. $x = t^3$, $y = t^2$, $-\infty < t < \infty$.

[1] The reader should try his hand at solving the pair of equations (23) for y in terms of x.

8. $x = t^3 - 3t$, $y = t$, $\qquad -\infty < t < \infty$.

*9. $x = t^3 - 3t$, $y = \tan\dfrac{\pi}{4}t$, $\quad -2 < t < 2$.

*10. $x = t^3 - 3t$, $y = 4 - t^2$, $\quad -\infty < t < \infty$.

11. Show that $\mathbf{R} = \mathbf{OP_1} + t\mathbf{P_1P_2}$ is a vector equation for the straight line through P_1 and P_2. Use this vector equation to find parametric equations for the line through (a_1, b_1) and (a_2, b_2).

12. Show that the equations

$$x = a + k \cos \theta$$
$$y = b + k \sin \theta$$

are parametric equations for a circle. What is the center and radius of this circle?

In Problems 13 through 15, a curve is described by some geometric condition on a moving point P. Obtain a vector equation for the position vector $\mathbf{R}$ of the point P. If possible, find a Cartesian equation for the curve.

13. From the origin a line OQ is drawn to an arbitrary point Q on a fixed vertical line $\mathscr{L}$ that is b units to the right of O (see Fig. 16). A line segment QP is then drawn parallel to the x-axis, and to the right of $\mathscr{L}$ and such that $|\mathbf{QP}| = |\mathbf{OQ}|^2$. Use the parametric equations to show that the point P describes a parabola.

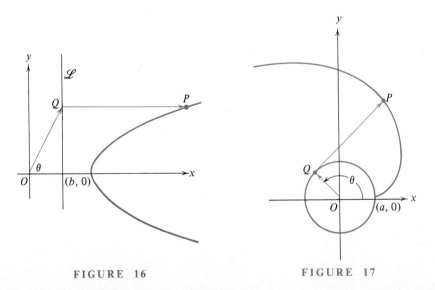

FIGURE 16   FIGURE 17

14. If in Problem 13, $|\mathbf{QP}| = 4|\mathbf{OQ}|$, prove that the point P describes one branch of a hyperbola.

*15. The *involute* of a circle is the curve described by the end point P of a thread as the thread is unwound from a fixed spool. Suppose that the radius of the spool is a, and for simplicity assume that when the spool is placed with its center at the origin, the point P starts at $(a, 0)$ (see Fig. 17). HINT: The length of $\mathbf{QP}$ is the amount of thread unwound, namely, $a\theta$.

*16. If a circle of radius b rolls on the inside of a second circle of radius a $(a > b)$ without slipping, the curve described by a fixed point P on the circumference of the first circle is called a *hypocycloid*. Show that if the fixed point P is initially at $(a, 0)$ as indicated in Fig. 18, then the vector equation of the hypocycloid is

$$\mathbf{R} = \left[(a - b)\cos\theta + b\cos\frac{a - b}{b}\theta\right]\mathbf{i} + \left[(a - b)\sin\theta - b\sin\frac{a - b}{b}\theta\right]\mathbf{j}.$$

HINT: If there is no slipping, then $a\theta = b\varphi$.

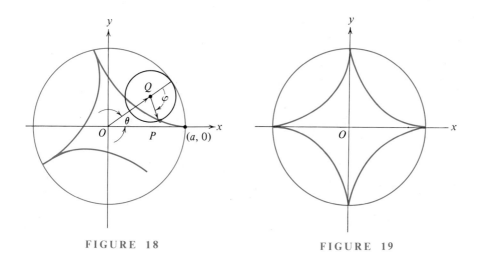

FIGURE 18   FIGURE 19

*17. If in Problem 16, the ratio a/b is an integer n, then the hypocycloid will have n cusps. Show that the hypocycloid of 4 cusps (see Fig. 19) has the Cartesian equation

$$x^{2/3} + y^{2/3} = a^{2/3}.$$

HINT: Use the relations $a - b = 3b$, $(a - b)/b = 3$, and the standard trigonometric identities $\cos 3\theta = 4\cos^3\theta - 3\cos\theta$ and $\sin 3\theta = 3\sin\theta - 4\sin^3\theta$.

*18. Show that the hypocycloid of two cusps obtained from Problem 16 when $a = 2b$ is just the straight-line segment between $(-a, 0)$ and $(a, 0)$.

19. Sketch the curve $\mathbf{R} = e^t\cos t\,\mathbf{i} + e^t\sin t\,\mathbf{j}$.

20. Show that the area of the region bounded by the x-axis and one arch of the cycloid (see Fig. 15) is three times the area of the rolling circle. HINT:

$$A = \int_{x=0}^{x=2\pi a} y\,dx = \int_{\varphi=0}^{\varphi=2\pi} y(\varphi)\frac{dx}{d\varphi}\,d\varphi.$$

21. Use parametric equations $x = a \cos t$, $y = a \sin t$ to compute the area of a circle.
22. Use parametric equations $x = a \cos t$, $y = b \sin t$ to prove that the area of the region bounded by this ellipse is πab.
*23. Show that the area of the region bounded by the hypocycloid of four cusps (see Problem 17) is $3\pi a^2/8$.
*24. If a circle of radius b rolls on the outside of a fixed circle of radius a without slipping, the curve described by a fixed point P on the circumference of the rolling circle is called an *epicycloid*. If the fixed circle has its center at the origin and if P is initially at $(a, 0)$, show that

$$x = (a + b)\cos\theta - b\cos\frac{a+b}{b}\theta, \qquad y = (a + b)\sin\theta - b\sin\frac{a+b}{b}\theta$$

are parametric equations for the epicycloid. Observe that these equations can be obtained from the equations for the hypocycloid (Problem 16) by replacing b by $-b$.

4 *Differentiation of Vector Functions*

If $\mathbf{R} = \mathbf{R}(t)$ is a vector function of a scalar t, it should be possible to differentiate this vector function with respect to t. Suppose that the vector function is given by means of its components $f(t)$ and $g(t)$:

(24) $$\mathbf{R}(t) = f(t)\mathbf{i} + g(t)\mathbf{j}.$$

It would be very nice if we could differentiate the vector function by just differentiating its scalar components; thus

(25) $$\frac{d\mathbf{R}(t)}{dt} = \frac{df(t)}{dt}\mathbf{i} + \frac{dg(t)}{dt}\mathbf{j}.$$

For example, if

$$\mathbf{R} = t \sin t\,\mathbf{i} + (t^3 - 3t + e^t)\mathbf{j},$$

then we should like to know that

$$\frac{d\mathbf{R}}{dt} = (t \cos t + \sin t)\mathbf{i} + (3t^2 - 3 + e^t)\mathbf{j}.$$

To prove that this is indeed the case, we must first agree on the definition of the derivative of a vector function. Since this involves a limit, we begin with

---

DEFINITION 8 (Limit). We say that $\mathbf{R}(t)$ approaches the vector $\mathbf{L}$ as t approaches t_0, and we write either

$$(26) \qquad \lim_{t \to t_0} \mathbf{R}(t) = \mathbf{L} \qquad \text{or} \qquad \lim_{h \to 0} \mathbf{R}(t_0 + h) = \mathbf{L}$$

if

$$(27) \qquad \lim_{t \to t_0} |\mathbf{R}(t) - \mathbf{L}| = 0.$$

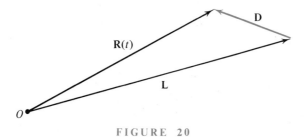

FIGURE 20

Observe that in (27) the quantity $\mathbf{D} = \mathbf{R}(t) - \mathbf{L}$ is a vector whose length approaches zero as t approaches t_0. These three vectors are shown in Fig. 20, where the variable vector $\mathbf{R}(t)$ and the limit vector $\mathbf{L}$ both have their initial points at the origin. Consequently, if $|\mathbf{D}| \to 0$, the end point of $\mathbf{R}(t)$ must approach the end point of $\mathbf{L}$, and this means that the x and y components of $\mathbf{R}(t)$ must approach the x and y components of $\mathbf{L}$. This gives

---

THEOREM 9. Suppose that $\mathbf{R}(t) = f(t)\mathbf{i} + g(t)\mathbf{j}$, and $\mathbf{L} = a\mathbf{i} + b\mathbf{j}$. Then

$$(28) \qquad \lim_{t \to t_0} \mathbf{R}(t) = \mathbf{L}$$

if and only if

$$(29) \qquad \lim_{t \to t_0} f(t) = a \qquad \text{and} \qquad \lim_{t \to t_0} g(t) = b.$$

Example 1. Find $\lim_{t \to 2} [(t^2 + 3)\mathbf{i} + e^t\mathbf{j}]$.

Solution. We take the limit of each component. Thus

$$\lim_{t \to 2} [(t^2 + 3)\mathbf{i} + e^t\mathbf{j}] = \lim_{t \to 2} (t^2 + 3)\mathbf{i} + \lim_{t \to 2} e^t\mathbf{j} = 7\mathbf{i} + e^2\mathbf{j}. \quad \bullet$$

DEFINITION 9 (Continuity). A vector function $\mathbf{R}(t)$ is said to be *continuous* at t_0 if

$$\lim_{t \to t_0} \mathbf{R}(t) = \mathbf{R}(t_0).$$

The vector function is continuous in an interval $\mathcal{I}$ if it is continuous at every point of $\mathcal{I}$.

From Theorem 9 we see that a vector function is continuous in $\mathcal{I}$ if and only if its components are continuous in $\mathcal{I}$. Thus it is a relatively easy matter to recognize a continuous vector function. We merely examine its components. The vector function in Example 1 is continuous for all t because each of its components $t^2 + 3$ and e^t is continuous for all t.

DEFINITION 10 (Derivative). The vector function $\mathbf{R}(t)$ is said to be *differentiable* at t_0 in $\mathcal{I}$ if the limit

(30)
$$\lim_{h \to 0} \frac{\mathbf{R}(t_0 + h) - \mathbf{R}(t_0)}{h}$$

exists. When this limit exists, it is called the derivative at t_0 and is denoted by $\mathbf{R}'(t_0)$. If $\mathbf{R}(t)$ is differentiable at every point in an interval $\mathcal{I}$, then $\mathbf{R}(t)$ is said to be differentiable in $\mathcal{I}$.

Note that vector subtraction has already been defined and that division by the scalar h means multiplication by the scalar $1/h$. Consequently, the expression

$$\frac{\mathbf{R}(t_0 + h) - \mathbf{R}(t_0)}{h}$$

is meaningful and is a vector function of h.

As one might expect, other notations for the derivative are available. If $\mathbf{R}(t)$ is differentiable at each point in $\mathcal{I}$, then we can drop the subscript and write $\mathbf{R}'(t)$ for the derivative in place of $\mathbf{R}'(t_0)$. At the same time, we may on occasion write

$$\frac{d\mathbf{R}}{dt} = \lim_{\Delta t \to 0} \frac{\Delta \mathbf{R}}{\Delta t} = \lim_{\Delta t \to 0} \frac{\mathbf{R}(t + \Delta t) - \mathbf{R}(t)}{\Delta t}$$

in place of

$$\mathbf{R}'(t) = \lim_{h \to 0} \frac{\mathbf{R}(t + h) - \mathbf{R}(t)}{h}.$$

THEOREM 10. Let $f(t)$ and $g(t)$ be differentiable in an interval $\mathscr{I}$. Then the vector function

(24) $$\mathbf{R}(t) \equiv f(t)\mathbf{i} + g(t)\mathbf{j}$$

is differentiable in $\mathscr{I}$ and its derivative is given by

(25) $$\mathbf{R}'(t) = f'(t)\mathbf{i} + g'(t)\mathbf{j}.$$

Proof. By the definition of $\Delta \mathbf{R}$,

$$\Delta \mathbf{R} = \mathbf{R}(t + h) - \mathbf{R}(t) = [f(t + h)\mathbf{i} + g(t + h)\mathbf{j}] - [f(t)\mathbf{i} + g(t)\mathbf{j}].$$

Using the algebraic properties of vectors, we can rearrange the right side, obtaining

$$\mathbf{R}(t + h) - \mathbf{R}(t) = [f(t + h) - f(t)]\mathbf{i} + [g(t + h) - g(t)]\mathbf{j}.$$

Dividing both sides by h and letting $h \to 0$, we obtain

$$\lim_{h \to 0} \frac{\mathbf{R}(t + h) - \mathbf{R}(t)}{h} = \lim_{h \to 0} \frac{f(t + h) - f(t)}{h}\mathbf{i} + \lim_{h \to 0} \frac{g(t + h) - g(t)}{h}\mathbf{j}$$

(31) $$\frac{d\mathbf{R}}{dt} = \frac{df}{dt}\mathbf{i} + \frac{dg}{dt}\mathbf{j}. \quad \blacksquare$$

In Theorem 10 we explicitly stated that $f(t)$ and $g(t)$ are assumed to be differentiable functions. We can avoid such distracting phrases if we agree that whenever a theorem, problem, and so on, involves differentiating a function, we naturally assume it is differentiable.

As in the scalar case, the derivative of a vector function has a geometric interpretation that is useful and beautiful.

THEOREM 11. Let $\mathbf{R}(t)$ be the position vector to a point P moving on a curve $\mathscr{C}$. If $\mathbf{R}'(t_0) \neq \mathbf{0}$, then the derivative $\mathbf{R}'(t_0)$ is a vector tangent to the curve $\mathscr{C}$ at the point $P = P(t_0)$, and points in the direction of motion of P along the curve $\mathscr{C}$ as t increases.

Proof. As shown in Fig. 21, the difference vector $\Delta \mathbf{R} = \mathbf{R}(t_0 + \Delta t) - \mathbf{R}(t_0)$ can be re-

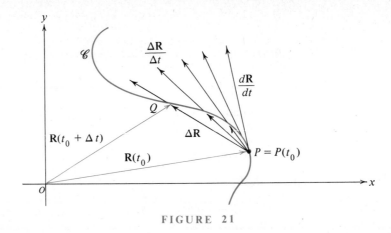

FIGURE 21

garded as a vector from the fixed point P on the curve $\mathscr{C}$ to a second point Q on the curve $\mathscr{C}$. Then

$$\frac{\Delta\mathbf{R}}{\Delta t}$$

is a vector pointing in the same direction if $\Delta t > 0$. In fact, if $\Delta t < 1$, then this vector will be longer than $\Delta\mathbf{R}$, as indicated in the figure. Now let $\Delta t \to 0$. The point Q slides along the curve toward the limiting position P, and at the same time the limiting position of the chord $\Delta\mathbf{R}$ is just the position of a tangent vector, pointing in the direction along $\mathscr{C}$ specified by increasing t. Figure 21 shows several intermediate vectors in this limit process. Since

$$\lim_{\Delta t \to 0^+} \frac{\Delta\mathbf{R}}{\Delta t} = \frac{d\mathbf{R}}{dt} = \mathbf{R}'(t_0) \neq \mathbf{0},$$

it follows that $\mathbf{R}'(t_0)$ is a vector with the properties described in the theorem.

The drawing in Fig. 21 and the proof have been given in the case that Δt is positive. We leave it to the student to make a drawing and consider the case that Δt is negative. ∎

DEFINITION 11 (Smooth Curve). A curve $\mathscr{C}$ is called a *smooth curve* if a parametrization can be selected so that $\mathbf{R}'(t)$ is continuous and never zero.

This means that it is possible to select a parametrization of $\mathscr{C}$ so that the tangent vector is a continuous function of t that is never zero. Geometrically, a smooth curve has no corners or cusps.

THEOREM 12. Let $x = f(t)$ and $y = g(t)$ be parametric equations for a smooth curve $\mathscr{C}$. Then $\dfrac{dy}{dx}$, the slope of the tangent line to the curve $\mathscr{C}$, is given by the formula

(32)
$$\frac{dy}{dx} = \frac{g'(t)}{f'(t)}$$

whenever $f'(t) \neq 0$.

Since $x = f(t)$ and $y = g(t)$, this formula is frequently written in the form

(33)
$$\frac{dy}{dx} = \frac{\dfrac{dy}{dt}}{\dfrac{dx}{dt}}.$$

In this form the equation is easy to memorize, for if we regard the right side of (33) as the ratio of a pair of fractions, we can invert the denominator and multiply. The formal cancellation of dt thus leads from (33) to

$$\frac{dy}{dx} = \frac{dy}{dt}\frac{dt}{dx} = \frac{dy}{dx},$$

an obviously true identity. In fact, this theorem is frequently proved by writing the identity

$$\frac{\Delta y}{\Delta x} = \frac{\dfrac{\Delta y}{\Delta t}}{\dfrac{\Delta x}{\Delta t}}$$

and then taking the limits on both sides as $\Delta t \to 0$.

Proof of Theorem 12. The position vector $\mathbf{R} = f(t)\mathbf{i} + g(t)\mathbf{j}$ describes the same curve $\mathscr{C}$ that the parametric equations describe. By Theorem 11,

$$\mathbf{R}'(t) = f'(t)\mathbf{i} + g'(t)\mathbf{j}$$

is a vector tangent to $\mathscr{C}$. Then the slope of the tangent line is just the slope of this vector, and this is just the ratio $g'(t)/f'(t)$ of its y-component to its x-component. On the other hand, this slope is also dy/dx. ∎

Example 2. Find $\dfrac{dy}{dx}$ for the curve $x = 2 \cos t$, $y = 1 + 3 \sin t$ **(a)** in general, **(b)** at $t = \pi/4$, and **(c)** at $t = \pi$.

Solution. **(a)** By Theorem 12,

$$\frac{dy}{dx} = \frac{dy/dt}{dx/dt} = \frac{3 \cos t}{-2 \sin t} = -\frac{3}{2} \cot t.$$

(b) If $t = \pi/4$, then $\sin t = \cos t = \sqrt{2}/2$, and hence

$$\frac{dy}{dx} = -\frac{3}{2}.$$

(c) If $t = \pi$, then $\cos \pi = -1$, $\sin \pi = 0$, and $\dfrac{dy}{dx} = -\dfrac{3(-1)}{2(0)}$ is undefined. For the vector equation of $\mathscr{C}$, we have

$$\mathbf{R}'(\pi) = -2 \sin \pi \mathbf{i} + 3 \cos \pi \mathbf{j} = 0 - 3\mathbf{j}$$

and hence the tangent line is vertical. ●

In general, if $f'(t) = 0$ and $g'(t) \neq 0$, then the tangent vector is vertical. We may symbolize this situation by writing $dy/dx = \pm\infty$. In case both of the components of $\mathbf{R}'(t) = f'(t)\mathbf{i} + g'(t)\mathbf{j}$ are simultaneously zero at $t = t_0$, the determination of the direction of the tangent vector to the curve is a little more complicated. A good working rule is to consider the slope given by equation (32) for t near t_0.

Once we have a nonzero tangent vector $\mathbf{R}'(t) = f'(t)\mathbf{i} + g'(t)\mathbf{j}$, we can always find a *unit* tangent vector $\mathbf{T}$. Indeed,

(34)
$$\mathbf{T} = \frac{\mathbf{R}'(t)}{|\mathbf{R}'(t)|} = \frac{f'(t)\mathbf{i} + g'(t)\mathbf{j}}{\sqrt{f'(t)^2 + g'(t)^2}}.$$

Further, we can also find $\mathbf{N}$, a unit normal (a vector perpendicular to $\mathbf{T}$). Indeed from Problem 14, Exercise 1,

(35)
$$\mathbf{N} = \frac{-g'(t)\mathbf{i} + f'(t)\mathbf{j}}{\sqrt{f'(t)^2 + g'(t)^2}}.$$

The unit normal vector given by (35) is always $90°$ in advance of the unit tangent $\mathbf{T}$ given by (34).

EXERCISE 3

In Problems 1 through 4, differentiate the given vector function with respect to the independent variable.

1. $\mathbf{R} = \tan t\mathbf{i} + \sec t\mathbf{j}.$
2. $\mathbf{R} = u \cos u\mathbf{i} + u \ln u\mathbf{j}.$
3. $\mathbf{R} = v^3 e^{2v}\mathbf{i} + v^2 e^{-3v}\mathbf{j}.$
4. $\mathbf{R} = \text{Sin}^{-1} w\mathbf{i} + \text{Tan}^{-1} w\mathbf{j}.$

In Problems 5 through 10, the positive vector of a moving particle is given for time t (where t is any real number). In each case find: (a) $\mathbf{R}'(t)$, *(b)* $\mathbf{R}''(t)$, *(c) the location of the particle when* $\mathbf{R}'(t) = 0$, *and (d) the location of the particle when the tangent to the path is either horizontal or vertical.*

5. $\mathbf{R} = 5 \cos^2 t\mathbf{i} + 5 \sin^2 t\mathbf{j}$ (see Problem 3 of Exercise 2).
6. $\mathbf{R} = 3 \cos t\mathbf{i} + 5 \sin t\mathbf{j}$ (see Problem 4 of Exercise 2).
7. $\mathbf{R} = \sin t\mathbf{i} + \cos 2t\mathbf{j}$ (see Problem 5 of Exercise 2).
8. $\mathbf{R} = t^3\mathbf{i} + t^2\mathbf{j}$ (see Problem 7 of Exercise 2).
9. $\mathbf{R} = (t^3 - 3t)\mathbf{i} + t\mathbf{j}$ (see Problem 8 of Exercise 2).
10. $\mathbf{R} = (t^3 - 3t)\mathbf{i} + (4 - t^2)\mathbf{j}$ (see Problem 10 of Exercise 2).

11. Locate all horizontal and vertical tangents on the cycloid $\mathbf{R}(\varphi) = a(\varphi - \sin \varphi)\mathbf{i} + a(1 - \cos \varphi)\mathbf{j}$ (see Example 2, Fig. 15, Section 3).

12. Prove that the vector $\mathbf{i} + f'(x)\mathbf{j}$ is a vector tangent to the curve $y = f(x)$.

13. If $\mathbf{R}(t) = f(t)\mathbf{i} + g(t)\mathbf{j}$, find a formula for $\dfrac{d^2 y}{dx^2}$ in terms of f', f'', g', and g''. HINT:

$$\frac{d^2 y}{dx^2} = \frac{d}{dx}\left(\frac{dy}{dx}\right) = \frac{d}{dt}\left(\frac{dy}{dx}\right)\frac{dt}{dx}.$$

14. Prove that the formula obtained in Problem 13 gives

$$\frac{d^2 y}{dx^2} = F''(t)$$

for the curve $\mathbf{R}(t) = t\mathbf{i} + F(t)\mathbf{j}$.

15. Use the formula from Problem 13 to compute $\dfrac{d^2 y}{dx^2}$ for the curve of: (a) Problem 5, (b) Problem 7, and (c) Problem 9.

16. Do Problem 15 for: (a) Problem 6, (b) Problem 8, and (c) Problem 10.

17. Find a unit normal vector 90° in advance of the unit tangent vector for the curve of: (a) Problem 5, (b) Problem 7, and (c) Problem 9.

18. Do Problem 17 for the curves of: (a) Problem 6, (b) Problem 8, and (c) Problem 10.

19. Find an equation for the line tangent to the curve $\mathbf{R}(t) = e^t\mathbf{i} + (t^2 + 1)\mathbf{j}$ at the point when (a) $t = 0$ and (b) $t = 3$.

20. Do Problem 19 for the curve $\mathbf{R}(\theta) = 3 \sin \theta\mathbf{i} + 4 \cos \theta\mathbf{j}$ when (a) $\theta = \pi/2$, and (b) $\theta = \pi/4$.

5 *Arc Length, Velocity, and Acceleration*

Suppose that we have a smooth curve $\mathscr{C}$ given by the usual parametric equations $x = f(t)$ and $y = g(t)$. We select some point P_0 as our initial point on $\mathscr{C}$ (see Fig. 22) and we can always change the parameter so that $t = 0$ gives the point P_0. The parameter t automatically assigns a direction to the curve and we use this direction to attach a sign to s, the length of arc from P_0 to P. Thus we agree that if $t > 0$ [$P(t)$ follows $P(0)$ on the curve], then $s > 0$. If $t < 0$ [$P(t)$ precedes $P(0)$ on the curve], then $s < 0$. Thus s (the arc length) and t (the parameter) increase together. With these preparations, we have

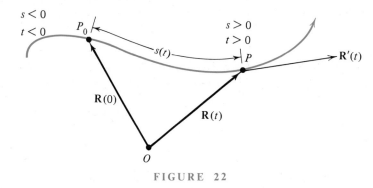

FIGURE 22

THEOREM 13. If $\mathbf{R}(t) = f(t)\mathbf{i} + g(t)\mathbf{j}$ is a vector equation for a smooth curve, for which s and t increase together, then

(36)
$$\frac{ds}{dt} = |\mathbf{R}'(t)|.$$

This result can be written in a variety of equivalent forms. Using Theorem 10 with equation (36), we have

(37)
$$\frac{ds}{dt} = \sqrt{[f'(t)]^2 + [g'(t)]^2} = \sqrt{\left(\frac{dx}{dt}\right)^2 + \left(\frac{dy}{dt}\right)^2}.$$

Proof. Returning to the proof of Theorem 11 and the associated Fig. 21, we can write

(38)
$$\frac{\Delta \mathbf{R}}{\Delta t} = \frac{\Delta \mathbf{R}}{\Delta s} \frac{\Delta s}{\Delta t}$$

merely by inserting any nonzero term Δs in two places where it will obviously cancel. Now let Δs denote the length of arc between the two points P and Q, the end points of the vector $\Delta \mathbf{R}$ (Fig. 21). Thus Δs is indeed the change in the arc length as the parameter t changes by an amount Δt. Taking the lengths of the vectors in (38), we have

$$(39) \qquad \left| \frac{\Delta \mathbf{R}}{\Delta t} \right| = \left| \frac{\Delta \mathbf{R}}{\Delta s} \right| \frac{\Delta s}{\Delta t},$$

where we have dropped the absolute value signs from the last term because s and t increase together, so the ratio $\Delta s/\Delta t$ is positive. Taking limits as $\Delta t \to 0$ in (39), we have

$$(40) \qquad |\mathbf{R}'(t)| = \left| \frac{d\mathbf{R}}{dt} \right| = \left| \frac{d\mathbf{R}}{ds} \right| \frac{ds}{dt},$$

provided that the various limits exist. Now

$$\left| \frac{\Delta \mathbf{R}}{\Delta s} \right| = \frac{\text{length of chord from } P \text{ to } Q}{\text{length of arc from } P \text{ to } Q},$$

and it is intuitively clear that this ratio tends to 1 as $\Delta s \to 0$. Since $|d\mathbf{R}/ds| = 1$, equation (40) gives (36). ∎

Now that we have a formula for ds/dt [see equations (36) or (37)], we can compute the length of any arc by integration.

Example 1. Find the length of the arc of $\mathbf{R}(t) = -5 \sin t\mathbf{i} + 5 \cos t\mathbf{j}$ for $0 \leqq t \leqq \pi$.

Solution. We have

$$\frac{dx}{dt} = -5 \cos t, \qquad \frac{dy}{dt} = -5 \sin t,$$

and hence, by equation (37),

$$\frac{ds}{dt} = \sqrt{(-5 \cos t)^2 + (-5 \sin t)^2} = \sqrt{25(\cos^2 t + \sin^2 t)} = 5.$$

Thus, on integration, we find that

$$s = \int_0^\pi 5 \, dt = 5\pi.$$

The alert reader will observe that in this example, $\mathbf{R}(t)$ is a vector equation for a circle of radius 5 and that 5π is precisely one half of the circumference of this circle. ●

If *t* is regarded as time, then $\mathbf{R}'(t)$ has a nice interpretation as the velocity of a moving particle. For this we need

---

DEFINITION 12 (Speed and Velocity). If a particle is moving on a smooth curve $\mathscr{C}$, and *s* denotes arc length along that curve measured from some fixed point on the curve $\mathscr{C}$ with *s* increasing in the direction of motion of the particle, then the *speed* of the particle is defined to be the instantaneous rate of change of arc length with respect to time; that is,

$$\text{speed} \equiv \frac{ds}{dt}.$$

If the speed is not zero, the *velocity* of the particle is defined to be a vector $\mathbf{V}$ that is tangent to the curve, points in the direction of motion of the particle, and has length equal to the speed. If the speed is zero, then the velocity is the zero vector.

THEOREM 14. If $\mathbf{R}(t)$ is the position vector of a moving particle and *t* is time, then the derivative $\mathbf{R}'(t)$ is the velocity $\mathbf{V}$ of the particle and $|\mathbf{R}'(t)|$ is its speed.

Proof. We have already proved in Theorem 11 that if $\mathbf{R}'(t) \neq \mathbf{0}$, then it is tangent to the curve and points in the proper direction. Further, from Theorem 13, $|\mathbf{R}'(t)|$ is the speed of the particle. Hence $\mathbf{R}'(t)$ has all the required properties of the velocity vector given in Definition 12, whence $\mathbf{R}'(t) = \mathbf{V}$. In case $\mathbf{R}'(t) = \mathbf{0}$, then by equation (36), the speed is also zero, and hence by Definition 12, we have $\mathbf{V} = \mathbf{0}$. Therefore, in both cases $\mathbf{R}'(t) = \mathbf{V}$. ∎

The speed can be zero when the moving particle stops momentarily and then resumes its motion. The curve may well have a tangent at such a point, but of course the zero vector does not give the direction of the tangent to the curve.

DEFINITION 13 (Acceleration). The acceleration vector $\mathbf{A}(t)$ of a moving particle is defined to be the derivative of the velocity vector; that is

(41) $$\mathbf{A}(t) \equiv \frac{d}{dt}\mathbf{V}(t) = \frac{d^2}{dt^2}\mathbf{R}(t).$$

Example 2. Suppose that a wheel on a fixed axis rotates counterclockwise steadily at a rate of ω (Greek lowercase letter omega) revolutions per second. Find the velocity, acceleration, and speed of a particle *P* on the wheel *r* cm from the center.

Solution. We introduce a coordinate system with the origin at the center of the wheel, and we let P be on the positive x-axis when $t = 0$. If θ is the angle (measured in radians) that the vector **OP** makes with the positive x-axis, then $\theta = 2\pi\omega t$. The position vector for P is

$$\mathbf{R}(t) = r\cos 2\pi\omega t\,\mathbf{i} + r\sin 2\pi\omega t\,\mathbf{j}.$$

Differentiating twice with respect to t, we obtain the velocity and acceleration:

(42) $$\mathbf{V}(t) = \mathbf{R}'(t) = r2\pi\omega(-\sin 2\pi\omega t\,\mathbf{i} + \cos 2\pi\omega t\,\mathbf{j}),$$

(43) $$\mathbf{A}(t) = \mathbf{R}''(t) = r4\pi^2\omega^2(-\cos 2\pi\omega t\,\mathbf{i} - \sin 2\pi\omega t\,\mathbf{j}).$$

From equation (43) it is clear that $\mathbf{A}(t) = -4\pi^2\omega^2\mathbf{R}(t)$. Hence the acceleration is always toward the center of the circle. For the speed, equation (42) gives

$$|\mathbf{V}| = V = 2\pi r\omega \text{ cm/sec.}$$

Naturally, the units for the speed depend on the units used for r and ω. For example, if ω is in revolutions per hour and r is in miles, then $V = 2\pi r\omega$ will be in miles/hr.   ●

Example 3. Use vector methods to find the equation of a projectile near the earth's surface when we neglect the resistance due to the air.

Solution. We assume that the projectile is moving in a vertical plane, and we take the x- and y-axes in their usual position (Fig. 23). Then the vector expression of the

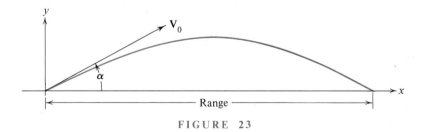

FIGURE 23

physical law, that the acceleration is constant, is

(44) $$\mathbf{A} = \frac{d\mathbf{V}}{dt} = -g\mathbf{j},$$

where g is the scalar acceleration, and the vector $-\mathbf{j}$ gives the direction toward the center of the earth. Experiment shows that near the earth's surface, g is a constant. In the British system, $g \approx 32$ ft/sec^2 (more accurately 32.17 ft/sec^2), and in the CGS system $g \approx 980$ cm/sec^2 (more accurately 980.5 cm/sec^2). If we integrate the vector equation (44) with respect to t, we obtain

$$\mathbf{V} = -gt\mathbf{j} + \mathbf{C}_1,$$

where $\mathbf{C}_1$ is some suitable vector constant of integration. But at $t = 0$, $\mathbf{V} = \mathbf{V}_0$, the initial vector velocity; hence $\mathbf{C}_1 = \mathbf{V}_0$. Therefore,

$$\mathbf{V} = \frac{d\mathbf{R}}{dt} = -gt\mathbf{j} + \mathbf{V}_0,$$

and integrating this equation with respect to t gives the position vector for the projectile,

$$\mathbf{R} = -\frac{gt^2}{2}\mathbf{j} + \mathbf{V}_0 t + \mathbf{C}_2,$$

where again $\mathbf{C}_2$ is a vector constant of integration. Putting $t = 0$, we find $\mathbf{C}_2 = \mathbf{R}_0$, the position vector giving the initial position of the projectile. Hence

(45)
$$\mathbf{R} = -\frac{gt^2}{2}\mathbf{j} + \mathbf{V}_0 t + \mathbf{R}_0,$$

and this is the equation that gives the motion of a body falling freely under gravity in terms of its initial position $\mathbf{R}_0$ and its initial velocity $\mathbf{V}_0$.

Of course, some simplification can be achieved if we select the origin of the coordinate system to be at the starting point of the projectile. Then $\mathbf{R}_0 = \mathbf{0}$. Further, let us write the initial velocity in terms of its components

$$\mathbf{V}_0 = V_1\mathbf{i} + V_2\mathbf{j}$$

and set

$$\mathbf{R} = x\mathbf{i} + y\mathbf{j}.$$

With these conventions, equation (45) can be written in terms of its components. This yields

(46)
$$x = V_1 t \quad \text{and} \quad y = -\frac{gt^2}{2} + V_2 t$$

as the parametric equations for the motion of a projectile starting at the origin. If α is the angle of $\mathbf{V}_0$, then $V_1 = V_0 \cos \alpha$ and $V_2 = V_0 \sin \alpha$, where $V_0 = |\mathbf{V}_0|$.

To obtain the Cartesian equation for the path of a projectile, we solve the first equation in (46) for t and use this in the second equation. Thus $t = x/V_1$ and hence

$$y = -\frac{g}{2}\left(\frac{x}{V_1}\right)^2 + V_2\left(\frac{x}{V_1}\right) = -\frac{g}{2V_1^2}x^2 + \frac{V_2}{V_1}x \equiv Ax^2 + Bx.$$

Hence the path is a parabola if $V_1 \neq 0$. ●

Example 4. The angle of elevation of a gun is $30°$. If the muzzle speed is 1600 ft/sec, what is the range? How long after firing does the projectile land?

Solution. We are assuming that the gun and the point of impact of the projectile are in the same horizontal plane (see Fig. 23). The range of the gun means the distance between these two points. We set the coordinate axes so that the origin is at the muzzle of the gun. Then $\mathbf{R}_0 = \mathbf{0}$, and equations (46) are applicable. The projectile hits the earth when $y = 0$. Under this condition the second equation in (46) gives

$$0 = -\frac{g}{2}t^2 + V_2 t = t\left(-\frac{g}{2}t + V_2\right),$$

or

(47) $$t = \frac{2V_2}{g} = \frac{2V_0 \sin \alpha}{32}$$

as the time of flight. Using this in equation (46), we find that

$$x = V_1 t = (V_0 \cos \alpha)t = \frac{2V_0^2 \sin \alpha \cos \alpha}{32} = \frac{V_0^2}{32} \sin 2\alpha.$$

This is a general expression for the range of a projectile fired with initial speed V_0 and angle of elevation α. Using the given data, we find that the range is

$$x = \frac{(1600)^2}{32} \sin 60° \approx 69{,}280 \text{ ft} \approx 13.1 \text{ miles.}$$

The time of flight is given by equation (47),

$$t = \frac{V_0 \sin \alpha}{16} = \frac{1600}{16} \cdot \frac{1}{2} = 50 \text{ sec.} \quad \bullet$$

EXERCISE 4

In Problems 1 through 4, find the length of arc of the given curve for the given interval.
1. $\mathbf{R}(t) = 2t^2\mathbf{i} + 5t^2\mathbf{j}$, $1 \leqq t \leqq 5$.
2. $\mathbf{R}(t) = 5 \cos^2 t\,\mathbf{i} + 5 \sin^2 t\,\mathbf{j}$, $0 \leqq t \leqq \pi/2$.
3. $\mathbf{R}(t) = (6t - 3t^2)\mathbf{i} + 8t^{3/2}\mathbf{j}$, $0 \leqq t \leqq 5$.
4. $\mathbf{R}(t) = (30t + 15t^2)\mathbf{i} + (20t^{3/2} + 12t^{5/2})\mathbf{j}$, $0 \leqq t \leqq 3$.

5–10. In Problems 5 through 10 of Exercise 3, suppose that $\mathbf{R}(t)$ is the vector equation of a moving particle and t is the time. As Problems 5 through 10 of this set, find: (a) the velocity, (b) the acceleration, and (c) the speed for $\mathbf{R}(t)$ given in Problems 5 through 10 of Exercise 3.

*11. For the cycloid $\mathbf{R} = a(\varphi - \sin\varphi)\mathbf{i} + a(1 - \cos\varphi)\mathbf{j}$ (see Example 2, Fig. 15, Section 3), find an expression for the speed in terms of $d\varphi/dt$. Assuming that $d\varphi/dt$ is a constant, find an expression for the maximum speed. Show that if a car is traveling 60 miles/hr, then the speed of a particle on the tire can be as high as 120 miles/hr.

12. Prove that if the acceleration vector of a particle is the zero vector for all time t, the motion of the particle is along a straight line.

13. What is the range of a gun if $V_0 = 800$ ft/sec and the angle of elevation is $15°$?

14. Find the range if $V_0 = 200$ m/sec and the angle of elevation is $20°$. HINT: Use $g = 9.8$ m/sec².

15. What is the maximum height reached by a projectile fired under the conditions of Problem 13?

16. Find the maximum height reached by a projectile fired under the conditions of Problem 14.

17. Prove that for a given fixed V_0 the maximum range for the gun is obtained by firing the gun at a $45°$ angle of elevation. Find a formula for this maximum range.

18. During World War I a German gun threw a shell approximately 64 miles. Assuming that this was the maximum range of the gun, find V_0.

19. Develop a general formula for the maximum height reached by a projectile fired with initial speed V_0 and angle of elevation α.

20. Show that doubling the initial speed of a projectile has the effect of multiplying both the range and the maximum height of the projectile by a factor of four.

21. An airplane drops a bomb while flying level at a height of 1500 m and at a speed of 360 km/hr. How long does it take for the bomb to land? How far does the bomb land in front of the position of the airplane at the time of release?

22. Do Problem 21 if the airplane is flying 240 miles/hr at an altitude of 2500 ft.

23. Tracy finds that no matter how hard she tries, she can throw a ball 200 ft but not farther. Find the maximum speed in miles per hour that she can throw a ball.

24. A man standing 30 ft from a tall building throws a ball with an initial speed of 80 ft/sec. If the angle of elevation of the throw is $60°$, how far up the building does the ball hit? Neglect the height of the man, and assume that he throws at the building. In what direction is the ball going when it hits the building?

6 Curvature

Our objective in this section is to obtain a means of measuring how fast a curve is turning. This amounts to finding the rate at which the tangent line is rotating as the point P of tangency travels along the curve. Let φ be the angle that the tangent line at P makes with the x-axis (see Fig. 24). It would seem at first glance that a satisfactory measure of the "curvature" of the curve should be

$$\frac{d\varphi}{dx}$$

because this gives a rate of turning of the tangent line. In Problems 17 and 18 of the next exercise, we will see that this definition of curvature is not satisfactory because it is not an invariant under a rigid motion ($d\varphi/dx$ may change as we move the curve).

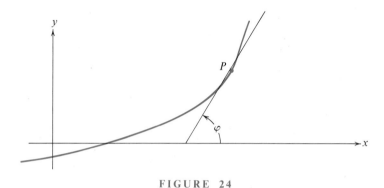

FIGURE 24

It turns out that if we replace x by the arc length s, then the quantity $d\varphi/ds$ is invariant (see Problem 21 of the next exercise).

DEFINITION 14 (Curvature). Let s denote the arc length on a curve, and suppose that a definite direction on the curve has been selected for the direction of increasing s. Let φ be the angle that the tangent line to the curve makes with the x-axis. Then the *curvature* of the curve, denoted by κ (Greek lowercase letter kappa), is defined by

(48) $$\kappa = \frac{d\varphi}{ds}.$$

To obtain a formula for computing κ, we suppose that the curve is given parametrically by

(15) $$x = f(t), \qquad y = g(t),$$

where the parameter t is selected so that s and t increase together. Then by Theorem 13, equation (37),

$$\frac{ds}{dt} = \sqrt{\left(\frac{dx}{dt}\right)^2 + \left(\frac{dy}{dt}\right)^2}.$$

Now φ is a function of t and t is a function of s, so we may write (48) in the form

(49)
$$\kappa = \frac{d\varphi}{dt}\frac{dt}{ds} = \frac{d\varphi}{dt}\bigg/\frac{ds}{dt} = \frac{d\varphi}{dt}\frac{1}{\sqrt{\left(\dfrac{dx}{dt}\right)^2 + \left(\dfrac{dy}{dt}\right)^2}}.$$

To compute $\dfrac{d\varphi}{dt}$, we observe that $\tan\varphi = m = \dfrac{dy}{dt}\bigg/\dfrac{dx}{dt}$, and on differentiating with respect to t, we obtain

$$\sec^2\varphi\,\frac{d\varphi}{dt} = \frac{\dfrac{dx}{dt}\dfrac{d^2y}{dt^2} - \dfrac{dy}{dt}\dfrac{d^2x}{dt^2}}{\left(\dfrac{dx}{dt}\right)^2}.$$

But

$$\sec^2\varphi = 1 + \tan^2\varphi = 1 + \frac{\left(\dfrac{dy}{dt}\right)^2}{\left(\dfrac{dx}{dt}\right)^2},$$

and hence

(50)
$$\frac{d\varphi}{dt} = \frac{\dfrac{dx}{dt}\dfrac{d^2y}{dt^2} - \dfrac{dy}{dt}\dfrac{d^2x}{dt^2}}{\left(\dfrac{dx}{dt}\right)^2 + \left(\dfrac{dy}{dt}\right)^2}.$$

Using (50) in (49), we arrive at

THEOREM 15. If the curve $\mathscr{C}$ is defined parametrically and if the arc length increases with increasing parameter t, then the curvature is given by

(51)
$$\kappa = \frac{\dfrac{dx}{dt}\dfrac{d^2y}{dt^2} - \dfrac{dy}{dt}\dfrac{d^2x}{dt^2}}{\left[\left(\dfrac{dx}{dt}\right)^2 + \left(\dfrac{dy}{dt}\right)^2\right]^{3/2}}.$$

Example 1. Compute the curvature of the parabola $y = x^2$ **(a)** in general and **(b)** at the point $(1, 1)$.

Solution. A convenient parameterization of this parabola is $x = t$ and $y = t^2$. Then (the arrows will help us remember the formula)

$$\frac{dx}{dt} = 1, \qquad \frac{dy}{dt} = 2t,$$

$$\frac{d^2x}{dt^2} = 0, \qquad \frac{d^2y}{dt^2} = 2,$$

$$\kappa = \frac{1(2) - 2t(0)}{(1 + 4t^2)^{3/2}} = \frac{2}{(1 + 4t^2)^{3/2}}.$$

At $(1, 1)$ we must have $t = 1$, and then $\kappa = 2/\sqrt{125} = 2/5\sqrt{5}$. ●

It can be proved that the formula (51) is invariant under a rotation of the curve, and also under a change of parameterization. Although these results are intuitively obvious, the actual computations are involved, so we reserve this important point for the starred problems in the next exercise.

Two special cases of formula (51) are worth noting. Frequently, our curve is given by a formula $y = f(x)$. In this case a convenient parameterization is obtained by setting $x = t$ and $y = f(t)$. Then

$$\frac{dx}{dt} = 1, \qquad \frac{d^2x}{dt^2} = 0, \qquad \frac{dy}{dt} = \frac{dy}{dx}, \qquad \frac{d^2y}{dt^2} = \frac{d^2y}{dx^2},$$

and formula (51) yields the following

COROLLARY. If $y = f(x)$, then the curvature is given by

(52)
$$\kappa = \frac{\dfrac{d^2y}{dx^2}}{\left[1 + \left(\dfrac{dy}{dx}\right)^2\right]^{3/2}} = \frac{f''(x)}{[1 + [f'(x)]^2]^{3/2}}.$$

If the curve is given by $x = g(y)$, a similar computation shows that

(53)
$$\kappa = \frac{-\dfrac{d^2x}{dy^2}}{\left[1 + \left(\dfrac{dx}{dy}\right)^2\right]^{3/2}} = \frac{-g''(y)}{[1 + [g'(y)]^2]^{3/2}}.$$

In each of these formulas the direction of increasing arc length is *assumed* to be the same as the direction of increasing parameter. Thus in (52) x and s are both increasing as a point P on the curve moves from left to right, and in (53) y and s are both increasing as a point P moves upward along the curve.

> **Example 2.** Show that for a circle the curvature is a constant, and the absolute value of this constant is the reciprocal of the radius of the circle.
>
> **Solution.** One is tempted to use $y = \sqrt{r^2 - x^2}$ and carry out the computations for the upper half of the circle. But it turns out that the computation is simpler if we use the parameterization $x = r \cos t$ and $y = r \sin t$. Then
>
> $$x' = -r \sin t \qquad y' = r \cos t$$
> $$x'' = -r \cos t \qquad y'' = -r \sin t$$
>
> and from (51),
>
> $$\kappa = \frac{r^2 \sin^2 t + r^2 \cos^2 t}{[r^2 \sin^2 t + r^2 \cos^2 t]^{3/2}} = \frac{1}{r}. \quad \bullet$$

Suppose that we fasten our attention upon a fixed point P on a curve $\mathscr{C}$ and try to draw through P a circle that most closely fits $\mathscr{C}$. We would first require that the circle and the curve be tangent at P, and we might next ask that they have the same curvature. When we do this, the circle is uniquely determined, and we call this circle the *circle of curvature*, or the *osculating circle*, of the curve $\mathscr{C}$ at P. The radius of this circle is denoted by ρ (Greek lowercase letter rho) and is called the *radius of curvature* of the curve $\mathscr{C}$ at P. From our example, we have $\rho = 1/\kappa$ for any circle and hence for any curve, $y = f(x)$,

(54)
$$\rho = \frac{1}{|\kappa|} = \frac{[1 + (y')^2]^{3/2}}{|y''|},$$

provided, of course, that the denominator y'' is not zero at P.

We did not really prove that this circle of curvature is the "closest" circle to the curve at P, nor could we do so, because we did not define what is meant by "closest" fitting circle. The definition of "closest" is complicated and the proof that the circle of curvature is "closest" is still more complicated. This topic is best reserved for the course in advanced calculus.

EXERCISE 5

In Problems 1 through 9, find the curvature for the given curve.

1. $y = mx + b.$
2. $y = x^2 + 2x.$
3. $y = e^x.$
4. $x = \sqrt{y}.$
5. $y = \sin 3x.$
6. $x = \ln (\cos y).$

7. $x = a(t - \sin t)$,   8. $x = b(\cos\theta + \theta\sin\theta)$,   9. $y = x^3$.
   $y = a(1 - \cos t)$, $a > 0$. $y = b(\sin\theta - \theta\cos\theta)$, $b > 0$.

10. Show that the curvature of the upper half-circle $y = \sqrt{r^2 - x^2}$ is $-1/r$. Note that this minus sign seems to be inconsistent with the result in Example 2, where it was proved that $\kappa = 1/r$. Explain this discrepancy.

11. At what point on the parabola $y = x^2/4$ is the radius of curvature a minimum?

12. Find the minimum value of the radius of curvature for the curve $y = \ln x$.

$\star$13. Find the minimum value of the radius of curvature for the curve $y = x^4/4$.

$\star$14. Suppose that a circle $\mathscr{C}_1$ and a curve $\mathscr{C}_2$ are tangent at a point P, and that for the equations which give these two curves d^2y/dx^2 is the same at P for both curves. Prove that both curves have the same curvature at P, and hence $\mathscr{C}_1$ is the circle of curvature of $\mathscr{C}_2$ at P.

$\star$15. Prove the converse of the theorem stated in Problem 14, namely, that if $\mathscr{C}_1$ is the circle of curvature of $\mathscr{C}_2$ at P, then d^2y/dx^2 is the same at P for both curves.

16. Prove that at the ends of the major and minor axes of an ellipse $x = a\cos t$, $y = b\sin t$, the two values for the curvature are a/b^2 and b/a^2. Notice that when $a = b$, the ellipse is a circle, and both of these formulas give $\kappa = 1/r$, as they should.

17. Compute $d\varphi/dx$ for the parabola $y = x^2$ at the point $P(1, 1)$. HINT: Use the fact that $\varphi = \text{Tan}^{-1} y' = \text{Tan}^{-1} 2x$.

$\star$18. If we rotate the right half of the parabola of Problem 17 backward (clockwise) through an angle of $90°$, we obtain the curve $y = -\sqrt{x}$, and the point $P(1, 1)$ goes into the point $Q(1, -1)$. Find $d\varphi/dx$ for $y = -\sqrt{x}$ at Q. Is this answer the same as the answer to Problem 17?

19. In Example 1 we found that the curvature of $y = x^2$ at $(1, 1)$ is $2/5\sqrt{5}$. How does this result compare with the answers to Problems 17 and 18.

20. Find the curvature of $y = -\sqrt{x}$ at $(1, -1)$.

$\star\star$21. Prove that the curvature is invariant under a rotation of the curve about the origin. Outline of solution: Let (x, y) be the coordinates in the original position and let (X, Y) be the new coordinates. Then (from Chapter 11, Exercise 6, Problem 19)

$$X = x\cos\alpha - y\sin\alpha,$$
$$Y = x\sin\alpha + y\cos\alpha.$$

If the curve $\mathscr{C}$ is given parametrically, x, y, X, and Y are all functions of t. Prove that

$$\frac{X'Y'' - Y'X''}{[(X')^2 + (Y')^2]^{3/2}} = \frac{x'y'' - y'x''}{[(x')^2 + (y')^2]^{3/2}}.$$

The computation can be simplified by considering the numerators and denominators separately because it turns out that these two pieces are each invariant under a rotation of the curve.

$\star\star$22. Prove that the curvature is invariant under a change of parametrization. Outline of solution: Let $x = f(t), y = g(t)$ be one parametrization. A new parameter T can be introduced by letting t be a function of T, $t = h(T)$. Prove first that

$$\frac{dx}{dT} = \frac{dx}{dt}\frac{dt}{dT} \qquad \frac{d^2x}{dT^2} = \frac{d^2x}{dt^2}\left(\frac{dt}{dT}\right)^2 + \frac{dx}{dt}\frac{d^2t}{dT^2},$$

with similar equations for y. Then use these to prove that

$$\frac{\dfrac{dx}{dT}\dfrac{d^2y}{dT^2} - \dfrac{dy}{dT}\dfrac{d^2x}{dT^2}}{\left[\left(\dfrac{dx}{dT}\right)^2 + \left(\dfrac{dy}{dT}\right)^2\right]^{3/2}} = \frac{\dfrac{dx}{dt}\dfrac{d^2y}{dt^2} - \dfrac{dy}{dt}\dfrac{d^2x}{dt^2}}{\left[\left(\dfrac{dx}{dt}\right)^2 + \left(\dfrac{dy}{dt}\right)^2\right]^{3/2}}.$$

7   Arc Length as a Parameter

In Section 5 we saw that the arc length s can be used as a parameter. To be specific, if $\mathscr{C}$ is a smooth curve and we fix an initial point P_0 on $\mathscr{C}$ (see Fig. 22), then for each point on $\mathscr{C}$ there is a unique number s (the directed length of arc from P_0 to P), and conversely for each s there is a unique point on the curve (if the curve is long enough).

Thus, in theory, we can always introduce the arc length as a parameter, but the actual computations are frequently quite messy, and sometimes impossible in terms of our elementary functions. It turns out that this computation is easy for the straight line and for the circle, but even for such an elementary curve as the ellipse, the integrals involved can *not* be expressed in finite terms.

Example 1. Find parametric equations for

(55)
$$\mathbf{R}(t) = \frac{t^2}{2}\mathbf{i} + \frac{t^3}{3}\mathbf{j}, \qquad t \geqq 0$$

using s as the parameter.

Solution. For convenience we take $(0, 0)$ as the fixed point from which we measure arc length. By Theorem 13,

(56)
$$\frac{ds}{dt} = |\mathbf{R}'(t)| = \sqrt{t^2 + t^4} = t\sqrt{1 + t^2},$$

and hence

$$s = \int_0^t u\sqrt{1 + u^2}\, du = \frac{1}{2}\int_0^t (1 + u^2)^{1/2} 2u\, du = \frac{1}{2}\frac{2}{3}(1 + u^2)^{3/2}\Big|_0^t,$$

(57) $$s = \frac{1}{3}(1 + t^2)^{3/2} - \frac{1}{3}.$$

Solving equation (57) for t in terms of s, we find $t = \sqrt{(3s + 1)^{2/3} - 1}$. Using this expression for t in (55), we obtain

(58) $$x = \frac{1}{2}((3s + 1)^{2/3} - 1), \qquad y = \frac{1}{3}((3s + 1)^{2/3} - 1)^{3/2}$$

as parametric equations for the curve (55), where s is now the parameter.   ●

    The arc length as a parameter is very helpful when applied to the unit tangent and the unit normal vectors. We recall from Section 4 that if $\mathbf{R}(t) = f(t)\mathbf{i} + g(t)\mathbf{j}$, then the unit tangent vector $\mathbf{T}$ is given by

(34) $$\mathbf{T} = \frac{f'(t)\mathbf{i} + g'(t)\mathbf{j}}{\sqrt{f'(t)^2 + g'(t)^2}}.$$

The unit normal $\mathbf{N}$ that is $90°$ in advance of $\mathbf{T}$ (see Figs. 25 and 26) is given by

(35) $$\mathbf{N} = \frac{-g'(t)\mathbf{i} + f'(t)\mathbf{j}}{\sqrt{f'(t)^2 + g'(t)^2}}.$$

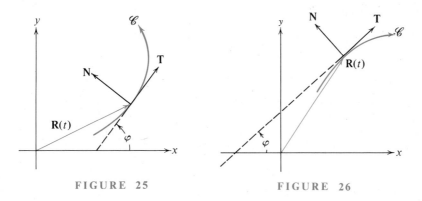

FIGURE 25   FIGURE 26

When the parameter is s rather than t, we have

THEOREM 16. If $\mathbf{R} = \mathbf{R}(s)$ is the vector equation of a smooth curve $\mathscr{C}$, and s is arc length on $\mathscr{C}$, then

(59) $$\frac{d\mathbf{R}}{ds} = \mathbf{T}.$$

Proof. By Theorem 13 we have $|\mathbf{R}'(t)| = ds/dt$. But now $s = t$, so the right side is just 1. Hence $\mathbf{R}'(s)$ is a unit vector. But $\mathbf{R}'$ is a tangent vector for any parameter, as long as $\mathbf{R}' \neq 0$. ∎

Written in component form, equation (59) states that

(60)
$$\mathbf{T} = \frac{dx}{ds}\mathbf{i} + \frac{dy}{ds}\mathbf{j},$$

On the other hand, if φ is the angle that $\mathbf{T}$ makes with the positive x axis, then

(61)
$$\mathbf{T} = \cos\varphi\,\mathbf{i} + \sin\varphi\,\mathbf{j}.$$

If we compare (60) and (61), we see that

(62)
$$\frac{dx}{ds} = \cos\varphi, \qquad \frac{dy}{ds} = \sin\varphi.$$

Consequently,

(63)
$$\left(\frac{dx}{ds}\right)^2 + \left(\frac{dy}{ds}\right)^2 = 1.$$

We can also find $\cos\varphi$ and $\sin\varphi$ when the parameter is t. Indeed, if we compare (61) and (34), we see that

(64)
$$\cos\varphi = \frac{f'(t)}{\sqrt{f'(t)^2 + g'(t)^2}} \qquad \text{and} \qquad \sin\varphi = \frac{g'(t)}{\sqrt{f'(t)^2 + g'(t)^2}}.$$

Example 2. Use equation (63) to check the results in Example 1.

Solution. The components $x(s)$ and $y(s)$ are given by equation set (58). Hence, from these equations,

$$\left(\frac{dx}{ds}\right)^2 = \left[\frac{1}{2}\cdot\frac{2}{3}\frac{1}{(3s+1)^{1/3}}\cdot 3\right]^2 = \frac{1}{(3s+1)^{2/3}},$$

$$\left(\frac{dy}{ds}\right)^2 = \left[\frac{1}{3}\cdot\frac{3}{2}((3s+1)^{2/3}-1)^{1/2}\cdot\left(\frac{2}{3}\right)\frac{3}{(3s+1)^{1/3}}\right]^2 = \frac{(3s+1)^{2/3}-1}{(3s+1)^{2/3}},$$

and the sum is 1, as it should be. ●

Example 3. For the curve of Example 1, find $\mathbf{T}$, $\mathbf{N}$, $\cos\varphi$, and $\sin\varphi$ as functions of t.

Solution. Differentiating gives $\mathbf{R}' = t\mathbf{i} + t^2\mathbf{j}$ and $|\mathbf{R}'| = \sqrt{t^2 + t^4}$. In this example $\mathbf{R}' = \mathbf{0}$ when $t = 0$, and we may infer that the point $(0, 0)$ is some sort of singular point on the curve at which **T** and **N** are not defined. For convenience we restrict ourselves to $t > 0$. Then $|\mathbf{R}'| = t\sqrt{1 + t^2}$, and by equation (34) the unit tangent is

(65)
$$\mathbf{T} = \frac{t\mathbf{i} + t^2\mathbf{j}}{t\sqrt{1 + t^2}} = \frac{\mathbf{i} + t\mathbf{j}}{\sqrt{1 + t^2}}.$$

Although **T** is not initially defined at $(0, 0)$, equation (65) shows that the unit tangent vector approaches the limit **i** as $t \to 0$, and thus **i** can be regarded as the unit tangent at $(0, 0)$. Comparing (34) and (35), we see that **N** can be obtained from **T** by a switch in the components, and an alteration in sign. Performing this operation on (65), we have

(66)
$$\mathbf{N} = \frac{-t\mathbf{i} + \mathbf{j}}{\sqrt{1 + t^2}}.$$

Finally, the components of **T** yield

(67)
$$\cos \varphi = \frac{1}{\sqrt{1 + t^2}} \quad \text{and} \quad \sin \varphi = \frac{t}{\sqrt{1 + t^2}}. \quad \bullet$$

THEOREM 17. For the unit tangent and normal vectors, we have the differentiation formulas

(68)
$$\frac{d\mathbf{T}}{d\varphi} = \mathbf{N}$$

and

(69)
$$\frac{d\mathbf{T}}{ds} = \kappa\mathbf{N},$$

where φ is the angle the tangent vector makes with the x-axis and κ is the curvature.

Proof. If we differentiate $\mathbf{T} = \cos \varphi \mathbf{i} + \sin \varphi \mathbf{j}$ with respect to φ, we obtain $\mathbf{T}'(\varphi) = -\sin \varphi \mathbf{i} + \cos \varphi \mathbf{j}$, and this is a unit vector $90°$ in advance of **T** and hence must be **N**. This gives (68). For (69) we have

$$\frac{d\mathbf{T}}{ds} = \frac{d\mathbf{T}}{d\varphi}\frac{d\varphi}{ds} = \mathbf{N}\kappa$$

by using (68), and the definition of the curvature κ, equation (48). ∎

EXERCISE 6

For each of the curves given in Problems 1 through 5, find parametric equations in which the parameter is the arc length measured from the given P_0. In each case check your answer by showing that $(dx/ds)^2 + (dy/ds)^2 = 1$.

1. The straight line $\mathbf{R} = (a + mt)\mathbf{i} + (b + nt)\mathbf{j}$, $P_0(a, b)$.
2. The circle $\mathbf{R} = (a + r\cos\theta)\mathbf{i} + (b + r\sin\theta)\mathbf{j}$, $P_0(a + r, b)$.
3. The involute $\mathbf{R} = 2(\cos\theta + \theta\sin\theta)\mathbf{i} + 2(\sin\theta - \theta\cos\theta)\mathbf{j}$, $P_0(2, 0)$.
4. The spiral $\mathbf{R} = \dfrac{e^t}{\sqrt{2}}(\cos t\,\mathbf{i} + \sin t\,\mathbf{j})$, $P_0(1/\sqrt{2}, 0)$.
5. One arch of the hypocyloid of four cusps $\mathbf{R} = (2/3)(\cos^3\theta\,\mathbf{i} + \sin^3\theta\,\mathbf{j})$, where $0 \le \theta \le \pi/2$ and $P_0(2/3, 0)$.

6. Find the length of one arch of the cycloid $\mathbf{R} = a(t - \sin t)\mathbf{i} + a(1 - \cos t)\mathbf{j}$.

7. Show that for the curve $x = 2e^t$, $y = \dfrac{1}{2}e^{2t} - t$, it is possible to find s as an elementary function of t. Observe that it seems to be difficult to solve for t in terms of s, so that an explicit formula for this curve in terms of elementary functions of s appears to be impossible.

8. Find an integral expression for the arc length of the ellipse $x = 4\cos t$, $y = 3\sin t$, $P_0(4, 0)$. It is a known fact that this integral cannot be evaluated in finite terms.

In Problems 9 through 16, find the unit normal $\mathbf{N}$ for the given curve.

9. $\mathbf{R} = (a + mt)\mathbf{i} + (b + nt)\mathbf{j}$.
10. $\mathbf{R} = (a + r\cos t)\mathbf{i} + (b + r\sin t)\mathbf{j}$, $r > 0$.
11. $\mathbf{R} = a\cos t\,\mathbf{i} + b\sin t\,\mathbf{j}$.
12. $\mathbf{R} = 2t\mathbf{i} + t^2\mathbf{j}$.
13. $\mathbf{R} = (t^3 - 3t)\mathbf{i} + (4 - 3t^2)\mathbf{j}$.
14. $\mathbf{R} = 2t\mathbf{i} + (e^t + e^{-t})\mathbf{j}$.
$\star$15. $\mathbf{R} = a(t - \sin t)\mathbf{i} + a(1 - \cos t)\mathbf{j}$, $a > 0$.
16. $\mathbf{R} = b(\cos u + u\sin u)\mathbf{i} + b(\sin u - u\cos u)\mathbf{j}$, $b > 0$, $u > 0$.

17. If φ is the angle that $\mathbf{T}$ makes with the positive x-axis, find $\cos\varphi$ for the curve in: (a) Problem 9, (b) Problem 11, (c) Problem 13, and (d) Problem 15.
18. Do Problem 17 for the curve in: (a) Problem 10, (b) Problem 12, (c) Problem 14, and (d) Problem 16.
19. Show that if $\kappa \ne 0$, then the vector $\kappa\mathbf{N}$ always points toward the concave side of the curve. There are four cases to consider. Two of the cases are shown in Fig. 27, and the other two arise from reversing the positive direction on the curve.

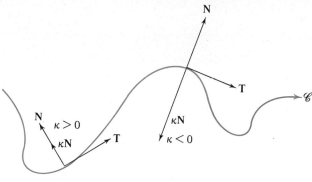

FIGURE 27

Tangential and Normal Components of Acceleration $8^\star$

The motion of a particle is completely determined when the x- and y-coordinates of the particle are given for any time t. This is done quite simply by giving its position vector

(70) $$\mathbf{R}(t) = f(t)\mathbf{i} + g(t)\mathbf{j},$$

from which we obtain at once, by differentiating, the velocity and acceleration vectors

(71) $$\mathbf{V}(t) = f'(t)\mathbf{i} + g'(t)\mathbf{j},$$

(72) $$\mathbf{A}(t) = f''(t)\mathbf{i} + g''(t)\mathbf{j}.$$

Now equation (72) gives the components of the acceleration in the direction of the x- and y-axes. Frequently, it is more important to have the components of $\mathbf{A}$ along the tangent to the curve and along the normal to the curve. If we use subscripts A_T and A_N to denote these two scalars, they are defined by the equation

(73) $$\mathbf{A} = A_T\mathbf{T} + A_N\mathbf{N}.$$

Formulas for these components are given in

THEOREM 18. The velocity and acceleration vectors of a particle moving on a curve $\mathbf{R}(t)$ are

(74) $$\mathbf{V} = \frac{ds}{dt}\mathbf{T} = V\mathbf{T}$$

and

(75)
$$A = \frac{d^2s}{dt^2}T + \left(\frac{ds}{dt}\right)^2 \kappa N,$$

where T and N are the unit tangent and unit normal vectors to the curve, s is the arc length measured in the direction of the motion, κ is the curvature, and t is the time.

Equation (74) states that the normal component of the velocity is zero. Equation (75) tells us that A_T and A_N are given by the formulas

(76)
$$A_T = \frac{d^2s}{dt^2} = \frac{dV}{dt} \quad \text{and} \quad A_N = \left(\frac{ds}{dt}\right)^2 \kappa = \frac{V^2}{\rho},$$

where of course V is the speed of the particle, and ρ, the radius of curvature, is the reciprocal of the curvature.

The importance of the acceleration vector lies in the fact that from all experimental evidence particles in nature move in accordance with Newton's Law,

$$F = mA,$$

where F is the vector force applied to the particle and m is the mass of the particle. Thus mA_T is the component of the force tangential to the path required to keep the particle moving with the desired speed, and mV^2/ρ is the component of the force normal to the curve required to keep the particle on the curve. This latter force is usually a frictional force, or a restraining force, as in the case of an automobile or a train going around a curve.

Proof of Theorem 18. By definition,

$$V = \frac{dR}{dt} = \frac{dR}{ds}\frac{ds}{dt} = T\frac{ds}{dt}.$$

Since $ds/dt = V$, the speed, this proves (74). Differentiating both sides of (74) with respect to t and using equation (69) of Theorem 17, we find that

$$A = \frac{dV}{dt} = \frac{d}{dt}\left(\frac{ds}{dt}T\right) = \frac{d^2s}{dt^2}T + \frac{ds}{dt}\frac{dT}{dt}$$

$$= \frac{d^2s}{dt^2}T + \frac{ds}{dt}\left(\frac{dT}{ds}\frac{ds}{dt}\right) = \frac{d^2s}{dt^2}T + \left(\frac{ds}{dt}\right)^2\frac{dT}{ds} = \frac{d^2s}{dt^2}T + \left(\frac{ds}{dt}\right)^2\kappa N. \quad \blacksquare$$

If $R(t) = f(t)i + g(t)j$, then the tangential and normal components of velocity and acceleration are given by

(77)
$$V = \frac{ds}{dt} = \sqrt{f'(t)^2 + g'(t)^2},$$

(78)
$$A_T = \frac{d^2s}{dt^2} = \frac{f'(t)f''(t) + g'(t)g''(t)}{\sqrt{f'(t)^2 + g'(t)^2}},$$

and

(79)
$$A_N = \left(\frac{ds}{dt}\right)^2 \kappa = \frac{f'(t)g''(t) - g'(t)f''(t)}{\sqrt{f'(t)^2 + g'(t)^2}}.$$

Example 1. A common amusement device found at large entertainment parks consists of a giant horizontal flat wheel. Volunteers climb onto this wheel while it is stationary. The operator starts the wheel rotating about the fixed center and the volunteers attempt to stay on the wheel as long as possible. Find a formula for the critical angular velocity beyond which the volunteer must slide off.

Solution. To be definite, let us assume that the wheel has a radius of 20 ft and that the coefficient of friction μ (Greek lowercase letter mu) is $1/10$. This means that if a person has weight W, the maximum frictional force that can be exerted between the wheel and the person on it is μW. If the person is pushed horizontally outward with a larger force, he will tend to slide off. Thus it is this frictional force μW which provides the person riding on the wheel with the necessary normal component of acceleration to stay in his place on the wheel.

Let us assume that the wheel has a steady motion of ω revolutions per second. Then the vector equation for the motion of a particle r feet from the center is $\mathbf{R}(t) = r(\cos 2\pi\omega t\, \mathbf{i} + \sin 2\pi\omega t\, \mathbf{j})$. Hence, for the speed, $V = |\mathbf{R}'(t)| = 2\pi r\omega$ ft/sec. For the volunteer to stay in place he needs a normal acceleration [equation (76)]

$$A_N = V^2\kappa = \frac{V^2}{r} = 4\pi^2 r\omega^2 \text{ ft/sec}^2.$$

In the equation $\mathbf{F} = m\mathbf{A}$, the mass in the British system is $m = W/g$, where W is the weight of the body and g is the acceleration due to gravity. Since the wheel is rotating steadily, the tangential component of acceleration is zero [see equation (76)], so the frictional force required to keep the volunteer in position is

(80)
$$F = mA_N = \frac{W}{g}4\pi^2 r\omega^2.$$

On the other hand, $F = \mu W$ is the maximum force that can be exerted by the volunteer in his effort to stay on. Hence the critical ω satisfies the equation

(81)
$$\mu W = \frac{W}{g}4\pi^2 r\omega^2,$$

or

(82)
$$\omega = \frac{1}{2\pi} \sqrt{\frac{\mu g}{r}}.$$

Any greater rate of turning, and the volunteer slides off. Note that W has canceled, so a heavy person and a light person have equal opportunity.

To be specific, suppose that our volunteer is at the outer edge of the wheel. Then

$$\omega = \frac{1}{2\pi} \sqrt{\frac{32 \times 0.1}{20}} = \frac{1}{2\pi} \sqrt{0.16}$$

$$= \frac{0.2}{\pi} \text{ rev/sec} = \frac{12}{\pi} \text{ rev/min} \approx 3.82 \text{ rev/min}.$$

If the wheel turns more rapidly, the person on the edge must slide off. ●

EXERCISE 7

1. What is the critical speed for the wheel described in the example if the volunteer is 5 ft from the center of the wheel?

2. Find the critical speed for the wheel described in the example if the volunteer is only 1 ft from the center of the wheel and $\mu = 1/4$ (he is wearing gym shoes and has rosin on his hands).

3. Show that if a person can sit right on the center of the wheel described in the example, then (barring physiological effects) he can stay on indefinitely no matter how fast the wheel turns.

4. A car weighing 3200 lb going steadily at 60 miles/hr makes a circular turn on a flat road. If the radius of the circular turn is 44 ft, what frictional force is required on the bottom of the tires to keep the car from skidding?

5. What is the least possible value of μ, the coefficient of friction between the tire and the road, that is sufficient to keep the car of Problem 4 from slipping?

6. Show that if the driver of the car in Problem 4 will cut his speed in half, the frictional force required to keep his car from slipping will be reduced by a factor of one fourth. Show that the same is true of the minimum value of the coefficient of friction.

7. If the metric (MKS) system is used in equation (82), we should obtain the same ω since revolutions per second is the same in all systems. Show that if $g = 9.8$ meters/sec^2 and r is measured in meters, equation (82) gives (approximately) the same answer. HINT: 1 ft is 0.3048 m.

8. Check that equations (77), (78), and (79) are correct.

In Problems 9 through 14, the motion of a particle is described by its vector equation $\mathbf{R}(t)$, *where t is time. In each case find:* (a) V, (b) A_T, *and* (c) A_N.

9. $\mathbf{R}(t) = (3 + 5t)\mathbf{i} + (2 - 12t)\mathbf{j}.$ 10. $\mathbf{R}(t) = 3\cos t\mathbf{i} + 4\sin t\mathbf{j}.$

11. $\mathbf{R}(t) = t\mathbf{i} + t^2\mathbf{j}.$

12. $\mathbf{R}(t) = t^2\mathbf{i} + 2\cos t\mathbf{j}.$

13. $\mathbf{R}(t) = \dfrac{t^2}{2}\mathbf{i} - \dfrac{t^3}{3}\mathbf{j}, \quad t > 0.$

14. $\mathbf{R}(t) = 2t\mathbf{i} + (e^t + e^{-t})\mathbf{j}.$

15. If the metric (MKS) system is used in the equation $\mathbf{F} = m\mathbf{a},$ where the mass is in kilograms and the acceleration is in m/sec^2, then the force is measured in newtons (nt). If a particle of 10 grams is moving on the curve of Problem 11, where t is time in seconds and the coordinates are given in centimeters, find the tangential and normal components of the force in newtons acting on the particle at the point $(2, 4)$.

16. Do Problem 15 for the curve of Problem 13 at the point $(1/2, -1/3)$.

17. Do Problem 15 for a mass of 500 grams, moving on the curve of Problem 14, when the coordinates are in meters and the mass is at the point $(2, e + 1/e)$.

18. A locomotive weighing 120,000 kg is moving at 36 km/hr on a circle of radius 400 m. Find the total force on the outer wheels exerted by the rails that is necessary to keep the locomotive on the tracks.

19. A locomotive weighing 120 tons is going steadily at 60 miles/hr along a level track that is at first straight and then takes a turn. The equation of the curve is $1760y = x^2$ (in feet), and the curved piece joins the straight piece at the vertex $(0, 0)$ of the parabola. Find (approximately) the horizontal radial thrust of the locomotive on the outer rail just after the locomotive enters on the parabolic turn.

20. A man holds onto a rope to which is tied a pail holding 5 lb of water. He swings the pail in a vertical circle with a radius of 4 ft. If the pail is making 60 rpm, what is the pressure of the water on the bottom of the pail at the high point and low point of the swing? Find the least number of rpm in order that the water will stay in the pail.

21. A popular amusement ride called the Round-Up consists of a flat circular ring with outer radius 16 ft. The ring is at first horizontal and the volunteers stand on the outer edge of the ring, each facing the center of the ring. Each person is supported in back by a wire fence that is roughly 7 ft high, mounted securely on the outer edge of the ring. When all is ready, the ring, fence, and volunteers begin rotating. The volunteers are thrown backward against the fence, and when the force is large enough to assure safety, the ring, fence, and volunteers are gradually lifted until the collection is rotating in a vertical plane. If the ring rotates in a vertical plane at 21 rpm, what force does a man of weight W exert against the wire fence: **(a)** when he is at the top of the circular path looking downward, and **(b)** when he is at the bottom of the path?

22. An astronaut is traveling in a circular orbit 440 miles above the surface of the earth. If the radius of the earth is 3960 miles, find his speed. How long does it take for the spaceship to make one circuit around the earth? HINT: The ship and the astronaut are both weightless when in a stable orbit; that is, the total force on the ship is zero. For simplicity, assume that at that height g is still 32 ft/sec^2, although actually it will be closer to $32(9/10)^2$ ft/sec^2.

23. An object weighing 10 lbs moves on the curve $y = x^2$ from left to right. At the mo-

ment that the object is at the origin, the total force on the object is $\mathbf{F} = 5\mathbf{i} + 10\mathbf{j}$, where $|\mathbf{F}|$ is in pounds. When the object is at the origin, find: (a) the speed and (b) A_T.

REVIEW PROBLEMS

In Problems 1 through 9, let $\mathbf{X} = 2\mathbf{i} - 3\mathbf{j}$, $\mathbf{Y} = 3\mathbf{i} + \mathbf{j}$, $\mathbf{Z} = 5\mathbf{i} + 6\mathbf{j}$, *and* $\mathbf{W} = \mathbf{i} - 2\mathbf{j}$. *Compute each of the specified vectors.*

1. $\mathbf{X} + 3\mathbf{Y}$.
2. $2\mathbf{X} + \mathbf{Y}$.
3. $\mathbf{X} + \mathbf{Y} - \mathbf{Z}$.
4. $10\mathbf{Y} + 5\mathbf{W}$.
5. $10\mathbf{X} + 2\mathbf{Z}$.
6. $\mathbf{Y} + \mathbf{Z} - 7\mathbf{W}$.
7. $2\mathbf{W} - \mathbf{X}$.
8. $3\mathbf{Z} - 5\mathbf{Y}$.
9. $3\mathbf{X} - 5\mathbf{Y} + 2\mathbf{Z} - \mathbf{W}$.

In Problems 10 through 15, use the vectors $\mathbf{X}, \mathbf{Y}, \mathbf{Z}$, *and* $\mathbf{W}$ *given above and solve the given equation.*

10. $a\mathbf{X} + b\mathbf{Y} = 5\mathbf{i} + 9\mathbf{j}$.
11. $a\mathbf{Z} + b\mathbf{W} = 17\mathbf{i} - 2\mathbf{j}$.
12. $c\mathbf{Y} + d\mathbf{Z} = -10\mathbf{i} + \mathbf{j}$.
13. $c\mathbf{X} + d\mathbf{W} = 2\mathbf{Y} - 9\mathbf{j}$.
14. $c\mathbf{X} + d\mathbf{W} = \mathbf{Z} - \mathbf{Y}$.
15. $a(\mathbf{Z} + 2\mathbf{X}) = b(\mathbf{Y} - 3\mathbf{W})$.

16. Let the vectors $\mathbf{X}, \mathbf{Y}, \mathbf{Z}$, and $\mathbf{W}$ given above be the position vectors of points X, Y, Z, and W, respectively. Find the vectors: (a) $\overrightarrow{XY}$, (b) $\overrightarrow{XZ}$, (c) $\overrightarrow{XW}$, (d) $\overrightarrow{YZ}$, and (e) $\overrightarrow{YW}$. Make a picture showing each vector.
17. For the points given in Problem 16, find the midpoint of the line segments XY, ZW, and WX.

In Problems 18 through 23, the vector function $\mathbf{R}(t)$ *gives the position vector of a point P. Sketch the curve generated by* $\mathbf{R}(t)$. *You may find it helpful to compute* $\mathbf{R}'(t)$ *and find the tangent vector at several points.*

18. $\mathbf{R}(t) = \cos^2 t\mathbf{i} + \sin t\mathbf{j}$, $0 \le t \le 2\pi$.
19. $\mathbf{R}(t) = t^2\mathbf{i} + (1 - t^2)\mathbf{j}$, $-4 \le t \le 4$.
20. $\mathbf{R}(t) = t^2\mathbf{i} + (1/t)\mathbf{j}$, $0 < t$.
21. $\mathbf{R}(t) = \sec t\mathbf{i} + \tan t\mathbf{j}$, $0 < t < \pi/2$.
22. $\mathbf{R}(t) = (3 + 2\sin t)\mathbf{i} + (2 - 3\cos t)\mathbf{j}$, $0 \le t \le 4\pi$.
23. $\mathbf{R}(t) = (3 + 2\sin^2 t)\mathbf{i} + (2 - 3\cos^2 t)\mathbf{j}$, $0 \le t \le 4\pi$.

24. For each of the curves given in Problems 18 through 23, find a general expression for $\mathbf{N}$, the unit normal vector.
25. For each of the curves given in Problems 18 through 23, find a general expression for the curvature.
26. Find the maximum value of the curvature for the curve: (a) of Problem 20, and (b) of Problem 22.

27. Find those points on the curve of Problem 20 where the tangent to the curve is parallel to the vector: (a) $\mathbf{i} - 4\mathbf{j}$, (b) $\mathbf{i} - \mathbf{j}$, and (c) $2\mathbf{i} - \mathbf{j}$.

28. Suppose that $\mathbf{R}(t)$ gives the position vector of a moving particle and t is the time. Find the maximum and minimum values for the speed when $\mathbf{R}(t)$ is the function given: (a) in Problem 22, and (b) in Problem 23.

29. For each of the motions given in Problems 18 through 23, find the normal component of the acceleration.

★30. Let $\mathscr{C}$ be a curve composed of two rays meeting at the origin and having slopes m and $-m$, respectively. Show that $\mathbf{R}(t) = |t^3|\mathbf{i} + mt^3\mathbf{j}$ gives a vector equation for this "angle." Prove that $\mathbf{R}'(t)$ is continuous. Does this curve seem like a "smooth" curve to you? Is there a point where $\mathbf{R}'(t) = \mathbf{0}$?

The integral of a vector function can be defined as a limit of a sum, but with only a little effort it can be proved that if $\mathbf{R}(t) = f(t)\mathbf{i} + g(t)\mathbf{j}$, then

$$\int_a^b \mathbf{R}(t)\, dt = \left(\int_a^b f(t)\, dt \right)\mathbf{i} + \left(\int_a^b g(t)\, dt \right)\mathbf{j}.$$

In Problems 31 through 35, compute the indicated integral.

31. $\mathbf{R}(t) = 2t\mathbf{i} + 3t^2\mathbf{j}$, $\qquad \displaystyle\int_1^3 \mathbf{R}(t)\, dt.$

32. $\mathbf{R}(t) = \sin t\mathbf{i} + \cos t\mathbf{j}$, $\qquad \displaystyle\int_0^{2\pi} \mathbf{R}(t)\, dt.$

33. $\mathbf{R}(t) = 2t\mathbf{i} + 3t^2\mathbf{j}$, $\qquad \displaystyle\int_1^t \mathbf{R}(u)\, du.$

34. $\mathbf{R}(t) = t(\sin t\mathbf{i} + \cos t\mathbf{j})$, $\qquad \displaystyle\int_0^t \mathbf{R}(u)\, du.$

35. $\mathbf{R}(u) = \dfrac{1}{4 + u^2}\mathbf{i} + \dfrac{u}{4 + u^2}\mathbf{j}$, $\quad \displaystyle\int_0^u \mathbf{R}(t)\, dt.$

36. Explain why a Ferris wheel of radius 20 ft should not rotate at more than 12 rpm. Actual operating speed is usually about 6 rpm.

37. The magnetic drum in the IBM 650 computer is 4 in. in diameter and rotates at about 12,000 rpm. Show that the normal acceleration of a particle on the surface of the drum is $(80,000\pi^2/3)$ ft/sec². Hence the adhesive force necessary to keep the particle from flying off is approximately $8000\, W$, where W is the weight of the particle.

13

Polar Coordinates

1 The Polar Coordinate System

So far, we have studied exclusively the rectangular coordinate system. But there are other coordinate systems that are frequently useful. Of these systems, the most important is the polar coordinate system. In this system a fixed point O is selected and from this fixed point a fixed ray (half-line) OA is drawn. The point O is called the *pole* or *origin,* and the ray OA is called the *polar axis,* or *polar line.* All points in the plane are located with respect to the point O and the ray OA. For convenience, the ray OA is always drawn horizontal and to the right, as shown in Fig. 1.

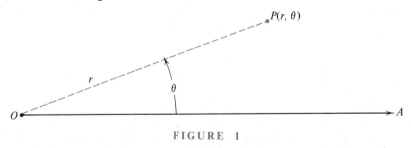

FIGURE 1

Now let P be any point in the plane other than O. Let r be the distance from P to O, and let θ be the angle from OA to OP. Then the numbers r and θ serve as *polar coordinates* for the point P and these coordinates are written (r, θ). In this determination of the polar coordinates for P it is obvious that $r > 0$ and $0 \leq \theta < 2\pi$. Except for the origin, each point in the plane has a unique pair of coordinates (r, θ) such that $r > 0$ and

522

$0 \leqq \theta < 2\pi$. Conversely, for each pair of numbers (r, θ) that satisfies these conditions, there is a unique point P in the plane that has the polar coordinates (r, θ).

Suppose that we wish to remove the restriction on r and θ and allow (r, θ) to be any pair of real numbers. Let (r, θ) be given. We first locate a ray OL by turning the polar line through an angle $|\theta|$, counterclockwise if θ is positive and clockwise if θ is negative. Then on the ray OL a point P is located so that $OP = r$ if r is positive or zero. If r is negative, then the ray is extended backward through O, and P is located on this extension so that $OP = |r|$. This process is illustrated in Fig. 2, where a number of points have been located from their given polar coordinates. Observe that θ may exceed 2π. For example, $\theta = 134.5\pi$ for the point C in Fig. 2. The reader should also note that $r = -2$ (negative) for the point E.

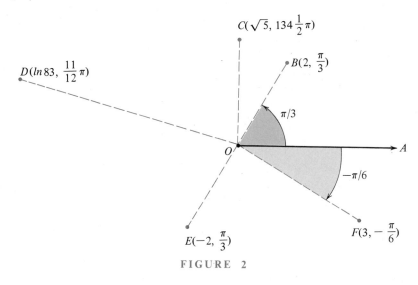

FIGURE 2

With this procedure it is clear that to each pair of real numbers (r, θ) there corresponds a uniquely determined point P in the plane. The numbers (r, θ) are also regarded as polar coordinates of P, but we can no longer say *the* polar coordinates of P because for any fixed point there will be infinitely many pairs (r, θ) that correspond to P. For example, if P is at the pole, then $r = 0$, but θ can be any real number. If $r \neq 0$, then θ can be changed by any multiple of 2π. Or we can replace r by $-r$ and add π to θ to obtain another set of polar coordinates for the same point P. For example, the point C in Fig. 2 has the polar coordinates $(\sqrt{5}, (2n + 1/2)\pi)$, where n is any integer. The same point C also has the coordinates of $(-\sqrt{5}, (2n + 3/2)\pi)$. This multiplicity of coordinates for a given point P may be disturbing at first, but it has many advantages. When we want the coordinates of P to be uniquely determined, we will use the phrase "*the* polar coordinates of P" and in this situation it is understood that $0 < r$ and $0 \leqq \theta < 2\pi$.

Let us superimpose a rectangular coordinate system on the polar coordinate system, making the origins in both systems coincide and making the positive *x*-axis fall on the polar line (see Fig. 3). Then each point *P* has two types of coordinates, a rectangular set (x, y) and a polar set (r, θ). We leave it to the student to prove that these coordinates are related by the equations

(1)
$$x = r \cos \theta,$$
$$y = r \sin \theta$$

and

(2)
$$r^2 = x^2 + y^2,$$
$$\tan \theta = \frac{y}{x}.$$

The set (1) allows us to pass from the polar coordinates to the rectangular coordinates, and the set (2) takes us in the reverse direction. For example, the point $B(2, \pi/3)$ of Fig. 2 has the rectangular coordinates $x = 2 \cos (\pi/3) = 1$ and $y = 2 \sin (\pi/3) = \sqrt{3}$.

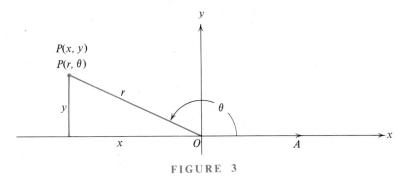

FIGURE 3

EXERCISE 1

In Problems 1 through 9, a point is given by a pair of polar coordinates. Plot the point and find its rectangular coordinates.

1. $(4, 0)$.   2. $(3, \pi)$.   3. $(-5, -\pi)$.
4. $(2, \pi/2)$.   5. $(4, 3\pi/2)$.   6. $(2, -\pi/6)$.
7. $(8, 3\pi/4)$.   8. $(-8, 5\pi/4)$.   9. $(6, -10\pi/3)$.

In Problems 10 through 18, the rectangular coordinates of a point are given. In each case find all possible sets of polar coordinates for the point.

10. $(1, 1)$.   11. $(-\sqrt{3}, \sqrt{3})$.   12. $(2, -2\sqrt{3})$.

13. $(0, -5)$.   14. $(-3, -3)$.   15. $(-4, 0)$.

16. $(3, 4)$.   17. $(-5, 12)$.   18. $(2, -1)$.

19. Prove that the points (r, θ) and $(r, -\theta)$ are symmetric with respect to the x-axis.

20. Prove that the points (r, θ) and $(-r, -\theta)$ are symmetric with respect to the y-axis.

21. What can you say about the symmetry of the pair of points (r, θ) and $(-r, \theta)$?

22. Do Problem 21 for the points (r, θ) and $(r, \pi - \theta)$.

23. Do Problem 21 for the points (r, θ) and $(-r, \pi - \theta)$.

24. Do Problem 21 for the points (r, θ) and $\left(r, \dfrac{\pi}{2} - \theta\right)$.

CALCULATOR PROBLEMS

In Problems C1 through C6, a point is given in polar coordinates. Find the rectangular coordinates. Note that the angle is given in radians.

C1. $(5, 0.5)$.   C2. $(2.34, 1.23)$.   C3. $(7.77, -0.333)$.

C4. $(-10, 2.12)$.   C5. $(-9, 5.55)$.   C6. $(25, 25)$.

In Problems C7 through C12, a point is given in rectangular coordinates. Find (r, θ) if $0 \leq \theta < 2\pi$ and θ is in radians.

C7. $(1.23, 2.34)$.   C8. $(1984, -1776)$.   C9. $(-606, 567)$.

C10. $(-2.54, -9.81)$.   C11. $(88.8, -77.7)$.   C12. $(-1945, 2001)$.

The Graph of a Polar Equation **2**

Just as in rectangular coordinates, the graph of an equation

$$(3) \qquad F(r, \theta) = 0$$

is by definition the collection of all points $P(r, \theta)$ whose polar coordinates satisfy the equation. Here the point P has many different pairs of coordinates, but P is in the graph if just *one* of its many different pairs of coordinates satisfies the equation.

    In many cases equation (3) can be solved explicitly for r or θ, giving either

$$(4) \qquad r = f(\theta)$$

or

$$(5) \qquad \theta = g(r).$$

In either of these cases it is easy to sketch the graph. For example, with equation (4), we merely select a sequence of values for θ and compute the associated value for r.

Example 1. Sketch the graph of $r = 2(1 + \cos \theta)$.

Solution. Since $\cos \theta$ is an even function, it is sufficient to make a table for $\theta \geqq 0$. Selecting the popular angles for θ and computing r from $r = 2(1 + \cos \theta)$ yields the following table of coordinates for points on the curve:

θ	0	$\pm\dfrac{\pi}{6}$	$\pm\dfrac{\pi}{4}$	$\pm\dfrac{\pi}{3}$	$\pm\dfrac{\pi}{2}$	$\pm\dfrac{2\pi}{3}$	$\pm\dfrac{3\pi}{4}$	$\pm\dfrac{5\pi}{6}$	$\pm\pi$
r	4	$2 + \sqrt{3}$	$2 + \sqrt{2}$	3	2	1	$2 - \sqrt{2}$	$2 - \sqrt{3}$	0

The graph of the equation $r = 2(1 + \cos \theta)$ is shown in Fig. 4. This type of curve is called a *cardioid* because it resembles a heart. ●

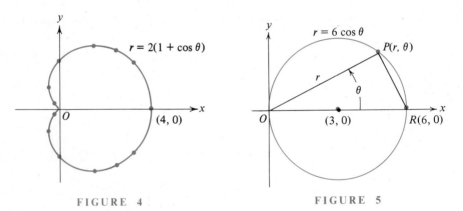

FIGURE 4   FIGURE 5

We can also pose the converse problem: Given a collection of points, find a polar equation whose graph is the given point set. When the graph is a circle with center at O, then its equation is just $r = r_0$, where r_0 is the radius. If the graph is a line through O with angle of inclination α, then its equation is $\theta = \alpha$ (keeping in mind that r can be negative).

Example 2. Find an equation (in polar coordinates) for the circle with center at $(3, 0)$ and radius 3.

Solution. This circle is shown in Fig. 5. One diameter of the circle will be the line joining the pole O and the point $R(6, 0)$. If P is on the circle, then $\angle OPR$ is a right angle (inscribed in a semicircle) and hence $r = |OR| \cos \theta$ or $r = 6 \cos \theta$. On the other hand, it is easy to see that if (r, θ) satisfy $r = 6 \cos \theta$, then P is on the circle. Therefore, this is an equation for the given circle. Note that as θ runs from 0 to π, the circle is described once. As θ runs from π to 2π, the circle is described again; and so on. ●

To find points of intersection of two given curves, we solve their polar coordinate equations simultaneously. But unfortunately, this may not give *all* the points of intersection. This peculiar behavior can occur because the polar coordinates of a point are not unique.

Example 3. Find all points of intersection of the two curves $r = 6\cos\theta$ and $r = 2(1 + \cos\theta)$.

Solution. These are the curves of Figs. 4 and 5, so the desired points could be found geometrically by superimposing the two sketches (see Fig. 6).

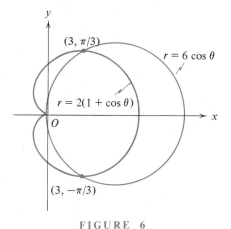

FIGURE 6

Suppose that we are to solve this problem analytically. Let (r_1, θ_1) be coordinates of a point P that is on both curves. If (r_1, θ_1) satisfy both of the equations $r = 2(1 + \cos\theta)$ and $r = 6\cos\theta$, then these two equations yield $2(1 + \cos\theta_1) = 6\cos\theta_1$ since both sides give r_1. Therefore, $2\cos\theta_1 = 1$, and hence $\cos\theta_1 = 1/2$, and $\theta_1 = \pi/3 + 2n\pi$ or $-\pi/3 + 2n\pi$. Using these values for θ in either $r = 2(1 + \cos\theta)$ or $r = 6\cos\theta$ yields $r_1 = 3$. Therefore, $(3, \pi/3)$ and $(3, -\pi/3)$ are intersection points of the two given curves. But if we superimpose the two curves as indicated in Fig. 6, we find that the curves also intersect at the origin. How did we miss this point of intersection?

The cardioid passes through the origin because the coordinates $(0, \pi)$ satisfy the equation $r = 2(1 + \cos\theta)$. The curve $r = 6\cos\theta$ passes through the origin because the coordinates $(0, \pi/2)$ satisfy this equation. Although the point is the same in both cases, the two coordinates $(0, \pi)$ and $(0, \pi/2)$ are different. Hence this point is missed when we solve the pair of equations simultaneously. ●

There are a number of general rules to handle this situation. One rule is to test the given equations to see if the origin is on both curves, by setting $r = 0$ in each of the equations. A second rule is to replace θ by $\theta + 2n\pi$ in one of the equations. Thus, if $r = f(\theta)$ and $r = g(\theta)$ are the equations for the two curves, we would solve

$$(6) \qquad\qquad f(\theta) = g(\theta + 2n\pi)$$

for θ. Such values of θ would lead to points of intersection of the two curves that may have been missed when we simply set $f(\theta) = g(\theta)$. In most cases, however, a sketch of the two curves will be helpful in determining the points of intersection.

EXERCISE 2

In Problems 1 through 9, sketch the curve and if possible give its name. Throughout this chapter a and b denote positive constants, and A and B denote arbitrary constants.

1. $r = 4$.
2. $r \cos \theta = 5$.
3. $r \sin \theta = 3$.
4. $r = 4(1 + \sin \theta)$.
5. $r = 2 \cos \theta - 1$.
6. $r\theta = \pi$.
7. $r = a \cos 2\theta$.
8. $r^2 = a^2 \sin 2\theta$.
9. $r = a \cos 3\theta$.

10. Prove that the vertical line $x = A$ has $r \cos \theta = A$ as an equation in polar coordinates.
11. Prove that the horizontal line $y = B$ has $r \sin \theta = B$ as an equation in polar coordinates.
12. Find a suitable equation in polar coordinates for the circle $x^2 + y^2 = 2By$.

In Problems 13 through 21, find all of the points of intersection of the given pair of curves.

13. $r \sin \theta = 2$, $\quad r = 4 \sin \theta$.
14. $r \cos \theta = 2$, $\quad r \sin \theta + 2\sqrt{3} = 0$.
15. $r = a$, $\quad r = 4a \cos \theta$.
16. $r = a \sin \theta$, $\quad r = a \cos \theta$.
17. $r = \cos \theta$, $\quad r = \dfrac{2}{3 + 2 \cos \theta}$.
18. $r \cos \theta = 1$, $\quad r = 2 \cos \theta + 1$.
★19. $r^2 = 2 \cos \theta$, $\quad r = 2(\cos \theta + 1)$.
★20. $r = a \cos 2\theta$, $\quad 4r \cos \theta = a\sqrt{3}$.
★★21. $r = \theta \ (\theta \geq 0)$, $r = 2\theta \ (\theta \geq 0)$. A curve $r = c\theta$ is called a *spiral of Archimedes*.

★22. A certain circle has its center on the polar line and passes through the origin. Find a suitable equation for the locus of the midpoints of the chords through the origin. Identify the curve.

★23. A line segment of fixed length $2a$ slides in such a way that one end is always on the x-axis and the other end is always on the y-axis. Find an equation in polar coordinates for the locus of points P in which a line from the origin perpendicular to the moving segment intersects the segment. Sketch the curve.

Curve Sketching in Polar Coordinates **3**

We can frequently shorten the labor of sketching a curve by testing the equation of the curve for symmetries of the curve. We leave it to the student to justify the rules that are given in Table 1. Here the variables r and θ are replaced as indicated in the first two columns, and if the equation $F(r, \theta) = 0$ remains unchanged or is transformed into an equivalent equation, then the curve has the symmetry indicated in the third column.

TABLE 1

Original	Replaced by:	Curve is symmetric with respect to:
r, θ	$r, -\theta$	the x-axis ($\theta = 0$)
r, θ	$-r, -\theta$	the y-axis ($\theta = \pi/2$)
r, θ	$-r, \theta$	the origin
r, θ	$r, \pi - \theta$	the y-axis
r, θ	$-r, \pi - \theta$	the x-axis
r, θ	$r, \pi/2 - \theta$	the line $y = x$ ($\theta = \pi/4$)

Example 1. Examine the curve $r = 4 \sin 2\theta$ for symmetry, and sketch the curve.

Solution. Since the sine function is an odd function, $\sin(-2\theta) = -\sin 2\theta$. Hence, applying the second test from Table 1, we obtain the equation

(7) $$-r = 4 \sin(-2\theta).$$

But this equation is equivalent to $r = 4 \sin 2\theta$, so the curve is symmetric with respect to the y-axis. Let us apply the fifth test from Table 1. The altered equation is $-r = 4 \sin 2(\pi - \theta)$. But $\sin 2(\pi - \theta) = \sin(2\pi - 2\theta) = \sin(-2\theta) = -\sin 2\theta$. Hence the equation $-r = 4 \sin 2(\pi - \theta)$ is equivalent to $r = 4 \sin 2\theta$, and the curve is symmetric with respect to the x-axis. Now any curve that is symmetric with respect to the x- and y-axis is also symmetric with respect to the origin. But please note that while the curve has all these symmetries, the first, third, and fourth tests from Table 1 fail to reveal them. Consequently, these tests are *sufficient* to ensure the symmetries stated but are not *necessary*. We leave it to the student to apply the sixth test from Table 1 and show that the curve is also symmetric with respect to the line $y = x$. As a result of all these symmetries, it is sufficient to sketch the curve, just in the sector between $\theta = 0$ and $\theta = \pi/4$, and then the rest of the curve can be obtained by reflections. The graph of $r = 4 \sin 2\theta$ is shown in Fig. 7, and for obvious reasons the curve is known as the *four-leafed rose*. ●

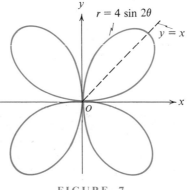

FIGURE 7

In sketching a curve, it is helpful to know something about the tangent lines to the curve. We shall consider the general problem in Section 4. However, when the curve passes through the origin, there is a very simple rule for the tangent line, given in

THEOREM 1. If $f'(\theta_0) \neq 0$ and $f(\theta_0) = 0$, then the curve $r = f(\theta)$ has the line $\theta = \theta_0$ as a tangent line at the origin.

Proof. The origin is on the curve, since $r = f(\theta_0) = 0$ by hypothesis. We change θ_0 by a small amount $\Delta\theta$ and consider the point $P(r_1, \theta_0 + \Delta\theta)$ on the curve where $r_1 = f(\theta_0 + \Delta\theta)$. Then the line joining O and P is a secant of the curve. But as $\Delta\theta \to 0$, this secant tends to the line $\theta = \theta_0$. On the other hand, since $f(\theta)$ is continuous at θ_0, r_1 tends to zero as $\Delta\theta \to 0$, so that the point P moves toward O. Thus the limiting position of the secant line $\theta = \theta_0 + \Delta\theta$ is the line $\theta = \theta_0$ as P approaches O. But then the line $\theta = \theta_0$ is a tangent line, by the definition of a tangent line as the limiting position of the secant line. ▮

In Fig. 8, where this proof is illustrated, the increment $\Delta\theta$ is negative.

As an example, consider the equation $r = 4 \sin 2\theta$, and its graph as shown in Fig. 7. We have that $r = 0$ whenever $\sin 2\theta = 0$, and this occurs for $\theta = 0, \pi/2, \pi, 3\pi/2, \ldots$. Hence each of these lines is tangent to the curve at the origin. In this case, however, there are only two distinct tangent lines at the origin, although the curve passes through the origin four times as θ runs from $-\epsilon$ to $2\pi - \epsilon, \epsilon > 0$.

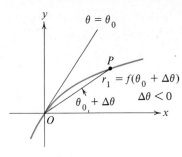

FIGURE 8

In sketching a curve, we can always seek help by transforming the given equation from one coordinate system to another. In changing the equation from polar coordinates to rectangular coordinates, or in the reverse direction, the equation set (1) $x = r \cos \theta$ and $y = r \sin \theta$ is very useful.

Example 2. Sketch the curve $r = 2(\sin \theta + \cos \theta)$.

Solution. We transform this equation into rectangular coordinates to see if it looks better in that system. If we multiply both sides by r, we introduce $r = 0$ as a solution, and hence add the origin to the curve. But the origin is already on the curve (set $\theta = 3\pi/4$ in the original equation), so no harm is done. After multiplying by r, we have

$$r^2 = 2(r \sin \theta + r \cos \theta).$$

But $r^2 = x^2 + y^2$, $r \sin \theta = y$, and $r \cos \theta = x$. Hence this equation is equivalent to

(8)
$$x^2 + y^2 = 2y + 2x$$
$$(x - 1)^2 + (y - 1)^2 = 2.$$

Consequently, we see that the graph of $r = 2(\sin \theta + \cos \theta)$ is a circle with center $(1, 1)$ (in Cartesian coordinates) and radius $\sqrt{2}$. ●

Example 3. Sketch the curve $(x^2 + y^2)^3 = 16x^2y^2$.

Solution. Clearly, it would be troublesome to make a table of values for points (x, y) on the curve. We try transforming the curve into polar coordinates. We have as equivalent equations

$$(r^2)^3 = 16(r \cos \theta)^2(r \sin \theta)^2$$
$$r^6 = 16r^4 \sin^2 \theta \cos^2 \theta$$

$$r^2 = 16 \sin^2 \theta \cos^2 \theta$$
$$r = \pm 4 \sin \theta \cos \theta = \pm 2 \sin 2\theta.$$

But this last is exactly the equation of Example 1 with the factor 4 replaced by ± 2. Since the curve $r = 4 \sin 2\theta$ is a four-leafed rose, the graph of $r = 2 \sin 2\theta$ is also a four-leafed rose, as shown in Fig. 7, except shrunk by a factor of $1/2$. Finally, the curve is already symmetric with respect to the origin, so the $\pm$ sign may be dropped in $r = \pm 2 \sin 2\theta$ without changing the curve.   ●

EXERCISE 3

In Problems 1 through 10, find the lines of symmetry and sketch the given curve.

1. $r \cos \theta = 5$.
2. $r \sin \theta = 3$.
3. $r = 4 + \sin \theta$.
4. $r = 1 + 4 \cos \theta$.
5. $r^2 = 16 \sin 2\theta$.
6. $r = 4 \sin 3\theta$.
7. $r = 4 \cos 3\theta$.
8. $r = 2 \tan \theta \sin \theta$.
9. $r = \dfrac{2}{1 - \sin \theta}$.
★10. $r = \dfrac{1}{\cos 2\theta}$.

11. Prove that the graph of $r(A \cos \theta + B \sin \theta) = C$ is always a straight line provided that A and B are not both zero.
12. Prove that the graph of $r = 2A \sin \theta + 2B \cos \theta$ is either a circle through the origin or a single point. Find the radius and center of the circle.

In Problems 13 through 18, transform the given equation into an equation in polar coordinates for the same curve.

13. $x^2 + y^2 - 6y = \sqrt{x^2 + y^2}$.
14. $x^4 + y^4 = 2xy(2 - xy)$.
15. $y^2(1 + x) = x^3$.
16. $x^3 + y^3 = 8xy$.
17. $y^2 = x^2 \dfrac{a + x}{a - x}$.
★18. $x^4 + 2x^2 y^2 + y^4 = 6x^2 y - 2y^3$.

In Problems 19 through 22, transform the given equation into an equation in rectangular coordinates for the same curve.

19. $r = 2 \tan \theta \sin \theta$.
20. $r = \dfrac{8}{1 - \cos \theta}$.
21. $r^2 = \tan \theta \sin^2 \theta$.
22. $r^2 = a^2 \cos 2\theta$.

★23. Sketch the curve $r = a \sin n\theta$ for $0 \le \theta \le \pi/n$. From this part of the graph deduce the fact that if n is an integer, the complete graph is a rose. If n is an even integer, the rose has $2n$ petals or loops, but if n is an odd integer, the rose has only n petals or loops. Prove the same assertion about the graph of $r = a \cos n\theta$, by proving that it is congruent to the graph of $r = a \sin n\theta$. HINT: Replace θ by $\theta + 3\pi/2n$ in $r = a \cos n\theta$.

$\star$24. If $\mathscr{C}_1$ is the curve $r = f(\theta)$ and $\mathscr{C}_2$ is the curve $r = kf(\theta)$, where k is a nonzero constant, the curve $\mathscr{C}_2$ is said to be *similar* to $\mathscr{C}_1$ with the origin as the *center of similitude*.

  (a) Prove that if $\mathscr{C}_1$ is a straight line, then $\mathscr{C}_2$ is also a straight line.
  (b) Prove that if $\mathscr{C}_1$ is a circle through the origin, then $\mathscr{C}_2$ is also a circle through the origin.
  (c) Show that **(b)** includes the result of Problem 22, Exercise 2, as a special case.

CONIC SECTIONS IN POLAR FORM

$\star\star$25. Let $\mathscr{C}$ be a conic section with one focus at the origin, the associated directrix a vertical line p units to the left of 0, and eccentricity e. Find an equation for $\mathscr{C}$ in the form $r = f(\theta)$.

26. Suppose that $0 < e < 1$. Use calculus and the formula from Problem 25 to find the points on $\mathscr{C}$: **(a)** closest to the origin, and **(b)** farthest from the origin.

In Problems 27 through 32, (a) identify the conic section, (b) give its eccentricity, and (c) find the distance of the closest directrix to the origin. HINT: *Use the formula developed in Problem 25.*

27. $r = \dfrac{7}{1 - \cos\theta}$.

28. $r = \dfrac{5}{1 - 3\cos\theta}$.

29. $r = \dfrac{10}{4 - 3\cos\theta}$.

30. $r = \dfrac{10}{3 - 4\cos\theta}$.

31. $r = \dfrac{8}{1 - 2\cos\theta}$.

32. $r = \dfrac{5}{4 - \cos\theta}$.

$\star$33. Derive a standard form similar to the one obtained in Problem 25 for a conic with one focus at the origin O, and its associated directrix: **(a)** vertical and p units to the right of O, **(b)** horizontal and p units above O, and **(c)** horizontal and p units below O.

Differentiation in Polar Coordinates *4*

Of course, the differentiation formulas for $r = f(\theta)$ are just the same as for $y = f(x)$; only the names of the variables have been altered. But the geometric interpretation of the derivative must be different, because we are now using a different coordinate system. Our first impulse is to search for a formula for the slope of a line tangent to the curve $r = f(\theta)$. This is given in

THEOREM 2. If m is the slope of the line tangent to the curve $r = f(\theta)$ at the point $P_1(r_1, \theta_1)$, then

(9)
$$m = \frac{f(\theta_1) \cos \theta_1 + f'(\theta_1) \sin \theta_1}{f'(\theta_1) \cos \theta_1 - f(\theta_1) \sin \theta_1},$$

whenever the denominator is not zero.

Equation (9) can be written in the alternative form

(10)
$$m = \frac{r \cos \theta + \dfrac{dr}{d\theta} \sin \theta}{\dfrac{dr}{d\theta} \cos \theta - r \sin \theta},$$

where the right side is computed at P_1.

Proof. The rectangular coordinates of a point on the curve can be obtained from the polar coordinates through the equation set

(11)
$$x = r \cos \theta = f(\theta) \cos \theta,$$
$$y = r \sin \theta = f(\theta) \sin \theta.$$

Looking at the extreme right side of equation set (11), we see that these equations can be regarded as parametric equations for the curve in rectangular coordinates, with θ as the parameter. Then from (11) we have

(12)
$$m = \frac{dy}{dx} = \frac{\dfrac{dy}{d\theta}}{\dfrac{dx}{d\theta}} = \frac{f(\theta) \cos \theta + f'(\theta) \sin \theta}{f'(\theta) \cos \theta - f(\theta) \sin \theta},$$

and this gives (9) at the point $P_1(r_1, \theta_1)$ whenever the denominator is not zero. If the denominator is zero and the numerator is not zero at P_1, then it is clear that the tangent line to the curve is vertical. ∎

Because formula (9) is a little complicated, we prefer to have some geometric quantity that is given by a simpler formula. Such a quantity is the angle ψ (Greek letter psi), shown in Fig. 9, and the formula for $\tan \psi$ is given in

THEOREM 3. Let P_1 be a point on the curve $r = f(\theta)$ and let ψ be the angle from the radial line OP_1 extended to the tangent line to the curve $r = f(\theta)$ at P_1 (see Fig. 9). If $r_1 \neq 0$ and $f'(\theta_1) \neq 0$, then

(13)
$$\tan \psi = \frac{f(\theta_1)}{f'(\theta_1)}.$$

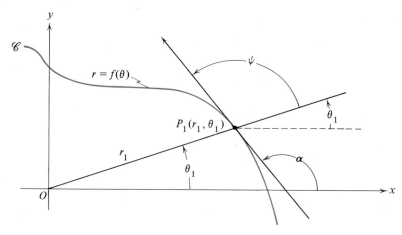

FIGURE 9

Equation (13) can be written in the alternative form

(14)
$$\tan \psi = \frac{r}{\dfrac{dr}{d\theta}},$$

where the right side is computed at P_1.

Proof of Theorem 3. If $\cos \theta_1 \neq 0$, then equation (9) can be put in the form

(15)
$$m = \tan \alpha = \frac{f(\theta_1) + f'(\theta_1) \tan \theta_1}{f'(\theta_1) - f(\theta_1) \tan \theta_1},$$

where α is the angle that the tangent line to $\mathscr{C}$ makes with the x-axis (see Fig. 9). If we solve equation (15) for $f(\theta_1)$, we find that

(16)
$$f(\theta_1) = \frac{\tan \alpha - \tan \theta_1}{1 + \tan \alpha \tan \theta_1} f'(\theta_1) = f'(\theta_1) \tan (\alpha - \theta_1).$$

Since $\tan (\alpha - \theta_1) = \tan \psi$ (see Fig. 9), equation (16) will give (13), when we divide both sides by $f'(\theta_1)$. ∎

Example 1. For the cardioid $r = a(1 - \cos \theta)$, find $\tan \alpha$ and $\tan \psi$ **(a)** in general, and **(b)** at $P(a, \pi/2)$ (see Fig. 10).

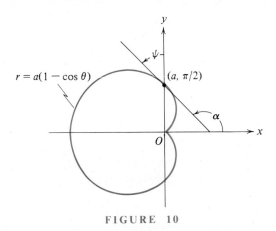

$$r = a(1 - \cos \theta) \qquad (a, \pi/2)$$

FIGURE 10

Solution. **(a)** From equation (10), we have

$$\tan \alpha = \frac{a(1 - \cos \theta) \cos \theta + a \sin^2 \theta}{a \sin \theta \cos \theta - a(1 - \cos \theta) \sin \theta}$$

$$= \frac{\cos \theta + \sin^2 \theta - \cos^2 \theta}{2 \sin \theta \cos \theta - \sin \theta}$$

$$= \frac{\cos \theta - \cos 2\theta}{\sin 2\theta - \sin \theta}$$

$$= \frac{2 \sin \dfrac{3\theta}{2} \sin \dfrac{\theta}{2}}{2 \cos \dfrac{3\theta}{2} \sin \dfrac{\theta}{2}} = \tan \frac{3\theta}{2}.$$

From equation (14),

$$\tan \psi = \frac{a(1 - \cos \theta)}{a \sin \theta} = \frac{2 \sin^2 \dfrac{\theta}{2}}{2 \sin \dfrac{\theta}{2} \cos \dfrac{\theta}{2}} = \tan \frac{\theta}{2}.$$

(b) At $\theta = \pi/2$, $\tan \alpha = \tan(3\pi/4) = -1$, and hence $\alpha = 3\pi/4$. Similarly, $\tan \psi = \tan(\pi/4) = 1$, so $\psi = \pi/4$. These angles are shown in Fig. 10. ●

The same procedure used for proving Theorem 2 will give us a formula for the length of an arc of a curve.

THEOREM 4. If s denotes the length of the arc of the curve $r = f(\theta)$ between the points $P_1(r_1, \theta_1)$ and $P_2(r_2, \theta_2)$, then

(17)
$$s = \int_{\theta_1}^{\theta_2} \sqrt{[f'(\theta)]^2 + [f(\theta)]^2}\, d\theta,$$

provided that $f'(\theta)$ is continuous for $\theta_1 \leqq \theta \leqq \theta_2$.

Proof. Once again we regard θ as the parameter and equation set (11) as parametric equations for the curve in rectangular coordinates. From equation (37) of Chapter 12 (page 498), with the parameter t replaced by θ, we have

(18)
$$s = \int_{\theta_1}^{\theta_2} \sqrt{\left(\frac{dx}{d\theta}\right)^2 + \left(\frac{dy}{d\theta}\right)^2}\, d\theta.$$

From equation set (11), we find that

$$\left(\frac{dx}{d\theta}\right)^2 + \left(\frac{dy}{d\theta}\right)^2 = [f'(\theta)\cos\theta - f(\theta)\sin\theta]^2 + [f'(\theta)\sin\theta + f(\theta)\cos\theta]^2$$

$$= [f'(\theta)]^2(\sin^2\theta + \cos^2\theta) + [f(\theta)]^2(\sin^2\theta + \cos^2\theta)$$

$$= [f'(\theta)]^2 + [f(\theta)]^2.$$

Using this in (18) gives (17). ∎

If we differentiate (17) with respect to θ_2, drop the subscript, and replace $f(\theta)$ by r, we obtain

$$\frac{ds}{d\theta} = \sqrt{\left(\frac{dr}{d\theta}\right)^2 + r^2}.$$

Thus (17) gives the equivalent differential form

(19)
$$ds^2 = dr^2 + r^2\, d\theta^2.$$

Once we have proved the formulas of Theorem 3 and 4, we may use any convenient device to assist in memorizing them. One such device is shown in Fig. 11. Here P and Q are neighboring points on the curve and the point R is the intersection of the circle with

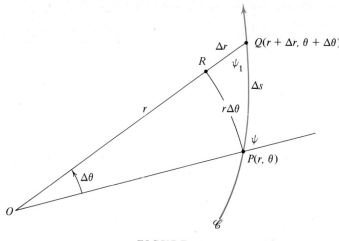

FIGURE 11

center O and radius r, with the ray OQ. Since the angle at R is a right angle, the figure PQR is a curvilinear right triangle. Then (approximately)

$$(\Delta s)^2 \approx |RQ|^2 + |PR|^2 \approx (\Delta r)^2 + (r\,\Delta\theta)^2$$

and

$$\tan\psi \approx \tan\psi_1 \approx \frac{PR}{RQ} \approx \frac{r\,\Delta\theta}{\Delta r} = \frac{r}{\dfrac{\Delta r}{\Delta\theta}}.$$

These approximate equations readily suggest the limit equations (19) and (14), respectively.

Example 2. Find the total length of the cardioid $r = a(1 - \cos\theta)$, where $a > 0$.

Solution. This curve is shown in Fig. 10. By Theorem 4, equation (17),

$$s = \int_0^{2\pi} \sqrt{[f'(\theta)]^2 + [f(\theta)]^2}\, d\theta = \int_0^{2\pi} \sqrt{a^2 \sin^2\theta + a^2(1 - \cos\theta)^2}\, d\theta$$

$$= \int_0^{2\pi} a\sqrt{2 - 2\cos\theta}\, d\theta = \int_0^{2\pi} 2a \left|\sin\frac{\theta}{2}\right| d\theta,$$

since $1 - \cos\theta = 2\sin^2(\theta/2)$. We are forced to use the absolute value signs in this last step because by definition $ds/d\theta \geqq 0$, while $\sin(\theta/2)$ might be negative. Fortunately, however, for $0 \leqq \theta \leqq 2\pi$ it is true that $\sin(\theta/2) \geqq 0$. Hence we can drop the absolute value signs and write

$$s = \int_0^{2\pi} 2a \sin \frac{\theta}{2} \, d\theta = -4a \cos \frac{\theta}{2} \Big|_0^{2\pi} = 4a - (-4a) = 8a. \quad \bullet$$

For some problems it is advantageous to express the arc length as an integral on r. Starting from equation (19), it is easy to obtain

(20)
$$s = \int_{r_1}^{r_2} \sqrt{1 + r^2 \left(\frac{d\theta}{dr} \right)^2} \, dr.$$

EXERCISE 4

1. Find those points on the cardioid $r = a(1 - \sin \theta)$ where the tangent line is horizontal.
2. Find all the points on the parabola $r = a/(1 - \cos \theta)$ where the tangent line has slope 1.
3. Find α and ψ for the circle $r = a \sin \theta$.
4. Prove that $\tan \psi = \theta$ for the spiral of Archimedes, $r = a\theta$.
5. Prove that for the *logarithmic spiral* $r = ae^{b\theta}$, the angle ψ is a constant. For this reason the curve is also called the *equiangular spiral*.
★6. For the parabola $r = a/(1 - \cos \theta)$, prove that if $0 \leq \theta < \pi$, then $\tan \psi = -\tan (\theta/2)$ and consequently $\psi = \pi - \theta/2$. Show that this establishes the following optical property. If a ray of light starts from the focus of a parabolic reflector, it is reflected from the walls in a line parallel to the axis of the parabola.
7. If φ is the angle of intersection between the curves $r_1 = f_1(\theta)$ and $r_2 = f_2(\theta)$, measured from the first curve to the second curve, show that

$$\tan \varphi = \frac{\tan \psi_2 - \tan \psi_1}{1 + \tan \psi_2 \tan \psi_1}.$$

Deduce from this equation a condition that a pair of curves intersect orthogonally.
8. Use the condition found in Problem 7 to show that the following pairs of curves intersect orthogonally. Recall that a and b denote positive constants.
   (a) $r = a \sin \theta,$ $\qquad\qquad r = b \cos \theta.$
   (b) $r = \dfrac{a}{1 - \cos \theta},$ $\qquad r = \dfrac{b}{1 + \cos \theta}.$
   (c) $r^2 = \dfrac{a^2}{\sin 2\theta},$ $\qquad r^2 = \dfrac{b^2}{\cos 2\theta}.$
   (d) $r = a(1 - \cos \theta),$ $\qquad r = a(1 + \cos \theta),$ except at the origin.

(e) $r^2 = a^2 \sin 2\theta$, $r^2 = b^2 \cos 2\theta$, except at the origin.

(f) $r = \dfrac{9}{2 - \cos \theta}$, $r = \dfrac{3}{\cos \theta}$.

In Problems 9 and 10, find the length of the indicated arc.

9. (a) $r = 2\theta^2$, $0 \leq \theta \leq 5$.   (b) $r = ae^{b\theta}$, $0 \leq \theta \leq \pi$, $b \neq 0$.

10. (a) $r = a \sin^3 (\theta/3)$, $0 \leq \theta \leq 3\pi$.   (b) $r = \dfrac{a}{\theta^3}$, $1 \leq \theta \leq 4$.

11. Use equation (20) to compute the length of the arc of the curve $r\theta = 1$ for $1/2 \leq \theta \leq 1$.

12. Show that the part of the curve $r\theta = 1$ that lies inside the circle $r = 1$ has infinite length.

13. An arc of the curve $r = f(\theta)$ that lies above the x-axis is rotated about the x-axis. Find a formula for the area of the surface generated.

14. The lemniscate $r^2 = a^2 \cos 2\theta$ is rotated about the x-axis. Find the area of the surface. HINT: Use symmetry and integrate from 0 to $\pi/4$.

*15. The cardioid $r = a(1 - \cos \theta)$ is rotated about the x-axis. Find the area of the surface.

*16. The arc of the curve $r = e^\theta$ for $0 \leq \theta \leq \pi$ is rotated about the x-axis. Find the area of the surface.

17. The lemniscate of Problem 14 is rotated about the y-axis. Find the area of the surface.

18. The circle $r = a \cos \theta$ is rotated about the y-axis. Find the area of the surface.

*19. Sketch the two curves $r = a/\theta$ and $r = a\theta$, and show that these two curves intersect infinitely often. Prove that they are orthogonal at infinitely many of these points, but at infinitely many (other) intersection points they are not orthogonal.

5 *Plane Areas in Polar Coordinates*

The formula for area in polar coordinates is given in

THEOREM 5. Let $r = f(\theta)$ be a positive continuous function for $\alpha \leq \theta \leq \beta$. Let $\mathscr{R}$ be the region bounded by the curve $r = f(\theta)$ and the rays $\theta = \alpha$ and $\theta = \beta$, $r \geq 0$ (see Fig. 12). Then the area of $\mathscr{R}$ is given by

(21)
$$A = \int_\alpha^\beta \frac{1}{2} r^2 \, d\theta = \int_\alpha^\beta \frac{1}{2} f^2(\theta) \, d\theta.$$

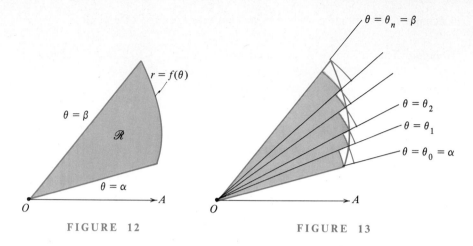

FIGURE 12   FIGURE 13

Proof. We partition the interval $\alpha \leqq \theta \leqq \beta$ into n subintervals and let $\theta_0, \theta_1, \ldots, \theta_n$ be the points of the partition, where $\alpha = \theta_0$ and $\beta = \theta_n$. We draw the rays $\theta = \theta_k$, $r \geqq 0$, for $k = 1, 2, 3, \ldots, n-1$. These rays divide the region $\mathcal{R}$ into n parts (see Fig. 13, where $n = 5$). Let R_k be the maximum of $r = f(\theta)$ in the interval $\theta_{k-1} \leqq \theta \leqq \theta_k$, and let r_k be the minimum of $r = f(\theta)$ in the same interval. Using the radial lines $\theta = \theta_{k-1}$ and $\theta = \theta_k$ and an arc of the circle $r = r_k$, we obtain a sector of a circle whose area is given by

(22) $$\frac{1}{2} r_k^2 (\theta_k - \theta_{k-1}) = \frac{1}{2} r_k^2 \, \Delta \theta_k, \qquad k = 1, 2, \ldots, n$$

[see equation (30) of Chapter 8]. The union of these sectors gives a set $\mathcal{R}_1$ that is contained in $\mathcal{R}$. This is the set shown shaded in Fig. 13. On the other hand, if the circular boundary of the sector has radius R_k, then the area of the sector is given by

(23) $$\frac{1}{2} R_k^2 (\theta_k - \theta_{k-1}) = \frac{1}{2} R_k^2 \, \Delta \theta_k, \qquad k = 1, 2, \ldots, n,$$

and the union of these sectors is a set $\mathcal{R}_2$ that contains $\mathcal{R}$.

Since $\mathcal{R}_1 \subset \mathcal{R} \subset \mathcal{R}_2$, we have $A(\mathcal{R}_1) \leqq A(\mathcal{R}) \leqq A(\mathcal{R}_2)$. Hence

(24) $$\sum_{k=1}^{n} \frac{1}{2} r_k^2 \, \Delta \theta_k \leqq A(\mathcal{R}) \leqq \sum_{k=1}^{n} \frac{1}{2} R_k^2 \, \Delta \theta_k.$$

The proof is completed by noting that $r = f(\theta)$ is a continuous function so that as the mesh of the partition tends to zero, both sums in (24) have the same limit, and consequently (24) gives (21). ∎

Theorem 5 is still true if $f(\theta)$ is sometimes zero or negative, as long as we replace the region $\mathcal{R}$ by a suitable set.

Example 1. Use polar coordinates to compute the area of a circle $r = a \cos \theta$.

Solution. The full circle is described as θ runs from $-\pi/2$ to $\pi/2$. Hence

$$A = \int_{-\pi/2}^{\pi/2} \frac{1}{2} r^2 \, d\theta = \int_{-\pi/2}^{\pi/2} \frac{1}{2} a^2 \cos^2 \theta \, d\theta$$

$$= \frac{a^2}{2} \int_{-\pi/2}^{\pi/2} \frac{1 + \cos 2\theta}{2} \, d\theta = \frac{a^2}{4} \left(\theta + \frac{\sin 2\theta}{2} \right) \Big|_{-\pi/2}^{\pi/2} = \frac{\pi a^2}{4}. \quad \bullet$$

We expected this answer because a is the diameter of the circle $r = a \cos \theta$.

Example 2. Find the area of the region that lies outside the cardioid $r = 2a(1 + \cos \theta)$ and inside the circle $r = 6a \cos \theta$.

Solution. The region in question is shown shaded in Fig. 14. The intersection points

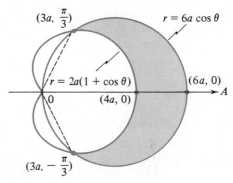

$$\left(3a, \frac{\pi}{3}\right) \qquad r = 6a \cos \theta$$
$$r = 2a(1 + \cos \theta) \qquad (6a, 0)$$
$$0 \qquad (4a, 0) \qquad A$$
$$\left(3a, -\frac{\pi}{3}\right)$$

FIGURE 14

of these two curves were already determined in Example 3 of Section 2 as the origin and the points $(3a, \pm\pi/3)$. The area is obviously the difference between two areas, one bounded by the circle, and the other bounded by the cardioid. Hence

$$A = \int_{-\pi/3}^{\pi/3} \frac{1}{2} r_2^2 \, d\theta - \int_{-\pi/3}^{\pi/3} \frac{1}{2} r_1^2 \, d\theta$$

$$= \int_{-\pi/3}^{\pi/3} \frac{1}{2} (6a \cos \theta)^2 \, d\theta - \int_{-\pi/3}^{\pi/3} \frac{1}{2} [2a(1 + \cos \theta)]^2 \, d\theta$$

$$= a^2 \int_{-\pi/3}^{\pi/3} [18 \cos^2 \theta - 2(1 + 2 \cos \theta + \cos^2 \theta)] \, d\theta$$

$$= a^2 \int_{-\pi/3}^{\pi/3} \left(16\frac{1 + \cos 2\theta}{2} - 2 - 4 \cos \theta \right) d\theta$$

$$= a^2(6\theta + 4 \sin 2\theta - 4 \sin \theta) \Big|_{-\pi/3}^{\pi/3} = 4\pi a^2. \quad \bullet$$

EXERCISE 5

In Problems 1 through 5, find the area of the region enclosed by the given curve.

1. $r = a \sin \theta$.   2. $r = 3 + 2 \cos \theta$.

3. $r = 3 + 2 \cos n\theta$.   4. $r = 1 + \sin^2 \theta$.

★5. $r = a + b \sin \theta + c \cos \theta$, $a > |b| + |c|$.

6. Find the area of one petal of the rose $r = a \sin n\theta$.

7. Find the total area of the regions enclosed by the lemniscate $r^2 = a^2 \cos 2\theta$.

8. Find the area of the region enclosed by the small loop of the limaçon $r = \sqrt{2} + 2 \cos \theta$.

In Problems 9 through 11, find the area of the region bounded by the given curve and the given rays.

9. $r = a \sec \theta$, $\theta = 0$ and $\theta = \pi/4$.

10. $r = a \tan \theta$, $\theta = 0$ and $\theta = \pi/4$.

11. $r = e^{\sin \theta} \sqrt{\cos \theta}$, $\theta = 0$ and $\theta = \pi/2$.

In Problems 12 through 15, calculate the area of the region that lies outside of the first curve and inside the second curve.

12. $r = a(1 - \cos \theta)$, $r = a$.   13. $r = a$, $r = 2a \cos \theta$.

★14. $r = a$, $r = 2a \sin n\theta$, n odd.   ★15. $r^2 = 2a^2 \cos 2\theta$, $r = a$.

★16. Find the area of the region common to the two circles $r = a \cos \theta$ and $r = b \sin \theta$.

★★17. Interpret geometrically as an area the following computation:

$$A = \frac{1}{2} \int_0^{4\pi} (e^\theta)^2 \, d\theta = \frac{1}{4}(e^{8\pi} - 1).$$

REVIEW PROBLEMS

In Problems 1 through 4, the rectangular coordinates of a point are given. In each case find (a) all possible sets of polar coordinates for the point and (b) the polar coordinates of the point with $r > 0$ and $0 \le \theta < 2\pi$.

1. $(10, 10)$.   2. $(-11, 0)$.   3. $(0, -19)$.   4. $(-1/4, -1/4)$.

In Problems 5 through 8, polar coordinates of a point are given. In each case find the rectangular coordinates of the point.

    5. $(5, \pi/4)$.   **6.** $(5, 5\pi/4)$.   **7.** $(5, 5\pi)$.   **8.** $(10, 40\pi/3)$.

9. Transform into polar coordinates the formula in rectangular coordinates for the distance between two points. Show that the resulting formula is merely a disguised form of the Law of Cosines from trigonometry.

In Problems 10 through 31, sketch the graph of the given equation.

    10. $r \cos \theta + 7 = 0$.   **11.** $r \sin \theta + 3 = 0$.   **12.** $r \cos \theta = 1 + r \sin \theta$.

    13. $r(2 \sin \theta + 3 \cos \theta) = 4$.   **14.** $r = 2 \sin \theta + 3 \cos \theta$.   **15.** $r(1 - \cos \theta) = 6$.

    16. $r(2 - \sin \theta) = 4$.   **17.** $r(1 + 2 \sin \theta) = 8$.   **18.** $r = 2 + \cos \theta$.

    19. $r \sin 2\theta = 4$.   **20.** $r = 4 \cos 4\theta$.   **21.** $r = 10 \sin 5\theta$.

    22. $r = 4 \sec \theta + 3$.   **23.** $r = 4 \cos \theta + 3$.   **24.** $r = 4 - 3 \cos \theta$.

    25. $r^2 = 16 \cos 2\theta$.   **26.** $r^2 = 16 \cos \theta$.   **27.** $r = \dfrac{4 \sin^2 \theta}{\cos \theta}$.

    28. $r = 4 + \sin^2 \theta$.   **29.** $r = 5 - \cos^2 \theta$.   **30.** $r = \cos \dfrac{\theta}{2}$.

    31. $r^2 = \cos \dfrac{\theta}{2}$.

32. Find a general expression for $\tan \psi$ and locate those points where the tangent line to the curve is normal to the radius vector for the curve in: **(a)** Problem 10, **(b)** Problem 12, **(c)** Problem 14, **(d)** Problem 20, **(e)** Problem 27, and **(f)** Problem 30.

In Problems 33 through 38, find all points of intersection of the given pair of curves.

    33. $r \cos \theta = 3$, $r = 6 \cos \theta$.   **34.** $r^2 = 9 \cos 2\theta$, $r = \sqrt{6} \cos \theta$.

    35. $r^2 = 8 \cos \theta$, $r^2 = 4 \sin \theta$.   **36.** $r = \dfrac{6}{1 - \cos \theta}$, $r = \dfrac{2}{1 + \cos \theta}$.

    37. $r = 2 \tan \theta \sin \theta$, $r = 6 \cos \theta$.   **38.** $r = 8 \sin^3 \theta$, $r^2 = 2 \sin \theta$.

39. Prove that the two curves $r = 6 + 3 \sin \theta + 4 \cos \theta$ and $r = 2 + 5 \sin \theta + \cos \theta$ never meet.

In Problems 40 and 41, find the length of the given arc.

    40. $r = 5 \sin^2 \dfrac{\theta}{2}$, $0 \le \theta \le \dfrac{\pi}{3}$.   **41.** $r = 3 \sin \theta + 7 \cos \theta$, $\dfrac{4\pi}{7} \le \theta \le \dfrac{11\pi}{7}$.

In Problems 42 through 45, find the area of the region bounded by the given curve, and the given rays.

42. $r = \sqrt{\cos \theta},$ $\theta = 0,$ $\theta = \dfrac{\pi}{4}.$

43. $r^2 = 4 \cos 3\theta,$ $\theta = -\dfrac{\pi}{6},$ $\theta = \dfrac{\pi}{6}.$

44. $r = 4 \cos 3\theta,$ $\theta = -\dfrac{\pi}{6},$ $\theta = \dfrac{\pi}{6}.$

$\star$45. $r = \dfrac{1}{1 - \cos \theta},$ $\theta = \dfrac{\pi}{4},$ $\theta = \dfrac{\pi}{2}.$

46. Find the area of the region that lies inside the curve $r^2 = 9 \cos 2\theta$ and inside the curve $r = \sqrt{6} \cos \theta$. (See Problem 34.)

47. Find the area of the region in the first quadrant that lies inside the curve $r^2 = 8 \cos \theta$ and inside the curve $r^2 = 4 \sin \theta$. (See Problem 35.)

14

Indeterminate Forms and Improper Integrals

1 Indeterminate Forms

Suppose that we are to compute

$$(1) \qquad \lim_{x \to 1} \frac{x^3 - 1}{e^{1-x} - 1}.$$

Clearly, both the numerator and the denominator are 0 at $x = 1$ so that the ratio in (1) has the form $0/0$. An expression of this type is called an *indeterminate form*. We have already met such indeterminate forms, and in the past we were able to determine the limit by some suitable algebraic manipulations. In the present case no such manipulations present themselves because of the presence of the exponential function in the denominator. What we need here is a systematic procedure for computing such limits as (1). In Section 3 we will prove a theorem, called L'Hospital's Rule, that gives just such a method. Briefly, the rule states that we should differentiate the numerator and the denominator, and find the limit of the ratio of these two derivatives. Applying this rule in (1), we find that

$$(2) \qquad \lim_{x \to 1} \frac{x^3 - 1}{e^{1-x} - 1} = \lim_{x \to 1} \frac{3x^2}{-e^{1-x}} = \frac{3}{-1} = -3,$$

a result that would have been hard to guess.

In order to prove this rule, we must first generalize the Mean Value Theorem (Theorem 8 of Chapter 5, page 199).

546

The Cauchy Mean Value Theorem $\boldsymbol{2}$

Let us recall Rolle's Theorem (Theorem 7 of Chapter 5, page 197).

ROLLE'S THEOREM. Suppose that $f(x)$ is a continuous function of x in the closed interval $a \leq x \leq b$, and a differentiable function in the open interval $a < x < b$. If $f(a) = f(b) = 0$, then there is some point ξ in the open interval $a < \xi < b$ such that $f'(\xi) = 0$.

We will use Rolle's Theorem to prove the more general result of

---

THEOREM 1 (The Cauchy Mean Value Theorem). Let $f(t)$ and $g(t)$ be two functions, each continuous in the closed interval $a \leq t \leq b$, and each differentiable in the open interval $a < t < b$. Suppose further that $f(b) \neq f(a)$, and that the derivatives $f'(t)$ and $g'(t)$ do not vanish simultaneously in $a < t < b$. Then there is some point ξ, with $a < \xi < b$, such that

(3)
$$\frac{g(b) - g(a)}{f(b) - f(a)} = \frac{g'(\xi)}{f'(\xi)}.$$

Before proving this theorem, let us give a geometric interpretation of this result. Let $x = f(t)$ and $y = g(t)$ be parametric equations for some curve $\mathscr{C}$. As t runs from $t = a$ to $t = b$, the point $P(x, y)$ describes a continuous curve joining the points $A(f(a), g(a))$ and $B(f(b), g(b))$, as indicated in Fig. 1. Now the left side of (3) is just $\Delta y/\Delta x$, the slope of the straight line AB. On the other hand, $g'(t)/f'(t)$ is the slope of the tangent line to the curve at the point $P(f(t), g(t))$ on the curve $\mathscr{C}$. Then Theorem 1 states that there is some point P on the curve between A and B such that the tangent line to the curve is parallel to the chord AB. Of course, there may be more than one such point, but there is always at least one. The theorem may fail if the curve has a cusp as shown in Fig. 2. This can occur if $f'(t)$ and $g'(t)$ vanish simultaneously, so the hypothesis that they do not vanish simultaneously is necessary in the statement of the theorem. Parametric equations for the curve of Fig. 2 are given in Problem 10 of Exercise 1.

Proof of Theorem 1. For the reader who is familiar with determinants, let $F(t)$ be defined

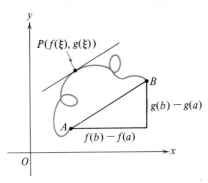

FIGURE 1

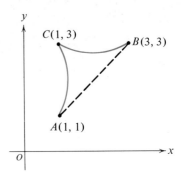

FIGURE 2

by the determinant

$$(4) \qquad F(t) = \begin{vmatrix} f(t) & g(t) & 1 \\ f(a) & g(a) & 1 \\ f(b) & g(b) & 1 \end{vmatrix}.$$

When $t = a$, the first and second rows are identical and hence $F(a) = 0$. When $t = b$, the first and third rows are identical and so $F(b) = 0$. It is easy to see that $F(t)$ satisfies the other conditions of Rolle's Theorem. Hence there is some suitable ξ for which $F'(\xi) = 0$. Expanding the determinant in (4) by minors of the first row gives

$$(5) \qquad F(t) = f(t)[g(a) - g(b)] - g(t)[f(a) - f(b)] + f(a)g(b) - g(a)f(b).$$

For the reader who is not familiar with determinants, the proof can start with the function $F(t)$ defined by equation (5). A brief computation from (5) will show that $F(a) = F(b) = 0$, and hence Rolle's Theorem can be applied to $F(t)$.

If we differentiate $F(t)$, we have from (5) that

$$(6) \qquad F'(t) = f'(t)[g(a) - g(b)] - g'(t)[f(a) - f(b)].$$

By Rolle's Theorem there is a ξ such that $F'(\xi) = 0$. For this ξ, equation (6) gives

$$(7) \qquad 0 = f'(\xi)[g(a) - g(b)] - g'(\xi)[f(a) - f(b)].$$

From (7), simple algebraic manipulations give equation (3). However, in doing the divisions involved we must be certain that we do not divide by zero. But this is assured by the hypotheses that $f(a) \neq f(b)$ and that $f'(t)$ and $g'(t)$ are not simultaneously zero. ∎

Example 1. Find a value for ξ as predicted by Theorem 1 for the pair of functions $f(t) = t^2 + 2t + 1$ and $g(t) = t^3 + 3t + 1$ for $0 \le t \le 1$.

Solution. The curve $x = t^2 + 2t + 1$, $y = t^3 + 3t + 1$ joins the points $(1, 1)$ and

$(4, 5)$ as t runs from $t = 0$ to $t = 1$. For this curve equation (3) leads to

$$\frac{5 - 1}{4 - 1} = \frac{3t^2 + 3}{2t + 2}.$$

Solving for t gives $8t + 8 = 9t^2 + 9$ or $9t^2 - 8t + 1 = 0$. Hence

$$t = \frac{8 \pm \sqrt{64 - 36}}{18} = \frac{4 \pm \sqrt{7}}{9}.$$

Since both of these numbers lie in the interval $0 < \xi < 1$, we have two values for ξ, namely $\xi_1 = (4 - \sqrt{7})/9$ and $\xi_2 = (4 + \sqrt{7})/9$. ●

EXERCISE 1

In Problems 1 through 8, find all values of ξ satisfying equation (3) (Theorem 1) for the given pair of functions and the given interval. Observe that for each of these problems you have already sketched the curve $x = f(t), y = g(t)$ (see Exercise 2 of Chapter 12, page 487).

1. $x = 5 \cos t, \quad y = 5 \sin t, \quad 0 \le t \le \pi/2.$
2. $x = 3 \cos t, \quad y = 5 \sin t, \quad 0 \le t \le \pi/2.$
3. $x = 3 \cos t, \quad y = 5 \sin t, \quad 0 \le t \le \pi.$
4. $x = \sin t, \quad y = \cos 2t, \quad 0 \le t \le \pi/2.$
5. $x = \cosh t, \quad y = \sinh t, \quad 0 \le t \le 6.$
6. $x = t^3, \quad y = t^2, \quad 1 \le t \le 2.$
7. $x = t^3 - 3t, \quad y = t, \quad -3 \le t \le 3.$
8. $x = t^3 - 3t, \quad y = 4 - t^2, \quad -3 \le t \le 2.$

★9. Sketch the curve $x = t^3 \equiv f(t)$, $y = t^2 \equiv g(t)$ for $-2 \le t \le 3$. Observe that this curve has a cusp at the origin. Show that for this curve $f'(t)$ and $g'(t)$ vanish simultaneously for a suitable value of t. Prove that there is no ξ in the interval $-2 < t < 3$ such that

$$\frac{g(3) - g(-2)}{f(3) - f(-2)} = \frac{g'(\xi)}{f'(\xi)}.$$

★10. Repeat Problem 9 for the curve

$$x = 1 + t(t - 1)^2, \quad y = 3 + (t - 2)(t - 1)^2$$

for the interval, $0 \le t \le 2$.

3 *The Form 0/0*

If $f(x)$ and $g(x)$ are continuous functions at $x = a$ with $f(a) = 0$ and $g(a) = 0$, then

$$(8) \qquad \lim_{x \to a} \frac{g(x)}{f(x)}$$

is referred to briefly as the *indeterminate form* $0/0$. In (8) x can approach a either from the left or from the right. For simplicity we will state our theorem just for the second case, since the first case is handled in exactly the same way. The method of finding a value for the indeterminate form $0/0$ is given by

---

THEOREM 2 (L'Hospital's Rule). Suppose that $f(a) = g(a) = 0$, $f(x)$ and $g(x)$ are each continuous in $a \leq x \leq b$, and differentiable in $a < x < b$, and $f'(x) \neq 0$ in $a < x < b$. Then

$$(9) \qquad \lim_{x \to a^+} \frac{g(x)}{f(x)} = \lim_{x \to a^+} \frac{g'(x)}{f'(x)}$$

whenever the latter limit exists. If the limit on the right side of equation (9) is $+\infty$ or $-\infty$, then the limit on the left side is $+\infty$ or $-\infty$, respectively.

We have already illustrated equations (8) and (9) by equations (1) and (2).

Proof. We apply Theorem 1, but instead of the interval $a \leq x \leq b$, we use t as the variable and x as the right end point. Thus we consider the interval $a \leq t \leq x$. Then the Cauchy Mean Value Theorem states that there is a ξ with $a < \xi < x$ such that

$$(10) \qquad \frac{g(x) - g(a)}{f(x) - f(a)} = \frac{g'(\xi)}{f'(\xi)}.$$

But $g(a) = f(a) = 0$, so (10) simplifies to

$$(11) \qquad \frac{g(x)}{f(x)} = \frac{g'(\xi)}{f'(\xi)}, \qquad a < \xi < x.$$

Now let $x \to a^+$. Since $a < \xi < x$, ξ must also approach a^+. Hence if

$$\lim_{\xi \to a^+} \frac{g'(\xi)}{f'(\xi)} = L,$$

then by (11) the left side of (9) has the same limit L. ∎

Example 1. Compute $\lim\limits_{x\to 0} \dfrac{x}{\ln^2 (1 + x)}$.

Solution. At $x = 0$ both the numerator and the denominator are zero, so Theorem 2 is applicable. By L'Hospital's Rule, equation (9),

$$\lim_{x\to 0} \frac{x}{\ln^2 (1 + x)} = \lim_{x\to 0} \frac{1}{\dfrac{2 \ln (1 + x)}{1 + x}} = \lim_{x\to 0} \frac{1 + x}{2 \ln (1 + x)}.$$

If $x > 0$ and $x \to 0$, the denominator of this last fraction is positive and tending to zero. Hence the fraction grows large without bound and therefore the limit is $+\infty$. When $x \to 0$ and $x < 0$, the denominator is negative and hence the limit is $-\infty$. This result is symbolized by writing

$$\lim_{x\to 0^+} \frac{x}{\ln^2 (1 + x)} = \infty \quad \text{and} \quad \lim_{x\to 0^-} \frac{x}{\ln^2 (1 + x)} = -\infty. \quad \bullet$$

Example 2. Compute $L = \lim\limits_{x\to 2} \dfrac{3\sqrt[3]{x - 1} - x - 1}{3(x - 2)^2}$.

Solution. At $x = 2$, both the numerator and denominator are zero, so Theorem 2 is applicable. Hence

$$(12) \qquad L = \lim_{x\to 2} \frac{3\sqrt[3]{x - 1} - x - 1}{3(x - 2)^2} = \lim_{x\to 2} \frac{\dfrac{1}{(x - 1)^{2/3}} - 1}{6(x - 2)}.$$

In this last fraction the numerator and denominator are both zero at $x = 2$, so we are faced with another indeterminate form $0/0$. But we can apply L'Hospital's Rule to this new indeterminate form. Differentiating the numerator and denominator, we find that

$$(13) \qquad \lim_{x\to 2} \frac{\dfrac{1}{(x - 1)^{2/3}} - 1}{6(x - 2)} = \lim_{x\to 2} \frac{-\dfrac{2}{3} \dfrac{1}{(x - 1)^{5/3}}}{6} = -\frac{2}{18} = -\frac{1}{9}.$$

Combining equations (12) and (13) yields $L = -1/9$. $\quad \bullet$

     If the original indeterminate form is sufficiently complicated, it may be necessary to use L'Hospital's Rule a large number of times to obtain a numerical answer.

     L'Hospital's Rule is exactly the same in case the independent variable x is tending to ∞ instead of some finite number a. Precisely stated, we have

---

THEOREM 3. Suppose that $f(x)$ and $g(x)$ are each differentiable in $M < x < \infty$, $f'(x) \neq 0$ in $M < x < \infty$, and

$$\lim_{x \to \infty} f(x) = 0 \quad \text{and} \quad \lim_{x \to \infty} g(x) = 0.$$

Then

(14)
$$\lim_{x \to \infty} \frac{g(x)}{f(x)} = \lim_{x \to \infty} \frac{g'(x)}{f'(x)},$$

whenever the latter limit exists.

Proof. We bring the point at infinity into the origin by the substitution $x = 1/t$ and apply L'Hospital's Rule with $a = 0$. Clearly, as $x \to \infty$, $t \to 0^+$ and conversely as $t \to 0^+$, $x \to \infty$. The details of this program are

$$\lim_{x \to \infty} \frac{g(x)}{f(x)} = \lim_{t \to 0^+} \frac{g(1/t)}{f(1/t)} = \lim_{t \to 0^+} \frac{g'(1/t)(-1/t^2)}{f'(1/t)(-1/t^2)} \quad \text{(Chain Rule)}$$

$$= \lim_{t \to 0^+} \frac{g'(1/t)}{f'(1/t)} = \lim_{x \to \infty} \frac{g'(x)}{f'(x)}. \quad \blacksquare$$

Example 3. Compute $\lim\limits_{x \to \infty} \dfrac{1/x}{\sin (\pi/x)}$.

Solution. We differentiate the numerator and denominator. By Theorem 3,

$$\lim_{x \to \infty} \frac{1/x}{\sin (\pi/x)} = \lim_{x \to \infty} \frac{-1/x^2}{\cos (\pi/x)(-\pi/x^2)} = \lim_{x \to \infty} \frac{1}{\pi \cos (\pi/x)} = \frac{1}{\pi}. \quad \bullet$$

EXERCISE 2

Evaluate each of the following limits.

1. $\lim\limits_{x \to 3} \dfrac{x^2 - 4x + 3}{x^2 + x - 12}$.

2. $\lim\limits_{x \to -1} \dfrac{x^2 + 6x + 5}{x^2 - x - 2}$.

3. $\lim\limits_{x \to 1} \dfrac{\sin \pi x}{x^2 - 1}$.

4. $\lim\limits_{x \to 1} \dfrac{\ln x}{x^2 - x}$.

5. $\lim\limits_{x \to 0} \dfrac{e^x - e^{-x}}{\sin 3x}$.

6. $\lim\limits_{x \to \pi} \dfrac{\ln \cos 2x}{(\pi - x)^2}$.

7. $\lim\limits_{x \to 0} \dfrac{e^x - 1 - x}{1 - \cos \pi x}$.

8. $\lim\limits_{x \to 0} \dfrac{\tan 2x - 2x}{x - \sin x}$.

9. $\lim\limits_{x \to 0} \dfrac{\sin x - \sinh x}{x^3}$.

10. $\displaystyle\lim_{x\to 0^+}\frac{\operatorname{Sin}^{-1}x}{\sin^2 3x}$.

11. $\displaystyle\lim_{x\to 0^+}\frac{\operatorname{Tan}^{-1}x}{1-\cos 2x}$.

12. $\displaystyle\lim_{x\to \pi}\frac{1+\cos x}{\sin 2x}$.

13. $\displaystyle\lim_{x\to 0}\frac{b^x-a^x}{x}$, $\quad b>a>0$.

14. $\displaystyle\lim_{x\to 4}\frac{4e^{4-x}-x}{\sin \pi x}$.

15. $\displaystyle\lim_{t\to 0}\frac{\sin^2 t-\sin t^2}{t^2}$.

16. $\displaystyle\lim_{t\to 0}\frac{e^{5t}\sin 2t}{\ln(1+t)}$.

$\star$**17.** $\displaystyle\lim_{x\to 0}\frac{2\cos x-2+x^2}{3x^4}$.

$\star$**18.** $\displaystyle\lim_{x\to 0}\frac{x\sin(\sin x)}{1-\cos(\sin x)}$.

19. $\displaystyle\lim_{x\to 1}\frac{x^3-3x+1}{x^4-x^2-2x}$.

20. $\displaystyle\lim_{x\to 2}\frac{e^{x-2}+2-x}{\cos^2 \pi x}$.

21. $\displaystyle\lim_{x\to \infty}\frac{e^{3/x}-1}{\sin(1/x)}$.

$\star$**22.** $\displaystyle\lim_{x\to \infty}\frac{\tan^2(1/x)}{\ln^2(1+4/x)}$.

$\star$**23.** Extend Theorem 12 of Chapter 12 as follows. Let $x=f(t)$ and $y=g(t)$ be parametric equations for a curve $\mathscr{C}$, and suppose that $f'(a)=g'(a)=0$. Prove that if

$$L=\lim_{t\to a}\frac{g''(t)}{f''(t)},$$

then L is the slope of the line tangent to $\mathscr{C}$ at the point $(f(a),g(a))$.

The Form ∞/∞ 4

One of the attractive features of L'Hospital's Rule is that it works for the indeterminate form ∞/∞ just as it works for $0/0$.

THEOREM 4 (PWO). Let $\displaystyle\lim_{x\to a^+}f(x)=\infty$ and $\displaystyle\lim_{x\to a^+}g(x)=\infty$, and suppose that $f'(x)\neq 0$ in $a<x<b$. Then

(15)
$$\lim_{x\to a^+}\frac{g(x)}{f(x)}=\lim_{x\to a^+}\frac{g'(x)}{f'(x)},$$

whenever the latter limit exists. Here a^+ may be replaced by ∞ if $f'(x)\neq 0$ in $M<x<\infty$.

In other words, we can evaluate the indeterminate form

$$\lim_{x \to a^+} \frac{g(x)}{f(x)} = \frac{\infty}{\infty}$$

by just differentiating the numerator and denominator and then taking the limit.

It turns out that a proof of Theorem 4 as stated is somewhat difficult, and so we omit it. The interested reader can find it in any book on advanced calculus. But if we add a little to the hypotheses, then the proof is easier.

THEOREM 5. Let f and g satisfy the conditions of Theorem 4, and suppose that in addition the two limits indicated in equation (15) both exist:

(16) $$\lim_{x \to a^+} \frac{g(x)}{f(x)} = L \quad \text{and} \quad \lim_{x \to a^+} \frac{g'(x)}{f'(x)} = M.$$

If $L \neq 0$ and $L \neq \infty$, then $L = M$.

Proof. We apply Theorem 2 (or Theorem 3 if $x \to \infty$) to the evaluation of

$$\lim_{x \to a^+} \frac{\dfrac{1}{g(x)}}{\dfrac{1}{f(x)}},$$

which is an indeterminant form $0/0$. Hence

$$\lim_{x \to a^+} \frac{\dfrac{1}{g(x)}}{\dfrac{1}{f(x)}} = \lim_{x \to a^+} \frac{-\dfrac{1}{g^2(x)} g'(x)}{-\dfrac{1}{f^2(x)} f'(x)}.$$

This is equivalent to

(17) $$\lim_{x \to a^+} \frac{f(x)}{g(x)} = \lim_{x \to a^+} \frac{f^2(x)}{g^2(x)} \frac{g'(x)}{f'(x)} = \left(\lim_{x \to a^+} \frac{f^2(x)}{g^2(x)} \right)\left(\lim_{x \to a^+} \frac{g'(x)}{f'(x)} \right),$$

where the "factorization" is legitimate because both limits exist by hypothesis. Using the conditions (16) in equation (17), we find that

$$\frac{1}{L} = \frac{1}{L^2} M,$$

and hence $L = M$. ∎

Example 1. Compute $\lim\limits_{x\to\infty}\dfrac{\ln^2 x}{x}$.

Solution. This has the form ∞/∞. Applying L'Hospital's Rule,

$$\lim_{x\to\infty}\frac{\ln^2 x}{x}=\lim_{x\to\infty}\frac{2(\ln x)\dfrac{1}{x}}{1}=\lim_{x\to\infty}\frac{2\ln x}{x}.$$

This last is still an indeterminate form ∞/∞. Applying our rule to this new form,

$$\lim_{x\to\infty}\frac{2\ln x}{x}=\lim_{x\to\infty}\frac{2\dfrac{1}{x}}{1}=0. \quad\bullet$$

EXERCISE 3

In Problems 1 through 8, evaluate each of the indeterminate forms.

1. $\lim\limits_{x\to\infty}\dfrac{x}{e^x}$.
2. $\lim\limits_{x\to\infty}\dfrac{\ln x}{\sqrt[3]{x}}$.
3. $\lim\limits_{x\to\infty}\dfrac{x^2+3x+2}{2x^2+x+3}$.
4. $\lim\limits_{x\to\infty}\dfrac{e^x}{x^2}$.

$\star$5. $\lim\limits_{x\to\pi/2}\dfrac{\tan x}{\tan 3x}$.
6. $\lim\limits_{x\to\infty}\dfrac{x+\ln x}{x\ln x}$.
7. $\lim\limits_{x\to\pi/2}\dfrac{\sec x+5}{\tan x}$.
8. $\lim\limits_{x\to0^+}\dfrac{\ln x}{\csc x}$.

9. Prove that if ϵ is any fixed positive number, no matter how small, then

$$\lim_{x\to\infty}\frac{\ln x}{x^\epsilon}=0.$$

10. Prove that if M is any fixed positive number, no matter how large, then

$$\lim_{x\to\infty}\frac{e^x}{x^M}=\infty.$$

HINT: First assume that M is an integer.

$\star$11. Without using Theorem 4, prove that

$$\lim_{x\to\infty}\frac{x+\sin x}{x}=1.$$

Observe that if we differentiate the numerator and denominator, we must consider

$$\lim_{x\to\infty}\frac{\cos x}{1}$$

and this ratio does not tend to any limit. Does this show that Theorem 4 is false?

CALCULATOR PROBLEM

C1. Use a calculator to check mechanically (by computation) the results obtained in some of Problems 1 through 8.

5 *Other Indeterminate Forms*

Aside from $0/0$ and ∞/∞, the types of indeterminate forms that occur most frequently are $0 \times \infty$, $\infty - \infty$, 0^0, ∞^0, and 1^∞. Here the meaning of each of these five symbols is obvious, and each is illustrated by an example below. Rather than prove a new theorem for each one of these five types, it is simpler to reduce them by suitable manipulations to a type already treated.

Example 1. Compute $L_1 = \lim\limits_{x \to 0^+} x^3 \ln x$.

Solution. This has the form $0 \times (-\infty)$. We transform this into a type ∞/∞ (the negative sign is unimportant here) by writing

$$L_1 = \lim_{x \to 0^+} x^3 \ln x = \lim_{x \to 0^+} \frac{\ln x}{\dfrac{1}{x^3}}.$$

We apply Theorem 4 to the second ratio (∞/∞) and find that

$$L_1 = \lim_{x \to 0^+} \frac{\dfrac{1}{x}}{-\dfrac{3}{x^4}} = \lim_{x \to 0^+} -\frac{x^4}{3x} = \lim_{x \to 0^+} -\frac{x^3}{3} = 0. \quad \bullet$$

Example 2. Compute $L_2 = \lim\limits_{x \to 0} \left(\dfrac{1}{x} - \dfrac{1}{\ln(1+x)} \right)$.

Solution. This has the form $\infty - \infty$. But on adding the two fractions, we obtain

$$L_2 = \lim_{x \to 0} \frac{\ln(1+x) - x}{x \ln(1+x)}$$

and this has the form $0/0$. L'Hospital's Rule gives us

$$L_2 = \lim_{x \to 0} \frac{\dfrac{1}{1+x} - 1}{\ln(1+x) + \dfrac{x}{1+x}} = \lim_{x \to 0} \frac{1 - 1 - x}{(1+x)\ln(1+x) + x}.$$

This still has the form 0/0. Using L'Hospital's Rule again yields

$$L_2 = \lim_{x \to 0} \frac{-1}{1 + \ln(1 + x) + 1} = -\frac{1}{2}. \quad \bullet$$

Example 3. Compute $L_3 = \lim_{x \to 0^+} x^{x^3}$.

Solution. This has the form 0^0. To simplify the work, we let $Q = x^{x^3}$. Then $\ln Q = \ln/x^{x^2} = x^3 \ln x$. Now as $x \to 0^+$, $x^3 \ln x$ has the form $0 \times (-\infty)$, and in fact this is just the one evaluated in Example 1. Hence we already know that

$$\lim_{x \to 0^+} \ln Q = \lim_{x \to 0^+} x^3 \ln x = 0.$$

Therefore, $L_3 = \lim_{x \to 0^+} Q = \lim_{x \to 0^+} e^{\ln Q} = e^0 = 1. \quad \bullet$

Example 4. Compute $L_4 = \lim_{x \to 0^+} \left(1 + \frac{5}{x}\right)^{2x}$.

Solution. This has the form ∞^0. We let $Q = \left(1 + \frac{5}{x}\right)^{2x}$. Then we find that $\ln Q = 2x \ln \left(1 + \frac{5}{x}\right)$, and as $x \to 0^+$, this has the form $0 \times \infty$. Now

$$\lim_{x \to 0^+} \ln Q = \lim_{x \to 0^+} 2x \ln \left(1 + \frac{5}{x}\right) = \lim_{x \to 0^+} \frac{2 \ln \left(1 + \frac{5}{x}\right)}{\frac{1}{x}}.$$

This last limit has the form ∞/∞. Hence

$$(18) \qquad \lim_{x \to 0^+} \ln Q = \lim_{x \to 0^+} \frac{2 \dfrac{1}{1 + \dfrac{5}{x}} \left(-\dfrac{5}{x^2}\right)}{-\dfrac{1}{x^2}} = \lim_{x \to 0^+} \frac{10}{1 + \dfrac{5}{x}} = 0.$$

Therefore, $L_4 = \lim_{x \to 0^+} Q = \lim_{x \to 0^+} e^{\ln Q} = e^0 = 1. \quad \bullet$

Example 5. Compute $L_5 = \lim_{x \to \infty} \left(1 + \frac{5}{x}\right)^{2x}$.

Solution. This has the form 1^∞. The function involved is the same as in Example 4.

The only difference is that now $x \to \infty$ instead of $x \to 0^+$. We follow the same pattern up to equation (18), but this time we have

$$\lim_{x \to \infty} \ln Q = \lim_{x \to \infty} \frac{10}{1 + \dfrac{5}{x}} = 10.$$

Hence $L_5 = e^{10}$. ●

EXERCISE 4

In Problems 1 through 18, compute each of the given indeterminate forms.

1. $\lim\limits_{x \to 0} \left(\dfrac{1}{x} - \dfrac{1}{\sin x} \right)$.

2. $\lim\limits_{x \to 0} \left(\dfrac{1}{x} - \dfrac{1}{e^x - 1} \right)$.

3. $\lim\limits_{x \to \pi/2^+} (\sec^3 x - \tan^3 x)$.

4. $\lim\limits_{x \to 0^+} \sqrt{x} \ln x^2$.

5. $\lim\limits_{x \to \infty} x^3 e^{-x}$.

6. $\lim\limits_{x \to \pi} \left(\dfrac{1}{\sin x} - \dfrac{1}{\pi - x} \right)$.

7. $\lim\limits_{x \to \pi/2} (\sin x)^{\tan x}$.

★8. $\lim\limits_{x \to 0^+} x^3 e^{1/x}$.

9. $\lim\limits_{x \to 0} (1 + x^3)^{4/x^3}$.

10. $\lim\limits_{x \to \infty} \left(\cos \dfrac{3}{x} \right)^x$.

★11. $\lim\limits_{x \to 0^+} x^{\ln(1+x)}$.

★12. $\lim\limits_{x \to 0^+} (e^x - 1)^{\sin x}$.

13. $\lim\limits_{x \to \infty} (e^x - 1)^{1/x}$.

14. $\lim\limits_{x \to 1} x^{1/(1-x)}$.

15. $\lim\limits_{x \to 0^+} (\tan x)^{\sin x}$.

16. $\lim\limits_{x \to 0} x^2 \csc (3 \sin^2 x)$.

17. $\lim\limits_{x \to \infty} (1 + 2x)^{e^{-x}}$.

18. $\lim\limits_{x \to 1^-} \left(\dfrac{1}{1 - x} \right)^{(1-x)^2}$.

19. The expressions in Problems 11, 12, and 15 all have the form 0^0 and in each case the limit is 1. Prove that for any positive A and n,

$$\lim_{x \to 0^+} (Ax)^{x^n} = 1.$$

20. Continuation of Problem 19. Our results seem to indicate that 0^0 is always 1. Explode this conjecture by finding: (a) $\lim\limits_{x \to 0^+} x^{-k/\ln x}$ and (b) $\lim\limits_{x \to 0^+} x^{-k/\sqrt{-\ln x}}$, where k is a positive constant.

21. Explain why 1^0, 0^1, and 0^∞ are not indeterminate forms.

22. Explain why the expression in Problem 8 is not indeterminate if $x \to 0^-$. What is the limit in this case?

23. Why is the expression

$$\lim_{x \to 2\pi} \left(\frac{1}{\sin x} - \frac{1}{2\pi - x} \right)$$

not an indeterminate form? Compare this with the form in Problem 6.

Improper Integrals 6

In the definition of a definite integral, $f(x)$ was assumed to be bounded in the interval $\mathcal{I}: a \leq x \leq b$. If $f(x)$ is not bounded in $a \leq x \leq b$, then the integral

$$\int_a^b f(x)\, dx$$

is said to be an *improper integral*. We can also replace the interval of integration by a ray $-\infty < x \leq b$, or by a ray $a \leq x < \infty$, or by the set $-\infty < x < \infty$ of all real numbers. It is convenient to extend our terminology and call these sets *infinite intervals*. An integral over an interval that is infinite is also called an *improper integral*. Thus an integral can be improper in two ways: (1) the interval of integration may be infinite, or (2) the integrand $f(x)$ may become infinite at one or possibly more points either inside the interval or at an end point.

Suppose that we wish to compute

$$(19) \qquad\qquad I_1 = \int_2^\infty \frac{1}{x^3}\, dx.$$

The natural thing to do is to replace the upper limit ∞, by M. Then we compute

$$I(M) \equiv \int_2^M \frac{1}{x^3}\, dx$$

and take the limit of $I(M)$ as $M \to \infty$. In fact, we make this our definition of (19). In other words, we agree by definition that

$$(20) \qquad\qquad \int_a^\infty f(x)\, dx = \lim_{M \to \infty} \int_a^M f(x)\, dx,$$

whenever the expression on the right side of (20) has a limit. If the expression on the right side has a finite limit, then the integral is said to be *convergent;* otherwise, the integral is said to be *divergent*.

Example 1. Compute the integral in (19).

Solution. By definition,

$$\int_2^\infty \frac{dx}{x^3} = \lim_{M \to \infty} \int_2^M \frac{dx}{x^3} = \lim_{M \to \infty} \frac{-1}{2x^2}\bigg|_2^M = \lim_{M \to \infty}\left(-\frac{1}{2M^2} + \frac{1}{8}\right) = \frac{1}{8}. \quad \bullet$$

If the interval of integration is $-\infty < x \leq b$, equation (20) must be modified in an obvious way. If the interval of integration is $-\infty < x < \infty$, equation (20) is replaced by

$$\int_{-\infty}^{\infty} f(x)\, dx \equiv \lim_{L \to -\infty} \int_{L}^{0} f(x)\, dx + \lim_{M \to \infty} \int_{0}^{M} f(x)\, dx,$$

provided that *both* limits on the right side exist (as finite numbers). Integrals of this type will be found in Exercise 5.

As an example of the second type of improper integral, consider

$$(21) \qquad\qquad I_2 = \int_{1}^{3} \frac{dx}{(3-x)^2}.$$

Here the integrand $1/(3-x)^2$ becomes infinite at the upper end point of the interval of integration. The natural thing to do is to compute

$$I(\epsilon) = \int_{1}^{3-\epsilon} \frac{dx}{(3-x)^2}$$

for $\epsilon > 0$, and then take the limit of $I(\epsilon)$ as $\epsilon \to 0^+$. In fact, if $\lim_{x \to b} f(x) = \infty$, we make the definition

$$(22) \qquad\qquad \int_{a}^{b} f(x)\, dx = \lim_{\epsilon \to 0^+} \int_{a}^{b-\epsilon} f(x)\, dx,$$

whenever the expression on the right side of (22) has a finite limit. When this occurs, the integral is said to be *convergent*. If the expression on the right side does not have a limit, then the integral is said to be *divergent*. We leave it to the reader to frame the proper definition, when $f(x)$ becomes infinite at the left-hand end point, or in the interior, of the interval of integration.

Example 2. Compute the integral in (21).

Solution. By definition,

$$\int_{1}^{3} \frac{dx}{(3-x)^2} = \lim_{\epsilon \to 0^+} \int_{1}^{3-\epsilon} \frac{dx}{(3-x)^2} = \lim_{\epsilon \to 0^+} \frac{1}{3-x} \Big|_{1}^{3-\epsilon} = \lim_{\epsilon \to 0^+} \left(\frac{1}{\epsilon} - \frac{1}{2} \right) = \infty.$$

In this case the integral diverges. ●

Example 3. Compute $I_3 = \int_{0}^{6} \frac{2x\, dx}{(x^2 - 4)^{2/3}}.$

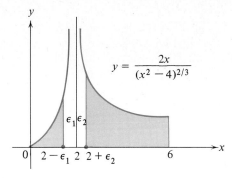

$$y = \frac{2x}{(x^2 - 4)^{2/3}}$$

FIGURE 3

Solution. The graph of $y = 2x/(x^2 - 4)^{2/3}$ is shown in Fig. 3. Obviously, the function tends to infinity as $x \to 2$, so the curve has a vertical asymptote at $x = 2$. Since $x = 2$ is inside the interval $0 \leq x \leq 6$, we must break the computation of I_3 into two parts. Indeed, by definition (if the integrals converge),

(23) $$I_3 = \lim_{\epsilon_1 \to 0^+} \int_0^{2-\epsilon_1} \frac{2x \, dx}{(x^2 - 4)^{2/3}} + \lim_{\epsilon_2 \to 0^+} \int_{2+\epsilon_2}^6 \frac{2x \, dx}{(x^2 - 4)^{2/3}}.$$

Thus, in (23), we are to compute the area of the shaded regions in Fig. 3 and then take the limit of that area as $\epsilon_1 \to 0^+$ and $\epsilon_2 \to 0^+$. For simplicity, let I' and I'' be the two integrals on the right side of (23). Then, on integrating, we obtain

$$I' = \lim_{\epsilon_1 \to 0^+} 3(x^2 - 4)^{1/3} \Big|_0^{2-\epsilon_1} = \lim_{\epsilon_1 \to 0^+} (3 \sqrt[3]{(2 - \epsilon_1)^2 - 4} + 3\sqrt[3]{4}) = 3\sqrt[3]{4}$$

and

$$I'' = \lim_{\epsilon_2 \to 0^+} 3(x^2 - 4)^{1/3} \Big|_{2+\epsilon_2}^6 = \lim_{\epsilon_2 \to 0^+} (3 \sqrt[3]{32} - 3\sqrt[3]{(2 + \epsilon_2)^2 - 4}) = 3\sqrt[3]{32}.$$

Thus $I_3 = I' + I'' = 3\sqrt[3]{4} + 3\sqrt[3]{32} = 9\sqrt[3]{4}$. Therefore, although the region below the curve $y = 2x/(x^2 - 4)^{2/3}$ between $x = 0$ and $x = 6$ is infinite in extent, it has a finite area, namely, $9\sqrt[3]{4}$. ●

Example 4. Compute $I_4 = \displaystyle\int_0^6 \frac{2x \, dx}{x^2 - 4}$.

Solution. Following the pattern of Example 3, we have $I_4 = I' + I''$, where

$$I' = \lim_{\epsilon_1 \to 0^+} \ln |x^2 - 4| \Big|_0^{2-\epsilon_1} = \lim_{\epsilon_1 \to 0^+} [\ln \epsilon_1(4 - \epsilon_1) - \ln 4]$$

and

$$I'' = \lim_{\epsilon_2 \to 0^+} \ln |x^2 - 4| \Big|_{2+\epsilon}^{6} = \lim_{\epsilon_2 \to 0^+} [\ln 32 - \ln \epsilon_2(4 + \epsilon_2)].$$

Clearly, $I' = -\infty$ and $I'' = +\infty$, so this integral is divergent.   ●

EXERCISE 5

In Problems 1 through 27, evaluate the given improper integral in case it converges. In each problem interpret the integral as an area under a suitable curve.

1. $\displaystyle\int_1^\infty \frac{dx}{x^2}.$

2. $\displaystyle\int_0^1 \frac{dx}{(1-x)^2}.$

3. $\displaystyle\int_0^1 \frac{dx}{\sqrt{x}}.$

4. $\displaystyle\int_1^\infty \frac{dx}{\sqrt[3]{x}}.$

5. $\displaystyle\int_0^2 \frac{dx}{4-x^2}.$

6. $\displaystyle\int_0^3 \frac{x\,dx}{\sqrt{9-x^2}}.$

7. $\displaystyle\int_1^\infty e^{-2x}\,dx.$

8. $\displaystyle\int_{-\infty}^\infty \frac{x\,dx}{1+2x^2}.$

9. $\displaystyle\int_{-\infty}^\infty \frac{dx}{1+x^2}.$

10. $\displaystyle\int_0^4 \frac{8\,dx}{\sqrt{16-x^2}}.$

11. $\displaystyle\int_8^\infty \frac{dx}{x^{4/3}}.$

12. $\displaystyle\int_0^\infty (x-1)e^{-x}\,dx.$

13. $\displaystyle\int_0^2 \frac{\ln x\,dx}{x}.$

14. $\displaystyle\int_0^2 \frac{\ln x\,dx}{\sqrt{x}}.$

15. $\displaystyle\int_0^{\pi/2} \csc x\,dx.$

16. $\displaystyle\int_{-\infty}^\infty \operatorname{sech} x\,dx.$

17. $\displaystyle\int_{-\infty}^0 \tanh x\,dx.$

18. $\displaystyle\int_0^{\pi/2} \tan x\,dx.$

19. $\displaystyle\int_0^\infty e^{-x} \sin x\,dx.$

20. $\displaystyle\int_{-1}^1 \frac{dx}{\sqrt[3]{x}}.$

21. $\displaystyle\int_{-1}^1 \frac{dx}{x}.$

22. $\displaystyle\int_0^7 \frac{dx}{x^2-5x+6}.$

23. $\displaystyle\int_0^\infty \frac{dx}{\sqrt[3]{e^x}}.$

24. $\displaystyle\int_0^\infty \frac{dx}{\sqrt[5]{x}}.$

25. $\displaystyle\int_0^{\pi/2} \frac{dx}{1-\sin x}.$

26. $\displaystyle\int_0^\infty \frac{x^2\,dx}{e^{x^3}}.$

27. $\displaystyle\int_1^\infty \frac{dx}{x \ln x}.$

28. Prove that if $k > 1$, then $\displaystyle\int_1^\infty \frac{dx}{x^k} = \frac{1}{k-1}$, and if $k \leqq 1$, then this integral diverges.

29. State and prove a result similar to that in Problem 28 for $\displaystyle\int_0^1 \frac{dx}{x^k}.$

★30. Use induction to prove that for each positive integer n,

$$\int_0^\infty x^n e^{-x}\,dx = n!.$$

What is the situation when $n = 0$?

*31. The region bounded by the x-axis and the curve $y = 4/(3x^{3/4})$ and to the right of the line $x = 1$ is rotated about the x-axis. Find the volume of the solid generated.

*32. Prove that the surface area of the solid described in Problem 31 is infinite. As a result we have a container that can be filled with paint (finite volume) but whose surface cannot be painted (infinite surface). HINT: Use the obvious inequality

$$\frac{1}{x^{3/4}} \sqrt{1 + \frac{1}{x^{7/2}}} > \frac{1}{x^{3/4}}$$

to prove that

$$\sigma \geqq \frac{8\pi}{3} \int_1^\infty \frac{dx}{x^{3/4}} = \infty.$$

**33. A formal computation indicates that the derivative of $-2\sqrt{1 - \sin x}$ is $\sqrt{1 + \sin x}$. Consequently,

(24) $$\int_0^M \sqrt{1 + \sin x}\, dx = -2\sqrt{1 - \sin x}\, \Big|_0^M = 2 - 2\sqrt{1 - \sin M}.$$

But as $M \to \infty$, the area under the curve $y = \sqrt{1 + \sin x}$ between $x = 0$ and $x = M$ becomes infinite (graph the function). On the other hand, the right side of (24) is never greater than 2. Where is the error?

34. Note that $\lim_{M \to \infty} \int_{-M}^M x\, dx = \lim_{M \to \infty} \frac{x^2}{2} \Big|_{-M}^M = 0$. Is it true that the improper integral $\int_{-\infty}^\infty x\, dx = 0$?

REVIEW PROBLEMS

In Problems 1 through 19, evaluate the given limit.

1. $\lim_{x \to 1} \dfrac{\sin (x - 1)}{\ln (3x - 2)}$.

2. $\lim_{x \to 0} \dfrac{e^x - 1}{x \cos^2 x}$.

3. $\lim_{x \to 0} \dfrac{\pi/2 - \text{Cos}^{-1} 3x}{\tan^{-1} 5x}$.

4. $\lim_{x \to 0} \dfrac{e^{ax} - e^{bx}}{\tan x}$, $a \neq b$.

5. $\lim_{x \to 0} \dfrac{e^x - 1 - \sin x}{x \sin x}$.

6. $\lim_{x \to 0} \dfrac{\cos^2 3x - \cos x^2}{3x^2 - 2x^3}$.

7. $\lim_{x \to \pi/4} \dfrac{\sin x - \cos x}{\ln \tan^2 x}$.

8. $\lim_{x \to 0} \dfrac{\cos 3x - \cosh 2x}{x^2}$.

9. $\lim_{x \to 0} \dfrac{x \cos x}{\sin x + \cos x}$.

10. $\lim_{x \to \infty} \dfrac{(\ln x)^4}{x}$.

11. $\lim\limits_{x \to 0^+} x^{\sin 2x}$.

12. $\lim\limits_{x \to 4^+} \left(\dfrac{1}{x-4} - \dfrac{1}{\sqrt{x-4}} \right)$.

13. $\lim\limits_{x \to \infty} (x^2 + e^{3x})^{4/\sqrt{x}}$.

14. $\lim\limits_{x \to \infty} (x^4 + e^{6x})^{8/x}$.

15. $\lim\limits_{x \to \infty} (\sin x + e^x)^{5/x^2}$.

16. $\lim\limits_{x \to \infty} (\sqrt{x^4 + 5x^2 + 7} - x^2)$.

17. $\lim\limits_{x \to 0} (\sinh x)^{x^2}$.

18. $\lim\limits_{x \to 0} (\cosh x)^{1/x^2}$.

19. Find $\lim\limits_{x \to \infty} (\sinh x)^{1/x^n}$ in the following three cases:

   (a) $n > 1$.   (b) $n = 1$.   (c) $0 < n < 1$.

In Problems 20 through 43, evaluate the given improper integral whenever it is convergent.

20. $\displaystyle\int_0^1 \dfrac{dx}{\sqrt[5]{x^4}}$.

21. $\displaystyle\int_1^\infty \dfrac{dx}{\sqrt[5]{x^4}}$.

22. $\displaystyle\int_0^1 \dfrac{x^2\,dx}{\sqrt{1-x^3}}$.

23. $\displaystyle\int_0^\infty \dfrac{dx}{4+x^2}$.

24. $\displaystyle\int_0^\infty \dfrac{x\,dx}{4+x^2}$.

25. $\displaystyle\int_0^\infty \dfrac{x^2}{4+x^3}\,dx$.

26. $\displaystyle\int_0^\infty \dfrac{x^2}{4+x^6}\,dx$.

27. $\displaystyle\int_0^{10} \dfrac{dx}{4-x^2}$.

28. $\displaystyle\int_0^{10} \dfrac{(2-x)\,dx}{4-x^2}$.

29. $\displaystyle\int_0^\infty e^{-ax}\,dx, \quad a > 0$.

30. $\displaystyle\int_0^\infty \cos x\, e^{\sin x}\,dx$.

31. $\displaystyle\int_{-\infty}^{+\infty} xe^{-ax^2}\,dx, \quad a > 0$.

★32. $\displaystyle\int_0^1 \dfrac{dx}{\sqrt{x(1-x)}}$.

★33. $\displaystyle\int_0^1 \sqrt{\dfrac{x}{1-x^3}}\,dx$.

34. $\displaystyle\int_0^{16} \dfrac{dx}{x(\sqrt{x}-4)}$.

35. $\displaystyle\int_0^{16} \dfrac{dx}{(x-8)^{2/3}}$.

36. $\displaystyle\int_0^1 \ln x\,dx$.

37. $\displaystyle\int_0^1 x^n \ln x\,dx, \quad n > 0$.

★38. $\displaystyle\int_1^\infty \dfrac{(x+1)e^{-x}}{x^2}\,dx$.

★39. $\displaystyle\int_0^{\pi/2} \dfrac{x\cos x - \sin x}{x^2}\,dx$.

★40. $\displaystyle\int_{1/2}^2 \dfrac{\ln x - 1}{\ln^2 x}\,dx$.

41. $\displaystyle\int_0^{\pi^3/8} \dfrac{\sin \sqrt[3]{x}}{\sqrt[3]{x}}\,dx$.

42. $\displaystyle\int_{-\infty}^0 xe^x\,dx$.

43. $\displaystyle\int_0^\infty e^{-x}\cos x\,dx$.

15
Infinite Series

An expression of the form

(1)
$$a_1 + a_2 + a_3 + \cdots + a_k + \cdots$$

is called an *infinite series*. The three dots at the end indicate that the terms go on indefinitely. In other words, for each positive integer k, there is a term a_k and in (1) we are instructed to add this infinite collection of terms. The sum in (1) may be written more briefly as

(2)
$$\sum_{k=1}^{\infty} a_k$$

(read "the sum from $k = 1$ to infinity of a_k").

Now at first glance it may seem impossible to form such a sum, and certainly it is physically impossible to perform the operation of addition an infinite number of times (there is not enough time). It is one of the real achievements of mathematics that a perfectly satisfactory meaning can be attached to the symbols in (1) and (2), a meaning that will allow us to work with such expressions just as though they involved only a finite number of terms. In fact, in many cases we will be able to find the number that is the sum of the infinite series.

The theory of infinite series is both useful and beautiful, and the manipulations involved are rather simple, in spite of the fact that we are dealing with infinite sets of numbers. The key to understanding series is a good grasp of the fundamentals of infinite sequences—a topic already discussed in Chapter 3, Section 4. Thus, as our first objective, we shall review old facts and present some new facts on the behavior of sequences.

2 *Sequences Revisited*

We recall that an infinite sequence is simply an unending string of numbers

(3) $S_1, S_2, S_3, \ldots, S_n, \ldots.$

For brevity, we denote the sequence (3) by

$$\{S_n\} \qquad \text{or} \qquad \{S_n\}_{n=1}^{\infty}.$$

For example,

$$\left\{\frac{n}{n+1}\right\}_{n=1}^{\infty} = \frac{1}{2}, \frac{2}{3}, \frac{3}{4}, \ldots, \frac{n}{n+1}, \ldots.$$

A sequence $\{S_n\}$ is said to *converge*[1] to the number L if its terms eventually stay arbitrarily close to L. In other words, the distance between S_n and L, given by $|S_n - L|$, becomes less than any prescribed positive number for all sufficiently large values of the subscript n. Under these circumstances, we write

(4) $\lim_{n \to \infty} S_n = L.$

Several examples of convergent sequences were presented in Chapter 3. Furthermore, in Chapter 4, we studied the related limit:

(5) $\lim_{x \to \infty} f(x) = L.$

The reader should observe that there is a strong similarity between (4) and (5). In (5) the function $f(x)$ is defined for all real numbers greater than some suitable x_0, while in (4) S_n is a function whose domain of definition is (usually) the set of all positive integers. Indeed, this is the only difference, and hence our knowledge of functions and their limits carries over immediately to sequences and their limits.

Of course, there are sequences that do not converge. *Divergence* will occur if, for example, the absolute values of the terms of the sequence grow without bound or if the terms oscillate between two (or more) different values.

> **Example 1.** Settle the convergence or divergence of the following sequences:
> (a) $S_n = (-2)^n$, (b) $S_n = 1/n^2$, (c) $S_n = \sqrt{2} + \sin(\pi/n)$, (d) $S_n = (\ln n)/n$, and
> (e) $S_n = (n^3 + 1)/(n^2 + 100n)$.
>
> **Solution.** (a) The sequence is $-2, 4, -8, 16, \ldots$, which has no limit because the

---

[1] See Appendix 1 for a further discussion of this definition.

absolute values of the terms grow large without bound. So the sequence diverges.

(b) The sequence is 1, 1/4, 1/9, 1/16, . . . and obviously $\lim_{n\to\infty} 1/n^2 = 0$. Hence the sequence converges and the limit is zero.

(c) As $n \to \infty$, we have $\pi/n \to 0$ and hence $\sin(\pi/n) \to 0$. Therefore,

$$\lim_{n\to\infty}\left[\sqrt{2} + \sin\left(\frac{\pi}{n}\right)\right] = \sqrt{2};$$

that is, the sequence converges to $\sqrt{2}$.

(d) Here it is helpful to consider the function $f(x) = (\ln x)/x$, which is defined for all $x > 0$. By L'Hospital's Rule (Chapter 14, Section 4) we know that

$$\lim_{x\to\infty}\frac{\ln x}{x} = \lim_{x\to\infty}\frac{1/x}{1} = 0.$$

In particular, if we let x approach infinity through positive integer values, we obtain

$$\lim_{n\to\infty}\frac{\ln n}{n} = 0.$$

Thus the sequence converges to 0.

(e) Clearly,

$$S_n = \frac{n^3 + 1}{n^2 + 100n} = \frac{n + \dfrac{1}{n^2}}{1 + \dfrac{100}{n}} \to \infty, \qquad \text{as } n \to \infty.$$

Hence the sequence is divergent. ●

Example 2. Prove that

(6) $$\lim_{n\to\infty} r^n = 0 \qquad \text{for } -1 < r < 1.$$

Solution. We shall make use of the fact that for any positive number a, the Binomial Theorem gives the expansion

$$(1 + a)^n = 1 + na + \text{positive terms}, \qquad n \geq 2,$$

and hence

(7) $$(1 + a)^n \geq 1 + na, \qquad a > 0.$$

Now (6) is certainly true for $r = 0$, so we assume that $0 < |r| < 1$. Since $1/|r| > 1$, we can write

$$\frac{1}{|r|} = 1 + a, \qquad \text{where } a \equiv \frac{1}{|r|} - 1 > 0.$$

Then, using (7), we have

$$\frac{1}{|r|^n} = (1 + a)^n \geqq 1 + na \to \infty \qquad \text{as } n \to \infty.$$

Because $1/|r^n| \to \infty$, we have $r^n \to 0$.   ●

DEFINITION 1 (Bounded Sequence). A sequence $\{S_n\}_{n=1}^{\infty}$ is *bounded above* if there is some number B that is greater than each term of the sequence, that is

$$(8) \qquad\qquad\qquad\qquad S_n \leqq B$$

for every positive integer n. A sequence is *bounded below* if there is a number b such that

$$(9) \qquad\qquad\qquad\qquad b \leqq S_n$$

for every positive integer n. If the sequence is bounded above and below, then the sequence is *bounded*.

The values for b and B in this definition are not unique.

Example 3. Which of the sequences in Example 1 are bounded above and which are bounded below?

Solution. **(a)** The sequence $-2, 4, -8, 16, -32, \ldots$ is neither bounded above nor bounded below.

(b) The sequence $1, 1/4, 1/9, 1/16, \ldots$ is bounded below and above. Take $b = 0$, $B = 1$.

(c) The terms $S_n = \sqrt{2} + \sin{(\pi/n)}$ satisfy the inequalities

$$\sqrt{2} \leqq \sqrt{2} + \sin{(\pi/n)} \leqq \sqrt{2} + 1 \qquad \text{for } n = 1, 2, 3, \ldots.$$

Hence the sequence is bounded from below by $b = \sqrt{2}$ and above by $B = \sqrt{2} + 1$.

(d) The terms $(\ln n)/n$, $n = 1, 2, 3, \ldots$ are certainly all nonnegative. Furthermore, the function $f(x) = (\ln x)/x$ for $x > 0$ attains its maximum when

$$f'(x) = \frac{x(1/x) - \ln x}{x^2} = \frac{1 - \ln x}{x^2} = 0$$

that is, when $x = e$. Thus

$$\frac{\ln n}{n} \leqq f(e) = \frac{\ln e}{e} = \frac{1}{e}, \qquad n = 1, 2, 3, \ldots.$$

Consequently, the sequence is bounded below by $b = 0$ and above by $B = 1/e$.

(e) The sequence

$$\left\{\frac{n^3 + 1}{n^2 + 100n}\right\}_{n=1}^{\infty} = \frac{2}{101}, \frac{9}{204}, \frac{28}{309}, \frac{65}{416}, \ldots$$

is bounded below (take $b = 0$) but is not bounded above (the terms approach plus infinity). ●

We remark that for a convergent sequence, the terms eventually cluster about the (finite) limit L, and therefore the terms cannot grow arbitrarily large in absolute value. Consequently, every *convergent* sequence is bounded above and below. This theoretical fact could have been used for the convergent sequences in parts (b), (c), and (d), but our purpose in Example 3 was to actually find numerical values for the bounds.

DEFINITION 2 (Monotonic Sequence). A sequence $\{S_n\}_{n=1}^{\infty}$, is said to be *increasing*[1] if

(10) $$S_1 \leqq S_2 \leqq S_3 \cdots \leqq S_n \leqq S_{n+1} \leqq \cdots,$$

that is, if each element is less than or equal to the following one. The sequence is said to be *decreasing* if

(11) $$S_1 \geqq S_2 \geqq S_3 \cdots \geqq S_n \geqq S_{n+1} \geqq \cdots.$$

In either case the sequence is said to be *monotonic* (or *monotone*).

An important property of monotone sequences is stated in

THEOREM 1 (PWO). An increasing sequence that is bounded above converges. Similarly, a decreasing sequence that is bounded below converges.

---

[1] Some authors prefer to have the strict inequality $S_n < S_{n+1}$ (for all n) in (10) and reserve the word "nondecreasing" for what we have called "increasing." There is no real gain to be had by making such a fine distinction. Our increasing sequences may be stationary; that is, adjacent terms may be equal. If $S_1 < S_2 < S_3 < \cdots < S_n < S_{n+1} < \cdots$ (equality does not occur), the sequence is said to be *strictly increasing*. Similarly, if $S_1 > S_2 > S_3 > \cdots > S_n > S_{n+1}, \ldots$ the sequence is said to be *strictly decreasing*.

A rigorous proof of this theorem is not difficult. But it does require a detailed examination of the definition of a number. This path leads us backward into a study of the foundations of mathematics. Instead, we wish to go forward to see how mathematics grows. So instead of presenting a rigorous proof, we will give the following convincing argument. We imagine a directed line with an origin, and corresponding to each number S_n in the sequence we mark on the line a point whose coordinate is S_n. The situation for an increasing sequence is illustrated in Fig. 1. The point whose coordinate is the upper bound B will lie to the right of each S_n. Hence it acts as a physical barrier to the sequence of points, and beyond this barrier the points of the sequence cannot penetrate. On the other hand, the sequence is increasing so each point is either to the right of its predecessor or perhaps coincides with it. Thus the points move steadily to the right toward B as n increases. Clearly, either these points approach B in the limit, or they approach some other point L to the left of B. In either case the sequence has a limit L, where $L \leq B$, and $\lim\limits_{n \to \infty} S_n = L$.

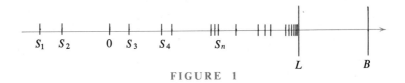

FIGURE 1

Example 4. Is the sequence with $S_n = \dfrac{2n - 3}{5n - 4}$ increasing? Is it bounded above? Does it have a limit?

Solution. We test the inequality $S_n \leq S_{n+1}$; that is,

(12)
$$\frac{2n - 3}{5n - 4} \leq \frac{2(n + 1) - 3}{5(n + 1) - 4} = \frac{2n - 1}{5n + 1}.$$

For $n \geq 1$, both denominators are positive, so we obtain an equivalent inequality by multiplying both sides of (12) by $(5n - 4)(5n + 1)$. This gives

$$10n^2 - 13n - 3 \leq 10n^2 - 13n + 4,$$

or

$$-3 \leq 4.$$

Since this last is true, retracing our steps we infer that (12) is true for $n \geq 1$, and in fact without the equality sign. Hence the given sequence is increasing.

We test that this sequence has the upper bound 2. The inequality $S_n < 2$ leads to

$$\frac{2n - 3}{5n - 4} < 2$$

$$2n - 3 < 10n - 8$$

$$5 < 8n.$$

But this last is obvious for $n \geq 1$. Since the steps are reversible, we have proved that $S_n < 2$. Now we have an increasing sequence that is bounded above, so by Theorem 1 this sequence has a limit.

Of course, we knew all along that this sequence has the limit 2/5, because

$$\lim_{n \to \infty} S_n = \lim_{n \to \infty} \frac{2n - 3}{5n - 4} = \lim_{n \to \infty} \frac{2 - \dfrac{3}{n}}{5 - \dfrac{4}{n}} = \frac{2 - 0}{5 - 0} = \frac{2}{5}. \quad \bullet$$

EXERCISE 1

In Problems 1 through 22, the nth term of a sequence is given. In each case write out the first four terms of the sequence and state whether the sequence is convergent or divergent. If it is convergent, give its limit.

1. $\dfrac{1}{\sqrt{n}}$.

2. $\dfrac{128}{2^n}$.

3. $\dfrac{1}{100} \ln n$.

4. $2 + \dfrac{(-1)^n}{n}$.

5. $\dfrac{3n}{\sqrt{n + 700}}$.

6. $\dfrac{5n^2 + 100}{n^3 + 1}$.

7. $\left(\dfrac{\pi}{4}\right)^n$.

8. $\dfrac{\sqrt{n + 5}}{\sqrt{2n + 3}}$.

9. $\left(50 + \dfrac{1}{\sqrt{n}}\right)^2$.

10. $\dfrac{3n}{n + 2} - \dfrac{n + 3}{2n}$.

11. $\dfrac{n^2}{2n + 5} - \dfrac{n^2}{2n + 1}$.

12. $\dfrac{\sqrt{n + 1}}{\sqrt[3]{n + 2}}$.

13. $\dfrac{n}{e^n}$.

14. $\dfrac{n^2}{e^n}$.

15. $n^{1/n}$.

16. $\mathrm{Tan}^{-1}\left(\dfrac{3}{n}\right)$.

17. $\mathrm{Tan}^{-1} n$.

18. $\left(1 + \dfrac{1}{n}\right)^n$.

19. $\dfrac{\cos(n\pi)}{n}$.

20. $\dfrac{[\ln n]^2}{n}$.

21. $\ln\left(\dfrac{1 + 2n}{n}\right)$.

22. S_n is the nth digit after the decimal point in the decimal representation for 131/150.

23. Prove that the sequence $S_n = r^n$ diverges if $|r| > 1$.

24. If all the terms of the sequence S_n are positive and $S_{n+1}/S_n \geq 1$ for each positive integer n, then the sequence is increasing. If the inequality is reversed, then the sequence is decreasing. Prove this statement.

In Problems 25 through 29, the nth term of a sequence is given. Use either the test of Problem 24 or the test $S_{n+1} - S_n \gtrless 0$ to determine whether the given sequence is increasing or decreasing.

25. $\dfrac{n}{2^n}$.

26. $\dfrac{n!}{2^n}$.

27. $n^2 + (-1)^n n$.

28. $\dfrac{(n+1)!}{1 \cdot 3 \cdot 5 \cdots (2n+1)}$.

29. $\dfrac{2^{2n}(n!)^2}{(2n)!}$.

30. Prove that the sequence $S_n = (2n-3)/(5n-4)$ is increasing by considering the derivative of $f(x) = (2x-3)/(5x-4)$.

In Problems 31 through 36, determine whether the given sequence is bounded above and whether it is bounded below.

31. $\left\{ \dfrac{(-1)^n}{n} \right\}_{n=1}^{\infty}$.

32. $\left\{ -3 - \dfrac{5}{n} \right\}_{n=1}^{\infty}$.

33. $\left\{ n^2 \left(\dfrac{1+n}{2+n} \right) \right\}_{n=1}^{\infty}$.

34. $\left\{ n - \dfrac{2}{n} \right\}_{n=2}^{\infty}$.

35. $\left\{ \dfrac{-n}{\ln n} \right\}_{n=2}^{\infty}$.

36. $\{(-e)^n\}_{n=0}^{\infty}$.

37. For the sequence $S_n = (\ln n)/n$, $n = 1, 2, 3, \ldots$, considered in part (d) of Example 3, prove that $B = (\ln 3)/3$ is the smallest upper bound.

38. Give an example of an increasing sequence that diverges.

39. Give an example of a bounded sequence (from above and below) that diverges.

CALCULATOR PROBLEMS

C1. Let $d > 0$ be a fixed number, and let S_1 be any number satisfying $0 < S_1 \leq 1/d$.
   (a) Prove, by induction, that the sequence $\{S_n\}_{n=1}^{\infty}$ defined recursively by

$$S_{n+1} = S_n(2 - dS_n), \qquad n = 1, 2, 3, \ldots$$

   is increasing and bounded above by $1/d$. (Therefore, by Theorem 1, the sequence has a limit L.) HINT: Consider the maximum value of the function $x(2 - dx)$ for $0 \leq x \leq 1/d$.
   (b) By taking the limit as $n \to \infty$ in the equation $S_{n+1} = S_n(2 - dS_n)$, prove that $L = 1/d$.
   (c) As a result of parts (a) and (b) we have found a method for computing the reciprocal of a number *without* using division. Apply this method to approximate the reciprocal of $d = 3.152483$ by choosing $S_1 = 0.25$ and computing S_6. Compare with your calculator's value for the reciprocal.

C2. If A is a constant and $|r| < 1$, then the sequence $S_n = A + r^n$ has limit A. Prove that this limit can be found exactly from any three consecutive terms of the sequence by verifying the formula

$$A = S_n - \frac{(S_{n+1} - S_n)^2}{S_{n+2} - 2S_{n+1} + S_n}.$$

C3. Aitken's Δ^2-Method. Problem C2 suggests that for *certain* convergent sequences $\{S_n\}$, the new sequence $\{\widehat{S}_n\}$ defined by

$$\widehat{S}_n = S_n - \frac{(S_{n+1} - S_n)^2}{S_{n+2} - 2S_{n+1} + S_n}$$

will have the same limit L and, moreover, will yield better approximations to L than the quantities S_n. This technique for speeding convergence is called Aitken's Δ^2-method. For the sequence $S_n = n2^n/(1 + n2^n)$ which has limit $L = 1$, compute $S_1, S_2, \ldots, S_6$ and use these values to compute the terms $\widehat{S}_1, \widehat{S}_2, \widehat{S}_3, \widehat{S}_4$ of the Aitken sequence. Does the Aitken sequence appear to converge more rapidly?

C4. Compute to three decimal places the solution of the equation $x = e^{-x}$ using the following procedure. Start with the guess $x_1 = 0.5$ and use the iteration $x_{n+1} = e^{-x_n}$ to compute x_2, x_3, x_4, x_5. From these values compute the terms $\widehat{x}_1, \widehat{x}_2, \widehat{x}_3$ of the accelerated Aitken sequence of Problem C3.

Convergence and Divergence of Series 3

Consider the specific series

(13)
$$\sum_{k=1}^{\infty} \frac{1}{2^k} = \frac{1}{2} + \frac{1}{4} + \frac{1}{8} + \cdots + \frac{1}{2^k} + \cdots.$$

We begin to add the terms, and we let S_n denote the sum of the first n terms. Then simple arithmetic gives

(14)

$$
\begin{aligned}
S_1 &= \frac{1}{2} & &= \frac{1}{2}, & S_5 &= \frac{31}{32}, \\
S_2 &= \frac{1}{2} + \frac{1}{4} & &= \frac{3}{4}, & S_6 &= \frac{63}{64}, \\
S_3 &= \frac{1}{2} + \frac{1}{4} + \frac{1}{8} & &= \frac{7}{8}, & S_7 &= \frac{127}{128}, \\
S_4 &= \frac{1}{2} + \frac{1}{4} + \frac{1}{8} + \frac{1}{16} &= \frac{15}{16}, & S_8 &= \frac{255}{256},
\end{aligned}
$$

and so on. But why continue? It is perfectly obvious that the sum S_n is always less than 1, but gets closer and closer to 1 as we take more and more terms. This is clear because at each stage the next term to be added is just $1/2$ of the difference between 1 and the sum of all the preceding terms. Therefore, it is reasonable to claim that 1 is the sum of the infinite series (13), and we have just found the sum of our first infinite series. All that remains is to give a *precise* definition of the sum of an infinite series, and see that the above work fits the definition.

For the present we assume that each term a_k of the infinite series

$$(15) \qquad \sum_{k=1}^{\infty} a_k = a_1 + a_2 + a_3 + a_4 + \cdots$$

is a number. (Later we shall consider series in which each term is a function.) Just as in our example, we let S_n denote the sum of the first n terms; that is,

$$(16) \qquad S_n = \sum_{k=1}^{n} a_k = a_1 + a_2 + a_3 + \cdots + a_n.$$

These finite sums, S_n, are called the *partial sums* of the series.

DEFINITION 3 (Convergence of Series). The series (15) is said to *converge* if the sequence of its partial sums $\{S_n\}$ converges. In this case, if $S = \lim_{n \to \infty} S_n$, we write

$$(17) \qquad \sum_{k=1}^{\infty} a_k = \lim_{n \to \infty} S_n = S$$

and say that the series converges (or sums) to the number S. If the sequence $\{S_n\}$ of partial sums diverges, then the series (15) is said to *diverge*.

Example 1. Settle the divergence or convergence of the following series:

(a) $\displaystyle\sum_{k=1}^{\infty} \frac{1}{2^k}.$

(b) $\displaystyle\sum_{k=1}^{\infty} 0 = 0 + 0 + 0 + \cdots.$

(c) $\displaystyle\sum_{k=1}^{\infty} 1 = 1 + 1 + 1 + \cdots.$

(d) $\displaystyle\sum_{k=1}^{\infty} (-1)^{k+1} = 1 - 1 + 1 - \cdots.$

Solution. (a) This is the series displayed in (13). We have already seen that for $n = 1, 2, \ldots, 8$, we have

$$(18) \qquad S_n = 1 - \frac{1}{2^n},$$

and it is easy to prove this for every $n > 1$, using mathematical induction. Since

$$\lim_{n\to\infty} S_n = \lim_{n\to\infty} \left(1 - \frac{1}{2^n}\right) = 1,$$

the series (13) converges and its sum is 1.

(b) Each partial sum S_n is zero because all the terms are zero. But then

$$\lim_{n\to\infty} S_n = \lim_{n\to\infty} 0 = 0.$$

Consequently, the series in **(b)** converges and the sum is zero.

(c) If we add n ones, the sum is n. Hence $S_n = n$, so $S_n \to \infty$ as $n \to \infty$. Since the sequence of partial sums does not converge to a finite number, the series diverges.

(d) If we take an even number of terms, the sum is zero. If we take an odd number, the sum is 1. In symbols, $S_{2n} = 0$ and $S_{2n-1} = 1$. Hence there is no number S to which the partial sums converge. The series is divergent. ●

The series $\displaystyle\sum_{k=1}^{\infty} 1/2^k$ of (13) is actually a special case of an important class of series given in

---

DEFINITION 4 (Geometric Series). A series $\displaystyle\sum_{k=1}^{\infty} a_k$ is called a *geometric series* if there is a fixed constant r such that $a_{k+1} = ra_k$ for each $k > 0$. In other words, the ratio of each term to its predecessor is r.

If we let a_1, the first term, be a, then clearly $a_2 = ar$, $a_3 = ar^2$, and in general $a_k = ar^{k-1}$. Note that the power on r is one less than the subscript. This is a nuisance but can easily be adjusted. Instead of writing

$$(19) \qquad \sum_{k=1}^{\infty} ar^{k-1} = a + ar + ar^2 + \cdots + ar^{k-1} + \cdots,$$

we will lower the index of the beginning term by 1 and write

$$(20) \qquad \sum_{k=0}^{\infty} ar^{k} = a + ar + ar^2 + \cdots + ar^{k} + \cdots,$$

which obviously gives the same series. The form (20) is more convenient, but in either series the nth term is ar^{n-1}.

THEOREM 2. If $a \neq 0$ and $-1 < r < 1$, the geometric series (20) converges and the sum is given by

(21)
$$\sum_{k=0}^{\infty} ar^k = \frac{a}{1-r}.$$

If $a \neq 0$, and $|r| \geq 1$, the series diverges. If $a = 0$, the series converges to the sum 0 for all r.

This theorem is extremely important and should be memorized. The proof is easy.

Proof. If $a = 0$, the convergence is obvious, since all terms are zero. If $a \neq 0$ and $r = 1$, the divergence is also obvious. Suppose now that $a \neq 0$ and $r \neq 1$. As usual, let S_n be the sum of the first n terms. Then

(22) $$S_n = a + ar + ar^2 + \cdots + ar^{n-2} + ar^{n-1}.$$

Multiplying both sides of (22) by r, we have

(23) $$rS_n = \quad ar + ar^2 + ar^3 + \cdots + ar^{n-1} + ar^n.$$

We next subtract equation (23) from equation (22). Most of the terms drop out on the right side, and we find that

$$S_n - rS_n = a - ar^n.$$

Since $r \neq 1$, we can divide both sides by $1 - r$ (not zero) and we obtain

(24) $$S_n = \frac{a(1-r^n)}{1-r}.$$

Now take the limit in (24) as $n \to \infty$. If $|r| < 1$, the term $r^n \to 0$, and hence

(25) $$\lim_{n \to \infty} S_n = \lim_{n \to \infty} \frac{a(1-r^n)}{1-r} = \frac{a}{1-r}.$$

This proves the convergence part of the theorem, and establishes formula (21). If $|r| > 1$, then $|r|^n \to \infty$ as $n \to \infty$. If r is negative, r^n oscillates in sign, giving negative numbers when n is odd and positive numbers when n is even. But they grow in absolute value without bound as n grows; hence the quantity S_n cannot tend to a limit. If $r = -1$, then $S_{2n} = 0$, and $S_{2n+1} = a$. In this case the series is also divergent. ∎

Example 2. Find the sum of each of the following series.

(a) $\displaystyle\sum_{k=0}^{\infty} 15\left(\frac{2}{7}\right)^k.$

(b) $\displaystyle\sum_{k=0}^{\infty} \frac{3^k}{5^{k+2}}.$

(c) $\displaystyle\sum_{k=2}^{\infty} \left(\frac{4}{7}\right)^k.$

(d) $\displaystyle\sum_{k=0}^{\infty} \left(\frac{-7}{11}\right)^k.$

Solution. Each of the series above is a geometric series with $-1 < r < 1$. Hence we can use formula (21).

(a) Here $a = 15$ and $r = \dfrac{2}{7}$. Hence

$$S = \frac{15}{1 - \dfrac{2}{7}} = 15 \cdot \frac{7}{5} = 21.$$

(b) If we factor 5^2 from the denominator, we see that **(b)** then fits the standard pattern with $a = 1/5^2$ and $r = 3/5$. Hence

$$S = \frac{1}{5^2} \frac{1}{1 - \dfrac{3}{5}} = \frac{1}{10}.$$

(c) Notice that here the sum starts with $k = 2$, so the first term is $(4/7)^2$. Since $r = 4/7$, we have

$$S = \left(\frac{4}{7}\right)^2 \frac{1}{1 - \dfrac{4}{7}} = \frac{16}{49} \cdot \frac{7}{3} = \frac{16}{21}.$$

(d) Here $a = 1$ and $r = -7/11$. Hence $S = \dfrac{1}{1 + \dfrac{7}{11}} = \dfrac{11}{18}.$ ●

One important area where the reader has already encountered infinite series is in the decimal expansion of numbers. Indeed, when we write

(26)
$$\frac{1}{3} = 0.3333\ldots,$$

the symbol $0.3333\ldots$ represents the infinite series

(27)
$$\frac{3}{10} + \frac{3}{10^2} + \frac{3}{10^3} + \frac{3}{10^4} + \cdots = \sum_{k=0}^{\infty} \frac{3}{10}\left(\frac{1}{10}\right)^k$$

and equation (26) states that this series sums to 1/3. This is consistent with Theorem 2 because for $a = 3/10$ and $r = 1/10$, we obtain

$$\sum_{k=0}^{\infty} \frac{3}{10}\left(\frac{1}{10}\right)^k = \frac{3}{10} \frac{1}{1 - \frac{1}{10}} = \frac{3}{10 - 1} = \frac{3}{9} = \frac{1}{3}.$$

Example 3. The decimal fraction $0.31555\ldots$ continues with infinitely many 5's, as indicated by the dots. Find the equivalent rational fraction in lowest terms.

Solution. The meaning of the decimal fraction is

$$0.31555\ldots = \frac{31}{100} + 5\left(\frac{1}{1000} + \frac{1}{10{,}000} + \frac{1}{100{,}000} + \cdots\right)$$

$$= \frac{31}{100} + \frac{5}{1000} \sum_{k=0}^{\infty} \frac{1}{10^k}.$$

Hence

$$0.31555\ldots = \frac{31}{100} + \frac{5}{1000} \cdot \frac{1}{1 - \frac{1}{10}} = \frac{31}{100} + \frac{5}{1000} \cdot \frac{10}{9}$$

$$= \frac{9 \cdot 31 + 5}{900} = \frac{279 + 5}{900} = \frac{284}{900} = \frac{71}{225}. \quad \bullet$$

EXERCISE 2

In Problems 1 through 6, find a simple formula for the nth partial sum S_n of the given series. Determine whether the series converges or diverges, and if it converges give the sum S.

1. $\displaystyle\sum_{k=1}^{\infty} 5.$

2. $\displaystyle\sum_{k=1}^{\infty} k.$

3. $\displaystyle\sum_{k=1}^{\infty} \frac{1 + (-1)^k}{2}.$

4. $\displaystyle\sum_{k=0}^{\infty} (0.53)^k.$

5. $\displaystyle\sum_{k=1}^{\infty} \left(\frac{1}{k+1} - \frac{1}{k}\right).$

6. $\displaystyle\sum_{k=1}^{\infty} (\sqrt{k+1} - \sqrt{k}).$

In Problems 7 through 10, prove that the given series has the partial sums indicated.

7. $\displaystyle\sum_{k=1}^{\infty} \frac{1}{2^{3k-1}},$ $\qquad S_n = \frac{2}{7}\left(1 - \frac{1}{2^{3n}}\right).$

8. $\displaystyle\sum_{k=1}^{\infty} \frac{4}{(k+1)(k+2)},$ $S_n = 2 - \dfrac{4}{n+2}.$

9. $\displaystyle\sum_{k=1}^{\infty} \frac{(-1)^{k+1}}{3^k},$ $S_n = \dfrac{1}{4}\left[1 - \dfrac{(-1)^n}{3^n}\right].$

10. $\displaystyle\sum_{k=1}^{\infty} \frac{2-k}{2^k},$ $S_n = \dfrac{n}{2^n}.$ HINT: Use mathematical induction.

11. Find the sum of the series given in Problems 7 through 10.

In Problems 12 through 20, state whether the geometric series is convergent or divergent. If the series is convergent, find its sum.

12. $\displaystyle\sum_{k=0}^{\infty} \left(\frac{3}{4}\right)^k.$

13. $\displaystyle\sum_{k=0}^{\infty} \left(\frac{4}{5}\right)^k.$

14. $\displaystyle\sum_{k=0}^{\infty} \left(-\frac{5}{6}\right)^k.$

15. $\displaystyle\sum_{k=14}^{\infty} \left(-\frac{3}{8}\right)^k.$

16. $\displaystyle\sum_{k=0}^{\infty} \left(\frac{5}{7}\sqrt{2}\right)^k.$

17. $\displaystyle\sum_{k=0}^{\infty} \left(\frac{8}{5\sqrt{3}}\right)^k.$

18. $\displaystyle\sum_{k=1}^{\infty} 3\left(\frac{2}{9}\right)^k.$

19. $\displaystyle\sum_{k=1}^{\infty} \frac{5^k}{100 \cdot 4^k}.$

20. $\displaystyle\sum_{k=1}^{\infty} \frac{1}{(4-\sqrt{10})^k}.$

In Problems 21 through 26, find the rational fraction in lowest terms that is equal to the given infinite repeating decimal fraction. The bar indicates the part that is repeated.

21. $0.13\overline{33}\ldots.$ 22. $0.27\overline{77}\ldots.$ 23. $0.61\overline{11}\ldots.$

24. $0.2727\overline{27}\ldots.$ 25. $0.848\overline{484}\ldots.$ 26. $0.9189\overline{1891}8\ldots.$

27. A ball is dropped from a height of 10 ft onto a concrete walk. Each time it bounces it rises to a height of $3h/4$, where h is the height attained after the previous bounce. Find the total distance that the ball travels before coming to a rest.

28. If the ball of Problem 27 is dropped from an initial height of H ft, and after each bounce rises to a height rh ft, where r is a constant, prove that the total distance traveled by the ball is $s = H(1+r)/(1-r)$.

**29. A ball dropped from a height of h ft will reach the ground in $\sqrt{h}/4$ sec. The same length of time is required for a ball to bounce upward to a maximum height of h ft. Find a formula for the length of time required for the ball of Problem 28 to come to a rest. Apply this formula to the ball of Problem 27.

30. For what values of x does the series $\displaystyle\sum_{k=0}^{\infty} 29\left(\frac{x}{3}\right)^k$ converge?

31. For what values of x does the series $\sum\limits_{k=0}^{\infty} \dfrac{x^{k+2}}{5^k}$ converge?

In Problems 32 through 37, the first few terms of an infinite series are given. Try to find a general expression (a function of k) for the kth term. In each such problem there are infinitely many functions that will do. Try to find a very simple function.

32. $1 + \dfrac{1}{4} + \dfrac{1}{9} + \dfrac{1}{16} + \cdots$

33. $\dfrac{1}{4} + \dfrac{1}{7} + \dfrac{1}{10} + \dfrac{1}{13} + \cdots$

34. $\dfrac{1}{4} + \dfrac{1}{7} + \dfrac{1}{12} + \dfrac{1}{19} + \cdots$

35. $\dfrac{1}{3} + \dfrac{1}{15} + \dfrac{1}{35} + \dfrac{1}{63} + \cdots$

36. $\dfrac{3}{2} + \dfrac{5}{4} + \dfrac{7}{8} + \dfrac{9}{16} + \cdots$

★37. $\dfrac{1}{3} + \dfrac{1}{3} + \dfrac{3}{11} + \dfrac{2}{9} + \dfrac{5}{27} + \cdots$

★38. Prove that *any* sequence $S_1, S_2, S_3, \ldots$ can be made to generate a series for which $\{S_n\}$ is the sequence of partial sums. HINT: Let $a_1 = S_1$, $a_2 = S_2 - S_1$, $a_3 = S_3 - S_2$, and in general set $a_k = S_k - S_{k-1}$ for $k \geq 2$.

CALCULATOR PROBLEMS

C1. Calculate the partial sums S_7, S_8, and S_9 of the series $\sum\limits_{k=0}^{\infty} 1/k!$ (note that $0! = 1$). This series converges; can you guess its sum?

C2. Does the series $\sum\limits_{k=0}^{\infty} \left(\dfrac{\pi^2 - 2.2}{e^2} \right)^k$ converge?

C3. For each of the given series, compute several of the partial sums and guess whether the series converges or diverges. If the series converges, try to guess an exact (or approximate) value for the sum.

(a) $\sum\limits_{k=1}^{\infty} \dfrac{k}{3^k}$.

(b) $\sum\limits_{k=1}^{\infty} \dfrac{k}{k+1}$.

(c) $\sum\limits_{k=0}^{\infty} \dfrac{(-1)^k}{(2k)!} \left(\dfrac{\pi}{3} \right)^{2k}$.

C4. Let $S_n = \sum\limits_{k=1}^{n} a_k$ be the nth partial sum of the infinite series $\sum\limits_{k=1}^{\infty} a_k$. Show that the corresponding Aitken sequence (see Problem C3, Exercise 1) is given by

$$\widehat{S}_n = S_n + \dfrac{a_{n+1}^2}{a_{n+1} - a_{n+2}}.$$

HINT: Use the fact that $S_{n+1} = S_n + a_{n+1}$.

C5. Evaluate the sum of the series

$$\sum\limits_{k=1}^{\infty} \dfrac{k^2 + 1}{2^k k^2}$$

to three decimal places using the following procedure. Compute $S_1, S_2, \ldots, S_5$ and use these values to compute the terms $\widehat{S}_1, \widehat{S}_2, \ldots, \widehat{S}_5$ of the corresponding Aitken sequence as given in Problem C4.

Some General Theorems 4

In a definite integral the variable of integration is a dummy variable. Similarly, in a sum, the index k is a dummy index, and any letter may be used in its place. Thus if $a_1, a_2, \ldots$ is a given sequence, the five series

$$\sum_{k=1}^{\infty} a_k, \quad \sum_{n=1}^{\infty} a_n, \quad \sum_{j=1}^{\infty} a_j, \quad \sum_{m=1}^{\infty} a_m, \quad \sum_{\alpha=1}^{\infty} a_\alpha$$

are all equal. Up to now we have been using the index k consistently. Henceforth we shall use any letter that seems suitable as a summation index. However, the letter n is the most popular one for this task.

As we have emphasized, the convergence of an infinite series $\sum_{n=1}^{\infty} a_n$ is determined by the behavior of its sequence of partial sums $\{S_n\}$. Thus any theorem about the convergence of sequences gives rise to a corresponding theorem concerning infinite series.

For example, the fact that the limit of the sum of two convergent sequences is the sum of the limits[1] easily gives

---

THEOREM 3. Convergent series can be added. If the two convergent series

$$S = \sum_{n=1}^{\infty} a_n \quad \text{and} \quad T = \sum_{n=1}^{\infty} b_n$$

have the sums indicated, then the series

$$\sum_{n=1}^{\infty} (a_n + b_n) = (a_1 + b_1) + (a_2 + b_2) + \cdots + (a_n + b_n) + \cdots$$

converges, and has the sum $S + T$.

---

[1] See Chapter 3, Section 4, Theorem 3, and Appendix 1.

Proof. Let S_n and T_n denote the sum of the first n terms of the two given series. Then by hypothesis $S_n \to S$ and $T_n \to T$ as $n \to \infty$. But

$$\sum_{k=1}^{n} (a_k + b_k) = (a_1 + b_1) + (a_2 + b_2) + \cdots + (a_n + b_n)$$

$$= (a_1 + a_2 + \cdots + a_n) + (b_1 + b_2 + \cdots + b_n) = S_n + T_n.$$

Thus from the limit properties of sequences, we have

$$\lim_{n \to \infty} \sum_{k=1}^{n} (a_k + b_k) = \lim_{n \to \infty} (S_n + T_n) = \lim_{n \to \infty} S_n + \lim_{n \to \infty} T_n = S + T. \quad \blacksquare$$

For example, by Theorem 3, we have

$$\sum_{n=0}^{\infty} \left(\frac{1}{2^n} + \frac{1}{3^n} \right) = \sum_{n=0}^{\infty} \frac{1}{2^n} + \sum_{n=0}^{n} \frac{1}{3^n} = \frac{1}{1 - \frac{1}{2}} + \frac{1}{1 - \frac{1}{3}} = 2 + \frac{3}{2} = \frac{7}{2}.$$

In a similar fashion, the reader can immediately prove

THEOREM 4 (PLE). A convergent series can be multiplied by a constant. If the convergent series $\displaystyle\sum_{n=1}^{\infty} a_n$ sums to S, then the series

$$\sum_{n=1}^{\infty} ca_n = ca_1 + ca_2 + \cdots + ca_n + \cdots$$

converges and has the sum cS.

The next result exploits the relationship between the partial sums and the individual terms of a series.

THEOREM 5. If the series $a_1 + a_2 + \cdots$ converges, then $a_n \to 0$ as $n \to \infty$. If a_n does not approach zero as $n \to \infty$, then the series diverges.

Proof. By definition, $S_n \to S$ and $S_{n-1} \to S$ as $n \to \infty$. Hence

$$\lim_{n \to \infty} (S_n - S_{n-1}) = \lim_{n \to \infty} S_n - \lim_{n \to \infty} S_{n-1} = S - S = 0.$$

But $S_n - S_{n-1} = (a_1 + a_2 + \cdots + a_n) - (a_1 + a_2 + \cdots + a_{n-1}) = a_n.$ Therefore, $a_n \to 0$ as $n \to \infty$.

The second part of the theorem follows immediately from the first part.   ∎

Thus at a glance we see that each of the series

(28)
$$\sum_{n=1}^{\infty} \sqrt{n}, \quad \sum_{n=1}^{\infty} \frac{n}{2n+1}, \quad \text{and} \quad \sum_{n=1}^{\infty} 2^{1/n}$$

diverges because the terms a_n do not tend to zero.

A word of warning is appropriate here. We have proved that if the series converges, then $a_n \to 0$. But the converse of this statement is *not* true. In other words, if the limit of the nth term of a series is zero, there is no guarantee that the series converges. This is illustrated in

Example 1. Show that the series

(29)
$$\sum_{n=1}^{\infty} \frac{1}{n} = 1 + \frac{1}{2} + \frac{1}{3} + \cdots + \frac{1}{n} + \cdots,$$

called the *harmonic series,* diverges even though the limit of the nth term is zero.

Solution. We will prove divergence by showing that the partial sums

$$S_n = 1 + \frac{1}{2} + \frac{1}{3} + \cdots + \frac{1}{n}$$

grow arbitrarily large. It is interesting to note that numerical calculation of the partial sums lends no insight into this behavior. In Table 1 we list the computed values (to six decimal places) of several of the partial sums, and while these sums are growing, it is by no means apparent that they tend to plus infinity.

TABLE 1

$S_{10} = 2.928968$	$S_{500} = 6.792823$
$S_{50} = 4.499205$	$S_{1000} = 7.485471$
$S_{100} = 5.187376$	$S_{5000} = 9.094509$

Nevertheless, we can give a proof of this fact by considering the particular partial sums $S_{10}, S_{100}, \ldots, S_{10^n}, \ldots$.

First, we note that S_{10} satisfies the inequality

(30) $S_{10} = 1 + \dfrac{1}{2} + \dfrac{1}{3} + \cdots + \dfrac{1}{10} \geqq 1 + \dfrac{1}{10} + \dfrac{1}{10} + \cdots + \dfrac{1}{10} = 1 + \dfrac{9}{10}.$

Next, since the last 90 terms of the partial sum S_{100} are each greater than or equal to $1/100$, we have

$$S_{100} = \sum_{n=1}^{10} \frac{1}{n} + \sum_{n=11}^{100} \frac{1}{n} \geqq S_{10} + \sum_{n=11}^{100} \frac{1}{100} = S_{10} + 90\left(\frac{1}{100}\right),$$

and so, using (30), we find that

(31) $$S_{100} \geqq S_{10} + \frac{9}{10} \geqq 1 + \frac{9}{10} + \frac{9}{10} = 1 + 2\left(\frac{9}{10}\right).$$

Similarly,

$$S_{1000} = \sum_{n=1}^{100} \frac{1}{n} + \sum_{n=101}^{1000} \frac{1}{n} \geqq S_{100} + \sum_{n=101}^{1000} \frac{1}{1000} \geqq S_{100} + 900\left(\frac{1}{1000}\right),$$

and hence from (31) we have

$$S_{1000} \geqq S_{100} + \frac{9}{10} \geqq 1 + 2\left(\frac{9}{10}\right) + \frac{9}{10} = 1 + 3\left(\frac{9}{10}\right).$$

Continuing in this manner, we find inductively that

(32) $$S_{10^n} \geqq 1 + n\left(\frac{9}{10}\right) \qquad \text{for each } n = 1, 2, 3, \ldots.$$

Since $1 + n(9/10) \to \infty$ and the partial sums S_n are increasing, we have that

$$\lim_{n \to \infty} S_n = \lim_{n \to \infty} S_{10^n} = \infty,$$

which means the harmonic series diverges. (Another proof of this fact is given in Section 5.) ●

If a series has nonnegative terms a_n for every $n \geqq 2$, then the sequence of partial sums

$$S_1 = a_1, \qquad S_2 = a_1 + a_2, \qquad S_3 = a_1 + a_2 + a_3, \qquad \cdots$$

is obviously increasing. So Theorem 1 (Section 2) gives the following very important result.

THEOREM 6.   If $a_n \geqq 0$ for $n \geqq 2$, and if the partial sums of the infinite series

$$\sum_{n=1}^{\infty} a_n = a_1 + a_2 + a_3 + \cdots$$

are bounded above, then the series converges.

Example 2. Show that if $0 \leq x < 1$, then the series $\sum\limits_{k=1}^{\infty} \dfrac{k}{k+5} x^k$ converges.

Solution. We apply Theorem 6. Each term of this series is obviously nonnegative. Further, since $k/(k+5) < 1$, the kth term is less than x^k. Thus

$$S_n = \sum_{k=1}^{n} \frac{k}{k+5} x^k \leq \sum_{k=0}^{n} x^k = \frac{1 - x^{n+1}}{1 - x} \leq \frac{1}{1 - x}.$$

Therefore, the partial sums are bounded above by $1/(1-x) \equiv B$. By Theorem 6, the infinite series converges. ●

We will learn in Section 7 that this series also converges for $-1 < x \leq 0$, and consequently it converges for all x in the interval $-1 < x < 1$.

THEOREM 7 (The Comparison Test). Suppose that for each positive integer n,

(33) $$0 \leq a_n \leq b_n.$$

(a) If $\sum b_n$ converges, then $\sum a_n$ converges.

(b) If $\sum a_n$ diverges, then $\sum b_n$ diverges.[1]

Proof. Assume that the series $\sum\limits_{n=1}^{\infty} b_n$ is convergent and let the sum be T. Then from (33) it follows that for each positive integer n, the partial sum

$$S_n = a_1 + a_2 + \cdots + a_n \leq b_1 + b_2 + \cdots + b_n \leq \sum_{n=1}^{\infty} b_n = T.$$

Hence the sequence of partial sums $\{S_n\}$ is bounded above by T. By hypothesis, $a_n \geq 0$. Hence by Theorem 6, the series with terms a_n converges. This proves **(a)**.

To prove **(b)**, assume that the series $\sum a_n$ diverges. There are only two possibilities for the series $\sum b_n$; namely it can converge or diverge. If the series $\sum b_n$ converges, then by part **(a)** of the theorem, $\sum a_n$ converges. But this contradicts the hypothesis that $\sum a_n$ diverges. Therefore, $\sum b_n$ cannot converge, and hence must diverge. ∎

[1] For convenience in printing we have written $\sum$ instead of $\sum\limits_{n=1}^{\infty}$.

Example 3. Does the series

(34)
$$\sum_{n=1}^{\infty} \frac{1}{\sqrt{n}}$$

converge or diverge?

Solution. For each $n = 1, 2, 3, \ldots$, we have $n \geq \sqrt{n}$, so

$$0 < \frac{1}{n} \leq \frac{1}{\sqrt{n}}.$$

Since the harmonic series $\sum_{n=1}^{\infty} 1/n$ diverges, so does the series (34), by part **(b)** of Theorem 7. ●

It is appropriate to remark at this point that the early terms in a series have no effect on the convergence or divergence of the series, although they do contribute to the sum. The first million or so terms can be quite "wild," but if the terms eventually "settle down and behave nicely," the series can still converge. Stated more precisely, the condition $0 \leq a_n \leq b_n$ in Theorem 7 need not be satisfied for all n. If $0 \leq a_n \leq b_n$ for all $n \geq N$, where N is some suitably selected integer, then the conclusions of the theorem are still true.

Example 4. Does the series

(35)
$$\sum_{n=1}^{\infty} \left(\frac{n + 47}{3n}\right)^n$$

converge or diverge?

Solution. For the quantity in parentheses, we have $(n + 47)/3n \to 1/3$ as $n \to \infty$. Thus if we let r_0 be a number such that $1/3 < r_0 < 1$ (say $r_0 = 2/3$), then the terms $(n + 47)/3n$ will eventually be less than r_0. That is,

$$\frac{n + 47}{3n} < r_0 \qquad \text{for } n \geq N,$$

where N is suitably selected. Consequently,

$$0 \leq \left(\frac{n + 47}{3n}\right)^n < r_0{}^n \qquad \text{for } n \geq N,$$

and since the geometric series $\sum\limits_{n=0}^{\infty} r_0{}^n$ converges (remember $r_0 < 1$), so does the series (35). ●

An important variation of Theorem 7 is given in

THEOREM 8 PLE (Limit Comparison Test). Suppose that for each integer $n > N$, we have $a_n > 0$ and $b_n > 0$, and suppose that

$$\lim_{n \to \infty} \frac{a_n}{b_n} = L \neq 0.$$

Then the two series $\sum b_n$ and $\sum a_n$ converge or diverge together.

This means that if either series converges, then the other one also converges, and if either diverges, then the other one also diverges.

Example 5. Does the series

$$\sum_{n=1}^{\infty} \frac{2}{5n + 14}$$

converge or diverge?

Solution. We compare the terms of the given series with those of the harmonic series. Setting

$$a_n \equiv \frac{2}{5n + 14} \qquad \text{and} \qquad b_n \equiv \frac{1}{n}$$

we find that

$$\lim_{n \to \infty} \frac{a_n}{b_n} = \lim_{n \to \infty} \frac{2/(5n + 14)}{1/n} = \lim_{n \to \infty} \frac{2n}{5n + 14} = \frac{2}{5}.$$

As this limiting value is a nonzero number, and the harmonic series diverges, Theorem 8 states that the given series also diverges. ●

EXERCISE 3

In Problems 1 through 9, use a comparison test (Theorems 7 or 8) to determine whether the given series converges or diverges.

1. $\sum_{n=1}^{\infty} \dfrac{5n}{n^2 + 4}$.

2. $\sum_{n=1}^{\infty} \dfrac{8n^3 + 7}{3n^4 + 9}$.

3. $\sum_{n=1}^{\infty} \left(\dfrac{1}{4}\right)^n \left(1 + \dfrac{7}{n}\right)$.

4. $\sum_{k=1}^{\infty} \dfrac{13}{6^k + 27}$.

5. $\sum_{k=1}^{\infty} \dfrac{1}{(2k)^k}$.

6. $\sum_{k=1}^{\infty} \dfrac{3 + \sin k}{7^k}$.

7. $\sum_{n=1}^{\infty} \left(\dfrac{n + 1}{2n}\right)^n$.

8. $\sum_{n=1}^{\infty} \dfrac{5}{\sqrt[3]{n}}$.

9. $\sum_{n=10}^{\infty} \sqrt{\dfrac{2n}{(n^2 + 1)(n - 4)}}$.

10. Use the Comparison Test (Theorem 7) to prove that the series $\sum_{n=1}^{\infty} 1/n^p$ diverges for each $p \leq 1$.

In Problems 11 through 22, determine whether the given series converges or diverges.

11. $\sum_{n=1}^{\infty} \dfrac{(0.84)^n}{n}$.

12. $\sum_{n=1}^{\infty} 5^{1/n}$.

13. $\sum_{n=1}^{\infty} \dfrac{n + 1}{3n + 4}$.

14. $\sum_{k=1}^{\infty} \dfrac{1}{k^k}$.

15. $\sum_{k=1}^{\infty} \dfrac{3^k}{5^k + 4}$.

16. $\sum_{k=1}^{\infty} \dfrac{7^k + 9}{5^k + 13}$.

17. $\sum_{n=1}^{\infty} \dfrac{8}{\sqrt[3]{n}}$.

18. $\sum_{n=3}^{\infty} \dfrac{5}{\sqrt{n - 2}}$.

19. $\sum_{n=1}^{\infty} \dfrac{n^2 + 4}{5n^3 + 9}$.

20. $\sum_{n=1}^{\infty} \left(\dfrac{n}{n + 3}\right) \left(\dfrac{3}{4}\right)^n$.

21. $\sum_{n=1}^{\infty} \left(\dfrac{n + 1}{2n + 3}\right)^n$.

22. $\sum_{n=1}^{\infty} \left(\dfrac{4 - 3n}{n + 1}\right)^n$.

*23. Prove Theorem 8. HINT: For all sufficiently large n,

$$\frac{1}{2} L b_n < a_n < 2 L b_n.$$

24. Prove that the series $\sum_{k=0}^{\infty} \dfrac{1}{k!}$ is convergent. HINT: First show that if $k \geq 4$, then $1/k! < 1/2^k$.

In Problems 25 through 28, prove that the given series converges for $0 \leq x < 1$.

25. $\displaystyle\sum_{k=1}^{\infty} \frac{x^k}{k}.$ **26.** $\displaystyle\sum_{n=1}^{\infty} \frac{5n+1}{n+3} x^n.$ ★**27.** $\displaystyle\sum_{k=1}^{\infty} \frac{3^k x^k}{k!}.$ ★**28.** $\displaystyle\sum_{k=1}^{\infty} \frac{\ln^2 k}{k} x^k.$

CALCULATOR PROBLEMS

C1. It was shown in Example 1 that the harmonic series diverges. The mathematician

Euler showed, in fact, that the partial sums $S_n = \displaystyle\sum_{k=1}^{n} 1/k$ grow large as fast as $\ln n$

grows. Specifically, Euler proved that there exists a constant γ (called *Euler's constant*) such that

$$\lim_{n \to \infty} (S_n - \ln n) = \gamma.$$

Use the values for S_n given in Table 1 to illustrate that this limit exists (although the convergence is quite slow) and obtain an approximate value for γ.

C2. If the partial sums of the harmonic series are computed on a calculator, eventually one would get a constant value for all successive partial sums. Explain why this happens.

Some Practical Tests for Convergence 5

So far in our work we have had only one type of series that is convergent, namely the geometric series. To use Theorem 7 or Theorem 8 more effectively, we need to enlarge our supply of convergent series. A very nice supply is provided below by Theorem 10 and its corollary. The next two theorems are the really useful ones for practical testing for convergence.

THEOREM 9 (The Ratio Test). Suppose that $a_n > 0$ for each n and that

(36) $$\lim_{n \to \infty} \frac{a_{n+1}}{a_n} = L.$$

(a) If $L < 1$, then the series $\displaystyle\sum_{n=1}^{\infty} a_n$ converges.

(b) If $L > 1$, then this series diverges.

(c) If $L = 1$, no conclusion can be drawn.

Proof. (a) Suppose first that $L < 1$. We can select a number R such that $L < R < 1$, and then for all n sufficiently large we have, from (36),

$$(37) \qquad \qquad \frac{a_{n+1}}{a_n} < R.$$

Suppose that N is some integer such that for all $n \geq N$ the inequality (37) holds. Then for each $k \geq N$,

$$a_k = \frac{a_k}{a_{k-1}} \cdot \frac{a_{k-1}}{a_{k-2}} \cdot \frac{a_{k-2}}{a_{k-3}} \cdots \frac{a_{N+1}}{a_N} \cdot a_N,$$

$$(38) \qquad \qquad a_k < R \cdot R \cdot R \cdots R \cdot a_N = R^{k-N} a_N.$$

Then, using (38), we see that

$$(39) \qquad \qquad \sum_{k=N}^{\infty} a_k = a_N + a_{N+1} + a_{N+2} + \cdots$$

is termwise less than

$$(40) \qquad \qquad \sum_{k=N}^{\infty} R^{k-N} a_N = a_N + a_N R + a_N R^2 + \cdots,$$

a convergent geometric series. This proves **(a)**, thanks to the Comparison Test.

(b) Suppose that $L > 1$. This time we select R such that $L > R > 1$. Then there is some N such that for all $n \geq N$ we have

$$\frac{a_{n+1}}{a_n} > R.$$

Notice that this is (37) with the inequality sign reversed. Then in (38) the inequality sign is also reversed. But $R > 1$ and hence we have

$$a_k > R^{k-N} a_N > a_N, \qquad k = N + 1, N + 2, \ldots.$$

Consequently, the general term cannot tend to zero, and by Theorem 5 the series is divergent.

(c) Consider the two series

$$(41) \qquad \qquad \sum_{n=1}^{\infty} \frac{1}{n} = 1 + \frac{1}{2} + \frac{1}{3} + \cdots + \frac{1}{n} + \cdots$$

and

(42)
$$\sum_{n=1}^{\infty} \frac{1}{n^2} = 1 + \frac{1}{4} + \frac{1}{9} + \cdots + \frac{1}{n^2} + \cdots.$$

The Ratio Test [equation (36)] applied to (41) gives

$$\lim_{n \to \infty} \frac{a_{n+1}}{a_n} = \lim_{n \to \infty} \frac{1/(n+1)}{1/n} = \lim_{n \to \infty} \frac{n}{n+1} = 1.$$

When applied to the series (42), the Ratio Test yields

$$\lim_{n \to \infty} \frac{a_{n+1}}{a_n} = \lim_{n \to \infty} \frac{1/(n+1)^2}{1/n^2} = \lim_{n \to \infty} \frac{n^2}{n^2 + 2n + 1} = 1.$$

We already know that the series (41) diverges and, as a corollary to the next theorem, we will see that the series in (42) converges. Hence, if $L = 1$, no conclusion can be drawn about the convergence of the series. ∎

Example 1. Does the series $\displaystyle\sum_{n=1}^{\infty} n^2 \left(\frac{3}{4}\right)^n$ converge?

Solution. Applying the Ratio Test, we find that

$$\lim_{n \to \infty} \frac{(n+1)^2 (3/4)^{n+1}}{n^2 (3/4)^n} = \lim_{n \to \infty} \frac{(n+1)^2}{n^2} \frac{3}{4} = \lim_{n \to \infty} \left(1 + \frac{2}{n} + \frac{1}{n^2}\right)\frac{3}{4} = \frac{3}{4} < 1.$$

Hence by Theorem 9, the given series converges. ●

THEOREM 10 (The Cauchy Integral Test). Suppose that $f(x)$ has the following properties:
(a) $f(n) = a_n$ for each integer $n > 0$.
(b) $f(x)$ is a decreasing function for $x \geq 1$.
(c) $f(x) > 0$, for $x \geq 1$.

Then the series $\displaystyle\sum_{n=1}^{\infty} a_n$ and the integral $\displaystyle\int_{1}^{\infty} f(x)\,dx$ converge or diverge together.

When the integral converges, the sum can be estimated by

(43)
$$\int_{1}^{\infty} f(x)\,dx \leq \sum_{n=1}^{\infty} a_n \leq a_1 + \int_{1}^{\infty} f(x)\,dx.$$

Proof. Since the function is decreasing, the area under the curve $y = f(x)$ between $x = k$

and $x = k + 1$ lies between the areas of two rectangles of unit width and height $f(k) = a_k$ and $f(k + 1) = a_{k+1}$ (see Fig. 2). In symbols,

$$(44) \qquad a_k \geqq \int_k^{k+1} f(x)\,dx \geqq a_{k+1}.$$

Forming the sum of n such terms $k = 1, 2, 3, \ldots, n$, we have

$$(45) \qquad a_1 + a_2 + \cdots + a_n \geqq \int_1^{n+1} f(x)\,dx \geqq a_2 + a_3 + \cdots + a_{n+1}.$$

This inequality is shown graphically in Fig. 3.

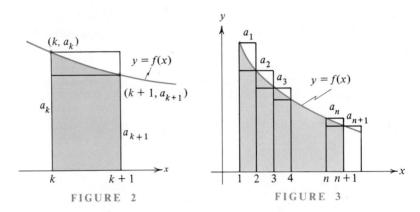

FIGURE 2   FIGURE 3

If the integral converges, then by the right side of (45) the partial sums are bounded, and hence by Theorem 6 the series converges. If the integral diverges, it must tend to infinity because $f(x)$ is positive. Then the left side of (45) shows that the series is also divergent. Finally, if the series converges, we can let $n \to \infty$ in (45). Then the left half of (45) gives the left half of (43). To obtain the right half of (43), just add a_1 to the middle and right side of (45) and let $n \to \infty$. ∎

> **COROLLARY.** The *p-series*
>
> $$(46) \qquad \sum_{n=1}^\infty \frac{1}{n^p} = 1 + \frac{1}{2^p} + \frac{1}{3^p} + \cdots + \frac{1}{n^p} + \cdots$$
>
> converges if $p > 1$, and diverges if $p \leqq 1$.

Proof. For $p > 0$, the function $f(x) = 1/x^p$ satisfies the three conditions of Theorem 10, so we only need to compute the integral. If $0 < p < 1$, then

(47)
$$\int_1^M f(x)\,dx = \int_1^M \frac{dx}{x^p} = \frac{x^{1-p}}{1-p}\Big|_1^M = \frac{M^{1-p}-1}{1-p},$$

and this quantity tends to infinity as $M \to \infty$. In this case the series diverges. The case $p = 1$ requires special treatment. In this case

$$\int_1^M f(x)\,dx = \int_1^M \frac{dx}{x} = \ln M,$$

and since $\ln M \to \infty$ as $M \to \infty$ the series again diverges.

If $p > 1$, then

$$\int_1^M \frac{dx}{x^p} = \frac{1}{(1-p)x^{p-1}}\Big|_1^M = \frac{1}{p-1}\left(1 - \frac{1}{M^{p-1}}\right).$$

This tends to $1/(p-1)$ as $M \to \infty$. In this case the series converges. Furthermore, by (43) we have the estimate

$$\frac{1}{p-1} \leqq \sum_{n=1}^{\infty} \frac{1}{n^p} \leqq 1 + \frac{1}{p-1}, \qquad p > 1.$$

If $p \leqq 0$, then $a_n = n^{|p|} \geqq 1$ for all n. Therefore, the series (46) diverges. ∎

Two special cases are worth noting. When $p = 1$ the p-series gives the harmonic series (41), so the Corollary furnishes another proof that this series diverges. When $p = 2$, we have the series (42) and hence that series converges. This completes the proof of Theorem 9.

Example 2. Does the series $\displaystyle\sum_{n=3}^{\infty} \frac{n}{(4+n^2)^{3/4}}$ converge?

Solution. We apply the integral test (Theorem 10), with $x \geqq 3$.

$$I_M = \int_3^M \frac{x\,dx}{(4+x^2)^{3/4}} = \frac{1}{2}\cdot\frac{(4+x^2)^{1/4}}{1/4}\Big|_3^M = 2\sqrt[4]{4+M^2} - 2\sqrt[4]{13}.$$

Since $I_M \to \infty$ as $M \to \infty$, the given series diverges. ●

Example 3. Does the series $\displaystyle\sum_{n=1}^{\infty} \frac{1}{\sqrt{n^3+1}}$ converge?

Solution. The student should try the Ratio Test on this series. He will find that the limit of the ratio is 1. More powerful methods are needed. The integral test is more powerful, but in this case the integration of $1/\sqrt{x^3 + 1}$ is difficult. But the function is approximately $1/\sqrt{x^3}$, so this suggests the use of the Comparison Test (Theorem 7). Indeed, $\sqrt{n^3 + 1} > \sqrt{n^3}$, and hence

$$\frac{1}{\sqrt{n^3 + 1}} < \frac{1}{\sqrt{n^3}} = \frac{1}{n^{3/2}}.$$

But on the right side we have the general term of the p-series with $p = 3/2$, and hence the series $\sum 1/n^{3/2}$ is convergent. Therefore, the given series is convergent. ●

The comparison tests are quite useful when used in conjunction with the p-series. Suppose that a_n is an algebraic function of n. By a careful inspection of a_n, we can usually guess at a power p such that $n^p a_n$ is essentially a constant $L > 0$, for large n. With this guess we then try to prove that

$$\lim_{n \to \infty} \frac{a_n}{1/n^p} = \lim_{n \to \infty} n^p a_n = L.$$

If $L > 0$ and $p > 1$, then by Theorem 8 and the convergence of the p-series, we see that $\sum_{k=1}^{\infty} a_k$ converges. If $L > 0$ and $p \leq 1$, then the series diverges.

Example 4*. Settle the convergence or divergence of the series

$$\sum_{n=1}^{\infty} \frac{\sqrt[3]{5n^2 + 6n + 7}}{n\sqrt[5]{n^4 + 11n - 3}}.$$

Solution. Examining the largest exponents on n, we observe that

$$\frac{2}{3} - 1 - \frac{4}{5} = -1 + \frac{10 - 12}{15} = -1 - \frac{2}{15} = -\frac{17}{15}.$$

We leave to the reader the labor of showing that

$$\lim_{n \to \infty} n^{17/15} a_n = \lim_{n \to \infty} n^{17/15} \frac{\sqrt[3]{5n^2 + 6n + 7}}{n\sqrt[5]{n^4 + 11n - 3}} = \sqrt[3]{5} > 0.$$

Since $17/15 > 1$ and $L = \sqrt[3]{5} > 0$, the given series converges. ●

EXERCISE 4

In Problems 1 through 6, determine whether the given series converges or diverges by using the Ratio Test, Theorem 9. (Use another method only if the test fails.)

1. $\displaystyle\sum_{n=1}^{\infty} \frac{2^n}{n!}$.

2. $\displaystyle\sum_{n=1}^{\infty} \frac{n+3}{n!}$.

3. $\displaystyle\sum_{n=1}^{\infty} \frac{5^n}{n^2 3^n}$.

4. $\displaystyle\sum_{n=1}^{\infty} \sqrt{n}\left(\frac{2}{3}\right)^n$.

5. $\displaystyle\sum_{n=1}^{\infty} \frac{n!}{(2n)!}$.

6. $\displaystyle\sum_{n=1}^{\infty} \frac{\sqrt{n}}{n+3}$.

In Problems 7 through 12, determine whether the given series converges or diverges by using the integral test, Theorem 10.

7. $\displaystyle\sum_{n=3}^{\infty} \frac{\text{Tan}^{-1} n}{n^2 + 1}$.

8. $\displaystyle\sum_{n=1}^{\infty} \frac{\ln n}{n}$.

9. $\displaystyle\sum_{n=2}^{\infty} \frac{1}{n \ln n}$.

10. $\displaystyle\sum_{k=1}^{\infty} \frac{1}{3k+1}$.

11. $\displaystyle\sum_{k=1}^{\infty} \text{Cot}^{-1} k$.

12. $\displaystyle\sum_{k=3}^{\infty} \frac{\ln k}{k^2}$.

In Problems 13 through 30, determine whether the given series converges or diverges. You may try any one of the tests discussed so far.

13. $\displaystyle\sum_{n=1}^{\infty} \frac{n^3}{2^n}$.

14. $\displaystyle\sum_{n=1}^{\infty} \frac{10^n}{n!}$.

15. $\displaystyle\sum_{n=4}^{\infty} \frac{n^2 + 2n - 9}{n^3 - 5n - 17}$.

16. $\displaystyle\sum_{n=5}^{\infty} \frac{43n + 51}{n^3 + n^2 - 11}$.

17. $\displaystyle\sum_{n=2}^{\infty} \frac{\sqrt{n}}{n^2 - 3}$.

18. $\displaystyle\sum_{n=1}^{\infty} \frac{n}{n^3 + 1}$.

19. $\displaystyle\sum_{n=2}^{\infty} \frac{3^n + n}{2^n - n}$.

20. $\displaystyle\sum_{n=2}^{\infty} \frac{1}{\ln n}$.

21. $\displaystyle\sum_{n=1}^{\infty} \frac{n^n}{n!}$.

22. $\displaystyle\sum_{n=2}^{\infty} \frac{(n+2)!}{(n-1)! 2^n}$.

23. $\displaystyle\sum_{n=1}^{\infty} \frac{(n!)^2 2^n}{(2n)!}$.

24. $\displaystyle\sum_{n=1}^{\infty} \frac{n^{1/3} + 6}{n^{5/4} + 13}$.

25. $\displaystyle\sum_{n=1}^{\infty} \frac{n^{1/4}}{n^{4/3} + 47}$.

26. $\displaystyle\sum_{n=1}^{\infty} \frac{|\sin n^3|}{n^2}$.

$\star$27. $\displaystyle\sum_{n=2}^{\infty} \frac{1}{n \ln^2 n}$.

$\star$28. $\displaystyle\sum_{n=1}^{\infty} \frac{\ln (n + 5)}{n \sqrt{7n - 3}}$.

$\star$29. $\displaystyle\sum_{n=1}^{\infty} \frac{1}{n^n \sqrt{n}}$.

$\star$30. $\displaystyle\sum_{n=1}^{\infty} \tan\left(\frac{\pi}{n^2}\right)$.

31. Show that the Ratio Test cannot be applied directly to the series

$$\sum_{n=1}^{\infty} \frac{1 + (-1)^{n+1}}{2^n} = 1 + 0 + \frac{1}{4} + 0 + \frac{1}{16} + 0 \cdots.$$

Prove that this series converges and has the sum 4/3.

32. Prove that $\dfrac{\pi}{4} < \displaystyle\sum_{k=1}^{\infty} \frac{1}{1 + k^2} < \dfrac{\pi}{4} + \dfrac{1}{2}$.

CALCULATOR PROBLEMS

C1. The mathematician Euler proved that for $p = 2$ and $p = 4$, the p-series have the sums

$$\sum_{n=1}^{\infty} \frac{1}{n^2} = \frac{\pi^2}{6}, \qquad \sum_{n=1}^{\infty} \frac{1}{n^4} = \frac{\pi^4}{90}.$$

Calculate the partial sums S_5, S_{10}, S_{15}, and S_{20} for each of these series and compare their values with the stated sums.

C2. If $\displaystyle\sum_{n=1}^{\infty} a_n$ is a convergent series with sum S, then the difference between S and the nth partial sum S_n is called the *remainder* or *truncation error* and is usually denoted by R_n (i.e., $R_n \equiv S - S_n$). Note that R_n is itself an infinite series, namely $R_n = \displaystyle\sum_{k=n+1}^{\infty} a_k$. Prove that for the p-series, with $p > 1$, we have the inequalities

$$\frac{1}{(p-1)(n+1)^{p-1}} \leqq R_n \leqq \frac{1}{(p-1)n^{p-1}}.$$

Illustrate these inequalities using your computed values from Problem C1 and the stated sums.

C3. For $p = 2$, the estimates for the truncation error found in Problem C2 give

$$\frac{1}{n+1} \leqq R_n = S - S_n \leqq \frac{1}{n}.$$

Let A_n be the average of these estimates and let B_n be half their difference; that is,

$$A_n = \frac{1}{2}\left(\frac{1}{n} + \frac{1}{n+1}\right) = \frac{2n+1}{2n^2 + 2n}, \qquad B_n = \frac{1}{2}\left(\frac{1}{n} - \frac{1}{n+1}\right) = \frac{1}{2n^2 + 2n}.$$

Prove that

$$|S - (S_n + A_n)| \leqq B_n = \frac{1}{2n^2 + 2n}.$$

Since $|S - S_n| \leq 1/n$ and $1/(2n^2 + 2n) < 1/n$, this suggests that the quantities $S_n + A_n$ may yield better approximations to the sum S than the partial sums S_n above. In other words, a more efficient way to approximate S is to compute S_n and add the *correction term* A_n. For this case when $p = 2$, compute $S_n + A_n$ for $n = 5, 10, 15$, and 20. Compare your answers with the actual sum $S = \pi^2/6$.

Alternating Series 6

In Section 5 all the terms of the series were greater than or equal to zero. We now allow negative terms to appear in the series in a regular way.

DEFINITION 5 (Alternating Series). A series of the form

(48) $$\sum_{n=1}^{\infty} (-1)^{n+1} a_n = a_1 - a_2 + a_3 - a_4 + a_5 - \cdots + (-1)^{n+1} a_n + \cdots$$

is called an *alternating series* if $a_n \geq 0$ for each integer $n > 0$.

A simple and beautiful test for the convergence of an alternating series is given in

---

THEOREM 11. If the terms of an alternating series (48) satisfy the following two conditions:

(a) The terms a_n are decreasing; that is, $a_1 \geq a_2 \geq a_3 \geq \cdots \geq a_n \geq \cdots$,
(b) The terms tend to zero; that is, $\lim_{n \to \infty} a_n = 0$,

then the series (48) converges. Further, if S is the sum of the series and S_n is the sum of the first n terms, then

(49) $$S_{2n} \leq S \leq S_{2n+1}, \qquad n = 1, 2, 3, \ldots.$$

Before proving this theorem, let us observe that the behavior of the partial sums is very similar to a swinging pendulum that is slowly coming to rest in a fixed position that is equivalent to the sum of the series. This is illustrated in Fig. 4. It may be helpful to keep this figure in view while reading the proof.

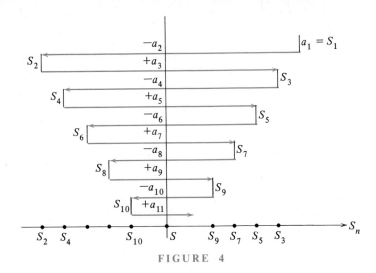

$$\text{FIGURE } 4$$

Proof. We first consider the partial sums with an even number of terms. These terms can be paired thus:

$$S_{2n} = a_1 - a_2 + a_3 - a_4 + \cdots + a_{2n-1} - a_{2n}$$
$$= (a_1 - a_2) + (a_3 - a_4) + \cdots + (a_{2n-1} - a_{2n}).$$

Since the individual terms are decreasing, the difference in each pair of parentheses gives a positive or zero result. More precisely,

$$a_{2n-1} - a_{2n} \geqq 0, \qquad n = 1, 2, 3, \ldots.$$

In going from S_{2n} to S_{2n+2}, we therefore add a quantity $a_{2n+1} - a_{2n+2}$ that is either positive or zero. Hence the sequence of partial sums with even subscripts is increasing, or in symbols

$$(50) \qquad\qquad S_2 \leqq S_4 \leqq S_6 \leqq \cdots \leqq S_{2n} \leqq S_{2n+2} \leqq \cdots.$$

For the partial sums with an odd number of terms, we may group the terms a little differently and write

$$S_{2n+1} = a_1 - a_2 + a_3 - a_4 + a_5 - \cdots - a_{2n} + a_{2n+1}$$
$$= a_1 - (a_2 - a_3) - (a_4 - a_5) - \cdots - (a_{2n} - a_{2n+1}).$$

Here again each grouped pair gives a positive or zero result, but now this result is to be *subtracted*. Hence the sequence of partial sums with odd subscripts is decreasing, or in symbols

$$(51) \qquad\qquad S_1 \geqq S_3 \geqq S_5 \geqq \cdots \geqq S_{2n-1} \geqq S_{2n+1} \geqq \cdots.$$

In order to combine the inequalities in (50) and (51), we observe that

$$(52) \qquad S_{2n+1} - S_{2n} = \sum_{k=1}^{2n+1} (-1)^{k+1} a_k - \sum_{k=1}^{2n} (-1)^{k+1} a_k$$

$$= (-1)^{2n+1+1} a_{2n+1} = a_{2n+1}.$$

But $a_{2n+1} \geq 0$, so $S_{2n} \leq S_{2n+1}$. Combining this with (50) and (51), we find that

$$S_2 \leq S_4 \leq S_6 \leq \cdots \leq S_{2n} \leq S_{2n+1} \leq \cdots \leq S_5 \leq S_3 \leq S_1.$$

Hence the sequence $S_2, S_4, \ldots$ is increasing and bounded above by S_1. So by Theorem 1 this sequence has a limit. Call it L. Similarly, the sequence $S_1, S_3, \ldots$ is decreasing and bounded below, so it has a limit which we call M. Now $a_n \to 0$ as $n \to \infty$, so using equation (52), we have

$$0 = \lim_{n \to \infty} a_{2n+1} = \lim_{n \to \infty} (S_{2n+1} - S_{2n}) = \lim_{n \to \infty} S_{2n+1} - \lim_{n \to \infty} S_{2n} = M - L.$$

Hence $M = L$. Both sequences converge to the same limit. Therefore, the series (48) converges and its sum S is the common limit. Since the one sequence is increasing to S and the other is decreasing to S, this also proves the inequality (49). ∎

It is evident from Fig. 4 and the preceding proof that for any alternating series satisfying the conditions of Theorem 11, we have

$$(53) \qquad |S - S_n| \leq a_{n+1}.$$

In other words, the error $|S - S_n|$ in approximating the sum S by the nth partial sum is at most the absolute value of the next term.

Example 1. Investigate the convergence of the series:

(a) $\displaystyle\sum_{n=1}^{\infty} \frac{(-1)^{n+1}}{n}$.   **(b)** $\displaystyle\sum_{n=1}^{\infty} \frac{(-1)^{n+1} n}{100 + 10n}$.   **(c)** $\displaystyle\sum_{n=0}^{\infty} \frac{\sin\left(\frac{\pi}{4} + \frac{n\pi}{2}\right)}{n+1}$.

Solution. The series in **(a)** is

$$(54) \qquad 1 - \frac{1}{2} + \frac{1}{3} - \frac{1}{4} + \cdots + \frac{1}{2n-1} - \frac{1}{2n} + \cdots.$$

This series is alternating and the general term $1/n$ decreases steadily to zero. Hence by Theorem 11, this series converges. Further, using the first three or four terms, (49) gives for the sum S, the bounds $0.5833 \ldots < S < 0.8333. \ldots$

(b) This series is alternating, but the absolute value of the general term, $n/(100 + 10n)$, tends to $1/10$ as $n \to \infty$. Hence, by Theorem 5, this series diverges.

(c) Each term has the common factor $\sqrt{2}/2$. Putting this out front, this series can be written

$$\frac{\sqrt{2}}{2}\left(1 + \frac{1}{2} - \frac{1}{3} - \frac{1}{4} + \frac{1}{5} + \frac{1}{6} - \frac{1}{7} - \frac{1}{8} + \cdots\right).$$

Now this series is not an alternating series, but if we pair the terms in an obvious way, we get an alternating series, namely,

$$\frac{\sqrt{2}}{2}\left(\frac{3}{2} - \frac{7}{12} + \frac{11}{30} - \frac{15}{56} + \cdots + \frac{(-1)^n(4n + 3)}{(2n + 1)(2n + 2)} + \cdots\right).$$

Since its terms satisfy the conditions of Theorem 11, this series converges. Consequently, the given series also converges, because its general term tends to zero. ●

★The general theory for grouping terms or rearranging terms in an infinite series is a little complicated and it is best to postpone this study until the advanced calculus course. Here are a few of the essential results.

A divergent series may become convergent when the terms are grouped. For example, the series

$$\sum_{n=0}^{\infty}(-1)^n = 1 - 1 + 1 - 1 + 1 - 1 + \cdots$$

is divergent, but when the terms are paired, we have

$$(1 - 1) + (1 - 1) + (1 - 1) + \cdots = 0 + 0 + 0 + \cdots,$$

a convergent series. However, if the general term of the original series tends to zero, and the new series obtained by grouping converges (where the number of terms in each group is less than some constant), then the original series also converges. This is the situation in Example 1(c).

When the terms of a series are rearranged (reordered), a convergent series can become divergent, or may remain convergent but have a different sum.

In the next section we discuss series that are *absolutely convergent*. It turns out that if a series is absolutely convergent, the pathological behavior described above cannot occur. Thus if the terms are rearranged in an absolutely convergent series, the new series is convergent and has the same sum as the original series.

EXERCISE 5

In Problems 1 through 15, determine whether the given series converges or diverges.

1. $\displaystyle\sum_{n=1}^{\infty} \frac{(-1)^{n+1}}{\sqrt{n}}$.

2. $\displaystyle\sum_{n=3}^{\infty} \frac{(-1)^{n}}{\sqrt[3]{n^2}}$.

3. $\displaystyle\sum_{n=1}^{\infty} (-1)^{n} \ln \frac{1}{n}$.

4. $\displaystyle\sum_{n=1}^{\infty} (-1)^{n} \sin n\pi$.

5. $\displaystyle\sum_{n=0}^{\infty} (-1)^{n} \sin \frac{\pi}{n+1}$.

6. $\displaystyle\sum_{n=2}^{\infty} \frac{(-1)^{n} \ln n}{n}$.

7. $\displaystyle\sum_{n=2}^{\infty} \frac{(-1)^{n} \sqrt[4]{n}}{\ln n}$.

8. $\displaystyle\sum_{n=2}^{\infty} \frac{(-1)^{n} 2^{n}}{n^{10}}$.

9. $\displaystyle\sum_{n=1}^{\infty} \frac{(-1)^{n} 3^{n}}{n!}$.

10. $\displaystyle\sum_{n=2}^{\infty} \frac{(-1)^{n}}{\ln n}$.

11. $\displaystyle\sum_{n=1}^{\infty} (-1)^{n} \ln \sqrt[n]{n}$.

12. $\displaystyle\sum_{n=1}^{\infty} \frac{(-1)^{n+1} n!}{2^{n}}$.

⋆13. $\displaystyle\sum_{n=1}^{\infty} \frac{(-1)^{n} \cos n\pi}{n}$.

⋆14. $\displaystyle\sum_{n=1}^{\infty} \frac{(-1)^{n} \ln^3 n^3}{n}$.

⋆15. $\displaystyle\sum_{n=7}^{\infty} \frac{(-1)^{n} \sqrt[n]{n+100}}{n-5}$.

CALCULATOR PROBLEMS

For each of the alternating series given in Problems C1 through C4, find the sum S to within ± 0.005 by computing the partial sum S_n, where n is the first positive integer for which $a_{n+1} < 0.005$. [According to inequality (53), this will give $|S - S_n| < 0.005$.]

C1. $\displaystyle\sum_{n=1}^{\infty} \frac{(-1)^{n+1}}{n 3^{n}}$.

C2. $\displaystyle\sum_{n=1}^{\infty} \frac{(-1)^{n+1}}{n^2 4^{n}}$.

C3. $\displaystyle\sum_{n=1}^{\infty} \frac{(-1)^{n+1}}{n!(1.03)^{n}}$.

C4. $\displaystyle\sum_{n=1}^{\infty} \frac{(-1)^{n+1}}{(n!)^2}$.

Absolute and Conditional Convergence 7

We have seen in Sections 4 and 5 that the harmonic series

(41)
$$1 + \frac{1}{2} + \frac{1}{3} + \frac{1}{4} + \cdots + \frac{1}{n} + \cdots$$

diverges. However, if we alter this series by subtracting instead of adding the terms $1/2n$, we have the series

(54)
$$1 - \frac{1}{2} + \frac{1}{3} - \frac{1}{4} + \cdots + \frac{(-1)^{n+1}}{n} + \cdots,$$

which converges (by Theorem 11). Thus a series [such as (54)] may converge, but when each term is replaced by its absolute value the new series may diverge. This suggests

---

DEFINITION 6 (Absolute Convergence). The series $\displaystyle\sum_{n=1}^{\infty} a_n$ is said to be *absolutely convergent* if the series formed by using the absolute values of the terms,

(55)
$$\sum_{n=1}^{\infty} |a_n| = |a_1| + |a_2| + \cdots + |a_n| + \cdots,$$

is convergent.

It can happen that a series is convergent and yet not be absolutely convergent. The prize example is the series (54). It is a convergent series, but when absolute values are taken, (54) gives the divergent series (41). Such a series is said to be *conditionally convergent*.

DEFINITION 7 (Conditional Convergence). If a series

(56)
$$\sum_{n=1}^{\infty} a_n = a_1 + a_2 + \cdots + a_n + \cdots$$

is convergent, and if its series of absolute values (55) diverges, then the series is said to be conditionally convergent.

A central result about absolutely convergent series is

THEOREM 12. An absolutely convergent series is convergent.

This states that if (55) converges, then so does (56).

Proof. Let p_n denote the positive and zero terms and let q_n denote the magnitude of the negative terms in (56). More precisely, set

(57)
$$\begin{cases} p_n = a_n \\ q_n = 0 \end{cases} \text{ if } a_n \geqq 0, \qquad \begin{cases} p_n = 0 \\ q_n = -a_n > 0 \end{cases} \text{ if } a_n < 0,$$

and consider the two new series

$$\sum_{n=1}^{\infty} p_n \qquad \text{and} \qquad \sum_{n=1}^{\infty} q_n.$$

For example, if our series is the series (54), then by (57), the two new series are

$$\sum_{n=1}^{\infty} p_n = 1 + 0 + \frac{1}{3} + 0 + \frac{1}{5} + \cdots + 0 + \frac{1}{2n+1} + 0 + \cdots$$

and

$$\sum_{n=1}^{\infty} q_n = 0 + \frac{1}{2} + 0 + \frac{1}{4} + 0 + \cdots + \frac{1}{2n} + 0 + \frac{1}{2n+2} + \cdots.$$

From the definition of p_n and q_n in (57) it is clear that

(58)
$$p_n + q_n = |a_n| \qquad \text{and} \qquad p_n - q_n = a_n,$$

for $n = 1, 2, \ldots$. Let A_n, S_n, P_n, and Q_n denote the sum of the first n terms of the series $\sum |a_n|$, $\sum a_n$, $\sum p_n$, and $\sum q_n$, respectively. Then from (58), we have

(59)
$$P_n + Q_n = (p_1 + q_1) + (p_2 + q_2) + \cdots + (p_n + q_n)$$
$$= |a_1| + |a_2| + \cdots + |a_n| = A_n$$

and

(60)
$$P_n - Q_n = (p_1 - q_1) + (p_2 - q_2) + \cdots + (p_n - q_n)$$
$$= a_1 + a_2 + \cdots + a_n = S_n.$$

Since the given series is absolutely convergent, the partial sums A_n form an increasing sequence that converges to a limit A. By (59), we have $P_n \leqq A_n \leqq A$ and $Q_n \leqq A_n \leqq A$ for $n = 1, 2, \ldots$. But the sequence of partial sums $P_1, P_2, \ldots$ is increasing and since it is bounded above, it converges to a limit, which we denote by P. Similarly, the sequence Q_1, $Q_2, \ldots$ is increasing and bounded above, so it has a limit Q. Finally, from (60),

$$\lim_{n \to \infty} S_n = \lim_{n \to \infty} (P_n - Q_n) = \lim_{n \to \infty} P_n - \lim_{n \to \infty} Q_n = P - Q.$$

Hence the given series converges and its sum is $P - Q$. ∎

Note that this theorem is completely reasonable because it states that when a series is absolutely convergent, the sum of the series is $P - Q$ (the sum of its positive terms minus the sum of the absolute values of its negative terms).

This theorem is also a comfortable one, for it allows us to "throw away the negative signs" in a preliminary investigation, because whenever the new series is convergent, then the one with the "negative signs restored" is also convergent.

Example 1. Find the convergence set for the series

(61)
$$\sum_{n=1}^{\infty} \frac{(-1)^{n+1}x^n}{n} = x - \frac{x^2}{2} + \frac{x^3}{3} - \frac{x^4}{4} + \cdots.$$

Solution. We are to find all values of x for which this series converges. We first consider the absolute convergence; that is, we consider

(62)
$$\sum_{n=1}^{\infty} \frac{|x|^n}{n} = |x| + \frac{|x|^2}{2} + \frac{|x|^3}{3} + \frac{|x|^4}{4} + \cdots.$$

Applying the Ratio Test, we have

$$\lim_{n \to \infty} \frac{a_{n+1}}{a_n} = \lim_{n \to \infty} \frac{|x|^{n+1}}{n+1} \frac{|x|^n}{n} = \lim_{n \to \infty} \frac{|x|^{n+1}}{n+1} \cdot \frac{n}{|x|^n} = |x|.$$

Therefore, if $-1 < x < 1$, the series converges absolutely and hence it converges.

Now suppose that $|x| > 1$. Then by the Ratio Test the series (62) of absolute values diverges. But what about our original series (61)? Could it be that (61) converges conditionally for $|x| > 1$? The answer is *no* and the reason is contained in the proof of part (b) of the Ratio Test (page 590). In that proof we showed that whenever the limit of a_{n+1}/a_n is greater than 1, the series $\sum a_n$ diverges because *its terms do not tend to zero*. But if the absolute values $a_n = |x|^n/n$ do not tend to zero, the same must be true of the original terms $(-1)^{n+1}x^n/n$. Hence the series (61) diverges for $|x| > 1$.

All that remains to consider are the end points of the interval $-1 < x < 1$. When $x = 1$, the series (61) becomes the series (54), which we already know converges. When $x = -1$, the series (61) becomes the negative of the harmonic series (41) and hence diverges. Summarizing, the given series converges for x in the half-open interval $-1 < x \leq 1$ and diverges for all other real values of x. ●

Example 2. Find the convergence set for the series

(63)
$$\sum_{n=1}^{\infty} \frac{3^n}{n^2}(x - 1)^n.$$

Solution. Again we test for absolute convergence by applying the Ratio Test. Setting $a_n = 3^n|x - 1|^n/n^2$, we find that

$$\lim_{n \to \infty} \frac{a_{n+1}}{a_n} = \lim_{n \to \infty} \frac{[3^{n+1}|x - 1|^{n+1}/(n + 1)^2]}{[3^n|x - 1|^n/n^2]} = \lim_{n \to \infty} 3|x - 1| \left(\frac{n}{n + 1}\right)^2 = 3|x - 1|.$$

Hence the series (63) converges absolutely if $3|x - 1| < 1$, that is, if

$$|x - 1| < \frac{1}{3} \quad \text{or} \quad \frac{2}{3} < x < \frac{4}{3}.$$

Furthermore, as in the previous example, the Ratio Test gives divergence of (63) for $|x - 1| > 1/3$.

Finally we consider the end points. When $x = 4/3$, the series (63) becomes

$$\sum_{n=1}^{\infty} \frac{3^n}{n^2} \left(\frac{4}{3} - 1\right)^n = \sum_{n=1}^{\infty} \frac{3^n}{n^2} \cdot \frac{1}{3^n} = \sum_{n=1}^{\infty} \frac{1}{n^2},$$

which we know converges. When $x = 2/3$, we obtain

$$\sum_{n=1}^{\infty} \frac{3^n}{n^2} \left(\frac{2}{3} - 1\right)^n = \sum_{n=1}^{\infty} \frac{3^n}{n^2} \left(-\frac{1}{3}\right)^n = \sum_{n=1}^{\infty} \frac{(-1)^n}{n^2},$$

which converges absolutely. Summarizing (or should we say "summing up"), we find that the series (63) converges absolutely for $2/3 \leq x \leq 4/3$ and diverges for all other real values of x. •

Example 3. Find the convergence set for the series

(64)
$$\sum_{n=1}^{\infty} \frac{(-1)^{n+1}}{n} \left(\frac{x}{3x + 8}\right)^n.$$

Solution. To simplify matters, we replace the quantity $x/(3x + 8)$ by t and consider the series

(65)
$$\sum_{n=1}^{\infty} \frac{(-1)^{n+1}}{n} t^n.$$

We recognize this as being the same series as (61) (with x replaced by t) and hence (65) converges for $-1 < t \leq 1$. Therefore, the convergence set for the given series (64) is the set of x for which

$$-1 < t = \frac{x}{3x + 8} \leq 1.$$

We leave it to the reader to show that solving these inequalities for x gives $x \leqq -4$ or $x > -2$. Hence the desired convergence set is $(-\infty, -4] \cup (-2, \infty)$, as depicted in Fig. 5. ●

FIGURE 5

EXERCISE 6

In Problems 1 through 14, find the convergence set for the given series. Observe that in each of these problems the convergence set is an interval. Be sure to test the series at the end points of the interval.

1. $\displaystyle\sum_{n=1}^{\infty} nx^n$.

2. $\displaystyle\sum_{n=1}^{\infty} \frac{2^n x^n}{n^2}$.

3. $\displaystyle\sum_{n=1}^{\infty} \frac{x^n}{n}$.

4. $\displaystyle\sum_{n=3}^{\infty} \frac{(-1)^n x^n}{n5^n}$.

5. $\displaystyle\sum_{n=0}^{\infty} \frac{x^n}{n!}$.

6. $\displaystyle\sum_{n=0}^{\infty} \frac{(-1)^n x^{2n+1}}{(2n+1)!}$.

7. $\displaystyle\sum_{n=0}^{\infty} n! x^n$.

8. $\displaystyle\sum_{n=1}^{\infty} \sqrt{n}(x-3)^n$.

9. $\displaystyle\sum_{n=2}^{\infty} \frac{(x+5)^n}{2^n \ln n}$.

10. $\displaystyle\sum_{n=2}^{\infty} \frac{2^n (x-4)^n}{n \ln n}$.

11. $\displaystyle\sum_{n=1}^{\infty} \frac{(x-2)^n}{3n+1}$.

12. $\displaystyle\sum_{n=1}^{\infty} \frac{(2x+11)^n}{3^n(2n-1)}$.

13. $\displaystyle\sum_{n=1}^{\infty} \frac{(-1)^n (3x+2)^n}{5^n n \sqrt{n}}$.

14. $\displaystyle\sum_{n=1}^{\infty} \frac{105 x^n}{n(n+3)}$.

In Problems 15 through 26, find the convergence set for the given series.

15. $\displaystyle\sum_{n=1}^{\infty} \frac{n}{x^n}$.

16. $\displaystyle\sum_{n=1}^{\infty} \frac{1}{nx^n}$.

17. $\displaystyle\sum_{n=2}^{\infty} \frac{1}{n^2} \left(\frac{3x-18}{x-2}\right)^n$.

18. $\displaystyle\sum_{n=0}^{\infty} n\left(\frac{3x-15}{x+3}\right)^n$.

19. $\displaystyle\sum_{n=1}^{\infty} \frac{3^n}{n} \left(\frac{x-3}{x+13}\right)^n$.

20. $\displaystyle\sum_{n=0}^{\infty} \left(\frac{3x-17}{2x+5}\right)^n$.

21. $\displaystyle\sum_{n=2}^{\infty} \frac{\ln n}{n} \left(\frac{2x+3}{3x+2}\right)^n$.

22. $\displaystyle\sum_{n=0}^{\infty} \left(\frac{x+5}{x-3}\right)^n$.

★23. $\displaystyle\sum_{n=1}^{\infty} \frac{(x-n)^n}{n}$.

$\star$**24.** $\displaystyle\sum_{n=1}^{\infty} \frac{x^{n!}}{n}$.

$\star$**25.** $\displaystyle\sum_{n=1}^{\infty} \frac{(-1)^n}{n + x^2}$.

$\star$**26.** $\displaystyle\sum_{n=1}^{\infty} \frac{1}{n^2} \sin n^3 x$.

$\star$**27.** Suppose that in the series $\sum a_n x^n$, we have $\lim |a_{n+1}|/|a_n| = L$ as $n \to \infty$. Prove that the series converges for $|x| < 1/L$.

$\star$**28.** Prove that if we differentiate term by term the series of Problem 27, then the new series $\sum n a_n x^{n-1}$ also converges in the same interval, $|x| < 1/L$.

$\star$**29.** Find the convergence set for $\displaystyle\sum_{n=1}^{\infty} \frac{(x^2 - 5)^n}{4^n}$ and the sum of the series.

30. Prove that $\lim_{n \to \infty} x^n/n! = 0$ for each fixed x no matter how large, by observing that the series of Problem 5 converges for all x.

$\star$**31.** Find the convergence set for $\displaystyle\sum_{n=1}^{\infty} \frac{x^n}{n + x^{2n}}$.

$\star\star$**32.** Prove that by rearranging the terms of a conditionally convergent series, we can obtain a new series that has for its sum any number S that we wish. HINT: Use the notation in the proof of Theorem 12. If (56) is *not* absolutely convergent, then $P_n \to \infty$ and $Q_n \to \infty$ as $n \to \infty$. Select terms from the set $\mathscr{P} \equiv \{p_1, p_2, \ldots\}$ so that the sum exceeds S. Then add terms from the set $\mathscr{Q} \equiv \{-q_1, -q_2, \ldots\}$ until the sum falls below S. Continue alternating between the sets $\mathscr{P}$ and $\mathscr{Q}$. If done properly, the new series will be a rearrangement of (56) and will have the sum S.

Power Series and Taylor Series 8

We now consider a special type of series whose terms are functions of x.

DEFINITION 8 (Power Series). A power series centered at $x = a$ is an infinite series of the form

(66) $$\sum_{n=0}^{\infty} a_n(x - a)^n = a_0 + a_1(x - a) + a_2(x - a)^2 + \cdots,$$

where x is a variable and the a_n are constants.[1]

---

[1] Notational difficulty arises in the first term $a_0(x - a)^0$ of (66) when $x = a$ for this term becomes $a_0 0^0$ and 0^0 is undefined. However, when $x \neq a$, we have $a_0(x - a)^0 = a_0$ and for the sake of consistency we assign the value a_0 to the first term even when $x = a$.

For example,

$$\sum_{n=0}^{\infty} 2^n(x-5)^n, \qquad \sum_{n=0}^{\infty} \frac{(x+3)^n}{n!}, \qquad \text{and} \qquad \sum_{n=0}^{\infty} \frac{(-1)^n}{n+1} x^n$$

are power series centered at 5, -3, and 0, respectively.

Observe that every power series converges at its center $x = a$, for then the series (66) is just

$$\sum_{n=0}^{\infty} a_n(a-a)^n = a_0 + 0 + 0 + 0 + \cdots = a_0.$$

But what about other values of x? The use of the Ratio Test in the examples and exercises of the previous section suggest that the convergence set of a power series is an interval centered about $x = a$. The next theorem states that this is always the case.

THEOREM 13. For each power series (66), there is a number $R \geq 0$ called the *radius of convergence* of the series, such that if $|x - a| < R$, then the series converges absolutely; and if $|x - a| > R$, the series diverges.

In other words, the series converges in the open interval $a - R < x < a + R$ and diverges outside the closed interval $a - R \leq x \leq a + R$. We remark that we may have $R = 0$, and then the series converges only at $x = a$ (see Problem 7 of Exercise 6). Or it may be that $R = \infty$, and then the series converges for all x (see Problems 5 and 6 of Exercise 6). Further, the behavior of the series at $x = a + R$ and $x = a - R$ (the end points of the interval of convergence) follows no general pattern, as Problems 1, 2, 3, and 4 of Exercise 6 clearly illustrate.

$\star$*Proof.* We first prove Theorem 13 for power series of the form

$$(67) \qquad \qquad \sum_{n=0}^{\infty} a_n x^n,$$

which are centered about the origin. The series (67) always converges when $x = 0$. If it diverges for all other values of x, then $R = 0$.

We can suppose now that the series (67) converges not only at $x = 0$ but for some other value of x, let us call it x_0. Then $a_n x_0^n \to 0$ as $n \to \infty$. Let x_1 be any number such that $|x_1| < |x_0|$. We can write

$$(68) \qquad \qquad \sum_{n=0}^{\infty} |a_n x_1^n| = \sum_{n=0}^{\infty} |a_n| \, |x_0|^n \frac{|x_1|^n}{|x_0|^n} = \sum_{n=0}^{\infty} A_n r^n,$$

where $A_n \equiv |a_n| \, |x_0|^n$ and $r \equiv |x_1|/|x_0| < 1$. Since $A_n \to 0$ as $n \to \infty$, these numbers have a bound M. Hence the series (68) is term by term less than the terms of the geometric series $\sum Mr^n$ with $r < 1$. By the Comparison Test, the series (68) converges. Thus (67) is absolutely convergent when $x = x_1$ and hence is convergent for $x = x_1$.

This proves that if the series converges for $x = x_0$, then it converges for all x in the interval $-|x_0| < x < |x_0|$. If the series diverges for all x outside this interval, then $|x_0|$ is the value for R mentioned in the theorem. Otherwise, we can find another x, say x_0', outside of this interval for which (67) converges. Then the same argument shows that the series (67) converges for all x in the larger interval $-|x_0'| < x < |x_0'|$.

Now consider the collection of all numbers ρ with the property that the series (67) converges for x in the interval $-\rho < x < \rho$. If this set contains all positive numbers, then the series (67) converges for all x and the radius of convergence of the series is infinite ($R = \infty$). If this set does not contain all the positive numbers, let R be the least upper bound of the set. Then R has just the properties ascribed to it in the theorem for the case $a = 0$.

Finally, for the general power series (66), we merely replace $x - a$ by t and obtain the power series $\sum a_n t^n$, which, from the first part of the proof, converges if $|t| < R$ and diverges if $|t| > R$. Then (66) converges if $|x - a| < R$ and diverges if $|x - a| > R$. ∎

Now for each x for which the series (66) converges, we get a number that is the sum of the series. It is appropriate to denote this sum by $f(x)$ since its value depends on the choice of x. Thus we write

$$f(x) = \sum_{n=0}^{\infty} a_n (x - a)^n$$

for all numbers x in the convergence interval. In particular, the power series $\sum_{n=0}^{\infty} x^n$ is a geometric series (with r replaced by x), so we know that it has the radius of convergence $R = 1$ and the sum function $f(x) = 1/(1 - x)$. That is,

$$(69) \qquad \frac{1}{1 - x} = \sum_{n=0}^{\infty} x^n \qquad \text{for } -1 < x < 1.$$

Now let us reverse the picture. Suppose that we start with a function $f(x)$ and wish to represent it by a power series about $x = a$. How should we choose the coefficients a_n so that (hopefully) we obtain the equation

$$(70) \qquad f(x) = a_0 + a_1(x - a) + a_2(x - a)^2 + a_3(x - a)^3 + a_4(x - a)^4 + \cdots?$$

To answer this, let us assume that (70) holds in an interval about $x = a$. Furthermore, we suppose that all derivatives of f exist and that term-by-term differentiation of (70) is valid in this interval. Then, if we differentiate both sides of (70), we obtain

(71) $$f'(x) = a_1 + 2a_2(x - a) + 3a_3(x - a)^2 + 4a_4(x - a)^3 + \cdots.$$

Differentiating (71) gives

(72) $$f''(x) = 1 \cdot 2a_2 + 2 \cdot 3a_3(x - a) + 3 \cdot 4a_4(x - a)^2 + 4 \cdot 5(x - a)^3 + \cdots,$$

and differentiating (72), we have

(73) $$f^{(3)}(x) = 1 \cdot 2 \cdot 3a_3 + 2 \cdot 3 \cdot 4a_4(x - a) + 3 \cdot 4 \cdot 5(x - a)^2 + \cdots,$$

and so on.

Now when we put $x = a$ in (70), all the terms of the series disappear except for the first, and we find that $f(a) = a_0$. When we put $x = a$ in equation (71), the only surviving term is a_1, and hence $f'(a) = a_1$. Similarly, by putting $x = a$ in (72), (73), and the subsequent equations, we find that

$$f''(a) = 1 \cdot 2a_2, \qquad f^{(3)}(a) = 1 \cdot 2 \cdot 3a_3,$$

and so on.

Thus

$$a_0 = f(a), \qquad a_1 = f'(a), \qquad a_2 = \frac{f''(a)}{2!}, \qquad a_3 = \frac{f^{(3)}(a)}{3!}, \qquad \cdots,$$

and so in general we have the formula

(74) $$a_n = \frac{f^{(n)}(a)}{n!}.$$

This suggests

DEFINITION 9 (Taylor and Maclaurin Series). If all the derivatives of a function f exist at $x = a$, then the series[1]

(75) $$\sum_{n=0}^{\infty} \frac{f^{(n)}(a)}{n!}(x - a)^n = f(a) + f'(a)(x - a)$$

$$+ \frac{f''(a)}{2!}(x - a)^2 + \frac{f^{(3)}(a)}{3!}(x - a)^3 + \cdots$$

is called the *Taylor series* for f about a. When $a = 0$, the series (75) becomes

---

[1] For notational consistency, we set $0! = 1$ and $f^{(0)}(x) = f(x)$.

(76)
$$\sum_{n=0}^{\infty} \frac{f^{(n)}(0)}{n!} x^n$$

and is called the *Maclaurin series* for f.

At this point we would like to write

(77)
$$f(x) = \sum_{n=0}^{\infty} \frac{f^{(n)}(a)}{n!} (x - a)^n.$$

Certainly, this is true for $x = a$, but for other values of x we have derived (77) only under certain assumptions on the existence and behavior of a power series expansion for f in an interval around a. Nevertheless, for most of the functions f encountered in this text, such as e^x, $\ln x$, $\sin x$, and $\cos x$, equation (77) is indeed valid for all x within the open interval of convergence of the Taylor series.[1] In the next sections we will treat this important question more carefully. Our purpose now is to give several examples of Taylor series, and in so doing we shall write (77) without any proof of when it is valid.

Example 1. Find the Maclaurin series expansion for e^x.

Solution. Computing the successive derivatives of e^x, we have

$$
\begin{aligned}
f(x) &= e^x & f(0) &= e^0 = 1 \\
f'(x) &= e^x & f'(0) &= e^0 = 1 \\
&\;\;\vdots & &\;\;\vdots \\
f^{(n)}(x) &= e^x & f^{(n)}(0) &= e^0 = 1.
\end{aligned}
$$

Thus $f^{(n)}(0)/n! = 1/n!$ for $n = 0, 1, 2, \ldots$, and the Maclaurin expansion is

(78)
$$e^x = \sum_{n=0}^{\infty} \frac{x^n}{n!} = 1 + x + \frac{x^2}{2!} + \frac{x^3}{3!} + \cdots + \frac{x^n}{n!} + \cdots. \qquad \bullet$$

If we apply the Ratio Test to the power series (78), we find that

$$\lim_{n\to\infty} \frac{[|x|^{n+1}/(n+1)!]}{[|x|^n/n!]} = \lim_{n\to\infty} \frac{|x|n!}{(n+1)!} = \lim_{n\to\infty} \frac{|x|}{n+1} = 0,$$

so the series $\sum_{n=0}^{\infty} x^n/n!$ converges for all x.

[1] There are pathological examples for which a Taylor series generated by a function f sums to a function quite different from f (see Problem 33 of Exercise 7).

Example 2. Find the Taylor series for $\ln x$ about $x = 1$.

Solution. The successive derivatives are

$$
\begin{aligned}
f(x) &= \ln x & f(1) &= 0 \\
f'(x) &= 1/x & f'(1) &= 1 \\
f''(x) &= -1/x^2 & f''(1) &= -1 \\
f^{(3)}(x) &= 2 \cdot 1/x^3 & f^{(3)}(1) &= 2! \\
f^{(4)}(x) &= -3 \cdot 2 \cdot 1/x^4 & f^{(4)}(1) &= -3! \\
f^{(5)}(x) &= 4 \cdot 3 \cdot 2 \cdot 1/x^5 & f^{(5)}(1) &= 4!,
\end{aligned}
$$

and so on. Here a pattern becomes evident after computing a few derivatives, and it is easy to guess (and one can prove by induction) that

$$
f^{(n)}(1) = (-1)^{n+1}(n-1)! \qquad \text{for } n = 1, 2, 3, \ldots.
$$

Hence the Taylor coefficients are

$$
\frac{f^{(0)}(1)}{0!} = f(1) = 0 \quad \text{and} \quad \frac{f^{(n)}(1)}{n!} = (-1)^{n+1}\frac{(n-1)!}{n!} = \frac{(-1)^{n+1}}{n} \qquad \text{for } n \geq 1,
$$

and the expansion for $\ln x$ about $x = 1$ is given by

$$
(79) \qquad \ln x = \sum_{n=1}^{\infty} \frac{(-1)^{n+1}}{n}(x-1)^n
$$

$$
= (x-1) - \frac{(x-1)^2}{2} + \frac{(x-1)^3}{3} - \frac{(x-1)^4}{4} + \cdots. \qquad \bullet
$$

Note that the series in (79) is essentially the same as the series (61) considered in Example 1 of Section 7, except that $x - 1$ appears in place of x. Thus the series (79) converges for $-1 < x - 1 \leq 1$, that is, for $0 < x \leq 2$.

Example 3. Find the Maclaurin expansion for $\sin x$.

Solution. Here

$$
\begin{aligned}
f(x) &= \sin x & f(0) &= 0 \\
f'(x) &= \cos x & f'(0) &= 1 \\
f''(x) &= -\sin x & f''(0) &= 0 \\
f^{(3)}(x) &= -\cos x & f^{(3)}(0) &= -1 \\
f^{(4)}(x) &= \sin x & f^{(4)}(0) &= 0,
\end{aligned}
$$

and so on. Thus all the even derivatives are zero and the odd derivatives alternate between 1 and -1. So the Maclaurin expansion for $\sin x$ is

$$0 + \frac{1}{1!}x + \frac{0}{2!}x^2 - \frac{1}{3!}x^3 + \frac{0}{4!}x^4 + \frac{1}{5!}x^5 - \cdots = x - \frac{x^3}{3!} + \frac{x^5}{5!} - \cdots,$$

which we can write as

(80)
$$\sin x = x - \frac{x^3}{3!} + \frac{x^5}{5!} - \frac{x^7}{7!} + \cdots = \sum_{n=0}^{\infty} \frac{(-1)^n x^{2n+1}}{(2n+1)!}. \quad \bullet$$

If we have the Taylor series

(75) $f(a) + f'(a)(x-a) + \dfrac{f''(a)}{2!}(x-a)^2 + \dfrac{f^{(3)}(a)}{3!}(x-a)^3 + \dfrac{f^{(4)}(a)}{4!}(x-a)^4 + \cdots$

for a function $f(x)$, then term-by-term differentiation of (75) gives the series

$$f'(a) + 2\frac{f''(a)}{2!}(x-a) + 3\frac{f^{(3)}(a)}{3!}(x-a)^2 + 4\frac{f^{(4)}(a)}{4!}(x-a)^3 + \cdots$$

$$= f'(a) + f''(a)(x-a) + \frac{f^{(3)}(a)}{2!}(x-a)^2 + \frac{f^{(4)}(a)}{3!}(x-a)^3 + \cdots.$$

But, referring to Definition 9, this is precisely the same as the Taylor series for the function $f'(x)$. Thus we have proved

---

THEOREM 14. The Taylor series expansion for the derivative $f'(x)$ can be obtained by termwise differentiation of the Taylor series expansion for $f(x)$.

Example 4. Find the Maclaurin expansion for $\cos x$.

Solution. We can simply differentiate the Maclaurin series (80) for $\sin x$ to find

$$\cos x = 1 - \frac{3x^2}{3!} + \frac{5x^4}{5!} - \frac{7x^6}{7!} + \cdots,$$

or

(81)
$$\cos x = 1 - \frac{x^2}{2!} + \frac{x^4}{4!} - \frac{x^6}{6!} + \cdots = \sum_{n=0}^{\infty} \frac{(-1)^n}{(2n)!} x^{2n}. \quad \bullet$$

In Section 10 we will discuss the fact that power series expansions for functions are *unique*. This means that if we can somehow produce a power series that converges to $f(x)$ in an open interval centered at a, this power series must be the Taylor series for f about a. For example, the power series expansion for $f(x) = 1/(1 - x)$ given in (69) is, in fact, the Maclaurin expansion for this function, as the reader can easily verify.

Example 5. Find the Maclaurin expansion for $1/(1 + x^3)$.

Solution. If we attempt to find a general formula for the nth derivative of this function, we will soon give up in despair, because the derivatives of higher order are very complicated. But if we use the geometric series

$$(82) \qquad \frac{1}{1 - u} = \sum_{n=0}^{\infty} u^n = 1 + u + u^2 + \cdots + u^n + \cdots,$$

and then replace u by $-x^3$, we obtain the desired Maclaurin series,

$$(83) \quad \frac{1}{1 - (-x^3)} = \frac{1}{1 + x^3} = 1 - x^3 + x^6 - x^9 + \cdots = \sum_{n=0}^{\infty} (-1)^n x^{3n}. \quad \bullet$$

Since the series in (82) converges for $-1 < u < 1$, the series in (83) will converge for $-1 < -x^3 < 1$, that is, for $-1 < x < 1$.

Example 6. Find the Maclaurin series for $x^2 \cosh x$.

Solution. Here the differentiations are not too bad, but we want another and shorter solution. We already have a series for e^x,

$$(78) \qquad e^x = \sum_{n=0}^{\infty} \frac{x^n}{n!} = 1 + x + \frac{x^2}{2!} + \frac{x^3}{3!} + \cdots.$$

On replacing x by $-x$ in (78), we find that

$$(84) \qquad e^{-x} = \sum_{n=0}^{\infty} \frac{(-1)^n x^n}{n!} = 1 - x + \frac{x^2}{2!} - \frac{x^3}{3!} + \cdots.$$

If we add (78) and (84), the odd powers of x drop out. Hence

$$\cosh x = \frac{e^x + e^{-x}}{2} = 1 + \frac{x^2}{2!} + \frac{x^4}{4!} + \cdots = \sum_{n=0}^{\infty} \frac{x^{2n}}{(2n)!}.$$

Finally, multiplying both sides by x^2, we have

$$x^2 \cosh x = x^2 + \frac{x^4}{2!} + \frac{x^6}{4!} + \cdots = \sum_{n=0}^{\infty} \frac{x^{2n+2}}{(2n)!} = \sum_{n=1}^{\infty} \frac{x^{2n}}{(2n-2)!}. \quad \bullet$$

Either one of the two forms on the right side is acceptable.

EXERCISE 7

1. Using Definition 9, find the Maclaurin series for $f(x) = 1/(2 - x)^2$.
2. Prove that (69) is actually the Maclaurin series for $f(x) = 1/(1 - x)$.
3. Starting with the Maclaurin series for $1/(1 - x)$, how would you obtain the Maclaurin series for $1/(1 - x)^2$?
4. Prove that the Taylor series about a for the integral $\int f(x)\,dx$ is given by termwise integration of the Taylor series about a for $f(x)$.
5. Starting with the Maclaurin series for $1/(1 - x)$, use the result of Problem 4 to find the Maclaurin series for $-\ln(1 - x)$.

In Problems 6 through 14, find the Maclaurin series for the given function, and determine the interval of convergence. The expansions can each be obtained using suitable operations (substitution, addition, multiplication, differentiation, and integration) on the series we already know $[e^x, \sin x, \cos x, \text{ and } (1 - x)^{-1}]$.

6. e^{2x}.

7. $\dfrac{x^4}{1 + x^2}$.

8. $\sin x^2$.

9. $\dfrac{1}{(1 - x)^4}$.

10. $\dfrac{1 + x}{1 - x}$.

11. $x \cos \sqrt{x}$.

12. $x \sinh x$.

13. $\mathrm{Tan}^{-1} x$.

14. $\dfrac{1 - 3x}{1 + 2x}$.

In Problems 15 through 23, find the Taylor series for the given function about the given point.

15. e^x, $a = 2$.
16. $1/x$, $a = 1$.
17. $1/x$, $a = 3$.
18. $\ln(1 + x)$, $a = 0$.
19. $\ln x$, $a = 3$.
20. $\cos x$, $a = \pi/2$.
21. $\sin x$, $a = \pi/2$.
22. $x^4 - 6x^2 + 9$, $a = 3$.
23. $7x^2 + 13x - 15$, $a = -5$.

24. Explain why the function $f(x) = x^{1/3}$ does not have a Maclaurin series expansion.

In some cases, it is not easy to find an explicit formula for the nth derivative. In Problems 25 through 27, find the first three nonzero terms of the Maclaurin series for the given function.

25. $e^{x(x-1)}$.
26. $e^x \sin x$.
27. $\tan x$.

28. Prove that the coefficients of a polynomial $P(x) = a_0 + a_1 x + a_2 x^2 + \cdots + a_n x^n$ are given by $a_k = P^{(k)}(0)/k!$ for $k = 0, 1, 2, \ldots, n$.

29. Find a third-degree polynomial $P(x)$ such that $P(0) = 1$, $P'(0) = 5$, $P''(0) = 6$, $P'''(0) = 18$.

30. Starting with the Maclaurin series for $\ln(1 + x)$ and $\ln(1 - x)$, find the Maclaurin series for $\ln\left(\dfrac{1 + x}{1 - x}\right)$.

*31. Use partial fractions to decompose the function $10/(x^2 - x - 6)$ and then find the Maclaurin series for this function.

32. The binomial series is frequently given without proof in algebra courses. This series for $(1 + x)^m$ is

$$(1 + x)^m = 1 + \frac{m}{1} x + \frac{m(m - 1)}{1 \cdot 2} x^2 + \frac{m(m - 1)(m - 2)}{1 \cdot 2 \cdot 3} x^3 + \cdots$$

$$+ \frac{m(m - 1)(m - 2) \cdots (m - k + 1)}{1 \cdot 2 \cdot 3 \cdots k} x^k + \cdots.$$

Prove that the binomial series is just the Maclaurin series for the function $(1 + x)^m$. Show that if m is an integer greater than or equal to zero, then the binomial series is just a polynomial. In all other cases the binomial series has infinitely many nonzero terms.

*33. Let

$$f(x) = \begin{cases} e^{-1/x^2}, & \text{if } x \neq 0, \\ 0, & \text{if } x = 0. \end{cases}$$

Show that $f^{(n)}(0) = 0$ for each $n = 0, 1, 2, \ldots$, so that the Maclaurin series for $f(x)$ is $0 + 0 + 0 + \cdots$. Thus the Maclaurin series converges for all x, but it does not converge to $f(x)$ except when $x = 0$. HINT: To find $f'(0)$, compute the limit of the difference quotient $(e^{-1/h^2} - 0)/h$ by writing it in the form $(1/h)/e^{1/h^2}$ and using L'Hospital's Rule.

9 Taylor's Theorem with Remainder

We know from the previous section that the series $\displaystyle\sum_{n=0}^{\infty} x^n/n!$ converges for all x, and we suspect that the sum of this series is e^x. Can we prove it?

In general, if we start with a function $f(x)$ and generate its Taylor series, we wish to investigate how well its partial sums approximate $f(x)$. First, we note that the partial sums of a Taylor series are polynomials in x, and we now denote them by $P_n(x)$:

(85) $\qquad P_n(x) = f(a) + f'(a)(x-a) + \dfrac{f''(a)}{2!}(x-a)^2 + \cdots + \dfrac{f^{(n)}(a)}{n!}(x-a)^n.$

We can also see that the nth-degree polynomial $P_n(x)$ agrees with the function $f(x)$ and its first n derivatives at $x = a$; indeed, on successively differentiating (85) and setting $x = a$, we find that $P_n(a) = f(a)$, $P_n'(a) = f'(a)$, $P_n''(a) = f''(a)$, ..., $P_n^{(n)}(a) = f^{(n)}(a)$.

To measure just how close the polynomial $P_n(x)$ is to $f(x)$, we introduce a new quantity $R_n(x)$, which is defined as the difference between $f(x)$ and $P_n(x)$; that is,

(86) $\qquad\qquad\qquad\qquad R_n(x) \equiv f(x) - P_n(x).$

The function $R_n(x)$ is called the *remainder*. From (85) and (86), we have

(87) $\quad f(x) = f(a) + f'(a)(x-a) + \dfrac{f''(a)}{2!}(x-a)^2 + \cdots + \dfrac{f^{(n)}(a)}{n!}(x-a)^n + R_n(x).$

To prove that the infinite series

(88) $\qquad \displaystyle\sum_{n=0}^{\infty} \dfrac{f^{(n)}(a)}{n!}(x-a)^n = f(a) + f'(a)(x-a) + \dfrac{f''(a)}{2!}(x-a)^2 + \cdots$

converges to $f(x)$, it is sufficient to prove that $R_n(x) \to 0$ as $n \to \infty$. For this we need some information about $R_n(x)$. A nice formula for $R_n(x)$ is given in

---

THEOREM 15. If $R_n(x)$ is the remainder, as defined by equations (85) and (86), then for $n = 0, 1, 2, \ldots,$

(89) $\qquad\qquad\qquad R_n(x) = \dfrac{1}{n!} \displaystyle\int_a^x f^{(n+1)}(t)(x-t)^n \, dt.$

Of course, we must assume that the function $f(x)$ has enough properties so that the proof is valid. For this purpose we will assume that x and a are interior points of some interval $\mathscr{I}$ and that $f^{(n+1)}(x)$ is continuous in $\mathscr{I}$.

Proof. We use mathematical induction. Clearly, for x and a in $\mathscr{I}$,

$$\int_a^x f'(t)\, dt = f(t)\Big|_a^x = f(x) - f(a)$$

or, on transposition,

$$f(x) = f(a) + \int_a^x f'(t)\, dt.$$

But this is (87) and (89) in the special case that $n = 0$.

We next assume that the theorem is true for index k. Thus we assume that

$$(90) \quad f(x) = f(a) + f'(a)(x - a) + \cdots + \frac{f^{(k)}(a)}{k!}(x - a)^k + \frac{1}{k!}\int_a^x f^{(k+1)}(t)(x - t)^k \, dt.$$

We now integrate the last term in (90) using integration by parts. Keeping in mind that t is the variable of integration and x is a constant, we have

$$u = f^{(k+1)}(t) \longrightarrow dv = (x - t)^k \, dt,$$

$$du = f^{(k+2)}(t) \, dt \longleftarrow v = \frac{-(x - t)^{k+1}}{k + 1},$$

$$\int_a^x u \, dv = \int_a^x f^{(k+1)}(t)(x - t)^k \, dt = \frac{-f^{(k+1)}(t)(x - t)^{k+1}}{k + 1} \Big|_{t=a}^{t=x}$$

$$+ \int_a^x \frac{f^{(k+2)}(t)(x - t)^{k+1} \, dt}{k + 1},$$

or

$$(91) \quad \int_a^x f^{(k+1)}(t)(x - t)^k \, dt = \frac{f^{(k+1)}(a)}{k + 1}(x - a)^{k+1} + \frac{1}{k + 1}\int_a^x f^{(k+2)}(t)(x - t)^{k+1} \, dt.$$

Using (91) in (90) and observing that $k!(k + 1) = (k + 1)!$, we find that

$$f(x) = f(a) + f'(a)(x - a) + \cdots + \frac{f^{(k)}(a)}{k!}(x - a)^k + \frac{f^{(k+1)}(a)}{(k + 1)!}(x - a)^{k+1}$$

$$+ \frac{1}{(k + 1)!}\int_a^x f^{(k+2)}(t)(x - t)^{k+1} \, dt.$$

But this is the statement of the theorem when the index n is $k + 1$. ∎

COROLLARY. Let M be the maximum of $|f^{(n+1)}(t)|$ for t in the interval between a and x. Then

$$(92) \qquad\qquad |R_n(x)| \leq \frac{M}{(n + 1)!}|x - a|^{n+1}.$$

Proof. We observe that there are two cases to consider, depending on whether $x \geqq a$ or $x < a$. Suppose first that $x \geqq a$. Then in the integral (89) we have $a \leqq t \leqq x$, or

$x - t \geqq 0$. Consequently,

$$|R_n(x)| = \left| \frac{1}{n!} \int_a^x f^{(n+1)}(t)(x-t)^n \, dt \right| \leqq \frac{1}{n!} \int_a^x M(x-t)^n \, dt$$

$$\leqq \frac{-M}{n!} \frac{(x-t)^{n+1}}{n+1} \bigg|_{t=a}^{t=x}.$$

Hence

$$|R_n(x)| \leqq \frac{M}{(n+1)!}(x-a)^{n+1} = \frac{M}{(n+1)!}|x-a|^{n+1}.$$

If $x < a$, then $x \leqq t \leqq a$ in (89) and hence $x - t \leqq 0$. In this case

$$|R_n(x)| = \left| \frac{1}{n!} \int_a^x f^{(n+1)}(t)(x-t)^n \, dt \right| \leqq \frac{1}{n!} \int_x^a M(t-x)^n \, dt$$

$$\leqq \frac{M}{n!} \frac{(t-x)^{n+1}}{n+1} \bigg|_{t=x}^{t=a} = \frac{M}{(n+1)!}(a-x)^{n+1} = \frac{M}{(n+1)!}|x-a|^{n+1}.$$

Hence in either case we get (92). ∎

Example 1. Prove that the series

$$(93) \qquad \sum_{n=0}^{\infty} \frac{x^n}{n!} = 1 + x + \frac{x^2}{2!} + \frac{x^3}{3!} + \cdots$$

converges to e^x for all x.

Solution. We already know that when $a = 0$, the first $n + 1$ terms of the series (93) is the approximating polynomial (85) when $f(x) = e^x$. In other words, we already have

$$e^x = 1 + x + \frac{x^2}{2!} + \frac{x^3}{3!} + \cdots + \frac{x^n}{n!} + R_n(x).$$

We apply formula (92) of the corollary. In this case $f^{(n+1)}(t) = e^t$. If $x > 0$, the maximum value of e^t for $0 \leqq t \leqq x$ is e^x. If $x < 0$, the maximum value of e^t for $x \leqq t \leqq 0$ is $e^0 = 1$. In the first case,

$$(94) \qquad |R_n(x)| \leqq \frac{e^x}{(n+1)!} x^{n+1}.$$

But we already know from Section 8 that the series (93) converges, so by Theorem 5 the general term $x^{n+1}/(n+1)! \to 0$ as $n \to \infty$. Then from (94) we see that $|R_n(x)| \to 0$ as $n \to \infty$, and consequently the series (93) converges to e^x. If $x < 0$, then (94) is replaced by $|R_n(x)| \leqq |x|^{n+1}/(n+1)!$ with the same conclusion. Hence *for all values of x,*

$$e^x = 1 + x + \frac{x^2}{2!} + \frac{x^3}{3!} + \cdots + \frac{x^n}{n!} + \cdots. \quad \bullet$$

Example 2. Compute $\sqrt{e}$ to three decimal places by using the partial sums of the Maclaurin series for e^x.

Solution. This means find q such that $|q - e^{1/2}| < 0.0005$. If we use the corollary, it suffices to find a value of n such that

$$|R_n(x)| \leq \frac{M}{(n+1)!} |x|^{n+1} < 0.0005$$

when $x = 1/2$. Now M is the maximum of e^t in the interval $0 \leq t \leq 1/2$. If we take as a conservative estimate $e < 4$, then $M = e^{1/2} < \sqrt{4} = 2$. Consequently,

$$|R_n(1/2)| < \frac{2}{(n+1)!} \left(\frac{1}{2}\right)^{n+1} = \frac{1}{(n+1)!2^n}.$$

For $n = 4$ the right side gives $1/1920 \approx 0.00052$ and this is not quite sufficient. For $n = 5$, the right side gives $1/23{,}040 \approx 0.000043$. So we need only the first six terms of the series, for the desired accuracy. We find that

$$\sqrt{e} \approx 1 + \frac{1}{2} + \frac{1}{2!2^2} + \frac{1}{3!2^3} + \frac{1}{4!2^4} + \frac{1}{5!2^5}$$

$$\approx 1.0000 + 0.5000 + 0.1250 + 0.0208 + 0.0026 + 0.0003$$

$$\approx 1.6487 \approx 1.649.^1 \quad \bullet$$

EXERCISE 8

In Problems 1 through 5, prove that the given series converges to the indicated sum, for x in the given set.

1. $\sin x = \displaystyle\sum_{n=0}^{\infty} \frac{(-1)^n x^{2n+1}}{(2n+1)!}$, $-\infty < x < \infty$.

2. $\cos x = \displaystyle\sum_{n=0}^{\infty} \frac{(-1)^n x^{2n}}{(2n)!}$, $-\infty < x < \infty$.

---

[1] The reader should keep in mind that if x and x^* are two numbers for which $|x - x^*| < 0.0005$, then x^* rounded to three decimal places may not equal x rounded to three decimal places. Indeed, let $x = 1.00666$ and $x^* = 1.00633$. Then $|x - x^*| < 0.0005$, and x rounded to three decimals is 1.007, but x^* rounded to three decimals is 1.006.

3. $\sinh x = \displaystyle\sum_{n=0}^{\infty} \frac{x^{2n+1}}{(2n+1)!}$, $-\infty < x < \infty.$

4. $\cosh x = \displaystyle\sum_{n=0}^{\infty} \frac{x^{2n}}{(2n)!}$, $-\infty < x < \infty.$

*5. $\sqrt{1+x} = 1 + \displaystyle\sum_{n=1}^{\infty} \frac{(-1)^{n+1}(2n-2)!\,x^n}{2^{2n-1}n!(n-1)!}$, $0 \leq x \leq 1.$

6. Use enough terms of the series (93) to prove that $e > 2.71$.

7. Show that for the partial sums of the Maclaurin expansion of $1/(1-x)$, we have

$$R_n(x) = (n+1) \int_0^x \frac{(x-t)^n}{(1-t)^{n+2}}\,dt = \frac{x^{n+1}}{1-x}.$$

HINT: Since this Maclaurin series is a geometric series, $R_n(x) = x^{n+1}/(1-x)$ follows from an earlier identity.

8. Using the result of Problem 7, show that for the partial sums of the Maclaurin expansion of $\ln(1+x)$, we have

$$R_n(x) = (-1)^n \int_0^x \frac{t^n}{1+t}\,dt.$$

Then prove that

$$\ln(1+x) = \sum_{n=1}^{\infty} \frac{(-1)^{n+1}x^n}{n} \quad\quad \text{for } -1 < x \leq 1.$$

In Problems 9 through 14, use inequality (92) to find the maximum error in approximating the given number by using the nth-degree partial sums of the Maclaurin expansion. Keep in mind that in the expansions for the trigonometric functions x is in radians.

9. e, $n = 2$.   10. $e^{0.1}$, $n = 3$.   11. $\sin 5°$, $n = 3$.

12. $\cos 5°$, $n = 4$.   13. $\sinh 1$, $n = 3$.   14. $\cosh 1$, $n = 4$.

CALCULATOR PROBLEMS

In Problems C1 through C8, compute the value of the given number to the indicated accuracy by using the Taylor series about the given value of a.

C1. $\sqrt[5]{e}$; $a = 0$; $|R_n| < 0.001$.   C2. e^2; $a = 0$; $|R_n| < 0.01$.

C3. $\ln(1.2)$; $a = 1$; $|R_n| < 0.005$.   C4. $\ln 0.8$; $a = 1$; $|R_n| < 0.005$.

C5. $\sin 4°$; $a = 0$; $|R_n| < 0.0005$.   C6. $\cos 15°$; $a = 0$; $|R_n| < 0.0005$.

C7. $\sin 65°$; $a = \pi/3$; $|R_n| < 0.0005$.   C8. $\sin 33°$; $a = \pi/6$; $|R_n| < 0.0005$.

10   *Differentiation and Integration of Power Series*

Some important applications of power series are based on the following fact: When a power series has a positive radius of convergence, it can be differentiated and integrated term by term to give the derivative or integral, respectively, of the function represented by the original series. This is stated precisely in

---

THEOREM 16 (PWO). If the series $\sum_{n=0}^{\infty} a_n(x-a)^n$ converges to $f(x)$ in the interval $|x-a| < R$, where $R > 0$, then $f'(x)$ exists for $|x-a| < R$ and the series

$$(95) \qquad \sum_{n=1}^{\infty} na_n(x-a)^{n-1}$$

converges to $f'(x)$ for $|x-a| < R$.

Furthermore, the series

$$(96) \qquad \sum_{n=0}^{\infty} \frac{a_n}{n+1}(x-a)^{n+1}$$

converges to the definite integral $\displaystyle\int_{a}^{x} f(t)\,dt$ for $|x-a| < R$.

We shall not take time here to prove Theorem 16. The interested reader can find the details in any advanced calculus text.

Note that Theorem 16 is stated only for an open interval and does not say anything about the end points. It sometimes happens that in using termwise differentiation we lose convergence at an end point, while with termwise integration we may in fact gain convergence at an end point.

From Theorem 16 we can easily prove

THEOREM 17. If the series $\sum_{n=0}^{\infty} a_n(x-a)^n$ converges to $f(x)$ for $|x-a| < R$, where $R > 0$, then

(A) All the derivatives $f^{(n)}(x)$ exist for $|x - a| < R$,

(B) $a_n = \dfrac{f^{(n)}(a)}{n!}$,

and

(C) $\displaystyle\sum_{n=0}^{\infty} a_n(x - a)^n$ is the Taylor series for $f(x)$ about $x = a$.

Proof. From Theorem 16 we know that $f'(x)$ exists and

$$f'(x) = \sum_{n=1}^{\infty} na_n(x - a)^{n-1} \qquad \text{for } |x - a| < R.$$

By applying Theorem 16 to this last series, it follows that $f''(x)$ exists and

$$f''(x) = \sum_{n=1}^{\infty} n(n - 1)a_n(x - a)^{n-2} = \sum_{n=2}^{\infty} n(n - 1)a_n(x - a)^{n-2}$$

for $|x - a| < R$. Continuing in this manner, we deduce that all derivatives of $f(x)$ exist for $|x - a| < R$ and that their power series expansions are given by successive differentiation of the power series for $f(x)$. Hence by the process described in Section 8, we find that

(74) $$a_n = \frac{f^{(n)}(a)}{n!} \qquad \text{for } n = 0, 1, 2, \ldots.$$

This proves parts **(A)** and **(B)** of the theorem. Part **(C)** follows at once from the definition of a Taylor series. ∎

We remark that Theorem 17 implies that the power series representation of a function about $x = a$ is unique. Indeed, if there are two expansions

$$f(x) = \sum_{n=0}^{\infty} a_n(x - a)^n \qquad \text{and} \qquad f(x) = \sum_{n=0}^{\infty} b_n(x - a)^n,$$

for $|x - a| < R$, then by part **(B)** of Theorem 17, we have $a_n = f^{(n)}(a)/n!$ and $b_n = f^{(n)}(a)/n!$, so that $a_n = b_n$ for all n.

Example 1. Use Theorem 16 to show that

(97) $$\frac{1}{(1 - x)^3} = \sum_{n=0}^{\infty} \frac{(n + 2)(n + 1)}{2} x^n \qquad \text{for } -1 < x < 1.$$

Solution. We begin with the equation for geometric series

$$\frac{1}{1-x} = \sum_{n=0}^{\infty} x^n = 1 + x + x^2 + \cdots + x^n + \cdots,$$

which we know holds for $-1 < x < 1$. By Theorem 16 we can differentiate both sides, twice with respect to x. Thus

$$\frac{1}{(1-x)^2} = \sum_{n=1}^{\infty} nx^{n-1} = 1 + 2x + 3x^2 + 4x^3 + \cdots$$

for $-1 < x < 1$, and

$$\frac{2}{(1-x)^3} = \sum_{n=2}^{\infty} n(n-1)x^{n-2} = 2 + 6x + 12x^2 + \cdots,$$

for $-1 < x < 1$. Hence, on dividing by 2, we obtain the desired result,

$$\frac{1}{(1-x)^3} = 1 + 3x + 6x^2 + \cdots = \sum_{n=2}^{\infty} \frac{n(n-1)}{2}x^{n-2} = \sum_{n=0}^{\infty} \frac{(n+2)(n+1)}{2}x^n,$$

which is valid for $-1 < x < 1$. ●

Of course, by Theorem 17, the series (97) is necessarily the Maclaurin series for $1/(1-x)^3$.

Example 2. Find $\displaystyle\int \sin x^2 \, dx$.

Solution. This indefinite integral cannot be expressed with a finite number of combinations of the elementary functions, but we can find an infinite series for it. We know (see Problem 1, Exercise 8) that the equation

$$\sin x = \sum_{n=0}^{\infty} \frac{(-1)^n x^{2n+1}}{(2n+1)!}$$

holds for all x, and so by substitution,

$$\sin t^2 = \sum_{n=0}^{\infty} \frac{(-1)^n (t^2)^{2n+1}}{(2n+1)!} = \sum_{n=0}^{\infty} \frac{(-1)^n t^{4n+2}}{(2n+1)!}$$

for all t. From Theorem 16 the function

$$F(x) = \int_0^x \sin t^2 \, dt$$

can be written as the infinite series

$$F(x) = \sum_{n=0}^{\infty} \int_0^x \frac{(-1)^n t^{4n+2}}{(2n+1)!} \, dt = \sum_{n=0}^{\infty} \frac{(-1)^n x^{4n+3}}{(2n+1)!(4n+3)}.$$

But $F(x)$ is an antiderivative for $\sin x^2$ (see Theorem 8, Chapter 6), so for all x,

(98) $$\int \sin x^2 \, dx = C + \sum_{n=0}^{\infty} \frac{(-1)^n x^{4n+3}}{(2n+1)!(4n+3)}. \quad \bullet$$

Example 3. Find $\displaystyle\int_0^1 \sin x^2 \, dx$ to four decimal places.

Solution. From equation (98),

$$\int_0^1 \sin x^2 \, dx = \sum_{n=0}^{\infty} \frac{(-1)^n x^{4n+3}}{(2n+1)!(4n+3)} \Big|_0^1$$

$$= \frac{1}{1!3} - \frac{1}{3!7} + \frac{1}{5!11} - \frac{1}{7!15} + \cdots.$$

Since this is an alternating series with terms decreasing in absolute value, we know from inequality (53), Section 6, that the error in using S_n in place of S does not exceed a_{n+1}. Hence we look for the smallest value of n for which

$$a_{n+1} = \frac{1}{(2n+1)!(4n+3)} < 0.00005.$$

Now when $n = 3$, we find that $1/7!15 \approx 0.0000132$; hence we need only the first three terms of the series:

$$\frac{1}{1!3} - \frac{1}{3!7} + \frac{1}{5!11} \approx 0.33333 - 0.02381 + 0.00076 = 0.31028.$$

Therefore,

$$\int_0^1 \sin x^2 \, dx \approx 0.3103 \qquad \text{(to four decimal places).} \quad \bullet$$

EXERCISE 9

Starting with the equation for geometric series prove that each of the series given in Problems 1 through 8 has the indicated sum for $-1 < x < 1$.

1. $\displaystyle\sum_{n=1}^{\infty} \frac{x^n}{n} = -\ln(1-x).$

2. $\displaystyle\sum_{n=1}^{\infty} \frac{(-1)^{n+1}x^n}{n} = \ln(1+x).$

3. $\displaystyle\sum_{n=1}^{\infty} \frac{x^{2n-1}}{2n-1} = \frac{1}{2}\ln\frac{1+x}{1-x}.$

4. $\displaystyle\sum_{n=0}^{\infty} \frac{(-1)^n x^{2n+1}}{2n+1} = \mathrm{Tan}^{-1}x.$

★5. $\displaystyle\sum_{n=1}^{\infty} \frac{x^n}{n^2} = -\int_0^x \frac{\ln(1-t)}{t}\,dt.$

6. $\displaystyle\sum_{n=0}^{\infty} (-1)^n (n+1)(n+2)x^n = \frac{2}{(x+1)^3}.$

7. $\displaystyle\sum_{n=0}^{\infty} \frac{(n+1)(n+2)(n+3)}{6}x^n = \frac{1}{(1-x)^4}.$

★8. $\displaystyle\sum_{n=1}^{\infty} nx^{2n-1} = \frac{x}{1-2x^2+x^4}.$

9. By differentiating the Maclaurin series for $1/(1-x)$, find the Maclaurin series for $1/(1-x)^k$ for any positive integer k.

10. Define R_n by writing the series of Problem 3 in the form

$$\ln\frac{1+x}{1-x} = 2\left(x + \frac{x^3}{3} + \cdots + \frac{x^{2n-1}}{2n-1}\right) + R_n.$$

Using a geometric series that is term by term greater than the terms of R_n, prove that

$$|R_n| \leqq \frac{2|x|^{2n+1}}{2n+1}\frac{1}{1-x^2}.$$

11. Let $f(x) = \displaystyle\sum_{n=0}^{\infty} \frac{n^3}{3^n}x^n$. Find $f'(0)$ and $f^{(6)}(0)$.

12. Let $f(x) = \begin{cases} \dfrac{\sin x}{x}, & \text{if } x \neq 0, \\ 1, & \text{if } x = 0. \end{cases}$

    (a) Use the Maclaurin series for $\sin x$ to prove that for *all* x,

$$f(x) = 1 - \frac{x^2}{3!} + \frac{x^4}{5!} - \frac{x^6}{7!} + \cdots.$$

    (b) Find $f^{(3)}(0)$ and $f^{(4)}(0)$.

★13. By completing the following steps, find the Maclaurin series expansion for the solution of the initial value problem

$$y'' - xy' - y = 0, \qquad y(0) = 1, \qquad y'(0) = 0.$$

(a) Write $y(x) = \sum_{n=0}^{\infty} a_n x^n$, with $a_0 = 1$, $a_1 = 0$, and substitute this power series into the differential equation to obtain

$$\sum_{n=2}^{\infty} n(n-1)a_n x^{n-2} - x \sum_{n=1}^{\infty} na_n x^{n-1} - \sum_{n=0}^{\infty} a_n x^n = 0.$$

(b) Show that on collecting like powers of x, the equation in part (a) can be written in the form

$$2a_2 - a_0 + \sum_{n=1}^{\infty} [(n+2)(n+1)a_{n+2} - (n+1)a_n]x^n = 0.$$

(c) Set the coefficients of the power series in part (b) equal to zero to obtain the recurrence relation

$$a_{n+2} = \frac{a_n}{n+2}, \qquad n = 0, 1, 2, \ldots.$$

(d) Use the fact that $a_0 = 1$, $a_1 = 0$ to prove from the recurrence relation that $a_{2n} = 1/2^n(n!)$, $a_{2n+1} = 0$, and hence

$$y(x) = \sum_{n=0}^{\infty} \frac{x^{2n}}{2^n(n!)}$$

is the desired solution.

14. Use the identity $\sin^2 x = \dfrac{1 - \cos 2x}{2}$ to find the Maclaurin series expansion for $\sin^2 x$.

In Problems 15 through 18, express the given definite integral as the sum of an infinite series of constants.

15. $\displaystyle\int_0^1 \cos x^2 \, dx.$

16. $\displaystyle\int_0^1 \frac{\sin x}{x} \, dx.$

17. $\displaystyle\int_0^{1/2} \frac{e^{-x} - 1}{x} \, dx.$

★18. $\displaystyle\int_0^1 \sqrt{1 - x^3} \, dx.$

In Problems 19 through 22, use series to evaluate the limit.

19. $\lim\limits_{x\to 0} \dfrac{\cos x - 1}{x}$.

20. $\lim\limits_{x\to 0} \dfrac{1 - 2x^2 - \cos(2x)}{x^4}$.

21. $\lim\limits_{x\to 0} \dfrac{\sin(3x)}{\sin(2x)}$.

22. $\lim\limits_{x\to 0} \dfrac{e^x - e^{-x} - 2x}{x^3}$.

CALCULATOR PROBLEMS

In Problems C1 through C5, use series to approximate the given integral to three decimal places.

C1. $\displaystyle\int_0^1 e^{-x^2}\,dx$.

C2. $\displaystyle\int_0^{0.8} \sin x^2\,dx$.

C3. $\displaystyle\int_0^1 \cos x^3\,dx$.

C4. $\displaystyle\int_{0.5}^1 \dfrac{\cos x}{x}\,dx$.

C5. $\displaystyle\int_0^{0.5} \dfrac{\sin x}{x}\,dx$.

C6. Compute $\ln 2$ to four decimal places by setting $x = 1/3$ in the series of Problem 10.

C7. Compute $\ln(3/2)$ to four decimal places by setting $x = 1/5$ in the series of Problem 10.

C8. Approximate $\ln 3$ using the results of Problems C6 and C7.

C9. Compute $\mathrm{Tan}^{-1}(1/2)$ to four decimal places using the series of Problem 4.

11 Some Concluding Remarks on Infinite Series

Although it may seem to the student that we have learned quite a bit about infinite series, the truth is that we have just scratched the surface. The reader who desires further information may consult any book on advanced calculus. The most complete single book on the topic is the one by Konrad Knopp, *Theory and Application of Infinite Series* (Blackie, Glasgow, 1951).

In closing this chapter, we remark that given two infinite series, we may form their product or their quotient under suitable conditions. The process is illustrated in the following examples.

Example 1. Find the Maclaurin series for $1/(1 - x)^2$ by squaring the series for $1/(1 - x)$.

Solution. The series for $1/(1 - x)$ is just the geometric series

$$\frac{1}{1 - x} = 1 + x + x^2 + \cdots + x^n + \cdots.$$

To square this series, we multiply first by 1, then by x, then by $x^2, \ldots$, and add the results. The computation can be arranged as follows.

$$1 + x + x^2 + x^3 + x^4 + x^5 + x^6 + x^7 + \cdots$$
$$1 + x + x^2 + x^3 + x^4 + x^5 + x^6 + x^7 + \cdots$$

$$
\begin{array}{l}
1 + x + x^2 + x^3 + x^4 + x^5 + x^6 + x^7 + \cdots \\
\quad\ x + x^2 + x^3 + x^4 + x^5 + x^6 + x^7 + \cdots \\
\qquad\quad x^2 + x^3 + x^4 + x^5 + x^6 + x^7 + \cdots \\
\qquad\qquad\quad x^3 + x^4 + x^5 + x^6 + x^7 + \cdots \\
\qquad\qquad\qquad\quad x^4 + x^5 + x^6 + x^7 + \cdots \\
\qquad\qquad\qquad\qquad\quad x^5 + x^6 + x^7 + \cdots \\
\qquad\qquad\qquad\qquad\qquad\quad x^6 + x^7 + \cdots \\
\qquad\qquad\qquad\qquad\qquad\qquad\quad x^7 + \cdots \\
\qquad\qquad\qquad\qquad\qquad\qquad\qquad\quad \cdots
\end{array}
$$

$$1 + 2x + 3x^2 + 4x^3 + 5x^4 + 6x^5 + 7x^6 + 8x^7 + \cdots$$

It is clear that in general the coefficient of x^{n-1} is n. Consequently,

$$\frac{1}{(1-x)^2} = \sum_{n=1}^{\infty} nx^{n-1}. \quad \bullet$$

Example 2. Find a rule for forming the product of two Maclaurin series.

Solution. Let

(99) $$f(x) = \sum_{n=0}^{\infty} a_n x^n \qquad \text{and} \qquad g(x) = \sum_{n=0}^{\infty} b_n x^n$$

be the given series. Let the product have the Maclaurin series

$$f(x)g(x) = \sum_{n=0}^{\infty} c_n x^n.$$

We are to find a rule that gives each c_n in terms of the coefficients a_k and b_k. A little reflection shows that each product of the form

$$a_j x^j b_k x^k = a_j b_k x^{j+k}, \qquad j = 0, 1, 2, \ldots, \qquad k = 0, 1, 2, \ldots$$

enters exactly once in the series for $f(x)g(x)$. We can group together those for which the exponent on x is the same. Thus let $j + k = n$; then as j runs through the integers $0, 1, 2, \ldots, n$, the index k runs through the same set in the reverse order. Hence c_n, the coefficient of x^n in $f(x)g(x)$, is just the sum of such terms, so

(100) $$c_n = \sum_{j=0}^{n} a_j b_{n-j} = a_0 b_n + a_1 b_{n-1} + a_2 b_{n-2} + \cdots + a_n b_0.$$

Thus the desired rule is expressed by the formula

(101) $$f(x)g(x) = \left(\sum_{n=0}^{\infty} a_n x^n \right)\left(\sum_{n=0}^{\infty} b_n x^n \right) = \sum_{n=0}^{\infty} \left(\sum_{j=0}^{n} a_j b_{n-j} \right) x^n.$$

The right side of (101) is known as the *Cauchy product* of the series in (99).   ●

Example 3. Find the Maclaurin series for tan x by dividing the series for sin x by the series for cos x.

Solution. We can arrange the work just as in the division of one polynomial by another, except that this time we write the two quantities with the exponents increasing. Now

$$\sin x = x - \frac{x^3}{6} + \frac{x^5}{120} - \cdots, \qquad \cos x = 1 - \frac{x^2}{2} + \frac{x^4}{24} - \cdots.$$

$$
\begin{array}{r}
x + \dfrac{x^3}{3} + \dfrac{2x^5}{15} + \cdots \\[1ex]
\hline
\end{array}
$$

$$1 - \frac{x^2}{2} + \frac{x^4}{24} - \cdots \bigg)\; x - \frac{x^3}{6} + \frac{x^5}{120} - \cdots$$

$$x - \frac{x^3}{2} + \frac{x^5}{24} - \cdots$$

$$\frac{x^3}{3} - \frac{x^5}{30} + \cdots$$

$$\frac{x^3}{3} - \frac{x^5}{6} + \cdots$$

$$\frac{2x^5}{15} + \cdots$$

Hence

(102) $$\tan x = x + \frac{x^3}{3} + \frac{2x^5}{15} + \cdots.$$

It is practically impossible to obtain a general formula for the nth term. It can be proved that this series converges for $|x| < \pi/2$, but the proof is not easy.   ●

There are two other methods for solving this type of problem. One such method is to write

$$\cos x = 1 - u, \qquad \text{where} \qquad u = \frac{x^2}{2} - \frac{x^4}{24} + \frac{x^6}{720} - \cdots$$

and expand $1/\cos x = 1/(1 - u)$ as a geometric series. Thus

$$\frac{\sin x}{\cos x} = \left(x - \frac{x^3}{6} + \frac{x^5}{120} - \cdots\right)\left(1 + u + u^2 + u^3 + \cdots\right)$$

$$= \left(x - \frac{x^3}{6} + \frac{x^5}{120} - \cdots\right)\left[1 + \left(\frac{x^2}{2} - \frac{x^4}{24} + \frac{x^6}{720} - \cdots\right)\right.$$

$$\left. + \left(\frac{x^2}{2} - \frac{x^4}{24} + \frac{x^6}{720} - \cdots\right)^2 + \left(\frac{x^2}{2} - \cdots\right)^3 + \cdots\right].$$

The reader can continue this computation and show that it also gives (102).

A third method is to write the quotient with unknown coefficients a_k, multiply both sides by the denominator using (101), and then equate coefficients of like powers of x on both sides and solve for the unknown coefficients. In this example the computation runs as follows.

$$x - \frac{x^3}{6} + \frac{x^5}{120} - \cdots = \left(1 - \frac{x^2}{2} + \frac{x^4}{24} - \cdots\right)(a_0 + a_1 x + a_2 x^2 + \cdots)$$

$$= a_0 + a_1 x + \left(a_2 - \frac{a_0}{2}\right)x^2 + \left(a_3 - \frac{a_1}{2}\right)x^3$$

$$+ \left(a_4 - \frac{a_2}{2} + \frac{a_0}{24}\right)x^4 + \cdots.$$

Equating coefficients of like powers of x on both sides gives

$$0 = a_0, \qquad 1 = a_1, \qquad 0 = a_2 - \frac{a_0}{2}, \qquad -\frac{1}{6} = a_3 - \frac{a_1}{2}, \qquad 0 = a_4 - \frac{a_2}{2} + \frac{a_0}{24}, \qquad \cdots.$$

The reader should find the next equation in this infinite set, and show that on solving we again obtain (102).

EXERCISE 10

In Problems 1 through 13, use the methods of this section to show that the given function has the Maclaurin series on the right, as far as the terms indicated.

1. $\ln^2 (1 - x) = x^2 + x^3 + \frac{11}{12}x^4 + \frac{5}{6}x^5 + \cdots.$

2. $\dfrac{\sin x}{1-x} = x + x^2 + \dfrac{5}{6}x^3 + \dfrac{5}{6}x^4 + \dfrac{101}{120}x^5 + \cdots.$

3. $\dfrac{e^x}{2+x} = \dfrac{1}{2} + \dfrac{1}{4}x + \dfrac{1}{8}x^2 + \dfrac{1}{48}x^3 + \dfrac{1}{96}x^4 + \cdots.$

4. $e^x \cos x = 1 + x - \dfrac{1}{3}x^3 - \dfrac{1}{6}x^4 - \dfrac{1}{30}x^5 + \cdots.$

5. $e^{x+x^2} = 1 + x + \dfrac{3}{2}x^2 + \dfrac{7}{6}x^3 + \dfrac{25}{24}x^4 + \cdots.$

6. $\dfrac{x}{\sin x} = 1 + \dfrac{1}{6}x^2 + \dfrac{7}{360}x^4 + \cdots.$

7. $\sec x = \dfrac{1}{\cos x} = 1 + \dfrac{1}{2}x^2 + \dfrac{5}{24}x^4 + \cdots.$

8. $\dfrac{1}{\displaystyle\sum_{n=0}^{\infty} x^n} = 1 - x.$

9. $\dfrac{\sin x}{\ln(1+x)} = 1 + \dfrac{1}{2}x - \dfrac{1}{4}x^2 - \dfrac{1}{24}x^3 + \cdots.$

10. $\dfrac{x+x^2}{1+x-x^2} = x + x^3 - x^4 + 2x^5 + \cdots.$

11. $e^{\sin x} = 1 + x + \dfrac{1}{2}x^2 - \dfrac{1}{8}x^4 + \cdots.$

12. $\ln \cos x = -\dfrac{1}{2}x^2 - \dfrac{1}{12}x^4 - \dfrac{1}{45}x^6 + \cdots.$

13. $\tanh x = x - \dfrac{1}{3}x^3 + \dfrac{2}{15}x^5 + \cdots.$

14. Prove that $\dfrac{1+x}{1+x+x^2} = 1 - x^2 + x^3 - x^5 + x^6 - x^8 + x^9 - \cdots.$

15. Prove that $\dfrac{1}{1-x+x^2-x^3} = 1 + x + x^4 + x^5 + x^8 + x^9 + x^{12} + x^{13} + \cdots.$

16. The Maclaurin series for $1/\sqrt{1-x}$ is

$$\frac{1}{\sqrt{1-x}} = 1 + \frac{1}{2}x + \frac{1\cdot 3}{2^2\cdot 2}x^2 + \frac{1\cdot 3\cdot 5}{2^3\cdot 3!}x^3 + \frac{1\cdot 3\cdot 5\cdot 7}{2^4\cdot 4!}x^4 + \cdots.$$

Check this result by squaring the series and showing that the result is $1 + x + x^2 + x^3 + x^4 + \cdots$, as far as the first five terms are concerned.

*17. By squaring the series for $\sin x$ and $\cos x$, show that $\sin^2 x + \cos^2 x = 1$, at least as far as the first four terms are concerned.

**18. By multiplying the two series for e^x and e^y, prove that $e^x e^y = e^{x+y}$.

REVIEW PROBLEMS

In Problems 1 through 10, test the given series for convergence or divergence.

1. $\displaystyle\sum_{n=1}^{\infty} \frac{1}{n^2 - \ln n}$.

2. $\displaystyle\sum_{n=1}^{\infty} \frac{(-1)^n}{2n - \sqrt{17n}}$.

3. $\displaystyle\sum_{n=1}^{\infty} \frac{(-1)^n}{2\sqrt{n} + 3}$.

4. $\displaystyle\sum_{n=1}^{\infty} \frac{1 \cdot 4 \cdot 7 \cdots (3n - 2)}{7 \cdot 9 \cdot 11 \cdots (2n + 5)}$.

5. $\displaystyle\sum_{n=2}^{\infty} \frac{1}{n \ln^p n}$, $p > 1$.

6. $\displaystyle\sum_{n=1}^{\infty} \frac{3^n (n!)^2}{(2n)!}$.

7. $\displaystyle\sum_{n=1}^{\infty} \frac{(n!)^{5/2}}{(2n)!}$.

8. $\displaystyle\sum_{n=1}^{\infty} \frac{2^n (n!)}{4 \cdot 7 \cdot 10 \cdots (3n + 1)}$.

★9. $\displaystyle\sum_{n=1}^{\infty} \frac{(n + 10)!}{n^n}$.

★10. $\displaystyle\sum_{n=1}^{\infty} \frac{1}{n \sqrt[n]{1 + n}}$.

★11. By finding an explicit expression for the sum of the first n terms, settle the convergence or divergence of the series:

(a) $\displaystyle\sum_{n=1}^{\infty} \ln\left(\frac{n}{n + 1}\right)$.

(b) $\displaystyle\sum_{n=2}^{\infty} \ln\left(1 - \frac{1}{n^2}\right)$.

12. By suitable substitution in the series for $\ln[(1 + x)/(1 - x)]$, derive the series

$$\ln\left(1 + \frac{1}{M}\right) = \sum_{n=1}^{\infty} \frac{2}{2n - 1} \left(\frac{1}{2M + 1}\right)^{2n-1}$$

$$= 2\left[\frac{1}{2M + 1} + \frac{1}{3(2M + 1)^3} + \frac{1}{5(2M + 1)^5} + \cdots\right].$$

13. Find the convergence set for $\displaystyle\sum_{n=1}^{\infty} x^{n^2} = x + x^4 + x^9 + x^{16} + \cdots$.

14. By integrating the series for $1/\sqrt{1 - x^2}$, derive the series

$$\mathrm{Sin}^{-1} x = x + \frac{1}{2}\frac{x^3}{3} + \frac{1 \cdot 3}{2 \cdot 4}\frac{x^5}{5} + \cdots + \frac{1 \cdot 3 \cdot 5 \cdots (2n - 1)}{2 \cdot 4 \cdot 6 \cdots 2n}\frac{x^{2n+1}}{2n + 1} + \cdots$$

$$= \sum_{n=0}^{\infty} \frac{(2n)!}{4^n (n!)^2}\frac{x^{2n+1}}{2n + 1}.$$

15. Find the Maclaurin series for $\ln(a + x)$, where $a > 0$.

16. Prove that for $|x| > 1$,

$$\frac{1}{1 - x} = -\sum_{n=1}^{\infty} \frac{1}{x^n}.$$

17. Expand $\ln x$ in a Taylor series about the point $x = 2$.

18. Find the Maclaurin series for $c/(ax + b)$, where $a \neq 0$ and $b \neq 0$.

19. Use partial fractions to find a Maclaurin series for $7x/(6 - 5x + x^2)$.

20. Expand $2/(1 - x)$ as a Taylor series about the point $x = 3$.

21. By multiplying the series for $1/(1 - x)$ by the series

$$\frac{1}{(1 - x)^2} = 1 + 2x + 3x^2 + \cdots + (n + 1)x^n + \cdots,$$

find the first five terms of the series for $1/(1 - x)^3$.

⋆22. By integrating a suitable series prove that

$$(1 + x) \ln(1 + x) = x + \sum_{n=2}^{\infty} \frac{(-1)^n x^n}{n(n - 1)}.$$

23. Find the convergence set for $\displaystyle\sum_{n=1}^{\infty} \frac{1}{n}\left(\frac{x - 1}{x}\right)^n$.

⋆24. By comparing areas show that $\ln n > \displaystyle\int_{n-1}^{n} \ln x \, dx$ for $n \geq 2$. Use this to prove that $\ln(n!) > n \ln n - n + 1$ for $n \geq 2$. Then show that for $n \geq 2$,

$$\frac{n!}{n^n} > e \frac{1}{e^n}.$$

⋆⋆25. Use the methods of Problem 24 to prove that for $n \geq 3$,

$$\frac{n!}{n^n} < \frac{ne^2}{4} \frac{1}{e^n}.$$

In Problems 26 through 28, show that the given function has the Maclaurin series on the right, as far as the terms indicated.

26. $(1 + x) \cos \sqrt{x} = 1 + \dfrac{1}{2}x - \dfrac{11}{24}x^2 + \dfrac{29}{720}x^3 + \cdots.$

27. $\dfrac{1}{1 + \sin x} = 1 - x + x^2 - \dfrac{5}{6}x^3 + \dfrac{2}{3}x^4 - \dfrac{61}{120}x^5 + \cdots.$

28. $\ln(1 + \sin x) = x - \dfrac{1}{2}x^2 + \dfrac{1}{6}x^3 - \dfrac{1}{12}x^4 + \dfrac{1}{24}x^5 + \cdots.$

In Problems 29 through 32, estimate the given integral to the fourth decimal place.

29. $\displaystyle\int_0^{1/4} \sqrt{x}\,\sin x\,dx.$

30. $\displaystyle\int_0^{0.1} \ln\,(1 + \sin x)\,dx.$

31. $\displaystyle\int_0^{1/2} e^{-x^2}\,dx.$

32. $\displaystyle\int_0^1 \cos x^3\,dx.$

In Problems 33 through 36, use infinite series to find the indicated limit.

33. $\displaystyle\lim_{x\to 0}\frac{1 - \sqrt{1 + x}}{x}.$

34. $\displaystyle\lim_{x\to 0}\frac{x^2 - \sinh x^2}{x^6}.$

35. $\displaystyle\lim_{x\to 0}\frac{x\,\sin x}{1 - \cos x}.$

$\star$36. $\displaystyle\lim_{x\to 0}\frac{\sin^2 x - \sin x^2}{x^3\,\sin x}.$

37. By rationalizing the denominator, show that

$$\sum_{k=1}^{n}\frac{1}{\sqrt{k+1} + \sqrt{k}} = \sqrt{n+1} - 1.$$

Deduce from this that $\displaystyle\sum_{k=1}^{\infty}\frac{1}{\sqrt{k}}$ diverges.

16

Vectors and Solid Analytic Geometry

1 The Rectangular Coordinate System

To locate points in three-dimensional space, we must have some fixed reference frame. We obtain such a frame by selecting a fixed point O for the origin and three directed lines that are mutually perpendicular at O (see Fig. 1). These three lines are called the x-axis, y-axis, and z-axis.

It is customary and convenient to have the x-axis and y-axis in a horizontal plane and the z-axis vertical. Suppose that we place our right hand so that the thumb points in the positive direction of the x-axis and the index finger points in the positive direction of the y-axis. If the middle finger points in the positive direction of the z-axis, then the coordinate system is called *right-handed*. If the middle finger points in the negative direction along the z-axis, the coordinate system is called *left-handed*. The system shown in Fig. 1 is right-handed, and in this book we will always use a right-handed system.

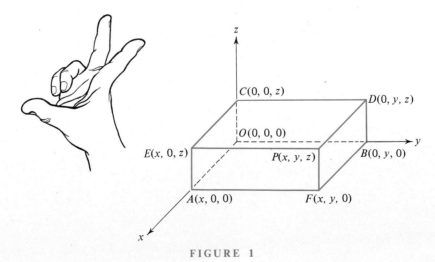

FIGURE 1

636

The *x*-axis and *y*-axis together determine a horizontal plane called the *xy*-plane. Similarly, the *xz*-plane is the vertical plane containing the *x*-axis and *z*-axis, and the *yz*-plane is the plane determined by the *y*-axis and *z*-axis.

If *P* is any point in space it has three coordinates with respect to this fixed frame of reference, and these coordinates are indicated by writing $P(x, y, z)$. These coordinates can be defined thus:

> *x* is the directed distance of *P* from the *yz*-plane,
> *y* is the directed distance of *P* from the *xz*-plane,
> *z* is the directed distance of *P* from the *xy*-plane.

Referring to Fig. 1, we see that these are the directed distances *DP*, *EP*, and *FP*, respectively. These line segments are the edges of a box,[1] with each face perpendicular to one of the coordinate axes. With the lettering of Fig. 1, *A* is the projection of *P* on the *x*-axis, *B* is the projection of *P* on the *y*-axis, and *C* is the projection of *P* on the *z*-axis. Clearly, an alternative definition for the coordinates of *P* is

> *x* is the directed distance *OA*,
> *y* is the directed distance *OB*,
> *z* is the directed distance *OC*.

The points $Q(-7, 4, 3)$, $R(-2, -8, 5)$, and $S(2, -5, -4)$ are shown in Fig. 2, along with their associated boxes. It is clear that if *x* is negative, the point (x, y, z) lies in back of the *yz*-plane. If *y* is negative, the point lies to the left of the *xz*-plane; and if *z* is negative, the point lies below the *xy*-plane. These three coordinate planes divide space into eight separate regions called *octants*. The octant in which all three coordinates are positive is called the *first octant*. The other octants could be numbered, but there is no real reason for doing so.

The preceding discussion suggests the obvious

THEOREM 1. With a fixed set of coordinate axes, there is a one-to-one correspondence between the set of points in space and the set of all ordered triples of real numbers (x, y, z). For each point *P* there is a uniquely determined ordered triple of real numbers, the coordinates (x, y, z) of *P*. Conversely, for each such triple of numbers there is a uniquely determined point in space with this triple for its coordinates.

Many of the theorems from plane analytic geometry have a simple and obvious extension in solid analytic geometry, and in fact their proofs can be relegated to the exercises. As a simple example we have

[1] This three-dimensional figure with vertices *O*, *A*, *F*, *B*, *C*, *E*, *P*, and *D* is usually called a rectangular parallelepiped. We will use the shorter word *box* for such figures.

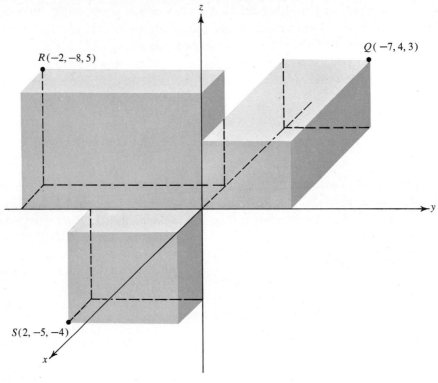

FIGURE 2

THEOREM 2 (PLE). The distance between the origin and the point $P(x, y, z)$ is given by the formula

$$r = \sqrt{x^2 + y^2 + z^2}.$$

Just as in the plane, the graph of an equation in the three variables x, y, and z is the collection of all points whose coordinates satisfy the given equation. Such an equation can be written symbolically as $F(x, y, z) = 0$, where $F(x, y, z)$ denotes, as usual, some suitable function of three variables. We recall that in plane analytic geometry the graph of an equation $F(x, y) = 0$ usually turns out to be some curve. By contrast we will see that in three-dimensional analytic geometry the graph of an equation $F(x, y, z) = 0$ is usually a surface. One way of describing a curve in three-dimensional space is as the intersection of two surfaces. Hence the points P whose coordinates satisfy simultaneously two given equations, $F(x, y, z) = 0$ and $G(x, y, z) = 0$, usually form a curve.

We will prove in Section 7 that the graph of $Ax + By + Cz - D = 0$ is always a plane if at least one of the coefficients A, B, and C is not zero. Conversely, each plane has an equation of this form. In the meantime let us assume these two facts, while we gain some experience with the material through examples.

Example 1. Sketch **(a)** the surface $2x + 3y = 6$, **(b)** the surface $5y + 2z = 10$, and **(c)** the curve of intersection of these two surfaces.

Solution. **(a)** In the xy-plane the graph of $2x + 3y = 6$ is a straight line with intercepts 3 and 2 on the x-axis and y-axis, respectively. Hence the straight line through the points $A(3, 0, 0)$ and $B(0, 2, 0)$ is a part of the three-dimensional graph of $2x + 3y = 6$ (see Fig. 3). Since z does not enter explicitly in the equation $2x + 3y = 6$, we see that z can assume any value as long as $x = x_0$ and $y = y_0$ satisfy the given equation. Hence $P(x_0, y_0, z)$ will be a point of the graph of $2x + 3y = 6$ if and only if the point (x_0, y_0) is on the line AB. Thus P is on the graph if and only if it lies directly above (or on, or below) the line AB. Hence the graph of $2x + 3y = 6$ is the plane containing the line AB and perpendicular to the xy-plane. A portion of this plane is shown in Fig. 3.

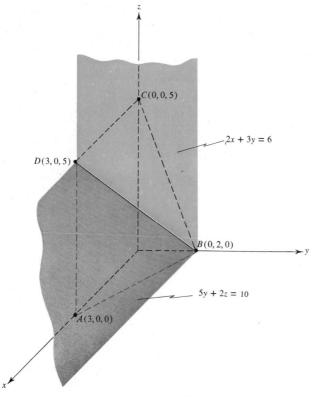

FIGURE 3

(b) The graph of $5y + 2z = 10$ in the yz-plane is just the straight line through the points $B(0, 2, 0)$ and $C(0, 0, 5)$. Since x may assume any value, the graph of $5y + 2z = 10$ in three-dimensional space is just the plane through the line BC and perpendicular to the yz-plane. A portion of this plane is shown in Fig. 3.

(c) The intersection of these two planes is a straight line. Since $B(0, 2, 0)$ and $D(3, 0, 5)$ lie on both planes, the graph of the pair of equations $2x + 3y = 6$ and $5y + 2z = 10$ is the line through the points B and D. ●

DEFINITION 1 (Cylinder). Let $\mathscr{C}$ be a plane curve, let $L^\star$ be a fixed line through some point on $\mathscr{C}$, and let $\mathscr{L}$ be the set consisting of $L^\star$ and each line L that is parallel to $L^\star$ and contains a point of $\mathscr{C}$. Then the union of all lines in $\mathscr{L}$ is a surface that is called a *cylinder*. A point is on the cylinder if and only if it is on some line in the set $\mathscr{L}$. The curve $\mathscr{C}$ is called a *directrix* of the cylinder and each line in $\mathscr{L}$ is called an element (or a generator) of the cylinder. The cylinder is called a *circular cylinder* if $\mathscr{C}$ is a circle. If, in addition, $L^\star$ is perpendicular to the plane of $\mathscr{C}$, then the cylinder is called a *right circular cylinder*.

A plane is a simple example of a cylinder in which the directrix is a straight line. In fact, any line in the plane can be regarded as a directrix for the plane.

A generator of a cylinder need not be parallel to a coordinate axis, but whenever it is parallel to an axis, the cylinder has an equation in which the corresponding variable is missing. Conversely, if one variable is missing in an equation, then the graph is a cylinder with generators that are parallel to the corresponding axis. For example, the graph of $y^2 + z^2 - 16 = 0$ in the yz-plane is a circle about the origin. Then, in three-dimensional space, the graph of the same equation is a right circular cylinder with generators parallel to the x-axis. A portion of the cylinder is shown in Fig. 4.

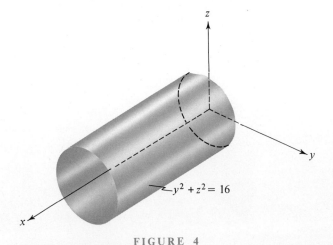

$-y^2 + z^2 = 16$

FIGURE 4

EXERCISE 1

1. Locate each of the following points, and sketch its associated box. Observe that there is exactly one point in each octant. $P_1(1, 2, 3)$, $P_2(-1, 2, 3)$, $P_3(5, -8, 1)$, $P_4(-5, -8, 1)$, $P_5(3, 4, -9)$, $P_6(-7, 4, -3.5)$, $P_7(\pi, -\sqrt{15}, -1.5)$, and $P_8(-1, -1, -1)$.

2. Sketch the three-dimensional graph of each of the equations (a) $z = 5$, (b) $x = 3$, and (c) $y = -6$. Observe that each of these graphs is a plane parallel to one of the coordinate planes.

3. Sketch the box bounded by the following six planes: $x = 2$, $x = 7$, $y = 3$, $y = 10$, $z = 4$, and $z = -3$. Find the coordinates for each of the eight vertices of this box.

In Problems 4 through 7, sketch the given pair of planes, and their line of intersection. In each case find the coordinates of the points in which the line of intersection meets the xz-plane and the yz-plane.

4. $y + 6z = 6$, $6x + 5y = 30$.   5. $2x + 7y = 14$, $x + z = 7$.

6. $x + y = 4$, $y + 2z = 16$.   ★7. $5x + z = 10$, $2y - z = 5$.

In Problems 8 through 11, sketch that portion of the graph of the given equation that lies in the first octant.

8. $x^2 + y^2 = 4$.   9. $4y = x^2$.

10. $y = 8 - z^2$.   11. $9x^2 + 25y^2 = 225$.

12. Prove Theorem 2. HINT: In the box of Fig. 1, draw the lines OF and OP and consider the right triangles OAF and OFP. First compute $|OF|$ and then $r = |OP|$.

13. Give an equation for the sphere of radius 5 and center at the origin.

14. Find the length of the diagonal of the box of Problem 3.

15. If $F(x, y, z) \equiv x + 2y - z^2$, find (a) $F(1, 2, 3)$, (b) $F(2, 4, -3)$, and (c) $F(2, -1, 0)$.

★16. Let F be defined as in Problem 15. Prove that if t is any real number, then $F(t^2 + 2t, -t, t) \equiv 0$. Thus each point with coordinates $(t^2 + 2t, -t, t)$ lies on the surface $x + 2y - z^2 = 0$.

★17. Prove that for any real t, the point $(2t^3, 8t^2 - t^3, 4t)$ lies on the surface defined in Problem 16.

Vectors in Three-Dimensional Space **2**

Solid analytic geometry can be presented without the use of vectors. However, when vectors are used the presentation can be simplified, and indeed to such an extent that we will be amply rewarded for the additional time and energy required to learn the necessary vector algebra.

Much of the material on vectors covered in Chapter 12 is valid for three-dimensional space and needs no detailed discussion. We ask the reader to quickly review the following items: (1) the definition of a vector, (2) the definition of addition of two vectors, (3) the definition of multiplication of a vector by a scalar, and (4) the properties of these two operations stated in Theorems 1, 2, 3, and 4 of Chapter 12. It is easy to see that in all these items, there is no need to suppose that the vectors lie in a plane.

By contrast, when we compute, using the two unit vectors **i** and **j**, we are able to handle only those vectors that lie in the plane of **i** and **j**. In order to compute with vectors in three-dimensional space, it is convenient to introduce a third unit vector, **k**. Accordingly, we let **i**, **j**, and **k** be three mutually perpendicular unit vectors with **i** pointing in the positive direction along the x-axis, **j** pointing in the positive direction along the y-axis, and **k** pointing in the positive direction along the z-axis (see Fig. 5). If $P(x, y, z)$ is any point, then the position vector $\mathbf{R} = \mathbf{OP}$ can always be written in the form[1]

$$(1) \qquad\qquad \mathbf{OP} = x\mathbf{i} + y\mathbf{j} + z\mathbf{k}.$$

For example, as indicated in Fig. 5, the vector from the origin to the point $(3, 7, 4)$ is $\mathbf{R} = 3\mathbf{i} + 7\mathbf{j} + 4\mathbf{k}.$

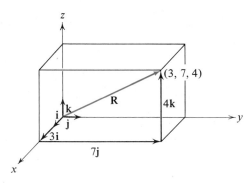

FIGURE 5

Since any vector may be shifted parallel to itself, we can always regard a vector as starting at the origin of the coordinate system, whenever it is convenient to do so. Thus any vector can be written in the form (1).

Given two vectors, $\mathbf{A} = a_1\mathbf{i} + a_2\mathbf{j} + a_3\mathbf{k}$ and $\mathbf{B} = b_1\mathbf{i} + b_2\mathbf{j} + b_3\mathbf{k}$, their sum is found by adding the corresponding components; thus

$$(2) \qquad\qquad \mathbf{A} + \mathbf{B} = (a_1 + b_1)\mathbf{i} + (a_2 + b_2)\mathbf{j} + (a_3 + b_3)\mathbf{k}.$$

---

[1] Some authors prefer the notation $[x, y, z]$ or $\langle x, y, z \rangle$ for the vector $x\mathbf{i} + y\mathbf{j} + z\mathbf{k}$. Although $[x, y, z]$ and $\langle x, y, z \rangle$ are shorter and easier to write than $x\mathbf{i} + y\mathbf{j} + z\mathbf{k}$, we feel that when studying vectors for the first time, the unit vectors **i**, **j**, and **k** are a real help in understanding the subject. The gain in understanding is well worth the extra time and energy necessary to write **i**, **j**, and **k**.

This is just the generalization of Theorem 5 of Chapter 12 to three-dimensional space. The proof is similar to the proof for plane vectors, and hence we omit it.

In the same way, it is easy to see that if c is any number, then

(3) $$\mathbf{A} = ca_1\mathbf{i} + ca_2\mathbf{j} + ca_3\mathbf{k}.$$

By Theorem 2 the length of the vector $\mathbf{R} = x\mathbf{i} + y\mathbf{j} + z\mathbf{k}$ is given by

(4) $$R = |\mathbf{R}| = \sqrt{x^2 + y^2 + z^2}.$$

The vector $\mathbf{R}$ is completely determined whenever we specify its three components x, y, and z along the three axes. But we could also specify the vector by giving its length and the angle that it makes with each of the coordinate axes. These angles are called the *direction angles* of the vector and are denoted by α, β, and γ. Thus, as indicated in Fig. 6:

> α is the angle between the vector and the positive x-axis.
> β is the angle between the vector and the positive y-axis.
> γ is the angle between the vector and the positive z-axis.

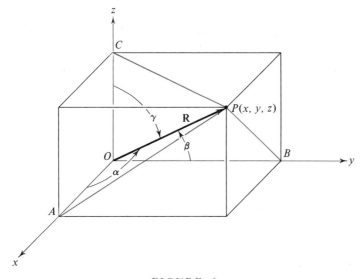

FIGURE 6

By definition, each of these angles lies between 0 and π. Actually, we are more interested in the cosines of these angles, and if P is in the first octant, it is obvious from the right triangles shown in Fig. 6 that

$$\cos \alpha = \frac{x}{\sqrt{x^2 + y^2 + z^2}}, \qquad \text{right triangle } OAP,$$

$$(5) \qquad \cos \beta = \frac{y}{\sqrt{x^2 + y^2 + z^2}}, \qquad \text{right triangle } OBP,$$

$$\cos \gamma = \frac{z}{\sqrt{x^2 + y^2 + z^2}}, \qquad \text{right triangle } OCP.$$

Of course, if P is not in the first octant, some of the numbers $\cos \alpha$, $\cos \beta$, $\cos \gamma$ may be negative. For example, if P lies to the left of the xz-plane, then β is a second quadrant angle and $\cos \beta$ is negative. But in this case y is also negative. A little reflection will show that the formulas (5) hold for any position of the point P, as long as $R \neq 0$.

The numbers ($\cos \alpha$, $\cos \beta$, $\cos \gamma$) are called the *direction cosines* of the vector **R**. Obviously, these specify the direction of **R**. However, these direction cosines cannot be assigned arbitrarily because they must satisfy the condition

$$(6) \qquad \cos^2 \alpha + \cos^2 \beta + \cos^2 \gamma = 1.$$

To prove this, we have from the equation set (5),

$$\cos^2 \alpha + \cos^2 \beta + \cos^2 \gamma = \frac{x^2}{x^2 + y^2 + z^2} + \frac{y^2}{x^2 + y^2 + z^2} + \frac{z^2}{x^2 + y^2 + z^2}$$

$$= \frac{x^2 + y^2 + z^2}{x^2 + y^2 + z^2} = 1.$$

If m is any positive constant, the set of numbers

$$(m \cos \alpha, \ m \cos \beta, \ m \cos \gamma)$$

also serves to specify the direction of **R**. Such a triple is called a triple of *direction numbers* for the vector **R**. Conversely, given some direction numbers, we can always find the direction cosines for the vector, as illustrated in the next example.

Finally, we remark that if we know the length of the vector and its direction cosines, then the vector is completely specified, and combining equation (4) with equation set (5), we have

$$(7) \qquad \mathbf{R} = |\mathbf{R}| \, (\cos \alpha \mathbf{i} + \cos \beta \mathbf{j} + \cos \gamma \mathbf{k}).$$

Since $x = |\mathbf{R}| \cos \alpha$, $y = |\mathbf{R}| \cos \beta$, and $z = |\mathbf{R}| \cos \gamma$, it is clear that the components of a vector form a triple of direction numbers. We obtain the components from the direction cosines by multiplying by $|\mathbf{R}|$. Conversely, dividing the components by $|\mathbf{R}|$ gives the direction cosines.

Example 1. A certain vector **R** has length 21 and direction numbers $(2, -3, 6)$. Find the direction cosines and the components of **R**.

Solution. From the statement of the problem, **R** is parallel to the vector **S** = 2**i** − 3**j** + 6**k**. Since 4 + 9 + 36 = 49, we have |**S**| = 7. Hence for the direction cosines of **S**, and also **R**, we have

$$\cos \alpha = \frac{2}{7}, \qquad \cos \beta = -\frac{3}{7}, \qquad \cos \gamma = \frac{6}{7}.$$

Then

$$\mathbf{R} = 21 \left(\frac{2}{7}\mathbf{i} - \frac{3}{7}\mathbf{j} + \frac{6}{7}\mathbf{k} \right) = 6\mathbf{i} - 9\mathbf{j} + 18\mathbf{k}. \quad \bullet$$

Example 2. Find the direction cosines of the vector from $A(4, 8, -3)$ to $B(-1, 6, 2)$.

Solution. We recall from Theorem 8 of Chapter 12 that for any two points A and B, we have **AB** = **OB** − **OA**. Hence

$$\mathbf{AB} = -\mathbf{i} + 6\mathbf{j} + 2\mathbf{k} - (4\mathbf{i} + 8\mathbf{j} - 3\mathbf{k}) = -5\mathbf{i} - 2\mathbf{j} + 5\mathbf{k}.$$

It follows that $|\mathbf{AB}| = \sqrt{5^2 + 2^2 + 5^2} = \sqrt{54} = 3\sqrt{6}$, and then from equation set (5) that

$$\cos \alpha = \frac{-5}{3\sqrt{6}}, \qquad \cos \beta = \frac{-2}{3\sqrt{6}}, \qquad \cos \gamma = \frac{5}{3\sqrt{6}}. \quad \bullet$$

This example suggests

THEOREM 3. The length of the vector from $A(a_1, a_2, a_3)$ to $B(b_1, b_2, b_3)$ is given by

(8)
$$|\mathbf{AB}| = \sqrt{(b_1 - a_1)^2 + (b_2 - a_2)^2 + (b_3 - a_3)^2}.$$

If A and B are distinct points, then the direction cosines of **AB** are given by

(9)
$$\cos \alpha = \frac{b_1 - a_1}{|\mathbf{AB}|}, \qquad \cos \beta = \frac{b_2 - a_2}{|\mathbf{AB}|}, \qquad \cos \gamma = \frac{b_3 - a_3}{|\mathbf{AB}|}.$$

Proof. Just as in the example,

$$\mathbf{AB} = \mathbf{OB} - \mathbf{OA} = b_1\mathbf{i} + b_2\mathbf{j} + b_3\mathbf{k} - (a_1\mathbf{i} + a_2\mathbf{j} + a_3\mathbf{k})$$

(10)
$$\mathbf{AB} = (b_1 - a_1)\mathbf{i} + (b_2 - a_2)\mathbf{j} + (b_3 - a_3)\mathbf{k}.$$

Then (8) follows by applying equation (4) to (10), and (9) follows from equation set (5) in the same way. ∎

COROLLARY. The distance between the points $A(a_1, a_2, a_3)$ and $B(b_1, b_2, b_3)$ is given by equation (8).

THEOREM 4 (PLE). If $r > 0$, then

$$(x - a)^2 + (y - b)^2 + (z - c)^2 = r^2$$

is an equation for a sphere with center at (a, b, c) and radius r.

Example 3. Describe the graph of the equation

$$x^2 + y^2 + z^2 - 2x + 4z = 0.$$

Solution. On completing the squares, we obtain

$$(x^2 - 2x + 1) + y^2 + (z^2 + 4z + 4) = 1 + 4$$

or

$$(x - 1)^2 + (y - 0)^2 + (z + 2)^2 = 5.$$

Hence by Theorem 4, the graph is a sphere with center $(1, 0, -2)$ and radius $\sqrt{5}$. ●

EXERCISE 2

1. Make a three-dimensional sketch showing each of the following vectors with its initial point at the origin. $\mathbf{A} = \mathbf{i} + 2\mathbf{j} + 3\mathbf{k}$, $\mathbf{B} = 4\mathbf{i} - 3\mathbf{j} - \mathbf{k}$, $\mathbf{C} = -5\mathbf{i} - 3\mathbf{j} + 5\mathbf{k}$, $\mathbf{D} = -7\mathbf{i} + \mathbf{j} - 15\mathbf{k}$, and $\mathbf{E} = 4\mathbf{i} - 7\mathbf{k}$.
2. Using the vectors of Problem 1, compute each of the following vectors.
   (a) $\mathbf{A} + \mathbf{B}$. (b) $2\mathbf{A} - \mathbf{C}$. (c) $\mathbf{C} + \mathbf{D} + \mathbf{E}$. (d) $3\mathbf{A} - 2\mathbf{B} + \mathbf{C} - 2\mathbf{D} + \mathbf{E}$.
3. Using the vectors of Problem 1, compute
   (a) $\mathbf{B} + \mathbf{C}$. (b) $2\mathbf{C} + \mathbf{A}$. (c) $2\mathbf{A} - \mathbf{C} - \mathbf{D}$. (d) $\mathbf{A} + \mathbf{B} + \mathbf{D} - \mathbf{E}$.
4. With the vectors of Problem 1, show that $4\mathbf{A} + 2\mathbf{B} + \mathbf{C} + \mathbf{D} = \mathbf{0}$.

5. Find the direction cosines for the vectors $\mathbf{F} = 2\mathbf{i} + \mathbf{j} - 2\mathbf{k}$, $\mathbf{G} = 6\mathbf{i} - 3\mathbf{j} + 2\mathbf{k}$, $\mathbf{H} = \mathbf{i} + \mathbf{j} + \mathbf{k}$, and $\mathbf{I} = -5\mathbf{i} + 6\mathbf{j} + 8\mathbf{k}$.

6. Find the direction cosines for the vectors $\mathbf{A} = \mathbf{i} - 2\mathbf{j} + 3\mathbf{k}$, $\mathbf{B} = 2\mathbf{i} - 6\mathbf{j} - 3\mathbf{k}$, and $\mathbf{C} = -\mathbf{i} - 3\mathbf{j} + 5\mathbf{k}$.

7. In each of the following, find the length of the vector from the first point to the second point.

    (a) $(3, 2, -2)$, $(7, 4, 2)$.   (b) $(5, -1, -6)$, $(-3, -5, 2)$.

8. Do Problem 7 for

    (a) $(-3, 11, -4)$, $(4, 10, -9)$.   (b) $(-1, 9, 11)$, $(-13, 22, 16)$.

*9. Equation (6) is the generalization to three-dimensional space of an important formula from plane trigonometry. What is the formula?

10. Prove that two vectors $\mathbf{A} = a_1\mathbf{i} + a_2\mathbf{j} + a_3\mathbf{k}$ and $\mathbf{B} = b_1\mathbf{i} + b_2\mathbf{j} + b_3\mathbf{k}$ are equal if and only if $a_1 = b_1$, $a_2 = b_2$, and $a_3 = b_3$.

*11. Given the vectors $\mathbf{A} = -\mathbf{i} + 3\mathbf{j} + \mathbf{k}$, $\mathbf{B} = 8\mathbf{i} + 2\mathbf{j} - 4\mathbf{k}$, $\mathbf{C} = \mathbf{i} + 2\mathbf{j} - \mathbf{k}$, and $\mathbf{D} = -\mathbf{i} + \mathbf{j} + 3\mathbf{k}$, find scalars m, n, and p such that

$$m\mathbf{A} + n\mathbf{B} + p\mathbf{C} = \mathbf{D}.$$

12. Prove that the points $A(2, -1, 6)$, $B(3, 2, 8)$, and $C(8, 7, -9)$ form the vertices of a right triangle.

13. Find an equation for the sphere with center $P_0(2, -3, 6)$ and radius 7.

14. Do Problem 13 if the center is at $(-1, 3, 5)$ and the radius is 8.

15. Find a simple equation for the set of all points that are equidistant from the points $A(1, 3, 5)$ and $B(2, -3, -4)$. Naturally, these points form a plane.

16. Repeat Problem 15 for the pairs of points: (a) $(0, 0, 1)$, $(1, 1, 0)$, and (b) $(0, 0, 0)$, $(2a, 2b, 2c)$.

*17. Let a, b, and c be any three real numbers. Prove that $P_1(a, b, c)$, $P_2(b, c, a)$, and $P_3(c, a, b)$ either all coincide or form the vertices of an equilateral triangle.

*18. Prove Theorem 4.

19. Find the center and radius for each of the following spheres.

    (a) $x^2 + y^2 + z^2 + 4x - 6z = 0$.

    (b) $x^2 + y^2 + z^2 + 12x - 6y + 4z = 0$.

20. Do Problem 19 for:

    (a) $x^2 + y^2 + z^2 - 10x + 6y + 8z + 14 = 0$.

    (b) $3x^2 + 3y^2 + 3z^2 - x - 5y - 4z = 2$.

21. A point P moves in three-dimensional space so that its distance from the point $A(0, 2, 0)$ is always twice its distance from the point $B(0, 5, 0)$. Prove that P always lies on a sphere. Find the center and radius of that sphere.

22. Find the distance of the point $P_1(x_1, y_1, z_1)$ from:

    (a) The x-axis.   (b) The y-axis.

    (c) The z-axis.   (d) The plane $x = 2$.

    (e) The line of intersection of the planes $y = -3$, $z = 5$.

3 *Equations of Lines in Space*

A straight line $\mathscr{L}$ in three-dimensional space is completely determined if we are given two points on the line, or if we are given one point on the line together with the direction of the line. The direction of the line can be specified by giving a vector parallel to the line. Suppose that the two points $P_0(x_0, y_0, z_0)$ and $P_1(x_1, y_1, z_1)$ are on the line $\mathscr{L}$. Then the vector

$$\mathbf{P_0P_1} = \mathbf{OP_1} - \mathbf{OP_0} = (x_1 - x_0)\mathbf{i} + (y_1 - y_0)\mathbf{j} + (z_1 - z_0)\mathbf{k}$$

that joins the two points on the line specifies the direction of the line.

To obtain a vector equation for the line, we observe that we can reach any point P on the line by proceeding first from O to P_0 and then traveling along the line a suitable multiple t of the vector $\mathbf{P_0P_1}$ to P, as illustrated in Fig. 7. Thus if $\mathbf{R}$ is the position vector of a point P on the line, then there is a scalar t such that

(11) $$\mathbf{R} = \mathbf{OP_0} + t\mathbf{P_0P_1}.$$

Conversely, for each real number t, the vector $\mathbf{R}$ of equation (11) is the position vector of

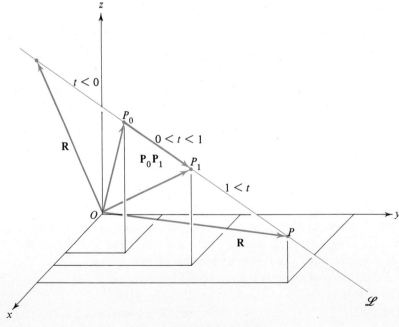

FIGURE 7

a point on the line through P_0 and P_1. Thus we have proved that (11) is a *vector equation* for the line through P_0 and P_1.

Observe that if $0 \leq t \leq 1$, then P is on the line segment joining P_0 and P_1. If $t > 1$, the points are in the order P_0, P_1, P on the line. Finally, if $t < 0$, the points are in the order P, P_0, P_1 on the line.

If we put $\mathbf{R} = x\mathbf{i} + y\mathbf{j} + z\mathbf{k}$, equation (11) can be written as

$$(12) \quad x\mathbf{i} + y\mathbf{j} + z\mathbf{k} = x_0\mathbf{i} + y_0\mathbf{j} + z_0\mathbf{k} + t(x_1 - x_0)\mathbf{i} + t(y_1 - y_0)\mathbf{j} + t(z_1 - z_0)\mathbf{k}.$$

Equating corresponding components on both sides of (12) leads to

$$(13) \quad \begin{aligned} x &= x_0 + t(x_1 - x_0), \\ y &= y_0 + t(y_1 - y_0), \\ z &= z_0 + t(z_1 - z_0). \end{aligned}$$

This set of equations forms a *set of parametric equations* for the line. For brevity we set $x_1 - x_0 = a$, $y_1 - y_0 = b$, and $z_1 - z_0 = c$. Then the vector $a\mathbf{i} + b\mathbf{j} + c\mathbf{k}$ is a vector parallel to the line and the set of numbers (a, b, c) is a set of direction numbers for the line. The equations (13) are equivalent to

$$(14) \quad x = x_0 + at, \qquad y = y_0 + bt, \qquad z = z_0 + ct.$$

Finally, since (14) is a set of simultaneous equations in t, we may eliminate t and arrive at

$$(15) \quad \frac{x - x_0}{a} = \frac{y - y_0}{b} = \frac{z - z_0}{c},$$

as long as a, b, and c are all different from zero. These equations are often called *symmetric equations* of the line.

To summarize, each of (11), (13), (14), and (15) determines the same straight line in space, and the vector $a\mathbf{i} + b\mathbf{j} + c\mathbf{k}$ is parallel to the straight line. We observe that (15) can be thought of as a pair of equations

$$\frac{x - x_0}{a} = \frac{y - y_0}{b} \qquad \text{and} \qquad \frac{y - y_0}{b} = \frac{z - z_0}{c}, \qquad a, b, c \neq 0.$$

The first equation is the equation of a plane perpendicular to the xy-plane. The second is the equation of a plane perpendicular to the yz-plane. The straight line in question is then the intersection of these two planes.

Example 1. Find a vector equation for the line through the points $(7, -3, 5)$ and $(-2, 8, 1)$. Where does this line pierce the xy-plane?

Solution. Let the two points be P_0 and P_1, respectively. Then we have $P_0P_1 = -9\mathbf{i} + 11\mathbf{j} - 4\mathbf{k}$, and a vector equation is

(11e) $$\mathbf{R} = 7\mathbf{i} - 3\mathbf{j} + 5\mathbf{k} + t(-9\mathbf{i} + 11\mathbf{j} - 4\mathbf{k}).$$

The set of parametric equations obtained from (11e) is

(14e) $$x = 7 - 9t, \qquad y = -3 + 11t, \qquad z = 5 - 4t.$$

Eliminating t from this set, we find that

(15e) $$\frac{x-7}{-9} = \frac{y+3}{11} = \frac{z-5}{-4}.$$

The numbers $(-9, 11, -4)$ form a set of direction numbers for this line.

This line meets the xy-plane when $z = 0$. Using this in the last equation of set (14e) gives $t = 5/4$. Then, from the other equations of this set, we have

$$x = 7 - 9 \cdot \frac{5}{4} = -\frac{17}{4} \quad \text{and} \quad y = -3 + 11 \cdot \frac{5}{4} = \frac{43}{4}.$$

Hence the pierce point is $(-17/4, 43/4, 0)$. ●

Example 2. Find a vector equation for $\mathscr{L}$, the line of intersection of the two planes

(16) $$\mathscr{P}_1: x + y - 2z = -3 \quad \text{and} \quad \mathscr{P}_2: x - 3y - 4z = -19.$$

First Solution. Our plan is to find two points on $\mathscr{L}$ and then use the method of Example 1. We can select one of the variables arbitrarily and then solve the system of two equations in the remaining two variables. If we set $z = 0$, then the system (16) becomes

$$x + y = -3$$
$$-x + 3y = 19,$$

and this system has the solution $x = -7$, $y = 4$. Hence the point $P_0(-7, 4, 0)$ is on $\mathscr{L}$.

If we set $y = 0$, then the system (16) becomes

$$x - 2z = -3$$
$$-x + 4z = 19,$$

and this system has the solution $x = 13$, $z = 8$. Hence the point $P_1(13, 0, 8)$ is also on $\mathscr{L}$. Following the method of Example 1, we find that a vector equation for $\mathscr{L}$ is

$$\mathbf{R} = \mathbf{OP_0} + t\mathbf{P_0P_1}$$

$$= -7\mathbf{i} + 4\mathbf{j} + t[13\mathbf{i} + 8\mathbf{k} - (-7\mathbf{i} + 4\mathbf{j})]$$

$$= -7\mathbf{i} + 4\mathbf{j} + t[20\mathbf{i} - 4\mathbf{j} + 8\mathbf{k}]$$

$$(17) \qquad \mathbf{R} = -7\mathbf{i} + 4\mathbf{j} + 4t[5\mathbf{i} - \mathbf{j} + 2\mathbf{k}]. \quad \bullet$$

In this method, an "unfortunate" set of equations, or an "unfortunate" selection of the variable, may lead to "unpleasant-looking numbers." For example, suppose that we set $x = 0$. Then the system (16) becomes

$$y - 2z = -3$$

$$3y + 4z = 19,$$

and this system has the solution $y = 13/5$, $z = 14/5$.

Second Solution. We first obtain a set of symmetric equations for $\mathscr{L}$ and use these to derive a vector equation. We let one of the variables act as a parameter; for example, we set $z = T$ (avoiding the letter t used in the first solution). Then the system (16) becomes

$$x + y = -3 + 2z = -3 + 2T$$

$$-x + 3y = 19 - 4z = 19 - 4T.$$

If we solve the system (18) in terms of T, we find that

$$(18) \qquad x = -7 + \frac{5}{2}T, \qquad y = 4 - \frac{1}{2}T.$$

This together with $z = T$ gives for $\mathscr{L}$ the symmetric equations

$$(19) \qquad T = \frac{x + 7}{5/2} = \frac{y - 4}{-1/2} = \frac{z - 0}{1}.$$

Hence $\mathscr{L}$ passes through the point $(-7, 4, 0)$ and has the direction numbers $(5/2, -1/2, 1)$. Consequently, a vector equation for $\mathscr{L}$ is

$$(20) \qquad \mathbf{R} = -7\mathbf{i} + 4\mathbf{j} + T\left(\frac{5}{2}\mathbf{i} - \frac{1}{2}\mathbf{j} + \mathbf{k}\right). \quad \bullet$$

We observe that equations (17) and (20) are both vector equations for the same line, although they look somewhat different. However, if we set $T = 8t$ in equation (20), then (20) becomes identical with (17).

More generally, if in (20) we set $T = C_1 t + C_2$, where $C_1 \neq 0$, then the new equation still gives the same line. Further, every equation for $\mathscr{L}$ can be obtained from (20) by selecting C_1 and C_2 properly in $T = C_1 t + C_2$. For example, if we set $T = 2t + 4$ in (20), we obtain the equation

$$\mathbf{R} = 3\mathbf{i} + 2\mathbf{j} + 4\mathbf{k} + t(5\mathbf{i} - \mathbf{j} + 2\mathbf{k}).$$

Example 3. Find a formula for the coordinates of the midpoint of the line segment joining the points P_0 and P_1.

Solution: Referring to Fig. 7, we see that the position vector of the midpoint is

$$\mathbf{R} = \mathbf{OP}_0 + \frac{1}{2}\mathbf{P}_0\mathbf{P}_1.$$

Setting $t = 1/2$ in equations (12) or (13) yields

$$x = x_0 + \frac{1}{2}(x_1 - x_0), \qquad y = y_0 + \frac{1}{2}(y_1 - y_0), \qquad z = z_0 + \frac{1}{2}(z_1 - z_0)$$

or

$$x = \frac{x_0 + x_1}{2}, \qquad y = \frac{y_0 + y_1}{2}, \qquad z = \frac{z_0 + z_1}{2}. \qquad \bullet$$

EXERCISE 3

In Problems 1 through 4, find symmetric equations for the line joining the two given points.
1. $(1, 2, 3)$, $(4, 6, -9)$.
2. $(0, -5, 8)$, $(1, 6, 2)$.
3. $(2, 1, -3)$, $(5, -4, 7)$.
4. $(-2, 6, 8)$, $(2, 6, -8)$.

5. Where does the line of Problem 1 meet the yz-plane?
6. Where does the line of Problem 2 meet the plane $y = -16$?
7. Is there a point on the line of Problem 3 at which all the coordinates are equal?
*8. Find all points on the line of Problem 3 in which two of the coordinates are equal.

In Problems 9 through 12, find a vector equation for the line of intersection of the two given planes.
9. $2x + 7y = 14$, $x + z = 7$.
10. $5x + z = 10$, $2y - z = 5$.
11. $3x - 4y = 9$, $4y - 5z = 20$.
12. $x + 7y - z = 7$, $2x + 21y - 4z = 14$.

In Problems 13 through 16, the points P_0, P_1, and P_2 are collinear and occur in that order on the line.

13. Find the coordinates of P_2 if $|\mathbf{P_0P_2}| = 2|\mathbf{P_0P_1}|$ and P_0 and P_1 are $(1, 2, 3)$ and $(4, 6, -9)$, respectively.

14. Find the coordinates of P_2 if $2|\mathbf{P_0P_1}| = 3|\mathbf{P_1P_2}|$ and P_0 and P_1 are $(-2, 7, 4)$ and $(7, -2, 1)$, respectively.

*15. Find the coordinates of P_0 if $|\mathbf{P_0P_1}| = 10|\mathbf{P_1P_2}|$ and P_1 and P_2 are $(2, 1, 3)$ and $(-3, 2, -1)$, respectively.

*16. Find the coordinates of P_1 if $4|\mathbf{P_0P_1}| = |\mathbf{P_1P_2}|$ and P_0 and P_2 are $(4, 6, 7)$ and $(9, -14, 2)$, respectively.

*17. Prove that the two sets of equations

$$\frac{x - 1}{3} = \frac{y - 2}{4} = \frac{z - 3}{-12} \quad \text{and} \quad \frac{x + 5}{-6} = \frac{y + 6}{-8} = \frac{z - 27}{24}$$

are equations for the same straight line.

18. Find symmetric equations of the line through the origin that makes the same angle with each of the three coordinate axes.

19. Find the coordinates of the point P in which the line

$$\frac{x - 2}{3} = \frac{y - 3}{4} = \frac{z + 4}{2}$$

intersects the plane $4x + 5y + 6z = 87$.

*20. Prove that the line

$$\frac{x - 1}{9} = \frac{y - 6}{-4} = \frac{z - 3}{-6}$$

lies in the plane $2x - 3y + 5z = -1$.

21. Find the point of intersection of the line through P and Q with the given plane in each of the following cases.
    (a) $P(-1, 5, 1)$, $Q(-2, 8, -1)$, $2x - 3y + z = 10$.
    (b) $P(-1, 0, 9)$, $Q(-3, 1, 14)$, $3x + 2y - z = 6$.
    (c) $P(0, 0, 0)$, $Q(A, B, C)$, $Ax + By + Cz = D$.

22. Give an appropriate interpretation for the equations (14) and (15) in case some or all of the numbers a, b, and c are zero.

The Scalar Product of Two Vectors 4

We want to formulate a definition for the product of two vectors. It turns out that there are two different ways of doing this, both of which give interesting and useful results. Rather than select one of these definitions in preference to the other, we keep both using the "dot" in the first case and the "cross" in the second, in order to distinguish between the two. The scalar product (or *dot product*) $\mathbf{A} \cdot \mathbf{B}$ of the two vectors $\mathbf{A}$ and $\mathbf{B}$ is a number. The *vector product* (or *cross product*) $\mathbf{A} \times \mathbf{B}$ is a vector. We will devote this section to a study of the scalar product, and postpone the vector product until Section 5.

The dot product of vectors $\mathbf{A}$ and $\mathbf{B}$ can be defined either algebraically or geometrically. It turns out that if we adopt the algebraic definition, (which may look artificial at first), then it is easier to obtain the results we wish, so we follow this course.

DEFINITION 2 (Dot Product). Suppose that $\mathbf{A} = a_1\mathbf{i} + a_2\mathbf{j} + a_3\mathbf{k}$ and $\mathbf{B} = b_1\mathbf{i} + b_2\mathbf{j} + b_3\mathbf{k}$. Then the dot product of $\mathbf{A}$ and $\mathbf{B}$ is defined by

(21)
$$\mathbf{A} \cdot \mathbf{B} = a_1b_1 + a_2b_2 + a_3b_3.$$

Example 1. Find $\mathbf{A} \cdot \mathbf{B}$ if $\mathbf{A} = \mathbf{i} - 2\mathbf{j} + 3\mathbf{k}$ and $\mathbf{B} = 5\mathbf{i} + 8\mathbf{j} + 6\mathbf{k}$.

Solution. By equation (21), we have

$$\mathbf{A} \cdot \mathbf{B} = 1(5) + (-2)(8) + 3(6) = 5 - 16 + 18 = 7.$$

Notice that the result is a *number*. ●

This definition of $\mathbf{A} \cdot \mathbf{B}$ leads immediately to

---

THEOREM 5 (PLE). For any three vectors $\mathbf{A}$, $\mathbf{B}$, and $\mathbf{C}$ and any scalar c, we have

(22) $(c\mathbf{A}) \cdot \mathbf{B} = \mathbf{A} \cdot (c\mathbf{B}) = c(\mathbf{A} \cdot \mathbf{B}),$

(23) $\mathbf{A} \cdot \mathbf{B} = \mathbf{B} \cdot \mathbf{A},$

(24) $\mathbf{A} \cdot (\mathbf{B} + \mathbf{C}) = \mathbf{A} \cdot \mathbf{B} + \mathbf{A} \cdot \mathbf{C},$

(25) $\mathbf{0} \cdot \mathbf{A} = \mathbf{A} \cdot \mathbf{0} = 0,$

(26) $\mathbf{A} \cdot \mathbf{A} > 0,$ if $\mathbf{A} \neq \mathbf{0}.$

As an example we prove the hardest one, namely (24). If $\mathbf{C} = c_1\mathbf{i} + c_2\mathbf{j} + c_3\mathbf{k}$, then

$$\mathbf{A} \cdot (\mathbf{B} + \mathbf{C}) = (a_1\mathbf{i} + a_2\mathbf{j} + a_3\mathbf{k}) \cdot ([b_1 + c_1]\mathbf{i} + [b_2 + c_2]\mathbf{j} + [b_3 + c_3]\mathbf{k})$$

$$= a_1(b_1 + c_1) + a_2(b_2 + c_2) + a_3(b_3 + c_3)$$

$$= a_1b_1 + a_1c_1 + a_2b_2 + a_2c_2 + a_3b_3 + a_3c_3$$

$$= (a_1 b_1 + a_2 b_2 + a_3 b_3) + (a_1 c_1 + a_2 c_2 + a_3 c_3)$$

$$= \mathbf{A} \cdot \mathbf{B} + \mathbf{A} \cdot \mathbf{C}.$$

We observe that

$$\mathbf{A} \cdot \mathbf{A} = (a_1 \mathbf{i} + a_2 \mathbf{j} + a_3 \mathbf{k}) \cdot (a_1 \mathbf{i} + a_2 \mathbf{j} + a_3 \mathbf{k}) = a_1^2 + a_2^2 + a_3^2.$$

Hence $\mathbf{A} \cdot \mathbf{A}$ is the square of the length of $\mathbf{A}$, or $\mathbf{A} \cdot \mathbf{A} = |\mathbf{A}|^2$.

The very important geometric property of the dot product is given in

---

THEOREM 6. The scalar product of two vectors is the product of their lengths and the cosine of the angle between them. In symbols,

(27)
$$\mathbf{A} \cdot \mathbf{B} = |\mathbf{A}| \, |\mathbf{B}| \cos \theta,$$

where θ is the angle between the two vectors.

Proof. As shown in Fig. 8, we place the given vectors $\mathbf{A}$ and $\mathbf{B}$ with their initial point at the origin. Let $A(a_1, a_2, a_3)$ and $B(b_1, b_2, b_3)$ be their end points, and consider the triangle OAB. We apply the law of cosines to this triangle to determine the length of the side opposite the angle θ. This gives us

$$|\mathbf{B} - \mathbf{A}|^2 = |\mathbf{A}|^2 + |\mathbf{B}|^2 - 2|\mathbf{A}| \, |\mathbf{B}| \cos \theta,$$

or

(28)
$$|\mathbf{A}| \, |\mathbf{B}| \cos \theta = \frac{1}{2} (|\mathbf{A}|^2 + |\mathbf{B}|^2 - |\mathbf{B} - \mathbf{A}|^2).$$

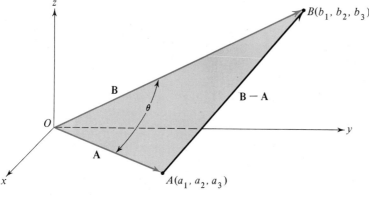

FIGURE 8

Using formulas (4) and (8) for the length of a vector in the right side of (28), we find that

$$|\mathbf{A}|\,|\mathbf{B}|\cos\theta = \frac{1}{2}\left(a_1^2 + a_2^2 + a_3^2 + b_1^2 + b_2^2 + b_3^2\right.$$

$$\left. - (b_1 - a_1)^2 - (b_2 - a_2)^2 - (b_3 - a_3)^2\right) = a_1b_1 + a_2b_2 + a_3b_3$$

$$= \mathbf{A}\cdot\mathbf{B}. \quad\blacksquare$$

Notice that if **A** and **B** are perpendicular, then $\cos\theta = 0$ and $\mathbf{A}\cdot\mathbf{B} = 0$. Conversely, if $\mathbf{A}\cdot\mathbf{B} = 0$ and both the vectors are nonzero, then equation (27) tells us that $\cos\theta = 0$. Hence $\theta = \pi/2$ and the vectors **A** and **B** are mutually perpendicular. We have therefore proved

---

THEOREM 7. Two nonzero vectors **A** and **B** are perpendicular if and only if

(29) $$\mathbf{A}\cdot\mathbf{B} = 0$$

Example 2. For the points $A(1, 1, 2)$, $B(3, 4, 1)$, $C(-1, 4, 8)$, and $D(-6, 5, 1)$, prove that the line through A and B is perpendicular to the line through C and D.

Solution. The vectors **AB** and **CD** specify the directions of the two lines. The lines are perpendicular if the vectors are. Now

$$\mathbf{AB} = \mathbf{OB} - \mathbf{OA} = \quad 3\mathbf{i} + 4\mathbf{j} + \mathbf{k} - (\ \mathbf{i} + \ \mathbf{j} + 2\mathbf{k}) = \quad 2\mathbf{i} + 3\mathbf{j} - \ \mathbf{k},$$

$$\mathbf{CD} = \mathbf{OD} - \mathbf{OC} = -6\mathbf{i} + 5\mathbf{j} + \mathbf{k} - (-\mathbf{i} + 4\mathbf{j} + 8\mathbf{k}) = -5\mathbf{i} + \ \mathbf{j} - 7\mathbf{k}.$$

By definition, the dot product is

$$\mathbf{AB}\cdot\mathbf{CD} = (2\mathbf{i} + 3\mathbf{j} - \mathbf{k})\cdot(-5\mathbf{i} + \mathbf{j} - 7\mathbf{k})$$
$$= 2(-5) + 3\cdot1 + (-1)(-7) = -10 + 3 + 7 = 0.$$

Since the dot product is zero, the two vectors are perpendicular. ●

Observe that we never proved that the two lines intersect, and in fact they do not. This step was omitted because the concept of an angle between two lines is independent of whether or not the lines intersect.

Example 3. Find the angle between the vector $\mathbf{A} = \mathbf{i} - 2\mathbf{j} + 3\mathbf{k}$ and the vector $\mathbf{B} = 5\mathbf{i} + 8\mathbf{j} + 6\mathbf{k}$.

Solution. From Theorem 6 and the definition of the scalar product,

$$|\mathbf{A}|\,|\mathbf{B}|\cos\theta = \mathbf{A}\cdot\mathbf{B} = (\mathbf{i}-2\mathbf{j}+3\mathbf{k})\cdot(5\mathbf{i}+8\mathbf{j}+6\mathbf{k}) = 5-16+18 = 7.$$

Therefore, on dividing by $|\mathbf{A}|\,|\mathbf{B}|$ and using formula (4), we have

$$\cos\theta = \frac{7}{|\mathbf{A}|\,|\mathbf{B}|} = \frac{7}{\sqrt{1^2+2^2+3^2}\sqrt{5^2+8^2+6^2}} = \frac{7}{\sqrt{14}\sqrt{125}} = \frac{\sqrt{7}}{5\sqrt{10}},$$

$$\theta = \mathrm{Cos}^{-1}\frac{\sqrt{7}}{5\sqrt{10}} \approx \mathrm{Cos}^{-1}0.1673 \approx 80°22'. \quad\bullet$$

EXERCISE 4

1. Compute the following scalar products.
   (a) $(3\mathbf{i}+2\mathbf{j}-4\mathbf{k})\cdot(3\mathbf{i}-2\mathbf{j}+7\mathbf{k})$.   (b) $(-\mathbf{i}+6\mathbf{j}+5\mathbf{k})\cdot(10\mathbf{i}+3\mathbf{j}-\mathbf{k})$.
   (c) $(2\mathbf{i}+5\mathbf{j}+6\mathbf{k})\cdot(6\mathbf{i}+6\mathbf{j}-7\mathbf{k})$.   (d) $(7\mathbf{i}+8\mathbf{j}+9\mathbf{k})\cdot(5\mathbf{i}-9\mathbf{j}+4\mathbf{k})$.

2. For each pair of vectors given in Problem 1, find $\cos\theta$, where θ is the angle between the vectors.

3. Find z so that the vectors $\mathbf{i}+2\mathbf{j}+3\mathbf{k}$ and $4\mathbf{i}+5\mathbf{j}+z\mathbf{k}$ are perpendicular.

4. Find t so that the vectors $2\mathbf{i}-3t\mathbf{j}+5\mathbf{k}$ and $t\mathbf{i}-4\mathbf{j}-6\mathbf{k}$ are perpendicular.

5. A triangle has vertices at $A(1,0,0)$, $B(0,2,0)$, and $C(0,0,3)$. Find $\cos\theta$ for each angle of the triangle.

6. Repeat Problem 5 for the points $D(1,1,1)$, $E(-1,-1,1)$, and $F(1,-1,-1)$.

7. Repeat Problem 5 for the points $G(3,1,-5)$, $H(-5,3,1)$, and $J(1,-5,3)$. Compare your result with that obtained in Exercise 2, Problem 17.

8. Find the cosine of the angle between the diagonal of a cube and the diagonal of one of its faces.

9. A right pyramid has a square base 2 ft on each side and a height of 3 ft. Find the cosine of the angle of intersection of two adjacent edges that meet at the vertex.

10. Suppose that $\mathbf{A}$ and $\mathbf{B}$ are vectors of unit length. Prove that $\mathbf{A}+\mathbf{B}$ is a vector that bisects the angle between $\mathbf{A}$ and $\mathbf{B}$.

11. Use the result of Problem 10 to find a vector that bisects the angle between $3\mathbf{i}+2\mathbf{j}+6\mathbf{k}$ and $9\mathbf{i}+6\mathbf{j}+2\mathbf{k}$.

★12. The Cauchy–Bunyiakowsky–Schwarz Inequality states that if $a_1, a_2, \ldots, a_n, b_1, b_2, \ldots, b_n$ are any real numbers, then

$$\left(\sum_{\alpha=1}^{n} a_\alpha b_\alpha\right)^2 \leqq \left(\sum_{\alpha=1}^{n} a_\alpha^2\right)\left(\sum_{\alpha=1}^{n} b_\alpha^2\right).$$

Use the dot product to prove this inequality when $n = 3$. When does the equality sign occur?

13. Complete the proof of Theorem 5.
14. Fill in the terms in the table

$A \cdot B$	i	j	k
i			
j			
k			

where each entry is the dot product of the vectors in that row and that column.

15. Prove that $A \cdot B / |B|$ is the scalar projection of A on a line that has the direction of B.
16. Use the dot product to derive formulas (8) and (9) of Theorem 3.
★17. Use the dot product to prove that an angle inscribed in a semicircle is a right angle. HINT: From Fig. 9,

$$(B - A) \cdot (B + A) = B \cdot B - A \cdot A = |B|^2 - |A|^2 = 0.$$

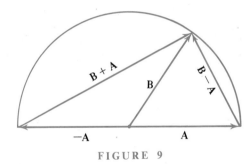

FIGURE 9

CALCULATOR PROBLEMS

C1. Find the angle in degrees to four significant figures, between the lines from the origin to $P_1(17, 18, 19)$ and $P_2(3, 2, 1)$.
C2. Do Problem C1 for $P_1 = (17, 18, 19)$ and $P_2(18, 19, 20)$.
C3. Check that the inequality of Problem 12 is true by computing both sides for the two ordered 5-tuples,

$$(a_1, a_2, a_3, a_4, a_5) = (10, 11, 12, 13, 14) \quad \text{and} \quad (b_1, b_2, b_3, b_4, b_5) = (11, 12, 13, 14, 15).$$

C4. Are the two vectors $A = 17.4i - 21.2j + 17.8k$ and $B = 13.9i + 25.6j + 16.9k$ perpendicular?

C5. Are the vectors $\mathbf{A} = 198\mathbf{i} + 743\mathbf{j} + 322\mathbf{k}$ and $\mathbf{B} = -597\mathbf{i} + 215\mathbf{j} - 129\mathbf{k}$ perpendicular? If not, find the angle (in degrees) between them.

The Vector Product of Two Vectors 5

This product can be defined in two different ways, (1) algebraically and (2) geometrically. Just as in the case of the dot product, the algebraic definition looks artificial, but it is easier to use so we adopt it here.

DEFINITION 3 (The Vector Product). Suppose that $\mathbf{A} = a_1\mathbf{i} + a_2\mathbf{j} + a_3\mathbf{k}$ and $\mathbf{B} = b_1\mathbf{i} + b_2\mathbf{j} + b_3\mathbf{k}$ are two vectors. The vector product (or cross product) is

(30) $$\mathbf{A} \times \mathbf{B} = (a_2b_3 - a_3b_2)\mathbf{i} + (a_3b_1 - a_1b_3)\mathbf{j} + (a_1b_2 - a_2b_1)\mathbf{k}.$$

At first glance, this definition seems rather arbitrary, and it may be hard to understand why one would define $\mathbf{A} \times \mathbf{B}$ in such a complicated way. But when $\mathbf{A} \times \mathbf{B}$ is defined in this way, then it has a very important geometric property. This is given in

---

THEOREM 8. For any two vectors $\mathbf{A}$ and $\mathbf{B}$:
  (i) the vector $\mathbf{A} \times \mathbf{B}$ has the length $|\mathbf{A}|\,|\mathbf{B}|\sin\theta$, where θ is the angle between the vectors $\mathbf{A}$ and $\mathbf{B}$.
 (ii) If the length is not zero, then $\mathbf{A} \times \mathbf{B}$ is perpendicular to the plane determined by $\mathbf{A}$ and $\mathbf{B}$.
(iii) If the length is not zero, $\mathbf{A}$, $\mathbf{B}$, and $\mathbf{A} \times \mathbf{B}$ form a right-handed set. We can write this result in symbols as

(31) $$\mathbf{A} \times \mathbf{B} = |\mathbf{A}|\,|\mathbf{B}|\sin\theta\,\mathbf{e},$$

where $\mathbf{e}$ is a unit vector perpendicular to the plane of $\mathbf{A}$ and $\mathbf{B}$ and such that $\mathbf{A}$, $\mathbf{B}$, and $\mathbf{e}$ form a right-handed set (see Figure 10).

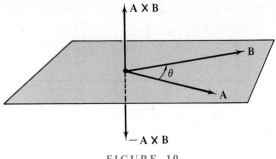

FIGURE 10

We will postpone the proof of this theorem until later. Let us first examine some simple consequences of Definition 3.

Equation (30) is not exactly easy to memorize, but if we are familiar with the expansions of third-order determinants, and not afraid to use vectors as elements of a determinant, then it is easy to see that the expansion of the determinant

(32)
$$\begin{vmatrix} \mathbf{i} & \mathbf{j} & \mathbf{k} \\ a_1 & a_2 & a_3 \\ b_1 & b_2 & b_3 \end{vmatrix} = \mathbf{i}\begin{vmatrix} a_2 & a_3 \\ b_2 & b_3 \end{vmatrix} - \mathbf{j}\begin{vmatrix} a_1 & a_3 \\ b_1 & b_3 \end{vmatrix} + \mathbf{k}\begin{vmatrix} a_1 & a_2 \\ b_1 & b_2 \end{vmatrix}$$

gives exactly the right side of (30) (see Appendix 3 for properties of determinants). Thus we have

---

THEOREM 9. If $\mathbf{A} = a_1\mathbf{i} + a_2\mathbf{j} + a_3\mathbf{k}$ and $\mathbf{B} = b_1\mathbf{i} + b_2\mathbf{j} + b_3\mathbf{k}$, then

(33)
$$\mathbf{A} \times \mathbf{B} = \begin{vmatrix} \mathbf{i} & \mathbf{j} & \mathbf{k} \\ a_1 & a_2 & a_3 \\ b_1 & b_2 & b_3 \end{vmatrix}.$$

Example 1. Find $\mathbf{A} \times \mathbf{B}$, where $\mathbf{A} = \mathbf{i} + 2\mathbf{j} - 3\mathbf{k}$ and $\mathbf{B} = 4\mathbf{i} - 5\mathbf{j} - 6\mathbf{k}$.

Solution. By Theorem 9 (or by Definition 3)

$$\mathbf{A} \times \mathbf{B} = \begin{vmatrix} \mathbf{i} & \mathbf{j} & \mathbf{k} \\ 1 & 2 & -3 \\ 4 & -5 & -6 \end{vmatrix} = (-12 - 15)\mathbf{i} + (-12 + 6)\mathbf{j} + (-5 - 8)\mathbf{k}$$

$$= -27\mathbf{i} - 6\mathbf{j} - 13\mathbf{k}. \quad \bullet$$

Using either Definition 3, or the properties of determinants and equation (33), we can prove

THEOREM 10. For any two vectors **A** and **B** and any scalar c:

(34) $$\mathbf{A} \times \mathbf{B} = -\mathbf{B} \times \mathbf{A} \qquad \text{(the anticommutative law)}$$

(35) $$c(\mathbf{A} \times \mathbf{B}) = (c\mathbf{A}) \times \mathbf{B} = \mathbf{A} \times (c\mathbf{B}),$$

(36) $$\mathbf{A} \times \mathbf{0} = \mathbf{0} \times \mathbf{A} = \mathbf{A} \times \mathbf{A} = \mathbf{0}.$$

For any three vectors **A**, **B**, and **C** we have the two distributive laws

(37) $$(\mathbf{A} + \mathbf{B}) \times \mathbf{C} = \mathbf{A} \times \mathbf{C} + \mathbf{B} \times \mathbf{C},$$

and

(38) $$\mathbf{A} \times (\mathbf{B} + \mathbf{C}) = \mathbf{A} \times \mathbf{B} + \mathbf{A} \times \mathbf{C}.$$

Proof. We appeal to certain properties of determinants. The reader who is not familiar with determinants must use Definition 3 and then some of the computations may be more complicated.

To prove (34), we interchange the second and third row of the determinant for **A** × **B**. Thus we have

$$\mathbf{A} \times \mathbf{B} = \begin{vmatrix} \mathbf{i} & \mathbf{j} & \mathbf{k} \\ a_1 & a_2 & a_3 \\ b_1 & b_2 & b_3 \end{vmatrix} = - \begin{vmatrix} \mathbf{i} & \mathbf{j} & \mathbf{k} \\ b_1 & b_2 & b_3 \\ a_1 & a_2 & a_3 \end{vmatrix} = -\mathbf{B} \times \mathbf{A}.$$

For (35), we multiply a determinant by c if we multiply any one row by c. Indeed,

$$c(\mathbf{A} \times \mathbf{B}) = c \begin{vmatrix} \mathbf{i} & \mathbf{j} & \mathbf{k} \\ a_1 & a_2 & a_3 \\ b_1 & b_2 & b_3 \end{vmatrix} = \begin{vmatrix} \mathbf{i} & \mathbf{j} & \mathbf{k} \\ ca_1 & ca_2 & ca_3 \\ b_1 & b_2 & b_3 \end{vmatrix} = (c\mathbf{A}) \times \mathbf{B},$$

with a similar computation for **A** × (c**B**).

The assertion in (36) is left for the reader.

To prove (37), let $\mathbf{C} = c_1\mathbf{i} + c_2\mathbf{j} + c_3\mathbf{k}$. Then

$$(\mathbf{A} + \mathbf{B}) \times \mathbf{C} = \begin{vmatrix} \mathbf{i} & \mathbf{j} & \mathbf{k} \\ a_1 + b_1 & a_2 + b_2 & a_3 + b_3 \\ c_1 & c_2 & c_3 \end{vmatrix}$$

$$= \begin{vmatrix} \mathbf{i} & \mathbf{j} & \mathbf{k} \\ a_1 & a_2 & a_3 \\ c_1 & c_2 & c_3 \end{vmatrix} + \begin{vmatrix} \mathbf{i} & \mathbf{j} & \mathbf{k} \\ b_1 & b_2 & b_3 \\ c_1 & c_2 & c_3 \end{vmatrix}$$

$$= \mathbf{A} \times \mathbf{C} + \mathbf{B} \times \mathbf{C}.$$

The proof of (38) is similar. ∎

We now prove part (i) of Theorem 8, namely

THEOREM 11. For any two vectors **A** and **B**,

(39) $$|\mathbf{A} \times \mathbf{B}| = \sqrt{|\mathbf{A}|^2|\mathbf{B}|^2 - (\mathbf{A} \cdot \mathbf{B})^2} = |\mathbf{A}|\,|\mathbf{B}|\sin\theta$$

where θ is the angle between the two vectors $(0 \leqq \theta \leqq \pi)$.

Proof. This is an exercise in computation. Using Definition 3, we see that

$$|\mathbf{A} \times \mathbf{B}|^2 = (a_2 b_3 - a_3 b_2)^2 + (a_3 b_1 - a_1 b_3)^2 + (a_1 b_2 - a_2 b_1)^2$$

(40) $$|\mathbf{A} \times \mathbf{B}|^2 = a_2^2 b_3^2 + a_3^2 b_2^2 + a_3^2 b_1^2 + a_1^2 b_3^2 + a_1^2 b_2^2 + a_2^2 b_1^2$$

$$- 2a_2 a_3 b_2 b_3 - 2a_1 a_3 b_1 b_3 - 2a_1 a_2 b_1 b_2.$$

On the other hand, in the expansion on the right side of

$$|\mathbf{A}|^2|\mathbf{B}|^2 - (\mathbf{A} \cdot \mathbf{B})^2 = (a_1^2 + a_2^2 + a_3^2)(b_1^2 + b_2^2 + b_3^2) - (a_1 b_1 + a_2 b_2 + a_3 b_3)^2,$$

the terms $a_1^2 b_1^2$, $a_2^2 b_2^2$, and $a_3^2 b_3^2$ will drop out and the remaining terms will be identical with the right side of (40). Therefore, we have

(41) $$|\mathbf{A} \times \mathbf{B}|^2 = |\mathbf{A}|^2|\mathbf{B}|^2 - (\mathbf{A} \cdot \mathbf{B})^2,$$

and this is the first equality in Theorem 11. Further, we know from Theorem 6 that $\mathbf{A} \cdot \mathbf{B} = |\mathbf{A}|\,|\mathbf{B}|\cos\theta$. Thus (41) yields

$$|\mathbf{A} \times \mathbf{B}|^2 = |\mathbf{A}|^2|\mathbf{B}|^2 - |\mathbf{A}|^2|\mathbf{B}|^2 \cos^2\theta$$

(42) $$|\mathbf{A} \times \mathbf{B}|^2 = |\mathbf{A}|^2|\mathbf{B}|^2(1 - \cos^2\theta) = |\mathbf{A}|^2|\mathbf{B}|^2 \sin^2\theta.$$

By agreement, $0 \leqq \theta \leqq \pi$, so $\sin\theta \geqq 0$. Thus we can take the square root of both sides of (42) to obtain $|\mathbf{A} \times \mathbf{B}| = |\mathbf{A}|\,|\mathbf{B}|\sin\theta$. ∎

Example 2. Use the cross product to find $\sin\theta$, where θ is the angle between the vectors $\mathbf{A} = \mathbf{i} + 2\mathbf{j} - 3\mathbf{k}$ and $\mathbf{B} = 4\mathbf{i} - 5\mathbf{j} - 6\mathbf{k}$.

Solution. These are the vectors used in Example 1 where we computed $\mathbf{A} \times \mathbf{B}$ and found that $\mathbf{A} \times \mathbf{B} = -27\mathbf{i} - 6\mathbf{j} - 13\mathbf{k}$. Then from Theorem 11, equation (39),

$$\sin\theta = \frac{|\mathbf{A} \times \mathbf{B}|}{|\mathbf{A}|\,|\mathbf{B}|} = \frac{\sqrt{27^2 + 6^2 + 13^2}}{\sqrt{1^2 + 2^2 + 3^2}\sqrt{4^2 + 5^2 + 6^2}}$$

$$= \frac{\sqrt{934}}{\sqrt{14}\sqrt{77}} = \frac{\sqrt{467}}{7\sqrt{11}}. \quad \bullet$$

We can check this result by computing $\cos\theta$ from the dot product. Indeed,

$$(43) \qquad \cos\theta = \frac{\mathbf{A}\cdot\mathbf{B}}{|\mathbf{A}|\,|\mathbf{B}|} = \frac{4-10+18}{\sqrt{14}\sqrt{77}} = \frac{12}{7\sqrt{2}\sqrt{11}} = \frac{6\sqrt{2}}{7\sqrt{11}}.$$

Hence

$$(44) \qquad \sin^2\theta + \cos^2\theta = \frac{467}{49\cdot 11} + \frac{72}{49\cdot 11} = \frac{539}{539} = 1.$$

Observe that if $0 < \theta < \pi$, then $\sin\theta > 0$. Hence, if we wish to determine whether θ is a first or second quadrant angle for two given vectors, the computation of $\sin\theta$ by the cross product is useless. For this purpose we should examine $\mathbf{A}\cdot\mathbf{B}$. Clearly, if $0 \leqq \theta < \pi/2$, then $\mathbf{A}\cdot\mathbf{B} > 0$, and if $\pi/2 < \theta \leqq \pi$, then $\mathbf{A}\cdot\mathbf{B} < 0$.

As a consequence of Theorem 11 we have the following two results.

THEOREM 12 (PLE). If $\mathbf{A}$ and $\mathbf{B}$ are coterminal sides of a parallelogram, then $|\mathbf{A}\times\mathbf{B}|$ is the area of the parallelogram.

COROLLARY. Let $\mathbf{A}$ and $\mathbf{B}$ be nonzero vectors. Then $\mathbf{A}\times\mathbf{B} = \mathbf{0}$ if and only if $\mathbf{A}$ and $\mathbf{B}$ are parallel.

Now that we know the magnitude of $\mathbf{A}\times\mathbf{B}$ (it is $|\mathbf{A}|\,|\mathbf{B}|\sin\theta$), we turn our attention to the direction of this vector. Of course, if $\mathbf{A} = \mathbf{0}$ or $\mathbf{B} = \mathbf{0}$ or $\sin\theta = 0$ (the trivial cases), then $\mathbf{A}\times\mathbf{B} = \mathbf{0}$ and the determination of a direction for $\mathbf{A}\times\mathbf{B}$ is not required.

THEOREM 13. In the nontrivial case, the vector $\mathbf{A}\times\mathbf{B}$ is perpendicular to $\mathbf{A}$ and perpendicular to $\mathbf{B}$.

Thus $\mathbf{A}\times\mathbf{B}$ is perpendicular to the plane determined by $\mathbf{A}$ and $\mathbf{B}$.

Proof. We examine $\mathbf{A}\cdot(\mathbf{A}\times\mathbf{B})$. By Definition 3, equation (30),

$$\mathbf{A}\cdot(\mathbf{A}\times\mathbf{B}) = a_1(a_2b_3 - a_3b_2) + a_2(a_3b_1 - a_1b_3) + a_3(a_1b_2 - a_2b_1)$$

$$= a_1a_2b_3 - a_1a_3b_2 + a_2a_3b_1 - a_2a_1b_3 + a_3a_1b_2 - a_3a_2b_1 = 0.$$

A similar computation shows that $\mathbf{B} \cdot (\mathbf{A} \times \mathbf{B}) = 0$. Hence $\mathbf{A}$ and $\mathbf{B}$ are both perpendicular to $\mathbf{A} \times \mathbf{B}$. ∎

★We can now complete the proof of Theorem 8. First observe that most of Theorem 8 has been settled. We know the length of $\mathbf{A} \times \mathbf{B}$ is $|\mathbf{A}|\,|\mathbf{B}| \sin \theta$ (see Theorem 11) and that this vector is the zero vector whenever any one of the three factors is zero.

We also know that if $\mathbf{A} \times \mathbf{B}$ is not the zero vector, then it is perpendicular to the plane of $\mathbf{A}$ and $\mathbf{B}$ (see Fig. 10). As indicated in Fig. 10, the only item in doubt is the direction of $\mathbf{A} \times \mathbf{B}$. We use a continuity argument to settle this point. We observe that in the definition of $\mathbf{A} \times \mathbf{B}$, this product is a continuous function of its coordinates. Thus we can gradually alter the coordinates (move the vectors), being careful that during this process $\mathbf{A} \times \mathbf{B}$ is never zero. Then the direction of $\mathbf{A} \times \mathbf{B}$ relative to $\mathbf{A}$ and $\mathbf{B}$ will remain the same. We first shrink or enlarge θ so that $\theta = \pi/2$ ($\mathbf{A}$ and $\mathbf{B}$ are perpendicular). Then we rotate the system until $\mathbf{A}$ has the direction of $\mathbf{i}$ and $\mathbf{B}$ has the direction of $\mathbf{j}$. Then for the new system, $\mathbf{A}^{\star} = a_1^{\star}\mathbf{i}$ and $\mathbf{B}^{\star} = b_1^{\star}\mathbf{j}$. Hence

$$(45) \qquad \mathbf{A}^{\star} \times \mathbf{B}^{\star} = \begin{vmatrix} \mathbf{i} & \mathbf{j} & \mathbf{k} \\ a_1^{\star} & 0 & 0 \\ 0 & b_1^{\star} & 0 \end{vmatrix} = a_1^{\star} b_1^{\star} \mathbf{k},$$

where $a_1^{\star} > 0$ and $b_1^{\star} > 0$. Since $\mathbf{i}$, $\mathbf{j}$, and $\mathbf{k}$ form a right-handed set, it follows that the original set of vectors $\mathbf{A}$, $\mathbf{B}$, and $\mathbf{e}$ form a right-handed set. ∎

Note that if we are using a left-handed system for x, y, and z, then $\mathbf{A}$, $\mathbf{B}$, and $\mathbf{A} \times \mathbf{B}$ will form a left-handed set.

EXERCISE 5

1. In Problem 1 of Exercise 4, replace the dot by a cross and compute the cross products.
2. Use the cross product to find $\sin \theta$, where θ is the angle between the two vectors, for each pair given in Problem 1 of Exercise 4. Check your answers by showing that they satisfy $\sin^2 \theta + \cos^2 \theta = 1$.
3. Find a unit vector that is perpendicular to both $\mathbf{i} + \mathbf{j}$ and $\mathbf{j} + \mathbf{k}$.
4. Prove Theorem 12 and its corollary.
5. Use the cross product to find the area of the triangle OP_1P_2 in each of the following cases.
   (a) $P_1(5, 1, 0)$, $P_2(2, 3, 0)$. (b) $P_1(4, 2, 0)$, $P_2(-3, 7, 0)$.
   (c) $P_1(1, 2, 3)$, $P_2(4, 5, 6)$. (d) $P_1(2, -1, 4)$, $P_2(-3, 5, -7)$.

*6. Prove that the area of the plane triangle with vertices at $A(a_1, a_2)$, $B(b_1, b_2)$, and $C(c_1, c_2)$ is $|D|/2$, where

$$D = \begin{vmatrix} a_1 & a_2 & 1 \\ b_1 & b_2 & 1 \\ c_1 & c_2 & 1 \end{vmatrix}.$$

Show that D is positive if the points A, B, and C are in counterclockwise order, and D is negative if A, B, and C are in clockwise order. When is $D = 0$?

7. Find a vector that is perpendicular to the plane through the points P_1, P_2, and P_3 for each of the following sets of points.
   (a) $P_1(1, 3, 5)$, $P_2(2, -1, 3)$, $P_3(-3, 2, -6)$.
   (b) $P_1(2, 4, 6)$, $P_2(-3, 1, -5)$, $P_3(2, -6, 1)$.

8. Fill in the terms in the table

$A \times B$	i	j	k
i			
j			
k			

where each entry is the cross product of the vectors in that row and that column.

9. Find three vectors **A**, **B**, and **C** such that

$$(A \times B) \times C \neq A \times (B \times C).$$

Hence the associative law is not true for the vector product.

10. Let **A**, **B**, and **C** be vectors that form the coterminal edges of a parallelepiped (a box whose vertex angles are not necessarily right angles). Prove that $|A \cdot B \times C|$ is the volume of the box. The product $A \cdot B \times C$ is called the *triple scalar product* of the three vectors, or the *box product*.

11. Use the result of Problem 10 to prove that

$$A \cdot B \times C = B \cdot C \times A = C \cdot A \times B.$$

12. Prove that $A \cdot B \times C = A \times B \cdot C$.

CALCULATOR PROBLEMS

C1. Find $A \times B$ if $A = 2.3i + 3.5j + 7.9k$ and $B = 1.4i - 2.6j + 3.8k$. Use the cross product to find $\sin \theta$. Check your work using the dot product.

C2. Do Problem C1 if $A = 81i + 72j - 65k$ and $B = 32i - 54j + 76k$.

6 *Computations with Products of Vectors*

The dot and cross products are quite useful in solving problems in three-dimensional geometry.

Example 1. Find the distance from the point $A(1, 2, 3)$ to the line through the points $B(-1, 2, 1)$ and $C(4, 3, 2)$.

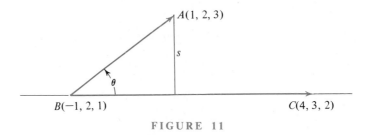

FIGURE 11

Solution. We pass a plane through the three points. From the plane diagram shown in Fig. 11, it is clear that the distance s is given by

(46) $s = |\mathbf{BA}| \sin \theta.$

But $\mathbf{BA} \times \mathbf{BC} = |\mathbf{BA}||\mathbf{BC}| \sin \theta \, \mathbf{e}$. Hence we can find s by taking the length of $\mathbf{BA} \times \mathbf{BC}$ and dividing by the length of $\mathbf{BC}$. Thus

(47) $s = \dfrac{|\mathbf{BA} \times \mathbf{BC}|}{|\mathbf{BC}|}.$

Carrying out these computations, we have

$$\mathbf{BA} = 2\mathbf{i} + 0\mathbf{j} + 2\mathbf{k}, \qquad \mathbf{BC} = 5\mathbf{i} + \mathbf{j} + \mathbf{k},$$

(48) $$\mathbf{BA} \times \mathbf{BC} = \begin{vmatrix} \mathbf{i} & \mathbf{j} & \mathbf{k} \\ 2 & 0 & 2 \\ 5 & 1 & 1 \end{vmatrix} = -2\mathbf{i} + 8\mathbf{j} + 2\mathbf{k},$$

$$s = \frac{|-2\mathbf{i} + 8\mathbf{j} + 2\mathbf{k}|}{|5\mathbf{i} + \mathbf{j} + \mathbf{k}|} = \frac{2\sqrt{1 + 16 + 1}}{\sqrt{25 + 1 + 1}} = \frac{2\sqrt{18}}{3\sqrt{3}} = \frac{2}{3}\sqrt{6}. \quad \bullet$$

Example 2. Find the distance from the point $P(2, 2, 9)$ to the plane through the three points $A(2, 1, 3)$, $B(3, 3, 5)$, and $C(1, 3, 6)$.

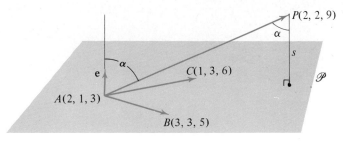

FIGURE 12

Solution. As indicated in Fig. 12, the vectors **AB** and **AC** lie in the plane $\mathscr{P}$ through the three given points. Hence **AB** × **AC** gives the direction of a unit vector **e** perpendicular to $\mathscr{P}$. Thus

(49) $$\mathbf{e} = \frac{\mathbf{AB} \times \mathbf{AC}}{|\mathbf{AB} \times \mathbf{AC}|}.$$

Then the distance s from P to the plane $\mathscr{P}$ is

(50) $$s = |\mathbf{AP}| \cos \alpha = \mathbf{AP} \cdot \mathbf{e} = \frac{\mathbf{AP} \cdot (\mathbf{AB} \times \mathbf{AC})}{|\mathbf{AB} \times \mathbf{AC}|}.$$

Carrying out the computations for the specific points given, we have

$$\mathbf{AB} = \mathbf{i} + 2\mathbf{j} + 2\mathbf{k}, \qquad \mathbf{AC} = -\mathbf{i} + 2\mathbf{j} + 3\mathbf{k}, \qquad \mathbf{AP} = \mathbf{j} + 6\mathbf{k},$$

(51) $$\mathbf{AB} \times \mathbf{AC} = \begin{vmatrix} \mathbf{i} & \mathbf{j} & \mathbf{k} \\ 1 & 2 & 2 \\ -1 & 2 & 3 \end{vmatrix} = 2\mathbf{i} - 5\mathbf{j} + 4\mathbf{k}$$

$$s = \frac{(\mathbf{j} + 6\mathbf{k}) \cdot (2\mathbf{i} - 5\mathbf{j} + 4\mathbf{k})}{|2\mathbf{i} - 5\mathbf{j} + 4\mathbf{k}|} = \frac{-5 + 24}{\sqrt{4 + 25 + 16}} = \frac{19}{3\sqrt{5}} = \frac{19}{15}\sqrt{5}. \quad \bullet$$

Example 3. Find the distance between the two lines $\mathscr{L}_1$ and $\mathscr{L}_2$, where $\mathscr{L}_1$ goes through the points $A(1, 2, 1)$ and $B(2, 7, 3)$, and $\mathscr{L}_2$ goes through the points $C(2, 3, 5)$ and $D(0, 6, 6)$.

Solution. We first observe that if $\mathbf{N} = \mathbf{AB} \times \mathbf{CD}$, then $\mathbf{N}$ is a vector that is simultaneously perpendicular to the two lines $\mathscr{L}_1$ and $\mathscr{L}_2$. Hence there is a pair of parallel planes $\mathscr{P}_1$ and $\mathscr{P}_2$ containing the lines $\mathscr{L}_1$ and $\mathscr{L}_2$, respectively, namely, two planes each perpendicular to the vector $\mathbf{N}$. These planes are shown in Fig. 13. If we take any two points, one in each plane, and project the line segment joining them onto the common perpendicular, we obtain s, the distance between the two planes. But this is

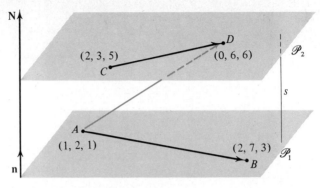

FIGURE 13

also the distance between the two lines. Thus if A and D are the two points chosen, and $\mathbf{n} = \mathbf{N}/|\mathbf{N}|$ is a unit normal, then

(52) $$s = |\mathbf{AD} \cdot \mathbf{n}| = \left| \mathbf{AD} \cdot \frac{\mathbf{AB} \times \mathbf{CD}}{|\mathbf{AB} \times \mathbf{CD}|} \right|.$$

In this formula we could replace $\mathbf{AD}$ by $\mathbf{AC}$ or $\mathbf{BC}$ or $\mathbf{BD}$ and obtain the same result.

Carrying out the computations for the specific points given, we find that

$$\mathbf{AB} = \mathbf{i} + 5\mathbf{j} + 2\mathbf{k}, \qquad \mathbf{CD} = -2\mathbf{i} + 3\mathbf{j} + \mathbf{k},$$

$$\mathbf{N} = \mathbf{AB} \times \mathbf{CD} = \begin{vmatrix} \mathbf{i} & \mathbf{j} & \mathbf{k} \\ 1 & 5 & 2 \\ -2 & 3 & 1 \end{vmatrix} = -\mathbf{i} - 5\mathbf{j} + 13\mathbf{k}.$$

To obtain a unit normal, we must divide $\mathbf{N}$ by $|\mathbf{N}| = \sqrt{195}$. Finally, we have $\mathbf{AD} = -\mathbf{i} + 4\mathbf{j} + 5\mathbf{k}$ so that

$$s = \frac{(-\mathbf{i} + 4\mathbf{j} + 5\mathbf{k}) \cdot (-\mathbf{i} - 5\mathbf{j} + 13\mathbf{k})}{\sqrt{195}} = \frac{1 - 20 + 65}{\sqrt{195}} = \frac{46}{\sqrt{195}}. \quad \bullet$$

As a check we use $\mathbf{CB}$ in place of $\mathbf{AD}.$ This gives us

$$s = \left| \frac{(0\mathbf{i} + 4\mathbf{j} - 2\mathbf{k}) \cdot (-\mathbf{i} - 5\mathbf{j} + 13\mathbf{k})}{\sqrt{195}} \right| = \left| \frac{-20 - 26}{\sqrt{195}} \right| = \frac{46}{\sqrt{195}}.$$

EXERCISE 6

1. Find the distance from the point A to the line through the points B and C for each of the following sets of points.
   (a) $A(1, 1, 7)$, $B(2, -1, 4)$, $C(3, 1, 6)$.
   (b) $A(1, 2, 2)$, $B(-1, 3, 5)$, $C(1, 6, 11)$.

2. Do Problem 1 for the points:
   (a) $A(3, 3, 4)$, $B(0, 0, 0)$, $C(6, 6, 7)$.
   (b) $A(2, 6, 5)$, $B(3, 1, 11)$, $C(5, -9, 23)$.

3. Find the distance from the point P to the plane through the points A, B, and C for each of the following sets of points.
   (a) $P(0, 0, 1)$, $A(-1, -2, -3)$, $B(0, 5, 1)$, $C(-2, 1, 0)$.
   (b) $P(2, 1, 7)$, $A(3, -1, 6)$, $B(1, 5, 5)$, $C(4, -6, 4)$.

4. Do Problem 3 for the points:
   (a) $P(3, 12, 17)$, $A(2, -1, 5)$, $B(4, -2, 2)$, $C(1, 4, 11)$.
   (b) $P(0, 0, -1)$, $A(-1, -1, 0)$, $B(0, -2, 0)$, $C(0, 0, 1)$.

5. Let $\mathbf{A} = a_1\mathbf{i} + a_2\mathbf{j} + a_3\mathbf{k}$, $\mathbf{B} = b_1\mathbf{i} + b_2\mathbf{j} + b_3\mathbf{k}$, and $\mathbf{C} = c_1\mathbf{i} + c_2\mathbf{j} + c_3\mathbf{k}$. Prove that

$$\mathbf{A} \cdot \mathbf{B} \times \mathbf{C} = \begin{vmatrix} a_1 & a_2 & a_3 \\ b_1 & b_2 & b_3 \\ c_1 & c_2 & c_3 \end{vmatrix}.$$

   HINT: Use Theorem 9.

*6. Using the result of Problem 5, interpret the result of Problem 10, Exercise 5 as a theorem on determinants.

7. Find the volume of the parallelepiped if three of the edges are the vectors $\mathbf{A}$, $\mathbf{B}$, and $\mathbf{C}$ for each of the following sets. (See Problem 10 of Exercise 5, page 665.)
   (a) $\mathbf{A} = \mathbf{i} + \mathbf{j}$, $\mathbf{B} = -2\mathbf{i} + 3\mathbf{j}$, $\mathbf{C} = \mathbf{i} + \mathbf{j} + \mathbf{k}$.
   (b) $\mathbf{A} = \mathbf{i} - \mathbf{j} - \mathbf{k}$, $\mathbf{B} = \mathbf{i} + 3\mathbf{j} + \mathbf{k}$, $\mathbf{C} = 2\mathbf{i} + 3\mathbf{j} + 5\mathbf{k}$.

8. Do Problem 7 for the vectors:
   (a) $\mathbf{A} = 2\mathbf{i} + \mathbf{j} + \mathbf{k}$, $\mathbf{B} = -\mathbf{i} + 4\mathbf{j} + 2\mathbf{k}$, $\mathbf{C} = 7\mathbf{i} - 10\mathbf{j} - 4\mathbf{k}$.
   (b) $\mathbf{A} = 2\mathbf{i} - \mathbf{j} + \mathbf{k}$, $\mathbf{B} = \mathbf{i} + 2\mathbf{j} + 3\mathbf{k}$, $\mathbf{C} = \mathbf{i} + \mathbf{j} - 2\mathbf{k}$.

9. Find the distance between the line through the points A and B and the line through the points C and D for each of the following sets of points.
   (a) $A(1, 0, 0)$, $B(0, 1, 1)$, $C(-1, 0, 0)$, $D(0, -1, 1)$.
   (b) $A(1, 1, 0)$, $B(-2, 3, 1)$, $C(1, -1, 3)$, $D(0, 0, 0)$.

10. Do Problem 9 for the points:
   (a) $A(-2, 3, -1)$, $B(2, 4, 4)$, $C(1, 2, 1)$, $D(-1, 5, 2)$.
   (b) $A(-2, 3, -1)$, $B(2, 4, 4)$, $C(-1, 0, 3)$, $D(-3, 3, 4)$.

11. Find the distance between the two lines

$$\frac{x-1}{2} = \frac{y-2}{3} = \frac{z+1}{-1} \qquad \text{and} \qquad \frac{x+1}{3} = \frac{y-1}{2} = \frac{z-2}{1}.$$

12. Find the distance of the origin from each of the lines in Problem 11.
13. Under what circumstances may formula (52) of Example 3 fail?
14. If A, B, and C are three noncollinear points and $\mathbf{A}$, $\mathbf{B}$, and $\mathbf{C}$ are their position vectors, prove that $\mathbf{A} \times \mathbf{B} + \mathbf{B} \times \mathbf{C} + \mathbf{C} \times \mathbf{A}$ is a vector perpendicular to the plane of the triangle ABC.
15. Let A, B, C, and D be the vertices of a proper tetrahedron (the points are not coplanar). On each face of the tetrahedron erect a vector normal to the face, pointing outward from the tetrahedron, and having length equal to the area of the face. Prove that the sum of these four vectors is zero.
16. Under certains conditions, formula (50) may give a negative answer. What are these conditions and how would you correct the formula?

7 *Equations of Planes*

Let $\mathscr{P}$ be the plane through the three noncollinear points P_0, P_1, and P_2, where P_0 has the coordinates (x_0, y_0, z_0). For simplicity, let

$$\mathbf{P_0P_1} = a_1\mathbf{i} + a_2\mathbf{j} + a_3\mathbf{k} \qquad \text{and} \qquad \mathbf{P_0P_2} = b_1\mathbf{i} + b_2\mathbf{j} + b_3\mathbf{k}.$$

To find an equation for this plane, we first observe that the vector

(53) $$\mathbf{N} = \mathbf{P_0P_1} \times \mathbf{P_0P_2}$$

is normal to the plane (see Fig. 14). Then (by the definition of a plane) the point $P(x, y, z)$ is on the plane if and only if $\mathbf{P_0P}$ is perpendicular to $\mathbf{N}$ (or is zero). Hence P is on the plane if and only if

(54) $$\mathbf{P_0P} \cdot \mathbf{N} = 0$$

and this is a vector equation of the plane $\mathscr{P}$. To obtain an equation in the usual form, we use the result of Problem 5 of Exercise 6. Since $\mathbf{P_0P} = (x - x_0)\mathbf{i} + (y - y_0)\mathbf{j} + (z - z_0)\mathbf{k}$, equations (53) and (54) give

(55) $$\begin{vmatrix} x - x_0 & y - y_0 & z - z_0 \\ a_1 & a_2 & a_3 \\ b_1 & b_2 & b_3 \end{vmatrix} = 0.$$

On expanding this determinant, we see that the plane has an equation of the form

(56)
$$Ax + By + Cz = D.$$

Of course, this equation is not unique because we can always multiply through by any nonzero constant. Nevertheless, we shall refer to (56) occasionally as *the* equation of the plane.

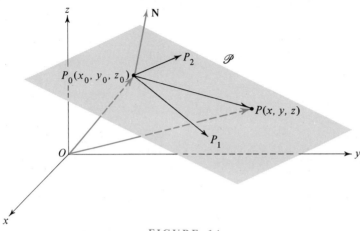

FIGURE 14

THEOREM 14. Any plane has an equation of the form (56). Conversely, if the numbers A, B, and C are not all zero, then equation (56) is an equation of some plane. Finally, the vector $\mathbf{N} = A\mathbf{i} + B\mathbf{j} + C\mathbf{k}$ is always perpendicular to the plane (56).

We have already proved the first part of this theorem. To prove the converse, suppose that we are given an equation of the form (56). If A, B, and C are not all zero, we can always find *one* point $P_0(x_0, y_0, z_0)$ whose coordinates satisfy (56); that is,

(57)
$$Ax_0 + By_0 + Cz_0 = D.$$

Suppose that $P(x, y, z)$ is any point whose coordinates satisfy equation (56). Subtracting equation (57) from equation (56) gives

$$A(x - x_0) + B(y - y_0) + C(z - z_0) = 0,$$

or

$$[A\mathbf{i} + B\mathbf{j} + C\mathbf{k}] \cdot [(x - x_0)\mathbf{i} + (y - y_0)\mathbf{j} + (z - z_0)\mathbf{k}] = 0,$$

or, in vector form,

$$(58) \qquad\qquad\qquad \mathbf{N} \cdot \mathbf{P_0P} = 0.$$

This shows that $\mathbf{P_0P}$ is either the zero vector or is perpendicular to the vector $\mathbf{N}$. Hence every point P whose coordinates satisfy (56) lies in the plane through P_0 perpendicular to $\mathbf{N}$. Further, each such point satisfies (58) and hence (56). Consequently, (56) is an equation of that plane. ▮

Example 1. Find an equation for the plane that passes through $P_0(1, 2, 2)$, $P_1(2, -1, 1)$, and $P_2(-1, 3, 0)$.

Solution. $\mathbf{P_0P_1} = \mathbf{i} - 3\mathbf{j} - \mathbf{k}$, and $\mathbf{P_0P_2} = -2\mathbf{i} + \mathbf{j} - 2\mathbf{k}$. Then (55) gives

$$\begin{vmatrix} x - 1 & y - 2 & z - 2 \\ 1 & -3 & -1 \\ -2 & 1 & -2 \end{vmatrix} = 7(x - 1) + 4(y - 2) - 5(z - 2) = 0,$$

or

$$(56e) \qquad\qquad\qquad 7x + 4y - 5z = 5.$$

The vector $7\mathbf{i} + 4\mathbf{j} - 5\mathbf{k}$ is normal to this plane. The student should check that each of the given points satisfies equation (56e). ●

Example 2. Find parametric equations for the line that contains the point $(3, -1, 2)$ and is perpendicular to the plane $x - 2y + 7z = 28$.

Solution. The vector $\mathbf{i} - 2\mathbf{j} + 7\mathbf{k}$ is perpendicular to the given plane and hence gives the direction of the line. Therefore, from equation (15), we have

$$\frac{x - 3}{1} = \frac{y + 1}{-2} = \frac{z - 2}{7}$$

as symmetric equations of the line. In parametric form we have

$$x = 3 + t, \qquad y = -1 - 2t, \qquad z = 2 + 7t. \quad ●$$

Example 3. Prove that if $P_1(x_1, y_1, z_1)$ is any point and d is the distance from P_1 to the plane $Ax + By + Cz = D$, then

$$(59) \qquad\qquad\qquad d = \frac{|Ax_1 + By_1 + Cz_1 - D|}{\sqrt{A^2 + B^2 + C^2}}.$$

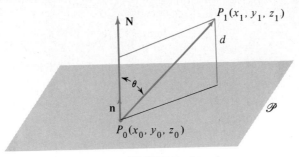

FIGURE 15

Solution. If **n** is a unit normal to the plane, then it is clear from Fig. 15 that

$$d = |\mathbf{P_0P_1}| \cos \theta = \mathbf{P_0P_1} \cdot \mathbf{n}$$

if $\mathbf{P_0P_1}$ and **n** lie on the same side of the plane. If the plane separates these vectors, then $|\mathbf{P_0P_1}| \cos \theta$ is negative. But in any case, $|\mathbf{P_0P_1}| \cos \theta = \mathbf{n} \cdot \mathbf{P_0P_1}$ and $d = |\mathbf{n} \cdot \mathbf{P_0P_1}|$. Since $\mathbf{N} = A\mathbf{i} + B\mathbf{j} + C\mathbf{k}$ is normal to the plane, we have

$$d = \left| \frac{[A\mathbf{i} + B\mathbf{j} + C\mathbf{k}] \cdot [(x_1 - x_0)\mathbf{i} + (y_1 - y_0)\mathbf{j} + (z_1 - z_0)\mathbf{k}]}{|\mathbf{N}|} \right|,$$

$$d = \frac{|A(x_1 - x_0) + B(y_1 - y_0) + C(z_1 - z_0)|}{\sqrt{A^2 + B^2 + C^2}},$$

$$(60) \qquad d = \frac{|Ax_1 + By_1 + Cz_1 - Ax_0 - By_0 - Cz_0|}{\sqrt{A^2 + B^2 + C^2}}.$$

But $P_0(x_0, y_0, z_0)$ lies on the plane, so that $Ax_0 + By_0 + Cz_0 = D$. Using this in (60), we obtain (59). ●

Example 4. Are the two planes

$$\mathscr{P}_1: x - 2y + 3z = 14, \qquad \text{and} \qquad \mathscr{P}_2: 8x + y - 2z = 37$$

perpendicular?

Solution. Since the vector $\mathbf{N}_1 = \mathbf{i} - 2\mathbf{j} + 3\mathbf{k}$ is normal to the plane $\mathscr{P}_1$ and the vector $\mathbf{N}_2 = 8\mathbf{i} + \mathbf{j} - 2\mathbf{k}$ is normal to the plane $\mathscr{P}_2$ we need only check to see if $\mathbf{N}_1$ is perpendicular to $\mathbf{N}_2$. But

$$\mathbf{N}_1 \cdot \mathbf{N}_2 = (\mathbf{i} - 2\mathbf{j} + 3\mathbf{k}) \cdot (8\mathbf{i} + \mathbf{j} - 2\mathbf{k}) = 8 - 2 - 6 = 0,$$

so $\mathscr{P}_1$ and $\mathscr{P}_2$ are indeed perpendicular. ●

EXERCISE 7

1. Find an equation of the plane through P_0, P_1, and P_2 for each of the following sets of points.
   (a) $P_0(2, 1, 6)$, $P_1(5, -2, 0)$, $P_2(4, -5, -2)$.
   (b) $P_0(1, 2, 17)$, $P_1(-1, -2, 3)$, $P_2(-4, 2, 2)$.
2. Do Problem 1 for the following sets of points.
   (a) $P_0(2, -2, 2)$, $P_1(1, -8, 6)$, $P_2(4, 3, -1)$.
   (b) $P_0(a, 0, 0)$, $P_1(0, b, 0)$, $P_2(0, 0, c)$.
   (c) $P_0(a, b, 0)$, $P_1(0, b, c)$, $P_2(a, 0, c)$.
3. Find an equation of the plane that passes through $(2, -3, 5)$ and is parallel to the plane $3x + 5y - 7z = 11$.
4. Do Problem 3 for the point $(3, 5, -9)$ and the plane $5x + 7y + 10z = \sqrt{37} - e^2$.
5. Find the distance from the origin to each of the planes in Problems 1 and 2.
6. Find the distance of the point $(1, -2, 3)$ from each of the planes in Problem 1.
7. Prove that the planes $Ax + By + Cz = D_1$ and $Ax + By + Cz = D_2$ are parallel.
8. Prove that the distance between the two planes of Problem 7 is

$$|D_1 - D_2| \big/ \sqrt{A^2 + B^2 + C^2}.$$

9. Find the distance between the two planes

$$2x - 3y - 6z = 5 \qquad \text{and} \qquad 4x - 6y - 12z = -11.$$

10. Find the distance between the two planes

$$-x + 4y + 2z = \pi^3 + 5\sqrt{19} \qquad \text{and} \qquad -x + 4y + 2z = \pi^3 + 2\sqrt{19}.$$

11. Find symmetric equations for the line that passes through $P(-9, 4, 3)$ and is perpendicular to the plane $2x + 6y + 9z = 0$. Find the point Q where this line intersects the plane.
12. Do Problem 11 for $P(2, 5, -7)$ and the line $x - 2y + 5z = 77$.
13. For the points of Problem 11, find the distance $|PQ|$ in two ways: (a) directly from coordinates, and (b) by the method of Example 3.
14. Do Problem 13 for the points P and Q of Problem 12.
15. Find symmetric equations for the straight line that passes through $(3, -1, 6)$ and is parallel to both of the planes $x - 2y + z = 2$ and $2x + y - 3z = 5$.
16. Find symmetric equations for the line of intersection for each of the following pairs of planes.
    (a) $3x + 4y - z = 10$, $2x + y + z = 0$.
    (b) $x + 5y + 3z = 14$, $x + y - 2z = 4$.
    (c) $2x + 3y + 5z = 21$, $3x - 2y + z = 12$.
17. Find an equation for the plane that passes through A and is perpendicular to the line through B and C for the points $A(0, 0, 0)$, $B(1, 2, 3)$, $C(3, 2, 1)$.

18. Do Problem 17 for the following sets of points.
    (a) $A(1, 5, 9)$, $B(2, 3, -4)$, $C(5, 1, -1)$.
    (b) $A(-3, -7, 11)$, $B(7, 5, 3)$, $C(8, -4, 2)$.

19. Prove that if $P(x, y)$ is any point in the xy-plane and d is the distance of P from the line $ax + by = c$, then $d = |ax + by - c|/\sqrt{a^2 + b^2}$.

20. Let $\mathscr{L}$ be the common edge of two half-planes $\mathscr{P}_1$ and $\mathscr{P}_2$ (see Fig. 16). Let P be a point on $\mathscr{L}$ and let $\mathscr{L}_1$ and $\mathscr{L}_2$ be lines in $\mathscr{P}_1$ and $\mathscr{P}_2$, respectively, each perpendicular to $\mathscr{L}$ at P. Then θ, the least positive angle between $\mathscr{L}_1$ and $\mathscr{L}_2$, is called the *angle between the two half-planes*, or the *dihedral angle* formed by the half-planes. Show that $\cos \theta$ can be found by considering the vectors normal to the half-planes. Compute $\cos \theta$ for the smaller dihedral angle for each pair of planes given in Problem 16.

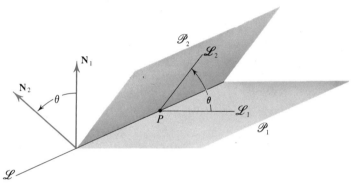

FIGURE 16

21. Let the line $\mathscr{L}$ intersect the plane $\mathscr{P}$ at a point P. If the line $\mathscr{L}$ is not normal to the plane, then different lines in $\mathscr{P}$ through P may make different angles with $\mathscr{L}$ (see Fig. 17). By definition *the angle θ between $\mathscr{L}$ and $\mathscr{P}$ is the smallest of these various angles.*

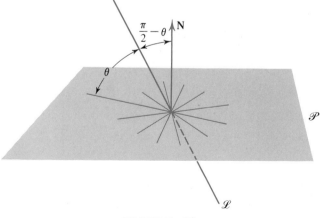

FIGURE 17

Find $\sin \theta$ for the angle between the line and the plane in each of the following.

(a) $\dfrac{x - 5}{2} = \dfrac{y + 17}{-3} = \dfrac{z - \ln 15}{6}$, $x + 2y + 2z = e^2$.

(b) $\dfrac{x + \sqrt{2}}{6} = \dfrac{y + \sqrt{3}}{7} = \dfrac{z + \sqrt{5}}{-6}$, $2x - 4y + 4z = \sqrt{11}$.

(c) $\dfrac{x}{5} = \dfrac{y}{6} = \dfrac{z}{2}$, $20x - 11y - 17z = 1$.

(d) $x - 1 = y - 2 = z - 3$, $5x + 5y + 5z = 1$.

★22. A unit cube lies in the first octant with three edges along the coordinate axes. Find the equation of a plane that intersects the surface of this cube in a regular hexagon.

23. Determine whether the given pair of planes meet at right angles.

(a) $2x - 3y + 5z = 14$, $-y + 18z = 0$.

(b) $-x + 3y - 2z = 0$, $4x - 2y - 5z = \pi$.

(c) $2x - y + 3z = 20$, $-4x + 2y - 6z = \sqrt{5}$.

24. Which of the pairs of planes in Problem 23 are parallel?

8 *Differentiation of Vector Functions*

Just as we can have vector functions where the vector $\mathbf{R}(t)$ lies in the xy-plane, so we can also consider vector functions in three-dimensional space. In this case $\mathbf{R} = \mathbf{R}(t)$ denotes a vector function, in which to every real value of t (perhaps t is restricted to some interval) there corresponds a three-dimensional vector $\mathbf{R}$. Such a function arises quite naturally if we consider some particle moving in space along a curve, and $\mathbf{R}$ is the vector from the origin to this particle. Then the symbol t would denote the time at which the particle had a given location and $\mathbf{R}(t)$ would be the position vector of the particle at that time. Of course, it is not necessary to regard t as representing time, nor $\mathbf{R}(t)$ as a position vector, but in most cases it is convenient to make these interpretations.

Most of the material in Section 2 of Chapter 12 is valid in three-dimensional space, and the student would do well to review that section at this time. The following three-dimensional generalizations will then appear to be obvious.

The vector function $\mathbf{R}(t)$ can always be written in terms of its components as

(61) $$\mathbf{R}(t) = f(t)\mathbf{i} + g(t)\mathbf{j} + h(t)\mathbf{k}.$$

If (x, y, z) denotes the coordinates of P, the end point of the position vector $\mathbf{R}(t)$, then (61) represents three scalar equations,

(62) $$x = f(t), \qquad y = g(t), \qquad z = h(t),$$

the parametric equations of the curve described by the point P. The vector function $\mathbf{R}(t)$ is continuous in an interval $\mathscr{I}$ if and only if each of its three components $f(t)$, $g(t)$, and $h(t)$ is continuous in $\mathscr{I}$. Certain elementary combinations of vector functions are covered in

> **THEOREM 15.** Let $\mathbf{U}(t)$ and $\mathbf{V}(t)$ be two vector functions and let $f(t)$ be a scalar function. If $\mathbf{U}$, $\mathbf{V}$, and f are continuous in a common interval $\mathscr{I}$, then each of the functions
>
> $$\mathbf{U} + \mathbf{V}, \qquad f\mathbf{U}, \qquad \mathbf{U} \cdot \mathbf{V}, \qquad \mathbf{U} \times \mathbf{V}$$
>
> is continuous in $\mathscr{I}$.

If the scalar functions in (62) are differentiable in $\mathscr{I}$, then (just as in the theory for plane vectors) the vector function $\mathbf{R}(t)$ is differentiable in $\mathscr{I}$ and indeed

$$(63) \qquad \mathbf{R}'(t) = \frac{d\mathbf{R}(t)}{dt} = \frac{df}{dt}\mathbf{i} + \frac{dg}{dt}\mathbf{j} + \frac{dh}{dt}\mathbf{k}.$$

This is frequently abbreviated by writing

$$(64) \qquad \frac{d\mathbf{R}}{dt} = \frac{dx}{dt}\mathbf{i} + \frac{dy}{dt}\mathbf{j} + \frac{dz}{dt}\mathbf{k}.$$

For example, if

$$(65) \qquad \mathbf{R} = \sin t \cos t\,\mathbf{i} + \cos^2 t\,\mathbf{j} + \sin t\,\mathbf{k},$$

then

$$\frac{d\mathbf{R}}{dt} = (\cos^2 t - \sin^2 t)\mathbf{i} - 2\sin t \cos t\,\mathbf{j} + \cos t\,\mathbf{k}.$$

If in (63), $\mathbf{R}'(t) \neq \mathbf{0}$, then $\mathbf{R}'(t)$ is a vector tangent to the curve. In fact, $\mathbf{R}'(t)$ is just the velocity of the moving particle and $|\mathbf{R}'(t)|$ is its speed. Similarly, the acceleration of the particle is just $\mathbf{R}''(t)$. For example, if a particle moves so that (65) is the position vector, then

$$\mathbf{V}(t) = (\cos^2 t - \sin^2 t)\mathbf{i} - 2\sin t \cos t\,\mathbf{j} + \cos t\,\mathbf{k} = \cos 2t\mathbf{i} - \sin 2t\mathbf{j} + \cos t\,\mathbf{k},$$

$$\mathbf{A}(t) = -2\sin 2t\mathbf{i} - 2\cos 2t\mathbf{j} - \sin t\,\mathbf{k},$$

give the velocity and acceleration, respectively, for the particle.

Whenever $\mathbf{R}'(t)$ is not zero, we can obtain a unit tangent vector $\mathbf{T}$, by dividing by $|\mathbf{R}'(t)|$. Thus

$$(66) \qquad \mathbf{T} = \frac{\dfrac{dx}{dt}\mathbf{i} + \dfrac{dy}{dt}\mathbf{j} + \dfrac{dz}{dt}\mathbf{k}}{\sqrt{\left(\dfrac{dx}{dt}\right)^2 + \left(\dfrac{dy}{dt}\right)^2 + \left(\dfrac{dz}{dt}\right)^2}}$$

and if α, β, and γ denote the direction angles for **T**, then

$$\cos \alpha = \frac{\dfrac{dx}{dt}}{\sqrt{\left(\dfrac{dx}{dt}\right)^2 + \left(\dfrac{dy}{dt}\right)^2 + \left(\dfrac{dz}{dt}\right)^2}},$$

$$(67)\qquad \cos \beta = \frac{\dfrac{dy}{dt}}{\sqrt{\left(\dfrac{dx}{dt}\right)^2 + \left(\dfrac{dy}{dt}\right)^2 + \left(\dfrac{dz}{dt}\right)^2}},$$

$$\cos \gamma = \frac{\dfrac{dz}{dt}}{\sqrt{\left(\dfrac{dx}{dt}\right)^2 + \left(\dfrac{dy}{dt}\right)^2 + \left(\dfrac{dz}{dt}\right)^2}}.$$

Example 1. For the curve defined by (65), prove that the tangent vector is always perpendicular to the position vector.

Solution. Since neither **R** nor **V** is zero, it suffices to prove that $\mathbf{R} \cdot \mathbf{V} = 0$. Clearly,

$$\mathbf{R} \cdot \mathbf{V} = [\sin t \cos t\mathbf{i} + \cos^2 t\mathbf{j} + \sin t\mathbf{k}] \cdot [(\cos^2 t - \sin^2 t)\mathbf{i} - 2 \sin t \cos t\mathbf{j} + \cos t\mathbf{k}]$$

$$= \sin t \cos^3 t - \sin^3 t \cos t - 2 \sin t \cos^3 t + \sin t \cos t$$

$$= \sin t \cos t(-\sin^2 t - \cos^2 t) + \sin t \cos t = 0. \quad \bullet$$

A few formulas for the differentiation of vector functions are covered in

---

THEOREM 16. If $\mathbf{U}(t)$ and $\mathbf{V}(t)$ are differentiable vector functions and $f(t)$ is a differentiable scalar function, then

$$(68)\quad \frac{d}{dt}(\mathbf{U} + \mathbf{V}) = \frac{d\mathbf{U}}{dt} + \frac{d\mathbf{V}}{dt}, \qquad (69)\quad \frac{d}{dt}f\mathbf{V} = f\frac{d\mathbf{V}}{dt} + \frac{df}{dt}\mathbf{V},$$

$$(70)\quad \frac{d}{dt}\mathbf{U} \cdot \mathbf{V} = \mathbf{U} \cdot \frac{d\mathbf{V}}{dt} + \frac{d\mathbf{U}}{dt} \cdot \mathbf{V}, \qquad (71)\quad \frac{d}{dt}\mathbf{U} \times \mathbf{V} = \mathbf{U} \times \frac{d\mathbf{V}}{dt} + \frac{d\mathbf{U}}{dt} \times \mathbf{V}.$$

The proof of (68) is sufficiently simple, so we omit it. The other three formulas all have the form of products, although the type of multiplication is different in each case. Since the method of proof is exactly the same in each case, we will give the proof only for (71).

Let $\mathbf{W} = \mathbf{U} \times \mathbf{V}$ and suppose that for a certain increment Δt, in the variable t, the vector functions $\mathbf{U}$, $\mathbf{V}$, and $\mathbf{W}$ change by amounts $\Delta\mathbf{U}$, $\Delta\mathbf{V}$, and $\Delta\mathbf{W}$, respectively. Then from the definition of $\Delta\mathbf{W}$, we have

$$\mathbf{W} + \Delta\mathbf{W} = (\mathbf{U} + \Delta\mathbf{U}) \times (\mathbf{V} + \Delta\mathbf{V})$$
$$= \mathbf{U} \times \mathbf{V} + \mathbf{U} \times \Delta\mathbf{V} + \Delta\mathbf{U} \times \mathbf{V} + \Delta\mathbf{U} \times \Delta\mathbf{V}.$$

Subtracting $\mathbf{W} = \mathbf{U} \times \mathbf{V}$ from both sides and then dividing by Δt yields

(72)
$$\frac{\Delta\mathbf{W}}{\Delta t} = \mathbf{U} \times \frac{\Delta\mathbf{V}}{\Delta t} + \frac{\Delta\mathbf{U}}{\Delta t} \times \mathbf{V} + \frac{\Delta\mathbf{U}}{\Delta t} \times \Delta\mathbf{V}.$$

If we let $\Delta t \to 0$, then $\Delta\mathbf{V} \to \mathbf{0}$, and the last term in (72) vanishes. The other three terms obviously give (71).

To obtain the proof of (70), just replace the crosses by dots in the preceding proof. To obtain (69), just suppress the crosses and replace $\mathbf{U}$ by f. ∎

Because the dot product of two vectors is commutative, we could reverse the order of the factors in the last term of (70) and write

$$\frac{d}{dt}\mathbf{U} \cdot \mathbf{V} = \mathbf{U} \cdot \frac{d\mathbf{V}}{dt} + \mathbf{V} \cdot \frac{d\mathbf{U}}{dt}.$$

But in (71) such a reversal will lead to an error unless a minus sign is also introduced. In deference to (71), it is customary to preserve the order of the factors in (69) and (70).

Example 2. Using the functions

$$f(t) = t^2, \qquad \mathbf{U}(t) = t\mathbf{i} + t^2\mathbf{j} + 2t\mathbf{k}, \qquad \mathbf{V}(t) = (1 + t^2)\mathbf{i} + (2 - t)\mathbf{j} + 3\mathbf{k},$$

compute the derivative with respect to t of $f\mathbf{V}$, $\mathbf{U} \cdot \mathbf{V}$, and $\mathbf{U} \times \mathbf{V}$.

Solution. Direct computation for $f\mathbf{V}$ gives

$$\frac{d}{dt}f\mathbf{V} = \frac{d}{dt}[(t^2 + t^4)\mathbf{i} + (2t^2 - t^3)\mathbf{j} + 3t^2\mathbf{k}]$$
$$= (2t + 4t^3)\mathbf{i} + (4t - 3t^2)\mathbf{j} + 6t\mathbf{k}.$$

If we use the right side of (69) for the same computation, we find that

$$f\frac{d\mathbf{V}}{dt} + \frac{df}{dt}\mathbf{V} = t^2[2t\mathbf{i} - \mathbf{j} + 0\mathbf{k}] + 2t[(1 + t^2)\mathbf{i} + (2 - t)\mathbf{j} + 3\mathbf{k}]$$

$$= (2t^3 + 2t + 2t^3)\mathbf{i} + (-t^2 + 4t - 2t^2)\mathbf{j} + 6t\mathbf{k},$$

and this agrees with the result obtained from the first computation, just as the theorem tells us it should.

Similarly, direct computation gives

$$\frac{d}{dt}\mathbf{U} \cdot \mathbf{V} = \frac{d}{dt}[t + t^3 + 2t^2 - t^3 + 6t] = \frac{d}{dt}[7t + 2t^2] = 7 + 4t.$$

If we use the right side of (70) for the same computation, we find that

$$\mathbf{U} \cdot \frac{d\mathbf{V}}{dt} + \frac{d\mathbf{U}}{dt} \cdot \mathbf{V} = [t\mathbf{i} + t^2\mathbf{j} + 2t\mathbf{k}] \cdot [2t\mathbf{i} - \mathbf{j} + 0\mathbf{k}]$$

$$+ [\mathbf{i} + 2t\mathbf{j} + 2\mathbf{k}] \cdot [(1 + t^2)\mathbf{i} + (2 - t)\mathbf{j} + 3\mathbf{k}]$$

$$= (2t^2 - t^2 + 0) + (1 + t^2 + 4t - 2t^2 + 6) = 7 + 4t.$$

We leave it for the student to compute $\dfrac{d}{dt}\mathbf{U} \times \mathbf{V}$ in two different ways and show that both ways give $(10t - 4)\mathbf{i} + (6t^2 - 1)\mathbf{j} + (2 - 4t - 4t^3)\mathbf{k}$. ●

Example 3. Prove that if the length of the vector function $\mathbf{R}(t)$ is a constant, then $\mathbf{R}$ and $d\mathbf{R}/dt$ are perpendicular whenever neither of the vectors is zero.

Solution. By hypothesis, $\mathbf{R} \cdot \mathbf{R} = |\mathbf{R}|^2 = $ constant. Then differentiating with respect to t, and using (70), we obtain

$$\mathbf{R} \cdot \frac{d\mathbf{R}}{dt} + \frac{d\mathbf{R}}{dt} \cdot \mathbf{R} = 0.$$

Consequently, $2\mathbf{R} \cdot \dfrac{d\mathbf{R}}{dt} = 0$ and the two vectors are perpendicular. ●

EXERCISE 8

In Problems 1 through 6, $\mathbf{R}$ is the position vector for a moving particle, and t denotes time. Find the velocity and acceleration.

1. $\mathbf{R} = a \sin 5t\mathbf{i} + a \cos 5t\mathbf{j} + 3t\mathbf{k}.$
2. $\mathbf{R} = a \sin t\mathbf{i} + a \cos t\mathbf{j} + 2 \sin 2t\mathbf{k}.$
3. $\mathbf{R} = (1 + 3t)\mathbf{i} + (2 - 5t)\mathbf{j} + (7 - t)\mathbf{k}.$
4. $\mathbf{R} = t\mathbf{i} + t^2\mathbf{j} + t^3\mathbf{k}.$
5. $\mathbf{R} = (t^2 - 1)\mathbf{i} + (t^3 - 3t^2)\mathbf{j} + 5t\mathbf{k}.$
6. $\mathbf{R} = (1 - te^{-t})\mathbf{i} + (t^{-1} + 5)\mathbf{j} + t^{-1} \ln t\mathbf{k}.$

7. For the motion of Problem 1 prove that: (a) **A** and **V** have constant length, (b) **A** and **V** are perpendicular, and (c) **A** is always parallel to the *xy*-plane.

*8. In Problem 3 the particle moves on a line and the acceleration vector is zero. Prove that whenever the acceleration vector is constantly zero, then the motion is along a line.

9. For the motion of Problem 4, show that if $t \neq 0$, then no two of the vectors **R**, **V**, and **A** are ever perpendicular.

10. For the motion of Problem 5, find where the particle is when the velocity vector is parallel to the *xz*-plane.

11. As $t \to \infty$, for the motion of Problem 6, what is the limiting position of the particle? What is the limiting velocity and acceleration vector?

**12. Suppose that for some motion, the particle tends to a limiting position as $t \to \infty$. Is it necessary for either the velocity vector or the acceleration vector to approach zero?

*13. Suppose that for a certain motion we always have $\mathbf{R} \cdot \mathbf{V} = 0$. Prove that the particle must at all times lie on the surface of some sphere.

*14. At a certain instant one airplane is 1 mile directly above another airplane. Both are flying level, the first going due north at 120 miles/hr and the second going due west at twice the speed. Find the rate at which the distance between them is changing 2 min later.

*15. If in Problem 14 one airplane is 2 miles above the other, and they are traveling as before but with speeds of 100 and 110 miles/hr, respectively, find the rate at which they are separating 6 min later.

Space Curves 9

Any curve in space can be described by giving parametric equations

(73) $$x = f(t), \qquad y = g(t), \qquad z = h(t)$$

for the coordinates (x, y, z) of a point P on the curve. The same curve is described by writing the equation for the position vector to the point on the curve

(74) $$\mathbf{R} = \mathbf{R}(t) = f(t)\mathbf{i} + g(t)\mathbf{j} + h(t)\mathbf{k}.$$

A number of results follow immediately from this vector equation for the curve. The proofs are completely similar to those given in the case of a plane curve in Chapter 12, so it will be sufficient to state the facts.

If $\mathbf{R}'(t)$ is not zero, it is a vector tangent to the curve, and pointing in the direction of increase of t on the curve. If the arc length s is taken as the parameter, then

$$\frac{d\mathbf{R}}{ds} = \mathbf{T},$$

a unit vector. Using the Chain Rule for differentiation, we have

$$\frac{d\mathbf{R}}{dt} = \frac{d\mathbf{R}}{ds}\frac{ds}{dt} = \mathbf{T}\frac{ds}{dt}.$$

Taking the dot product of each side with itself, we have

$$(75) \qquad \frac{d\mathbf{R}}{dt} \cdot \frac{d\mathbf{R}}{dt} = \mathbf{T} \cdot \mathbf{T}\left(\frac{ds}{dt}\right)^2 = \left(\frac{ds}{dt}\right)^2$$

since $\mathbf{T}$ is a unit vector. Consequently, using (74) and (75), we obtain

$$\left(\frac{ds}{dt}\right)^2 = \left(\frac{dx}{dt}\right)^2 + \left(\frac{dy}{dt}\right)^2 + \left(\frac{dz}{dt}\right)^2 = [f'(t)]^2 + [g'(t)]^2 + [h'(t)]^2,$$

and if s and t increase together,

$$(76) \qquad \frac{ds}{dt} = \sqrt{[f'(t)]^2 + [g'(t)]^2 + [h'(t)]^2} = |\mathbf{R}'(t)|.$$

The curvature for a space curve is closely related to the curvature for a plane curve, but there is an *essential difference,* which may cause trouble for the careless reader. We recall (Chapter 12, Theorem 17) that for a plane curve

$$(77) \qquad \boxed{\frac{d\mathbf{T}}{ds} = \kappa\mathbf{N}.}$$

In this equation κ is the curvature of $\mathscr{C}$ and $\mathbf{N}$ is a unit vector, 90° in *advance* of $\mathbf{T}$. Under these conditions we may have $\kappa > 0$ for some points on $\mathscr{C}$ and $\kappa < 0$ for other points on $\mathscr{C}$. In space there is no way of specifying a vector 90° in advance of $\mathbf{T}$, so the unit vector $\mathbf{N}$ must be defined in some other manner.

For a space curve we use (77) as a basis for the definition of κ and $\mathbf{N}$. At any point of $\mathscr{C}$ at which the left side of (77) is not the zero vector, we let the curvature κ be positive, and let $\mathbf{N}$ be a unit vector. Then both κ and $\mathbf{N}$ are uniquely determined by (77), and indeed

$$(78) \qquad \kappa = \left|\frac{d\mathbf{T}}{ds}\right| = \sqrt{\frac{d\mathbf{T}}{ds} \cdot \frac{d\mathbf{T}}{ds}}.$$

If the left side of (77) is the zero vector, then we set $\kappa = 0$, and we leave $\mathbf{N}$ undefined.

For a plane curve we recall that $\mathbf{N}$ was defined as a vector perpendicular to $\mathbf{T}$, and the relation (77) appeared as a theorem. By contrast, (77) is now a definition for $\mathbf{N}$, and hence we should prove that **N** and **T** are mutually perpendicular. Since **T** is a unit vector, $\mathbf{T} \cdot \mathbf{T} = 1$. Differentiating both sides of this identity with respect to s and using (70) and (77), we have

$$0 = \mathbf{T} \cdot \frac{d\mathbf{T}}{ds} + \frac{d\mathbf{T}}{ds} \cdot \mathbf{T} = 2\mathbf{T} \cdot \frac{d\mathbf{T}}{ds} = 2\kappa\mathbf{T} \cdot \mathbf{N}.$$

Consequently, if $\kappa \neq 0$, then **T** and **N** are mutually perpendicular.

Once **T** and **N** have been found, it is convenient to define a third unit vector **B**, which is perpendicular to both **T** and **N**, by the equation

(79)
$$\mathbf{B} = \mathbf{T} \times \mathbf{N}.$$

The vector **N** is called the *principal normal* to the curve, and the vector **B** is called the *binormal* to the curve. These three unit vectors are extremely useful in the differential geometry of space curves.

Example 1. Find κ, **T**, **N**, and **B** for the *circular helix*

(80)
$$\mathbf{R} = a \cos t\mathbf{i} + a \sin t\mathbf{j} + bt\mathbf{k},$$

where a and b are positive constants.

Solution. The projection of this curve on the xy-plane is obtained by setting the z-component equal to zero. This gives $\mathbf{R} = a \cos t\mathbf{i} + a \sin t\mathbf{j}$, so the projection of this space curve is a circle with center at the origin and radius a. Since $z = bt$ is steadily increasing with t, it is easy to see that this curve has the appearance indicated in Fig. 18 (next page).

Differentiating $\mathbf{R}(t)$, we find that

(81)
$$\mathbf{R}'(t) = \frac{d\mathbf{R}}{dt} = -a \sin t\mathbf{i} + a \cos t\mathbf{j} + b\mathbf{k}$$

is a tangent vector. To obtain a unit tangent vector, we divide by

(82)
$$|\mathbf{R}'(t)| = \sqrt{a^2 \sin^2 t + a^2 \cos^2 t + b^2} = \sqrt{a^2 + b^2}.$$

Consequently,

(83)
$$\mathbf{T} = \frac{1}{\sqrt{a^2 + b^2}}[-a \sin t\mathbf{i} + a \cos t\mathbf{j} + b\mathbf{k}].$$

If we decide to measure arc length along this curve from the point $(a, 0, 0)$ corre-

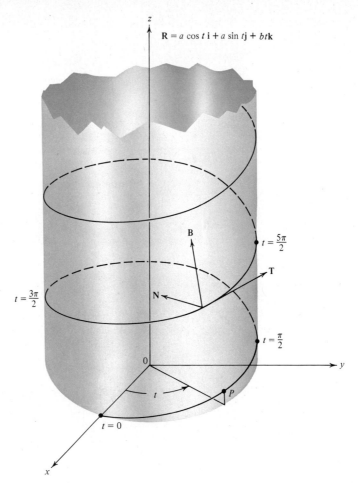

$$\mathbf{R} = a\cos t\,\mathbf{i} + a\sin t\mathbf{j} + bt\mathbf{k}$$

FIGURE 18

sponding to $t = 0$, we use $s'(t) = |\mathbf{R}'(t)|$ and integrate from 0 to t with u as the dummy variable of integration. Then equations (76), (81), and (82) yield

$$s = \int_0^t \sqrt{(-a\sin u)^2 + (a\cos u)^2 + b^2}\,du = \int_0^t \sqrt{a^2 + b^2}\,du = t\sqrt{a^2 + b^2}.$$

In this case it is a simple matter to introduce the arc length s as a parameter for the curve. Indeed, using $t = s/\sqrt{a^2 + b^2}$ in (80), we have

$$\mathbf{R} = a\cos\frac{s}{\sqrt{a^2 + b^2}}\mathbf{i} + a\sin\frac{s}{\sqrt{a^2 + b^2}}\mathbf{j} + \frac{bs}{\sqrt{a^2 + b^2}}\mathbf{k},$$

the parametric equations for the circular helix with the arc length s as parameter.

To find κ, we use equation (77). Indeed, differentiating equation (83) and using

$$\frac{dt}{ds} = \frac{1}{|\mathbf{R}'(t)|} = \frac{1}{\sqrt{a^2 + b^2}},$$

we have

$$\kappa\mathbf{N} = \frac{d\mathbf{T}}{ds} = \frac{d\mathbf{T}}{dt}\frac{dt}{ds} = \frac{1}{\sqrt{a^2 + b^2}}[-a\cos t\mathbf{i} - a\sin t\mathbf{j}]\frac{dt}{ds}$$

$$= \frac{-a}{\sqrt{a^2 + b^2}}[\cos t\mathbf{i} + \sin t\mathbf{j}]\frac{1}{\sqrt{a^2 + b^2}} = \frac{a}{a^2 + b^2}[-\cos t\mathbf{i} - \sin t\mathbf{j}].$$

Since $-\cos t\mathbf{i} - \sin t\mathbf{j}$ is a unit vector, it is clear that

$$\kappa = \frac{a}{a^2 + b^2} \quad \text{and} \quad \mathbf{N} = -\cos t\mathbf{i} - \sin t\mathbf{j}.$$

Finally, $\mathbf{B} = \mathbf{T} \times \mathbf{N}$, so

$$\mathbf{B} = \frac{1}{\sqrt{a^2 + b^2}}\begin{vmatrix} \mathbf{i} & \mathbf{j} & \mathbf{k} \\ -a\sin t & a\cos t & b \\ -\cos t & -\sin t & 0 \end{vmatrix} = \frac{1}{\sqrt{a^2 + b^2}}[b\sin t\mathbf{i} - b\cos t\mathbf{j} + a\mathbf{k}].$$

As a check, we observe that $\mathbf{B} \cdot \mathbf{T} = 0$, $\mathbf{B} \cdot \mathbf{N} = 0$, and $\mathbf{N} \cdot \mathbf{T} = 0$. ●

Example 2. Find a formula for κ in terms of $\mathbf{R}(t)$ and its derivatives.

Solution. Using the Chain Rule for differentiation, we have

(84)
$$\frac{d\mathbf{R}}{dt} = \frac{d\mathbf{R}}{ds}\frac{ds}{dt} = \mathbf{T}\frac{ds}{dt},$$

$$\frac{d^2\mathbf{R}}{dt^2} = \mathbf{T}\frac{d^2s}{dt^2} + \frac{d\mathbf{T}}{dt}\frac{ds}{dt} = \mathbf{T}\frac{d^2s}{dt^2} + \left(\frac{d\mathbf{T}}{ds}\frac{ds}{dt}\right)\frac{ds}{dt},$$

(85)
$$\frac{d^2\mathbf{R}}{dt^2} = \frac{d^2s}{dt^2}\mathbf{T} + \left(\frac{ds}{dt}\right)^2\kappa\mathbf{N}.$$

[Compare this equation with equation (75) of Chapter 12, obtained there for plane curves.] Taking the cross product of the vectors in equation (84) with those in (85), and noting that $\mathbf{T} \times \mathbf{T} = \mathbf{0}$, we have

(86)
$$\frac{d\mathbf{R}}{dt} \times \frac{d^2\mathbf{R}}{dt^2} = \mathbf{T}\frac{ds}{dt} \times \left[\frac{d^2s}{dt^2}\mathbf{T} + \left(\frac{ds}{dt}\right)^2\kappa\mathbf{N}\right] = \kappa\left(\frac{ds}{dt}\right)^3\mathbf{B}.$$

But $\mathbf{B}$ is a unit vector, so taking lengths in (86), we obtain

$$\kappa\left(\frac{ds}{dt}\right)^3 = \left|\frac{d\mathbf{R}}{dt} \times \frac{d^2\mathbf{R}}{dt^2}\right|,$$

or

$$(87) \qquad \kappa = \frac{|\mathbf{R}'(t) \times \mathbf{R}''(t)|}{|\mathbf{R}'(t)|^3}. \qquad \bullet$$

We leave it to the reader to apply this formula to the curve of Example 1 and show that $\kappa = a/(a^2 + b^2)$ for that curve.

EXERCISE 9

In Problems 1 through 5, find an equation for the line tangent to the given curve at the given point, and an equation for the plane normal to the given curve at that point.

1. $\mathbf{R} = 6t\mathbf{i} + 3t^2\mathbf{j} + t^3\mathbf{k},$ $P(0, 0, 0).$
2. $\mathbf{R} = \sqrt{2}\,t\mathbf{i} + e^t\mathbf{j} + e^{-t}\mathbf{k},$ $P(0, 1, 1).$
3. $\mathbf{R} = t\sin t\mathbf{i} + t\cos t\mathbf{j} + \sqrt{3}\,t\mathbf{k},$ $P(0, 0, 0).$
4. $\mathbf{R} = \sin 3t\mathbf{i} + \cos 3t\mathbf{j} + 2t^{3/2}\mathbf{k},$ $P(0, 1, 0).$
5. $\mathbf{R} = t\mathbf{i} + t\mathbf{j} + \dfrac{2}{3}t^{3/2}\mathbf{k},$ $P(9, 9, 18).$

6. For each of the curves of Problems 1 through 5, find the arc length as a function of t, assuming that $s = 0$ when $t = 0$, and that they increase together.

In Problems 7 through 9, find the curvature for the given curve. Use the formula from Example 2.

7. $\mathbf{R} = e^t\mathbf{i} + \sqrt{2}\,t\mathbf{j} + e^{-t}\mathbf{k}.$ 8. $\mathbf{R} = 6t\mathbf{i} + 3\sqrt{2}\,t^2\mathbf{j} + 2t^3\mathbf{k}.$
*9. $\mathbf{R} = 3at^2\mathbf{i} + a(3t + t^3)\mathbf{j} + a(3t - t^3)\mathbf{k}.$

10. Find $\cos\theta$ for the angle of intersection of the two curves

$$\mathbf{R}_1 = (1 + t^4)\mathbf{i} + 2\cos\pi t\mathbf{j} + t^3\mathbf{k} \qquad \text{and} \qquad \mathbf{R}_2 = (t + t^2)\mathbf{i} + (t - 3t^2)\mathbf{j} + te^{t-1}\mathbf{k}$$

at the point $(2, -2, 1).$

11. Find $\cos\theta$ for the angle of intersection of the curve $\mathbf{R}_1$ of Problem 10 and the curve

$$\mathbf{R}_3 = \frac{1}{4}t^3\mathbf{i} + (6 - t^3)\mathbf{j} + (t^2 - 3)\mathbf{k} \text{ at the point } (2, -2, 1).$$

12. Show that for the straight line, $d\mathbf{T}/ds = \mathbf{0}$. Consequently, for the straight line the vectors $\mathbf{N}$ and $\mathbf{B}$ are not defined.

In Problems 13 through 15, find the principal normal vector $\mathbf{N}$ and check that $\mathbf{N} \cdot \mathbf{T} = 0$ in each case.

13. The conical helix $\mathbf{R} = e^t\mathbf{i} + e^t\cos t\mathbf{j} + e^t\sin t\mathbf{k}.$
14. $\mathbf{R} = 4\sin t\mathbf{i} + (2t - \sin 2t)\mathbf{j} + \cos 2t\mathbf{k}.$
15. The curve of Problem 1.

16. Prove that for a curve

$$\frac{dx}{ds} = \cos\alpha, \qquad \frac{dy}{ds} = \cos\beta, \qquad \frac{dz}{ds} = \cos\gamma,$$

where α, β, and γ are the direction angles of the unit tangent vector.

★17. Assuming that the binormal vector $\mathbf{B}$, defined by $\mathbf{B} = \mathbf{T} \times \mathbf{N}$, is a differentiable function and that $d\mathbf{B}/ds \neq \mathbf{0}$, prove that it is a vector perpendicular to both $\mathbf{T}$ and $\mathbf{B}$. Hence the equation

$$\frac{d\mathbf{B}}{ds} = -\tau\mathbf{N}$$

defines a quantity τ (Greek lowercase letter tau), known as the *torsion of* $\mathscr{C}$. HINT: Differentiate the identities $\mathbf{B} \cdot \mathbf{T} = 0$ and $\mathbf{B} \cdot \mathbf{B} = 1$.

★18. Prove that $d\mathbf{N}/ds = -\kappa\mathbf{T} + \tau\mathbf{B}$. HINT: Differentiate the identity $\mathbf{N} = \mathbf{B} \times \mathbf{T}$. The three formulas [see equation (77) and Problem 17]

$$\frac{d\mathbf{T}}{ds} = \kappa\mathbf{N},$$

$$\frac{d\mathbf{N}}{ds} = -\kappa\mathbf{T} + \tau\mathbf{B},$$

$$\frac{d\mathbf{B}}{ds} = -\tau\mathbf{N},$$

are known as the *Frenet–Serret Formulas,* and they are fundamental in the study of the differential geometry of space curves.

★19. Starting from equation (85), prove that

$$\mathbf{R}'''(t) = (s''' - (s')^3\kappa^2)\mathbf{T} + (3s''s'\kappa + (s')^2\kappa')\mathbf{N} + (s')^3\kappa\tau\mathbf{B},$$

where primes denote differentiation with respect to t.

★20. Use the result in Problem 19 to prove that

$$(\mathbf{R}' \times \mathbf{R}'') \cdot \mathbf{R}''' = \kappa^2\tau(s')^6.$$

★21. Combine the results of Problem 20 with equation (87) to prove that

$$\tau = \frac{(\mathbf{R}' \times \mathbf{R}'') \cdot \mathbf{R}'''}{|\mathbf{R}' \times \mathbf{R}''|^2}.$$

★22. Find the torsion at an arbitrary point of the curve:
   (a) $\mathbf{R}(t) = t\mathbf{i} + t^2\mathbf{j} + t^3\mathbf{k}$.   (b) $\mathbf{R}(t) = a\cos t\,\mathbf{i} + a\sin t\,\mathbf{j} + bt\mathbf{k}$.
   (c) $\mathbf{R}(t) = e^t\mathbf{i} + \sin t\,\mathbf{j} + t\mathbf{k}$.

★23. Suppose that a curve lies in the xy-plane. Prove that at every point on the curve where τ is defined, we have $\tau = 0$.

10 *Surfaces*

The set of all points $P(x, y, z)$ whose coordinates satisfy an equation

(88) $$F(x, y, z) = 0$$

usually constitutes a surface, and will always do so for any function $F(x, y, z)$ that is of practical importance. For example,

$$Ax + By + Cz - D = 0$$

represents a plane if not all of the coefficients A, B, and C are zero. The equation

$$(x - a)^2 + (y - b)^2 + (z - c)^2 - r^2 = 0$$

is the equation of a sphere of radius r and center (a, b, c).

In some cases (88) can be explicitly solved for z, and we may write

(89) $$z = f(x, y)$$

as the equation for a surface. Thus to each point (x, y) in the base plane (or in some suitable region of the plane) there corresponds a point on the surface, z units directly above (or below if $z < 0$). The situation is illustrated in Fig. 19, where (x, y) is supposedly restricted to lie in a rectangle

As a specific illustration of (89), consider the surface

(90) $$z = 4 - x^2 - y^2.$$

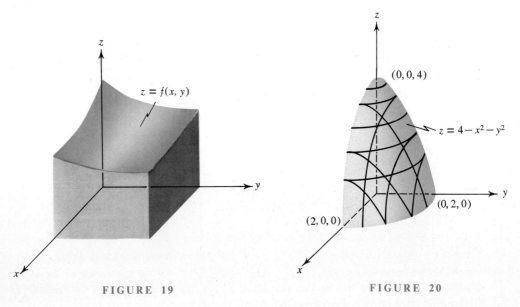

FIGURE 19   FIGURE 20

We will prove in the discussion of Example 1 below that this surface can be obtained by rotating the parabola $z = 4 - y^2$ (lying in the yz-plane) about the z-axis. The surface is called a *paraboloid of revolution*. That portion of the surface that lies in the first octant is shown in Fig. 20.

Just as the vector function $\mathbf{R}(t)$, of a single variable, describes a curve, so we may expect a vector function $\mathbf{R}(u, v)$, of two variables, to describe a surface. Thus in (89) we could set $x = u$, $y = v$, and $z = f(x, y) = f(u, v)$, and then the position vector $\mathbf{R} = x\mathbf{i} + y\mathbf{j} + z\mathbf{k}$ to a point on the surface would take the form

$$(91) \qquad \mathbf{R}(u, v) = u\mathbf{i} + v\mathbf{j} + f(u, v)\mathbf{k},$$

a vector equation for the surface (89).

More generally, any three functions of two variables

$$(92) \qquad x = f(u, v), \qquad y = g(u, v), \qquad z = h(u, v)$$

give *parametric equations* for a surface, and the vector function

$$(93) \qquad \mathbf{R} = f(u, v)\mathbf{i} + g(u, v)\mathbf{j} + h(u, v)\mathbf{k}$$

describes the same surface.

There are always infinitely many ways in which a given surface can be described by a vector equation. Although perhaps not obvious, the four vector equations

$$\mathbf{R} = u\mathbf{i} + v\mathbf{j} + (4 - u^2 - v^2)\mathbf{k},$$

$$\mathbf{R} = (u - 2)\mathbf{i} + (v - 1)\mathbf{j} + (4u + 2v - u^2 - v^2 - 1)\mathbf{k},$$

$$\mathbf{R} = u \cos v\mathbf{i} + u \sin v\mathbf{j} + (4 - u^2)\mathbf{k},$$

$$\mathbf{R} = u^2 \cos 3v\mathbf{i} + u^2 \sin 3v\mathbf{j} + (4 - u^4)\mathbf{k},$$

all describe the paraboloid of revolution (90) (see Fig. 20).

We will postpone the study of the vector representation of a surface until Chapter 18, and for the present devote our attention to the graph of $F(x, y, z) = 0$. Here it is best to proceed by examples, and then at the end to summarize with some general principles on sketching a surface from its equation.

Example 1. Sketch the surface $z = 4 - x^2 - y^2$.

Solution. A portion of this surface is already shown in Fig. 20, but how did we arrive at this sketch? If we want to find the curve of intersection of the surface with some plane parallel to one of the coordinate planes, we merely regard the appropriate variable as a constant. Now the plane parallel to the xy-plane and 1 unit above it is just the collection of points on which $z = 1$; that is, it has the equation $z = 1$. Setting $z = 1$ in $z = 4 - x^2 - y^2$ gives the equation in x and y for the points on the curve of intersection. In this case we find $1 = 4 - x^2 - y^2$ or $x^2 + y^2 = 3$. Hence the curve of intersection is a circle, with center on the z-axis and radius $\sqrt{3}$. If we write the equation of the surface in the form $x^2 + y^2 = 4 - z$, we see that if z_0 is any fixed number with $z_0 < 4$, then the intersection of the plane $z = z_0$ with the surface is a circle with center on the z-axis and radius $\sqrt{4 - z_0}$. If $z_0 = 4$, the intersection is the point $(0, 0, 4)$, and if $z_0 > 4$, the intersection is empty (the plane and the surface do not meet). Since each section of the surface obtained by cutting with a plane perpendicular to the z-axis is a circle (or a point, or empty) with center on the z-axis, the surface itself can be generated by rotating a suitable curve around the z-axis. To find this suitable curve, we take the intersection of the surface with the yz-plane, by setting $x = 0$ in the equation of the surface. We find $z = 4 - y^2$, the equation of a parabola.

It is of interest to find the intersection of this surface with other planes, for example, a plane parallel to the yz-plane. This is done by setting $x = x_0$, a constant. We obtain $z = (4 - x_0^2) - y^2$, or $z = A - y^2$, the equation of a parabola. Similarly, any plane parallel to the xz-plane also intersects this surface in a parabola. Portions of these parabolas are shown in Fig. 20. ●

Example 2. Sketch the *ellipsoid* $\dfrac{x^2}{a^2} + \dfrac{y^2}{b^2} + \dfrac{z^2}{c^2} = 1$.

Solution. A portion of this surface is shown in Fig. 21. Obviously, the complete surface is symmetric with respect to each of the coordinate planes, and consequently, it is symmetric with respect to each of the three coordinate axes and with respect to the origin. The full surface resembles an egg except that the egg has only one axis of symmetry. Solving for x gives

$$x = \pm a \sqrt{1 - \frac{y^2}{b^2} - \frac{z^2}{c^2}}$$

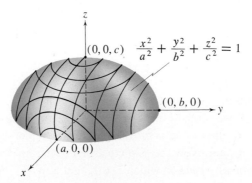

FIGURE 21

and hence for points on the surface with real coordinates we must have $-a \leq x \leq a$. Similarly, $-b \leq y \leq b$ and $-c \leq z \leq c$ and therefore the surface must lie inside the box bounded by the six planes $x = \pm a, y = \pm b, z = \pm c$. The intersection of this surface with a plane $z = z_0, |z_0| < c$, is the curve

$$\frac{x^2}{a^2} + \frac{y^2}{b^2} = 1 - \frac{z_0^2}{c^2}.$$

This is a circle if $a = b$ and an ellipse if $a \neq b$. Consequently, if $a \neq b, b \neq c$, and $a \neq c$, this surface is *not* a surface of revolution. ●

Example 3. Sketch the surface $\dfrac{z}{c} = \dfrac{y^2}{a^2} - \dfrac{x^2}{b^2}$, $c > 0$.

Solution. This surface is called the *hyperbolic paraboloid,* and a portion is shown in Fig. 22. For obvious reasons, it is also called a *saddle surface,* and the origin in this case is called a *saddle point*. This surface is important in further theoretical studies, and the student should convince himself that the graph of this equation does indeed have the form indicated in the picture. We leave it for the student to examine the symmetry and to prove that every plane parallel to the *xy*-plane intersects this sur-

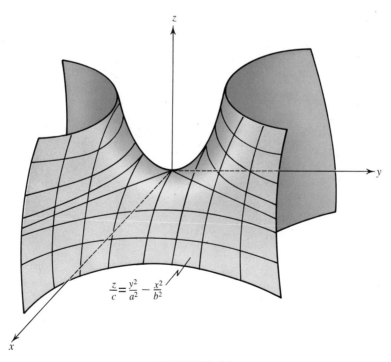

$$\frac{z}{c} = \frac{y^2}{a^2} - \frac{x^2}{b^2}$$

FIGURE 22

face in a hyperbola, except in one case where the hyperbola degenerates into two straight lines. Each plane parallel to the xz-plane or yz-plane intersects the surface in a parabola, but in the first case the parabola opens downward, while in the second case it opens upward. ●

The following statements (really theorems) are useful in sketching surfaces or finding their equations. We leave the proofs to the student.

1. If $F(x, y, -z) \equiv F(x, y, z)$, then the surface $F(x, y, z) = 0$ is symmetric with respect to the xy-plane.
2. If z is missing in $F(x, y, z) = 0$, then the surface is a cylinder perpendicular to the xy-plane.
3. If $F(x, y) = 0$ is the equation of a curve in the xy-plane, then $F(x, \pm\sqrt{y^2 + z^2}) = 0$ is an equation for the surface generated by rotating this curve about the x-axis.
4. If $F(x, y, z) = 0$ and $G(x, y, z) = 0$ are the equations of two surfaces, $\mathcal{S}_1$ and $\mathcal{S}_2$, and if elimination of the variable z yields $\Phi(x, y) = 0$, then the graph of this latter equation contains (possibly as a proper subset) the projection of the intersection of $\mathcal{S}_1$ and $\mathcal{S}_2$ on the xy-plane.
5. If $F(x, y, z)$ is a quadratic in x, y, and z, then the intersection of any plane with the surface $F(x, y, z) = 0$ projects onto the xy-plane (or xz-plane, or yz-plane), giving a conic section, or a circle, or one or two straight lines, or a point, or no points.
6. A curve defined by $x = f(t)$, $y = g(t)$, $z = h(t)$ for t in $\mathcal{I}$ lies on the surface $F(x, y, z) = 0$ if $F(f(t), g(t), h(t)) \equiv 0$ for every t in $\mathcal{I}$.

Example 4. The straight line $y = 2z$ is rotated about the z-axis. Find an equation for the cone generated, and prove that each of the straight lines

$$\mathbf{R}(t) = 2at\mathbf{i} + 2bt\mathbf{j} + \sqrt{a^2 + b^2}\, t\mathbf{k}$$

lies on the surface.

Solution. By Statement 3, with a suitable change of letters, an equation for the cone is $2z = \pm\sqrt{y^2 + x^2}$ or $4z^2 - x^2 - y^2 = 0$. Substituting $x = 2at$, $y = 2bt$, and $z = \sqrt{a^2 + b^2}\,t$ in this equation yields

$$4(a^2 + b^2)t^2 - 4a^2t^2 - 4b^2t^2 = 0.$$

But this is true for any t, so the lines $\mathbf{R}(t)$ lie on the surface. ●

EXERCISE 10

1. Give a condition on the function $F(x, y, z)$ so that the surface $F(x, y, z) = 0$ is
   (a) symmetric with respect to the xz-plane, (b) symmetric with respect to the yz-

plane, **(c)** symmetric with respect to the *x*-axis, **(d)** symmetric with respect to the *z*-axis, and **(e)** symmetric with respect to the plane $x = y$.

2. Without computing any points, what can you say about the surface

$$x \sin z + zx^5 = 7?$$

3. Sketch the surface

$$\frac{x^2}{a^2} + \frac{y^2}{b^2} - \frac{z^2}{c^2} = 1.$$

This surface is called a *hyperboloid of one sheet*. Find the values of x_0 such that the plane $x = x_0$ intersects this surface in two straight lines.

4. For what value of z_0 does the plane $z = z_0$ intersect the saddle surface of Example 3 in two straight lines?

In Problems 5 through 12, sketch enough of the given surface to indicate clearly what the surface looks like.

5. $z + 9 - y^2 = 0$.   6. $4x^2 + z^2 - 24x + 32 = 0$.

7. $x^2 - y^3 = 0$.   8. $4z - x^2 - y^2 = 0$.

9. $4z^2 - x^2 - y^2 = 0$.   10. $\dfrac{z}{c^2} - \dfrac{x^2}{a^2} - \dfrac{y^2}{b^2} = 0$.

11. $y - x^2 - z^2 = 0$.   12. $x^2 - 2y^2 + z^2 - 1 = 0$.

13. The ellipse $\dfrac{z^2}{a^2} + \dfrac{y^2}{b^2} = 1$ is rotated about the *z*-axis. Find a formula for the surface generated. If $a > b$, the surface is called a *prolate ellipsoid*. If $a < b$, the surface is called an *oblate ellipsoid*. Sketch the surface in both cases.

14. The surface $\dfrac{z^2}{c^2} - \dfrac{x^2}{a^2} - \dfrac{y^2}{b^2} = 1$ is called a *hyperboloid of two sheets*. Sketch this surface, and explain the "two sheets." Compare this surface with the one in Problem 3.

In Problems 15 through 21, find an equation for the projection onto the xy-plane of the curve of intersection of the two given surfaces. In each case sketch the surfaces.

15. $z = x^2 + y^2$, $z = 4y$.   16. $z = 8 - x^2 - y^2$, $z = 2x$.

17. $x^2 + z^2 = 4$, $y^2 + z^2 = 4$.   18. $x^2 + z^2 = 9$, $y^2 + z^2 = 4$.

19. $z = y^2 + 4x^2$, $z = 4xy$.   20. $z = y^2 + 4x^2$, $z = Axy$, $A > 4$.

21. $x^2 + y^2 + z^2 = 4A^2$, $x + y + z = 2A$.

22. Show that the curve $\mathbf{R} = a \cos t\mathbf{i} + a \sin t\mathbf{j} + a \cos t\mathbf{k}$ lies on both of the cylinders $x^2 + y^2 = a^2$ and $y^2 + z^2 = a^2$.

23. Prove that for any pair of numbers a and b the straight line

$$\mathbf{R} = (t + a)\mathbf{i} + (t + b)\mathbf{j} + [2(b - a)t + b^2 - a^2]\mathbf{k}$$

lies on the saddle surface $z = y^2 - x^2$. Show further that through each point on this surface there passes at least one such line. This shows that the saddle surface can be obtained as a union of straight lines.

24. Show that the curve $\mathbf{R} = e^t \cos 3t\mathbf{i} + e^t \sin 3t\mathbf{j} + (4 - e^{2t})\mathbf{k}$ lies on the surface $z = 4 - x^2 - y^2$.

$\star$25. Find a condition on a, b, and c so that the curve $\mathbf{R} = at^4\mathbf{i} + bt^3\mathbf{j} + ct^6\mathbf{k}$ lies on the surface $z^2 = x^3 + 2y^4$. Prove that through each point on the surface there is at least one curve from this family. Is there a point on the surface through which every curve of the family runs?

26. Consider the circle $(x - A)^2 + z^2 = a^2$ (with $A > a > 0$) lying in the xz-plane. If this circle is rotated about the z-axis, it generates a surface called the *torus*, or *anchor ring* (an idealized doughnut). Find an equation for this surface.

11 The Cylindrical Coordinate System

In addition to the rectangular coordinate system, there are two other coordinate systems that are useful in solving problems in three-dimensional space. These are the spherical coordinate system, which we will study in the next section, and the cylindrical coordinate system, which we consider now.

The cylindrical coordinate system is obtained by putting a z-axis "on top" of a polar coordinate system. Thus, as indicated in Fig. 23, if O is the pole of a plane polar coordinate system, a z-axis is erected at O perpendicular to that plane. Then the coordinates (r, θ, z) for a point in space describe its location.

In many applications of cylindrical coordinates it is convenient to require that if P is not on the z-axis, then P has a unique set of cylindrical coordinates. This can be done by imposing the conditions $r \geq 0$ and $0 \leq \theta < 2\pi$. However, in this text we will leave r and θ unrestricted unless the contrary is explicitly stated.

We always consider the cylindrical coordinate system as superimposed on a rectangular coordinate system in such a way that the two origins and the two z-axes coincide, and the positive x-axis falls on the polar line. Then the coordinates in the two different systems, for a given point, are related by the equation set

(94)
$$x = r \cos \theta, \qquad y = r \sin \theta, \qquad z = z.$$

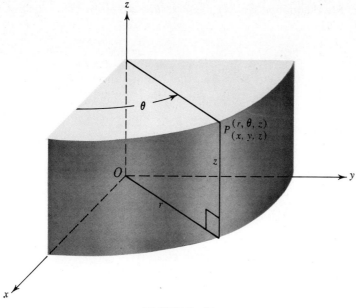

FIGURE 23

Example 1. Describe each of the surfaces: (a) $\theta = \pi/3$, (b) $r = 5$, (c) $z + r = 7$, and (d) $r(2 \sin \theta + 3 \cos \theta) + 4z = 0$.

Solution. (a) The collection of all points $P(r, \theta, z)$ for which $\theta = \pi/3$ fills a plane that contains the z-axis and that intersects the xy-plane in a line that makes an angle $\pi/3$ with the polar line. At first glance it might seem that we get only a half-plane, but we recall that r may be negative, and those values for r give the points in back of the yz-plane.

(b) The surface $r = 5$ is just a right circular cylinder of radius 5, with the z-axis as a center line.

(c) For the surface $z + r = 7$, think of its intersection with the yz-plane ($\theta = \pi/2$). Then $r = y$ and we have just the straight line $z + y = 7$. But the original equation does not contain θ; hence the surface is just the cone obtained by rotating the line $z + y = 7$ about the z-axis.

(d) Using the equation set (94), the equation $r(2 \sin \theta + 3 \cos \theta) + 4z = 0$ can be transformed into $2y + 3x + 4z = 0$. So the surface is just a plane through the origin with the normal vector $3\mathbf{i} + 2\mathbf{j} + 4\mathbf{k}$. ●

Example 2. Find an equation for the saddle surface $z = x^2 - y^2$ in cylindrical coordinates.

Solution. From equation set (94) $x = r \cos \theta$, $y = r \sin \theta$; hence the given equation is transformed into

$$z = x^2 - y^2 = r^2 \cos^2 \theta - r^2 \sin^2 \theta = r^2(\cos^2 \theta - \sin^2 \theta) = r^2 \cos 2\theta.$$

Consequently, $z = r^2 \cos 2\theta$ is an equation for the saddle surface in cylindrical coordinates. ●

EXERCISE 11

In Problems 1 through 6, change from the given cylindrical coordinates of a point to the set of rectangular coordinates for the same point.

1. $(3, \pi/2, 5)$.
2. $(-3, \pi/2, -5)$.
3. $(4, -4\pi/3, 1)$.
4. $(-1, 25\pi, 6)$.
5. $(6, 7\pi/4, 19)$.
6. $(4, 2, 1)$.

In Problems 7 through 12, change from the given rectangular coordinates of a point to a suitable set of cylindrical coordinates for the same point.

7. $(1, 1, 1)$.
8. $(2, -2, -2)$.
9. $(-3\sqrt{3}, 3, 6)$.
10. $(-4, 4, -7)$.
11. $(-8, -8\sqrt{3}, \pi)$.
12. $(10, 0, -10)$.

In Problems 13 through 16, translate the given equation into an equation in cylindrical coordinates.

13. $x^2 + y^2 + z^2 = 16$.
14. $z = x^3 - 3xy^2$.
15. $z^2(x^2 - y^2) = 2xy$.
16. $Ax + By + Cz = D$.

In Problems 17 through 20, translate the given equation into an equation in rectangular coordinates.

17. $r = 4 \cos \theta$.
18. $r^3 = z^2 \sin^3 \theta$.
19. $r^3 = 2z \sin 2\theta$.
20. $r^2 \cos 2\theta = z^3$.

21. Sketch a portion of the surfaces (a) $z = \sin \theta$, $0 \leq \theta \leq \pi/2$, (b) $z = \tan \theta$, $0 \leq \theta < \pi/2$, (c) $z = r$, and (d) $z = \theta$, $0 \leq \theta \leq 4\pi$, where the equations are given in cylindrical coordinates.

22. If ds denotes the differential of arc length for a curve $\mathscr{C}$, find an expression for $ds^2 = |\mathbf{R}'(t)|^2 \, dt^2$ when $\mathscr{C}$ is given in cylindrical coordinates.

12 The Spherical Coordinate System

In Fig. 24 we show a spherical coordinate system superimposed on a rectangular coordinate system. The spherical coordinates of a point P are (ρ, φ, θ). Here ρ is the distance of the point P from O, the common origin in both systems. Hence by agreement, $\rho \geq 0$. The angle φ is the angle from the positive z-axis to the radial line OP. Here it is convenient to restrict φ to the interval $0 \le \varphi \le \pi$. Finally, θ is the angle from the positive x-axis to the projection OP' of the ray OP on the xy-plane. In cylindrical coordinates θ may be any real number, but in spherical coordinates θ must lie in the interval $0 \le \theta < 2\pi$. Consequently, the spherical coordinates (ρ, φ, θ) of a point P always satisfy the conditions

(95) $$0 \le \rho, \qquad 0 \le \varphi \le \pi, \qquad 0 \le \theta < 2\pi.$$

The angle φ is often called the *colatitude* of P, and θ is often called the *longitude* of P.

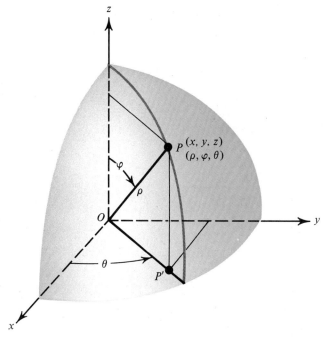

FIGURE 24

If we refer to Fig. 24, it is clear that $|OP'| = \rho \sin \varphi = r \ge 0$, where (r, θ) are the polar coordinates of P'. Consequently, we have $x = r \cos \theta = (\rho \sin \varphi) \cos \theta$, and $y = r \sin \theta = (\rho \sin \varphi) \sin \theta$. Therefore,

(96) $$x = \rho \sin \varphi \cos \theta, \qquad y = \rho \sin \varphi \sin \theta, \qquad z = \rho \cos \varphi,$$

are the equations of transformation from the spherical coordinate system to the rectangular coordinate system.

Example 1. Describe each of the surfaces **(a)** $\rho = 5$, **(b)** $\varphi = 2\pi/3$, **(c)** $\theta = \pi/2$, and **(d)** $\rho = 2 \sin \varphi$, where the equations are given in spherical coordinates.

Solution. **(a)** The collection of all points five units from the origin forms a sphere of radius 5 with center at the origin.

(b) The graph of $\varphi = 2\pi/3$ consists of one nappe of the cone with the z-axis as an axis of the cone and with angle $\pi/3$ between the axis and any one of its elements. The graph is only a half-cone because only points on or below the xy-plane can be on this surface.

(c) The graph of $\theta = \pi/2$ is that half of the yz-plane that lies to the right of the z-axis together with the points on the z-axis.

(d) The equation $\rho = 2 \sin \varphi$ is independent of θ, so we have a surface of revolution with the z-axis as the axis of revolution. In the yz-plane the equation $\rho = 2 \sin \varphi$ gives a circle of unit radius, as indicated in Fig. 25, whence the full surface is the one obtained by rotating this circle about the z-axis. This surface is a degenerate torus in which the hole has radius zero.   ●

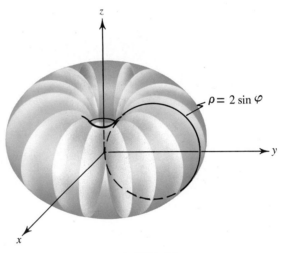

$\rho = 2 \sin \varphi$

FIGURE 25

Example 2. Find an equation in rectangular coordinates for the surface $\rho = 2 \sin \varphi$ shown in Fig. 25.

Solution. From $\rho = 2 \sin \varphi$ we have $\rho^2 = 2\rho \sin \varphi$ and, on squaring,

$$\rho^4 = 4\rho^2 \sin^2 \varphi = 4\rho^2(1 - \cos^2 \varphi) = 4\rho^2 - 4\rho^2 \cos^2 \varphi.$$

Since $\rho^2 = x^2 + y^2 + z^2$, we have

$$(x^2 + y^2 + z^2)^2 = 4(x^2 + y^2 + z^2) - 4z^2,$$

or

(97) $$(x^2 + y^2 + z^2)^2 = 4(x^2 + y^2).$$ ●

EXERCISE 12

In Problems 1 through 6, change from the given spherical coordinates of a point to the set of rectangular coordinates for the same point.

1. $(8, \pi/6, \pi/4)$.   2. $(4, \pi/2, \pi/3)$.   3. $(16, \pi/6, 5\pi/4)$.

4. $(7, 0, 7\pi/13)$.   5. $(\pi, \pi, 4)$.   6. $(0, 5, 9)$.

In Problems 7 through 10, the equation of a surface is given in rectangular coordinates. Transform the equation into an equation in spherical coordinates and describe or sketch the surface.

7. $x^2 + y^2 + z^2 - 8z = 0$.   8. $z = 10 - x^2 - y^2$.

9. $(x^2 + y^2 + z^2)^3 = (x^2 + y^2)^2$.   ★10. $(x^2 + y^2)\sqrt{x^2 + y^2 + z^2} = x^2$.

11. Deduce the equation set

$$\sin\varphi = \frac{\sqrt{x^2 + y^2}}{\sqrt{x^2 + y^2 + z^2}}, \qquad \sin\theta = \frac{y}{\sqrt{x^2 + y^2}},$$

$$\cos\varphi = \frac{z}{\sqrt{x^2 + y^2 + z^2}}, \qquad \cos\theta = \frac{x}{\sqrt{x^2 + y^2}}.$$

These equations can be helpful in transforming an equation in spherical coordinates into an equivalent equation in rectangular coordinates.

In Problems 12 through 15, the equation of a surface is given in spherical coordinates. Transform the equation into an equivalent equation in rectangular coordinates, and describe the surface.

12. $\rho \cos\varphi = -7$.   13. $\rho = 2\cos\varphi + 4\sin\varphi \cos\theta$.

14. $\rho \sin\varphi = 10$.   15. $\rho = 2\tan\theta$.

16. Obtain equations for transforming from spherical coordinates to cylindrical coordinates.

17. For the torus of Problem 26, Exercise 10, find an equation: (a) in cylindrical coordinates, and (b) in spherical coordinates.

18. Transform each of the following equations for a sphere of radius A into spherical coordinates and compare the results for simplicity:

    (a) $(x - A)^2 + y^2 + z^2 = A^2$.   (b) $x^2 + y^2 + (z - A)^2 = A^2$.

19. Find ds^2 in spherical coordinates. (See Exercise 11, Problem 22.)

REVIEW PROBLEMS

In Problems 1 through 4, prove that the points A, B, and C are vertices of a right triangle:
(a) using the Pythagorean Theorem, and (b) using a dot product.

1. $A(-2, -1, 1)$, $B(1, -2, 2)$, $C(-1, -3, 2)$.
2. $A(2, 2, 2)$, $B(-3, -2, -1)$, $C(-4, -3, 2)$.
3. $A(1, 2, 3)$, $B(3, 5, 9)$, $C(4, 4, 1)$.
4. $A(2, 0, -1)$, $B(-1, 3, 3)$, $C(5, -1, 2)$.

5. For each of the right triangles in Problems 1 through 4, find the coordinates of the midpoint of the hypotenuse.
6. Prove that the points $A(1, 3, -1)$, $B(4, 6, -1)$, $C(1, 6, 2)$, and $D(4, 3, 2)$ are the vertices of a regular tetrahedron.
★7. Prove that the six points $A(1, 0, 1)$, $B(0, 1, 1)$, $C(-1, 1, 0)$, $D(-1, 0, -1)$, $E(0, -1, -1)$, and $F(1, -1, 0)$ all lie in a plane and form the vertices of a regular hexagon with center at $(0, 0, 0)$.

In Problems 8 through 13, a condition is given for the set of all points P of a surface $\mathscr{S}$. Find a simple equation for $\mathscr{S}$. In each case name the surface.

8. $|PF_1| = |PF_2|$, where F_1 and F_2 are the points $(3, 1, -2)$ and $(5, -3, 0)$, respectively.
9. $|PD_1| = |PD_2|$, where these are the distances from P to the x-axis and y-axis, respectively.
10. $|PF| = |PD|$, where F is the point $(0, 4, 0)$ and $|PD|$ is the distance from P to the xz-plane.
11. $|PF|^2 = |PD|$, where the symbols have the same meaning as in Problem 10.
12. $|PF| = |PD|$, where F is the point $(0, 4, 0)$ and $|PD|$ is the distance from P to the x-axis.
13. $|PF_1| = \sqrt{2}|PF_2|$, where F_1 and F_2 are the points $(3, 0, 2)$ and $(1, 2, 3)$, respectively.

14. Find a vector normal to the plane of Problem 8 in two different ways. Where does the line through F_1 and F_2 meet the plane?

In Problems 15 through 19, sketch that portion of the given surface that lies in the first octant. Also sketch the intersection of the surface with the given planes.

15. $z = x^2 + 2y^2 + 1$, $x = 1, y = 1, z = 5$.
16. $z = 8 - 2x^2 - y^2$, $x = 1, y = 2, z = 2$.
17. $9x^2 + 4y^2 = 36(1 + z^2)$, $x = 2, y = 3, z = \sqrt{3}$.
18. $z = y^2 - 4x^2$, $x = 1, y = 2, z = 2$.
19. $z = 4 - xy$, $z = 2, z = 3, y = x$.

20. Prove that the curve $\mathbf{R} = t\mathbf{i} + (4 - t)\mathbf{j} + (2 - t)^2\mathbf{k}$ lies on the surface of Problem 19 and sketch that part of the curve that lies in the first octant.

In Problems 21 and 22, determine the point where the line through A and B meets the indicated plane.

21. $A(1, -1, -2)$, $B(-1, 2, 3)$, yz-plane.
22. $A(0, 0, 0)$, $B(-2, -3, -4)$, $4x + 3y + 2z = 75$.

In Problems 23 through 25, determine if the given pair of lines meet, and, if so, where.

23. $\mathbf{R}_1 = (1 + 2t)\mathbf{i} + (2 + 3t)\mathbf{j} + (3 - t)\mathbf{k}$,
    $\mathbf{R}_2 = (9 + 3T)\mathbf{i} + (1 - 2T)\mathbf{j} + (6 + 2T)\mathbf{k}$,
24. $\mathbf{R}_1 = (6 + 4t)\mathbf{i} + (-3 + 5t)\mathbf{j} + (11 - 3t)\mathbf{k}$,
    $\mathbf{R}_2 = (19 - T)\mathbf{i} + (-8 + 3T)\mathbf{j} - (1 + 2T)\mathbf{k}$.
25. $\mathbf{R}_1 = (5 - t)\mathbf{i} + (6 + 2t)\mathbf{j} + (1 + 3t)\mathbf{k}$,
    $\mathbf{R}_2 = (1 + 2t)\mathbf{i} + (-13 + 5t)\mathbf{j} + (7 - 4t)\mathbf{k}$.

In Problems 26, 27, and 28, find the distance from the given point to the given plane.

26. $P(-5, 3, 4)$, $\qquad 2x - y + 2z + 4 = 0$.
27. $P(2, -3, 0)$, $\qquad 2x + 3y - 4z = 1$.
28. $P(3\sqrt{2}, \sqrt{8}, 3)$, $\qquad 2x - 3y + 6z = 11$.

29. Where does the line $\mathbf{R} = (2 - t)\mathbf{i} + t\mathbf{j} + (3 + 4t)\mathbf{k}$ meet the surface

$$z = 4 + x^2 + 2y^2?$$

30. Where does the line $\mathbf{R} = (1 + t)\mathbf{i} + (2 - t)\mathbf{j} - (2 - t)\mathbf{k}$ meet the surface

$$z = 8 - x^3 - y^3?$$

31. Find $\cos B$ for each of the triangles in Problems 1, 3, and 4.
32. Suppose that $\mathbf{A} = a_1\mathbf{i} + a_2\mathbf{j} + a_3\mathbf{k}$, $\mathbf{B} = b_1\mathbf{i} + b_2\mathbf{j} + b_3\mathbf{k}$, and $\mathbf{C} = c_1\mathbf{i} + c_2\mathbf{j} + c_3\mathbf{k}$ are three coplanar vectors. Use a theorem on determinants to prove that the vectors $\mathbf{U} = a_1\mathbf{i} + b_1\mathbf{j} + c_1\mathbf{k}$, $\mathbf{V} = a_2\mathbf{i} + b_2\mathbf{j} + c_2\mathbf{k}$, and $\mathbf{W} = a_3\mathbf{i} + b_3\mathbf{j} + c_3\mathbf{k}$ are also coplanar.
33. Find the distance from the point $(1, 2, 3)$ to the line $\mathcal{L}$ in each of the following cases.
    (a) $\mathbf{R} = (2 + t)\mathbf{i} + (3 - t)\mathbf{j} + (4 + 3t)\mathbf{k}$.
    (b) $\mathbf{R} = (-1 + t)\mathbf{i} + 2t\mathbf{j} + (4 - 3t)\mathbf{k}$.
    (c) $\mathbf{R} = (22 + 7t)\mathbf{i} - (13 + 5t)\mathbf{j} + (36 + 11t)\mathbf{k}$.
34. Find the distance from the origin to the plane containing the points A, B, and C for the points given in (a) Problem 1, (b) Problem 2, (c) Problem 3, and (d) Problem 4.
35. Find the distance from the point C to the plane containing the origin, A, and B for the points given in (a) Problem 1, (b) Problem 2, (c) Problem 3, and (d) Problem 4.

36. Find the distance between the two lines given in: (a) Problem 23, (b) Problem 24, and (c) Problem 25.

37. The points $A(1, 1, 1)$, $B(1, 2, 3)$, $C(2, 2, 4)$, and $D(-1, 2, 3)$ form the vertices of a tetrahedron. Find $\cos \theta$ for the dihedral angle θ at the edge: (a) AB, (b) AC, and (c) AD.

38. Find an equation for the plane that is perpendicular to the given vector $\mathbf{R}$ and contains the given point Q.
   (a) $\mathbf{R} = 2\mathbf{i} - 3\mathbf{j} + 5\mathbf{k}$, $Q(-1, -1, -1)$.
   (b) $\mathbf{R} = 3\mathbf{i} + 13\mathbf{j} - 2\mathbf{k}$, $Q(-7, 3, 9)$.
   (c) $\mathbf{R} = 2\mathbf{i} - \mathbf{j} + 3\mathbf{k}$, $Q(\sqrt{3} + 5, 3\sqrt{2} + 2\sqrt{3}, \sqrt{2})$.

39. Find an equation for the plane that passes through the origin and is perpendicular to the two planes determined in (a) and (b) of Problem 38.

40. Find an equation for the plane that is parallel to the vector $\mathbf{R} = 2\mathbf{i} + \mathbf{j} - \mathbf{k}$ and contains the points $(1, 1, 0)$ and $(0, 2, 3)$.

41. For each of the following curves, find those points where the tangent to the curve is parallel to the xy-plane. At each such point find a tangent vector.
   (a) $\mathbf{R} = (2 + t)\mathbf{i} + (t^2 - t)\mathbf{j} + (3t - t^3)\mathbf{k}$.
   (b) $\mathbf{R} = \sin t\mathbf{i} + \cos t\mathbf{j} + \sinh t\mathbf{k}$.
   (c) $\mathbf{R} = t\mathbf{i} + (t + t^2)\mathbf{j} + \cos \pi t\mathbf{k}$.
   (d) $\mathbf{R} = t^2\mathbf{i} + t^3\mathbf{j} + te^t\mathbf{k}$.

42. Find the curvature for the curves given in Problem 41(a) and 41(b).

43. Transform into an equation in rectangular coordinates, and (if possible) identify the surface:
   (a) $\rho = 4 \cos \varphi + 6 \sin \varphi \cos \theta$,
   (b) $\rho = 2 \sin \varphi (\sin \theta + \cos \theta) - 6 \cos \varphi$,
   (c) $\cos \varphi = \sin \varphi (2 \sin \theta + 5 \cos \theta)$.

44. Transform into an equation in cylindrical coordinates:
   (a) $1 = \sin^2 \theta + \cos^2 \varphi$,
   (b) $1 = \cos^2 \theta + \sin^2 \varphi$.

45. For the surfaces given in Problems 15 through 17, transform the equation into an equation: (a) in cylindrical coordinates, and (b) in spherical coordinates.

17

Moments and Centroids

Objective 1

In this chapter we apply the calculus to the computation of certain quantities that are important in physics and mechanics. Unfortunately the numerical results depend on the system of measurements used. In Appendix 4 we discuss units of measurement rather thoroughly. This appendix is essentially arithmetic and does not properly belong in a calculus text. However, it should be helpful to the person who wishes to apply the calculus to problems of the real world that involve the quantities treated in this chapter.

The Moment of a System of Particles 2

In many physical problems where the bodies under consideration are small in comparison with the distances between them, it is convenient to think of all the material of each body as concentrated at a single point, presumably the center of that body. The best example of this is the system of planets revolving about the sun. Here the diameter of the earth is roughly 8000 miles, while its distance from the sun is 92,000,000 miles, so that indeed the earth is but a small particle in the solar system.

Henceforth the term *system of particles* will mean a collection of objects in which the mass of each object is regarded as concentrated at a point. We are interested in finding the turning effect of a collection of particles about an axis. For simplicity we will assume first that all the particles lie in a plane, and in fact we will regard the plane as horizontal and use it as our *xy*-plane (see Fig. 1).

703

     Simple observations of two children on a seesaw indicate that the seesaw will balance when $m_1 d_1 = m_2 d_2$, where naturally m_1 and m_2 are the masses of the two children and d_1 and d_2 are their respective distances from the fulcrum, which is located between the children. This suggests that we regard the product ml as the turning effect of a particle of mass m placed at directed distance l from an axis, assigning a positive sign to l for those particles on one side of the axis (fulcrum) and a negative sign to l for those particles on the other side. Then the condition for balance is that $m_1 l_1 + m_2 l_2 = 0$; that is, the total turning effect or total moment of the system is zero. Abstracting the essentials from the above discussion, we are led to

---

DEFINITION 1 (Moment of a System About an Axis). Let $\mathscr{P}$ be a plane system of particles of masses $m_1, m_2, \ldots, m_n$ located at the points $P_1, P_2, \ldots, P_n$, respectively. Let the line $\mathscr{L}$ in the plane be taken as an axis and let $l_1, l_2, \ldots, l_n$ be the directed distances from the line $\mathscr{L}$ of the points $P_1, P_2, \ldots, P_n$, respectively. Then the moment of this system of particles about the axis $\mathscr{L}$ is denoted by $M_{\mathscr{L}}$ and is given by

$$(1) \qquad M_{\mathscr{L}} = \sum_{k=1}^{n} m_k l_k = m_1 l_1 + m_2 l_2 + \cdots + m_n l_n.$$

     For the most part we are interested in M_x and M_y, the moments about the x- and y-axes, respectively. In these two cases the positive direction for measuring directed distances is the usual one, and hence the positive moments have the rotational effect indicated by the arrows in Fig. 1.

     If each point P_k has coordinates (x_k, y_k), then equation (1) gives

$$(2) \qquad M_x = \sum_{k=1}^{n} m_k y_k = m_1 y_1 + m_2 y_2 + \cdots + m_n y_n$$

and

$$(3) \qquad M_y = \sum_{k=1}^{n} m_k x_k = m_1 x_1 + m_2 x_2 + \cdots + m_n x_n.$$

The reader should note carefully the interchange of x and y in formulas (2) and (3). Thus, to compute M_x we use y_k, and to compute M_y we use x_k.

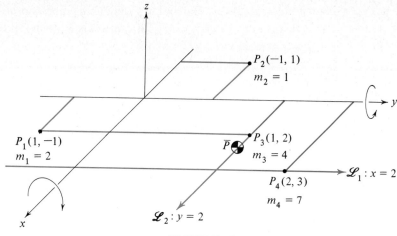

FIGURE 1

Example 1. For the system of particles P_1, P_2, P_3, P_4 shown in Fig. 1, compute M_x, M_y, $M_{\mathscr{L}_1}$, and $M_{\mathscr{L}_2}$, where $\mathscr{L}_1$ is the line $x = 2$ and $\mathscr{L}_2$ is the line $y = 2$.

Solution. From equations (2) and (3), we have

$$M_x = \sum_{k=1}^{4} m_k y_k = 2 \cdot (-1) + 1 \cdot (1) + 4 \cdot (2) + 7 \cdot (3) = -2 + 1 + 8 + 21 = 28,$$

$$M_y = \sum_{k=1}^{4} m_k x_k = 2 \cdot (1) + 1 \cdot (-1) + 4 \cdot (1) + 7 \cdot (2) = 2 - 1 + 4 + 14 = 19.$$

A similar computation for the axes $\mathscr{L}_1$ and $\mathscr{L}_2$ gives

$$M_{\mathscr{L}_1} = 2 \cdot (-1) + 1 \cdot (-3) + 4 \cdot (-1) + 7 \cdot (0) = -2 - 3 - 4 = -9,$$

$$M_{\mathscr{L}_2} = 2 \cdot (-3) + 1 \cdot (-1) + 4 \cdot (0) + 7 \cdot (1) = -6 - 1 + 7 = 0. \quad \bullet$$

Notice that $M_{\mathscr{L}_1}$ is negative. This is to be expected, since all the points are on the axis $\mathscr{L}_1$ or behind it, so the turning effect is negative. Finally, observe that $M_{\mathscr{L}_2} = 0$, so the given system is in balance with respect to this axis.

DEFINITION 2 (Center of Mass of a Plane System). Let $\mathscr{P}$ be a plane system of particles and let $\bar{\mathscr{P}}$ be the system obtained by concentrating the total mass of the system $\mathscr{P}$ at a single point $\bar{P}(\bar{x}, \bar{y})$. Let $M_{\mathscr{L}}$ and $\bar{M}_{\mathscr{L}}$ denote the moments about $\mathscr{L}$ of the systems $\mathscr{P}$ and $\bar{\mathscr{P}}$, respectively. If $M_{\mathscr{L}} = \bar{M}_{\mathscr{L}}$ for every axis $\mathscr{L}$, then $\bar{P}$ is called the *center of mass*[1] of the system $\mathscr{P}$.

We will not prove that there is always such a center of mass because the proof is a little complicated. However, assuming that there is such a point, we can obtain simple formulas for its coordinates. Let m denote the total mass of the system,

$$(4) \qquad m = m_1 + m_2 + \cdots + m_n.$$

If the moment of the system $\bar{\mathscr{P}}$ is to equal the moment of the original system about the two coordinate axes, then we must have

$$(5) \qquad m\bar{x} = M_y = m_1 x_1 + m_2 x_2 + \cdots + m_n x_n$$

and

$$(6) \qquad m\bar{y} = M_x = m_1 y_1 + m_2 y_2 + \cdots + m_n y_n.$$

Division by m gives

THEOREM 1. If $\bar{P}(\bar{x}, \bar{y})$ is the center of mass of a system of particles of mass m_k at $P_k(x_k, y_k)$, $k = 1, 2, \ldots, n$, then

$$(7) \qquad \bar{x} = \frac{m_1 x_1 + m_2 x_2 + \cdots + m_n x_n}{m} = \frac{\displaystyle\sum_{k=1}^{n} m_k x_k}{\displaystyle\sum_{k=1}^{n} m_k}$$

and

$$(8) \qquad \bar{y} = \frac{m_1 y_1 + m_2 y_2 + \cdots + m_n y_n}{m} = \frac{\displaystyle\sum_{k=1}^{n} m_k y_k}{\displaystyle\sum_{k=1}^{n} m_k}.$$

[1] The center of mass is often called the center of gravity. However, the title "center of mass" is preferred because the system of particles will have a "center" $\bar{P}$ even if the system is located in some remote region of space where the gravitational forces are zero.

Example 2. Find the center of mass of the system described in Example 1.

Solution. Since $m = 2 + 1 + 4 + 7 = 14$, then $\bar{x} = M_y/m = 19/14$ and $\bar{y} = M_x/m = 28/14 = 2$. Then $\bar{P}$ is the point $(19/14, 2)$. ●

As a check we compute the moment about the axis $\mathscr{L}_1$ of the system in which the total mass 14 is concentrated at $(\bar{x}, \bar{y})$. This point is $2 - 19/14$ units behind $\mathscr{L}_1$ and hence $M_{\mathscr{L}_1} = -14(2 - 19/14) = -28 + 19 = -9$, the same moment that we found before.

Clearly, the moment about any axis through $\bar{P}$ must be zero, so $\bar{P}$ represents a point of balance for the original system of particles.

EXERCISE 1

In Problems 1 through 5, a system of particles is given. In each case find M_x, M_y, $\bar{P}$, and the moment about the line $y = 3$. For each problem make a plane diagram showing the points and the center of mass.

1. $P_1(2, 3)$, $m_1 = 7$; $P_2(5, 3)$, $m_2 = 14$.
2. $P_1(6, -2)$, $m_1 = 10$; $P_2(-2, 6)$, $m_2 = 10$.
3. $P_1(7, 1)$, $m_1 = 5$; $P_2(3, 5)$, $m_2 = 5$; $P_3(-2, -4)$, $m_3 = 2$.
4. $P_1(1, 5)$, $m_1 = 2$; $P_2(3, 6)$, $m_2 = 1$; $P_3(4, 2)$, $m_3 = 1$;
      $P_4(2, -2)$, $m_4 = 2$; $P_5(-7, 3)$, $m_5 = 4$.
5. $P_1(0, 2)$, $m_1 = 2$; $P_2(1, 7)$, $m_2 = 5$; $P_3(8, 4)$, $m_3 = 2$;
      $P_4(7, -1)$, $m_4 = 5$; $P_5(4, 3)$, $m_5 = 6$.

6. What mass should be placed at $(4, -7)$ in addition to the particles described in Problem 1 so that the new system will have its center of mass at $(4, 0)$?

★7. Suppose that each particle of a given system is moved parallel to the x-axis h units to the right. Prove that the center of mass of the new system is h units to the right of the center of mass of the original system.

Observe that in your proof h could be negative (particles moved to the left). Also observe that the same type of proof will give a similar result for a translation of the particles parallel to the y-axis. Thus you have proved that for a translation of a system of particles, the center of mass undergoes the same translation.

★★8. Suppose that each particle of a given system is rotated about the origin through an angle α. Prove that the center of mass of this new system of particles can be obtained by rotating the center of mass of the original system about the origin through an angle α. (See Chapter 11, Exercise 6, Problem 19.)

Observe that in Problems 7 and 8 we could just as easily regard the points as fixed, and the axes as being translated or rotated. Thus you have proved that the center of mass of a system *of particles is independent of the selection of the axes used to compute it. For this reason we may be content to compute M_x and M_y. In practical work we can select our axes in a convenient way. Thus in the design of ships or airplanes, one axis is taken along the longitudinal center line, and the other is taken perpendicular to this line at the front end of the ship or airplane.*

★9. Prove that if P_1 and P_2 are two given points, then positive masses can be assigned at these two points in such a way that the center of mass of the system will be at any preassigned point in the interior of the line segment joining P_1 and P_2. HINT: By Problems 7 and 8 you may assume that P_1 is at the origin and P_2 is on the x-axis.

★10. Generalize the statement of Problem 9 to the case of three points P_1, P_2, and P_3.

11. A particle of mass $1/2^k$ is placed on the x-axis at $x = 1/2^k$ for $k = 0, 1, 2, \ldots$. Find $\bar{x}$ for this system of infinitely many particles.

★12. **A Paradox.** Consider the set of infinitely many particles on the positive part of the x-axis distributed as follows. Mass $m_1 = 1$ located at $x_1 = 1/2$; $m_2 = 1/2$ at $x_2 = 1/3$; $m_3 = 1/3$ at $x_3 = 1/4$; .... In general, for each positive integer n, there is a mass $m_n = 1/n$ located at $x_n = 1/(n + 1)$. Show that this system has no center of mass; that is, there is no axis parallel to the y-axis about which the moment of the system is zero. HINT: For this system

$$M_y = \sum_{n=1}^{\infty} \frac{1}{n(n + 1)} = 1.$$

13. Suppose that a system consisting of a finite number of particles is symmetric with respect to the y-axis. This means that if at $P_k(x_k, y_k)$ there is a particle of mass m_k in the system, then there is a second particle with the same mass at $(-x_k, y_k)$ that is also in the system. Prove that such a system, symmetric with respect to the y-axis, has its center of mass on the y-axis.

★14. Consider the system of particles with mass $m_n = 1/n^2$ placed at $(\pm n, 0)$ for $n = 1, 2, 3, \ldots$. This system has infinitely many particles and is symmetric with respect to the y-axis. Prove that the total mass of the system is finite but that the system does not have a uniquely determined center of mass. This explains the need for the hypothesis of finitely many particles in Problem 13.

3   *Systems of Particles in Space*

Our problem is to find the center of mass of a system of particles when they do not lie in a plane. Our first impulse would be to compute M_x and M_y as before and also M_z, the moment about the z-axis. To see that this is incorrect, consider the system of two particles of unit mass, one at $A(3, 3, 0)$ and the other at $B(3, 3, 4)$, as indicated in Fig. 2.

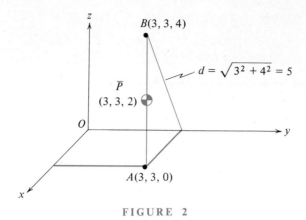

FIGURE 2

On the one hand, we already know that the center of mass must lie on the midpoint of the line segment AB, and this is the point $\overline{P}(3, 3, 2)$. On the other hand, if we compute $\bar{x}$ using M_y, the moment about the y-axis, we obtain

$$\bar{x} = \frac{M_y}{m} = \frac{1(3) + 1\sqrt{3^2 + 4^2}}{1 + 1} = \frac{3 + 5}{2} = 4 \neq 3.$$

Consequently, this procedure must be *wrong*.

To find the correct procedure, we observe that the force of gravity acting upon the particle at B acts along the line BA and the turning effect about the x-axis is really the product of the mass and the distance of the line of action of the force from the x-axis. This distance is not $\sqrt{3^2 + 4^2} = 5$ but is just 3, the y-coordinate of B. We observe that this is the distance of B from the yz-plane. For convenience we regard the product $m_1 x_1$ as a moment about the yz-plane, and we use M_{yz} to denote the sum of such moments for a system of particles.

DEFINITION 3 (Moment of a System About a Plane). The moments of a system of particles about the coordinate planes are denoted by M_{yz}, M_{zx}, and M_{xy} and are given by

(9)
$$M_{yz} = \sum_{k=1}^{n} m_k x_k = m_1 x_1 + m_2 x_2 + \cdots + m_n x_n,$$

(10)
$$M_{zx} = \sum_{k=1}^{n} m_k y_k = m_1 y_1 + m_2 y_2 + \cdots + m_n y_n,$$

(11) $$M_{xy} = \sum_{k=1}^{n} m_k z_k = m_1 z_1 + m_2 z_2 + \cdots + m_n z_n.$$

DEFINITION 4 (Center of Mass of a System in 3-Space). The coordinates of $\overline{P}(\overline{x}, \overline{y}, \overline{z})$, the center of mass of a system of particles in three-dimensional space, are given by

(12a, b, c) $$\overline{x} = \frac{M_{yz}}{m}, \qquad \overline{y} = \frac{M_{zx}}{m}, \qquad \overline{z} = \frac{M_{xy}}{m}.$$

It may seem unreasonable to speak about the moment of our system about the *xy*-plane, because gravity acts perpendicular to this plane. But keep in mind that the system of particles together with the *y*- and *z*-coordinate axes could be rotated 90° about the *x*-axis and then the *xy*-plane would be a vertical plane, with the force of gravity acting parallel to it.

Example 1. Find the center of mass of the system of particles shown in Fig. 2.

Solution. By Definitions 3 and 4,

$$M_{yz} = 1 \cdot (3) + 1 \cdot (3) = 3 + 3 = 6, \quad \overline{x} = \frac{6}{2} = 3,$$

$$M_{zx} = 1 \cdot (3) + 1 \cdot (3) = 3 + 3 = 6, \quad \overline{y} = \frac{6}{2} = 3,$$

$$M_{xy} = 1 \cdot (0) + 1 \cdot (4) = 0 + 4 = 4, \quad \overline{z} = \frac{4}{2} = 2.$$

Thus we find that $\overline{P}$ is at $(3, 3, 2)$, where it should be, as we knew all along. ●

EXERCISE 2

In Problems 1 through 5, a system of particles is given. In each case find $\overline{P}$, the center of mass. Make a sketch of the given system.
  1. $m = 1$ at each of the points $(0, 0, 0)$, $(2, 0, 0)$, $(0, 2, 0)$, $(2, 2, 0)$, and $(1, 1, 5)$.
  2. $m = 1$ at $(3a, 0, 0)$, $(0, 3b, 0)$, and $(0, 0, 3c)$.

3. $m = 1$ at $(0, 0, 0)$, $(4a, 0, 0)$, $(0, 4b, 0)$, and $(0, 0, 4c)$.
4. $m = 1$ at $(1, 0, 0)$, $m = 2$ at $(2, 1, 1)$, $m = 3$ at $(0, 1, 1)$, $m = 4$ at $(0, 1, -1)$, $m = 5$ at $(2, 1, -1)$, $m = 6$ at $(1, 3, 1)$.
5. $m = 2$ at $(-1, 0, 5)$, $m = 1$ at $(0, 1, 4)$, $m = 1$ at $(2, 3, 2)$, $m = 2$ at $(3, 4, 1)$.

*6. Prove that if all the particles of a system lie in a fixed plane $Ax + By + Cz = D$, then the center of mass also lies in that plane.
*7. Prove that if all the particles of a system lie on a straight line, then the center of mass also lies on that straight line.
*8. Suppose that in a given system each mass is multiplied by the same positive constant c to form a new system. Prove that the center of mass of the new system coincides with the center of mass of the original system.

Density 4

The density[1] of a material can be specified either by giving the weight per unit volume or the mass per unit volume. In the first case we obtain δ, the weight density of the body. In the second case we obtain ρ, the mass density of the body. We recall that $w = mg$ gives the weight of a body as the product of its mass and the acceleration of gravity. Consequently, the weight density δ and the mass density are related by

$$\delta = \rho g$$

as long as we stay in the same system of units.[2] Consequently, it is an easy matter to convert from either type of density to the other whenever the need arises. Henceforth the term *density* will mean mass density.

In some cases we are interested in flat sheets of material, where the thickness of the sheet is the same throughout. Under such circumstances it is natural to divide the mass of the material by the area of the sheet (instead of the volume) and to call the quotient the *surface density* (or the *area density*) of the material. For example, suppose that a sheet of copper 0.25 cm thick, 90 cm wide, and 120 cm long has a mass of 24,300 grams. Then the surface density of this particular sheet of copper is $24{,}300/(90 \times 120) = 2.25$ grams/cm^2.

In the case of a wire of uniform thickness we would divide the mass by the length of the wire to obtain the *linear density*. For example, suppose that a certain aluminum wire is 3 meters long and has a mass of 12 grams. Then the linear density of this wire is $12/(3 \times 100) = 0.04$ grams/cm.

---

[1] We have already touched on this concept in Chapter 7, Section 7, page 317.
[2] For a discussion of the British and metric units, see Appendix 4.

In these examples we are assuming that the material is homogeneous. This means that the density is the same throughout the material. If the material is not homogeneous, then the computations above give the *average density*. We define the *density at a point* as the limit of the average density. We use ρ (Greek letter rho) for this quantity, and whenever it is necessary to distinguish among *linear, surface,* and *solid* density, we use ρ_1, ρ_2, and ρ_3, respectively.

DEFINITION[1] **5 (Mass Density at a Point).** Let $P_0(x_0, y_0, z_0)$ be a fixed point inside a solid body and let C be a cube of side r with P_0 as center. Let $m(r)$ denote the mass of the material inside the cube C and let $V(r) = r^3$ be the volume of C. Then $\rho_3(x_0, y_0, z_0)$, *the density of the solid at P_0,* is given by

(13)
$$\rho_3(x_0, y_0, z_0) = \lim_{r \to 0} \frac{m(r)}{V(r)}.$$

For the *surface density,* (13) is replaced by

(14)
$$\rho_2(x_0, y_0) = \lim_{r \to 0} \frac{m(r)}{A(r)},$$

where the sheet of material is now considered as lying in the *xy*-plane and $A(r) = r^2$ denotes the area of a square of side r and center at (x_0, y_0).

For *linear density,* we take a segment of the arc of length s with P_0 as center, and then by definition

---

[1] There are some logical difficulties with this definition. We naturally think of the cube C as having its faces parallel to the coordinate planes. But if we select our cubes so that this is not true, will we get the same number as a limit in (13)? Again, suppose that we use spheres, or ellipsoids, or some other closed surface and that $m(r)$ and $V(r)$ denote the mass and volume enclosed by the surface. Will we still obtain the same number as a limit in (13)? At this stage, it is much simpler to just ignore such questions.

Another block that may occur to the reader at this point lies in the atomic theory of matter. If P_0 happens to be at some point not occupied by an elementary particle (electron, proton, neutron, etc.), then the limit in (13) is zero. For other locations, we need to have a knowledge of the internal composition of these elementary particles in order to evaluate the limit in (13). Actually, this causes no practical difficulty because in the applications the solids will be extremely large compared to the dimensions of the atoms. On the other hand, we can regard the work of this chapter as purely mathematical, in which we assume that matter is continuously distributed (not discrete as in the atomic theory), and our results are completely independent of the true nature of the physical world.

(15)
$$\rho_1(x_0, y_0) = \lim_{s \to 0} \frac{m(s)}{s},$$

where $m(s)$ denotes the mass of the segment.

Although these limiting ratios are not derivatives in the usual sense, it is convenient to use derivative notation and write

(16a, b, c)
$$\rho_1 = \frac{dm}{ds}, \qquad \rho_2 = \frac{dm}{dA}, \qquad \rho_3 = \frac{dm}{dV}.$$

Equation set (16) suggests the approximate relations

(17a, b, c)
$$\Delta m \approx \rho_1 \Delta s, \qquad \Delta m \approx \rho_2 \Delta A, \qquad \Delta m \approx \rho_3 \Delta V,$$

when the quantities Δs, ΔA, and ΔV are small, and these last are easily justified on the basis of the definitions (13), (14), and (15).

If we are considering weight density, then equation set (16) is replaced by

(18a, b, c)
$$\delta_1 = \frac{dw}{ds}, \qquad \delta_2 = \frac{dw}{dA}, \qquad \delta_3 = \frac{dw}{dV},$$

and the approximations (17a, b, c) are replaced by

(19a, b, c)
$$\Delta w \approx \delta_1 \Delta s, \qquad \Delta w \approx \delta_2 \Delta A, \qquad \Delta w \approx \delta_3 \Delta V.$$

Example 1. A piece of wire bent in the shape of the parabola $y = x^2$ runs from the point $(1, 1)$ to the point $(2, 4)$. The density is variable and when the wire is placed in the position described above, $\rho_1 = 30x$ grams/meter. If x and y are in meters, find the mass of the wire.

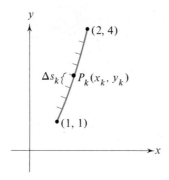

FIGURE 3

Solution. As indicated in Fig. 3, we divide the wire into n pieces by a partition of the interval $1 \le x \le 2$. From equation (17a) a good approximation to Δm_k, the mass of the kth piece, is

$$\Delta m_k \approx \rho_1(x_k)\Delta s_k \approx 30x_k \Delta s_k,$$

and hence for the total mass m,

$$m = \sum_{k=1}^{n} \Delta m_k \approx \sum_{k=1}^{n} 30x_k \, \Delta s_k.$$

In the limit as $n \to \infty$ and $\mu(\mathscr{P}) \to 0$,

$$m = \int_{x=1}^{x=2} 30x \, ds = \int_{1}^{2} 30x \sqrt{1 + (2x)^2} \, dx = \frac{30}{8} \frac{(1 + 4x^2)^{3/2}}{3/2} \Big|_{1}^{2}$$

$$= \frac{10}{4}[(\sqrt{17})^3 - (\sqrt{5})^3] \approx 147 \text{ grams.} \quad \bullet$$

EXERCISE 3

In Problems 1 through 6, density means weight density. In Problems 9 through 20, density means mass density.

1. Aluminum has a density of 168 lb/ft³. Find the surface density of a homogeneous sheet 0.05 in. thick.

2. Copper has a density of 540 lb/ft³. Find the surface density of a homogeneous sheet 0.03 in. thick.

3. Find the linear density of a homogeneous copper wire if the area of the cross section is 0.012 in².

4. Find the linear density of a homogeneous aluminum wire if the cross-sectional area is 0.006 in².

5. Find the surface density in lb/in.² of a sheet of copper 0.01 in. thick.

6. A homogeneous piece of copper wire 6 ft long and of constant cross section weighs 0.09 lb. Find the cross-sectional area.

7. Show that if δ_1 is the linear density in pounds per foot, then $\rho_1 = 14.88\delta_1$ grams/cm.

8. Show that if δ_2 is the surface density in pounds per square foot, then $\rho_2 = 0.4882\delta_2$ grams/cm².

9. If aluminum has a density of 2.7 grams/cm³, find the surface density of a homogeneous sheet 0.09 cm thick.

10. If copper has a density of 8.7 grams/cm³, find the surface density of a homogeneous sheet 0.3 cm thick.

11. Find the linear density of a homogeneous copper wire if the area of the cross section is 0.04 cm².

12. Find the linear density of a homogeneous aluminum wire if the area of the cross section is 0.22 cm².

13. A certain piece of wire is placed on the x-axis from $x = 1$ to $x = 4$ (units in cm) and in this position has a variable density of x grams/cm. Find the mass of the wire.

14. Find the mass of the wire if the density in Problem 13 is $\sqrt{x}$ grams/cm.

★15. A piece of wire of variable density is placed on the x-axis and in this position has the property that for each $a > 0$, the mass of the wire lying between $x = -a$ and $x = a$ is a^n. Show that if $n > 1$, the linear density of the wire at $x = 0$ is 0, and if $n < 1$, the linear density at $x = 0$ is infinite.

★★16. Assuming that the mass of the wire in Problem 15 is distributed symmetrically with respect to $x = 0$, find the linear density at $x = 3$.

★17. A rectangular plate 10 cm wide and 12 cm high has a variable surface density $\rho_2 = c(3 + y)$ grams/cm², where y is the distance from the base. Find the mass of this plate.

★18. A sphere of radius 2 cm has density $1/(r + 4)$ grams/cm³, where r is the distance from the center. Find the mass of this sphere. HINT: Consider the solid as built of spherical shells so that $dV = 4\pi r^2 \, dr$.

★19. Suppose that the solid of Problem 18 has density $1/r$ gram/cm³, so that at the center the density is infinite. Is the mass of the sphere finite?

★20. Find the mass of a wire bent in the form of the catenary $y = \cosh x$, $-a \leq x \leq a$, if $\rho_1 = cy$ grams/cm and y and x are in centimeters.

The Centroid of a Plane Region 5

Example 1. A sheet of metal has the form of the region in the first quadrant bounded by the axes and the curve $y = 4 - x^2$, as shown in Fig. 4. If the sheet is homogeneous and has the surface[1] density ρ, find the center of mass.

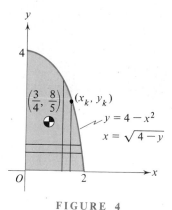

FIGURE 4

[1] From here on it will always be obvious whether the density is linear, surface, or volume, and hence we can drop the subscript and use ρ without decorations.

Solution. We first compute the mass of the sheet. Dividing the sheet into vertical strips by lines parallel to the y-axis, we find that each strip has mass

$$\Delta m_k \approx \rho \, \Delta A_k \approx \rho y_k \, \Delta x_k.$$

Hence $m \approx \sum_{k=1}^{n} \rho y_k \, \Delta x_k$ and taking the limit as $n \to \infty$ and $\mu(\mathcal{P}) \to 0$,

$$m = \int_0^2 \rho y \, dx = \rho \int_0^2 y \, dx = \rho \int_0^2 (4 - x^2) \, dx = \rho\left(4x - \frac{x^3}{3}\right)\Big|_0^2 = \frac{16\rho}{3}.$$

To find $\bar{x}$, we must compute M_y. Each particle in the kth vertical strip is approximately at distance x_k from the y-axis, and this approximation becomes better as $\Delta x_k \to 0$. Hence the moment of each strip about the y-axis is approximately

$$x_k \, \Delta m_k \approx x_k \rho \, \Delta A_k \approx \rho x_k y_k \, \Delta x_k;$$

and, on summing,

(20)
$$M_y \approx \sum_{k=1}^{n} \rho x_k y_k \, \Delta x_k.$$

Then, on taking the limit as $n \to \infty$ and $\mu(\mathcal{P}) \to 0$,

$$M_y = \int_0^2 \rho x y \, dx = \rho \int_0^2 x y \, dx = \rho \int_0^2 x(4 - x^2) \, dx$$

$$= \rho\left(2x^2 - \frac{x^4}{4}\right)\Big|_0^2 = \rho(8 - 4) = 4\rho.$$

Hence

(21)
$$\bar{x} = \frac{M_y}{m} = \frac{4\rho}{16\rho/3} = \frac{3}{4}.$$

To find M_x, we divide the sheet into horizontal strips by lines parallel to the x-axis. The reasoning is similar, except that now the roles played by x and y are interchanged. We find that

(22) $$M_x = \int_0^4 y \, dm = \int_0^4 y \, \rho \, dA = \int_0^4 y \, \rho x \, dy = \rho \int_0^4 y \sqrt{4 - y} \, dy.$$

We make the substitution $u = 4 - y$. Then $y = 4 - u$ and

$$M_x = \rho \int_4^0 (4 - u) \sqrt{u}(-du) = \rho \int_0^4 (4\sqrt{u} - u\sqrt{u}) \, du$$

$$= \rho\left(\frac{8}{3}u^{3/2} - \frac{2}{5}u^{5/2}\right)\Big|_0^4 = \rho\left(\frac{64}{3} - \frac{64}{5}\right) = \frac{128\rho}{15}.$$

Finally,

(23)
$$\bar{y} = \frac{M_x}{m} = \frac{128\rho/15}{16\rho/3} = \frac{128}{15} \times \frac{3}{16} = \frac{8}{5}.$$

Then the center of mass is at $(3/4, 8/5)$. ●

Note that this point lies inside the shaded region of Fig. 4, as our intuition tells us it must.

We observe that in finding M_x the integration in (22) was a little difficult. In some cases the computation may be quite involved. Therefore, it pays to have an alternative method. We now compute M_x using *vertical* strips. In this case each strip is approximately a rectangle, but some of the particles are near the x-axis while others are far away. The key idea is that the center of mass of a homogeneous rectangle is its geometric center. Hence for the near rectangle the center of mass is very close to the point $(x_k, y_k/2)$. The moment of the kth strip about the x-axis is approximately $\Delta m_k y_k/2$. Summing over all such strips and taking the limit as $n \to \infty$ and $\mu(\mathscr{P}) \to 0$, we have

$$M_x = \int_0^2 \frac{y}{2}\, dm = \int_0^2 \frac{y}{2} \rho\, dA = \rho \int_0^2 \frac{y}{2} y\, dx = \frac{\rho}{2} \int_0^2 (4 - x^2)^2\, dx$$

(24)
$$= \frac{\rho}{2} \int_0^2 (16 - 8x^2 + x^4)\, dx = \frac{\rho}{2}\left(16x - \frac{8}{3}x^3 + \frac{1}{5}x^5\right)\Big|_0^2$$

$$= \frac{\rho}{2}\left(32 - \frac{64}{3} + \frac{32}{5}\right) = \frac{128\rho}{15}.$$

Similarly, we can compute M_y using horizontal strips.

We now abstract from this example the essential features. We observe that in equations (21) and (23) the density ρ canceled. Consequently, the center of mass $(3/4, 8/5)$ did not depend on ρ but only on the shape of the metal sheet. Indeed, whenever a solid is homogeneous, the center of mass is independent of the density of the solid. Hence as a matter of convenience we may take $\rho = 1$. When this is done, we drop the term *metal sheet,* and instead refer to a *region.* The center of mass is then called the *centroid* of the region. This is stated precisely in

DEFINITION 6 (Moments and Centroid of a Plane Region). Let $\mathscr{R}$ be a plane region, lying in the rectangle $a \leq x \leq b$, $c \leq y \leq d$ (see Fig. 5). Let the line $x = x_0$ intersect this region in a segment of length $h(x_0)$ for each x_0 in $a \leq x_0 \leq b$, and let the line $y = y_0$ intersect this region in a line segment of length $l(y_0)$ for each y_0 in $c \leq y_0 \leq d$. Then M_x and M_y, the moments of $\mathscr{R}$ about the x- and y-axes, respectively, are given by

(25a, b)
$$M_x = \int_c^d yl(y)\, dy, \qquad M_y = \int_a^b xh(x)\, dx.$$

The coordinates $(\overline{x}, \overline{y})$ of the centroid $\overline{P}$ of the region are given by

(26a, b)
$$\overline{x} = \frac{M_y}{A}, \qquad \overline{y} = \frac{M_x}{A},$$

where A is the area of the region $\mathcal{R}$.

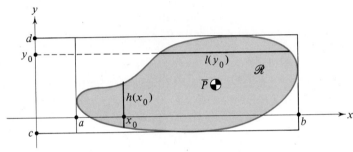

FIGURE 5

Example 2. Find the centroid of the shaded region of Fig. 4.

Solution. We have already solved this problem in Example 1. Just repeat all of the computations, omitting the factor ρ. The centroid is $(3/4, 8/5)$. ●

Example 3. Find the centroid of the region bounded above by the line $y = 1$ and bounded below by the curve $y = x^2/4$.

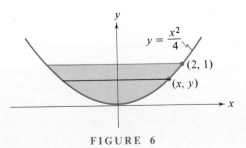

FIGURE 6

Solution. The region is shown shaded in Fig. 6. It is immediately obvious that the region is symmetric with respect to the y-axis, and hence (see Problem 21 in the next exercise) the centroid lies on the y-axis. Therefore, $\bar{x} = 0$. Whenever such a symmetry presents itself, we can and should use it to shorten the labor of locating the centroid.

To find $\bar{y}$, we need M_x. Using horizontal strips as indicated in the figure, we see that $l(y) = 2x$, where x is the coordinate of the appropriate point on the curve. Since $x = 2\sqrt{y}$, equation (25a) gives

$$M_x = \int_0^1 yl(y)\,dy = \int_0^1 y2(2\sqrt{y})\,dy = 4 \times \frac{2}{5}y^{5/2}\Big|_0^1 = \frac{8}{5}.$$

To find the area, we can just drop the first y in the computation above:

$$A = \int_0^1 l(y)\,dy = \int_0^1 2(2\sqrt{y})\,dy = 4 \times \frac{2}{3}y^{3/2}\Big|_0^1 = \frac{8}{3}.$$

Then $\bar{y} = \dfrac{M_x}{A} = \dfrac{8}{5} \times \dfrac{3}{8} = \dfrac{3}{5}$. The centroid is at $\left(0, \dfrac{3}{5}\right)$. ●

Example 4. Find the centroid of the region bounded by the parabola $y = 4x - x^2$ and the line $y = x$.

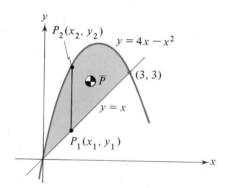

FIGURE 7

Solution. The region is shown shaded in Fig. 7. For the area we have, as usual,

$$A = \int_0^3 (y_2 - y_1)\,dx = \int_0^3 (4x - x^2 - x)\,dx = \frac{3x^2}{2} - \frac{x^3}{3}\Big|_0^3 = \frac{27}{2} - 9 = \frac{9}{2}.$$

Since the length of each vertical strip is given by

$$h(x) = y_2 - y_1 = 4x - x^2 - x = 3x - x^2,$$

equation (25b) gives

$$M_y = \int_0^3 xh(x)\, dx = \int_0^3 (3x^2 - x^3)\, dx = x^3 - \frac{x^4}{4}\Big|_0^3 = \frac{27}{4}.$$

Hence $\bar{x} = \dfrac{M_y}{A} = \dfrac{27}{4} \times \dfrac{2}{9} = \dfrac{3}{2}$. It is not easy to find M_x using formula (25a) because $l(y)$, the length of the horizontal segment, is given by two different formulas: one if $0 \le y \le 3$, and a second one if $3 \le y \le 4$. The curious reader should attempt to find M_x using (25a). To avoid this difficulty, we stay with the vertical strips and observe that the midpoint of the line segment P_1P_2 has y-coordinate $(y_1 + y_2)/2$. Then the vertical strip of width Δx_k will have a moment about the x-axis approximately equal to

$$\frac{y_1 + y_2}{2}\Delta A_k = \frac{y_1 + y_2}{2}(y_2 - y_1)\,\Delta x_k.$$

Taking a sum of such strips, and then the limit as $\mu(\mathscr{P}) \to 0$ in the usual way, we have the general formula

(27)
$$M_x = \int_a^b \frac{y_1 + y_2}{2}(y_2 - y_1)\, dx.$$

In our specific case, equation (27) yields

$$M_x = \int_0^3 \left(\frac{5x - x^2}{2}\right)(3x - x^2)\, dx = \frac{1}{2}\int_0^3 (15x^2 - 8x^3 + x^4)\, dx$$

$$= \frac{1}{2}\left(5x^3 - 2x^4 + \frac{1}{5}x^5\right)\Big|_0^3 = \frac{x^3}{2}\left(5 - 2x + \frac{1}{5}x^2\right)\Big|_0^3 = \frac{54}{5}.$$

Hence $\bar{y} = \dfrac{M_x}{A} = \dfrac{54}{5} \times \dfrac{2}{9} = \dfrac{12}{5}$. The reader should prove that the centroid $(3/2, 12/5)$ lies inside the region. ●

EXERCISE 4

In these problems a and b are positive constants.

1. Prove by integration that the rectangle bounded by the coordinate axes and the lines $x = a$ and $y = b$ has its centroid at $(a/2, b/2)$, the geometric center of the rectangle.

In Problems 2 through 15, find the centroid of the region bounded by the given curves.

2. $bx + ay = ab$, $y = 0$, and $x = 0$ (a triangle).

3. $y = 3 - x$, $y = 0$, and $x = 0$ (a triangle).

4. $y = 8 - x$, $y = 0$, $y = 6$, and $x = 0$ (a trapezoid).

5. $y = x^2$, $x = 2$, and $y = 0$.

6. $y = \sqrt[3]{x}$, $x = 8$, and $y = 0$.

7. $y = \sqrt{x}$, $x = 1$, $x = 4$, and $y = 0$.

8. $y = 2 + x^2$, $x = -1$, $x = 1$, and $y = 0$.

9. $y = x - x^4$, and $y = 0$.

10. $y = x^3$, and $y = 4x$ $(x \geq 0)$.

★11. $x = y^2 - 2y$, and $x = 6y - y^2$.

12. $y = \sqrt{a^2 - x^2}$, and $y = 0$ (a semicircle).

13. $y = \sin x$, and $y = 0$ $(0 \leq x \leq \pi)$.

14. $y = e^x$, $y = 0$, $x = 0$, and $x = 2$.

★15. $y = e^x$, and $y = 0$ $(-\infty < x \leq 2)$.

★16. The region below the curve $y = 1/x$, above the x-axis, and to the right of the line $x = 1$ has infinite extent. Prove that the area and M_y are infinite, but M_x is finite. Find M_x.

★17. The region below the curve $y = 1/x^2$, above the x-axis, and to the right of the line $x = 1$ has infinite extent. Prove that the area and M_x are finite, but M_y is infinite. Hence we can compute $\bar{y}$, but $\bar{x} = \infty$. Find $\bar{y}$.

★18. In Problem 17, replace $y = 1/x^2$ by the curve $y = 1/x^3$ and find the centroid of the region. Notice that $\bar{y}$ for this region is *greater* than $\bar{y}$ for the region of Problem 17.

★19. A rectangular sheet of metal is placed in the first quadrant of a rectangular coordinate system in such a way that one corner is at the origin, and another corner is at the point $(4, 6)$. In this position the surface density is $(3 + x)$. Find M_x, M_y, and the center of mass.

★20. The sheet of Problem 19 is cut along the diagonal from $(4, 0)$ to $(0, 6)$ and the upper part is rejected. Find the center of mass of the remaining piece.

★21. Prove that if a region is bounded and symmetric with respect to the y-axis, then the centroid is on the y-axis.

22. The *First Theorem of Pappus* states that if a plane region $\mathcal{R}$ lies entirely on one side of a line, then the volume V of the solid generated by revolving $\mathcal{R}$ about that line is given by the formula $V = AL$, where A is the area of $\mathcal{R}$ and L is the distance traveled by the centroid of $\mathcal{R}$ during one revolution. Verify this for the region $\mathcal{R}$ of Example 1 (see Fig. 4) revolved about the x-axis.

23. Use the First Theorem of Pappus (Problem 22) to find the volume of the torus generated when the region bounded by the circle $(x - R)^2 + y^2 = 1 (R > 1)$ is rotated about the y-axis.

24. **Fluid Pressure.** Suppose that a plate of area A is submerged vertically in a liquid of weight density δ, and that the centroid of the plate has a distance $\bar{y}$ from the surface of the liquid. Prove that the total force F exerted by the liquid on one side of the plate

is given by $F = \delta A\bar{y}$. HINT: Set the x-axis at the surface of the liquid, and take the positive direction on the y-axis *downward*.

25. A tank car is full of crude oil of weight density 50 lb/ft^3. Using the result of Problem 24, find the force on one end if the tank is a cylinder of radius 3 ft.

6 The Moment of Inertia of a Plane Region

In Section 2 we defined the moment of a particle about an axis as ml, where m is the mass of the particle and l is its distance from the axis. There is nothing to prevent us from considering other types of moments in which we use different powers on l. We call ml^2 the *second moment,* ml^3 the *third moment,* and so on. All of these quantities are mathematically of interest, but we will consider only the second moment ml^2, because it is this one that is important in dynamics and in the mechanics of materials. Because of the importance of the second moment, it is given a special name *moment of inertia,* and a special symbol I. Just as in the case of the first moment, whenever the body is homogeneous, ρ is a constant, and it is convenient to assume that $\rho = 1$. For the present we consider just the moment of inertia of a flat homogeneous sheet of material, and from a mathematical point of view this amounts to a plane region with surface density $\rho = 1$.

DEFINITION 7 (Moments of Inertia). With the notation and the conditions of Definition 6 (as illustrated in Fig. 5), the moment of inertia of the region $\mathcal{R}$ about the x-axis and y-axis is given by

(28a, b)
$$I_x = \int_c^d y^2 l(y)\,dy, \qquad I_y = \int_a^b x^2 h(x)\,dx.$$

It is natural to seek some ideal point such that if all the material were concentrated at that point, it would give the same second moment. For the first moment, this ideal point is the center of mass. For the moment of inertia, *there is no such point,* because the location of such a point is found to change as the axis changes. However, if we *fix the axis,* we can define a distance r such that $mr^2 = I$, and r is called the *radius of gyration.*

In the simple case of a plane region of area A (with $\rho = 1$), the two radii of gyration with respect to the x- and y-axes are defined by the equations $Ar_x^2 = I_x$ and $Ar_y^2 = I_y$. Consequently,

(29a, b)
$$r_x = \sqrt{\frac{I_x}{A}}, \qquad r_y = \sqrt{\frac{I_y}{A}}.$$

Example 1. Find the moment of inertia of a rectangle about one of its edges. Find the radius of gyration with respect to that edge.

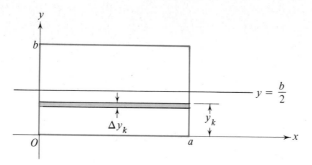

FIGURE 8

Solution. We place the rectangle in a coordinate system as indicated in Fig. 8, and compute I_x and r_x. We divide the rectangle into horizontal strips, the kth strip having all its points approximately y_k units from the x-axis. Forming a sum and taking the limit in the usual way (or appealing to Definition 7), we find that

(30) $$I_x = \int_0^b y^2 l(y)\, dy = \int_0^b y^2 a\, dy = \frac{ay^3}{3}\bigg|_0^b = \frac{ab^3}{3},$$

$$r_x = \sqrt{\frac{I_x}{A}} = \sqrt{\frac{ab^3}{3ab}} = \frac{b}{\sqrt{3}}.$$

By interchanging the role of x and y, it is easy to see that

(31) $$I_y = \frac{ba^3}{3}, \qquad r_y = \frac{a}{\sqrt{3}}. \quad \bullet$$

Example 2. For the rectangle of Fig. 8, find I about the axis $y = b/2$.

Solution. Since this line runs through the centroid, one might guess that the answer is zero, but this is wrong. Why?

We divide the rectangle into two pieces by the axis, and now each piece has dimensions a and $b/2$. Then applying (30) to each piece and adding, we have

$$I_{y=b/2} = \frac{1}{3}a\left(\frac{b}{2}\right)^3 + \frac{1}{3}a\left(\frac{b}{2}\right)^3 = \frac{1}{12}ab^3. \quad \bullet$$

Second Solution. Each horizontal line has distance $|y - b/2|$ from the axis. Since $|y - b/2|^2 = (y - b/2)^2$, formula (28a) gives

$$I_{y=b/2} = \int_0^b \left(y - \frac{b}{2}\right)^2 l(y)\,dy = \int_0^b \left(y - \frac{b}{2}\right)^2 a\,dy$$

$$= \frac{a}{3}\left(y - \frac{b}{2}\right)^3 \Big|_0^b = \frac{a}{3}\left(\frac{b}{2}\right)^3 - \frac{a}{3}\left(-\frac{b}{2}\right)^3 = \frac{ab^3}{12}. \quad \bullet$$

Example 3. Find I_x and I_y for the region shaded in Fig. 4.

Solution. Using vertical strips, equation (28b) gives

$$I_y = \int_0^2 x^2 y\,dx = \int_0^2 x^2(4 - x^2)\,dx = \frac{4x^3}{3} - \frac{x^5}{5}\Big|_0^2 = 2^5\left(\frac{1}{3} - \frac{1}{5}\right) = \frac{64}{15}.$$

Using horizontal strips, equation (28a) gives

(32) $$I_x = \int_0^4 y^2 x\,dy = \int_0^4 y^2 \sqrt{4 - y}\,dy.$$

This integral can be evaluated using the substitution $4 - y = u$. Our purpose is to show that we can also find I_x using vertical strips. The kth strip is approximately a rectangle of width Δx_k and height y_k, so formula (30), derived in Example 1, gives $y_k^3 \Delta x_k/3$ as an approximation to ΔI_x for that strip. Summing and taking a limit in the usual way yields

$$I_x = \int_0^2 \frac{y^3}{3}\,dx = \frac{1}{3}\int_0^2 (4 - x^2)^3\,dx = \frac{1}{3}\int_0^2 (64 - 48x^2 + 12x^4 - x^6)\,dx$$

$$= \frac{1}{3}\left(64x - 16x^3 + \frac{12x^5}{5} - \frac{x^7}{7}\right)\Big|_0^2$$

$$= \frac{2^7}{3}\left(1 - 1 + \frac{3}{5} - \frac{1}{7}\right) = \frac{2^{11}}{3 \cdot 5 \cdot 7}. \quad \bullet$$

The reader should complete the computation of (32) and show that he obtains the same answer. He should also compute I_y using horizontal strips.

EXERCISE 5

1. Find I_x for the rectangle with vertices at $(0, b_1)$, $(0, b_2)$, (a, b_1), and (a, b_2), where $a > 0$ and $b_2 > b_1 > 0$. Is the restriction $b_1 > 0$ necessary?

As Problems 2 through 9, find I_x and I_y for the regions described in Problems 2 through 9 of Exercise 4 of this chapter (see pages 720–721).

*10. Find I_x and I_y for the region under the curve $y = \sin x$ and above the x-axis for $0 \leq x \leq \pi$.

*11. Find I_x and I_y for the region under the curve $y = e^x$ and above the x-axis for $0 \leq x \leq 2$.

**12. Consider the region of infinite extent below the curve $y = 1/x^p$, above the x-axis, and to the right of the line $x = 1$. Find all values of p such that M_x is infinite and I_x is finite. Find I_x in terms of p, when I_x is finite.

*13. For the region of Problem 12, find all values of p for which I_y is finite.

*14. Find I_x for the region bounded by the curve $y = x^2$ and the straight line $y = 4$. HINT: Use the results of Problem 1.

Three-Dimensional Regions 7

We have already seen that in computing the moments for a three-dimensional set of particles it is the directed distance from a *plane* that is of interest. Hence for a solid body we have, by definition,

(33a, b, c) $$M_{yz} = \int x\,dm, \qquad M_{zx} = \int y\,dm, \qquad M_{xy} = \int z\,dm,$$

where the integration is taken over the figure under consideration, and M denotes the moment with respect to the plane indicated by the subscripts.

In contrast, the applications to be made in dynamics of the moment of inertia dictate that these be computed with respect to an axis. Hence by definition, the moments of inertia about the coordinate axes are

(34a, b, c) $$I_x = \int (y^2 + z^2)\,dm, \qquad I_y = \int (x^2 + z^2)\,dm, \qquad I_z = \int (x^2 + y^2)\,dm.$$

If ρ denotes the density, then we can substitute $dm = \rho\,dV$ in (33) and (34). For simplicity we assume that our body is homogeneous and $\rho = 1$. Then (33) and (34) give the moments, and moments of inertia of a three-dimensional region, and we have

(35a, b, c) $$M_{yz} = \int x\,dV, \qquad M_{zx} = \int y\,dV, \qquad M_{xy} = \int z\,dV,$$

and

(36a, b, c) $$I_x = \int (y^2 + z^2)\,dV, \qquad I_y = \int (x^2 + z^2)\,dV, \qquad I_z = \int (x^2 + y^2)\,dV.$$

At present we can carry through the detailed computations only for certain simple cases, such as figures of revolution. After we have mastered the technique of multiple integration, covered in Chapter 19, we will be able to make our definitions of M and I more precise, and we will be able to compute these quantities for a much greater selection of regions.

Example 1. A cone of height H and radius of base R is placed with its vertex at the origin and its axis on the positive y-axis. Find M_{zx} for this cone. Find the centroid of the cone.

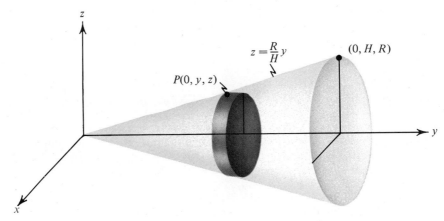

FIGURE 9

Solution. As indicated in Fig. 9, the cone is cut into elementary disks by planes perpendicular to the y-axis. Each such disk has volume

$$dV = \pi r^2 h = \pi z^2 \, dy.$$

All the points of such a disk are roughly the same distance y from the xz-plane. Then (35b) gives

$$M_{zx} = \int y \, dV = \int_0^H y \pi z^2 \, dy = \int_0^H \pi y \left(\frac{R}{H} y\right)^2 dy$$

$$= \frac{\pi R^2}{H^2} \int_0^H y^3 \, dy = \frac{\pi R^2}{H^2} \frac{H^4}{4} = \frac{\pi R^2 H^2}{4}.$$

We can locate the centroid in the usual way. By symmetry, $\bar{x} = \bar{z} = 0$. Since the volume of the cone is $\pi R^2 H/3$, we have

$$\bar{y} = \frac{M_{zx}}{V} = \frac{\pi R^2 H^2}{4} \cdot \frac{3}{\pi R^2 H} = \frac{3}{4} H. \quad \bullet$$

Example 2. Compute the moment of inertia of a solid right circular cylinder about its axis if the radius of the cylinder is R and its height is H.

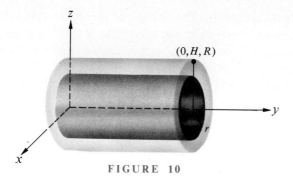

FIGURE 10

Solution. We place the cylinder as indicated in Fig. 10, and this time we consider the cylinder as built up of hollow cylindrical shells. The surface area of each shell of radius r is $2\pi rH$. Therefore, $dV = 2\pi rH\,dr$. Since all of the points in any one shell are roughly the same distance r from the axis, formula (36b) gives

$$(37) \qquad I_y = \int (x^2 + z^2)\,dV = \int_0^R r^2 2\pi rH\,dr = 2\pi H \int_0^R r^3\,dr$$

$$= \frac{\pi R^4 H}{2} = \frac{1}{2}R^2 V. \quad \bullet$$

Example 3. The region in the xz-plane bounded by the curve $z = 4 - x^2$ and the x-axis is rotated about the z-axis. For the figure generated, find I_z by two different methods.

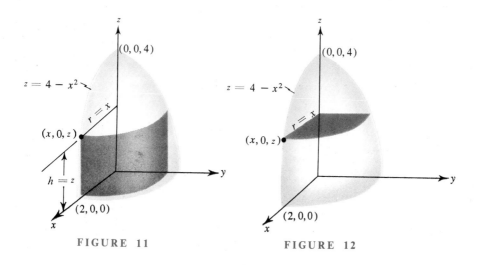

FIGURE 11　　　　　　　　　　　　**FIGURE 12**

Solution. SHELL METHOD. The portion of the figure that lies in the first octant is shown in Fig. 11. We consider the solid as built up from cylindrical shells with axes coinciding with the *z*-axis. One fourth of such a shell is indicated in the figure. The full shell has surface area $2\pi rh = 2\pi rz$. Consequently, $dV = 2\pi rz\,dr$. Since all points on the shell have the same distance $r = \sqrt{x^2 + y^2}$ from the *z*-axis, formula (36c) gives

$$I_z = \int (x^2 + y^2)\,dV = \int_0^2 r^2 2\pi rz\,dr = 2\pi \int_0^2 r^3(4 - r^2)\,dr$$

$$= 2\pi \int_0^2 (4r^3 - r^5)\,dr = 2\pi \left(r^4 - \frac{r^6}{6}\right)\Big|_0^2 = \frac{32}{3}\pi.$$

DISK METHOD. We consider the solid as built up from disks, obtained by slicing with planes perpendicular to the *z*-axis. One fourth of a typical disk is shown in Fig. 12. Now each such disk can be regarded as a solid cylinder of the type considered in Example 2, except that the height *H* of the disk is now *dz* and the radius *R* is $x = \sqrt{4 - z}$. Hence by the formula (37) obtained in Example 2, the moment of inertia of each such elementary disk is

$$\frac{1}{2}\pi R^4 H = \frac{1}{2}\pi(4 - z)^2\,dz.$$

Consequently,

$$I_z = \int_0^4 \frac{1}{2}\pi(4 - z)^2\,dz = -\frac{\pi}{2}\frac{(4 - z)^3}{3}\Big|_0^4 = \frac{64\pi}{6} = \frac{32\pi}{3}. \quad \bullet$$

The problem posed in Example 3 is easy to solve using either the shell method or the disk method. For a different problem, one of the two methods may be far superior to the other.

EXERCISE 6

In Problems 1 through 6, the region in the xy-plane bounded by the given curves is rotated about the x-axis. Find the centroid of the figure generated.

1. $y = x^2$, $y = 0$, and $x = 1$.
2. $y = x^3$, $y = 0$, and $x = 1$.
3. $x = y^2 - 4y$ and $x = 0$.
4. $y = x^2 + 3x$ and $y = 0$.
5. $y = e^x$, $x = 0$, and $x = 1$.
6. $x = \sqrt{1 + y^2}$ and $x = 2$.

In Problems 7 through 12, the region in the yz-plane bounded by the given curves is rotated about the z-axis. Find the centroid of the figure generated.

7. $z = \sqrt{R^2 - y^2}$ and $z = 0$.   8. $z = y^2$ and $z = 1$.

9. $z = 8 - y^3$, $z = 0$, and $y = 0$.   *10. $z = y^2$, $z = 0$, and $y = 1$.

*11. $z = \sqrt{y}$, $z = 0$, and $y = 4$.   **12. $z = \sin y$, $z = 0$, and $y = \pi/2$.

13. Find I_z for the solids of revolution described in each of Problems 7 through 12.

14. Consider the region of infinite extent below the curve $y = 1/x^n$, above the x-axis, and to the right of the line $x = 1$. This region is rotated about the x-axis to form a solid of infinite extent. For what values of n is I_x finite for this solid? Find I_x in those cases.

15. For the solid of Problem 14, find those values of n for which M_{yz} is finite. Find M_{yz} in those cases.

Curves and Surfaces   8

The definitions of moment and moment of inertia given in equations (33) and (34) can be applied also to curves and surfaces. In the first case $dm = \rho\, ds$, where ρ is the linear density and ds is the differential of arc length. For surfaces, $dm = \rho\, d\sigma$, where now ρ is the surface density and $d\sigma$ is the differential of surface area. Here a curve is the mathematical idealization of a bent wire, and a surface is the mathematical idealization of a curved sheet of metal. In most applications the wire or the sheet is homogeneous and ρ is a constant. We can then take $\rho = 1$ for simplicity.

Example 1. Find the centroid of a semicircular arc of radius R.

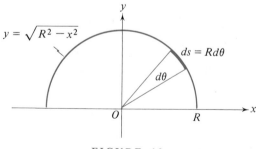

FIGURE 13

Solution. We consider the semicircle placed in an xy-plane as shown in Fig. 13. By symmetry, $\bar{x} = 0$. To find $\bar{y}$, we need M_x. Now

$$M_x = \int_{-R}^{R} y \, ds = 2 \int_{0}^{R} y \sqrt{1 + \left(\frac{dy}{dx}\right)^2} \, dx$$

$$= 2 \int_{0}^{R} \sqrt{R^2 - x^2} \sqrt{1 + \frac{x^2}{R^2 - x^2}} \, dx$$

$$= 2 \int_{0}^{R} \sqrt{R^2 - x^2 + x^2} \, dx = 2 \, Rx \Big|_{0}^{R} = 2R^2.$$

Sometimes it is convenient to use polar coordinates to shorten the labor. In this case we have $ds = R \, d\theta$ and $y = R \sin \theta$. Then

$$M_x = \int y \, ds = \int_{0}^{\pi} R \sin \theta \, R \, d\theta = R^2 \int_{0}^{\pi} \sin \theta \, d\theta = -R^2 \cos \theta \Big|_{0}^{\pi} = 2R^2.$$

The semicircle has length πR. Hence $\bar{y} = M_x/L = 2R^2/\pi R = 2R/\pi$. ●

Example 2. Find the centroid of a hemisphere of radius R.

Solution. We can consider the hemisphere as generated by rotating the semicircle of Fig. 13 about the y-axis. Then, by symmetry, $\bar{x} = \bar{z} = 0$.

$$M_{zx} = \int y \, d\sigma = \int y 2\pi x \, ds = \int_{0}^{\pi/2} (R \sin \theta) 2\pi (R \cos \theta) R \, d\theta$$

$$= \pi R^3 \int_{0}^{\pi/2} 2 \sin \theta \cos \theta \, d\theta = \pi R^3 \sin^2 \theta \Big|_{0}^{\pi/2} = \pi R^3.$$

Since the surface area is $2\pi R^2$, we have $\bar{y} = \dfrac{\pi R^3}{2\pi R^2} = \dfrac{R}{2}$. ●

EXERCISE 7

1. Find the centroid of the quarter circle $y = \sqrt{R^2 - x^2}$, $0 \leq x \leq R$.
2. Find the centroid of the portion of the spherical surface $x^2 + y^2 + z^2 = R^2$ that lies in the first octant.
3. Find I_x for the curve of Problem 1.
4. Find I_x for the surface of Problem 2.
5. The segment of the straight line $Bx + Ay = AB(A > 0, B > 0)$ that lies in the first quadrant is rotated about the x-axis. Find the centroid of the surface generated (right circular cone).
6. Find I_x for the surface of Problem 5.

7. Find the centroid of that portion of the hypocycloid $x = A \cos^3 t$, $y = A \sin^3 t$ that lies in the first quadrant.

*8. The curve of Problem 7 is rotated about the x-axis. Find $\bar{x}$ for the surface generated.

9. Find the centroid for the arch of the cycloid $x = A(t - \sin t)$, $y = A(1 - \cos t)$ for which $0 \leqq t \leqq 2\pi$.

10. Find $\bar{y}$ for the portion of the catenary $y = \cosh x$ between $x = -a$ and $x = a$.

*11. The curve of Problem 10 is rotated about the y-axis. Find $\bar{y}$ for the surface generated.

12. A wire has the form of the quarter circle $x^2 + y^2 = R^2$ lying in the first quadrant, and in that position has density $\rho = R + x$. Find the center of mass of the wire.

13. Find the centroid of the arc $y = \dfrac{x^3}{6} + \dfrac{1}{2x}$, $1 \leqq x \leqq 3$.

14. The *Second Theorem of Pappus* states that if a plane curve of length s lies entirely on one side of a line, then the area σ of the surface generated when the curve is rotated about the line is given by $\sigma = sL$, where L is the distance traveled by the centroid of the curve. Verify this for the semicircular arc of Example 1 (see Fig. 13) rotated about the x-axis.

15. Use the Second Theorem of Pappus (see Problem 14) to find the surface area of the torus generated when the circle $(x - R)^2 + y^2 = 1$ $(R > 1)$ is rotated about the y-axis.

REVIEW PROBLEMS

In Problems 1 through 4, find the center of mass of the given system of particles.

1. $m_k = k$ at the point $P_k(1, k)$, $k = 1, 2, 3, 4$.
2. $m_k = \sqrt{k}$ at the point $P_k(k, \sqrt{k})$, $k = 1, 4, 9, 16$.
3. $m_k = k$ at the point $P_k(1, k, k^2)$, $k = 1, 2, 3, 4, 5$.
4. $m_k = 1$ at the point $P_k(k^2, k, 3)$, $k = -2, -1, 0, 1, 2, 3$.

5. Is it possible to add a particle of mass m at the point $(-1, -1)$ to the system of Problem 1 so that the enlarged system has center of mass at the origin?

6. Is it possible to add two particles, one of mass m_1 at $(-1, 0)$ and one of mass m_2 at $(0, -2)$ to the system of Problem 1 so that the new system has center of mass at the origin?

*7. Find the masses of the two particles that must be added at the points $(-4, 0)$ and $(0, -3)$ to the system of Problem 1 so that the enlarged system has center of mass at $(-1, -1)$.

*8. Do Problem 7 for the system of Problem 2.

9. A piece of wire of length $b - a$ is placed on the interval $a \leqq x \leqq b$, where $a > 0$. If $\rho = x^n$, $n \neq -1$, find the center of mass of the wire.

10. A wire bent in the form of a semicircle, $y = \sqrt{R^2 - x^2}$, $-R \leqq x \leqq R$, has density $\rho = kx$. Find M_x and the center of mass.

In Problems 11 through 15, find the centroid of the region in the first quadrant bounded by the given curves.

11. $y = 1 - x^2$, $y = 0$, $x = 0$.
12. $y = x^2 - 1$, $y = 0$, $x = 3$.
13. $y = x + 4/x^2$, $y = 0$, $x = 1$, $x = 2$.
14. $y = x + 2/x$, $y = 0$, $x = 1$, $x = 2$.
15. $y = 1/(1 + x)$, $y = 0$, $x = 0$, $x = M > 0$.

16. Find I_y for the region given in Problem 12.
17. Find I_x and I_y for the region given in: (a) Problem 11, and (b) Problem 13.
18. Find I_x and I_y for the region given in: (a) Problem 14, and (b) Problem 15.
19. Find the centroid of the three-dimensional region generated when we rotate about the x-axis the region given in: (a) Problem 11, (b) Problem 14, and (c) Problem 15.
20. Prove that the centroid of a triangle is at the point where its three medians intersect.
21. Find the centroid of the surface formed by rotating about the y-axis the portion of $x^2 + y^2 = 1$ in the first quadrant.

18

Partial Differentiation

The concept of a function and its derivative extend in a natural way to functions of several variables. Formulas such as

(1) $$z = x^2 - 5y^2,$$

(2) $$w = \ln(5x + y - z),$$

and

(3) $$u = \frac{w}{x}(y^2 z^3 + v \sin wz^2),$$

define functions of 2, 3, and 5 variables, respectively.

In this chapter we begin the systematic study of the differential calculus of functions of several variables. The integral calculus for such functions will be considered in the next chapter.

If $z = f(x, y)$ is a function of the two variables x and y, it is natural to regard the function as defining a surface in three-dimensional space, and to use this surface as an aid in studying the properties of the function (see Chapter 16, Sections 10, 11, and 12). If f is a function of more than two variables, for example if $w = f(x, y, z)$, then such a geometric interpretation is lacking, although we often carry over the geometrical language. Thus we may describe the graph of $w = \ln(5x + y - z)$ as a surface in four-dimensional space, even though we may find it difficult (or impossible) to visualize the surface, and no one has ever made a picture of the surface.

733

Most of the interesting and distinctive features of the theory of functions of several variables appear when f is a function of just two independent variables. Consequently, we concentrate our attention on this simpler situation, and state our definitions and theorems for functions of two variables. However, occasionally we consider functions of more than two variables, to remind the reader that such functions are also interesting and useful.

2 The Domain of a Function

Strictly speaking, a function is not completely specified unless the domain is also given. However, we avoid this time-consuming duty by recalling and renewing the agreement made in Chapter 3.

> **AGREEMENT.** Whenever a function f is defined by a formula, the domain of f is the largest set for which the formula gives a real-valued function of real variables.

Example 1. Find the domain of $w = \ln(5x + y - z)$.

Solution. We recall that $\ln u$ is defined and real if and only if $u > 0$. Hence the domain of this function is the set of all triples (x, y, z) for which $5x + y - z > 0$. This is the set of points for which $z < 5x + y$, and this is the set of points that lie below the plane $z = 5x + y$. ●

3 Limits and Continuity

> **DEFINITION 1 (Limit).** The function $f(x, y)$ is said to have the limit L at the point $P_0(x_0, y_0)$, and we write
>
> (4)
> $$\lim_{\substack{x \to x_0 \\ y \to y_0}} f(x, y) = L$$
>
> if the function values $f(x, y)$ get closer and closer to L as the point $P(x, y)$ gets closer and closer to P_0 (but P does not coincide with P_0).[1]

[1] An analytic definition of a limit using ϵ and δ (Greek lowercase letters epsilon and delta) is given in Appendix 2.

Just as in the theory of functions of a single variable, the definition states that we can force $f(x, y)$ to be as close as we please to L if we require that (x, y) be sufficiently close to (x_0, y_0).

The notation in (4) can be condensed by writing

$$(5) \qquad \lim_{P \to P_0} f(x, y) = L,$$

where naturally P is the point (x, y) and P_0 is the point (x_0, y_0). We also write $f(x, y) \to L$ as $(x, y) \to (x_0, y_0)$, with exactly the same meaning.

We observe that the definition of a limit says nothing about the value of f at P_0, and in fact f need not even be defined at P_0. If $f(x_0, y_0) = L$, then f is *continuous* at P_0.

DEFINITION 2 (Continuity). The function $f(x, y)$ is said to be continuous at $P_0(x_0, y_0)$ if **(A)** $f(x, y) \to L$ as $(x, y) \to (x_0, y_0)$, and **(B)** $f(x_0, y_0) = L$. The function f is continuous in a set $\mathbb{S}$ if it is continuous at every point of $\mathbb{S}$.

A thorough analysis of these two definitions is best postponed because intuitively the meaning is clear. Indeed, if we consider the surface defined by $z = f(x, y)$, then the function f is continuous if and only if the surface has no breaks or vertical cliffs.

All the basic theorems on limits and continuity for functions of a single variable can be generalized to functions of two or more variables. We combine a few of these to obtain

---

THEOREM 1 (PWO). If $f(x, y)$ and $g(x, y)$ are continuous in a common set $\mathbb{S}$ and c is any constant, then

(I) $f(x, y) + g(x, y)$ is continuous in $\mathbb{S}$.
(II) $cf(x, y)$ is continuous in $\mathbb{S}$.
(III) $f(x, y)g(x, y)$ is continuous in $\mathbb{S}$.
(IV) $f(x, y)/g(x, y)$ is continuous at each point P of $\mathbb{S}$, where $g(x, y) \neq 0$.

It is also easy to see that any *polynomial in two variables*,[1]

$$(6) \qquad \sum_{j=0}^{m} \sum_{k=0}^{n} a_{jk} x^j y^k \equiv a_{00} + a_{10}x + a_{01}y + \cdots + a_{mn} x^m y^m,$$

---

[1] The double subscript jk on a_{jk} is not a product of j and k but merely serves to indicate that particular coefficient of the term $x^j y^k$. Thus the coefficient of $x^2 y^3$ in equation (6) is a_{23} (and this is *not* a_6 and it is *not* a sub-twenty-three).

is continuous at every point of the plane. It follows from (IV) that a *rational function of two variables,*

$$(7) \qquad R(x, y) \equiv \frac{N(x, y)}{D(x, y)} \equiv \frac{\displaystyle\sum_{j=0}^{m} \sum_{k=0}^{n} a_{jk} x^j y^k}{\displaystyle\sum_{j=0}^{p} \sum_{k=0}^{q} b_{jk} x^j y^k},$$

is continuous at each point where $D \neq 0$. The reader should be warned that $R(x, y)$ can exhibit some peculiar properties near a point at which $D = 0$.

The composition of two continuous functions again leads to continuous functions, whenever the new function is meaningful.

As examples we cite:

(A) If $f(x, y)$ is continuous, then $e^{f(x, y)}$, $\sin f(x, y)$, and $\cos f(x, y)$ are continuous.
(B) If $f(x, y)$ is continuous, then $\ln f(x, y)$ is continuous wherever $f(x, y) > 0$.
(C) If $f(x, y)$ is continuous, then $\tan f(x, y)$ is continuous wherever $f(x, y) \neq \pi/2 + n\pi$.

The reader may easily formulate his own rules for other combinations as the need arises.

Although the above discussion has emphasized the similarity between the limit concept for a function of a single variable and the limit concept for a function of two variables, there is an important distinction to be kept in mind. Whereas in the one-variable case it suffices to study the behavior of $f(x)$ as x approaches x_0 from "either side," in the two-variable case (x, y) can approach (x_0, y_0) from *any* direction in the plane, or even along a parabola, spiral, or more complicated curve. And in order for the limit of $f(x, y)$ to exist and equal L, all these possible approaches of (x, y) to (x_0, y_0) must yield the same limiting value L for $f(x, y)$.

Example 1. Let

$$(8) \qquad f(x, y) = \frac{xy}{x^2 + y^2}, \qquad (x, y) \neq (0, 0).$$

Does this function have a limit as $P \to (0, 0)$?

Solution. For brevity we write $f(P)$ for $f(x, y)$. We let P approach $(0, 0)$ along the line $y = mx$. Using this relation in (8), we see that for points on this line

$$(9) \qquad f(P) = \frac{x(mx)}{x^2 + (mx)^2} = \frac{x^2 m}{x^2(1 + m^2)} = \frac{m}{1 + m^2}, \qquad x \neq 0.$$

Hence on each line through the origin $f(P)$ is constant. For the particular line $y = x$, we have $m = 1$ and from (9),

$$f(P) \to \frac{1}{1+1} = \frac{1}{2}$$

as $P \to (0,0)$ on this line. If $P \to (0,0)$ along the x-axis, then $m = 0$ and from (9) we see that

$$f(P) \to \frac{0}{1+0} = 0.$$

Since $0 \neq 1/2$, it is clear that no L exists that will satisfy Definition 1 at $(0,0)$. Hence the function defined by (8) does not have a limit at $(0,0)$. ●

EXERCISE 1

In Problems 1 through 10, find the domain of the given function in accordance with our agreement.

1. $f(x, y) = \sqrt{1 - x^2 - y^2}$.

2. $g(x, y) = \sqrt{x^2 - y}$.

3. $h(x, y) = \ln(5 - x - y)$.

4. $F(x, y) = \dfrac{3 + e^x}{4 - x^2 - y^2}$.

5. $G(x, y) = \sqrt{1 - |x| - |y|}$.

6. $H(x, y) = \ln(1 - x^2 + y)$.

7. $w = \ln(1 - \cosh xy^2)$.

8. $z = \dfrac{\sinh x - \sinh y}{xy}$.

9. $z = \dfrac{x + y}{x - y}$.

10. $v = \ln\left(1 + \dfrac{2x}{x^2 + y^2}\right)$.

In Problems 11 through 20, find the indicated limit.

11. $\displaystyle\lim_{(x,y)\to(1,2)} (2x + 3y)$.

12. $\displaystyle\lim_{(x,y)\to(2,3)} (4x - 5y)$.

13. $\displaystyle\lim_{P\to(1,-1)} (x^2 + 4y^2)$.

14. $\displaystyle\lim_{P\to(1,-1)} (2x^2 - 3y^2)$.

15. $\displaystyle\lim_{P\to(0,1,2)} (x^2 + y - 2z)$.

16. $\displaystyle\lim_{P\to(1,2,-3)} (x + 2y^2 + 3z)$.

17. $\displaystyle\lim_{P\to(3,4)} \dfrac{x + y}{x - y}$.

18. $\displaystyle\lim_{P\to(3,5)} \dfrac{x^2 + y^2}{x^3 - y^2}$.

19. $\displaystyle\lim_{P\to(3,2,0)} \sin(x^2 y^3 z^{10})$.

20. $\displaystyle\lim_{P\to(2,0)} e^{x^2 + \cos y}$.

Discuss $\lim\limits_{P\to(0,0)} f(P)$ *for each of the functions given in Problems 21 through 28.*

21. $f(P) = \dfrac{x}{x+y}$.

22. $f(P) = \dfrac{x^2 y^2}{x^2 - y^2}$.

23. $f(P) = \dfrac{x^2 y^2}{x^2 + y^2}$.

24. $f(P) = \dfrac{x^3 + y^3}{x^2 + y^2}$.

25. $f(P) = \dfrac{x + y^5}{x^2 + y^4}$.

26. $f(P) = \dfrac{\sin(2x^2 + 2y^2)}{x^2 + y^2}$.

27. $f(P) = \dfrac{\sin(x^2 + 2y^2)}{x^2 + y^2}$.

28. $f(P) = \dfrac{x^2 + y}{x^2 + y^2}$.

★29. Let

$$f(P) = \frac{x^2 y}{x^4 + y^2} \qquad \text{if } P \neq (0, 0).$$

Prove that along every straight line through $(0, 0)$ we have $f(P) \to 0$ as $P \to (0, 0)$. Prove that if $P \to (0, 0)$ along the parabola $y = x^2$, then $f(P) \to 1/2$.

In Problems 30 through 36, a composite function is given. Make the indicated substitution and simplify the resulting expression (if possible).

30. $f(x, y) = 5x^2 - 7xy + 9y^2$, $x = t$, $y = 2t$.

31. $g(x, y) = x^2 + 2xy + 3y^2$, $x = t^2$, $y = -t^3$.

32. $h(x, y) = 3x + 5y + 1$, $x = 5t - 7$, $y = 4 - 3t$.

33. $F(x, y) = x^2 + 6xy + y^2$, $x = \cos t$, $y = \sin t$.

34. $G(x, y) = \dfrac{2xy}{x^2 + y^2}$, $x = 2u$, $y = u^2$.

35. $H(x, y) = \ln(x^2 - 3xy^2 + 4y^4)$, $x = 2v^2$, $y = v$.

36. $f(x, y, z) = x^2 y^3 z^4$, $x = w^3$, $y = w^4$, $z = 1/w^5$.

4 Partial Derivatives

Just as in the case of a function of one variable, the derivative gives the rate of change of the function. But now that we have many independent variables a difficulty arises in deciding what increment to give each of the independent variables. Actually, this complicated situation will be easy to handle if we first consider the simple case in which we let just *one* of the independent variables change while keeping all of the others constant. When we do this, we obtain the *partial derivative*. If $z = f(x, y)$ and x is varying while y

is constant, the partial derivative is symbolized by writing[1]

$$\frac{\partial z}{\partial x} \quad \text{or} \quad \frac{\partial f}{\partial x}$$

and is read "the partial of z with respect to x" or "the partial of f with respect to x." Stated precisely, we have

---

DEFINITION 3. If $z = f(x, y)$, then

(10)
$$\frac{\partial f}{\partial x} = \lim_{\Delta x \to 0} \frac{f(x + \Delta x, y) - f(x, y)}{\Delta x} = \lim_{\Delta x \to 0} \frac{\Delta z}{\Delta x}$$

and

(11)
$$\frac{\partial f}{\partial y} = \lim_{\Delta y \to 0} \frac{f(x, y + \Delta y) - f(x, y)}{\Delta y} = \lim_{\Delta y \to 0} \frac{\Delta z}{\Delta y},$$

whenever the limits exist.

Observe that in equation (10), y is held constant and x is changed by an amount Δx. In equation (11), x is held constant and y is given the increment Δy.

Naturally, these concepts extend to functions of more than two variables. All the formulas for differentiation that we have learned so far are still valid for functions of several variables. The only difficulty is keeping in mind which variables are held constant during the differentiation.

Example 1. Find $\dfrac{\partial f}{\partial x}$ and $\dfrac{\partial f}{\partial y}$ for each of the functions

(a) $f(x, y) = x^2 + y^3 + 5xy.$ **(b)** $f(x, y) = x^3 \sin y.$

Solution. **(a)** $\dfrac{\partial f}{\partial x} = 2x + 5y,$ $\dfrac{\partial f}{\partial y} = 3y^2 + 5x.$

(b) $\dfrac{\partial f}{\partial x} = 3x^2 \sin y,$ $\dfrac{\partial f}{\partial y} = x^3 \cos y.$ ●

The functions may be more complicated and involve more than two independent variables.

---

[1] The symbol ∂ is frequently called a "roundback d." It is the italic d from the Russian alphabet.

Example 2. If u is the function

$$u = \frac{w}{x}(y^2 z^3 + v \sin wz^2),$$

find $\partial u/\partial x$, $\partial u/\partial y$, $\partial u/\partial z$, and $\partial u/\partial v$.

Solution. We use the standard differentiation formulas.

$$\frac{\partial u}{\partial x} = -\frac{w}{x^2}(y^2 z^3 + v \sin wz^2),$$

$$\frac{\partial u}{\partial y} = \frac{w}{x}(2yz^3 + 0) = \frac{2wyz^3}{x},$$

$$\frac{\partial u}{\partial z} = \frac{w}{x}(3y^2 z^2 + 2vwz \cos wz^2),$$

$$\frac{\partial u}{\partial v} = \frac{w}{x}(0 + \sin wz^2) = \frac{w}{x}\sin wz^2. \quad \bullet$$

Higher-order partial derivatives are defined just as in the case of one variable. For example,

$$\frac{\partial^2 z}{\partial x^2} \equiv \frac{\partial}{\partial x}\left(\frac{\partial z}{\partial x}\right), \qquad \frac{\partial^2 z}{\partial y\,\partial x} \equiv \frac{\partial}{\partial y}\left(\frac{\partial z}{\partial x}\right),$$

$$\frac{\partial^2 z}{\partial x\,\partial y} \equiv \frac{\partial}{\partial x}\left(\frac{\partial z}{\partial y}\right), \qquad \frac{\partial^4 z}{\partial y^4} \equiv \frac{\partial}{\partial y}\left(\frac{\partial^3 z}{\partial y^3}\right),$$

where in each equation the quantity on the left is defined by the expression on the right.

Example 3. Find each of the four partial derivatives listed above for the function $z = xy^3 + x \sin xy$.

Solution. First we must find $\dfrac{\partial z}{\partial x}$ and $\dfrac{\partial z}{\partial y}$.

(12) $$\frac{\partial z}{\partial x} = y^3 + \sin xy + x(\cos xy)y = y^3 + \sin xy + xy \cos xy,$$

(13) $$\frac{\partial z}{\partial y} = 3xy^2 + x(\cos xy)x = 3xy^2 + x^2 \cos xy.$$

Using (12), we find that

$$\frac{\partial^2 z}{\partial x^2} = \frac{\partial}{\partial x}(y^3 + \sin xy + xy \cos xy)$$

$$\frac{\partial^2 z}{\partial x^2} = (\cos xy)y + y \cos xy + xy(-\sin xy)y = 2y \cos xy - xy^2 \sin xy,$$

$$\frac{\partial^2 z}{\partial y \, \partial x} = \frac{\partial}{\partial y}(y^3 + \sin xy + xy \cos xy)$$

$$= 3y^2 + (\cos xy)x + x \cos xy + xy(-\sin xy)x,$$

(14)
$$\frac{\partial^2 z}{\partial y \, \partial x} = 3y^2 + 2x \cos xy - x^2 y \sin xy.$$

Using (13), we find that

(15)
$$\frac{\partial^2 z}{\partial x \, \partial y} = \frac{\partial}{\partial x}(3xy^2 + x^2 \cos xy)$$

$$= 3y^2 + 2x \cos xy - x^2 y \sin xy.$$

To find $\dfrac{\partial^4 z}{\partial y^4}$, we have, from (13),

$$\frac{\partial^2 z}{\partial y^2} = \frac{\partial}{\partial y}(3xy^2 + x^2 \cos xy) = 6xy - x^3 \sin xy,$$

$$\frac{\partial^3 z}{\partial y^3} = \frac{\partial}{\partial y}(6xy - x^3 \sin xy) = 6x - x^4 \cos xy,$$

$$\frac{\partial^4 z}{\partial y^4} = \frac{\partial}{\partial y}(6x - x^4 \cos xy) = x^5 \sin xy. \quad \bullet$$

Observe that from (14) and (15),

(16)
$$\frac{\partial^2 z}{\partial y \, \partial x} = \frac{\partial^2 z}{\partial x \, \partial y}$$

for the particular function $z = xy^3 + x \sin xy$, and hence the order in which the partial derivatives are taken seems to be unimportant. Actually, equation (16) is not true for every function of two variables, but in order to find a function for which (16) is false, one must work very hard. In all practical cases, equation (16) is true, and we assume this fact henceforth. The reader who wishes more details about equation (16) is referred to any good book on advanced calculus. The partial derivatives that occur in (16) are called *mixed partial derivatives,* for obvious reasons.

EXERCISE 2

In Problems 1 through 10, find the first partial derivative of the given function with respect to each of the independent variables. In each of Problems 1 through 8, compute the two mixed *partial derivatives of second order and show that for each of the given functions the two mixed partials are equal* [*see equation* (16)].

1. $z = x^2y - xy^3$.
2. $z = e^{xy} \sin(x + 2y)$.
3. $z = x \sec 2y \tan 3x$.
4. $z = \ln(x \cot y^2)$.

5. $v = \operatorname{Sin}^{-1} \dfrac{y}{\sqrt{x^2 + y^2}}$.
6. $w = \operatorname{Tan}^{-1} \dfrac{y - x}{y + x}$.

7. $x = r \cos \theta$.
8. $u = (s^{1/2} + t^{1/2})^{1/2}$.

9. $w = (x^2 + y^2 + z^2) \ln \sqrt{x^2 + y^2 + z^2}$.
10. $Z = \dfrac{x}{y^2} + \dfrac{y^2}{z^3} + \dfrac{z^3}{t^4} + \dfrac{t^4}{x}$.

11. Prove that if $z = Cx^ny^m$, then equation (16) is satisfied. Then observe that (16) is satisfied whenever z is a sum of such terms. This proves that (16) is satisfied whenever z is a polynomial in the two variables.

12. If $u = xz^2 + yx^2 + zy^2$, show that

$$\frac{\partial u}{\partial x} + \frac{\partial u}{\partial y} + \frac{\partial u}{\partial z} = (x + y + z)^2.$$

13. If $u = A \cos m(x + at) + B \sin n(x - at)$, prove that

$$\frac{\partial^2 u}{\partial t^2} = a^2 \frac{\partial^2 u}{\partial x^2}$$

for all values of the constants A, B, m, n, and a.

14. If $z = x \sin(x/y) + ye^{y/x}$, prove that

$$x \frac{\partial z}{\partial x} + y \frac{\partial z}{\partial y} = z.$$

15. A function $u(x, y)$ that satisfies *Laplace's equation*

$$\frac{\partial^2 u}{\partial x^2} + \frac{\partial^2 u}{\partial y^2} = 0.$$

is called a *harmonic function*. Prove that each of the following functions is a harmonic function.

(a) $u = \operatorname{Tan}^{-1} \dfrac{y}{x}$.
(b) $u = x^4 - 6x^2y^2 + y^4$.

★(c) $u = \dfrac{x + y}{x^2 + y^2}$.
★(d) $u = e^{x^2 - y^2} \sin 2xy$.

★(e) $u = e^x \sin y + \ln(x^2 + y^2) + x^3 - 3xy^2$.

★16. Find $\dfrac{\partial^8 z}{\partial y^5 \partial x^3}$ if $z = x^2y^9 + 2x^5y^3 - 9x^7y + y^2e^x \sin^3 x$.

Various Notations for Partial Derivatives 5

Suppose that we are to find $\partial z/\partial x$ at the point $(1, 2)$ when $z = x^2 y + xy^3$. Naturally, we compute

$$\frac{\partial z}{\partial x} = 2xy + y^3$$

and on putting $x = 1$ and $y = 2$, we obtain $4 + 8 = 12$. What we really want is a symbol to indicate this process. One reasonable suggestion is to write

(17)
$$\frac{\partial z}{\partial x}\Big|_{\substack{x=1 \\ y=2}}$$

where the bar indicates that we are to evaluate the quantity using the indicated values. Such a symbol is indeed frequently used, but it is somewhat awkward. A perfectly satisfactory alternative is to use subscripts to denote partial differentiation. Thus by the meaning of the symbols, we have

$$f_x(x, y) \equiv \frac{\partial f}{\partial x}, \qquad z_x \equiv \frac{\partial z}{\partial x}, \qquad f_y(x, y) \equiv \frac{\partial f}{\partial y}, \qquad z_y \equiv \frac{\partial z}{\partial y}.$$

If we use the subscript for partial differentiation, our problem would be stated: If $f(x, y) = x^2 y + xy^3$, find

(18)
$$f_x(1, 2).$$

For this purpose the notation (18) is superior to (17). However, if we are to replace x and y by functions rather than numbers, then the notation (18) may be ambiguous. To be specific, suppose that $f(x, y) = x^2 + 4y^2$, and we are asked to compute

(19)
$$f_x(x + y, x - y).$$

This can be interpreted in two different ways:

(I) First differentiate $f(x, y)$, then replace x by $x + y$ and y by $x - y$.
(II) First replace x by $x + y$ and y by $x - y$, and then differentiate the (new) function.

The interpretation **(I)** gives

$$\frac{\partial f(x, y)}{\partial x} = 2x,$$

$$f_x(x + y, x - y) = 2(x + y) = 2x + 2y.$$

The interpretation **(II)** gives

$$f(x + y, x - y) = (x + y)^2 + 4(x - y)^2,$$

$$f_x(x + y, x - y) = 2(x + y) + 8(x - y) = 10x - 6y.$$

As may be expected, the different interpretations give different results. In this book *we will adopt the first interpretation*. However, it is better to avoid the ambiguity, as described below.

We let f_1 denote the partial derivative of $f(\ ,\ )$ with respect to the first variable (whatever the letter may be). Then the first interpretation for (19) would be indicated clearly by writing $f_1(x + y, x - y)$. With this convention, the ambiguity has disappeared. As illustrations of the various possibilities with this new notation,[1] we have

$$f_1(x, y) \equiv f_x(x, y) = \frac{\partial f}{\partial x}, \qquad\qquad f_2(x, y) \equiv f_y(x, y) = \frac{\partial f}{\partial y},$$

$$f_{21}(x, y) \equiv f_{yx}(x, y) \equiv \frac{\partial}{\partial x}\left(\frac{\partial f}{\partial y}\right), \qquad f_{12}(x, y) \equiv f_{xy}(x, y) \equiv \frac{\partial}{\partial y}\left(\frac{\partial f}{\partial x}\right),$$

$$f_{123}(x, y, z) \equiv f_{xyz}(x, y, z) \equiv \frac{\partial}{\partial z}\left(\frac{\partial}{\partial y}\left(\frac{\partial f}{\partial x}\right)\right).$$

Observe that with the subscript notation, f_{xy} means that we differentiate f first with respect to x and then with respect to y.

For the second interpretation of (19) it is convenient to introduce a new symbol for the new function. Thus if we set $F(x, y) \equiv f(x + y, x - y)$, then $F_x(x, y)$ expresses clearly the second interpretation for (19).

Example 1. If $f(x, y) = 3x^3y - 3xy + xy^3$, find:

$$f_2(1, 2), \qquad f_2(a, b), \qquad f_2(\cos\theta, \sin\theta), \qquad \text{and} \qquad \frac{\partial}{\partial y}f(x + y, x - y).$$

Solution. $f_2(x, y) = \dfrac{\partial}{\partial y}(3x^3y - 3xy + xy^3) = 3x^3 - 3x + 3xy^2$. Therefore,

$$f_2(1, 2) = 3 - 3 + 12 = 12,$$

$$f_2(a, b) = 3a^3 - 3a + 3ab^2 = 3a(a^2 + b^2 - 1),$$

$$f_2(\cos\theta, \sin\theta) = 3\cos^3\theta - 3\cos\theta + 3\cos\theta\sin^2\theta$$

$$= 3\cos\theta(\cos^2\theta + \sin^2\theta - 1) \equiv 0.$$

[1] Here, multiple subscripts, such as f_{21} and f_{123}, enter in a natural way. See the footnote on page 735.

With our agreement, the first interpretation, we have

$$\frac{\partial}{\partial y} f(x + y, x - y) = f_2(x + y, x - y) = 3x^3 - 3x + 3xy^2 \Big|_{\substack{x \to x+y \\ y \to x-y}}$$

$$= 3(x + y)^3 - 3(x + y) + 3(x + y)(x - y)^2$$

$$= 3(x + y)(2x^2 + 2y^2 - 1). \quad \bullet$$

Example 2. If $f(x, y, z) = (x + yz^2)^4$, find $f_{13}(-2, -1, 1)$.

Solution. We have

$$f_1(x, y, z) = \frac{\partial}{\partial x}(x + yz^2)^4 = 4(x + yz^2)^3,$$

$$f_{13}(x, y, z) = \frac{\partial}{\partial z} f_1(x, y, z) = \frac{\partial}{\partial z} 4(x + yz^2)^3$$

$$= 12(x + yz^2)^2 \, 2yz = 24yz(x + yz^2)^2.$$

Hence

$$f_{13}(-2, -1, 1) = 24yz(x + yz^2)^2 \Big|_{\substack{x = -2 \\ y = -1 \\ z = 1}} = 24(-1)(1)[-2 + (-1) \cdot (1)]^2$$

$$= -24(-3)^2 = -216.$$

$$\bullet$$

EXERCISE 3

For the functions in Problems 1, 2, 3, and 4, compute $f_1(1, -2)$ and $f_2(2, 3)$.

1. $f(x, y) = x^2 y + y^2$.

2. $f(x, y) = x^2 y^3 \sin \pi xy$.

3. $f(x, y) = \dfrac{x + y^2}{x - y^2}$.

4. $f(x, y) = \sqrt{8 + \dfrac{x}{y}}$.

5. For each of the functions of Problems 1 through 4, let $F(x, y) = f(y^2, x)$ and compute $F_1(x, y)$.

6. For the functions $F(x, y)$ in Problem 5, find $F_2(x, y)$.

7. If $f(x, y) = y^2 e^{-x}$, compute $f_{xy}(0, 2)$ and $f_{yx}(0, 2)$.

8. If $g(x, y) = 3x^2 y^2 + 2y$, compute $g_{xx}(1, -2)$.

9. If $h(x, y) = \ln(x - 3y)$, compute $h_{xxy}(2, -1)$.

10. If $G(x, y) = (2 + xy)^7$, compute $G_{22}(-1, 1)$.

11. If $H(x, y) = -x^2(3 + \sin y)^4$, compute $H_{11}(0, \pi/2)$.

12. Find $f_{23}(-3, -2, 0)$ if $f(x, y, z) = xy + e^{xyz}$.

13. Find $f_{yx}(2, 0, 3)$ if $f(x, y, z) = \sin(xyz) + 3z^2$.

14. Find $g_{13}(-1, 1, 0)$ if $g(x, y, z) = 2x^2 + 3y - z^2$.

15. Find $g_{123}(-1, 1, 2)$ if $g(x, y, z) = (x + yz)^3$.

16. The volume of a right circular cone of height h and radius r is $V(r, h) = \pi r^2 h/3$. Find the rate of change of the volume with respect to the radius.

★17. Given that the function $f(x, y)$ satisfies

$$f_1(x, y) = 6x^2y^2 + y \qquad \text{and} \qquad f_2(x, y) = 4x^3y + x + 2y,$$

find $f(x, y)$.

6 Tangent Planes and Normal Lines to a Surface

Let $z = f(x, y)$ be a function, and let $\mathcal{S}$ be the surface represented by this function. Suppose that we hold x constant by setting $x = x_0$. Then we are selecting those points on the surface for which $x = x_0$. But geometrically, those points are just the points of intersection of the plane $x = x_0$ with $\mathcal{S}$. These points form the curve CP_0D in Fig. 1. On this curve z changes with y while x remains constant. Since dz/dy is just the slope of a line tangent to the curve CP_0D when z is a function of y alone, we conclude that at the point P_0

$$(20) \qquad \frac{\partial z}{\partial y} = \text{slope of the line } P_0N = \tan \beta,$$

where β is the angle indicated in the figure. Similarly, the plane $y = y_0$ cuts the surface in a curve AP_0B and on that curve y is a constant so that at P_0

$$(21) \qquad \frac{\partial z}{\partial x} = \text{slope of the line } P_0M = \tan \alpha.$$

The above work suggests that the plane containing the lines P_0M and P_0N is the tangent plane to the surface $\mathcal{S}$ at the point P_0. But we must first define a tangent plane.

DEFINITION 4. Let $\mathcal{T}$ be a plane through a point P_0 on the surface $\mathcal{S}: z = f(x, y)$, and let P be any other point on $\mathcal{S}$. If, as P approaches P_0, the angle between the line segment P_0P and the plane $\mathcal{T}$ tends to zero, then $\mathcal{T}$ is called the *tangent plane* to the surface $\mathcal{S}$ at P_0 (see Fig. 2).

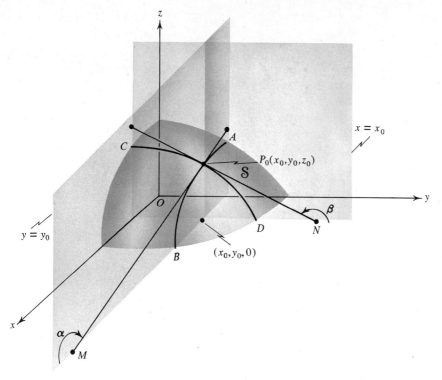

Of course, a surface need not have a tangent plane at a point P_0. The simplest example is the half-cone $z = a\sqrt{x^2 + y^2}$ shown in Fig. 3. Here it is clear that there is no plane that is tangent to this surface at the origin.

If the surface has a tangent plane at P_0, then it is obvious that the plane must contain the lines P_0M and P_0N, and this fact will provide us with a very easy way to obtain an equation for the tangent plane. Indeed, if $z = f(x, y)$ is the equation of the surface and $P_0(x_0, y_0, z_0)$ is the point under consideration, then $f_x(x_0, y_0)$ is the rate of change of z as x changes along the line P_0M. A unit change in x produces a change of $f_x(x_0, y_0)$ in z, and y does not change along the line P_0M. Consequently, the vector

$$\mathbf{V} = \mathbf{i} + 0\mathbf{j} + f_x(x_0, y_0)\mathbf{k}$$

is parallel to the line P_0M. Similarly, the vector

$$\mathbf{U} = 0\mathbf{i} + \mathbf{j} + f_y(x_0, y_0)\mathbf{k}$$

is parallel to the line P_0N. Then the cross product of $\mathbf{U}$ and $\mathbf{V}$ determines a vector $\mathbf{N}$ that is perpendicular to the lines P_0M and P_0N. The vector $\mathbf{N}$ is therefore perpendicular to the plane that contains these two lines. The vector $\mathbf{N}$ is said to be *normal* to the plane containing P_0M and P_0N. If this plane is tangent to the surface, then $\mathbf{N}$ is said to be *normal* to the

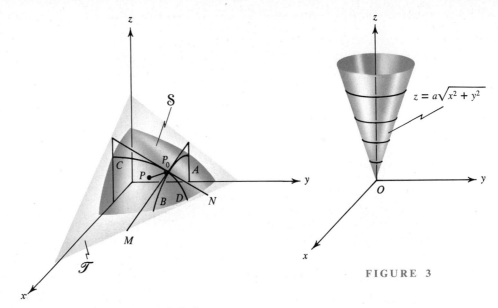

FIGURE 2

FIGURE 3

surface at P_0. A simple computation gives

$$\mathbf{N} = \mathbf{U} \times \mathbf{V} = \begin{vmatrix} \mathbf{i} & \mathbf{j} & \mathbf{k} \\ 0 & 1 & f_y(x_0, y_0) \\ 1 & 0 & f_x(x_0, y_0) \end{vmatrix}$$

or

(22)
$$\mathbf{N} = f_x(x_0, y_0)\mathbf{i} + f_y(x_0, y_0)\mathbf{j} - \mathbf{k}.$$

If $R(x, y, z)$ is any point on the tangent plane, then the line $P_0 R$ is perpendicular to $\mathbf{N}$; consequently, $\mathbf{P_0 R} \cdot \mathbf{N} = 0$. Using equation (22), we have

$$0 = [(x - x_0)\mathbf{i} + (y - y_0)\mathbf{j} + (z - z_0)\mathbf{k}] \cdot [f_x(x_0, y_0)\mathbf{i} + f_y(x_0, y_0)\mathbf{j} - \mathbf{k}]$$

$$0 = f_x(x_0, y_0)(x - x_0) + f_y(x_0, y_0)(y - y_0) - (z - z_0),$$

or

(23)
$$z - z_0 = f_x(x_0, y_0)(x - x_0) + f_y(x_0, y_0)(y - y_0).$$

We have proved

THEOREM 2. If the surface $z = f(x, y)$ has a tangent plane at $P_0(x_0, y_0, z_0)$, then the partial derivatives $f_x(x, y)$ and $f_y(x, y)$ exist at (x_0, y_0). Further, equation (23) is an equation of the tangent plane, and (22) gives a vector $\mathbf{N}$ that is normal to the surface at P_0.

From (22) it is easy to see that

(24) $$\mathbf{R} = [x_0 + f_x(x_0, y_0)t]\mathbf{i} + [y_0 + f_y(x_0, y_0)t]\mathbf{j} + [z_0 - t]\mathbf{k}$$

is a vector equation for the line normal to the surface $z = f(x, y)$ at P_0. This vector equation yields the set of equations

$$\frac{x - x_0}{f_x(x_0, y_0)} = \frac{y - y_0}{f_y(x_0, y_0)} = \frac{z - z_0}{-1}$$

for the same line, provided that $f_x(x_0, y_0) \neq 0$ and $f_y(x_0, y_0) \neq 0$.

We know that there are surfaces that do not have a tangent plane at certain points (see Fig. 3). It can be proved that whenever f_x and f_y are continuous at every point inside a small circle with center at $P_0(x_0, y_0)$, then the surface $z = f(x, y)$ has a tangent plane at the corresponding point (x_0, y_0, z_0).

Example 1. Find a normal vector and the tangent plane to the surface

$$z = 4 - x^2 - y^2/4$$

at the point $P_0(1, 2, 2)$. Where does the line normal to the surface meet the xy-plane? Find the intercepts of the tangent plane on the three axes. Make a sketch showing the intersection of the tangent plane with the coordinate axes.

Solution. We first check that the given point $(1, 2, 2)$ is on the surface. Using $x = 1$ and $y = 2$, we find that $z = 4 - 1^2 - 2^2/4 = 4 - 1 - 1 = 2$ and hence P_0 is indeed on the surface. Next we compute the partial derivatives and find that at P_0,

(25) $$f_x = -2x = -2 \quad \text{and} \quad f_y = \frac{-y}{2} = \frac{-2}{2} = -1.$$

From equation (23) an equation for the tangent plane is

$$z - 2 = -2(x - 1) - 1(y - 2)$$

or

$$z = -2x - y + 6.$$

The intercepts of this plane on the three axes are $x = 3$, $y = 6$, and $z = 6$. The vector $\mathbf{N}_1 = f_x\mathbf{i} + f_y\mathbf{j} - \mathbf{k} = -2\mathbf{i} - \mathbf{j} - \mathbf{k}$ is normal at P_0. But for convenience we can replace this vector by $\mathbf{N} = -\mathbf{N}_1 = 2\mathbf{i} + \mathbf{j} + \mathbf{k}$. For the line normal to the surface at P_0, we use equation (24) and find that

(26) $$\mathbf{R} = (1 - 2t)\mathbf{i} + (2 - t)\mathbf{j} + (2 - t)\mathbf{k}.$$

This line meets the xy-plane when $z = 2 - t = 0$ or when $t = 2$. This gives the point $(-3, 0, 0)$. All these items (except for the normal line) are shown in Fig. 4. ●

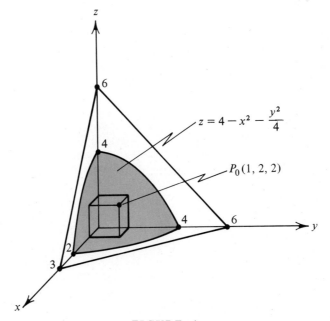

FIGURE 4

Example 2. At each point of intersection of the line $\mathbf{R} = (2 - t)\mathbf{i} + t\mathbf{j} + 2t\mathbf{k}$ and the surface $z = x^2 + y^2$, find $\cos\theta$, where θ is the angle between the line and the normal to the surface.

Solution. The line has parametric equations $x = 2 - t$, $y = t$, $z = 2t$. To find the points of intersection, we substitute these in $z = x^2 + y^2$, the equation of the surface, and obtain

$$2t = (2 - t)^2 + t^2,$$

$$0 = 2t^2 - 6t + 4 = 2(t - 1)(t - 2).$$

Hence $t = 1$ or $t = 2$ and the points are $P_1(1, 1, 2)$ and $P_2(0, 2, 4)$. In general, a normal vector to the surface is $f_1\mathbf{i} + f_2\mathbf{j} - \mathbf{k}$. For the surface $z = x^2 + y^2$, this gives $\mathbf{N} = 2x\mathbf{i} + 2y\mathbf{j} - \mathbf{k}$. At the points in question, $\mathbf{N}_1 = 2\mathbf{i} + 2\mathbf{j} - \mathbf{k}$ and $\mathbf{N}_2 = 4\mathbf{j} - \mathbf{k}$. A vector parallel to the given line is $\mathbf{R}'(t) = -\mathbf{i} + \mathbf{j} + 2\mathbf{k}$. The dot product gives

$$\text{At } P_1, \quad \cos \theta = \frac{\mathbf{N}_1 \cdot \mathbf{R}'(t)}{|\mathbf{N}_1|\,|\mathbf{R}'(t)|} = \frac{-2 + 2 - 2}{\sqrt{9}\sqrt{6}} = \frac{-2}{3\sqrt{6}}.$$

$$\text{At } P_2, \quad \cos \theta = \frac{\mathbf{N}_2 \cdot \mathbf{R}'(t)}{|\mathbf{N}_2|\,|\mathbf{R}'(t)|} = \frac{4 - 2}{\sqrt{17}\sqrt{6}} = \frac{2}{\sqrt{17}\sqrt{6}}. \quad \bullet$$

The student should make a sketch, showing the surface, the line through the two points, the normals, and the angle θ at each point.

EXERCISE 4

In Problems 1 through 6, find an equation for the tangent plane and the normal line at the indicated point on the given surface.

1. $z = 10 - x^2 - y^2$, $P_0(1, 2, 5)$.
2. $z = 2x^2 - 3y^2$, $P_0(3, 2, 6)$.
3. $z = 6/xy$, $P_0(1, 2, 3)$.
4. $z = e^x \sin \pi y$, $P_0(2, 1, 0)$.
5. $z = x + y + 2 \ln xy$, $P_0(1, 1, 2)$.
6. $x^2 + y^2 + z^2 = 121$, $P_0(6, 7, 6)$.

7. Find the angle between the line $\mathbf{R} = (-2 + 4t)\mathbf{i} + (5 + t)\mathbf{j} + (12 - 3t)\mathbf{k}$ and the normal to the sphere $x^2 + y^2 + z^2 = 121$ at the points of intersection of the line and the sphere.

*8. Show that at each point P_0 of the cone $z^2 = A(x^2 + y^2)$, other than the vertex, the tangent plane has an equation $z_0 z = A(x_0 x + y_0 y)$. Hence each such tangent plane passes through the vertex of the cone. Show that

$$\mathbf{R} = x_0(1 + At)\mathbf{i} + y_0(1 + At)\mathbf{j} + z_0(1 - t)\mathbf{k}$$

is an equation for the line normal to the cone at P_0.

*9. Suppose that on the cone of Problem 8 we take all points of fixed height H above the xy-plane and erect normals to the cone at these points. Prove that the set of points in which these normals intersect the xy-plane forms a circle. Find the radius of the circle.

**10. Find an equation of the curve of intersection with the xy-plane of the normals to the surface $z = ax^2 + by^2$ when all the normals are erected at the same height $z = H$ on the surface. Here assume that $a > 0$, $b > 0$, and $H > 0$.

*11. Since the half-cone $z = \sqrt{x^2 + y^2}$ does not have a tangent plane at $(0, 0, 0)$ (see Fig. 3), we expect that the partial derivatives z_x and z_y do not exist at $x = 0$, $y = 0$

[the limits in equations (10) and (11) do not exist]. Prove this fact. HINT: Recall that $\sqrt{(\Delta x)^2} = |\Delta x|$.

*12. Show that if $z = \sqrt{x^2 + y^2}$, then the partial derivatives z_x and z_y do not have a limit as $P \to (0, 0)$.

7 The Increment of a Function of Two Variables

We recall that if $z = f(x)$ is a differentiable function of a single independent variable, then

$$(27) \qquad \Delta z = \frac{df}{dx} \Delta x + \epsilon \, \Delta x$$

where $\epsilon \to 0$ as $\Delta x \to 0$. In fact, this is merely a restatement of the definition of a derivative. For if we divide both sides of (27) by Δx and take the limit, we have

$$(28) \qquad \lim_{\Delta x \to 0} \frac{\Delta z}{\Delta x} = \lim_{\Delta x \to 0} \left(\frac{df}{dx} + \epsilon \right) = \frac{df}{dx} + \lim_{\Delta x \to 0} \epsilon,$$

since at a given point df/dx is a constant. It is clear from (28) that $\epsilon \to 0$ as $\Delta x \to 0$.

Similarly, if $z = f(y)$ is a differentiable function of y (perhaps a different function, but we use the same letter f), then

$$(29) \qquad \Delta z = \frac{df}{dy} \Delta y + \epsilon \, \Delta y,$$

where $\epsilon \to 0$ as $\Delta y \to 0$.

What is the analogue of (27) and (29) when $z = f(x, y)$ is a function of two variables? Can we just add the two expressions on the right side of (27) and (29) using partial derivatives? The affirmative answer is contained in

THEOREM 3. Let $z = f(x, y)$ and suppose that the partial derivatives $f_x(x, y)$ and $f_y(x, y)$ are continuous in some neighborhood[1] of (x_0, y_0). Let

$$(30) \qquad \Delta z \equiv f(x_0 + \Delta x, y_0 + \Delta y) - f(x_0, y_0).$$

Then

---

[1] By a neighborhood of a point (x_0, y_0) we mean the set of all points (x, y) that lie inside a circle of radius r with center at (x_0, y_0). These are the points for which $(x - x_0)^2 + (y - y_0)^2 < r^2$. The phrase "some neighborhood" means that there is some $r > 0$ (r may be small).

(31)
$$\Delta z = f_x(x_0, y_0)\, \Delta x + f_y(x_0, y_0)\, \Delta y + \epsilon_1\, \Delta x + \epsilon_2\, \Delta y,$$

where $\epsilon_1 \to 0$ and $\epsilon_2 \to 0$ as both $\Delta x \to 0$ and $\Delta y \to 0$.

DISCUSSION. The similarity between (31) and (27) or (29) becomes obvious if we write (31) in the form

(32)
$$\Delta z = \frac{\partial f}{\partial x}\, \Delta x + \frac{\partial f}{\partial y}\, \Delta y + \epsilon_1\, \Delta x + \epsilon_2\, \Delta y.$$

From equation (30), Δz is the change in the function as the point $P(x, y)$ moves from (x_0, y_0) to a neighboring point $(x_0 + \Delta x, y_0 + \Delta y)$, as indicated in Fig. 5. The theorem asserts that this change can be approximated by

(33)
$$\Delta z \approx f_x(x_0, y_0)\, \Delta x + f_y(x_0, y_0)\, \Delta y,$$

because the terms $\epsilon_1\, \Delta x + \epsilon_2\, \Delta y$ will tend to zero much more rapidly than $\sqrt{(\Delta x)^2 + (\Delta y)^2}$. However, the nature of this approximation is a little complicated, and we will discuss it further in Example 1.

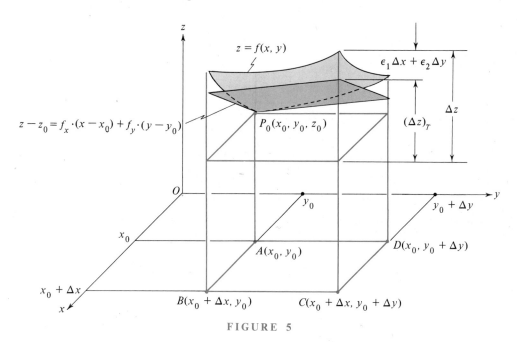

FIGURE 5

If we compare (33) with the equation for a tangent plane developed in Section 6,

(23)
$$z - z_0 = f_x(x_0, y_0)(x - x_0) + f_y(x_0, y_0)(y - y_0),$$

and identify Δx with $x - x_0$ and Δy with $y - y_0$, it is clear that Δz is approximated by $z - z_0$, where in (23), z is the corresponding point on the tangent plane (see Fig. 5). Using $(\Delta z)_T$ to denote the change of z on the tangent plane, and reserving Δz for the change in z on the surface, equation (31) states that

$$\Delta z = (\Delta z)_T + \epsilon_1 \Delta x + \epsilon_2 \Delta y.$$

Proof of Theorem 3. On the right side of (30) we subtract the quantity $f(x_0 + \Delta x, y_0)$ and then add the same quantity so that the equal sign is not disturbed. Thus we can write

$$\Delta z = [f(x_0 + \Delta x, y_0 + \Delta y) - f(x_0 + \Delta x, y_0)] + [f(x_0 + \Delta x, y_0) - f(x_0, y_0)]$$

$$(34) \qquad \Delta z \equiv \qquad \Delta_1 \qquad + \qquad \Delta_2.$$

Observe that in Δ_1 the first variable x is held constant at $x_0 + \Delta x$, so Δ_1 represents a change due to the change in the second variable only. Similarly, in Δ_2, y is held constant at y_0, and only x is changing. Thus the introduction of the terms $\pm f(x_0 + \Delta x, y_0)$ allows us to decompose Δz into the sum of two increments, in each of which only one variable actually undergoes a change. Referring to Fig. 5, we see that the changes Δ_1 and Δ_2 are the changes in z along the two line segments BC and AB, respectively.

We can apply the Mean Value Theorem for one variable to each of the expressions for Δ_1 and Δ_2. Thus for Δ_2 we can write

$$(35) \qquad \Delta_2 \equiv f(x_0 + \Delta x, y_0) - f(x_0, y_0) = f_x(\xi, y_0) \Delta x,$$

where ξ is some suitably selected point between x_0 and $x_0 + \Delta x$. Similarly,

$$(36) \qquad \Delta_1 = f(x_0 + \Delta x, y_0 + \Delta y) - f(x_0 + \Delta x, y_0) = f_y(x_0 + \Delta x, \eta) \Delta y,$$

where η (Greek lowercase letter eta) is some suitably selected point between y_0 and $y_0 + \Delta y$. Now the points (ξ, y_0) and $(x_0 + \Delta x, \eta)$ are in a neighborhood of (x_0, y_0), where the radius $r = |\Delta x| + |\Delta y|$. Since f_x and f_y are continuous in a neighborhood of (x_0, y_0), we can write

$$(37) \qquad f_x(\xi, y_0) = f_x(x_0, y_0) + \epsilon_1, \qquad f_y(x_0 + \Delta x, \eta) = f_y(x_0, y_0) + \epsilon_2,$$

where $\epsilon_1 \to 0$ and $\epsilon_2 \to 0$ as $|\Delta x| + |\Delta y| \to 0$. Using (37) in (36) and (35) and substituting the result in (34), we obtain (31). ∎

With considerably more writing but no new ideas, this proof can be extended to the case where $f(x, y, \ldots, s, t, u)$ is a function of any finite number of variables. There is no need to burden ourselves with the details. For three variables the result is stated in

THEOREM 4 (PWO). Let $w = f(x, y, z)$ and suppose that the partial derivatives $f_x(x, y, z)$, $f_y(x, y, z)$, and $f_z(x, y, z)$ are continuous in some neighborhood of (x_0, y_0, z_0). Let

(38) $\Delta w \equiv f(x_0 + \Delta x, y_0 + \Delta y, z_0 + \Delta z) - f(x_0, y_0, z_0).$

Then

(39) $\Delta w = f_x(x_0, y_0, z_0)\, \Delta x + f_y(x_0, y_0, z_0)\, \Delta y + f_z(x_0, y_0, z_0)\, \Delta z$
$$+ \epsilon_1 \Delta x + \epsilon_2 \Delta y + \epsilon_3 \Delta z,$$

where $\epsilon_1 \to 0$, $\epsilon_2 \to 0$, and $\epsilon_3 \to 0$ as Δx, Δy, and $\Delta z \to 0$.

Example 1. Illustrate Theorem 3 by finding an explicit expression for each of the quantities mentioned in that theorem, for the particular function $z = y^2 - x^2$ at the point $(1, 2)$. Discuss the nature of the approximation $\Delta z \approx (\Delta z)_T$ at the point $(1, 2)$ for various choices of Δx and Δy.

Solution. For this function and any point (x_0, y_0), equation (30) becomes

$$\Delta z = (y_0 + \Delta y)^2 - (x_0 + \Delta x)^2 - (y_0^2 - x_0^2)$$

$$= y_0^2 + 2y_0\, \Delta y + (\Delta y)^2 - x_0^2 - 2x_0\, \Delta x - (\Delta x)^2 - y_0^2 + x_0^2$$

$$= -2x_0\, \Delta x + 2y_0\, \Delta y - (\Delta x)^2 + (\Delta y)^2.$$

At the particular point $(1, 2)$ we have

(40) $\Delta z = -2\, \Delta x + 4\, \Delta y - (\Delta x)(\Delta x) + (\Delta y)(\Delta y).$

Comparing (40) with (31), we see that $f_x(1, 2)$ should be -2 and $f_y(1, 2)$ should be 4, and this is indeed the case for the function $f(x, y) = y^2 - x^2$. Further, $\epsilon_1 = -\Delta x$ and $\epsilon_2 = \Delta y$, so that obviously $\epsilon_1 \to 0$ and $\epsilon_2 \to 0$ as $(\Delta x, \Delta y) \to (0, 0)$. Comparing (40) and (33), we have $\Delta z \approx -2\, \Delta x + 4\, \Delta y$, and the difference, which we denote by R (the remainder), is given by $R = -(\Delta x)^2 + (\Delta y)^2$. This certainly tends to zero more rapidly than $\sqrt{(\Delta x)^2 + (\Delta y)^2}$ because

$$\lim_{(\Delta x, \Delta y) \to (0,0)} \frac{-(\Delta x)^2 + (\Delta y)^2}{\sqrt{(\Delta x)^2 + (\Delta y)^2}} = 0.$$

At the point $(1, 2)$ we have $z_0 = 4 - 1 = 3$, and hence the approximation $(\Delta z)_T = -2\, \Delta x + 4\, \Delta y$ can be written

$$z - 3 = -2(x - 1) + 4(y - 2).$$

This is the equation for the plane tangent to the surface $z = y^2 - x^2$ at the point $(1, 2, 3)$.

To investigate the way in which $(\Delta z)_T$ approximates Δz, we select a few values for Δx and Δy and compute the various quantities involved. The results are given in Table 1. We first observe that in every case R is either 0 or involves t^2, while $\sqrt{(\Delta x)^2 + (\Delta y)^2}$ involves only the first power of t. Thus if $t \to 0$, then $R/\sqrt{(\Delta x)^2 + (\Delta y)^2} \to 0$. In almost every case it is also true that the ratio $R/(\Delta z)_T \to 0$ as $\Delta t \to 0$. This is the relation that suggests the phrase "$(\Delta z)_T$ is a good approximation to Δz." But in case 1 we see that $(\Delta z)_T = 0$ and $\Delta z = R$. So $(\Delta z)_T$ is not really a good approximation to Δz in this one exceptional case. Because $(\Delta z)_T$ does give the major portion of Δz in most cases, we will continue to refer to $(\Delta z)_T$ as a first approximation to Δz, even though there are exceptional cases. ●

TABLE 1

Case	Δx	Δy	$\Delta z = (\Delta z)_T + R$	$(\Delta z)_T$	R	$\sqrt{(\Delta x)^2 + (\Delta y)^2}$
1	$2t$	t	$-3t^2$	0	$-3t^2$	$\sqrt{5}\,|t|$
2	t	t	$2t$	$2t$	0	$\sqrt{2}\,|t|$
3	0	t	$4t + t^2$	$4t$	t^2	$|t|$
4	t	0	$-2t - t^2$	$-2t$	$-t^2$	$|t|$
5	$-t$	t	$6t$	$6t$	0	$\sqrt{2}\,|t|$
6	$-2t$	t	$8t - 3t^2$	$8t$	$-3t^2$	$\sqrt{5}\,|t|$

Example 2. In a certain survey the two sides of a triangle were measured and found to be 160 and 300 ft, respectively, with an error at most of ± 0.5 ft. The included angle was found to be $60°$ with an error of at most 5 min. The third side is computed from these measurements. Find an approximate value for the maximum error in the third side, and for the percent error.

Solution. This amounts to finding the change in c, where

(41) $$c^2 = a^2 + b^2 - 2ab \cos C$$

when a, b, and C change as indicated. With the given measurements

$$c^2 = (160)^2 + (300)^2 - 2(160)(300)\frac{1}{2} = 67{,}600 = (260)^2,$$

so the computed value of $c = 260$. By Theorem 4,

(42) $$\Delta c \approx \frac{\partial c}{\partial a} \Delta a + \frac{\partial c}{\partial b} \Delta b + \frac{\partial c}{\partial C} \Delta C.$$

To compute the partial derivatives in equation (41), we may use $c = \sqrt{a^2 + b^2 - 2ab \cos C}$. However, it is simpler (and leads to the same result) to differentiate equation (41) implicitly. Indeed, we find that

$$2c \frac{\partial c}{\partial a} = 2a - 2b \cos C, \qquad 2c \frac{\partial c}{\partial b} = 2b - 2a \cos C, \qquad 2c \frac{\partial c}{\partial C} = 2ab \sin C,$$

and consequently,

$$\frac{\partial c}{\partial a} = \frac{a - b \cos C}{c}, \qquad \frac{\partial c}{\partial b} = \frac{b - a \cos C}{c}, \qquad \frac{\partial c}{\partial C} = \frac{ab \sin C}{c}.$$

Using these expressions for the partial derivatives in (42) together with $a = 160$, $b = 300$, $c = 260$, and $C = 60°$, we find that

$$(43) \qquad \Delta c \approx \frac{160 - 150}{260} \Delta a + \frac{300 - 80}{260} \Delta b + \frac{160 \times 300 \sqrt{3}/2}{260} \Delta C.$$

We want to maximize this approximate value for Δc, under the conditions $-1/2 \leq \Delta a \leq 1/2$, $-1/2 \leq \Delta b \leq 1/2$, and (converting 5 min to radian measure) $-5\pi/10{,}800 \leq \Delta C \leq 5\pi/10{,}800$. Let $(\Delta c)_{max}$ denote the maximum error. To maximize the right side of (43), we must give Δa, Δb, and ΔC their maximum values. We then find that

$$(\Delta c)_{max} \approx \frac{5 + 110 + 60.5}{260} \approx 0.675 \text{ ft.}$$

A close approximation for the maximum percent error is

$$\frac{(\Delta c)_{max}}{c} \times 100 \approx \frac{0.675}{260} \times 100 \approx 0.26 \text{ percent.} \quad \bullet$$

Returning to the notation $z = f(x, y)$ and Theorem 3, we observe that if

$$\frac{\partial f}{\partial x} \Delta x + \frac{\partial f}{\partial y} \Delta y$$

represents a good approximation to Δz, then it is reasonable to define dz, the differential of z, by the expression

$$(44) \qquad dz = \frac{\partial f}{\partial x} dx + \frac{\partial f}{\partial y} dy.$$

Similarly, if $w = f(x, y, z)$, then dw, the differential of w, is defined by the right side of

$$(45) \qquad dw = \frac{\partial w}{\partial x} dx + \frac{\partial w}{\partial y} dy + \frac{\partial w}{\partial z} dz.$$

With this notation and terminology, equation (42) takes the form

$$dc = \frac{\partial c}{\partial a} \, da + \frac{\partial c}{\partial b} \, db + \frac{\partial c}{\partial C} \, dC.$$

In Example 2, we used differentials to estimate the maximum error. If a calculator is available, we can obtain a more accurate estimate. Thus if $a = 160.5$ ft, $b = 300.5$ ft, and $C = 60°5'$, then (to five decimal places)

$$c^2 = a^2 + b^2 - 2ab \cos C = (160.5)^2 + (300.5)^2 - 2(160.5)(300.5) \cos 60°5'$$

$$= 25{,}760.25 + 90{,}300.25 - 48{,}108.69904 = 67{,}951.80096$$

$$c = 260.67566.$$

Therefore, the value given by differentials is at most only 0.00066 ft away from the true value for the maximum error.

EXERCISE 5

In Problems 1 through 4, find an explicit expression for

$$\Delta z - (\Delta z)_T = \Delta z - \left\{ \frac{\partial f}{\partial x} \Delta x + \frac{\partial f}{\partial y} \Delta y \right\}.$$

Observe that each term in your answer has at least one of the factors $(\Delta x)^2$, $(\Delta x)(\Delta y)$, *or* $(\Delta y)^2$.

1. $z = 2x^2 + 3y^2$.
                2. $z = x^3 + xy^2 - y^3$.
*3. $z = x/y$.
                4. $z = 3x^2y^2 + 5x - 7y$.

In Problems 5 through 16, use differentials to obtain a first approximation. Then if a calculator is available, use it to obtain a more accurate result.

5. Find an approximate value for the change in z, on the surface $z = 2x^2 - 3y^2$ when x changes from 4 to 4.3 and y changes from 5 to 4.8.
6. Repeat Problem 5 for the surface $z = 2x^2 + 3y^2$.
7. The legs of a right triangle were measured and found to be 120 and 160 ft with an error of at most 1 ft. Find an approximation for the maximum error when the area and the hypotenuse are computed from these measurements.
8. A certain tin can is supposed to be 10 in. high and have a base radius of 2 in. If each of these measurements may be in error by as much as 0.1 in., find an approximate value for the maximum error when the volume is computed from these dimensions.
9. In a certain survey the two sides of a triangle were measured and found to be 160 and 210 ft with an error of at most 0.1 ft. The included angle was measured to be 60° with

an error of at most 1 min. The third side is computed from these measurements. Find an approximate value for the maximum error in the third side and for the percent error.

10. Repeat Problem 9 if the sides were measured to be 550 and 160 ft, the angle was measured to be 60°, and each measurement was accurate within 1 percent.

11. The electrical resistance of a certain wire can be computed from the formula $V = IR$, where V is the voltage drop across the ends of the wire, I is the current flowing through the wire, and R is the resistance of the wire. If V and I are measured with an error of at most 1 percent, find an approximate value for the maximum percent error in the computed value of R.

12. The focal length f of a lens is given by

$$\frac{1}{f} = \frac{1}{p} + \frac{1}{q},$$

where p and q are the distances of the lens from the object and image, respectively. For a certain lens p and q are each 20 cm, with a possible error of at most 0.5 cm. Find an approximate value for the maximum error in the computed value of f.

13. The period of a pendulum is given by $P = 2\pi \sqrt{l/g}$, where l is the length of the pendulum and g is the acceleration due to gravity. But if P and l are measured accurately, this equation can be used to compute g. Suppose that in a certain pendulum $l = 5.1$ ft with an error of at most 0.1 ft and $P = 2.5$ sec with an error of at most 0.05 sec. Find an approximate value for the maximum error in the computed value of g.

14. The eccentricity of the ellipse $b^2x^2 + a^2y^2 = a^2b^2$ is given by $e = \sqrt{a^2 - b^2}/a$. In a certain ellipse a and b were measured and found to be 25 and 24, respectively, with an accuracy of ± 0.2. Find an approximate value for the maximum error in the computed value of e.

15. If three electrical resistances are connected in parallel, the circuit resistance R is given by

$$\frac{1}{R} = \frac{1}{r_1} + \frac{1}{r_2} + \frac{1}{r_3}.$$

Suppose that R is computed using $r_1 = 200$, $r_2 = 200$, and $r_3 = 100$ (in ohms). Estimate upper and lower bounds for R, assuming that r_1, r_2, and r_3 are correct within 2 ohms.

16. In 1965, Craig Breedlove drove his special auto over a 2-mile course at slightly over 600 miles/hr. For simplicity assume that the measured time was 12.00 sec so that the computed speed was exactly 600 miles/hr. Find approximate upper and lower bounds for the true speed if the distance was correct within 0.01 mile and the time was correct within 0.03 sec.

CALCULATOR PROBLEMS

In Problems C1 through C4, find: (a) Δz, (b) $(\Delta z)_T$, and (c) $\Delta z - (\Delta z)_T$ as P changes from P_0 to P_1. Give answers to five decimal places.

C1. $z = x^3 + 3y^3$, $P_0(2, 1)$, $P_1(2.01, 1.03)$.

C2. $z = \dfrac{x^2}{y^3}$, $P_0(4, 3)$, $P_1(4.01, 2.95)$.

C3. $z = x \sin y$, $P_0(5, \pi/2)$, $P_1(4.96, 1.55)$.

C4. $z = x^2 y u^{1/2}$, $P_0(-1, 5, 7)$, $P_1(-.99, 5.02, 7.03)$, in (x, y, u) space.

8 *The Chain Rule*

Suppose that $z = f(x, y)$ and that x and y are each in turn functions of a third variable t; that is, $x = x(t)$ and $y = y(t)$. Then z is a function of t, a composite function denoted by $z = f(x(t), y(t))$. We now derive a formula for the derivative of this composite function. Let Δt be a change in t and Δx and Δy be the changes induced in x and y, respectively, by this change in t. If the conditions of Theorem 3 are satisfied, then we can write for Δz

(32)
$$\Delta z = \frac{\partial z}{\partial x} \Delta x + \frac{\partial z}{\partial y} \Delta y + \epsilon_1 \Delta x + \epsilon_2 \Delta y.$$

Dividing by Δt gives

(46)
$$\frac{\Delta z}{\Delta t} = \frac{\partial z}{\partial x} \frac{\Delta x}{\Delta t} + \frac{\partial z}{\partial y} \frac{\Delta y}{\Delta t} + \epsilon_1 \frac{\Delta x}{\Delta t} + \epsilon_2 \frac{\Delta y}{\Delta t}.$$

If we take the limit as $\Delta t \to 0$ and recall that $\epsilon_1 \to 0$ and $\epsilon_2 \to 0$, we obtain the desired formula,

(47)
$$\frac{dz}{dt} = \frac{\partial z}{\partial x} \frac{dx}{dt} + \frac{\partial z}{\partial y} \frac{dy}{dt}.$$

We have proved

THEOREM 5. If $x(t)$ and $y(t)$ are differentiable functions of t and if the partial derivatives $f_1(x, y)$ and $f_2(x, y)$ are continuous in a neighborhood of the point $(x(t), y(t))$, then the function

$$z = f(x(t), y(t))$$

is differentiable, and its derivative is given by (47).

This theorem is a natural generalization of the Chain Rule (Chapter 4, Theorem 17, page 160). Observe that we can obtain (47) mechanically if we divide both sides of (44), the expression for dz, by dt.

Example 1. If $z = 2x^2 + 3xy - 4y^2$, where $x = \cos t$ and $y = \sin t$, find dz/dt in two different ways.

Solution. By (47) we have

$$\frac{dz}{dt} = (4x + 3y)(-\sin t) + (3x - 8y)\cos t.$$

We can use $x = \cos t$, $y = \sin t$, to express everything in terms of t,

$$\frac{dz}{dt} = (4\cos t + 3\sin t)(-\sin t) + (3\cos t - 8\sin t)\cos t$$

$$= 3\cos^2 t - 12\sin t \cos t - 3\sin^2 t = 3\cos 2t - 6\sin 2t.$$

We could also first substitute and then differentiate, thus:

$$\frac{dz}{dt} = \frac{d}{dt}(2\cos^2 t + 3\cos t \sin t - 4\sin^2 t)$$

$$= 4\cos t(-\sin t) + 3\cos^2 t - 3\sin^2 t - 8\sin t \cos t$$

$$= 3\cos 2t - 6\sin 2t. \quad \bullet$$

Theorem 5 and its associated formula generalize in an obvious way to any number of variables. For example, if $w = f(x, y, z)$ and x, y, and z are each functions of t, then

(48)
$$\frac{dw}{dt} = \frac{\partial w}{\partial x}\frac{dx}{dt} + \frac{\partial w}{\partial y}\frac{dy}{dt} + \frac{\partial w}{\partial z}\frac{dz}{dt}.$$

Suppose that x, y, and z are each functions of *two* variables, for instance, $x = x(t, u)$, $y = y(t, u)$, and $z = z(t, u)$. Then the partial derivatives are given by

(49)
$$\frac{\partial w}{\partial t} = \frac{\partial w}{\partial x}\frac{\partial x}{\partial t} + \frac{\partial w}{\partial y}\frac{\partial y}{\partial t} + \frac{\partial w}{\partial z}\frac{\partial z}{\partial t}$$

and

(50)
$$\frac{\partial w}{\partial u} = \frac{\partial w}{\partial x}\frac{\partial x}{\partial u} + \frac{\partial w}{\partial y}\frac{\partial y}{\partial u} + \frac{\partial w}{\partial z}\frac{\partial z}{\partial u}.$$

If we use the subscript notation for partial derivatives, these two formulas would be written

$$w_t = w_x x_t + w_y y_t + w_z z_t$$

and

$$w_u = w_x x_u + w_y y_u + w_z z_u.$$

We can also consider the much simpler case in which $w = f(u)$, a single variable, where u in turn depends on several variables. This is illustrated in

Example 2. Prove that if f is any differentiable function, then $z = f(x^3 - y^2)$ is a solution of the partial differential equation

$$2y \frac{\partial z}{\partial x} + 3x^2 \frac{\partial z}{\partial y} = 0.$$

Solution. The notation $z = f(x^3 - y^2)$ means that $z = f(u)$, where $u = x^3 - y^2$. In this case, the Chain Rule gives

(51) $$\frac{\partial z}{\partial x} = \frac{\partial f}{\partial u} \frac{\partial u}{\partial x} = \frac{df}{du} \frac{\partial u}{\partial x} = f'(x^3 - y^2)3x^2, \quad \bigg| \quad 2y$$

(52) $$\frac{\partial z}{\partial y} = \frac{\partial f}{\partial u} \frac{\partial u}{\partial y} = \frac{df}{du} \frac{\partial u}{\partial y} = f'(x^3 - y^2)(-2y). \quad \bigg| \quad 3x^2$$

We multiply equation (51) by $2y$, and multiply equation (52) by $3x^2$ (as indicated schematically), and add. Obviously, the sum is

$$f'(x^3 - y^2)[3x^2(2y) + (-2y)(3x^2)] = f'(x^3 - y^2)(0) \equiv 0. \quad \bullet$$

To illustrate the meaning of this result, let us select for $f(u)$ the function

$$f(u) = \ln[u^2 + u^4] + \sinh[\text{Tan}^{-1} u].$$

Then our work shows that the function

$$z \equiv f(x^3 - y^2) \equiv \ln[(x^3 - y^2)^2 + (x^3 - y^2)^4] + \sinh[\text{Tan}^{-1}(x^3 - y^2)]$$

is a solution of the given partial differential equation.

EXERCISE 6

In Problems 1 through 5, find dw/dt in two ways: (a) by using the Chain Rule and then expressing everything in terms of t, and (b) first expressing w as a function of t alone and then differentiating.

1. $w = e^{x^2 + y^2}$, $x = \sin t, \quad y = \cos t.$

2. $w = \text{Tan}^{-1} xyz$, $x = t^2, \quad y = t^3, \quad z = 1/t^4.$

3. $w = xy + yz + zx$, $x = e^t, \quad y = 2t^3, \quad z = e^{-t}.$

4. $w = \dfrac{2xy}{x^2 + y^2}$, $x = 2t, \quad y = t^2.$

5. $w = \ln (x^2 + 3xy^2 + 4y^4)$, $x = 2t^2, \quad y = 3t.$

In Problems 6 and 7, find $\partial w/\partial t$ and $\partial w/\partial u$ in two ways.

6. $w = x \ln (x^2 + y^2)$, $x = t + u, \quad y = t - u.$

7. $w = e^{x + 2y} \sin (2x - y)$, $x = t^2 + 2u^2, \quad y = 2t^2 - u^2.$

8. The area of a rectangle is given by $A = xy$, where x is the base and y is the altitude. Let r be the length of a diagonal and θ the angle it makes with the base. Compute A_r and A_θ in two ways.

9. The volume of a right circular cone is given by $V = \frac{1}{3}\pi r^2 h$. Compute in two ways V_θ and V_l, where θ is the angle between the axis and an element of the surface of the cone and l is the slant height. Here the notation V_θ means that l is regarded as constant, and V_l means that θ is regarded as constant.

10. With the notation of Problem 9, prove that $lV_l = 3hV_h$ and that $V_\theta = hV_r - rV_h$.

★11. With θ and h as in Problem 9, prove that the volume is given by $V = \frac{1}{3}\pi h^3 \tan^2 \theta$. Then $V_h = \pi h^2 \tan^2 \theta$ and $V_\theta = \frac{2}{3}\pi h^3 \tan \theta \sec^2 \theta$. Prove that these expressions for V_h and V_θ are in general not equal to those obtained in Problems 9 and 10, and explain why.

12. Given $z = f(x, y)$ where $x = a + ht$ and $y = b + kt$, show that

$$\frac{dz}{dt} = hf_x(a + ht, b + kt) + kf_y(a + ht, b + kt),$$

and if $f_{xy} = f_{yx}$ show that

$$\frac{d^2z}{dt^2} = h^2 f_{xx}(a + ht, b + kt) + 2hk f_{xy}(a + ht, b + kt) + k^2 f_{yy}(a + ht, b + kt).$$

13. Prove that if f is any differentiable function, then $z = xf(y/x)$ is a solution of the partial differential equation

$$x \frac{\partial z}{\partial x} + y \frac{\partial z}{\partial y} = z.$$

14. Prove that if f is any differentiable function, then $z = f(x^2 - y^2)$ is a solution of the partial differential equation

$$x \frac{\partial z}{\partial y} + y \frac{\partial z}{\partial x} = 0.$$

15. Prove that $z = x^2 + xf(xy)$ is a solution of $x\dfrac{\partial z}{\partial x} - y\dfrac{\partial z}{\partial y} = x^2 + z$.

16. If $u = f(x, y)$ and $x = r\cos\theta$, $y = r\sin\theta$ (transformation equations from polar coordinates to rectangular coordinates), prove that

(53)
$$\left(\frac{\partial u}{\partial x}\right)^2 + \left(\frac{\partial u}{\partial y}\right)^2 = \left(\frac{\partial u}{\partial r}\right)^2 + \frac{1}{r^2}\left(\frac{\partial u}{\partial \theta}\right)^2.$$

*17. With the notation of Problem 16, prove that

(54)
$$\frac{\partial^2 u}{\partial x^2} + \frac{\partial^2 u}{\partial y^2} = \frac{\partial^2 u}{\partial r^2} + \frac{1}{r}\frac{\partial u}{\partial r} + \frac{1}{r^2}\frac{\partial^2 u}{\partial \theta^2}.$$

18. Use the result in Problem 17 to show that each of the functions $u_1 = r^n \cos n\theta$ and $u_2 = r^n \sin n\theta$ satisfies Laplace's Equation,

$$\frac{\partial^2 u}{\partial x^2} + \frac{\partial^2 u}{\partial y^2} = 0.$$

*19. Prove the following theorem. *Let $f(x, y)$ have first partial derivatives that are continuous in a region that contains the line segment joining $P_0(a, b)$ and $P_1(a + h, b + k)$. Then there is a point $P^{\star}(x^{\star}, y^{\star})$ on this segment such that*

(55)
$$f(a + h, b + k) - f(a, b) = hf_x(x^{\star}, y^{\star}) + kf_y(x^{\star}, y^{\star}).$$

This is the Mean Value Theorem for functions of two variables. HINT: Apply the Mean Value Theorem (Chapter 5, Theorem 8, page 99) to the function of a single variable t, $f(x_0 + ht, y_0 + kt)$, and use Problem 12.

*20. A *domain* $\mathscr{R}$ in the plane is a set of points with the following two properties: (a) For each point P there is a small circle with center at P such that every point inside the circle is in $\mathscr{R}$, and (b) any two points in $\mathscr{R}$ can be connected by a polygonal path all of whose points lie in $\mathscr{R}$. Suppose that $f_x(x, y) = 0$ and $f_y(x, y) = 0$ at every point of a domain $\mathscr{R}$. Prove that $f(x, y)$ is a constant in $\mathscr{R}$. HINT: Use Problem 19 and part (b) of the definition of a domain.

9   *The Directional Derivative*

Let $w = f(x, y)$, and let $\mathscr{C}$ be a directed curve in the xy-plane that has a tangent at the point P_0 on $\mathscr{C}$. We want to develop a formula for the rate of change of the function $f(x, y)$ along the curve $\mathscr{C}$. For this purpose, we use s, the arc length of the curve, and define the *directional derivative of $f(x, y)$ along the curve $\mathscr{C}$* to be the rate of change of $f(x, y)$ with respect to s along $\mathscr{C}$. In symbols,

$$(56) \qquad \frac{df}{ds} = \lim_{\Delta s \to 0} \frac{\Delta f}{\Delta s},$$

where P approaches P_0 on the curve $\mathscr{C}$ and Δs is the length of the arc $P_0 P$. We observe that the notation for the directional derivative is defective because the curve itself does not appear in (56), although the value of df/ds depends upon the curve. In other words, if $f(x, y)$ is a fixed function, two different curves through P_0 may well give rise to two different values for df/ds.

To obtain a formula for computing the directional derivative along the curve $\mathscr{C}$, we use the arc length as a parameter and suppose that the curve has the equations $x = x(s)$, $y = y(s)$. If the partial derivatives of $f(x, y)$ are continuous in a neighborhood of P_0, then the Chain Rule (Theorem 5) is applicable, since along $\mathscr{C}$, $f = f(x(s), y(s))$ is now a function of s alone. Hence

$$(57) \qquad \frac{df}{ds} = \frac{\partial f}{\partial x} \frac{dx}{ds} + \frac{\partial f}{\partial y} \frac{dy}{ds}.$$

If the tangent vector to the curve $\mathscr{C}$ at P_0 makes an angle α with the positive x-axis (see Fig. 6), then $dx/ds = \cos \alpha$, $dy/ds = \sin \alpha$ [see equations (62), Chapter 12] and (57) becomes

$$(58) \qquad \frac{df}{ds} = \frac{\partial f}{\partial x} \cos \alpha + \frac{\partial f}{\partial y} \sin \alpha,$$

where f_x and f_y are evaluated at P_0.

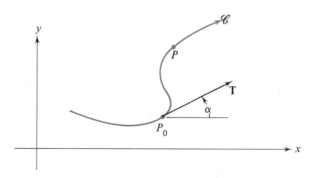

FIGURE 6

It is clear from (58) that if two curves through P_0 have the same direction (α is the same), then the directional derivative of f on the two curves is the same. Thus the directional derivative does not depend on the curve, but only on the direction of the curve at P_0.

Example 1. Find the directional derivative of $f(x, y) = x + 2xy - 3y^2$ in the direction of the vector $3\mathbf{i} + 4\mathbf{j}$ at the point $(1, 2)$.

Solution. For this function (58) gives

$$\frac{df}{ds} = (1 + 2y) \cos \alpha + (2x - 6y) \sin \alpha.$$

Using $\cos \alpha = 3/5$, $\sin \alpha = 4/5$, $x = 1$, and $y = 2$, we find that

$$\frac{df}{ds} = (1 + 4)\frac{3}{5} + (2 - 12)\frac{4}{5} = 3 - 8 = -5. \quad \bullet$$

Example 2. Find the directional derivative of the function defined in Example 1 at $(1, 2)$ along the curve $\mathscr{C}_1 : x = 1 + 3t + t^3$, $y = 2 + 4t + t^4$, and along the curve $\mathscr{C}_2 : x = 1 + 6t + \sin^3 t + \tan t^5$, $y = 2 + 8t + t^2 e^t \sinh^3 t$.

Solution. Each curve passes through the point $(1, 2)$, as is easily checked by setting $t = 0$. When $t = 0$, we have:

$$\text{For } \mathscr{C}_1: \quad \frac{dx}{dt} = 3, \quad \frac{dy}{dt} = 4. \quad \text{For } \mathscr{C}_2: \quad \frac{dx}{dt} = 6, \quad \frac{dy}{dt} = 8.$$

Hence for both curves the unit tangent vector is $(3\mathbf{i} + 4\mathbf{j})/5$, the same as in Example 1. Hence along both curves the directional derivative of

$$f(x, y) = x + 2xy - 3y^2$$

at $(1, 2)$ is -5, and the computation is identical with that performed in Example 1. $\bullet$

The concept of a directional derivative and the proof of formula (58) generalize immediately to functions of three variables. Stated precisely, we have

---

THEOREM 6. If f_x, f_y, and f_z are continuous in a neighborhood of the point P_0, and if $\mathbf{u} = \cos \alpha \mathbf{i} + \cos \beta \mathbf{j} + \cos \gamma \mathbf{k}$ is a unit vector, then the directional derivative of $f(x, y, z)$ along a directed curve whose tangent at P_0 has the direction of $\mathbf{u}$ is given by

$$(59) \qquad \frac{df}{ds} = \frac{\partial f}{\partial x} \cos \alpha + \frac{\partial f}{\partial y} \cos \beta + \frac{\partial f}{\partial z} \cos \gamma,$$

where f_x, f_y, and f_z are evaluated at P_0.

Example 3. Find the directional derivative of $f = ze^x \cos \pi y$ at the point $(0, -1, 2)$ in the direction of the vector $\mathbf{U} = -\mathbf{i} + 2\mathbf{j} + 2\mathbf{k}$.

Solution. Formula (59) gives

$$\frac{df}{ds} = ze^x \cos \pi y \cos \alpha - \pi ze^x \sin \pi y \cos \beta + e^x \cos \pi y \cos \gamma.$$

To obtain a unit vector, we write

$$\mathbf{u} = \frac{\mathbf{U}}{|\mathbf{U}|} = \frac{-\mathbf{i} + 2\mathbf{j} + 2\mathbf{k}}{\sqrt{1 + 4 + 4}} = -\frac{1}{3}\mathbf{i} + \frac{2}{3}\mathbf{j} + \frac{2}{3}\mathbf{k}.$$

For the point $(0, -1, 2)$, and the direction $\mathbf{u}$, we find that

$$\frac{df}{ds} = \frac{-2e^0 \cos(-\pi)}{3} - \frac{2\pi 2e^0 \sin(-\pi)}{3} + \frac{2e^0 \cos(-\pi)}{3}$$

$$= \frac{2 - 0 - 2}{3} = 0. \quad \bullet$$

EXERCISE 7

In Problems 1 through 6, compute the directional derivative of the given function at P in the direction specified by the angle α.

1. $x^2 + y^3$, (1, 2), $\alpha = \pi/4$. 2. $\ln x^2 y^3$, (1, 2), $\alpha = 3\pi/4$.
3. $\text{Tan}^{-1} xy^2$, (2, -1), $\alpha = \pi$. 4. $ax + by$, (-1, 2), $\alpha = 0$.
5. $ax^2 + by^2$, (-1, -1), $\alpha = \pi/2$. 6. $\sin(x + y)$, (0, $\pi/2$), $\alpha = \pi/3$.

7. As a special case of equation (58), show that $f_x(x, y)$ is just the directional derivative in the direction of the positive x-axis, and $f_y(x, y)$ is the directional derivative in the direction of the positive y-axis. How would you interpret $-f_x(x, y)$ and $-f_y(x, y)$?

8. We can use the notation $(df/ds)_\alpha$ to denote the directional derivative in equation (58), where the subscript α gives the direction. With this notation prove that

$$\left(\frac{df}{ds}\right)_\alpha = -\left(\frac{df}{ds}\right)_{\alpha+\pi} \quad \text{and} \quad \left(\frac{df}{ds}\right)_\alpha^2 + \left(\frac{df}{ds}\right)_{\alpha+\pi/2}^2 = f_x^2 + f_y^2.$$

Consequently, in two mutually perpendicular directions, the sum of the squares of the directional derivatives is a constant at a given point.

9. Prove that if either $f_x \neq 0$ or $f_y \neq 0$, at P_0, then there is always a curve leading from P_0 along which $f(x, y)$ is increasing, and a second curve along which $f(x, y)$ is decreasing.

10. Use the result of Problem 9 to prove that if $P_0(x_0, y_0, z_0)$ is a high point or a low point on the surface $z = f(x, y)$, then $f_x = 0$ and $f_y = 0$ at P_0.

11. Prove that the surface $z = x^2 + 4xy + 4y^2 + 7y - 13$ has neither a high point nor a low point.

10 *The Gradient*

The symmetric form of equation (59) suggests immediately that it can be regarded as the dot product of two vectors. Indeed, we can write

$$\text{(59)} \qquad \frac{df}{ds} = \frac{\partial f}{\partial x} \cos \alpha + \frac{\partial f}{\partial y} \cos \beta + \frac{\partial f}{\partial z} \cos \gamma$$

in the form

$$\text{(60)} \qquad \frac{df}{ds} = \left(\frac{\partial f}{\partial x} \mathbf{i} + \frac{\partial f}{\partial y} \mathbf{j} + \frac{\partial f}{\partial z} \mathbf{k} \right) \cdot (\cos \alpha \mathbf{i} + \cos \beta \mathbf{j} + \cos \gamma \mathbf{k}).$$

This representation as a dot product is so important that the first vector is given a special symbol and name.

DEFINITION 5. The gradient of $f(x, y, z)$, written ***grad** f* or $\boldsymbol{\nabla} f$ (read "del f") is by definition the vector

$$\text{(61)} \qquad \boldsymbol{grad\ f = \nabla f = f_x \mathbf{i} + f_y \mathbf{j} + f_z \mathbf{k}.}$$

Returning to equation (60), we recall that $\mathbf{u} = \cos \alpha \mathbf{i} + \cos \beta \mathbf{j} + \cos \gamma \mathbf{k}$ is a unit vector that specifies the direction for the directional derivative. Consequently, (60) can be written in the compact form

$$\text{(62)} \qquad \frac{df}{ds} = (\boldsymbol{\nabla} f) \cdot \mathbf{u} = |\boldsymbol{\nabla} f| \cos \varphi,$$

where φ is the angle between the two vectors $\boldsymbol{\nabla} f$ and $\mathbf{u}$. But (62) is more than a shorthand notation. We can derive very valuable information from it and with great ease. Suppose that $f_x = f_y = f_z = 0$. Then $\boldsymbol{\nabla} f = \mathbf{0}$ and the directional derivative is zero in all directions. In all other cases, $|\boldsymbol{\nabla} f| > 0$. It follows from (62) that the directional derivative ranges between $-|\boldsymbol{\nabla} f|$ and $|\boldsymbol{\nabla} f|$, and attains these two extreme values at $\varphi = \pi$ and $\varphi = 0$, respectively. Consequently, the maximum value of the directional derivative is attained by selecting for **u** the direction of the vector ∇f, and the minimum is attained by selecting the opposite direction. We summarize in

---

THEOREM 7. If ∇f is a continuous vector function in a neighborhood of P_0, and if $\nabla f \neq 0$ at P_0, then at P_0 the vector ∇f points in the direction in which $f(x, y, z)$ has its maximum directional derivative and this maximum is just $|\nabla f|$. Further, if **u** is any unit vector, then in the direction of **u**,

(63)
$$\frac{df}{ds} = (\nabla f) \cdot \mathbf{u}.$$

Example 1. Find the direction at the point $(2, -1, 5)$ for which the function $f(x, y, z) = x^2 y (z - 4)^3$ has its maximum directional derivative, and find this maximum.

Solution. By definition, equation (61),

$$\nabla f = \nabla[x^2 y (z - 4)^3] = 2xy(z - 4)^3 \mathbf{i} + x^2(z - 4)^3 \mathbf{j} + 3x^2 y (z - 4)^2 \mathbf{k}.$$

At the point $(2, -1, 5)$, we have $\nabla f = -4\mathbf{i} + 4\mathbf{j} - 12\mathbf{k}$. The maximum for df/ds occurs in the direction of this vector. Further, by Theorem 7,

$$\max \frac{df}{ds} = |\nabla f| = |-4\mathbf{i} + 4\mathbf{j} - 12\mathbf{k}| = 4\sqrt{1 + 1 + 9} = 4\sqrt{11}. \quad \bullet$$

EXERCISE 8

1. The material of Definition 5 and Theorem 7 is also valid if $f(x, y)$ is just a function of two variables. In this plane case, write the analogue of: **(a)** equation (60), **(b)** Definition 5, and **(c)** Theorem 7.

In Problems 2 through 7, find the directional derivative of the given function at the given point, in the given direction.

2. e^{x+y}, $(0, 0)$, $\mathbf{i} + \mathbf{j}$.   3. $e^{x^2 + y^3}$, $(0, 0)$,   any direction.

4. $\sin \pi x \cos \pi y^2$, $(1, 2)$, $\mathbf{i} - 2\mathbf{j}$.   5. $xy^2 z^3$, $(-3, 2, 1)$, $6\mathbf{i} - 2\mathbf{j} + 3\mathbf{k}$.

6. $xy^2 + y^2 z^3 + z^3 x$, $(4, -2, -1)$, $\mathbf{i} + 3\mathbf{j} + 2\mathbf{k}$.

7. $x \sin \pi yz + zy \tan \pi x$, $(1, 2, 3)$, $2\mathbf{i} + 6\mathbf{j} - 9\mathbf{k}$.

8. Find those points for which $\nabla(xy + \cos x + zy^2)$ is parallel to the *yz*-plane.

9. Is there any point at which $\nabla(x^3 + y^2 - 5z)$ makes the same angle with each of the three coordinate axes?

10. Find those points for which the gradient of $f = x^2 + xy - z^2 + 4y - 3z$ makes the same angle with each of the three coordinate axes.

11. Repeat Problem 10 for $f = xy^2z^3$. Observe that points in the xy-plane or xz-plane are exceptional.

★12. Repeat Problem 10 for $f = 2x \cos \pi y + 3yz$.

13. Find those points in the xy-plane for which the length of the vector $\nabla(x^2 + xy - y^2)$ is a constant.

★14. At each point in space, a line is drawn with the direction of $\nabla(x^3 + y^2 - 5z)$. Find those points for which the line passes through the origin.

15. Prove that the gradient of $F(x, y, z) = e^{ax+by+cz}$ is always parallel to the gradient of $f(x, y, z) = ax + by + cz$ and hence always points in the direction of the constant vector $a\mathbf{i} + b\mathbf{j} + c\mathbf{k}$.

★16. (A generalization of Problem 15.) If $F(x, y, z) = G(u)$, where $u = f(x, y, z)$, prove that ∇F and ∇f are parallel vectors at each point where neither one of the vectors is zero.

★17. If u and v are functions of x, y, and z, and if a and b are constants, prove that

$$\nabla(au + bv) = a\nabla u + b\nabla v, \qquad \nabla(uv) = u\nabla v + v\nabla u, \qquad \text{and} \qquad \nabla u^n = nu^{n-1}\nabla u.$$

★18. Suppose that $\nabla f = \mathbf{0}$ for all values of x, y, and z in a certain domain $\mathscr{R}$. What can you say about f in $\mathscr{R}$? HINT: See Exercise 6, Problem 20.

★19. Suppose that $\nabla f = a\mathbf{i} + b\mathbf{j} + c\mathbf{k}$ throughout $\mathscr{R}$. What can you say about f in $\mathscr{R}$?

★★20. Prove that if in Theorem 7 we omit the condition that ∇f is continuous in a neighborhood of P_0, then the resulting "theorem" is false. HINT: It is sufficient to consider a function of two variables. Let $f(x, y) = x + y - 3\sqrt{|xy|}$. Then at $P_0 = (0, 0)$ we have $\nabla f = \mathbf{i} + \mathbf{j}$. Suppose that P is on the ray $y = x$, with $x \geq 0$. Then we have $f(x, y) = x + x - 3\sqrt{|x^2|} = -x$ so that f is *decreasing*. But along the x-axis f is *increasing*.

11 *Implicit Functions*

Let $F(x, y)$ be a function of two variables and let $\mathscr{C}(c)$ be those points of the plane for which

(64) $$F(x, y) = c,$$

where c is a constant. In most cases (but not in all) the set $\mathscr{C}(c)$ is a curve, and whenever it is a curve, it is called a *level curve* for the function F. As c varies, we obtain a collection of level curves that together fill out the domain of F, which is usually the entire plane (see Fig. 7). Under certain very mild conditions it can be proved that $\mathscr{C}(c)$ can be decomposed into a finite number of smooth curves each with an equation of the form $y = f(x)$. This means that if we replace y by $f(x)$ in equation (64), we obtain an identity

(65) $$F(x, f(x)) \equiv c.$$

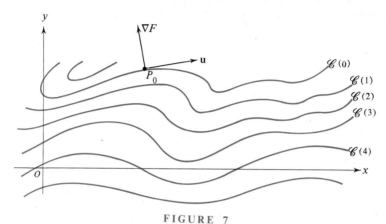

FIGURE 7

Any theorem that asserts the existence of an $f(x)$ with the properties described is called an *implicit function theorem,* because $F(x, y) = c$ is regarded as defining $f(x)$ implicitly. We will not prove an implicit function theorem,[1] because it would take us too far astray. For the present we assume that $f(x)$ exists (as it does in all practical cases) and proceed to obtain some related results.

We differentiate (65) with respect to x. The right side gives 0, and if we apply the Chain Rule (Theorem 5 with t replaced by x and f replaced by F), we obtain

(66) $$\frac{\partial F}{\partial x} \frac{dx}{dx} + \frac{\partial F}{\partial y} \frac{dy}{dx} = 0.$$

Here $dx/dx = 1$. If we solve (66) for dy/dx, we complete the proof of

---

THEOREM 8. Suppose that $\nabla F(x, y)$ is continuous in $\mathcal{N}$, a neighborhood of $P_0(x_0, y_0)$ and that $F_2(x, y) \neq 0$ in $\mathcal{N}$. Let $y = f(x)$ be the equation of $\mathscr{C}(c)$, a level curve that contains P_0. Then, for points of $\mathscr{C}(c)$ in $\mathcal{N}$,

(67) $$\frac{dy}{dx} = -\frac{\dfrac{\partial F}{\partial x}}{\dfrac{\partial F}{\partial y}} = -\frac{F_1(x, y)}{F_2(x, y)}.$$

---

[1] The reader can find these in any advanced calculus text.

If the curve $\mathscr{C}(c)$ is a smooth curve, we can introduce the arc length s as a parameter. If $x = x(s)$ and $y = y(s)$ are the parametric equations for $\mathscr{C}(c)$, then (64) yields the identity

(68)
$$F(x(s), y(s)) \equiv c.$$

The Chain Rule applied to (68) gives

(69)
$$\frac{\partial F}{\partial x}\frac{dx}{ds} + \frac{\partial F}{\partial y}\frac{dy}{ds} = 0.$$

This can be written in the vector form as

(70)
$$0 = \left[\frac{\partial F}{\partial x}\mathbf{i} + \frac{\partial F}{\partial y}\mathbf{j}\right] \cdot \left[\frac{dx}{ds}\mathbf{i} + \frac{dy}{ds}\mathbf{j}\right] = (\nabla F) \cdot \mathbf{u},$$

where $\mathbf{u} = (dx/ds)\mathbf{i} + (dy/ds)\mathbf{j}$. Since $\mathbf{u}$ is a tangent vector to the level curve $\mathscr{C}(c)$ of the function $F(x, y) = c$, equation (70) assures us that ∇F is perpendicular to $\mathbf{u}$ and hence is normal to $\mathscr{C}(c)$. We have proved

---

THEOREM 9. Suppose that the level curve $\mathscr{C}(c)$ is a smooth curve that contains the point P_0. If $\nabla F \neq 0$ at P_0 and is continuous in a neighborhood of P_0, then ∇F is normal to $\mathscr{C}(c)$ at P_0.

Further, we remark that from equation (69) we have for this curve

(71)
$$\frac{dy}{dx} = \frac{\dfrac{dy}{ds}}{\dfrac{dx}{ds}} = -\frac{F_1(x_0, y_0)}{F_2(x_0, y_0)}$$

as long as the denominators are not zero. Thus we obtain again formula (67) of Theorem 8.

Example 1. Sketch the three level curves of $F(x, y) = y^2 - 4x$ that pass through the points $P_1(-2, 2)$, $P_2(4, 4)$, and $P_3(7, 4)$, respectively. At each of these points, compute and sketch the vector ∇F and the unit vector $\mathbf{u}$ tangent to the level curve.

Solution. Each of the level curves of $y^2 - 4x$ is a parabola. For the curve through $P_1(-2, 2)$, we have $c = y^2 - 4x = 4 - 4(-2) = 12$. For the other two points the values of c are $16 - 4(4) = 0$ and $16 - 4(7) = -12$, respectively. The three level curves are shown in Fig. 8.

For this function $\nabla F = -4\mathbf{i} + 2y\mathbf{j}$ and at the points P_1, P_2, and P_3, ∇F is $-4\mathbf{i} + 4\mathbf{j}$, $-4\mathbf{i} + 8\mathbf{j}$, and $-4\mathbf{i} + 8\mathbf{j}$, respectively. Given any nonzero vector $\mathbf{v} =$

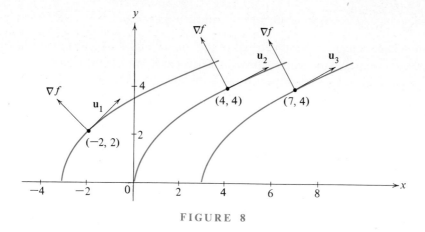

FIGURE 8

$a\mathbf{i} + b\mathbf{j}$, it is easy to see (from the dot product) that the vector $b\mathbf{i} - a\mathbf{j}$ is perpendicular to $\mathbf{v}$. Hence unit vectors tangent to the level curves at the given points can be found from ∇F. These are

$$\mathbf{u}_1 = (\mathbf{i} + \mathbf{j})/\sqrt{2}, \qquad \mathbf{u}_2 = (2\mathbf{i} + \mathbf{j})/\sqrt{5}, \qquad \mathbf{u}_3 = (2\mathbf{i} + \mathbf{j})/\sqrt{5}.$$

The three tangent vectors and the three vectors ∇F (not to scale) are shown in Fig. 8. ●

The above considerations generalize immediately to three-dimensional space. Let $F(x, y, z)$ be a function of three variables and let $\mathcal{S}(c)$ be the set of points for which

(72) $$F(x, y, z) = c.$$

In most cases (but not in all) the set $\mathcal{S}(c)$ is a surface, and whenever it is a surface it is called a *level surface* for the function F. As c varies we obtain a collection of level surfaces which together fill out the domain of F. Under certain very mild conditions it can be proved that $\mathcal{S}(c)$ can be decomposed into a finite number of smooth surfaces, each with an equation of the form $z = f(x, y)$ [or $x = g(y, z)$, or $y = h(z, x)$]. Again we will not stop to prove an implicit function theorem for this situation. For the present we assume that there is a function $f(x, y)$ such that when we replace z by $f(x, y)$ in (72), we obtain the identity

(73) $$F(x, y, f(x, y)) \equiv c.$$

As a familiar example, the reader may consider the function

(74) $$F(x, y, z) \equiv x^2 + y^2 + z^2.$$

Here, if $c > 0$, the set $\mathcal{S}(c)$ is a sphere of radius $\sqrt{c}$ with center at the origin. This surface can be decomposed into two pieces with equations

(75) $$z = \sqrt{c - x^2 - y^2} \quad \text{and} \quad z = -\sqrt{c - x^2 - y^2}.$$

The first equation gives the upper half of the sphere, and the second gives the lower half.

We now parallel the work in proving Theorem 8. We use the Chain Rule to compute the partial derivative with respect to x of the identity (73). This gives formally

(76) $$\frac{\partial F}{\partial x} \frac{\partial x}{\partial x} + \frac{\partial F}{\partial y} \frac{\partial y}{\partial x} + \frac{\partial F}{\partial z} \frac{\partial z}{\partial x} = 0.$$

But $\partial x/\partial x = 1$, and since y is constant in this process, $\partial y/\partial x = 0$. Consequently, if $F_z \neq 0$, then

(77) $$\frac{\partial z}{\partial x} \equiv \frac{\partial f(x, y)}{\partial x} = -\frac{\dfrac{\partial F}{\partial x}}{\dfrac{\partial F}{\partial z}} = -\frac{F_1(x, y, z)}{F_3(x, y, z)}.$$

Similarly, differentiating the identity (73) with respect to y gives

(78) $$\frac{\partial z}{\partial y} \equiv \frac{\partial f(x, y)}{\partial y} = -\frac{\dfrac{\partial F}{\partial y}}{\dfrac{\partial F}{\partial z}} = -\frac{F_2(x, y, z)}{F_3(x, y, z)}.$$

We have proved

> **THEOREM 10.** Suppose that $\nabla F(x, y, z)$ is continuous in $\mathcal{N}$, a neighborhood of $P_0(x_0, y_0, z_0)$, and that $F_3(x, y, z) \neq 0$ in $\mathcal{N}$. Let $z = f(x, y)$ be the equation of $\mathcal{S}(c)$, a level surface of F that contains P_0. Then for points of $\mathcal{S}(c)$ in $\mathcal{N}$, equations (77) and (78) hold.

Example 2. For the level surface of the function

(79) $$F(x, y, z) \equiv xy^2z^3 + (x - 3)^5(y + 2)^7(z - 1)^{11}$$

that contains the point $P_0(3, -2, 1)$, find $\partial z/\partial x$ and $\partial z/\partial y$.

Solution. We observe that the second term in the right side of (79), namely, $F^\star(x, y, z) \equiv (x - 3)^5(y + 2)^7(z - 1)^{11}$, has been included as part of the function merely to prevent us from solving $F(x, y, z) = 12$ for z. This forces us to use formulas (77) and (78). At P_0, $F_1^\star = F_2^\star = F_3^\star = 0$. Then formulas (77) and (78) give

$$\frac{\partial z}{\partial x} = -\frac{F_1(P_0)}{F_3(P_0)} = -\left.\frac{y^2z^3 + F_1^\star}{3xy^2z^2 + F_3^\star}\right|_{P_0} = \frac{-4}{3 \cdot 3 \cdot 4} = -\frac{1}{9}$$

and

$$\frac{\partial z}{\partial y} = -\frac{F_2(P_0)}{F_3(P_0)} = -\frac{2xyz^3 + F_2^\star}{3xy^2z^2 + F_3^\star}\Bigg|_{P_0} = -\frac{2\cdot 3(-2)}{3\cdot 3\cdot 4} = \frac{1}{3}. \quad \bullet$$

To generalize Theorem 9 to three-dimensional space, we suppose that

$$x = x(s), \qquad y = y(s), \qquad z = z(s)$$

are parametric equations (where s is the arc length) for a smooth curve that lies on $\mathcal{S}(c)$ and has a tangent at P_0 (see Fig. 9). Since the curve lies on $\mathcal{S}(c)$, we have the identity

(80) $$F(x(s), y(s), z(s)) \equiv c.$$

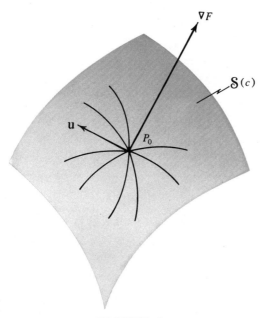

FIGURE 9

The Chain Rule applied to (80) gives

$$\frac{\partial F}{\partial x}\frac{dx}{ds} + \frac{\partial F}{\partial y}\frac{dy}{ds} + \frac{\partial F}{\partial z}\frac{dz}{ds} = 0.$$

This can be put in the vector form

(81) $$0 = \left[\frac{\partial F}{\partial x}\mathbf{i} + \frac{\partial F}{\partial y}\mathbf{j} + \frac{\partial F}{\partial z}\mathbf{k}\right]\cdot\left[\frac{dx}{ds}\mathbf{i} + \frac{dy}{ds}\mathbf{j} + \frac{dz}{ds}\mathbf{k}\right] = (\nabla F)\cdot\mathbf{u},$$

where $\mathbf{u} = (dx/ds)\mathbf{i} + (dy/ds)\mathbf{j} + (dz/ds)\mathbf{k}$ and is a unit vector tangent to the curve. Thus if ∇F is not zero, it is perpendicular to **u,** and hence it is normal to the curve. But this is true for any such curve through P_0, and hence ∇F is normal to the surface $\mathcal{S}(c)$ at P_0. This gives

---

THEOREM 11. Suppose that the level surface $\mathcal{S}(c)$ has a tangent plane at the point P_0. If $\nabla F \neq \mathbf{0}$ at P_0 and is continuous in a neighborhood of P_0, then ∇F is normal to $\mathcal{S}(c)$ at P_0.

From this we deduce immediately an equation for the tangent plane.

THEOREM 12. Under the conditions of Theorem 11,

(82)
$$(x - x_0)F_x(x_0, y_0, z_0) + (y - y_0)F_y(x_0, y_0, z_0)$$
$$+ (z - z_0)F_z(x_0, y_0, z_0) = 0$$

is an equation for the plane tangent to the level surface $F(x, y, z) = c$, at P_0.

Proof. The point $P(x, y, z)$ is on the tangent plane if and only if the vector $\mathbf{P_0P}$ is perpendicular to ∇F. Clearly, this is the case if and only if $\mathbf{P_0P} \cdot \nabla F = 0$. But when this orthogonality condition is written in component form, it gives (82). ∎

Equations (23) and (82) are both equations for a plane tangent to a surface at a point. But the equations look somewhat different. The reader should note that equation (23) applies to the surface $z = f(x, y)$. Equation (82) applies to a level surface, $F(x, y, z) = c$.

THEOREM 13 (PLE). Under the conditions of Theorem 11 the straight line normal to the surface $F(x, y, z) = c$ at P_0 has the set of equations

(83)
$$\frac{x - x_0}{F_x(x_0, y_0, z_0)} = \frac{y - y_0}{F_y(x_0, y_0, z_0)} = \frac{z - z_0}{F_z(x_0, y_0, z_0)},$$

provided that the denominators are all different from zero.

Example 3. For the surface and point given in Example 2, find equations for the tangent plane and the normal line.

Solution. For any point we have

$$\nabla F = (y^2z^3 + F_1^\star)\mathbf{i} + (2xyz^3 + F_2^\star)\mathbf{j} + (3xy^2z^2 + F_3^\star)\mathbf{k}.$$

At the given point $(3, -2, 1)$, $F_1^\star = F_2^\star = F_3^\star = 0$. Hence

$$\nabla F = (-2)^2 \times 1^3\mathbf{i} + 2 \times 3 \times (-2) \times 1^3\mathbf{j} + 3 \times 3 \times (-2)^2 \times 1^2\mathbf{k}$$

$$= 4\mathbf{i} - 12\mathbf{j} + 36\mathbf{k}.$$

Setting $\mathbf{P_0P} \cdot \nabla F = 0$, where $\mathbf{P_0P} = (x - 3)\mathbf{i} + (y + 2)\mathbf{j} + (z - 1)\mathbf{k}$, we have

$$4(x - 3) - 12(y + 2) + 36(z - 1) = 0.$$

On simplifying, the equation of the tangent plane is $x - 3y + 9z = 18$. For the normal line at P_0, equation (83) gives

(84) $$\frac{x - 3}{1} = \frac{y + 2}{-3} = \frac{z - 1}{9}. \quad \bullet$$

EXERCISE 9

In Problems 1 through 8, sketch the level curves for the given function through the given points. At each of these points compute and sketch ∇F and a unit vector $\mathbf{u}$ tangent to the level curve.

1. $F(x, y) = 2x + 5y$, $(1, 1)$, $(4, 7)$.
2. $F(x, y) = x^2 - y$, $(1, 3)$, $(-1, -5)$.
3. $F(x, y) = x^2 + y^2$, $(3, -4)$, $(-5, 12)$.
4. $F(x, y) = 4x^2 + y^2$, $(1, 4)$, $(0, -3)$.
5. $F(x, y) = y + 4x - x^3$, $(1, -3)$, $(0, 5)$.
$\star$6. $F(x, y) = \cos(x + y)$, $(0, 0)$, $(\pi, 0)$.
$\star$7. $F(x, y) = y^2 - \sin x$, $(\pi, 0)$, $(\pi, -2)$.
$\star$8. $F(x, y) = x^2 - y^2$, $(0, 0)$, $(4, 0)$.

In Problems 9 through 12, find an equation for the plane tangent to the given surface at the given point.

9. $x^2 + y^2 + z^2 = 30$, $(2, 1, -5)$.
10. $x^3 + y^3 + z^3 = 6xyz$, $(1, 2, 3)$.
11. $x \sin \pi y + ze^{x^2} = e - y \tan \pi z$, $(1, 1, 1)$.
12. $\ln(x + y) + x \cos z + \text{Tan}^{-1}(y + z) = 1$, $(1, 0, 0)$.

13. Prove that the plane tangent to the quadric surface $ax^2 + by^2 + cz^2 = r^2$ at (x_0, y_0, z_0) has the equation $ax_0x + by_0y + cz_0z = r^2$.

14. Prove that the parametric equations for a line normal to the surface $F(x, y, z) = c$ at P_0 are

$$x = x_0 + tF_x, \qquad y = y_0 + tF_y, \qquad z = z_0 + tF_z,$$

where F_x, F_y, and F_z are computed at P_0. From this, prove Theorem 13.

In Problems 15 and 16, find symmetric equations for the line normal to the given surface at the given point.

15. $xy^3 + yz^3 + zx^3 = 5$, $(1, 2, -1)$.

16. $\dfrac{x}{yz} + \dfrac{4z}{xy} = xz - 2$, $(2, 1, -1)$.

*17. Give an example of a function $F(x, y, z)$ together with a particular constant such that $F_x \neq 0$, $F_y \neq 0$, and $F_z \neq 0$, for any point, and yet the set $F(x, y, z) = c$ is not a surface (as we usually understand the term).

18. Show that equation (23) of Theorem 2 can be obtained as a special case of equation (82) of Theorem 12.

19. If two surfaces $F(x, y, z) = c_1$ and $G(x, y, z) = c_2$ intersect in a curve, the tangent to the curve at a point P_0 must lie in each of the planes tangent to the surface at P_0. Prove that therefore the tangent line must be parallel to $\nabla F \times \nabla G$.

20. Use the result of Problem 19 to find equations for the tangent line to the curve of intersection of the two surfaces $x^2 + 2y^2 = 3z^2$ and $x + 3y + 4z = 8$ at the point $(1, 1, 1)$.

21. Repeat Problem 20 for the two surfaces $y = 1 + \sin \pi xz$ and $z = 2 + \cos \pi xy$ at $(2, 1, 3)$.

*22. Suppose that under the conditions of Problem 19 the curve of intersection has an equation of the form $y = f(x)$, $z = g(x)$. Prove that

$$\frac{dy}{dx} = \frac{F_3G_1 - F_1G_3}{F_2G_3 - F_3G_2} \qquad \text{and} \qquad \frac{dz}{dx} = \frac{F_1G_2 - G_1F_2}{F_2G_3 - F_3G_2},$$

provided, of course, that $F_2G_3 - F_3G_2 \neq 0$.

23. Apply the results of Problem 22 to find $f'(x)$ and $g'(x)$ for the curve formed by the intersection of the plane $Ax + By + Cz = D$, $D \neq 0$, with the cone $z^2 = x^2 + y^2$. Explain the restriction $D \neq 0$.

24. Repeat Problem 23 for the curve of intersection of the surfaces $x^2 + y^2 + z^2 = 14$ and $x^3 + y^3 + z^3 + 3xyz = 54$.

25. Prove that the surfaces in Problem 24 actually intersect in more than just one point, by showing that $(1, 2, 3)$ is on both surfaces and that ∇F is not parallel to ∇G at that point.

*26. Show that the two surfaces

$$x^2 + y^2 + z^2 = 18 \qquad \text{and} \qquad 72x^2 + 13y^2 + 13z^2 - 10yz = 144$$

meet in just two points $P_1(0, 3, 3)$ and $P_2(0, -3, -3)$. HINT: Consider the rotation of axes $x = X$, $\sqrt{2}y = Y + Z$, and $\sqrt{2}z = Y - Z$. Show that for these surfaces at P_1, the computations of Problem 22 yield the indeterminate forms $0/0$. Prove that at P_1, these two surfaces have the same tangent plane by showing that $\nabla G = 8\nabla F$.

★27. Prove that if at a common point P on the surfaces $F = c$, $G = c$, the tangent planes coincide, then $\nabla F = k\nabla G$ at P, and as a result the computations of Problem 22 will always yield indeterminate forms $0/0$.

★28. Show that at every point common to the two surfaces $z = x^2 + y^2$ and $z = x^2 + 2y^2$, the two tangent planes coincide. Prove that on the curve of intersection of these two surfaces, $dz/dx = 2x$ and $dz/dy = 0$. Make a sketch of these two surfaces.

Maxima and Minima of Functions of Several Variables   12

Suppose that $z = f(x, y)$ is a continuous function for some closed and bounded[1] region $\mathscr{R}$ in the xy-plane, that is, for some domain plus its boundary points. We want to find the maximum and minimum values for z as the point (x, y) varies over $\mathscr{R}$. It is convenient to think of $z = f(x, y)$ as the equation of some surface, and then the maximum value of z corresponds to the highest point on the surface and the minimum value of z corresponds to the lowest point on the surface. The situation is illustrated in Fig. 10, where $\mathscr{R}$ is a certain closed rectangular region with two sides on the axes. It is clear from the picture that for this particular function the maximum occurs at the point (a, b) in the interior of the region, while the minimum occurs at $(c, 0)$ on the boundary. To obtain some necessary conditions for a relative maximum at an interior point, let us pass the plane $x = a$ through this surface and let $\mathscr{C}_1$ be the intersection curve. If z_M is the maximum value of z on the surface it is also the maximum value of z on the curve $\mathscr{C}_1$. On this curve x is constant and y is the variable; hence from our work on plane curves it is obvious that at a maximum point we must have

(85)
$$\frac{\partial f}{\partial y} = 0 \quad \text{and} \quad \frac{\partial^2 f}{\partial y^2} \leq 0.$$

Similar considerations about $\mathscr{C}_2$, the intersection curve of the surface with the plane $y = b$, show that at a maximum point we must also have

(86)
$$\frac{\partial f}{\partial x} = 0 \quad \text{and} \quad \frac{\partial^2 f}{\partial x^2} \leq 0.$$

---

[1] A region is said to be *bounded* if there is a large circle $x^2 + y^2 = M$ such that every point of the region is inside the circle. It is *closed* if it contains all its boundary points.

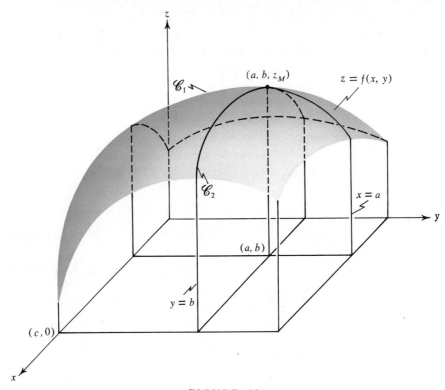

FIGURE 10

Thus (85) and (86) must be simultaneously satisfied for a maximum point. Unfortunately, these necessary conditions are not sufficient, as we shall see in the next section.

   To find a minimum point for $f(x, y)$, we may modify the above considerations with a different surface in mind, or we may apply the results already obtained to the function $z = -f(x, y)$. In either case we will find that for $f(x, y)$ to have a relative minimum value at an interior point (a, b) it is necessary that at (a, b)

$$(87) \qquad \frac{\partial f}{\partial x} = 0, \qquad \frac{\partial f}{\partial y} = 0, \qquad \frac{\partial^2 f}{\partial x^2} \geq 0, \qquad \text{and} \qquad \frac{\partial^2 f}{\partial y^2} \geq 0.$$

We summarize in

THEOREM 14. Suppose that an extreme value for the function $f(x, y)$ occurs at an interior point $P_0(a, b)$ of a region, and suppose that the first and second partial derivatives exist in a neighborhood of P_0. If P_0 is a relative minimum point, then the conditions (87) must be satisfied at P_0.

If P_0 is a relative maximum point, then at P_0,

(88) $$\frac{\partial f}{\partial x} = 0, \qquad \frac{\partial f}{\partial y} = 0, \qquad \frac{\partial^2 f}{\partial x^2} \leq 0, \qquad \text{and} \qquad \frac{\partial^2 f}{\partial y^2} \leq 0.$$

In looking for extreme values for z, the first step is to solve simultaneously the pair of equations

(89) $$\frac{\partial f}{\partial x} = 0, \qquad \frac{\partial f}{\partial y} = 0.$$

Each point at which (89) is satisfied is called a *critical point*. If the maximum or minimum value of z occurs at an interior point, these equations must be satisfied and the point must be a critical point. If it occurs on the boundary of the region under consideration, then the situation is more complicated. Thus for the function pictured in Fig. 10, the minimum of z occurs at $(c, 0)$ but neither partial derivative is zero at that point. The simultaneous solution of the equations in (89) thus furnishes the candidates for *interior* extreme points.

Example 1. Find the extreme values for $z = x^2 + xy + y^2 - 6x + 2$ inside the closed region $\mathscr{R}$ bounded by a circle of radius 6 and center at the origin.

Solution. We first find all the critical points. Solving

$$\frac{\partial f}{\partial x} = 2x + y - 6 = 0, \qquad \frac{\partial f}{\partial y} = x + 2y = 0$$

simultaneously yields $x = 4$, $y = -2$. At the point $(4, -2)$ we find $z = -10$,

$$\frac{\partial^2 f}{\partial x^2} = 2 > 0, \qquad \frac{\partial^2 f}{\partial y^2} = 2 > 0,$$

and consequently -10 appears to be a minimum value for z.

We next investigate the behavior of the function on the boundary of our region, namely, on the circle $x^2 + y^2 = 36$. Using $x = 6 \cos \theta$, $y = 6 \sin \theta$, for points on the boundary, we see that

$$z = (6 \cos \theta)^2 + (6 \cos \theta)(6 \sin \theta) + (6 \sin \theta)^2 - 36 \cos \theta + 2$$

$$= 38 + 36 \sin \theta \cos \theta - 36 \cos \theta = 38 + 36(\sin \theta \cos \theta - \cos \theta).$$

Then

$$\frac{dz}{d\theta} = 36(\cos^2\theta - \sin^2\theta + \sin\theta) = 36(1 - 2\sin^2\theta + \sin\theta)$$

$$= 36(1 - \sin\theta)(1 + 2\sin\theta).$$

Hence the relative extreme values of z on the boundary must be among the points corresponding to $\sin\theta = 1, \theta = \pi/2$, or $\sin\theta = -1/2, \theta = 7\pi/6$ or $11\pi/6$. For these values of θ we find $z = 38, 38 + 27\sqrt{3}$, and $38 - 27\sqrt{3}$, respectively. Consequently, it appears that $38 + 27\sqrt{3}$ is the maximum value of z and -10 is the minimum value of z in the given closed region.

A rigorous proof that this conclusion is correct, is long and tedious, but not difficult. A basic step is the proof of

---

THEOREM 15 (PWO). If $f(x, y)$ is a continuous function in a closed and bounded region $\mathcal{R}$, then it has a maximum value at some point of $\mathcal{R}$, and a minimum value at some (other) point of $\mathcal{R}$.

Our intuition tells us that this theorem must be true, and indeed it is. We apply Theorem 15 to our example as follows. The first computation proved that $z = -10$ is the only possible contender for the title of an extreme value for interior points of $\mathcal{R}$. The second computation gave $z = 38, 38 + 27\sqrt{3}$, and $38 - 27\sqrt{3} \approx -8.765$ as the only possible contenders on the boundary of $\mathcal{R}$. But by Theorem 15 there is a largest and smallest value for z. From these contenders it is easy to select $38 + 27\sqrt{3}$ as the largest and -10 as the smallest. ●

Example 2. Find three positive numbers whose product is as large as possible, and such that the first plus twice the second plus three times the third is 54.

Solution. Restated in symbols, we are to maximize $Q = xyz$ subject to the side conditions that $x + 2y + 3z = 54, x > 0, y > 0$, and $z > 0$. Using $x = 54 - 2y - 3z$ in Q, we find that we are to maximize

$$Q = yz(54 - 2y - 3z),$$

where the point (y, z) must lie in a certain triangular region in the yz-plane. We leave it for the student to determine this region. Searching for the critical points for the function Q, we set

$$Q_y = z(54 - 2y - 3z) - 2yz = 0,$$

$$Q_z = y(54 - 2y - 3z) - 3yz = 0.$$

Solving this pair of equations simultaneously for *positive* y and z, we find that $y = 9$, $z = 6$, and consequently $x = 54 - 2 \times 9 - 3 \times 6 = 18$. Thus it appears that the maximum for the product is $Q = 18 \times 9 \times 6 = 972$. We leave it for the student to show that at the point $y = 9$, $z = 6$,

$$Q_{yy} < 0, \qquad Q_{zz} < 0,$$

and that on the boundary of our triangular region, $Q = 0$. ●

A Sufficient Condition for a Relative Extreme **13**

Our object is to sort out from among the critical points those that are relative maximum or minimum points and those that are not. If $P_0(a, b)$ is a critical point of $f(x, y)$, then $f_x(a, b) = f_y(a, b) = 0$, and consequently the tangent plane to the surface $z = f(x, y)$ at P_0 is horizontal. If near P_0 the surface lies above or on the tangent plane, then P_0 is a *relative minimum point*. If the surface lies below or on the tangent plane, then P_0 is a *relative maximum point*. If the surface lies partly above and partly below, then P_0 is called a *saddle point*. The simplest example of a saddle point is the origin on the surface $z = y^2 - x^2$, shown in Fig. 11. Here it is clear that at $(0, 0)$

$$\frac{\partial z}{\partial x} = -2x = 0, \qquad \frac{\partial z}{\partial y} = 2y = 0.$$

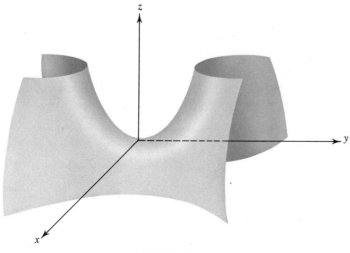

FIGURE 11

In this case it is easy to spot the origin as a saddle point by looking at the second derivatives. Indeed, at $(0, 0)$,

$$\frac{\partial^2 z}{\partial x^2} = -2 < 0, \qquad \frac{\partial^2 z}{\partial y^2} = 2 > 0,$$

so that along the x-axis, z has a local maximum, while along the y-axis z has a local minimum.

But even if both second partial derivatives have the same sign, the point may still be a saddle point. Consider, for example, the surface

(90) $$z = x^2 + y^2 - 4xy.$$

It is easy to see that at $(0, 0)$ this function gives

$$\frac{\partial z}{\partial x} = 2x - 4y = 0, \qquad \frac{\partial z}{\partial y} = 2y - 4x = 0, \qquad \frac{\partial^2 z}{\partial x^2} = 2 > 0, \qquad \frac{\partial^2 z}{\partial y^2} = 2 > 0,$$

and hence the conditions (87) are satisfied. Along the x- and y-axes we do have $z \geq 0$ so that at first glance $z = 0$ seems to be a relative minimum value for z. But for points along the line $y = x$, we find that

$$z = x^2 + x^2 - 4x^2 = -2x^2 < 0, \qquad \text{when} \qquad x \neq 0.$$

Consequently, $z = 0$ is not the smallest value for z in any neighborhood of the origin. Hence for the surface defined by (90), the origin is a saddle point. A method for classifying the type of a critical point is given in

---

THEOREM 16 (PWO). Let $P_0(a, b)$ be a critical point of $z = f(x, y)$ and suppose that f_{xx}, f_{xy}, and f_{yy} are continuous in some neighborhood $\mathcal{N}(P_0)$. For simplicity let

(91) $$A \equiv f_{xx}(a, b), \qquad B \equiv f_{xy}(a, b), \qquad C \equiv f_{yy}(a, b).$$

(I) If $AC - B^2 > 0$, and $A > 0$, then P_0 is a relative minimum point for f.
(II) If $AC - B^2 > 0$, and $A < 0$, then P_0 is a relative maximum point for f.
(III) If $AC - B^2 < 0$, then P_0 is a saddle point for f.

If $AC - B^2 = 0$, then the nature of P_0 requires a more elaborate investigation, but in most practical problems Theorem 16 will supply the needed information.

Example 1. Use Theorem 16 to determine the nature of the critical point for the function $f(x, y) = x^2 + y^2 - 4xy$.

Solution. Solving

$$f_x = 2x - 4y = 0 \quad \text{and} \quad f_y = 2y - 4x = 0$$

simultaneously gives $(0, 0)$ as the only critical point. Now $A = f_{xx} = 2$, $B = f_{xy} = -4$, and $C = f_{yy} = 2$; consequently, $AC - B^2 = 4 - 16 < 0$. So $(0, 0)$ is a saddle point. ●

Example 2. Find all the relative minimum and relative maximum points of $z = x^3 + y^3 - 9xy$.

Solution. Setting $f(x, y) = x^3 + y^3 - 9xy$, we solve simultaneously

$$f_x = 3x^2 - 9y = 0 \quad \text{and} \quad f_y = 3y^2 - 9x = 0$$

to find the critical points. These equations can be written in the equivalent form

$$y = \frac{1}{3} x^2 \quad \text{and} \quad x = \frac{1}{3} y^2,$$

so that on substituting for x in the first equation, we obtain

$$y = \frac{1}{3} \left(\frac{1}{3} y^2 \right)^2 = \frac{1}{27} y^4.$$

Hence $27y - y^4 = 0$ or $y(27 - y^3) = 0$, which gives the values $y = 0$ and $y = 3$. Thus there are two critical points: $(0, 0)$ and $(3, 3)$. Now

$$f_{xx} = 6x, \quad f_{xy} = -9, \quad f_{yy} = 6y.$$

At $(0, 0)$ we have $A = 0$, $B = -9$, $C = 0$; consequently, $AC - B^2 < 0$ and $(0, 0)$ is a saddle point. At $(3, 3)$ we have $A = 18$, $B = -9$, and $C = 18$. Consequently, $A > 0$ and $AC - B^2 = 18(18) - 81 = 243 > 0$ and so by Theorem 16 the point $(3, 3)$ is a relative minimum. ●

EXERCISE 10

In Problems 1 through 6, find the maximum and minimum values for the given function in the given closed region.

1. $z = 8x^2 + 4y^2 + 4y + 5$, $x^2 + y^2 \leq 1$.

2. $z = 8x^2 - 24xy + y^2$, $x^2 + y^2 \leq 25$.
   HINT: Observe that if $\tan 2\theta = -24/7$, then $\tan \theta = 4/3$ or $-3/4$.

3. $z = 8x^2 - 24xy + y^2$, $x^2 + y^2 \leq r^2$, r fixed.

4. $z = 6x^2 + y^3 + 6y^2$, $x^2 + y^2 \leq 25$.

⋆5. $z = x^3 - 6xy + y^3,$ $-8 \le x \le 8,$ $-8 \le y \le 8.$

⋆6. $z = \dfrac{2x + 2y + 1}{x^2 + y^2 + 1},$ $x^2 + y^2 \le 4.$

7. Show that on the circle $x^2 + y^2 = r^2$ the maximum and minimum values of z for the function of Problem 6 both tend to zero as $r \to \infty$.

8. Find the maximum value of $\sin x + \sin y + \sin (x + y)$.

9. Find the maximum value of xyz when x, y, and z are positive numbers such that $x + 3y + 4z = 108$.

10. Find the point on the surface $xyz = 25$ in the first octant that makes $Q = 3x + 5y + 9z$ a minimum.

11. Find the box of maximum volume that has three faces in the coordinate planes and one vertex in the plane $ax + by + cz = d$, where a, b, c, and d are positive constants. Observe that Problem 9 is a special case of this problem.

12. Show that the cube is the largest box that can be placed inside a sphere.

13. Find the volume of the largest box with sides parallel to the coordinate planes that can be inscribed in the ellipsoid

$$\frac{x^2}{a^2} + \frac{y^2}{b^2} + \frac{z^2}{c^2} = 1.$$

14. Use the methods of this section to find the distance from the origin to the given planes:
    (a) $x - y + z = 7.$ (b) $3x + 2y - z + 10 = 0.$

15. Use the methods of this section to find the distance between the two given lines:

    (a) $\begin{cases} x = 1 - t \\ y = t \\ z = t \end{cases}$ $\begin{cases} x = -1 + u \\ y = -u \\ z = u. \end{cases}$ (b) $\begin{cases} x = -2 + 4t \\ y = 3 + t \\ z = -1 + 5t \end{cases}$ $\begin{cases} x = -1 - 2t \\ y = 3t \\ z = 3 + t. \end{cases}$

    (c) $\begin{cases} x = 2 + 4t \\ y = 4 + t \\ z = 4 + 5t \end{cases}$ $\begin{cases} x = 5 + 2t \\ y = -4 - 3t \\ z = -1 - t. \end{cases}$ ⋆(d) $\begin{cases} x = 12 + 3t \\ y = 13 + t \\ z = 15 + 8t \end{cases}$ $\begin{cases} x = 20 + 6t \\ y = -10 - 3t \\ z = -10 - 4t. \end{cases}$

16. Find the point (x, y, z) in the first octant for which $x + 2y + 3z = 24$ and the function $f(x, y, z) \equiv xyz^2$ is a maximum.

17. Find the point (x, y, z) in the first octant for which $2x + y + 5z = 40$ and the function $f(x, y, z) \equiv xy^3z^2$ is a maximum.

18. Find the maximum and minimum values of the function $Q = x^4 + y^4 + z^4$ for points on the surface of the sphere $x^2 + y^2 + z^2 = R^2$.

19. Find the points on the surface $x^2y^2z^3 = 972$ that are closest to the origin. Prove that at each such point the line normal to the surface passes through the origin.

20. What are the dimensions of a box that has a volume of 125 in.^{3} and has the least possible surface area?

21. A cage for a snake is to be built in the shape of a box with one glass face, the remaining five sides of wood, and should have a volume of 12 ft^3. If the glass costs twice as much per square foot as the wood, what are the dimensions of the cage that will minimize the total cost for materials?

The methods of finding extreme values for a function $F(x, y)$ can be extended to functions of three or more independent variables. This is illustrated in Problems 22 through 26.

22. Find the maximum value of $F = xyze^{-(2x+3y+5z)}$ for points (x, y, z) in the first octant.

23. Assuming that the function

$$F = 2x^2 + 6y^2 + 45z^2 - 4xy + 6yz + 12xz - 6y + 14$$

has a minimum, find it. Prove that this function has no maximum value.

*24. Show that the function

$$G = x^2 + 10y^2 - 27z^2 - 8xy + 6yz - 12xz - 6y + 14$$

has the same critical point as the function F defined in Problem 23. Prove that G has neither a maximum nor a minimum value.

*25. Let $P_k(x_k, y_k)$, $k = 1, 2, \ldots, n$, be n fixed points in the plane, let $P(x, y)$ be a variable point, and let s_k be the distance from P to P_k. Prove that

$$F = s_1^2 + s_2^2 + \cdots + s_n^2$$

is a minimum when P is at the center of mass of the system obtained by placing the same mass at each of the points P_k.

**26. Let $a_1, a_2, \ldots, a_n$ be n fixed positive constants. Find the maximum and minimum values of

$$F = \sum_{k=1}^{n} a_k x_k = a_1 x_1 + a_2 x_2 + \cdots + a_n x_n,$$

where $x_1, x_2, \ldots, x_n$ is any set of numbers such that $x_1^2 + x_2^2 + \cdots + x_n^2 = 1$.

*27. Theorem 16 makes no assertion if $AC - B^2 = 0$. Prove that if $F = x^4 + y^2$ and $G = x^4 - y^2$, then both F and G have a critical point at $(0, 0)$ and that for both F and G we have $AC - B^2 = 0$. Prove further that F has a minimum point at $(0, 0)$ while G has a saddle point at $(0, 0)$. Hence it is impossible to give a simple extension of Theorem 16 that will cover the case $AC - B^2 = 0$.

**28. Find the set $\mathcal{S}$ of points for which the function

$$f(x, y) = \sqrt{x^2 + y^2} + \sqrt{(x - 1)^2 + y^2}$$

is a minimum.

⋆**29.** Prove that the function $f(x, y) = x^2 y^3 (6 - x - y)$ has infinitely many saddle points and infinitely many relative minimum points.

CALCULATOR PROBLEMS

(Newton's Method, Generalized): Finding the critical points of a function $f(x, y)$ requires that we solve the simultaneous equations

$$f_x(x, y) = 0$$

$$f_y(x, y) = 0.$$

Newton's method (recall Chapter 5, Section 12) can be adapted for this purpose, namely to give a procedure for the numerical solution of two equations in the two unknowns x and y. To describe the method, we consider the more general system

$$g(x, y) = 0$$

$$h(x, y) = 0,$$

and seek approximations to a point $(x^\star, y^\star)$ for which $g(x^\star, y^\star) = 0$, $h(x^\star, y^\star) = 0$. Here is the procedure.

Step 1. Make a guess (x_1, y_1) at the solution.

Step 2. Replace each of the equations in the system by the corresponding equations for the tangent planes to the surfaces at $(x_1, y_1, g(x_1, y_1))$ and $(x_1, y_1, h(x_1, y_1))$, respectively. This gives the equations

$$g(x_1, y_1) + g_x(x_1, y_1)(x - x_1) + g_y(x_1, y_1)(y - y_1) = 0$$

$$h(x_1, y_1) + h_x(x_1, y_1)(x - x_1) + h_y(x_1, y_1)(y - y_1) = 0,$$

which are *linear* in x and y. For convenience we set $\Delta x \equiv x - x_1$, $\Delta y \equiv y - y_1$ so that the last system becomes

$$g_x(x_1, y_1) \Delta x + g_y(x_1, y_1) \Delta y = -g(x_1, y_1)$$

$$h_x(x_1, y_1) \Delta x + h_y(x_1, y_1) \Delta y = -h(x_1, y_1).$$

Step 3. Solve the linear system for $\Delta x, \Delta y$ using elimination or Cramer's Rule. Put $x_2 = x_1 + \Delta x$, $y_2 = y_1 + \Delta y$. This gives a second approximation to the solution.

Step 4. Repeat Steps 1 to 3 using (x_2, y_2) in place of (x_1, y_1) to obtain the next approximation (x_3, y_3), and so on. With a "good" initial guess the process should converge (to the accuracy of your calculator) after just a few iterations.

For example, by applying the method to the system

$$g(x, y) = x^2 + 2y^2 - 4 = 0$$

$$h(x, y) = 4xy - 1 = 0,$$

the equations in Step 2 become

$$(2x_1) \Delta x + (4y_1) \Delta y = -(x_1^2 + 2y_1^2 - 4)$$

$$(4y_1) \Delta x + (4x_1) \Delta y = -(4x_1 y_1 - 1).$$

With the initial guess $x_1 = 2$, $y_1 = 0$, this reduces to

$$4 \Delta x = 0$$

$$8 \Delta y = 1$$

so that $\Delta x = 0$ and $\Delta y = 0.125$. Thus $x_2 = 2 + 0 = 2$ and $y_2 = 0 + 0.125 = 0.125$. This and further steps in the process are recorded in Table 2 to six significant figures.

TABLE 2

n	x_n	y_n	$g(x_n, y_n)$	$h(x_n, y_n)$
1	2	0	0	-1
2	2	0.125	0.03125	0
3	1.99213	0.125492	7.84210×10^{-5}	-1.44880×10^{-5}
4	1.99211	0.125495	2.42×10^{-7}	-6.22×10^{-7}
5	1.99211	0.125495	2.42×10^{-7}	-6.22×10^{-7}

In Problems C1 through C3, apply Newton's Method and give answers to five significant figures.

C1. Find a solution of the system

$$\begin{cases} x^2 + 2y^2 - 4 = 0 \\ \quad 4xy - 1 = 0, \end{cases}$$

starting with the guess $x_1 = 0.2$, $y_1 = 1.5$.

C2. Find a critical point of $f(x, y) = \dfrac{x^3}{3} + xy^3 - 2x - y$ near $x = 1.5$, $y = 0.5$. Determine if it is a local maximum, a local minimum, or neither.

C3. Find a critical point of $f(x, y) = \dfrac{x^4}{4} + xy^3 - x - y$ near (a) $x = 0$, $y = 1$, and (b) $x = 1$, $y = 0.5$. Determine if a local maximum, a local minimum, or neither occurs at these points.

★14 *The Regression Line*

Suppose that an experiment gives us n points $P_1(x_1, y_1)$, $P_2(x_2, y_2)$, ..., $P_n(x_n, y_n)$, and that a graph shows that they *almost* lie on a straight line. We wish to determine the constants m and b such that the straight line $y = mx + b$ lies "closest" to the points P_k given by the experiment. Before we continue the theory, we consider

Example 1. Find the straight line that lies "closest" to the points $P_1(1, 1)$, $P_2(2, 1)$, $P_3(3, 2)$, $P_4(4, 2.5)$, $P_5(5, 3.1)$.

Solution. These points are shown in Fig. 12. The reader is invited to add to Fig. 12 the straight line that seems "by eye" to be closest to these points. ●

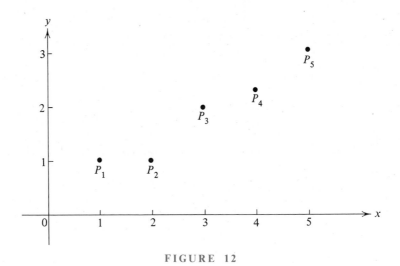

FIGURE 12

We now proceed with the theory. There are many ways to define "closest," but the theory of least squares gives a method for finding m and b that is simple, practical, and has a reasonably good theoretical foundation. According to this theory, we let $y = mx_k + b$ be the computed value of y corresponding to x_k, and we let y_k be the experimental value.

Then we determine m and b so that $\sum_{k=1}^{n} (y - y_k)^2$ is a minimum. Since we set $y = mx + b$, our task is to find m and b such that

(92)
$$f(m, b) = \sum_{k=1}^{n} (mx_k + b - y_k)^2$$

is a minimum.

DEFINITION 6 (Regression Line). For a set of points $P_1, P_2, \ldots, P_n$, determine m and b so that $f(m, b)$, given by (92), is a minimum. Then $y = mx + b$ is called the regression line for that set of points.

To find m and b, we set $f_m = 0$ and $f_b = 0$ and solve. The condition $f_m = 0$ yields

$$\sum_{k=1}^{n} 2(mx_k + b - y_k)x_k = 0,$$

or

(93)
$$\left(\sum_{k=1}^{n} x_k^2\right)m + \left(\sum_{k=1}^{n} x_k\right)b = \sum_{k=1}^{n} x_k y_k.$$

In a similar manner, the condition $f_b = 0$ yields

$$\sum_{k=1}^{n} 2(mx_k + b - y_k) = 0,$$

or

(94)
$$\left(\sum_{k=1}^{n} x_k\right)m + nb = \sum_{k=1}^{n} y_k.$$

We must solve the pair of equations (93) and (94) for m and b. The student may find it helpful to introduce the new symbols

$$A \equiv \sum_{k=1}^{n} x_k^2, \qquad B \equiv \sum_{k=1}^{n} x_k, \qquad C \equiv \sum_{k=1}^{n} x_k y_k, \qquad D \equiv \sum_{k=1}^{n} y_k.$$

Then the set (93) and (94) take on the simpler-looking form

$$Am + Bb = C$$

$$Bm + nb = D.$$

Standard methods of elimination (or determinants) lead to

$$m = \frac{nC - BD}{nA - B^2}, \qquad b = \frac{AD - BC}{nA - B^2}.$$

Restoring the meaning of A, B, C and D, we obtain

THEOREM 17. If $y = mx + b$ is the equation of the regression line, then

$$(95) \quad m = \frac{n \sum x_k y_k - \left(\sum x_k\right)\left(\sum y_k\right)}{n \sum x_k^2 - \left(\sum x_k\right)^2}, \qquad b = \frac{\sum x_k^2 \sum y_k - \sum x_k \sum x_k y_k}{n \sum x_k^2 - \left(\sum x_k\right)^2},$$

where all sums run from 1 to n.

Solution to Example 1. To compute the sums in Theorem 17, we arrange the data as in Table 3.

TABLE 3

k	x_k	y_k	$x_k y_k$	x_k^2
1	1	1	1	1
2	2	1	2	4
3	3	2	6	9
4	4	2.5	10	16
5	5	3.1	15.5	25
Sums	15	9.6	34.5	55

If we use these sums in equation set (95), we find that

$$m = \frac{5(34.5) - 15(9.6)}{5(55) - (15)^2} = \frac{28.5}{50} = 0.57,$$

$$b = \frac{55(9.6) - 15(34.5)}{5(55) - (15)^2} = \frac{10.5}{50} = 0.21.$$

The equation of the regression line for the points in Fig. 12 is $y = 0.57x + 0.21$. ●

The reader is invited to add this line to Fig. 12 and see how close it comes to the one drawn by eye.

EXERCISE 11

In Problems 1 through 4, coordinates are given for a set of points. In each case, make a graph graph showing the points and draw a straight line that seems to best fit the points. Then compute the equation for the regression line. In each case give the coefficients to the nearest three significant figures.

1.
x	0	2	3	5
y	1	2	6	6

2.
x	2	6	8	14
y	6	7	8	10

3.
x	8	10	12	14	16
y	4	6	7	8	10

4.
x	3	6	9	12	15	18
y	23	18	17	12	11	6

CALCULATOR PROBLEMS

C1. Turnover's Television Store keeps a careful record of their weekly advertising expenses and the number of television sets they sell the following week. Some of their data are given below.

x = weekly advertising expenditure	100	300	500	700
y = sales the following week	16	31	53	75

Use the regression line to predict the sales: **(a)** if they quit advertising, and **(b)** if they spend \$800 on advertising in a special week.

C2. The number of grams of copper sulfate, $CuSO_4$, that will dissolve in 100 grams of water at various temperatures is given in the table on page 794.

Use the regression line to predict the solubility of copper sulfate at **(a)** 5°C, **(b)** 40°C, and **(c)** 95°C.

Centigrade temperature	10°	30°	50°	70°
Grams of $CuSO_4$	17	25	35	47

C3. The following table gives the populations for certain states in 1970 and the motor-vehicle traffic deaths for the same year.

State	Nevada	Georgia	Florida	Illinois	New York
P: Approx. population (10^3)	489	4,590	6,790	11,100	18,200
D: Motor-vehicle deaths	260	1,800	2,170	2,350	3,100

Use the regression line to predict the number of auto deaths in: **(a)** Colorado, $P = 2,210,000$, **(b)** Michigan, $P = 8,860,000$, **(c)** California, $P = 20,000,000$, and **(d)** a desert state in which nobody is living.

C4. The following table gives R, the gross yearly income of American Telephone and Telegraph Company in billions of dollars (to the nearest hundred million), and P, the profit per share of stock to the nearest 10 cents. Using the year 1965 as the zero point, find the regression line for $R(t)$ and $P(t)$. Use the equations to predict R and P in 1990.

Year	1965	1966	1967	1968	1969	1970	1971	1972	1973	1974
R	11.1	12.1	13.0	14.1	15.7	17.0	18.4	20.9	23.5	26.2
P	3.40	3.70	3.80	3.80	4.00	4.00	3.90	4.30	5.00	5.30

★15 *Lagrange Multipliers*

We illustrate the method of Lagrange Multipliers by reworking Example 2 of Section 12. In that example we were asked to find the maximum value of the product $Q = xyz$ when x, y, and z are positive numbers such that

$$(96) \qquad x + 2y + 3z = 54.$$

An equation such as (96) which puts conditions on the variables is called a *constraint*. That is, the variables x, y, and z cannot wander freely but are constrained by the condition (96). We had no difficulty solving the problem in Section 12, but the reader must have noticed that our method was "unsymmetrical"—we selected one variable, x, at random and solved equation (96) for x.

The Lagrange Method avoids this unsymmetrical approach, and at the same time it often simplifies the manipulations. Of course, if the method is correct (and it is), it must give the same solution, but usually much more quickly. Here is the method.

Let $f(x, y, z)$ be the function to be maximized (or minimized). We write the constraint in the form $g(x, y, z) = 0$. Then, using a new variable λ (Greek lowercase letter lambda), we form a new function

$$(97) \qquad F(x, y, z, \lambda) \equiv f(x, y, z) + \lambda g(x, y, z),$$

of the four variables x, y, z, and λ. Here λ is called the *Lagrange multiplier*. We next look for the critical points of this new function $F(x, y, z, \lambda)$. Thus we solve the set of equations

$$(98) \qquad \begin{array}{ll} F_x(x, y, z, \lambda) = 0, & F_y(x, y, z, \lambda) = 0, \\ F_z(x, y, z, \lambda) = 0, & F_\lambda(x, y, z, \lambda) = 0. \end{array}$$

The last equation in this set is always the constraint equation, $g(x, y, z) = 0$. According to the method, the solution of the set (98) will lead to the extreme values of $f(x, y, z)$.

To apply the Lagrange Method in our example, we write (96) in the required form

$$g(x, y, z) = x + 2y + 3z - 54 = 0.$$

With $Q = xyz$ to be maximized, equation (97) becomes

$$(99) \qquad F(x, y, z, \lambda) = xyz + \lambda(x + 2y + 3z - 54).$$

Following the instructions to differentiate [see equation set (98)], we obtain

$$(100) \qquad F_x = yz + \lambda = 0 \qquad \text{or} \qquad yz = -\lambda.$$

$$(101) \qquad F_y = xz + 2\lambda = 0 \qquad \text{or} \qquad xz = -2\lambda.$$

$$(102) \qquad F_z = xy + 3\lambda = 0 \qquad \text{or} \qquad xy = -3\lambda.$$

$$(103) \qquad F_\lambda = x + 2y + 3z - 54 = 0.$$

We know that x, y, and z are positive, and hence $\lambda \neq 0$. Looking at the equations on the extreme right, we divide equation (100) by (101) and (102), respectively, obtaining

$$(104) \qquad \frac{y}{x} = \frac{1}{2} \qquad \text{and} \qquad \frac{z}{x} = \frac{1}{3}.$$

Using these relations, $2y = x$ and $3z = x$, in (103), we find that

$$x + x + x - 54 = 0.$$

Thus $3x = 54$, or $x = 18$. Then (104) yields $y = 9$ and $z = 6$. Consequently, the maximum value of $Q = xyz$ is $(18)(9)(6) = 972$ when x, y, z are restricted (constrained) by

$x + 2y + 3z = 54$. Although we used the Lagrange multiplier λ, we were able to solve the problem without finding λ explicitly.

We have just considered a function of three variables with one constraint, but the Lagrange Multiplier Method applies, more generally, to a function of any number of variables and with any number of constraints. The reader is cautioned to keep in mind that while the Lagrange Method gives us the critical points for the original problem, it does not tell us whether a relative maximum, relative minimum, or neither occurs at these points. Quite often, however, these matters can be decided based on the physical or geometrical context of the stated problem. For a justification of the Lagrange Multiplier technique, see Problem 4, Exercise 12.

Example 1. Find the (shortest) distance from the plane $3x + 2y + 6z = 56$ to the origin.

Solution. We are to minimize $D \equiv (x^2 + y^2 + z^2)^{1/2}$ subject to the constraint $g(x, y, z) = 56 - 3x - 2y - 6z = 0$. But the extreme occurs for both D and D^2 at exactly the same point. Hence we can minimize $x^2 + y^2 + z^2$, and this function is easier to handle. Our new method tells us to examine

$$(105) \qquad F(x, y, z, \lambda) = x^2 + y^2 + z^2 + \lambda(56 - 3x - 2y - 6z).$$

According to the method we differentiate (105) obtaining,

$$F_x = 2x - 3\lambda = 0 \qquad \text{or} \qquad x = \frac{3\lambda}{2},$$

$$F_y = 2y - 2\lambda = 0 \qquad \text{or} \qquad y = \lambda,$$

$$F_z = 2z - 6\lambda = 0 \qquad \text{or} \qquad z = 3\lambda,$$

$$(106) \qquad F_\lambda = 56 - 3x - 2y - 6z = 0.$$

Substituting these relations for x, y, z in the constraint equation (106), we find that

$$0 = 56 - 3\left(\frac{3\lambda}{2}\right) - 2\lambda - 6(3\lambda) = 56 - \frac{49}{2}\lambda,$$

so $\lambda = (56)2/49 = 16/7$. This value of λ gives $x = 3\lambda/2 = 24/7$, $y = \lambda = 16/7$, and $z = 3\lambda = 48/7$. Hence $(24/7, 16/7, 48/7)$ is the point on the given plane that is nearest to the origin, and the desired distance is given by

$$\left[\left(\frac{24}{7}\right)^2 + \left(\frac{16}{7}\right)^2 + \left(\frac{48}{7}\right)^2\right]^{1/2} = \frac{8}{7}[3^2 + 2^2 + 6^2]^{1/2} = \frac{8}{7}\sqrt{49} = 8. \quad \bullet$$

We remark that the answer to Example 1 is easily corroborated using the analytic geometry technique of Chapter 16, Section 6, page 666.

Example 2. Find the minimum of the function $w = 2x^2 + y^2 + z^2$ subject to the two conditions

(107) $$2x - y + z = 2 \quad \text{and} \quad 2x + y - z = -4.$$

Solution. Here we have *two* constraints, so we introduce the Lagrange multipliers λ_1 and λ_2, and look for the critical points of

(108) $\quad F(x, y, z, \lambda_1, \lambda_2) = 2x^2 + y^2 + z^2 + \lambda_1(2x - y + z - 2)$
$$+ \lambda_2(2x + y - z + 4).$$

Differentiating (108) with respect to each of its five variables gives

(109) $$F_x = 4x + 2\lambda_1 + 2\lambda_2 = 0, \qquad F_y = 2y - \lambda_1 + \lambda_2 = 0,$$
$$F_z = 2z + \lambda_1 - \lambda_2 = 0,$$

(110) $\quad F_{\lambda_1} = 2x - y + z - 2 = 0, \qquad F_{\lambda_2} = 2x + y - z + 4 = 0.$

Solving (109) for x, y, and z, we find that

$$x = -\frac{\lambda_1 + \lambda_2}{2}, \qquad y = \frac{\lambda_1 - \lambda_2}{2}, \qquad z = -\frac{\lambda_1 - \lambda_2}{2}.$$

Inserting these expressions in (110) leads to the equations

$$-2\lambda_1 - 2 = 0, \qquad -2\lambda_2 + 4 = 0$$

so that $\lambda_1 = -1$, $\lambda_2 = 2$. These values, in turn, yield

$$x = -\frac{1}{2}, \qquad y = -\frac{3}{2}, \qquad z = \frac{3}{2},$$

and the desired minimum value of w is

$$2\left(-\frac{1}{2}\right)^2 + \left(-\frac{3}{2}\right)^2 + \left(\frac{3}{2}\right)^2 = \frac{1}{4}(2 + 9 + 9) = 5. \quad \bullet$$

EXERCISE 12

1. Use Lagrange's Method to find the distance from the origin to each of the following planes.
   (a) $2x + y + 2z = 12$.   (b) $6x - 3y + 2z = 35$.
2. Use Lagrange's Method to prove that the distance from the origin to the plane with equation $ax + by + cz = d$ (where $d > 0$) is given by $d/\sqrt{a^2 + b^2 + c^2}$. Use this result to check your answers to Problem 1.
3. Find the volume and the dimensions of the largest box that can be made if the total

cost is $48, the bottom and sides cost $1 per square foot, and the top costs $3 per square foot.

★4. **Justification of Lagrange's Method.** Assume that $f(x, y, z)$ and $g(x, y, z)$ are differentiable and that the surface $\mathcal{S}$ given by $g(x, y, z) = 0$ has a tangent plane at each point.

(a) Show that the equations (98) can be written, using the gradient notation, as
$$\nabla f + \lambda \nabla g = \mathbf{0}, \quad g(x, y, z) = 0.$$

(b) Suppose that a relative maximum (or minimum) of $f(x, y, z)$ on $\mathcal{S}$ occurs at the point $P_0(x_0, y_0, z_0)$. Let $\mathcal{C}$ denote any smooth curve on $\mathcal{S}$ containing P_0 which is parametrized by $\mathcal{C}: x = x(s), \; y = y(s), \; z = z(s)$, where s denotes arc length. Using the Chain Rule, differentiate $f(x(s), y(s), z(s))$ with respect to s to prove that at P_0, we must have $\nabla f \cdot \mathbf{T} = 0$, where $\mathbf{T}$ is an arbitrary tangent vector to $\mathcal{S}$ at P_0.

(c) By Theorem 11, Section 11, ∇g is normal to $\mathcal{S}$ at P_0, and from part (b), so is ∇f. Argue that, at P_0, we must have $\nabla f = -\lambda \nabla g$ for some scalar λ. From part (a), this gives Lagrange's Method.

5. Find the point in the first octant that is closest to the origin and on the surface $xyz^2 = 32$.

6. Find the point in the first octant that is closest to the origin and on the surface $xy^2z^2 = 128$.

7. Find the point in the first octant that is on the surface $z = xy$ and closest to the point $(0, 0, 17)$.

In Problems 8 through 13, find the maximum value of $f(x, y, z)$ subject to the given constraint.

8. $f(x, y, z) = x + 2y + 3z, \quad 2x^2 + y^2 + 2z^2 = 36.$
9. $f(x, y, z) = 3x + 2y + 4z, \quad x^2 + 2y^2 + 8z^2 = 52.$
10. $f(x, y, z) = xyz, \qquad\qquad 2x^2 + y^2 + z^2 = 2.$
11. $f(x, y, z) = xyz, \qquad\qquad x^2 + 2y^2 + 4z^2 = 4.$
12. $f(x, y, z) = xy^2 + z^3, \qquad x^2 + y^2 + z^2 = 1.$
★13. $f(x, y, z) = x + 2y + 3z, \quad x^{1.5} + y^{1.5} + z^{1.5} = 288.$

14. Solve Example 2 by first eliminating λ_1 and λ_2 from the equations in (109) to deduce that $y = -z$, and then substituting this fact in the equations (110).

15. Find the point on the line of intersection of the two planes $6x + y - z = 2$, $3x - y - z = -4$ that is closest to the origin.

16. Find the minimum value of $w = x^2 + y^2 + 4z^2$ subject to the constraints $x + y + z = 1$ and $x + 3y - 2z = 6$.

17. Find the point on the intersection of $x + y - z = 2$ and $x^2 + 2y^2 + z^2 = 34$ that is closest to the origin.

18. Do Problem 13, Exercise 10, using Lagrange's Method.

19. Find the maximum value of $w = xyz$ subject to the constraints $x - 2y - 2z = 1$ and $x + 2y + 2z = 0$.

REVIEW PROBLEMS

In Problems 1 through 6, find the domain of the given function in accordance with our agreement.

1. $f(x, y) = \sqrt{\dfrac{x + y}{x - y}}$.

2. $g(x, y) = \text{Sin}^{-1}\dfrac{x}{y}$.

3. $F(x, y) = \ln(x^2 + y^2 - 1)$.

4. $G(x, y) = \ln(x^2 - 2xy + y^2 - 1)$.

5. $H(x, y) = \sqrt{\sin(x + y)}$.

6. $\varphi(x, y) = \ln(\sin(y - 4x^2))$.

In Problems 7 through 12, determine whether the given function has a limit as $P \to (0, 0)$.

7. $\dfrac{xy^2}{x^2 + y^2}$.

8. $\dfrac{x^3 + y^2}{x^2 + y}$.

9. $\dfrac{xy^3}{x^4 + y^4}$.

10. $\dfrac{x^2 y^3}{x^4 + y^4}$.

11. $(x + y) \sin \dfrac{1}{x^2 + y^2}$.

12. $\dfrac{\sin(x + y)}{x^2 + y^2}$.

In Problems 13 through 18, find f_x and f_y for the given function.

13. $f(x, y) = \ln(x^2 + 2y)$.

14. $f(x, y) = \sin xy^2$.

15. $f(x, y) = \tan x^2 \sec y^3$.

16. $f(x, y) = \text{sech}(x^2 + y^2)$.

17. $f(x, y) = xy^2 \cos(x + y)$.

18. $f(x, y) = \tan(\ln(1 + x^2 y^2))$.

19. A function $u(x, y, z)$ is said to be *harmonic* in a domain D if in D it satisfies *Laplace's equation*

$$\frac{\partial^2 u}{\partial x^2} + \frac{\partial^2 u}{\partial y^2} + \frac{\partial^2 u}{\partial z^2} = 0.$$

Prove that $u(x, y, z) = (x^2 + y^2 + z^2)^{-1/2}$ is a harmonic function.

20. Prove that for any set of constants a, b, c, d, e, f, and g, the function

$$u = axy + byz + czx + dx + ey + fz + g$$

is harmonic.

21. If $ax^2 + by^2 + cz^2$ is a harmonic function, what relation must the constants $a, b,$ and c satisfy? Is this necessary condition also a sufficient condition?

22. Prove that if u and v are harmonic functions of three-variables, then $au + bv$ is also a harmonic function.

23. Under the conditions of Problem 22, is it true that the product uv is always harmonic?

24. The equation for the flow of heat in a homogeneous wire (one-dimensional heat equation) is

$$\frac{\partial u}{\partial t} = k^2 \frac{\partial^2 u}{\partial x^2}.$$

Let $f(x)$ and $g(t)$ be arbitrary solutions of the equations

$$\frac{d^2 f}{dx^2} + \lambda f = 0 \qquad \text{and} \qquad \frac{dg}{dt} + k^2 \lambda g = 0,$$

respectively. Prove that $u(x, t) \equiv Cf(x)g(t)$ is a solution of the one-dimensional heat equation.

25. Prove that if u_1 and u_2 are two solutions of the one-dimensional heat equation, then $u \equiv Au_1 + Bu_2$ is also a solution of the same equation.

26. The equation for the displacement u in a vibrating string is

$$\frac{\partial^2 u}{\partial t^2} = a^2 \frac{\partial^2 u}{\partial x^2}.$$

Let $f(x)$ and $g(t)$ be arbitrary solutions of the equations

$$\frac{d^2 f}{dx^2} + \lambda f = 0 \qquad \text{and} \qquad \frac{d^2 g}{dt^2} + a^2 \lambda g = 0,$$

respectively. Prove that $u(x, t) \equiv Cf(x)g(t)$ is a solution of the vibrating string equation.

27. Do Problem 25 for the vibrating string equation.

In Problems 28 through 34, find an equation for the plane tangent to the given surface at P_0. Find a vector equation for the line normal to the surface at P_0.

28. $z = 4y^2 - x^2$, $P_0(3, 2, 7)$.   29. $z = x^2 + 2y^2 - 5$, $P_0(1, 2, 4)$.

30. $z = xy - 3\sqrt{xy}$, $P_0(4, 1, -2)$.   31. $z = x \cos \pi y - y \cos \pi x$, $P_0(2, 2, 0)$.

32. $xy^2 + yz^3 + zx^4 = 1$, $P_0(1, 2, -1)$.   33. $\sqrt{xyz} + x^3 y^3 - z^3 = 2$, $P_0(1, 2, 2)$.

34. $y \operatorname{Sin}^{-1}(x - 2) + x \operatorname{Cos}^{-1}(z - 3) + z \operatorname{Tan}^{-1} y = 7\pi/4$, $P_0(2, 1, 3)$.

35. Each one of the measurements of a box 2 by 3 by 12 in. may be off by as much as 0.02 in. Use differentials to find an approximate value for the maximum error that may be made if the volume is computed from these data. Find the maximum error, using a calculator.

36. Do Problem 35, for the surface area of the box.

37. Do Problem 35, for the length of the longest diagonal of the box.

38. A concrete base for a statue honoring Peace is to be in the shape of an ellipse. The

amount of concrete ordered is based on the assumption that the base is 2 ft thick, the minor axis is 10 ft, and the major axis is 20 ft. Using differentials, find an approximate value for the maximum error in the computed value of the volume if the dimensions may be off by 1, 2, and 3 in., respectively. Compare with the exact value of the maximum error.

39. Use differentials to find an approximate value for $u = xy^2 - y^2z^3 + 2x^2z$ at the point $P_0(0.99, 1.01, 2.02)$. Use a calculator to find the true value of u.

40. Find an approximate value for $u = \ln(2x + y - z)$ at the point $P_0(2.01, 3.02, 5.95)$. What value does your calculator give?

In Problems 41 through 44, find the directional derivative of the given function at P_0 in the direction of **U**.

41. $xy + 2yz + 3z^2$, $P_0(3, -2, 1)$, $\mathbf{U} = 3\mathbf{i} + 12\mathbf{j} - 4\mathbf{k}$.
42. $\ln(x + 3y + 2z)$, $P_0(-3, 2, 1)$, $\mathbf{U} = 6\mathbf{i} + 2\mathbf{j} + 9\mathbf{k}$.
43. $xy^2 + 4x^2z - y^4$, $P_0(-1, \sqrt{3}, 1)$, $\mathbf{U} = 2\mathbf{i} - \sqrt{3}\mathbf{j} + 3\mathbf{k}$.
44. $xyze^{x+2y-4z}$, $P_0(-1, 5, 3)$, $\mathbf{U} = \sqrt{7}\mathbf{i} + 5\mathbf{j} + 7\mathbf{k}$.

45. For the function $xy + 2yz + 3z^2$, prove that $\nabla f = \mathbf{0}$ only at the origin. Find the set of points for which ∇f is parallel to the vector $\mathbf{i} + \mathbf{j} + \mathbf{k}$.

46. For the function $\ln(x + 3y + 2z)$ prove that wherever the function is defined, ∇f always has the same direction. Explain why (without computations) we should expect this.

47. For the function $xy^2 + 4x^2z - y^4$ find the set of points for which: (a) $\nabla f = \mathbf{0}$, (b) ∇f is parallel to the xz-plane, and (c) ∇f is parallel to the yz-plane.

48. Find the set of points for which $\nabla xyze^{x+2y-4z} = \mathbf{0}$.

49. Prove that if $F(x, y, z) = C$ defines each variable as a differentiable function of the other two, then

$$\frac{\partial y}{\partial x} \frac{\partial z}{\partial y} \frac{\partial x}{\partial z} = -1.$$

50. Suppose that $F(x, y, z, w) = C$ defines each variable as a differentiable function of the other three. What can you say about the product

$$\frac{\partial y}{\partial x} \frac{\partial z}{\partial y} \frac{\partial u}{\partial z} \frac{\partial x}{\partial u}?$$

Check your assertion by considering $ax + by + cz + du = 1$.

In Problems 51 through 56, let a and b be constants and let $f(u)$ be a differentiable function of u. Prove that the given function is a solution of the given partial differential equation.

51. $z = f(ax + by)$, $b\dfrac{\partial z}{\partial x} = a\dfrac{\partial z}{\partial y}$.

52. $z = f(ax^2 + by^2)$, $by \dfrac{\partial z}{\partial x} = ax \dfrac{\partial z}{\partial y}$.

53. $z = f(xy)$, $x \dfrac{\partial z}{\partial x} = y \dfrac{\partial z}{\partial y}$.

54. $z = f(x^a y^b)$, $bx \dfrac{\partial z}{\partial x} = ay \dfrac{\partial z}{\partial y}$.

55. $z = f\left(\dfrac{x + y}{x - y}\right)$, $x \dfrac{\partial z}{\partial x} + y \dfrac{\partial z}{\partial y} = 0$.

56. $z = f\left(\dfrac{y}{x} + \dfrac{x}{y}\right)$, $x \dfrac{\partial z}{\partial x} + y \dfrac{\partial z}{\partial y} = 0$.

In Problems 57 through 64, locate all the critical points of the given function. Find each relative maximum value, each relative minimum value, and each saddle point.

57. $f(x, y) = x^2 + 6xy + 2y^2 + 16x + 6y$.

58. $f(x, y) = x^2 + 5xy - y^2 - 19x + 25y$.

59. $f(x, y) = 6xy - 2x^2 - 5y^2 + 10x - 18y$.

60. $f(x, y) = x^2 y - 2xy^2 + 10x$.

61. $f(x, y) = 8x^3 - 24xy + y^3$.

62. $f(x, y) = \dfrac{16}{x} + \dfrac{32}{y} + xy$.

*63. $f(x, y) = (x^2 - 4x) \cos y$.

*64. $f(x, y) = xye^{-x-y}$.

*65. Prove that $AC - B^2 = 0$ at the critical point of $(3x - y)^4 + x^2 - 8x$. Then prove directly that the point is an absolute minimum for the function.

*66. Let $Ax + By + Cz = D$ be the equation of a plane, and set $R = \sqrt{A^2 + B^2 + C^2}$. Use the methods of this chapter to prove that the distance from the origin to the plane is $|D|/R$. Prove that the point on the plane closest to the origin has coordinates $(AD/R^2, BD/R^2, CD/R^2)$.

67. Find the point on the plane $2x - 3y + 6z = 17$ that is closest to the origin.

68. Find the distance between the two lines

$$\mathbf{R}_1 = (1 + t)\mathbf{i} + (2 + t)\mathbf{j} + (3 + 2t)\mathbf{k} \quad \text{and} \quad \mathbf{R}_2 = (3 + 2u)\mathbf{i} + u\mathbf{j} + (1 - u)\mathbf{k}.$$

69. Find the distance from the origin to the surface $z = xy + 2$.

70. Find the distance from the origin to the surface $xyz^2 = 32$.

71. Find the regression line for the points $(1, 1)$, $(3, 2)$, $(4, 6)$, and $(6, 6)$. Make a graph showing the points and the regression line.

72. Repeat Problem 71 for the points $(-2, 1)$, $(-1, 0)$, $(0, 0)$, $(1, 0)$, and $(2, 2)$.

**73. Extend the theory of least squares to find the parabola $y = Ax^2 + Bx + C$ that is "closest" to n given points. In other words, find three equations in A, B, and C that must be satisfied if the quantity

$$f(A, B, C) \equiv \sum_{k=1}^{n} (Ax_k^2 + Bx_k + C - y_k)^2$$

is to be a minimum.

74. Use the formulas developed in Problem 73 to find the parabola that best fits the data given in Problem 72. Plot the points and sketch the graph of the parabola.

19

Multiple Integrals

1 *Objective*

In the preceding chapter we studied the differential calculus for functions of several variables. We now begin the integral calculus for several variables. In this theory two types of integrals appear: the iterated integral and the multiple integral. These mathematical cousins resemble each other, and, to add to the confusion, they always give the same number when the two integrals are computed for the same continuous function over the same closed and bounded region. However, the iterated integral and the multiple integral are conceptually quite different.

2 *Regions Described by Inequalities*

One simple and convenient way to describe a region in the plane is to give an inequality or a system of inequalities which the coordinates of P must satisfy in order that P belong to the region. Of course, if the region is extremely complicated, then we may expect the system of inequalities to be correspondingly complicated. However, in most practical cases the inequalities will be rather elementary. Sometimes a region can be described by two different sets of inequalities which at first glance appear to have no connection. This is illustrated in

Example 1. Sketch the region in the plane described by the inequalities

$$(1) \qquad\qquad 0 < x < 4, \qquad 0 < y < \sqrt{x}.$$

804

Give a second set of inequalities that describe the same region.

Solution. We notice that the upper limit on y depends on x. Consequently, we sketch the two curves $y = 0$ and $y = \sqrt{x}$, and observe that (1) states that a point in the region must lie above the curve $y = 0$, and below the curve $y = \sqrt{x}$. Further, the point must lie between the vertical lines $x = 0$ and $x = 4$. This gives the shaded region shown in Fig. 1.

An alternative set of inequalities is obtained by allowing $y = y_0$ to be any number in the interval $0 < y < 2$, and then finding restrictions on x. A glance at Fig. 1 shows that we must have $y_0^2 < x < 4$. Consequently, the set

(2) $$0 < y < 2, \qquad y^2 < x < 4$$

describes the same region that (1) does.   ●

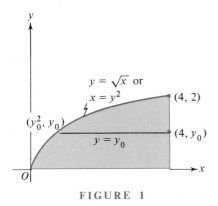

FIGURE 1   FIGURE 2

Example 2. Repeat the instructions of Example 1 for the region defined by

(3) $$0 < x < 3, \qquad 0 < y < 3 + 2x - x^2.$$

Solution. A sketch of the curve $y = 3 + 2x - x^2$ shows that the inequalities (3) describe the shaded region of Fig. 2.

To obtain a set of inequalities in which the bounds on x are given as functions of y, we solve the equation $y = 3 + 2x - x^2$ for x, obtaining

$$x = 1 \pm \sqrt{4 - y}.$$

Here the plus sign gives x for the right-hand branch of the parabola (i.e., $x > 1$) and the minus sign gives x for the left branch of the parabola. Then, if y is given, the bounds on x depend on whether $y \leq 3$, or $y \geq 3$. The two cases yield

(4) $$0 < y \leq 3, \qquad 0 < x < 1 + \sqrt{4 - y},$$

(5) $$3 \leq y < 4, \qquad 1 - \sqrt{4 - y} < x < 1 + \sqrt{4 - y}.$$

From the figure it is clear that if the coordinates of $P(x, y)$ satisfy (3), they satsify either (4) or (5). Conversely, if they satisfy either (4) or (5), then they satisfy (3). Hence (3) and the system (4) and (5) describe the same region. ●

The same techniques can be used for three-dimensional regions.

Example 3. Sketch the region $\mathscr{R}$ described by the inequalities

(6) $$0 < y < 4, \qquad 0 < x < y, \qquad \sqrt{y} < z < 2\sqrt{y}.$$

Give a second set of inequalities that describe $\mathscr{R}$.

Solution. The first inequality in (6) tells us that $\mathscr{R}$ lies between the xz-plane and the plane $y = 4$. The middle inequality tells us that $\mathscr{R}$ lies in a triangular-shaped cylinder generated by a triangle in the xy-plane. The last inequality in (6) describes the upper and lower boundary surface for $\mathscr{R}$. The region is shown in Fig. 3.

It is easy to see that the same region is described by the inequalities

(7) $$0 < x < 4, \qquad x < y < 4, \qquad \sqrt{y} < z < 2\sqrt{y}.$$

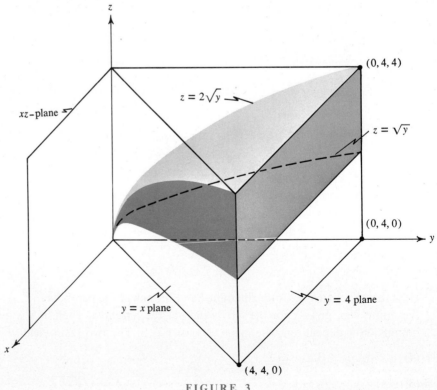

FIGURE 3

We can obtain a third set of inequalities if we start with any z such that $0 < z < 4$. Here we must provide for two cases:

$$(8) \qquad\qquad 0 < z \leqq 2, \qquad 0 < x < y, \qquad \frac{z^2}{4} < y < z^2$$

and

$$(9) \qquad\qquad 2 < z < 4, \qquad 0 < x < y, \qquad \frac{z^2}{4} < y < 4.$$

Thus P is in $\mathscr{R}$ if and only if (x, y, z) satisfies either the set (8) or the set (9).   ●

Each of the sets in Examples 1, 2, and 3 is an *open set*, by which we mean that none of its boundary points belong to the set. If we add *all* the boundary points to an open set, we obtain a *closed region*. To describe the closed regions corresponding to the sets in Examples 1, 2, and 3, we merely include the possibility of equality in the inequalities that describe the region. For example, the set of points (x, y, z) for which

$$0 \leqq y \leqq 4, \qquad 0 \leqq x \leqq y, \qquad \sqrt{y} \leqq z \leqq 2\sqrt{y}$$

is just the closure of the region $\mathscr{R}$ of Example 3.

The regions sketched in Figs. 1, 2, and 3 have a common character that facilitates their descriptions by inequalities. This character is described accurately in Definition 2. But first we need

---

DEFINITION 1 (Convex in the Direction of a Line). A region $\mathscr{R}$ is said to be *convex in the direction of a line* $\mathscr{L}^\star$ if for every line $\mathscr{L}$ with the direction of $\mathscr{L}^\star$, we have $\mathscr{L} \cap \mathscr{R}$ is either a line segment, or a point, or the empty set.

The region in the plane between two concentric circles is not convex in any direction. The regions shown in Figs. 1 and 2 are both convex in the direction of the x-axis and in the direction of the y-axis. They are also convex in certain other directions, but this does not concern us now. The region shown in Fig. 3 is convex in the direction of each of the three coordinate axes. It is this directional convexity that makes these regions easy to describe by inequalities. For convenience we introduce

---

DEFINITION 2 (Normal Region). A closed and bounded region that is convex in the direction of each of the coordinate axes is called a *normal region*.

The closed region shown in Fig. 4 is obviously not convex in any direction. However, such a region can be represented as the union of a finite number of normal regions. Consequently, we concentrate our attention on normal regions with the understanding that we can extend our results to the union of a finite number of normal regions if the need should ever arise.

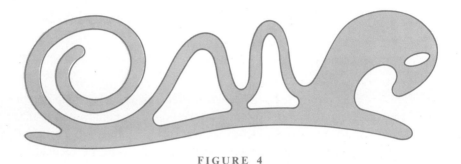

FIGURE 4

EXERCISE 1

In Problems 1 through 12, sketch the region in the plane determined by the given inequalities. In each case give an alternative set of inequalities that describes the same region.

1. $0 < x < 2,$ $0 < y < 4 - x^2.$
2. $-2 < x < 2,$ $0 < y < 4 - x^2.$
3. $0 < x < 4,$ $0 < y < e^x - 1.$
4. $\pi/6 < x < \pi/2,$ $1/2 < y < \sin x.$
5. $0 < x < \ln 5,$ $2/5 < y < 2e^{-x}.$
6. $-1 < x < 1,$ $-3 < y < x^4.$
7. $16x^2 + 9y^2 < 144.$
8. $4y > x^2.$
*9. $2x + 8 > 4y > x^2.$
*10. $x^4 < y < 20 - x^2.$
*11. $x^4 < y < 12 + x^2.$
*12. $x^2 + y^2 < 25,$ $x^2 + y^2 < 20x - 35.$

13. Let $\mathcal{R}$ be the closed region between the two concentric circles of radius 1 and 2 with center at the origin. Thus $\mathcal{R}$ consists of the points (x, y) for which $1 \leqq x^2 + y^2 \leqq 4$. Show that $\mathcal{R}$ is not a normal region.

In Problems 14 through 17, a three-dimensional region is described by a set of inequalities. In each case sketch the given region and try to determine at least one alternative set of inequalities that describes the same region.

14. $0 < x,$ $0 < y,$ $0 < z,$ $6x + 3y + 2z < 6.$
15. $0 < x < 4,$ $0 < y < x/2,$ $y < z < x - y.$
16. $0 < x < 1,$ $0 < y < x,$ $0 < z < xy.$
17. $0 < y,$ $4 - 2x < z < 4 - x^2 - 4y^2.$

*18. A region in the first octant is bounded above by the surface $z = y(1 - x)$, below by the surface $z = y^2(1 + x)$, and in back by the yz-plane. Describe this region by two different sets of inequalities.

Iterated Integrals 3

The expression

(10)
$$I = \int_a^b \int_{y_1(x)}^{y_2(x)} f(x, y) \, dy \, dx$$

is called an *iterated integral* or a *repeated integral*. By definition the symbol on the right side of (10) means that we are to integrate first with respect to y, regarding x as a constant. The result is a function of x which we call $g(x)$. Thus

(11)
$$g(x) \equiv \int_{y_1(x)}^{y_2(x)} f(x, y) \, dy,$$

where x is held constant, during the integration. We then follow by integrating $g(x)$, obtaining

(12)
$$I = \int_a^b g(x) \, dx.$$

So by definition, the computation of I in (10) is done in the two steps, indicated by equations (11) and (12). Alternative forms for writing (10) are

$$\int_a^b \left[\int_{y_1(x)}^{y_2(x)} f(x, y) \, dy \right] dx \qquad \text{and} \qquad \int_a^b dx \int_{y_1(x)}^{y_2(x)} f(x, y) \, dy.$$

Naturally, the functions $y_1(x)$ and $y_2(x)$ are functions of x that in some cases may be constant.

If in the iterated integral we wish to integrate first with respect to x and then with respect to y, we would write

(13)
$$J = \int_c^d \int_{x_1(y)}^{x_2(y)} f(x, y) \, dx \, dy$$

with suitable changes in (11) and (12). In Section 6 we will see that these iterated integrals have a very nice interpretation as the volume of a certain solid.

Example 1. Compute the two iterated integrals

$$I = \int_0^2 \int_0^{4-x^2} (4 - x^2 - y)\, dy\, dx, \qquad J = \int_0^4 \int_0^{\sqrt{4-y}} (4 - x^2 - y)\, dx\, dy.$$

Solution. To compute I, we have for the first integral

$$g(x) = \int_0^{4-x^2} (4 - x^2 - y)\, dy = \left((4 - x^2)y - \frac{y^2}{2} \right)\Big|_{y=0}^{y=4-x^2} = \frac{(4 - x^2)^2}{2}.$$

$$I = \int_0^2 g(x)\, dx = \int_0^2 \left(8 - 4x^2 + \frac{x^4}{2} \right) dx = \left(8x - \frac{4}{3}x^3 + \frac{x^5}{10} \right)\Big|_0^2 = \frac{128}{15}.$$

Similarly for J, the first integration gives

$$\int_0^{\sqrt{4-y}} (4 - x^2 - y)\, dx = \left(4x - \frac{x^3}{3} - yx \right)\Big|_{x=0}^{x=\sqrt{4-y}}$$

$$= 4\sqrt{4-y} - \frac{(4-y)\sqrt{4-y}}{3} - y\sqrt{4-y}$$

$$= \frac{2}{3}(4 - y)\sqrt{4-y}.$$

Integrating this function over the interval $0 \le y \le 4$, we have

$$J = \int_0^4 \frac{2}{3}(4 - y)^{3/2}\, dy = -\frac{2}{3} \times \frac{2}{5}(4 - y)^{5/2}\Big|_0^4 = \frac{128}{15}. \quad \bullet$$

The fact that $I = 128/15 = J$ is not an accident. In Section 6 we will see that both I and J give the volume of the same figure and hence we must have $I = J$.

Any number of integrations may occur in an iterated integral, and the notation extends in an obvious way. This is illustrated in

Example 2. Compute each of the iterated integrals:

$$A \equiv \int_0^1 \int_0^{z^2} \int_0^{yz} xy^2z^3\, dx\, dy\, dz,$$

$$B \equiv \int_0^{x_{n+1}} \int_0^{x_n} \cdots \int_0^{x_3} \int_0^{x_2} x_1 x_2 \cdots x_n\, dx_1\, dx_2 \cdots dx_n.$$

Solution. $A = \dfrac{1}{2} \displaystyle\int_0^1 \int_0^{z^2} x^2 y^2 z^3 \Big|_{x=0}^{x=yz}\, dy\, dz = \dfrac{1}{2} \displaystyle\int_0^1 \int_0^{z^2} y^4 z^5\, dy\, dz$

$$= \frac{1}{10} \int_0^1 y^5 z^5 \Big|_{y=0}^{y=z^2} dz = \frac{1}{10} \int_0^1 z^{15} dz = \frac{1}{160} z^{16} \Big|_0^1 = \frac{1}{160}.$$

Of course, B is not clearly defined, but the notation seems to indicate that there are n integral signs, with the lower limit always zero and the subscripts on the upper limit decreasing in the obvious way. We assume that this is what the proposer intended. Mathematical induction seems to be indicated, so we replace B on the left by $B_n(x_{n+1})$. The first two cases give

$$B_1(x_2) = \int_0^{x_2} x_1 \, dx_1 = \frac{x_1^2}{2} \Big|_0^{x_2} = \frac{1}{2} x_2^2,$$

$$B_2(x_3) = \int_0^{x_3} B_1(x_2) x_2 \, dx_2 = \frac{1}{2} \int_0^{x_3} x_2^3 \, dx_2 = \frac{1}{2 \cdot 4} x_3^4.$$

The pattern may now be clear. We leave it for the reader to carry out the formal proof and show that

$$B_n(x_{n+1}) = \frac{x_{n+1}^{2n}}{2 \cdot 4 \cdot 6 \cdots 2n} = \frac{x_{n+1}^{2n}}{2^n n!}. \quad \bullet$$

EXERCISE 2

In Problems 1 through 9, evaluate the given iterated integral.

1. $\int_0^3 \int_1^5 dy \, dx.$ **2.** $\int_0^4 \int_2^5 xy^2 \, dy \, dx.$ **3.** $\int_2^5 \int_0^4 xy^2 \, dx \, dy.$

4. $\int_0^1 \int_0^2 (x + y^2) \, dy \, dx.$ **5.** $\int_0^1 \int_0^2 (x + y^2) \, dx \, dy.$ **6.** $\int_0^1 \int_0^x (x + y^3) \, dy \, dx.$

7. $\int_0^1 \int_0^y (x + y^3) \, dx \, dy.$ **8.** $\int_1^2 \int_0^{x^2-1} xy^3 \, dy \, dx.$ **9.** $\int_0^\pi \int_0^{\cos x} y \sin x \, dy \, dx.$

In Problems 10 through 17, evaluate the given iterated integral. Observe that in this set, the answer to problem 2n is the same as the answer to problem 2n + 1 (n = 5, 6, 7, 8).

10. $\int_0^1 \int_0^{2-2x} (5 - x - 2y) \, dy \, dx.$ **11.** $\int_0^2 \int_0^{1-y/2} (5 - x - 2y) \, dx \, dy.$

12. $\int_0^2 \int_0^{4-x^2} y \, dy \, dx.$ **13.** $\int_0^4 \int_0^{\sqrt{4-y}} y \, dx \, dy.$

14. $\int_{-2}^{2} \int_{0}^{4-x^2} (3+x)\,dy\,dx.$

15. $\int_{0}^{4} \int_{-\sqrt{4-y}}^{\sqrt{4-y}} (3+x)\,dx\,dy.$

16. $\int_{0}^{1} \int_{0}^{2-2x} (4-4x^2-y^2)\,dy\,dx.$

17. $\int_{0}^{2} \int_{0}^{1-y/2} (4-4x^2-y^2)\,dx\,dy.$

In Problems 18 through 27, compute the indicated iterated integral.

18. $\int_{0}^{2} \int_{0}^{z^3} \int_{0}^{y^2} xy^2z^3\,dx\,dy\,dz.$

19. $\int_{0}^{1} \int_{0}^{x} \int_{0}^{x+y} (x+y+z)\,dz\,dy\,dx.$

20. $\int_{0}^{1} \int_{0}^{1} \int_{0}^{1} (x+y+z)^2\,dy\,dx\,dz.$

21. $\int_{1}^{2} \int_{0}^{z^2} \int_{0}^{xz} z^2\,dy\,dx\,dz.$

22. $\int_{0}^{1} \int_{0}^{x^2} \int_{0}^{x} y^2\,dz\,dy\,dx.$

23. $\int_{0}^{1} \int_{0}^{x^2} \int_{z}^{x^2} dy\,dz\,dx.$

*24. $\int_{0}^{1} \int_{0}^{z} \int_{y-z}^{y+z} \sqrt{x+y+z}\,dx\,dy\,dz.$

*25. $\int_{0}^{1} \int_{0}^{x_n} \cdots \int_{0}^{x_3} \int_{0}^{x_2} x_1^2\, x_2^2 \cdots x_n^2\,dx_1\,dx_2 \cdots dx_n.$

*26. $\int_{0}^{1} \int_{0}^{1} \cdots \int_{0}^{1} \int_{0}^{1} (x_1 + 2x_2 + \cdots + nx_n)\,dx_1\,dx_2 \cdots dx_n.$

*27. $\int_{0}^{n} \int_{0}^{n-1} \cdots \int_{0}^{2} \int_{0}^{1} (x_1 + x_2 + \cdots + x_n)\,dx_1\,dx_2 \cdots dx_n.$

4 The Definition of a Double Integral

We recall that in Chapter 6, Section 5, the integral of $f(x)$ is defined as the limit of a certain sum. A multiple integral is the natural extension of this concept to functions of several variables. For simplicity we begin with a function $f(x, y)$ of two variables, but it will be clear that the definitions, theorems, and methods can be extended to functions of three or more variables without difficulty.

Our aim is to define the integral of $f(x, y)$ over a closed and bounded region $\mathscr{R}$ in the xy-plane (see Fig. 5), when $f(x, y)$ is continuous on $\mathscr{R}$. Now the concepts are rather simple, but if we proceed directly toward our objective, we will encounter some difficulty with the

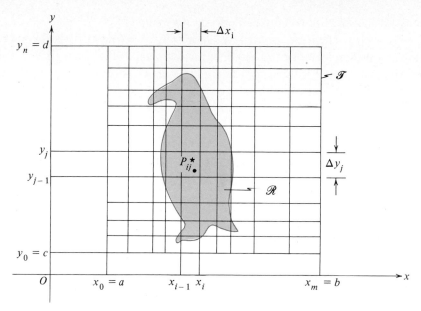

FIGURE 5

notation for sums. We can avoid this difficulty by the following device. We let $\mathcal{T}$ be a closed rectangle that contains the region $\mathcal{R}$ (see Fig. 5). We extend the definition of $f(x, y)$ to points of $\mathcal{T}$ by setting it equal to zero for every point of $\mathcal{T}$ that is not in $\mathcal{R}$ (the unshaded part of $\mathcal{T}$ in Fig. 5). If we are rather casual about matters, we may still call the new function $f(x, y)$. However, clarity demands that we give it a new name. Following this edict, we define $F(x, y)$ by

(14)
$$F(x, y) = \begin{cases} f(x, y), & \text{if } (x, y) \text{ is in } \mathcal{R}, \\ 0, & \text{if } (x, y) \text{ is in } \mathcal{T} - \mathcal{R}. \end{cases}$$

In most cases this new function $F(x, y)$ will not be continuous in $\mathcal{T}$, but if the boundary of $\mathcal{R}$ is sufficiently smooth, then the discontinuities of $F(x, y)$ on the boundary will do no harm.

Suppose now that $\mathcal{T}$ is the closed rectangle: $a \leq x \leq b$, $c \leq y \leq d$. Let $a = x_0 < x_1 < \cdots < x_m = b$ be a partition of the interval $a \leq x \leq b$, and let $c = y_0 < y_1 < \cdots < y_n = d$ be a partition of the interval $c \leq y \leq d$. Then (as indicated in Fig. 5) the vertical lines $x = x_i$, $i = 0, 1, 2, \ldots, m$, and the horizontal lines $y = y_j$, $j = 0, 1, 2, \ldots, n$, divide the rectangle $\mathcal{T}$ into mn rectangles[1] $\mathcal{T}_{ij}: x_{i-1} \leq x \leq x_i$,

---

[1] The double subscript ij on $\mathcal{T}_{ij}$ is not a product of i and j but is a convenient method of indicating the particular rectangle consisting of points such that $x_{i-1} \leq x \leq x_i$ and $y_{j-1} \leq y \leq y_j$. See the footnotes on pages 735 and 744.

$y_{j-1} \leqq y \leqq y_j$. The set $\mathscr{P}$ of rectangles $\mathscr{T}_{ij}$ forms a *partition* of $\mathscr{T}$. We let $\Delta x_i = x_i - x_{i-1}$ and $\Delta y_j = y_j - y_{j-1}$. Then $\Delta A_{ij} \equiv (x_i - x_{i-1})(y_j - y_{j-1}) = (\Delta x_i)(\Delta y_j)$ is the area of the rectangle $\mathscr{T}_{ij}$. Further, let d_{ij} be the length of the diagonal of the rectangle $\mathscr{T}_{ij}$. The *mesh* or *norm* of the partition $\mathscr{P}$ is the largest of the numbers $d_{11}, d_{12}, \ldots, d_{mn}$, and is denoted by $\mu(\mathscr{P})$. As the reader may anticipate from the definition of the definite integral of a function of a single variable, our intention is to take the limit of a certain sum as $\mu(\mathscr{P}) \to 0$. Clearly, if $\mu(\mathscr{P}) \to 0$, then $m \to \infty$ and $n \to \infty$.

Now select a point $P_{ij}^\star(x_{ij}^\star, y_{ij}^\star)$ in the rectangle $\mathscr{T}_{ij}$ for each rectangle of the partition, and let $F(P_{ij}^\star)$ be the value of the function at the point $(x_{ij}^\star, y_{ij}^\star)$. We form the sum

$$(15) \qquad S(\mathscr{P}) = \sum_{i=1}^{m} \sum_{j=1}^{n} F(P_{ij}^\star) \, \Delta A_{ij},$$

which is merely a condensed version of

$$(16) \qquad S(\mathscr{P}) = \sum_{i=1}^{m} \sum_{j=1}^{n} F(x_{ij}^\star, y_{ij}^\star)(\Delta x_i)(\Delta y_j).$$

The double sum $\sum \sum$ in (15) means that we are to add all the mn terms of the form indicated, where i takes on all integer values from 1 to m, and j takes on all integer values from 1 to n. Thus the sum in (15) is over all the mn rectangles $\mathscr{T}_{ij}$ of the partition (see Fig. 6).

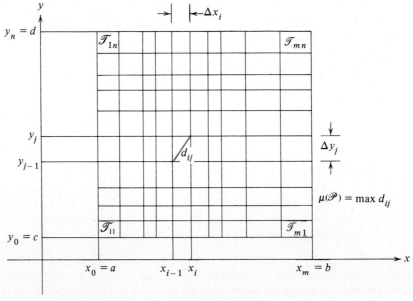

FIGURE 6

The sum $S(\mathscr{P})$ depends on the way in which $\mathscr{T}$ is partitioned. For example, a shift in any one of the lines may change the sum. Further, a change in the selection of any one of the points $P_{ij}^{\star}$ may change $S(\mathscr{P})$. Despite the wide variety of possible values for $S(\mathscr{P})$, we may expect that as we add more lines and make the rectangles progressively smaller, the sum may tend to a limit. This will certainly occur if $f(x, y)$ is continuous in $\mathscr{T}$. Whether the function is continuous or not, we make this limit the definition of the double integral and we denote this limit by the symbol

$$(17) \qquad\qquad \int\int_{\mathscr{T}} F(x, y)\, dA,$$

[read "the double integral of $F(x, y)$ over $\mathscr{T}$"].

DEFINITION 3 (Double Integral over a Rectangle). The double integral of $F(x, y)$ over $\mathscr{T}$ is defined by

$$(18) \qquad \int\int_{\mathscr{T}} F(x, y)\, dA = \lim_{\mu(\mathscr{P})\to 0} \sum_{i=1}^{m} \sum_{j=1}^{n} F(P_{ij}^{\star})\, \Delta A_{ij},$$

whenever the limit on the right side of (18) exists.[1]

Of course, the computation of the limit directly from the definition seems to involve a tremendous amount of effort, but as we will soon see, in many cases a double integral can be converted into an iterated integral, and these are relatively easy to compute.

What is the situation if we want the double integral over some closed and bounded region $\mathscr{R}$ that is not a rectangle?

DEFINITION 4 (Double Integral over a Region). Suppose that $f(x, y)$ is defined on a bounded region $\mathscr{R}$. Let $\mathscr{T}$ be a rectangle that contains $\mathscr{R}$. Let $F(x, y)$ be defined by

$$(14) \qquad F(x, y) = \begin{cases} f(x, y), & \text{if } (x, y) \text{ is in } \mathscr{R}, \\ 0 & \text{if } (x, y) \text{ is in } \mathscr{T} - \mathscr{R}. \end{cases}$$

Then

---

[1] This means that the limit is the same for every sequence of partitions with $\mu(\mathscr{P}) \to 0$, and every choice of $P_{ij}^{\star}$ in $\mathscr{T}_{ij}$.

(19)
$$\iint\limits_{\mathscr{R}} f(x, y)\, dA \equiv \int \int\limits_{\mathscr{T}} F(x, y)\, dA,$$

whenever the integral on the right side exists.

Since the area of a rectangle can be written as $\Delta A = \Delta x\, \Delta y$, this suggests that

(20)
$$\int \int\limits_{\mathscr{R}} f(x, y)\, dx\, dy$$

is a good alternative notation for the double integral on the left side of (19). In this form the double integral looks like an iterated integral, and in fact we will see, in Theorem 3, that when $f(x, y)$ is continuous in a normal region, then the double integral in (19) is always equal to a suitably chosen iterated integral. Because of this equality, students sometimes regard the double integral and the iterated integral as being the same. But they are quite different in concept, and in order to emphasize the difference, the notation (19) is frequently used rather than the notation (20).

When can we be certain that the limit in (18) does exist? Suppose first that $F(x, y)$ is continuous in $\mathscr{T}$. For each rectangle $\mathscr{T}_{ij}$ in the partition of $\mathscr{T}$, we set

(21)
$$M_{ij} = \text{maximum of } F(x, y) \text{ in } \mathscr{T}_{ij},$$

$$m_{ij} = \text{minimum of } F(x, y) \text{ in } \mathscr{T}_{ij}.$$

Then in each rectangle $\mathscr{T}_{ij}$,

$$m_{ij}\, \Delta A_{ij} \leqq F(P_{ij}^{\star})\, \Delta A_{ij} \leqq M_{ij}\, \Delta A_{ij}.$$

Consequently, for the sum $S(\mathscr{P})$ defined in equation (15) we have the two bounds

(22)
$$\sum_{i=1}^{m} \sum_{j=1}^{n} m_{ij}\, \Delta A_{ij} \leqq S(\mathscr{P}) \leqq \sum_{i=1}^{m} \sum_{j=1}^{n} M_{ij}\, \Delta A_{ij}.$$

Now let s_{mn} and S_{mn} denote the two extreme sums in (22). Then

$$S_{mn} - s_{mn} = \sum_{i=1}^{m} \sum_{j=1}^{n} M_{ij}\, \Delta A_{ij} - \sum_{i=1}^{m} \sum_{j=1}^{n} m_{ij}\, \Delta A_{ij}$$

(23)
$$S_{mn} - s_{mn} = \sum_{i=1}^{m} \sum_{j=1}^{n} (M_{ij} - m_{ij})\, \Delta A_{ij}.$$

It can be proved that if $F(x, y)$ is continuous in $\mathcal{T}$, then for each $\epsilon > 0$, there[1] is a partition of $\mathcal{T}$ such that $M_{ij} - m_{ij} < \epsilon$ for every rectangle of the partition. Consequently, we have $S_{mn} - s_{mn} < \epsilon A$, where A is the area of $\mathcal{T}$. Hence $S_{mn} - s_{mn} \to 0$ as $\mu(\mathcal{P}) \to 0$. Thus we expect

---

THEOREM 1 (PWO). If $F(x, y)$ is continuous in the rectangle $\mathcal{T}$, then the limit in (18) exists.

    Actually, we have given an outline of a proof, and the numerous details that must be supplied to convert the outline into a rigorous proof add very little to the main idea.

    Suppose now that $f(x, y)$ is continuous in a bounded region $\mathcal{R}$ and we extend the definition in the usual way to form $F(x, y)$ defined in $\mathcal{T}$ [see equation (14) in Definition 4]. In most cases this extended function $F(x, y)$ is not continuous in $\mathcal{T}$, and hence for those rectangles $\mathcal{T}_{ij}$ which contain points of the boundary of $\mathcal{R}$ it may happen that $M_{ij} - m_{ij}$ is large. Despite this, if the boundary of $\mathcal{R}$ is nice, then the effect of these terms on the sum $S(\mathcal{P})$ will be negligible. Indeed, we have

---

THEOREM 2 (PWO). Let $\mathcal{R}$ be a closed and bounded region and suppose that the boundary of $\mathcal{R}$ consists of a finite number of smooth curves. If $f(x, y)$ is continuous in $\mathcal{R}$, then the double integral

(24)
$$\int\int_{\mathcal{R}} f(x, y)\, dA \equiv \int\int_{\mathcal{T}} F(x, y)\, dA$$

exists.

EXERCISE 3

In Problems 1 through 7, we give a number of statements about multiple integrals. In each statement the regions are normal regions, and each function is continuous in the region involved. At least one of the statements is false, and the others are theorems that can be

---

[1] We have tried to keep the exposition simple, but we really need much more. It can be proved that for each $\epsilon > 0$, there is a μ_0 such that for every partition for which $\mu(\mathcal{P}) < \mu_0$ and every rectangle of the partition, we have $M_{ij} - m_{ij} < \epsilon$.

proved from the definition. Find all the false statements and in each case prove that the statement is false by a suitable example.

1. If c is a constant, then $\displaystyle\iint_R cf(x, y)\, dA = c \iint_R f(x, y)\, dA$.

2. $\displaystyle\iint_R [f(x, y) + g(x, y)]\, dA = \iint_R f(x, y)\, dA + \iint_R g(x, y)\, dA$.

3. If $f(x, y) \geqq g(x, y)$ in R, then $\displaystyle\iint_R f(x, y)\, dA \geqq \iint_R g(x, y)\, dA$.

4. If R_1 and R_2 have some boundary points in common but no interior points in common, and if R_3 is the union of R_1 and R_2, then

$$\iint_{R_1} f(x, y)\, dA + \iint_{R_2} f(x, y)\, dA = \iint_{R_3} f(x, y)\, dA.$$

5. If $f(x, y) \geqq 0$ in R_2 and if R_1 is contained in R_2, then

$$\iint_{R_1} f(x, y)\, dA \leqq \iint_{R_2} f(x, y)\, dA.$$

6. If R is the rectangle $a \leqq x \leqq b,\ c \leqq y \leqq d$, then

$$\iint_R f(x)g(y)\, dA = \left(\int_a^b f(x)\, dx \right)\left(\int_c^d g(y)\, dy \right).$$

7. $\displaystyle\iint_R f(x)g(y)\, dA = \left(\iint_R f(x)\, dA \right)\left(\iint_R g(y)\, dA \right).$

8. Let $\mathcal{T}$ be the rectangle $0 \leqq x \leqq 2,\ 0 \leqq y \leqq 1$. Let $f(x, y) = 1$ if x and y are both irrational, and let $f(x, y) = 0$ otherwise (at least one coordinate is rational). Compute the two extremes, s_{mn} and S_{mn}, in the inequality (22). Does the double integral over $\mathcal{T}$ exist for this function?

In Problems 9 through 12, the given function is defined over the rectangle $0 \leqq x \leqq 2$, $0 \leqq y \leqq 2$. In each case find a μ_0 such that for every partition of $\mathcal{T}$ with $\mu(\mathcal{P}) < \mu_0$ we have the inequality (a) $M_{ij} - m_{ij} < 1/10$ for every rectangle of $\mathcal{P}$ and (b) $M_{ij} - m_{ij} < \epsilon\ (\epsilon > 0)$ for every rectangle of $\mathcal{P}$.

⋆9. $f(x, y) \equiv 5$.

⋆10. $f(x, y) = x + y$.

⋆11. $f(x, y) = x^2 + y^2$.

⋆12. $f(x, y) = xy$.

The Volume of a Solid $\boldsymbol{5}$

The volume of a solid was treated in an intuitive manner in Chapter 7. We now consider the concept of volume a little more carefully.

Suppose that $f(x, y) \geqq 0$ for (x, y) in some closed and bounded region $\mathscr{R}$ in the xy-plane and we want to compute the volume of the solid[1] bounded above by the surface $z = f(x, y)$, below by the xy-plane, and on the sides by the cylinder generated by erecting at each point on the boundary of $\mathscr{R}$, a line parallel to the z-axis. Such a solid is shown in Fig. 7. Henceforth, for brevity we shall speak of such a solid as the solid under the surface $z = f(x, y)$ and above $\mathscr{R}$.

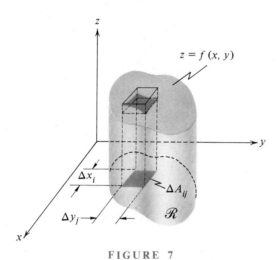

FIGURE 7

To estimate the volume of the solid, we partition a rectangle $\mathscr{T}$ that contains $\mathscr{R}$, and we consider first some rectangle $\mathscr{T}_{ij}$ of the partition that is contained entirely in $\mathscr{R}$. The product $m_{ij} \Delta A_{ij}$ is just the volume of a prism with height m_{ij} and base $\mathscr{T}_{ij}$. Similarly, $M_{ij} \Delta A_{ij}$ is the volume of a prism with the same base and height M_{ij} (see Fig. 7). Hence, (recalling (21)) if ΔV_{ij} is the volume of the solid under the surface $z = f(x, y)$ and above the rectangle $\mathscr{T}_{ij}$, then

(25) $$m_{ij} \Delta A_{ij} \leqq \Delta V_{ij} \leqq M_{ij} \Delta A_{ij}.$$

---

[1] We are really computing the volume of a three-dimensional *region* but it seems to be more natural and more intuitive to refer to the volume of a *solid*.

From the definition of m_{ij} and M_{ij} we also have

(26) $$m_{ij}\,\Delta A_{ij} \leqq f(P_{ij}^{\star})\,\Delta A_{ij} \leqq M_{ij}\,\Delta A_{ij}.$$

If $\mathcal{T}_{ij}$ lies partly outside $\mathcal{R}$ or completely outside $\mathcal{R}$, then $m_{ij} = 0$, but the inequalities (25) and (26) still hold. We next take the sum of such inequalities over all rectangles of the partition. The inequality (25) yields

(27) $$\sum_{i=1}^{m}\sum_{j=1}^{n} m_{ij}\,\Delta A_{ij} \leqq V \leqq \sum_{i=1}^{m}\sum_{j=1}^{n} M_{ij}\,\Delta A_{ij},$$

where V is the volume of the solid under the surface $z = f(x, y)$ and above the region $\mathcal{R}$. The inequality (26) yields

(28) $$\sum_{i=1}^{m}\sum_{j=1}^{n} m_{ij}\,\Delta A_{ij} \leqq \sum_{i=1}^{m}\sum_{j=1}^{n} f(P_{ij}^{\star})\,\Delta A_{ij} \leqq \sum_{i=1}^{m}\sum_{j=1}^{n} M_{ij}\,\Delta A_{ij}.$$

Suppose now that the two extremes in (27) and (28) approach the same limit as $\mu(\mathscr{P}) \to 0$. Then clearly the double integral of $f(x, y)$ over $\mathcal{R}$ exists and moreover

(29) $$V = \iint_{\mathcal{R}} f(x, y)\, dA.$$

One may expect us to formulate (29) as a theorem, and indeed on an intuitive basis we have proved that the volume of the solid is given by the double integral of $f(x, y)$ over $\mathcal{R}$. But a close look at the argument shows one slight flaw: We do not have a definition of volume on which to base a proof. Our efforts have not been wasted, because our discussion leads in a natural way to

DEFINITION 5 (Volume). If $f(x, y) \geqq 0$ in a bounded region $\mathcal{R}$ of the plane and if the double integral over $\mathcal{R}$ of $f(x, y)$ exists, then the volume of the solid under the surface $z = f(x, y)$ and above $\mathcal{R}$ is given by equation (29).

If $f(x, y) \leqq 0$ in $\mathcal{R}$, then it is clear that the double integral gives the negative of the volume of the solid that lies *above* the surface $z = f(x, y)$ and *below* the closed region $\mathcal{R}$. Finally, suppose that $f(x, y)$ changes sign in $\mathcal{R}$. Let $\mathcal{R}_2$ be the set of points at which $f(x, y) \geqq 0$, and let $\mathcal{R}_1$ be the remaining points of $\mathcal{R}$, namely, the points at which $f(x, y) < 0$. Let V_2 be the volume of the solid under the surface $z = f(x, y)$ and over $\mathcal{R}_2$,

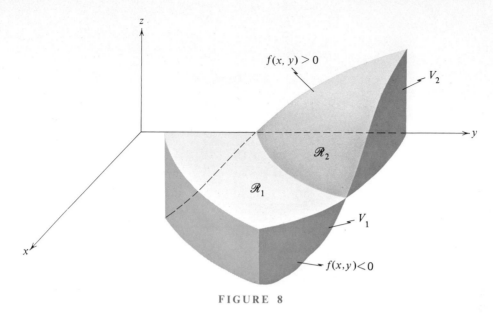

FIGURE 8

and let V_1 be the volume of the solid above that surface and below $\mathcal{R}_1$ (see Fig. 8). Then it is clear that

(30)
$$\int \int_{\mathcal{R}} f(x, y)\, dA = V_2 - V_1.$$

To compute the quantities given in equations (29) and (30), we will use the fact that these same volumes are given by an iterated integral when $f(x, y)$ is continuous and $\mathcal{R}$ is a normal region. This is covered in the next section.

Volume as an Iterated Integral 6

As in Section 5, we suppose that $f(x, y) \geqq 0$ in $\mathcal{R}$, and we are to compute V, the volume of the solid under the surface $z = f(x, y)$ and above $\mathcal{R}$. We now make one additional assumption: that $\mathcal{R}$ is convex in the direction of the y-axis. This means that the region $\mathcal{R}$ can be described by a set of inequalities of the form

(31)
$$a \leqq x \leqq b, \qquad y_1(x) \leqq y \leqq y_2(x),$$

as indicated in Fig. 9.

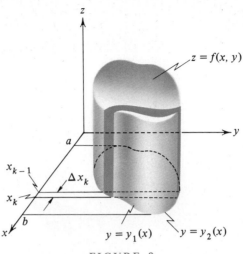

FIGURE 9

To compute V, we first cut the solid with a plane perpendicular to the x-axis at $x = x_k$. Then the area of the face of the solid cut by this plane is

$$(32) \qquad A(x_k) = \int_{y_1(x_k)}^{y_2(x_k)} f(x_k, y)\, dy$$

for each fixed x_k in the interval $a \leqq x \leqq b$ (see Fig. 9). If we use two such planes at distance Δx_k apart, then the volume ΔV_k of the slab cut out from our solid is given approximately by

$$(33) \qquad \Delta V_k \approx A(x_k)\, \Delta x_k = \int_{y_1(x_k)}^{y_2(x_k)} f(x_k, y)\, dy\, \Delta x_k.$$

Forming a sum of n such terms in the usual way, we have

$$(34) \qquad V = \sum_{k=1}^{n} \Delta V_k \approx \sum_{k=1}^{n} \left[\int_{y_1(x_k)}^{y_2(x_k)} f(x_k, y)\, dy \right] \Delta x_k.$$

If now we let $n \to \infty$ and $\mu(\mathscr{P}) \to 0$, the right side of (34) becomes the iterated integral (by definition) and hence we may expect that

$$(35) \qquad V = \int_a^b \int_{y_1(x)}^{y_2(x)} f(x, y)\, dy\, dx.$$

The proof that V is indeed given by the iterated integral in (35) is not difficult, but merely long, and hence we omit it.

In a similar way, we can first cut our solid with planes perpendicular to the y-axis. If $\mathscr{R}$ is also specified by a set of inequalities of the form $c \leq y \leq d$, $x_1(y) \leq x \leq x_2(y)$, then the volume is also given by the iterated integral

(36)
$$V = \int_c^d \int_{x_1(y)}^{x_2(y)} f(x, y)\, dx\, dy.$$

Now the two volumes given by (35) and (36) must be equal, so the two types of iterated integrals (35) and (36) will give the the same result, provided that the limits chosen determine the same closed and bounded region $\mathscr{R}$, and $f(x, y)$ is the same in both integrals.

Example 1. Find the volume of the solid under the surface $z = 4 - x^2 - y$ and in the first octant.

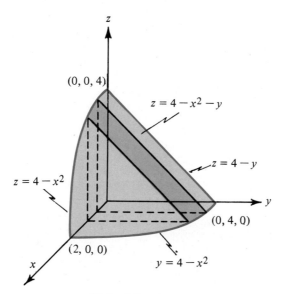

FIGURE 10

Solution. This solid is shown in Fig. 10. If we set $z = 0$, it becomes clear that the bottom of our solid is the region bounded by the x-axis, the y-axis, and the curve $y = 4 - x^2$ (with $x \geq 0$). Thus our solid is above the closed region $\mathscr{R}$, where $\mathscr{R}$ is described by the two inequalities $0 \leq x \leq 2$ and $0 \leq y \leq 4 - x^2$. Hence by equation (35),

$$V = \int_0^2 \int_0^{4-x^2} (4 - x^2 - y)\, dy\, dx.$$

The base $\mathscr{R}$ is also described by the two inequalities

$$0 \leqq y \leqq 4 \quad \text{and} \quad 0 \leqq x \leqq \sqrt{4 - y}.$$

Hence by equation (36),

$$V = \int_0^4 \int_0^{\sqrt{4-y}} (4 - x^2 - y) \, dx \, dy.$$

But these are just the integrals evaluated in Example 1 of Section 3. Hence $V = 128/15$, and we now see why the two integrals in that example gave the same number. ●

If $f(x, y) \geqq 0$ in $\mathscr{R}$, then the two iterated integrals in (35) and (36) give the volume of the solid under $z = f(x, y)$ and above $\mathscr{R}$, and hence each one is equal to the double integral of $f(x, y)$ over $\mathscr{R}$. Suppose that $f(x, y)$ changes sign in $\mathscr{R}$. Then a reconsideration of our argument shows that each of the iterated integrals gives $V_2 - V_1$, where the symbols have the meaning used in equation (30). In this way we arrive at

---

THEOREM 3 (PWO). Suppose that the normal region $\mathscr{R}$ in the plane is described by the pair of inequalities

(31) $$a \leqq x \leqq b, \quad y_1(x) \leqq y \leqq y_2(x),$$

and is also described by the pair of inequalities

(37) $$x_1(y) \leqq x \leqq x_2(y), \quad c \leqq y \leqq d.$$

If $f(x, y)$ is continuous in $\mathscr{R}$, then

(38) $$\iint_{\mathscr{R}} f(x, y) \, dA = \int_a^b \int_{y_1(x)}^{y_2(x)} f(x, y) \, dy \, dx = \int_c^d \int_{x_1(y)}^{x_2(y)} f(x, y) \, dx \, dy.$$

This is the theorem that permits us to compute double integrals by converting them to iterated integrals, and as we have seen, these are frequently easy to compute. If $\mathscr{R}$ is not a normal region but is the union of a finite number of normal regions $\mathscr{R}_i$, then we apply equation (38) to each $\mathscr{R}_i$ and add the results to obtain the double integral over $\mathscr{R}$.

Example 2. Find $I = \iint_{\mathscr{R}} (y^2 - x^2) \, dA$, where $\mathscr{R}$ is the closed rectangle

$$0 \leqq x \leqq 3, \quad 0 \leqq y \leqq 3.$$

Solution. If we recall the shape of the saddle surface $z = y^2 - x^2$ (Fig. 11, page 783), we suspect that this double integral is zero, because "the surface lies as much above the plane as it does below the plane." By Theorem 3, I is equal to an appropriate iterated integral. Thus

$$I = \int_0^3 \int_0^3 (y^2 - x^2)\, dy\, dx = \int_0^3 \left(\frac{y^3}{3} - x^2 y\right)\Big|_{y=0}^{y=3} dx$$

$$= \int_0^3 (9 - 3x^2)\, dx = (9x - x^3)\Big|_0^3 = 27 - 27 = 0,$$

as predicted. ●

Example 3. Find the volume of the solid bounded by the surface $z = y^2 - x^2$ and the planes $z = 0$, $x = 0$, $x = 3$, $y = 0$, and $y = 3$.

Solution. The plane $z = 0$ divides this solid into two pieces of volumes V_1 and V_2, respectively. By the result of Example 2, we see that $V_1 = V_2$. Hence we can compute V_2 and double the result. Now $z = y^2 - x^2 \geq 0$ in $\mathcal{R}$ if and only if $0 \leq x \leq y$. Hence

$$V = 2V_2 = 2 \int_0^3 \int_0^y (y^2 - x^2)\, dx\, dy = 2 \int_0^3 \left(y^2 x - \frac{x^3}{3}\right)\Big|_{x=0}^{x=y} dy$$

$$= 2 \int_0^3 \left(y^3 - \frac{y^3}{3}\right) dy = 2 \left(\frac{2}{3}\frac{y^4}{4}\right)\Big|_0^3 = 27. \quad ●$$

We return for a moment to Theorem 3. We arrived at this theorem by geometrical considerations: namely if the volume of a fixed solid is computed in two different ways, then the two computations must give the same number. In years past such as argument would have been accepted as a proof, but today this type of reasoning is not allowed in a rigorous proof. We may use the invariance of a physical object, such as volume, mass, moment, and so on, to suggest certain theorems, but the proofs must be purely analytic.

    *We have no intention of giving a detailed proof of Theorem 3, but we will give a brief outline of the proof of the first part. Indeed, consider the double sum

(39)
$$S(\mathscr{P}) = \sum_{i=1}^m \sum_{j=1}^n f(P_{ij}^{\star})\, \Delta A_{ij}.$$

In each rectangle $\mathscr{T}_{ij}$ we select $P_{ij}^{\star}$ to be the upper right-hand corner. Thus $P_{ij}^{\star}$ is the point (x_i, y_j) and (39) takes the form

(40)
$$S(\mathscr{P}) = \sum_{i=1}^m \sum_{j=1}^n f(x_i, y_j)(\Delta x_i)(\Delta y_j).$$

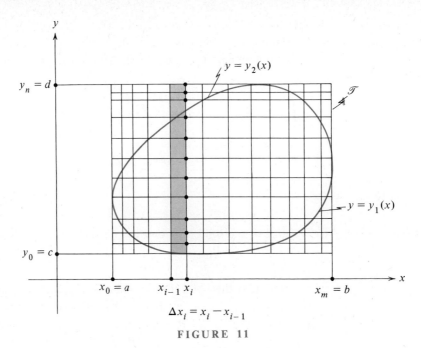

FIGURE 11

We next rearrange the terms in (40) so that we group those terms that correspond to rectangles in the same vertical column (see Fig. 11). Thus for each fixed i the terms that correspond to $\mathcal{T}_{ij}$ are

(41)
$$\left[\sum_{j=1}^{n} f(x_i, y_j)\, \Delta y_j\right] \Delta x_i,$$

where we can "factor out" Δx_i because it is the same for each term in (41). The sum $S(\mathcal{P})$ is a "sum of sums" and this "sum of sums" is indicated by writing

(42)
$$S(\mathcal{P}) = \sum_{i=1}^{m} \left[\sum_{j=1}^{n} f(x_i, y_j)\, \Delta y_j\right] \Delta x_i.$$

Now as $\mu(\mathcal{P}) \to 0$, the sum inside the bracket in (42) approaches

$$\int_{y=y_1(x_i)}^{y=y_2(x_i)} f(x_i, y)\, dy$$

because if (x_i, y_j) is outside $\mathcal{R}$, then $f(x_i, y_j) = 0$ and the corresponding term can be dropped from the sum in (42). Consequently, $S(\mathcal{P})$ differs only slightly from

$$S_m(\mathcal{P}) \equiv \sum_{i=1}^{m} \left[\int_{y=y_1(x_i)}^{y=y_2(x_i)} f(x_i, y)\, dy\right] \Delta x_i.$$

Finally, as $\mu(\mathscr{P}) \to 0$, this last sum approaches the integral of the function in the bracket, namely,

(43)
$$\lim_{\mu(\mathscr{P}) \to 0} S_m(\mathscr{P}) = \int_{x=a}^{x=b} \left[\int_{y=y_1(x)}^{y=y_2(x)} f(x, y)\, dy \right] dx.$$

From equation (43) we see that the double integral of $f(x, y)$ over $\mathscr{R}$ is the iterated integral on the right side.

EXERCISE 4

In Problems 1 through 9, sketch the solid under the given surface and above the given closed region in the xy-plane. Compute the volume of the solid. In Problems 1, 2, 3, 4, and 5, check your work by computing the volume in two different ways.

1. $z = 5 - x - 2y,$ $0 \le x \le 1,$ $0 \le y \le 2.$
2. $z = 5 - x - 2y,$ $0 \le x \le 1,$ $0 \le y \le 2(1 - x).$
3. $z = y,$ $0 \le x \le 2,$ $0 \le y \le 4 - x^2.$
4. $z = 3 + x,$ $-2 \le x \le 2,$ $0 \le y \le 4 - x^2.$
5. $z = 4 - 4x^2 - y^2,$ $0 \le x \le 1,$ $0 \le y \le 2 - 2x.$
6. $z = e^y - x,$ $0 \le x \le e^y - 1,$ $0 \le y \le 4.$
7. $z = 6,$ $\pi/6 \le x \le \pi/2,$ $1/2 \le y \le \sin x.$
8. $z = 1 - x^2,$ $0 \le y \le 1,$ $0 \le x \le y.$
9. $z = \sin(x + y),$ $0 \le y \le \pi,$ $0 \le x \le \pi - y.$

10. Find the volume of the solid bounded by the four planes $x = 0, y = 0, z = 0,$ and $\dfrac{x}{a} + \dfrac{y}{b} + \dfrac{z}{c} = 1,$ where $a, b,$ and c are positive.

11. Find the volume of the solid bounded by the planes $x = 0, x = 2, y = 0, y = 1,$ $z = 0,$ and the surface $z = \dfrac{16}{(x + 2)^2(y + 1)^2}.$

*12. The solid in the first octant bounded by the three coordinate planes and the surface $z = \dfrac{16}{(x - 2)^2(y + 1)^2}$ has infinite extent, but finite volume. Find the volume.

*13. Find the volume of the solid bounded by the planes $x = 0, y = 0, z = 0,$ $x + y = 1,$ and the surface of Problem 12.

14. Find the volume of the solid in the first octant bounded by the cylinders $y = x^2,$ $y = x^3$ and the planes $z = 0$ and $z = 1 + 3x + 2y.$

15. Find the volume of the solid bounded by the surfaces $y = e^x, y = z, z = 0, x = 0,$ and $x = 2.$

16. Find the volume of the solid lying in the first octant and bounded by the coordinate planes and the cylinders $x^2 + z^2 = 16$ and $x^2 + y^2 = 16$.

17. The surfaces $z = 1 + y^2$, $3x + 2y = 12$, and $x = 2$ divide the first octant into a number of pieces, two of which are bounded. Find the volume of the piece that contains the point $(1, 2, 3)$.

**18. Equations (35) and (36) seem to suggest that we always have

$$(44) \qquad \int_a^b \int_c^d f(x, y) \, dy \, dx = \int_c^d \int_a^b f(x, y) \, dx \, dy.$$

Give an argument to show that (44) is always true. Now consider (44) when $a = c = 0$ and $b = d = 1$ and $f(x, y) = (x - y)/(x + y)^3$. Show that the right side of (44) gives $-1/2$ and (by symmetry) the left side gives $1/2$. Hence (44) is *not* always true. What seems to be the trouble?

7 Applications of the Double Integral

The double integral

$$(45) \qquad \int\int_{\mathscr{R}} f(x, y) \, dA$$

has many useful interpretations that arise by making special selections for the function $f(x, y)$. If $f(x, y) \equiv 1$, then (45) gives the area of the closed region $\mathscr{R}$. Indeed, in the sum

$$(46) \qquad s_{mn} = \sum_{i=1}^m \sum_{j=1}^n m_{ij} \, \Delta A_{ij},$$

$m_{ij} = 1$ for each rectangle $\mathscr{T}_{ij}$ contained in $\mathscr{R}$ and $m_{ij} = 0$ otherwise. Thus s_{mn} is the total area of the rectangles contained in $\mathscr{R}$ (see Fig. 12). But as $\mu(\mathscr{P}) \to 0$, these rectangles fill out the interior of $\mathscr{R}$ and hence $\lim s_{mn} = A(\mathscr{R})$.

If $f(x, y) = \rho_2(x, y)$, the surface density of a sheet of material having the form of the closed region $\mathscr{R}$, then (45) gives the mass of the material. This is clear because the terms in the sum (46) give a close approximation to the mass of the material in those rectangles contained in $\mathscr{R}$.

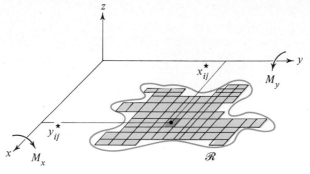

FIGURE 12

In the same way, an inspection of Fig. 12 immediately suggests the following formulas for the moments (see Chapter 17):

$$M_x = \int\int_{\mathscr{R}} y\rho_2(x,y)\,dA, \qquad M_y = \int\int_{\mathscr{R}} x\rho_2(x,y)\,dA,$$

$$I_x = \int\int_{\mathscr{R}} y^2\rho_2(x,y)\,dA, \qquad I_y = \int\int_{\mathscr{R}} x^2\rho_2(x,y)\,dA.$$

Finally, if we set $f(x,y) = r^2\rho_2(x,y)$, where $r^2 = x^2 + y^2$ is the distance of the point (x,y) from the origin, then we obtain the *polar moment of inertia* (by definition). Using J to denote this new quantity, we have

(47)
$$J = \int\int_{\mathscr{R}} (x^2 + y^2)\rho_2(x,y)\,dx\,dy.$$

In all these formulas we may set $\rho = 1$ and obtain the moments and moments of inertia of the closed region $\mathscr{R}$.

Example 1. A triangular sheet of metal is placed as shown in Fig. 13 with its vertices at the points $(0,0)$, $(1,0)$, and $(0,2)$, and in this position has surface density $\rho = (1 + x + y)$. Find its center of mass and I_x.

Solution. To find the mass, we have

$$m = \int\int_{\mathscr{R}} (1 + x + y)\,dA.$$

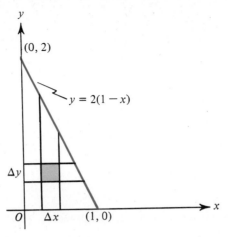

FIGURE 13

By Theorem 3 this can be computed as an iterated integral,

$$m = \int_0^1 \int_0^{2(1-x)} (1 + x + y)\,dy\,dx = \int_0^1 \left[y(1+x) + \frac{y^2}{2} \right]\Big|_0^{2(1-x)} dx$$

$$= \int_0^1 [2(1 - x^2) + 2(1 - x)^2]\,dx$$

$$= 4 \int_0^1 (1 - x)\,dx = 4\left(x - \frac{x^2}{2} \right)\Big|_0^1 = 2.$$

Similarly, an iterated integral for M_x gives

$$M_x = \int_0^1 \int_0^{2(1-x)} y(1 + x + y)\,dy\,dx = \int_0^1 \left[(1 + x)\frac{y^2}{2} + \frac{y^3}{3} \right]\Big|_0^{2(1-x)} dx$$

$$= \int_0^1 \left[2(1 + x)(1 - x)^2 + \frac{8}{3}(1 - x)^3 \right] dx$$

$$= \int_0^1 \left(\frac{14}{3} - 10x + 6x^2 - \frac{2}{3}x^3 \right) dx$$

$$= \left(\frac{14}{3}x - 5x^2 + 2x^3 - \frac{1}{6}x^4 \right)\Big|_0^1 = \frac{14}{3} - 5 + 2 - \frac{1}{6} = \frac{3}{2}.$$

Therefore, $\bar{y} = M_x/m = 3/4$. A similar computation gives $M_y = 2/3$ and hence $\bar{x} = M_y/m = 1/3$. The center of mass is at $(1/3, 3/4)$.

For the moment of inertia I_x we have

$$I_x = \int_0^1 \int_0^{2(1-x)} y^2(1 + x + y) \, dy \, dx$$

$$I_x = \int_0^1 \left[(1+x)\frac{y^3}{3} + \frac{y^4}{4} \right]\Big|_0^{2(1-x)} dx = \int_0^1 8(1-x)^3 \left[\frac{1+x}{3} + \frac{1-x}{2} \right] dx$$

$$= \int_0^1 8(1-x)^3 \frac{5-x}{6} \, dx = \int_0^1 \left[\frac{16}{3}(1-x)^3 + \frac{4}{3}(1-x)^4 \right] dx$$

$$= \left(-\frac{4}{3}(1-x)^4 - \frac{4}{15}(1-x)^5 \right)\Big|_0^1 = \frac{4}{3} + \frac{4}{15} = \frac{8}{5}. \quad \bullet$$

We leave it to the reader to reverse the order of integration and show that

$$I_x = \int_0^2 \int_0^{1-y/2} y^2(1 + x + y) \, dx \, dy = \frac{8}{5}.$$

Up to this point we have been careful to distinguish between a domain (an open connected set) and the closed region obtained by adding all its boundary points. However, it is intuitively clear that such physical quantities as area, mass, moment of inertia, and so on, will be the same, whether computed for a domain or for the closed region obtained by adjoining its boundary points, as long as the boundary consists of a finite number of smooth curves. Hence, whenever it is convenient to do so, we may drop the requirement that $\mathscr{R}$ be closed.

EXERCISE 5

In Problems 1 through 4, use double integrals to find the area of the region enclosed by the given curves.

1. $y = x$, $y = 4x - x^2$.
2. $xy = 4$, $x + y = 5$.
3. $x = y^2$, $x = 4 + 2y - y^2$.
4. $x + y = 8$, $xy - x^2 = 6$.

In Problems 5 through 7, a metal sheet has the shape of the region bounded by the given lines, and has the given surface density. In each case find the center of mass, I_x, and I_y.

5. $x = 1$, $x = 2$, $y = 0$, $y = 3$, $\rho = x + y$.
6. $x = 1$, $y = 2$, $2x + y = 2$, $\rho = 2x + 4y$.
7. $x = 0$, $y = 0$, $x + y = 1$, $\rho = 2xy$.

*8. Prove that $J = I_x + I_y$ and use this to compute J for the metal sheets of Problems 5, 6, and 7.

9. Find the centroid of the region in the first quadrant bounded by the curves $y = x^3$ and $y = 2x^2$. Prove that the centroid is a point of the region.

10. Find I_x and I_y for the region of Problem 9.

★11. Find the moment of inertia about the line $y = x$ for the region of Problem 9. HINT: Recall the formula for the distance from a point to a line.

12. Find the centroid of the region in the first quadrant bounded by the curve $y = \sin x$ and the line $y = 2x/\pi$. Show that the centroid lies in the given region.

★13. Find $\bar{y}$ for the region bounded by the two curves $y = x^4$ and $y = (Ax^2 + 1)/(A + 1)$ for $A > 0$.

★14. Prove that if $A > 2$, then the centroid of the region of Problem 13 does *not* lie in the region. Why is this possible?

8 Area of a Surface

Our objective is to find a formula for the area of that part of the surface $z = f(x, y)$ that lies over a given region $\mathcal{R}$ in the xy-plane. To obtain such a formula, we first take the simplest case in which the surface is a plane $z = ax + by$ and $\mathcal{R}$ is a rectangle of sides Δx and Δy (see Fig. 14). Let γ denote the angle between the vector $\mathbf{n}$ normal to the plane and the z-axis, and let $\Delta \sigma$ be the area of that part of the plane that lies over the rectangle. Since

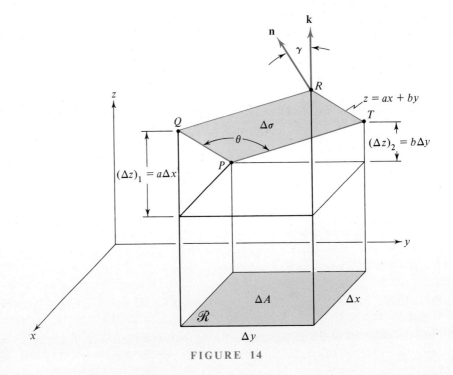

FIGURE 14

γ is also the angle between the given plane and the xy-plane, we might conjecture, by projection, that

$$\Delta A = \Delta\sigma \cos\gamma.$$

Then on division by $\cos\gamma$, we have $\Delta\sigma = \sec\gamma\,\Delta A$. This suggests

---

THEOREM 4. With the meanings of the symbols as shown in Fig. 14,

(48) $$\Delta\sigma = \sec\gamma\,\Delta x\,\Delta y.$$

Proof. Equation (48) has been obtained purely by intuition. To give a proof, we first observe that the area of the parallelogram $PQRT$ is the product $|\mathbf{PQ}|\,|\mathbf{PT}|\sin\theta$ and hence is the length of $\mathbf{V}$, where $\mathbf{V} = \mathbf{PQ}\times\mathbf{PT}$. But

$$\mathbf{PQ} = \Delta x\mathbf{i} + (\Delta z)_1\mathbf{k} = \Delta x\mathbf{i} + a\,\Delta x\mathbf{k}$$

and

$$\mathbf{PT} = \Delta y\mathbf{j} + (\Delta z)_2\mathbf{k} = \Delta y\mathbf{j} + b\,\Delta y\mathbf{k}.$$

Then an easy computation shows that $\mathbf{V} = (-a\mathbf{i} - b\mathbf{j} + \mathbf{k})\Delta x\,\Delta y$ and consequently,

(49) $$\Delta\sigma = |\mathbf{V}| = \sqrt{a^2 + b^2 + 1}\,\Delta x\,\Delta y.$$

But the vector $\mathbf{V}$ is also normal to the plane $z = ax + by$, so that it is easy to find $\sec\gamma$. Indeed,

$$\cos\gamma = \frac{\mathbf{V}\cdot\mathbf{k}}{|\mathbf{V}|} = \frac{\Delta x\,\Delta y}{\sqrt{a^2 + b^2 + 1}\,\Delta x\,\Delta y} = \frac{1}{\sqrt{a^2 + b^2 + 1}},$$

or

$$\sec\gamma = \sqrt{a^2 + b^2 + 1}.$$

Using this in (49) gives (48). ■

We are now prepared to find the area of a curved surface. We select a rectangle $\mathcal{T}$ that contains $\mathcal{R}$ and we partition $\mathcal{T}$ in the usual way. (Note that it is no longer necessary to picture $\mathcal{T}$ in Fig. 15.) We select an arbitrary point $P_{ij}^{\star}(x_{ij}^{\star}, y_{ij}^{\star})$ in each rectangle $\mathcal{T}_{ij}$ that is contained in $\mathcal{R}$ and at the corresponding point P_{ij} on the surface $z = f(x, y)$ we pass a plane tangent to the surface (see Fig. 15). Let $\Delta\sigma_{ij}$ be the area of that part of the tangent plane at P_{ij} that lies over the rectangle $\mathcal{T}_{ij}$. Then $\Delta\sigma_{ij}$ represents a good approximation for

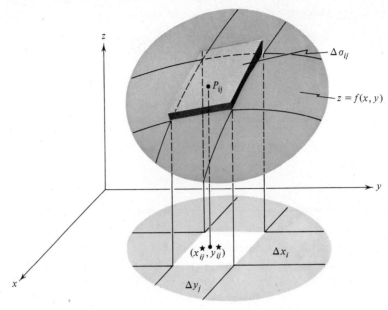

FIGURE 15

the area of the surface that lies over $\mathcal{T}_{ij}$, and we expect that $\sum\sum \Delta\sigma_{ij}$ will be a good approximation for the area under consideration. This suggests

---

DEFINITION 6. For each $\mathcal{T}_{ij}$ that lies on the boundary or outside $\mathcal{R}$, we set $\sigma_{ij} = 0$. If σ denotes the area of the surface $z = f(x, y)$ that lies over the region $\mathcal{R}$, then

$$(50) \qquad \sigma = \lim_{\mu(\mathcal{P})\to 0} \sum_{i=1}^{m} \sum_{j=1}^{n} \Delta\sigma_{ij}.$$

By Theorem 4 we can replace $\Delta\sigma_{ij}$ by $\sec\gamma_{ij}\,\Delta x_i\,\Delta y_j$ in (50). To obtain $\sec\gamma_{ij}$, we recall that the vector

$$(51) \qquad \mathbf{N} = -f_x(P_{ij}^{\star})\mathbf{i} - f_y(P_{ij}^{\star})\mathbf{j} + \mathbf{k}$$

is normal to the surface $z = f(x, y)$ at P_{ij} and consequently,

$$(52) \qquad \sec\gamma_{ij} = \sqrt{f_x^{\,2}(P_{ij}^{\star}) + f_y^{\,2}(P_{ij}^{\star}) + 1}.$$

Using (52) and $\Delta\sigma_{ij} = \sec\gamma_{ij}\,\Delta A_{ij}$ in (50), we obtain

$$(53) \qquad \sigma = \lim_{\mu(\mathcal{P})\to 0} \sum_{i=1}^{m} \sum_{j=1}^{n} \sqrt{f_x^{\,2}(P_{ij}^{\star}) + f_y^{\,2}(P_{ij}^{\star}) + 1}\,\Delta A_{ij},$$

where it is understood that we replace the radical in (53) by zero for those rectangles of the partition that are not contained in $\mathcal{R}$. Then, by the definition of a double integral, we have

---

THEOREM 5. If σ is the area of the part of the surface $z = f(x, y)$ that lies over the closed and bounded region $\mathcal{R}$ and if f_x and f_y are continuous in $\mathcal{R}$, then

(54)
$$\sigma = \int\int_{\mathcal{R}} \sqrt{1 + f_x^2 + f_y^2}\, dA.$$

If the surface is given by an equation of the form $x = f(y, z)$ or $y = f(x, z)$, then certain obvious changes must be made in (54). If the surface is given by an equation $F(x, y, z) = c$, then the vector $\mathbf{V} = F_x\mathbf{i} + F_y\mathbf{j} + F_z\mathbf{k}$ is normal to the surface. In this case equation (54) is replaced by

$$\sigma = \int\int_{\mathcal{R}} \frac{\sqrt{F_x^2 + F_y^2 + F_z^2}}{|F_z|}\, dA.$$

Example 1. Find the area of the part of the surface $z = \dfrac{2}{3}(x^{3/2} + y^{3/2})$ that lies over the square $0 \leqq x \leqq 3,\ 0 \leqq y \leqq 3$.

Solution. Here $f_x = x^{1/2}, f_y = y^{1/2}$, and (54) gives

$$\sigma = \int_0^3\int_0^3 \sqrt{1 + x + y}\, dy\, dx = \int_0^3 \frac{2}{3}(1 + x + y)^{3/2}\Big|_0^3 dx$$

$$= \frac{2}{3}\int_0^3 [(x + 4)^{3/2} - (x + 1)^{3/2}]\, dx = \frac{2}{3}\frac{2}{5}[(x + 4)^{5/2} - (x + 1)^{5/2}]\Big|_0^3$$

$$= \frac{4}{15}[7^{5/2} - 32 - 32 + 1] = \frac{4}{15}[7^{5/2} - 63] \approx 17.771. \quad \bullet$$

Observe that $\sigma > 9$, the area of the square base, as it should be.

EXERCISE 6

In Problems 1 through 8, find the area of that portion of the given surface that lies over the given region.

1. $z = a + bx + cy,\quad 0 \leqq y \leqq x^2,\quad 0 \leqq x \leqq 3.$

2. $z = x + \frac{2}{3}y^{3/2}$, $1 \leq x \leq 4$, $2 \leq y \leq 7$.

3. $z = x^2 + \sqrt{3}y$, $0 \leq y \leq 2x$, $0 \leq x \leq 2\sqrt{2}$.

4. $z = e^{-y} + \sqrt{7}x$, $0 \leq x \leq e^{-2y}$, $0 \leq y \leq 3$.

5. $z = \sqrt{x^2 + y^2}$, $0 \leq x \leq 2$, $0 \leq y \leq 5$.

6. $z = \sqrt{a^2 - y^2}$, $0 \leq x \leq 2y$, $0 \leq y \leq a$.

$\star$7. $z = -\ln y$, $0 \leq x \leq y^2$, $0 \leq y \leq 1$.

$\star$8. $z = 1/y$, $0 \leq x \leq y$, $0 \leq y \leq 1$.

9. Prove that the points $A(0, 0, 1)$, $B(2, 0, 4)$, and $C(5, 6, 2)$ form the vertices of a right triangle. Find the area of this triangle **(a)** by elementary means and **(b)** by double integration.

10. Find the area of that part of the cylinder $x^2 + z^2 = a^2$ lying in the first octant and between the planes $y = x$ and $y = 3x$.

11. Find the area of that part of the cylinder $y^2 + z^2 = a^2$ that lies inside the cylinder $x^2 + y^2 = a^2$.

12. Prove that if $\mathscr{R}$ is any region in the xy-plane, then the area of the portion of the cone $z = a\sqrt{x^2 + y^2}$ that lies above $\mathscr{R}$ is $A\sqrt{1 + a^2}$, where A is the area of $\mathscr{R}$.

13. Consider that portion of the cylinder $y^2 + z^2 = 1$ that lies in the first octant and above the unit square $0 \leq x \leq 1, 0 \leq y \leq 1$. A rough sketch seems to indicate that the plane $y = x$ bisects this surface into two congruent parts. Prove that the parts are not congruent by finding the area of each part.

$\star$14. We have previously used the formula

$$2\pi \int_a^b y\sqrt{1 + \left(\frac{dy}{dx}\right)^2}\, dx$$

for the area of the surface of revolution obtained when the curve $y = f(x)$ is rotated about the x-axis. Prove that our new definition is consistent, by deriving this formula from (54). HINT: The surface of revolution has the equation $y^2 + z^2 = f^2(x)$.

15. Prove that if $\mathscr{R}$ is any region in the xy-plane, then the area of the portion of the paraboloid $z = ax^2 + by^2$ that lies above $\mathscr{R}$ is equal to the area of the portion of the saddle surface $z = ax^2 - by^2$ that lies above (or below) $\mathscr{R}$. Here $a > 0$ and $b > 0$.

16. Prove that the statement of Problem 15 is also true for each of the following pairs of surfaces.
    (a) $z = x^2 + y^2$, $z = 2xy$.   (b) $z = \ln(x^2 + y^2)$, $z = 2\,\text{Tan}^{-1}(y/x)$.
    (c) $z = e^x(x\cos y - y\sin y)$, $z = e^x(y\cos y + x\sin y)$.

Double Integrals in Polar Coordinates 9

In certain problems, the evaluation of a given double integral becomes simpler if polar coordinates are used. Now x and y are replaced by $r\cos\theta$ and $r\sin\theta$, respectively, and $f(x, y)$ becomes $f(r\cos\theta, r\sin\theta) \equiv F(r, \theta)$, a function of r and θ. Instead of using a network of horizontal and vertical lines, the region $\mathscr{R}$ is partitioned by a finite number of circles with center at the origin, and a finite number of radial lines, as indicated in Fig. 16.

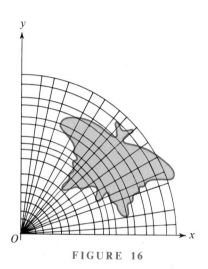

FIGURE 16

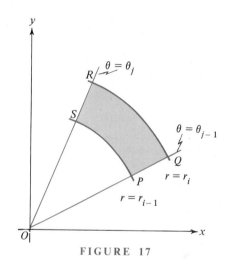

FIGURE 17

These circles and lines form the boundaries of a finite number of regions called *polar rectangles*. A typical polar rectangle is shown in Fig. 17. The *diameter* of a polar rectangle is the length of the longest line segment that has its end points on the boundary of the polar rectangle. The *mesh* of the partition $\mu(\mathscr{P})$ is the maximum of the diameters of the polar rectangles that form the partition $\mathscr{P}$.

To find an expression for ΔA_{ij}, the area of a polar rectangle, let r_{i-1} and r_i be the radii of the smaller and larger circles, respectively, with a similar meaning for θ_{j-1} and θ_j. Then referring to the lettering of Fig. 17, the area of the sector SOP is $(1/2)r_{i-1}^2(\theta_j - \theta_{j-1})$, the area of the sector ROQ is $(1/2)r_i^2(\theta_j - \theta_{j-1})$, and consequently the difference gives

$$\Delta A_{ij} = \frac{1}{2}r_i^2\,\Delta\theta_j - \frac{1}{2}r_{i-1}^2\,\Delta\theta_j = \frac{1}{2}(r_i^2 - r_{i-1}^2)\,\Delta\theta_j,$$

(55)
$$\Delta A_{ij} = \frac{r_i + r_{i-1}}{2}(r_i - r_{i-1})\,\Delta\theta_j = r_{ij}^\star\,\Delta r_i\,\Delta\theta_j.$$

Here $r_{ij}^\star \equiv (r_i + r_{i-1})/2$ and represents the radius of a circle that is midway between the smaller and larger circles forming the boundary of the polar rectangle.

Following the pattern used for double integrals in rectangular coordinates, we let $\mathcal{T}$ be a polar rectangle that contains $\mathcal{R}$ and we extend our partition of $\mathcal{R}$ to the polar rectangle $\mathcal{T}$. We let $P_{ij}^\star$ be the point $(r_{ij}^\star, \theta_{ij}^\star)$, where $r_{ij}^\star = (r_i + r_{i-1})/2$ and $\theta_{j-1} \leqq \theta_{ij}^\star \leqq \theta_j$.

Just as before, we set $F(r, \theta) = 0$ for points of $\mathcal{T}$ that are not in $\mathcal{R}$. Then the sum

$$\sum_{i=1}^{m} \sum_{j=1}^{n} F(P_{ij}^\star)\, \Delta A_{ij} \equiv \sum_{i=1}^{m} \sum_{j=1}^{n} F(r_{ij}^\star, \theta_{ij}^\star) r_{ij}^\star\, \Delta r_i\, \Delta \theta_j$$

is a good approximation to the double integral of $f(x, y)$ over $\mathcal{R}$. If we take the limit as $\mu(\mathcal{P}) \to 0$, we have

$$(56) \qquad \iint\limits_{\mathcal{R}} f(x, y)\, dx\, dy = \iint\limits_{\mathcal{R}} F(r, \theta) r\, dr\, d\theta,$$

where $F(r, \theta) \equiv f(r \cos \theta, r \sin \theta)$.

To evaluate a double integral in polar coordinates we convert it to an iterated integral (see Theorem 3) as described in

---

THEOREM 6 (PWO). Suppose that the region $\mathcal{R}$ is described by the pair of inequalities for polar coordinates

$$r_1(\theta) \leqq r \leqq r_2(\theta), \qquad \alpha \leqq \theta \leqq \beta,$$

and is also described by the pair of inequalities

$$a \leqq r \leqq b, \qquad \theta_1(r) \leqq \theta \leqq \theta_2(r).$$

If $F(r, \theta)$ is continuous in $\mathcal{R}$, then

$$(57) \qquad \iint\limits_{\mathcal{R}} F(r, \theta) r\, dr\, d\theta = \int_{\alpha}^{\beta} \int_{r_1(\theta)}^{r_2(\theta)} F(r, \theta) r\, dr\, d\theta$$

$$= \int_{a}^{b} \int_{\theta_1(r)}^{\theta_2(r)} F(r, \theta) r\, d\theta\, dr.$$

Just as before, various physical interpretations can be assigned to (57) in accordance with the selection of the function $F(r, \theta)$. If $z = F(r, \theta)$ is the equation of a surface in cylindrical coordinates, then (57) gives the volume of the solid under the surface and over $\mathcal{R}$. If $F(r, \theta) = 1$, then (57) gives the area of $\mathcal{R}$, and so on.

It is convenient to refer to the expression $r\,dr\,d\theta$ as the *differential element of area* in polar coordinates. This expression is easy to recall if we observe that in Fig. 17 the arc PS has length $r\,\Delta\theta$ and the segment PQ has length Δr so that the polar rectangle has area (approximately) equal to their product $r\,\Delta r\,\Delta\theta$.

Example 1. For the solid bounded by the xy-plane, the cylinder $x^2 + y^2 = a^2$ and the paraboloid $z = b(x^2 + y^2)$ with $b > 0$, find **(a)** the volume, **(b)** its centroid, **(c)** I_z, and **(d)** the area of its upper surface.

Solution. Since $x^2 + y^2 = r^2$, the equation for the paraboloid in cylindrical coordinates is $z = br^2$. For the volume, we have

$$V = \int\!\!\int_{\mathscr{R}} z\,dx\,dy = \int\!\!\int_{\mathscr{R}} zr\,dr\,d\theta = \int_0^{2\pi}\!\!\int_0^a br^2 r\,dr\,d\theta$$

$$= \int_0^{2\pi} b\frac{r^4}{4}\bigg|_{r=0}^{r=a} d\theta = \frac{ba^4}{4}\int_0^{2\pi} d\theta = \frac{\pi ba^4}{2}.$$

By symmetry the centroid is on the z-axis, so that $\bar{x} = \bar{y} = 0$. To find $\bar{z}$, we must compute M_{xy}. A partition of the plane region $\mathscr{R}$, of the type shown in Fig. 16, will divide the solid into elements with cylindrical sides. A typical element is shown in Fig. 18. If the sides of the polar rectangle are sufficiently small, then $zr\,\Delta r\,\Delta\theta$ represents a good approximation to the volume of the element, and $z/2$ is a good approximation for the z-coordinate of its centroid. Hence the sum

$$\sum_{i=1}^{m}\sum_{j=1}^{n} \frac{z_{ij}^{\star}}{2} z_{ij}^{\star} r_{ij}^{\star} \Delta r_i\,\Delta\theta_j$$

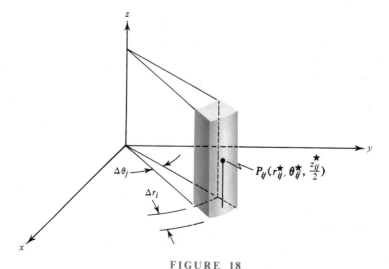

FIGURE 18

should give a good approximation to M_{xy}. Taking the limit as $\mu(\mathscr{P}) \to 0$, we can write

$$M_{xy} = \int\int_{\mathscr{R}} \int \frac{z^2}{2} r \, dr \, d\theta = \int_0^{2\pi} \int_0^a \frac{b^2 r^4}{2} r \, dr \, d\theta = \int_0^{2\pi} \frac{b^2 r^6}{12}\Big|_0^a \, d\theta = \frac{\pi b^2 a^6}{6}.$$

Then

$$\bar{z} = \frac{M_{xy}}{V} = \frac{\pi b^2 a^6}{6} \times \frac{2}{\pi b a^4} = \frac{ba^2}{3}.$$

To find I_z, we note that each point in the element shown in Fig. 18 has roughly the same distance $r_{ij}^\star$ from the z-axis. Then forming a sum and taking the usual limit, we find that

$$I_z = \int\int_{\mathscr{R}} \int r^2 \, dV = \int\int_{\mathscr{R}} r^2 z \, dA = \int\int_{\mathscr{R}} r^2 z r \, dr \, d\theta$$

$$= \int_0^{2\pi} \int_0^a br^5 \, dr \, d\theta = \int_0^{2\pi} \frac{br^6}{6}\Big|_0^a \, d\theta = \frac{\pi ba^6}{3}.$$

For the surface area, we have

$$\sigma = \int\int_{\mathscr{R}} \sqrt{1 + f_x^2 + f_y^2} \, dx \, dy = \int\int_{\mathscr{R}} \sqrt{1 + 4b^2(x^2 + y^2)} \, dx \, dy$$

$$= \int_0^{2\pi} \int_0^a \sqrt{1 + 4b^2 r^2} \, r \, dr \, d\theta = \frac{1}{8b^2} \int_0^{2\pi} \frac{2}{3}(1 + 4b^2 r^2)^{3/2}\Big|_0^a \, d\theta$$

$$= \frac{\pi}{6b^2}[(1 + 4b^2 a^2)^{3/2} - 1]. \quad\bullet$$

Example 2. Use double integration to find the area enclosed by the circle $r = a \cos \theta$.

Solution. The circle is described once as θ runs from $-\pi/2$ to $\pi/2$. Thus

$$A = \int_{-\pi/2}^{\pi/2} \int_0^{a \cos \theta} r \, dr \, d\theta = \int_{-\pi/2}^{\pi/2} \frac{r^2}{2}\Big|_0^{a \cos \theta} \, d\theta = \frac{a^2}{2} \int_{-\pi/2}^{\pi/2} \cos^2 \theta \, d\theta$$

$$= \frac{a^2}{4} \int_{-\pi/2}^{\pi/2} (1 + \cos 2\theta) \, d\theta = \frac{a^2}{4}\left(\theta + \frac{1}{2}\sin 2\theta\right)\Big|_{-\pi/2}^{\pi/2} = \frac{\pi a^2}{4}. \quad\bullet$$

EXERCISE 7

In Problems 1 through 6, use double integrals to find the area of the region enclosed by the given curve or curves.

1. $r^2 = a^2 \cos 2\theta$, $\theta = 0$, and $\theta = \pi/6$.   2. One loop of the curve $r = 6 \cos 3\theta$.

3. $r = 3 + \sin 4\theta$.   4. $r = \tan \theta$, $\theta = 0$, and $\theta = \pi/4$.

5. Outside the circle $r = 3a$ and inside the cardioid $r = 2a(1 + \cos \theta)$.

*6. $\theta = r$, $\theta = 2r$, $r = \pi/4$, $r = \pi/2$ ($\theta \geqq 0$).

In Problems 7 through 10, a solid is bounded by the given surfaces. Find the indicated quantity. Here σ denotes the area of the upper surface.

7. $z = 4 - x^2 - y^2$, $z = 0$.   Find V, I_z, and σ.

8. $z = a + x$, $z = 0$, $x^2 + y^2 = a^2$.   Find V, $\bar{x}$, $\bar{z}$, and I_z.

9. $z = 0$, $z = r$ ($r \geqq 0$), $r = a + b \cos \theta$ ($a > |b|$).   Find V.

10. $z = 0$, $z = 1/r$ ($r > 0$), $r = a \sec \theta$, $r = b \sec \theta$,   ($b > a > 0$), $\theta = 0$, $\theta = \pi/4$.
    Find V, M_{yz}, M_{zx}, and M_{xy}.

11. Find the volume of the solid in the first octant bounded by the planes $z = 0$, $z = y$, and the cylinder $r = \sin 2\theta$.

12. Find the volume of that portion of the sphere $x^2 + y^2 + z^2 = a^2$ that is also inside the cylinder $x^2 + y^2 = ax$.

13. Find the area of the top surface of the solid of Problem 12.

*14. Show that the volume of the solid bounded by $z = 0$, $z = 1/r$, and the cylinder $(x - 2)^2 + y^2 = 1$ is given by the elliptic integral

$$V = 4 \int_0^{\pi/6} \sqrt{4 \cos^2 \theta - 3} \, d\theta.$$

15. A sheet of material has the shape of a plane region $\mathscr{R}$ and has surface density ρ. Give the specific function $F(r, \theta)$ to be used in equation (57) to compute M_x, M_y, I_x, I_y, and J.

16. For the circular region bounded by $r = a$ find I_x and I_y by first finding J and dividing by 2.

In Problems 17 through 21, find the centroid of the region described.

17. The half-circle $0 < r < a$, $-\pi/2 < \theta < \pi/2$.

18. $0 < r < \sqrt[3]{\theta}$, $0 < \theta < \pi$.

19. Inside the cardioid $r = a(1 + \cos \theta)$.

20. Inside the circle $r = 2a \cos \theta$ and outside the circle $r = \sqrt{2}a$.

*21. The smaller of the two regions bounded by the circle $r = a$ and the straight line $r = b \sec \theta$, $0 < b < a$.

22. Obtain the answer to Problem 17, by putting $b = 0$ in the answer to Problem 21.
23. Find $\bar{x}$ for a sheet of material having the shape of the region bounded by the curve $r = a + b \cos \theta$ $(0 < b < a)$ with $0 \leq \theta \leq \pi$, and the x-axis, where the material has the surface density $\rho = \sin \theta$.
*24. Find $\bar{x}$ for the material of Problem 23 if $\rho = r \sin \theta$.
25. Find I_x and I_y for the region enclosed by the circle $r = \sin \theta$.
26. Check your answers to Problem 25 by finding J for that circle in two different ways.
27. Find I_x and I_y for the semicircle $0 < r < a$, $0 < \theta < \pi$, if the surface density is $\rho = \sin \theta$. As in Problem 26, check your answer by finding J.
*28. We call the region on the surface of a sphere between two circles of latitude and two circles of longitude a *spherical rectangle*. Suppose that in cylindrical coordinates the sphere has the equation $z^2 + r^2 = \rho^2$. Then the spherical rectangle can be described by the inequalities

$$\rho \sin \varphi_0 < r < \rho \sin (\varphi_0 + \Delta\varphi) \qquad \text{and} \qquad \theta_0 < \theta < \theta_0 + \Delta\theta$$

(here φ_0 has the meaning that is standard in spherical coordinates). Prove that the area of this spherical rectangle is $\sigma = \rho^2[\cos \varphi_0 - \cos (\varphi_0 + \Delta\varphi)] \Delta\theta$. Prove further that if $\sin \varphi_0 \neq 0$, then

$$\lim_{\substack{\Delta\theta \to 0 \\ \Delta\varphi \to 0}} \frac{\sigma}{\rho^2 \sin \varphi_0 \, \Delta\theta \, \Delta\varphi} = 1.$$

Hence the denominator furnishes a good approximation to σ.
*29. The region in the first octant bounded by the coordinate planes and the surface $z = e^{-(x^2+y^2)}$ has infinite extent but finite volume. Find the volume.
*30. It can be proved that under suitable conditions

$$\int_0^a f(x)\, dx \int_0^b g(y)\, dy = \int\int_{\mathscr{R}} f(x)g(y)\, dA,$$

where $\mathscr{R}$ is the rectangle $0 \leq x \leq a$, $0 \leq y \leq b$. Use this result, together with the result of Problem 29, to prove that

$$\int_0^\infty e^{-x^2}\, dx = \frac{\sqrt{\pi}}{2}.$$

This is curious, because if the upper limit a in $\int_0^a e^{-x^2}\, dx$ is not infinity or zero, the integral cannot be evaluated by elementary means.

Triple Integrals   *10*

The double integral was obtained by dividing a plane region into little rectangles and taking the limit of a certain sum. A triple integral is the natural extension of this concept to three-dimensional regions.

Suppose that $f(x, y, z)$ is defined for all points in a three-dimensional closed region $\mathcal{R}$ contained in a box $\mathcal{T}$. We partition $\mathcal{T}$ by a finite number of planes parallel to the coordinate planes: $x = x_i, i = 0, 1, \ldots, m$; $y = y_j, j = 0, 1, \ldots, n$; $z = z_k, k = 0, 1, \ldots, p$. These planes form a finite number of boxes[1]

$$\mathcal{T}_{ijk} : x_{i-1} \leq x \leq x_i, \quad y_{j-1} \leq y \leq y_j, \quad z_{k-1} \leq z \leq z_k.$$

We let $\Delta V_{ijk} \equiv \Delta x_i \, \Delta y_j \, \Delta z_k$ denote the volume of the box $\mathcal{T}_{ijk}$, and we let d_{ijk} be the diameter of $\mathcal{T}_{ijk}$. The mesh of the partition $\mathcal{P}$ is the largest of these diameters, and as usual we denote the mesh by $\mu(\mathcal{P})$.

We extend the definition of $f(x, y, z)$ to all of $\mathcal{T}$ in the usual way obtaining the new function $F(x, y, z)$, where

(58)
$$F(x, y, z) = \begin{cases} f(x, y, z), & \text{if } (x, y, z) \text{ is in } \mathcal{R}, \\ 0, & \text{if } (x, y, z) \text{ is in } \mathcal{T} - \mathcal{R}. \end{cases}$$

In each box $\mathcal{T}_{ijk}$ we select a point $P^{\star}_{ijk}(x^{\star}_{ijk}, y^{\star}_{ijk}, z^{\star}_{ijk})$ and we let $F(P^{\star}_{ijk})$ denote the value of F at that point. Finally, we form the triple sum

(59)
$$S(\mathcal{P}) \equiv \sum_{i=1}^{m} \sum_{j=1}^{n} \sum_{k=1}^{p} F(P^{\star}_{ijk}) \, \Delta V_{ijk},$$

where the sum includes each of the *mnp* boxes in $\mathcal{T}$. If this sum has a limit as $\mu(\mathcal{P}) \to 0$, we denote the limit by either one of the two symbols

(60)
$$\iiint_{\mathcal{T}} F(x, y, z) \, dV, \qquad \iiint_{\mathcal{R}} f(x, y, z) \, dV$$

[read "the triple integral of $F(x, y, z)$ over $\mathcal{T}$" or "the triple integral of $f(x, y, z)$ over $\mathcal{R}$"]. Stated accurately, we have

---

[1] Here triple subscripts enter in a natural way. See the footnotes on double subscripts on pages 813, 744, and 735.

DEFINITION 7. Let $f(x, y, z)$ be defined in a bounded region $\mathscr{R}$ contained in a box $\mathscr{T}$ and extend the definition of $f(x, y, z)$ to $\mathscr{T}$ by equation (58). Then the triple integral of $f(x, y, z)$ over $\mathscr{R}$ is defined by

$$(61) \qquad \iiint\limits_{\mathscr{R}} f(x, y, z)\, dV = \lim_{\mu(\mathscr{P}) \to 0} \sum_{i=1}^{m} \sum_{j=1}^{n} \sum_{k=1}^{p} F(P_{ijk}^{\star})\, \Delta V_{ijk},$$

whenever the limit on the right side exists.

The theory of the triple integral parallels that of the double integral. Hence we may omit the details and merely state the results.

A surface $z = f(x, y)$ over a closed region $\mathscr{R}_{xy}$ in the xy-plane is said to be *smooth* if the partial derivatives f_x and f_y are continuous in $\mathscr{R}_{xy}$. A similar definition holds for surfaces of the form $y = g(x, z)$ and $x = h(y, z)$. With this terminology we have

THEOREM 7 (PWO). If $f(x, y, z)$ is continuous in a closed and bounded region $\mathscr{R}$, and if the boundary of $\mathscr{R}$ consists of a finite number of smooth surfaces, then the triple integral (61) exists.

We next consider some of the physical interpretations of the triple integral.

If $f(x, y, z) \equiv 1$ in $\mathscr{R}$, then the triple integral (61) gives the volume of $\mathscr{R}$.

Suppose next that the closed region $\mathscr{R}$ is occupied by some solid of variable mass density $\rho(x, y, z)$, and we set $f(x, y, z) = \rho(x, y, z)$. Then $f(P_{ijk}^{\star})\, \Delta V_{ijk} = \rho(P_{ijk}^{\star})\, \Delta V_{ijk}$ is a good approximation for m_{ijk}, the mass of the material in the box $\mathscr{T}_{ijk}$. In the limit as $\mu(\mathscr{P}) \to 0$ we obtain

$$(62) \qquad m = \iiint\limits_{\mathscr{R}} \rho(x, y, z)\, dV,$$

where m is the mass of the solid.

Similar considerations lead to formulas for M_{yz}, M_{zx}, and M_{xy}, the moments with respect to the various coordinate planes; and to formulas for I_z, I_x, and I_y, the moments of inertia about the various axes. These formulas are

$$M_{yz} = \iiint\limits_{\mathcal{R}} x\rho \, dV, \qquad M_{zx} = \iiint\limits_{\mathcal{R}} y\rho \, dV, \qquad M_{xy} = \iiint\limits_{\mathcal{R}} z\rho \, dV,$$

$$(63) \qquad I_z = \iiint\limits_{\mathcal{R}} (x^2 + y^2)\rho \, dV, \qquad I_x = \iiint\limits_{\mathcal{R}} (y^2 + z^2)\rho \, dV,$$

$$I_y = \iiint\limits_{\mathcal{R}} (z^2 + x^2)\rho \, dV.$$

As usual, when we speak of the moment, or moment of inertia of a closed region (instead of a solid) we mean the above quantities computed with $\rho = 1$.

To compute a triple integral, we convert it to an iterated integral. One practical difficulty is that of assigning limits for the iterated integral, and we now discuss this problem. To simplify matters we select $P_{ijk}^{\star}$ to be the corner point (x_i, y_j, z_k) for each box $\mathcal{T}_{ijk}$ (see Fig. 19). Further, we assume that $\mathcal{R}$ is a normal three-dimensional region and that the projection of $\mathcal{R}$ on the xy-plane forms a region $\mathcal{S}$ that is also normal. Any partition of $\mathcal{T}$ by planes simultaneously effects a partition of $\mathcal{S}$ by lines parallel to the x- and y-axis.

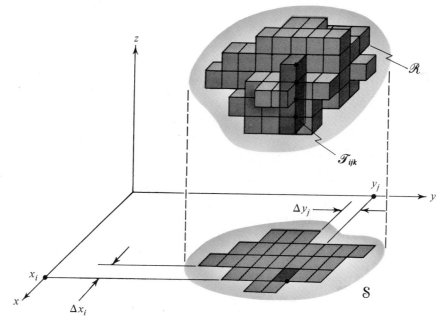

FIGURE 19

We now rearrange the terms in the triple sum (59) into a "sum of sums of sums." We first fix i and j and select only those terms in (59) for which i and j have those fixed values. This amounts to summing over those boxes that are in a vertical stack above a fixed rectangle in $\mathbb{S}$ (see Fig. 19). This sum can be put in the form

$$
(64) \qquad \sum_{k=1}^{p} f(P_{ijk})\,\Delta V_{ijk} = \left[\sum_{k=1}^{p} f(x_i, y_j, z_k)\,\Delta z_k\right]\Delta y_j\,\Delta x_i,
$$

where we can "factor out" Δy_j and Δx_i because i and j are constant for the terms of this sum.

We next form a sum of sums of the type (64). Thus with i fixed and j running from 1 to n, we obtain

$$
(65) \qquad \left\{\sum_{j=1}^{n}\left[\sum_{k=1}^{p} f(x_i, y_j, z_k)\,\Delta z_k\right]\Delta y_j\right\}\Delta x_i,
$$

where we can "factor out" Δx_i because i is a constant for the terms of this sum. The sum in (65) is the sum over all rectangles of the partition that have their "front face" on the plane $x = x_i$ (see Fig. 20).

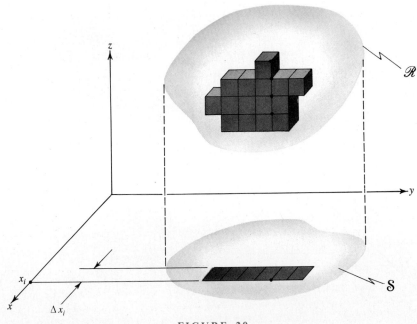

FIGURE 20

Finally, we form the sum of the sums in (65) as i runs from 1 to m. This gives all of the terms in (59) and hence

$$(66) \qquad S(\mathscr{P}) = \sum_{i=1}^{m} \left\{ \sum_{j=1}^{n} \left[\sum_{k=1}^{p} f(x_i, y_j, z_k)\, \Delta z_k \right] \Delta y_j \right\} \Delta x_i.$$

Now consider the limit in this rearranged form of (66) as $\mu(\mathscr{P}) \to 0$. Since $\mathscr{R}$ is convex in the direction of the z-axis the upper and lower bounding surfaces are given by functions $z = z_2(x, y)$ and $z = z_1(x, y)$. If we recall that $f(x, y, z) = 0$ for P in $\mathscr{T} - \mathscr{R}$ [for $z > z_2(x, y)$ and $z < z_1(x, y)$], then it is clear that for fixed i and j the "inner sum" in (66) (the sum in brackets) approaches

$$(67) \qquad \int_{z_1(x_i, y_j)}^{z_2(x_i, y_j)} f(x_i, y_j, z)\, dz.$$

Suppose next that the boundary of $\mathcal{S}$ is given by two equations $y = y_2(x)$ and $y = y_1(x)$ for $a \leq x \leq b$. We multiply (67) by Δy_j and form a sum of such terms to obtain a good approximation for the sum in braces in (66). Then as $\mu(\mathscr{P}) \to 0$,

$$(68) \qquad \sum_{j=1}^{n} \left[\int_{z_1(x_i, y_j)}^{z_2(x_i, y_j)} f(x_i, y_j, z)\, dz \right] \Delta y_j \to \int_{y_1(x_i)}^{y_2(x_i)} \left[\int_{z_1(x_i, y)}^{z_2(x_i, y)} f(x_i, y, z)\, dz \right] dy.$$

Finally, we multiply both sides of (68) by Δx_i and for the last time form a sum and consider the limit as $\mu(\mathscr{P}) \to 0$. Since we started with $S(\mathscr{P})$, an approximating sum for a triple integral, and arrived at an iterated integral, it is intuitively clear that we have

---

THEOREM 8 (PWO). Let $\mathscr{R}$ be a closed and bounded three-dimensional region that is described by the inequalities

$$(69) \qquad a \leq x \leq b, \qquad y_1(x) \leq y \leq y_2(x), \qquad z_1(x, y) \leq z \leq z_2(x, y).$$

Suppose further that the two surfaces $z = z_1(x, y)$ and $z = z_2(x, y)$ are smooth surfaces, and the two curves $y = y_1(x)$ and $y = y_2(x)$ are smooth curves. Then

$$(70) \qquad \iiint_{\mathscr{R}} f(x, y, z)\, dV = \int_{a}^{b} \left\{ \int_{y_1(x)}^{y_2(x)} \left[\int_{z_1(x, y)}^{z_2(x, y)} f(x, y, z)\, dz \right] dy \right\} dx.$$

In (70) the integral from $z_1(x, y)$ to $z_2(x, y)$ corresponds to the summation over the boxes in a vertical stack [see equation (64)] and is performed with x and y fixed. The integral from $y_1(x)$ to $y_2(x)$ corresponds to the summation over the rectangles, with forward edge on a fixed line parallel to the y-axis in the xy-plane [see (65)] and is performed with x fixed. Finally, the integral from a to b corresponds to a summation over all of the rectangles in the partition of $\mathcal{S}$.

It should be clear that $\mathcal{R}$ could also be projected on the xz-plane or on the yz-plane. Further, in each case the computation of the resulting double integral over the shadow region can be done in two different ways. Hence there is a total of *six* different ways of writing the right side of (70), one way for each of the six permutations of the symbols dx, dy, and dz. With each of these six different ways, the limits of integration must be selected accordingly. This is illustrated in

Example 1. Compute the triple integral of $f(x, y, z) = 2x + 4y$ over the region in the first octant bounded by the coordinate planes and the plane

$$6x + 3y + 2z = 6.$$

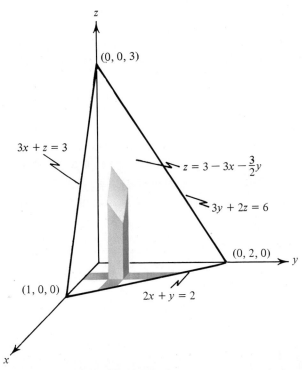

FIGURE 21

Solution. The region is shown in Fig. 21, together with the equations of the boundary lines. For this region (70) becomes

$$I = \int_0^1 \int_0^{2-2x} \int_0^{3-3x-3y/2} (2x + 4y)\, dz\, dy\, dx$$

$$= \int_0^1 \int_0^{2-2x} (2x + 4y)z \Big|_{z=0}^{z=3-3x-3y/2} dy\, dx$$

$$= \int_0^1 \int_0^{2-2x} (2x + 4y)\left(3 - 3x - \frac{3}{2}y\right) dy\, dx$$

$$= \int_0^1 \int_0^{2(1-x)} (6x + 12y - 6x^2 - 15xy - 6y^2)\, dy\, dx$$

$$= \int_0^1 [12x(1 - x) + 24(1 - x)^2 - 12x^2(1 - x)$$

$$- 30x(1 - x)^2 - 16(1 - x)^3]\, dx$$

$$= \int_0^1 (8 - 18x + 12x^2 - 2x^3)\, dx = 8x - 9x^2 + 4x^3 - \frac{x^4}{2} \Big|_0^1 = \frac{5}{2}. \quad \bullet$$

Other iterated integrals that give this same triple integral are

$$\int_0^2 \int_0^{1-y/2} \int_0^{3-3x-3y/2} (2x + 4y)\, dz\, dx\, dy,$$

$$\int_0^1 \int_0^{3-3x} \int_0^{2-2x-2z/3} (2x + 4y)\, dy\, dz\, dx,$$

$$\int_0^3 \int_0^{1-z/3} \int_0^{2-2x-2z/3} (2x + 4y)\, dy\, dx\, dz,$$

$$\int_0^3 \int_0^{2-2z/3} \int_0^{1-y/2-z/3} (2x + 4y)\, dx\, dy\, dz,$$

$$\int_0^2 \int_0^{3-3y/2} \int_0^{1-y/2-z/3} (2x + 4y)\, dx\, dz\, dy.$$

EXERCISE 8

1. Evaluate the triple integral of $f = 24xy^2z^3$ over the box $0 \leq x \leq a$, $0 \leq y \leq b$, $0 \leq z \leq c$.
2. Evaluate the triple integral of $f = 24xy^2z^3$ over the region bounded by the planes $x = 0$, $x = 1$, $y = 0$, $z = y$, and $z = 2$.
3. Check the answer to Example 1 of this section by evaluating at least two of the other five iterated integrals given at the end of the example.

In Problems 4 through 9, use triple integration to find the volume of the region bounded by the given surfaces.

    4. $x = 0$, $y = 0$, $z = 0$, and $6x + 4y + 3z = 12$.
    5. $z = 6\sqrt{y}$, $z = \sqrt{y}$, $y = x$, $y = 4$, and $x = 0$.
    6. $y = 0$, $x = 4$, $z = y$, and $z = x - y$.
  ★**7.** $z = x^2 + 2y^2$, and $z = 16 - x^2 - 2y^2$.
  ★**8.** $z = x^2 + y^2$, and $z = 2y$.
  ★**9.** $y = x^2 + 2x$, and $y = 4 - z^2 + 2x$.

In Problems 10 through 15, find the mass of the solid bounded by the given surfaces, and having the given mass density. In these problems a, b, c, and k are positive constants.

  10. $x = 0$, $x = 1$, $y = 0$, $y = 1$, $z = 0$, $z = 1$, $\rho = ky$.
  11. $x = 0$, $x = a$, $y = 0$, $y = b$, $z = 0$, $z = c$, $\rho = ky$.
  12. The solid of Problem 11, $\rho = 1 + 24kxy^2z^3$.
  13. $z = x^2 + y^2$, $z = 4$, $x = 0$, $y = 0$,   first octant, $\rho = 4ky$.
  14. $z = xy$, $x = 1$, $y = x$, $z = 0$, $\rho = 1 + 2z$.
  15. $z = e^{x+y}$, $z = 4$, $x = 0$, $y = 0$, $\rho = 1/z$.

  16. Obtain the answer to Problem 12 from the solution to Problem 1.
  17. Find the center of mass for the solid of: **(a)** Problem 10, **(b)** Problem 11, **(c)** Problem 12, and **(d)** Problem 14. In **(d)** prove that the center of mass lies inside the solid.
  18. Find I_x and I_y for the solid of: **(a)** Problem 10, **(b)** Problem 11, **(c)** Problem 12, and **(d)** Problem 14.
  19. Find the centroid of the tetrahedron bounded by the coordinate planes and the plane $bcx + acy + abz = abc$, where a, b, and c are positive.
  20. Find I_z for the region of Problem 19.
★★**21.** The Parallel Axis Theorem. For a given solid let I_p and I_g be the moments of inertia about axes p and g, respectively, where p is parallel to g and g runs through the center of mass of the solid. Then

$$I_p = I_g + ms^2,$$

where m is the mass of the solid and s is the distance between p and g. Prove this theorem. HINT: Select the origin of the coordinate system at the center of mass of the solid, let the y-axis coincide with g, and then rotate the xz-plane so that the x-axis intersects the line p at the point $(s, 0, 0)$. Then

$$I_p = \iiint_{\mathscr{R}} [(s - x)^2 + z^2]\rho \, dx \, dy \, dz$$

$$I_p = \int\!\!\int\!\!\int_{\mathscr{R}} (x^2 + z^2)\,\rho\,dx\,dy\,dz - 2s \int\!\!\int\!\!\int_{\mathscr{R}} x\rho\,dx\,dy\,dz + s^2 \int\!\!\int\!\!\int_{\mathscr{R}} \rho\,dx\,dy\,dz$$

$$= I_y - 2sM_{yz} + s^2 m = I_g - 0 + ms^2.$$

Triple Integrals in Cylindrical Coordinates *11*

In many cases the triple integral over a region $\mathscr{R}$ can be evaluated more easily if cylindrical coordinates are used in place of rectangular coordinates. As usual, the region is partitioned into parts, but this time the partitioning is done by cylinders with the z-axis for an axis, together with horizontal planes and planes containing the z-axis. These surfaces form a number of cylindrical boxes, and a typical one is shown in Fig. 22.

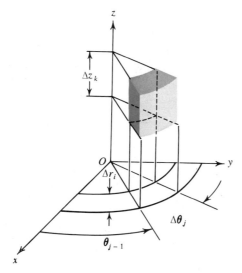

FIGURE 22

We have already seen in Section 9 that in polar coordinates the differential element of area is $dA = r\,dr\,d\theta$. Consequently, it follows that in cylindrical coordinates the differential element of volume is $dV = r\,dr\,d\theta\,dz$. Then

(71)
$$\int\!\!\int\!\!\int_{\mathscr{R}} f(x, y, z)\,dV = \int\!\!\int\!\!\int_{\mathscr{R}} f(r\cos\theta, r\sin\theta, z)\,r\,dr\,d\theta\,dz.$$

As usual, $f = 1$ gives the volume of $\mathcal{R}$, $f = r^2$ gives I_z for $\mathcal{R}$, and so on.

Example 1. For the region inside both the sphere $x^2 + y^2 + z^2 = 4a^2$ and the cylinder $(x - a)^2 + y^2 = a^2$, find **(a)** the volume, and **(b)** I_z.

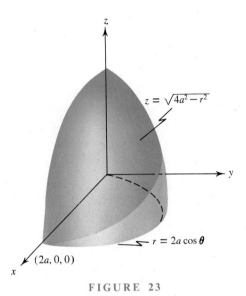

FIGURE 23

Solution. The portion of the region in the first octant is shown in Fig. 23. In cylindrical coordinates the equation of the sphere is $r^2 + z^2 = 4a^2$, and the equation of the cylinder is $r = 2a \cos \theta$. Using the symmetry, (71) gives

$$V = 4 \int_0^{\pi/2} \int_0^{2a \cos \theta} \int_0^{\sqrt{4a^2 - r^2}} r \, dz \, dr \, d\theta$$

$$= 4 \int_0^{\pi/2} \int_0^{2a \cos \theta} (4a^2 - r^2)^{1/2} r \, dr \, d\theta$$

$$= -2 \int_0^{\pi/2} \frac{2}{3} (4a^2 - r^2)^{3/2} \Big|_{r=0}^{r=2a \cos \theta} d\theta$$

$$= \frac{4}{3} \int_0^{\pi/2} [8a^3 - (4a^2 - 4a^2 \cos^2 \theta)^{3/2}] \, d\theta$$

$$= \frac{32a^3}{3} \int_0^{\pi/2} (1 - \sin^3 \theta) \, d\theta = \frac{16a^3}{9} (3\pi - 4).$$

Similarly, for I_z we have

$$I_z = 4 \int_0^{\pi/2} \int_0^{2a \cos \theta} \int_0^{\sqrt{4a^2 - r^2}} r^2 r \, dz \, dr \, d\theta$$

$$= 4 \int_0^{\pi/2} \int_0^{2a \cos \theta} (4a^2 - r^2)^{1/2} (r^2 - 4a^2 + 4a^2) r \, dr \, d\theta$$

$$= 4 \int_0^{\pi/2} \int_0^{2a \cos \theta} [4a^2 (4a^2 - r^2)^{1/2} - (4a^2 - r^2)^{3/2}] r \, dr \, d\theta$$

$$= -2 \int_0^{\pi/2} \left[\frac{8}{3} a^2 (4a^2 - r^2)^{3/2} - \frac{2}{5} (4a^2 - r^2)^{5/2} \right] \Big|_{r=0}^{r=2a \cos \theta} d\theta$$

$$= \frac{128 a^5}{15} \int_0^{\pi/2} (2 + 3 \sin^5 \theta - 5 \sin^3 \theta) \, d\theta$$

$$= \frac{128 a^5}{15} \left[\pi + \int_0^{\pi/2} (-2 - \cos^2 \theta + 3 \cos^4 \theta) \sin \theta \, d\theta \right]$$

$$= \frac{128 a^5}{15} \left(\pi - \frac{26}{15} \right). \quad \bullet$$

EXERCISE 9

1. Write the expression to be used for f in the right side of equation (71) when computing (a) M_{xy}, (b) M_{yz}, (c) I_x, and (d) I_y, assuming unit density.

2. Find the volume of the region bounded above by the sphere $z^2 + r^2 = \rho_0^2$ and below by the cone $z = r \cot \varphi_0$. Use this result to find the volume of a hemisphere.

3. Consider the portion of the region of Problem 2 that lies between the half-planes $\theta = \theta_0$ and $\theta = \theta_0 + \Delta\theta$ $(r \geq 0)$. Prove that the volume is given by

$$V = \frac{\rho_0^3}{3} (1 - \cos \varphi_0) \, \Delta\theta.$$

This result will be used in the next section, on spherical coordinates.

4. Find the centroid of the region bounded by the cone $z = m \sqrt{x^2 + y^2}$ and the plane $z = H$, where $m > 0$ and $H > 0$.

5. Find the volume of the region inside the sphere $r^2 + z^2 = 8$ and above the paraboloid $2z = r^2$.

6. Find $\bar{z}$ for the region of Problem 5.

7. Find the volume of the region below the plane $z = y$ and above the paraboloid $z = r^2$.

8. Find the centroid of the region of Problem 7.

9. A right circular cylinder has base radius R and height H. Find the moment of inertia about a generator (a line lying in its lateral surface). Find the moment of inertia about a diameter of the base.

10. A right circular cone has base radius R and height H. Find the moment of inertia about its axis, and about a diameter of the base.

★11. Find the volume of the region in the first octant above the surface $z = (x - y)^2$, below the surface $z = 4 - 2xy$, and between the planes $y = 0$ and $y = x$.

★12. A solid has the form of the region of Problem 11. If the density is proportional to the distance from the xz-plane, find the mass, $\bar{x}$, and $\bar{y}$.

★★13. Find the volume of the region bounded by the planes $z = H + my$ and $z = 0$ and the cylinder $r = a + b \sin \theta$. Here $a > b > 0$, $m > 0$, and $H > m(a - b)$.

★14. Find the volume and the centroid of the region in the first octant bounded by the coordinate planes, the cylinder $r = R$, and the surface $z = e^{-r}$.

★15. As $R \to \infty$, the region of Problem 14 becomes infinite in extent, but this infinite region has a centroid. Find it.

12   *Triple Integrals in Spherical Coordinates*

For triple integrals in spherical coordinates, the region is partitioned[1] by the surfaces $\rho = \rho_i$, $\varphi = \varphi_j$, and $\theta = \theta_k$. The surfaces $\rho = \rho_i$ form a set of spheres with center at the origin, the surfaces $\varphi = \varphi_j$ form a set of cones with vertex at the origin, and the surfaces $\theta = \theta_k$ form a set of half-planes, each one containing the z-axis. These surfaces are the boundaries of a finite number of regions called *spherical boxes*. A typical spherical box is shown in Fig. 24. Our purpose is to find an expression for the volume of such a spherical box, and to deduce from it the proper form for dV when spherical coordinates are used.

    Let us proceed first on the basis of our intuition. The three coordinate arcs PQ, PR, and PS that meet at $P(\rho_{i-1}, \varphi_{j-1}, \theta_{k-1})$ are mutually perpendicular (see Fig. 24). We therefore expect that ΔV, the volume of the spherical box, will be approximately the product of the lengths of the three arcs meeting at P. These lengths are $\Delta \rho_i$, $\rho_{i-1} \Delta \varphi_j$, and $\rho_{i-1} (\sin \varphi_{j-1}) \Delta \theta_k$, respectively, and consequently,

(72)
$$\Delta V \approx \rho_{i-1}^2 \sin \varphi_{j-1} \, \Delta \rho_i \, \Delta \varphi_j \, \Delta \theta_k.$$

---

[1] The letter ρ is traditionally used as one of the spherical coordinates. It is also the standard symbol for mass density. To avoid confusion, we use μ for mass density in this section and the next.

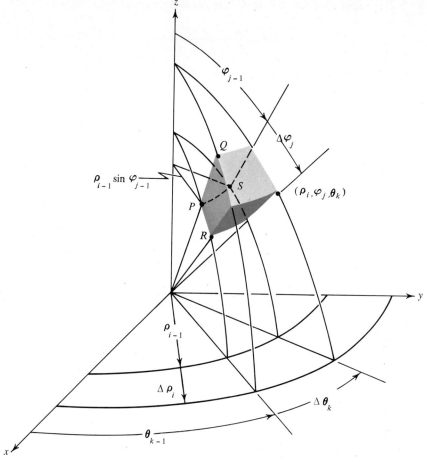

FIGURE 24

If this approximation is close enough, then we can write

(73) $$dV = \rho^2 \sin \varphi \, d\rho \, d\varphi \, d\theta,$$

and to evaluate a triple integral in spherical coordinates, we can use

(74) $$\iiint_{\mathcal{R}} f(x, y, z) \, dV = \iiint_{\mathcal{R}} f(\rho \sin \varphi \cos \theta, \rho \sin \varphi \sin \theta, \rho \cos \varphi)\rho^2 \sin \varphi \, d\rho \, d\varphi \, d\theta.$$

To prove that the approximation for ΔV given in (72) is sufficiently close, we will obtain a precise expression for ΔV using the formula

$$V = \frac{\rho_0^3(1 - \cos \varphi_0)\, \Delta\theta}{3}$$

obtained in Problem 3 of Exercise 9. First, let V_1 denote the volume of the region inside the sphere $\rho = \rho_i$, between the half-planes $\theta = \theta_{k-1}$ and $\theta = \theta_k$, and between the cones $\varphi = \varphi_{j-1}$ and $\varphi = \varphi_j$. This region is shaped like a pyramid, with its vertex at O and one corner of its spherical base at Q (Fig. 24). But then V_1 is just the difference

$$V_1 = \frac{\rho_i^3}{3}[1 - \cos \varphi_j]\, \Delta\theta_k - \frac{\rho_i^3}{3}[1 - \cos \varphi_{j-1}]\, \Delta\theta_k$$

$$= \frac{\rho_i^3}{3}[\cos \varphi_{j-1} - \cos \varphi_j]\, \Delta\theta_k.$$

By the Mean Value Theorem (with φ as the variable), we can write

(75) $$V_1 = \frac{\rho_i^3}{3}\sin \varphi_j^\star\, \Delta\varphi_j\, \Delta\theta_k, \qquad \varphi_{j-1} < \varphi_j^\star < \varphi_j.$$

Finally, the spherical box is just the region of the pyramid described above that lies inside a sphere of radius ρ_i and outside a sphere of radius ρ_{i-1} and hence its volume is just the difference of two expressions of the form (75). This gives

(76) $$\Delta V = \frac{\rho_i^3}{3}\sin \varphi_j^\star\, \Delta\varphi_j\, \Delta\theta_k - \frac{\rho_{i-1}^3}{3}\sin \varphi_j^\star\, \Delta\varphi_j\, \Delta\theta_k,$$

where $\varphi_j^\star$ is the same in both terms. Using the Mean Value Theorem again (this time with ρ as the variable) (76) gives

(77) $$\Delta V = (\rho_i^\star)^2 \sin \varphi_j^\star\, \Delta\rho_i\, \Delta\varphi_j\, \Delta\theta_k, \qquad \rho_{i-1} < \rho_i^\star < \rho_i.$$

But this is (72) computed at $(\rho_i^\star, \varphi_j^\star, \theta_k)$ instead of at $(\rho_{i-1}, \varphi_{j-1}, \theta_k)$. Since the point $(\rho_i^\star, \varphi_j^\star, \theta_k)$ lies in the closed spherical box, it follows that the approximation (72) is sufficiently close, and this proves (74).

> **Example 1.** Find the mass of a sphere of radius R if the density at each point is inversely proportional to its distance from the center of the sphere.
>
> **Solution.** Naturally, we place the sphere so that its center is at the origin of a spherical coordinate system. Then $\mu = k/\rho$ is the density and (74) gives for the mass
>
> $$m = \iiint_{\mathcal{R}} \frac{k}{\rho}\, dV = \int_0^{2\pi} \int_0^\pi \int_0^R \frac{k}{\rho}\rho^2 \sin \varphi\, d\rho\, d\varphi\, d\theta$$

$$= \int_0^{2\pi} \int_0^{\pi} k\frac{R^2}{2} \sin \varphi \, d\varphi \, d\theta = \int_0^{2\pi} k\frac{R^2}{2} 2 \, d\theta = 2k\pi R^2. \quad \bullet$$

This result is interesting, because the mass is finite, although at the center of the sphere the density is infinite.

EXERCISE 10

1. Prove that the volume of the spherical box shown in Fig. 24 is given by the formula

$$\Delta V = 2\left(\rho_{i-1}^2 + \rho_{i-1}\Delta\rho_i + \frac{(\Delta\rho_i)^2}{3}\right) \sin\left(\varphi_{j-1} + \frac{\Delta\varphi_j}{2}\right) \sin\frac{\Delta\varphi_j}{2} \Delta\rho_i \, \Delta\theta_k.$$

2. Use the result of Problem 1 to prove that if $\rho_{i-1} \sin \varphi_{j-1} \neq 0$, then as $\Delta\rho_i$, $\Delta\varphi_j$, $\Delta\theta_k$ all approach zero,

$$\lim \frac{\Delta V}{\rho_{i-1}^2 \sin \varphi_{j-1} \Delta\rho_i \, \Delta\varphi_j \, \Delta\theta_k} = 1.$$

This permits an alternative approach to the proof of (74).

3. Assuming unit density, write the expression to be used for f in the right side of (74) when computing: **(a)** M_{xy}, **(b)** M_{zx}, **(c)** I_x, **(d)** I_y, and **(e)** I_z.

4. Use triple integration to find **(a)** the volume of a sphere and **(b)** I_z, where the sphere has radius R and center at the origin.

5. Find the centroid of the hemispherical shell $A \leq \rho \leq B$, $0 \leq \varphi \leq \pi/2$.

6. For the hemispherical shell of Problem 5, find I_x and I_z.

7. For the region inside the sphere $\rho = R$ and above the cone $\varphi = \gamma$ (a constant), find the volume, the centroid, I_z, and I_y.

*8. As $\gamma \to 0$, the closed region of Problem 7 tends to the line segment $0 \leq z \leq R$. The centroid of this line segment has $\bar{z} = R/2$. But for the $\bar{z}$ of the region of Problem 7, we have $\lim_{r \to 0} \bar{z} = 3R/4 \neq R/2$. Explain this apparent inconsistency.

9. Find the volume of the torus $\rho = A \sin \varphi$.

10. Find the volume of the region bounded by the surface $\rho = A(\sin \varphi)^{1/3}$.

11. Find $\bar{z}$ for the region between two spheres of radii A and $B (A < B)$ that are tangent to the xy-plane at the origin.

12. Discuss the limit of $\bar{z}$ in Problem 11 as A approaches zero, and as A approaches B.

*13. Find the volume of the region inside of both of the spheres $\rho = B \cos \varphi$ and $\rho = A(A < B)$. Check your answer by considering the special case $A = B$.

*14. Find the volume of the region bounded by the surface $\rho = A \sin \varphi \sin \theta$, with $0 \leq \theta \leq \pi$.

15. A hemispherical solid is bounded above by the sphere $\rho = R$ and below by the xy-plane, and has density $\mu = k/\rho^n$. For what values of n is the mass finite? Find the mass when it is finite.

16. For the solid of Problem 15, find M_{xy} when it is finite. Find $\bar{z}$. For what values of n is M_{xy} finite and m infinite?

17. For the solid of Problem 15, find I_z when it is finite.

18. Suppose that in Problems 15, 16, and 17, the solid is a sphere tangent to the xy-plane at the origin. Without doing any computation, state whether m, M_{xy}, and I_z are finite for exactly the same values of n, as found in those problems.

*19. Consider a solid sphere of radius R with density $\mu = c/\rho^n$ when the sphere is placed with its center at the origin of a spherical coordinate system. Prove that if $3 \leq n < 5$, then the moment of inertia about any axis through the center of mass is finite, but for any other axis the moment of inertia is infinite. HINT. Use the Parallel Axis Theorem. (See Exercise 8, Problem 21, page 850.)

*13 *Gravitational Attraction*

According to Newton's Law of universal gravitation, any two particles attract each other with a force that acts along the line joining them. The magnitude of the force is given by the formula

(78)
$$F = \gamma \frac{Mm}{r^2},$$

where M and m are the masses of the two particles, r is the distance between them, and γ is a constant that depends on the units used. In the CGS system, experimental determinations give $\gamma \approx 6.675 \times 10^{-8}$ cm^3/gram sec^2, but we will have no need for the value of the constant γ in this book.

Equation (78) can be put into the vector form

(79)
$$\mathbf{F} = \gamma \frac{Mm}{r^2} \mathbf{e}$$

merely by introducing a unit vector $\mathbf{e}$ that has the proper direction. Equation (79) is convenient for theoretical discussions, but in practical work we decompose the vectors into components, and obtain from (79) three scalar equations which we use for computation.

The *gravitational attraction* of a certain body $\mathscr{B}$ at a point P is the attractive force that the body would exert on a particle of *unit* mass placed at P. If the body $\mathscr{B}$ is small compared to its distance from P, we can regard all the mass of $\mathscr{B}$ as being concentrated at a point. Then $m = 1$, and (79) gives

(80)
$$\mathbf{F} \approx \gamma \frac{M}{r^2} \mathbf{e}$$

for the gravitational attraction. If the body is large in comparison with its distance from P, then (80) is not a good approximation. We must then resort to integration. Dividing the body up into pieces of small diameter, forming a sum and taking the limit in the usual way, we arrive at the vector integral

(81)
$$\mathbf{F} = \gamma \iiint_{\mathcal{B}} \frac{\mu}{r^2} \mathbf{e} \, dV$$

for the gravitational attraction due to $\mathcal{B}$. Here μ is the mass density of the body, and it, together with r and $\mathbf{e}$, may vary over the region of integration. In contrast, γ has been placed in front of the integral sign, because according to Newton's Law (and in conformity with all experimental evidence) it is a universal constant. Again we observe that for computation, equation (81) yields three scalar equations by taking components.

We mention in passing that the laws for electrical attraction and magnetic attraction have the same form as (79), but with slightly different meanings attached to the symbols. Hence a careful study of gravitational attraction automatically gives useful information about the other two phenomena.

Example 1. Find the gravitational force exerted by a homogeneous solid sphere at a point Q outside the sphere.

Solution. We place the sphere so that its center is at the origin of a rectangular coordinate system. By the symmetry of the sphere, we may assume that Q is on the z-axis at $(0, 0, a)$, where $a > R$ the radius of the sphere. If we write $\mathbf{F} = F_1\mathbf{i} + F_2\mathbf{j} + F_3\mathbf{k}$ for the force of attraction, then by the symmetry of the sphere, it follows that $F_1 = F_2 = 0$, and it only remains to compute F_3.

We now introduce a spherical coordinate system along with the rectangular coordinate system in the usual manner, as indicated in Fig. 25. Then

$$|\mathbf{QP}|^2 = r^2 = a^2 + \rho^2 - 2a\rho \cos \varphi$$

and the component of the vector $\mathbf{QP}$ on the z-axis is $-(a - \rho \cos \varphi)$.

For the integrand in equation (81), we have

$$\frac{\mu}{r^2} \mathbf{e} = \frac{\mu}{r^2} \frac{\mathbf{QP}}{|\mathbf{QP}|} = \frac{\mu \mathbf{QP}}{r^3}$$

and for the z-component of this vector, we have

$$\frac{-\mu(a - \rho \cos \varphi)}{(a^2 + \rho^2 - 2a\rho \cos \varphi)^{3/2}}.$$

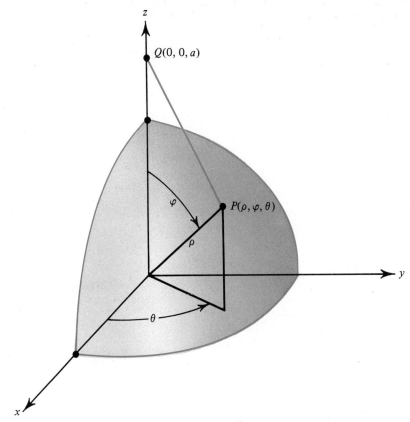

FIGURE 25

If we use spherical coordinates for the integration, (81) yields

$$F_3 = \gamma\mu \int_0^{2\pi} \int_0^{\pi} \int_0^R \frac{\rho\cos\varphi - a}{(a^2 + \rho^2 - 2a\rho\cos\varphi)^{3/2}} \rho^2 \sin\varphi \, d\rho \, d\varphi \, d\theta.$$

Since the integrand is independent of θ, we carry out this integration first and obtain

(82) $$F_3 = 2\pi\gamma\mu \int_0^R \int_0^{\pi} \frac{(\rho\cos\varphi - a)\rho^2 \sin\varphi}{(a^2 + \rho^2 - 2a\rho\cos\varphi)^{3/2}} \, d\varphi \, d\rho.$$

To carry out the integration on φ, we make the substitutions

$$u^2 = a^2 + \rho^2 - 2a\rho\cos\varphi > 0,$$

$$2u \, du = 2a\rho \sin\varphi \, d\varphi,$$

$$\rho\cos\varphi = \frac{a^2 + \rho^2 - u^2}{2a}.$$

Then

$$\int_0^\pi \frac{(\rho \cos \varphi - a)\rho^2 \sin \varphi \, d\varphi}{(a^2 + \rho^2 - 2a\rho \cos \varphi)^{3/2}} = \int_{a-\rho}^{a+\rho} \frac{\left(\dfrac{a^2 + \rho^2 - u^2}{2a} - a\right)\rho \dfrac{u}{a}}{u^3} \, du$$

$$= \frac{-\rho}{2a^2} \int_{a-\rho}^{a+\rho} \frac{u^2 + a^2 - \rho^2}{u^2} \, du$$

$$= \frac{-\rho}{2a^2} \left(u - \frac{a^2 - \rho^2}{u}\right)\Bigg|_{a-\rho}^{a+\rho} = \frac{-2\rho^2}{a^2}.$$

Using this result in (82), we have

(83) $$F_3 = 2\pi\gamma\mu \int_0^R \frac{-2\rho^2}{a^2} \, d\rho = -\frac{\gamma}{a^2} \frac{4}{3} \pi R^3 \mu = -\gamma \frac{M}{a^2},$$

where M is the mass of the sphere.   ●

We see from (83) that the gravitational attraction due to a homogeneous sphere at a point outside the sphere is the same as the attraction that would result if all the mass were concentrated at the center of the sphere.

EXERCISE 11

1. Why is F_3 negative in equation (83)?

In Problems 2, 3, and 5, express the answer in terms of the mass of the given body.

2. A homogeneous wire of length $2a$ is placed on the x-axis with its midpoint at the origin. Find the gravitational attraction at $(b, 0)$, where $b > a$.

3. For the wire of Problem 2, find the attraction at $(0, b)$, where $b > 0$.

4. Find the attraction if the wire has infinite length in both directions, but all other conditions in Problem 3 are the same.

5. A homogeneous circular disk of radius R is placed on the xy-plane with its center at the origin. Find the attraction at the point $(0, 0, a)$, $a > 0$.

*6. Let $\mathcal{R}$ be the region between two concentric spheres of radii R_1 and R_2, respectively ($R_1 < R_2$), and suppose that this region is filled with a homogeneous material. Prove that the gravitational attraction at any point inside the smaller sphere is zero. HINT: We may set up the integral just as in the example problem of this section, except that the point $Q(0, 0, a)$ is now inside the smaller circle and hence $a < R_1$. The integrand is exactly the same, and the same substitution $u^2 = a^2 + \rho^2 - 2a\rho \cos \varphi$

permits the same evaluation. The only difference is that now as φ runs from 0 to π, the new variable u runs from $\rho - a$ to $\rho + a$ (previously it ran from $a - \rho$ to $a + \rho$).

★7. Find the gravitational attraction at the vertex of a homogeneous right circular cone of base radius R and height H. HINT: Place the cone with its vertex at the origin and its axis on the z-axis. Then use spherical coordinates.

★8. Find the gravitational attraction for the cone of Problem 7 if the density is $k\rho$, where k is a constant.

★9. Repeat Problem 8 if $\mu = k\rho^n$, $n > 1$.

10. Obtain the answer to Problem 8 from the answer to Problem 9 by letting $n \to 1$.

REVIEW PROBLEMS

In Problems 1 through 11, evaluate the given iterated integral.

1. $\displaystyle\int_0^2 \int_0^x x^2 y^3 \, dy \, dx.$

2. $\displaystyle\int_0^2 \int_0^y x^2 y^3 \, dx \, dy.$

3. $\displaystyle\int_1^3 \int_2^x (x - 1)(y - 2) \, dy \, dx.$

4. $\displaystyle\int_1^2 \int_{x^2}^{x^3} \frac{x^{13}}{y^5} \, dy \, dx.$

5. $\displaystyle\int_0^{\pi/4} \int_0^\beta \sin(\alpha + \beta) \, d\alpha \, d\beta.$

6. $\displaystyle\int_1^2 \int_{v^2}^{v^5} e^{u/v^2} \, du \, dv.$

7. $\displaystyle\int_0^2 \int_0^x \int_0^y (x + y + z)^2 \, dz \, dy \, dx.$

8. $\displaystyle\int_0^1 \int_0^z \int_0^y e^{x+y+z} \, dx \, dy \, dz.$

9. $\displaystyle\int_n^{n+1} \int_{n-1}^n \cdots \int_2^3 \int_1^2 x_1 x_2 \cdots x_{n-1} x_n \, dx_1 \, dx_2 \cdots dx_{n-1} \, dx_n.$

10. $\displaystyle\int_0^1 \int_0^1 \cdots \int_0^1 \int_0^1 x_1 x_2^2 \cdots x_{n-1}^{n-1} x_n^n \, dx_1 \, dx_2 \cdots dx_n.$

★★11. $\displaystyle\int_0^1 \int_0^1 \cdots \int_0^1 \int_0^1 (x_1 + x_2 + \cdots + x_n)^2 \, dx_1 \, dx_2 \cdots dx_n.$

In Problems 12 through 19, find the volume under the given surface and above the given region in the xy-plane. In each case try to make a sketch of the solid.

12. $z = 8 - 2x - y, \quad 0 \le x \le 4, \quad 0 \le y \le (4 - x)/4.$

13. $z = 8 - 2x - y, \quad 0 \le x \le 2, \quad 0 \le y \le 4 - x^2.$

14. $z = 6 - x^2 - y^2, \quad 0 \le x \le 2, \quad 0 \le y \le 2 - x.$

15. $z = 6 - x^2 - y^2, \quad 0 \le x \le 1, \quad 0 \le y \le 1 - x^2.$

★16. $z = e^{-(x+2y)}, \quad 0 \le x \le 1, \quad 0 \le y \le 3 - 3x.$

*17. $z = xe^{-(x^2+y)}$, $\quad 0 \le x \le 1$, $\quad 0 \le y \le 1 - x^2$.

18. $z = \sin(x + 2y)$, $\quad 0 \le y \le \pi/2$, $\quad 0 \le x \le \pi - 2y$.

19. $z = 12/x^2 y$, $\quad\quad 1 \le y \le 2$, $\quad\quad 1 \le x \le 3 - y$.

In Problems 20 through 23, find the volume of the solid that lies above the first surface, below the second surface, and above the given region in the xy-plane.

*20. $z = 3x^2 + y^3$, $\quad z = 7 - 2x - y^2$, $\quad 0 \le y \le 1$, $\quad 0 \le x \le 1 - y$.

21. $z = \sin^2 \sqrt{xy}$, $\quad z = 4 - \cos^2 \sqrt{xy}$, $\quad 0 \le y \le 2$, $\quad 0 \le x \le 4 - y^2$.

*22. $z = \sin(x + 2y)$, $\quad z = 6 - \sin(x - 2y)$, $\quad 0 \le y \le \pi$, $\quad 0 \le x \le \pi$.

23. $z = x^3 - y^2$, $\quad\quad z = x^3 + 2y^2$, $\quad\quad 0 \le y \le x^3$, $\quad 0 \le x \le 1$.

24. In Problems 20, 21, and 22, prove that for the specified region the second-named surface is above the first-named surface.

25. Prove that for all (x, y) the surface $z = x^3 + 2y^2$ is above or touches the surface $z = x^3 - y^2$. (See Problem 23.)

*26. Find the centroid for the region given in Problem 12.

27. Find $\bar{x}$ and $\bar{y}$ for the region given in Problem 14.

28. Find $\bar{y}$ for the region given in Problem 19.

29. Find $\bar{x}$ and $\bar{y}$ for the region given in Problem 21. Note that the computation of $\bar{z}$ is difficult.

30. Find I_z for the region given in (a) Problem 19 and (b) Problem 23.

In Problems 31 through 34, a flat sheet of material is bounded by the given curves and has the given mass density. In each case find the center of mass.

31. $y = x$, $\quad\quad\quad y = 0$, $\quad x = 2$, $\quad \rho = 1 + x$.

32. $y = x$, $\quad\quad\quad y = 0$, $\quad x = 2$, $\quad \rho = 1/(1 + x)$.

33. $y = x - x^2$, $\quad\quad y = 0$, $\quad\quad\quad\quad \rho = x$.

34. $y = \pm(2 + \sin x)$, $\quad x = 0$, $\quad x = 2\pi$, $\quad \rho = x$.

In Problems 35 through 38, find the area of that portion of the given surface that lies over the given region in the xy-plane.

35. $z = 2x^{3/2} + 2\sqrt{2}\,y$, $\quad 0 \le x \le 8$, $\quad\quad 0 \le y \le 5$.

36. $z = \sqrt{3}x + y^2$, $\quad\quad -y \le x \le y$, $\quad\quad 1 \le y \le 3$.

37. $z = 2y + x^3$, $\quad\quad\quad -x^3 \le y \le x^3$, $\quad\quad 0 \le x \le 1$.

38. $z = \sin x + 3y$, $\quad\quad 0 \le y \le \sin 2x$, $\quad 0 \le x \le \pi/2$.

39. A flat sheet of material is bounded by the curve $r = c + d \sin \theta$ (in polar coordinates) with $c - 1 > d > 0$. Find the mass if $\rho = a + br$, $a > b > 0$. HINT: Recall that

$$\int_0^{2\pi} \sin \theta \, d\theta = \int_0^{2\pi} \sin^3 \theta \, d\theta = 0, \qquad \int_0^{2\pi} \sin^2 \theta \, d\theta = \pi.$$

40. Find the mass of the material described in Problem 39 if: (a) $\rho = a - br$, and
   (b) $\rho = a + b/r$.

*41. Show that if $z = f(r,\theta)$ is the equation of a surface in polar coordinates, then
   equation (54) of Theorem 5 for the surface area becomes

$$\sigma = \iint_{\mathcal{R}} \sqrt{1 + z_r^2 + \frac{1}{r^2}z_\theta^2}\; r\, dr\, d\theta.$$

42. Find the area of that portion of the paraboloid $z = a(x^2 + y^2)$ that lies above the
   region $0 \leqq r \leqq \sqrt{\theta},\, 0 \leqq \theta \leqq \pi$.

43. Find the area of that portion of the surface $z = ar^2 \sin 2\theta$ that lies over the region
   $0 \leqq r \leqq 2,\, 0 \leqq \theta \leqq 2\pi$.

*44. Find the area of that portion of the surface $z = a \ln r$ that lies over the region
   $0 \leqq \theta \leqq \pi,\, \epsilon \leqq r \leqq 2,\, \epsilon > 0$. Find the limit of the area as $\epsilon \to 0^+$.

*In Problems 45 through 48, find the mass of the solid bounded by the coordinate planes and
the given surface (or surfaces), and having the given density.*

45. $x = 3,\quad y = 2,\quad z = 1,\quad \rho = a + bx^n,\quad n > 0$.
46. $x = 3,\quad y = 2,\quad z = 1,\quad \rho = a + bxy$.
47. $x + 2y + 4z = 4,\qquad \rho = a + bz^2$.
48. $z = e^{x+y},\quad x + y = 3,\qquad \rho = a + 2by$.

49. For the solid of Problem 45, find the center of mass and I_x.
50. For the solid of Problem 46, find the center of mass, I_x, and I_z.
*51. For the solid of Problem 48, find M_{xy} and M_{yz}.
*52. Use spherical coordinates to find the mass of the hemispherical solid
   $0 \leqq z \leqq \sqrt{R^2 - x^2 - y^2}$ if the mass density is $\mu = 2a + 3b\rho + 4c\rho^2,\, a, b, c > 0$.
   Find M_{xy} and I_z for this solid.

53. Find the mass of the solid torus bounded by $\rho = A \sin \varphi,\, A > 0$ if $\mu = 3a + b/\rho$,
   $a > 0,\, b > 0$.

20

Line and Surface Integrals

Objective **1**

The integral of a function of a single variable is defined over some interval $a \leqq x \leqq b$. A natural suggestion is to replace this segment of the x-axis by a curve in the xy-plane or a curve in space. When we do this we obtain an integral over a curve, or a curvilinear integral. Even though the integral is taken over a curve, custom decrees that it be called a *line integral,* and we follow the standard terminology.

A double integral of a function of two variables is defined over a region $\mathcal{R}$ in the plane. A natural suggestion is to replace $\mathcal{R}$ by a surface in three-dimensional space. When we do this, we obtain an integral over a surface or, briefly, a *surface integral.* We begin to investigate these new concepts.

Vector Fields and Line Integrals **2**

Let us consider a fluid flowing through a region. At each point the fluid has a velocity that can be represented by a vector. This gives rise to a vector function because at each point of the region there is now a well-defined vector. Such a vector function is also called a *vector field.* If the velocity changes with time, then the function (or field) is called a *dynamic function* (or field). If the vector function does not change with time, then we have a *static*[1] *vector function* (or field). For the present we consider only a plane region and static fields. Then the vector function $\mathbf{F}(P)$ can be put in the form

$$\mathbf{F} = \mathbf{F}(P) = M(x, y)\mathbf{i} + N(x, y)\mathbf{j},$$

[1] This type of vector field is also called a "steady state."

where M and N, the components of **F**, are well-defined scalar functions in some region. For example, the equation

(2)
$$\mathbf{F} = 7y^2\mathbf{i} - 4xy\mathbf{j}$$

gives a vector function (or field) for the entire plane because for each point (x, y) it specifies the vector associated with that point.

A vector field also arises when we consider a field of forces. The forces can be gravitational, electrical, or magnetic, but in any case, there is at each point a well-defined vector that represents the force exerted at the point on a unit mass (or unit electrical charge, or unit magnet).

Suppose now that we have a gravitational field and a particle moving along some smooth directed curve $\mathscr{C}$:

(3)
$$\mathbf{R}(t) = f(t)\mathbf{i} + g(t)\mathbf{j}$$

that lies in the field (see Fig. 1).

As the particle moves along a very small portion of $\mathscr{C}$, that portion is nearly a straight line so the amount of work done on the particle is very close to $\mathbf{F} \cdot \Delta \mathbf{R}$, where the dot represents, as usual, scalar multiplication, and enters here because only the component of **F** in the direction of $\Delta \mathbf{R}$ is pushing the particle along the path $\mathscr{C}$.

If we subdivide $\mathscr{C}$ into small arcs and take a sum of $\mathbf{F} \cdot \Delta \mathbf{R}$ for each subarc, we obtain

(4)
$$\mathbf{F}(P_1) \cdot \Delta \mathbf{R}_1 + \mathbf{F}(P_2) \cdot \Delta \mathbf{R}_2 + \cdots + \mathbf{F}(P_n) \cdot \Delta \mathbf{R}_n$$

as a close approximation to the work done, and hence we arrive at

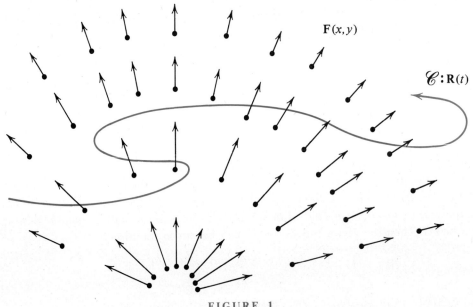

FIGURE 1

DEFINITION 1 (Work). The work done by a force field on a particle as it moves along a curve $\mathscr{C}$ is given by

(5)
$$W = \int_{\mathscr{C}} \mathbf{F} \cdot d\mathbf{R},$$

where the integral on the right is the limit of the expression (4) as $n \to \infty$ and the length of the largest subarc approaches zero.

The integral (5) is called a *line integral* (even though it is computed over a curve). We now turn our attention to the real meaning of (5) and how to compute the integral when $\mathbf{F}$ and $\mathscr{C}$ are given explicitly.

Let us first indulge in manipulations with $\mathbf{R}(t)$, the parametric equation for the curve, and $\mathbf{F}(P)$, the vector function that gives the force. From $\mathbf{R}(t) = f(t)\mathbf{i} + g(t)\mathbf{j}$, we obtain

(6)
$$d\mathbf{R} = \frac{d\mathbf{R}}{dt}\,dt = [f'(t)\mathbf{i} + g'(t)\mathbf{j}]\,dt.$$

We evaluate the vector function $\mathbf{F} = M(x, y)\mathbf{i} + N(x, y)\mathbf{j}$ for points on the curve by setting $x = f(t)$ and $y = g(t)$ so that on $\mathscr{C}$

(7)
$$\mathbf{F} = M(f(t), g(t))\mathbf{i} + N(f(t), g(t))\mathbf{j}.$$

Finally, suppose that the curve goes from the initial point P to the terminal point Q as t runs from a to b. If we take the dot product $\mathbf{F} \cdot d\mathbf{R}$ and use (6) and (7), the integral in (5) becomes

(8)
$$\int_{\mathscr{C}} \mathbf{F} \cdot d\mathbf{R} = \int_a^b [M(f(t), g(t))f'(t) + N(f(t), g(t))g'(t)]\,dt,$$

and this is a well-defined integral of a scalar function over the interval $a \leq t \leq b$.

Example 1. Evaluate the line integral

(9)
$$W = \int_{\mathscr{C}} (7y^2\mathbf{i} - 4xy\mathbf{j}) \cdot d\mathbf{R}$$

over the arc that joins the points $(0, 0)$ and $(2, 4)$: **(a)** for $\mathscr{C}_1 : \mathbf{R}(t) = t\mathbf{i} + 2t\mathbf{j}$, and **(b)** for $\mathscr{C}_2 : \mathbf{R}(t) = t\mathbf{i} + 2\sqrt{2t}\mathbf{j}$.

Solution. The reader should check that as t runs from 0 to 2, both curves have the same initial point $(0, 0)$ and the same terminal point $(2, 4)$. Further, $\mathcal{C}_1$ is a straight-line segment and $\mathcal{C}_2$ is a piece of a parabola.

(a) For $\mathcal{C}_1 : f(t) = t,\ g(t) = 2t$, and equation (8) gives

$$\int_{\mathcal{C}_1} \mathbf{F} \cdot d\mathbf{R} = \int_0^2 (7y^2 \cdot 1 - 4xy \cdot 2)\, dt$$

$$= \int_0^2 [7(2t)^2 - 8t(2t)]\, dt = \int_0^2 (28 - 16)t^2\, dt$$

$$= 4t^3 \Big|_0^2 = 32.$$

(b) For $\mathcal{C}_2 : f(t) = t,\ g(t) = 2\sqrt{2t},\ f'(t) = 1$, and $g'(t) = \sqrt{2}/\sqrt{t}$. Then

$$\int_{\mathcal{C}_2} \mathbf{F} \cdot d\mathbf{R} = \int_0^2 \left(7y^2 \cdot 1 - 4xy\, \frac{\sqrt{2}}{\sqrt{t}}\right) dt$$

$$= \int_0^2 \left[7(2\sqrt{2t})^2 - 4t(2\sqrt{2t})\, \frac{\sqrt{2}}{\sqrt{t}}\right] dt$$

$$= \int_0^2 (56 - 16)t\, dt = 20t^2 \Big|_0^2 = 80. \quad \bullet$$

Observe that the two different curves give two different values to W although the force field is the same in both problems.

We return a moment to the integral in (5) and observe that we can put it in the form

$$\int_{\mathcal{C}} \mathbf{F} \cdot d\mathbf{R} = \int_{\mathcal{C}} [M\mathbf{i} + N\mathbf{j}] \cdot [dx\mathbf{i} + dy\mathbf{j}],$$

or

(10)
$$\int_{\mathcal{C}} \mathbf{F} \cdot d\mathbf{R} = \int_{\mathcal{C}} M\, dx + N\, dy = \int_{\mathcal{C}} M\, dx + \int_{\mathcal{C}} N\, dy.$$

Thus the integral in (10) can be evaluated in two pieces. Let us look at the first piece. Each partition of the interval $a \leqq t \leqq b$ automatically induces a partition of the curve $\mathcal{C}$, as indicated in Fig. 2. Further, each selection of $t_k^\star$ with $t_{k-1} \leqq t_k^\star \leqq t_k$ gives a point $P_k^\star(x_k^\star, y_k^\star)$ on the curve. We let $M(P_k^\star) = M(x_k^\star, y_k^\star)$ and form the sum

$$S_n^\star = M(P_1^\star)(x_1 - x_0) + M(P_2^\star)(x_2 - x_1) + \cdots + M(P_n^\star)(x_n - x_{n-1}),$$

or

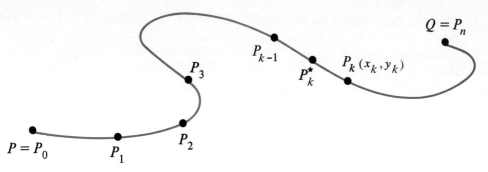

FIGURE 2

(11) $$S_n^\star \equiv \sum_{k=1}^{n} M(P_k^\star)\, \Delta x_k.$$

Then

(12) $$\int_{\mathscr{C}} M\, dx = \lim_{\substack{n\to\infty \\ \mu(\mathscr{P})\to 0}} \sum_{k=1}^{n} M(P_k^\star)\, \Delta x_k.$$

The same partition leads to a similar expression for the second integral in (10):

(13) $$\int_{\mathscr{C}} N\, dy = \lim_{\substack{n\to\infty \\ \mu(\mathscr{P})\to 0}} \sum_{k=1}^{n} N(P_k^\star)\, \Delta y_k.$$

Suppose that $\mathscr{C}$ is a vertical line segment. Then in (12) each $\Delta x_k = 0$ and the sum is also 0. Similarly, in (13) each $\Delta y_k = 0$ on any horizontal line segment. Hence we have

---

THEOREM 1. Let $\mathscr{V}$ be a vertical line segment and $\mathscr{H}$ be a horizontal line segment. Then

(14) $$\int_{\mathscr{V}} M(x, y)\, dx = 0 \qquad \text{and} \qquad \int_{\mathscr{H}} N(x, y)\, dy = 0.$$

The reader will also find it easy to prove

THEOREM 2. Let $\mathscr{C}$ be a smooth curve and let $-\mathscr{C}$ denote a curve that coincides with $\mathscr{C}$ as a point set but has the opposite direction. Then

(15) $$\int_{-\mathscr{C}} M(x, y)\, dx + N(x, y)\, dy = -\int_{\mathscr{C}} M(x, y)\, dx + N(x, y)\, dy.$$

Consequently, reversing the direction of a curve reverses the sign of the line integral (5).

It is often important in physics to consider the work done as the particle moves over a *simple closed curve*. A curve is said to be *closed* if its initial and terminal points coincide. A closed curve is called *simple* if it has no self-intersections (does not cross itself; see Fig. 3).

Simple closed curves   Not a simple curve

FIGURE 3

Simple closed curves in the plane can be directed in two possible ways—clockwise or counterclockwise. The standard agreement is to take the counterclockwise direction as the positive direction. For elementary curves as a circle, ellipse, or square, the meaning of "counterclockwise" is reasonably clear, but for a curve of the type shown in Fig. 4a one might demand a precise definition of "counterclockwise." Such a definition lies outside the realm of the calculus, so we omit it and instead offer the following. According to the Jordan Closed Curve Theorem, each simple closed curve $\mathscr{C}$ divides the plane into two regions, a bounded one, called the *inside,* or the region enclosed by $\mathscr{C}$; and an unbounded one. Then at each point on $\mathscr{C}$, the *counterclockwise* (or *positive*) *direction* is that direction for which the enclosed region is on the *left*. This is indicated by the arrows in Fig. 4a. This definition applies also to a region whose boundary consists of several simple closed curves.

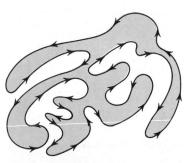 

FIGURE 4a   FIGURE 4b

This is illustrated in Fig. 4b where, by definition, the positive direction on the boundary of the region is that direction for which the region is on the left. Again this is indicated by arrows. Henceforth, when we speak of integrating around the boundary of a region, we will mean in the positive direction unless the contrary is explicitly stated.

A line integral over a simple closed curve is indicated by a variety of notations:

(16)
$$\oint \mathbf{F} \cdot d\mathbf{R}, \qquad \oint \mathbf{F} \cdot d\mathbf{R}, \qquad \int_{\circ} \mathbf{F} \cdot d\mathbf{R},$$

or just the notation in (5) with the statement that $\mathscr{C}$ is a simple closed curve.

Example 2. Evaluate $\oint x\, dy$, over the boundary of the rectangle $\mathscr{T}$ bounded by the lines $x = x_0$, $x = x_0 + a$, $y = y_0$, and $y = y_0 + b$, where $a > 0$, $b > 0$.

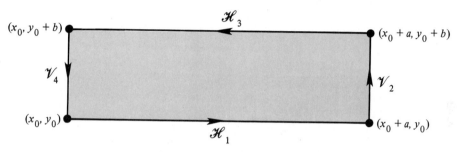

FIGURE 5

Solution. Let $\mathscr{H}_1$, $\mathscr{V}_2$, $\mathscr{H}_3$, and $\mathscr{V}_4$ be the horizontal and vertical line segments that form the boundary of $\mathscr{T}$ (see Fig. 5). Obviously,

$$\oint x\, dy = \int_{\mathscr{H}_1} x\, dy + \int_{\mathscr{V}_2} x\, dy + \int_{\mathscr{H}_3} x\, dy + \int_{\mathscr{V}_4} x\, dy.$$

By Theorem 1, the first and third integrals are zero. Notice that the positive direction on $\mathscr{V}_4$ is downward. Then

$$\oint x\, dy = \int_{y_0}^{y_0+b} (x_0 + a)\, dy + \int_{y_0+b}^{y_0} x_0\, dy = (x_0 + a)b - x_0 b = ab = A(\mathscr{T}),$$

the area of the rectangle $\mathscr{T}$. ●

Most of this material extends to three-dimensional space without any difficulty. In this case we write $\mathbf{F} = L\mathbf{i} + M\mathbf{j} + N\mathbf{k}$ and

(17)
$$\int_{\mathscr{C}} \mathbf{F} \cdot d\mathbf{R} = \int_{\mathscr{C}} (L\mathbf{i} + M\mathbf{j} + N\mathbf{k}) \cdot (dx\mathbf{i} + dy\mathbf{j} + dz\mathbf{k})$$

$$= \int_{\mathscr{C}} L\, dx + M\, dy + N\, dz.$$

With suitable modifications, Theorems 1 and 2 are still true in three-dimensional space. However, there is no preferred direction for a curve in three-dimensional space; the positive direction is just the one assigned to the curve by its parameter.

EXERCISE 1

In Problems 1 through 6, evaluate the given line integral along the indicated curve from P to Q.

1. $\int_{\mathscr{C}} 3y^2 \, dx + 2xy \, dy$, $\mathscr{C} : y^2 = 9x$, $P(0,0), \, Q(1,3)$.

2. $\int_{\mathscr{C}} 9x^2y \, dx - 11xy^2 \, dy$, $\mathscr{C} : \mathbf{R} = t^2\mathbf{i} + t^3\mathbf{j}$, $P(0,0), \, Q(1,1)$.

3. Same integrand and points as Problem 2, $\mathscr{C}$ a straight line.

4. Same integrand and points as Problem 2, $\mathscr{C} : \mathbf{R} = t\mathbf{i} + t^n\mathbf{j}, \, n > 1$.

5. $\int_{\mathscr{C}} y \, dx + x \, dy, \, P(A,0), \, Q(-A,0), \, A > 0, \, \mathscr{C}$ the upper half of a circle.

6. $\int_{\mathscr{C}} y^2 \, dx + x^2 \, dy$, same curve as in Problem 5.

In Problems 7 through 9, compute the indicated integral.

7. $\oint \dfrac{x \, dy - y \, dx}{x^2 + y^2}$, $\mathscr{C}$ a circle with center at 0.

8. $\oint y \, dx$, $\mathscr{C}$ the curve of Example 2.

9. $\dfrac{1}{2} \oint x \, dy - y \, dx$, $\mathscr{C}$ the curve of Example 2.

If $\mathscr{C}$ is a simple closed curve, then the area of the region enclosed by the curve is given by

$$A = \frac{1}{2} \oint x \, dy - y \, dx.$$

In Problems 10 through 15, use this formula to find the area of the region enclosed by the given curve. A proof of this formula will be indicated in Exercise 3, Problem 7.

10. The ellipse $x = a \cos t$, $y = b \sin t$.

11. The epicycloid of $n - 1$ cusps,

$$x = a(n \cos t - \cos nt), \qquad y = a(n \sin t - \sin nt).$$

12. The hypocycloid of $n + 1$ cusps,

$$x = a(n \cos t + \cos nt), \qquad y = a(n \sin t - \sin nt).$$

13. The loop of the strophoid,

$$x = a \cos 2t, \qquad y = a \tan t \cos 2t, \qquad -\frac{\pi}{4} \le t \le \frac{\pi}{4}.$$

14. The curve $x = t(1 - t)$, $y = t^2(1 - t)$, $0 \le t \le 1$.

15. The curve $x = \sin^2 t$, $y = \cos^2 t$, $0 \le t \le \pi$.

*16. Compute $\oint x \, dy - y \, dx$ over the closed curve

$$\mathbf{R} = \sin t\mathbf{i} + \sin 2t\mathbf{j}, \qquad 0 \le t \le 2\pi,$$

and explain your answer.

In Problems 17 through 21, evaluate the given line integral along the indicated curve from P to Q.

17. $\displaystyle\int_{\mathscr{C}} x \, dx + y \, dy + z \, dz$, $\qquad \mathbf{R} = t\mathbf{i} + 2t\mathbf{j} + 3t\mathbf{k}$, $P(0, 0, 0)$, $Q(2, 4, 6)$.

18. $\displaystyle\int_{\mathscr{C}} xz \, dx + yz \, dy + (3z^2 - 2x) \, dz$, $\mathscr{C}$ a straight line, $P(0, 0, 0)$, $Q(1, 2, 3)$.

19. Same integrand and points as in Problem 18, $\mathscr{C} : \mathbf{R} = t\mathbf{i} + 2t^3\mathbf{j} + 3t^2\mathbf{k}$.

20. $\displaystyle\int_{\mathscr{C}} yz^2 \, dx + xz^2 \, dy + 2xyz \, dz$, $\mathscr{C}$ a straight line, $P(0, 0, 0)$, $Q(1, 2, 3)$.

21. Same integrand and points as in Problem 20, $\mathscr{C} : \mathbf{R} = t\mathbf{i} + 2t^4\mathbf{j} + 3t^5\mathbf{k}$.

*22. Let M be the maximum value of $|\mathbf{F}(x, y, z)|$ for P on a curve $\mathscr{C}$, and let L be the length of $\mathscr{C}$. Prove that

$$\left| \int_{\mathscr{C}} \mathbf{F} \cdot d\mathbf{R} \right| \le ML.$$

3 *Independence of Path*

The student may have noticed that in Problems 20 and 21 of Exercise 1 the computation of

(18) $$\int_{\mathscr{C}} y z^2 \, dx + x z^2 \, dy + 2xyz \, dz$$

on two different curves (paths) from P to Q gave the same result. We investigate this "accident" more closely. To clarify the issue, let us introduce the symbol $I(\mathbf{F}, P, Q, \mathscr{C})$ to denote the integral

(17) $$\int_{\mathscr{C}} \mathbf{F} \cdot d\mathbf{R} \equiv \int_{\mathscr{C}} L \, dx + M \, dy + N \, dz$$

over the curve $\mathscr{C}$ from P to Q.

DEFINITION 2 (Independence of Path). Let $\mathbf{F}$ be defined in a region $\mathscr{R}$. The integral of $\mathbf{F}$ is said to be independent of the path in $\mathscr{R}$ if for every pair of points P and Q in $\mathscr{R}$ and every pair of curves $\mathscr{C}_1$ and $\mathscr{C}_2$ from P to Q and lying in $\mathscr{R}$ we have

(19) $$I(\mathbf{F}, P, Q, \mathscr{C}_1) = I(\mathbf{F}, P, Q, \mathscr{C}_2).$$

This requirement (19) seems to be so severe that one might expect that no such vector function $\mathbf{F}$ exists. We will see shortly that it is easy to find such functions and to recognize them. We first dispose of

---

THEOREM 3. The integral of $\mathbf{F}$ is independent of path in a region $\mathscr{R}$ if and only if for every simple closed curve lying in $\mathscr{R}$

(20) $$\oint \mathbf{F} \cdot d\mathbf{R} = 0.$$

Proof. As indicated in Fig. 6, let $\mathscr{C}_1$ and $\mathscr{C}_2$ be two curves from P to Q and let $\mathscr{C}$ be the closed curve formed from $\mathscr{C}_1 \cup (-\mathscr{C}_2)$. Thus $\mathscr{C}$ starts at P, goes to Q, and returns to P. Now

$$\oint_{\mathscr{C}} \mathbf{F} \cdot d\mathbf{R} = \int_{\mathscr{C}_1} \mathbf{F} \cdot d\mathbf{R} + \int_{-\mathscr{C}_2} \mathbf{F} \cdot d\mathbf{R},$$

or

(21)
$$\oint_{\mathscr{C}} \mathbf{F} \cdot d\mathbf{R} = \int_{\mathscr{C}_1} \mathbf{F} \cdot d\mathbf{R} - \int_{\mathscr{C}_2} \mathbf{F} \cdot d\mathbf{R}.$$

If the integral is independent of path, then the right side of (21) is zero, and hence (20) is satisfied. Conversely, if (20) is satisfied, then the right side of (21) is zero and the integral is independent of path. ∎

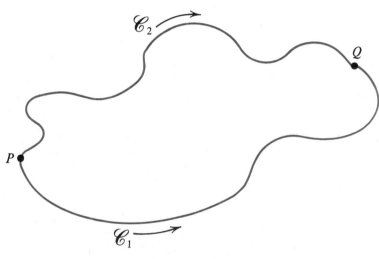

FIGURE 6

Returning now to the particular integral in (18), a sharp reader may notice that the integrand is an "exact" differential. For, if we set $\varphi(x, y, z) = xyz^2$, then we have

$$d\varphi = \frac{\partial \varphi}{\partial x} dx + \frac{\partial \varphi}{\partial y} dy + \frac{\partial \varphi}{\partial z} dz = yz^2 dx + xz^2 dy + 2xyz \, dz,$$

and this is just the integrand in (18). Consequently, we may expect that

$$\int_{\mathscr{C}} yz^2 dx + xz^2 dy + 2xyz \, dz = \int_{\mathscr{C}} d\varphi = \varphi(Q) - \varphi(P) = xyz^2 \bigg|_{(0,0,0)}^{(1,2,3)} = 18.$$

The path has disappeared in the manipulations, and we have also obtained the right answer! We put this material on a firm footing.

DEFINITION 3 (Exact Differential). The expression

(22) $$M(x, y) \, dx + N(x, y) \, dy$$

is called an *exact differential* in a region $\mathscr{R}$ if there is a $\varphi(x, y)$, differentiable in $\mathscr{R}$, such that in $\mathscr{R}$,

(23) $$\frac{\partial \varphi}{\partial x} = M(x, y) \quad \text{and} \quad \frac{\partial \varphi}{\partial y} = N(x, y).$$

The expression

(24) $$L(x, y, z) \, dx + M(x, y, z) \, dy + N(x, y, z) \, dz$$

is called an exact differential in a region $\mathscr{R}$ if there is a $\varphi(x, y, z)$ such that in $\mathscr{R}$,

(25) $$\frac{\partial \varphi}{\partial x} = L(x, y, z), \quad \frac{\partial \varphi}{\partial y} = M(x, y, z), \quad \frac{\partial \varphi}{\partial z} = N(x, y, z).$$

This concept can be phrased efficiently by using the gradient.[1] Thus (24) is an exact differential if the vector

(26) $$\mathbf{F} = L(x, y, z)\mathbf{i} + M(x, y, z)\mathbf{j} + N(x, y, z)\mathbf{k},$$

is the gradient of some scalar function φ; that is, if

$$\nabla \varphi = \mathbf{F}.$$

When such a φ exists, it is called a *potential function,* or a *potential* of $\mathbf{F}$. With these conventions the differential (24) can be condensed to $\mathbf{F} \cdot d\mathbf{R}$, and it is an exact differential if it can be expressed as $(\nabla \varphi) \cdot d\mathbf{R}$.

THEOREM 4. If $\mathbf{F} \cdot d\mathbf{R}$ is an exact differential in $\mathscr{R}$, then (17) is independent of the path in $\mathscr{R}$. Further, if φ is a potential of $\mathbf{F}$, then for any curve in $\mathscr{R}$, from P to Q,

(27) $$\int_{\mathscr{C}} \mathbf{F} \cdot d\mathbf{R} = \varphi(Q) - \varphi(P).$$

Proof. Let $\mathbf{R} = f(t)\mathbf{i} + g(t)\mathbf{j} + h(t)\mathbf{k}$ be a vector equation that generates $\mathscr{C}$ for t in $[a, b]$.

---

[1] The gradient was introduced and studied in Chapter 18, Section 10. The student might benefit by reviewing that section.

Then by definition,

$$\int_{\mathscr{C}} \mathbf{F} \cdot d\mathbf{R} = \int_a^b L(f(t), g(t), h(t)) f'(t) \, dt$$

$$+ \int_a^b M(f(t), g(t), h(t)) g'(t) \, dt + \int_a^b N(f(t), g(t), h(t)) h'(t) \, dt.$$

If φ is a potential for **F**, then equations (25) give

$$\int_{\mathscr{C}} \mathbf{F} \cdot d\mathbf{R} = \int_a^b \frac{\partial \varphi}{\partial x} \frac{dx}{dt} \, dt + \int_a^b \frac{\partial \varphi}{\partial y} \frac{dy}{dt} \, dt + \int_a^b \frac{\partial \varphi}{\partial z} \frac{dz}{dt} \, dt$$

$$= \int_a^b \frac{d\varphi(\mathbf{R}(t))}{dt} \, dt = \varphi(\mathbf{R}(b)) - \varphi(\mathbf{R}(a)) = \varphi(Q) - \varphi(P).$$

This is (27), and since the right side of (27) is independent of $\mathscr{C}$, the integral is independent of the path. ∎

The manipulations involved in the proof can be greatly condensed by writing

$$\int_{\mathscr{C}} \mathbf{F} \cdot d\mathbf{R} = \int_{\mathscr{C}} \nabla\varphi \cdot d\mathbf{R} = \int_a^b \left(\nabla\varphi \cdot \frac{d\mathbf{R}}{dt} \right) dt = \int_a^b \frac{d\varphi}{dt} \, dt = \varphi(Q) - \varphi(P).$$

Example 1. Investigate the line integral

(28)
$$\int_{\mathscr{C}} \frac{y^2 + z^2}{(z - xy)^2} \, dx + \frac{z(1 + x^2)}{(z - xy)^2} \, dy - \frac{y(1 + x^2)}{(z - xy)^2} \, dz$$

for independence of the path.

Solution. It is not at all obvious, but a little labor will show that if

$$\varphi = \frac{xz + y}{z - xy},$$

then the integrand in (28) is just $\nabla\varphi \cdot d\mathbf{R}$. Consequently, wherever φ is differentiable, the integral is independent of the path. The only crucial points are those for which the denominator in φ vanishes. If $z - xy = 0$, then P lies on the surface $z = xy$.

If P and Q lie on the same side of this surface and $\mathscr{C}$ does not meet the surface $z = xy$, then (28) is independent of path. However, no assertion is made for a path that contains a point of the surface $z = xy$. ●

The converse of Theorem 4 is not difficult to prove, but the proof does occupy time, so we are content to state

---

THEOREM 5 (PWO). If the integral (17) is independent of path in a region $\mathscr{R}$ and $\mathbf{F}$ is continuous in $\mathscr{R}$, then there is a φ such that $\nabla\varphi = \mathbf{F}$, that is,

$$(29) \qquad \frac{\partial\varphi}{\partial x} = L, \qquad \frac{\partial\varphi}{\partial y} = M, \qquad \frac{\partial\varphi}{\partial z} = N.$$

To aid in the search for exact differentials, we can use

THEOREM 6. Suppose that the first partial derivatives of L, M, and N are continuous in a region $\mathscr{R}$. If

$$(22) \qquad M(x, y)\, dx + N(x, y)\, dy$$

is an exact differential in $\mathscr{R}$, then in $\mathscr{R}$,

$$(30) \qquad \frac{\partial M}{\partial y} = \frac{\partial N}{\partial x}.$$

If

$$(24) \qquad L(x, y, z)\, dx + M(x, y, z)\, dy + N(x, y, z)\, dz$$

is an exact differential in $\mathscr{R}$, then in $\mathscr{R}$,

$$(31) \qquad \frac{\partial L}{\partial y} = \frac{\partial M}{\partial x}, \qquad \frac{\partial M}{\partial z} = \frac{\partial N}{\partial y}, \qquad \frac{\partial N}{\partial x} = \frac{\partial L}{\partial z}.$$

Conversely, if the conditions (30) hold for all (x, y) in the plane or the conditions (31) hold for all (x, y, z) in three-dimensional space, then the corresponding expressions (22) or (24) are exact differentials.[1]

The proof of the direct part of this theorem follows from the equality of the mixed partial derivatives of φ. The details are left as an exercise for the student. The converse part is a little more complicated and will be omitted.

Example 2. Test the differential

$$(32) \qquad (2xy^3 + 5\cos x)\, dx + (3x^2 y^2 - 4e^y)\, dy$$

to see if it may be exact, and try to find a suitable φ.

---

[1] More generally, the converse is true if (30) or (31) hold throughout a region $\mathscr{R}$ that is *simply connected* as defined in the next section.

Solution. The test described in equation (30) gives

(33) $$\frac{\partial M}{\partial y} = \frac{\partial}{\partial y}(2xy^3 + 5\cos x) = 6xy^2 = \frac{\partial}{\partial x}(3y^2x^2 - 4e^y) = \frac{\partial N}{\partial x}.$$

Hence the expression in (32) is an exact differential. To find φ, we integrate along an arbitrary path from a fixed point P to a variable point $Q(x, y)$. For convenience we select the "elbow path" composed of two line segments as indicated in Fig. 7.

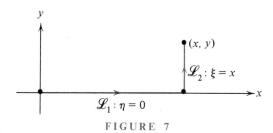

FIGURE 7

On $\mathscr{L}_1$ we can set $x = \xi$ and $y = 0$ with $0 \leq \xi \leq x$. On $\mathscr{L}_2$ we have $0 \leq \eta \leq y$, while x is fixed. Hence with $\mathbf{F} = (2xy^3 + 5\cos x)\mathbf{i} + (3x^2y^2 - 4e^y)\mathbf{j}$, we have

$$\varphi(x, y) = \int_{\mathscr{L}_1} \mathbf{F} \cdot d\mathbf{R} + \int_{\mathscr{L}_2} \mathbf{F} \cdot d\mathbf{R}$$

$$= \int_0^x (2\xi \cdot 0^3 + 5\cos \xi)\, d\xi + \int_0^y (3\eta^2 x^2 - 4e^\eta)\, d\eta$$

$$\varphi(x, y) = 5\sin x + y^3 x^2 - 4e^y + 4.$$

To obtain the most general potential Φ, we can replace 4 with an arbitrary constant. This gives

(34) $$\Phi = 5\sin x + x^2 y^3 - 4e^y + C.$$

This amounts to replacing $P(0, 0)$ by some arbitrary initial point.   ●

Integration along an elbow path is a systematic way of finding a potential function φ, but it is not the only way or the best way. Frequently, one is able to guess φ, and one can then prove the guess is correct by computing φ_x and φ_y.

EXERCISE 2

In Problems 1 through 7, use the test of Theorem 6 to see if the given differential may be exact. Whenever a potential function exists, find it.

1. $(6x^2 + 3y^2)\,dx + 6xy\,dy$.

2. $(3x^2y^4 + 5x^4y^2 + 4)\,dx + (2x^5y + 4x^3y^3 + 6)\,dy$.

3. $(3y^2 + 6xy)\,dx + (3x^2 + 6y)\,dy$.

4. $3(x^2 \sin y + \cos^2 y)\,dx - x(3 \sin 2y - x^2 \cos y)\,dy$.

★5. $(4x^3 - 2xy)e^{y/x^2}\,dx + x^2 e^{y/x^2}\,dy$.

★6. $\dfrac{2xy^2\,dx - (x^3 + x^2y)\,dy}{(x - y)^3}$.

7. $e^{2x} \sin y\,dx + e^{2x} \cos y\,dy$.

8. A force field in $\mathscr{R}$ is said to be *conservative* if the integral (17) is independent of the path in $\mathscr{R}$. Prove that if $\mathbf{F}$ is a constant, then the force field is conservative. Physically, this is the situation near the earth's surface, where the gravitational attraction is essentially constant.

9. Let $\mathbf{F}$ be the gravitational force due to a mass concentrated at a point, located at the origin for convenience. Then $\mathbf{F} = -k\mathbf{e}/r^2$ at P, where $\mathbf{e}$ is a unit vector having the direction of $\mathbf{OP}$ and $r = |\mathbf{OP}|$. Prove that $\mathbf{F}$ is a conservative field.

★10. Let $g(r^2)$ be any differentiable scalar function of $r^2 = x^2 + y^2 + z^2$. Prove that the field $\mathbf{F} \equiv g(r^2)(x\mathbf{i} + y\mathbf{j} + z\mathbf{k})$ is a conservative field.

★11. Prove that if $\mathbf{F}$ and $\mathbf{G}$ are two conservative fields, then $\mathbf{F} + \mathbf{G}$ is a conservative field.

In Problems 12 through 15, show that $\mathbf{F}$ is conservative by finding a suitable potential.

12. $(y - 3z + 2z^2)\mathbf{i} + (x + 2z)\mathbf{j} + (2y - 3x + 4xz)\mathbf{k}$.

13. $\dfrac{xy^2}{z^5}[2yz\mathbf{i} + 3xz\mathbf{j} - 4xy\mathbf{k}]$.

14. $(e^{y^2} + 2xze^{x^2})\mathbf{i} + (e^{z^2} + 2xye^{y^2})\mathbf{j} + (e^{x^2} + 2yze^{z^2})\mathbf{k}$.

15. $ze^{xz} \sin yz\mathbf{i} + ze^{xz} \cos yz\mathbf{j} + e^{xz}[x \sin yz + y \cos yz]\mathbf{k}$.

4 Green's Theorem

Let $\mathscr{R}$ be a closed and bounded region that is convex in the direction of the x-axis and y-axis (see Fig. 8). We recall from Chapter 19, Definition 2, page 807, that a region of this type is called a *normal* region. Our objective is to express the line integral

$$(35) \qquad \oint \mathbf{F} \cdot d\mathbf{R} = \oint M\,dx + N\,dy$$

taken over the boundary of $\mathscr{R}$ in terms of a double integral over $\mathscr{R}$ of a suitable function.

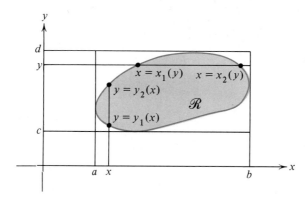

FIGURE 8

We first examine the iterated integral

$$I_1 \equiv \int_c^d \int_{x_1(y)}^{x_2(y)} \frac{\partial N}{\partial x}\,dx\,dy,$$

where $x_1(y)$ and $x_2(y)$ for $c \leq y \leq d$ are just those functions which give the boundary curves of $\mathscr{R}$ on the left and right sides, respectively. Then

$$(36) \qquad I_1 = \int_c^d [N(x_2(y), y) - N(x_1(y), y)]\,dy$$

$$= \int_c^d N(x_2(y), y)\,dy + \int_d^c N(x_1(y), y)\,dy$$

$$I_1 = \oint N(x, y)\,dy.$$

A similar iterated integral gives

$$(37) \qquad I_2 = \int_a^b \int_{y_1(x)}^{y_2(x)} \frac{\partial M}{\partial y}\,dy\,dx$$

$$= \int_a^b [M(x, y_2(x)) - M(x, y_1(x))]\,dx$$

$$= -\left[\int_a^b M(x, y_1(x))\, dx + \int_b^a M(x, y_2(x))\, dx \right]$$

$$I_2 = -\oint M(x, y)\, dx.$$

If N_x and M_y are continuous in $\mathcal{R}$, then the two iterated integrals I_1 and I_2 can be written as double integrals over $\mathcal{R}$. The combination $I_1 - I_2$ yields

(38)
$$\oint M\, dx + N\, dy = \iint\limits_{\mathcal{R}} \left(\frac{\partial N}{\partial x} - \frac{\partial M}{\partial y} \right) dA.$$

THEOREM 7 (Green's Theorem). Let $\mathcal{R}$ be a closed region that can be obtained as the union of a finite number of normal regions. If M, N, M_y, and N_x are continuous in $\mathcal{R}$, then equation (38) holds, where the contour on the left is over the boundary of $\mathcal{R}$ in the positive sense.

Proof. We have already proved Green's Theorem in the simple case that $\mathcal{R}$ is itself a normal region. Now let $\mathcal{R} = \mathcal{R}_1 \cup \mathcal{R}_2 \cup \cdots \cup \mathcal{R}_n$, where each region $\mathcal{R}_k$ is a normal region, and, if $j \neq k$, then $\mathcal{R}_j \cap \mathcal{R}_k$ is either empty or contains only boundary points of $\mathcal{R}_j$ and $\mathcal{R}_k$. In Fig. 9 we see a region of this type, where $\mathcal{R}$ is the union of six normal regions.

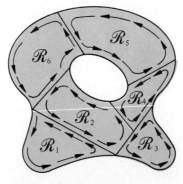

FIGURE 9

The positive direction along the boundaries of $\mathcal{R}_k$ is indicated by the arrows. Let $\mathcal{C}_k$ be the boundary of $\mathcal{R}_k$, $k = 1, 2, \ldots, n$. Then for each of the normal regions, we already know

that

(39)
$$\int_{\mathscr{C}_k} M\,dx + N\,dy = \int\int_{\mathscr{R}_k} \left(\frac{\partial N}{\partial x} - \frac{\partial M}{\partial y}\right) dA.$$

We form the sum of n such equations (39), for $k = 1, 2, \ldots, n$. The right side of the sum gives the right side of (38). The left side of the sum also gives the left side of (38), because the line integrals over common boundary curves of adjacent regions always occur in pairs with opposing directions and hence cancel (see Theorem 2), leaving only the line integrals over the boundary of $\mathscr{R}$. ∎

The composition and cancellation described in the proof is shown graphically in the transition from Fig. 9 to Fig. 10.

FIGURE 10

Example 1. Let $\mathscr{C}$ be the ellipse $x^2 + 4y^2 = 4$. Compute

$$I = \int_{\mathscr{C}} (\cos x + 4xy)\,dx + (e^y + 2x^2 + 5x)\,dy.$$

Solution. We apply Green's Theorem, equation (38). Then

$$\frac{\partial N}{\partial x} - \frac{\partial M}{\partial y} = \frac{\partial}{\partial x}(e^y + 2x^2 + 5x) - \frac{\partial}{\partial y}(\cos x + 4xy)$$

$$= 4x + 5 - 4x = 5.$$

Therefore,

$$I = \int\int_{\mathcal{R}} \left(\frac{\partial N}{\partial x} - \frac{\partial M}{\partial y} \right) dx\, dy = \int\int_{\mathcal{R}} 5\, dx\, dy$$

$$= 5(\text{area of } \mathcal{R}).$$

For this ellipse, $A = \pi ab = \pi(2)(1)$. Hence $I = 10\pi$. ●

We observe that the region $\mathcal{R}$ shown in Figs. 9 and 10 has a "hole." This certainly marks $\mathcal{R}$ as essentially different from a circular disk or a rectangle. A region without "holes" is called *simply connected*. This notion is made precise in

---

DEFINITION 4 (Simple Connectivity). A region $\mathcal{R}$ is said to be *simply connected* if for every simple closed curve $\mathcal{C}$ contained in $\mathcal{R}$, the region enclosed by $\mathcal{C}$ is also contained in $\mathcal{R}$.

A region that is *not* simply connected is called *multiply connected*. Any region with holes is multiply connected. In this regard the reader should keep in mind that holes are not necessarily visible. The deletion of a single point from a simply connected region is sufficient to destroy the simple connectivity. Thus the punctured disk consisting of (x, y) such that $0 < x^2 + y^2 < 1$ is multiply connected. The relation between connectivity and Green's Theorem is exposed in

Example 2. Let $\mathcal{C}$ be a simple closed curve that bounds a normal region. Compute the line integrals

$$I_1 \equiv \int_{\mathcal{C}} (x^3 - 3xy^2)\, dx - 3x^2y\, dy \qquad \text{and} \qquad I_2 \equiv \int_{\mathcal{C}} \frac{-y\, dx + x\, dy}{x^2 + y^2}.$$

Solution. We apply Green's Theorem to I_1. We find that

$$\frac{\partial N}{\partial x} - \frac{\partial M}{\partial y} = \frac{\partial}{\partial x}(-3x^2y) - \frac{\partial}{\partial y}(x^3 - 3xy^2) = -6xy - (-6xy) = 0.$$

Consequently, the integral on the right side of (38) is zero, and hence $I_1 = 0$. The same manipulation for I_2 yields

$$\frac{\partial N}{\partial x} - \frac{\partial M}{\partial y} = \frac{\partial}{\partial x}\frac{x}{x^2 + y^2} - \frac{\partial}{\partial y}\frac{-y}{x^2 + y^2} = \frac{y^2 - x^2}{(x^2 + y^2)^2} - \frac{y^2 - x^2}{(x^2 + y^2)^2} = 0.$$

But the parallel conclusion, that $I_2 = 0$, may be *false*. According to Green's Theorem, we assume that M, N, M_y and N_x are continuous in $\mathscr{R}$, and conclude (38). For I_1 these hypotheses are always met, but in I_2 all four of M, N, M_y and N_x are discontinuous at $(0,0)$. If $(0,0)$ is not on $\mathscr{C}$ or in the region $\mathscr{R}$ enclosed by $\mathscr{C}$, then Green's Theorem applies and indeed $I_2 = 0$. If $(0,0)$ is in $\mathscr{R}$, then $I_2 = 2\pi$. To see this, we first select $\mathscr{C}$ to be $\mathscr{C}_r$, a circle of radius $r > 0$ and center at the origin. Then $x = r \cos t$, $y = r \sin t$, and the computation of I_2 yields

$$(40) \quad I_2 = \int_{\mathscr{C}_r} \frac{-y\,dx + x\,dy}{x^2 + y^2} = \int_0^{2\pi} \frac{-r\sin t(-r\sin t) + r\cos t(r\cos t)}{r^2}\,dt = 2\pi.$$

More generally, if $\mathscr{C}$ is a closed curve as indicated in Fig. 11, we introduce a circle $\mathscr{C}_r$

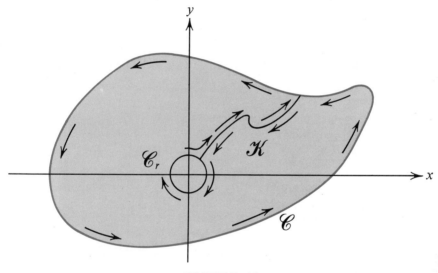

FIGURE 11

that lies in the region enclosed by $\mathscr{C}$. Next join $\mathscr{C}$ and $\mathscr{C}_r$ with an arc $\mathscr{K}$ as indicated in Fig. 11. Now Green's Theorem applied to the closed curve $\mathscr{C} \cup \mathscr{K} \cup [-\mathscr{C}_r] \cup [-\mathscr{K}]$ gives

$$\oint \frac{-y\,dx + x\,dy}{x^2 + y^2} = 0.$$

The reader can combine this result with equation (40) and Theorem 2, to prove that under these conditions $I_2 = 2\pi$. ●

How does it happen that on some curves $I_2 = 0$ and on others $I_2 \neq 0$? In the first case $\mathscr{C}$ is contained in a *simply connected* region in which M, N, M_y and N_x are continuous, and in the second case it does not.

We gather up some of the pieces in

THEOREM 8. Let $\mathscr{R}$ be a closed region in the plane which can be obtained as the union of a finite number of normal regions, and suppose that M, N, M_y, and N_x are continuous in $\mathscr{R}$. If in addition $\mathscr{R}$ is simply connected, then any one of the following five properties implies the remaining four.

(a) On every simple closed curve in $\mathscr{R}$,

$$(41) \qquad\qquad \oint M\,dx + N\,dy = 0.$$

(b) The integral

$$(35) \qquad\qquad \int_{\mathscr{C}} \mathbf{F} \cdot d\mathbf{R} \equiv \int_{\mathscr{C}} M\,dx + N\,dy$$

is independent of path in $\mathscr{R}$.
(c) There is a scalar function φ such that $\nabla\varphi = \mathbf{F}$ in $\mathscr{R}$.
(d) $M\,dx + N\,dy$ is an exact differential in $\mathscr{R}$.
(e) Throughout $\mathscr{R}$ we have

$$(30) \qquad\qquad \frac{\partial M}{\partial y} = \frac{\partial N}{\partial x}.$$

Proof. We will prove the following sequence of implications: **(a)** $\Longrightarrow$ **(b)** $\Longrightarrow$ **(c)** $\Longrightarrow$ **(d)** $\Longrightarrow$ **(e)** $\Longrightarrow$ **(a)**.

First, we note that Theorem 3 (in the plane) gives **(a)** $\Longrightarrow$ **(b)**. Next, Theorem 5 (also specialized to the plane) gives **(b)** $\Longrightarrow$ **(c)**. Items **(c)** and **(d)** are equivalent by definition. They are merely two different ways of stating that there is a φ such that

$$(23) \qquad\qquad \frac{\partial \varphi}{\partial x} = M \quad \text{and} \quad \frac{\partial \varphi}{\partial y} = N.$$

The first part of Theorem 6 gives **(d)** $\Longrightarrow$ **(e)**.

Finally, the crucial step is the assertion that **(e)** $\Longrightarrow$ **(a)**. This follows immediately from Green's Theorem, because whenever the equation (30) holds, then the right side of (38) is zero. This gives (41). ∎

Example 3. Let $\mathscr{C}$ be a simple closed curve. Compute

$$I_3 \equiv \int_{\mathscr{C}} \frac{(y^2 - x^2)\,dx - 2xy\,dy}{(x^2 + y^2)^2}.$$

Solution. The reader can show that for this integral $M_y = N_x$ everywhere except at the origin. But our experience from Example 2 warns us *not* to conclude that $I_3 = 0$. Further we observe that N and M are both discontinuous at the origin. To compute I_3 on the circle $\mathscr{C}_r$, we set $x = r \cos t$, $y = r \sin t$, and we find that

$$I_3 = \int_0^{2\pi} \frac{r^3(\sin^2 t - \cos^2 t)(-\sin t) - 2r^3 \sin t \cos t \cos t}{r^4}\, dt$$

$$= \frac{1}{r} \int_0^{2\pi} (\cos 2t \sin t - \sin 2t \cos t)\, dt = \frac{1}{r} \cos t \,\Big|_0^{2\pi} = 0.$$

We leave it for the reader to show that $I_3 = 0$ for every simple closed curve that does not pass through the origin. ●

EXERCISE 3

In Problems 1 through 6, use Green's Theorem to evaluate the given line integral over the given curve.

1. $\oint (x + y)\, dx + (x - y)\, dy$, $\mathscr{C}$: any circle.

2. $\oint (y^3 + 1)\, dx + 3y^2 x\, dy$, $\mathscr{C}$: $x^2 + y^2 = 10$.

3. $\oint (2xy^3 + \cos x)\, dx + (3x^2 y^2 + 5x)\, dy$, $\mathscr{C}$: $x^2 + y^2 = 10$.

4. $\oint e^x \sin y\, dx + e^x \cos y\, dy$, $\mathscr{C}$: $7x^2 + 11y^2 = 13$.

5. $\oint (e^x - 4y \sin^2 x)\, dx + (2x + \sin 2x)\, dy$, $\mathscr{C}$: square $x = \pm 2$, $y = \pm 2$.

6. $\oint e^{\sin^5 x^3}\, dx + y(\cos^4 e^y)^{\sin^2 y}\, dy$, $\mathscr{C}$: $13x^2 + 17y^2 = 99$.

7. Use Green's Theorem to prove the formula for the area of a region that was stated in Exercise 1.

8. Explain the result in equation (40) on the basis of

$$d \tan^{-1} \frac{y}{x} = \frac{-y\, dx + x\, dy}{x^2 + y^2}.$$

9. Explain the result in Example 3 on the basis of

$$d\frac{x}{x^2 + y^2} = \frac{(y^2 - x^2)\, dx - 2xy\, dy}{(x^2 + y^2)^2}.$$

In Problems 10 through 13, assume that the functions involved have continuous second-order partial derivatives in a simply connected region $\mathcal{R}$ and that the curves involved are smooth curves that lie in $\mathcal{R}$.

$\star$**10.** Prove that $\displaystyle\oint (u\nabla v + v\nabla u) \cdot d\mathbf{R} = 0.$

$\star$**11.** Let $\mathbf{R}(t) = f(t)\mathbf{i} + g(t)\mathbf{j}$. If $\mathbf{R}'(t) = f'(t)\mathbf{i} + g'(t)\mathbf{j} \neq \mathbf{0}$, then it is tangent to $\mathcal{C}$. The vector

$$\mathbf{n} = \mathbf{n}(t) \equiv \frac{1}{|\mathbf{R}'(t)|}\left(g'(t)\mathbf{i} - f'(t)\mathbf{j}\right)$$

is a unit vector normal to $\mathbf{R}'(t)$ and is called the *unit outer normal* of $\mathcal{C}$. Prove that

$$\oint (M\mathbf{i} + N\mathbf{j}) \cdot \mathbf{n}\, ds = \oint M\, dy - N\, dx.$$

This is the integral of the normal component of $\mathbf{F} = M\mathbf{i} + N\mathbf{j}$ over $\mathcal{C}$. If $\mathbf{F}$ is a vector that represents the velocity of a fluid, the integral gives the total flow across $\mathcal{C}$ per unit of time. HINT: $ds = |\mathbf{R}'(t)|\, dt.$

$\star$**12.** The *normal derivative* of φ on $\mathcal{C}$ is defined by

$$\frac{\partial \varphi}{\partial n} = \nabla\varphi \cdot \mathbf{n}$$

and is just the derivative of φ in the direction of $\mathbf{n}$. Prove that

$$\oint \frac{\partial \varphi}{\partial n}\, ds = \iint_{\mathcal{R}} \nabla^2\varphi\, dA,$$

where $\nabla^2\varphi \equiv \varphi_{xx} + \varphi_{yy}$. The reason for this notation will appear in the next section.

$\star$**13.** Prove that

$$\oint u\frac{\partial v}{\partial n}\, ds = \iint_{\mathcal{R}} (u\nabla^2 v + \nabla u \cdot \nabla v)\, dA,$$

and

$$\oint \left(u \frac{\partial v}{\partial n} - v \frac{\partial u}{\partial n} \right) ds = \int \int_{\mathscr{R}} (u \nabla^2 v - v \nabla^2 u) \, dA.$$

HINT: Let $M = -u \dfrac{\partial v}{\partial y}$ and $N = u \dfrac{\partial v}{\partial x}$.

Divergence and Curl 5

We recall that the gradient of a scalar function $\varphi(x, y, z)$ is defined by

$$\nabla \varphi = \frac{\partial \varphi}{\partial x} \mathbf{i} + \frac{\partial \varphi}{\partial y} \mathbf{j} + \frac{\partial \varphi}{\partial z} \mathbf{k}. \tag{42}$$

In equation (42) the symbol ∇ by itself has no meaning, but if we try to "factor" (42) we might arrive at

$$\nabla \equiv \frac{\partial}{\partial x} \mathbf{i} + \frac{\partial}{\partial y} \mathbf{j} + \frac{\partial}{\partial z} \mathbf{k}. \tag{43}$$

We still have no meaning for ∇ by itself, other than the collection of symbols on the right side of (43), but this collection of symbols can be combined in a meaningful way with others. Thus ∇ combined with a scalar function φ gives the meaningful quantity in (42). We can combine ∇ with a vector function $\mathbf{F} = L\mathbf{i} + M\mathbf{j} + N\mathbf{k}$ in two ways, using (an analogue of) the dot and cross product. The dot product of ∇ and $\mathbf{F}$ gives

$$\nabla \cdot \mathbf{F} = \left(\frac{\partial}{\partial x} \mathbf{i} + \frac{\partial}{\partial y} \mathbf{j} + \frac{\partial}{\partial z} \mathbf{k} \right) \cdot (L\mathbf{i} + M\mathbf{j} + N\mathbf{k}),$$

or

$$\nabla \cdot \mathbf{F} = \frac{\partial L}{\partial x} + \frac{\partial M}{\partial y} + \frac{\partial N}{\partial z}. \tag{44}$$

This quantity is called the *divergence* of the vector function $\mathbf{F}$ and is often written

$$\operatorname{div} \mathbf{F} = \frac{\partial L}{\partial x} + \frac{\partial M}{\partial y} + \frac{\partial N}{\partial z}.$$

The cross product of ∇ and $\mathbf{F}$ gives

$$\nabla \times \mathbf{F} \equiv \begin{vmatrix} \mathbf{i} & \mathbf{j} & \mathbf{k} \\ \dfrac{\partial}{\partial x} & \dfrac{\partial}{\partial y} & \dfrac{\partial}{\partial z} \\ L & M & N \end{vmatrix} = \left(\frac{\partial N}{\partial y} - \frac{\partial M}{\partial z} \right) \mathbf{i} + \left(\frac{\partial L}{\partial z} - \frac{\partial N}{\partial x} \right) \mathbf{j} + \left(\frac{\partial M}{\partial x} - \frac{\partial L}{\partial y} \right) \mathbf{k}. \tag{45}$$

This quantity is called the *curl* of **F**, and sometimes is called the *rotation* of **F**. It is often denoted by **curl F** or **rot F**. Both the divergence and the curl are very important in various branches of physics and applied mathematics.

We frequently refer to ∇ as an *operator* because in the combinations $\nabla\varphi$, $\nabla \cdot \mathbf{F}$, and $\nabla \times \mathbf{F}$ it "operates" on φ or **F** to give a new function.

Example 1. For the vector function $\mathbf{F} = 3xy\mathbf{i} + 5xy^2z\mathbf{j} - (yz + x)\mathbf{k}$, compute $\nabla \cdot \mathbf{F}$ and $\nabla \times \mathbf{F}$.

Solution. Using equations (44) and (45), we obtain

$$\nabla \cdot \mathbf{F} = 3y + 10xyz - y = 2y(1 + 5xz),$$

$$\nabla \times \mathbf{F} = \begin{vmatrix} \mathbf{i} & \mathbf{j} & \mathbf{k} \\ \dfrac{\partial}{\partial x} & \dfrac{\partial}{\partial y} & \dfrac{\partial}{\partial z} \\ 3xy & 5xy^2z & -yz - x \end{vmatrix} = (-z - 5xy^2)\mathbf{i} + \mathbf{j} + (5y^2z - 3x)\mathbf{k}. \quad \bullet$$

Example 2. Find a reasonable meaning for $\nabla \cdot \nabla\varphi$.

Solution. From (43) we may infer that (symbolically)

$$\nabla \cdot \nabla = \left(\frac{\partial}{\partial x}\mathbf{i} + \frac{\partial}{\partial y}\mathbf{j} + \frac{\partial}{\partial z}\mathbf{k}\right) \cdot \left(\frac{\partial}{\partial x}\mathbf{i} + \frac{\partial}{\partial y}\mathbf{j} + \frac{\partial}{\partial z}\mathbf{k}\right)$$

$$= \frac{\partial^2}{\partial x^2} + \frac{\partial^2}{\partial y^2} + \frac{\partial^2}{\partial z^2}.$$

Consequently, we might guess that

$$(46) \qquad \nabla^2\varphi \equiv \nabla \cdot \nabla\varphi \equiv \frac{\partial^2\varphi}{\partial x^2} + \frac{\partial^2\varphi}{\partial y^2} + \frac{\partial^2\varphi}{\partial z^2},$$

and indeed the right side of (46) is the definition of the first two items in (46). The expression $\nabla^2\varphi$ is called the *Laplacian* of φ and the equation $\nabla^2\varphi = 0$ is called *Laplace's differential equation*. Any solution φ for which the second partial derivatives are continuous is called a *harmonic* function. $\bullet$

Example 3. Derive formulas for the expansion of $\nabla^2(fg)$ and $\nabla \times (\varphi\mathbf{F})$.

Solution

$$\nabla^2(fg) = \frac{\partial^2(fg)}{\partial x^2} + \frac{\partial^2(fg)}{\partial y^2} + \frac{\partial^2(fg)}{\partial z^2}$$

$$= f_{xx}g + 2f_x g_x + fg_{xx} + \text{similar terms in } y \text{ and } z$$
$$= g(f_{xx} + f_{yy} + f_{zz}) + 2(f_x g_x + f_y g_y + f_z g_z) + f(g_{xx} + g_{yy} + g_{zz})$$
$$= g\nabla^2 f + f\nabla^2 g + 2\nabla f \cdot \nabla g.$$

We leave it for the reader to make the laborious computations and show that

$$\nabla \times (\varphi \mathbf{F}) = \begin{vmatrix} \mathbf{i} & \mathbf{j} & \mathbf{k} \\ \dfrac{\partial}{\partial x} & \dfrac{\partial}{\partial y} & \dfrac{\partial}{\partial z} \\ \varphi L & \varphi M & \varphi N \end{vmatrix} = \varphi \nabla \times \mathbf{F} - \mathbf{F} \times \nabla \varphi. \quad \bullet$$

EXERCISE 4

In Problems 1 through 7, compute the divergence and curl of the given vector function.

1. $\mathbf{R} = x\mathbf{i} + y\mathbf{j} + z\mathbf{k}$.
2. $\mathbf{F} = 2xy^3 z\mathbf{i} + 3x^2 y^2 z\mathbf{j} + x^2 y^3 \mathbf{k}$.
3. $\mathbf{F} = (2y^2 - 2z)\mathbf{i} + xy\mathbf{j} - xz\mathbf{k}$.
4. $\mathbf{F} = e^x \sin y\mathbf{i} + e^y \sin z\mathbf{j} + e^x \cos z\mathbf{k}$.
5. $\mathbf{F} = x(3y^2 - 2z)\mathbf{i} + y^2(1 - y)\mathbf{j} + z(z - 2y)\mathbf{k}$.
6. $\mathbf{F} = e^y(\cos x \cos z\mathbf{i} + \sin x \cos z\mathbf{j} - \sin x \sin z\mathbf{k})$.
7. $\mathbf{F} = \varphi(x, y, z)(x\mathbf{i} + y\mathbf{j} + z\mathbf{k}) \equiv \varphi\mathbf{R}$.

In Problems 8 through 14, prove the given differentiation formulas, assuming the continuity of all the partial derivatives involved.

8. $\nabla \cdot (\mathbf{F} + \mathbf{G}) = \nabla \cdot \mathbf{F} + \nabla \cdot \mathbf{G}$.
9. $\nabla \times (\mathbf{F} + \mathbf{G}) = \nabla \times \mathbf{F} + \nabla \times \mathbf{G}$.
10. $\nabla \times (\nabla \varphi) = 0$.
11. $\nabla \cdot (\nabla \times \mathbf{F}) = 0$.
12. $\nabla \cdot (\varphi \mathbf{F}) = \varphi(\nabla \cdot \mathbf{F}) + \mathbf{F} \cdot (\nabla \varphi)$.
★13. $\nabla \cdot (\mathbf{F} \times \mathbf{G}) = \mathbf{G} \cdot (\nabla \times \mathbf{F}) - \mathbf{F} \cdot (\nabla \times \mathbf{G})$.
14. $\nabla \times (\varphi(r)\mathbf{R}) = 0$, where $r = |\mathbf{R}| = |x\mathbf{i} + y\mathbf{j} + z\mathbf{k}|$.

15. Prove that if $\mathbf{A}$ is a constant vector, then

$$\nabla \times (\mathbf{A} \times \mathbf{R}) = 2\mathbf{A}, \qquad \text{and} \qquad \nabla \cdot (\mathbf{A} \times \mathbf{R}) = 0.$$

16. For what value of n is $\nabla \cdot (r^n \mathbf{R}) = 0$?

In Problems 17 through 20, find the indicated quantity: (a) in space where $r = \sqrt{x^2 + y^2 + z^2}$, and (b) in the plane where $r = \sqrt{x^2 + y^2}$.

17. $\nabla^2 r^n$.
18. $\nabla^2 \ln r$.
19. $\nabla \cdot [r\nabla r^{-3}]$.
20. $\nabla \times (\varphi \nabla \varphi)$.

6 The Area of a Surface and Surface Integrals

In Chapter 19, Section 8, we learned how to find σ, the area of a surface $\mathcal{S}$ if the surface is given by an equation of the form $z = f(x, y)$.

We now suppose that the surface is given in the parametric form

$$(47) \qquad \mathbf{R} = \mathbf{R}(u, v) = f(u, v)\mathbf{i} + g(u, v)\mathbf{j} + h(u, v)\mathbf{k},$$

where $\mathbf{R}$ is the position vector to points on the surface and u and v are parameters. As u and v vary over some region in the uv-plane, the vector $\mathbf{R}$ will usually generate some nice surface, and will always do so in any case of practical importance (see Figs. 12 and 13). For examples of surfaces with this type of equation, see Chapter 16, Section 10, page 689.

If the surface is described by an equation of the form $z = f(x, y)$, then $\sigma(\mathcal{S})$, the surface area, is given by

$$(48) \qquad \sigma(\mathcal{S}) = \iint\limits_{\mathcal{A}} \sqrt{1 + f_x{}^2 + f_y{}^2} \, dx \, dy,$$

where (x, y) varies over some suitable base region $\mathcal{A}$ in the xy-plane.

To arrive at a formula when the surface is described by a vector function $\mathbf{R}(u, v) = f(u, v)\mathbf{i} + g(u, v)\mathbf{j} + h(u, v)\mathbf{k}$, we consider a partition of $\mathcal{A}$ into little rectangles as indicated in Fig. 12. The vector function $\mathbf{R}(u, v)$ maps each little rectangle $\mathcal{T}_\alpha$ onto a piece of

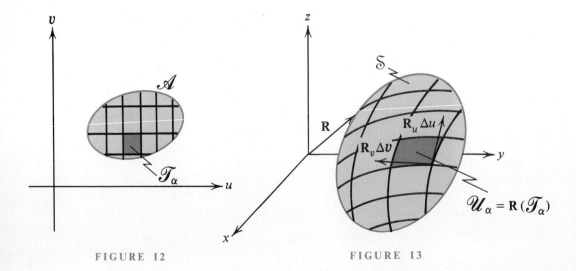

FIGURE 12   FIGURE 13

the surface [which we denote by $\mathbf{R}(\mathcal{T}_\alpha)$]. If $\mathcal{T}_\alpha$ is small enough, then the image $\mathcal{U}_\alpha$ will be very nearly a parallelogram (a curvilinear parallelogram for which the curvature of the sides is very small). We use the approximations $|\mathbf{R}_u \, \Delta u|$ and $|\mathbf{R}_v \, \Delta v|$ for the length of two coterminal sides, and we observe that these sides may meet at an angle that is not a right angle so that we must introduce $\sin \theta$, where θ is the angle of intersection. Thus a "good" approximation for the area of $\mathcal{U}_\alpha$ is

(49)
$$A(\mathcal{U}_\alpha) \approx |\mathbf{R}_u \times \mathbf{R}_v| \, \nabla u \, \nabla v$$

computed at any point in $\mathcal{T}_\alpha$. Taking a limit of sums of terms of the form (49), we arrive at

---

DEFINITION 5 (Surface Area). If $\mathcal{S}$ is given parametrically and $\sigma(\mathcal{S})$ denotes the area of the surface $\mathcal{S}$, then

(50)
$$\sigma(\mathcal{S}) \equiv \int\!\!\int_{\mathscr{A}} |\mathbf{R}_u \times \mathbf{R}_v| \, du \, dv.$$

If we write the right side of (50) in full, the expression becomes rather complicated, so some abbreviations are helpful. We let J_1, J_2, and J_3 be defined by

(51)
$$\mathbf{R}_u \times \mathbf{R}_v = \begin{vmatrix} \mathbf{i} & \mathbf{j} & \mathbf{k} \\ \dfrac{\partial x}{\partial u} & \dfrac{\partial y}{\partial u} & \dfrac{\partial z}{\partial u} \\ \dfrac{\partial x}{\partial v} & \dfrac{\partial y}{\partial v} & \dfrac{\partial z}{\partial v} \end{vmatrix} \equiv J_1 \mathbf{i} + J_2 \mathbf{j} + J_3 \mathbf{k},$$

where (using subscript notation for the partial derivatives)

(52)
$$J_1 = \begin{vmatrix} y_u & z_u \\ y_v & z_v \end{vmatrix} = y_u z_v - z_u y_v,$$

(53)
$$J_2 = \begin{vmatrix} z_u & x_u \\ z_v & x_v \end{vmatrix} = z_u x_v - x_u z_v,$$

and

(54)
$$J_3 = \begin{vmatrix} x_u & y_u \\ x_v & y_v \end{vmatrix} = x_u y_v - y_u x_v.$$

All the determinants are called "*Jacobians*" and are often abbreviated thus:

(55)
$$J_1 = \frac{\partial(y, z)}{\partial(u, v)}, \qquad J_2 = \frac{\partial(z, x)}{\partial(u, v)}, \qquad J_3 = \frac{\partial(x, y)}{\partial(u, v)}.$$

The Jacobian notation and definition generalize to n functions of n variables and is very important in advanced mathematics.

If we combine (50) and (51), we have for the area of our surface,

$$(56) \qquad \sigma(\mathbb{S}) = \int\!\!\!\int_{\mathscr{A}} \sqrt{J_1^2 + J_2^2 + J_3^2}\, du\, dv.$$

Example 1. The triangular region $\mathscr{A}: 0 \leq u \leq 1 - v,\, 0 \leq v \leq 1$, is mapped onto a surface $\mathbb{S}$ by the vector function $\mathbf{R} = u^2\mathbf{i} + 2v^2\mathbf{j} + (3u^2 - v^2)\mathbf{k}$. Find the area of the surface.

Solution. We first find $|\mathbf{R}_u \times \mathbf{R}_v|$. Indeed,

$$(57) \qquad \mathbf{R}_u \times \mathbf{R}_v = \begin{vmatrix} \mathbf{i} & \mathbf{j} & \mathbf{k} \\ 2u & 0 & 6u \\ 0 & 4v & -2v \end{vmatrix} = -24uv\mathbf{i} + 4uv\mathbf{j} + 8uv\mathbf{k}.$$

Hence

$$(58) \qquad |\mathbf{R}_u \times \mathbf{R}_v| = \sqrt{(576 + 16 + 64)u^2 v^2} = \sqrt{656}\,|uv| = 4\sqrt{41}\,uv,$$

where in the last equation we can write $|uv| = uv$ because in $\mathscr{A}$ both u and v are positive. Then, for the area, equation (50) gives

$$(59) \quad \sigma(\mathbb{S}) = \int_0^1 \int_0^{1-v} 4\sqrt{41}\, uv\, du\, dv = 2\sqrt{41} \int_0^1 vu^2 \Big|_0^{1-v} dv$$

$$= 2\sqrt{41} \int_0^1 v(1 - 2v + v^2)\, dv = 2\sqrt{41}\left(\frac{v^2}{2} - \frac{2}{3}v^3 + \frac{v^4}{4}\right)\Big|_0^1 = \frac{\sqrt{41}}{6}. \quad \bullet$$

The reader may have noticed that $\mathbb{S}$ is really a piece of a plane. We select such a simple surface because for a surface selected at random, the integral of $|\mathbf{R}_u \times \mathbf{R}_v|$ may be very difficult to evaluate, and in many cases we must settle for some numerical approximation of the integral.

We now consider the integral of an arbitrary function over a surface. First, suppose that we have a scalar function $\varphi(x, y, z)$. We partition $\mathbb{S}$ in the usual way, and in each curvilinear parallelogram $\mathscr{T}_\alpha$ we select a point P_α and compute the sum

$$(60) \qquad \sum_{\alpha=1}^n \varphi(P_\alpha)\, \Delta\sigma_\alpha,$$

where $\Delta\sigma_\alpha$ is the area of $\mathscr{T}_\alpha$. Taking the limit in the usual way [and always assuming that the limit in (60) exists for every selection of P_α and every sequence of partitions], we arrive at the definition of the integral of φ over the surface $\mathcal{S}$, denoted by

(61)
$$\iint_{\mathcal{S}} \varphi(x, y, z)\, d\sigma.$$

If $\mathcal{S}$ is parametrized by $\mathbf{R}(u, v)$, then

$$d\sigma = |\mathbf{R}_u \times \mathbf{R}_v|\, du\, dv$$

and the integral in (61) can be computed as a double integral:

(62)
$$\iint_{\mathcal{A}} \varphi(f(u, v), g(u, v), h(u, v))|\mathbf{R}_u \times \mathbf{R}_v|\, du\, dv.$$

If the surface is given in pieces (e.g., the surface of a cube), then we compute (62) for each piece and add the results to obtain the integral over the entire surface.

We may consider the surface integral of a vector function over $\mathcal{S}$, but this merely reduces to three surface integrals of scalar functions. Indeed, if

(63)
$$\mathbf{F}(x, y, z) = L(x, y, z)\mathbf{i} + M(x, y, z)\mathbf{j} + N(x, y, z)\mathbf{k},$$

then, by definition,

(64)
$$\iint_{\mathcal{S}} \mathbf{F}\, d\sigma = \mathbf{i} \iint_{\mathcal{S}} L\, d\sigma + \mathbf{j} \iint_{\mathcal{S}} M\, d\sigma + \mathbf{k} \iint_{\mathcal{S}} N\, d\sigma.$$

Various other possibilities may occur, such as

$$\iint_{\mathcal{S}} \mathbf{F} \cdot \mathbf{n}\, d\sigma \quad \text{or} \quad \iint_{\mathcal{S}} \mathbf{F} \times \mathbf{n}\, d\sigma,$$

where $\mathbf{n}$ is a unit vector normal to the surface. However, the computation of these more complicated integrals is ultimately reduced to the computation of a number of integrals of the form (62).

Example 2. Compute the surface integral

(65)
$$\iint_{\mathcal{S}} \mathbf{F} \cdot \mathbf{n}\, d\sigma,$$

where $\mathbf{F} = y\mathbf{i} + (y - x)\mathbf{j} + e^x\mathbf{k}$ over that portion of the paraboloid $z = x^2 + y^2$ that lies over $\mathcal{A}: 0 \leq x \leq 1,\ 0 \leq y \leq 3$.

Solution. We must select a parametrization for the surface, and the simplest procedure is to use

(66) $x = u, \quad y = v, \quad z = u^2 + v^2$

with $\mathscr{A} : 0 \leq u \leq 1$ and $0 \leq v \leq 3$. A suitable normal is

$$N = R_u \times R_v = \begin{vmatrix} i & j & k \\ 1 & 0 & 2u \\ 0 & 1 & 2v \end{vmatrix} = -2u i - 2v j + k.$$

We can convert to a unit normal $\mathbf{n}$ by dividing by $|R_u \times R_v|$. Then

(67) $$\iint_S F \cdot n \, d\sigma = \iint_{\mathscr{A}} F \cdot \frac{R_u \times R_v}{|R_u \times R_v|} |R_u \times R_v| \, dA = \iint_{\mathscr{A}} F \cdot (R_u \times R_v) \, dA$$

$$= \iint_{\mathscr{A}} (v i + (v - u) j + e^u k) \cdot (-2u i - 2v j + k) \, dA$$

$$= \int_0^3 \int_0^1 (-2uv - 2v^2 + 2uv + e^u) \, du \, dv$$

$$= \int_0^3 (-2v^2 u + e^u) \Big|_0^1 \, dv = \frac{-2v^3}{3} + (e - 1)v \Big|_0^3 = 3e - 21.$$

The reader should observe that the statement of the problem did not specify the
direction of the normal **n,** and consequently the problem was not clearly stated. In the
solution we arbitrarily selected the parametrization (66) and found a normal vector
for which the z-component is always positive. If we use $x = v$, $y = u$, and
$z = u^2 + v^2$, we obtain $N = 2u i + 2v j - k$, and the answer is $21 - 3e$.   ●

Suppose now that **F** represents the velocity of some fluid at each point of a region that
contains S. Then $F \cdot n$ is the component normal to S, and it seems reasonably clear that
the integral in (65) measures the total amount of fluid that flows across the surface S per
unit of time. Consequently, this type of integral is very important in hydrodynamics and
aerodynamics.

EXERCISE 5

In Problems 1 through 4, find the area of the surface $R(\mathscr{A})$ *for the given function* **R** *and the
given closed region* $\mathscr{A}$.

1. $R = u \cos v\, i + u \sin v\, j + v k,$ $\mathscr{A} : 0 \leq u \leq \sqrt{3}, \qquad 0 \leq v \leq 2\pi.$

2. $\mathbf{R} = 8u^2\mathbf{i} + v^2\mathbf{j} + 4uv\mathbf{k}$, $\mathcal{A}: 0 \le u \le 3$, $0 \le v \le 1$.

3. $\mathbf{R} = (u^3 + v)\mathbf{i} + u^2\mathbf{j} + v\mathbf{k}$, $\mathcal{A}: 1/3 \le u \le \sqrt{8}/3$, $2 \le v \le 5$.

4. $\mathbf{R} = (u^2 + v)\mathbf{i} + (u^2 - v)\mathbf{j} + \sqrt{u^4 + v^2}\mathbf{k}$, $\mathcal{A}: 0 \le v \le \sqrt{u}$, $0 \le u \le 2$.

*5. In the study of surfaces it is customary to introduce the fundamental quantities

$$E \equiv \mathbf{R}_u \cdot \mathbf{R}_u, \qquad F \equiv \mathbf{R}_u \cdot \mathbf{R}_v, \qquad G \equiv \mathbf{R}_v \cdot \mathbf{R}_v.$$

Prove that with this notation (50) is equivalent to

$$\sigma(\mathcal{S}) = \iint_{\mathcal{A}} \sqrt{EG - F^2} \, du \, dv.$$

6. Prove that the surface in Example 1 is a piece of a plane.

7. If the equation of $\mathcal{S}$ has the form $z = f(x, y)$, prove that equation (50) for $\sigma(\mathcal{S})$ will give equation (48), our old formula for the area of a surface. HINT: Set $x = u$ and $y = v$.

8. State formulas similar to (48) when the surface is given by **(a)** $y = g(x, z)$ and **(b)** $x = h(y, z)$.

In Problems 9 and 10, set up integrals for the area of the indicated surface but do not evaluate the integrals.

9. $\mathbf{R} = uv\mathbf{i} + (u + 2v)\mathbf{j} + (u^2 + v^2)\mathbf{k}$.

10. $\mathbf{R} = u \cos v\mathbf{i} + u^2 v\mathbf{j} + uv^2\mathbf{k}$.

In Problems 11 through 14, compute the surface integral (61) for the given φ over the given surface $\mathcal{S}$.

11. $\varphi = y^2$; **(a)** $\mathcal{S}$ is the portion of the plane $x + y + z = 1$ that lies in the first octant, and **(b)** $\mathcal{S}$ is the portion of the cylinder $y^2 + z^2 = 1$ that lies between $x = 0$ and $x = 5$.

12. $\varphi = y^2 + z^2$; surfaces $\mathcal{S}$ as in Problem 11.

13. $\varphi = \sqrt{1 + 4x^2 + 4y^2}$; $\mathcal{S}$ is that portion of the paraboloid $z = x^2 + y^2$ that lies over the square $0 \le x \le 3$, $0 \le y \le 3$.

14. $\varphi = xy$; $\mathcal{S}$ is the portion of the sphere $x^2 + y^2 + z^2 = r^2$ in the first octant. HINT: Let $\mathbf{R}(\varphi, \theta) = r \sin\varphi \cos\theta\mathbf{i} + r \sin\varphi \sin\theta\mathbf{j} + r \cos\varphi\mathbf{k}$.

In Problems 15 through 21, compute $\displaystyle\iint_{\mathcal{S}} \mathbf{R} \, d\sigma$, *where* $\mathbf{R} = x\mathbf{i} + y\mathbf{j} + z\mathbf{k}$.

15. $\mathcal{S}$ is the rectangle $0 \le x \le 2$, $0 \le z \le 4$, $y = 5$.

16. $\mathcal{S}$ is the triangle, $0 \le x + y \le 1$, $0 \le x$, $0 \le y$, $z = 7$.

17. $\mathcal{S}$ is the surface described in Problem 14.

18. $\mathcal{S}$ is the sphere $x^2 + y^2 + z^2 = r^2$. HINT: You should obtain the answer without any computation.

19. $\mathcal{S}$ is the upper half of the sphere mentioned in Problem 18. HINT: Same as in Problem 18.
20. $\mathcal{S}$ is the square $-1 \leqq x \leqq 1$, $-1 \leqq y \leqq 1$, $z = 1$.
21. $\mathcal{S}$ is the surface of the cube $-1 \leqq x \leqq 1$, $-1 \leqq y \leqq 1$, $-1 \leqq z \leqq 1$.

In Problems 22 through 26, compute $\displaystyle\iint_{\mathcal{S}} \mathbf{F} \cdot \mathbf{n} \, d\sigma$, *where* $\mathbf{n}$ *is a unit normal as indicated and*

$\mathbf{R}$ *is the position vector to a point on the given surface.*

22. $\mathbf{F} = \mathbf{R}$; $\mathcal{S}$ is the rectangle described in Problem 15, $\mathbf{n} = \mathbf{j}$.
23. $\mathbf{F} = \mathbf{R}$; $\mathcal{S}$ is the sphere $x^2 + y^2 + z^2 = r^2$, and $\mathbf{n}$ is the outer normal.
24. $\mathbf{F} = \mathbf{R} \times \mathbf{k}$; $\mathcal{S}$ and $\mathbf{n}$ as in Problem 23.
25. $\mathbf{F} = \mathbf{R} \times \mathbf{k}$; $\mathcal{S}$ and $\mathbf{n}$ as in Example 2.
26. $\mathbf{F} = \mathbf{R} \times \mathbf{j}$; $\mathcal{S}$ and $\mathbf{n}$ as in Example 2.

7 The Divergence Theorem

This theorem asserts that under very general conditions

(68)
$$\iint_{\mathcal{S}} \mathbf{F} \cdot \mathbf{n} \, d\sigma = \iiint_{\mathcal{R}} \nabla \cdot \mathbf{F} \, dV,$$

where $\mathcal{S}$ is a closed surface, $\mathcal{R}$ is the region bounded by $\mathcal{S}$, and $\mathbf{n}$ is the unit outer normal (points away from $\mathcal{R}$). In the proof we will impose conditions on $\mathbf{F}$, $\mathcal{S}$, and $\mathcal{R}$, which reduce the labor required. Relaxing these conditions is a task that is best left for the specialist.

Suppose now that $\mathcal{R}$ is a normal region in three-dimensional space, that the boundary of $\mathcal{R}$ is piecewise smooth,[1] and that the first partial derivatives of $\mathbf{F}$ are continuous in $\mathcal{R}$. We decompose (68) into three parts:

$$\iint_{\mathcal{S}} L\mathbf{i} \cdot \mathbf{n} \, d\sigma = \iiint_{\mathcal{R}} \frac{\partial L}{\partial x} \, dV,$$

$$\iint_{\mathcal{S}} M\mathbf{j} \cdot \mathbf{n} \, d\sigma = \iiint_{\mathcal{R}} \frac{\partial M}{\partial y} \, dV,$$

---

[1] A surface is *smooth* if the unit normal is a continuous vector function on the surface. A surface is *piecewise smooth* if it can be decomposed into a finite number of smooth pieces. A cube furnishes a good example of a piecewise smooth surface.

and

(69)
$$\iint_{\mathcal{S}} N\mathbf{k} \cdot \mathbf{n}\, d\sigma = \iiint_{\mathcal{R}} \frac{\partial N}{\partial z}\, dV.$$

If we can prove these three relations, then (68) will follow by addition. Further, it suffices to prove (69) because the other two relations will follow by a cyclic interchange of letters.

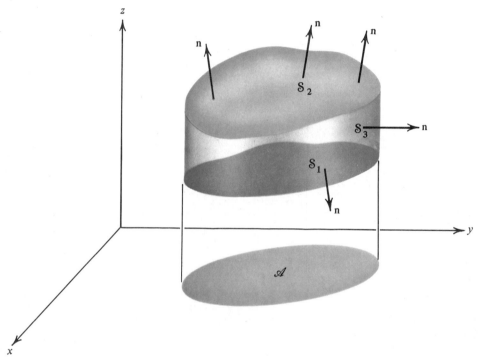

FIGURE 14

We let $\mathcal{A}$ be the projection of $\mathcal{R}$ on the xy-plane (see Fig. 14) and, since $\mathcal{R}$ is convex in the direction of the z-axis, we have $\mathcal{S} = \mathcal{S}_1 \cup \mathcal{S}_2 \cup \mathcal{S}_3$, where $\mathcal{S}_1$ is the lower portion, $\mathcal{S}_2$ is the upper portion, and $\mathcal{S}_3$ (which may degenerate into a common boundary curve) is a lateral surface on which $\mathbf{n}$ is parallel to the xy-plane. Since $\mathbf{k} \cdot \mathbf{n} = 0$ on $\mathcal{S}_3$, the surface integral (69) over $\mathcal{S}_3$ is zero, and we have

(70)
$$\iint_{\mathcal{S}} N\mathbf{k} \cdot \mathbf{n}\, d\sigma = \iint_{\mathcal{S}_1} N\mathbf{k} \cdot \mathbf{n}\, d\sigma + \iint_{\mathcal{S}_2} N\mathbf{k} \cdot \mathbf{n}\, d\sigma.$$

We can use x and y as the parameters u and v for the surfaces $\mathcal{S}_1$ and $\mathcal{S}_2$. Let $z = z_\alpha(x, y)$ be the equation for S_α, $\alpha = 1, 2$. Then $\mathbf{R} = x\mathbf{i} + y\mathbf{j} + z_\alpha(x, y)\mathbf{k}$ and

(71)
$$\mathbf{R}_u \times \mathbf{R}_v \equiv \mathbf{R}_x \times \mathbf{R}_y = -\frac{\partial z_\alpha}{\partial x}\mathbf{i} - \frac{\partial z_\alpha}{\partial y}\mathbf{j} + \mathbf{k}.$$

Since $\mathbf{n}$ has a positive z-component on S_2, the normal defined by (71) has the "right" direction and can be used in the last term of (70). From (67) we find that

(72)
$$\iint_{S_2} N\mathbf{k} \cdot \mathbf{n}\, d\sigma = \iint_{\mathscr{A}} N\mathbf{k} \cdot (\mathbf{R}_x \times \mathbf{R}_y)\, dx\, dy$$

$$= \iint_{\mathscr{A}} N\mathbf{k} \cdot \mathbf{k}\, dx\, dy = \iint_{\mathscr{A}} N(x, y, z_2(x, y))\, dx\, dy.$$

On S_1 the normal vector has a negative z-component, and hence the normal in (71) has the "wrong" direction. This can be corrected by using $-\mathbf{R}_x \times \mathbf{R}_y$ in (72) and we find that

(73)
$$\iint_{S_1} N\mathbf{k} \cdot \mathbf{n}\, d\sigma = -\iint_{\mathscr{A}} N(x, y, z_1(x, y))\, dx\, dy.$$

Using (73) and (72) in (70) gives

(74)
$$\iint_{S} N\mathbf{k} \cdot \mathbf{n}\, d\sigma = \iint_{\mathscr{A}} [N(x, y, z_2(x, y)) - N(x, y, z_1(x, y))]\, dx\, dy.$$

On the other hand, if we convert the right side of (69) into an iterated integral, we have

(75)
$$\iiint_{\mathscr{R}} \frac{\partial N}{\partial z}\, dV = \iint_{\mathscr{A}} \left[\int_{z_1(x,y)}^{z_2(x,y)} \left(\frac{\partial N}{\partial z}\, dz \right) \right] dx\, dy$$

$$= \iint_{\mathscr{A}} [N(x, y, z_2(x, y)) - N(x, y, z_1(x, y))]\, dx\, dy,$$

which is identical to (74). This proves (69) and consequently (68) under the conditions previously mentioned.

Suppose now that $\mathscr{R} = \mathscr{R}_1 \cup \mathscr{R}_2$, where $\mathscr{R}_1$ and $\mathscr{R}_2$ are normal and $\mathscr{R}_1 \cap \mathscr{R}_2$ is a smooth surface. We can apply (68) to $\mathscr{R}_1$ and $\mathscr{R}_2$ and, when we add the two equations, we again obtain (68) for $\mathscr{R}$, because in the two surface integrals over the common smooth boundary $\mathscr{R}_1 \cap \mathscr{R}_2$, the outer normal for $\mathscr{R}_1$ is the negative of the outer normal for $\mathscr{R}_2$, and the two integrals will cancel. We have proved

---

THEOREM 9 (The Divergence Theorem). Let $\mathcal{R} = \mathcal{R}_1 \cup \mathcal{R}_2 \cup \cdots \cup \mathcal{R}_k$, where each $\mathcal{R}_\alpha$, $\alpha = 1, 2, \ldots, k$, is a normal region and, if $\alpha \neq \beta$, then $\mathcal{R}_\alpha \cap \mathcal{R}_\beta$ is either empty or a smooth surface. Let $\mathbf{F}$ be a vector function that has continuous first-order partial derivatives in $\mathcal{R}$, and suppose that $\mathcal{S}$, the boundary of $\mathcal{R}$, is a piecewise smooth surface. Then equation (68) holds.

It is easy to sense the importance of this theorem. Suppose that at each point of a region that contains $\mathcal{R}$, the vector $\mathbf{F}$ represents the velocity of a fluid and that the fluid is in a steady state. Then it is easy to argue that the integral on the left side of (68) represents the total amount of fluid that crosses the surface $\mathcal{S}$ in a unit of time. Consequently, $\nabla \cdot \mathbf{F}$ represents in some sense what is happening to the fluid in $\mathcal{R}$. For example, if the fluid is incompressible, and is neither being created nor destroyed in $\mathcal{R}$, then for each sphere in $\mathcal{R}$, the total flow across the boundary [left side of (68)] must be zero. Hence under these conditions $\nabla \cdot \mathbf{F} = 0$ throughout the region $\mathcal{R}$.

Example 1. Check the Divergence Theorem by computing both sides of (68) when $\mathbf{F} = 2y\mathbf{i} + (z - y)\mathbf{j} + (xy + z)\mathbf{k}$ and $\mathcal{R}$ is the box $0 \leqq x \leqq 1, 0 \leqq y \leqq 2, 0 \leqq z \leqq 3$.

Solution. Since $\nabla \cdot \mathbf{F} = -1 + 1 = 0$, the volume integral on the right side of (68) is zero. For the left side we write $I = I_1 + I_2 + \cdots + I_6$, where I_α refers to a surface integral over a face of the box. For the top, $\mathbf{n} = \mathbf{k}$, $z = 3$, and we have

$$I_1(\text{top}) = \iint_{\mathcal{S}_1} \mathbf{F} \cdot \mathbf{n}\, d\sigma = \int_0^2 \int_0^1 (xy + 3)\, dx\, dy = 1 + 6 = 7.$$

We leave it for the reader to sketch the box in a standard position and prove that

$$I_2(\text{bottom}) = -1, \qquad I_3(\text{front}) = 12, \qquad I_4(\text{back}) = -12,$$

$$I_5(\text{right}) = -3/2, \qquad I_6(\text{left}) = -9/2.$$

Consequently, $I_1 + I_2 + \cdots + I_6 = 0$, as assured by Theorem 9.   ●

Example 2. Prove that the volume of a cone is $A(\mathcal{B})h/3$, where $A(\mathcal{B})$ is the area of the base and h is the altitude.

Solution. Let $\mathcal{B}$ be a simply connected plane region with a piecewise smooth boundary $\mathcal{C}$, and let Q be a point not in the plane of $\mathcal{B}$ (see Fig. 15). The collection of all line segments that join Q to a point of $\mathcal{C}$ forms a surface $\mathcal{S}$. By definition the solid bounded by $\mathcal{S} \cup \mathcal{B}$ is called a *cone*. To compute the volume of this solid, we apply

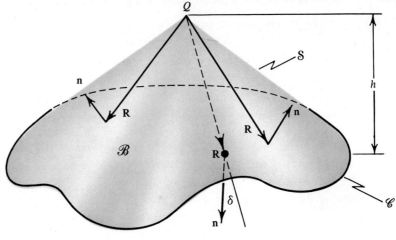

FIGURE 15

the divergence theorem, using Q as the origin and $\mathbf{F} \equiv \mathbf{R} = x\mathbf{i} + y\mathbf{j} + z\mathbf{k}$. Since $\nabla \cdot \mathbf{F} \equiv \nabla \cdot \mathbf{R} \equiv 3$, the right side of (68) gives $3V$, where V is the volume of the cone. For the left side we observe that on $\mathbb{S}$ we have $\mathbf{R} \cdot \mathbf{n} = 0$. Over the region $\mathscr{B}$ we have $\mathbf{R} \cdot \mathbf{n} = |\mathbf{R}| \cos \delta = h$, a constant. Hence the surface integral yields $hA(\mathscr{B})$ and, consequently, $V = A(\mathscr{B})h/3$.   ●

EXERCISE 6

In Problems 1 through 4, use the Divergence Theorem to compute $\displaystyle\iint_{\mathbb{S}} \mathbf{F} \cdot \mathbf{n}\, d\sigma$ *over the given surface.*

1. $\mathbb{S}$ is the surface of the box $0 \leq x \leq 1$, $0 \leq y \leq 2$, $0 \leq z \leq 3$, and
   (a) $\mathbf{F} = x\mathbf{i} + y\mathbf{j} + z\mathbf{k}$.   (b) $\mathbf{F} = x^2\mathbf{i} + y^2\mathbf{j} + z^2\mathbf{k}$.   (c) $\mathbf{F} = yz\mathbf{i} + xz\mathbf{j} + xy\mathbf{k}$.
2. $\mathbb{S}$ as in Problem 1 and
   (a) $\mathbf{F} = xy\mathbf{i} + yz\mathbf{j} + xz\mathbf{k}$.   (b) $\mathbf{F} = (x^2 + y^2)\mathbf{R}$.
3. $\mathbb{S}$ is the surface of the cylindrical solid; $x^2 + y^2 = 4$, $0 \leq z \leq 5$, and

$$\mathbf{F} = x^2\mathbf{i} + (3y^4 - 2z)\mathbf{j} + (4z - \sin x)\mathbf{k}.$$

4. $\mathbb{S}$ is the boundary of the tetrahedron bounded by the coordinate planes and the plane $x + 2y + z = 4$, and

$$\mathbf{F} = (4x + e^y)\mathbf{i} + (xy + \sin z)\mathbf{j} - (2yz + \sqrt{xy})\mathbf{k}.$$

5. Let $\mathscr{R}$ be a homogeneous solid of unit density, and let M_{xy}, M_{yz}, M_{zx}, I_x, I_y, and I_z denote, as usual, moments and moments of inertia. Let $\mathbb{S}$ be the surface of $\mathscr{R}$ and let

n be the outer normal to $\mathcal{S}$. Prove that

(a) $\displaystyle\iint_{\mathcal{S}} x^2\mathbf{i}\cdot\mathbf{n}\,d\sigma = 2M_{yz}.$ (b) $\displaystyle\iint_{\mathcal{S}} (y^3\mathbf{j}+z^3\mathbf{k})\cdot\mathbf{n}\,d\sigma = 3I_x.$

Develop similar formulas for M_{zx}, M_{xy}, I_y, and I_z.

6. Use the result of Problem 5 to evaluate (without integration)

$$\iint_{\mathcal{S}} (x^2\mathbf{i}+y^2\mathbf{j}+z^2\mathbf{k})\cdot\mathbf{n}\,d\sigma,$$

where $\mathcal{S}$ is the surface of a sphere of radius r and center at $(a, 0, 0)$.

In Problems 7 through 13, prove the given identities. Assume that the regions and functions involved satisfy the conditions of Theorem 9 and that $r = |\mathbf{R}| = |x\mathbf{i} + y\mathbf{j} + z\mathbf{k}|$.

7. $\displaystyle\iint_{\mathcal{S}} (\boldsymbol{\nabla}\times\mathbf{F})\cdot\mathbf{n}\,d\sigma = 0.$

8. $\displaystyle\iint_{\mathcal{S}} \frac{\partial\varphi}{\partial n}\,d\sigma = \iint_{\mathcal{S}} \boldsymbol{\nabla}\varphi\cdot\mathbf{n}\,d\sigma \equiv \iiint_{\mathcal{R}} \nabla^2\varphi\,dV.$

*9. $\displaystyle\iint_{\mathcal{S}} \varphi\mathbf{n}\,d\sigma = \iiint_{\mathcal{R}} \boldsymbol{\nabla}\varphi\,dV.$ HINT: Let $\mathbf{F} = \varphi\mathbf{A}$, where $\mathbf{A}$ is a constant vector.

10. $\displaystyle\iint_{\mathcal{S}} r^k\mathbf{n}\,d\sigma = \iiint_{\mathcal{R}} kr^{k-2}\mathbf{R}\,dV, \quad k \geq 2.$

11. $\displaystyle\iint_{\mathcal{S}} \mathbf{n}\,d\sigma = \mathbf{0}.$

12. $\displaystyle\iint_{\mathcal{S}} \varphi\frac{\partial\psi}{\partial n}\,d\sigma = \iiint_{\mathcal{R}} (\boldsymbol{\nabla}\varphi\cdot\boldsymbol{\nabla}\psi + \varphi\nabla^2\psi)\,dV.$

13. $\displaystyle\iint_{\mathcal{S}} \left(\varphi\frac{\partial\psi}{\partial n} - \psi\frac{\partial\varphi}{\partial n}\right)d\sigma = \iiint_{\mathcal{R}} (\varphi\nabla^2\psi - \psi\nabla^2\varphi)\,dV.$

14. In Problem 11, replace **n** by $-$**n** and interpret the result in terms of the total force exerted on a body by a surrounding fluid pressing uniformly on the surface.

*15. Alter the integral in Problem 11 to obtain a proof of Archimedes' principle on the buoyancy of a solid immersed in a fluid. HINT: Use the result of Problem 9.

16. Prove that a region may be normal for one set of rectangular coordinate axes but not for another.

17. We recall that if φ is harmonic in a region $\mathscr{R}$, then (by definition) $\nabla^2\varphi = 0$ in $\mathscr{R}$. Prove that for every closed piecewise smooth surface in $\mathscr{R}$,

$$\iint\limits_{\mathcal{S}} \frac{\partial \varphi}{\partial n}\, d\sigma = 0$$

if φ is a harmonic function in $\mathscr{R}$.

8 Stokes' Theorem

We have just learned that the Divergence Theorem relates an integral over a closed surface $\mathcal{S}$, and an integral over the region $\mathscr{R}$, which has $\mathcal{S}$ as its boundary. We naturally look for a relation between an integral over a closed curve $\mathscr{C}$ and an integral over a surface $\mathcal{S}$, which has $\mathscr{C}$ as its boundary. Indeed, if $\mathscr{C}$ is a plane curve, Green's Theorem does relate two such integrals. In three-dimensional space the relation we seek is called *Stokes' Theorem*. This theorem asserts that under suitable conditions

(76)
$$\int_{\mathscr{C}} \mathbf{F} \cdot \mathbf{T}\, ds = \iint\limits_{\mathcal{S}} (\nabla \times \mathbf{F}) \cdot \mathbf{n}\, d\sigma,$$

where $\mathcal{S}$ is a surface, $\mathscr{C}$ is the boundary of $\mathcal{S}$, the unit vector $\mathbf{T}$ is tangent to $\mathscr{C}$, and s is arc length on $\mathscr{C}$. Equation (76) may be put in various forms. If $\mathbf{F} = L\mathbf{i} + M\mathbf{j} + N\mathbf{k}$, we can write

$$\mathbf{F} \cdot \mathbf{T} = (L\mathbf{i} + M\mathbf{j} + N\mathbf{k}) \cdot \left(\frac{dx}{ds}\mathbf{i} + \frac{dy}{ds}\mathbf{j} + \frac{dz}{ds}\mathbf{k}\right),$$

so that the left side of (76) gives

(77)
$$\int_{\mathscr{C}} \mathbf{F} \cdot \mathbf{T}\, ds = \int_{\mathscr{C}} L\, dx + M\, dy + N\, dz = \int_{\mathscr{C}} \mathbf{F} \cdot d\mathbf{R},$$

an integral that we have already studied extensively in Sections 2 and 3.

If α, β, and γ are the direction angles for the unit normal vector $\mathbf{n}$, then we can write

$$(\nabla \times \mathbf{F}) \cdot \mathbf{n} = \begin{vmatrix} \mathbf{i} & \mathbf{j} & \mathbf{k} \\ \dfrac{\partial}{\partial x} & \dfrac{\partial}{\partial y} & \dfrac{\partial}{\partial z} \\ L & M & N \end{vmatrix} \cdot (\cos\alpha\, \mathbf{i} + \cos\beta\, \mathbf{j} + \cos\gamma\, \mathbf{k}),$$

and the right side of (76) gives

$$(78) \quad \iint\limits_{S} (\nabla \times \mathbf{F}) \cdot \mathbf{n} \, d\sigma = \iint\limits_{S} [(N_y - M_z) \cos \alpha + (L_z - N_x) \cos \beta + (M_x - L_y) \cos \gamma] \, d\sigma.$$

The surface S must be orientable (two-sided) and one side of S is selected arbitrarily as the positive side. The direction of the curve $\mathscr{C}$ (and the tangent vector $\mathbf{T}$) is then assigned so that a person walking along $\mathscr{C}$ in the positive direction and on the positive side of S always has the surface on his left. The normal vector $\mathbf{n}$ is selected so that it points away from S on the positive side of S (see Fig. 16).

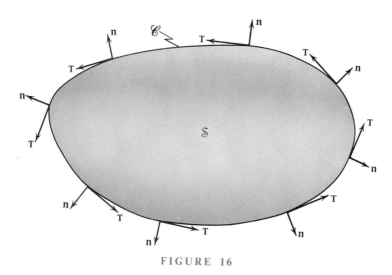

FIGURE 16

The proof of Stokes' Theorem is a little more complicated than the proof of the Divergence Theorem, so we omit it. A proof can be found in any good book on vector analysis or advanced calculus.

Example 1. Evaluate the line integral

$$I \equiv \int_{\mathscr{C}} y^2 z^3 \, dx + 2xyz^3 \, dy + 3xy^2 z^2 \, dz$$

over the curve $\mathscr{C} : \mathbf{R} = a \sin t \mathbf{i} + b \cos t \mathbf{j} + c \cos t \mathbf{k}$, $0 \le t \le 2\pi$, $abc \ne 0$.

Solution. We can use Stokes' Theorem with $\mathbf{F} = y^2 z^3 \mathbf{i} + 2xyz^3 \mathbf{j} + 3xy^2 z^2 \mathbf{k}$. An easy computation shows that $\nabla \times \mathbf{F} = \mathbf{0}$. Hence for any surface with boundary $\mathscr{C}$, the right side of (76) gives zero. Hence $I = 0$.

We can check Stokes' Theorem by computing I directly. We find that

$$I = \int_0^{2\pi} [ab^2c^3 \cos^6 t - 2ab^2c^3 \sin^2 t \cos^4 t - 3ab^2c^3 \sin^2 t \cos^4 t]\, dt$$

$$= \int_0^{2\pi} ab^2c^3[\cos^6 t - 5 \sin^2 t \cos^4 t]\, dt = ab^2c^3 \sin t \cos^5 t \Big|_0^{2\pi} = 0. \quad \bullet$$

Stokes' Theorem paves the way for a natural extension of Theorem 8 from the plane to three-dimensional space.

THEOREM 10 (PWO). Let $\mathcal{R}$ be a simply connected region in space, and suppose that all operations below take place in $\mathcal{R}$ ($\mathcal{C}$ is a closed curve in $\mathcal{R}$, etc.). Further suppose that $\nabla \times \mathbf{F}$ is continuous in $\mathcal{R}$. Then any one of the following five properties implies the remaining four.

(a) The integral

$$(77) \qquad \int_{\mathcal{C}} \mathbf{F} \cdot d\mathbf{R} \equiv \int_{\mathcal{C}} L\, dx + M\, dy + N\, dz$$

is zero for every simple closed curve in $\mathcal{R}$.

(b) The integral (77) is independent of path in $\mathcal{R}$.

(c) There is a scalar function φ such that $\nabla \varphi = \mathbf{F}$ in $\mathcal{R}$.

(d) $L\, dx + M\, dy + N\, dz$ is an exact differential in $\mathcal{R}$.

(e) $\nabla \times \mathbf{F} \equiv \nabla \times (L\mathbf{i} + M\mathbf{j} + N\mathbf{k}) = \mathbf{0}$ in $\mathcal{R}$.

The crucial step is the proof that **(e)** implies **(a)**. If $\nabla \times \mathbf{F} = \mathbf{0}$ in $\mathcal{R}$, then the right side of equation (76) is always zero. Hence the left side of the same equation gives

$$\oint \mathbf{F} \cdot d\mathbf{R} = 0$$

for every simple closed curve in $\mathcal{R}$.

For a physical interpretation of Stokes' Theorem, we again suppose that the vector $\mathbf{F}$ represents the velocity of a fluid. Then $\mathbf{F} \cdot \mathbf{T}$ is the component of $\mathbf{F}$ that is tangent to $\mathcal{C}$, and the left side of (76) measures the circulation or rotation of the fluid around $\mathcal{C}$. Suppose that we select a particular point P, and a plane through P. Let $\mathcal{C}$ be a circle of radius r and center P lying in this plane. Then from (76) we deduce that

$$(79) \qquad (\nabla \times \mathbf{F}) \cdot \mathbf{n} \Big|_P = \lim_{r \to 0} \frac{1}{\pi r^2} \int_{\mathcal{C}} \mathbf{F} \cdot \mathbf{T}\, ds.$$

The right side of (79) gives a physical interpretation for $\nabla \times \mathbf{F}$ at P in terms of an average circulation of the fluid. Further, the right side is clearly an invariant under a rotation or translation of the coordinate axes, and hence $\nabla \times \mathbf{F}$ is also an invariant, although the definition of $\nabla \times \mathbf{F}$ in equation (45) leaned heavily on the coordinate system.

If $\nabla \times \mathbf{F} = \mathbf{0}$, the vector field is said to be *irrotational*. If $\nabla \cdot \mathbf{F} = 0$, the field is said to be *solenoidal*. The reader who is interested in learning more about irrotational or solenoidal vector fields and about the applications of Stokes' Theorem and the divergence theorem should consult the books devoted to vector analysis or to the many applications; electrodynamics, hydrodynamics, mechanics, and so on.

EXERCISE 7

1. Sketch a portion of the vector field $\mathbf{F} = -y\mathbf{i} + x\mathbf{j} + 0\mathbf{k}$ that lies in the xy-plane. Find $\nabla \times \mathbf{F}$. Evaluate the integral (77) on the circle $\mathscr{C}:\mathbf{R} = r\cos\theta\mathbf{i} + r\sin\theta\mathbf{j} + 0\mathbf{k}$, $0 \leq \theta \leq 2\pi$. Does this example help with the interpretation of $\nabla \times \mathbf{F}$?

2. Show that Green's Theorem is a special case of Stokes' Theorem.

3. Repeat Problem 1 for the vector fields **(a)** $\mathbf{F} = Ax\mathbf{j}$, $A > 0$ and **(b)** $\mathbf{F} = \varphi(r)\mathbf{R}$, where φ is an arbitrary differentiable function.

4. Let $\mathbf{F} = (-y\mathbf{i} + x\mathbf{j})/(x^2 + y^2) + z^2\mathbf{k}$. Prove that $\nabla \times \mathbf{F} = \mathbf{0}$ except on the z-axis, where it is undefined. Evaluate the integral (77) on the circle defined in Problem 1. Does this example contradict Stokes' Theorem?

In Problems 5 through 7, use Stokes' Theorem to compute $\displaystyle\int_{\mathscr{C}} \mathbf{F} \cdot d\mathbf{R}$ *for the given* $\mathbf{F}$ *and the given curve.*

5. $\mathbf{F} = (3y + z)\mathbf{i} + (e^y - 3x)\mathbf{j} + (z^2 + x)\mathbf{k}$,
$$\mathscr{C}:\mathbf{R} = \cos\theta\mathbf{i} + \sin\theta\mathbf{j} - 7\mathbf{k},\ 0 \leq \theta \leq 2\pi.$$

6. $\mathbf{F} = y(x - z)\mathbf{i} + (2x^2 + z^2)\mathbf{j} + y^3\cos xz\mathbf{k}$,
$$\mathscr{C}:\text{The edge of the square } 0 \leq x,\, y \leq 2,\, z = 5.$$

7. $\mathbf{F} = e^{x^2}\mathbf{i} + (x + z)\sin y^3\mathbf{j} + (y^2 - x^2 + 2yz)\mathbf{k}$, $\mathscr{C}:$ The equilateral triangle formed by the intersection of the plane $x + y + z = 3$ with the three coordinate planes.

*★8. Let $\mathscr{R}$ be a closed convex region with smooth boundary $\mathscr{S}$, and let $\mathbf{F}$ have continuous first partial derivatives in $\mathscr{R}$. Prove that

$$\iint_{\mathscr{S}} \nabla \times \mathbf{F} \cdot \mathbf{n}\, d\sigma = 0.$$

*★9. Prove that if $\mathbf{F}$ is normal to a smooth surface $\mathscr{S}$ at each point of $\mathscr{S}$, then $\nabla \times \mathbf{F}$ is zero or tangent to $\mathscr{S}$ at each point of $\mathscr{S}$.

REVIEW PROBLEMS

1. Compute $\displaystyle\int \mathbf{F}\cdot d\mathbf{R}$ if

   (a) $\mathbf{F} = (x^2 + y)\mathbf{i} + x^2 y\mathbf{j} + z\mathbf{k}$, $\mathbf{R} = (3t\mathbf{i} + t^2\mathbf{j} + 2t\mathbf{k})$, $0 \le t \le 2$.

   (b) $\mathbf{F}$ as in part (a), $\mathbf{R} = (e^t\mathbf{i} + e^{-t}\mathbf{j} + 4t\mathbf{k})$, $0 \le t \le 2$.

2. Compute $\displaystyle\oint \frac{Ax\,dy - y\,dx}{x^2 + y^2}$, where $\mathscr{C}$ is the circle with center at $(0,0)$ and radius R.

3. Compute $\displaystyle\oint \frac{dx + dy}{x^2 + y^2}$ for the curve in Problem 2.

4. Is the integral in Problem 3 independent of path?

5. Find the work done on a particle if the force field is $\mathbf{F} = x^2\mathbf{i} + xy\mathbf{j} + xz\mathbf{k}$ and it moves along the curve $\mathbf{R} = \cos t\mathbf{i} + \sin t\mathbf{j} + 5t\mathbf{k}$ from $Q(1,0,0)$ to $P(1,0,10\pi)$.

6. Find the work done if the particle moves in the force field of Problem 5 on a straight line from $Q(1,0,0)$ to $P(1,0,10\pi)$.

7. Apply the test of Theorem 6 [equation (31)] for independence of path to the integrals in (a) Problem 1 and (b) Problem 5.

8. Prove that any line integral of the form

   $$\int_{\mathscr{C}} f(x)\,dx + g(y)\,dy$$

   is independent of path as long as $f(x)$ and $g(y)$ are continuous functions.

In Problems 9 through 11, use Green's Theorem to evaluate the given integral over the boundary of the rectangle with sides $x = 0$, $y = 0$, $x = 2$, $y = 3$.

9. $\displaystyle\oint xy^2\,dx - x^2 y\,dy$.   10. $\displaystyle\oint xe^y\,dx + e^x\,dy$.

11. $\displaystyle\oint (x^3 + 3x^2 y + 2y^2)\,dx + (x^3 + 2xy + y^4)\,dy$.

In Problems 12 through 16, find: (a) the divergence and (b) the curl of the given vector function.

12. $\mathbf{F} = 3x\mathbf{i} + 2y\mathbf{j} + 5z\mathbf{k}$.   13. $\mathbf{F} = 3y\mathbf{i} + 2z\mathbf{j} + 5x\mathbf{k}$.

14. $\mathbf{F} = xy^2\mathbf{i} + yz^2\mathbf{j} + zx^2\mathbf{k}$.   15. $\mathbf{F} = yz^2\mathbf{i} + zx^2\mathbf{j} + xy^2\mathbf{k}$.

16. $\mathbf{F} = xye^z\mathbf{i} + yze^x\mathbf{j} + zxe^y\mathbf{k}$.

17. Find the area of the surface described by $\mathbf{R} = 2uv\mathbf{i} + 2u^2\mathbf{j} + v^2\mathbf{k}$, $0 \le v \le u^2$, $0 \le u \le 1$.

18. Let $\mathbf{R} = Auv\mathbf{i} + Bu^2\mathbf{j} + Cv^2\mathbf{k}$. Prove that if $A^2 = 2BC$, then $|\mathbf{R}_u \times \mathbf{R}_v|$ is a polynomial in u and v.

19. The upper half of the sphere $x^2 + y^2 + z^2 = 1$ is described by

$$\mathbf{R} = u \cos v\mathbf{i} + u \sin v\mathbf{j} + \sqrt{1 - u^2}\mathbf{k},$$

where $0 \leq u \leq 1$ and $0 \leq v \leq 2\pi$. With this parametrization, find $|\mathbf{R}_u \times \mathbf{R}_v|$ and the area.

20. For the surface given in Problem 19, compute

(a) $\displaystyle\iint_S z \, d\sigma.$

(b) $\displaystyle\iint_S z^2 \, d\sigma.$

(c) $\displaystyle\iint_S \mathbf{R} \cdot \mathbf{k} \, d\sigma.$

(d) $\displaystyle\iint_S \mathbf{R} \cdot (y\mathbf{i} - x\mathbf{j}) \, d\sigma.$

21. Use the Divergence Theorem to compute $\displaystyle\iint_S \mathbf{F} \cdot \mathbf{n} \, d\sigma$, where S is the surface of the box $0 \leq x \leq 1$, $0 \leq y \leq 2$, $0 \leq z \leq 3$, and $\mathbf{F}$ is given by

(a) $\mathbf{F} = yz\mathbf{i} + xz\mathbf{j} + xy\mathbf{k}.$
(b) $\mathbf{F} = xz\mathbf{i} + yx\mathbf{j} + zy\mathbf{k}.$
(c) $\mathbf{F} = (1 - x)(x\mathbf{i} - y\mathbf{j} - z\mathbf{k}).$

22. Let S be a smooth closed surface bounding a region R. Prove that the volume of R is given by the surface integral

$$\iint_S \frac{1}{3}\mathbf{R} \cdot \mathbf{n} \, d\sigma,$$

where, as usual, $\mathbf{R} = x\mathbf{i} + y\mathbf{j} + z\mathbf{k}.$

23. Let $\mathscr{C}$ be the boundary of the square $0 \leq x, y \leq 3$, $z = 5$. Use Stokes' Theorem to compute

$$\int_{\mathscr{C}} \mathbf{F} \cdot d\mathbf{R}$$

for each of the following vector functions.
(a) $\mathbf{F} = (x + y)\mathbf{i} + (y + z)\mathbf{j} + (z + x)\mathbf{k}.$ (b) $\mathbf{F} = e^{xy}\mathbf{i} + e^{yz}\mathbf{j} + e^{zx}\mathbf{k}.$
(c) $\mathbf{F} = Ax^3\mathbf{i} + By^3\mathbf{j} + Cz^3\mathbf{k}.$ (d) $\mathbf{F} = Ay^3\mathbf{i} + Bz^3\mathbf{j} + Cx^3\mathbf{k}.$

24. Let $\mathscr{C}$ be the triangle ABC, where $A(1, 0, 0)$, $B(0, 1, 0)$, and $C(0, 0, 1)$. Compute the line integral in Problem 23 for each of the following functions.
(a) $\mathbf{F} = Az\mathbf{i} + Bx\mathbf{j} + Cy\mathbf{k}.$
(b) $\mathbf{F} = (z + e^x)\mathbf{i} + (x + e^y)\mathbf{j} + (y + \text{Tan}^{-1} z)\mathbf{k}.$
(c) $\mathbf{F} = Ay^2\mathbf{i} + Bz^2\mathbf{j} + Cx^2\mathbf{k}.$

Appendix 1
Sequences: The ϵ, N Definition of Limit

In Section 4 of Chapter 3 we stated that a sequence $S_1, S_2, S_3, \ldots$ has limit L if its terms eventually come arbitrarily close to L. For the purpose of giving rigorous proofs of the basic limit theorems as well as answering deeper theoretical questions about limits, we need to make this definition more precise.

First, we remark that the distance between the number S_n and the number L is given by $|S_n - L|$. Thus, for example, the inequality

$$|S_n - L| < 0.05$$

means that S_n is within 0.05 units of L; in other words, S_n lies in the open interval $(L - 0.05, L + 0.05)$ of radius 0.05 about L. In the rigorous definition of a limit we will use ϵ, the Greek lowercase letter epsilon, to denote the radius of an interval about L.

DEFINITION 1 (Limit of a Sequence). A sequence $S_1, S_2, S_3, \ldots$ is said to have limit L if for each positive number ϵ (no matter how small) there exists an integer N (which depends on ϵ) such that

(1) $$|S_n - L| < \epsilon \qquad \text{whenever } n > N.$$

Under these conditions, we write

(2) $$\lim_{n \to \infty} S_n = L,$$

or

$$S_n \to L \qquad \text{as } n \to \infty.$$

Geometrically, this definition states that for each open interval $(L - \epsilon, L + \epsilon)$ of radius $\epsilon > 0$ about L, it is possible to find an integer N so that all of the terms $S_{N+1}, S_{N+2}, S_{N+3}, \ldots$ lie in this interval.

According to Definition 1, whenever we claim that

$$\lim_{n \to \infty} S_n = L,$$

we must be prepared to defend our statement against skeptics in the following manner. A doubting Thomas gives us a small open interval about L, say $(L - 0.001, L + 0.001)$. Our burden is then to produce an integer N so that S_{N+1} and all succeeding terms of the sequence lie within that interval. To defend against all possible skeptics, we imagine a general interval $(L - \epsilon, L + \epsilon)$ as being given, and produce a formula for the appropriate N in terms of the radius ϵ.

Example 1. Use Definition 1 to prove that

(3)
$$\lim_{n \to \infty} \frac{n}{n + 1} = 1.$$

Solution. Here $S_n = \dfrac{n}{n + 1}$ and $L = 1$. For any given $\epsilon > 0$, we must explain how to produce an integer N so that the inequality

(4)
$$|S_n - L| = \left| \frac{n}{n + 1} - 1 \right| < \epsilon$$

holds whenever $n > N$.

Let us examine the desired inequality (4) more closely. Since

$$\left| \frac{n}{n + 1} - 1 \right| = \left| \frac{n - n - 1}{n + 1} \right| = \frac{1}{n + 1},$$

inequality (4) is equivalent to

$$\frac{1}{n + 1} < \epsilon,$$

or

$$n + 1 > \frac{1}{\epsilon},$$

or

$$n > \frac{1}{\epsilon} - 1.$$

Thus (4) holds for every n greater than $1/\epsilon - 1$. We therefore specify N to be the first positive integer greater than or equal to $1/\epsilon - 1$.

In particular, if $\epsilon = 0.001$, then we take

$$N = \frac{1}{\epsilon} - 1 = \frac{1}{0.001} - 1 = 1000 - 1 = 999,$$

and we have

(5)
$$\left| \frac{n}{n+1} - 1 \right| < 0.001 \qquad \text{for } n > N = 999.$$

If $\epsilon = 0.0005$, then

$$N = \frac{1}{\epsilon} - 1 = \frac{1}{0.0005} - 1 = 2000 - 1 = 1999,$$

and we have

(6)
$$\left| \frac{n}{n+1} - 1 \right| < 0.005 \qquad \text{for } n > N = 1999.$$

If $\epsilon = 3/8$, then $1/\epsilon - 1 = 8/3 - 1 = 5/3$ and we may take $N = 2$, since 2 is the first positive integer greater than $5/3$. ●

It is important to keep in mind that there is no *unique* formula for the integer N in Definition 1. Indeed, if for a given ϵ we can produce one value for N that satisfies the conditions of Definition 1, then any *larger* value for N will also satisfy the definition. For example, the inequality (5) certainly holds for $n > 1000$, $n > 8547$, $n > 10^8$, and so on.

Example 2. Prove that

(7)
$$\lim_{n \to \infty} \frac{2}{(n+3)^{1/2}} = 0.$$

Solution. For each given $\epsilon > 0$ we must produce a positive integer N so that

(8)
$$|S_n - L| = \left| \frac{2}{(n+3)^{1/2}} - 0 \right| < \epsilon$$

whenever $n > N$. Since inequality (8) can be written in the equivalent forms

$$\frac{2}{(n+3)^{1/2}} < \epsilon, \qquad \frac{(n+3)^{1/2}}{2} > \frac{1}{\epsilon}, \qquad (n+3) > \left(\frac{2}{\epsilon}\right)^2, \qquad n > \left(\frac{2}{\epsilon}\right)^2 - 3,$$

we specify N to be the first positive integer greater than or equal to $(2/\epsilon)^2 - 3$. For then, if $n > N$, we have $n > (2/\epsilon)^2 - 3$, so that

$$\left| \frac{2}{(n+3)^{1/2}} - 0 \right| < \epsilon. \quad \bullet$$

We now give the proofs of the basic limit properties mentioned in Theorem 3 of Chapter 3.

THEOREM 1. Let $S_1, S_2, S_3, \ldots$ and $T_1, T_2, T_3, \ldots$ be two convergent sequences with

(9) $$\lim_{n \to \infty} S_n = S \quad \text{and} \quad \lim_{n \to \infty} T_n = T.$$

Then

(10) $$\lim_{n \to \infty} (S_n + T_n) = S + T.$$

Proof. Let ϵ be a given positive number. Since $S_n \to S$ as $n \to \infty$, we know from the definition of limit that for the positive number $\epsilon/2$, there is an N_1 such that if $n > N_1$, then

(11) $$|S_n - S| < \frac{\epsilon}{2}.$$

Similarly, since $T_n \to T$ as $n \to \infty$, there is an N_2 such that if $n > N_2$, then

(12) $$|T_n - T| < \frac{\epsilon}{2}.$$

We now specify N to be the maximum of the two numbers N_1 and N_2. If $n > N$, then both (11) and (12) are satisfied simultaneously. Consequently, if $n > N$, then by the Triangle Inequality

$$|(S_n + T_n) - (S + T)| = |(S_n - S) + (T_n - T)|$$
$$\leqq |S_n - S| + |T_n - T|$$
$$< \frac{\epsilon}{2} + \frac{\epsilon}{2} = \epsilon.$$

But this is the meaning of equation (10). ∎

THEOREM 2. If $S_n \to S$ as $n \to \infty$, then the sequence of terms $S_1 - S$, $S_2 - S$, $S_3 - S, \ldots, S_n - S, \ldots$ approaches zero as $n \to \infty$. Conversely, if $(S_n - S) \to 0$ as $n \to \infty$, then $S_n \to S$ as $n \to \infty$.

Proof. Both of the statements $S_n \to S$ and $(S_n - S) \to 0$ flow from inequality

$$(13) \qquad |S_n - S| < \epsilon$$

for n sufficiently large. Hence $S_n \to S$ implies that $(S_n - S) \to 0$, and conversely. ∎

THEOREM 3. If $S_n \to S$ and c is any constant, then

$$(14) \qquad \lim_{n \to \infty} cS_n = cS.$$

Proof. If $c = 0$, we are to prove that $0 \cdot S_n \to 0$, and this is clear. Assume next that $c \neq 0$. For each $\epsilon > 0$, we apply the definition of a limit with $\epsilon/|c|$ in place of ϵ. Then for $\epsilon/|c| > 0$ there is an N such that if $n > N$, then

$$(15) \qquad |S_n - S| < \frac{\epsilon}{|c|}.$$

Multiplying both sides of (15) by $|c|$, we obtain

$$(16) \qquad |cS_n - cS| < \epsilon \qquad \text{for } n > N. ∎$$

THEOREM 4. If $\lim_{n \to \infty} S_n = 0$ and $\lim_{n \to \infty} T_n = 0$, then

$$(17) \qquad \lim_{n \to \infty} S_n T_n = 0.$$

Proof. For each ϵ such that $0 < \epsilon < 1$, there is an N_1 and an N_2 such that

$$(18) \qquad |S_n| < \epsilon \qquad \text{for } n > N_1$$

$$(19) \qquad |T_n| < \epsilon \qquad \text{for } n > N_2.$$

Now choose N to be the maximum of the two numbers N_1 and N_2. Then if $n > N$, we have from (18) and (19) that

$$|S_n T_n| < \epsilon^2.$$

But if $0 < \epsilon < 1$, then $\epsilon^2 < \epsilon$ so that

(20) $$|S_n T_n - 0| < \epsilon \qquad \text{for } n > N.$$

In Definition 1 there is no loss of generality in assuming that $\epsilon < 1$. (Why?) Hence our proof is complete. ∎

THEOREM 5. If

(21) $$\lim_{n \to \infty} S_n = S \qquad \text{and} \qquad \lim_{n \to \infty} T_n = T.$$

then

(22) $$\lim_{n \to \infty} S_n T_n = ST.$$

Proof. By Theorem 2 each of the sequences $(S_n - S)$ and $(T_n - T)$ approaches zero as $n \to \infty$. Hence, by Theorem 4, we have

(23) $$\lim_{n \to \infty} (S_n - S)(T_n - T) = 0.$$

Since $(S_n - S)(T_n - T) = S_n T_n - ST_n - TS_n + ST$, it follows from Theorems 1 and 3 that

$$0 = \lim_{n \to \infty} S_n T_n - S(\lim_{n \to \infty} T_n) - T(\lim_{n \to \infty} S_n) + \lim_{n \to \infty} ST,$$

$$0 = \lim_{n \to \infty} S_n T_n - ST - TS + ST,$$

(24) $$0 = \lim_{n \to \infty} S_n T_n - ST.$$

From (24) we obtain (22). ∎

THEOREM 6. If $\lim_{n \to \infty} T_n = T$ and $T \neq 0$, then

(25) $$\lim_{n \to \infty} \frac{1}{T_n} = \frac{1}{T}.$$

Proof. Given $\epsilon > 0$, we need to find an N such that

(26) $$\left| \frac{1}{T_n} - \frac{1}{T} \right| < \epsilon$$

whenever $n > N$. Inequality (26) is equivalent to

(27)
$$\left| \frac{T - T_n}{T_n T} \right| < \epsilon.$$

Since $T \neq 0$, there is an N_1 such that

(28)
$$|T_n - T| < \frac{|T|}{2} \qquad \text{for } n > N_1.$$

As $|T| - |T_n| < |T_n - T|$, inequality (28) implies that

$$|T| - |T_n| < \frac{|T|}{2},$$

or

$$\frac{|T|}{2} < |T_n| \qquad \text{for } n > N_1.$$

Consequently, on multiplying this last inequality by $|T|$, we obtain

(29)
$$\frac{|T|^2}{2} < |T_n T| \qquad \text{for } n > N_1.$$

Further, there is an integer N_2 such that

(30)
$$|T - T_n| < \frac{|T|^2}{2}\epsilon \qquad \text{for } n > N_2.$$

Let $N = \max \{N_1, N_2\}$. Then both (29) and (30) hold for $n > N$, and we have

$$\left| \frac{T - T_n}{T_n T} \right| = \frac{1}{|T_n T|}|T - T_n| < \frac{2}{|T|^2}\frac{|T|^2}{2}\epsilon = \epsilon. \quad \blacksquare$$

THEOREM 7. If $\lim_{n \to \infty} S_n = S$ and $\lim_{n \to \infty} T_n = T$, and $T \neq 0$, then

(31)
$$\lim_{n \to \infty} \frac{S_n}{T_n} = \frac{S}{T}.$$

Proof. We use Theorems 5 and 6. These give

$$\lim_{n \to \infty} \frac{S_n}{T_n} = \lim_{n \to \infty} \left(S_n \frac{1}{T_n} \right) = \left(\lim_{n \to \infty} S_n \right)\left(\lim_{n \to \infty} \frac{1}{T_n} \right) = S\frac{1}{T} = \frac{S}{T}. \quad \blacksquare$$

EXERCISE 1

In Problems 1 through 4, use Definition 1 to prove the stated limit.

1. $\lim\limits_{n\to\infty} \dfrac{2n}{n+3} = 2.$

2. $\lim\limits_{n\to\infty} \dfrac{1}{n^p} = 0$ for $p > 0.$

3. $\lim\limits_{n\to\infty} \dfrac{(-1)^n}{n} = 0.$

4. $\lim\limits_{n\to\infty} \dfrac{n^2+5}{3n^2+8} = \dfrac{1}{3}.$

5. Prove that if $\lim\limits_{n\to\infty} S_n = S$, then $\lim\limits_{n\to\infty} |S_n| = |S|.$

6. Prove that if $\lim\limits_{n\to\infty} S_n = S$ and $S > 0$, then there is an integer N such that $S_n > 0$ for

    each $n > N$. HINT: Take $\epsilon = S/2$ in Definition 1.

7. (Squeeze Theorem). Suppose that

$$S_n \leqq A_n \leqq T_n \qquad \text{for } n = 1, 2, \ldots .$$

    Prove that if $\lim\limits_{n\to\infty} S_n = L = \lim\limits_{n\to\infty} T_n$, then $\lim\limits_{n\to\infty} A_n = L.$

8. Use the Squeeze Theorem (Problem 7) to prove that $\lim\limits_{n\to\infty} 2^{1/n} = 1.$ HINT: Use the

    Binomial Theorem to show that $\left(1 + \dfrac{1}{n}\right)^n \geqq 2$, and then deduce the inequalities

    $1 \leqq 2^{1/n} \leqq 1 + \dfrac{1}{n}.$

9. Use the Squeeze Theorem (Problem 7) to prove that

$$\lim\limits_{n\to\infty} \sqrt[n]{4^n + 5^n} = 5.$$

10. If $0 < a < b < c$, find

$$\lim\limits_{n\to\infty} \sqrt[n]{a^n + b^n + c^n}.$$

★11. If all the a_j are positive, find $\lim\limits_{n\to\infty} \sqrt[n]{a_1^n + a_2^n + \cdots + a_k^n}.$

Appendix 2
Functions: The ϵ, δ Definition of Limit

In Section 3 of Chapter 4 we described the meaning of the limit

(1) $$\lim_{x \to a} f(x) = L$$

by saying that the function values $f(x)$ stay arbitrarily close to L whenever x is sufficiently close (but not equal to) a. To make this definition more precise, we use the two Greek letters δ (delta) and ϵ (epsilon) to specify the "closeness" of x to a and the "closeness" of $f(x)$ to L, respectively.

Before giving the more precise definition of (1), we recall that the condition $|x - a| < \delta$ is satisfied if and only if

(2) $$a - \delta < x < a + \delta.$$

It is convenient to call (2) a δ-*neighborhood* of a. When we want to consider x near a, but want x to be different from a, we write $0 < |x - a| < \delta$. This gives the set obtained by removing the center point $x = a$ from the interval (2), and is called a *deleted* (or *punctured*) δ-neighborhood of a.

DEFINITION 1 (Limit of a Function). Let $f(x)$ be a function defined in some deleted neighborhood of $x = a$. We say that the limit of $f(x)$, as x approaches a, is the number L if for each positive ϵ, no matter how small, there is a corresponding positive number δ such that the inequality

(3) $$|f(x) - L| < \epsilon$$

holds for all x with

918

(4) $$0 < |x - a| < \delta.$$

Under these conditions we write

$$\lim_{x \to a} f(x) = L,$$

or

$$f(x) \to L \qquad \text{as} \qquad x \to a.$$

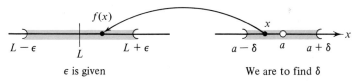

ϵ is given   We are to find δ

FIGURE 1

The definition states that $\lim_{x \to a} f(x) = L$ if for each given $\epsilon > 0$ (no matter how small) we can force $f(x)$ to be in an ϵ-neighborhood of L by restricting x to be in a sufficiently small deleted δ-neighborhood of a (see Fig. 1). We use a deleted neighborhood of a because we are interested in how $f(x)$ behaves for x *near* a, but we are not interested in how $f(x)$ behaves *at* $x = a$. Observe that δ depends on ϵ. The quantity ϵ is given first, and then the value assigned to δ depends on the particular value given to ϵ.

Another way to picture the situation is by drawing the graph of the function $y = f(x)$, as shown in Fig. 2. Let $P(x, y)$ be a point of the graph. If condition (3) is satisfied, then the

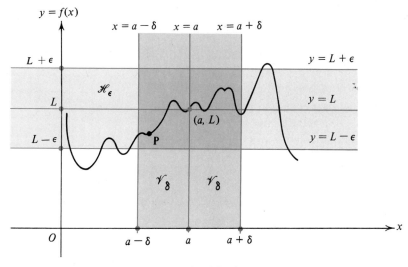

FIGURE 2

point P must lie in the horizontal strip $\mathcal{H}_\epsilon$ bounded by the lines $y = L + \epsilon$ and $y = L - \epsilon$. The condition (4) means that P must lie in the set $\mathcal{V}_\delta$ consisting of the two vertical strips bounded by the lines $x = a - \delta$, $x = a$, and $x = a + \delta$. The entire definition states that if the width of $\mathcal{H}_\epsilon$ is specified, no matter how small (i.e., if ϵ is given), then there is a set $\mathcal{V}_\delta$ ($\delta > 0$ can be found) such that when P is in the set $\mathcal{V}_\delta$, then it must also be in the horizontal strip $\mathcal{H}_\epsilon$. Thus near the point (a, L) the curve must lie in the little box shown shaded in the figure.

One should observe that Definition 1 does not depend on the picture shown in Fig. 2. Rather the picture merely illustrates the definition and aids our understanding. Further, the value of f at $x = a$ is not necessarily L. In fact, condition (4) specifically excludes $x = a$ from our considerations. We may have $f(a) = L$, or we may have $f(a) \neq L$, or indeed the function f may not even be defined at $x = a$.

Example 1. Start the proof that $\lim_{x \to 2} (3x + 7) = 13$ by finding a suitable δ **(a)** when $\epsilon = 0.1$ and **(b)** when $\epsilon = 0.01$.

Solution. **(a)** Here $|f(x) - L| = |3x + 7 - 13| = |3x - 6|$. To force the inequality (3),

$$|3x - 6| < \epsilon = 0.1,$$

we observe that

$$|3x - 6| = 3|x - 2|.$$

Clearly, if $|x - 2| < 1/30$, then $3|x - 2| < 1/10$. Hence any $\delta \leq 1/30$ will do. We might set $\delta = 0.025$ for simplicity. Then the condition on x that $|x - 2| < 0.025$ gives us that

$$|f(x) - L| = |3x + 7 - 13| = |3x - 6|$$
$$= 3|x - 2| < 3 \times 0.025 = 0.075 < 0.1.$$

(b) It is clear from the analysis above that for this particular problem we can just divide the previous δ by 10. Therefore, $\delta = 0.0025$ will ensure that $|3x + 7 - 13| < 0.01$. ●

Example 2. Use Definition 1 to prove that $\lim_{x \to 2} (3x + 7) = 13$.

Solution. The analysis is just as before. Let $\epsilon > 0$ be given. Set $\delta = \epsilon/3$. Now if $|x - 2| < \epsilon/3$, then

$$|f(x) - L| = |3x + 7 - 13| = |3x - 6| = 3|x - 2| < 3 \times \epsilon/3 = \epsilon.$$

Thus if $|x - 2| < \epsilon/3$, then $|f(x) - L| < \epsilon$. ●

In the solution of Example 2, the rule $\delta = \epsilon/3$ gives the largest possible δ for each given ϵ. However, any rule that prescribes a smaller value for δ will also satisfy the conditions of Definition 1. For example, taking $\delta = \epsilon/7$ yields another valid solution to Example 2.

We now present the proofs of the fundamental limit theorems for functions. Both the statements and the proofs are similar to those given in Appendix 1 for sequences.

THEOREM 1. If

(5) $$\lim_{x \to a} f(x) = L \quad \text{and} \quad \lim_{x \to a} g(x) = M,$$

then

(6) $$\lim_{x \to a} (f(x) + g(x)) = L + M.$$

Proof. By the definition of a limit we are to prove that for each $\epsilon > 0$, there is a $\delta > 0$ such that if $0 < |x - a| < \delta$, then

$$|f(x) + g(x) - (L + M)| < \epsilon.$$

This is the meaning of equation (6). By hypothesis, $\lim_{x \to a} f(x) = L$. Hence there is a $\delta_1 > 0$ such that if $0 < |x - a| < \delta_1$, then

(7) $$|f(x) - L| < \frac{\epsilon}{2}.$$

Similarly, by hypothesis, $\lim_{x \to a} g(x) = M$. Hence there is a $\delta_2 > 0$ such that whenever $0 < |x - a| < \delta_2$, then

(8) $$|g(x) - M| < \frac{\epsilon}{2}.$$

Now let δ be the minimum of the two numbers δ_1 and δ_2. If $0 < |x - a| < \delta$, then both (7) and (8) are satisfied. Consequently, if $0 < |x - a| < \delta$, then

$$|f(x) + g(x) - (L + M)| = |f(x) - L + g(x) - M|$$

$$\leq |f(x) - L| + |g(x) - M| < \frac{\epsilon}{2} + \frac{\epsilon}{2} = \epsilon. \quad \blacksquare$$

THEOREM 2. If $\lim_{x \to a} f(x) = L$, then $\lim_{x \to a} (f(x) - L) = 0$, and conversely.

Proof. Assume that $\lim_{x \to a} f(x) = L$. Then for each $\epsilon > 0$, there is a $\delta > 0$ such that if $0 < |x - a| < \delta$, then

$$(9) \qquad\qquad\qquad |f(x) - L| < \epsilon.$$

But the inequality (9) is just the condition that must be satisfied to prove that $\lim_{x \to a} (f(x) - L) = 0$. Conversely, if $\lim_{x \to a} (f(x) - L) = 0$, then (9) is satisfied for $0 < |x - a| < \delta$, and hence $\lim_{x \to a} f(x) = L$. ∎

THEOREM 3. If $\lim_{x \to a} f(x) = L$ and c is any constant, then

$$\lim_{x \to a} cf(x) = cL.$$

Proof. If $c = 0$, we are to prove that $\lim_{x \to a} 0 = 0$ and this is clear. Assume that $c \neq 0$. For each $\epsilon > 0$, we apply the definition of a limit using $\epsilon/|c|$ in place of ϵ. Then for each $\epsilon/|c| > 0$, there is a $\delta > 0$ such that if $0 < |x - a| < \delta$, then

$$(10) \qquad\qquad\qquad |f(x) - L| < \frac{\epsilon}{|c|}.$$

Multiplying both sides of (10) by $|c|$, we obtain

$$(11) \qquad\qquad\qquad |cf(x) - cL| < \epsilon. \quad ∎$$

If we apply this same technique to the proofs of Theorems 4, 5, 6, and 7 of Appendix 1, we obtain the proofs of Theorems 4, 5, 6, and 7 of this appendix. We merely make the replacements indicated by the arrows: $S_n \to f(x)$, $T_n \to g(x)$, $S \to L$, $T \to M$, $n > N \to 0 < |x - a| < \delta$. We leave it for the energetic student to give the details.

THEOREM 4. If $\lim_{x \to a} f(x) = 0$ and $\lim_{x \to a} g(x) = 0$, then

$$\lim_{x \to a} f(x)g(x) = 0.$$

THEOREM 5. If $\lim_{x \to a} f(x) = L$ and $\lim_{x \to a} g(x) = M$, then

$$\lim_{x \to a} f(x)g(x) = LM.$$

THEOREM 6. If $\lim\limits_{x \to a} g(x) = M$ and $M \neq 0$, then

$$\lim_{x \to a} \frac{1}{g(x)} = \frac{1}{M}.$$

THEOREM 7. If $\lim\limits_{x \to a} f(x) = L$ and $\lim\limits_{x \to a} g(x) = M$, and $M \neq 0$, then

$$\lim_{x \to a} \frac{f(x)}{g(x)} = \frac{L}{M}.$$

Now let us recall the definition of a continuous function given in Chapter 4 (page 131).

DEFINITION 2. A function $y = f(x)$ is said to be continuous at $x = a$ if
(A) $y = f(x)$ is defined at $x = a$.
(B) $\lim\limits_{x \to a} f(x)$ exists (as a real number).
(C) $\lim\limits_{x \to a} f(x) = f(a)$.

As we mentioned in Chapter 4, only (C) is really necessary because the symbol $f(a)$ already implies that the function $f(x)$ is defined at $x = a$, and the form of writing (C) already implies that there is a limit.

Now the definition of limit applied to (C) states that (C) is satisfied if and only if for each given $\epsilon > 0$, there is a $\delta > 0$ such that if

(12) $$0 < |x - a| < \delta,$$

then

(13) $$|f(x) - f(a)| < \epsilon.$$

The effect of the condition $0 < |x - a|$ is to prevent x from assuming the value $x = a$. But here no harm is done if $x = a$, because $f(x)$ is defined at $x = a$. Further, when $x = a$ the left side of (13) becomes $|f(a) - f(a)| = 0$, and this is certainly less than ϵ. Consequently, in defining a continuous function, we can drop the zero on the left side of (12). This gives the following alternative definition of a continuous function in terms of ϵ and δ.

DEFINITION 2★. A function $f(x)$ is said to be continuous at $x = a$ if for each given $\epsilon > 0$ there is a $\delta > 0$ such that if

(14) $$|x - a| < \delta,$$

then

(15) $$|f(x) - f(a)| < \epsilon.$$

The reader should note carefully the very slight difference between (12) and (14). We have proved that if $f(x)$ satisfies the conditions of Definition 2, it satisfies the conditions of Definition 2★. We leave it to the student to prove conversely that if $f(x)$ satisfies the conditions of Definition 2★, then it satisfies the conditions of Definition 2. Consequently, the two definitions are equivalent.

It is a simple matter to prove the following theorem concerning continuous functions.

THEOREM 8. If $f(x)$ and $g(x)$ are each continuous at $x = a$, then $f(x) + g(x)$, $cf(x)$, and $f(x)g(x)$ are continuous at $x = a$. Further, $f(x)/g(x)$ is continuous at $x = a$, provided that $g(a) \neq 0$.

Proof. Let us consider the sum $f(x) + g(x)$. Since $f(x)$ and $g(x)$ are continuous at $x = a$, we know from (C) of Definition 2 that

$$\lim_{x \to a} f(x) = f(a) \quad \text{and} \quad \lim_{x \to a} g(x) = g(a).$$

Hence, by Theorem 1, we have

$$\lim_{x \to a} [f(x) + g(x)] = \lim_{x \to a} f(x) + \lim_{x \to a} g(x) = f(a) + g(a).$$

Consequently, $f(x) + g(x)$ is continuous at $x = a$. Utilizing in the same manner Theorems 3, 5, and 7, we establish the continuity of the functions $cf(x)$, $f(x)g(x)$, and $f(x)/g(x)$ [if $g(a) \neq 0$]. ∎

EXERCISE 1

In Problems 1 through 4, use Definition 1 to prove the stated limits.

1. $\lim\limits_{x \to 2} (2x - 5) = -1.$

2. $\lim\limits_{x \to -1} (\pi x + 8) = 8 - \pi.$

★3. $\lim\limits_{x \to 3} x^2 = 9.$

★4. $\lim\limits_{x \to 4} \dfrac{8}{x} = 2.$

5. How would you modify Definition 1 to define the *one-sided* limits $\lim\limits_{x \to a^+} f(x)$ and $\lim\limits_{x \to a^-} f(x)$?

6. Using the definition of limit of a sequence given in Appendix 1, prove that if $\lim\limits_{x \to a} f(x) = L$, then for any sequence $x_1, x_2, x_3, \ldots$ with limit a $(x_n \neq a)$ we have
$$\lim\limits_{n \to \infty} f(x_n) = L.$$

7. Prove that if $f(x)$ is continuous at $x = a$ and $f(a) > 0$, then there exists a $\delta > 0$ such that $f(x) > 0$ for all x such that $|x - a| < \delta$. HINT: Use Definition 2★ with $\epsilon = f(a)/2$.

8. When $L = \infty$, we modify Definition 1 in the following manner. *The limit of $f(x)$ as x approaches a is ∞ if for each positive number M there is a positive number δ such that $f(x) > M$ for all x with $0 < |x - a| < \delta$.* Use this definition to prove that
$$\lim\limits_{x \to 0} \dfrac{1}{x^2} = \infty.$$

9. (Squeeze Theorem for Functions) Suppose that $f(x) \leq g(x) \leq h(x)$ for all x in some deleted neighborhood of a. Prove that if
$$\lim\limits_{x \to a} f(x) = \lim\limits_{x \to a} h(x) = L,$$
then $\lim\limits_{x \to a} g(x) = L$.

10. Use the Squeeze Theorem (Problem 9) to prove that
$$\lim\limits_{x \to 0} x \sin\left(\dfrac{1}{x}\right) = 0.$$

11. Suppose that $\lim\limits_{x \to a} F(x) = 0$ and that $|G(x)| \leq 25$ for all $x \neq a$. Prove that $\lim\limits_{x \to a} F(x)G(x) = 0.$

Appendix 3
Determinants

Determinants are often helpful in the solution of systems of linear equations. They are also useful in many other situations, for example, in the computation of the vector product (see Chapter 16, page 660). Further, the theory of determinants is in itself rather attractive and certainly deserves study on its own merits. Here we give a brief summary of the facts. The reader who wants to see the proofs of the theorems stated here, or who wishes to dig deeper into the subject, should consult some of the older algebra books.

1 Matrices and Determinants

A rectangular array of numbers is called a *matrix*. As examples of matrices we cite

$$A = [1, 3, 7], \quad B = \begin{bmatrix} 3 \\ -4 \\ 0 \end{bmatrix}, \quad C = \begin{bmatrix} 1 & 3 \\ -2 & -11 \end{bmatrix}, \quad D = \begin{bmatrix} a, & b, & c \\ x, & y, & z \end{bmatrix},$$

$$E = \begin{bmatrix} 3 & 7/2 \\ -5 & 9 \\ \pi & \sqrt{17} \\ -\sqrt{2} & 4^5 \end{bmatrix}, \quad F = \begin{bmatrix} -1 & -2 & -3 & -4 & -5 \\ 2 & 5 & 8 & 11 & 14 \\ \sqrt{2} & \sqrt[3]{3} & \sqrt[5]{5} & \sqrt[7]{7} & \sqrt{11} \end{bmatrix}.$$

It is customary to use square brackets to indicate that the array is a matrix. The numbers in the array are called the *elements* of the matrix. If a matrix M has m rows and n columns, it is called an $m \times n$ (m by n) matrix. If $m = n$, then M is said to be a *square matrix of order n*, or an *nth-order matrix*.

To each square matrix M we associate a number $D(M)$, called the *determinant of M*. We also indicate the determinant of M by replacing the brackets with vertical lines. For example, if we have the matrix

$$M = \begin{bmatrix} 1 & 3 \\ 2 & 9 \end{bmatrix},$$

then for the determinant of M, we write

$$D(M) = \begin{vmatrix} 1 & 3 \\ 2 & 9 \end{vmatrix}.$$

If M is an nth-order matrix (n rows and n columns), then $D(M)$ is called an *nth-order determinant*.

DEFINITION 1 (2 × 2 Determinants). The value of a second-order determinant is given by the formula

(1)
$$\begin{vmatrix} a_1 & b_1 \\ a_2 & b_2 \end{vmatrix} = a_1 b_2 - b_1 a_2.$$

The expression on the right is called the expansion of the determinant.

This formula is easy to remember if we observe that we multiply elements on the diagonals and use a plus sign if the diagonal descends from left to right, and a minus sign if the diagonal descends from right to left. For example,

$$\begin{vmatrix} 1 & 3 \\ 2 & 9 \end{vmatrix} = 1(9) - 3(2) = 9 - 6 = 3, \qquad \begin{vmatrix} 2 & -5 \\ 3 & 7 \end{vmatrix} = 2(7) - (-5)(3) = 14 + 15 = 29.$$

DEFINITION 2 (3 × 3 Determinants). The value of a third-order determinant is given by the formula

(2)
$$D = \begin{vmatrix} a_1 & b_1 & c_1 \\ a_2 & b_2 & c_2 \\ a_3 & b_3 & c_3 \end{vmatrix} = a_1 \begin{vmatrix} b_2 & c_2 \\ b_3 & c_3 \end{vmatrix} - b_1 \begin{vmatrix} a_2 & c_2 \\ a_3 & c_3 \end{vmatrix} + c_1 \begin{vmatrix} a_2 & b_2 \\ a_3 & b_3 \end{vmatrix}.$$

Equation (2) is equivalent to

(3)
$$D = \begin{vmatrix} a_1 & b_1 & c_1 \\ a_2 & b_2 & c_2 \\ a_3 & b_3 & c_3 \end{vmatrix} = a_1 b_2 c_3 + a_3 b_1 c_2 + a_2 b_3 c_1 - a_3 b_2 c_1 - a_1 b_3 c_2 - a_2 b_1 c_3.$$

Although both of the formulas (2) and (3) look somewhat complicated, there are several devices that make it easy to obtain the expansion (the right side of (3)). In one such device we repeat the first two columns to the right of the determinant, and then take products along the diagonals as indicated in the following diagram, using a plus sign if the diagonal descends from left to right, and a minus sign if the diagonal descends from right to left.

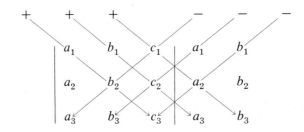

For example, to compute

$$D = \begin{vmatrix} 3 & 2 & 7 \\ -1 & 5 & 3 \\ 2 & -3 & -6 \end{vmatrix},$$

we have

$$\begin{vmatrix} 3 & 2 & 7 \\ -1 & 5 & 3 \\ 2 & -3 & -6 \end{vmatrix} \begin{matrix} 3 & 2 \\ -1 & 5 \\ 2 & -3 \end{matrix}$$

$$D = 3 \cdot 5(-6) + 2 \cdot 3 \cdot 2 + 7(-1)(-3) - 2 \cdot 5 \cdot 7 - (-3)3 \cdot 3 - (-6)(-1)2$$

$$= -90 + 12 + 21 - 70 + 27 - 12 = -112.$$

This method of evaluating a third-order determinant does *not* work when the order is greater than three. However, in this book we have little need for determinants of order greater than three.

EXERCISE 1

In Problems 1 through 12, find the value of the given determinant.

1. $\begin{vmatrix} 1 & 2 \\ 3 & 4 \end{vmatrix}$.

2. $\begin{vmatrix} 3 & 6 \\ 9 & 12 \end{vmatrix}$.

3. $\begin{vmatrix} -1 & 2 \\ -4 & 7 \end{vmatrix}$.

4. $\begin{vmatrix} 10 & -9 \\ 8 & 7 \end{vmatrix}.$ **5.** $\begin{vmatrix} \sqrt{2} & 3 \\ 2 & \sqrt{18} \end{vmatrix}.$ **6.** $\begin{vmatrix} 1 & 3 \\ -5 & 2 \end{vmatrix}.$

7. $\begin{vmatrix} 1 & 2 & 1 \\ 2 & 4 & 2 \\ 1 & -1 & 1 \end{vmatrix}.$ **8.** $\begin{vmatrix} 1 & 1 & -1 \\ -1 & 1 & 1 \\ 1 & -1 & 2 \end{vmatrix}.$ **9.** $\begin{vmatrix} 2 & 1 & 3 \\ -4 & -5 & 6 \\ 1 & -9 & 5 \end{vmatrix}.$

10. $\begin{vmatrix} -3 & 1 & 2 \\ 5 & 0 & -1 \\ 0 & 3 & 0 \end{vmatrix}.$ **11.** $\begin{vmatrix} \frac{1}{2} & -1 & \frac{1}{3} \\ -6 & 3 & 1 \\ 5 & 2 & -2 \end{vmatrix}.$ **12.** $\begin{vmatrix} 1 & 2 & 3 \\ 4 & 5 & 6 \\ 7 & 8 & 9 \end{vmatrix}.$

13. In the following scheme we multiply the numbers on the same "broken diagonal" attaching a sign $+$ or $-$ in accordance with the direction as indicated. Show that this scheme also gives the expansion of the determinant [the right side of equation (3)].

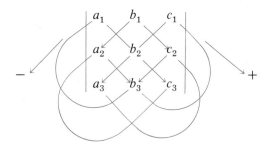

Some Theorems on Determinants 2

We state a number of theorems concerning the columns of a determinant. However, each theorem in this section is still true if the word *column* is replaced by the word *row*, everywhere it occurs. We omit the proofs—but each theorem follows from Definitions 1 and 2 for second-order and third-order determinants. For an *n*th-order determinant, the definition and the proofs are more complicated.

THEOREM 1 (PLE). If in one column of a matrix M every element is zero, then $D(M) = 0$.

THEOREM 2 (PLE). Suppose that the matrix P is obtained from the matrix M by the interchange of two adjacent columns. Then, $D(P) = -D(M)$.

THEOREM 3 (PLE). If two columns of the matrix M are identical, then $D(M) = 0$.

THEOREM 4 (PLE). If the matrix P is obtained from the matrix M by multiplying each element in a fixed column by the constant k, then $D(P) = kD(M)$.

In practice, we use this theorem to simplify determinants by "factoring out" common factors in a column or row. For example, if we factor 3 from the third column, we have

$$\begin{vmatrix} 5 & 4 & 9 \\ 2 & -1 & 3 \\ 0 & 7 & -12 \end{vmatrix} = 3 \begin{vmatrix} 5 & 4 & 3 \\ 2 & -1 & 1 \\ 0 & 7 & -4 \end{vmatrix}.$$

THEOREM 5 (PLE). If two columns of a matrix M are proportional, then $D(M) = 0$.

For example,

(4)
$$\begin{vmatrix} 11 & -5 & 15 \\ 19 & 1 & -3 \\ 65 & -7 & 21 \end{vmatrix} = 0,$$

because the last two columns are proportional, with -3 as the constant of proportionality. To see this, observe that each element in the third column is -3 times the element in the same row and in the second column.

THEOREM 6 (PWO). If the matrix P is obtained from the matrix M by adding to any one column (termwise) k times some other column, then $D(P) = D(M)$.

For example,

(5)
$$\begin{vmatrix} 7 & 2 & 17 \\ -15 & -5 & 6 \\ 9 & 3 & 49 \end{vmatrix} = \begin{vmatrix} 1 & 2 & 17 \\ 0 & -5 & 6 \\ 0 & 3 & 49 \end{vmatrix},$$

because the second determinant can be obtained from the first one by multiplying each element in the second column by -3 and adding it to the first column. In this example the work of evaluating the second determinant is much simpler than the work for the first determinant because of the zeros that have been introduced.

Example 1. Find $D(M)$ for the matrix

(6)
$$M = \begin{bmatrix} 21 & 31 & 41 \\ 51 & 62 & 73 \\ 5 & 6 & 8 \end{bmatrix}.$$

Solution. We use our theorems to replace the large numbers in M by smaller ones and then eventually by zeros. To simplify the work we introduce the symbol $\text{R2} - k \,\text{R3}$ to mean that we subtract k times the third row from the second row. A similar notation with C indicates the corresponding operation on columns. Then for $D(M)$, we have

$$\begin{vmatrix} 21 & 31 & 41 \\ 51 & 62 & 73 \\ 5 & 6 & 8 \end{vmatrix} \underset{\text{R1}}{\overset{\text{R2}}{=}} \begin{vmatrix} 21 & 31 & 41 \\ 30 & 31 & 32 \\ 5 & 6 & 8 \end{vmatrix} \begin{array}{c} \text{C2} \overset{=}{} \text{C1} \\ \text{C3} - \text{C1} \end{array} \begin{vmatrix} 21 & 10 & 20 \\ 30 & 1 & 2 \\ 5 & 1 & 3 \end{vmatrix}$$

$$\text{C3} \overset{=}{} 2\,\text{C2} \begin{vmatrix} 21 & 10 & 0 \\ 30 & 1 & 0 \\ 5 & 1 & 1 \end{vmatrix} = 21(1)(1) + 0 + 0 - 0 - 0 - 1(10)(30)$$

$$= 21 - 300 = -279. \quad \bullet$$

For those who prefer brute force, we can apply equation (3) to the matrix (6). This gives us

$$a_1 b_2 c_3 = 21(62)(8) = 10{,}416 \qquad -a_3 b_2 c_1 = -5(62)(41) = -12{,}710$$

$$a_3 b_1 c_2 = 5(31)(73) = 11{,}315 \qquad -a_1 b_3 c_2 = -21(6)(73) = -\ 9{,}198$$

$$\underline{a_2 b_3 c_1 = 51(6)(41) = 12{,}546} \qquad \underline{-a_2 b_1 c_3 = -51(31)(8) = -12{,}648}$$

$$\text{sum} \quad = 34{,}277 \qquad\qquad\quad \text{sum} \quad = -34{,}556$$

Hence $D(M) = 34{,}277 - 34{,}556 = -279$, just as before.

There is one more theorem that is helpful in evaluating a determinant. We need two new terms. The *minor* of an element in an nth-order determinant is the $(n-1)$th-order determinant obtained by striking out the row and column of that element. Thus the minors of b_2 and c_2 in

$$\begin{vmatrix} a_1 & b_1 & c_1 \\ a_2 & b_2 & c_2 \\ a_3 & b_3 & c_3 \end{vmatrix} \quad \text{are} \quad \begin{vmatrix} a_1 & c_1 \\ a_3 & c_3 \end{vmatrix} \quad \text{and} \quad \begin{vmatrix} a_1 & b_1 \\ a_3 & b_3 \end{vmatrix},$$

respectively. The *cofactor* of an element is obtained from the minor by adjoining the factor $(-1)^{i+j}$, where the element is in the ith row and the jth column. For b_2 and c_2 these factors are $(-1)^{2+2} = 1$ and $(-1)^{2+3} = -1$, respectively. Thus the cofactors of b_2 and c_2 are

$$\begin{vmatrix} a_1 & c_1 \\ a_3 & c_3 \end{vmatrix} \quad \text{and} \quad -\begin{vmatrix} a_1 & b_1 \\ a_3 & b_3 \end{vmatrix},$$

respectively.

If we examine Definition 2 carefully, we see that equation (2) is really an expansion by using cofactors of the first row. It is easy to prove that we can also evaluate a determinant by using the cofactors of any fixed row, or the cofactors of any fixed column. For example, if we expand by minors of the third row, we find that

$$(7) \quad \begin{vmatrix} 3 & 5 & 6 \\ -1 & 7 & 8 \\ 2 & 1 & 4 \end{vmatrix} = 2\begin{vmatrix} 5 & 6 \\ 7 & 8 \end{vmatrix} - \begin{vmatrix} 3 & 6 \\ -1 & 8 \end{vmatrix} + 4\begin{vmatrix} 3 & 5 \\ -1 & 7 \end{vmatrix}$$

$$= 2(40 - 42) - (24 + 6) + 4(21 + 5) = -4 - 30 + 104 = 70.$$

If we expand (7) by minors of the third column, we find that

$$(8) \quad \begin{vmatrix} 3 & 5 & 6 \\ -1 & 7 & 8 \\ 2 & 1 & 4 \end{vmatrix} = 6\begin{vmatrix} -1 & 7 \\ 2 & 1 \end{vmatrix} - 8\begin{vmatrix} 3 & 5 \\ 2 & 1 \end{vmatrix} + 4\begin{vmatrix} 3 & 5 \\ -1 & 7 \end{vmatrix}$$

$$= 6(-1 - 14) - 8(3 - 10) + 4(21 + 5) = -90 + 56 + 104 = 70.$$

The reader is invited to expand this determinant by cofactors of any row or column and obtain the same result: 70. He can also use the method of Example 1. This should also give 70.

We remark that this method of expansion by using cofactors applies to any $n \times n$ determinant. Thus by expanding along a row or column a 4×4 determinant can be reduced to computing four 3×3 determinants. More generally, an $n \times n$ determinant can be evaluated by computing n determinants of order $n - 1$.

EXERCISE 2

In Problems 1 through 6, evaluate the given determinant. HINT: *Use the methods of Example 1 to obtain two zeros in some row or column.*

1. $\begin{vmatrix} 1 & 4 & 9 \\ 4 & 9 & 16 \\ 9 & 16 & 25 \end{vmatrix}.$

2. $\begin{vmatrix} 12 & 15 & 11 \\ 6 & 2 & 7 \\ 13 & 16 & 12 \end{vmatrix}.$

3. $\begin{vmatrix} 17 & 4 & 5 \\ 16 & 10 & 9 \\ 8 & 1 & 3 \end{vmatrix}.$

4. $\begin{vmatrix} 2 & 3 & 5 \\ 7 & 11 & 13 \\ 17 & 19 & 23 \end{vmatrix}.$

5. $\begin{vmatrix} 171 & 110 & 60 \\ 141 & 71 & 70 \\ 344 & 221 & 121 \end{vmatrix}.$

6. $\begin{vmatrix} 28 & 30 & -35 \\ -15 & -14 & 19 \\ 6 & 11 & -5 \end{vmatrix}.$

7. Use the methods of this section to evaluate the determinants in Problems 7 through 12 of Exercise 1.

In Problems 8 through 11, prove that the given expansion is correct.

8. $\begin{vmatrix} a & b & c \\ c & a & b \\ b & c & a \end{vmatrix} = a^3 + b^3 + c^3 - 3abc.$

9. $\begin{vmatrix} 1 & 1 & 1 \\ x & y & z \\ x^2 & y^2 & z^2 \end{vmatrix} = (z - x)(z - y)(y - x).$

10. $\begin{vmatrix} b + c & c & b \\ c & c + a & a \\ b & a & a + b \end{vmatrix} = 4abc.$

11. $\begin{vmatrix} b + c & a - c & a - b \\ b - c & c + a & b - a \\ c - b & c - a & a + b \end{vmatrix} = 8abc.$

Cramer's Rule 3

This rule tells us how to use determinants to solve systems of simultaneous equations. The rule is good for n linear equations in n unknowns, but for simplicity we state it for three linear equations in three unknowns.

THEOREM 7 Cramer's Rule (PWO). If for the system of simultaneous equations

(9)
$$a_1 x + b_1 y + c_1 z = k_1$$
$$a_2 x + b_2 y + c_2 z = k_2$$
$$a_3 x + b_3 y + c_3 z = k_3,$$

the determinant of the coefficients

(10)
$$D = \begin{vmatrix} a_1 & b_1 & c_1 \\ a_2 & b_2 & c_2 \\ a_3 & b_3 & c_3 \end{vmatrix}$$

is not zero, then the solution is given by

(11) $$x = \frac{\begin{vmatrix} k_1 & b_1 & c_1 \\ k_2 & b_2 & c_2 \\ k_3 & b_3 & c_3 \end{vmatrix}}{D}, \qquad y = \frac{\begin{vmatrix} a_1 & k_1 & c_1 \\ a_2 & k_2 & c_2 \\ a_3 & k_3 & c_3 \end{vmatrix}}{D}, \qquad z = \frac{\begin{vmatrix} a_1 & b_1 & k_1 \\ a_2 & b_2 & k_2 \\ a_3 & b_3 & k_3 \end{vmatrix}}{D}.$$

Example 1. Solve the system of simultaneous equations

(12)
$$\begin{aligned} 14x + 2y - 6z &= 9 \\ -4x + y + 9z &= 3 \\ 6x - 4y + 3z &= -4. \end{aligned}$$

Solution. The denominator D is formed using the coefficients of the unknowns x, y, and z. We find that

$$D = \begin{vmatrix} 14 & 2 & -6 \\ -4 & 1 & 9 \\ 6 & -4 & 3 \end{vmatrix} = \begin{vmatrix} 26 & -6 & 0 \\ -22 & 13 & 0 \\ 6 & -4 & 3 \end{vmatrix} = 3 \begin{vmatrix} 26 & -6 \\ -22 & 13 \end{vmatrix} = 3 \begin{vmatrix} 4 & 7 \\ -22 & 13 \end{vmatrix}$$

$$= 3(52 + 154) = 618.$$

For the numerator of x, we replace the coefficients of x by the constants. Thus

(13) $$xD = \begin{vmatrix} 9 & 2 & -6 \\ 3 & 1 & 9 \\ -4 & -4 & 3 \end{vmatrix} = \begin{vmatrix} 1 & -6 & 0 \\ 15 & 13 & 0 \\ -4 & -4 & 3 \end{vmatrix} = 3(13 + 90) = 309.$$

Hence

$$x = \frac{309}{D} = \frac{309}{618} = \frac{1}{2}.$$

Similarly, for the numerator of y, we find that

(14) $$yD = \begin{vmatrix} 14 & 9 & -6 \\ -4 & 3 & 9 \\ 6 & -4 & 3 \end{vmatrix} = \begin{vmatrix} 26 & 1 & 0 \\ -22 & 15 & 0 \\ 6 & -4 & 3 \end{vmatrix} = 3 \begin{vmatrix} 26 & 1 \\ -22 & 15 \end{vmatrix} = 1236.$$

Hence

$$y = \frac{1236}{D} = \frac{1236}{618} = 2.$$

Finally, for the numerator of z, we have

$$zD = \begin{vmatrix} 14 & 2 & 9 \\ -4 & 1 & 3 \\ 6 & -4 & -4 \end{vmatrix} = 2 \begin{vmatrix} 14 & 2 & 9 \\ -4 & 1 & 3 \\ 3 & -2 & -2 \end{vmatrix} = 2 \begin{vmatrix} 22 & 0 & 3 \\ -4 & 1 & 3 \\ -5 & 0 & 4 \end{vmatrix} = 2 \begin{vmatrix} 22 & 3 \\ -5 & 4 \end{vmatrix}$$

$$= 2(88 + 15) = 206,$$

$$z = \frac{206}{618} = \frac{1}{3}. \quad \bullet$$

Cramer's Rule for a system of two linear equations in two unknowns,

(15)
$$\begin{cases} a_1 x + b_1 y = c_1 \\ a_2 x + b_2 y = c_2, \end{cases}$$

is given in

THEOREM 8 (PLE). If the denominator is not zero, then the solution of the system (15) is given by

(16)
$$x = \frac{\begin{vmatrix} c_1 & b_1 \\ c_2 & b_2 \end{vmatrix}}{\begin{vmatrix} a_1 & b_1 \\ a_2 & b_2 \end{vmatrix}} \quad \text{and} \quad y = \frac{\begin{vmatrix} a_1 & c_1 \\ a_2 & c_2 \end{vmatrix}}{\begin{vmatrix} a_1 & b_1 \\ a_2 & b_2 \end{vmatrix}}.$$

EXERCISE 3

In Problems 1 through 12, use Cramer's Rule to solve the given system of equations.

1. $x + y = 6$
    $3x - 2y = 8.$

2. $2x - y = 5$
    $5x + 3y = 18.$

3. $2x + 5y = 12$
    $5x - 3y = -1.$

4. $2x + 3y = 9$
    $11x + 7y = 2.$

5. $11x + 3 = -3y$
    $5x + 2y = 5.$

6. $10x + 6y = 1$
    $14x = 3y + 9.$

7. $x + y + z = 6$
    $2x - y + 3z = 9$
    $3x - 2y - z = -4.$

8. $6x + 5y + 4z = 5$
    $5x + 4y + 3z = 5$
    $4x + 3y + z = 7.$

9. $x - 2y + 3z = 15$
    $5x + 7y - 11z = -29$
    $-13x + 17y + 19z = 37.$

10. $x - 3y + 6z = -8$
    $3x - 2y - 10z = 11$
    $-5x + 6y + 2z = -7.$

11. $5u + 3v + 5w = 3$
    $3u + 5v + w = -5$
    $2u + 2v + 3w = 7.$

12. $6A + 5B = 1$
    $7B + 4C = 13$
    $8A + 3C = 23.$

Appendix 4
Units: The British and Metric Systems

The first step in measuring any physical object is to select a unit. The length of an object is expressed as a multiple of the length of some standard object selected as having unit length, such as 1 foot, 1 yard, 1 meter, and so on. Unfortunately, different civilizations have selected different units, and this may cause some confusion. The three important systems still in use today are listed in Table 1, together with the names for some of the units.

TABLE 1

System	Force	Mass	Acceleration
MKS meter, kilogram, second	newton (N)	kilogram (kg)	meters/sec^2
CGS centimeter, gram, second	dyne	gram (g)	cm/sec^2
British engineering	pound (lb)	slug	ft/sec^2

The unit of time is the same in all three systems. The units of length are related as indicated[1] in Table 2. Since 1 meter is 100 centimeters and 1 kilogram is 1000 grams, it is a simple matter to change from the CGS system to the MKS system. Consequently, we may drop the latter from further consideration.

---

[1] These relations are known with much greater accuracy. For simplicity we have rounded off all numbers to four significant figures.

936

TABLE 2

Centimeters	Feet	Inches
1	0.03281	0.3937
30.48	1	12
2.540	0.08333	1

From the entries in Table 2, it is easy to derive relations for units of area, volume, velocity, and acceleration. Some of these are given in Table 3 and others are left to the exercises.

TABLE 3

	British	CGS
Area	1 in.^{2}	6.452 cm^2
Volume	1 in.^{3}	16.39 cm^3
Velocity	1 ft/sec	30.48 cm/sec
Velocity	1 mile/hr	1.467 ft/sec
Velocity	1 mile/hr	44.70 cm/sec
Acceleration	32.17 ft/sec^2	980.5 cm/sec^2

The last item in Table 3 is the acceleration due to gravity at the earth's surface. Since this item may be different at different points on the earth's surface, one often replaces the table values by 32.2 ft/sec^2 and 980 cm/sec^2.

In converting from one system of units to another, we can cancel units in the same way that we cancel common factors in algebra. This technique is often useful in guiding us to the correct result.

Example 1. Check the fifth item in Table 3.

Solution. Using the second entry in Table 2 we have

$$1\,\frac{\text{mile}}{\text{hr}} = 1\,\frac{\text{mile}}{\text{hr}} \times \frac{5{,}280\;\text{ft}}{1\,\text{mile}} \times \frac{1\;\text{hr}}{3600\;\text{sec}} \times \frac{30.48\;\text{cm}}{1\;\text{ft}}$$

$$= 1 \times \frac{5280}{3600} \times 30.48\,\frac{\text{cm}}{\text{sec}} \approx 44.70\;\text{cm/sec.} \quad \bullet$$

The conversion factors for mass and force are a little more complex because the fundamental concepts are different in the two systems. Both systems make use of the basic relation

(1) $$F = mA,$$

which is Newton's famous law relating the force applied to a body and its acceleration, but the two systems (CGS and British engineering) proceed differently.

We note first that the *mass* of a body is regarded as an intrinsic quantity that is the same wherever the body may be placed. In contrast, the *weight* of a body varies slightly as it is moved about the earth's surface and (as we know) is much less on the surface of the moon. This can be checked by a spring balance or by more delicate weighing devices.

In both systems the weight of an object is measured by the force exerted on it by gravity. Since this varies, the weight is a variable, but this variation over the earth's surface is less than 0.4 percent, and for practical purposes this does little harm.

In the British system, equation (1) gives

(2) $$F[\text{lb}] = m \,[\text{slugs}] \cdot g \,[\text{ft/sec}^2],$$

and hence the mass is computed by setting $m = F/g$.

In the CGS system, equation (2) is replaced by

(3) $$F[\text{dynes}] = m \,[\text{gm}] \cdot g \,[\text{cm/sec}^2].$$

Note that in equation (2), the "pound" unit is on the left side, and in equation (3) the "gram" unit is on the right side. Thus the pound is a unit of force and the gram is a unit of mass.

When we write that 1 lb = 453.6 grams we are equating a weight on the left side (which may be variable) with a mass on the right side (which is constant). This equation makes sense only if the value of g is specified. Consequently, in Table 4 the equivalences are valid only at those places where g has the value specified.

TABLE 4. $g = 32.17 \text{ ft/sec}^2$ or 980.5 cm/sec^2

	British	CGS
Weight-mass	1 lb	453.6 g
Weight-mass	0.002205 lb	1 g
Pressure	1 lb/ft^2	0.4882 g/cm^2
Pressure	2.048 lb/ft^2	1 g/cm^2
Density	1 lb/ft^3	0.01602 g/cm^3
Density	62.42 lb/ft^3	1 g/cm^3
Weight-force	1 lb	4.448×10^5 dynes
Weight-force	2.248×10^{-6} lb	1 dyne

The first two entries in Table 4 are obtained by comparing the standard bodies in the two systems. The remaining entries in Table 4 can be derived from the first two, and the appropriate entries from Tables 2 and 3.

EXERCISE 1

1. Let us regard the second row of entries in Table 2 as definitions. Use these to derive the other entries in Table 2.
2. Derive the entries in Table 3 from those in Table 2.
3. Use the first entry in Table 4 (and any previous results that you need) to derive the other entries in the table.

In Problems 4 through 7, derive the indicated equivalence.
4. 1 cm^2 is 0.1550 in.^{2}.
5. 1 cm/sec is 0.03281 ft/sec.
6. 1 ft-lb of work is 0.1383 kg-meters of work.
7. 1 ft/sec is 0.6818 mile/hr.

Appendix 5
Evaluation of Polynomial Functions

A polynomial function is a function of the special form

(1) $$P(x) = a_n x^n + a_{n-1} x^{n-1} + \cdots + a_1 x + a_0,$$

where a_n, a_{n-1}, $\ldots$, a_1, a_0 represent constants and n is a nonnegative integer. For example,

$$P_1(x) = -2x^3 + 4x^2 - 3x + 1 \quad \text{and} \quad P_2(x) = \pi x^9 - 8x^2 + x + \sqrt{7}$$

are polynomial functions.

The degree of a polynomial is the highest power of x that appears. Thus the polynomial $x^3 - 2x^2 + 5$ is of degree 3, and the polynomial $3x^{72} - 6x^{40}$ is of degree 72.

The evaluation of polynomial functions simply involves the operations of addition and multiplication. This can be done in a straightforward manner on the calculator by first computing each term in the expression, storing the values of these terms in memory (or recording them on a scratch pad), and finally adding all the terms together. For polynomials of high degree, however, this procedure is rather tedious and requires a large number of key strokes. The purpose of this section is to present a time-saving device for polynomial evaluations called *Horner's Method*. We first illustrate the method for quadratic polynomials.

To compute $a_2 x^2 + a_1 x + a_0$, we write the polynomial in the equivalent form

(2) $$a_2 x^2 + a_1 x + a_0 = (a_2 x + a_1) x + a_0.$$

940

The evaluation of the right side is then done on the calculator by working from the inside of the parentheses outward:

$$a_2 \boxed{\times}\, x \boxed{+}\, a_1 \boxed{=} \boxed{\times} x \boxed{+}\, a_0 \boxed{=}.^{1}$$

For cubic polynomials we use a similar parenthetical representation,

(3) $$a_3 x^3 + a_2 x^2 + a_1 x + a_0 = ((a_3 x + a_2)x + a_1)x + a_0$$

and again calculate from the inside out thus:

$$a_3 \boxed{\times}\, x \boxed{+}\, a_2 \boxed{=} \boxed{\times} x \boxed{+}\, a_1 \boxed{=} \boxed{\times} x \boxed{+}\, a_0 \boxed{=}.$$

Note that the first $\boxed{=}$ gives the value of $a_3 x + a_2$; the second $\boxed{=}$ gives the value of $((a_3 x + a_2)x + a_1)$; and the third $\boxed{=}$ gives the value of the whole polynomial in (3). In general, the polynomial $P(x)$ in (1) can be written using nested parentheses:

(4) $$P(x) = ((\cdots((a_n x + a_{n-1})x + a_{n-2})x + \cdots + a_2)x + a_1)x + a_0,$$

and the right side can be computed according to the instructions:

(5) $$a_n \boxed{\times}\, x \boxed{+}\, a_{n-1} \boxed{=} \boxed{\times} x \boxed{+}\, a_{n-2} \boxed{=} \cdots \boxed{+}\, a_1 \boxed{=} \boxed{\times} x \boxed{+}\, a_0 \boxed{=}.$$

Although (4), with its multitude of parentheses, looks far more complex than (1), it is far superior to (1) when computing $P(x)$. To see this, it is convenient to introduce a notation for the intermediate values in (5). We let b_{n-1} denote the value obtained for the first $\boxed{=}$, b_{n-2} denote the value obtained for the second $\boxed{=}$, and so on until we reach the last $\boxed{=}$, which is denoted by b_0. The procedure of Horner can now be summarized as follows:

$$\text{SET} \quad b_n = a_n.$$

$$\text{COMPUTE} \quad b_{n-1} = b_n x + a_{n-1},$$

(6)
$$b_{n-2} = b_{n-1} x + a_{n-2},$$
$$\vdots$$
$$b_1 = b_2 x + a_1,$$
$$b_0 = b_1 x + a_0 = P(x).$$

Note that the innermost pair of parentheses in (4) gives b_{n-1}, the next larger pair of parentheses gives b_{n-2}, and so on, until the whole polynomial equals b_0.

[1] For calculators using Reverse-Polish notation, the instructions are

$$a_2 \boxed{\text{ENTER}}\, x \boxed{\times}\, a_1 \boxed{+}\, x \boxed{\times}\, a_0 \boxed{+}.$$

Example 1. Evaluate

(7) $$P(x) = 2x^3 - 7x^2 + 3$$

for $x = 5$.

Solution. Here $a_3 = 2$, $a_2 = -7$, $a_1 = 0$, and $a_0 = 3$. So, from (6) with $n = 3$ and $x = 5$ we compute

(8)
$$b_3 = 2,$$
$$b_2 = b_3 \cdot 5 - 7 = 2 \cdot 5 - 7 = 3,$$
$$b_1 = b_2 \cdot 5 + 0 = 3 \cdot 5 + 0 = 15,$$
$$b_0 = b_1 \cdot 5 + 3 = 15 \cdot 5 + 3 = 78.$$

Thus $P(5) = 78$. ●

Observe that the set of equations (8) requires three multiplications, three additions, and no storage. The conventional method of computing $2x^3 - 7x^2 + 3$ for $x = 5$ requires seven multiplications, two additions, and one storage. The difference is even larger for higher-degree polynomials. Of course, in Horner's Method it is not necessary to record the intermediate values $b_n, b_{n-1}, \ldots, b_1$, but these quantities, which depend on x, have an interesting application.

Suppose that we wish to divide a polynomial $P(x)$ by $x - c$, where c is a given number. In general, we will obtain some quotient polynomial $Q(x)$ plus a constant remainder term R such that

(9) $$P(x) = (x - c)Q(x) + R.$$

For example, if $P(x) = 2x^3 - 7x^2 + 3$ and $c = 5$, then we can use the usual process of division to find

$$
\begin{array}{r}
2x^2 + 3x + 15 \\
x - 5 \overline{\smash{)}2x^3 - 7x^2 + 0x + 3} \\
\underline{2x^3 - 10x^2} \\
3x^2 + 0x \\
\underline{3x^2 - 15x} \\
15x + 3 \\
\underline{15x - 75} \\
78.
\end{array}
$$

Hence $Q(x) = 2x^2 + 3x + 15$ and $R = 78$. Thus we have

(10) $$2x^3 - 7x^2 + 3 = (x - 5)(2x^2 + 3x + 15) + 78.$$

Note that the coefficients 2, 3, and 15 of the quotient polynomial $Q(x)$ are precisely the same as the values obtained in (8) for b_3, b_2, and b_1. Furthermore, $b_0 = 78$ is the remainder R. In other words, for this cubic polynomial,

$$P(x) = (x - c)(b_3 x^2 + b_2 x + b_1) + b_0,$$

where the numbers b_3, b_2, b_1, and b_0 are given by Horner's Method (6) for $x = c$. This is no mere coincidence, as the following theorem asserts.

THEOREM 1. If the polynomial

(1)
$$P(x) = a_n x^n + a_{n-1} x^{n-1} + \cdots + a_1 x + a_0$$

is divided by $x - c$ to obtain

(9)
$$P(x) = (x - c)Q(x) + R,$$

then

(11)
$$Q(x) = b_n x^{n-1} + b_{n-1} x^{n-2} + \cdots + b_2 x + b_1$$

and

(12)
$$R = b_0,$$

where b_n, b_{n-1}, $\ldots$, b_1, b_0 are the values given in (6) for $x = c$.

Proof. Since the general case is similar, we only give the proof for polynomials $P(x)$ of degree 4. Let

(13)
$$P(x) = a_4 x^4 + a_3 x^3 + a_2 x^2 + a_1 x + a_0,$$

and denote the unknown quotient polynomial $Q(x)$ in (9) by

$$Q(x) = e_3 x^3 + e_2 x^2 + e_1 x + e_0.$$

Our goal is to show that

(14)
$$e_3 = b_4, \quad e_2 = b_3, \quad e_1 = b_2, \quad e_0 = b_1, \quad \text{and } R = b_0,$$

where b_4, b_3, b_2, b_1, and b_0 are the numbers defined in (6) when $n = 4$ and $x = c$.
  Multiplying out the right-hand side of (9), we obtain

$$(x - c)Q(x) + R = (x - c)(e_3 x^3 + e_2 x^2 + e_1 x + e_0) + R$$

$$= e_3 x^4 + e_2 x^3 + e_1 x^2 + e_0 x - ce_3 x^3 - ce_2 x^2 - ce_1 x - ce_0 + R$$

(15)
$$= e_3 x^4 + (e_2 - ce_3)x^3 + (e_1 - ce_2)x^2 + (e_0 - ce_1)x + (R - ce_0).$$

On the other hand, the right side of (9) equals $P(x)$. Thus the polynomials in (13) and (15) are identical, and equating the coefficients of like powers of x gives:

$$a_4 = e_3 \qquad \text{or} \qquad e_3 = a_4,$$

$$a_3 = e_2 - ce_3 \qquad \text{or} \qquad e_2 = e_3 c + a_3,$$

$$a_2 = e_1 - ce_2 \qquad \text{or} \qquad e_1 = e_2 c + a_2,$$

$$a_1 = e_0 - ce_1 \qquad \text{or} \qquad e_0 = e_1 c + a_1,$$

$$a_0 = R - ce_0 \qquad \text{or} \qquad R = e_0 c + a_0.$$

But these equations are identical with the defining equations (6) for $n = 4$ and $x = c$. Hence (14) is true. ∎

We remark that this procedure of using the equations (6) to find the quotient polynomial is called *synthetic division*.

Example 2. Divide the polynomial

$$(16) \qquad\qquad P(x) = 4x^4 - 12x^3 + 11x^2 - 15x + 6$$

by $x - 0.5$. Also find $P(0.5)$.

Solution. Here

$$a_4 = 4, \quad a_3 = -12, \quad a_2 = 11, \quad a_1 = -15, \quad a_0 = 6, \quad \text{and } c = 0.5.$$

Thus, from the equations (6) with $n = 4$ and $x = 0.5$, we compute

$$b_4 = 4,$$

$$b_3 = b_4(0.5) - 12 = 4(0.5) - 12 = -10,$$

$$b_2 = b_3(0.5) + 11 = (-10)(0.5) + 11 = 6,$$

$$b_1 = b_2(0.5) - 15 = 6(0.5) - 15 = -12,$$

$$b_0 = b_1(0.5) + 6 = (-12)(0.5) + 6 = 0.$$

Therefore, $P(0.5) = 0$, and so the division comes out exactly:

$$P(x) = (x - 0.5)(4x^3 - 10x^2 + 6x - 12). \qquad \bullet$$

EXERCISE 1

CALCULATOR PROBLEMS

C1. Use Horner's method to evaluate $P(x) = 1.25x^3 - 2.8x^2 + x + 1$ for **(a)** $x = 3$, **(b)** $x = 5.3$, and **(c)** $x = 0.2$.

C2. Use Horner's method to evaluate $P(x) = 5.6x^4 - 2.3x^3 + 8x - 1$ for **(a)** $x = 2$, **(b)** $x = 0.4$, and **(c)** $x = -1.3$.

C3. Compute $P(1.85)$, where $P(x) = x^5 + 2.39x^4 - 7.26x^3 + 5.33x + 0.68$.

C4. Verify the nested parenthetical representation

$$a_4x^4 + a_3x^3 + a_2x^2 + a_1x + a_0 = (((a_4x + a_3)x + a_2)x + a_1)x + a_0$$

by multiplying out the right-hand side.

C5. Sketch the graph of $P(x) = 4x^3 + 4x^2 - 7x + 2$ on the interval $[0, 1]$ by first evaluating the polynomial for $x = 0, 0.1, 0.2, \ldots, 0.9, 1$.

C6. Use synthetic division to divide the polynomial

$$P(x) = 25x^5 - 338x^4 + 186x^3 - 240x^2 + 263x - 208 \quad \text{by} \quad x - 13.$$

Also find $P(13)$.

In Problems C7 through C11, divide the given polynomial $P(x)$ by $x - 1.5$. Also find $P(1.5)$.

C7. $P(x) = x^3 + 2.3x^2 + 4x - 3$.   **C8.** $P(x) = 8x^4 + 6x^3 + 2x^2 + 3x + 2$.

C9. $P(x) = -2x^4 + 3x^2 - 8.2x + 0.9$.   **C10.** $P(x) = x^5 + x^4 + x^3 + x^2 + x + 1$.

C11. $P(x) = 6x^6 + 5x^5 + 4x^4 + 3x^3 + 2x^2 + x + 1$.

Table A
Values of the Trigonometric Functions

Radians	Degrees	Sin	Tan	Cot	Cos		
.0000	0° 00′	.0000	.0000	—	1.0000	90° 00′	1.5708
.0175	1° 00′	.0175	.0175	57.29	.9998	89° 00′	1.5533
.0349	2° 00′	.0349	.0349	28.64	.9994	88° 00′	1.5359
.0524	3° 00′	.0523	.0524	19.08	.9986	87° 00′	1.5184
.0698	4° 00′	.0698	.0699	14.30	.9976	86° 00′	1.5010
.0873	5° 00′	.0872	.0875	11.43	.9962	85° 00′	1.4835
.1047	6° 00′	.1045	.1051	9.514	.9945	84° 00′	1.4661
.1222	7° 00′	.1219	.1228	8.144	.9925	83° 00′	1.4486
.1396	8° 00′	.1392	.1405	7.115	.9903	82° 00′	1.4312
.1571	9° 00′	.1564	.1584	6.314	.9877	81° 00′	1.4137
.1745	10° 00′	.1736	.1763	5.671	.9848	80° 00′	1.3963
.1920	11° 00′	.1908	.1944	5.145	.9816	79° 00′	1.3788
.2094	12° 00′	.2079	.2126	4.705	.9781	78° 00′	1.3614
.2269	13° 00′	.2250	.2309	4.331	.9744	77° 00′	1.3439
.2443	14° 00′	.2419	.2493	4.011	.9703	76° 00′	1.3265
.2618	15° 00′	.2588	.2679	3.732	.9659	75° 00′	1.3090
.2793	16° 00′	.2756	.2867	3.487	.9613	74° 00′	1.2915
.2967	17° 00′	.2924	.3057	3.271	.9563	73° 00′	1.2741
.3142	18° 00′	.3090	.3249	3.078	.9511	72° 00′	1.2566
.3316	19° 00′	.3256	.3443	2.904	.9455	71° 00′	1.2392
		Cos	Cot	Tan	Sin	Degrees	Radians

Radians	Degrees	Sin	Tan	Cot	Cos		
.3491	20° 00′	.3420	.3640	2.747	.9397	70° 00′	1.2217
.3665	21° 00′	.3584	.3839	2.605	.9336	69° 00′	1.2043
.3840	22° 00′	.3746	.4040	2.475	.9272	68° 00′	1.1868
.4014	23° 00′	.3907	.4245	2.356	.9205	67° 00′	1.1694
.4189	24° 00′	.4067	.4452	2.246	.9135	66° 00′	1.1519
.4363	25° 00′	.4226	.4663	2.145	.9063	65° 00′	1.1345
.4538	26° 00′	.4384	.4877	2.050	.8988	64° 00′	1.1170
.4712	27° 00′	.4540	.5095	1.963	.8910	63° 00′	1.0996
.4887	28° 00′	.4695	.5317	1.881	.8829	62° 00′	1.0821
.5061	29° 00′	.4848	.5543	1.804	.8746	61° 00′	1.0647
.5236	30° 00′	.5000	.5774	1.732	.8660	60° 00′	1.0472
.5411	31° 00′	.5150	.6009	1.664	.8572	59° 00′	1.0297
.5585	32° 00′	.5299	.6249	1.600	.8480	58° 00′	1.0123
.5760	33° 00′	.5446	.6494	1.540	.8387	57° 00′	.9948
.5934	34° 00′	.5592	.6745	1.483	.8290	56° 00′	.9774
.6109	35° 00′	.5736	.7002	1.428	.8192	55° 00′	.9599
.6283	36° 00′	.5878	.7265	1.376	.8090	54° 00′	.9425
.6458	37° 00′	.6018	.7536	1.327	.7986	53° 00′	.9250
.6632	38° 00′	.6157	.7813	1.280	.7880	52° 00′	.9076
.6807	39° 00′	.6293	.8098	1.235	.7771	51° 00′	.8901
.6981	40° 00′	.6428	.8391	1.192	.7660	50° 00′	.8727
.7156	41° 00′	.6561	.8693	1.150	.7547	49° 00′	.8552
.7330	42° 00′	.6691	.9004	1.111	.7431	48° 00′	.8378
.7505	43° 00′	.6820	.9325	1.072	.7314	47° 00′	.8203
.7679	44° 00′	.6947	.9657	1.036	.7193	46° 00′	.8029
.7854	45° 00′	.7071	1.000	1.000	.7071	45° 00′	.7854
		Cos	Cot	Tan	Sin	Degrees	Radians

Table B
Napierian or Natural Logarithms

N	ln N	N	ln N	N	ln N
		1.5	0.40547	**3.0**	1.09861
0.1	−2.30259	1.6	0.47000	3.1	1.13140
0.2	−1.60944	1.7	0.53063	3.2	1.16315
0.3	−1.20397	1.8	0.58779	3.3	1.19392
0.4	−0.91629	1.9	0.64185	3.4	1.22378
0.5	−0.69315	**2.0**	0.69315	3.5	1.25276
0.6	−0.51083	2.1	0.74194	3.6	1.28093
0.7	−0.35667	2.2	0.78846	3.7	1.30833
0.8	−0.22314	2.3	0.83291	3.8	1.33500
0.9	−0.10536	2.4	0.87547	3.9	1.36098
1.0	0.00000	2.5	0.91629	**4.0**	1.38629
1.1	0.09531	2.6	0.95551	4.1	1.41099
1.2	0.18232	2.7	0.99325	4.2	1.43508
1.3	0.26236	2.8	1.02962	4.3	1.45862
1.4	0.33647	2.9	1.06471	4.4	1.48160

N	$\ln N$	N	$\ln N$	N	$\ln N$
4.5	1.50408	7.5	2.01490	10.5	2.35138
4.6	1.52606	7.6	2.02815	10.6	2.36085
4.7	1.54756	7.7	2.04122	10.7	2.37024
4.8	1.56862	7.8	2.05412	10.8	2.37955
4.9	1.58924	7.9	2.06686	10.9	2.38876
5.0	1.60944	**8.0**	2.07944	**11.0**	2.39790
5.1	1.62924	8.1	2.09186	11.1	2.40695
5.2	1.64866	8.2	2.10413	11.2	2.41591
5.3	1.66771	8.3	2.11626	11.3	2.42480
5.4	1.68640	8.4	2.12823	11.4	2.43361
5.5	1.70475	8.5	2.14007	11.5	2.44235
5.6	1.72277	8.6	2.15176	11.6	2.45101
5.7	1.74047	8.7	2.16332	11.7	2.45959
5.8	1.75786	8.8	2.17475	11.8	2.46810
5.9	1.77495	8.9	2.18605	11.9	2.47654
6.0	1.79176	**9.0**	2.19722	**12.0**	2.48491
6.1	1.80829	9.1	2.20827	12.1	2.49321
6.2	1.82455	9.2	2.21920	12.2	2.50144
6.3	1.84055	9.3	2.23001	12.3	2.50960
6.4	1.85630	9.4	2.24071	12.4	2.51770
6.5	1.87180	9.5	2.25129	12.5	2.52573
6.6	1.88707	9.6	2.26176	12.6	2.53370
6.7	1.90211	9.7	2.27213	12.7	2.54160
6.8	1.91692	9.8	2.28238	12.8	2.54945
6.9	1.93152	9.9	2.29253	12.9	2.55723
7.0	1.94591	**10.0**	2.30259	**13.0**	2.56495
7.1	1.96009	10.1	2.31254	13.1	2.57261
7.2	1.97408	10.2	2.32239	13.2	2.58022
7.3	1.98787	10.3	2.33214	13.3	2.58776
7.4	2.00148	10.4	2.34181	13.4	2.59525

Table C
Exponential and Hyperbolic Functions

x	e^x	e^{-x}	sinh x	cosh x	tanh x
0	1.0000	1.0000	.00000	1.0000	.00000
0.1	1.1052	.90484	.10017	1.0050	.09967
0.2	1.2214	.81873	.20134	1.0201	.19738
0.3	1.3499	.74082	.30452	1.0453	.29131
0.4	1.4918	.67032	.41075	1.0811	.37995
0.5	1.6487	.60653	.52110	1.1276	.46212
0.6	1.8221	.54881	.63665	1.1855	.53705
0.7	2.0138	.49659	.75858	1.2552	.60437
0.8	2.2255	.44933	.88811	1.3374	.66404
0.9	2.4596	.40657	1.0265	1.4331	.71630
1.0	2.7183	.36788	1.1752	1.5431	.76159
1.1	3.0042	.33287	1.3356	1.6685	.80050
1.2	3.3201	.30119	1.5095	1.8107	.83365
1.3	3.6693	.27253	1.6984	1.9709	.86172
1.4	4.0552	.24660	1.9043	2.1509	.88535
1.5	4.4817	.22313	2.1293	2.3524	.90515
1.6	4.9530	.20190	2.3756	2.5775	.92167
1.7	5.4739	.18268	2.6456	2.8283	.93541
1.8	6.0496	.16530	2.9422	3.1075	.94681
1.9	6.6859	.14957	3.2682	3.4177	.95624

x	e^x	e^{-x}	$\sinh x$	$\cosh x$	$\tanh x$
2.0	7.3891	.13534	3.6269	3.7622	.96403
2.1	8.1662	.12246	4.0219	4.1443	.97045
2.2	9.0250	.11080	4.4571	4.5679	.97574
2.3	9.9742	.10026	4.9370	5.0372	.98010
2.4	11.023	.09072	5.4662	5.5569	.98367
2.5	12.182	.08208	6.0502	6.1323	.98661
2.6	13.464	.07427	6.6947	6.7690	.98903
2.7	14.880	.06721	7.4063	7.4735	.99101
2.8	16.445	.06081	8.1919	8.2527	.99263
2.9	18.174	.05502	9.0596	9.1146	.99396
3.0	20.086	.04979	10.018	10.068	.99505
3.1	22.198	.04505	11.076	11.122	.99595
3.2	24.533	.04076	12.246	12.287	.99668
3.3	27.113	.03688	13.538	13.575	.99728
3.4	29.964	.03337	14.965	14.999	.99777
3.5	33.115	.03020	16.543	16.573	.99818
3.6	36.598	.02732	18.285	18.313	.99851
3.7	40.447	.02472	20.211	20.236	.99878
3.8	44.701	.02237	22.339	22.362	.99900
3.9	49.402	.02024	24.691	24.711	.99918
4.0	54.598	.01832	27.290	27.308	.99933
4.1	60.340	.01657	30.162	30.178	.99945
4.2	66.686	.01500	33.336	33.351	.99955
4.3	73.700	.01357	36.843	36.857	.99963
4.4	81.451	.01228	40.719	40.732	.99970
4.5	90.017	.01111	45.003	45.014	.99975
4.6	99.484	.01005	49.737	49.747	.99980
4.7	109.95	.00910	54.969	54.978	.99983
4.8	121.51	.00823	60.751	60.759	.99986
4.9	134.29	.00745	67.141	67.149	.99989
5.0	148.41	.00674	74.203	74.210	.99991

Table D
A Brief Table of Integrals

Note: An arbitrary constant is to be added to the formula for each indefinite integral.

Elementary Integrals

1. $\displaystyle \int cf(u)\,du = c \int f(u)\,du.$

2. $\displaystyle \int (f(u) + g(u))\,du = \int f(u)\,du + \int g(u)\,du.$

3. $\displaystyle \int u^n\,du = \frac{u^{n+1}}{n+1}, \qquad n \neq -1.$

4. $\displaystyle \int \frac{du}{u} = \ln|u|.$

5. $\displaystyle \int e^u\,du = e^u.$

6. $\displaystyle \int a^u\,du = \frac{a^u}{\ln a}, \qquad a > 0, a \neq 1.$

7. $\displaystyle\int \sin u \, du = -\cos u.$

8. $\displaystyle\int \cos u \, du = \sin u.$

9. $\displaystyle\int \tan u \, du = -\ln |\cos u|.$

10. $\displaystyle\int \cot u \, du = \ln |\sin u|.$

11. $\displaystyle\int \sec u \, du = \ln |\sec u + \tan u|.$

12. $\displaystyle\int \csc u \, du = -\ln |\csc u + \cot u| = \ln |\csc u - \cot u|.$

13. $\displaystyle\int \sec^2 u \, du = \tan u.$

14. $\displaystyle\int \csc^2 u \, du = -\cot u.$

15. $\displaystyle\int \sec u \tan u \, du = \sec u.$

16. $\displaystyle\int \csc u \cot u \, du = -\csc u.$

17. $\displaystyle\int \sinh u \, du = \cosh u.$

18. $\displaystyle\int \cosh u \, du = \sinh u.$

19. $\displaystyle\int \frac{du}{a^2 + u^2} = \frac{1}{a} \operatorname{Tan}^{-1} \frac{u}{a}.$

20. $\displaystyle\int \frac{du}{a^2 - u^2} = \frac{1}{2a} \ln \left| \frac{a + u}{a - u} \right|.$

21. $\displaystyle\int \frac{du}{u^2 - a^2} = \frac{1}{2a} \ln \left| \frac{u - a}{u + a} \right|.$

22. $\displaystyle\int \frac{du}{\sqrt{a^2 - u^2}} = \text{Sin}^{-1}\frac{u}{a}, \qquad a^2 \geqq u^2.$

23. $\displaystyle\int \frac{du}{\sqrt{u^2 - a^2}} = \ln|u + \sqrt{u^2 - a^2}|, \qquad u^2 \geqq a^2.$

24. $\displaystyle\int \frac{du}{\sqrt{u^2 + a^2}} = \ln(u + \sqrt{u^2 + a^2}).$

Integrand Containing $a + bu$

25. $\displaystyle\int \frac{u\,du}{a + bu} = \frac{1}{b^2}(a + bu - a\ln|a + bu|).$

26. $\displaystyle\int \frac{u\,du}{(a + bu)^2} = \frac{1}{b^2}\left(\frac{a}{a + bu} + \ln|a + bu|\right).$

27. $\displaystyle\int \frac{u^2\,du}{(a + bu)^2} = \frac{1}{b^3}\left(a + bu - \frac{a^2}{a + bu} - 2a\ln|a + bu|\right).$

28. $\displaystyle\int \frac{u\,du}{(a + bu)^3} = \frac{1}{b^2}\left(-\frac{1}{a + bu} + \frac{a}{2(a + bu)^2}\right).$

29. $\displaystyle\int \frac{du}{u(a + bu)} = \frac{1}{a}\ln\left|\frac{u}{a + bu}\right|.$

30. $\displaystyle\int \frac{du}{u^2(a + bu)} = -\frac{1}{au} + \frac{b}{a^2}\ln\left|\frac{a + bu}{u}\right|.$

31. $\displaystyle\int \frac{du}{u(a + bu)^2} = \frac{1}{a(a + bu)} - \frac{1}{a^2}\ln\left|\frac{a + bu}{u}\right|.$

32. $\displaystyle\int u^m(a + bu)^n\,du = \frac{u^{m+1}(a + bu)^n}{m + n + 1} + \frac{an}{m + n + 1}\int u^m(a + bu)^{n-1}\,du.$

Integrand Containing $\sqrt{a + bu}$

33. $\displaystyle\int u\sqrt{a + bu}\,du = -\frac{2(2a - 3bu)(a + bu)^{3/2}}{15b^2}.$

34. $\displaystyle\int u^n \sqrt{a + bu}\, du = \frac{2u^n(a + bu)^{3/2}}{b(2n + 3)} - \frac{2an}{b(2n + 3)} \int u^{n-1}\sqrt{a + bu}\, du.$

35. $\displaystyle\int \frac{u\, du}{\sqrt{a + bu}} = \frac{2(bu - 2a)\sqrt{a + bu}}{3b^2}.$

36. $\displaystyle\int \frac{u^n\, du}{\sqrt{a + bu}} = \frac{2u^n\sqrt{a + bu}}{b(2n + 1)} - \frac{2an}{b(2n + 1)} \int \frac{u^{n-1}\, du}{\sqrt{a + bu}}.$

37. $\displaystyle\int \frac{du}{u\sqrt{a + bu}} = \frac{1}{\sqrt{a}} \ln \left| \frac{\sqrt{a + bu} - \sqrt{a}}{\sqrt{a + bu} + \sqrt{a}} \right|, \qquad a > 0.$

38. $\displaystyle\int \frac{du}{u\sqrt{a + bu}} = \frac{2}{\sqrt{-a}} \mathrm{Tan}^{-1} \sqrt{\frac{a + bu}{-a}}, \qquad a < 0.$

39. $\displaystyle\int \frac{du}{u^n\sqrt{a + bu}} = -\frac{\sqrt{a + bu}}{a(n - 1)u^{n-1}} - \frac{b(2n - 3)}{2a(n - 1)} \int \frac{du}{u^{n-1}\sqrt{a + bu}}.$

40. $\displaystyle\int \frac{\sqrt{a + bu}}{u}\, du = 2\sqrt{a + bu} + a \int \frac{du}{u\sqrt{a + bu}}.$

Integrand Containing $\sqrt{a^2 - u^2}$

See also formulas (20) and (22).

41. $\displaystyle\int \sqrt{a^2 - u^2}\, du = \frac{u}{2}\sqrt{a^2 - u^2} + \frac{a^2}{2} \mathrm{Sin}^{-1} \frac{u}{a}.$

42. $\displaystyle\int \frac{\sqrt{a^2 - u^2}}{u}\, du = \sqrt{a^2 - u^2} - a \ln \left| \frac{a + \sqrt{a^2 - u^2}}{u} \right|.$

43. $\displaystyle\int \frac{\sqrt{a^2 - u^2}}{u^2}\, du = -\frac{\sqrt{a^2 - u^2}}{u} - \mathrm{Sin}^{-1} \frac{u}{a}.$

44. $\displaystyle\int \frac{du}{u\sqrt{a^2 - u^2}} = -\frac{1}{a} \ln \left| \frac{a + \sqrt{a^2 - u^2}}{u} \right|.$

45. $\displaystyle\int \frac{du}{u^2\sqrt{a^2 - u^2}} = -\frac{\sqrt{a^2 - u^2}}{a^2 u}.$

46. $\displaystyle\int \frac{du}{(a^2 - u^2)^{3/2}} = \frac{u}{a^2\sqrt{a^2 - u^2}}.$

47. $\displaystyle\int \frac{u^2\,du}{\sqrt{a^2 - u^2}} = -\frac{u}{2}\sqrt{a^2 - u^2} + \frac{a^2}{2}\,\mathrm{Sin}^{-1}\frac{u}{a}.$

Integrand Containing $\sqrt{u^2 \pm a^2}$

See also formulas (19), (21), (23), and (24).

48. $\displaystyle\int \sqrt{u^2 \pm a^2}\,du = \frac{u}{2}\sqrt{u^2 \pm a^2} \pm \frac{a^2}{2}\ln|u + \sqrt{u^2 \pm a^2}|.$

49. $\displaystyle\int \sqrt{(u^2 \pm a^2)^n}\,du = \frac{u\sqrt{(u^2 \pm a^2)^n}}{n + 1} \pm \frac{na^2}{n + 1}\int \sqrt{(u^2 \pm a^2)^{n-2}}\,du, \qquad n \neq -1.$

50. $\displaystyle\int \frac{du}{\sqrt{u^2 \pm a^2}} = \ln|u + \sqrt{u^2 \pm a^2}|.$

51. $\displaystyle\int \frac{du}{(u^2 \pm a^2)^{3/2}} = \frac{\pm u}{a^2\sqrt{u^2 \pm a^2}}.$

52. $\displaystyle\int \frac{u^2\,du}{\sqrt{u^2 \pm a^2}} = \frac{u}{2}\sqrt{u^2 \pm a^2} \mp \frac{a^2}{2}\ln|u + \sqrt{u^2 \pm a^2}|.$

53. $\displaystyle\int \frac{du}{u\sqrt{u^2 + a^2}} = -\frac{1}{a}\ln\left|\frac{a + \sqrt{u^2 + a^2}}{u}\right|.$

54. $\displaystyle\int \frac{du}{u\sqrt{u^2 - a^2}} = \frac{1}{a}\,\mathrm{Cos}^{-1}\frac{a}{u}, \qquad u > a > 0.$

55. $\displaystyle\int \frac{du}{u^2\sqrt{u^2 \pm a^2}} = \mp\frac{\sqrt{u^2 \pm a^2}}{a^2 u}.$

56. $\displaystyle\int \frac{du}{u^3\sqrt{u^2 + a^2}} = -\frac{\sqrt{u^2 + a^2}}{2a^2 u^2} + \frac{1}{2a^3}\ln\left|\frac{a + \sqrt{u^2 + a^2}}{u}\right|.$

57. $\displaystyle\int \frac{du}{u^3\sqrt{u^2 - a^2}} = \frac{\sqrt{u^2 - a^2}}{2a^2 u^2} + \frac{1}{2a^3}\,\mathrm{Cos}^{-1}\frac{a}{u}, \qquad u > a > 0.$

58. $\displaystyle\int \frac{\sqrt{u^2 + a^2}}{u} \, du = \sqrt{u^2 + a^2} - a \ln \left| \frac{a + \sqrt{u^2 + a^2}}{u} \right|.$

59. $\displaystyle\int \frac{\sqrt{u^2 - a^2}}{u} \, du = \sqrt{u^2 - a^2} - a \operatorname{Cos}^{-1} \frac{a}{u}, \qquad u > a > 0.$

60. $\displaystyle\int \frac{\sqrt{u^2 \pm a^2}}{u^2} \, du = -\frac{\sqrt{u^2 \pm a^2}}{u} + \ln |u + \sqrt{u^2 \pm a^2}|.$

Integrand Containing $\sqrt{Ax^2 + Bx + C}$

These integrals can be transformed into types considered in formulas (41) through (60) by the following substitutions.

Let $D = B^2 - 4AC$.

CASE 1. If $A > 0$ and $D > 0$, let $u = \sqrt{A}x + \dfrac{B}{2\sqrt{A}}$, and let $a = \sqrt{D/4A}$.

Then $Ax^2 + Bx + C = u^2 - a^2$.

CASE 2. If $A > 0$ and $D < 0$, let $u = \sqrt{A}x + \dfrac{B}{2\sqrt{A}}$, and let $a = \sqrt{-D/4A}$.

Then $Ax^2 + Bx + C = u^2 + a^2$.

CASE 3. If $A < 0$ and $D > 0$, let $u = \sqrt{-A}x - \dfrac{B}{2\sqrt{-A}}$, and let $a = \sqrt{-D/4A}$.

Then $Ax^2 + Bx + C = a^2 - u^2$.

CASE 4. If $A < 0$ and $D < 0$ then $\sqrt{Ax^2 + Bx + C}$ is imaginary for all real values of x. This case cannot arise in a real integrand.

Integrand Containing Trigonometric Functions

See also formulas (7) through (16).

61. $\displaystyle\int \sin^2 u \, du = \frac{1}{2} u - \frac{1}{4} \sin 2u.$

62. $\int \cos^2 u \, du = \dfrac{1}{2} u + \dfrac{1}{4} \sin 2u.$

63. $\int \tan^2 u \, du = \tan u - u.$

64. $\int \cot^2 u \, du = -\cot u - u.$

65. $\int \tan^3 u \, du = \dfrac{1}{2} \tan^2 u + \ln |\cos u|.$

66. $\int \cot^3 u \, du = -\dfrac{1}{2} \cot^2 u - \ln |\sin u|.$

67. $\int \sec^3 u \, du = \dfrac{1}{2} \sec u \tan u + \dfrac{1}{2} \ln |\sec u + \tan u|.$

68. $\int \csc^3 u \, du = -\dfrac{1}{2} \csc u \cot u + \dfrac{1}{2} \ln |\csc u - \cot u|.$

69. $\int \sin^n u \, du = -\dfrac{\sin^{n-1} u \cos u}{n} + \dfrac{n-1}{n} \int \sin^{n-2} u \, du.$

70. $\int \cos^n u \, du = \dfrac{\cos^{n-1} u \sin u}{n} + \dfrac{n-1}{n} \int \cos^{n-2} u \, du.$

71. $\int \tan^n u \, du = \dfrac{\tan^{n-1} u}{n-1} - \int \tan^{n-2} u \, du.$

72. $\int \cot^n u \, du = -\dfrac{\cot^{n-1} u}{n-1} - \int \cot^{n-2} u \, du.$

73. $\int \sec^n u \, du = \dfrac{\tan u \sec^{n-2} u}{n-1} + \dfrac{n-2}{n-1} \int \sec^{n-2} u \, du.$

74. $\int \csc^n u \, du = -\dfrac{\cot u \csc^{n-2}}{n-1} + \dfrac{n-2}{n-1} \int \csc^{n-2} u \, du.$

75. $\int \sin au \sin bu \, du = -\dfrac{\sin (a+b)u}{2(a+b)} + \dfrac{\sin (a-b)u}{2(a-b)}, \qquad a^2 \neq b^2.$

76. $\displaystyle\int \cos au \cos bu \, du = \frac{\sin (a + b)u}{2(a + b)} + \frac{\sin (a - b)u}{2(a - b)}, \qquad a^2 \neq b^2.$

77. $\displaystyle\int \sin au \cos bu \, du = -\frac{\cos (a + b)u}{2(a + b)} - \frac{\cos (a - b)u}{2(a - b)}, \qquad a^2 \neq b^2.$

78. $\displaystyle\int u \sin u \, du = \sin u - u \cos u.$

79. $\displaystyle\int u \cos u \, du = \cos u + u \sin u.$

80. $\displaystyle\int u^n \sin u \, du = -u^n \cos u + n \int u^{n-1} \cos u \, du.$

81. $\displaystyle\int u^n \cos u \, du = u^n \sin u - n \int u^{n-1} \sin u \, du.$

Integrand Containing the Inverse Trigonometric Functions

82. $\displaystyle\int \mathrm{Sin}^{-1} u \, du = u \, \mathrm{Sin}^{-1} u + \sqrt{1 - u^2}.$

83. $\displaystyle\int \mathrm{Cos}^{-1} u \, du = u \, \mathrm{Cos}^{-1} u - \sqrt{1 - u^2}.$

84. $\displaystyle\int \mathrm{Tan}^{-1} u \, du = u \, \mathrm{Tan}^{-1} u - \frac{1}{2} \ln (1 + u^2).$

85. $\displaystyle\int u \, \mathrm{Sin}^{-1} u \, du = \frac{2u^2 - 1}{4} \mathrm{Sin}^{-1} u + \frac{u \sqrt{1 - u^2}}{4}.$

86. $\displaystyle\int u \, \mathrm{Cos}^{-1} u \, du = \frac{2u^2 - 1}{4} \mathrm{Cos}^{-1} u - \frac{u \sqrt{1 - u^2}}{4}.$

87. $\displaystyle\int u \, \mathrm{Tan}^{-1} u \, du = \frac{u^2 + 1}{2} \mathrm{Tan}^{-1} u - \frac{u}{2}.$

88. $\displaystyle\int \frac{\mathrm{Sin}^{-1} u}{u^2} \, du = \ln \left| \frac{1 - \sqrt{1 - u^2}}{u} \right| - \frac{\mathrm{Sin}^{-1} u}{u}.$

89. $\displaystyle\int \frac{\text{Tan}^{-1} u}{u^2}\, du = \ln |u| - \frac{1}{2}\ln (1 + u^2) - \frac{\text{Tan}^{-1} u}{u}.$

Integrand Containing the Exponential Function

See also formulas (5) and (6).

90. $\displaystyle\int u e^u\, du = (u - 1)e^u.$

91. $\displaystyle\int u^n e^u\, du = u^n e^u - n \int u^{n-1} e^u\, du.$

92. $\displaystyle\int \frac{e^u}{u^n}\, du = -\frac{e^u}{(n - 1)u^{n-1}} + \frac{1}{n - 1}\int \frac{e^u\, du}{u^{n-1}}, \qquad n \neq 1.$

93. $\displaystyle\int e^u \ln u\, du = e^u \ln u - \int \frac{e^u}{u}\, du, \qquad u > 0.$

94. $\displaystyle\int e^{au} \sin nu\, du = \frac{e^{au}(a \sin nu - n \cos nu)}{a^2 + n^2}.$

95. $\displaystyle\int e^{au} \cos nu\, du = \frac{e^{au}(a \cos nu + n \sin nu)}{a^2 + n^2}.$

Integrand Containing the Logarithmic Function

Assume $u > 0$ in formulas (96) through (101).

96. $\displaystyle\int \ln u\, du = u \ln u - u.$

97. $\displaystyle\int u^n \ln u\, du = u^{n+1}\left(\frac{\ln u}{n + 1} - \frac{1}{(n + 1)^2}\right), \qquad n \neq -1.$

98. $\displaystyle\int \frac{du}{u \ln u} = \ln |\ln u|.$

99. $\int u^m (\ln u)^n \, du = \dfrac{u^{m+1}}{m+1} (\ln u)^n - \dfrac{n}{m+1} \int u^m (\ln u)^{n-1} \, du, \qquad m \neq -1.$

100. $\int \sin (\ln u) \, du = \dfrac{u}{2} [\sin (\ln u) - \cos (\ln u)].$

101. $\int \cos (\ln u) \, du = \dfrac{u}{2} [\sin (\ln u) + \cos (\ln u)].$

Some Definite Integrals

102. $\displaystyle\int_0^\infty x^{n-1} e^{-x} \, dx = \int_0^1 \left(\ln \dfrac{1}{x} \right)^{n-1} dx = \Gamma(n), \qquad n > 0,$

and $\Gamma(n) = (n-1)!$, if n is a positive integer.

103. $\displaystyle\int_0^1 x^{m-1} (1-x)^{n-1} \, dx = \int_0^\infty \dfrac{x^{m-1} \, dx}{(1+x)^{m+n}} = \dfrac{\Gamma(m)\Gamma(n)}{\Gamma(m+n)}, \qquad m, n > 0.$

104. $\displaystyle\int_0^\infty \dfrac{x^{p-1} \, dx}{1+x} = \dfrac{\pi}{\sin p\pi}, \qquad 0 < p < 1.$

105. $\displaystyle\int_0^\infty \dfrac{\sin ax}{x} \, dx = \dfrac{\pi}{2}, \qquad a > 0.$

106. $\displaystyle\int_0^\infty \dfrac{\cos mx}{1+x^2} \, dx = \dfrac{\pi}{2} e^{-m}, \qquad 0 < m.$

107. $\displaystyle\int_0^\infty \sin (x^2) \, dx = \int_0^\infty \cos (x^2) \, dx = \dfrac{1}{2} \sqrt{\dfrac{\pi}{2}}.$

108. $\displaystyle\int_0^\infty e^{-a^2 x^2} \, dx = \dfrac{\sqrt{\pi}}{2a}.$

109. $\displaystyle\int_0^{\pi/2} \sin^n x \, dx = \int_0^{\pi/2} \cos^n x \, dx$

$= \dfrac{1 \cdot 3 \cdot 5 \cdots (n-1)}{2 \cdot 4 \cdot 6 \cdots n} \dfrac{\pi}{2},$ if n is an even positive integer,

$= \dfrac{2 \cdot 4 \cdot 6 \cdots (n-1)}{3 \cdot 5 \cdot 7 \cdots n},$ if n is an odd positive integer.

110. $\displaystyle\int_0^{\pi/2} \sin^m x \cos^n x \, dx,$ m and n positive integers,

$$= \frac{2 \cdot 4 \cdot 6 \cdots (m-1)}{(n+1)(n+3) \cdots (n+m)}, \qquad \text{if } m \text{ is odd, } m > 1,$$

$$= \frac{2 \cdot 4 \cdot 6 \cdots (n-1)}{(m+1)(m+3) \cdots (m+n)}, \qquad \text{if } n \text{ is odd, } n > 1,$$

$$= \frac{[1 \cdot 3 \cdots (m-1)][1 \cdot 3 \cdots (n-1)]}{2 \cdot 4 \cdot 6 \cdots (m+n)} \frac{\pi}{2}, \qquad m \text{ and } n \text{ both even.}$$

Answers to
Odd-Numbered Problems

CHAPTER 1

Exercise 1, page 6

1. (a) $1.06\overline{6}\ldots,$ (b) $1.6818\overline{18}\ldots,$ (c) $0.0002,$ (d) $-0.314\overline{314}\ldots.$
3. (a), (c), (d).
5. $6.85983478.$
7. If $a = b$ and $c = d$, then $ac = bd$.
9. If $a < c$ and $b < d$, then $a + b < c + d$.
11. If n is an integer greater than 1, then there is some prime number that lies between n and $2n$.
13. If $a + c < b + c$, then $a < b$ and, conversely, if $a < b$, then $a + c < b + c$.
C1. (a) $0.1831,$ (b) $1.4679,$ (c) $2.2361,$ (d) $9.9184.$
C3. (a) $7.53 \times 10^{10},$ (b) $6.95 \times 10^{-6},$ (c) $-1.71 \times 10^2,$ (d) $9.32 \times 10^6.$

Exercise 2, page 13

1. $\sqrt{19} + \sqrt{21}.$ 3. $5\sqrt{7}.$ 5. $-0.08351.$ 7. $a = 1.$ 9. $c = d.$ 11. $a = b.$
13. $x = 2y.$ 17. $x < -3/2.$ 19. $x > 4/3.$ 21. $x \geq 5.$ 23. All real numbers.
25. $-5 < x < 1.$ 27. $0 \leq x \leq 3/2.$ 29. $-1/4 < x < 5/2.$ 31. $4/3 < x.$
33. $-3 < x < 1/4.$ 35. $-2 < x < 1.$ 37. $2 < x < 23/6.$ 39. $-6 < x < 0.$
41. $x < -8$ or $x > 1.$ 43. $x \leq -2$ or $x \geq 0.$ 45. $3 \leq x < 39.$
C1. $159/135.$ C3. $\sqrt{2} + \sqrt{5}.$ C5. $(1.023)^3.$ C7. $x > 0.042.$
C9. $x < -14.579.$ C11. $7.289 < x < 7.877.$ C13. $x > 2.236$ or $x < -0.349.$
C15. All $n \geq 11.$

Exercise 3, page 18

1. $-3, 5.$ 3. $2/3.$ 5. $-8/3, -2.$ 7. $-1/2, 3/8.$ 9. $-3, -1, 0.$ 11. $-1.$
13. $1/2, 3.$ 15. $7 < x < 9.$ 17. $1/7 < x < 1.$ 19. $x \leq 1$ or $7/3 \leq x.$
21. $x \leq 1/2$ or $x \geq 1.$

C1. 0.485, 2.596. **C3.** −0.144, 0.781. **C5.** 1.296, 3.023. **C7.** 1.478 $< x <$ 14.632. **C9.** $x \leqq 0.559$ or $x \geqq 8.043$.

Exercise 4, page 21

1. 0.017453 radian.
3. (a) $\pi/3$, (b) $4\pi/3$, (c) 4π, (d) $-3\pi/4$, (e) $\pi/15$, (f) $11\pi/15$, (g) $\pi/5$, (h) $-5\pi/6$, (i) $17\pi/36$.
5. (a) $1/2$, (b) $-1/2$, (c) -1, (d) -1, (e) 0, (f) ∞, or does not exist, (g) -1, (h) -1.
C1. (a) 0.5299, (b) 0.9010, (c) -2.1850, (d) 0.9897, (e) -1.7034, (f) -0.6089.

CHAPTER 2

Exercise 1, page 26

1.
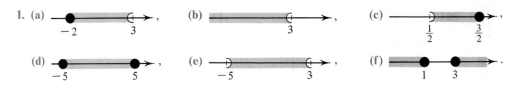

3. $x < 0$ or $1 < x$. 5. $-3 \leqq x \leqq 1/2$. 7. $3 \leqq x \leqq 5$. 9. $-4 \leqq x \leqq 0$ or $4 \leqq x$.
11. $x \leqq -3$, or $-2 \leqq x \leqq 2$, or $3 \leqq x$.
13. (a) 5, (b) 4, (c) -6, (d) 109/91, (e) 15/2, (f) $10 - \sqrt{2}$.
15. (a) 7, 10, (b) 6, 19, (c) 11/10, 9/5, (d) -11, $-8 + 3\sqrt{2}$.
17. $-20/3$, $\sqrt{7}$. 19. x is at most 5 units from -2; $-7 \leqq x \leqq 3$.
21. x is closer to 3 than to 4; $x < 3.5$. 23. $x \leqq 2$ or $x > 4$.
25. $1.99 < x < 2.01$, $x \neq 2$. 27. $-1 < x < 3$ or $4 < x$.
C1. (a) 0.494, (b) 1.297, (c) 0.829. **C3.** 1.6714, 1.9286, 2.1858, 2.4429, 2.7001, 2.9573.

Exercise 2, page 31

1. 5, 11, 13, 5, 3, 7, $\sqrt{5}$.
3. (a) Horizontal line 6 units above the x-axis, (b) vertical line 3 units to the left of the y-axis.
5. Yes, $5 + 45 = 50$. 7. No, $13 + 29 \neq 34$. 9. No, $41 + 145 \neq 194$.
11. $x + 2y = 3$. 13. $x^2 + y^2 = 6x - 8y$. 15. $6y = x^2 - 2x + 10$.
17. $3x^2 + 3y^2 + 38x - 4y + 55 = 0$.
C1. 0.7 hr.
C3. *ABDCA* is shortest, but *ACDBA* is the same length, namely 1630 km (rounded to the nearest 10 km).

Exercise 3, page 37

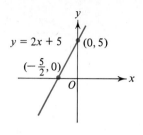

$y = 2x + 5$

$(0, 5)$

$(-\frac{5}{2}, 0)$

1.

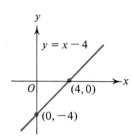

$y = x - 4$

$(4, 0)$

$(0, -4)$

3.

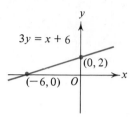

$3y = x + 6$

$(0, 2)$

$(-6, 0)$

5.

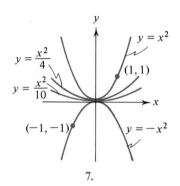

$y = x^2$

$y = \frac{x^2}{4}$

$y = \frac{x^2}{10}$

$(1, 1)$

$(-1, -1)$

$y = -x^2$

7.

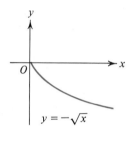

$y = \sqrt{x}$

9. (a)

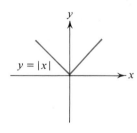

$y = -\sqrt{x}$

9. (b)

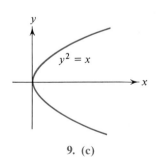

$y^2 = x$

9. (c)

$(0, 5)$

$(-5, 0)$

$(5, 0)$

$(0, -5)$

11.

$y = |x|$

13.

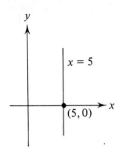

15.

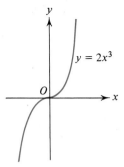

17.

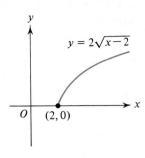

19.

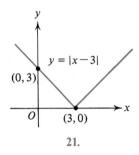

21.

23. Graph is the single point $(0, 0)$.   25. (a) $y = 0$,   (b) $x = 0$,   (c) $xy = 0$.
27. $x^2 + y^2 = 9$.   29. $x = 3$.   31. $y = 3x + 2$.   33. $y = x^2 - 3x + 2$.

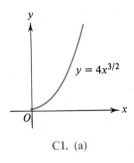

C1. (a)

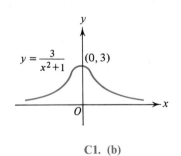

C1. (b)

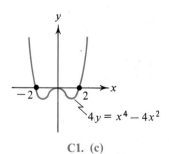

C1. (c)

Exercise 4, page 41
   1. 1.   3. -3.   5. $3b/a$.   7. Yes, $m = -1/2$.   9. No, $-3/5 \neq -11/18$.
   11. (a) $(1, 2)$,   (b) $(-1, 6)$,   (c) $(-1/2, 2)$,   (d) $(-7/2, -7/2)$.   13. $(1, -4)$.
   21. $\tan \varphi = \dfrac{m_2 - m_1}{1 + m_1 m_2}$.
   C1. $53.020°$.

Exercise 5, page 46

1. $y = 3x + 5$.
3. (a) $y = x + 1$,   (b) $2y + x = 11$,   (c) $2x - y = 2$,   (d) $3\sqrt{y} = x + 4\sqrt{3} - 1$.
5. $6y + 2x = 7$.
7. (a) $-2/3, -4/3$,   (b) $5, -7$,   (c) $1/3, -3$,   (d) $-3, 6$,   (e) $0, 10$,
   (f) $\sqrt{3}, 4\sqrt{3}$.
9. (a) 2,   (b) $14/3$,   (c) $7/2$.

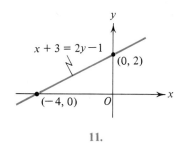

11.

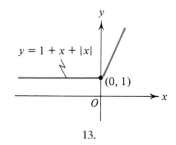

13.

15. $(-12, -17)$.   17. $(3, 1)$.
19. $C = (5/9)(F - 32)$,   (a) $35°C$,   (b) $257°F$,   (c) $37°C$,   (d) $-17.8°C$,
   (e) $140°F$.
21. (a) and (b) above,   (c) and (d) below.
23. $E = 56 - 2A/3$,   (a) $A = 42$, the table gives 44,   (b) $E = 56$, the table gives 67.5,
   (c) $A = 84$, the table does not give an age at which $E = 0$, but from 80 to 85 the table
   gives $E = 5.8$.
C1. $I = 85 + (0.0575)S$, $\$273.83$.   C3. $y = 2.747x - 2.523$.

Exercise 6, page 51

1. (a) $y = x - 10$,   (b) $y = -x$.   3. (a) $3y = -x$,   (b) $y = 3x$.
5. (a) $x = 100$,   (b) $y = 200$.   7. Yes, $3/11$ and $9/5$.   9. Yes, $(4/3)(-3/4) = -1$.
11. $y = -5x + 3$.   13. (a) $x + 2y = 5$,   (b) $\sqrt{5}/5$.   15. $5\sqrt{5}$.   17. 5.
23. $2(c - a)x + 2(d - b)y + a^2 + b^2 - c^2 - d^2 = 0$.

Exercise 7, page 54

1. $x^2 + y^2 = 10x + 24y$.   3. $x^2 + y^2 - 2x + 2y = 2$.
5. $x^2 + y^2 - 2ax - 4ay = 0$.   7. $x^2 + y^2 - 6x - 4y + 9 = 0$.
9. $x^2 + y^2 - 10x + 4y = 52$.   11. Circle, center $(2, -1)$, $r = 5$.
13. The point $(3, 8)$.   15. Circle, center $(4/5, 2/5)$, $r = 5$.
17. No points.   19. $(x - 6)^2 + (y - 1)^2 = 18$.   21. $y = x - 2$, $7y + x = 10$.
23. $x^2 + y^2 - 6x - 4y = 12$.   25. $(2, 0), (2/5, 4/5)$.
29. $(2x_0 - 1)^2 + (2y_0)^2 = 1$, circle, center $(1/2, 0)$, $r = 1/2$.
33. (a) 3,   (b) 5,   (c) 6.
C1. 24.53 (rounded to two decimals).

Exercise 8, page 66

1. (a) $(0, 1)$, $y = -1$,   (b) $(0, 1/4)$, $y = -1/4$,   (c) $(0, 1/16)$, $y = -1/16$,   (d) $(0, 8)$,

$y = -8,$ (e) $(1/2, 0),$ $x = -1/2,$ (f) $(1/64, 0),$ $x = -1/64,$ (g) $(1/12, 0),$
$x = -1/12,$ (h) $(9/4, 0),$ $x = -9/4.$

3. (a) $(-1/2, 0),$ $x = 1/2,$ (b) $(0, -1),$ $y = 1,$ (c) $(-1, 0),$ $x = 1,$
    (d) $(0, -2),$ $y = 2,$ (e) $(-1/4, 0),$ $x = 1/4,$ (f) $(0, -7/20),$ $y = 7/20.$

5. (a) $x^2 = 12y,$ (b) $y^2 = -8x,$ (c) $x^2 = -16y,$ (d) $y^2 = 32x.$

7. $4x = -(y^2 + 14y + 53).$

11. (a) $x^2/25 + y^2/16 = 1,$ (b) $x^2/64 + y^2/100 = 1,$ (c) $x^2/25 + y^2 = 1,$
    (d) $x^2/24 + y^2/25 = 1.$

13. (a) $(\pm 4, 0),$ $2a = 10,$ (b) $(\pm 1, 0),$ $2a = 10,$ (c) $(0, \pm 4),$ $2a = 10,$ (d) $(0, \pm \sqrt{5}),$
$2a = 6,$ (e) $(0, \pm 1),$ $2a = 4,$ (f) $(\pm 3\sqrt{11}/2, 0),$ $2a = 10.$

15. $x^2/144 + y^2/169 = 1.$

17. (a) $x^2/16 - y^2/9 = 1,$ (b) $x^2/9 - y^2/16 = 1,$ (c) $x^2/4 - y^2/21 = 1,$
    (d) $y^2 - x^2/8 = 1,$ (e) $4y^2 - 4x^2/15 = 1,$ (f) $y^2/16 - x^2/9 = 1.$

19. (a) $(\pm 13, 0),$ (b) $(\pm 2, 0),$ (c) $(\pm 13, 0),$ (d) $(0, \pm 3),$ (e) $(0, \pm 13),$
    (f) $(\pm 3, 0).$

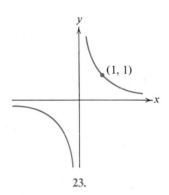

23.

25. $y^2 - 2x^2 = 4.$ 27. $(6, -9),$ $(-1, -2).$ 29. $(0, 1),$ $(2, 5).$ 31. $(1, 0),$ $(0, 1).$

Review Problems, page 69

1. (a) $1 < x < 2,$ (b) $x \leq 0$ or $x > 1,$ (c) $x \leq -5$ or $x \geq 3,$
    (d) $0 \leq x \leq 1$ or $x \geq 2,$ (e) $x \leq 4,$ (f) $x \leq -1/2$ or $x \geq 1.$ 3. *PR.*

5. (a) $m = 3/2,$ intercept $-3,$ (b) $m = -1/3,$ intercept $3,$ (c) $m = 2,$ intercept $-4.$

7. (a) $y = -2x - 4,$ (b) $x - 3y = 19,$ (c) $20x + 9y = 39,$ (d) $y = x + 3.$

9. (a) $2x + 3y = -5,$ (b) $4x + 5y = -9.$

11. (a) $(1/2, 1),$ (b) $(x - 1/2)^2 + (y - 1)^2 = 73/4.$

13. (a) $x^2 + y^2 + 8y - 48 = 0,$ (b) $x^2 + y^2 + 3y - 18 = 0.$

15. Parabola, $F(0, -4),$ directrix $y = 4.$ 17. Circle, center $(-5, 2),$ $r = 6.$

19. Ellipse, $F_1(-\sqrt{3}, 0),$ $F_2(\sqrt{3}, 0).$ 21. $60y^2 - 4x^2 = 135.$ 23. $(3, -4).$ 25. None.

CHAPTER 3

Exercise 1, page 76

3. $g(1) = 1, g(-2) = -4/5, g(1/2) = 4/5.$
5. $f(0) = 0, f(-2) = 4, f(11) = 3.$
7. $f(3 + h) = 2(3 + h) - (3 + h)^2 = -3 - 4h - h^2.$ 9. All reals except -2 and 1.
11. $(-\infty, 4]$.   13. $[-3, 3]$.   15. $[-1, 1] \cup [4, \infty)$.   17. $(-\infty, \infty)$.
19. $[1, \infty)$.   21. All reals except 0.   23. $[-1, 1]$.   25. $[4, 18]$.
27. $S = 6x^2$.   29. $L = 50 - w$, domain $= (0, 50)$, range $= (0, 50)$.   31. $d = \sqrt{16.84t}$.
33. $d_2(1/2) = 0, d_2(\sqrt{2}) = 1, d_2(\pi) = 4, d_2(4) = 0, d_2(1/6) = 6$. Range $= \{0, 1, 2, 3, 4, 5, 6, 7, 8, 9\}$.
39. (a) 9,   (b) 8,   (c) 28.
41. (a) 0,   (b) 2/3,   (c) 4/3.
43. (a) 0,   (b) $-1/5$,   (c) 5,   (d) 3 if $t \neq 0$,   (e) y/x if $x \neq 0$.
45. (a) -2,   (b) -2,   (c) -30,   (d) $-12t^3$.
C3. Minimum value $= f(1) = 10$.

Exercise 2, page 84

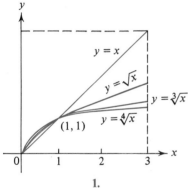

1.

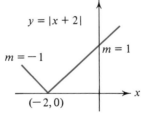

3. (a)

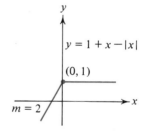

3. (b)

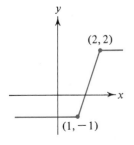

3. (c)

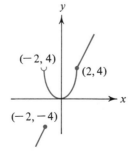

3. (d)   5. (d) $x = -2$.

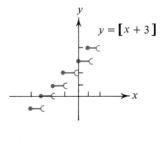

7.

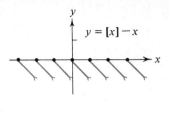

9.

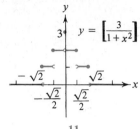

11.

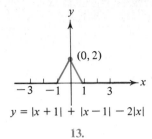

13.

19. (a) Even, **(b)** odd, **(c)** neither, **(d)** neither, **(e)** even.

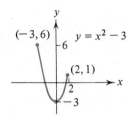

21. (a)

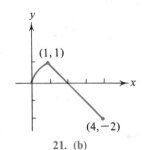

21. (b)

21. (a) (i) $f(-3) = 6$, **(ii)** $f(0) = -3$. **21. (b) (i)** $f(1) = 1$, **(ii)** $f(4) = -2$.
C1. (a) -12, **(b)** 27, **(c)** 43, **(d)** -3.

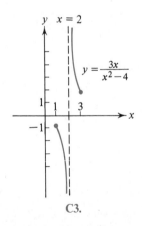

C3.

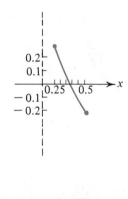

C5. (a).

C5. (b).

Exercise 3, page 95
1. (a) 1/8, 2/9, 3/10, 4/11, **(b)** -2, 4, -8, 16, **(c)** 0, 2, 6, 12.
3. $-1/30$. **5.** 0, 2, 7, 53. **7.** 0, 1, 3, 11, 52. **9.** 1, 2, 3/2, 5/3, 8/5, 13/8, 21/13.
17. Terms grow large without bound. **19.** Terms oscillate between 3 and 1.
21. Terms grow large without bound. **23.** $L = 6$.

25. $L = 9$. **27.** $L = 0$. **29.** $L = 1/2$. **31.** Diverges. **33.** $L = 2$.
35. $L = 5$. **37.** Diverges. **39. (a)** $L = 7$, **(b)** $L = 0$. **41. (a)** 0, **(b)** 28,
  (c) 2/7. **43. (a)** 0, **(b)** 2.
C3. $n \geqq 11$. **C5.** The absolute values of the terms grow large without bound.
C7. $L = \sqrt{2} = 1.414\ldots$. **C9.** $L = 0$. **C11.** $L = 2.718\ldots$. **C13.** $L = 1$.
C15. $(0.98)^n$, $75/n^2$, $1/2n$, $1/\log_{10} n$.

Exercise 4, page 106

C1. $2.366\ldots$. **C3.** $0.322\ldots$. **C5.** $-0.18\ldots$, $0.65\ldots$. **C7.** $-0.75\ldots$.
C9. 1.8, 1.733333334, 1.732051282, 1.732050808.
C11. (a) 1.5, 1.416666667, 1.414215687, 1.414213563,
  (b) 6, 5.720833333, 5.714021910, 5.714017850.
C13. 0.450183611 radian. **C15.** -0.144140603.
C17. If one uses a starting value greater than 3, then the successive approximations get larger, while if one uses a starting value between 1 and 3, the approximations converge to $x = 1$.

CHAPTER 4

Exercise 1, page 110

7. $x_1 = 4$, $y_1 = 4$. **9.** $y = 0$. **11.** $y = -2x + 9/4$. **13.** $y = 4cx - 4c$.
15. $y = 3cx - 2c$. **17.** $m = cnx^{n-1}$.

Exercise 2, page 117

1. -6. **3.** -6. **5.** 0. **7.** -7. **9.** 4. **11. (a)** 1, **(b)** 2.
13. (a) 2, **(b)** 6. **15. (a)** 1, **(b)** -3, **(c)** does not exist, **(d)** -1.
17. (a) -4, **(b)** -4, **(c)** -4, **(d)** does not exist. **19. (a)** 2, **(b)** 2, **(c)** 3.
C3. -256. **C5.** 1. **C7.** 6.

Exercise 3, page 123

7. 23. **9.** -47. **11.** -1. **13.** 8. **15.** 8/27. **17.** 1/3. **19.** 6.
21. (a) -13, **(b)** $-5/11$, **(c)** 6. **23.** 2.
C3. 12.545. **C5.** 0.522.

Exercise 4, page 129

1. 3. **3.** $-\infty$. **5.** 1/4. **7.** 0. **9.** $-\infty$. **11.** ∞. **13.** -8.
15. $-\infty$. **17.** $-\infty$. **19.** ∞. **21.** 9. **23.** 0. **25.** $A/C - B/D$.
27. $\lim\limits_{x \to 0^+} f(1/x) = 5$.

Exercise 5, page 134

1. C. **3.** D at $x = 5$ and $x = -5$. **5.** D at $x = 0$ and $x = 4$.
7. D at $x = 2$. **9.** C. **11.** No. **13.** $A = 3/4$. **15.** $B = 1/8$.
17. $a = 3$, $b = -8$. **21.** No. $P(x) = x^2 + 1 = 0$ has no real solutions.
C3. -1.69.

Exercise 6, page 142

1. 2.   3. $1 + 2x$.   5. $-1/x^2$.   7. $-3x^2$.   9. $\sqrt{5}/2\sqrt{x}$.
11. $-7/(x-5)^2$.   13. -2.   15. Yes.
C1. 0.25.   C3. 0.   C5. 0.5.   C7. $1/[(1 + \Delta x)^{1/4} + 1][(1 + \Delta x)^{1/2} + 1]$.

Exercise 7, page 146

1. Crit. $x = 0$; high point $(0, 2)$.
3. Crit. $x = 2, -2$; low point $(2, -14)$, high point $(-2, 18)$.
5. Crit. $x = 0$; high point $(0, 6)$.   7. $y + 8x = 12$.
9. $y - 2x = -3$.   11. $6y - x = -45$.

Exercise 8, page 151

1. $24x^7 - 24x^5$.   3. $25x^4 + 3x^2 + 1$.   5. $-1/3$.   7. $84t^{83} - 72t^{35}$.
9. $4\pi r^2$.   11. $24x^2 - 72x + 54$.   13. 48.   15. -13.
17. $y' = 2x + 6$; crit. at $x = -3$; low point $(-3, -4)$.
19. $y' = 3x^2 - 9$, crit. at $\sqrt{3}$ and $-\sqrt{3}$; low point $(\sqrt{3}, -6\sqrt{3})$, high point $(-\sqrt{3}, 6\sqrt{3})$.
21. $y = 11x - 8$.   23. $30y + 5x = 26$.   25. $(2, -34), (-3, 91)$.   27. $\pi/4$ radian.
C1. 1.3941 radians.   C3. $x = 1.4830, x = 0.21464$.   C5. $x = -0.92371$.

Exercise 9, page 156

1. $5x^4 - 3x^2 + 6x$.   3. $4x(x^2 - 1)$.   5. $-4x^{-3} + 3x^{-2} = (3x - 4)/x^3$.
7. $-x^{-2} - 15x^{-4} = -(15 + x^2)/x^4$.   9. $2/(x + 1)^2$.   11. $(x^2 - 8x + 5)/(x - 4)^2$.
13. $-4(x + 1)/(x - 1)^3$.   15. $-12x - 30x^{-3} = -(12x^4 + 30)/x^3$.   17. $(1 - \theta^2)/(1 + \theta^2)^2$.
19. $2(1 - 2u^2 - u^4)/u^3(u^2 - 1)^2$.   21. Crit. at $x = 1$; low point $(1, 3)$.   23. None.
25. Crit. at $\sqrt{2}/2$ and $-\sqrt{2}/2$; high $(\sqrt{2}/2, 3\sqrt{2}/4)$, low $(-\sqrt{2}/2, -3\sqrt{2}/4)$.   29. 11.
31. 1.   33. $y' = t'uvw + tu'vw + tuv'w + tuvw'$.   35. $(0, 2), (\pm 1, 1)$.

Exercise 10, page 163

1. (a) $(2 - x)/(2 + x)$, $\mathscr{D}$ = all reals except $0, -2$.
   (b) $2(u + 1)/(u - 1)$, $\mathscr{D}$ = all reals except ± 1.   (c) x, $\mathscr{D}$ = all reals except 0.
3. $2(x^3 + x)(3x^2 + 1)$.   5. $2x/(x^2 + 1)^2$.   7. $30(3x + 5)^9$.
9. $5(2x^3 - 3x^2 + 6x)^4(6x^2 - 6x + 6)$.   11. $-2x(3x + 2)/(x^3 + x^2 - 1)^3$.
13. $-15(x + 2)^2/(x - 3)^4$.   15. $28(11x + 4)(7x + 3)^3(4x + 1)^6$.
17. $-2(3x + 4)/(2x + 1)^2(x + 3)^3$.   19. $-2x(x^3 + 15x + 2)(x^2 + 5)/(x^3 - 1)^3$.
21. $-6x(x^2 + 1)^2(x + 1)/(x^3 - 1)^3$.   23. $(x + 1)(x - 3)/(x - 1)^2$.   25. $36v^2(v^3 + 17)^{11}$.
27. $2(7\theta + 1)(\theta + 1)^5(\theta - 1)^7$.   29. $-4/x^3$.   31. $2x$.   33. -60.   35. -36.

Exercise 11, page 169

1. $(3x + 1)/2\sqrt{x}$.   3. $8x/7(x^2 - 1)^{3/7}$.   5. $(1 - 2x^2)/\sqrt{1 - x^2}$.
7. $x(2 - x^2)/(1 - x^2)^{3/2}$.   9. $1/(9 - x^2)^{3/2}$.   11. $-30(x^2 + 9)^{2/3}/(x - 9)^{8/3}$.
13. $(8x^{3/2} - 1)/4x\sqrt{4x^2 + \sqrt{x}}$.   15. $-x/y$.   17. $(x^2 - 2y)/(2x - y^2)$.
19. $-y^5/x^5$.   21. $8x\sqrt{y}/(2\sqrt{y} + 1)$.   23. $(4 - 2y^{3/2})/(3xy^{1/2} - 2)$.
25. $[x^3 - (x + y)^3]/[(x + y)^3 - y^3] = y/x$.   27. $4y - 3x + 4 = 0$.

Exercise 12, page 173

1. $90(x^8 + x^4)$.   3. $2(2 + x)/(1 - x)^4$.   5. $2(t - 1)^3(21t^2 - 12t + 1)$.
7. $-2/9x^{4/3}$.   9. $480(2x - 1)^2$.   11. $(18 + 6x)/(1 - x)^5$.   13. $24/(1 + t)^5$.
15. $-16/9y^3$.   17. $-2x/y^5$.   19. $a^{1/2}/2x^{3/2}$.   21. $a = 3, b = -2$.
23. $(-1)^n n!/(2 + x)^{n+1}$.   25. $(-1)^n 3^n n!/(2 + 3x)^{n+1}$.
27. $(-1)^n ac^n n!/(b + cx)^{n+1}$.   29. $-1 \cdot 3 \cdot 5 \cdots (2n - 3)/2^n(1 - x)^{(2n-1)/2}$ for $n \geqq 2$.

C3.

0.2	0.1	0.01	0.001
-27.512	-29.679	-31.760799	-31.97601
-32.32	-32.08	-32.0008	-32.00001

C5. 0.303177, 0.302885, 0.302882.
C7. 0.937331, 0.937458, 0.937498, $f''(2) = 15/16 = 0.9375$.

Review Problems, page 175

1. $6(x - 1)$.   3. $12(v^3 - 1)$.   5. $(1 + z)/(1 - z)^3$.   7. $420x(7x + 1)^4(5x - 1)^6$.
9. $5(2u - 1/u^2)(u^2 + 1/u)^4$.   11. $(x^2 - 1)^{1/2}(x^4 + 5)^{1/4}(8x^5 - 5x^3 + 15x)$.
13. $(s^4 + 6s^2 - 3)/(s^2 + 1)^2$.   15. $\sqrt{x}(x - 1)^{3/2}(x + 3)^{5/2}(15x^2 + 14x - 9)/2$.
17. $(2x + 9)/2y$.   19. $(2xy^3 - 12x^3)/(10y^4 - 3x^2y^2)$.   21. $(a - dy - 2cx)/(dx - b)$.

$$y = \frac{x^3 + x^2 - 12x}{12}$$

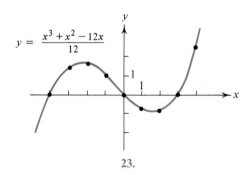

23.

25. (a) $x + 16y = 16$,   (b) $3x + 2y = 26$,   (c) $y = 2x$,   (d) $10y = 27x - 31$,
(e) $7x + 15y = 36$, $15x + 7y = 36$,   (f) $13x + 14y = 40$, $14x + 13y = 40$.
27. (a) $(-2, -11)$,   (b) $(0, 5), (2, 1)$,   (c) $(3, -92), (-2, 33)$,
(d) $(-3, 0), (5, 0), (9/5, (24)^3(16)^2/5^5)$,   (e) $(2, 1/4), (-2, -1/4)$,   (f) no points.
29. (a) $4/(x^2 + 4)^{3/2}$,   (b) $2 + 4/u^3$,   (c) $(x^3 - 1)^{-3/2}(15x^5/4 - 6x^2)$,
(d) $-3t/(t^2 + 1)^{5/2}$,   (e) $(3t^2 + 6t - 1 + 8t\sqrt{t})/[2t\sqrt{t}(t - 1)^3]$,
(f) $3(ad - bc)(ad - 4act - 5cb)/[4(ct + d)^{7/2}(at + b)^{1/2}]$.
31. (a) $f'(x) = 2$ if $x > 0$, $f'(x) = 0$ if $x < 0$, $f'(0)$ does not exist,   (b) $f'(x) = 1$ for
$x \neq \pm n$, $n = 0, 1, 2, \ldots, f'(x)$ does not exist for $x = \pm n$.

CHAPTER 5

Exercise 1, page 183

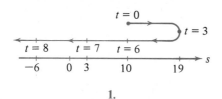

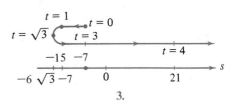

1. 3.

5. 64 ft, 3 sec, just about to hit the ground.
7. 30 sec, 294 m/sec. 9. $H = v_0^2/19.6$, 28 m/sec.
11. 40 ft/sec. 15. (a) $t = \sqrt{H/4.9} = \sqrt{10H}/7$ sec, (b) $t = (\sqrt{v_0^2 + 19.6H} - |v_0|)/9.8$ sec.
17. Yes, if $s = 6t^5 - 15t^4 + 10t^3$, then $v = 30t^2(1 - t)^2$.
19. (a) 84 ft/sec, (b) 36 ft/sec, (c) 16.5 ft/sec.
21. (a) 19.6 m/sec, (b) 17.06 m/sec, (c) 7.32 m/sec, (d) 3.36 m/sec.
23. $x = 5$, 13. 25. (a) $s = -33$, $v = -36$, (b) $s = -60$, $v = 27$.

Exercise 2, page 187

1. $4/45\pi \approx 0.02829$ in./min ≈ 1.698 in./hr.
3. $[(35 - 14t)^2 + 400t^2]^{1/2}$, $106/29 \approx 3.655$ km/hr.
5. $1/4\pi \approx 0.07958$ m/min. 7. 5 ft/sec. 9. $2\sqrt{10}$ cm³/sec.
11. (a) -3, (b) $3/5$, (c) $51/13$, (d) $3\sqrt{2}$.
13. (a) $4x(2x^2 - 9)/\sqrt{x^4 - 9x^2 + 25}$, (b) $-8\sqrt{5}/5$. 15. $(1/2, 1/4)$.

Exercise 3, page 195

1. $x > -1$, rel. min. $(-1, -6)$. 3. $x < 3/2$, rel. max. $(3/2, 33/4)$.
5. $-\infty < x < -2$ and $3 < x < \infty$, rel. max. $(-2, 51)$, rel. min. $(3, -74)$.
7. $0 < x < 2$, rel. min. $(0, 3)$, rel. max. $(2, 11)$.
9. $1 < x < 3$, rel. max. $(3, 5)$, rel. min. $(1, -103)$.
11. Nowhere, no rel. min. or max. 13. Nowhere, no rel. min. or max. 15. $-3, 0$.
17. $-1, -2/3$. 19. 0, 1/2. 21. 11. 23. $x^2 - 2x + 3$.
25. (a) Min. $(0, 0)$, (b) Max. $(0, 1)$, (c) Min. $(1, 0)$.

Exercise 4, page 203

1. 13/2, 3. $1 - \sqrt{3}/3$.
5. $\xi = -B/2A$ and is the midpoint of the interval $[a, b]$.
7. $2x^3 + C$. 9. $-3/x + C$. 11. $5x^4 - 2x^{10} + C$. 15. 30.25.
17. $7 + \sqrt{2}/10 \approx 7.141$. 19. 4. 23. $3x^2 + a > 0$ for all x.
25. $f(x)$ is not differentiable at $x = 1$.
C1. $\xi = 8.1937$. C3. $\xi_1 = 15.0002$.

Exercise 5, page 210

1. Concave upward all x, no inflection points.
3. C.U. $x \geq 0$, infl. pt. $(0, 0)$. 5. C.U. nowhere, no infl. pts.

7. C.U. $x \geqq 0$, infl. pt. $(0, 0)$.   9. C.U. $x \geqq 1$, infl. pt. $(1, -11)$.
11. C.U. $x \leqq -\sqrt{3}/3$ and $x \geqq \sqrt{3}/3$, infl. pts. $(\pm\sqrt{3}/3, 3)$.
13. C.U. $x > 0$, no infl. pts.
15. C.U. $2 \leqq x \leqq 4$, and $x \geqq 6$, and $x \leqq 0$, infl. pts. $(0, 0)$, $(2, -2^{15})$, $(4, -2^{15})$, $(6, 0)$.
17. C.U. $x \leqq -3$ and $0 \leqq x \leqq 3$, infl. pts. $(-3, -9/4)$, $(0, 0)$, $(3, 9/4)$.
19. C.U. nowhere, no infl. pts.

Exercise 6, page 212
1. Min. $(-3, -22)$.   3. R. max. $(-\sqrt{3}, 6\sqrt{3})$, r. min. $(\sqrt{3}, -6\sqrt{3})$.
5. Min. $(0, 0)$.   7. None.   9. R. min. $(3, -27)$, r. max. $(-1, 5)$.
11. Max. $(0, 4)$.   13. R. max. $(-1, -2)$, r. min. $(1, 2)$.   15. Min. $(3, -3^{10})$.
17. No max. or min.   19. R. max. $(-2, -3)$.

Exercise 7, page 217
1. 3 in.   7. 100 cm.   9. 10 and 10.   11. 1.   13. 4.   15. $2R/\sqrt{3}$.
19. $x = 10$, $y = 25$.   21. No solution.   23. $x = 2$, $y = 16$.
25. $(\pm 2\sqrt{2}, -4)$, $R = 2\sqrt{6}$.   27. 15 by 60 m.   29. $m = 1$.
31. $H = P/(4 + \pi)$, $R = P/(4 + \pi)$.   35. 32 kms.   37. Min. $= 6$ at $x = -1$.
39. Max. $= 45$ at $x = 5$.   41. Max. $= 24$ at $x = -5/2$.
C1. (a) $(1.61, 2.59)$,   (b) $(2.54, 6.45)$,   (c) $(1.29, 1.66)$.   C3. $x = 10.566$ cms.

Exercise 8, page 227
1. (a) $6x^2 \Delta x + 6x(\Delta x)^2 + 2(\Delta x)^3$,   (b) $(6x + 2\Delta x)(\Delta x)^2$.
3. (a) $3x^2 \Delta x + 3x(\Delta x)^2 + (\Delta x)^3 - 2(\Delta x)$,   (b) $(3x + \Delta x)(\Delta x)^2$.
5. (a) $\dfrac{-(2x + \Delta x)\,\Delta x}{x^2(x + \Delta x)^2}$,   (b) $\dfrac{(3x + 2\Delta x)(\Delta x)^2}{x^3(x + \Delta x)^2}$.
7. 8.063.   9. 2.005.   11. 11.88.   13. 2.030.   15. 10.550, 10.545.   17. 9.6 cm^3, 0.01875.
19. Range $= 6400$ ft, max. error $= 128$ ft, rel error $= 0.02$.   21. 12.6 cm^2.
C1. Range $= 5903$ m, max. error $= 118$ m, rel error $= 0.02$.   C3. 0.7096 sec.   C5. 0.002 sec.

Exercise 9, page 233
C1. 2.6458.   C3. 1.1795.   C5. 2.3129.   C7. -2.7214.
C11. 36.400895. . . .   C13. Max. $= 6.150100.\ldots$, Min. $= 3.8125$.
C15. 1.1795.   C17. 2.3129.   C19. $f(x) = \sqrt{60 - 1/x^3}$, $7 \leqq x \leqq 10$, $x^\star = 7.745827.\ldots$

Review Problems, page 234
1. Rel. max. $(3, 37)$, rel. min. $(-1, 5)$, infl. pt. $(1, 21)$.
3. Min. $(0, 0)$ and $(5, 0)$, rel. max. $(4, 16)$,
   infl. pt. $(4 - 2\sqrt{6}/3, [(56 - 16\sqrt{6})\sqrt{1 + 2\sqrt{6}/3}]/3)$.
5. Inc. on $(-1, 1)$ and $(3, \infty)$, c.u. on $(-\infty, 1 - 2\sqrt{3}/3]$ and $[1 + 2\sqrt{3}/3, \infty)$.
7. $v_0 = 32$ ft/sec, $s_0 = 240$ ft.   9. 2 ft.   11. $x = 6$, $y = 14$.
13. (a) 2.04,   (b) 2.03,   (c) 1.96,   (d) 1.97,   (e) 3.17.   15. 22/5 by 28/5.
17. 2.   19. $x = 20\sqrt{5}$.   21. $(4, 0)$.   23. $-7xy/4\sqrt{x^2 + y^2}$.
25. (a) $(3, 15/7)$, $(-3, -15/7)$,   (b) $(-2, 2)$,   (c) $(5, 36)$,   (d) no solution.

CHAPTER 6

Exercise 1, page 242

1. $3x^6 + C$.   3. $2y^{3/2}/3 + C$.   5. $-6t^{-7/2} + C$.

7. $500x^2 - x^5 + C$.   9. $\dfrac{\sqrt{2}}{8}x^8 - \dfrac{1}{2}x^3 - \dfrac{1}{4x^4} + C$.   11. $\dfrac{1}{1028}(2x + 7)^{514} + C$.

13. $\dfrac{1}{77}(3 + 11x^2)^{7/2} + C$.   15. $\dfrac{-2}{9(1 + u^3)^{3/2}} + C$.   17. $\dfrac{2}{9}(y^3 + y + 55)^{9/2} + C$.

19. $y = x^3 + x^2$.   21. $15y = (4 + 5x^2)^{3/2} - 24$.

C1. $0.2159 (1 + 14x)^{8.531} + C$.   C3. $138{,}700 (x + 1)^{0.0187} + C$.

Exercise 2, page 245

1. $s = 16t^2 + 5t + 100$.   3. $y = \sqrt{(2x^2 - 17)/(19 - 2x^2)}$.

5. $9u = v^3 + 6v^{3/2} + 9$.   7. $y = \sqrt{(7x^2 + 2)/(3 - 2x^2)}$.

Exercise 3, page 249

3. T.   5. F.   7. F.   9. F.   11. T.   13. T.   15. T.

C1. 10.832.   C3. 5.0210.   C5. 1.5241.   C7. 1.9883.

Exercise 4, page 261

1. (a) $3/8, 5/8$,   (b) $7/16, 9/16$,   (c) $9/20, 11/20$.

3. $1/2$.   7. $(b^3 - a^3)/3$.   9. $(b^{n+1} - a^{n+1})/(n + 1)$, yes.   11. No.

C1. (a) 1.1981,   (b) 1.2395,   (c) 0.0414.

C3. (a) 0.66877,   (b) 0.71877,   (c) 0.05000.

C5. (a) 0.75998,   (b) 0.80998,   (c) 0.05000.

C7. (a) 0.15333,   (b) 0.25333,   (c) 0.10000.

Exercise 5, page 268

1. 20.   3. 0.   5. 1.   7. $\dfrac{2}{3}a + 2c$.   9. $1/3$.   11. 70.   13. $62/9$.   15. $\sqrt{2}$.

C1. (a) 0.68080,   (b) 0.70580.   C3. (a) 0.68080,   (b) 0.70580.

Exercise 6, page 273

1. 0.   3. Same as Problem 1.   5. $4\sqrt{2}$.   7. 0.

Exercise 7, page 279

1. (a) $25/3$,   (b) 13.   3. (a) 6,   (b) $22/3$.   5. (a) $4/3$,   (b) 4.

7. (a) $112/15$,   (b) 8.   9. (a) $79\frac{1}{2}$,   (b) $96\frac{1}{2}$.   11. 8.

13. $b/2$.   15. 0.   17. $1/2$.   19. 0.

23. (a) $-10°C$,   (b) minimum is $-30°C$ at midnight, maximum is $6°C$ at noon.

C1. (a) -0.87812,   (b) 1.7111.   C3. -0.35580.

Exercise 8, page 283

3. $\sin x$.   5. $\sin t$.   7. $2x^3 \sin x^2$.   9. $2t^3 \sin t^2$.

11. 0.   13. $2x\sqrt{1 + x^{10}} - \sqrt{1 + x^5}$.   15. $-2 \leqq x \leqq 2$, $y'' > 0$ in the open interval.

Exercise 9, page 290

C1. (a) 3.75, (b) 3.75. C3. (a) 63.75, (b) 63.75.
C5. (a) 2.797, (b) $2\sqrt{3} - 2/3 \approx 2.797$. C7. $\pi \approx 3.1416$. C9. 1.51 miles^2.
C11. $A = 16,000$ ft$^2 >$ one-third acre $= 43,560/3 = 14,520$ ft^2.

Review Problems, page 292

1. $\frac{4}{9}(1 + 2x^2 + x^5)^{9/4} + C$. 3. $-\frac{2}{15}(1 - 3r^5)^{1/2} + C$. 5. $y = [x^{2/3} + C]^{3/2}$, where
   $C = 1 - 2\sqrt[3]{2}$. 7. $v = -2u^2/(u^4 - 4u^2 + 1)$.
11. $3n(n + 1)$. 13. $\frac{n}{3}(16n^2 + 36n + 23)$. 15. $\frac{1}{6}n(n + 1)(2n - 23)$.
19. $(2\sqrt{2} - 2)/\sqrt{5}$. 21. 9/2. 23. 119/6. 25. 16/3. 27. $11c^3/6$. 29. $b^n/(n + 1)$.
31. $2(\sqrt{1000 + b} - 10\sqrt{10})/b$. 33. $4(\sqrt[4]{(1 + b^3)^3} - 1)/9b$.
35. (a) 0, (b) 1, (c) ∞, (d) ∞.

CHAPTER 7

Exercise 1, page 299

1. 32/3. 3. 32/3. 5. 4/15. 7. 9/2. 9. 27/4. 11. 9/2.
13. 37/12. 15. 4. 17. 1/12. 19. 37/6. 21. 14/3.

Exercise 2, page 306

1. $500\pi/3$. 3. $4\pi/3$. 5. $392\pi/3$. 7. $128\pi/3$. 9. $64\pi/15$.
11. $192\pi/55$. 13. $4\pi ab^2/3$. 15. $28\pi/3$. 17. $16\pi/7$. 19. $19\pi/3$.

Exercise 3, page 309

1. $4\sqrt{3}/3$. 3. 648. 5. $4 \cdot 3^7/7$. 7. $\pi\sqrt{2}/15$. 9. 2000/3 cm^3.

Exercise 4, page 312

1. $8(10^{3/2} - 1)/27$. 3. 12. 5. 14/3. 7. 146/27.
11. See Chapter 14, Section 6. 13. $\int_0^4 \sqrt{1 + 9x^4}\, dx$. 15. $\int_4^9 \sqrt{1 + 25y^3}\, dy$.
C1. (a) 64.124878..., (b) 64.281914..., (c) 64.525090....

Exercise 5, page 316

1. $56\pi/3$. 3. 7π. 5. $515\pi/64$. 7. $\pi B^2/4$. 9. 168π.
11. $\left(\frac{515}{64} + \frac{59}{12}C\right)\pi$.

Exercise 6, page 320

1. 4500 lb. 3. 4125 lb. 5. 12,500 lb. 7. 14.06 lb. 9. 83,330 tons.
11. $\sqrt{2} \cdot 10^3$ lb.

Exercise 7, page 325

1. 930 ton-ft.   3. (a) 10 in.-lb,   (b) 30 in.-lb,   (c) 50 in.-lb.   5. 28/9 dynes.
7. 200 ft-lb.   9. $56{,}000\pi$ ft-lb.   11. $576\delta\pi$ ft-lb.   13. 399 ft-lb.

Review Problems, page 326

1. 9.   3. 9/2.   5. 6.   7. 4/3.   9. 27/4.   11. $32\pi/5$.   13. $a^2\pi(1 - b^{-1})$.
15. (a) 8π,   (b) $a^4\pi/6$,   (c) $2a\pi(b - 1)$,   (d) $a^5\pi/10$.   17. $b/\sqrt[5]{2}$.
19. $8\pi/5$.   21. $\pi/3$.   23. (a) 150,000 lb,   (b) 60,000 lb,   (c) 20,000 lb.
25. $kA^2/2$.   27. 150 ft-lb.   31. 33/8.   33. $208\pi/9$.

CHAPTER 8

Exercise 1, page 334

Problem Number	1	3	5	7	9	11
θ	$9\pi/4$	$-2\pi/3$	$11\pi/6$	$7\pi/3$	$0°$	$-420°$
$\sin \theta$	$\sqrt{2}/2$	$-\sqrt{3}/2$	$-1/2$	$\sqrt{3}/2$	0	$-\sqrt{3}/2$
$\cos \theta$	$\sqrt{2}/2$	$-1/2$	$\sqrt{3}/2$	$1/2$	1	$1/2$
$\tan \theta$	1	$\sqrt{3}$	$-\sqrt{3}/3$	$\sqrt{3}$	0	$-\sqrt{3}$

23. (a) $(\sqrt{6} - \sqrt{2})/4 \approx 0.2588$,   (b) $(\sqrt{6} + \sqrt{2})/4 \approx 0.9659$,   (c) $2 - \sqrt{3} \approx 0.2679$.

C1.

θ	1°	1 radian	3°	3 radians	6.5°	6.5 radians
$\sin \theta$	0.01745	0.8415	0.05234	0.1411	0.1132	0.2151
$\cos \theta$	0.9998	0.5403	0.9986	-0.9900	0.9936	0.9766
$\tan \theta$	0.01746	1.557	0.05241	-0.1425	0.1139	0.2203

C3. 1.   C5. 1.   C7. 0.

Exercise 2, page 340

1. $3 \cos 3x$.   3. $2 \sin x \cos x = \sin 2x$.
5. $-6x^2 \cos x^3 \sin x^3 = -3x^2 \sin 2x^3$.   7. $2 \cos 2x \cos 3x - 3 \sin 2x \sin 3x$.
9. $6(\sin 2x + \cos 2x)^2(\cos 2x - \sin 2x) = 6(\sin 2x + \cos 2x) \cos 4x$.
11. $6 \sin 3x(\cos 2x \cos 3x + \sin 2x \sin 3x)/\cos^4 2x = 6 \sin 3x \cos x/\cos^4 2x$.
13. $\sin^2(3t^2 + 5) + 6t^2 \sin (6t^2 + 10)$.   15. $-\left(u^2 \sin \dfrac{1}{u^2} + 2 \cos \dfrac{1}{u^2}\right)\Big/u^4$.

19. $(-1)^{n+1}5^{2n-1}\cos 5x$.   21. Max. $(\pi/4, 1)$, $(5\pi/4, 1)$, min. $(3\pi/4, -1)$, $(7\pi/4, -1)$.
23. Max. $(\pi/2, 2)$, r. max. $(3\pi/2, 0)$, min. $(7\pi/6, -1/4)$, $(11\pi/6, -1/4)$.
25. $((2n + 1)\pi, (2n + 1)\pi)$, $n = 0, \pm1, \pm2, \ldots$.   29. Max. $(n\pi + \pi/2, 1)$, min. $(n\pi, 0)$.
31. Min. $(n\pi, 0)$, max. $(n\pi + \pi/2, 1)$.   33. (a) 0.530,   (b) 0.695,   (c) 0.515.
C1. (a) 1.90 radians,   (b) 109°.   C3. (a) 0.680 radians,   (b) 39.0°.   C5. 1.5.

Exercise 3, page 342

1. $-5\cos 2x + C$.   3. $4\sin^2 x + C$.   5. $5\sin t^4 + C$.   7. $\sqrt{3}/10$.

9. $\sin(t^2 + t + 5) + C$.   11. $\frac{2}{3}(\sin^3 \theta - \cos^3 \theta) + \sin \theta - \cos \theta + C$.

13. $-5\cos 2\sqrt{x} + C$.   15. $-1/bc(n - 1)(a + b \sin cx)^{n-1} + C$.
17. $(x^2 + \cos 6x)^{10}/20 + C$.   19. $2/3$.   21. $\pi^2/4$.

Exercise 4, page 346

1. $2\tan x \sec^2 x$.   3. $-6\csc^3(2x + 1)\cot(2x + 1)$.   5. $\sin x(1 + \sec^2 x)$.
7. $3\sec^2 t(\tan^2 t + 1) = 3\sec^4 t$.   9. $-3\sec x \csc^4 x + 2\sec^3 x \csc^2 x$.
11. $-[2\sec 2x + (2x \sec 2x + 1)(\sec 2x - \tan 2x)]/(x + \tan 2x)^2$.
13. $2\sec^2 x(3 + 6x \tan x + 2x^2 \tan^2 x + x^2 \sec^2 x)$.

15. $8\sin 2x(6\sec^4 2x - \sec^2 2x - 1)$.   19. $-\frac{1}{30}\csc^5 6x + C$.

21. $2\tan x - x + C$.   23. $\frac{1}{2}\tan^2 y + C$.   25. $\frac{1}{3}\tan^3 \theta + \tan \theta + C$.

29. (a) 1.063,   (b) 0.969,   (c) 1.940,   (d) 1.143.   31. π.   33. $4\pi/3$.
35. $(\sin x \sin y + 1)/(\cos x \cos y - 1)$.   37. $(\sin x + y)/(\cos y - x)$.
39. None.   41. R. min. $(\pi/3, -3\sqrt{3})$, r. max. $(-\pi/3, 3\sqrt{3})$.
45. Neither $\sec x$ nor $\tan x$ is defined at $x = \pi/2$.
C1. Minimum $= -0.275906516\ldots$ at $x = 0.5530300792\ldots$ C3. 1.87628065.

Exercise 5, page 354

1. 0.   3. $\pi/2$.   5. $\pi/3$.   7. $2\pi/3$.   9. $-\pi/3$.
11. (a) $(9\sqrt{3} - 8\sqrt{5})/11$,   (b) $-(3\sqrt{2} + \sqrt{14})/8$,   (c) $41\sqrt{2}/58$.   13. T.
15. F. Set $x = -1/2$.   17. T.   19. T.   21. F. Set $y = -1/2$.   23. $0 \leq y \leq 1$.
C1. 0.20136.   C3. -0.30469.   C5. 2.4981.   C7. 0.079830.
C9. -1.5608.   C11. -0.86901.   C13. 1.4862.   C15. 1.8703.

Exercise 6, page 359

1. $-5/\sqrt{1 - 25x^2}$.   3. $1/2\sqrt{x(1 - x)}$.   5. $-(1 + y^2)/(1 - y^2 + y^4)$.
7. $-2t/\sqrt{1 - t^4}$.   9. $2x \sin^{-1}(x/2)$.   11. $\pi/2$.   13. No.   15. 4 ft.

17. (a) 0.132 rad/sec,   (b) 0.066 rad/sec,   (c) $\theta = 0$.   21. $\frac{1}{12}\tan^{-1}(y/3) + C$.

23. $-\sin^{-1}\dfrac{\cos x}{\sqrt{10}} + C$.   25. $\dfrac{7}{8}\tan^{-1}\dfrac{2x - 3}{4} + C$.

27. $\sin^{-1}\dfrac{3y + 2}{3} + C$.   29. $\sin^{-1} M, \pi/2$.

C1. (c) 1.1656,   (d) 4.6042.   C3. 1.083.   C5. 0.4551.
C7. $3.561096338\ldots$ at $x = 0.860333589\ldots$

Review Problems, page 362

1. $-2 \csc^2 2x$.   3. $-3x^2 \sqrt{\csc x^3} (\cot x^3)/2$.   5. $2z/\sqrt{1-z^4}$.
7. $-2 \csc^2 t/(1 - \cot t)^2$.   9. $(-t \, \mathrm{Cos}^{-1} t + \sqrt{1 - t^2})/\sqrt{1 - t^2}(\mathrm{Cos}^{-1} t)^2$.
11. $1/\sqrt{x}(x + 1)$.   13. $-(\cos 5x^2)/10 + C$.   15. $-(\csc^4 \theta)/4 + C$.
17. $\mathrm{Sin}^{-1} \dfrac{x - 3}{2\sqrt{2}} + C$.
19. (a) $2/\pi$,   (b) 0,   (c) $2/3\pi$,   (d) If m is even AV $= 0$. If m is odd AV $= 2/m\pi$.
  (e) $3\sqrt{3}/\pi$.
21. Max. $\left(\mathrm{Cos}^{-1} \left(-\dfrac{1}{3} \right), 2\sqrt{2} + \mathrm{Cos}^{-1} \left(-\dfrac{1}{3} \right) \right)$, min. $(0, 0)$.
23. R. max. $(-2, -2 - 5 \, \mathrm{Tan}^{-1}(-2))$, r. min. $(2, 2 - 5 \, \mathrm{Tan}^{-1} 2)$.
25. Max. $(\pm\pi/4, 1)$ and $(\pm 3\pi/4, 1)$, r. min. $(0, 0)$.
27. (a) $\sqrt{2}$,   (b) $(2 + \pi)/4$,   (c) $\sqrt{2} - \sqrt{2}/6$. $\sqrt{2}$ is the largest.

CHAPTER 9

Exercise 1, page 366

3. If $N \neq 1$ and $a = 1$, there is no L such that $1^L = N$. Hence there is no $L = \log_1 N$.
5. (a) 2,   (b) -2,   (c) -3,   (d) π,   (e) 42.   7. 0.   15. $2, -1$.
17. $3, -4, 0$.   19. 12.   21. 7.   23. 30 db.
C1. (a) 1.8226,   (b) 2.1746,   (c) 3.5672,   (d) 0.5479.
C3. (a) $(\sqrt{2})^{\sqrt{3} + \sqrt{5}} \approx (\sqrt{2})^{3.9681} \approx 3.9560$ and $(1.8226) \cdot (2.1705) \approx 3.9560$,
  (b) $(3.5672) \cdot (0.2803) = 0.99988616$,
  (c) $(0.5479) \cdot (1.8253) = 1.00008187$.
C5. (a) 0,   (b) 0,   (c) 2.5.

Exercise 2, page 369

1. (a) e,   (b) $\sqrt{e}$,   (c) $1/e$.
C1. (a) 2.7048,   (b) 2.7320,   (c) 2.7181,   (d) 2.7183.

Exercise 3, page 374

1. $6x(x^4 + 1)/(x^6 + 3x^2 + 1)$.   3. $3/(x + 1)$.   5. $\dfrac{2}{x} \ln x$.   7. $x(1 + 2 \ln x)$.
9. $(2x^2 + 4)/x(x^2 + 4)$.   11. 0.   13. $4 \, \mathrm{Tan}^{-1} 2x$.   15. $2\sqrt{x^2 - 5}$.
17. $(x^2 + 4x + 1)/((x - 1)(x + 2)(x + 5))^{2/3}$.   19. $6(35 - x^4)/x^7(x^2 - 5)^{1/2}(x^2 + 7)^{3/2}$.
23. $\dfrac{1}{3} \ln (5 - 3 \cos x) + C$.   25. $\dfrac{1}{5} \ln |\sec 5x + \tan 5x| + C$.   27. $2 \, \mathrm{Tan}^{-1} \sqrt{x} + C$.
29. $\dfrac{1}{4} \ln |\sin x^4| + C$.   31. $\ln |\ln x| + C$.   33. $\dfrac{1}{2} \ln 2 \approx 0.347$.
35. $\pi \ln 7$.   37. No infl. pts., min. $(2, 4 - 8 \ln 2) \approx (2, -1.545)$.
39. 0.   43. $1/x (\ln x) \ln (\ln x)$.   45. $14 \csc 2x$.   47. 0.
C1. (a) 0.48790,   (b) 0.49875,   (c) 0.49988.
C3. Computed value 0.69315. Correct value $0.693147180. \ldots$
C5. 1.8572 and 4.5364.

Exercise 4, page 378

1. $(2 - 3x)xe^{-3x}$.   3. $5e^{\sin^2 5x} \sin 10x$.   5. $(x^2 + 2xe^x - x^2 e^x)/(x + e^x)^2$.

7. $x^{\sin x}\left(\cos x \ln x + \dfrac{1}{x}\sin x\right)$.   9. $(1 + 3x)^{1/x}\left(\dfrac{3}{x(1 + 3x)} - \dfrac{\ln(1 + 3x)}{x^2}\right)$.

11. $-x^4 e^{-x}$.   13. $(-1)^{n-1}5(n - 1)!/(1 + x)^n$.   15. $7^n e^{7x}$.

17. $y' = x + 2x \ln x$, $y'' = 3 + 2 \ln x$, $y^{(n)} = (-1)^{n+1}2(n - 3)!/x^{n-2}$ for $n > 2$.

19. $7e^{x^2} + C$.   21. $6 \operatorname{Tan}^{-1} e^x + C$.   23. $\ln|e^x - e^{-x}| + C = x + \ln|1 - e^{-2x}| + C$.

25. Min $(0, 2)$. There are no infl. pts.

27. Min $(-3, -3/e) \approx (-3, -1.104)$, infl. pt. $(-6, -6/e^2) \approx (-6, -0.812)$.

29. Max $(1/\sqrt{2}, 1/\sqrt{2e})$, min $(-1/\sqrt{2}, -1/\sqrt{2e})$, infl. pts. $(0, 0)$, $(\pm\sqrt{3/2}, \pm\sqrt{3/2e^3})$.

31. $a + e^{2a}$.   33. $(e^a - e^{-a})/2$.   39. $\ln 2$.

C1. 3.1416.   C3. 1.6722.

Exercise 5, page 384

C1. (a) 0.958,   (b) 0.807,   (c) 0.652.   C3. 107 years.

C5. (a) -0.0001216,   (b) 8224.   C7. (a) 4.69 grams,   (b) 2.84 grams,
 (c) 1.71 grams.   C9. (a) $T = M + Ce^{kt}$,   (b) $68.4°$.   C11. 5958 B.C. ≈ 5960 B.C.

C13. (a) $\$115.76$,   (b) $\$116.18$.   C15. (a) $\$301.19$,   (b) $\$90.72$.

C17. If we take $k = 0.05$, then $Q \approx 24.5$ grams. To find the true value of k, set $1.05 = e^k$ and
then $k \approx 0.04879$. With this k, $Q \approx 22.7$ grams.

C19. (a) 6.93 years ≈ 6 years 11 months,   (b) 16.09 years ≈ 16 years 1 month.

C21. (a) 0.50,   (b) 0.25,   (c) 0.061,   (d) 0.0000008.   C23. $x^\star = \dfrac{\ln 20}{\ln 2}x_0 \approx 4.322x_0$.

Exercise 6, page 392

11. $3x^2 \coth x^3$.   13. $x^{-3/4}\cosh^3(x^{1/4})\sinh(x^{1/4})$.   15. $-e^{-x}\operatorname{sech}^2(e^{-x})$.

17. $\cosh x_0 = 5/3$, $\tanh x_0 = 4/5$, $\coth x_0 = 5/4$, $\operatorname{sech} x_0 = 3/5$, $\operatorname{csch} x_0 = 3/4$.

19. $\dfrac{1}{5}\cosh(5x + 1) + C$.   21. $(\sinh^3 x)/3 + C$.   23. $\ln(\cosh x) + C$.

25. $\dfrac{1}{2}x + \dfrac{1}{4}\sinh 2x + C$.   29. $1/\sqrt{x^2 - 1}$.   31. $1/(1 - x^2)$.

33. Max $= -\sqrt{5}$ at $x = \ln\sqrt{5}$.

C1. (a) 7.789,   (b) 1.128,   (c) 1.195.   C3. ± 1.244.

Review Problems, page 394

1. -2.   3. -3, $5/2$.   5. $\dfrac{2}{x} + \dfrac{2x}{x^2 - 4}$.   7. $\dfrac{5}{x} - \dfrac{3}{x - 2}$.

9. $2^{x^2+1}(1 + 2x^2 \ln 2)$.   11. $(2x + 1)^x\left(\dfrac{2x}{2x + 1} + \ln(2x + 1)\right)$.

13. $12 \sinh(4x - 1)\cosh^2(4x - 1)$.   15. $\cot x$.   17. $-e^{\cos x} + C$.

19. $\dfrac{1}{4}\ln|x^4 + 8x| + C$.   21. $2 \ln|\sec x^2 + \tan x^2| + C$.   23. $-\dfrac{1}{3}\operatorname{sech} 3x + C$.

25. $(e^3 - 3e + 2)/3$.   27. (a) $\pi \ln M$,   (b) ∞.   29. $\pi(2 + \sinh 2)$.

31. (a) Max $(1, 1/e)$, infl. pt. $(2, 2/e^2)$, increasing in $[0, 1]$,   (b) Max $(2, 4/e^2)$, infl. pt.
 $(2 + \sqrt{2}, (6 + 4\sqrt{2})/e^{2+\sqrt{2}})$, and $(2 - \sqrt{2}, (6 - 4\sqrt{2})/e^{2-\sqrt{2}})$, increasing in $[0, 2]$,
 (c) Max $(3, 27/e^3)$, infl. pts. $(3 + \sqrt{3}, (3 + \sqrt{3})^3/e^{3+\sqrt{3}})$ and $(3 - \sqrt{3}, (3 - \sqrt{3})^3/e^{3-\sqrt{3}})$,
 increasing in $[0, 3]$.

35. $(\ln 10)/(\ln 2)$ years.   37. $y = 10 - 5e^{-x/2}$.

CHAPTER 10

Exercise 1, page 402

Add a constant of integration to the answer to each indefinite integral in this chapter.

1. $2(x + 1)^{3/2}(3x - 2)/15.$ **3.** $(2x - 1)^{3/2}(3x + 1)/15.$

5. $2\sqrt{x - 3}(x + 6)/3.$ **7.** $2\sqrt{x + 1}(3x^2 - 4x + 8)/15.$

9. $-2(1 - x)^{3/2}(x + 4)/5.$ **11.** $3(1 + x)^{4/3}(4x - 3)/28.$

13. $(2x^2 + 1)^{3/2}(3x^2 - 1)/30.$ **15.** $4\sqrt{x} - \ln|1 + 4\sqrt{x}|.$

17. $(2x - 3x^{2/3} + 6x^{1/3})/2 - 3\ln|1 + x^{1/3}|.$ **19.** $2x^{1/2} - 3x^{1/3} + 6x^{1/6} - 6\ln|1 + x^{1/6}|.$

21. $2\sqrt{x + 5}(x^2 - 24).$ **23.** $\dfrac{4}{3}x^{3/2} + 2x + 10x^{1/2} + 10\ln|\sqrt{x} - 1|.$

25. $928/135 \approx 6.874.$ **27.** $1209/28 \approx 43.18.$

29. $34/3 \approx 11.33.$ **31.** $-18 + 48\ln(3/2) \approx 1.46.$

33. $(14\sqrt{2} - 16)/45 \approx 0.08442.$ **35.** $2\sqrt{e^x - 1}(e^x + 2)/3.$

37. $2\pi\ln(\sqrt{M} + 1) - 2\pi\sqrt{M}/(\sqrt{M} + 1).$ **39.** $2\pi M/(1 + \sqrt{M}).$

Exercise 2, page 407

1. $(3x^3\ln|x| - x^3)/9.$ **3.** $(3\theta\sin 3\theta + \cos 3\theta)/9.$

5. $e^{3x}(3x - 1)/9.$ **7.** $\dfrac{-1}{4x^2}(2\ln|x| + 1).$

9. $2(x\ln|x| - x).$ **11.** $x\,\text{Sin}^{-1}2x + \dfrac{1}{2}(1 - 4x^2)^{1/2}.$

13. $\dfrac{x^2 - 9}{2}\ln|x + 3| - \dfrac{x^2}{4} + \dfrac{3x}{2}.$ **15.** $e^{5y}(25y^2 - 10y + 2)/125.$

17. $(2y\sin 2y + \cos 2y - 2y^2\cos 2y)/4.$ **19.** $(2x^3\,\text{Tan}^{-1}x + \ln(1 + x^2) - x^2)/6.$

21. $-e^{-3x}(3\sin x + \cos x)/10.$ **23.** $(32x^3\sin 4x + 24x^2\cos 4x - 12x\sin 4x - 3\cos 4x)/128.$

25. $\dfrac{3}{8}(\ln|\sec x + \tan x| + \sec x\tan x) + \dfrac{1}{4}\tan x\sec^3 x.$ **27.** $\dfrac{1}{a}x\sinh ax - \dfrac{1}{a^2}\cosh ax.$

29. $2\pi^2.$

Exercise 3, page 410

The formula used is in parentheses.

1. $(30),\ -\dfrac{3}{x} - \dfrac{3}{2}\ln\left|\dfrac{2 - x}{x}\right|.$ **3.** $(84),\ 10(2x + 1)\,\text{Tan}^{-1}(2x + 1) - 5\ln(4x^2 + 4x + 2).$

5. $(57),\ \dfrac{\sqrt{t^2 - 4}}{8t^2} + \dfrac{1}{16}\text{Cos}^{-1}\dfrac{2}{t}.$ **7.** $(29),\ 2\ln|z| - 2\ln|4 + 13z|.$

9. $(47),\ -6x\sqrt{5 - x^2} + 30\,\text{Sin}^{-1}(x/\sqrt{5}).$ **11.** $(61),\ (8x - \sin 8x)/16.$

13. $(76),\ (\sin 8v + 4\sin 2v)/16.$ **15.** $(46),\ \dfrac{x - 3}{4\sqrt{6x - x^2 - 5}}.$

17. $(81, 80, 79),\ (x^3 - 6x)\sin x + (3x^2 - 6)\cos x.$ **19.** $(95),\ 4e^{2x}(2\cos 3x + 3\sin 3x).$

21. $(59),\ \sqrt{x^2 + 4x - 6} - \sqrt{10}\,\text{Cos}^{-1}\dfrac{\sqrt{10}}{x + 2}.$ **23.** $(35),\ \dfrac{1}{3}(x + 3)\sqrt{2x - 3}.$

25. $(69),\ -\cos x(3\sin^4 x + 4\sin^2 x + 8)/15.$ **27.** $(71, 63)\ \dfrac{1}{5}\tan^5 x - \dfrac{1}{3}\tan^3 x + \tan x - x.$

29. $(91)\ e^{5x}(125x^3 - 75x^2 + 30x - 6)/625.$

Exercise 4, page 415

1. $-\cos x + \dfrac{1}{3}\cos^3 x.$ 3. $\sin^3 y(5 - 3\sin^2 y)/15.$

5. $(12x - 3\sin 4x + 4\sin^3 2x)/192.$ 7. $-2(3 + \sin^2 x)/3\sqrt{\sin x}.$
9. $-\cot^3 3x(3\cot^2 3x + 5)/45.$ 11. $-(\cos 5x + 5\cos x)/10.$

13. $\dfrac{1}{3}\tan^3(x - 5) - \tan(x - 5) + x.$ 15. $(\tan 5x - \cot 5x - 2\ln|\csc 10x + \cot 10x|)/5.$

17. $-\dfrac{1}{2}\cot^2 x - \ln|\sin x|.$ 19. $\pi^2/2.$

21. $\pi(4 - \pi)/8.$ 27. $16\sqrt{2}/3.$

Exercise 5, page 420

1. $\sqrt{x^2 + 9}\,(x^2 - 18)/3.$ 3. $\text{Sin}^{-1}\dfrac{y}{\sqrt{2}} + \dfrac{y\sqrt{2 - y^2}}{2}.$

5. $\sqrt{y^2 - 6}/6y.$ 7. $\sqrt{9 - x^2} - 3\ln\left|\dfrac{3 + \sqrt{9 - x^2}}{x}\right|.$

9. $-\sqrt{4 - y^2}/y - \text{Sin}^{-1}(y/2).$ 11. $\sqrt{u^2 - a^2}\,(a^2 + 2u^2)/3a^4u^3.$

13. $3\sqrt{x^2 + 4x + 5} + \ln|x + 2 + \sqrt{x^2 + 4x + 5}|.$ 15. $\dfrac{x\sqrt{x^2 + 4}}{2} + 2\ln|x + \sqrt{x^2 + 4}|.$

17. $3 - \sqrt{2} + \ln(1 + \sqrt{2}/2).$ 19. $\pi r^2/2 - r^2\,\text{Sin}^{-1}(b/r) - b\sqrt{r^2 - b^2}.$
21. $4\pi^2/3 + \pi\sqrt{3}/2.$

Exercise 6, page 426

1. $\ln|(x + 1)(x + 2)^3|.$ 3. $\ln|(x - 1)/(x + 1)|.$
5. $\ln|(x + 3)/(x + 5)|.$ 7. $\ln|(x - 3)^3(x + 4)^4|.$
9. $x + \ln|(x - 2)/(x + 2)|.$ 11. $\ln[x^2(x + 2)^4(x - 2)^6].$
13. $\dfrac{1}{2}\ln|(x + 1)^3(x + 3)^{13}/(x + 2)^{14}|.$ 15. $\dfrac{1}{x} + \ln\left(\dfrac{x - 4}{x}\right)^2.$

17. $\dfrac{4}{x - 2} + \ln\dfrac{|x - 2|^3}{x^2}.$ 19. $\dfrac{6}{x + 2} - \dfrac{3}{(x + 2)^2} + \ln|x - 1|.$

21. $\dfrac{5}{2}\ln|2x - 1| - \dfrac{3}{2}\ln|2x + 1|.$ 23. $\dfrac{1}{4}\ln((2 + \sin x)/(2 - \sin x)).$

25. $x^2 - 3x + \ln|(x + 1)(x + 3)/(x + 2)(x + 4)|.$ 27. $\ln(3/2).$
29. $\pi\ln(3/2).$ 31. The denominator is zero at $x = 2$ and $x = 3$.

Exercise 7, page 430

1. $\ln|x| - \dfrac{1}{2}\ln(x^2 + 1).$ 3. $\dfrac{1}{2}\ln[(x^2 + 4)/(x^2 + 1)] + 2\,\text{Tan}^{-1}x.$

5. $\dfrac{1}{8}\ln\left|\dfrac{(2x - 1)^3}{(2x + 1)(4x^2 + 1)}\right|.$ 7. $-2/(x^2 + 9).$ 9. $3\ln(x^2 + 9) + 27/(x^2 + 9).$

11. $2\ln(x^2 + 1) + 7\,\text{Tan}^{-1}x.$ 13. $\ln x^2 - \dfrac{13}{3}\,\text{Tan}^{-1}(x/3).$

15. $\dfrac{1}{2}\ln(x^2 + 2x + 2) + 3\,\text{Tan}^{-1}(x + 1) + \dfrac{4}{x}.$

Review Problems, page 431

1. $-3 \ln |x| - \dfrac{1}{x} + \dfrac{3}{2} \ln (1 + x^2) + 4 \operatorname{Tan}^{-1} x$.

3. $-\ln |\csc 2x + \cot 2x|$.   5. $\dfrac{1}{2} \ln |1 - e^{-2x}|$.   7. $-\operatorname{Tan}^{-1} (\cos x)$.

9. $-2 \cos \sqrt{x}$.   11. $-\operatorname{Tan}^{-1} (\cos \theta)$.   13. $x \operatorname{Cot}^{-1} 2x + \dfrac{1}{4} \ln (1 + 4x^2)$.

15. $\dfrac{1}{8} (\ln |x|)^8$.   17. $-\ln (1 + e^{-x})$.   19. $\dfrac{1}{24} \cosh 6x - \dfrac{3}{8} \cosh 2x$.

21. $-\dfrac{1}{6} \cos 3x + \dfrac{1}{4} \cos 2x$.   23. $e^{\tan y}$.   25. $\dfrac{1}{3} \operatorname{Tan}^{-1} e^{3x}$.

27. $-2\sqrt{1 - \sin \theta}$.   29. $5x \ln (x^2 + a^2) - 10x + 10a \operatorname{Tan}^{-1} \dfrac{x}{a}$.

31. $2\sqrt{\sec y}$.   33. $\dfrac{1}{20} \sin 10x + \dfrac{1}{8} \sin 4x$.

35. $-(2e^{3x} + 3e^{2x} + 6e^x + 6 \ln |e^x - 1|)$.   37. $\theta \tan \theta + \ln |\cos \theta|$.

39. $-\ln |\ln |\cos x||$.   41. $\dfrac{1}{2} (\cosh x \sin x - \sinh x \cos x)$.

43. $\tan \theta - \cot \theta$.   45. $\ln \left| \dfrac{x^6 (x - 2)(x + 2)}{(x - 1)^4 (x + 1)^4} \right|$.   47. $\dfrac{x}{2} [\cos (\ln |x|) + \sin (\ln |x|)]$.

49. $x - 2 \operatorname{Tan}^{-1} x$.   51. $-x + 4\sqrt{x} - 4 \ln (\sqrt{x} + 1)$.

53. $\dfrac{1}{12} \ln \left| \dfrac{3 \cos \theta + 2 \sin \theta}{3 \cos \theta - 2 \sin \theta} \right|$.   55. $\ln |\sin \theta + 2 \cos \theta| + 3\theta$.

57. $t + 4\sqrt{t + 1} + 4 \ln |\sqrt{t + 1} - 1|$.   59. $-\dfrac{1}{44} \cos 11\theta + \dfrac{1}{28} \cos 7\theta - \dfrac{1}{12} \cos 3\theta - \dfrac{1}{4} \cos \theta$.

CHAPTER 11

Exercise 1, page 437

1. $x^2 = 8(y - 3)$, or $8y = x^2 + 24$.   3. $4x = -(y - 3)^2$.

5. $\dfrac{(x - 6)^2}{25} + \dfrac{y^2}{16} = 1$, or $16x^2 + 25y^2 - 192x + 176 = 0$.

7. $(x + 2)^2 + (y - 2)^2/2 = 1$ or $2x^2 + y^2 + 8x - 4y + 10 = 0$.

9. $y^2 - (x - 3)^2 = 2$ or $y^2 = x^2 - 6x + 11$.   11. P. $(-1, -6), (-1, -2), y = 2$.

13. P. $(3, 2), (1, 2), x = -1$.   15. E. $(1, 3), (9, 3), (0, 3), (10, 3), x = 5, y = 3$.

17. E. $(1, -2 \pm \sqrt{15}/2), (1, 0), (1, -4), x = 1, y = -2$.

19. H. $(-2, 5 \pm \sqrt{2}), (-2, 6), (-2, 4), x = -2, y = 5$.

21. $h = 3, k = -5, XY = 2$.   23. $h = -5, k = -2, XY = 19$.

25. $h = 4, k = -100, Y = X^3 - 41X$.

27. $\left(-\dfrac{B}{2A}, C + \dfrac{1 - B^2}{4A} \right)$.   29. $1/B^2 \sqrt{B^2 - 1}, (0, \pm\infty), (\pm 1, 0)$.

Exercise 2, page 446

1. $x = 1, y = 1$.   3. $x = 5/2, y = 1/2$.   5. $x = 0, y = 1$.

7. $x = 0, y = x$.   9. $x = 0, y = -3, y = 2$.   11. $x = 0, y = 1/2$.

13. $y = \pm\pi/2$.   15. $y = x/2$.   17. $x = 2, x = -5, y = 4$.

19. $x = 3/2$, $y = x + 1$.　　**21.** $x = 0$, $y = x/2 - 3$.　　**23.** None.
29. Yes. A line is its own asymptote.　　**33.** Min. $(0, 0)$.
C1. $(\sqrt[3]{12}, (3\sqrt[3]{12} - 12)/4) \approx (2.289, -1.283)$, R. min.

Exercise 3, page 450

3. (a) (h, k),　　(b) (h, k),　　(c) no,　　(d) no.　　**5.** x-axis.　　**7.** x-axis.
9. x-axis.　　**11.** (c) is not symmetrical about $y = x$, (a), (b), and (d) are.
15. Symmetrical with respect to $(b, 0)$.

Exercise 4, page 456

1. $e = 3/5$, $(\pm3, 0)$, $x = \pm25/3$.　　**3.** $e = 2/\sqrt{11}$, $(0, \pm2)$, $y = \pm11/2$.
5. $e = \sqrt{2}$, $(0, \pm\sqrt{2})$, $y = \pm\sqrt{2}/2$, $y = \pm x$.　　**7.** $e = 2$, $(\pm2, 0)$, $x = \pm1/2$, $y = \pm\sqrt{3}x$.
9. $e = 4/5$, $(0, \pm4)$, $y = \pm25/4$.　　**11.** $e = 1/2$, $(\pm3, 2)$, $x = \pm12$.
13. $e = 3$, $(1, -1)$, $(-5, -1)$, $x = -7/3$, $x = -5/3$, $y = \pm2\sqrt{2}(x + 2) - 1$.
15. $e = \sqrt{3}$, $(-2, 3 \pm 2\sqrt{3})$, $y = 3 \pm 2\sqrt{3}/3$, $\sqrt{2}(y - 3) = \pm(x + 2)$.
17. At $(-4, 0)$, $(0, 5)$.　　**21.** $80x^2 + 81y^2 = 6480$.　　**23.** $24y^2 - x^2 = 2400$.
25. $169x^2 + 25y^2 = 3600$.　　**27.** $8x^2 + 3y^2 = 35$.
C1. $b = 92,900,000$ miles, max. dist. $= 94,460,000$ miles, min. dist. $= 91,360,000$ miles.
C3. $e = 0.09349$.

Exercise 5, page 460

1. $1/3$.　　**3.** Same as Problem 1.　　**5.** $1/7$.　　**7.** $-3/4$.
9. -2 at $(0, 0)$, $2/25$ at $(2, 4)$.　　**11.** -3 at $(-4, -2)$, $1/5$ at $(1, 8)$, $-3/11$ at $(2, 4)$.

Exercise 6, page 466

1. $e = 2/\sqrt{6}$, $(2\sqrt{2}, 2\sqrt{2})$, $(-2\sqrt{2}, -2\sqrt{2})$, $x + y = \pm6\sqrt{2}$.
3. $e = \sqrt{6}$, $(3\sqrt{2}, 3\sqrt{2})$, $(-3\sqrt{2}, -3\sqrt{2})$, $x + y = \pm\sqrt{2}$, $y = -(3 \pm \sqrt{5})x/2$.
5. $e = 1$, $(-1, 1)$, $y = x - 2$.　　**7.** $e = 1/2$, $(\sqrt{3}, 1)$, $(-\sqrt{3}, -1)$, $\sqrt{3}x + y = \pm16$.
9. $e = \sqrt{10}/2$, $(3, 1)$, $(-3, -1)$, $3x + y = \pm4$, $y = (1 \pm 2\sqrt{6}/3)x$.

Review Problems, page 467

1. Sym. w.r. to y-axis, asymptote $y = 1$.
3. No sym., asymptote $y = 0$, max pt. $(1, 9)$, min pt. $(-9, -1)$.
5. Sym. w.r. to origin, asymptotes $x = \pm3$, $y = -x$, rel. max. $(3\sqrt{3}, -9\sqrt{3}/2)$, rel. min. $(-3\sqrt{3}, 9\sqrt{3}/2)$.
7. No sym. asymptote $x = -4$, $x = 0$, $x = 4$.　　**9.** Sym. w.r. to x-axis, asymptotes $x = \pm3$.
11. All three symmetries, asymptotes $x = 0$, $x = 1$, $x = -1$, rel. min at $(\pm\sqrt{2}/2, 2)$, rel. max at $(\pm\sqrt{2}/2, -2)$.
13. Yes, if $a = 0$ and $c = 0$.
17. Rel. max $(\pm\sqrt{2}/2, 1/2)$, rel. min $(\pm\sqrt{2}/2, -1/2)$, $m = \pm1$.
19. At $(1, 2)$, $\tan \varphi = -3$, at $(1, -2)$, $\tan \varphi = 3$.
21. At $(2, \sqrt{6})$, $\tan \varphi = 7\sqrt{6}/6$. At $(2, -\sqrt{6})$, $\tan \varphi = -7\sqrt{6}/6$.
23. At $(1, 3)$, $\tan \varphi = -16/7$, at $(-1, -3)$, $\tan \varphi = -16/7$.　　**31.** $(x')^2 + 4(y')^2 = 16$.
33. $9(x')^2 + 19(y')^2 = 36$.　　**35.** $(x')^2 + 5(y')^2 = 100$.

986

CHAPTER 12

Exercise 1, page 480

1. (a) $4\mathbf{i} - 3\mathbf{j}$, (b) $-7\mathbf{i} + 9\mathbf{j}$, (c) $-2\mathbf{i} - 7\mathbf{j}$, (d) $-4\mathbf{i} + 4\mathbf{j}$, (e) $7\mathbf{i} - 8\mathbf{j}$, (f) $25\mathbf{i} - 19\mathbf{j}$.

3. $\sqrt{5}, \sqrt{34}, \sqrt{113}, 3\sqrt{10}$.

5. (a) $-8\mathbf{i} + 6\mathbf{j}$, 10, (b) $30\mathbf{i} + 40\mathbf{j}$, 50, (c) $-12\mathbf{i} - 5\mathbf{j}$, 13, (d) $12\mathbf{i} - 8\mathbf{j}$, $4\sqrt{13}$.

7. (a) $\mathbf{i} + 3\mathbf{j}$ and $-\mathbf{i} - 3\mathbf{j}$, (b) $\mathbf{i} - 2\mathbf{j}$ and $-\mathbf{i} + 2\mathbf{j}$, (c) $\mathbf{i} + \sqrt{3}\mathbf{j}$ and $-\mathbf{i} - \sqrt{3}\mathbf{j}$.

11. (a) $(7, 11)$, (b) $(7, -7)$, (c) $(1, 2)$, (d) $(\sqrt{2}, \pi + e)$. 13. $\left(\dfrac{a_1 + 2b_1}{3}, \dfrac{a_2 + 2b_2}{3}\right)$.

15. $x = -1/2, y = 5/2$. 17. $x = 2n, y = 4n, z = -n$, where n is any integer.

19. $\sqrt{(a_1 + b_1)^2 + (a_2 + b_2)^2} \leq \sqrt{a_1^2 + a_2^2} + \sqrt{b_1^2 + b_2^2}$.

21. (a) $5\mathbf{i} + 5\mathbf{j}$, (b) $-5\mathbf{i} - 5\mathbf{j}$. 23. (a) $80\mathbf{i} + 10\mathbf{j}$, (b) $10\sqrt{65} \approx 80.62$ km/hr.

C1. (a) $2.880\mathbf{i} + 2.170\mathbf{j}$, (b) $-2.743\mathbf{i} + 3.388\mathbf{j}$, (c) $-0.137\mathbf{i} + 5.558\mathbf{j}$.

C3. (a) At an angle $131.08°$ with the positive x-axis (upstream $41.08°$ with the vertical), (b) 1.126 hours or 1 hour 7.6 min.

Exercise 2, page 487

1. The line segment joining $(0, 1)$ and $(1, 0)$.

3. The line segment joining $(0, 5)$ and $(5, 0)$ covered four times.

5. The part of the parabola $y = 1 - 2x^2$ between $(0, 1)$ and $(1, -1)$ doubly covered.

7.   9.

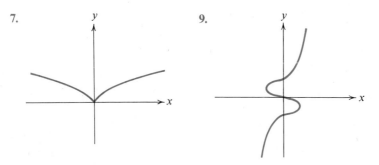

11. $x = a_1 + t(a_2 - a_1), y = b_1 + t(b_2 - b_1)$.

13. $\mathbf{R} = (b + b^2 \sec^2 \theta)\mathbf{i} + b \tan \theta \mathbf{j}, y^2 = x - (b + b^2)$.

15. $\mathbf{R} = a(\cos \theta + \theta \sin \theta)\mathbf{i} + a(\sin \theta - \theta \cos \theta)\mathbf{j}$.   19. A spiral.

Exercise 3, page 496

1. $\sec^2 t\mathbf{i} + \sec t \tan t\mathbf{j}$.   3. $(2v^3 + 3v^2)e^{2v}\mathbf{i} + (2v - 3v^2)e^{-3v}\mathbf{j}$.

5. (a) $\mathbf{R}'(t) = -10 \sin t \cos t\mathbf{i} + 10 \sin t \cos t\mathbf{j}$, (b) $\mathbf{R}''(t) = 10 \cos 2t (-\mathbf{i} + \mathbf{j})$, (c) $(5, 0), (0, 5)$, (d) none.

7. (a) $\mathbf{R}'(t) = \cos t\mathbf{i} - 2 \sin 2t\mathbf{j}$, (b) $\mathbf{R}''(t) = -\sin t\mathbf{i} - 4 \cos 2t\mathbf{j}$, (c) $(\pm 1, -1)$ (d) Horiz. tangent at $(0, 1)$.

9. (a) $\mathbf{R}'(t) = 3(t^2 - 1)\mathbf{i} + \mathbf{j}$, (b) $\mathbf{R}''(t) = 6t\mathbf{i}$, (c) $\mathbf{R}'(t) \neq \mathbf{0}$, (d) Vert. tangent at $(-2, 1), (2, -1)$.

11. Horiz. tangents at $((2n + 1)\pi a, 2a)$; vert. tangents at $(2n\pi a, 0), n = 0, \pm 1, \pm 2, \ldots$.

13. $\dfrac{d^2y}{dx^2} = \dfrac{f'(t)g''(t) - g'(t)f''(t)}{f'(t)^3}$.

15. (a) Since $f' = -g'$, $d^2y/dx^2 = 0$ except at $(5, 0)$ and $(0, 5)$,

   (b) $\dfrac{d^2y}{dx^2} = \dfrac{-4\cos t\cos 2t - 2\sin t\sin 2t}{\cos^3 t} = -4$ except at $(\pm 1, 1)$,

   (c) $\dfrac{d^2y}{dx^2} = \dfrac{-2t}{9(t^2 - 1)^3}$, except at $(-2, 1)$ and $(2, -1)$.

17. (a) $\pm(\mathbf{i} + \mathbf{j})/\sqrt{2}$,   (b) $(2\sin 2t\mathbf{i} + \cos t\mathbf{j})/\sqrt{4\sin^2 2t + \cos^2 t}$,
   (c) $(-\mathbf{i} + 3(t^2 - 1)\mathbf{j})/\sqrt{1 + 9(t^2 - 1)^2}$.

19. (a) $y = 1$,   (b) $y = 6e^{-3x} + 4$.

Exercise 4, page 503

1. $24\sqrt{29}$.   3. 105.

5. For $\mathbf{V}(t)$ and $\mathbf{A}(t)$ see $\mathbf{R}'$ and $\mathbf{R}''$ in Exercise 3,   (c) $|\mathbf{V}| = 5\sqrt{2}|\sin 2t|$.

7. (c) $|\mathbf{V}| = |\cos t|\sqrt{1 + 16\sin^2 t}$.   9. (c) $|\mathbf{V}| = \sqrt{9t^4 - 18t^2 + 10}$.

11. $|\mathbf{V}| = a\left|\dfrac{d\varphi}{dt}\right|\sqrt{2 - 2\cos\varphi}$. Max. $|\mathbf{V}| = 2a\left|\dfrac{d\varphi}{dt}\right|$.

13. $10{,}000$ ft.   15. 670 ft.

17. Range $= V_0^2(\sin 2\alpha)/g$, maximum range $= V_0^2/g$ when $\alpha = 45°$.

19. Maximum height $= V_0^2(\sin^2\alpha)/2g$.

21. (a) 17.5 sec,   (b) 1750 m.   23. 54.5 miles/hr.

Exercise 5, page 508

1. 0.   3. $e^x/(1 + e^{2x})^{3/2}$.   5. $-9(\sin 3x)/(1 + 9\cos^2 3x)^{3/2}$.

7. $-1/4a\,|\sin(t/2)|$.   9. $6x/(1 + 9x^4)^{3/2}$.   11. $\rho = 2$ at $(0, 0)$.

13. $9/7\sqrt[6]{28}$ at $x = \pm(2/7)^{1/6}$.   17. $2/5$.   19. $2/5\sqrt{5}$ is different from both.

Exercise 6, page 514

1. $x = a + ms/\sqrt{m^2 + n^2}$, $y = b + ns/\sqrt{m^2 + n^2}$.

3. $x = 2(\cos\sqrt{s} + \sqrt{s}\sin\sqrt{s})$, $y = 2(\sin\sqrt{s} - \sqrt{s}\cos\sqrt{s})$, $s \geq 0$.

5. $x = \dfrac{2}{3}(1 - s)^{3/2}$, $y = \dfrac{2}{3}s^{3/2}$.   7. $s = \dfrac{1}{2}e^{2t} + t - \dfrac{1}{2}$, $P_0(2, 1/2)$.

9. $(-n\mathbf{i} + m\mathbf{j})/\sqrt{m^2 + n^2}$.   11. $(-b\cos t\mathbf{i} - a\sin t\mathbf{j})/\sqrt{a^2\sin^2 t + b^2\cos^2 t}$.

13. $(2t\mathbf{i} + (t^2 - 1)\mathbf{j})/(1 + t^2)$.   15. $\sin(t/2)(-\cos(t/2)\mathbf{i} + \sin(t/2)\mathbf{j})/|\sin(t/2)|$.

17. (a) $m/\sqrt{m^2 + n^2}$,   (b) $-a\sin t/\sqrt{a^2\sin^2 t + b^2\cos^2 t}$,   (c) $(t^2 - 1)/(t^2 + 1)$,
   (d) $|\sin(t/2)|$.

Exercise 7, page 518

1. 7.64 rpm.   5. 5.5.   7. $9.805/0.3048 \approx 32$ ft/sec^2.   9. (a) 13,   (b) 0,   (c) 0.

11. (a) $\sqrt{1 + 4t^2}$,   (b) $4t/\sqrt{1 + 4t^2}$,   (c) $2/\sqrt{1 + 4t^2}$.

13. (a) $t\sqrt{1 + t^2}$,   (b) $(1 + 2t^2)/\sqrt{1 + t^2}$,   (c) $-t/\sqrt{1 + t^2}$.

15. (a) $F_T = 8 \times 10^{-4}/\sqrt{17} \approx 0.000194$ newton,   (b) $F_N = F_T/4 \approx 0.0000485$ newton.

17. (a) $F_T = (e - e^{-1})/2 \approx 1.175$ newtons,   (b) $F_N = 1$ newton.

19. 33 tons.   21. (a) 1.42 W,   (b) 3.42 W.   23. (a) 4 ft/sec,   (b) 16 ft/sec^2.

Review Problems, page 520

1. 11i.   3. $-8\mathbf{j}$.   5. $30\mathbf{i} - 18\mathbf{j}$.   7. $-\mathbf{j}$.   9. 0.   11. $a = 2$, $b = 7$.

13. $c = 5$, $d = -4$.   15. $a = 0$, $b = 0$.   17. (a) $(5/2, -1)$,   (b) $(3, 2)$,

   (c) $(3/2, -5/2)$.

19. Straight-line segment $x + y = 1$, $0 \leqq x \leqq 16$.

21. Part of the hyperbola $x^2 - y^2 = 1$ in the first quadrant.

23. Segment of the straight line $3x - 2y = 11$, $3 \leqq x \leqq 5$, covered eight times.

25. (a) $(\sin t \sin 2t + 2 \cos t \cos 2t)/(\sin^2 2t + \cos^2 t)^{3/2} = 2 \cos^3 t/(\sin^2 2t + \cos^2 t)^{3/2}$,

   (b) 0,   (c) $6t^4/(1 + 4t^6)^{3/2}$,   (d) $-\cos^3 t/(1 + \sin^2 t)^{3/2}$,   (e) $6/(4 + 5 \sin^2 t)^{3/2}$,

   (f) 0.

27. (a) $(1/4, 2)$,   (b) $(\sqrt[3]{2}/2, \sqrt[3]{2})$,   (c) $(1, 1)$.

29. (a) $2 \cos^3 t/\sqrt{\sin^2 2t + \cos^2 t}$,   (b) 0,   (c) $6/\sqrt{1 + 4t^6}$,

   (d) $-\sec^2 t/\sqrt{\tan^2 t + \sec^2 t}$,   (e) $6/\sqrt{4 + 5 \sin^2 t}$,   (f) 0.

31. $8\mathbf{i} + 26\mathbf{j}$.   33. $(t^2 - 1)\mathbf{i} + (t^3 - 1)\mathbf{j}$.   35. $\left(\frac{1}{2} \text{Tan}^{-1} \frac{u}{2}\right)\mathbf{i} + \left(\frac{1}{2} \ln \frac{4 + u^2}{4}\right)\mathbf{j}$.

CHAPTER 13

Exercise 1, page 524

1. $(4, 0)$.   3. $(5, 0)$.   5. $(0, -4)$.   7. $(-4\sqrt{2}, 4\sqrt{2})$.   9. $(-3, 3\sqrt{3})$.

In Problems 11 through 17, n is any integer.

11. $(\sqrt{6}, 2n\pi + 3\pi/4)$, $(-\sqrt{6}, 2n\pi - \pi/4)$.   13. $(-5, 2n\pi + \pi/2)$, $(5, 2n\pi - \pi/2)$.

15. $(4, (2n + 1)\pi)$, $(-4, 2n\pi)$.

17. $(13, 2n\pi + \beta)$, $(-13, (2n + 1)\pi + \beta)$, where $\beta = \text{Cos}^{-1}(-5/13)$.

21. Sym. w.r. to origin.   23. Sym w.r. to x-axis.

C1. $(4.388, 2.397)$.   C3. $(7.343, -2.540)$.

C5. $(-6.687, 6.023)$.   C7. $(2.644, 1.087)$.

C9. $(829.9, 2.389)$.   C11. $(118.0, 5.564)$.

Exercise 2, page 528

1. Circle.   3. Line $y = 3$.

5. Limaçon.

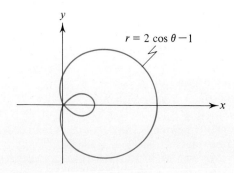

$r = 2 \cos \theta - 1$

7. Four-leafed rose.

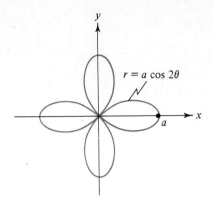

$r = a \cos 2\theta$

9. Three-leafed rose.

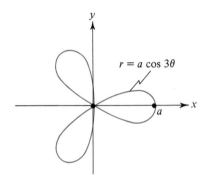

$r = a \cos 3\theta$

13. $(2\sqrt{2}, \pi/4)$, $(2\sqrt{2}, 3\pi/4)$.   15. $(a, \pm\text{Cos}^{-1}(1/4))$.
17. $(1/2, \pm\pi/3)$.   19. origin, $(-1, \pm\pi/3)$.
21. $(4m\pi, 4m\pi)$(first curve), $(4m\pi, 2m\pi)$(second curve), $m = 0, 1, 2, 3, \ldots$.
23. $r = a \sin 2\theta$, a four-leafed rose (see Fig. 7).

Exercise 3, page 532
1. $\theta = 0$.   3. $\theta = \pi/2$.   5. $\theta = \pi/4$.   7. $\theta = 0$.   9. $\theta = \pi/2$.
13. $r = 1 + 6 \sin \theta$.   15. $r = \sin^2 \theta/\cos \theta \cos 2\theta$.
17. $r = a \cos \theta (\tan^2 \theta - 1)$.   19. $y^2 = x^3/(2 - x)$.   21. $y^3 = x(x^2 + y^2)^2$.
25. $r = ep/(1 - e \cos \theta)$.   27. (a) Par.,   (b) 1,   (c) 7.
29. (a) Ellipse,   (b) 3/4,   (c) 10/3.   31. (a) Hyper.,   (b) 2,   (c) 4.
33. (a) $r = ep/(1 + e \cos \theta)$,   (b) $r = ep/(1 + e \sin \theta)$,   (c) $r = ep/(1 - e \sin \theta)$.

Exercise 4, page 539
1. $(a/2, \pi/6)$, $(a/2, 5\pi/6)$, $(2a, 3\pi/2)$.   3. $\alpha = 2\theta$ if $0 \leqq \theta < \pi/2$, $\psi = \theta$ if $0 \leqq \theta < \pi$.

5. $\tan \psi = 1/b$.   7. $r_1 r_2 = -\dfrac{dr_1}{d\theta} \dfrac{dr_2}{d\theta}$.

9. (a) $(58\sqrt{29} - 16)/3 \approx 98.78$,   (b) $a\sqrt{1 + b^2}(e^{b\pi} - 1)/b$.
11. $\sqrt{5} - \sqrt{2} + \ln(2\sqrt{2} + 2) - \ln(1 + \sqrt{5}) \approx 1.222$.

13. $\displaystyle\int 2\pi r \sin\theta \sqrt{r^2 + (dr/d\theta)^2}\, d\theta.$ 15. $32\pi a^2/5.$ 17. $2\pi a^2 \sqrt{2}.$

Exercise 5, page 543

1. $\pi a^2/4.$ 3. $11\pi.$ 5. $(2a^2 + b^2 + c^2)\pi/2.$ 7. $a^2.$
9. $a^2/2.$ 11. $(e^2 - 1)/4.$ 13. $a^2(2\pi + 3\sqrt{3})/6.$ 15. $a^2(2\pi - 6 + 3\sqrt{3})/3.$
17. The area swept out by the radial line $r = e^\theta$ as θ runs from 0 to 2π is counted twice in the answer.

Review Problems, page 543

1. (a) $(10\sqrt{2}, \pi/4 + 2n\pi),\ (-10\sqrt{2}, 5\pi/4 + 2n\pi),$ (b) $(10\sqrt{2}, \pi/4).$
3. (a) $(19, 3\pi/2 + 2n\pi),\ (-19, \pi/2 + 2n\pi),$ (b) $(19, 3\pi/2).$
5. $(5\sqrt{2}/2, 5\sqrt{2}/2).$ 7. $(-5, 0).$ 9. $d^2 = r_1^2 + r_2^2 - 2r_1 r_2 \cos(\theta_1 - \theta_2).$
11. The line $y = -3.$ 13. The line $3x + 2y = 4.$
15. The parabola, focus at the origin, directrix $x = -6.$
17. The hyperbola $x^2 - 3y^2 + 32y = 64,$ vertices at $(0, 8)$ and $(0, \tfrac{8}{3}),$ center at $(0, \tfrac{16}{3}).$
19. The rectangular equation is $y^2 = 4x^2/(x^2 - 4).$

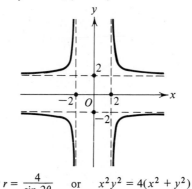

$$r = \frac{4}{\sin 2\theta} \quad \text{or} \quad x^2 y^2 = 4(x^2 + y^2)$$

21. A five-leafed rose.

23. A limaçon.

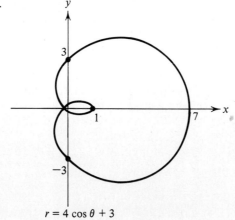

$r = 4\cos\theta + 3$

25. A lemniscate.

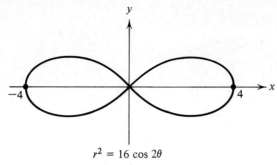

$$r^2 = 16 \cos 2\theta$$

27. The rectangular equation is $y^2 = x^3/(4 - x)$.

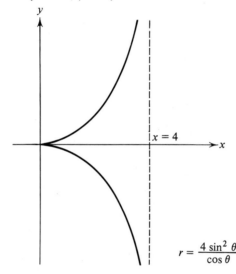

$x = 4$

$$r = \frac{4 \sin^2 \theta}{\cos \theta}$$

29. The graph resembles an ellipse through the points (± 4, 0) and (0, ± 5) but it is not an ellipse.

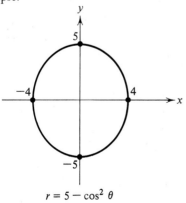

$$r = 5 - \cos^2 \theta$$

31.

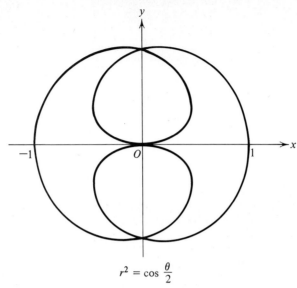

$$r^2 = \cos \frac{\theta}{2}$$

33. $(3\sqrt{2}, \pi/4)$, $(3\sqrt{2}, 7\pi/4)$.   35. $(\pm 2\sqrt[4]{500}/5, \text{Tan}^{-1} 2)$, origin.
37. $(3, \pi/3)$, $(3, 5\pi/3)$, origin.   41. $\sqrt{58\pi}$.   43. $4/3$.
45. $2\sqrt{2}/3 + 1/2$.   47. $6 - 2\sqrt{5}$.

CHAPTER 14

Exercise 1, page 549
1. $\pi/4$.   3. $\pi/2$.   5. 3.   7. $\pm\sqrt{3}$.

Exercise 2, page 552
1. $2/7$.   3. $-\pi/2$.   5. $2/3$.   7. $1/\pi^2$.   9. $-1/3$.   11. ∞.
13. $\ln(b/a)$.   15. 0.   17. $1/36$.   19. $1/2$.   21. 3.

Exercise 3, page 555
1. 0.   3. $1/2$.   5. 3.   7. 1.   11. No.

Exercise 4, page 558
1. 0.   3. ∞.   5. 0.   7. 1.   9. e^4.   11. 1.   13. e.   15. 1.   17. 1.

Exercise 5, page 562
1. 1.   3. 2.   5. Div.   7. $1/2e^2$.   9. π.   11. $3/2$.
13. Div.   15. Div.   17. Div.   19. $1/2$.   21. Div.   23. 3.
25. Div.   27. Div.   29. $1/(1 - k)$ if $k < 1$, div. if $k \geq 1$.   31. $32\pi/9$.
33. The derivative of $-2\sqrt{1 - \sin x}$ is $\cos x \sqrt{1 + \sin x}/|\cos x| = \epsilon \sqrt{1 + \sin x}$, where $\epsilon = \pm 1$ according as $\cos x > 0$ or $\cos x < 0$. At $x = 2n\pi + \pi/2$ the function $-2\sqrt{1 - \sin x}$ does not have a derivative.

Review Problems, page 563

1. $1/3$.   3. $3/5$.   5. $1/2$.   7. $\sqrt{2}/4$.   9. 0.   11. 1.

13. ∞.   15. 1.   17. 1.   19. (a) 1,   (b) e,   (c) ∞.   21. Div.

23. $\pi/4$.   25. Div.   27. Div.   29. $1/a$.   31. 0.   33. $\pi/3$.   35. 12.

37. $-1/(n+1)^2$.   39. $-1 + 2/\pi$.   41. 3.   43. $1/2$.

CHAPTER 15

Exercise 1, page 571

1. 0.   3. D.   5. D.   7. 0.   9. 2,500.   11. -1.   13. 0.

15. 1.   17. $\pi/2$.   19. 0.   21. ln 2.   25. Dec.   27. Inc.

29. Inc.   31. Bounded above and below.   33. Bounded below, but not above.

35. Bounded above, but not below.   39. $S_n = (-1)^n$ is one example.

C1. $S_6 = 0.31721028$.

C3. $\hat{S}_1 = 0.99346405$, $\hat{S}_2 = 0.99764706$, $\hat{S}_3 = 0.99923840$, $\hat{S}_4 = 0.99975155$.

Exercise 2, page 578

1. $S_n = 5n$, D.   3. $S_{2n-1} = n - 1$, $S_{2n} = n$, D.   5. $S_n = -n/(n+1)$, $S = -1$.

11. $2/7$, 2, $1/4$, 0.   13. 5.   15. $\dfrac{8}{11}\left(\dfrac{3}{8}\right)^{14}$.   17. $(75 + 40\sqrt{3})/11$.

19. D.   21. $2/15$.   23. $11/18$.   25. $28/33$.   27. 70 ft.

29. $\sqrt{H}(1 + \sqrt{r})/4(1 - \sqrt{r})$ sec, 11.01 sec.   31. $|x| < 5$.

33. $1/(3k + 1)$.   35. $1/(2k - 1)(2k + 1) = 1/(4k^2 - 1)$.   37. $k/(k^2 + 2)$.

C1. $S_7 = 2.71805556$, $S_8 = 2.71825397$, $S_9 = 2.71827877$; $S = e$.

C3 (a) $S = 0.75$,   (b) D,   (c) $S = 0.5$.   C5. 1.582.

Exercise 3, page 588

D.   3. C.   5. C.   7. C.   9. D.   11. C.   13. D.   15. C.

D.   19. D.   21. C.

$\gamma \approx 0.577$.

Exercise 4, page 595

1. C.   3. D.   5. C.   7. C.   9. D.   11. D.   13. C.   15. D.

17. C.   19. D.   21. D.   23. C.   25. C.   27. C.   29. D.

C1. For $p = 2$, $S_5 = 1.46361111$, $S_{10} = 1.54976773$, $S_{15} = 1.58044028$, $S_{20} = 1.59616324$, for
$p = 4$, $S_5 = 1.08035193$, $S_{10} = 1.08203658$, $S_{15} = 1.08223391$, $S_{20} = 1.08228459$.

C3. 1.64694444, 1.64522228, 1.64502362, 1.64497277.

Exercise 5, page 601

1. C.   3. D.   5. C.   7. D.   9. C.   11. C.   13. D.   15. C.

C1. $S \approx S_3 = 0.2901\ldots$ C3. $S \approx S_5 = 0.6222\ldots$

Exercise 6, page 606

1. $-1 < x < 1$.   3. $-1 \leq x < 1$.   5. $-\infty < x < \infty$.   7. $x = 0$.

9. $-7 \leq x < -3$.   11. $1 \leq x < 3$.   13. $-7/3 \leq x \leq 1$.   15. $x > 1$ or $x < -1$.

17. $5 \leq x \leq 8$.   **19.** $-1 \leq x < 11$.   **21.** $x \leq -1$ or $x > 1$.   **23.** D for all x.
25. C for all x.   **29.** $1 < |x| < 3$, $(x^2 - 5)/(9 - x^2)$.   **31.** All x except $x = 1$.

Exercise 7, page 615

1. $\displaystyle\sum_{n=0}^{\infty} \frac{n+1}{2^{n+2}} x^n$.   **3.** Differentiate, $1/(1 - x)^2 = \displaystyle\sum_{n=1}^{\infty} n x^{n-1}$.

5. $\displaystyle\sum_{n=0}^{\infty} \frac{x^{n+1}}{n+1} = \sum_{n=1}^{\infty} \frac{x^n}{n}$.   **7.** $\displaystyle\sum_{n=0}^{\infty} (-1)^n x^{2n+4}$, $-1 < x < 1$.

9. $\displaystyle\sum_{n=0}^{\infty} \frac{(n+3)(n+2)(n+1)}{6} x^n$, $-1 < x < 1$.   **11.** $\displaystyle\sum_{n=0}^{\infty} \frac{(-1)^n x^{n+1}}{(2n)!}$, all x.

13. $\displaystyle\sum_{n=0}^{\infty} \frac{(-1)^n x^{2n+1}}{2n+1}$, $-1 \leq x \leq 1$.   **15.** $\displaystyle\sum_{n=0}^{\infty} \frac{e^2 (x-2)^n}{n!}$.

17. $\displaystyle\sum_{n=0}^{\infty} \frac{(-1)^n (x-3)^n}{3^{n+1}}$.   **19.** $\ln 3 + \displaystyle\sum_{n=1}^{\infty} \frac{(-1)^{n+1}(x-3)^n}{3^n n}$.

21. $\displaystyle\sum_{n=0}^{\infty} \frac{(-1)^n (x - \pi/2)^{2n}}{(2n)!}$.   **23.** $95 - 57(x+5) + 7(x+5)^2$.

25. $1 - x + \dfrac{3}{2} x^2$.   **27.** $x + \dfrac{x^3}{3} + \dfrac{2}{15} x^5$.   **29.** $1 + 5x + 3x^2 + 3x^3$.

31. $\displaystyle\sum_{n=0}^{\infty} \left[\frac{(-1)^{n+1}}{2^n} - \frac{2}{3^{n+1}} \right] x^n$.

Exercise 8, page 620

9. $e/6$.   **11.** $\left(\dfrac{5\pi}{180}\right)^5 \Big/ 5!$, since $P_3(x) = P_4(x)$.   **13.** $(\cosh 1)/5!$, since $P_3(x) = P_4(x)$.
C1. 1.221.   **C3.** 0.182.   **C5.** 0.0698.   **C7.** 0.9063.

Exercise 9, page 626

9. $\displaystyle\sum_{n=0}^{\infty} \frac{(n+1)(n+2)\cdots(n+k-1)}{(k-1)!} x^n$.   **11.** $f'(0) = 1/3$, $f^{(6)}(0) = 6! \, 6^3/3^6$.

15. $\displaystyle\sum_{n=0}^{\infty} \frac{(-1)^n}{(4n+1)\cdot(2n)!}$.   **17.** $\displaystyle\sum_{n=1}^{\infty} \frac{(-1)^n}{n 2^n (n!)}$.   **19.** 0.   **21.** 3/2.
C1. 0.747.   **C3.** 0.932.   **C5.** 0.493.   **C7.** 0.4055.   **C9.** 0.4636 radian.

Review Problems, page 633

1. C.   **3.** C.   **5.** C.   **7.** D.   **9.** C.   **11.** (a) D,   (b) C.

13. $-1 < x < 1$.   **15.** $\ln a + \displaystyle\sum_{n=1}^{\infty} \frac{(-1)^{n+1} x^n}{n a^n}$.

17. $\ln 2 + \displaystyle\sum_{n=1}^{\infty} \frac{(-1)^{n+1}(x-2)^n}{n 2^n}$.   **19.** $\displaystyle\sum_{n=0}^{\infty} 7\left(\frac{1}{2^n} - \frac{1}{3^n} \right) x^n$.

21. $1 + 3x + 6x^2 + 10x^3 + 15x^4$.　**23.** $x \geqq 1/2$.　**29.** 0.0124.　**31.** 0.4613.
33. $-1/2$.　**35.** 2.

CHAPTER 16

Exercise 1, page 641

3. All possible combinations with $x = 2$ or 7, $y = 3$ or 10, $z = 4$ or -3.
5. $(7, 0, 0)$, $(0, 2, 7)$.　**7.** $(3, 0, -5)$, $(0, 7.5, 10)$.
9.　　　　　　　　　　　　　　　　**11.**

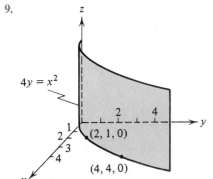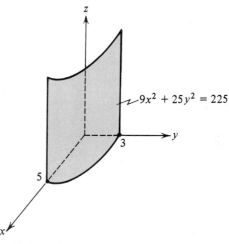

13. $x^2 + y^2 + z^2 = 25$.　**15.** (a) -4,　(b) 1,　(c) 0.

Exercise 2, page 646

1.

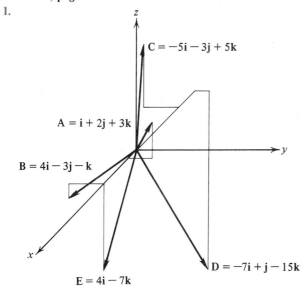

3. (a) $-i - 6j + 4k$, (b) $-9i - 4j + 13k$, (c) $14i + 6j + 16k$, (d) $-6i - 6k$.
5. (a) $(2/3, 1/3, -2/3)$, (b) $(6/7, -3/7, 2/7)$, (c) $(1/\sqrt{3}, 1/\sqrt{3}, 1/\sqrt{3})$,
   (d) $(-1/\sqrt{5}, 6/5\sqrt{5}, 8/5\sqrt{5})$.
7. (a) 6, (b) 12. 9. $\sin^2\theta + \cos^2\theta = 1$. 11. $m = 2, n = 1/2, p = -3$.
13. $x^2 + y^2 + z^2 - 4x + 6y - 12z = 0$. 15. $x - 6y - 9z + 3 = 0$.
19. (a) $(-2, 0, 3)$, $\sqrt{13}$, (b) $(-6, 3, -2)$, 7. 21. $(0, 6, 0)$, 2.

Exercise 3, page 652

1. $\dfrac{x-1}{3} = \dfrac{y-2}{4} = \dfrac{z-3}{-12}$. 3. $\dfrac{x-2}{3} = \dfrac{y-1}{-5} = \dfrac{z+3}{10}$. 5. $(0, 2/3, 7)$. 7. No.
9. $\mathbf{R} = 2j + 7k + t(-7i + 2j + 7k)$. 11. $\mathbf{R} = 3i - 4k + t(20i + 15j + 12k)$.
13. $(7, 10, -21)$. 15. $(52, -9, 43)$. 19. $(8, 11, 0)$.
21. (a) $(1, -1, 5)$, (b) $(3, -2, -1)$, (c) $(AD/R^2, BD/R^2, CD/R^2)$, where
   $R^2 = A^2 + B^2 + C^2$.

Exercise 4, page 657

1. (a) -23, (b) 3, (c) 0, (d) -1. 3. $-14/3$. 5. $1/5\sqrt{2}, 4/\sqrt{65}, 9/\sqrt{130}$.
7. $1/2, 1/2, 1/2$. 9. $9/11$. 11. $6i + 4j + 5k$.
C1. $24.80°$. C3. $624100 \leqq 624150$. C5. No, $\mathbf{A}\cdot\mathbf{B} = 1$, $\theta = 89.999894°$.

Exercise 5, page 664

1. (a) $3(2i - 11j - 4k)$, (b) $7(-3i + 7j - 9k)$, (c) $-71i + 50j - 18k$,
   (d) $113i + 17j - 103k$.
3. $(i - j + k)/\sqrt{3}$. 5. (a) $13/2$, (b) 17, (c) $3\sqrt{6}/2$, (d) $\sqrt{222}/2$.
7. (a) $42i + 19j - 17k$, (b) $19i + 5j - 10k$. 9. $(\mathbf{i} \times \mathbf{j}) \times \mathbf{j} = -\mathbf{i}$ but $\mathbf{i} \times (\mathbf{j} \times \mathbf{j}) = \mathbf{0}$.
C1. $\mathbf{A} \times \mathbf{B} = 33.84i + 2.32j - 10.88k$, $\sin\theta \approx 0.827819$.

Exercise 6, page 669

1. (a) $\sqrt{5}$, (b) $\sqrt{397}/7$. 3. (a) $35/\sqrt{230}$, (b) $\sqrt{11/30}$. 7. (a) 5, (b) 18.
9. (a) $\sqrt{2}$, (b) $17/5\sqrt{6}$. 11. $4/\sqrt{3}$. 13. If AB and CD are parallel.

Exercise 7, page 674

1. (a) $x - y + z = 7$, (b) $3x + 2y - z = -10$. 3. $3x + 5y - 7z = -44$.
5. (1a) $7/\sqrt{3}$, (1b) $10/\sqrt{14}$, (2a) 0,
   (2b) $|abc|/\sqrt{b^2c^2 + c^2a^2 + a^2b^2} = \left(\dfrac{1}{a^2} + \dfrac{1}{b^2} + \dfrac{1}{c^2}\right)^{-1/2}$, (2c) $2\left(\dfrac{1}{a^2} + \dfrac{1}{b^2} + \dfrac{1}{c^2}\right)^{-1/2}$.
9. $3/2$. 11. $\dfrac{x+9}{2} = \dfrac{y-4}{6} = \dfrac{z-3}{9}$, $Q(-105/11, 26/11, 6/11)$. 13. 3.
15. $x - 3 = y + 1 = z - 6$. 17. $x - z = 0$. 21. (a) $8/21$, (b) $20/33$, (c) 0, (d) 1.
23. Only the planes in (b) are perpendicular.

Exercise 8, page 680

1. $5a\cos 5ti - 5a\sin 5tj + 3k$, $-25a(\sin 5ti + \cos 5tj)$. 3. $3i - 5j - k$, $\mathbf{0}$.
5. $2ti + (3t^2 - 6t)j + 5k$, $2i + 6(t - 1)j$. 11. $P_\infty(1, 5, 0)$, $\mathbf{V}_\infty = \mathbf{0}$, $\mathbf{A}_\infty = \mathbf{0}$.
15. 147 miles/hr.

Exercise 9, page 686

1. (a) $y = z = 0$,   (b) $x = 0$.   3. (a) $\mathbf{R} = t\mathbf{j} + \sqrt{3}t\mathbf{k}$,   (b) $y + \sqrt{3}z = 0$.
5. (a) $3x = 3y = z + 9$,   (b) $x + y + 3z = 72$.   7. $\sqrt{2}/(e^t + e^{-t})^2$.
9. $1/3|a|(1 + t^2)^2$.   11. $24/65$.   13. $[-(\sin t + \cos t)\mathbf{j} + (\cos t - \sin t)\mathbf{k}]/\sqrt{2}$.
15. $[-2t\mathbf{i} + (2 - t^2)\mathbf{j} + 2t\mathbf{k}]/(2 + t^2)$.

Exercise 10, page 692

1. (a) $F(x, -y, z) \equiv F(x, y, z)$,   (b) $F(-x, y, z) \equiv F(x, y, z)$,
   (c) $F(x, -y, -z) \equiv F(x, y, z)$,   (d) $F(-x, -y, z) \equiv F(x, y, z)$,
   (e) $F(y, x, z) \equiv F(x, y, z)$.
3. $x_0 = \pm a$.

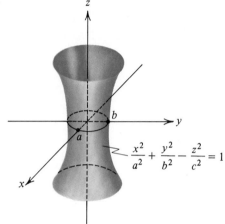

$$\frac{x^2}{a^2} + \frac{y^2}{b^2} - \frac{z^2}{c^2} = 1$$

5.

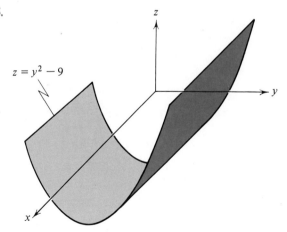

$z = y^2 - 9$

7.

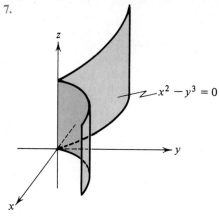

$x^2 - y^3 = 0$

9.

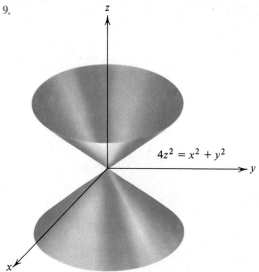

$4z^2 = x^2 + y^2$

11.

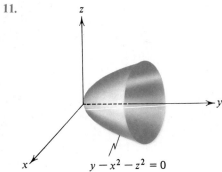

$y - x^2 - z^2 = 0$

13. $\dfrac{z^2}{a^2} + \dfrac{x^2}{b^2} + \dfrac{y^2}{b^2} = 1.$

15. $x^2 + (y-2)^2 = 4.$

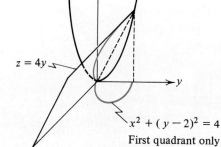

$z = x^2 + y^2$
First octant only

$z = 4y$

$x^2 + (y-2)^2 = 4$
First quadrant only

17. $x = \pm y,\ |x| \leqq 2.$

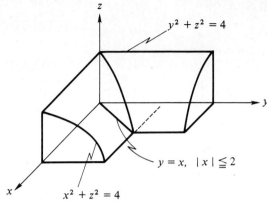

$y^2 + z^2 = 4$

$y = x,\ \ |x| \leqq 2$

$x^2 + z^2 = 4$

19. $y = 2x.$

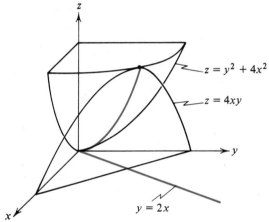

$z = y^2 + 4x^2$

$z = 4xy$

$y = 2x$

21. $x^2 + y^2 + xy = 2A(x + y).$

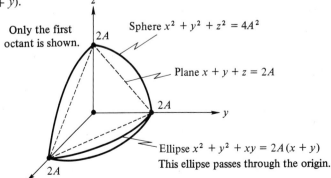

Only the first
octant is shown.

$2A$

Sphere $x^2 + y^2 + z^2 = 4A^2$

Plane $x + y + z = 2A$

$2A$

Ellipse $x^2 + y^2 + xy = 2A(x + y)$
This ellipse passes through the origin.

$2A$

25. $c^2 = a^3 + 2b^4$, yes $(0, 0, 0)$.

Exercise 11, page 696

1. $(0, 3, 5)$.　　3. $(-2, 2\sqrt{3}, 1)$.　　5. $(3\sqrt{2}, -3\sqrt{2}, 19)$.　　7. $(\sqrt{2}, \pi/4, 1)$.
9. $(6, 5\pi/6, 6)$.　　11. $(16, 4\pi/3, \pi)$.　　13. $r^2 + z^2 = 16$.
15. $z^2 = \tan 2\theta$.　　17. $x^2 + y^2 = 4x$.　　19. $(x^2 + y^2)^{5/2} = 4xyz$.

21. (a)

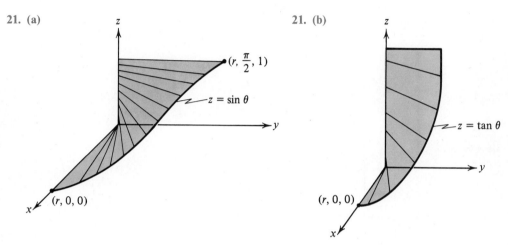

21. (b)

21. (c)

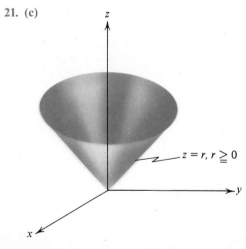

21. (d)

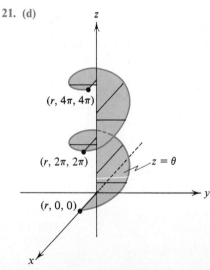

Exercise 12, page 699

1. $(2\sqrt{2}, 2\sqrt{2}, 4\sqrt{3})$.　　3. $(-4\sqrt{2}, -4\sqrt{2}, 8\sqrt{3})$.　　5. $(0, 0, -\pi)$.
7. Sphere $\rho = 8\cos\varphi$.　　9. A distorted torus $\rho = \sin^2\varphi$.
13. Sphere $(x - 2)^2 + y^2 + (z - 1)^2 = 5$.

15. $x^2(x^2 + y^2 + z^2) = 4y^2$. Each plane containing the z-axis intersects the surface in a circle of radius $2 \tan \theta$, except the plane $x = 0$, and except for points on the z-axis.

17. (a) $r^2 + z^2 + A^2 - a^2 = 2rA$, (b) $a^2 = \rho^2 + A^2 - 2\rho A \sin \varphi$.

19. $ds^2 = d\rho^2 + \rho^2 \, d\varphi^2 + \rho^2 \sin^2 \varphi \, d\theta^2$.

Review Problems, page 700

5. (a) $(-1/2, -3/2, 3/2)$, (b) $(-1, -1/2, 2)$, (c) $(7/2, 9/2, 5)$, (d) $(2, 1, 5/2)$.

9. $y = \pm x$, two planes.

11. $x^2 + (y - 9/2)^2 + z^2 = 17/4$, sphere, center $(0, 9/2, 0)$, radius $\sqrt{17}/2$.

13. $(x + 1)^2 + (y - 4)^2 + (z - 4)^2 = 18$, sphere, center $(-1, 4, 4)$, radius $3\sqrt{2}$.

15.

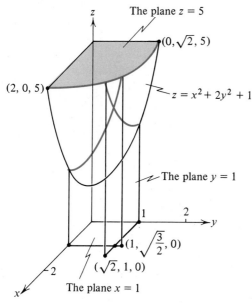

17.

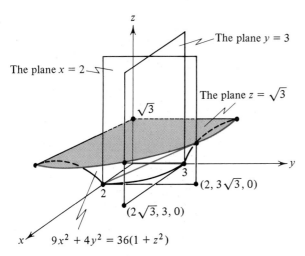

19.

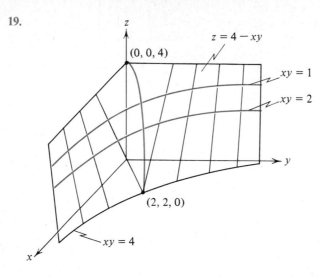

21. $(0, 1/2, 1/2)$.   **23.** $(3, 5, 2)$.   **25.** $(7, 2, -5)$.   **27.** $6/\sqrt{29}$.   **29.** $(1/3, 5/3, 29/3)$, $(1, 1, 7)$.
31. (a) $\sqrt{5/11}$,   (b) $7/\sqrt{66}$,   (c) $\sqrt{34/53}$.   **33.** (a) $2\sqrt{6/11}$,   (b) $3\sqrt{5/14}$,   (c) 0.
35. (a) $\sqrt{2}/2$,   (b) $2\sqrt{6}/3$,   (c) $11/\sqrt{10}$,   (d) $16\sqrt{70}/35$.
37. (a) $-\sqrt{5/6}$,   (b) $10/\sqrt{111}$,   (c) $19/\sqrt{370}$.   **39.** $-59x + 19y + 35z = 0$.
41. (a) $\mathbf{i} + \mathbf{j}$ at $(3, 0, 2)$, $\mathbf{i} - 3\mathbf{j}$ at $(1, 2, -2)$,   (b) nowhere,   (c) $\mathbf{i} + (1 \pm 2n)\mathbf{j}$ at
$(\pm n, \pm n + n^2, (-1)^n)$,   (d) $-2\mathbf{i} + 3\mathbf{j}$ at $(1, -1, -1/e)$.
43. (a) $(x - 3)^2 + y^2 + (z - 2)^2 = 13$, a sphere,   (b) $(x - 1)^2 + (y - 1)^2 + (z + 3)^2 = 11$,
a sphere,   (c) $z = 2y + 5x$, a plane.
45. (15a) $z = r^2 + r^2 \sin^2 \theta + 1$,   (15b) $\rho \cos \varphi = \rho^2 \sin^2 \varphi (1 + \sin^2 \theta) + 1$,
(16a) $z = 8 - r^2 \cos^2 \theta - r^2$,   (16b) $\rho \cos \varphi = 8 - \rho^2 \sin^2 \varphi(\cos^2 \theta + 1)$,
(17a) $5r^2 \cos^2 \theta + 4r^2 = 36 + 36z^2$,
(17b) $5\rho^2 \sin^2 \varphi \cos^2 \theta + 4\rho^2 \sin^2 \varphi = 36 + 36\rho^2 \cos^2 \varphi$.

CHAPTER 17

Exercise 1, page 707
 1. 63, 84, $(4, 3)$, 0.   **3.** 22, 46, $(23/6, 11/6)$, -14.   **5.** 60, 80, $(4, 3)$, 0.
11. $2/3$.

Exercise 2, page 710
1. $(1, 1, 1)$.   **3.** (a, b, c).   **5.** $(1, 2, 3)$.

Exercise 3, page 714
 1. $0.70 \, \text{lb}/\text{ft}^2$.   **3.** $0.045 \, \text{lb}/\text{ft}$.   **5.** $0.0031 \, \text{lb}/\text{in}^2$.   **9.** $0.243 \, \text{gram}/\text{cm}^2$.
11. $0.348 \, \text{gram}/\text{cm}$.   **13.** 7.5 grams.   **17.** $1080c$ grams.   **19.** Yes, 8π grams.

Exercise 4, page 720
 3. $(1, 1)$.   **5.** $(3/2, 6/5)$.   **7.** $(93/35, 45/56)$.   **9.** $(5/9, 5/27)$.   **11.** $(4, 2)$.

13. $(\pi/2, \pi/8)$.　　**15.** $(1, e^2/4)$.　　**17.** $1/6$.　　**19.** $360, 272, (34/15, 3)$.
23. $2\pi^2 R$.　　**25.** 4240 lb.

Exercise 5, page 724
1. $a(b_2^3 - b_1^3)/3$, no.　　**3.** $27/4, 27/4$.　　**5.** $2^7/21, 2^5/5$.
7. $62/15, 254/7$.　　**9.** $27/1820, 3/28$.　　**11.** $(e^6 - 1)/9, 2e^2 - 2$.　　**13.** $p > 3$.

Exercise 6, page 728
1. $(5/6, 0, 0)$.　　**3.** $(-8/5, 0, 0)$.　　**5.** $((e^2 + 1)/(2e^2 - 2), 0, 0)$.
7. $(0, 0, 3R/8)$.　　**9.** $(0, 0, 3)$.　　**11.** $(0, 0, 5/6)$.
13. $4\pi R^5/15, \pi/6, 192\pi/7, \pi/3, 2^{11}\pi/9, 3\pi(\pi^2 - 8)/2$.　　**15.** $n > 1, \pi/2(n - 1)$.

Exercise 7, page 730
1. $(2R/\pi, 2R/\pi)$.　　**3.** $\pi R^3/4$.　　**5.** $(A/3, 0, 0)$.　　**7.** $(2A/5, 2A/5)$.
9. $(\pi A, 4A/3)$.　　**11.** $(a^2 + a \sinh 2a - \sinh^2 a)/4(a \sinh a - \cosh a + 1)$.
13. $\left(\dfrac{15}{7} + \dfrac{3}{28}\ln 3, \dfrac{52}{21}\right)$.　　**15.** $4\pi^2 R$.

Review Problems, page 731
1. $(1, 3)$.　　**3.** $(1, 11/3, 15)$.　　**5.** No.
7. $m_1 = 16$ at $(-4, 0)$, $m_2 = 28$ at $(0, -3)$.
9. $\bar{x} = [(n + 1)(b^{n+2} - a^{n+2})]/[(n + 2)(b^{n+1} - a^{n+1})]$, $\bar{y} = 0$.　　**11.** $(3/8, 2/5)$.
13. $\left(\dfrac{2}{3} + \dfrac{8}{7}\ln 2, 1 + \dfrac{8}{7}\ln 2\right)$.　　**15.** $\bar{x} = \dfrac{M}{\ln(1 + M)} - 1$, $\bar{y} = \dfrac{M}{2(1 + M)\ln(1 + M)}$.
17. (a) $16/105, 2/15$,　　**(b)** $923/60, 31/4$.
19. (a) $(5/16, 0, 0)$,　　**(b)** $(3(39/4 + 4\ln 2)/25, 0, 0)$,　　**(c)** $\left(\dfrac{(1 + M)\ln(1 + M) - M}{M}, 0, 0\right)$.
21. $(0, 1/2, 0)$.

CHAPTER 18

Exercise 1, page 737
1. Inside and on the circle $x^2 + y^2 = 1$.　　**3.** Below the straight line $y = 5 - x$.
5. Inside and on the square with vertices $(\pm 1, 0)$ and $(0, \pm 1)$.
7. The empty set.　　**9.** Everywhere except the line $y = x$.　　**11.** 8.　　**13.** 5.
15. -3.　　**17.** -7.　　**19.** 0.　　**21.** No limit.　　**23.** $L = 0$.　　**25.** No limit.
27. No limit.　　**31.** $t^4(1 - 2t + 3t^2)$.　　**33.** $1 + 3\sin 2t$.　　**35.** $\ln 2v^4$.

Exercise 2, page 741
1. $2xy - y^3, x^2 - 3xy^2, 2x - 3y^2$.
3. $\sec 2y(\tan 3x + 3x \sec^2 3x)$,　 $2x \sec 2y \tan 2y \tan 3x$,　 $2 \sec 2y \tan 2y(\tan 3x + 3x \sec^2 3x)$.
5. $-y/(x^2 + y^2)$,　 $x/(x^2 + y^2)$,　 $(y^2 - x^2)/(x^2 + y^2)^2$.
7. $\cos \theta$,　 $-r \sin \theta$,　 $-\sin \theta$.
9. $x + x \ln(x^2 + y^2 + z^2)$,　 $y + y \ln(x^2 + y^2 + z^2)$,　 $z + z \ln(x^2 + y^2 + z^2)$.

Exercise 3, page 745

1. $-4, 10$.   3. $-8/9, 24/49$.
5. $y^4 + 2x$, $3x^2y^4 \sin \pi xy^2 + \pi x^3 y^6 \cos \pi xy^2$, $4xy^2/(y^2 - x^2)^2$, $-y^2/2x \sqrt{8x^2 + xy^2}$.
7. $f_{xy}(0, 2) = f_{yx}(0, 2) = -4$.   9. $-6/125$.   11. -512.   13. 3.   15. 18.
17. $2x^3y^2 + xy + y^2 + C$.

Exercise 4, page 751

1. $2x + 4y + z = 15$, $\mathbf{R} = (1 - 2t)\mathbf{i} + (2 - 4t)\mathbf{j} + (5 - t)\mathbf{k}$.
3. $6x + 3y + 2z = 18$, $\mathbf{R} = (1 + 6t)\mathbf{i} + (2 + 3t)\mathbf{j} + (3 + 2t)\mathbf{k}$.
5. $3x + 3y - z = 4$, $\mathbf{R} = (1 + 3t)\mathbf{i} + (1 + 3t)\mathbf{j} + (2 - t)\mathbf{k}$.   7. $\cos \theta = \pm \sqrt{26}/22$.
9. $H(1 + A)/\sqrt{A}$.

Exercise 5, page 758

1. $2(\Delta x)^2 + 3(\Delta y)^2$.   3. $\Delta y(-y \Delta x + x \Delta y)/y^2(y + \Delta y)$.
5. (a) 10.8,   (b) 10.86.   7. (a) 140 ft^2, 1.4 ft,   (b) 140.5 ft^2, 1.40010 ft.
9. (a) 0.142 ft, 0.075%,   (b) 0.14194 ft, 0.0747%.   11. (a) 2%,   (b) 2.0202%.
13. (a) 1.92 ft/sec^2,   (b) 1.98599 ft/sec^2.
15. (a) $49.25 \leqq R \leqq 50.75$,   (b) $49.24873 \leqq R \leqq 50.74877$.
C1. (a) 0.39878,   (b) 0.39000,   (c) 0.00878.
C3. (a) -0.04107,   (b) -0.04000,   (c) -0.00107.

Exercise 6, page 762

1. 0.   3. $(6t^2 + 2t^3)e^t + (6t^2 - 2t^3)e^{-t}$.   5. $4/t$.
7. $10te^{5t^2} \sin 5u^2$, $10ue^{5t^2} \cos 5u^2$.   9. $\pi l^3(2 \sin \theta \cos^2 \theta - \sin^3 \theta)/3$, $\pi l^2 \sin^2 \theta \cos \theta$.
11. In (10), V_h is computed with r held constant, and in (11), V_h is computed with θ held constant. In (9), V_θ is computed with l held constant and in (11), V_θ is computed with h held constant. A drawing will show that there is no reason to expect equality among these quantities.

Exercise 7, page 767

1. $7\sqrt{2}$.   3. $-1/5$.   5. $-2b$.
7. Directional derivative of f with $\alpha = \pi$ and $\alpha = 3\pi/2$, respectively.

Exercise 8, page 769

1. (a) $(f_x\mathbf{i} + f_y\mathbf{j}) \cdot (\cos \alpha\mathbf{i} + \sin \alpha\mathbf{j})$,   (b) $\nabla f = f_x\mathbf{i} + f_y\mathbf{j}$,
   (c) In Theorem 7 replace $f(x, y, z)$ by $f(x, y)$.
3. 0.   5. $-60/7$.   7. $12\pi/11$.   9. No.   11. The line $y = 2x, z = 3x$.
13. The circles $x^2 + y^2 = |\nabla f|^2/5$.   19. $f = ax + by + cz + d$ in $\mathscr{R}$.

Exercise 9, page 777

1. $2\mathbf{i} + 5\mathbf{j}$, $\mathbf{u} = (5\mathbf{i} - 2\mathbf{j})/\sqrt{29}$.
3. $6\mathbf{i} - 8\mathbf{j}$, $\mathbf{u} = (4\mathbf{i} + 3\mathbf{j})/5$; $-10\mathbf{i} + 24\mathbf{j}$, $\mathbf{u} = (12\mathbf{i} + 5\mathbf{j})/13$.
5. $\mathbf{i} + \mathbf{j}$, $\mathbf{u} = (\mathbf{i} - \mathbf{j})/\sqrt{2}$; $4\mathbf{i} + \mathbf{j}$, $\mathbf{u} = (\mathbf{i} - 4\mathbf{j})/\sqrt{17}$.
7. $\mathbf{i}$, $\mathbf{u} = \mathbf{j}$; $\mathbf{i} - 4\mathbf{j}$, $\mathbf{u} = (4\mathbf{i} + \mathbf{j})/\sqrt{17}$.   9. $2x + y - 5z = 30$.
11. $2ex - \pi y + (e + \pi)z = 3e$.   15. $\dfrac{x - 1}{5} = \dfrac{y - 2}{11} = \dfrac{z + 1}{7}$.

17. If $F(x, y, z) \equiv e^{x+y+z}$, then $\mathcal{S}(-1)$ is the empty set, and this could hardly be called a surface. **21.** $3\pi(x - 2) = y - 1, z = 3$.
23. $(-Az - Cx)/(Cy + Bz), (Bx - Ay)/(Cy + Bz)$.

Exercise 10, page 785

1. 14 at $(\pm\sqrt{3}/2, 1/2)$, 4 at $(0, -1/2)$.
3. $17r^2$ at $(-4r/5, 3r/5)$ and $(4r/5, -3r/5)$, $-8r^2$ at $(3r/5, 4r/5)$ and $(-3r/5, -4r/5)$, saddle pt. at $(0, 0)$.
5. 640 at $(8, 8)$, $(8, -4)$, and $(-4, 8)$, -1408 at $(-8, -8)$, rel. min. -8 at $(2, 2)$, saddle pt. at $(0, 0)$.
7. Max. $= (1 + 2\sqrt{2}\,r)/(1 + r^2)$, min. $= (1 - 2\sqrt{2}\,r)/(1 + r^2)$. **9.** 3888.
11. $V = d^3/27abc$ at $(d/3a, d/3b, d/3c)$. **13.** $V = 8abc/3\sqrt{3}$.
15. (a) $\sqrt{2}$, (b) $2\sqrt{3}$, (c) 0, (d) 13. **17.** $(10/3, 20, 8/3)$. **19.** $(\pm\sqrt{6}, \pm\sqrt{6}, 3)$.
21. 2 ft by 2 ft glass face, 3 ft deep. **23.** 5 at $(6, 3, -1)$.
C1. $x = 0.17748, y = 1.4086$.
C3. (a) $x = 0.34258, y = 0.98641$, neither; (b) $x = 0.92147, y = 0.60145$, r. min.

Exercise 11, page 793

1. $y = 1.12x + 0.962$. **3.** $y = 0.700x - 1.40$.
C1. $y = 0.0995x + 3.95$, (a) $3.95 \approx 4$, (b) $83.55 \approx 84$.
C3. $D = 0.142P/10^3 + 769$. (a) Computed 1083, real 690,
(b) computed 2027, real 2180, (c) computed 3609, real 4900,
(d) computed 769, real 0. The computed value probably refers to tourists.

Exercise 12, page 797

1. (a) 4, (b) 5. **3.** $2 \times 2 \times 4, V = 16$ ft^3. **5.** $(2, 2, 2\sqrt{2})$.
7. $(4, 4, 16)$. **9.** 26 at $(6, 2, 1)$.
11. $2\sqrt{6}/9$ at $(2\sqrt{3}/3, \sqrt{6}/3, \sqrt{3}/3), (-2\sqrt{3}/3, -\sqrt{6}/3, \sqrt{3}/3)$, and so on.
13. 144 at $(4, 16, 36)$. **15.** $(0, 3, 1)$. **17.** $(-1, 4, 1)$. **19.** $1/128$.

Review Problems, page 799

1. The shaded area except $(0, 0)$, and $y = x$.

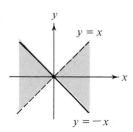

3. The region outside the circle $x^2 + y^2 = 1$.
5. The strip regions $2n\pi - x \leq y \leq (2n + 1)\pi - x, n = 0, \pm1, \pm2, \ldots$. **7.** 0. **9.** No
limit. **11.** 0. **13.** $f_x = 2x/(x^2 + 2y), f_y = 2/(x^2 + 2y)$.
15. $f_x = 2x \sec^2 x^2 \sec y^3, f_y = 3y^2 \tan x^2 \sec y^3 \tan y^3$.
17. $f_x = y^2[\cos(x + y) - x \sin(x + y)], f_y = xy[2 \cos(x + y) - y \sin(x + y)]$.

21. $a + b + c = 0$, yes.   **23.** No. Set $u = v = x + y + z$.

29. $2x + 8y - z = 14$. $\mathbf{R} = (1 + 2t)\mathbf{i} + (2 + 8t)\mathbf{j} + (4 - t)\mathbf{k}$.

31. $x - y - z = 0$. $\mathbf{R} = (2 + t)\mathbf{i} + (2 - t)\mathbf{j} - t\mathbf{k}$.

33. $50x + 25y - 23z = 54$, $\mathbf{R} = (1 + 50t)\mathbf{i} + (2 + 25t)\mathbf{j} + (2 - 23t)\mathbf{k}$.

35. (a) 1.32 in.^{3},   (b) 1.326808 in^3.

37. (a) $17/50\sqrt{157} \approx 0.0271350$,   (b) $\sqrt{157.6812} - \sqrt{157} \approx 0.0271534$.

39. (a) -3.43,   (b) -3.438577401.   **41.** $46/13$.   **43.** 11.

45. The points $(8t, 6t, -t)$.

47. (a) The z-axis,   (b) the points $y = 0$ and the points $x = 2y^2$,   (c) the points on the z-axis and the points on the surface $z = -y^2/8x$, $x \neq 0$.

57. $(1, -3)$ is a saddle point.   **59.** Relative max. 17 at $(-2, -3)$.

61. Saddle point at $(0, 0)$, Relative min. -64 at $(2, 4)$.

63. Relative max. 4 at $(2, (2n + 1)\pi)$, Relative min. -4 at $(2, 2n\pi)$, saddle point at $(0, (n + 1/2)\pi)$ and at $(4, (n + 1/2)\pi)$.

67. $(34/49, -51/49, 102/49)$.   **69.** $D = \sqrt{3}$ at $(1, -1, 1)$ and $(-1, 1, 1)$.

71. $y = (29/26)x - 2/13 \approx 1.1154x - 0.1538$.

73. All sums run from 1 to n.

$$\left(\sum x_k^4\right)A + \left(\sum x_k^3\right)B + \left(\sum x_k^2\right)C = \sum x_k^2 y_k,$$

$$\left(\sum x_k^3\right)A + \left(\sum x_k^2\right)B + \left(\sum x_k\right)C = \sum x_k y_k,$$

$$\left(\sum x_k^2\right)A + \left(\sum x_k\right)B + nC = \sum y_k.$$

CHAPTER 19

Exercise 1, page 808

1. $0 < y < 4$, $0 < x < \sqrt{4 - y}$.   **3.** $0 < y < e^4 - 1$, $\ln(y + 1) < x < 4$.

5. $2/5 < y < 2$, $0 < x < \ln(2/y)$.   **7.** $|x| < 3$, $|y| < 4\sqrt{1 - x^2/9}$.

9. If $0 < y \leq 1$, then $|x| < 2\sqrt{y}$; if $1 \leq y < 4$, then $2y - 4 < x < 2\sqrt{y}$.

11. If $0 < y < 12$, then $|x| < y^{1/4}$; if $12 \leq y < 16$, then $\sqrt{y - 12} < |x| < y^{1/4}$.

15. $0 < y < 2$, $y < z < 4 - y$, $y + z < x < 4$.

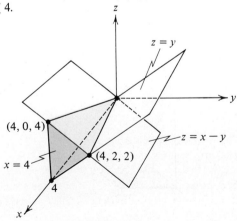

17. $0 < z < 4$, $(4-z)/2 < x < \sqrt{4-z}$, $0 < y < \dfrac{\sqrt{4-x^2-z}}{2}$.

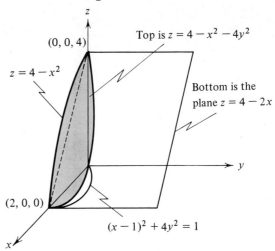

Top is $z = 4 - x^2 - 4y^2$

$(0, 0, 4)$

$z = 4 - x^2$

Bottom is the plane $z = 4 - 2x$

$(2, 0, 0)$

$(x-1)^2 + 4y^2 = 1$

Exercise 2, page 811
1. 12.   **3.** 312.   **5.** 8/3.   **7.** 11/30.   **9.** 1/3.   **11.** 10/3.
13. 128/15.   **15.** 32.   **17.** 8/3.   **19.** 7/8.   **21.** 255/16.   **23.** 1/10.
25. $1/3^n(n!)$.   **27.** $n!n(n+1)/4$.

Exercise 3, page 817
1. T.   **3.** T.   **5.** T.
7. False. Let $\mathcal{R}$ be the square $0 \leqq x \leqq 2$, $0 \leqq y \leqq 2$, and set $f(x)g(y) = xy$. The left side is 4 and the right side is 16.
9. (a), (b), μ_0 can be any positive number.   **11.** (a) $\sqrt{2}/80$,   (b) $\epsilon\sqrt{2}/8$.

Exercise 4, page 827
1. 5.

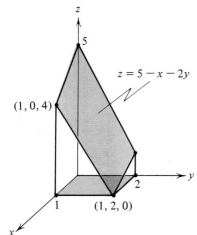

$z = 5 - x - 2y$

$(1, 0, 4)$

$(1, 2, 0)$

3. 128/15.

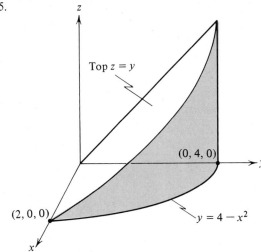

Top $z = y$

$(0, 4, 0)$

$(2, 0, 0)$

$y = 4 - x^2$

5. 8/3.

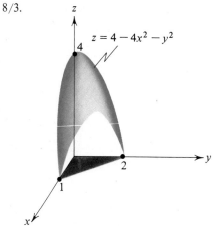

$z = 4 - 4x^2 - y^2$

7. $3\sqrt{3} - \pi$.

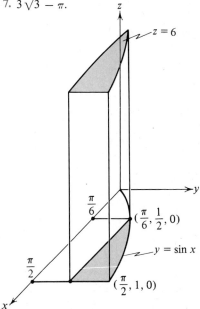

$z = 6$

$\dfrac{\pi}{6}$

$\left(\dfrac{\pi}{6}, \dfrac{1}{2}, 0\right)$

$y = \sin x$

$\dfrac{\pi}{2}$

$\left(\dfrac{\pi}{2}, 1, 0\right)$

9. π.

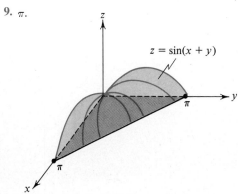

$z = \sin(x + y)$

π

π

11. 2.   **13.** $2 - \ln 3$.   **15.** $(e^4 - 1)/4$.   **17.** 153/2.

Exercise 5, page 831
1. 9/2.   **3.** 9.   **5.** (55/36, 7/4), 135/4, 87/4.   **7.** (2/5, 2/5), 1/60, 1/60.
9. (6/5, 96/35).   **11.** 64/21.   **13.** $(A^2 + 5A + 10)/3(A + 1)(A + 6)$.

Exercise 6, page 835
1. $9(1 + b^2 + c^2)^{1/2}$.   **3.** 104/3.   **5.** $10\sqrt{2}$.   **7.** $(2\sqrt{2} - 1)/3$.
9. $7\sqrt{13}/2$.   **11.** $8a^2$.   **13.** 1, $\pi/2 - 1$.

Exercise 7, page 841
1. $a^2\sqrt{3}/8$.   **3.** $19\pi/2$.   **5.** $\left(\dfrac{9\sqrt{3}}{2} - \pi\right)a^2$.   **7.** 8π, $32\pi/3$, $\pi(17\sqrt{17} - 1)/6$.

9. $\pi a(2a^2 + 3b^2)/3$.   **11.** 16/105.   **13.** $a^2(\pi - 2)$.

15. $\rho r \sin\theta$, $\rho r \cos\theta$, $\rho r^2 \sin^2\theta$, $\rho r^2 \cos^2\theta$, ρr^2.   **17.** $(4a/3\pi, 0)$.   **19.** $(5a/6, 0)$.

21. $\left(2(a^2 - b^2)^{3/2}/3(a^2 \cos^{-1}(b/a) - b\sqrt{a^2 - b^2}), 0\right)$.   **23.** $2b(5a^2 + b^2)/5(3a^2 + b^2)$.

25. $5\pi/64$, $\pi/64$.   **27.** $a^4/3$, $a^4/6$, $a^4/2$.   **29.** $\pi/4$.

Exercise 8, page 849
1. $a^2b^3c^4$.   **5.** 64.   **7.** $32\sqrt{2}\pi$.   **9.** 8π.   **11.** $kab^2c/2$.   **13.** $256k/15$.
15. $(\ln 4)^3/6$.   **17. (a)** (1/2, 2/3, 1/2),   **(b)** (a/2, 2b/3, c/2),
(c) $\left(\dfrac{a}{d}\left(\dfrac{1}{2} + \dfrac{2}{3}q\right), \dfrac{b}{d}\left(\dfrac{1}{2} + \dfrac{3}{4}q\right), \dfrac{c}{d}\left(\dfrac{1}{2} + \dfrac{4}{5}q\right)\right)$, where $q = kab^2c^3$, and $d = 1 + q$,
(d) (372/455, 258/455, 7/26).   **19.** (a/4, b/4, c/4).

Exercise 9, page 853
1. (a) z,   **(b)** $r\cos\theta$,   **(c)** $z^2 + r^2\sin^2\theta$,   **(d)** $z^2 + r^2\cos^2\theta$.
5. $4\pi(8\sqrt{2} - 7)/3$.   **7.** $\pi/32$.   **9.** $3\pi R^4 H/2$, $\pi R^2 H(3R^2 + 4H^2)/12$.
11. π.   **13.** $\pi(4Ha^2 + 2Hb^2 + 4ma^2b + mb^3)/4$.   **15.** $(4/\pi, 4/\pi, 1/8)$.

Exercise 10, page 857
3. (a) $\rho\cos\varphi$,   **(b)** $\rho\sin\varphi\sin\theta$,   **(c)** $\rho^2(\sin^2\varphi\sin^2\theta + \cos^2\varphi)$,
(d) $\rho^2(\sin^2\varphi\cos^2\theta + \cos^2\varphi)$,   **(e)** $\rho^2\sin^2\varphi$.
5. $(0, 0, 3(B^4 - A^4)/8(B^3 - A^3))$.
7. $2\pi R^3(1 - \cos\gamma)/3$, $(0, 0, 3R(1 + \cos\gamma)/8)$, $2\pi R^5(2 - 3\cos\gamma + \cos^3\gamma)/15$,
     $\pi R^5(4 - 3\cos\gamma - \cos^3\gamma)/15$.
9. $\pi^2 A^3/4$.   **11.** $(B^4 - A^4)/(B^3 - A^3)$.   **13.** $\pi A^3(4B - 3A)/6B$.
15. $2\pi kR^{3-n}/(3 - n)$ if $n < 3$.   **17.** $4\pi kR^{5-n}/3(5 - n)$, $n < 5$.

Exercise 11, page 861
3. $F_2 = -\gamma M/b\sqrt{a^2 + b^2}$.   **5.** $F_3 = -2\gamma M(\sqrt{a^2 + R^2} - a)/R^2\sqrt{a^2 + R^2}$.
7. $F_3 = 2\pi H\gamma\mu(1 - \cos\alpha)$, where $\tan\alpha = R/H$.   **9.** $F_3 = 2\pi\gamma kH^{n+1}(\sec^{n-1}\alpha - 1)/(n^2 - 1)$.

Review Problems, page 862

1. 32/7.   3. 1/3.   5. $(\sqrt{2} - 1)/2$.   7. 40/3.   9. $[3 \cdot 5 \cdots (2n + 1)]/2^n$.

11. $n(3n + 1)/12$.   13. 392/15.

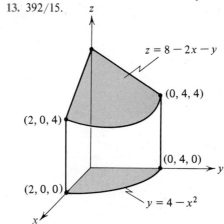

15. 26/7.

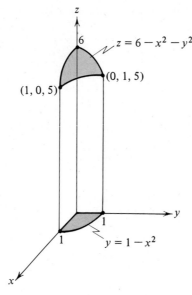

17. $(1 - 2/e)/2$.

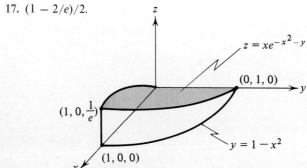

19. 4 ln 2.

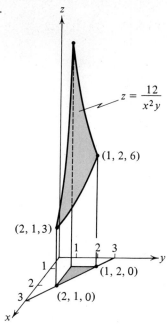

$$z = \frac{12}{x^2 y}$$

21. 16.   **23.** 1/10.   **27.** $\bar{x} = 22/35$, $\bar{y} = 22/35$.   **29.** $\bar{x} = 8/5$, $\bar{y} = 3/4$.
31. $\bar{x} = 10/7$, $\bar{y} = 5/7$.   **33.** $\bar{x} = 3/5$, $\bar{y} = 1/10$.   **35.** 260.   **37.** $(14\sqrt{14} - 5\sqrt{5})/27$.
39. $\pi(6ac^2 + 3ad^2 + 4bc^3 + 6bcd^2)/6$.   **43.** $\pi[(1 + 16a^2)^{3/2} - 1]/6a^2$.
45. $6a + 2b3^{n+1}/(n + 1)$.   **47.** $2(10a + b)/15$.

49. $\bar{x} = \left(\dfrac{n + 1}{n + 2}\right)\dfrac{3a(n + 2) + 2b3^{n+1}}{2a(n + 1) + 2b3^n}$, $\bar{y} = 1$, $\bar{z} = 1/2$, $I_x = 10(a + b3^n/(n + 1))$.

51. $M_{xy} = [a - b + e^6(5a + 13b)]/8$, $M_{yz} = \dfrac{e^3}{2}(5a + 8b) - a + 2b$.

53. $\pi A^2(9aA\pi + 16b)/12$.

CHAPTER 20

Exercise 1, page 872

1. 18.   **3.** $-1/2$.   **5.** 0.   **7.** 2π.   **9.** ab.   **11.** $n(n + 1)\pi a^3$.   **13.** $(4 - \pi)a^2/2$.
15. 0, the curve is a line segment covered twice.   **17.** 28.   **19.** $28\frac{1}{4}$.   **21.** 18.

Exercise 2, page 880

1. $2x^3 + 3xy^2 + C$.   **3.** Not exact.   **5.** $x^4 e^{y/x^2} + C$.   **7.** Not exact.
9. $\varphi = k/\sqrt{x^2 + y^2 + z^2} + C$.   **13.** $\varphi = x^2 y^3/z^4 + C$.   **15.** $\varphi = e^{xz} \sin yz + C$.

Exercise 3, page 887

1. 0.   3. 50π.   5. 64.

Exercise 4, page 891

1. $\mathbf{\nabla} \cdot \mathbf{R} = 3$, $\mathbf{\nabla} \times \mathbf{R} = 0$.   3. 0, $(z - 2)\mathbf{j} - 3y\mathbf{k}$.   5. 0, $-2(z\mathbf{i} + x\mathbf{j} + 3xy\mathbf{k})$.
7. $3\varphi + \mathbf{R} \cdot \mathbf{\nabla}\varphi$, $-\mathbf{R} \times \mathbf{\nabla}\varphi$.   17. (a) $n(n + 1)r^{n-2}$,   (b) $n^2 r^{n-2}$.
19. (a) $3/r^4$,   (b) $6/r^4$.

Exercise 5, page 896

1. $\pi[2\sqrt{3} + \ln(2 + \sqrt{3})]$.   3. 37/9.

9. $\displaystyle\iint_{\mathcal{R}} Q \, du \, dv$, where $Q^2 = 17u^2 - 20uv + 8v^2 + 4(u^2 - v^2)^2$.

11. (a) $\sqrt{3}/12$,   (b) 5π.   13. 225.   15. $8\mathbf{i} + 40\mathbf{j} + 16\mathbf{k}$.   17. $(\mathbf{i} + \mathbf{j} + \mathbf{k})\pi r^3/4$.
19. $\pi r^3 \mathbf{k}$.   21. 0.   23. $4\pi r^3$.   25. 0.

Exercise 6, page 902

1. (a) 18,   (b) 36,   (c) 0.   3. 80π.
15. In Problem 11, let the integrand be $-(a + \delta z)\mathbf{n}$, where δ is the density of the fluid. Then use Problem 9 with $\varphi = a + \delta z$. This gives $\delta V \mathbf{k}$, an upward force equal to the weight of the liquid displaced.

Exercise 7, page 907

1. $2\mathbf{k}$, $2\pi r^2 = r \times$ (length of $\mathcal{C}$).

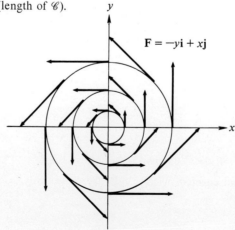

3. (a) $A\mathbf{k}$, $\pi r^2 A$,

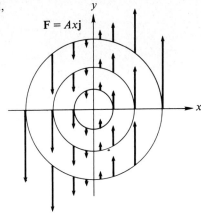

(b) **0**, 0,

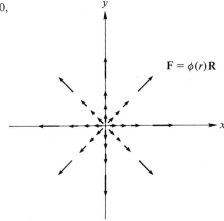

5. -6π.   7. ± 27, depending on the orientation.

Review Problems, page 908

1. (a) 280,   (b) $\frac{1}{3}(e^6 - 1) + 32$.   3. 0.   5. 0.

7. (a) and (b) are not independent of path.   9. -36.   11. -18.

13. (a) 0,   (b) $-2\mathbf{i} - 5\mathbf{j} - 3\mathbf{k}$.   15. (a) 0,   (b) $x(2y - x)\mathbf{i} + y(2z - y)\mathbf{j} + z(2x - z)\mathbf{k}$.

17. $188/105$.   19. (a) $u/\sqrt{1 - u^2}$,   (b) 2π.   21. (a) 0,   (b) 18,   (c) -6.

23. (a) -9,   (b) $-\frac{1}{3}(e^9 - 1) + 3$,   (c) 0,   (d) $-81A$.

APPENDIX 1

Exercise 1, page 917

11. $\max\{a_1, a_2, \ldots, a_k\}$.

APPENDIX 2

Exercise 1, page 924

5. For $\lim\limits_{x \to a^+} f(x) = L$, replace condition (4) of Definition 1 by $0 < x - a < \delta$.

For $\lim\limits_{x \to a^-} f(x) = L$, replace condition (4) by $-\delta < x - a < 0$.

APPENDIX 3

Exercise 1, page 928

1. -2.   3. 1.   5. 0.   7. 0.   9. 207.   11. -6.

Exercise 2, page 932

1. -8.   3. 133.   5. 1.

Exercise 3, page 935

1. $x = 4, y = 2$.   3. $x = 1, y = 2$.   5. $x = -3, y = 10$.
7. $x = 1, y = 2, z = 3$.   9. $x = 3, y = 0, z = 4$.
11. $u = -5, v = 1, w = 5$.

APPENDIX 5

Exercise 1, page 945

C1. (a) 12.55.   (b) 113.74425,   (c) 1.098.   C3. 14.238169.
C5.

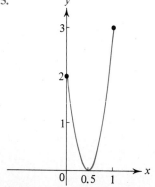

C7. $P(x) = (x - 1.5)(x^2 + 3.8x + 9.7) + 11.55$, $P(1.5) = 11.55$.
C9. $P(x) = (x - 1.5)(-2x^3 - 3x^2 - 1.5x - 10.45) - 14.775$, $P(1.5) = -14.775$.
C11. $P(x) = (x - 1.5)(6x^5 + 14x^4 + 25x^3 + 40.5x^2 + 62.75x + 95.125) + 143.6875$,
     $P(1.5) = 143.6875$.

Index

Selected Integration Formulas

$$\int (f(u) + g(u))\, du = \int f(u)\, du + \int g(u)\, du.$$

$$\int cf(u)\, du = c \int f(u)\, du.$$

$$\int u^n\, du = \frac{u^{n+1}}{n+1}, \quad n \neq -1.$$

$$\int \frac{du}{u} = \ln |u|.$$

$$\int e^u\, du = e^u.$$

$$\int a^u\, du = \frac{a^u}{\ln a}, \quad a > 0, a \neq 1.$$

$$\int \sin u\, du = -\cos u.$$

$$\int \cos u\, du = \sin u.$$

$$\int \tan u\, du = -\ln |\cos u|.$$

$$\int \cot u\, du = \ln |\sin u|.$$

$$\int \sec u\, du = \ln |\sec u + \tan u|.$$

$$\int \csc u\, du = \ln |\csc u - \cot u|.$$

$$\int \sec^2 u\, du = \tan u.$$

$$\int \csc^2 u\, du = -\cot u.$$

$$\int \sec u \tan u\, du = \sec u.$$

$$\int \csc u \cot u\, du = -\csc u.$$

$$\int \frac{du}{u \ln u} = \ln |\ln u|.$$

$$\int \ln u\, du = u \ln u - u.$$

$$\int u e^u\, du = (u - 1)e^u.$$

$$\int \frac{du}{a^2 + u^2} = \frac{1}{a} \operatorname{Tan}^{-1} \frac{u}{a}.$$

$$\int \frac{du}{a^2 - u^2} = \frac{1}{2a} \ln \left| \frac{a+u}{a-u} \right|.$$

$$\int \frac{du}{u^2 - a^2} = \frac{1}{2a} \ln \left| \frac{u-a}{u+a} \right|.$$

$$\int \frac{du}{\sqrt{a^2 - u^2}} = \operatorname{Sin}^{-1} \frac{u}{a}.$$

$$\int \sin^2 u\, du = \frac{1}{2}u - \frac{1}{4}\sin 2u.$$

$$\int \cos^2 u\, du = \frac{1}{2}u + \frac{1}{4}\sin 2u.$$

$$\int \tan^2 u\, du = \tan u - u.$$

$$\int \cot^2 u\, du = -\cot u - u.$$

$$\int u\, dv = uv - \int v\, du.$$

$$\int \operatorname{Sin}^{-1} u\, du = u \operatorname{Sin}^{-1} u + \sqrt{1 - u^2}.$$

$$\int \operatorname{Cos}^{-1} u\, du = u \operatorname{Cos}^{-1} u - \sqrt{1 - u^2}.$$

$$\int u^n \ln u\, du = u^{n+1} \left(\frac{\ln u}{n+1} - \frac{1}{(n+1)^2} \right), \quad n \neq -1.$$

$$\int u^n (\ln u)^m\, du = \frac{u^{n+1}}{n+1} (\ln u)^m - \frac{m}{n+1} \int u^n (\ln u)^{m-1}\, du.$$

$$\int \frac{du}{\sqrt{u^2 - a^2}} = \ln |u + \sqrt{u^2 - a^2}|.$$

$$\int \frac{du}{\sqrt{u^2 + a^2}} = \ln (u + \sqrt{u^2 + a^2}).$$

$$\int \frac{du}{u\sqrt{u^2 - a^2}} = \frac{1}{a} \operatorname{Cos}^{-1} \frac{a}{u}.$$

$$\int \frac{du}{u^2 \sqrt{u^2 \pm a^2}} = \mp \frac{\sqrt{u^2 \pm a^2}}{a^2 u}.$$

$$\int \frac{du}{(u^2 \pm a^2)^{3/2}} = \frac{\pm u}{a^2 \sqrt{u^2 \pm a^2}}.$$

$$\int u \sin u\, du = \sin u - u \cos u.$$

$$\int u \cos u\, du = \cos u + u \sin u.$$